CCME

单墫

数学与教育文选

单 墫◎著

华东师范大学出版社·上海

图书在版编目(CIP)数据

单墫数学与教育文选/单墫著. —上海:华东师范大学出版社,2021
(当代中国数学教育名家文选)
ISBN 978-7-5760-1762-5

Ⅰ.①单… Ⅱ.①单… Ⅲ.①数学教学-文集
Ⅳ.①O1-53

中国版本图书馆 CIP 数据核字(2021)第 094327 号

当代中国数学教育名家文选
单墫数学与教育文选

著 者 单 墫
策划编辑 刘祖希
责任编辑 刘祖希
特约审读 钟劲松 宋书华
责任校对 李琳琳
装帧设计 卢晓红

出版发行 华东师范大学出版社
社 址 上海市中山北路 3663 号 邮编 200062
网 址 www.ecnupress.com.cn
电 话 021-60821666 行政传真 021-62572105
客服电话 021-62865537 门市(邮购)电话 021-62869887
地 址 上海市中山北路 3663 号华东师范大学校内先锋路口
网 店 http://hdsdcbs.tmall.com

印 刷 者 上海雅昌艺术印刷有限公司
开 本 787×1092 16 开
印 张 65.75
插 页 4
字 数 965 千字
版 次 2021 年 7 月第 1 版
印 次 2021 年 7 月第 1 次
书 号 ISBN 978-7-5760-1762-5
定 价 198.00 元

出 版 人 王 焰

1964 年，大学毕业

1983 年 5 月，中国科学技术大学首批博士步出人民大会堂
左起白衣六男：赵林城，苏淳，白志东，李尚志，范洪义，单墫

1986 年 9 月，在北京一零一中学为第一届国家 IMO 集训队备课

1988 年，在加拿大作访问学者，摄于多伦多大学的菲尔兹研究所

1989 年，与中国 IMO 队员在德国高斯铜像前留影，当时中国首次取得 IMO 总分第一

左起：单墫，蒋步星，罗华章，唐若曦，俞杨，霍晓明，颜华菲，王寿仁

1993 年 6 月，与南京师范大学数学系毕业生合影，时任系主任

2005 年 11 月，南京师范大学数学教育博士点导师与部分博士合影
前排左起：李渺，常春艳，顾继玲，龙银美，韩龙淑
中排左起：王光明，杨骞，喻平，单墫，涂荣豹，李善良，曹一鸣
后排左起：钟志华，宋晓平，宁连华，葛军，黄晓学，郑庆全，李鹏，温建红，唐剑岚

师友家人

1980 年，全家福

1982 年，摄于无锡梅园
左起：杨劲根，肖刚，李克正，单墫

1990 年 7 月，摄于安徽屯溪
前排左起五人：苏淳，李炯生，史济怀，单墫，严镇军

1990 年，与张景中先生（中）、刘鸿坤先生（右）游北京香山公园

1992 年，陪同王元先生（左二）在南京参观《小学生数学报》等编辑部

1993 年，与熊斌先生合影

1995 年，与华罗庚先生之女华苏女士合影

1995 年，在江苏金坛参加“华杯赛”活动
前排左起：单墫，钱伟长，周春荔

1994 年，参加首届全国几何不等式会议（南京师范大学）
左起：杨路，谈祥柏，单墫

2009 年，在家中与孙儿下棋

2021 年 2 月，在家中讨论本书出版等事宜
左起：宋书华，单墫，刘祖希

2013 年 10 月，七十岁生日会合影

前排左起：熊　斌，王巧林，余红兵，杭顺清，刘鸿坤，单　墫，胡大同，吴建平，朱华伟，冷岗松，李　炘

中排左起：李　祎，葛　军，黄晓学，宁连华，倪　明，宋晓平，叶中豪，李善良，潘小明，崔恒兵，钟志华

后排左起：李　红，房剑平，郑庆全，李　鹏，闫伟峰，李　潜，李　忠，刘祖希，张新华，刘守军，黄志军

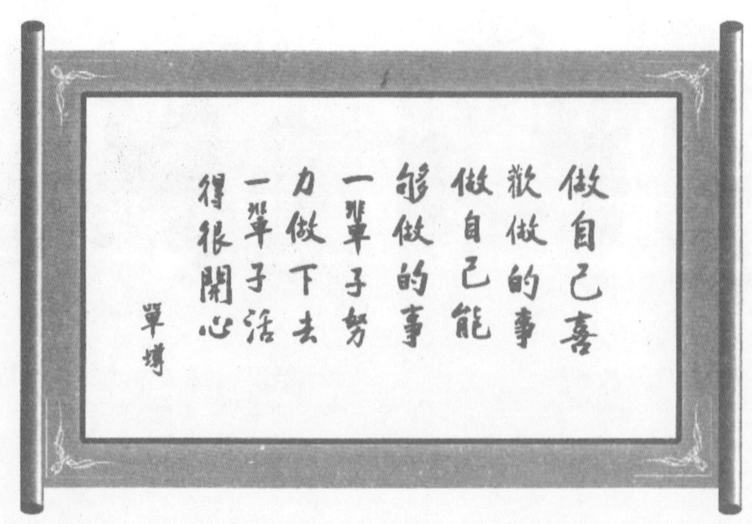

总序

数学教育具有悠久的历史.从一定程度上来讲,有数学就有数学教育.据记载,中国周代典章制度《礼记·内则》就有明确的对数学教学的内容要求:“六年教之数与方名……九年教之数日,十年出就外傅,居宿于外,学书计.”又据《周礼·地官》:“保氏掌谏王恶,而养国子以道,乃教之六艺,一曰五礼,二曰六乐,三曰五射,四曰五驭,五曰六书,六曰九数.”尽管周代就有关于数学教育的记载,但长期以来我国数学教学的规模很小,效果也不太好,大多数数学人才不是正规的官学(数学)教育培养出来的.中国古代的数学教育作为官方教育的一个组成部分,用现在话语体系来讲,其目标主要是培养管理型和技术型人才,既不是“精英”教育,也不是“大众”教育.

1582年,意大利传教士利玛窦来到中国.1600年,徐光启和李之藻向利玛窦学习西方的科学文化知识,翻译了欧几里得《几何原本》,对中国的数学与数学教育产生了一定的影响.1920年以后,在学习模仿和探索的基础上,中国人编写的数学教学法著作逐渐增多,内容不断扩展,水平也逐步提高,但主要还只是小学数学教育研究,大多数只是对前人或外国的教学法根据教学实践进行修补、总结而成的经验,并没有形成成熟的教育理论.1949年新中国成立后,通过苏联教育文献的引入,数学教学法得到系统的发展.如“中学数学教学法”就是从苏联伯拉基斯的《中学数学教学法》翻译而来,主要内容也只是介绍中学数学教学大纲的内容和体系,以及中学数学中的主要课题的教学法.

从国际范围来看,数学教育学科的形成、理论体系的建立时间也不长.在相当长一段时间内,数学教育主要是由数学家在从事数学研究的同时兼教数学,并没形成专职数学教师队伍.在社会经济、科学技术不发达的时代,能够有机会(需要)学习数学的人也只是少数,自然对数学教育(学)进行系统的研究就没有太多的需求.数学教师除了需要掌握数学还要懂得教学法才能胜任数学教学工作,这一点直到19世纪末才被人们充分认识到.“会数学不一定会教数学”“数学教师是有别于数学家的另一种职业”这样的观念开始逐渐被认同.最早提出把数学教育过程从教育过程中分离出来,作为一门独立的科学加以研究的是瑞士教育家别斯塔洛齐(J. H. Pestalozzi). 1911 年,哥廷根大学的鲁道夫 · 斯马克(Rudolf Schimmack)成为第一个数学教育的博士,其导师便是赫赫有名的德国数学家、数学教育学家菲利克斯 · 克莱因(Felix Klein). 数学家一直是数学教育与研究的中坚力量. 随着数学教育队伍的不断发展,教育学家、心理学家、哲学家、社会学家的不断融入,数学教育学术共同体不断走向了多元化,其中有些学者本身就是出自于数学界.

我国的数学教育系统深入的研究,总体上来讲起步则更晚. 1977 年恢复高考后,我国的教育开始走上了正规化的道路. 进入 21 世纪以后,随着我国经济的发展,教育进入了一个飞速发展的新时代.

(1) "数学教育学"的提出

随着 20 世纪以来对数学教育学科建设的探讨,人们逐渐认识到"数学教材教法"这一提法的局限性:相关研究主要集中在中小学数学内容如何教、教学大纲(课程标准)及教材如何编写等方面,而且以经验性的总结为主,从而提出了建立"数学教育学"学科的设想,在很大程度上赋予了这一领域更为广泛的学术内涵,并将其进一步细分为:数学教学论、数学课程论、数学教育心理学、数学教育哲学、数学教育测量与评价等相关研究领域,使得数学教育学科建设逐步走向深入.

(2) 数学教育学术共同体的形成

数学教育内涵的明晰与发展,伴随数学教育学术共同体的形成. 一方面,一批长期致力于数学与数学教育研究的专家学者,对我国数学教育研究领域的问题进行了深入的思考与研究,取得了丰硕成果,引领着我国数学教育的研究与实践. 另一方面,随着数学教育研究生培养体系的形成与完善,数学教育方向博士、硕士毕业生成为数学教育研究队伍中新生力量的主体. 更为重要的是,随着近年数学课程改革的不断深入,广大的一线教师成为新课程理念与实践的探索者、研究者,在数学课程改革中发挥了重要的作用. 一批长期致力于数学与数学教育的专家学者,以及广大的一线教师、教研员,形成了老中青数学教育工作者多维度梯队,为我国数学教育理论体系的建设作出了重要贡献.

(3) 国际数学教育学术交流与合作研究

随着我国改革开放的推进与社会经济发展,数学教育国际合作交流活动日渐频繁,逐步走向深层次、平等对话交流与合作研究. 20 世纪八九十年代,数学教育国际合作交流的形式主要是邀请国外专家来华访问、做学术报告,中国的研究者向国外学者请教、学习. 这对我国的数学教育研究走向国际起到了非常重要的作用. 这一阶段的主要特点是介绍国外先进的教育理论、数学教育理论,经常提到的话题是"与国际接轨". 进入 21 世纪,国内学者出国访问、参加学术会议、博士研究生联合培养,以及国外博士生毕业回国工作等人数爆发式增长. 通过参加国际学术交流,反思我国的数学教育研究,我国学者的数学教育研究水平得到了极大的提高. 这一时期我国的数学教育界经常提到的话题则是"要让中国的数学教育走向世界". 近年来,数学教育国际合作交流进入了的新发展时期. 人们逐渐认识到,听讲座报告、参加学术会议,已经不能满足我国数学教育发展的需求. 我国学者通过上述交流平台,与国外学者开展合作项目研究,针对中国以及国际数学教育共同关注的问题,形成中国特色数学教育理论. 通过举办、承办重大学术会议(如第 14 届国际数学教育大会)等让国际数学教育界更好地了解中国,在国际数学教育舞台上开展平等的对话交流、合作研究. 这一时期常常提到的是"在国际数学教育舞台上发出中国的声音". 我国数学教育国际化程度的不断提升,在很大程度上促进和提升了我国数学教育研究的水平.

(4) 数学教育研究成果的不断丰富

近年来,随着数学教育研究水平的提升、数学教育研究方法的不断完善,我国数学教育的成果不断丰富. 数学教育研究不仅在国内的学科教育研究领域独领风骚,而且在国际上的影响力不断提升. 这里特别需要提及的是,进入本世纪,数学教育方向的博士研究生的学位论文以及他们后续的相关研究,在某种程度上对整体拉升数学教育研究的水平起到了关键性作用,而《数学教育学报》则为此提供了最主要的阵地. 不言自明的是,我国数学教育博士点开创者,对中国的数学教育理论与实践逐步走向世界舞台,起到了关键的决定性的作用. 我们需要很好地学习、总结他们的研究成果.

华东师范大学出版社计划出版"当代中国数学教育名家文选"(丛书),开放式地逐步邀请对数学教育有系统深入研究的资深数学家、数学教育家,将他们的研究成果汇集在一起,供大家学习、研究. 本套丛书策划编辑刘祖希副编审代表出版社约请我担任丛书主编,虽然我一直有这样的朦胧念头,但未曾深入思考,我深感责任重大,担心不能很好完成这一历史使命. 然而,这一具有重大意义的工作机缘既然已到,就不应该推辞,必须责无旁贷全力去完成. 特别值得一提的是,正当丛书(第一批)即将正式出版之际,传来该丛书入选上海市重点图书出版项目的喜讯,这更增加了我们的信心和使命感.

当然,所收录的数学教育名家文选作者只是当代中国数学教育研究各领域的资深学者中的一部分,由于各种原因以及条件限制,并不是全部. 真诚欢迎数学教育同仁与我或刘祖希副编审联系,推荐(自荐)加入作者队伍.

北京师范大学特聘教授、博士生导师

义务教育数学课程标准修订组组长

2021 年春节完成初稿

"五一"劳动节定稿

前言

这本《文选》分为五个部分：

一、数学论文；

二、数学竞赛；

三、数学教育；

四、数学普及；

五、数学师友.

先说说数学论文.

1964 年，我毕业于扬州师范学院数学系，分配到南京四女中(现人民中学)任教至 1978 年. 1977 年中国科学技术大学常庚哲老师来南京对报名考研的人进行面试. 他问我：

“有没有什么工作？”

开始不明白这“工作”的含义，后来知道是问我有没有研究论文. 我未受过科研的训练，大学毕业后前两年专注于本职的教学工作，后来经历十年“文化大革命”，根本不可能从事科研.

1978 年 4 月到科大后，才开始做研究工作，其时已 35 岁.

常庚哲老师研究计算几何，我在这方向上完成一篇《平面 Bézier 曲线的凸性定理的一个证明》. 这是我第一篇数学论文(但发表时间不是最早的一篇).

我的兴趣主要在数论方面，其时科大数学系有冯克勤领导的代数数论讨论班(成员有杨劲根、张贤科等)与陆鸣皋领导的解析数论讨论班(成员有谢盛刚、杨照华等). 两个讨论班都欢迎我参加，代数数论要学的东西太多，难以写文章. 解析数论较易入门. 在陆鸣皋老师指导下，我与他合作完成了《A problem of Waring-Goldbach's type》. 接着又用圆法做变形的华林-歌德巴赫问题，陆老师做整数幂，将素数幂让给我做，我的结果《On a problem of the sums of powers of primes》即博士论文的主要部分. 正值王元先生(我们称他元老)来合肥指导，他见了陆老师与我的工作，表示很满意，后来又将我收入门下. 元老说每一个学生,他都要与之合作一篇文章,于是给了我一个题目,后来完成了,即《A conditional result on Goldbach problem》.

我做的论文中,最好的应是 Erdös 关于除数函数的猜测.

设 $d(n)=\sum_{k|n}1$ 为 n 的除数函数(正因数个数). 1952 年 Erdös 与 Mirsky 猜测 $d(n)=d(n+1)$ 有无穷多解? 这个问题直到 1984 年才为 Heath-Brown 解决. 但 Erdös 随即提出一个更难的猜测:"$\frac{d(n)}{d(n+1)}$ 的极限点应当在 $(0,+\infty)$ 上稠密."然而 Erdös 失望地说:"我们只知道 0 与∞为极限点."

不少人研究这个问题. 阚家海写过有关文章,并将这个问题告诉我. 我们合作写了一篇文章. 1995 年我在加拿大访问时,一天早晨,空气新鲜,心中澄澈,忽然觉得这个问题可以解决,于是又与阚家海合作写了有关文章. 数学家 A. J. Hildebrand 在综述文章《Erdös' Problems on Consective Integers》中对我们的工作作了评价,指出"... they give a full proof of the conjecture, and a far-reaching generalization of the Erdös-Mirsky conjecture, by showing that, for every positive rational number r, there are infinitely many n with $d(n)/d(n+1)=r$."

现在回顾我的科研生涯,成绩虽有一些,但只研究了一些问题,缺乏理论上的建树. 起步既晚,又未能全力以赴,在竞赛与普及上分心太多. 不过,我也尽了自己的努力,只能达到这样的高度.

数学竞赛,我是中学生时即很有兴趣. 但当时只有北京、上海组织过市一级的竞赛. "文化大革命"末期,我在中学任教,已经由朋友处见到一些国外的赛题,并做了一些题(如美国数学奥林匹克).

1978 年,我在中国科学技术大学时,与李克正、杜锡录合作,出了一本《1~20 届国际数学奥林匹克题解》,可能是国内第一本比较全面介绍 IMO(国际数学奥林匹克)的书,其中解答都是我们自己做的(因为未见到标准答案).

后来,数学奥林匹克在中国日益兴旺,我也在其中参加了一些工作,写了很多文章. 葛军等朋友帮我搜集了一部分,收在这本书里.

奥数(奥林匹克数学)这个词也随之传播开去. 1987 年,我在《曲阜师范大学学报》上发表了一篇文章《数学奥林匹克与奥林匹克数学》,或许奥林匹克数学这个词最早就出现在这篇文章中.

数学教育,又分为四个方面:(1) 我对于数学教育的一些看法;(2) 一些书的序与前言;(3) 一些评论;(4) 高考题的评析与解题漫谈.

数学普及重要.

我对数学普及饶有兴趣.因为当了一辈子教师,而且“好为人师”,所以写了不少普及的文章,这里选了一部分.

因为平时不注意搜集、整理,很多文章散失了,只好付之阙如.有的文章已经写到别的书里,本书不再重复.有的文章当时尚有新义,现在已成老生常谈,也不收集.

普及不容易,华罗庚先生说要“深入浅出”.

在深入、浅出这两方面,我均做得不够,希望有人比我做得更好.

若有来世,我倒希望仍做这项工作.

数学师友这部分,我写了几首打油诗.我爱“打油”,写了不少,大多与数学无关.或许有一天能出一本打油诗的集子.

2021年2月

目录

第一章　　数学论文

On a Problem of the Sums of Powers of Primes

中国科学技术大学学报,1981(4)

一些与置换群有关的不等式　　中国科学技术大学学报,1981(3)

一类不等式　　数学的实践与认识,1981(3)

一个华林-哥德巴赫型问题(与陆鸣皋合作)　中国科学技术大学学报,1982(S1)

一个 Waring-Гольдбах 型问题(与陆鸣皋合作)　　科学通报,1982(2)

A Problem of Waring-Goldbach's Type(与陆鸣皋合作)

科学通报(英文版),1982(3)

一个整数幂和问题　　中国科学技术大学学报,1982(2)

表自然数 n 为 $p_1^2+p_2^4+\cdots+p_s^{2^s}+p_{s+1}^{2^s}+p_{s+2}^k$

扬州师院学报(自然科学版),1982(2)

A Simple Proof for a Theorem of Kelisky and Rivlin (与常庚哲合作)

数学研究与评论,1983(1)

平面 Bézier 曲线的凸性定理的一个证明　　计算数学,1983(3)

初等数论中的一个猜测　　数学进展,1983(4)

$\sum e(\alpha p^k)$ 的估值　　扬州师院学报(自然科学版),1983(2)

On the Estimate for $\sum e(\alpha p^k)$　　中国科学技术大学学报,1983(S1)

关于 Gupta 定理的注记　　扬州师院学报(自然科学版),1984(2)

On Composite n for which $\varphi(n)|n-1$　　中国科学技术大学学报,1985(1)

A Conditional Result on Goldbach Problem(与王元合作)

数学学报(英文版),1985(1)

Шнирельман 常数的估计　　中国科学技术大学学报,1985(S2)

Hilbert 不等式　　科学通报,1984(1)

Hilbert 不等式　　中国科学技术大学学报,1985(S2)

初等数论中的一个猜测(Ⅱ)　　绍兴文理学院学报,1986(2)

On the Diophantine Equation $\sum_{i=0}^{k} \frac{1}{x_i} = \frac{a}{n}$ 数学年刊 B 辑(英文版),1986(2)

一个丢番图不等式 数学学报,1987(5)

A Note on Irrationality of Some Numbers

Journal of Number Theory, 1987(2)

An Application of Number Theory to Tournaments (与 Edward T. H. Wang 合作) *Graph Theory*, *Combinatorics*, *Algorithms and Applications* (published by SIAM), 1991, Chapter 54.

含有全部 k 元排列的短数列 中国科学技术大学学报,1989(4)

自然数集的(a, b, k)型可加划分(与朱平天合作)

四川大学学报(自然科学版),1989(89)

On (a, b, k)-Partitions of Positive Integers(与朱平天合作)

Southeast Asia Bulletin of Mathematics, 1993(1)

A Diophantine Inequality (Ⅱ) (与 Edward T. H. Wang 合作)

数学年刊 B 辑(英文版),1991(3)

Z_n 中的 D. F 集 南京师范大学学报(自然科学版),1991(2)

Mutual Multiples in Z_n(与 Edward T. H. Wang 合作)

Mathematics Magazine, 1999(72)

若干数论问题的注记 南京师范大学学报(自然科学版),1991(4)

关于哥德巴赫猜想、孪生素数猜想和余新河猜想的若干新结果(与阚家海合作) 南京邮电学院学报,1995(3)

哥德巴赫问题的一个推广(与阚家海合作) 中国科学技术大学学报,1997(2)

On the Divisor Function d(n) (与阚家海合作) *Mathematika*, 1996(43)

On the Divisor Function d(n) (Ⅱ) (与阚家海合作) *Mathematika*, 1999(46)

组合数论中的一个猜测(与李嘉昆合作)

扬州大学学报(自然科学版),1999(3)

A Simple Proof of a Curious Congruence by Sun (与 Edward T. H. Wang 合作) *Proceedings of the American Mathematical Society*, 1999(5)

模一个理想的 d 阶元的个数(与李嘉昆、张新华合作)

南京师范大学学报(自然科学版),2000(1)

A Remarkable Class of Congruences (与 P. Bundschuh,纪春岗合作)

Acta Sci. Math. (*Szeged*), 2001(67)

On a Problem of the Sums of Powers of Primes*

1 Introduction

In 1952, K. F. Roth[3] showed that every sufficiently large integer n might be represented in the form

$$n=\sum_{i=1}^{50} x_i^{i+1},$$

where $x_i's$ are natural numbers.

This result was improved upon by subsequent papers. In 1969, R. C. Vaughan[6] proved that every large odd integer n might be represented in the form

$$n=\sum_{i=1}^{31} p_i^{i+1},$$

where $p_i's$ are primes.

In 1971, Vaughan[7] improved upon his own result by showing that every large even number n might be represented in the form

$$n=\sum_{i=1}^{30} p_i^{i+1},$$

where $p_i's$ are primes. This statement depends mainly on the following theorem:

Let $k_1 \geqslant k_2 > k_3 > \cdots > k_s$ be positive integers, then the exponent density of the set $\{x_1^{k_1}+x_2^{k_2}+x_3^{k_3}+\cdots+x_s^{k_s}\}$ is

$$v \geqslant \frac{1}{k_1}+\frac{1}{k_2}+(1+\delta)\left(1-\frac{1}{k_2}\right)\left\{1-\prod_{j=3}^{s}\left(1-\frac{1}{k_j}\right)\right\},$$

* This paper was first submitted to 《*Acta Mathematica Sinica*》 in April 24, 1981.

where δ satisfies

$$\frac{1}{k_1}+\left(1-\frac{1}{k_2}\right)\delta\leqslant\frac{1}{k_3}\left(1-\frac{1}{k_2}\right)(1+\delta),$$

and

$$\delta\leqslant\frac{1}{k_2-1}.$$

In this paper, we improve upon Vaughan's result and obtain the following theorems:

Theorem 1. Let $k_1\geqslant k_2>k_3>\cdots>k_s$ be positive integers, then the exponent two density of the set $\{x_1^{k_1}+x_2^{k_2}+x_3^{k_3}+\cdots+x_s^{k_s}\}$ is

$$v\geqslant\frac{\theta_1}{k_1}+\frac{\theta_2}{k_2}+\cdots+\frac{\theta_s}{k_s},$$

where

$$\theta_1=\theta_2=1,$$

$$\theta_{i+1}=\begin{cases}\min\left(\theta_i,\ \left(1-\dfrac{2}{k_i}\right)\theta_i\Big/\left(1-\dfrac{1}{k_i+1}\right)\right), & \text{if } \theta_i\neq\theta_{i-1},\\ \min\left(\theta_i,\ \dfrac{\left(1-\dfrac{1}{k_i}-\dfrac{1}{k_{i-1}}-\cdots-\dfrac{1}{k_{i-j+1}}-\dfrac{2}{k_{i-j}}\right)\theta_i}{1-\dfrac{1}{k_{i+1}}}\right), & \\ \qquad \text{if } \theta_i=\theta_{i-1}=\cdots=\theta_{i-j}<\theta_{i-j-1}\,(1\leqslant j<i-2), & \\ \min\left(\theta_i,\ \dfrac{\left(1-\dfrac{1}{k_i}-\dfrac{1}{k_{i-1}}-\cdots-\dfrac{1}{k_2}-\dfrac{1}{k_1}\right)\theta_i}{1-\dfrac{1}{k_{i+1}}}\right), & \\ \text{if } \theta_i=\theta_{i-1}=\cdots=\theta_2\,(i=2,\ 3,\ \cdots,\ s-1). & \end{cases}$$

Theorem 1 is better than Vaughan's theorem. By Theorem 1, we find that the exponent densities of the sets $\{x_4^5+x_6^7+z_8^9+x_{21}^{22}+x_{22}^{23}\}$ and $\left\{x_5^6+x_7^8+\sum_{i=9}^{20}x_i^{i+1}\right\}$, where $x_i's$ are natural numbers, are greater than 0.500027448 and 0.724988662 respectively, while using Vaughan's theorem, we only obtain 0.479408997 and 0.72113922.

Theorem 2. Every large odd number n can be represented in the form

$$n=\sum_{i=1}^{23} p_i^{i+1},$$

where p_i are primes.

I am deeply indebted to Mr. Lu Ming-gao for his constant encouragement and kind advice.

2 Proof of Theorem 1

Let x_i, y_i be positive integers, p_i primes and θ_i the same as in § 1 ($i=1, 2, \cdots, s$).

Lemma 1. The number of the solutions of the equation

$$x_1^{k_1}+x_2^{k_2}+\cdots+x_s^{k_s}=y_1^{k_1}+y_2^{k_2}+\cdots+y_s^{k_s} \tag{1}$$

satisfying

$$\frac{1}{2}N^{\frac{\theta_i}{k_i}}\leqslant x_t,\ y_t\leqslant N^{\frac{\theta_i}{k_i}}\quad (i=1, 2, \cdots, s) \tag{2}$$

is

$$M\ll N^{\frac{\theta_1}{k_1}+\frac{\theta_2}{k_2}+\cdots+\frac{\theta_s}{k_s}+\varepsilon},$$

where ε is any small positive number.

Proof. We use induction on s. The cases $s=1$ and 2 are obvious. Suppose the conclusion is ture with $s-1$. Then there are two cases for x_s and y_s.

(1) $x_s=y_s$.

From the hypothesis, the number of the solutions of Equation (1) satisfying (2) is

$$M_1\ll N^{\frac{\theta_1}{k_1}+\frac{\theta_2}{k_2}+\cdots+\frac{\theta_{s-1}}{k_{s-1}}+\varepsilon+\frac{\theta_s}{k_s}}.$$

(2) $x_s \neq y_s$

First, we prove

$$\theta_i\left(1-\frac{1}{k_i}\right)\leqslant \theta_{i+1}, \quad (i=2, 3, \cdots, s-1). \tag{3}$$

There are four cases:

(a) $\theta_i=\theta_{i+1}$. Obviously

$$\theta_{i+1}=\theta_i>\theta_t\left(1-\frac{1}{k_i}\right).$$

(b) $\theta_{i+1}<\theta_i<\theta_{i-1}$. Since $k_{i+1}<k_i$,

$$\theta_{i+1}=\frac{\left(1-\frac{2}{k_i}\right)\theta_i}{1-\frac{1}{k_{i+1}}}\geqslant\left(1-\frac{1}{k_i}\right)\theta_i.$$

(c) $\theta_{i+1}<\theta_i=\theta_{i-1}=\cdots=\theta_{i-j}<\theta_{i-j-1} \quad (1\leqslant j<i-2)$.

In this case,

$$\theta_i\leqslant\frac{\left(1-\frac{1}{k_{i-1}}-\frac{1}{k_{i-2}}-\cdots-\frac{1}{k_{i-j+1}}-\frac{2}{k_{i-1}}\right)\theta_{i-1}}{1-\frac{1}{k_i}}$$

Hence,

$$1-\frac{1}{k_{i-1}}-\frac{1}{k_{i-2}}-\cdots-\frac{1}{k_{i-j+1}}-\frac{2}{k_{i-j}}\geqslant 1-\frac{1}{k_i},$$

$$\theta_{i+1}=\frac{\left(1-\frac{1}{k_i}-\frac{1}{k_{i-1}}-\cdots-\frac{1}{k_{i-j+1}}-\frac{2}{k_{i-j}}\right)\theta_i}{1-\frac{1}{k_{i+1}}}\geqslant\frac{\left(1-\frac{2}{k_i}\right)\theta_i}{1-\frac{1}{k_{i+1}}}\geqslant\left(1-\frac{1}{k_i}\right)\theta_i.$$

(d) $\theta_{i+1}<\theta_i=\theta_{i-1}=\cdots=\theta_2$. In this case,

$$\theta_i\leqslant\frac{\left(1-\frac{1}{k_{i-1}}-\cdots-\frac{1}{k_2}-\frac{1}{k_1}\right)\theta_{i-1}}{1-\frac{1}{k_i}}.$$

Hence,

$$1-\frac{1}{k_{i-1}}-\cdots-\frac{1}{k_2}-\frac{1}{k_1}\geqslant 1-\frac{1}{k_i},$$

$$\theta_{i+1}=\frac{\left(1-\frac{1}{k_i}-\frac{1}{k_{i-1}}-\cdots-\frac{1}{k_2}-\frac{1}{k_1}\right)\theta_i}{1-\frac{1}{k_{i+1}}}\geqslant\frac{\left(1-\frac{2}{k_i}\right)\theta_i}{1-\frac{1}{k_{i+1}}}\geqslant\left(1-\frac{1}{k_i}\right)\theta_i.$$

Since $\theta_1\geqslant\theta_2\geqslant\cdots\geqslant\theta_s$, we have

$$x_1^{k_1}+x_2^{k_2}+\cdots+x_{s-1}^{k_{s-1}}-y_1^{k_1}=y_2^{k_2}+y_3^{k_3}+\cdots+y_s^{k_s}-x_s^{k_s}=y_2^{k_2}+O(N^{\theta_3}).\tag{4}$$

The number of the possible selections of x_1, x_2, $\cdots$ x_{s-1} and y_1 is

$$O(N^{\frac{\theta_1}{k_1}+\frac{\theta_2}{k_2}+\cdots+\frac{\theta_{s-1}}{k_{s-1}}+\frac{\theta_1}{k_1}}).$$

Since

$$(y_2+1)^{k_2}-y_2^{k_2}\gg N^{\left(1-\frac{1}{k_2}\right)\theta_2},$$

if the value of the left side of (4) is fixed, say A, then the number of the possible selections of y_2 is $O(N^{\theta_3-\left(1-\frac{1}{k_2}\right)\theta_2})$.

Similarly, since

$$A-y_2^{k_2}=y_3^{k_3}+O(N^{\theta_4})$$

and

$$(y_3+1)^{k_3}-y_3^{k_3}\gg N^{\left(1-\frac{1}{k_j}\right)\theta_3},$$

if x_1, x_2, $\cdots$, x_{s-1}, y_1 and y_2 are fixed, then the number of the possible selections of y_3 is $O(N^{\theta_4-\left(1-\frac{1}{k_3}\right)\theta_3})$, etc. Finally, since

$$x_1^{k_1}+\cdots+x_{s-1}^{k_{s-1}}-y_1^{k_1}-\cdots-y_{s-1}^{k_{s-1}}=y_s^{k_s}-x_s^{k_s}\neq 0,$$

if x_1, $\cdots$, x_{s-1}, y_1, $\cdots$, y_{s-1} are fixed, then the number of the possible selections of x_s and y_s is $O(N^2)$.

Hence in this case, the number of the solutions of Equation (1) satisfying (2) is

$$M_2 \ll N^{\frac{\theta_1}{k_1}+\cdots+\frac{\theta_{s-1}}{k_{s-1}}+\frac{\theta_1}{k_1}+\theta_3-\left(1-\frac{1}{k_2}\right)\theta_2+\cdots+\theta_s-\left(1-\frac{1}{k_{s-1}}\right)\theta_{s-1+\varepsilon}}$$

Suppose $\theta_2=\theta_3=\cdots=\theta_{i_1}>\theta_{i_1+1}=\cdots=\theta_{i_2}>\theta_{i_2+1}=\cdots$

$$=\theta_{i_n}>\cdots>\theta_{i_n+1}=\cdots=\theta_{s-1},\quad (2\leqslant i_1<i_2<\cdots<i_h<s-1).$$

From the definition of θ_i, we have

$$\frac{\theta_1}{k_1}+\theta_3-\left(1-\frac{1}{k_2}\right)\theta_2+\cdots+\theta_{i_1+1}-\left(1-\frac{1}{k_{i_1}}\right)\theta_{i_1}=\theta_{i_1+1}$$

$$-\left(1-\frac{1}{k_{i_1}}-\cdots-\frac{1}{k_2}-\frac{1}{k_1}\right)\theta_{i_1}\leqslant\frac{\theta_{i_1+1}}{k_{i_1+1}},$$

$$\frac{\theta_{i_1+1}}{k_{i_1+1}}+\theta_{i_1+2}-\left(1-\frac{1}{k_{i_1+1}}\right)\theta_{i_1+1}+\cdots+\theta_{i_2+1}-\left(1-\frac{1}{k_{i_2}}\right)\theta_{i_2}=\theta_{i_2+1}$$

$$-\left(1-\frac{1}{k_{i_2}}-\cdots-\frac{1}{k_{i_1+2}}-\frac{2}{k_{i_1+1}}\right)\theta_{i_2}\leqslant\frac{\theta_{i_2+1}}{k_{i_2+1}},$$

……

$$\frac{\theta_{i_h+1}}{k_{i_h+1}}+\theta_{i_h+2}-\left(1-\frac{1}{k_{i_h+1}}\right)\theta_{i_h+1}+\cdots+\theta_s$$

$$-\left(1-\frac{1}{k_{s-1}}\right)\theta_{s-1}=\theta_s-\left(1-\frac{1}{k_{s-1}}-\cdots-\frac{1}{k_{i_h+2}}-\frac{2}{k_{i_h+1}}\right)\theta_{s-1}\leqslant\frac{\theta_s}{k_s}.$$

Hence

$$M_2\ll N^{\frac{\theta_1}{k_1}+\cdots+\frac{\theta_s}{k_s}+\varepsilon}$$

The proof of Lemma 1 is complete.

Now, we are in a position to prove Theorem 1. Let R (n) be the number of the solutions of the equation

$$x_1^{k_1}+x_2^{k_2}+\cdots+x_s^{k_s}=n\leqslant sN$$

satisfying the condition

$$\frac{1}{2}N^{\frac{\theta_i}{k_i}} \leqslant x_i \leqslant N^{\frac{\theta_i}{k_i}} \quad (i=1, 2, \cdots, s).$$

From Schwarz's inequality,

$$\sum_{R(n)>0} 1 \geqslant \left(\sum R(n)\right)^2 / \sum R^2(n).$$

By Lemma 1,

$$\sum R^2(n) \ll N^{\frac{\theta_1}{k_1}+\cdots+\frac{\theta_s}{k_s}+\varepsilon}$$

and obviously,

$$\sum R(n) \gg N^{\frac{\theta_1}{k_1}+\cdots+\frac{\theta_s}{k_s}},$$

Hence the result is obtained.

Corollary 1. The exponent density of the set $\{p_1^{k_1}+p_2^{k_2}+\cdots+p_s^{k_s}\}$ is

$$v \geqslant \frac{\theta_1}{k_1}+\frac{\theta_2}{k_2}+\cdots+\frac{\theta_s}{k_s}.$$

3 Notation

Let h, q, x, y, k, m, n (with or without suffix) be natural numbers and $h<q$, $(h, q)=1$.

c, c_1, c_2, $\cdots$ be positive constants.

p, p_1, p_2, $\cdots$ be primes.

N is a sufficiently large positive number. $L=\log N$.

ε is any small positive number. α is a fixed sufficiently small positive number.

$r=\frac{1}{4}-\alpha$.

$\theta \in [0, 1]$, $\beta=\theta-\frac{h}{q}$.

$$S_{h,q}^{(k)}=\sum_{x=1}^{q}e_q(hx^k),\ W_{h,q}^{(k)}=\sum_{\substack{x=1\\(x,q)=1}}^{q}e_q(hx^k).$$

$$g_k=\frac{1}{k}\sum_{\frac{1}{2^k}N\leqslant n\leqslant N}n^{\frac{1}{k}-1}e(n\beta),\quad (k=3,4).$$

$$\theta_k=1\ (k=2,3,4,9,22,23,24),\ \frac{\theta_7}{7}=\frac{1}{9}\times\frac{7}{6},\ \frac{\theta_5}{5}=\frac{1}{7}\times\frac{5}{4}.$$

$$\mu_1=\frac{1}{23}+\frac{1}{22}+\frac{1}{9}+\frac{\theta_7}{7}+\frac{\theta_5}{5}=0.500027488,\ \mu=\mu_1+\frac{1}{2}+\frac{1}{3}+\frac{1}{4}+\frac{1}{24}.$$

$$f_k=\sum_{\frac{1}{2}N^{\frac{\theta_k}{k}}\leqslant x\leqslant N^{\frac{\theta_k}{k}}}e(x^k\theta),\ k=2,3,4,5,7,9,22,23,24.$$

$$f_k^*=\frac{1}{q}S_{h,q}^{(k)}g_k(k=3,4).$$

$$V_k=\sum_{\frac{1}{2}N^{\frac{\theta_k}{k}}\leqslant r\leqslant N^{\frac{\theta_k}{k}}}e(p^k\theta),\ (k=2,3,4,5,7,9,22,23,24).$$

$$V_k^*=\frac{1}{\phi(q)}W_{h,q}^{(k)}\sum_{\frac{1}{2^k}N^{\theta_k}\leqslant n\leqslant N^{\theta_k}}n^{\frac{1}{k}-1}(\log n)^{-1}e(n\beta),\ (k=2,3,4,5,7,9,22,$$

23, 24).

$$J=\left\{p_5^6+p_7^8+\sum_{i=9}^{20}p_i^{i+1}\right\}.$$

$$Q(\theta)=\sum_{\substack{i\leqslant\frac{1}{2^\delta}N^{\frac{\delta}{4}}\\ j\in J}}e(j\theta),$$

where $\delta=3+\dfrac{3(1-v)}{3+v}$ and v is the exponent density of the set J.

$$A_q(m)=\{\phi(q)\}^{-1}\sum_{n}W_{p,q}^{(2)}W_{h,q}^{(3)}W_{p,q}^{(4)}W_{h,q}^{(5)}W_{h,q}^{(7)}W_{h,q}^{(9)}W_{h,q}^{(22)}W_{h,q}^{(23)}W_{h,q}^{(24)}e_q(-\mathrm{m}h).$$

$$\mathscr{S}(m)=\sum_{q=1}^{\infty}A_q(m).$$

$$\psi(m)=\sum_{\substack{\frac{1}{2^k}N^{\theta_k}\leqslant n_k\leqslant N^{\theta_k}\\ k=2,3,4,5,7,9,22,23,24\\ n_2+n_3+n_4+n_5+n_7+n_9+n_{21}+n_{23}+n_{24}=m}}\left(\prod_k n_k^{1-\frac{1}{k}}\log n_k\right)^{-1}.$$

$$V=V_2V_3V_4V_5V_7V_9V_{22}V_{23}V_{24}.$$

$$V^*=V_2^*V_3^*V_4^*V_5^*V_7^*V_9^*V_{22}^*V_{23}^*V_{24}^*.$$

$$r(N)=\int_0^1 VQ(\theta)e(-N\theta)\mathrm{d}\theta.$$

We divide the interval $\left[-\frac{1}{\tau},\ 1-\frac{1}{\tau}\right]$ into Farey arcs with $q\leqslant\tau=NL^{-\sigma_\theta}$. Let $\mathcal{M}_{h,q}$ be the are $\left[\frac{h}{q}-\frac{1}{q\tau},\ \frac{h}{q}+\frac{1}{q\tau}\right]$ with $q\leqslant L^{\sigma_\theta}$, $\mathcal{M}=\bigcup\limits_{h,q}\mathcal{M}_{h,q}$, $E=\left[-\frac{1}{\tau},\ 1-\frac{1}{\tau}\right]-\mathcal{M}$.

We also divide the interval $\left[-\frac{1}{\tau_1},\ 1-\frac{1}{\tau_1}\right]$ into Farey arcs with $q\leqslant\tau_1=N^{1-r}$, Let $\mathcal{M}'_{h,q}$ be the are $\left[\frac{h}{q}-\frac{1}{q\tau_1},\ \frac{h}{q}+\frac{1}{q\tau_1}\right]$ with $q\leqslant N^r$, $E'=\left[-\frac{1}{\tau_1},\ 1-\frac{1}{\tau_1}\right]-\bigcup\limits_{h,q}\mathcal{M}'_{h,q}$

4 Some Lemmas

The following lemmas are well known (e. g. see [1] and [4]).

Lemma 2. If $|\beta|\leqslant\frac{1}{2}$, then

$$g_k\ll\min(N^{\frac{1}{k}},\ N^{\frac{1}{k}-1}|\beta|^{-1}),$$

Lemma 3. For a given $\sigma>0$, there exists $\sigma_0>0$, such that in E

$$V_{24}=O(N^{\frac{1}{24}}L^{-\sigma}).$$

Lemma 4. In $\mathcal{M}$,

$$V_k-V_k^*=O(N^{\frac{\theta_k}{k}}e^{-c\sqrt{L}}).$$

Lemma 5. If $|\beta|\leqslant\frac{1}{2}$, then

$$V_k^*=O(q^{-\frac{1}{2}+\varepsilon}\min(N^{\frac{\theta_k}{k}}L^{-1},\ |\beta|^{-\frac{1}{k}})).$$

Lemma 6. If $\frac{1}{2}N \leqslant m \leqslant N$, then

$$c_1 L^{-9} N^{\mu-1} \leqslant \psi(m) \leqslant c_2 L^{-9} N^{\mu-1}.$$

5 Proof of Theorem 2

Lemma 7. The exponent density of the set J is

$$v \geqslant 0.724988662,$$

and

$$\frac{\delta v}{4} \geqslant 0.583885263 > \frac{7}{12} + 2\alpha.$$

Proof. It follows from Corollary 1.

Lemma 8. $\int_{E'} | f_3 f_4 Q_{(\theta)} |^2 \mathrm{d}\theta \ll N^{\frac{2}{3}+\frac{2}{4}-1} Q^2(0)$.

Proof. From Lemma 8.2 of Vaughan[7], for $\theta \in E'$,

$$f_3 \ll N^{\frac{1}{3}-\frac{1}{12}+\frac{\alpha}{3}}$$

By Davenport's lemma (Lemma 9.3 of Hua[1]),

$$\int_0^1 | f_4 Q |^2 \mathrm{d}\theta \ll N^{\frac{1}{4}+\varepsilon} Q(0) \ll N^{\frac{2}{4}-\frac{1}{4}-\frac{\delta v}{4}+\varepsilon} Q^2(0).$$

Hence

$$\int_{E'} | f_3 f_4 Q(\theta) |^2 \mathrm{d}\theta \ll N^{\frac{2}{3}-\frac{2}{12}+\frac{2\alpha}{3}} \int_0^1 | f_4 Q |^2 \mathrm{d}\theta \ll$$

$$N^{\frac{2}{3}+\frac{2}{4}-\frac{1}{4}-\frac{\delta v}{4}-\frac{1}{6}+\frac{2\alpha}{3}+\varepsilon} Q^2(0) \ll N^{\frac{2}{3}+\frac{2}{4}-1} Q^2(0)).$$

Let $S(\theta) = \begin{cases} f_3^* f_4^*, & if\ \theta \in \mathcal{M}'_{h,q}, \\ 0, & if\ \theta \overline{\in} \mathcal{M}'_{h,q}. \end{cases}$

$$T(\theta) = \begin{cases} f_3 f_4 - f_3^* f_4^*, & if\ \theta \in \mathcal{M}'_{h,q}, \\ 0, & if\ \theta \overline{\in} \mathcal{M}'_{h,q}. \end{cases}$$

Lemma 9. $\int_0^1 |S(\theta)|^2 d\theta \ll N^{\frac{2}{3}+\frac{2}{4}-1} L^{c_3}$.

Proof. $\int_0^1 |S(\theta)|^2 d\theta \ll \sum_{q \leqslant N^r} B(q) \int_{-\frac{1}{2}}^{\frac{1}{2}} |g_3 g_4|^2 d\beta$, where $B(q) = \sum_h q^{-4} |S_{h,q}^{(3)} S_{h,q}^{(4)}|^2$.

From the proof of Lemma 10.4 of Vaughan[7],

$$\sum_{q \leqslant N^r} B(q) \ll L^{c_3},$$

and from Lemma 2,

$$\int_{-\frac{1}{2}}^{\frac{1}{2}} |g_3 g_4|^2 d\beta \ll N^{\frac{2}{3}+\frac{2}{4}-1},$$

the conclusion is obtained.

Lemma 10. $\int_0^1 |T(\theta) Q(\theta)|^2 d\theta \ll N^{\frac{2}{3}+\frac{2}{4}-1} Q^2(0)$.

Proof. For $\theta \in \mathscr{M}'_{h,q}$, $f_k - f_k^* \ll q^{\frac{1}{2}+\varepsilon}$ (see [2]),

$$|T(\theta)| \leqslant |f_3 - f_3^*| |f_4^*| + |f_3| |f_4 - f_4^*| \ll$$

$$N^{\frac{1}{3}} q^{\frac{1}{2}-\frac{1}{3}+\varepsilon} + N^{\frac{1}{4}} q^{\frac{1}{2}-\frac{1}{4}+\varepsilon} \ll N^{\frac{1}{3}+\frac{1}{6}\times\frac{1}{4}} = N^{\frac{3}{8}}$$

Hence

$$\int_0^1 |T(\theta) Q(\theta)|^2 d\theta \ll N^{\frac{3}{4}} \int_0^1 |Q_{(\theta)}|^2 d\theta \ll N^{\frac{3}{4}} Q(0) \ll$$

$$N^{\frac{3}{4}-\frac{\delta v}{4}+\varepsilon} Q^2(0) \ll N^{\frac{2}{3}+\frac{2}{4}-1} Q^2(0).$$

Lemma 11. $\int_0^1 |f_3 f_4 Q_{(\theta)}|^2 d\theta \ll N^{\frac{2}{3}+\frac{2}{4}-1} Q^2(0) L^{c_3}$.

Proof. $\int_0^1 |f_3 f_4 Q_{(\theta)}|^2 d\theta = \left(\int_{E'} + \sum_{q \leqslant N^r} \sum_h \int_{M'_{h,q}}\right) |f_3 f_4 Q_{(\theta)}|^2 d\theta$

$$\ll \int_{E'} |f_3 f_4 Q_{(\theta)}|^2 d\theta + \int_0^1 |S_{(\theta)} Q_{(\theta)}|^2 d\theta + \int_0^1 |T_{(\theta)} Q_{(\theta)}|^2 d\theta.$$

The result follows from Lemmas 8, 9 and 10.

Lemma 12. Let d (n) be the divisor function and $\frac{\theta_1}{k_1} + \frac{\theta_2}{k_2} \cdots + \frac{\theta_s}{k_s} \geqslant \frac{s}{k}$, $k=$

$\max\limits_{1\leqslant i\leqslant s} k_i$, then

$$\sum_{\substack{\frac{1}{2}N^{\frac{\theta_i}{k_i}}\leqslant x_i,\, y_i\leqslant N^{\frac{\theta^i}{K_i}} \\ i=1,2,\cdots,s \\ x_1^{k_1}+\cdots+x_s^{k_s}-y_1^{k_1}-\cdots-y_s^{k_s}\neq 0}} d(|x_1^{k_1}+\cdots+x_s^{k_s}-y_1^{k_1}-\cdots-y_s^{k_s}|)\ll N^{2\left(\frac{\theta_1}{k_1}+\cdots+\frac{\theta_s}{k_s}\right)}L^c.$$

Proof. This can be proved in the same way as Theorem 3 of Hua[1].

Lemma 13. $\int_0^1 |f_2f_5f_7f_9f_{22}f_{23}|^2\mathrm{d}\theta\ll N^{2\mu_1}L^{c_5}$.

Proof. We consider the number of the solutions of the equation

$$x_1^2+x_4^5+x_6^7+x_8^9+x_{21}^{22}+x_{22}^{23}=y_1^1+y_4^5+y_6^7+y_8^9+y_{21}^{22}+y_{22}^{23}$$

satisfying

$$\frac{1}{2}N^{\frac{\theta_k}{k}}\leqslant x_k,\ y_k\leqslant N^{\frac{\theta_k}{k}}\ (k=2,\ 5,\ 7,\ 9,\ 22,\ 23).$$

There are two cases:

(1) $x_1\neq y_1$. From Lemma 12 the number of the solutions is

$$M_1\ll N^{2\mu_1}L^{c_6}.$$

(2) $x_1=y_1$. From Lemma 1 and Corollary 1, the number of the solutions is

$$M_2\ll N^{\frac{1}{2}}N^{\mu_1-1\varepsilon}\ll N^{2\mu_1}.$$

Now we take $\sigma>c_3+c_s+9$ and $\sigma_0\left(\mu-1-\frac{\theta_7}{7}-\frac{\theta_s}{5}\right)>\sigma-c_3-c_5=\sigma_1>9$.

Lemma 14. $\int_E |V_{(\theta)}Q_{(\theta)}|\mathrm{d}\theta\ll N^{\mu-1}L^{-\sigma_1}Q(0)$.

Proof. From Lemma 3, for $\theta\in E$,

$$V_{24}=O(N^{\frac{1}{24}}L^{-\sigma}).$$

Hence

$$\int_E |V(\theta)Q(\theta)|\,\mathrm{d}\theta\ll N^{\frac{1}{24}}L^{-\sigma}\int_0^1 |V_2V_3V_4V_5V_7V_9V_{22}V_{23}Q(\theta)|\,\mathrm{d}\theta$$

$$\ll N^{\frac{1}{24}}L^{-\sigma}\left(\int_0^1 |V_2V_5V_7V_9V_{22}V_{23}|^2 d\theta\right)^{1/2}\left(\int_0^1 |V_3V_4Q(\theta)|^2 d\theta\right)^{1/2}$$

$$\ll N^{\frac{1}{24}}L^{-\sigma}\left(\int_0^1 |f_2f_5f_7f_9f_{22}f_{23}|^2 d\theta\right)^{1/2}\left(\int_0^1 |f_3f_4Q(\theta)|^2 d\theta\right)^{1/2}.$$

From Lemmas 11 and 13, we have

$$\int_E |V(\theta)Q(\theta)| d\theta \ll N^{\frac{1}{24}}L^{-\sigma}N^{\mu_1}L^{c_s}N^{\frac{1}{3}+\frac{1}{4}-\frac{1}{2}}L^{c_3}Q(0) \ll N^{\mu-1}L^{-\sigma_1}Q(0).$$

Lemma 15. $\int_M |(V-V^*)Q(\theta)| d\theta \ll N^{\mu-1}L^{-\sigma_1}Q(0)$.

Proof. From Lemmas 4 and 5,

$$\int_M = \sum_{q\leqslant L^{\sigma_\theta}}\sum_h \int_{m_{h,q}} |(V-V^*)Q_{(\theta)}| d\theta$$

$$= O\Big(\sum_{q\leqslant L^{\sigma_\theta}} \phi(c)\int_0^{q^{-1}N^{-1}L^{\sigma_0}} N^{\mu}e^{-c\sqrt{L}} d\beta \times Q(0)\Big)$$

$$= O(N^{\mu-1}L^{-\sigma_1}Q(0)).$$

Lemma 16. $\sum_{q\leqslant L^{\sigma_\theta}}\sum_h \left(\int_0^1 - \int_{M_{h,q}}\right) |V^*Q_{(\theta)}| d\theta \ll N^{\mu-1}L^{-\sigma_1}Q(0)$.

Proof. From Lemma 5,

$$\sum_{q\leqslant L^{\sigma_\theta}}\sum_h \left(\int_0^1 - \int_{M_{h,q}}\right) |V^*_{(\theta)}Q_{(\theta)}| d\theta$$

$$\ll \sum_{q\leqslant L^{\sigma_\theta}} \phi(q)\int_{q^{-1}N^{-1}L^{\sigma_0}}^{\infty} (q^{-\frac{1}{2}+\epsilon})^9 N^{\frac{\theta_7}{7}+\frac{\theta_5}{5}} |\beta|^{-\mu+\frac{\theta_7}{7}+\frac{\theta_5}{5}} d\beta \times Q(0)$$

$$\ll N^{\mu-1}L^{-\sigma_0\left(\mu-1-\frac{\theta_7}{7}-\frac{\theta_5}{5}\right)}Q(0) \ll N^{\mu-1}L^{-\sigma_1}Q(0).$$

Lemma 17. Let $j \leqslant \left(\frac{1}{2}N^{\frac{1}{4}}\right)^{\delta}$, $j \in J$ and N be an odd number, then

$$\mathscr{S}(N-j) \geqslant c_7 > 0.$$

Proof.
$$1 + A_p(N-j) = \frac{p}{(p-1)^9}M_p,$$

where M_p is number of solutions of the equation

$$x_1^2+x_2^3+x_3^4+x_4^5+x_6^7+x_8^9+x_{20}^{22}+x_{22}^{23}+x_{23}^{24}\equiv N-j\,(\mathrm{mod}\ \mathrm{p})$$
$$p\nmid x_1x_2x_3x_4x_6x_8x_{21}x_{22}x_{23}$$

From the Lemma 13 of Prachar[5], we have $M_p>0$ and

$$A_p(N-j)\ll\frac{1}{(p-1)^9}(p^{\frac{1}{2}+\varepsilon})^9\ll\frac{1}{p^2}.$$

Hence

$$\mathscr{S}(N-j)=\prod_p(1+A_p(N-j))>c_8\prod_{p>c_9}\left(1-\frac{c_9}{p^2}\right)\geqslant c_7>0.$$

Lemma 18. $r(N)=\sum\limits_{\substack{i\leqslant(\frac{1}{2}N^{\frac{1}{4}})^{\delta}\\ i\in J}}\mathscr{S}(N-j)\psi(N-j)+O(N^{\mu-1}L^{-\sigma_1}Q(0)).$

Proof. From Lemmas 14, 15 and 16,

$$\begin{aligned}r(N)&=\left(\int_E+\int_M\right)VQ(\theta)e(-N\theta)\mathrm{d}\theta\\&=\int_M V^*Q(\theta)e(-N\theta)\mathrm{d}\theta+O(N^{\mu-1}L^{-\sigma_1}Q(0))\\&=\sum_{q\leqslant L^{\sigma_\theta}}\sum_h\int_0^1V^*Q(\theta)e(-N\theta)\mathrm{d}\theta+O(N^{\mu-1}L^{-\sigma_1}Q(0)\\&=\sum_{\substack{i\leqslant(\frac{1}{2}N^{\frac{1}{4}})^{\delta}\\ i\in J}}\sum_{q\leqslant L^{\sigma_\theta}}A_q(N-j)\psi(N-j)+O(N^{\mu-1}L^{-\sigma_1}Q(0)).\end{aligned}$$

Since $\sigma_0>\sigma_1-9$ and

$$\sum_{q>L^{\sigma_\theta}}|A_q(N-j)|=O(\sum_{q>L^{\sigma_\theta}}\{\phi(q)\}^{-9}(q^{\frac{1}{2}+\varepsilon})^9q)=O(L^{-\sigma_0}),$$

it follows from Lemma 7,

$$r(N)=\sum_{\substack{i\leqslant(\frac{1}{2}N^{\frac{1}{4}})^{\delta}\\ i\in J}}\mathscr{S}(N-j)\psi(N-j)+O(N^{\mu-1}L^{-\sigma_1}Q(0)).$$

From Lemmas 6, 17 and 18, we have

$$r(N)\ll N^{\mu-1}L^{-9}Q(0)>0.$$

This proves Theorem 2.

References

[1] Hua, L. K., *Additive Primzahltheorie*, Leipzig, 1959.
[2] Hua, L. K., *Sci. Record* (*N. S.*) 1(1957), No. 3, 15-16.
[3] Roth, K. F., *Proc. London Math. Soc.*, 53(1951), No. 2, 381-395.
[4] Prachar, K., *Monatsh. Math.*, 57(1953), 66-74.
[5] Prachar, K., *Monatsh. Math.*, 57(1953), 113-116.
[6] Vaughan, R. C., Ph. D. Thesis, London, 1969.
[7] Vaughan, R. C., *Proc. London Math. Soc.*, 21(1970), 160-180.
[8] Vaughan, R. C., *J. London Math. Soc.*, 3(1971), 677-688.

附本文中文摘要：

华林问题是解析数论的一个重要问题. 1952 年，Roth 证明了每个充分大的整数 $n=\sum_{i=1}^{50} x_i^{i+1}$，其中 x_i 为非负整数，Vaughan 改进了 Roth 的结果，并进一步考虑了素数幂和的问题，于 1971 年证明每个充分大的正偶数 $n=\sum_{i=1}^{30} p_i^{i+1}$，其中 p_i 为素数.

本文对 Vaughan 的结果作了较重大改进，先用最优化的思想改进了计算指数密率的方法，即证明了下列

定理 1　设自然数 $k_1 \geqslant k_2 > k_3 > \cdots > k_s$，则集合 $\{x_1^{k_1}+x_2^{k_2}+\cdots+x_s^{k_s}\}$ 的指数密率

$$v \geqslant \frac{\theta_1}{k_1}+\frac{\theta_2}{k_2}+\frac{\theta_3}{k_3}+\cdots+\frac{\theta_s}{k_s},$$

其中，$\theta_1=\theta_2=1$，

$$\theta_{i+1}=\begin{cases}\min\left(\theta_i,\ \left(1-\dfrac{2}{k_i}\right)\theta_i\Big/\left(1-\dfrac{1}{k_{i+1}}\right)\right), 若\ \theta_i\neq\theta_{i-1};\\[2ex] \min\left(\theta_i,\ \dfrac{\left(1-\dfrac{1}{k_i}-\dfrac{1}{k_{i-1}}-\cdots-\dfrac{1}{k_{i-j+1}}-\dfrac{2}{k_{i-j}}\right)\theta_i}{1-\dfrac{1}{k_{i+1}}}\right),\\ \qquad 若\ \theta_i=\theta_{i-1}=\cdots=\theta_{i-j}<\theta_{i-j-1}\quad(1\leqslant j<i-2);\\[2ex] \min\left(\theta_i,\ \dfrac{\left(1-\dfrac{1}{k_i}-\dfrac{1}{k_{i-1}}-\cdots-\dfrac{1}{k_2}-\dfrac{1}{k_1}\right)\theta_i}{1-\dfrac{1}{k_{i+1}}}\right),\\ \qquad 若\ \theta_i=\theta_{i-1}=\cdots=\theta_2(i=2,\ 3,\ \cdots,\ s-1).\end{cases}$$

运用定理1,采取新的分组方法并利用 Davenport 引理、华罗庚对优弧部分的估计及堆垒素数论方面的一些结果,得到下列

定理2 每一个充分大的正奇数 $n=\sum\limits_{i=1}^{23}p_i^{i+1}$,其中 p_i 为素数.

一些与置换群有关的不等式

本文证明了下面的定理 1，并应用置换群给出 Karamata 不等式[1]、[2]、[3]，Muirhead 不等式的一种新的证明.

设 $x=(x_1, x_2, \cdots, x_n)$ 为 n 维空间中的点. G 为集合$\{1, 2, \cdots, n\}$上的 n 元置换群. G 的元素用 ρ、σ、τ 等表示，对 $\rho \in G$, $\rho x=(x_{\rho_1}, x_{\rho_2}, \cdots, x_{\rho_n})$，其中 $\rho_k=\rho(k)$. 记 x 的 G 轨道为 Gx, Gx 的凸包为 $H(Gx)$.

定理 1　设 ϕ_1、ϕ_2、…、ϕ_n 为 $\mathbf{R}\to\mathbf{R}$ 的连续、凸函数，如果

$$y=(y_1, y_2, \cdots, y_n) \in H(Gx), \tag{1}$$

则

$$\sum_{\rho}\sum_{v}\phi_{\rho_v}(y_v) \leqslant \sum_{\rho}\sum_{v}\phi_{\rho_v}(x_v). \tag{2}$$

证　因为 $y \in H(Gx)$，所以存在一组 $t_\sigma \geqslant 0$ $(\sigma \in G)$, $\sum_{\sigma} t_\sigma=1$，使得

$$y_v=\sum_{\sigma} t_\sigma x_{\sigma}(v=1, 2, \cdots, n),$$

因此

$$\begin{aligned}\sum_{\rho}\sum_{v}\phi_{\rho_v}(y_v) &= \sum_{\rho}\sum_{v}\phi_{\rho_v}\left(\sum_{\sigma} t_\sigma x_{\sigma_\rho}\right)\\ &\leqslant \sum_{\rho}\sum_{v}\sum_{\sigma} t_\sigma \phi_{\rho_v}(x_{\sigma_v})\\ &= \sum_{\sigma} t_\sigma \sum_{\rho}\sum_{v}\phi_{\rho_v}(x_{\sigma_v})\\ &= \sum_{\sigma} t_\sigma \sum_{\rho}\sum_{v}\phi_{\rho\sigma^{-1}{}_v}(x_v)\\ &= \sum_{\sigma} t_\sigma \sum_{\tau}\sum_{v}\phi_{\tau_v}(x_v)\\ &= \sum_{\tau}\sum_{v}\phi_{\tau_v}(x_v).\end{aligned}$$

推论 1　设 $Z_v>0$ $(v=1, 2, \cdots, n)$，且(1)式成立，则

$$\sum_{\rho}\sum_{v} Z_{\rho_v}^{y_v} \leqslant \sum_{\rho}\sum_{v} Z_{\rho_v}^{x_v}.$$

推论 2 设 $Z_v>1$ $(v=1,2,\cdots,n)$，且(1)式成立，则

$$\sum_{\rho}\sum_{v}y_v^{Z_{\rho_v}}\leqslant\sum_{\rho}\sum_{v}x_v^{Z_{\rho_v}}.$$

由定理 1，我们可以得到 Karamata 不等式的一个新的简单证明

定理 2 (Karamata)设 ϕ 为 **R→R** 的连续凸函数，若

$$x_1\geqslant x_2\geqslant\cdots\geqslant x_n,\ y_1\geqslant y_2\geqslant\cdots\geqslant y_n, \tag{3}$$

$$\begin{aligned}&x_1\geqslant y_1,\\&x_1+x_2\geqslant y_1+y_2,\\&x_1+x_2+x_3\geqslant y_1+y_2+y_3,\\&\cdots\cdots\\&x_1+x_2+\cdots+x_n=y_1+y_2+\cdots+y_n,\end{aligned} \tag{4}$$

则

$$\phi_{(y_1)}+\phi_{(y_2)}+\cdots+\phi_{(y_n)}\leqslant\phi_{(x_1)}+\phi_{(x_2)}+\cdots+\phi_{(x_n)}.$$

证 在定理 1 中令 $\phi_1=\phi_2=\cdots=\phi_n=\phi$, $G=S_n$ (n 阶对称群)，只要证明在(3)式成立的前提下(1)与(4)等价就可以了.

假定(1)成立，那么存在一组 $t_\sigma\geqslant 0$, $\sum\limits_{\sigma}t_\sigma=1$，使得 $y_v=\sum\limits_{\sigma}t_\sigma x_{\sigma_v}$，因此对

$$1\leqslant m\leqslant n,$$

$$\begin{aligned}\sum_{v\leqslant m}y_v&=\sum_{v\leqslant m}\sum_{\sigma}t_\sigma x_{\sigma_v}=\sum_{\sigma}t_\sigma\sum_{v\leqslant m}x_{\sigma_v}\\&\leqslant\sum_{\sigma}t_\sigma\sum_{v\leqslant m}x_v=\sum_{v\leqslant m}x_v.\end{aligned}$$

当 $m=n$ 时等号成立.

假定(4)成立，如果 $y\in H(Gx)$，那么可以作一个超平面 M，y 到 M 的距离大于 $H(Gx)$中任一点到 M 的距离，即存在一组数 a_1, a_2, $\cdots$, a_n，使得

$$\sum_{v}a_vy_v>\sum_{v}a_vx_{\rho_v},$$

对所有的 $\rho\in G$ 成立. 换言之，对所有 $\sigma\in G$，

$$\sum_{v}a_{\sigma_v}y_{\sigma_v}>\sum_{v}a_{\sigma_v}x_v.$$

取 σ,使 $a_{\sigma_1}\geqslant a_{\sigma_2}\geqslant\cdots\geqslant a_{\sigma_n}$,则

$$\sum_{v<n}(a_{\sigma_v}-a_{\sigma_{v+1}})(y_1+y_2+\cdots+y_v)+a_{\sigma_n}(y_1+\cdots+y_n)$$

$$\geqslant\sum_{v<n}(a_{\sigma_v}-a_{\sigma_{v+1}})(y_{\sigma_1}+y_{\sigma_2}+\cdots+y_{\sigma_n})+a_{\sigma_n}(y_{\sigma_1}+\cdots+y_{\sigma_n})$$

$$=\sum_v a_{\sigma_v}y_{\sigma_v}>\sum_v a_{\sigma_v}x_v$$

$$=\sum_{v<n}(a_{\sigma_v}-a_{\sigma_{v+1}})(x_1+x_2+\cdots+x_v)+a_{\sigma_v}(x_1+\cdots+x_n)$$

与(4)矛盾.

由定理 2 的证明可知推论 2 是钟开莱 1954 年提出的一个问题[5]的推广.

定理 3　Z_1、Z_2、…、Z_n 为一组正数,如果(1)成立,即

$$\sum_\rho\prod_v Z_{\rho_v}^{y_v}\leqslant\sum_\rho\prod_v Z_{\rho_v}^{x_v}$$

证　t_σ 意义同定理一的证明,我们有

$$\sum_\rho\prod_v Z_{\rho_v}^{y_v}=\sum_\rho\prod_v Z_{\rho_v}^{\sum_\sigma t_\sigma x_{\sigma_v}}$$

$$=\sum_\rho\prod_\sigma\left(\prod_v Z_{\rho_v}^{x_{\sigma_v}}\right)^{t_\sigma}\leqslant\sum_\rho\sum_\sigma t_\sigma\prod_v Z_{\rho_v}^{x_{\sigma_v}}$$

$$=\sum_\sigma t_\sigma\sum_\rho\prod_v Z_{\rho_v}^{x_v}$$

$$=\sum_\rho\prod_v Z_{\rho_v}^{x_v}.$$

在定理 3 中令 $G=S_n$,使得到 Muirhead 不等式.

假定(1)成立,应用上面的方法可以推出下列结论.

设 Z_1, Z_2, …, Z_n 为正数,则

(i) $\prod_\rho\prod_v y_v^{Z_{\rho_v}}\geqslant\prod_\rho\prod_v x_v^{Z_{\rho_v}}$

(ii) $\prod_\rho\prod_v Z_{\rho_v}^{y_v}=\prod_\rho\prod_v Z_{\rho_v}^{x_v}$

(iii) $\prod_\rho\sum_v Z_{\rho_v}^{y_v}\leqslant\prod_\rho\sum_v Z_{\rho_v}^{x_v}$

(iv) $\sum_v\prod_\rho Z_{\rho_v}^{y_v}\leqslant\sum_v\prod_\rho Z_{\rho_v}^{x_v}$

(v) $\prod_v\sum_\rho Z_{\rho_v}^{y_v}\leqslant\prod_v\sum_\rho Z_{\rho_v}^{x_v}$

(iv) 若 $Z_v \leqslant 1\ (v=1, 2, \cdots, n)$, 则

$$\prod_v \sum_\rho y_v^{Z_{\rho_v}} \geqslant \prod_v \sum_\rho x_v^{Z_{\rho_v}}$$

(vii) 若 ϕ 为 $\mathbf{R}\to\mathbf{R}$ 的正值连续凹函数,则

$$\prod_\rho \prod_v \phi_{\rho_v}(y_v) \geqslant \prod_\rho \prod_v \phi_{\rho_v}(x_v)$$

参考文献

[1] Ostrowski, A. , *J* , *Math* , *Pure Appl.* , **31**(1952), 253 - 292.

[2] Karamata, J. , *Publ* , *Math* , *Univ* , *Belgrade.* , **1**(1932), 145 - 148.

[3] Beckenbach, E. F. and Bellman, R. , *Inequalities*, Springer-Verlag, Berlin, 1961, 30 - 31.

[4] Hardy, G. H. , Littlewood, J. E. and Pblya, G. , *Inequalities*, Cambridge university Press, 1952, 46 - 50.

[5] Kung, K. L. , *Am*, *Math*, *Monthly*. **60**(1953), 122.

Some Inequalities Associatad with the Permutation Group

Supposing $x=(x_1, x_2, \cdots, x_n)$ to be a point in $\mathbf{R}^n$. G a permutation group of $\{1, 2, \cdots, n\}$, Gx the G-orbit of x and $H(Gx)$ the convex hull of Gx, we have proved the following theorem.

Let ϕ_1、ϕ_2、$\cdots$、ϕ_n: $\mathbf{R}\to\mathbf{R}$ be continuous, convex functions. If

$$y=(y_1, y_2, \cdots, y_n)\in H(Gx),$$

then

$$\sum_\rho \sum_v \phi_{\rho_v}(y_v) \leqslant \sum_\rho \sum_v \phi_{\rho_v}(x_v).$$

We have also discussed some inequalities such as Karamata inequality Muirhead inequality, etc.

一类不等式

§1

钟开莱(Chung K. L.)曾提出过下面的问题[1]：

设正数 $\alpha_1, \alpha_2, \cdots, \alpha_n$ 及 $\beta_1, \beta_2, \cdots, \beta_n$ 依大小顺序排列，即

$$\begin{aligned} &\alpha_1 \geqslant \alpha_2 \geqslant \cdots \geqslant \alpha_n, \\ &\beta_1 \geqslant \beta_2 \geqslant \cdots \geqslant \beta_n, \end{aligned} \tag{1}$$

且

$$\begin{aligned} &\alpha_1 + \alpha_2 + \cdots + \alpha_k \leqslant \beta_1 + \beta_2 + \cdots + \beta_k (1 \leqslant k < n), \\ &\alpha_1 + \alpha_2 + \cdots + \alpha_n = \beta_1 + \beta_2 + \cdots + \beta_n, \end{aligned} \tag{2}$$

则

$$\sum_{i=1}^{n} \alpha_i^2 \leqslant \sum_{i=1}^{n} \beta_i^2, \tag{3}$$

当且仅当 $\alpha_i = \beta_i (i = 1, 2, \cdots, n)$ 时等号成立.

本文首先将(2)换为下面的(4)，然后将(3)推广，导出一类不等式.

§2

本文采用记号如下：

S_n 为 n 元素 $\{1, 2, \cdots, n\}$ 上的全体置换所组成的置换群，G 为 S_n 的一个子群.

$$x = (x_1, x_2, \cdots, x_n),\ \alpha = (\alpha_1, \alpha_2, \cdots, \alpha_n),\ \beta = (\beta_1, \beta_2, \cdots, \beta_n)$$

等均为 n 维欧氏空间中的点，并且不作特别申明时约定各个分量为正.

对 $\sigma, \tau \in G$，定义

$$\sigma x=(\sigma x_1,\ \sigma x_2,\ \cdots,\ \sigma x_n)=(x_{\sigma_1},\ x_{\sigma_2},\ \cdots,\ x_{\sigma_n})=x_\sigma,$$

其中 σ_i 即 $\sigma(i)$ 的简写.

显然 $(\tau\sigma)x=\tau(\sigma x)$.

记 β^G 为 β 在 G 作用下的象的集合, $H(\beta^G)$ 为 β^G 的凸闭包,即由满足

$$\alpha=\sum_{\tau\in G}t_\tau\beta_\tau,\ t_\tau\geqslant 0,\ \sum_\tau t_\tau=1$$

的 α 所成的点集.

显然, $H(\beta^G)\subset H(\beta^{S_n})$.

对 $i\in\{1,2,\cdots,n\}$,称 $\{\sigma i\mid\sigma\in G\}$ 为 i 在 G 作用下的轨道. 显然, $\{1,2,\cdots,n\}$ 在 G 的作用下分解为一个或几个轨道 $O_1,O_2,\cdots,O_h$.

用 $\sum\limits_i$ 或 $\prod\limits_i$ 表示对一个或几个轨道的并 $O_{j_1}\cup O_{j_2}\cup\cdots\cup O_{j_m}$ 求和或求积.

§3

本文主要结论如下:

定理 1 对于依(1)排列的 $\alpha_1,\alpha_2,\cdots,\alpha_n$ 与 $\beta_1,\beta_2,\cdots,\beta_n$,(2)与

$$\alpha\in H(\beta^{S_n}) \tag{4}$$

等价.

以下考虑更一般的条件

$$\alpha\in H(\beta^G). \tag{5}$$

定理 2 设 $x=(x_1,x_2,\cdots,x_n)$ 的各个分量 $x_i\geqslant 1$ 或为 $0(i=1,2,\cdots,n)$, 则在(5)成立时,有

$$\sum_{\sigma\in G}\sum_i\alpha_i^{x_{\sigma_i}}\leqslant\sum_{\sigma\in G}\sum_i\beta_i^{x_{\sigma_i}}. \tag{6}$$

定理 2′ 设 $\phi_1(z),\phi_2(z),\cdots,\phi_n(z)$ 为单变量 z 的一组凸函数,则在(5)成立时,

$$\sum_{\sigma\in G}\sum_i\phi_{\sigma_i}(\alpha_i)\leqslant\sum_{\sigma\in G}\sum_i\phi_{\sigma_i}(\beta_i). \tag{7}$$

系 1　设 $c \geqslant 1$ 或为 0,则在(5)成立时,有

$$\sum_i \alpha_i^c \leqslant \sum_i \beta_i^c.$$

系 2　(5)成立时,有

$$\sum_{i=1}^n \phi_i(\alpha_i) \leqslant \sum_{i=1}^n \phi_i(\beta_i).$$

系 3　如果 $\sum\limits_\sigma x_{\sigma_k} \geqslant 1$ 或为 0,则

$$\sum_i{}' \prod_{\sigma\in G} \alpha_i^{x_{\sigma_i}} \leqslant \sum_i{}' \prod_{\sigma\in G} \beta_i^{x_{\sigma_i}},$$

其中和号 $\sum\limits_i{}'$ 遍及 k 所在轨道, α, β 满足(5).

定理 3　若 $x_i > 0$ $(i=1, 2, \cdots, n)$, 且

$$\sum_{i=1}^n x_i = A \leqslant 1,$$

则在(5)成立时,有

$$\sum_\sigma \prod_i \alpha_i^{x_{\sigma_i}} \geqslant \sum_\sigma \prod_i \beta_i^{x_{\sigma_i}}. \tag{8}$$

系 4　(5)成立时,有

$$\prod_i \alpha_i \geqslant \prod_i \beta_i.$$

定理 4　(5)成立时,有

$$\prod_\sigma \prod_i \alpha_i^{x_{\sigma_i}} \geqslant \prod_\sigma \prod_i \beta_i^{x_{\sigma_i}}.$$

定理 5　(5)成立时,有

$$\sum_\sigma \prod_i x_{\sigma_i}^{\alpha_i} \leqslant \sum_\sigma \prod_i x_{\sigma_i}^{\beta_i}. \tag{9}$$

反之若

$$\sum_\sigma \prod_{i=1}^n x_{\sigma_i}^{\alpha_i} \leqslant \sum_\sigma \prod_{i=1}^n x_{\sigma_i}^{\beta_i} \tag{10}$$

对一切 $x_i > 0$ 均成立时,(5)也是必要的.

系 5　(5)成立时,有

$$\prod_i x^{\alpha_i} = \prod_i x^{\beta_i}.$$

系 6　(5)成立时,有

$$\prod_\sigma \prod_i x_{\sigma_i}^{\alpha_i} = \prod_\sigma \prod_i x_{\sigma_i}^{\beta_i}.$$

定理 6　(5)成立时,有

$$\sum_\sigma \sum_i x_{\sigma_i}^{\alpha_i} \leqslant \sum_\sigma \sum_i x_{\sigma_i}^{\beta_i}.$$

系 7　(5)成立时,有

$$\sum_i x^{\alpha_i} \leqslant \sum_i x^{\beta_i}.$$

系 8　(5)成立时,有

$$\sum_i \prod_\sigma x_{\sigma_i}^{\alpha_i} \leqslant \sum_i \prod_\sigma x_{\sigma_i}^{\beta_i}.$$

系 9　(5)成立时,有

$$\prod_\sigma \sum_i x_{\sigma_i}^{\alpha_i} \leqslant \prod_\sigma \sum_i x_{\sigma_i}^{\beta_i}.$$

§4

本节证明§3中的结论.

引理　设$\alpha \overline{\in} H(\beta^{S_n})$,则存在一组实数$\lambda_i (1 \leqslant i \leqslant n)$及正数$\delta$,对所有$\sigma \in S_n$,

$$\sum_{i=1}^n \lambda_i \alpha_i \geqslant \sum_{i=1}^n \lambda_i \beta_{\sigma_i} + \delta.$$

证　因为$H(\beta^{S_n})$是闭集,$\alpha \overline{\in} H(\beta^{S_n})$,所以$\alpha$到$H(\beta^{S_n})$有一最小距离$d > 0$,设$\gamma \in H(\beta^{S_n})$与$\alpha$的距离为$d$,$\delta = d^2 > 0$.

作与α,γ连线垂直的超平面

$$(\alpha - \gamma) \cdot \left(x - \frac{\alpha + \gamma}{2}\right) = 0,$$

则 $\lambda=\alpha-\gamma$ 及 δ 即为所求. 解析地说, 对任一点 $\gamma'\in H(\beta^{S_n})$, 由于 $H(\beta^{S_n})$ 是凸集, 所以

$$(1-t)\gamma+t\gamma'=\gamma+t(\gamma'-\gamma)\in H(\beta^{S_n}),$$

其中 $t\geqslant 0$.

由 γ 的定义,

$$\|\gamma+t(\gamma'-\gamma)-\alpha\|\geqslant\|\alpha-\gamma\|,$$

即

$$2t(\gamma-\gamma')(\alpha-\gamma)+t^2(\gamma-\gamma')^2\geqslant 0,$$

约去 $2t$, 并令 $t\to 0$ 得

$$(\alpha-\gamma)(\gamma-\gamma')\geqslant 0.$$

因此

$$\begin{aligned}\lambda(\alpha-\beta)&=(\alpha-\gamma)(\alpha-\gamma+\gamma-\beta)\\&=(\alpha-\gamma)^2+(\alpha-\gamma)(\gamma-\beta)\\&\geqslant(\alpha-\gamma)^2=\delta.\end{aligned}$$

定理 1 的证明

设(4)成立, 则存在 $t_\tau\geqslant 0$, $\sum t_\tau=1$, 使 $\alpha=\sum t_\tau\beta_\tau$.

对于 $1\leqslant k\leqslant n$,

$$\begin{aligned}\sum_{i\leqslant k}\alpha_i&=\sum_{i\leqslant k}\sum_\tau t_\tau\beta_{\tau_i}=\sum_\tau t_\tau\sum_{i\leqslant k}\beta_{\tau_i}\\&\leqslant\sum_\tau t_\tau\sum_{i\leqslant k}\beta_i=\sum_{i\leqslant k}\beta_i,\end{aligned}$$

并且当 $k=n$ 时, 符号成立.

反过来, 假定(2)成立而(4)不成立, 则由引理, 存在一组实数 $\lambda_i(1\leqslant i\leqslant n)$, 对所有 $\sigma\in S_n$,

$$\sum_{i=1}^n\lambda_i\alpha_i>\sum_{i=1}^n\lambda_i\beta_{\sigma_i},$$

亦即对所有 $\tau\in S_n$,

$$\sum_{i=1}^{n}\lambda_{\tau_i}\alpha_{\tau_i}>\sum_{i=1}^{n}\lambda_{\tau_i}\beta_i,$$

取 τ,使 $\lambda_{\tau_1}\geqslant\lambda_{\tau_2}\geqslant\cdots\geqslant\lambda_{\tau_n}$,则

$$\begin{aligned}&\sum_{i<n}(\lambda_{\tau_i}-\lambda_{\tau_{i+1}})(\alpha_1+\alpha_2+\cdots+\alpha_i)+\lambda_{\tau_n}(\alpha_1+\alpha_2+\cdots+\alpha_n)\\ \geqslant&\sum_{i<n}(\lambda_{\tau_i}-\lambda_{\tau_{i+1}})(\alpha_{\tau_1}+\alpha_{\tau_2}+\cdots+\alpha_{\tau_i})+\lambda_{\tau_n}(\alpha_{\tau_1}+\alpha_{\tau_2}+\cdots+\alpha_{\tau_n})\\ =&\sum_{i=1}^{n}\lambda_{\tau_i}\alpha_{\tau_i}>\sum_{i=1}^{n}\lambda_{\tau_i}\beta_i\\ =&\sum_{i<n}(\lambda_{\tau_i}-\lambda_{\tau_{i+1}})(\beta_1+\beta_2+\cdots+\beta_i)+\lambda_{\tau_n}(\beta_1+\beta_2+\cdots+\beta_n),\end{aligned}$$

与(2)矛盾.因此(2)⇒(4).

定理 2 的证明

$$\begin{aligned}\sum_{\sigma}\sum_{i}\alpha_i^{x_{\sigma_i}}&=\sum_{\sigma}\sum_{i}\Big(\sum_{r}t_{\tau}\beta_{\tau_i}\Big)^{x_{\sigma_i}}\\ &\leqslant\sum_{\sigma}\sum_{i}\sum_{\tau}t_{\tau}\beta_{\tau_i}^{x_{\sigma_i}}\ (\text{[2] 定理 16})\\ &=\sum_{\tau}t_{\tau}\sum_{\sigma}\sum_{i}\beta_{\tau_i}^{x_{\sigma_i}}\\ &=\sum_{\tau}t_{\tau}\sum_{\rho}\sum_{j}\beta_j^{x_{\rho_j}}\quad(j=\tau_i,\ \rho=\sigma\tau^{-1})\\ &=\sum_{\rho}\sum_{j}\beta_j^{x_{\rho_j}}.\end{aligned}$$

从上面的证明可以推知(6)式在 x_i 中有负数时仍然成立,而在 $0<x_i<1$ $(i=1,2\cdots,n)$ 时,(6)应改为方向相反的不等式.并且,(5)为等式时 x_i 全为 0 或 1,否则

$$\alpha_i=\beta_i\quad(i=1,2,\cdots,n).$$

定理 2′的证明(与定理 2 类似)

$$\begin{aligned}\sum_{\sigma}\sum_{i}\phi_{\sigma_i}(\alpha_i)&=\sum_{\sigma}\sum_{i}\phi_{\sigma_i}\Big(\sum_{\tau}t_{\tau}\beta_{\tau_i}\Big)\\ &\leqslant\sum_{\tau}t_{\tau}\sum_{\sigma}\sum_{i}\phi_{\sigma_i}(\beta_{\tau_i})\\ &=\sum_{\tau}t_{\tau}\sum_{\rho}\sum_{j}\phi_{\rho_j}(\beta_j)=\sum_{\rho}\sum_{j}\phi_{\rho_j}(\beta_i).\end{aligned}$$

系 1 的证明

在(6)中令 $x_1=x_2=\cdots=x_n=c$，则得

$$|G|\cdot\sum_i\alpha_i^c\leqslant|G|\cdot\sum_i\beta_i^c$$

即

$$\sum_i\alpha_i^c\leqslant\sum_i\beta_i^c.$$

当 $G=S_n$ 时，得

$$\sum_{i=1}^n\alpha_i^c\leqslant\sum_{i=1}^n\beta_i^c.$$

$c=2$ 时，即钟开莱所提的不等式(3).

系 2 的证明

在(7)中，令 $G=S_n$，$\phi_1=\phi_2=\cdots=\phi_n$ 即得. 这一结果属于 Baramate([3], p. 30)，我们这里的证法与它不同.

系 3 的证明

由系 1，

$$\sum_i{}'\prod_\sigma\alpha_i^{x_{\sigma_i}}=\sum_i{}'\alpha_i^{\sum\limits_\sigma x_{\sigma_i}}=\sum_i{}'\alpha_i^{\sum\limits_\sigma x_{\sigma_k}}$$

$$\leqslant\sum_i{}'\beta_i^{\sum\limits_\sigma x_{\sigma_k}}=\sum_i{}'\prod_\sigma\beta_i^{x_{\sigma_i}}$$

如果 $\sum\limits_\sigma x_{\sigma_k}\geqslant1$ 或为 0 对一切 $1\leqslant k\leqslant n$ 成立，则

$$\sum_i\prod_\sigma\alpha_i^{x_{\sigma_i}}\leqslant\sum_i\prod_\sigma\beta_i^{x_{\sigma_i}}.$$

定理 3 的证明

对任一 $\sigma\in G$，

$$A_\sigma=\sum_i x_{\sigma_i}\leqslant\sum_{i=1}^n x_i=A\leqslant1,$$

$$\sum_\sigma\prod_i\alpha_i^{x_{\sigma_i}}=\sum_\sigma\Big(\prod_i\alpha_i^{\frac{x_{\sigma_i}}{A_\sigma}}\Big)^{A_\sigma}=\sum_\sigma\Big(\prod_i\Big(\sum_\tau t_\tau\beta_{\tau_i}\Big)^{\frac{x_{\sigma_i}}{A_\sigma}}\Big)^{A_\sigma}$$

$$\geqslant\sum_\sigma\Big(\sum_\tau\prod_i(t_\tau\beta_{\tau_i})^{\frac{x_{\sigma_i}}{A_\sigma}}\Big)^{A_\sigma}\quad([2]\text{ 定理 }11)$$

$$=\sum_{\sigma}\Big(\sum_{\tau}t_{\tau}\prod_{i}\beta_{\tau_i}^{\frac{x_{\sigma_i}}{A_{\sigma}}}\Big)^{A_{\sigma}}\geqslant\sum_{\sigma}\sum_{\tau}t_{\tau}\prod_{i}\beta_{\tau_i}^{x_{\sigma_i}}\quad([2]\text{定理 }16)$$

$$=\sum_{\tau}t_{\tau}\sum_{\sigma}\prod_{i}\beta_{\tau_i}^{x_{\sigma_i}}=\sum_{\rho}\sum_{j}\beta_{j}^{x_{\rho_j}}.$$

系 4 的证明

取 $x_1=x_2=\cdots=x_n=\dfrac{1}{n}$，则

$$\sum_{\sigma}\prod_{i}\alpha_i^{x_{\sigma_i}}=|G|\cdot\prod_{i}\alpha_i^{\frac{1}{n}}\geqslant|G|\cdot\prod_{i}\beta_i^{\frac{1}{n}},$$

所以

$$\prod_{i}\alpha_i\geqslant\prod_{i}\beta_i.$$

定理 4 的证明

在每个轨道 O_j 上，$\sum\limits_{\sigma}x_{\sigma_i}\ (i\in O_j)$ 为一定值 A_j，所以

$$\prod_{\sigma}\prod_{i}\alpha_i^{x_{\sigma_i}}=\prod_{i}\alpha_i^{\sum\limits_{\sigma}x_{\sigma_i}}=\prod_{i\in O_{j_1}}\alpha_i^{A_{j_1}}\cdot\prod_{i\in O_{j_2}}\alpha_i^{A_{j_2}}\cdots\prod_{i\in O_{j_m}}\alpha_i^{A_{j_m}}$$

$$\geqslant\prod_{i\in O_{j_1}}\beta_i^{A_{j_1}}\prod_{i\in O_{j_2}}\beta_i^{A_{j_2}}\cdots\prod_{i\in O_{j_m}}\beta_i^{A_{j_m}}$$

$$=\prod_{i}\beta_i^{\sum\limits_{\sigma}x_{\sigma_i}}=\prod_{\sigma}\prod_{i}\beta_i^{x_{\sigma_i}}.$$

定理 5 的证明

$$\sum_{\sigma}\prod_{i}x_{\sigma_j}^{\alpha_i}=\sum_{\sigma}\prod_{i}x_{\sigma_i}^{\sum\limits_{\tau}t_{\tau}\beta_{\tau_i}}$$

$$=\sum_{\sigma}\prod_{\tau}\Big(\prod_{i}x_{\sigma_i}^{\beta_{\tau_i}}\Big)^{t_{\tau}}$$

$$\leqslant\sum_{\sigma}\sum_{\tau}t_{\tau}\prod_{i}x_{\sigma_i}^{\beta_{\tau_i}}\quad([2]\text{定理 }9)$$

$$=\sum_{\tau}t_{\tau}\sum_{\sigma}\prod_{i}x_{\sigma_i}^{\beta_{\tau_i}}=\sum_{\rho}\prod_{j}x_j^{\beta_{\rho_j}}.$$

反过来，设(10)对一切 $x_i>0$ 均成立. 若 $\alpha\,\overline{\in}\,H(\beta^G)$，则由引理，存在 $\lambda=(\lambda_1,\lambda_2,\cdots,\lambda_n)$ 及 $\delta>0$，对一切 $\sigma\in G$，

$$\sum_i \lambda_i \alpha_i \geqslant \sum_i \lambda_i \beta_{\sigma_i} + \delta$$

令 $x_i = N^{\lambda_i}$，$N > 1$ 为一个大正数，则

$$\begin{aligned}\sum_\sigma \prod_i x_{\sigma_i}^{\beta_i} &= \sum_\sigma N^{\sum_i \lambda_{\sigma_i}\beta_i} = \sum_\sigma N^{\sum_i \lambda_i \beta_{\sigma_i}} \\ &\leqslant \sum_\sigma N^{\sum_i \lambda_i \alpha_i - \delta} \leqslant n! \cdot N^{-\delta} \cdot N^{\sum_i \lambda_i \alpha_i} \\ &\leqslant n! \cdot N^{-\delta} \sum_\sigma \prod_i x_{\sigma_i}^{\sigma_i} < \sum_\sigma \prod_i x_{\sigma_i}^{\sigma_i} \quad (N\text{ 充分大}),\end{aligned}$$

与(10)矛盾. 故 $\alpha \in H(\beta^G)$.

特别地，令 $G = S_n$，便得到 Muirhead 的不等式([2](2.18.4))，我们这里的证法与它不同，并且 Muirhead 不等式原来的条件，即(1)、(2)，不能保证(10)或(9)对所有的 x 均成立，这只要取 $n=3$，$\beta=(4, 2, 0)$，$\alpha=(4, 1, 1)$，G 为三阶交错群$\langle(123)\rangle$，则 α、β 满足(1)、(2)，但

$$\alpha \in H(\beta^{S_n}) - H(\beta^G),$$

故(10)不成立.

系 5 的证明

在定理 5 的证明中，$x_1 = x_2 = \cdots = x_n$ 时，显然等号成立，故得证.

系 6 的证明

$$\begin{aligned}\prod_\sigma \prod_i x_{\sigma_i}^{\alpha_i} &= \prod_i \Big(\prod_{\substack{i \\ \sigma_i = j}} \prod_\sigma x_j^{\alpha_i}\Big) = \prod_j \prod_i \Big(\prod_{\sigma_i = j} x_j\Big)^{\alpha_i} \\ &= \prod_j \prod_i \Big(\prod_{\sigma_i = j} x_j\Big)^{\beta_i} = \prod_\sigma \prod_i x_{\sigma_i}^{\beta_i}.\end{aligned}$$

定理 6 的证明

在定理 2′中令 $\phi_i(x) = x_i^x$ 即得，或仿照定理 5 的证明.

系 7 的证明

在定理 6 中令 $x_1 = x_2 = \cdots = x_n = x$ 即得.

系 8 的证明

由系 7，

$$\sum_i \prod_\sigma x_{\sigma_i}^{\alpha_i} = \sum_i \left(\prod_\sigma x_{\sigma_i}\right)^{\alpha_i} = \sum_{h=1}^m \sum_{O_{j_n}} \left(\prod_\sigma x_{\sigma_i}\right)^{\alpha_i}$$

$$\leqslant \sum_{h=1}^m \sum_{O_{j_n}} \left(\prod_\sigma x_{\sigma_i}\right)^{\beta_i} = \sum_i \prod_\sigma x_{\sigma_i}^{\beta_i}.$$

又由系 7 可以推出系 9.

参考文献

[1] Chung K. L., Problem E1025. *Amer. Math. Monthly*, **59**(1952), 107.

[2] Hardy G. H., Littlewood J. E. and Pólya G., *Inequalities*, London Cambridge University Press, 1951.

[3] Beckenbach E. F. and Bellman R., *Inequalities*, Springer, Berlin, 1961.

[4] Bonnesen T. and Fenchel W., *Theorie Konvexen Körper*, Springer, Berlin, 1934.

一个华林-哥德巴赫型问题

1　引言

在 K. F. Roth[1] 证明了几乎所有的正整数都能表成一个自然数的平方，一个自然数的立方和一个自然数的四方之和以后，K. Prachar[2] 在 1952 年得到，几乎所有的偶数 n 都能表成

$$n=p_1^2+p_2^3+p_3^4+p_4^5,$$

其中 $p_i(1\leqslant i\leqslant 4)$ 是素数.

1974 年，K. Thanigasalam[3] 指出，序列 $\{x_1^2+x_2^3+x_3^5+x_4^k\}$ $(k\geqslant 5)$ 和 $\{x_1^2+x_2^3+x_3^6+x_4^k\}$ $(k\geqslant 6)$ 具有正的 Щнирельман 密率；其后，1978 年，Roger Cook[4] 证明了，几乎所有的正整数 n 都能表成

$$n=x_1^2+x_2^3+x_3^5+x_4^k \quad (k\geqslant 5)$$

及

$$n=x_1^2+x_2^3+x_3^6+x_4^k. \ (k\geqslant 6)$$

这里 $x_i(1\leqslant i\leqslant 4)$ 为正整数. 现在，作者将整幂问题改成素的问题，从而得到

定理 1　对任何的正整数 k，几乎所有的偶数 n 都能表成

$$n=p_1^2+p_2^3+p_3^5+p_4^k, \tag{1}$$

其中 $p_i(1\leqslant i\leqslant 4)$ 是素数.

定理 2　除了 $k\equiv 0(\bmod 2)$ 或 $k\equiv 0\ (\bmod 3)$ 的正整数 k 以外，几乎所有的偶数 n 都能表成

$$n=p_1^2+p_2^3+p_3^6+p_4^k, \tag{2}$$

其中 $p_i(1\leqslant i\leqslant 4)$ 是素数.

以下我们仅证定理 1，定理 2 可由完全类似的方法得到.

2 奇异级数

命 $e(x)=e^{2\pi lx}$,

$$e_q(x)=e(x/q),$$

$$W_{k,q}^{(r)}=\sum_{\substack{l=1\\(l,q)=1}}^{q}e_q(kl^r),\tag{3}$$

$$A_q(n)=(\varphi(q))^{-4}\sum_{\substack{h=1\\(h,q)=1}}^{q}W_{h,q}^{(2)}W_{h,q}^{(3)}W_{h,q}^{(5)}W_{h,q}^{(k)}e_q(-nh).\tag{4}$$

再命 N 是一个充分大的整数, $L=\log N$, c_1, c_2, c_3, …是与 N 无关的正常数, σ_0 是一待定的正实数,

$$S(n)=\sum_{q\leqslant L\sigma_0}A_q(n),\tag{5}$$

$$S_1(n)=\prod_{2\leqslant p<\frac{1}{5}L}(1+A_p(n)).\tag{6}$$

引理 1 [2]设 M_q 为同余方程

$$x^2+y^3+z^5+w^k\equiv n(\bmod q)\quad (xyzw,q)=1$$

的解数,则对素数 p

$$1+A_p(n)=\frac{p}{(p-1)^4}M_p\tag{7}$$

引理 2 对 $p\geqslant 23$ 及任何的自然数 u,同余方程

$$x^2+y^3+z^5\equiv u(\bmod p)p\nmid xyz\tag{8}$$

必有解.

证 对 $p>2$, 设 $B_1(p)$为存在 x, z, $p\nmid xz$ 使

$$x^2+z^5\equiv u_1(\bmod p),\tag{9}$$

且 $1\leqslant u_1\leqslant p$ 的 u_1 的个数, $B_2(p)$为存在 y, $p\nmid y$ 使

$$y^3\equiv u_2(\bmod p),\tag{10}$$

且 $1\leqslant u_2\leqslant p$ 的 u_2 的个数. 首先易知,

$$B_2(p)\geqslant\frac{p-1}{3},\tag{11}$$

存在 x, z, $p+xz$ 使(10)成立的 u_1 至少有如下互不包含的两类:(i) u_1-1 是 $\bmod p$ 的二次剩余;(ii) u_1-1 是 $\bmod p$ 的二次非剩余,但 u_1+1 是 $\bmod p$ 的二次剩余. 因此,

$$B_1(p)\geqslant\frac{p-1}{2}+b_1(p),$$

其中 $b_1(p)$是适合 $2\leqslant l\leqslant p$, $l\neq p-1$, $\left(\frac{l-1}{p}\right)=-1$, $\left(\frac{l+1}{p}\right)=1$ 的 l 的个数. 现有

$$\begin{aligned}b_1(p)&=\frac{1}{4}\sum_{\substack{l=2\\ l\neq p-1}}^{p}\left(1+\left(\frac{l+1}{p}\right)\right)\left(1-\left(\frac{l-1}{p}\right)\right)\\&=\frac{1}{4}\left(p-1-\left(\frac{2}{p}\right)+\left(\frac{-2}{p}\right)\right)\geqslant\frac{p-3}{4},\end{aligned}$$

即

$$B_1(p)\geqslant\frac{p-1}{2}+\frac{p-3}{4}.$$

于是当 $p>19$ 时,

$$B_1(p)+B_2(p)\geqslant\frac{p-1}{2}+\frac{p-1}{3}+\frac{p-3}{4}>p$$

由抽屉原则引理得证.

引理 3　对 $p>2$,

$$1+A_p(n)>0\tag{12}$$

及

$$1+A_2(n)=\begin{cases}0 & 若\ n\not\equiv 0(\bmod 2),\\ 2 & 若\ n\equiv 0(\bmod 2).\end{cases}\tag{13}$$

证　先证 $p>2$ 时的情形. 由引理 1,

$$1+A_p(n)=\frac{p}{(p-1)^4}M_p$$

再由引理 2,当 $p\geqslant 23$ 时,对任何的自然数 n,同余方程

$$x^2+y^3+z^5\equiv n-1(\mathrm{mad}\,p)\quad p\nmid xyz$$

至少有一解,故得 $M_p>0$. 又依[2]的引理 13 及[5]的引理 8.4,当

$$\frac{p-1}{(2,\ p-1)}+\frac{p-1}{(3,\ p-1)}+\frac{p-1}{(5,\ p-1)}+\frac{p-1}{(k,\ p-1)}-2>p \tag{14}$$

时,$M_p>0$. 现对素数 3, 5, 7, 11, 13, 17, 19 逐一验证,关系式(14)都成立,故得(12). 而关系式(13)显然成立.

引理 4[2] $A_p(n)=O(p^{-1}),\ A_q(n)=O(q^{-1+\varepsilon}).$ (15)

引理 5[2] $\sum\limits_{n=1}^{N}\{S(n)-S_1(n)\}^2=O(NL^{-1+\varepsilon}).$ (16)

引理 6 若 n 是偶数,则

$$S_1(n)>c_2(\log\log N)^{-\varepsilon_1}. \tag{17}$$

证 由引理 3,引理 4 及[2]的引理 16 即得.

3 基本引理

设

$$S_r(\alpha)=\sum_{p\leqslant N^{1/r}}e(\alpha p^r), \tag{18}$$

对 $\alpha=\dfrac{k}{q}+\beta,\ (h,\ q)=1$, 设

$$T_r(\alpha,\ h,\ q)=\frac{W_{h,\ q}^{(r)}}{\varphi(q)}\sum_{2\leqslant x\leqslant N}n^{\frac{1}{r}-1}(\log n)^{-1}e(n\beta), \tag{19}$$

$$\mu=\frac{1}{2}+\frac{1}{3}+\frac{1}{5}+\frac{1}{k},\ \bar{\mu}=\frac{1}{2}+\frac{1}{3}+\frac{1}{5},$$

引理 7 设 σ_0, σ_1 是任意给定的两个正实数,则对适合

$$|\beta|=\left|\alpha-\frac{h}{q}\right|\leqslant N^{-1}L^{\sigma_1},\ q\leqslant L^{\sigma_0} \tag{20}$$

的 α,

$$S_r(\alpha)-T_r(\alpha,h,q)=O(Ne^{-\sigma_1\sqrt{L}}), \tag{21}$$

这与[5]的引理 7.15 类似地证得.

引理 8[5]　当 $|\beta|\leqslant\frac{1}{2}$ 时,

$$T_r(\alpha,h,q)=O(q^{-\frac{1}{r}+\varepsilon}\min(N^{1/r},|\beta|^{-1/r})). \tag{22}$$

特别,当 α 适合引理 7 的条件时,

$$S_r(\alpha)=O(q^{-1/r+\varepsilon}\min(N^{1/r},|\beta|^{-1/r})).$$

引理 9　设 $r(n)$为(1)的解组数,

$$\psi(n)=\sum_{\substack{n_1+n_2+n_3+n_4=n\\ 2\leqslant n_i\leqslant N}}\frac{1}{n_1^{1-1/2}n_2^{1-1/3}n_3^{1-1/5}n_4^{1-1/k}\log n_1\log n_2\log n_3\log n_4} \tag{23}$$

并置

$$\prod_j S_l(\alpha)=S_2(\alpha)S_3(\alpha)S_5(\alpha)S_k(\alpha),$$

$$\prod_j T_l(\alpha,h,q)=T_2(\alpha,h,q)T_3(\alpha,h,q)T_5(\alpha,h,q)T_h(\alpha,h,q),$$

则

$$\begin{aligned}J(N)=\int_0^1\Big|\prod_j S_l(\alpha)-\sum_{q\leqslant L^{\alpha_0}}\sum_{\substack{h=1\\(h,q)=1}}^{q}\prod_j T_l(\alpha,h,q)\Big|^2 d\alpha\geqslant\\ \geqslant\sum_{n=3}^{N}\{r(n)-S(n)\psi(n)\}^2,\end{aligned} \tag{24}$$

且对 $c_4<n\leqslant N$

$$c_5n^{\mu-1}\log^{-4}n<\psi(n)<c_5n^{\mu-1}\log^{-4}n. \tag{25}$$

证　可与[2]中相应的结果类似地证得.

4 J(N)的估计

设 $\tau=NL^{-\sigma_1}$，以 $M_{h,q}$ 表示区间

$$\left|\alpha-\frac{h}{q}\right|<\tau^{-1},\ (h,q)=1,\ 0\leqslant h<q\leqslant L^{\sigma_0}. \tag{26}$$

熟知，当 N 充分大时这些小区间互不相交，而命

$$E=(-\tau^{-1},\ 1-\tau^{-1})-\sum M_{h,q}, \tag{27}$$

其中 $\sum M_{h,q}$ 表示所有适合(27)的区间的和集. 由(24),

$$\begin{aligned}J(N)&=\left(\sum\int_{M_{h,q}}+\int_E\right)\left|\prod_j S_j(\alpha)-\sum_{q\leqslant L^{\alpha_0}}\sum_{\substack{h=1\\(h,q)=1}}^{q}\prod_j T_j(\alpha,h,q)\right|^2\mathrm{d}\alpha\\&=J_2(N)+J_2(N).\end{aligned} \tag{28}$$

现先估计 $J_2(N)$. 对 $J_2(N)$,有

$$\begin{aligned}J_2(N)&\ll\int_E\left|\prod_j S_l(\alpha)\right|^2\mathrm{d}\alpha+\int_E\left|\sum_{q\leqslant L^{\alpha_0}}\sum_{\substack{h=1\\(h,q)=1}}^{q}\prod_j T_j(\alpha,h,q)\right|^2\mathrm{d}\alpha\\&=J_2'(N)+J_2''(N)\end{aligned} \tag{29}$$

引理 10[5] (Н. М. Виноградов)对任给的正实数 σ_2, σ_3, $\sigma_3\geqslant 2^{6r}(\sigma_2+1)$. 如果

$$\alpha=\frac{h}{q}+\beta(h,q)=1,\ L^{\sigma_1}<q\leqslant NL^{-\sigma_2},\ |\beta|\leqslant q^{-1}N^{-1}L^{\sigma_3},$$

那么

$$S_r(\alpha)\ll N^{1/r}L^{-\sigma_2}. \tag{30}$$

引理 11 存在正数 γ,使

$$\sum_{x_1=1}^{(3N)^{1/3}} \sum_{\substack{x_2=1 \\ x_1^3+x_2^5-y_1^3-y_2^5\neq 0}}^{(3N)^{1/5}} \sum_{y_1=1}^{(3N)^{1/3}} \sum_{y_2=1}^{(3N)^{1/5}} \mathrm{d}(|x_1^3+x_2^5-y_1^3-y_2^5|) \ll N^{2-\bar{\mu}_1} L^{\gamma}. \tag{31}$$

证　这可与[5]的定理 3 完全类似证得.

引理 12　设

$$F_r(\alpha)=\sum_{x\leqslant N^{1/r}} e(\alpha x^r), \tag{32}$$

则

$$\int_0^1 |F_2(\alpha)F_3(\alpha)F_5(\alpha)|^2 \mathrm{d}\alpha \ll N^{2\bar{\mu}-1} L^{\gamma}. \tag{33}$$

证　若设 $R(u)$为方程

$$x^2+y^3+z^5=u$$

的正整解组数,那么易知

$$\int_0^1 |F_2(\alpha)F_3(\alpha)F_5(\alpha)|^2 \mathrm{d}\alpha \leqslant \sum_{n\leqslant 3N} R^2(u),$$

但 $\sum_{n\leqslant 3N} R^2(u)$ 为不定方程

$$x_2^2+x_3^3+x_5^5=y_2^2+y_3^3+y_5^5\leqslant 3N \tag{34}$$

的解数,且其每组解必适合下面三个条件之一

$$x_2^2-y_2^2=y_3^3+y_5^5-x_3^3-x_5^5\neq 0, \tag{35}$$

$$x_2=y_2,\ y_3^3-x_3^3=x_5^5-y_5^5\neq 0, \tag{36}$$

$$x_2=y_2,\ x_3=y_3,\ x_5=y_5, \tag{37}$$

因由(34), $x_k\leqslant(3N)^{1/k}$, $y_k\leqslant(3N)^{1/k}$ $(k=2, 3, 5)$, 又不定方程 $|x_2^2-y_2^2|=u$, $|x_3^3-y_3^3|=u$ 之解数是$O(d(u))$,所以从引理 11 直接推得本引理.

引理 13　设 $\sigma\geqslant 9$ 是一任意给定的实数,那末

$$J_2'(N)\ll N^{2\mu-1}L^{-\sigma}. \tag{38}$$

证　取 $\sigma_2=\frac{1}{2}(\sigma+\gamma)$, 则对适合引理 10 之 σ_3 及 α,有

$$S_k(\alpha) \ll N^{1/k} L^{-\sigma_2}.$$

若取 $\sigma_0 = \sigma_2 = \max\left\{2^{\sigma r}(\sigma_2+1),\ \frac{k}{2}(\sigma+\gamma)\right\}$，对 $q \leqslant L^{\sigma_0}$，$|\beta| > N^{-1}L^{\sigma_3}$，由引理 8，

$$S_k(\alpha) \ll q^{-1/k+\varepsilon} |\beta|^{-1/k} \ll N^{1/k} L^{-\sigma_3/k},$$

于是由引理 12，

$$\begin{aligned} J_2'(N) &\ll N^{2/k} L^{-\sigma-\gamma} \int_0^1 |S_2(\alpha) S_3(\alpha) S_5(\alpha)|^2 \mathrm{d}\alpha \leqslant \\ &\leqslant N^{2/k} L^{-\sigma-\gamma} \int_0^1 |F_2(\alpha) F_3(\alpha) F_5(\alpha)|^2 \mathrm{d}\alpha \ll \\ &\ll N^{2\mu-1} L^{-\sigma}. \end{aligned}$$

引理 14[2] 取 $\sigma_1 = (2\mu-1)^{-1}(2\sigma_0+\sigma)$，则

$$\int_0^1 \left| \sum_{q \leqslant L^{\sigma_0}} \sum_{\substack{h=1 \\ (h,\,q)=1}}^{q}{}' \prod_j T_j(\alpha,\,h,\,q) \right|^2 \mathrm{d}\alpha = O(N^{2\mu-1} L^{-\sigma}), \tag{39}$$

其中求和号 $\sum\limits_{q \leqslant L^{\sigma_0}} \sum\limits_{\substack{h=1 \\ (h,\,q)=1}}^{q}{}'$ 表示当 $\alpha \in M_{h,\,q}$ 时去掉 $\prod\limits_j T_j(\alpha,\,h,\,q)$ 这一项.

对于 $J_1(N)$，由(28)，

$$\begin{aligned} J_1(N) &\ll \sum_{q \leqslant L^{\sigma_0}} \sum_{\substack{h=1 \\ (h,\,q)=1}}^{q} \int_{M_{h,\,q}} \left| \prod_j S_j(\alpha) - \prod_j T_j(\alpha,\,h,\,q) \right|^2 \mathrm{d}\alpha + \\ &+ \sum_{q \leqslant L^{\sigma_0}} \sum_{\substack{h=1 \\ (h,\,q)=1}}^{q} \int_{M_{h,\,q}} \left| \sum_{q' \leqslant L^{\sigma_0}} \sum_{\substack{h'=1 \\ (h',\,q')=1 \\ h',\,q' \neq hq}}^{q'} \prod_j T_j(\alpha,\,h',\,q') \right|^2 \mathrm{d}\alpha = \\ &= J_1'(N) + J_1''(N). \end{aligned} \tag{40}$$

引理 15

$$J_1''(N) + J_2''(N) \ll N^{2\mu-1} L^{-\sigma}. \tag{41}$$

证 因当 $\alpha \in E$ 时，

$$\begin{aligned} J_1''(N) + J_2''(N) &= \int_0^1 \left| \sum_{q \leqslant L^{\sigma_0}} \sum_{\substack{h=1 \\ (h,\,q)=1}}^{q}{}' \prod_j T_j(\alpha,\,h,\,q) \right|^2 \mathrm{d}\alpha = \\ &= O(N^{2\mu-1} L^{-\sigma}) \end{aligned}$$

引理 16

$$J_1'(N)=O(N^{2\mu-1}e^{-c_6L^{1/2}}). \tag{42}$$

证　由引理 7,引理 8 及[2]的引理 9 类似地推导可得.

引理 17

$$J(N)=O(N^{2\mu-1}L^{-\sigma}). \tag{43}$$

证　由(28),(29),(38),(41),(42)直接推得.

定理 1 的证　由(16),(24),(25),(43),

$$\sum_{n=3}^{N}\{r(n)-S_1(n)\psi(n)\}^2\leqslant$$

$$\leqslant 2\Big[\sum_{n=3}^{N}\{r(n)-S(n)\psi(n)\}^2+(\psi(n))^2\{S(n)-S_1(n)\}^2\Big]\ll$$

$$\ll N^{2\mu-1}L^{-9+\varepsilon}$$

于是由(17),(25)

$$\sum_{\substack{r(n)=0\\2|n\\n\leqslant N}}n^{2(\mu-1)}\ll N^{2\mu-1}L^{-1+\varepsilon}.$$

现设 $Q(N)$为 $1\leqslant n\leqslant N$,且使 $r(n)=0,\ 2\mid n$ 成立的 n 的个数,那么

$$Q(N)^{2\mu-1}\ll\sum_{r=1}^{c(N)}r^{2(\mu-1)}\leqslant\sum_{\substack{r(n)=0\\2|n\\n\leqslant N}}n^{2(\mu-1)}\ll N^{2\mu-1}L^{-1+\varepsilon},$$

此即

$$Q(N)\ll NL^{-(1-\varepsilon)/(2\mu-1)}$$

定理得证.

最后,我们指出用同样办法可得:

若 k 是奇数,$s\geqslant 3$,则几乎所有的满足 $n\equiv s,\ s+2(\bmod 6)$ 的正整数 n 都可以表为

$$n=p_1^2+p_2^4+\cdots+p_s^{2^s}+p_{s+1}^{2^s}+p_{s+2}^k,$$

其中 $p_i(1\leqslant i\leqslant s+2)$ 为素数.

参考文献

[1] K. F. Roth, *J. of London Math. Soc.*, **24**(1949), 4 - 13.
[2] K. Prachar, *Monatsh Math.*, **57**(1953), 66 - 74.
[3] K. Thanigasalam, *Trans. Amer. Math. Soc.*, **200**(1974), 199 - 205.
[4] Roger Cook, *J. Number Theory*, **4**(1979), 516 - 528.
[5] 华罗庚,堆垒素数论,科学出版社,1957.

一个 Waring-Гольдбах 型问题

在 Roth[1] 证明了几乎所有的正整数皆能表成一个整数的平方、一个自然数的立方和一个整数的四次方之和以后，Prachar[2] 在 1952 年得到，几乎所有的偶数 n 都能表成

$$n=p_1^2+p_2^3+p_3^4+p_4^5,$$

其中 $p_i(1\leqslant i\leqslant 4)$ 为素数.

1974 年，Thanigasalam[3] 指出，叙列 $\{x_1^2+x_2^3+x_3^5+x_4^k\}$, $(k\geqslant 5)$ 和 $\{x_1^2+x_2^3+x_3^6+x_4^k\}$, $(k\geqslant 6)$ 具有正的 Шнирельман 密率；其后，1978 年，Roger Cook[4] 证明，几乎所有的正整数 n 都能表成 $n=x_1^2+x_2^3+x_3^5+x_4^k$, $(k\geqslant 5)$ 及 $n=x_1^2+x_2^3+x_3^6+x_4^k$, $(k\geqslant 6)$. 这里 x_i 为正整数. 现在，作者将整幂问题改成素幂问题，从而得到

定理 1 对任何正整数 k，几乎所有的偶数 n 都能表成

$$n=p_1^2+p_2^3+p_3^5+p_4^k, \tag{1}$$

其中 $p_i(1\leqslant i\leqslant 4)$ 是素数.

定理 2 除了 $k\equiv 0(\mathrm{mod}2)$ 或 $k\equiv 0\ (\mathrm{mod}3)$ 的正整数 k 以外，几乎所有的偶数 n 都能表成

$$n=p_1^2+p_2^3+p_3^6+p_4^k, \tag{2}$$

其中 $p_i(1\leqslant i\leqslant 4)$ 是素数.

定理 1 是 Prachar[2] 的结果的一种推广. 以下我们仅证定理 1，定理 2 可完全类似地得到.

引理 1 对 $p\geqslant 23$，及任何自然数 u，同余方程

$$x^2+y^3+z^5\equiv u,\ (\mathrm{mod}\,p)\,p\nmid xyz \tag{3}$$

必有解.

由此引理可推出

引理 2 设

$$W_{h,q}^{(r)}=\sum_{\substack{l=1\\(l,q)=1}}^{q}e_q(hl^r),\tag{4}$$

$$A_q(n)=((\varphi(q))^{-4}\sum_{\substack{h=1\\(h,q)=1}}^{q}W_{h,q}^{(2)}W_{h,q}^{(3)}W_{h,q}^{(5)}W_{h,q}^{(k)}e_q(-nh),\tag{5}$$

则对 $p>2$,

$$1+A_p(n)>0,\tag{6}$$

及

$$1+A_2(n)=\begin{cases}0 & 若\ n\not\equiv 0(\bmod 2),\\ 2 & 若\ n\equiv 0(\bmod 2).\end{cases}\tag{7}$$

以下设 N 是一个充分大的整数, $L=\log N$.

引理 3[2] 设 $\sigma\geqslant 9$ 是一任意给定的实数, σ_0 是与 σ 有关的一个正数,

$$S(n)=\sum_{q\leqslant L^{\sigma_0}}A_q(n),\tag{8}$$

$$S_0(n)=\sum_{2\leqslant p<\frac{1}{5}L}(1+A_p(n)),\tag{9}$$

则

$$\sum_{n=3}^{N}\{S(N)-S_0(N)\}^2=O(NL^{-1+\varepsilon}).\tag{10}$$

由引理 2 可得到

引理 4 若 n 是偶数,则

$$S_0(n)>c_2(\log\log N)^{-c_1},\tag{11}$$

其中 c_1, c_2 是正常数.

引理 5 设 $r(n)$为(1)式的解数,及

$$\psi(n)=\sum_{\substack{n_1+n_2+n_3+n_4=n\\2\leqslant n_i\leqslant N}}\frac{1}{n_1^{1-(1/2)}n_2^{1-(1/3)}n_3^{1-(1/5)}n_4^{1-(1/k)}\log n_1\log n_2\log n_3\log n_4},\tag{12}$$

$$S_r(\alpha)=\sum_{p\leqslant N^{1/r}} e(\alpha p^r), \tag{13}$$

$$\prod_j S_j(\alpha)=S_2(\alpha)S_3(\alpha)S_5(\alpha)S_k(\alpha). \tag{14}$$

又对 $\alpha=\frac{h}{q}+\beta$, $(h, q)=1$, 设

$$T_r(\alpha, h, q)=\frac{W_{h,q}^{(r)}}{\varphi(q)}\sum_{2\leqslant n\leqslant N} n^{\frac{1}{r}-1}(\log n)^{-1}e(n\beta), \tag{15}$$

$$\prod_j T_1(\alpha, h, q)=T_2(\alpha, h, q)T_3(\alpha, h, q)T_5(\alpha, h, q)T_k(\alpha, h, q), \tag{16}$$

那么

$$\begin{aligned} J(N)&=\int_0^1\Big|\prod_j S_j(\alpha)-\sum_{q\leqslant L^{\sigma_0}}\sum_{\substack{h=1\\(h, q)=1}}^{q}\prod_j T_j(\alpha, h, q)\Big|^2 d\alpha \\ &\geqslant\sum_{n=3}^{N}\{r(n)-S(n)\psi(n)\}^2, \end{aligned} \tag{17}$$

且对 $0<c_3<n\leqslant N$,

$$c_4 n^{\mu-1}\log^{-4} n<\psi(n)<c_5 n^{\mu-1}\log^{-4} n, \tag{18}$$

其中 c_3, c_4, c_5 为正常数, $\mu=\frac{1}{2}+\frac{1}{3}+\frac{1}{5}+\frac{1}{k}$.

现设 $\tau=NL^{-\sigma_1}$, 以 $M_{h,q}$ 表示区间

$$\left|\alpha-\frac{h}{q}\right|<\tau^{-1}, (h, q)=1, 0\leqslant h<q\leqslant L^{\sigma_0}, \tag{19}$$

而命

$$E=(-\tau^{-1}, 1-\tau^{-1})-\sum_{h,q} M_{h,q}. \tag{20}$$

其中 $\sum\limits_{h,q} M_{h,q}$ 表示所有适合(19)式的区间的和集. 由(17)式,

$$\begin{aligned} J(N)&=\Big(\sum_{h,q}\int_{M_{h,q}}+\int_E\Big)\Big|\prod_j S_j(\alpha)-\sum_{q\leqslant L^{\sigma_0}}\sum_{\substack{h=1\\(h, q)=1}}^{q}\prod_j T_j(\alpha, h, q)\Big|^2\cdot d\alpha \\ &=J_1(N)+J_2(N). \end{aligned} \tag{21}$$

$$
\begin{aligned}
J_1(N) = &\sum_{q \leqslant L^{\sigma_0}} \sum_{\substack{h=1 \\ (h,q)=1}}^{q} \int_{M_{h,q}} \Big| \prod_j S_j(\alpha) - \prod_j T_j(\alpha, h, q) \\
&- \sum_{q' \leqslant L^{\sigma_0}} \sum_{\substack{h'=1 \\ (h',q')=1 \\ (h',q') \neq (h,q)}}^{q'} \prod_j T_j(\alpha, h', q') \Big|^2 \mathrm{d}\alpha \\
\leqslant &\sum_{q \leqslant L^{\sigma_0}} \sum_{\substack{h=1 \\ (h,q)=1}}^{q} \int_{M_{h,q}} \Big| \prod_j S_j(\alpha) - \prod_j T_j(\alpha, h, q) \Big|^2 \mathrm{d}\alpha \\
&+ \sum_{q \leqslant L^{\sigma_0}} \sum_{\substack{h=1 \\ (h,q)=1}}^{q} \int_{M_{h,q}} \Big| \sum_{q' \leqslant L^{\sigma_0}} \sum_{\substack{h'=1 \\ (h',q')=1 \\ (h',q') \neq (h,q)}}^{q'} \prod_j T_j(\alpha, h', q') \Big|^2 \mathrm{d}\alpha \\
= &J_1'(N) + J_1''(N).
\end{aligned} \tag{22}
$$

$$
\begin{aligned}
J_2(N) \ll &\int_E \Big| \prod_j S_j(\alpha) \Big|^2 \mathrm{d}\alpha + \int_E \Big| \sum_{q \leqslant L^{\sigma_0}} \sum_{\substack{h=1 \\ (h,q)=1}}^{q} \prod_j T_j(\alpha, h, q) \Big|^2 \mathrm{d}\alpha \\
= &J_2'(N) + J_2''(N).
\end{aligned} \tag{23}
$$

引理 6 存在正数 γ，使

$$
\sum_{x_1=1}^{(3N)^{1/3}} \sum_{\substack{x_2=1 \\ x_1^3+x_2^5-y_1^3-y_2^5 \neq 0}}^{(3N)^{1/5}} \sum_{y_1=1}^{(3N)^{1/3}} \sum_{y_2=1}^{(3N)^{1/5}} \mathrm{d}(|x_1^3 + x_2^5 - y_1^3 - y_2^5|) \ll N^{2\bar{\mu}-1} L^{\gamma}, \tag{24}
$$

其中 $\bar{\mu} = \frac{1}{2} + \frac{1}{3} + \frac{1}{5}$.

引理 7 设

$$
F_r(\alpha) = \sum_{x \leqslant N^{1/r}} e(\alpha x^r), \tag{25}
$$

则

$$
\int_0^1 |F_2(\alpha) F_3(\alpha) F_5(\alpha)|^2 \mathrm{d}\alpha \ll N^{2\bar{\mu}-1} L^{\gamma}. \tag{26}
$$

由此引理及熟知的推导过程可得

引理 8 $J_2'(N) \ll N^{2\mu-1} L^{-\sigma}$. (27)

引理 9

$$
J_1''(N) + J_2''(N) = \int_0^1 \Big| \sum_{q \leqslant L^{\sigma_0}} \sum_{\substack{h=1 \\ (h,q)=1}}^{q}{}' \prod_j T_j(\alpha, h, q) \Big|^2 \mathrm{d}\alpha
$$

$$=O(N^{2\mu-1}L^{-\sigma}), \tag{28}$$

其中求和号 $\sum\limits_{q\leqslant L^{\sigma_0}}\sum\limits_{\substack{h=1\\(h,q)=1}}^{q}{}'$ 表示当 $\alpha\in M_{h,q}$ 时去掉 $\prod\limits_{j}T_j(\alpha,h,q)$ 这一项,及存在正常数 c,

$$J_1'(N)\ll N^{2\mu-1}e^{-c\sqrt{L}}. \tag{29}$$

由此 $J(N)\ll N^{2\mu-1}L^{-\sigma}$. 再由(10)、(17)、(18)式

$$\begin{aligned}&\sum_{n=3}^{N}\{r(n)-S_0(n)\psi(n)\}^2\\ \leqslant\ &2\sum_{n=3}^{N}\{[r(n)-S(n)\psi(n)]^2+(\psi(n))^2[S(n)-S_0(n)]^2\}\\ \ll\ &N^{2\mu-1}L^{-9+\varepsilon}.\end{aligned}$$

于是由(11)、(18)式, $\sum\limits_{\substack{r(n)=0\\2\mid n\\n\leqslant N}}n^{2(\mu-1)}\ll N^{2\mu-1}L^{-1+\varepsilon}$.

设 $Q(N)$ 为使 $1\leqslant n\leqslant N$, $r(n)=0$, $2\mid n$ 成立的 n 的个数,那末

$$(Q(N))^{2\mu-1}\ll\sum_{r=1}^{Q(N)}r^{2(\mu-1)}\ll\sum_{\substack{r(n)=0\\2\mid n\\n\leqslant N}}n^{2(\mu-1)}\ll N^{2\mu-1}L^{-1+\varepsilon},$$

此即 $Q(N)\ll NL^{\frac{-1+\varepsilon}{2\mu-1}}$. 定理得证.

最后,我们指出用同样的方法可得:

若 k 是奇数, $s\geqslant3$, 则几乎所有的满足 $n\equiv s$, $s+2(\mathrm{mod}6)$ 的正整数 n 都可以表为

$$n=p_1^2+p_2^4+\cdots+p_s^{2^s}+p_{s+1}^{2^s}+p_{s+2}^k,$$

其中 $p_i(1\leqslant i\leqslant s+2)$ 为素数.

参考文献

[1] Roth, K. F., *J. London Math. Soc.*, **24**(1949), 4-13.
[2] Prachar, K., Typ. I. *Monatsh. Math.*, **57**(1953), 66-74.
[3] Thanigasalam, K., *Trams. Amer. Math. Soc.*, **200**(1974), 199-205.
[4] Roger Cook, *J. Number Theory*, **4**(1979), 516-528.

A Problem of Waring-Goldbach's Type

After K. F. Roth[1] had proved that almost all integers might be expressed as the sum of a square, a cube and a fourth power of positive integers, in 1952 K. Prachar[2] obtained that almost all positive even integers n may be represented by

$$n=p_1^2+p_2^3+p_3^4+p_4^5,$$

where p_i's $(1\leqslant i\leqslant 4)$ are primes.

In 1974, K. Thanigasalam[3] pointed out that the sequence of integers involved in the form $x_1^2+x_2^3+x_3^5+x_4^k$ or $x_1^2+x_2^3+x_3^6+x_4^k$, where x_i's are non-negative integers, had positive Schnirelmann density. Recently, R. Cook[4] indicated that almost all positive integers n might be expressed by

$$n=x_1^2+x_2^3+x_3^5+x_4^k$$

and

$$n=x_1^2+x_2^3+x_3^6+x_4^k,$$

where x_i's are positive integers.

In this paper, considering the corresponding problem for prime powers, we obtained the following

Theorem 1. *For any positive integer k, almost all even integers n may be expressed as*

$$n=p_1^2+p_2^3+p_3^5+p_4^k, \tag{1}$$

where p_i's are primes.

Theorem 2. *For any positive integer k, except $k\equiv 0 \pmod 2$ or $k\equiv 0$ $(\bmod 3)$, almost all even integers n may be expressed as*

$$n=p_1^2+p_2^3+p_3^6+p_4^k, \tag{2}$$

where p_i's are primes.

We need only to prove Theorem 1 and can complete the proof of Theorem 2 in the same way.

Lemma 1. *For any positive integer u and $p\geqslant 23$, there exist x, y and z satisfying*

$$x^2+y^3+z^5\equiv u\ (\bmod p),\ p\nmid xyz. \tag{3}$$

By this lemma we can obtain

Lemma 2. *Let*

$$W_{h,q}^{(r)}=\sum_{\substack{l=1\\(l,q)=1}}^{q} e_q(hl^r), \tag{4}$$

$$A_q(n)=(\phi(q))^{-4}\sum_{\substack{h=1\\(h,q)=1}}^{q} W_{h,q}^{(2)}W_{h,q}^{(3)}W_{h,q}^{(5)}W_{h,q}^{(k)}e_q(-nh), \tag{5}$$

then for $p>2$,

$$1+A_p(n)>0 \tag{6}$$

and

$$1+A_2(n)=\begin{cases}0, & if\ n\equiv 0\ (\bmod 2),\\ 2, & if\ n\not\equiv 0\ (\bmod 2).\end{cases} \tag{7}$$

Not let N be a sufficiently large integer, $L=\log N$.

Lemma 3[2]. *Let $\sigma\geqslant 9$ be any fixed positive number, σ_0 is a positive number that depends on σ,*

$$S(n)=\sum_{q\leqslant L^{\sigma_0}} A_q(n). \tag{8}$$

$$S_0(n)=\prod_{z\leqslant p<\frac{1}{5}L}(1+A_p(n)), \tag{9}$$

then

$$\sum_{n=3}^{N}\{S(n)-S_0(n)\}^2=O(NL^{-1+\varepsilon}). \tag{10}$$

From Lemma 2 we have

Lemma 4. *If n is an even integer, then*

$$S_0(n)>c_2(\log\log N)^{-c_1}, \tag{11}$$

where c_1, c_2 are positive constants.

Lemma 5. *Let $r(n)$ be the number of solutions of (1),*

$$\psi(n)=\sum_{\substack{n_1+n_2+n_3+n_4=n\\ 2\leqslant n_i\leqslant N}}\frac{1}{n_1^{1-\frac{1}{2}}n_2^{1-\frac{1}{3}}n_3^{1-\frac{1}{5}}n_4^{1-\frac{1}{k}}\log n_1\log n_2\log n_3\log n_4}, \tag{12}$$

$$S_r(\alpha)=\sum_{p\leqslant N^{\frac{1}{r}}}e^{(\alpha p^r)}. \tag{13}$$

$$\prod_j S_j(\alpha)=S_2(\alpha)S_3(\alpha)S_5(\alpha)S_k(\alpha) \tag{14}$$

and for $\alpha=\frac{h}{q}+\beta,\ (h,q)=1.$

let

$$T_r(\alpha,h,q)=\frac{W_{h,q}^{(r)}}{\phi(q)}\sum_{2\leqslant n\leqslant N}n^{\frac{1}{r}-1}(\log n)^{-1}e(n\beta), \tag{15}$$

$$\prod_j T_j(\alpha,h,q)=T_2(\alpha,h,q)T_3(\alpha,h,q)T_5(\alpha,h,q)T_k(\alpha,h,q), \tag{16}$$

then

$$J(N)=\int_0^1\Big|\prod_j S_j(\alpha)-\sum_{q\leqslant L^{\sigma_0}}\sum_{\substack{h=1\\(h,q)=1}}^{q}\prod_j T_j(\alpha,h,q)\Big|^2\mathrm{d}\alpha$$

$$\geqslant\sum_{n=3}^{N}\{r_{(n)}-S_{(n)}\psi_{(n)}\}^2, \tag{17}$$

and for $0<c_3<n\leqslant N$,

$$c_4n^{\mu-1}\log^{-4}n<\psi(n)<c_5n^{\mu-1}\log^{-4}n, \tag{18}$$

where c_3, c_4, c_5 are positive constants,

$$\mu = \frac{1}{2} + \frac{1}{3} + \frac{1}{5} + \frac{1}{k}.$$

Now let $\sigma_1 = \dfrac{2\sigma_0 + \sigma}{2\mu - 1}$ and $\tau = NL^{-\sigma_1}$, $M_{h,q}$ denotes the interval

$$\left| \alpha - \frac{h}{q} \right| < \tau^{-1},\ (h, q) = 1,\ 0 \leqslant h < q \leqslant L^{\sigma_0}, \tag{19}$$

and let

$$E = (-\tau^{-1},\ 1 - \tau^{-1}) - \sum_{h,q} M_{h,q}, \tag{20}$$

where $\sum\limits_{h,q} M_{h,q}$ is the union of all intervals satisfying (19).

By (17) we have

$$\begin{aligned} J(N) &= \Big(\sum_{h,q} \int_{M_{h,q}} + \int_E \Big) \Big| \prod_j S_j(\alpha) - \sum_{q \leqslant L^{\sigma_0}} \sum_{\substack{h=1 \\ (h,q)=1}}^{q} \prod_j T_j(\alpha, h, q) \Big|^2 \mathrm{d}\alpha \\ &= J_1(N) + J_2(N), \end{aligned} \tag{21}$$

$$\begin{aligned} J_1(N) &\ll \sum_{q \leqslant L^{\sigma_0}} \sum_{\substack{h=1 \\ (h,q)=1}}^{q} \int_{M_{h,q}} \Big| \prod_j S_j(\alpha) - \prod_j T_j(\alpha, h, q) \Big|^2 \mathrm{d}\alpha \\ &\quad + \sum_{q \leqslant L^{\sigma_0}} \sum_{\substack{h=1 \\ (h,q)=1}}^{q} \int_{M_{h,q}} \Big| \sum_{q' \leqslant L^{\sigma_0}} \sum_{\substack{h'=1 \\ (h',q')=1}}^{q'} \prod_j T_j(\alpha, h', q') \Big|^2 \mathrm{d}\alpha \\ &= J_1'(N) + J_1''(N), \end{aligned} \tag{22}$$

$$\begin{aligned} J_2(N) &\ll \int_E \Big| \prod_j S_{j(\alpha)} \Big|^2 \mathrm{d}\alpha + \int_E \Big| \sum_{q \leqslant L^{\sigma_0}} \sum_{\substack{h=1 \\ (h,q)=1}}^{q} \prod_j T_j(\alpha, h, q) \Big|^2 \mathrm{d}\alpha \\ &= J_2'(N) + J_2''(N). \end{aligned} \tag{23}$$

Lemma 6. *There is a positive number* γ *such that*

$$\sum_{x_1=1}^{(3N)^{1/3}} \sum_{x_2=1}^{(3N)^{1/5}} \sum_{y_1=1}^{(3N)^{1/3}} \sum_{y_2=1}^{(3N)^{1/5}} \mathrm{d}(|x_1^3 + x_2^5 - y_1^3 - y_2^5|) \ll N^{2\bar{\mu}-1} L^{\gamma}, \tag{24}$$

where

$$\bar{\mu} = \frac{1}{2} + \frac{1}{3} + \frac{1}{5}.$$

Lemma 7. *Let*

$$F_r(\alpha)=\sum_{x\leqslant N^{\frac{1}{r}}} e(\alpha x^r), \tag{25}$$

then

$$\int_0^1 |F_2(\alpha)F_3(\alpha)F_5(\alpha)|^2 \mathrm{d}\alpha \ll N^{2\bar{\mu}-1}L^{\gamma}. \tag{26}$$

By this Lemma we can obtain

Lemma 8.

$$J'_2(N) \ll N^{2\mu-1}L^{-\sigma}. \tag{27}$$

Lemma 9.

$$J'_1(N) \ll N^{2\mu-1}e^{-c\sqrt{L}} \tag{28}$$

and

$$J''_1(N)+J''_2(N)=\int_0^1 \Big|\sum_{q\leqslant L^{\sigma_0}} \sum_{\substack{h=1\\(h,q)=1}}^{q}{}' \prod_j T_j(\alpha,h,q)\Big|^2 \mathrm{d}\alpha = O(N^{2\mu-1}L^{-\sigma}), \tag{29}$$

where c is a positive number and the summation sign $\sum_{q\leqslant L^{\sigma_0}} \sum_{\substack{h=1\\(h,q)=1}}^{q}{}'$ *means that, if* α *lies on* $M_{h,q}$, *the term* $\prod_j T_j(\alpha,h,q)$ *is omitted from the sum.*

From these it follows that

$$J(N) \ll N^{2\mu-1}L^{-\sigma},$$

also by (10), (17), (18) we obtained

$$\sum_{n=3}^{N}\{r(n)-s_0(n)\psi(n)\}^2 \leqslant 2\sum_{n=3}^{N}\{[r(n)-s(n)\psi(n)]^2 + (\psi(n))^2[s(n)-s_0(n)]^2\} \ll N^{2\mu-1}L^{-q+\varepsilon}.$$

Therefore, by (11), (18) we have

$$\sum_{\substack{r(n)=0\\2|n\\n\leqslant N}} n^{2(\mu-1)} \ll N^{2\mu-1}L^{-1+\varepsilon}.$$

Let $Q(N)$ be the number of n satisfying

$$1 \leqslant n \leqslant N,\ r(n)=0,\ 2 \mid n,$$

then

$$(Q(N))^{2\mu-1} \ll \sum_{r=1}^{Q(N)} r^{2(\mu-1)} \ll \sum_{\substack{r(n)=0 \\ 2\mid n \\ n\leqslant N}} n^{2(\mu-1)} \ll N^{2\mu-1} L^{-1+\varepsilon},$$

i. e.

$$Q(N) \ll NL^{\frac{-1+\varepsilon}{2\mu-1}}.$$

The proof of Theorem 1 is complete.

Meanwhile, we point out that a similar result can be obtained.

If k is an odd integer, $s \geqslant 3$, then almost all positive integers n satisfying $n \equiv s,\ s+2 \pmod 6$ may be represented by

$$n = p_1^2 + p_2^4 + \cdots + p_s^{2^s} + p_{s+1}^{2^s} + p_{s+2}^k,$$

where p_i's are primes.

References

[1] Roth, K. F., *J. London Math. Soc.*, **24**(1949), 4 - 13.

[2] Prachar, K., *Monatsh. Math.*, **57**(1953), 66 - 74.

[3] Thanigasalam, K., *Trans. Amer. Math. Soc.*, **200**(1974), 199 - 205.

[4] Cook, R., *J. Number Theory*, **4**(1979), 516 - 528.

[5] 华罗庚,堆垒素数论,科学出版社,1957.

一个整数幂和问题

摘　要

萨利赫萨拉姆[16]证明了对任一正整数 k,由

$$x_1^2+x_2^4+\cdots+x_s^{2^s}+x_{s+1}^{2^s}+x_{s+2}^k \tag{$*$}$$

所表示的正整数数列具有正密率.

本文改进了他的结果,证明了下面的

定理　若整数 $s\geqslant 6$,则对一正整数 k,几乎所有的正整数可以表成($*$)的形式.

1　引言

K. Thanigasalam[16]在 1974 年证明了对任意的正整数 k,可以表示成

$$x_1^2+x_2^3+x_3^5+x_4^k \tag{1}$$

(其中 x_1, x_2, x_3, x_4 为非负整数)的正整数及可以表示成

$$x_1^2+x_2^3+x_3^6+x_4^k \tag{2}$$

(其中 x_1, x_2, x_3, x_4 为非负整数)的正整数都具有正的 Щнирельман 密率. 他又证明了对任意的正整数 s, k 可以表示成

$$x_1^2+x_2^{2^2}+\cdots+x_s^{2^s}+x_{s+1}^{2^s}+x_{s+2}^k \tag{3}$$

(其中 x_1, x_2, $\cdots x_{s+2}$ 为非负整数)的正整数也具有正的 Щнирельман 密率.

不久前,R. Cook[1]证明了对任意正整数 k,几乎所有的正整数 n 可以表示成(1)及几乎所有的正整数 n 可以表示成(2).

本文证明对于 $s\geqslant 6$ 及任意的正整数 k,几乎所有的正整数 n 可以表示成(3).即

定理 1 在正整数 $s \geqslant 6$ 时，对于任意的正整数 k，几乎所有的正整数 n 可以表示成一个平方数，一个 4 次幂，……，一个 2^{s-1} 次幂、两个 2^s 次幂及一个 k 次幂的和.

当 $s \leqslant 5$ 时，对于满足同余可解条件的 n，也就是使得同余方程

$$n-1 \equiv x_1^2+x_2^4+\cdots+x_s^{2^s}+x_{s+1}^{2^s}(\bmod 16) \tag{4}$$

有解的 n，亦有类似的结论成立. 例如，我们有

系 对于任意的正整数 k，几乎所有的满足

$$n \equiv 1,\ 2,\ 3,\ 4,\ 5,\ 6,\ 7,\ 8,\ 10,\ 11,\ 12,\ 13\ (\bmod 16)$$

的正整数 n 可以表示成一个平方数，一个 4 次幂，两个 8 次幂及一个 k 次幂的和.

显然，在证明定理 1 时，不妨假设 $k>2^s$.

2 记号

p 素数，l、n、x、v、r、q、h 表示非负整数（不论带不带下标），并且 $(q, h)=1$, c, c_1, c_2, $\cdots$ 表示正的常数，仅与 k、s 有关. N 表示大正数，$X=\frac{\delta}{2}\log N$，其中正数 δ 将在下面确定. ε 表示任意小的正数. θ、β 为实数

$$f_r=f_r(\theta)=\sum_{1\leqslant x\leqslant N^{\frac{1}{r}}} e(\theta x^r)$$

$$V(\theta)=f_2f_4\cdots f_{2^{s-1}}f_{2^s}^2f_k=\sum_{1\leqslant n\leqslant N}\rho(n)e(\theta n)+\sum_{N<n\leqslant(s+2)N}t(n)e(n\theta).$$

其中 $\rho(n)$ 表示使得

$$n=x_1^2+x_2^4+\cdots+x_s^{2^s}+x_{s+1}^{2^s}+x_{s+2}^k$$

成立的满足 $1\leqslant x_1\leqslant N^{1/2}$, $1\leqslant x_2\leqslant N^{1/4}$, $\cdots$, $1\leqslant x_s\leqslant N^{\frac{1}{2^s}}$, $1\leqslant x_{s+1}\leqslant N^{\frac{1}{2^s}}$, $1\leqslant x_{s+2}\leqslant N^{1/k}$ 的 $(x_1, \cdots, x_{s+2})$ 的个数，$t(n)$ 与 θ 无关.

$$I_r(\beta)=\frac{1}{r}\sum_{1\leqslant n\leqslant N}n^{\frac{1}{r}-1}e(\beta n),$$

$$S_{h,q}^{(r)}=\sum_{x=1}^{q}e_q(hx^r),$$

$$F_r=F_r(\theta,h,g)=\frac{1}{q}S_{h,q}^{(r)}I_{r\left(\theta-\frac{h}{q}\right)},$$

$$U(\theta,h,q)=F_2F_4\cdots F_{2^{s-1}}F_{2^s}^2F_k,$$

$$A(n,q)=\frac{1}{q^{s+2}}\sum_{\substack{h=1\\(h,q)=1}}^{q}S_{h,q}^{(2)}\cdot S_{h,q}^{(4)}\cdots(S_{h,q}^{(2^s)})^2\cdot S_{h,q}^{(k)}e_q(-hn),$$

$$\mathscr{S}(N,q)=\prod_{p\leqslant X}\sum_{\substack{v\geqslant 0\\p^v\leqslant X}}A(n,q^v),$$

$$\psi(n)=\sum_{n_1+n_2+\cdots+n_{s+2}=n}\frac{n_1^{1/2-1}n_2^{1/4-1}\cdots n_s^{1/2^s-1}n_{s+1}^{1/2^s-1}n_{s+2}^{1/k-1}}{2\times4\times\cdots\times2^{s-1}\times2^s\times2^s\times k}.$$

3 Farey 分割

取正数 $\alpha<\frac{c_1}{k}$，c_1 为适当小的正数，将区间 $0\leqslant\theta\leqslant1$ 分割为以有理点 $\frac{h}{q}$ 为中心的 $N^{1-\alpha}$ 阶的 Farey 弧（端点按通常约定），于是对每个 $\theta\in(0,1)$，存在既约分数 h/q，使得

$$\left|\theta-\frac{h}{q}\right|<q^{-1}N^{\alpha-1}.$$

对于 $q\leqslant N^\alpha$ 的弧用 $\mathscr{M}(h,q)$ 表示，$\mathscr{M}(h,q)$ 的全体记为 $\mathscr{M}$.

4 在 (0, 1) − $\mathscr{M}$ 上的估计

引理 1(Weyl) 设 $\theta\in(0,1)-\mathscr{M}$，则对任意 $\varepsilon>0$，$|f_k(\theta)|\ll N^{\frac{1}{k}-\frac{\alpha}{2^{k-1}}+\varepsilon}$.

证 见[2]引理 1 或[19]引理 3.5.

引理 2 对任意 $\varepsilon>0$，有

$$\int_0^1|f_2f_4\cdots f_{2^{s-1}}f_{2^s}^2|^2\mathrm{d}\theta\ll N^{1+\varepsilon}.$$

证 考虑方程

$$x_1^2+x_2^4+\cdots+x_s^{2^s}+x_{s+1}^{2^s}=y_1^2+y_2^4+\cdots+y_s^{2^s}+y_{s+1}^{2^s}$$

的满足 $x_1, y_1\leqslant N^{1/2}$, x_2、$y_2\leqslant N^{1/4}$, $\cdots$, x_s、$y_s\leqslant N^{\frac{1}{2^s}}$, x_{s+1}、$y_{s+1}\leqslant N^{\frac{1}{2^s}}$ 的解数 M.

在 $x_1\neq y_1$ 时,方程

$$x_1^2-y_1^2=y_2^4+\cdots+y_{s+1}^{2^s}-x_2^4-\cdots-x_{s+1}^{2^s}$$

的解数 $$M_1\ll N^{2\left(\frac{1}{4}+\cdots+\frac{1}{2^s}+\frac{1}{2^s}\right)+\varepsilon}=N^{1+\varepsilon};$$

在 $x_1=y_1$ 时,考虑方程

$$x_2^4+\cdots+x_{s+1}^{2^s}=y_2^4+\cdots+y_{s+1}^{2^s},$$

其解数

$$M_2\ll N^{\frac{1}{2}+\varepsilon}, M\leqslant M_1+N^{\frac{1}{2}}M_2\ll N^{1+\varepsilon}+N^{\frac{1}{2}}\cdot N^{\frac{1}{2}+\varepsilon}\ll N^{1+\varepsilon}.$$

引理 3 $\int_{(0,1)-m}|V(\theta)|^2\mathrm{d}\theta\ll N^{1+\frac{2}{k}-\varepsilon}$.

证 由引理 1、2 立得.

5 在 $\mathscr{M}$ 上的估计

引理 4 若 $\theta\in\mathscr{M}(h, q)$, 则

$$f_r(\theta)=F_r(\theta, h, q)+O(q^{1-\frac{1}{r}+\varepsilon}).$$

证 参见[2]引理 8 的证明或[19]引理 7.11.

引理 5 $S_{h,q}^{(r)}=O(q^{1-\frac{1}{r}})$.

证 参见[9]引理 1.

引理 6 若 $|\beta|\leqslant\frac{1}{2}$, 则

$$I_r(\beta)=O(\min(N^{\frac{1}{r}}, N^{\frac{1}{r}-1}|\beta|^{-1})).$$

证 非平凡部分可由 $\sum_{n_1 \leqslant n \leqslant n_2} e(\beta n) = O(|\beta|^{-1})$ 及 Abel 引理得出.

引理 7 $F_r(\theta, h, q) \ll q^{-\frac{1}{r}} \min\left(N^{\frac{1}{r}}, N^{\frac{1}{r}-1}\left\|\theta - \frac{h}{q}\right\|^{-1}\right)$，其中 $\left\|\theta - \frac{h}{q}\right\|$ 表示 $\theta - \frac{h}{q}$ 与最近整数之间的距离.

证 由引理 5、6 立得.

引理 8 如果 $\theta \in \mathscr{M}(h, q)$，则对任予 $\varepsilon > 0$，$V(\theta) - U(\theta, h, q) \ll q^{1-\frac{1}{k}+\varepsilon} N$.

证 因为 $q \leqslant N^{\alpha}$，$\alpha < \frac{c_1}{k}$，

$$f_r = F_r(\theta, h, q) + O(q^{1-\frac{1}{r}+\varepsilon}) \ll q^{-\frac{1}{r}} N^{\frac{1}{r}},$$

$$f_2 f_4 - F_2 F_4 = f_2(f_4 - F_4) +_4 (f_2 - F_2) \ll N^{\frac{1}{2}} q^{\frac{1}{4}+\varepsilon},$$

$$f_2 f_4 f_8 - F_2 F_4 F_8 \ll N^{\frac{3}{4}} q^{\frac{1}{8}+\varepsilon}, \text{等等.}$$

引理 9 $\int_M |V(\theta) - U(\theta, h, \varepsilon)|^2 \mathrm{d}\theta = \sum_{h, q} \int_{M(h, \varepsilon)} |V(\theta) - U(\theta, h, \varepsilon)|^2 \mathrm{d}\theta \ll N^{1+\frac{2}{k}-\varepsilon}$.

证

$$\int_M |V(\theta) - U(\theta, h, q)|^2 \mathrm{d}\theta \ll \sum_{q \leqslant N^{\alpha}} \sum_h q^{-1} N^{\alpha-1} (q^{1-\frac{1}{k}+\varepsilon} N)^2 \ll \sum_{q \leqslant N^{\alpha}} N^{\alpha+1} q^{2-\frac{2}{k}+\varepsilon}$$

$$\ll N^{1+\alpha+\left(3-\frac{2}{k}+\varepsilon\right)\alpha} \ll N^{1+\frac{2}{k}-\varepsilon} \text{(取 } c_1 \text{ 足够小).}$$

6 解析论证的完成

取 $\delta < \frac{\alpha}{3}$，设 $\mathscr{E}(X)$ 为在

$$\prod_{p \leqslant X} \sum_{\substack{v \geqslant 0 \\ p^v \leqslant X}} p^v$$

的展开式中出现的整数 $q = p_1^{v_1} \cdots p_l^{v_l}$ 的集合，则因为

$$X^{\pi(X)} < X^{2X/\log X} = e^{2X} = N^{\delta},$$

所以 $\mathscr{E}(X)$中的整数 $q \leqslant N^{\delta}$. 另一方面,显然一切$\leqslant X$ 的整数q 均在 $\mathscr{E}(X)$中.

记 $$T(\theta) = \sum_{q \in \mathscr{E}(X)} \sum_{\substack{h=1 \\ (h,\, q)=1}}^{q} U(\theta,\, h,\, q),$$

$$T'(\theta) = \sum_{q \in \mathscr{E}(X)} \sum_{\substack{h=1 \\ (h,\, q)=1}}^{q} U'(\theta,\, h,\, q),$$

其中,

$$U'(\theta,\, h,\, q) = \begin{cases} 0, & 若\ \theta \in \mathscr{M}(h,\, q); \\ U(\theta,\, h,\, q), & 若\ \theta \in \mathscr{M}(h,\, q). \end{cases}$$

引理 10　η 为一正数,则有

$$\sum_{q \geqslant \eta} \sum_{h} \int_0^1 |U(\theta,\, h,\, q)|^2 \mathrm{d}\theta \ll N^{1+\frac{2}{k}} \eta^{-\frac{2}{k}}.$$

证　由引理 7,

$$\begin{aligned} & \sum_{q \geqslant \eta} \sum_{h} \int_0^1 |U(\theta,\, h,\, q)|^2 \mathrm{d}\theta \\ \ll & \sum_{q \geqslant \eta} \sum_{h} \int_0^1 q^{-2-\frac{2}{k}} N^{2+\frac{2}{k}} \min\left(1,\ N^{-2} \left\| \theta - \frac{h}{q} \right\|^{-2}\right) \mathrm{d}\theta \\ \ll & \sum_{q \geqslant \eta} q^{-1-\frac{2}{k}} N^{1+\frac{2}{k}} \ll N^{1+\frac{2}{k}} \eta^{-\frac{2}{k}}. \end{aligned}$$

引理 11　$\sum_{q \leqslant \eta} \sum_{h} \int_{(0,\, 1)-\mathscr{M}(h,\, q)} |U(\theta,\, h,\, q)|^2 \mathrm{d}\theta \ll N^{1+\frac{2}{k}-\alpha} \eta$, 其中 η 为正数.

证　由引理 7,

$$\begin{aligned} & \sum_{q \leqslant \eta} \sum_{h} \int_{(0,\, 1)-\mathscr{M}(h,\, q)} |U(\theta,\, h,\, q)|^2 \mathrm{d}\theta \\ \ll & \sum_{q \leqslant \eta} \sum_{h} \int_{q^{-1} N^{\alpha-1}}^{\infty} q^{-2-\frac{2}{k}} N^{\frac{2}{k}} \beta^{-2} \mathrm{d}\beta \\ \ll & \sum_{q \leqslant \eta} q^{-1-\frac{2}{k}} N^{\frac{2}{k}} q N^{1-\alpha} \\ \ll & N^{1+\frac{2}{k}-\alpha} \eta \end{aligned}$$

引理 12 $\int_0^1 |T'(\theta)|^2 d\theta \ll N^{1+\frac{2}{k}-\varepsilon}$.

证 由 Cauchy 不等式

$$\left(\sum_{q\in\varepsilon(X)}\sum_{h}|U'(\theta, hq)|\right)^2 \leqslant \sum_{q\in\varepsilon(z)}\sum_{h}1^2\cdot\sum_{q\in\varepsilon(X)}\sum_{h}|U'(\theta, h, q)|^2$$
$$\leqslant N^{2\delta}\sum_{q\in E(X)}\sum_{h}|U'(\theta, h,)|^2,$$

由引理 11(取 $\eta=N^{\delta}$),

$$\int_0^1 |T'(\theta)|^2 d\theta \ll N^{2\delta}\sum_{q\leqslant N^{\delta}}\sum_{h}\int_{(0,1)-\mathscr{M}(h, q)}|U(\theta, h, q)|^2 d\theta$$
$$\ll N^{2\delta}\cdot N^{1+\frac{2}{k}-\alpha}N^{\delta} \ll N^{1+\frac{2}{k}-5}.$$

引理 13 $\int_0^1 |V(\theta)-T(\theta)|^2 d\theta \ll N^{1+\frac{2}{k}}(\log N)^{-\frac{2}{k}}$.

证 若 $\theta\in\mathscr{M}(h, q)$, $q\in\mathscr{E}(X)$, 则

$$\frac{1}{2}|V(\theta)-T(\theta)|^2 \leqslant |V(\theta)-U(\theta, h, q)|^2+|T'(\theta)|^2;$$

若 $\theta\in\mathscr{M}(h, q)$, $q\in\mathscr{E}(X)$,则 $q>X$,

$$\frac{1}{3}|V(\theta)-T(\theta)|^2 \leqslant |V(\theta)-U(\theta, h, q)|^2+|T'(\theta)|^2+|U(\theta, h, q)|^2;$$

若 $\theta\in(0, 1)-\mathscr{M}$, 则

$$\frac{1}{2}|V(\theta)-T(\theta)|^2 \leqslant |V(\theta)|^2+|T'(\theta)|^2,$$

所以

$$\int_0^1 |V(\theta)-T(\theta)|^2 d\theta \ll \sum_{q, h}\int_{M(h, q)}|V(\theta)-U(\theta, h, q)|^2 d\theta+\int_0^1|T'(\theta)|^2 d\theta$$
$$+\int_{(0,1)-M}|V(\theta)|^2 d\theta+\sum_{q\geqslant X}\sum_{h}\int_{M(h, q)}|U(\theta, h, q)|^2 d\theta,$$

分别引用引理 3、9、10(取 $\eta=X$)、12 即得.

引理 14 $\mathscr{S}(N, n)=\prod_{q\leqslant X}\sum_{\substack{v\geqslant 0\\ p^v\leqslant X}}A(n, p^v)=\sum_{q\in\varepsilon(X)}A(n, q)$.

证 注意 $A(n, q)$是 q 的积性函数.

引理 15 $\sum_{n\leqslant N}|\rho(n)-\psi(n)\mathscr{S}(N, n)|^2 \ll N^{\frac{2}{k}+1}(\log N)^{-\frac{2}{k}}$, 并且 $\psi(n) > cn^{\frac{1}{k}}$.

证 参见[19]或[12]引理 10.

7 小素数对奇异级数的贡献

考虑所有的 $p \leqslant ck$, 其中 c 为一适当的大于 1 的常数. 取 N 充分大,可假定 $p^4 < X$, 因此对这样的 p,

$$\sum_{\substack{v\geqslant 0\\ p^v\leqslant X}} A(n, p^v) = \sum_{v=0}^{l} A(n, p^v),$$

其中 $l \geqslant 4$. 设 $N(n, p^l)$为同余式

$$x_1^2 + x_2^4 + \cdots + x_s^{2^s} + x_{s+1}^{2^s} + x_{s+2}^k \equiv n(\bmod p^l), \ x_i \leqslant p^l,$$

的解数.

引理 16 $\sum_{v=0}^{l} A(n, p^v) = p^{-l(s+1)} N(n, p^l)$.

证 参见[19]引理 8.6.

引理 17 $p > 2$, $\left(\frac{x}{p}\right)$ 为 Legendre 符号,则

$$\sum_{l=1}^{r-2}\left(\frac{l(l+1)}{p}\right) = -1.$$

证 参见[20]定理 7.8.2.

引理 18 对于任意的 n、p,存在 $x_1, x_2, \cdots, x_{s+1}$,满足

$$x_1^2 + x_2^4 + \cdots + x_s^{2^s} + x_{s+1}^{2^s} \equiv n(\bmod p). \tag{5}$$

证 $p=2$ 时结论显然. 设 $p > 2$. 将

$$1^4, 2^4, \cdots, (p-1)^4$$

归入模 p 的剩余类中,每类至多含 4 个这样的 4 次幂,因此使

$$x_2^4 \equiv n_1 \pmod{p} \tag{6}$$

有解的、满足 $0 \leqslant n_1 \leqslant p-1$ 的 n_1 至少有 $1+\frac{1}{4}(p-1)$ 个.

当 n_2 为平方剩余或者 n_2-1 为平方剩余时,同余方程

$$x_1^2 + x_3^8 + \cdots + x_{s+1}^{2^s} \equiv n_2 \tag{7}$$

有解. 因此,使(7)有解的 n_2 的个数

$$\begin{aligned}
&\geqslant 1+\frac{1}{2}(p-1)+\frac{1}{4}\sum_{l=1}^{p-2}\left(1+\left(\frac{l}{p}\right)\right)\left(1-\left(\frac{l+1}{p}\right)\right) \\
&=1+\frac{1}{2}(p-1)+\frac{1}{4}(p-2)+\frac{1}{4}\sum_{l=1}^{r-2}\left(\frac{l}{p}\right)-\frac{1}{4}\sum_{l=1}^{p-2}\left(\frac{l+1}{p}\right)-\frac{1}{4}\sum_{l=1}^{p-2}\left(\frac{l(l+1)}{p}\right) \\
&=1+\frac{1}{2}(p-1)+\frac{1}{4}(p-1)-\frac{1}{4}\left(\frac{-1}{p}\right)+\frac{1}{4} \geqslant 1+\frac{1}{2}(p-1)+\frac{1}{4}(p-1).
\end{aligned}$$

因为

$$1+\frac{1}{4}(p-1)+1+\frac{1}{2}(p-1)+\frac{1}{4}(p-1) > p,$$

所以,根据 Dirichlet 抽屉原则,存在 n_1、n_2 分别使(6)、(7)有解,并且 $n_1+n_2=n$. 从而(5)有解.

$$\text{记 } b(p)=\begin{cases}1, & p>2, \\ 4, & p=2.\end{cases}$$

引理 19 若 $n \equiv 1 \ (\bmod p^{b(p)})$, $l \geqslant b(p)$, 则

$$x^X \equiv n \pmod{p^l}$$

有解.

证 参见[11]定理 36.

引理 20 存在正的常数 c_0 仅依赖于 K,使

$$\prod_{p \leqslant ck} \sum_{\substack{v \geqslant 0 \\ p^v \leqslant X}} A(u,\ p^v) > c_0.$$

证 由引理 18,对任意正整数 n,

$$x_1^2+x_2^4+\cdots+x_s^{2^s}+x_{s+1}^{2^s}\equiv n-1\ (\mathrm{mod}\,p)$$

有解.

又由试验，当 $s\geqslant 6$ 时，

$$x_1^2+x_2^4+\cdots+x_s^{2^s}+x_{s+1}^{2^s}\equiv n-1\ (\mathrm{mod}16)$$

有解.因此对所有的有 p，

$$x_1^2+x_2^4+\cdots+x_s^{2^s}+x_{s+1}^{2^s}\equiv n-1\ (\mathrm{mod}\,p^{b(p)}).$$

有解.

由引理 19，

$$x_1^2+x_2^4+\cdots+x_{s+1}^{2^s}+x_{s+2}^{K}\equiv n\ (\mathrm{mod}\,p^l)$$

有解，即 $N(u,\ p^l)>0$.

由引理 16 即得结论.

8　奇异级数的尾巴

设 $S_{h,\ p^v}^{(k)*}=\sum\limits_{\substack{x=1\\ p\nmid x}}^{p^v}e_{p^v}(hx^K)$，

$$B(n,\ p^v)=p^{-(s+2)v}\sum_{\substack{h=1\\ p\nmid h}}^{p^v}S_{h,\ p^v}^{(2)}\cdots(S_{h,\ p^v}^{(2^s)})^2S_{h,\ p^v}^{(k)*},$$

$N^*(n,\ p^v)$为

$$x_1^2+x_2^4+\cdots+x_{s+2}^{k}\equiv n\ (\mathrm{mod}\,p^l),\ x_i\leqslant p^l,\ p\nmid x_{s|2},$$

的解数.

引理 21　$1-p^{-1}+\sum\limits_{v=1}^{l}B(n,\ p^v)=N^*(n,\ p^v)$.

证　$$\sum_{v=1}^{l}B(n,\ p^v)=\sum_{v=1}^{l}\sum_{\substack{h=1\\ p\nmid h}}^{p^v}\sum_{\substack{x_i\leqslant p^v\\ p\nmid x_{s+2}}}e_{p^v}(h(x_1^2+\cdots+x_{s+2}^k-n))\cdot p^{-v(s+2)}$$

$$=p^{-l(s+2)}\sum_{\substack{x_i\leqslant p^l\\ p\nmid x_{s+2}}}\sum_{v=1}^{l}\sum_{\substack{h=1\\ p\nmid h}}^{p^v}e_{p^v}(h(x_1^2+\cdots+x_{s+2}^k-n))$$

$$=p^{-l(s+2)}\sum_{\substack{x_i\leqslant p^l\\ p\nmid x_{s+2}}}\left(\sum_{h=0}^{p^l}e_{p^v}(h(x_1^2+\cdots+x_{s+2}^k-n))-1\right)$$

$$=p^{-l(s+1)}N^*(n,\ p^l)-1+\frac{1}{p}.$$

考虑满足 $\chi^r\equiv\chi_0(\mathrm{mod}\,p)$ 的特征 χ,这样的特征共有 $d=(r,\ p-1)$ 个,记之为 $\chi_0^{(r)}=\chi_0,\ \chi_1^{(r)},\ \cdots,\ \chi_{d-1}^{(r)}$. 又设特征和 $\tau(\chi)=\sum\limits_{l=1}^{p-1}e_p(l)\chi(l)$.

引理 22 $\sum\limits_{i=1}^{d-1}\tau(\chi_i^{(r)})\chi_i^{(r)}(h)=S_{h,\ i}^{(r)}$.

证 参看[20]第七章.

设 $\mathrm{d}(p)=\prod\limits_{r}\{(p-1,\ r)-1\}$, 其中 r 遍及 $2,\ 4,\ \cdots,\ 2^s,\ 2^s,\ k$. 显然 $\mathrm{d}(p)$ 以一个与 p 无关的常数 c_1 为其上界.

引理 23 $|A(n,\ p)|\leqslant\mathrm{d}(p)\cdot p^{-1}<c_1p^{-1}$.

证 当 $\mathrm{d}(p)=0$ 时,必有某个 r,使得 $(r,\ p-1)=1$,从而 $x^r\equiv l(\mathrm{mod}\,p)$ 对于每一个 l 有且仅有一解,因此

$$S_{h,\ p}^{(r)}=\sum_{x=1}^{p}e_p(hx^r)=\sum_{l=1}^{p}e_p(hl)=0,$$

$$A(n,\ p)=0.$$

当 $\mathrm{d}(p)\neq0$ 时,由引理 22,

$$A(n,\ p)=p^{-(s+2)}\sum_{h=1}^{p-1}e_p(-hn)\prod_{r}\left\{\sum_{\tau=1}^{d-1}\tau(\chi_i^{(r)})\bar{\chi}_i^{(r)}\right\}$$

$$=p^{-(s+2)}\sum_{x^{(r)}\neq x_0}\tau(\chi^{(2)})\cdots\tau(\chi^{(k)})\sum_{h=1}^{p-1}ep(-hn)\bar{\chi}^{(2)}(h)\cdots\bar{\chi}^{(k)}(h),$$

其中各特征的下标省略掉了.

因为 $|\tau(\chi)|\leqslant p^{\frac{1}{2}}$ (华罗庚[20] § 7.4.),所以,

$$|\tau(\chi^{(2)})\cdots\tau(\chi^{(k)})|\leqslant p^{\frac{s+2}{2}},$$

从而不难得出

$$|A(n,\ p)|\leqslant p^{-(s+2)}\cdot p^{\frac{s+2}{2}}\cdot p\cdot\mathrm{d}(p)=\mathrm{d}(p)\cdot p^{-\frac{s}{2}}$$

$$\leqslant \mathrm{d}(p)\cdot p^{-1} < c_1 p^{-1}.$$

引理 24　$p > k$, $v \geqslant 2$ 时, $B(n,\ p^v) = 0$.

证　因为 $(p,\ k) = 1$, $v \geqslant 2$ 时, $S_{h,\ p^v}^{(k)*} = 0$.

引理 25　$|S_{h,\ p}^{(2)}\cdots S_{h,\ p}^{(k)}| < c_2 p^{\frac{s+2}{2}}$.

证　$|S_{h,\ p}^{(r)}| \leqslant (k-1)p^{\frac{1}{2}}$　([10]定理 311).

引理 26　$|B(n,\ p)| < c_3 p^{-1}$.

证　因为 $S_{h,\ p}^{(k)*} = S_{h,\ p}^{(k)} - 1$, 所以由引理 23、25,

$$|B(n,\ p)| = \left| A(n,\ p) - p^{-(s+2)} \sum_{h=1}^{p-1} S_{h,\ p}^{(2)} \cdots (S_{h,\ p}^{(2^s)})^2 e_p(-hn) \right|$$

$$< c_1 p^{-1} + c_2 p^{-\frac{s+2}{2}} < c_3 p^{-1}.$$

引理 27　对 $p > k$, 存在正的常数 c_4, 使

$$\sum_{v=0}^{l} A(n,\ p^l) > 1 - \frac{c_4}{p}.$$

证　由引理 16、21、24、26,

$$\sum_{v=0}^{l} A(n,\ p^l) = p^{-l(s+1)} N(n,\ p^l) > p^{-l(s+1)} N^*(n,\ p^l)$$

$$= 1 - p^{-1} + \sum_{v=1}^{l} B(n,\ p^l)$$

$$= 1 - p^{-1} + B(n,\ p^l) > 1 - p^{-1} - c_3 p^{-1} = 1 - c_4 p^{-1}.$$

引理 28　$\mathscr{S}(N,\ n) > c_5 (\log\log N)^{-c_6}$.

证　取 $c > c_4$, 则由引理 20、27

$$\mathscr{S}(N,\ n) = \prod_{p \leqslant X} \sum_{\substack{v \geqslant 0 \\ p^v \leqslant X}} A(n,\ p^v) \geqslant c_0 \sum_{ck < p \leqslant X} \left(1 - \frac{c_4}{p}\right) > c_5 (\log\log N)^{-c_6}.$$

9　定理的证明

由引理 15,

$$\sum_{\substack{n<N\\ \rho(n)=0}} n^{\frac{2}{k}} \ll N^{1+\frac{2}{k}}(\log N)^{-\frac{1}{k}+\varepsilon},$$

所以,

$$\begin{aligned}\sum_{\substack{n\leqslant N\\ \rho(n)=0}} 1 &\leqslant N(\log N)^{-\frac{1}{2k}} + \sum_{\substack{\rho(n)=0\\ N(\log N)^{-\frac{1}{2k}}\leqslant n\leqslant N}} 1\\ &\leqslant N(\log N)^{-\frac{1}{2k}} + \sum_{\substack{n\leqslant N\\ \rho(n)=0}}\left(\frac{n}{N(\log N)^{-\frac{1}{2k}}}\right)^{\frac{2}{k}}\\ &\ll N(\log N)^{-\frac{1}{2k}} + N^{-\frac{2}{k}}(\log N)^{\frac{1}{k^2}} N^{1+\frac{2}{k}}(\log N)^{-\frac{1}{k}+\varepsilon}\\ &\ll N(\log N)^{-\frac{1}{2k}+\varepsilon},\end{aligned}$$

因此定理 1 成立.

作者谨对陆鸣皋同志的指导与帮助表示衷心的感谢.

参考文献

[1] R. Cook, *J. of Number Theory*, **11**(1979), 516 - 528.
[2] H. Davenport, *Ann. of Math.*, **40**(1939), 731 - 747.
[3] H. Davenport, *Ann. Arbon*, 1962.
[4] H. Davenport, H. Heilbronn. *Proc. London Math. Soc.*, **43**(1937), 73 - 104.
[5] H. Davenport, H. Heilbronn. *Proc. London Math. Soc.*, **43**(1937), 142 - 151.
[6] G. A. Freiman, *Uspehi Mat. Nauk.*, **4**(1949), No. 1, 193.
[7] Loo-keng Hua(华罗庚), *Quart. J. of Math.* (Oxford), **9**(1938), 68 - 69.
[8] H. Halberstam, *Proc. London Math. Soc.*, **52**(1950), 455 - 466.
[9] H. Halberstam, *J. of London Math. Soc.*, **25**(1950), 127 - 140.
[10] E. Landau, Vorlesungen über Zahlentheorie, I. Leipzig, 1927.
[11] E. Landau, Über einige neuere Fortschritte der additiven Zahlentheorie, Cambridge, 1937.
[12] K. F. Roth, *J. London Math. Soc.*, **24**(1949), 4 - 13.
[13] K. F. Roth, *Proc. London Math. Soc.*, **53**(1951).
[14] E. J. Scourfield, *J London Math. Soc.*, **35**(1960), 98 - 116.
[15] K. Thanigasalam. *Acta Arith.*, **13**(1968), 237 - 258.
[16] K. Thanigasalam, *Trans. Amer. Soc.*, **200**(1974), 199 - 205.
[17] R. C. Vaughan, *Proc. London Math. Soc.*, **21**(1970), 160 - 180.

[18] R. C. Vaughan, *J. London Math. Soc.*, **3**(1971), 677-688.
[19] 华罗庚,堆垒素数论,科学出版社,1957.
[20] 华罗庚,数论导引,科学出版社,1957.

A Question of Sums of Powers of Integers

Abstract

K. Thanigasalam[16] has shown that for any positive integer k, the sequence of positive integers represented by

$$x_1^2 + x_2^4 + \cdots + x_s^{2^s} + x_{s+1}^{2^s} + x_{s+2}^k \tag{1}$$

has positive density.

We improve his results by showing the following

Theorem. If integer $s \geqslant 6$, then for any positive integer k almost all positive integers may be expressed in form (1).

表自然数 n 为 $p_1^2+p_2^4+\cdots+p_s^{2^s}+p_{s+1}^{2^s}+p_{s+2}^k$

在[3]中证明了，当 $S\geqslant 6$ 时，几乎所有的正整数 n 可以表为

$$n=x_1^2+x_2^4+\cdots+x_s^{2^s}+x_{s+1}^{2^s}+x_{s+2}^k.$$

其中 k 为给定的正整数，x_1，x_2，…，x_{s+2} 为非负整数.

本文将把这一结果推广到素数幂.

定理 在 k 为奇数，$s\geqslant 3$ 时，几乎所有的满足 $n\equiv s$，$s+2\pmod 6$ 的正整数 n 可以表示为

$$n=p_1^2+p_2^4+\cdots+p_s^{2^s}+p_{s+1}^{2^s}+p_{s+2}^k \tag{1}$$

其中 p_1，p_2，…，p_{s+2} 为素数.

我们采用如下记号：

n（带或不带下标）、r、v、h、q 为正整数，并且 $(h,q)=1$. p（带或不带下标）为素数. c（带或不带下标）为正的常数. N 为大正数，$L=\log N$，θ 为实数，

$$f_r=f_{r(\theta)}=\sum_{1\leqslant p\leqslant N^{\frac{1}{r}}}e_{(\theta p^r)}$$

$V_{(\theta)}=f_2f_4\cdots f_{2^s}f_{2^s}f_k=\sum\limits_{n\leqslant N}\rho_{(n)}e_{(n\theta)}+\sum\limits_{N<n}t_{(n)}e_{(n\theta)}$，其中 $\rho_{(n)}$ 为(1)的满足 $p_1\leqslant N^{\frac{1}{2}}$，$p_2\leqslant N^{\frac{1}{4}}$，…，$p$、$p_{s+1}\leqslant N^{\frac{1}{2^s}}$，$p_{s+2}\leqslant N^{\frac{1}{k}}$ 的解数，$t_{(n)}$ 与 θ 无关.

$$S_{h,q}^{(r)}=\sum_{\substack{x=1\\(q,x)=1}}^{q}e_q(hx^r)$$

$$A_{(n,q)}=\frac{1}{\phi_{(q)}^{s+2}}\sum_{\substack{h=1\\(h,q)=1}}^{q}S_{h,q}^{(2)}S_{h,q}^{(4)}\cdots S_{h,q}^{(2^s)}S_{h,q}^{(2^s)}S_{h,q}^{(k)}e_q(-nh)$$

$$S_{(n)}=\prod_{p\leqslant cL}\sum_{\substack{v\geqslant 0\\p^v\leqslant cL}}A(n,p^v)$$

$$\psi_{(n)}=\sum_{\substack{n_1+n_2+\cdots+n_{s+2}=n\\n_1,n_2,\cdots,n_{s+2}>1}}\frac{1}{n_1^{1-\frac{1}{2}}n_2^{1-\frac{1}{4}}\cdots n_s^{1-\frac{1}{2^s}}n_{s+1}^{1-\frac{1}{2^s}}n_{s+2}^{1-\frac{1}{k}}\log n_1\cdots\log n_{s+2}}$$

引理 1　设 $d_{(n)}$ 为除数函数，则

$$\sum_{x_1=1}^{N^{\frac{1}{n_1}}}\cdots\sum_{x_r=1}^{N^{\frac{1}{n_r}}}\sum_{y_1=1}^{N^{\frac{1}{n_1}}}\cdots\sum_{y_r=1}^{N^{\frac{1}{n_r}}}d(|x_1^{n_1}+\cdots+x_r^{n_r}-y_1^{n_1}-\cdots-y_r^{n_r}|)\ll N^{2\left(\frac{1}{n_1}+\cdots+\frac{1}{n_r}\right)}L^c.$$

$$x_1^{n_1}+\cdots+x_r^{n_r}-y_1^{n_1}-\cdots\cdots-y_r^{n_r}\neq 0$$

证明　取 $f(x_1,\cdots,x_r,y_1,\cdots,y_r)=x_1^{n_1}+\cdots+x_r^{n_r}-y_1^{n_1}-\cdots-y_r^{n_r}$. 仿照[2]定理 3 的证明即得.

引理 2　$\int_0^1|f_2f_4\cdots f_{2^r}^2|^2\mathrm{d}\theta\ll NL^c$.

证明　左边的积分值即方程

$$x_1^2+x_2^4+\cdots+x_s^{2^s}+x_{s+1}^{2^s}=y_1^2+y_2^4+\cdots+y_s^{2^s}+y_{s+1}^{2^s}\tag{2}$$

的满足

$$x_1y_1\leqslant N^{\frac{1}{2}},\ x_2y_2\leqslant N^{\frac{1}{4}},\ \cdots,\ x_s、y_s、x_{s+1}、y_{s+1}\leqslant N^{\frac{1}{2^s}}\tag{3}$$

的解数.

在 $x_1\neq y_1$ 时，这时的解满足

$$x_2^4+\cdots+x_s^{2^s}+x_{s+1}^{2^s}-y_2^4-\cdots-y_s^{2^s}-y_{s+1}^{2^s}=y_1^2-x_1^2\neq 0,$$

故由引理 1，解数

$$M_1\ll N^{2\left(\frac{1}{4}+\cdots+\frac{1}{2^s}+\frac{1}{2^s}\right)}L^{c_1}=NL^{c_1}.$$

同样，满足 $x_1=y_1,\ \cdots,\ x_{i-1}=y_{i-1},\ x_i\neq y_i(i\leqslant S)$ 的解的个数

$$M_i\ll N^{\frac{1}{2}+\cdots+\frac{1}{2^{i-1}}+2\left(\frac{1}{2^{i+1}}+\cdots+\frac{1}{2^s}+\frac{1}{2^s}\right)}L^{c_i}=NL^{c_i}.$$

又满足 $x_1=y_1,\ \cdots,\ x_s=y_s,\ x_{s+1}=y_{s+1}$ 的解数为

$$M_5\ll N^{\frac{1}{2}+\frac{1}{4}+\cdots+\frac{1}{2^s}\frac{1}{2^s}}=N.$$

因此方程(2)的满足(3)的解数 $\ll NL^c$. 引理 2 证毕.

引理 3　$\sum_{n\leqslant N}|\rho_{(n)}-\psi_{(n)}S_{(n)}|^2\ll N^{1+\frac{2}{k}}L^{-2(s+2)-1+\varepsilon}$,

并且

$$\psi_{(n)} > c_1 n^{\frac{1}{k}} (\log n)^{-s+2},$$

其中 ε 为任意的小正数.

证明 参见[1],用本文的引理 2 代替该文引理 3.

引理 4 k 为奇数, $v \geqslant 2$ 时, $A_{(n, p^v)} = 0$.

证明 参见[2]引理 8.3.

引理 5 $|A_{(n, p)}| < c_2 p^{-1}$.

证明 参见[3].

引理 6 $1 + A_{(n, p)} = \phi_{(p)^{-s+2}} p N_{(n, p)}$,

这里 $N_{(n, p)}$ 为

$$x_1^2 + x_2^4 + \cdots + x_s^{2^s} + x_{s+1}^{2^s} + x_{s+2}^k \equiv n (\bmod p) \tag{4}$$

的满足

$$1 \leqslant x_1, x_2, \cdots, x_{s+2} < p \tag{5}$$

的解数.

证明 参见[2]引理 8.6.

引理 7 在 k 为奇数, $S \geqslant 3$, $n \equiv S$, $S + 2 (\bmod 6)$ 时, $N_{(n, p)} > 0$.

证明 先设 $S \geqslant 4$. 当 $p \neq 2$、3、5、17 时,

$$2^8 = 256 \not\equiv 1 (\bmod p).$$

因此可表为

$$x_1^2 + x_3^8 \equiv n + 1 (\bmod p), \ 1 \leqslant x_1, x_3 < p \tag{6}$$

的 n 至少有以下两种:

(1) n 为平方剩余;

(2) n 非平方剩余,但 $n - 255$ 为平方剩余. 另外在 -255 为平方剩余时, $n = p$ 也能表成(6),从而在 x_1、x_3 经过 1, 2, $\cdots$, $p-1$ 时, $x_1^2 + x_3^8$ 经过的剩余类的个数为

$$M_1 \geqslant \frac{1}{2}(p-1) + \frac{1}{4} \sum_{\substack{n=1 \\ n \neq 255}}^{p-1} \left(1 + \left(\frac{n-255}{p}\right)\right)\left(1 - \left(\frac{n}{p}\right)\right) + \frac{1}{2}\left(1 + \left(\frac{-255}{p}\right)\right),$$

其中$\left(\frac{n}{p}\right)$为 Legendre 符号.

因为$\sum\limits_{\substack{n=1\\ n\neq 255}}^{p-1}\left(\frac{n(n-255)}{p}\right)=-1$，所以

$$\begin{aligned} M_1 &\geqslant \frac{1}{2}(p-1)+\frac{1}{4}(p'-1)+\frac{1}{4}\left(\left(\frac{255}{p}\right)-\left(\frac{-255}{p}\right)\right) \\ &\quad +\frac{1}{2}\left(1+\left(\frac{-255}{p}\right)\right) \\ &\geqslant \frac{1}{2}(p-1)+\frac{1}{4}(p-1)=\frac{3}{4}(p-1). \end{aligned} \tag{7}$$

由于$2^{16}-1=65535=3\times5\times17\times257$，所以在$p\neq257$时，$x_4^{16}$在$x_4=1$、2 时表示不同的剩余类，而在$p=257$时，$x_4^{16}$在$x_4=1,\ 2,\ \cdots,\ p-1$时表示$\frac{p-1}{(p-1,\ 16)}=\frac{256}{16}=16$个不同的剩余类. 而在$x_{s+2}$经过$1,\ 2,\ \cdots,\ p-1$时，$x_{s+2}^k$至少表示两个不同的剩余类$\pm1$. 在$x_2$经过$1,\ 2,\ \cdots,\ p-1$时，$x_2^4$经过$\frac{p-1}{(4,\ p-1)}$个不同的剩余类. 因此由 Cauchy 定理，当$x_1,\ x_2,\ \cdots,\ x_{s+2}$经过$1,\ 2,\ \cdots,\ p-1$时，$x_1^2+\cdots+x_{s+2}^k$经过的剩余类个数为

$$M_2\geqslant\frac{3}{4}(p-1)+\frac{p-1}{(4,\ p-1)}-1+(2-1)+(2-1)\geqslant p.$$

因而(4)有解，$N_{(n,\ p)}>0$.

不难直接验证当$p=2,\ 3,\ 5,\ 17$时，亦有$N_{(n,\ p)}>0$.

在$s=3$时，对于$p\neq2$、3、5、17，$2^8\not\equiv1(\bmod p)$，所以x_4^8在$x_4=1$、2 时表示两个不同的剩余类，从而在x_1、x_2、x_3、x_4、x_5经过$1,\ 2,\ \cdots,\ p-1$时，$x_1^2+x_2^4+x_3^8+x_4^8+x_5^k$经过的剩余类个数为$M_2\geqslant p$. 于是$N_{(n,\ p)}>0$. 当$p=2,\ 3,\ 5,\ 17$时，同样可以验证$N_{(n,\ p)}>0$.

引理 8　在k为奇数，$s\geqslant3$，$n\equiv s,\ s+2(\bmod 6)$时，$S_{(n)}>c_3(\log L)^{-c_4}$.

证明　由引理 4、5、6、7，

$$S_{(n)}>c_5\prod_{c_6<p<cL}\left(1-\frac{c_6}{p}\right)>c_3(\log L)-c^4$$

定理的证明 由引理3、8，

$$\sum_{\substack{\rho(n)=0\\ n\leqslant N\\ n=s,\ s+2(\mathrm{mod}\,6)}} \frac{n^{\frac{2}{k}}(\log L)^{-c}}{(\log n)^{2(s+2)}} \ll N^{1+\frac{2}{k}} L^{-2、s+2,\ -1+\varepsilon}.$$

记

$$R=\sum_{\substack{\rho(n)=0\\ n\leqslant N\\ n=s,\ s+2(\mathrm{mod}\,6)}} 1,\ T=\sum_{\substack{\rho(n)=0\\ n\leqslant N\\ n=s,\ s+2(\mathrm{mod}\,6)}} n^{\frac{2}{k}}.$$

则

$$R^{1+\frac{2}{k}} \ll \sum_{n=1}^{R} n^{\frac{2}{k}} < T \ll N^{1+\frac{2}{k}} L^{-1+\varepsilon}$$

从而

$$R \ll NL^{-\frac{1-\varepsilon}{1+\frac{2}{k}}}.$$

定理证毕.

参考文献

[1] K. Prachar, *Monatsh Math.*, 57(1953), 66 - 74.
[2] 华罗庚,堆垒素数论,科学出版社,1957.
[3] 单墫,一个整数幂和问题,中国科学技术大学学报,2(1982).

A Simple Proof for a Theorem of Kelisky and Rivlin

Theorem (Kelisky and Rivlin) Let $f(x)$ be a function defined in $[0, 1]$ and $B_n(f(x))=\sum_{k=0}^{n} f\left(\frac{k}{n}\right)\binom{n}{k} x^k(1-x)^{n-k}$ be the nth Bernstein polynomial of $f(x)$. Then $\lim_{l\to+\infty} B^l(f(x))=f(0)+(f(1)-f(0))x$.

Proof We can assume $f(0)=0$. Let $\phi_i(x)$ and $\psi_i(x)$ $(i=1, 2, \cdots, n)$ be Bernstein basis polynomials and Bezier basis polynomials respectively: Let $n\times n$ matrices

$$T=\begin{bmatrix} 1-1 & & & \\ & 1-1 & & \\ & & \ddots & \\ & & & 1-1 \\ & & & & 1 \end{bmatrix} \text{ and } B=\left(\phi_i\left(\frac{j}{n}\right)\right)=\begin{bmatrix} C & \begin{matrix} 0 \\ \vdots \\ 0 \end{matrix} \\ * & 1 \end{bmatrix}=V^{-1}\begin{pmatrix} 1 & \\ & J_{n-1} \end{pmatrix}V,$$

where V is a matrix which reduces B to Jordan form.

Each column sum of the matrix C is less than 1. Hence the absolute values of the characteristic roots of B are less than 1 except the simple root $\lambda=1$; Thus

$$\lim_{l\to+\infty} T^{-1}B^lT=T^{-1}V^{-1}\lim_{l\to+\infty}\begin{pmatrix} 1 & \\ & J_{n-1}^l \end{pmatrix}VT=T^{-1}V^{-1}\begin{pmatrix} 1 & \\ & O_{n-1} \end{pmatrix}VT=(a_ib_j)$$

where $(a_1, a_2, \cdots, a_n)'$ is the first column of $T^{-1}V^{-1}$ and $(b_1, b_2, \cdots, b_n)$ is the first row of VT:

From the properties of Bezier basis polynomials we know that $T^{-1}BT=\left(\psi_i\left(\frac{j}{n}\right)-\psi_i\left(\frac{j-1}{n}\right)\right)$ is a d. s. (doubly stochastic) matrix. Hence the matrix

(a_ib_j) is also a d. s. matrix. It follows that $a_ib_j=\frac{1}{n}$. Thus

$$\lim_{l\to+\infty}B^l(f(x))$$

$$=\lim_{l\to+\infty}\left(f\left(\frac{1}{n}\right),\ f\left(\frac{2}{n}\right),\ \cdots,\ f\left(\frac{n}{n}\right)\right)B^{l-1}(\phi_1(x),\ \phi_2(x),\ \cdots,\ \phi_n(x))'$$

$$=\left(f\left(\frac{1}{n}\right),\ f\left(\frac{2}{n}\right),\ \cdots,\ f\left(\frac{n}{n}\right)\right)T\lim_{l\to+\infty}(T^{-1}\ B^{-1}\ T)T^{-1}(\phi_1(x),\ \phi_2(x),\ \cdots,\ \phi_n(x))'$$

$$=\left(f\left(\frac{1}{n}\right),\ f\left(\frac{2}{n}\right),\ \cdots,\ f\left(\frac{n}{n}\right)\right)\begin{bmatrix}1-1 & & & \\ & 1-1 & & \\ & & \ddots & \\ & & & 1-1 \\ & & & 1\end{bmatrix}\begin{bmatrix}\frac{1}{n} & \frac{1}{n} & \cdots & \frac{1}{n} \\ \frac{1}{n} & \frac{1}{n} & \cdots & \frac{1}{n} \\ \cdots & & \cdots & \\ \frac{1}{n} & \frac{1}{n} & \cdots & \frac{1}{n}\end{bmatrix}\begin{bmatrix}1 & 1\cdots 1 \\ & 1\cdots 1 \\ & \ddots \\ & 1\end{bmatrix}\begin{bmatrix}\phi_1(x) \\ \phi_2(x) \\ \vdots \\ \phi_n(x)\end{bmatrix}$$

$$=\frac{1}{n}\left(f\left(\frac{1}{n}\right),\ \cdots,\ f\left(\frac{n}{n}\right)\right)\begin{bmatrix} & & \\ & O & \\ 1 & 1\cdots 1 & \end{bmatrix}\begin{bmatrix}\psi_1(x) \\ \vdots \\ \psi_n(x)\end{bmatrix}$$

$$=\frac{1}{n}\left(f\left(\frac{1}{n}\right),\ f\left(\frac{2}{n}\right),\ \cdots,\ f\left(\frac{n}{n}\right)\right)(0,\ \cdots,\ 0,\ nx)'=f(1)x.$$

References

[1] Kelisky, R. P. and Rivlin, T. T., *Pacific J. of Math.*, Vol. **21**(1967), No. 3, 511 - 520.

[2] Nielson, G. M. and others, *J, of Appro. Th*, Vol. **17**(1976), No. 4, 321 - 331.

平面 Bézier 曲线的凸性定理的一个证明

§1　引言

平面 Bézier 曲线的凸性定理是计算几何中一个重要的定理. 最近,苏步青、刘鼎元在[1, 2]中给出了凸性定理的证明.

本文的目的是从 Bézier 曲线的一阶导矢与二阶导矢的几何作图出发,给出凸性定理的另一个证明.

凸性定理　若特征多边形 $S_0S_1\cdots S_n(n>2)$ 是平面凸多边形,则对应的 Bézier 曲线

$$\boldsymbol{P}(u)=\sum_{i=0}^{n}\boldsymbol{S}_iB_{n,i}(u),\ 0\leqslant u\leqslant 1 \tag{1}$$

也是凸的. 其中 Bernstein 基函数

$$B_{n,i}(u)=\binom{n}{i}u^i(1-u)^{n-i},\ i=0,\ 1,\ \cdots,\ n.$$

§2　导矢的几何作图

Bézier 建议过一个有趣的作图方法:在边 S_iS_{i+1} 上取点 $S_i^{(1)}$,使

$$\frac{|S_iS_i^{(1)}|}{|S_iS_{i+1}|}=u,\ i=0,\ 1,\ \cdots,\ n-1.$$

这样,得到 n 个点:

$$S_0^{(1)},\ S_1^{(1)},\ \cdots,\ S_{n-1}^{(1)}.$$

把它们顺次联结起来,得到一个 $n-1$ 条边的折线 $S_0^{(1)}S_1^{(1)}\cdots S_{n-1}^{(1)}$. 这个作图我们称之为一次处理. 再在 $S_i^{(1)}S_{i+1}^{(1)}$ 上取点 $S_i^{(2)}$,使

$$\frac{|S_i^{(1)}S_i^{(2)}|}{|S_i^{(1)}S_{i+1}^{(1)}|}=u,\ i=0,1,\cdots,n-2.$$

得到折线 $S_0^{(2)}S_1^{(2)}\cdots S_{n-2}^{(2)}$. 这样连续作 $n-1$ 次处理后,得到一条直线(段) $S_0^{(n-1)}S_1^{(n-1)}$. 再作第 n 次处理,得到一个点 $S_0^{(n)}$,适合

$$\frac{|S_0^{(n-1)}S_0^{(n)}|}{|S_0^{(n-1)}S_1^{(n-1)}|}=u,$$

则 $S_0^{(n)}$ 就是曲线(1)上对应于参数 u 的点,并且 $\boldsymbol{S}_0^{(n-1)}\boldsymbol{S}_1^{(n-1)}$ 就是曲线(1)在该点的切线矢量(一阶导矢),至多相差一个正的常数因子(见图 1).

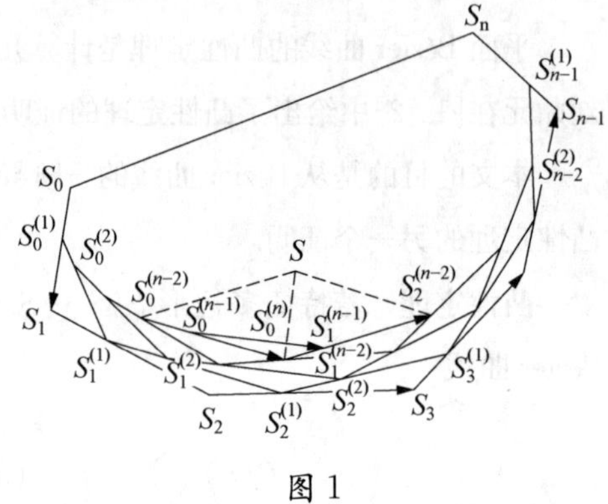

图 1

Bézier 本人并未给出上述结论的证明. 常庚哲、吴骏恒在[3]中用矩阵方法给出了一个证明. 我们现在利用移位算子 ε,给这个作图方法以一个简洁的证明,并且还得到了 Bézier 曲线二阶导矢的作图方法.

经过一次处理后,所得顶点 $S_i^{(1)}$ 的位置矢量

$$\boldsymbol{S}_i^{(1)}=\boldsymbol{S}_i\cdot(1-u)+\boldsymbol{S}_{i+1}\cdot u=[(1-u)I+u\varepsilon]\boldsymbol{S}_i,\ i=0,1,2,\cdots,n-1,$$

其中 I 为单位算子,ε 为移位算子,即

$$I\boldsymbol{S}_i=S_i,\ \varepsilon\boldsymbol{S}_i=\boldsymbol{S}_{i+1}.$$

因此, n 次处理后得到的点 $S_0^{(n)}$ 的位置矢量

$$\boldsymbol{S}_0^{(n)}=[(1-u)I+u\varepsilon]^n\boldsymbol{S}_0=\sum_{i=0}^{n}\binom{n}{i}u^i(1-u)^{n-i}\varepsilon^i\boldsymbol{S}_0$$

$$=\sum_{i=0}^{n}\binom{n}{i}u^i(1-u)^{n-i}\boldsymbol{S}_i=\sum_{i=0}^{n}B_{n,i}(u)\boldsymbol{S}_i,$$

即 $S_0^{(n)}$ 为曲线(1)上与参数 u 相对应的点. 由于

$$B'_{n,i}(u)=n[B_{n-1,i-1}(u)-B_{n-1,i}(u)],$$

引入差分算子 $\Delta=\varepsilon-I$ 之后,可将上式表为

$$B'_{n,i}(u)=-n\Delta B_{n-1,i-1}(u),$$

所以

$$\begin{aligned}\boldsymbol{S}_0^{(n-1)}\boldsymbol{S}_1^{(n-1)}&=\boldsymbol{S}_1^{(n-1)}-\boldsymbol{S}_0^{(n-1)}\\&=[(1-u)I+u\varepsilon]^{n-1}\boldsymbol{S}_1-[(1-u)I+u\varepsilon]^{n-1}\boldsymbol{S}_0\\&=[(1-u)I+u\varepsilon]^{n-1}\Delta\boldsymbol{S}_0=\sum_{i=0}^{n-1}B_{n-1,i}(u)\varepsilon^i\Delta\boldsymbol{S}_0\\&=\sum_{i=0}^{n-1}B_{n-1,i}(u)\Delta\boldsymbol{S}_i.\end{aligned}\tag{2}$$

但是,(2)中的最后一式又可以改写为

$$\begin{aligned}&-\sum_{i=1}^{n-1}\boldsymbol{S}_i\Delta B_{n-1,i-1}(u)+B_{n-1,n-1}(u)\boldsymbol{S}_n-B_{n-1,0}(u)\boldsymbol{S}_0\\&\qquad=\frac{1}{n}\sum_{i=0}^{n}B'_{n,i}(u)\boldsymbol{S}_i=\frac{1}{n}\boldsymbol{P}'(u),\end{aligned}\tag{3}$$

即 $\boldsymbol{S}_0^{(n-1)}\boldsymbol{S}_1^{(n-1)}$ 与曲线(1)在点 u 处的切线矢量只相差一个正的常数因子$\frac{1}{n}$.

与此类似,我们有如下的结论:以第 $n-2$ 次处理后所得的矢量 $\boldsymbol{S}_0^{(n-2)}\boldsymbol{S}_1^{(n-2)}$ 与 $\boldsymbol{S}_1^{(n-2)}\boldsymbol{S}_2^{(n-2)}$ 为边作平行四边形 $S_0^{(n-2)}S_1^{(n-2)}S_2^{(n-2)}S$(见图 1),则 $\boldsymbol{S}_1^{(n-2)}\boldsymbol{S}$ 与曲线(1)在点 $\boldsymbol{P}(u)$处的二阶导矢只相差一个正的常数因子 $\frac{1}{n(n-1)}$. 证明如下:

$$\begin{aligned}\boldsymbol{S}_1^{(n-2)}\boldsymbol{S}&=\boldsymbol{S}_1^{(n-2)}\boldsymbol{S}_2^{(n-2)}-\boldsymbol{S}_0^{(n-2)}\boldsymbol{S}_1^{(n-2)}\\&=[\boldsymbol{S}_2^{(n-2)}-\boldsymbol{S}_1^{(n-2)}]-[\boldsymbol{S}_1^{(n-2)}-\boldsymbol{S}_0^{(n-2)}]\\&=[(1-u)I+u\varepsilon]^{(n-2)}\Delta^2\boldsymbol{S}_0=\sum_{i=0}^{n-2}B_{n-2,i}(u)\varepsilon^i\Delta^2\boldsymbol{S}_0\\&=\sum_{i=0}^{n-2}B_{n-2,i}(u)\Delta^2\boldsymbol{S}_i=\frac{1}{n-1}\Big(\sum_{i=0}^{n-1}B_{n-1,i}(u)\Delta\boldsymbol{S}_i\Big)'\\&=\frac{1}{n(n-1)}\boldsymbol{P}(u)'',\end{aligned}$$

其中最后两步是利用了上面所得到的(2)式与(3)式相等的结果.

§3 凸性定理的证明

要证明曲线(1)是凸的,只需证明以下两点(见[1, 2]):

(i) $\boldsymbol{P}'(u)\times\boldsymbol{P}''(u)\neq\boldsymbol{0}$ $(u\in(0,1))$, 即曲线(1)无尖点与拐点.

(ii) $\boldsymbol{P}(u_1)\neq\boldsymbol{P}(u_2)$ $(0<u_1<u_2<1)$, 即曲线(1)无重点.

先证明(i). 连结 S_0S_n,形成一个封闭的凸多边形(如果 S_n 与 S_0 重合,那么特征多边形本身就是封闭的),我们可以把坐标原点取在 S_0 处(即令 $\boldsymbol{S}_0=\boldsymbol{0}$),并且不妨假设沿 $\boldsymbol{OS}_1=\boldsymbol{S}_0\boldsymbol{S}_1$, $\boldsymbol{S}_1\boldsymbol{S}_2$, $\cdots$, $\boldsymbol{S}_{n-1}\boldsymbol{S}_n$, $\boldsymbol{S}_n\boldsymbol{S}_0$ 前进时,多边形 $S_0S_1\cdots S_n$ 的内部恒在左侧(见图 1). 根据多边形的凸性, §2 中每次处理后所得的多边形也都是凸的并且具有同样的性质.

如果 $n+1$ 个顶点全部落在一条直线上,那么它们所决定的 Bézier 曲线是一段直线. 因此,无需讨论凸性. 我们不去讨论 $n+1$ 个顶点共线的情况. 特别, $n=1$ 已无需讨论.

当 $n=2$, 由于已设 $\boldsymbol{S}_0=\boldsymbol{0}$, 这时曲线(1)便是 $\boldsymbol{P}(u)=2(u-u^2)\boldsymbol{S}_1+u^2\boldsymbol{S}_2$. 直接计算可知 $\boldsymbol{P}'(u)\times\boldsymbol{P}''(u)=4\boldsymbol{S}_1\times\boldsymbol{S}_2$. 由于已经除去三个顶点共线的情形,所以 $\boldsymbol{S}_1\times\boldsymbol{S}_2\neq 0$. 这样,只需讨论 $n\geqslant 3$ 的情形.

我们指出,在 $n\geqslant 3$ 时,如果原来的 $n+1$ 个顶点不全在一条直线上,那么经过一次处理后,所得到的 n 个点也不全在一条直线上. 这是因为从原来的 $n+1$ 个点中总可以找出连续的四个点 S_i, S_{i+1}, S_{i+2}, S_{i+3} 不在同一条直线上,而这四个点的分布情况仅有以下三种,如图 2 所示:

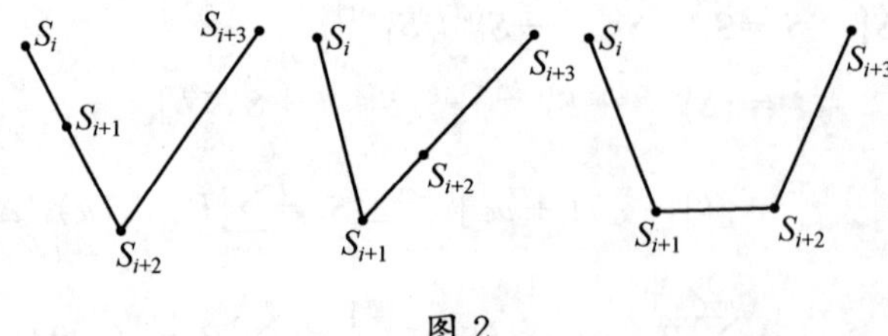

图 2

由于多边形的凸性,不论在图 2 的哪一种情况中, $S_i^{(1)}$, $S_{i+1}^{(1)}$, $S_{i+2}^{(1)}$ 都不在同一条直线上.

因此,经过 $n-2$ 次处理后,所得的三个点 $S_0^{(n-2)}$, $S_1^{(n-2)}$, $S_2^{(n-2)}$ 不在同一

条直线上，从而 $\boldsymbol{S}_1^{(n-2)}\boldsymbol{S}$ 与 $\boldsymbol{S}_0^{(n-1)}\boldsymbol{S}_1^{(n-1)}$ 都不是零矢量，即 $\boldsymbol{P}''(u)$ 与 $\boldsymbol{P}'(u)$ 都不等于零矢量. 又有

$$\begin{aligned}\frac{1}{n}\boldsymbol{P}'(u)&=\boldsymbol{S}_0^{(n-1)}\boldsymbol{S}_1^{(n-1)}=\boldsymbol{S}_1^{(n-1)}-\boldsymbol{S}_0^{(n-1)}\\&=[(1-u)\boldsymbol{S}_1^{(n-2)}+u\boldsymbol{S}_2^{(n-2)}]-[(1-u)\boldsymbol{S}_0^{(n-2)}+u\boldsymbol{S}_1^{(n-2)}]\\&=(1-u)\boldsymbol{S}_0^{(n-2)}\boldsymbol{S}_1^{(n-2)}+u\boldsymbol{S}_4^{(n-2)}\boldsymbol{S}_2^{(n-2)},\end{aligned}$$

$$\frac{1}{n(n-1)}\boldsymbol{P}''(u)=\boldsymbol{S}_1^{(n-2)}\boldsymbol{S}=\boldsymbol{S}_1^{(n-2)}\boldsymbol{S}_2^{(n-2)}-\boldsymbol{S}_0^{(n-2)}\boldsymbol{S}_1^{(n-2)},$$

所以，由 $S_0^{(n-2)}$, $S_1^{(n-2)}$, $S_2^{(n-2)}$ 不共线得

$$\boldsymbol{P}'(u)\times\boldsymbol{P}''(u)=n^2(n-1)\cdot\boldsymbol{S}_0^{(n-2)}\boldsymbol{S}_1^{(n-2)}\times\boldsymbol{S}_1^{(n-2)}\boldsymbol{S}_2^{(n-2)}\neq\mathbf{0}.$$

这就证明了(i).

现在证明(ii). 对于任意的 u_1, $u_2(0<u_1<u_2<1)$，不等式

$$\binom{n}{i}u_1^i(1-u_1)^{n-i}\leqslant\binom{n}{i}u_2^i(1-u_2)^{n-i}\tag{4}$$

的解为

$$i\geqslant\frac{n\log\dfrac{1-u_1}{1-u_2}}{\log\dfrac{u_2}{u_1}+\log\dfrac{1-u_1}{1-u_2}}=m+\theta,\ 0\leqslant\theta<1,$$

其中 m 是整数，并且 $0\leqslant m<n$. (4)中等式仅在 $\theta=0$, $i=m$ 时才能成立. 于是

$$\begin{aligned}\boldsymbol{P}(u_2)-\boldsymbol{P}(u_1)&=\sum_{i=0}^{n}\left[\binom{n}{i}u_2^i(1-u_2)^{n-i}-\binom{n}{i}u_1^i(1-u_1)^{n-i}\right]\boldsymbol{S}_i\\&=-\sum_{i\leqslant m}v_i\boldsymbol{S}_i+\sum_{i>m}v_i\boldsymbol{S}_i,\end{aligned}$$

其中 $v_i=\left|\binom{n}{i}u_2^i(1-u_2)^{n-i}-\binom{n}{i}u_1^i(1-u_1)^{n-i}\right|\geqslant 0$，并且等号仅在 $\theta=0$, $i=m$ 时才能成立.

由于 v_i 均非负及多边形 $S_0S_1\cdots S_n$ 的凸性，$\sum\limits_{i\leqslant m}v_i\boldsymbol{S}_i$ 在 $\angle S_1S_0S_m$ 内，

$\sum_{i>m} v_i \boldsymbol{S}_i$ 在$\angle S_{m+1}S_0S_n$ 内(当 S_n 与 S_0 重合时, $\sum_{i>m} v_i \boldsymbol{S}_i$ 在$\angle S_{m+1}S_0S_{n-1}$ 内. 某些顶点重合的退化情形,亦可仿此处理),如图 3 所示. 因为仅在 $\theta=0$, $i=m$ 时才有 $v_i=0$,所以$\sum_{i\leqslant m} v_i \boldsymbol{S}_i$ 与$\sum_{i>m} v_i \boldsymbol{S}_i$ 中至多有一个为零,并且不能同时有 $\sum_{i\leqslant m} v_i \boldsymbol{S}_i$ 在射线S_0S_m 上及 $\sum_{i>m} v_i \boldsymbol{S}_i$ 在射线S_0S_{m+1} 上. 于是,由于$\angle S_1S_0S_m$ 与$\angle S_{m+1}S_0S_n$ 无公共内部,所以 $\sum_{i\leqslant m} v_i \boldsymbol{S}_i \neq \sum_{i>m} v_i \boldsymbol{S}_i$. 从而 $\boldsymbol{P}(u_1) \neq \boldsymbol{P}(u_2)$, 即(ii)成立. 因此,曲线(1)是凸曲线.

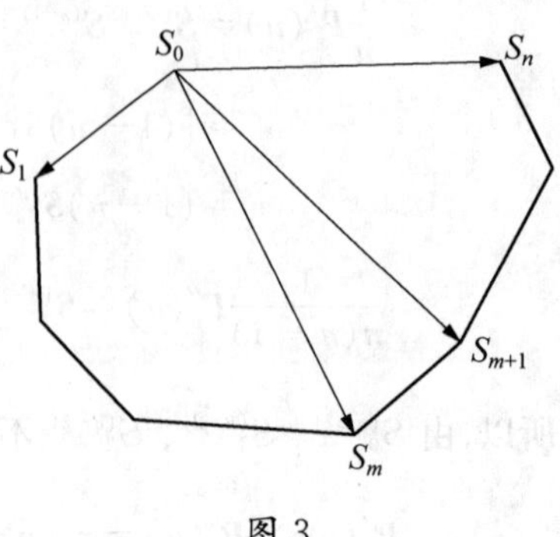

图 3

作者对常庚哲老师的指导表示衷心感谢.

参考文献

[1] 苏步青,刘鼎元,中国科学,10(1982), 867 - 877.
[2] 刘鼎元,数学年刊,3(1982), 45 - 55.
[3] 常庚哲,吴骏恒,计算数学,1(1980), 41 - 49.

A New Proof for A Theorem on the Convexity of Planar Bézier's Curves

Abstract

In this paper, we give a new proof of the following theorem. If the planar polygon is convex, then so is the Bézier's curve associated with the polygon.

初等数论中的一个猜测

任意的三个数(本文中数均指整数)中必有两个数的和为 2 整除,五个数中必有三个数的和为 3 整除,柯召、孙琦[1]证明了任意的七个数中必有四个数的和为 4 整除,并猜测任意的 $2n-1$ 个数中必有 n 个数的和为 n 整除. 本文证明这一猜测是正确的,即有

定理　设 n 为自然数,则任意的 $2n-1$ 个数中,必有 n 个数的和为 n 整除.

证明分三步进行.

(一) 几个引理.

引理 1　(Cauchy)设 p 为素数, x_1, x_2, $\cdots$, x_m 代表 m 个不同的剩余类 $(\bmod p)$, y_1, y_2, $\cdots$, y_n 代表 n 个不同的剩余类 $(\bmod p)$,则 $x_\mu+y_v(1\leqslant\mu\leqslant m,\ 1\leqslant v\leqslant n)$ 所代表的不同的剩余类 $(\bmod p)$ 的数目 $\geqslant\min(m+n-1,\ p)$.

证明　见华罗庚[2]引理 8.7.

引理 2　在任意的 n 个数 x_1, x_2, $\cdots$, x_n 中一定能选出若干个数,它们的和(包括一个数的情况,下同)为 n 整除.

证　n 个数 x_1, x_1+x_2, $\cdots$, $x_1+x_2+\cdots+x_n$,如果都不为 n 整除,即均不属于零类 $(\bmod n)$,那么根据 Dirichlet 的抽屉原则,其中必有两个数同余,它们的差为 n 整除.

引理 3　设 p 为素数. 任意 p 个数,按照 $\bmod p$ 的同余类分类,如果每一类中至多有 k 个数,那么可以从这 p 个数中取出 h 个数,它们的和为 p 整除,并且 $h\leqslant k$.

证　将这 p 个数,列成下表:

$$
\begin{matrix}
x_1 & x_2 & \cdots & x_{l_1} \\
x_1 & x_2 & \cdots & x_{l_2} \\
 & & \cdots & \\
x_1 & x_2 & \cdots & x_{l_k}
\end{matrix}
$$

其中每一列表示一个剩余类，并且同一类中的元素(至多 k 个)用同一字母表示. $l_1 \geqslant l_\tau \geqslant \cdots \geqslant l_k$. 显然 $l_1 + l_2 + \cdots + l_k = p$.

若 $x_1, x_2, \cdots, x_{l_1}$ 中有数属于零类，结论显然，故可设 $x_1, x_2, \cdots, x_{l_1}$ 均不属于零类.

因为 $x_1, x_2, \cdots, x_{l_1}$ 中剩余类个数为 l_1，而 $0, x_1, x_2, \cdots, x_{l_2}$ 中剩余类的个数为 l_2+1, 故由引理 1, $x_{i_1}, x_{i_1}+x_{i_2}$ $(1 \leqslant i_1 \leqslant l_1, 1 \leqslant i_2 \leqslant l_2)$ 中至少有 $l_1 + l_2$ 个不同的剩余类. 同样 $x_{i_1}, x_{i_1}+x_{i_2}, x_{i_1}+x_{i_2}+x_{i_3}$ $(i_j \leqslant l_j, j=1, 2, 3)$ 中至少有 $l_1 + l_2 + l_3$ 个不同的剩余类，依此类推, $x_{i_1}, x_{i_1}+x_{i_2}, \cdots, x_{i_1}+x_{i_2}+\cdots+x_{i_K}$ $(i_j \leqslant l_j, j=1, 2, \cdots, k)$ 中至少有 $l_1+l_2+\cdots+l_k=p$ 个剩余类，故引理 3 成立.

(二) 证明在 $n=$ 素数 p 时，定理成立.

将 $2p-1$ 个数，按 p 的剩余类分类，设其中最多的一类含 K_0 个数，不妨设这类为零类(如果这一类为 a，将每一个数减去 a，即化为这种情形). 如果 $K_0 \geqslant p$, 问题显然. 设 $K_0 < p$.

考虑其余的 $(2p-1)-K_0 \geqslant p$ 个数，由引理 2，其中可以取出个数 $\leqslant p$ 的若干个数，它们的和为 p 整除. 设在这样的和中，加数最多的为 $x_1+x_2+\cdots+x_{K_1}$ $(K_1 \leqslant p)$. 如果 $p-K_0 \leqslant K_1$, 结论显然成立. 设

$$p-K_0 > K_1, \tag{1}$$

我们证明这将导致矛盾.

考虑剩下的 $2p-1-K_0-K_1 \geqslant p$ 个数. 根据引理 2，其中可以选出若干个数，其和为 p 整除，设这样的和中，加数最少的为 $y_1+y_2+\cdots+y_{K_2}$.

如果 $K_1+K_2 \leqslant p$，那么 $x_1+x_2+\cdots+x_{K_1}+y_1+y_2+\cdots+y_{K_1} \equiv 0 \pmod p$，并且 $K_1+K_2 > K_1$, 这与 K_1 的定义矛盾. 所以有

$$K_1+K_2 > p. \tag{2}$$

由(1)、(2)得

$$K_2 > K_0. \tag{3}$$

而将上述 $2p-1-K_0-K_1$ 个数按 p 的剩余类分类，每一类中至多有 $K \leqslant K_0$ 个数. 由引理 3,

$$K_2 \leqslant K \leqslant K_0, \tag{4}$$

(3)、(4)矛盾！

故在 $n=$素数 p 时，定理成立.

（三）我们证明，如果定理在 $n=a>1$ 及 $n=b>1$ 时成立，则定理对于 $n=ab$ 也成立.

由 $2ab-1 \geqslant 2b-1$ 个数中可选出 b 个数满足

$$x_1^{(1)}+x_2^{(1)}+\cdots+x_b^{(1)}=K_1 b, \quad (K_1\text{ 为整数})$$

如果 $a>1$，剩下的数的个数为

$$2ab-1-b=(2a-1)b-1 \geqslant 2b-1,$$

再从中选出 b 个数满足

$$x_1^{(2)}+x_2^{(2)}+\cdots+x_b^{(2)}=K_2 b, \quad (K_2\text{ 为整数})$$

如此继续进行，直至 $2a-2$ 次，得

$$x_1^{(2a-2)}+\cdots+x_b^{(2a-2)}=K_{2a-2} b, \quad (K_{2a-2}\text{ 为整数})$$

这时 $2ab-1-(2a-2)b=2b-1$，还可以选出

$$x_1^{(2a-1)}+\cdots+x_b^{(2a-1)}=K_{2a-1} b, \quad (K_{2a-1}\text{ 为整数})$$

然后从 $2a-1$ 个数 K_1、K_2、…、K_{2a-1} 中取出 a 个数，不妨设为 $K_1, \cdots, K_a$，满足

$$K_1+K_2+\cdots+K_a=Ka, \quad (K\text{ 为整数})$$

于是

$$x_1^{(1)}+\cdots+x_b^{(1)}+\cdots+x_1^{(a)}+\cdots+x_b^{(a)}=(K_1+\cdots+K_a)b=Kab,$$

即 $n=ab$ 个数 $x_1^{(1)}, \cdots, x_b^{(1)}, \cdots, x_1^{(a)}, \cdots, x_b^{(a)}$ 的和为 n 整除.

由(二)、(三)即知定理对于一切自然数 n 成立.

参考文献

[1]　柯召、孙琦，初等数论 100 例，上海教育出版社，1979.

[2]　华罗庚，堆垒素数论，科学出版社，1957.

$\sum e(\alpha p^k)$ 的估值

§1 引言

关于三角和 $\sum e(\alpha p^k)$，过去常用维诺格拉朵夫的方法进行估值. 1977 年，R. C. Vaughan[1] 用新的方法给出 $\sum e(\alpha p)$ 的估值. 1981 年，A. Ghosh[2] 用 Vaughan 的方法给出 $\sum e(\alpha p^2)$ 及 $\sum e(\alpha p^3)$ 的估值. 在一定意义上，这些估值优于以往文献中的结果. 例如对于 $\sum e(\alpha p^3)$，Ghosh 得到

$$\sum_{n\leqslant N} \Lambda_{(n)} e(\alpha n^3) \ll N^{1+\varepsilon}(q^{-1}+N^{-5/8}+N^{-3}q)^{1/20} \tag{1}$$

而维诺格拉朵夫在条件 $\log q \geqslant 7\cdot 2^{16}(\log\log N)$ 下得到的结果为

$$\sum_{n\leqslant N} \Lambda_{(n)} e(\alpha n^3) \ll N(q^{-1}+N^{-2/3}+qN^{-3})^{1/1024}.$$

本文采用 Vaughan 的方法给出 $\sum e(\alpha p^k)$ 的估值得到下面的定理.

定理 设 α 为实数，正整数 $k>1$，正整数 a、q 满足条件 $(a, q)=1$ 及 $\left|\alpha-\frac{a}{q}\right|<q^{-2}$，则对任意给定的 $\varepsilon>0$，

$$\sum_{n\leqslant N} \Lambda_{(n)} e(\alpha n^k) \ll N^{1+\varepsilon}(q^{-1}+N^{-1/2}+N^{-k}q)^{2^{\frac{1}{2k-2}}} \tag{2}$$

其中与记号≪相关联的常数至多与 ε 及 k 有关.

在 $k=2$ 时，(2)与 Ghosh 的结果相同. 在 $k=3$ 时，(2)即

$$\sum_{n\leqslant N} \Lambda_{(n)} e(\alpha n^3) \ll N^{1+\varepsilon}(q^{-1}+N^{-1/2}+N^{-3}q)^{1/16} \tag{3}$$

显然优于 Ghosh 的结果.

§2　记号

X、Y、N 等大写字母为大正数.

$\|\beta\|$ 表示实数 β 与最近的整数之间的距离.

$$d_k(l)=\sum_{n_1\cdots n_k=l}1$$

$$\underset{y}{\triangle}f(x)=\frac{1}{y}(f(x+y)-f(x)).$$

$\sum\limits_{x}^{X}$ 表示一和,此和可分为 c_1 段,每段中变数 x 跑过 $\leqslant c_2$ 个连续整数. 这里 c_1、c_2 为绝对常数.

$\sum\limits_{m_1}{}'\cdots\sum\limits_{m_s}{}'\sum\limits_{n_1}{}'\cdots\sum\limits_{n_t}{}'$ 表示其中变数满足条件 $m_in_j\leqslant N$ $(1\leqslant i\leqslant s,\ 1\leqslant j\leqslant t)$.

α、a、q、k 的意义见定理.

§3　引理

引理 1　$\left|\sum\limits_{n}^{V}e(n\alpha)\right|\ll\min(V,\ \|\alpha\|^{-1}).$

证明　即维诺格拉朵夫[3]引理 6.

引理 2　$\sum\limits_{\substack{x\\x\neq0}}^{X}\min(Y/x,\ \|\alpha x\|^{-1})\ll(Yq^{-1}+X+q)\log XYq.$

证明　即维诺格拉朵夫[3]引理 8b.

引理 3　设 $\mu=1,\ 2,\ \cdots,\ k$, 则

$$\left|\sum_{x}^{X}e(f(x))\right|^{2^\mu}\ll X^{2^\mu-\mu-1}\sum_{y_1}^{X}\cdots\sum_{y_\mu}^{X}\sum_{x_{\mu+1}}^{X}e(y_1\cdots y_\mu\underset{y_1}{\triangle}\cdots\underset{y_\mu}{\triangle}f(x_{\mu+1}))$$

证明　即华罗庚[4]引理 3.3.

引理 4　设 (a_m) 为复数数列, $a_m\ll A$. M_1, M_2, N_1, N_2 为正整数,满足

$$M_1<M_2,\ N_1<N_2\text{ 及 }M_1N_1\leqslant N.$$

又 $\log q \ll \log N$,

$$S_1 = \sum_{M_1<m\leqslant M_2}{}' a_m \sum_{N_1<n\leqslant N_2}{}' e(\alpha m^k n^k),$$

则对任意 $\varepsilon > 0$, 有

$$S_1 \ll AN^{1+\varepsilon}\left(M_2^{\frac{k-1}{2^{k-1}}} q^{-\frac{1}{2^{k-1}}} + M_2^{\frac{k}{2^{k-1}}} N^{-\frac{1}{2^{k-1}}} + M_2^{\frac{k-1}{2^{k-1}}} N^{-\frac{k}{2^{k-1}}} q^{\frac{1}{2^{k-1}}}\right) \tag{4}$$

证明 采用对数分和,将$(M_1, M_2]$(同样地,将$(N_1, N_2]$)分为 $O(\log N)$ 个形如$(U, U']$($(V, V']$)的子区间,满足 $M_1 \leqslant U < U' \leqslant 2U \leqslant M_2$($N_1 \leqslant V < V' \leqslant 2V \leqslant N_2$). 考虑

$$S_1(U, V) = \sum_{\substack{m \\ m\neq 0}}^{U}{}' \sum_{n}^{V}{}' a_m e(\alpha m^k n^k).$$

由 Hölder 不等式,

$$S_1(U, V) \ll \left(\sum_{m}^{U} |a_m|^{\frac{2^{k-1}}{2^{k-1}-1}}\right)^{1-\frac{1}{2^{k-1}}} \left(\sum_{\substack{m \\ m\neq 0}}^{U}{}' \left|\sum_{n}^{V}{}' e(\alpha m^k n^k)\right|^{2^{k-1}}\right)^{\frac{1}{2^{k-1}}}$$

$$\ll AU^{1-\frac{1}{2^{k-1}}} \left(\sum_{\substack{m \\ m\neq 0}}^{U}{}' \left|\sum_{n}^{V}{}' e(\alpha m^k n^k)\right|^{2^{k-1}}\right)^{\frac{1}{2^{k-1}}}.$$

由引理 3,

$$S_1(U, V) \ll AU^{1-\frac{1}{2^{k-1}}} \left(\sum_{\substack{m \\ m\neq 0}}^{U}{}' V^{2^{k-1}-k} \sum_{n_1}^{V}{}' \cdots \sum_{n_k}^{V}{}' e(\alpha m^k n_1 \cdots n_k \cdot k!)\right)^{\frac{1}{2^{k-1}}}$$

$$\ll AU^{1-\frac{1}{2^{k-1}}} V^{1-\frac{k}{2^{k-1}}} \left(\sum_{\substack{m \\ m\neq 0}}^{U}{}' \sum_{n_1}^{V}{}' \cdots \sum_{n_k}^{V}{}' e(\alpha m^k n_1 \cdots n_k \cdot k!)\right)^{\frac{1}{2^{k-1}}}$$

$$\ll AU^{1-\frac{1}{2^{k-1}}} V^{1-\frac{k}{2^{k-1}}} \left(UV^{k-1} + \sum_{\substack{m \\ m\neq 0}}^{U}{}' \sum_{\substack{n_1 \\ n_1\neq 0}}^{V}{}' \cdots \sum_{\substack{n_k \\ n_k\neq 0}}^{V}{}' e(\alpha m^k n_1 \cdots n_k \cdot k!)\right)^{\frac{1}{2^{k-1}}}$$

$$\ll AUV^{1-\frac{1}{2^{k-1}}} + AU^{1-\frac{1}{2^{k-1}}} V^{1-\frac{k}{2^{k-1}}} (\Sigma_1)^{\frac{1}{2^{k-1}}} \tag{5}$$

由引理 1、2 得

$$\Sigma_1 = \sum_{\substack{m \\ m\neq 0}}^{U}{}' \sum_{\substack{n_1 \\ n_1\neq 0}}^{V}{}' \cdots \sum_{\substack{n_k \\ n_k\neq 0}}^{V}{}' e(\alpha m^k n_1 \cdots n_k \cdot k!)$$

$$\ll \sum_{\substack{m \\ m\neq 0}}^{U}{}' \sum_{\substack{n_1 \\ n_1\neq 0}}^{V}{}' \cdots \sum_{\substack{n_{k-1} \\ n_{k-1}\neq 0}}^{V}{}' \min(N/m,\ \| \alpha m^k n_1 \cdots n_{k-1} \cdot k!\ \|^{-1})$$

$$\ll \sum_{\substack{l \\ l\neq 0}}^{N^{k-1}U} d_k(l) \min(N^k/l,\ \| l\alpha \|^{-1}) \tag{6}$$

$$\ll N^{\varepsilon}(N^k q^{-1} + N^{k-1}U + q) \tag{7}$$

因此,由(5)、(7)得

$$S_1(U,\ V) \ll AN^{1+\varepsilon}(U^{\frac{k-1}{2^{k-1}}} q^{-\frac{1}{2^{k-1}}} + U^{\frac{1}{2^{k-1}}} N^{-\frac{1}{2^{k-1}}} + U^{\frac{k-1}{2^{k-1}}} N^{-\frac{k}{2^{k-1}}} q^{\frac{1}{2^{k-1}}}).$$

从而(4)式成立.

引理 5　设(b_n)为复数数列, $b_n \ll B$. (a_m)、q、M_1、M_2、N_1、N_2 同引理 4,

$$S_2 = \sum_{M_1 < m \leqslant M_2}{}' \sum_{N_1 < n \leqslant N_2}{}' a_m b_n e(\alpha m^k n^k).$$

则对任意 $\varepsilon > 0$, 有

$$S_2 \ll ABN^{1\pm\varepsilon}(q^{-\frac{1}{2^{2k-2}}} + N^{-\frac{1}{2^{k-1}}} M_2^{\frac{1}{2^{k-1}}} + N^{-\frac{1}{2^{2k-2}}} N_2^{\frac{1}{2^{2k-2}}} + N^{-\frac{k}{2^{2k-2}}} q^{\frac{1}{2^{2k-2}}}) \tag{8}$$

证明　采取与引理 4 的证明相同的对数分和. 考虑

$$S_2(U,\ V) = \sum_{m}^{U}{}' \sum_{n}^{V}{}' a_m b_n e(\alpha m^k n^k).$$

由 Hölder 不等式,

$$S_2(U,\ V) \ll (\sum_{m}^{U} |a_m|^{\frac{2^{k-1}}{2^{k-1}-1}})^{1-\frac{1}{2^{k-1}}} (\sum_{m}^{U}{}' \Big| \sum_{n}^{V}{}' b_n e(\alpha m^k n^k) \Big|^{2^{k-1}})^{\frac{1}{2^{k-1}}}$$

$$\ll AU^{1-\frac{1}{2^{k-1}}} (\sum_{m}^{U}{}' \Big| \sum_{n}^{V}{}' b_n e(\alpha m^k n^k) \Big|^{2^{k-1}})^{\frac{1}{2^{k-1}}}.$$

仿照引理 3 的证明可知

$$\Big| \sum_{n}^{V} b_n e(\alpha m^k n^k) \Big|^{2^{k-1}} \ll V^{2^{k-1}-k} \sum_{n_1}^{V} \cdots \sum_{n_k}^{V} b_{n_1 \cdots n_k} e(\alpha m^k n_1 \cdots n_k \cdot k!\ ),$$

其中 $b_{n_1 \cdots n_k} \ll B^{2^{k-1}}$.

因此，

$$
\begin{aligned}
& S_2(U, V) \\
& \ll AU^{1-\frac{1}{2^{k-1}}} V^{1-\frac{k}{2^{k-1}}} \Big(\sum_{m}^{U}{}' \sum_{n_1}^{V}{}' \cdots \sum_{n_k}^{V}{}' b_{n_1 \cdots n_k} e(\alpha m^k n_1 \cdots n_k \cdot k!) \Big)^{\frac{1}{2^{k-1}}} \\
& \ll AU^{1-\frac{1}{2^{k-1}}} V^{1-\frac{k}{2^{k-1}}} \Big(B^{2^{k-1}} UV^{k-1} + \sum_{m}^{U}{}' \sum_{\substack{n_1 \\ n_1 \neq 0}}^{V}{}' \cdots \sum_{\substack{n_k \\ n_k \neq 0}}^{V}{}' b_{n_1 \cdots n_k} e(\alpha m^k n_1 \cdots n_k \cdot k!) \Big)^{\frac{1}{2^{k-1}}} \\
& \ll ABUV^{1-\frac{1}{2^{k-1}}} + AU^{1-\frac{1}{2^{k-1}}} V^{1-\frac{k}{2^{k-1}}} (\Sigma_2)^{\frac{1}{2^{k-1}}} \qquad (9)
\end{aligned}
$$

由 Hölder 不等式及引理 3，

$$
\begin{aligned}
& \Sigma_2 \\
& = \sum_{m}^{U}{}' \sum_{\substack{n_1 \\ n_1 \neq 0}}^{V}{}' \cdots \sum_{\substack{n_k \\ n_k \neq 0}}^{V}{}' b_{n_1 \cdots n_k} e(\alpha m^k n_1 \cdots n_k \cdot k!) \\
& \ll \Big(\sum_{n_1}^{V} \cdots \sum_{n_k}^{V} |b_{n_1 \cdots n_k}|^{\frac{2^{k-1}}{2^{k-1}-1}} \Big)^{1-\frac{1}{2^{k-1}}} \Big(\sum_{\substack{n_1 \\ n_1 \neq 0}}^{V}{}' \cdots \sum_{\substack{n_k \\ n_k \neq 0}}^{V}{}' \Big| \sum_{m}^{U}{}' e(\alpha m^k n_1 \cdots n_k \cdot k!) \Big|^{2^{k-1}} \Big)^{\frac{1}{2^{k-1}}} \\
& \ll BV^{k-\frac{k}{2^{k-1}}} \Big(\sum_{\substack{n_1 \\ n_1 \neq 0}}^{V}{}' \cdots \sum_{\substack{n_k \\ n_k \neq 0}}^{V}{}' U^{2^{k-1}-k} \sum_{m_1}^{U}{}' \cdots \sum_{m_k}^{U}{}' e(\alpha (k!)^2 m_1 \cdots m_k n_1 \cdots n_k) \Big)^{\frac{1}{2^{k-1}}} \\
& \ll BV^{k-\frac{k}{2^{k-1}}} U^{1-\frac{k}{2^{k-1}}} (\Sigma_3)^{\frac{1}{2^{k-1}}} \qquad (10)
\end{aligned}
$$

由引理 1、2，

$$
\begin{aligned}
& \Sigma_3 \\
& = \sum_{\substack{n_1 \\ n_1 \neq 0}}^{V}{}' \cdots \sum_{\substack{n_k \\ n_k \neq 0}}^{V}{}' \sum_{m_1}^{U}{}' \cdots \sum_{m_k}^{U}{}' e(\alpha (k!)^2 m_1 \cdots m_k n_1 \cdots n_k) \\
& \ll N^{k-1} V + \sum_{\substack{n_1 \\ n_1 \neq 0}}^{V}{}' \cdots \sum_{\substack{n_k \\ n_k \neq 0}}^{V}{}' \sum_{\substack{m_1 \\ m_1 \neq 0}}^{U}{}' \cdots \sum_{\substack{m_k \\ m_k \neq 0}}^{U}{}' e(\alpha (k!)^2 m_1 \cdots m_k n_1 \cdots n_k) \\
& \ll N^{k-1} V + \sum_{\substack{n_1 \\ n_1 \neq 0}}^{V}{}' \cdots \sum_{\substack{n_k \\ n_k \neq 0}}^{V}{}' \sum_{\substack{m_1 \\ m_1 \neq 0}}^{U}{}' \cdots \sum_{\substack{m_{k-1} \\ m_{k-1} \neq 0}}^{U}{}' \min(N/n_k,\ 1/\| \alpha (k!)^2 m_1 \cdots m_{k-1} n_1 \cdots n_k \|) \\
& \ll N^{k-1} V + \sum_{\substack{l \\ l \neq 0}}^{N^{k-1} V} d_{2k-1}(l) \min(N^k/l,\ \| l\alpha \|^{-1}) \qquad (11)
\end{aligned}
$$

$$\ll N^{\varepsilon}(N^{k}q^{-1}+N^{k-1}V+q) \tag{12}$$

因此，由(9)、(10)、(12)得

$$S_2(U,V)\ll ABN^{1\pm\varepsilon}\left(q^{-\frac{1}{2^{2k-2}}}+N^{-\frac{1}{2^{k-1}}}U^{\frac{1}{2^{k-1}}}+N^{-\frac{1}{2^{2k-2}}}V^{\frac{1}{2^{2k-2}}}+N^{-\frac{k}{2^{2k-2}}}q^{\frac{1}{2^{2k-2}}}\right)$$

从而(8)式成立.

§4　定理的证明

当 $q>N^k$ 时，结论显然成立，因此，我们假定 $q\leqslant N^k$，从而 $\log q\ll\log N$.

采用 Vaughan 的方法，在恒等式

$$\sum_{u<n<N}f(1,n)=\sum_{d<u}\sum_{r}\sum_{n}\mu(d)f(rd,n)-\sum_{m>u}\sum_{n}\sum_{\substack{d\mid m\\ d\leqslant u}}\mu(d)f(m,n)$$

中，令

$$f(m,n)=\begin{cases}\Lambda_{(n)}e(\alpha m^k n^k), & \text{若 } u<n\leqslant Nm^{-1}.\\ 0, & \text{其他.}\end{cases}$$

其中正数 $u\leqslant N^{\frac{1}{4}}$ 将在后面确定.

于是，我们有

$$\sum_{n\leqslant N}\Lambda_{(n)}e(\alpha n^k)=T_1-T_2-T_3+O(N^{\frac{1}{4}}), \tag{13}$$

其中

$$\begin{aligned}T_1&=\sum_{d\leqslant u}\mu(d)\sum_{r\leqslant Nd^{-1}}\sum_{n\leqslant Nd^{-1}r^{-1}}\Lambda_{(n)}e(\alpha d^k r^k n^k)\\&=\sum_{d\leqslant u}\mu(d)\sum_{m\leqslant Nd^{-1}}\log m\,e(\alpha d^k m^k)\\&=\int_1^N T_1(\beta)\frac{\mathrm{d}\beta}{\beta},\end{aligned}$$

而

$$T_1(\beta)=\sum_{d\leqslant\min(uN\beta^{-1})}\mu(d)\sum_{\beta\leqslant n\leqslant Nd^{-1}}e(\alpha d^k n^k);$$

$$T_2 = \sum_{d \leqslant u} \mu(d) \sum_{n \leqslant u} \wedge_{(n)} \sum_{r \leqslant Nd^{-1}n^{-1}} e(\alpha d^k n^k r^k)$$

$$= \sum_{m \leqslant u^2} c_m \sum_{r \leqslant Nm^{-1}} e(\alpha m^k r^k),$$

而

$$c_m = \sum_{\substack{d<m \\ dn=m}} \mu(d) \wedge_{(n)} \ll N^{\varepsilon};$$

$$T_3 = \sum_{u<m \leqslant Nu^{-1}} t_m \sum_{u<n \leqslant Nm^{-1}} \wedge_{(n)} e(\alpha m^k n^k),$$

而

$$t_m = \sum_{\substack{d \mid m \\ d \leqslant u}} \mu(d) \ll N^{\varepsilon}.$$

因为

$$T_2 = \Big(\sum_{m \leqslant u} \sum_{r \leqslant Nm^{-1}} + \sum_{n \leqslant m \leqslant u^2} \sum_{r \leqslant N^{\frac{1}{2}}} + \sum_{N^{\frac{1}{2}} < r \leqslant Nu^{-2}} \sum_{u<m<u^2} + \sum_{Nu^{-2}<r<Nu^{-1}} \sum_{u<m \leqslant N r^{-1}}\Big) c_m e(\alpha m^k r^k)$$

$$= \sum_{j=1}^{4} T_2^{(j)},$$

$$T_3 = \Big(\sum_{u<m \leqslant N^{\frac{1}{2}}} \sum_{u<n \leqslant N^{\frac{1}{2}}} + \sum_{N^{\frac{1}{2}}<n \leqslant Nu^{-1}} \sum_{u<m \leqslant Nm^{-1}} + \sum_{N^{\frac{1}{2}}<m \leqslant Nu^{-1}} \sum_{u<n \leqslant Nm^{-1}}\Big) t_m \wedge_{(n)} e(\alpha m^n n^k)$$

$$= \sum_{j=1}^{3} T_3^{(j)}$$

所以

$$\sum_{n \leqslant N} \wedge_{(n)} e(\alpha n^k) \ll \int_1^N |T_1(\beta)| \frac{\mathrm{d}\beta}{\beta} + \sum_{j=1}^{4} |T_2^{(j)}| + \sum_{j=1}^{3} |T_3^{(j)}| + N^{\frac{1}{4}} \qquad (14)$$

对 $T_1(\beta)$及 $T_2^{(1)}$,用引理 4 估计,取 $M_2 = u$、$N_2 = N$. 对 $T_2^{(2)}$、$T_2^{(3)}$、$T_2^{(4)}$、$T_3^{(1)}$、$T_3^{(2)}$、$T_3^{(3)}$,用引理 5 估计,取 $M_2 = Nu^{-1}$, $N_2 = N^{\frac{1}{2}}$. 在所有的情况中, $A \ll N^{\varepsilon}$, $B \ll N^{\varepsilon}$.

于是由(14)式得

$$\sum_{n\leqslant N}\Lambda(n)e(\alpha n^k)\ll N^{1+\varepsilon}\Big(u^{\frac{k-1}{2^{k-1}}}q^{-\frac{1}{2^{k-1}}}+u^{\frac{k}{2^{k-1}}}N^{-\frac{1}{2^{k-1}}}u^{\frac{k-1}{2^{k-1}}}N^{-\frac{k}{2^{k-1}}}q^{\frac{1}{2^{k-1}}}+$$

$$q^{-\frac{1}{2^{2k-2}}}+N^{-\frac{1}{2^{2k-1}}}+u^{-\frac{1}{2^{k-1}}}+N^{-\frac{k}{2^{2k-2}}}q^{\frac{1}{2^{2k-2}}}\Big).$$

取 $u=\min(N^{\frac{1}{k+1}},q^{\frac{1}{k}},Nq^{-\frac{1}{k}})$，则得 $K\geqslant 3$ 时,

$$\sum_{n\leqslant N}\Lambda(n)e(\alpha n^k)\ll N^{1+\varepsilon}(q^{-1}+N^{-\frac{1}{2}}+N^{-k}q)^{\frac{1}{2^{2k-2}}}. \tag{15}$$

在 $k=2$ 时,改 $u\leqslant N^{\frac{1}{4}}$ 为 $u\leqslant N^{\frac{1}{3}}$，仍可证(15)成立,这一点 Ghosh 已经做过.定理证毕.

(2)式中出现 N^{ε}，主要是由于在引理 4、5 的证明中出现了 $d_k(l)$ 与 $d_{2k-1}(l)$. 如果在得到(6)及(11)后采用 Hölder 不等式,并将定理的证明略加修改,便不难将定理改进为

$$\sum_{n\leqslant N}\Lambda(n)e(\alpha n^k)\ll N(\log N)^{c(e)}(q^{-1}+N^{-\frac{1}{2}}+N^{-k}q)^{\frac{1}{2^{2k-2}}-\varepsilon}.$$

参考文献

[1] Vaughan, R. C., *Mathematika*, **24**, 136 - 41(1977).
[2] Ghosh, A., *Proc. London Math. Soc.*, (3)**42**, 252 - 259(1981).
[3] 维诺格拉朵夫,数学进展,**1**, 3 - 106(1955).
[4] 华罗庚,堆垒素数论,科学出版社,1957.

On the Estimate for $\sum e(\alpha p^k)$

In 1977, R. C. Vaughan introduced an elementary method which enables him to give the estimate for $\sum e(\alpha p)$. By means of the method of Vaughan, A. Ghosh gave new estimates for $\sum e(\alpha p^2)$ and $\sum e(\alpha p^3)$ in 1981. These results were better, in a certain sense, than the known results in the literature.

In this paper, using the method of Vaughan we obtain the following theorem.

Theorem. Suppose α is a real number and $k > 1$, a and q are positive integers satisfying $(a, q)=1$ and $\left|\alpha - \frac{a}{q}\right| < q^{-2}$. Then, given any real number $\varepsilon > 0$, we have

$$\sum_{n\leqslant N} e(\alpha p^k) \ll N^{1+\varepsilon}(q^{-1} + N^{-1/2} + N^{-k}q)^{\frac{1}{2^{2k-2}}} \tag{1}$$

where the constant implied by the $\ll$ notation depends at most on ε and k.

When $k=2$, the estimate (1) is just the same result obtained by Ghosh. In the case $k=3$, our estimate (1) is better than Ghosh's result.

References

[1] Vaughan, R. C., *Mathematatika*, **24**(1977), 136 - 141.

[2] Ghosh, A., *Proc. London Math. Soc.*, (3) **42**(1981), 252 - 259.

[3] Vinogradov, I, M., *The Method of Trigonometrical Sum in the Theory of Numbers*, Interscience Publishers, 1954.

[4] 华罗庚,堆垒素数论,科学出版社,1957.

关于 Gupta 定理的注记

1

Hansraj Gupta 所著 *Selected Topics in Number Theory* 第十二章有如下的定理(p. 335 定理 2).

定理 A　设

$$a_1, a_2, \cdots, a_n \tag{1}$$

为前 n 个自然数 1, 2, …, n 的一个排列, g_k 表示序列(1)中 a_k 左面(即 a_1, a_2, …, a_{k-1} 中) $>a_k$ 的数的个数, f_k 表示序列(1)中 a_k 右面(即 a_{k+1}, a_{k+2}, …, a_n 中) $<a_k$ 的数的个数,则有

$$\sum_{k=1}^{n} f_k = \sum_{k=1}^{n} g_k. \tag{2}$$

不难看出,如果用 g_k 表示序列(1)中 a_k 左面 $<a_k$ 的数的个数, f_k 表示序列(1)中 a_k 右面 $>a_k$ 的数的个数,则采用与该书同样的证法可知(2)式仍然成立.

有趣的是,下面的定理也成立.

定理 B　设序列(1)为前 n 个自然数的一个排列, g_k 表示序列(1)中 a_k 左面 $<k$ 的数的个数, f_k 表示序列(1)中 a_k 右面 $>k$ 的数的个数,则(2)成立.

这个定理是不能用 Gupta 的方法来证明的,必须另辟途径.

我们在§2 中证明这个定理. 在§3 中对于 Gupta 原来的定理给出两种新的证明.

2

我们用归纳法证明§1 中叙述的定理 B. $n=1$ 时,定理是显然成立的. 设将 n 换为 $n-1$ 时,定理成立. 考虑下面的两种情况:

(1) $a_n=n$. 这时序列为

$$a_1, a_2, \cdots, a_{n-1}, n, \tag{3}$$

又记序列

$$a_1, a_2, \cdots, a_{n-1} \tag{4}$$

中相应于 g_i, f_i 的量为 g'_k, $f'_k(k=1, 2, \cdots, n-1)$. 则与(3)比较可知

$$g'_k=g_k(1\leqslant k\leqslant n-1),\ g_n=n-1,$$
$$f'_k=f_k-1\ (1\leqslant k\leqslant n-1),\ f_n=0.$$

由归纳假设

$$\sum_{k=1}^{n-1}g'_k=\sum_{k=1}^{n-1}f'_k,$$

所以

$$\sum_{k=1}^{n}g_k=\sum_{k=1}^{n-1}g'_k+(n-1)=\sum_{k=1}^{n-1}f'_k+(n-1)=\sum_{k=1}^{n-1}(f'_k+1)=\sum_{k=1}^{n-1}f_k=\sum_{k=1}^{n}f_k,$$

即这时定理成立.

(2) $a_1=n\ (1\leqslant j<n)$. 这时序列为

$$a_1, a_2, \cdots, a_{j-1}, n, a_{j\mp1}, \cdots, a_n, \tag{5}$$

又记序列

$$a_1, a_2, \cdots, a_{j-1}, a_{j\mp1}, n, a_{j\mp2}, \cdots, a_n \tag{6}$$

中相应于 g_k, f_k 的量为 g'_k, f'_k.

这时又分为两种情况:

1° $a_{j+1}>j$

比较(5)与(6)可知

$$g'_k=g_k,\ f'_k=f_k\quad(k=1, 2, \cdots, n).$$

2° $a_{j+1}\leqslant j$

比较(5)与(6)可知

$$g'_k=g_k(k\neq j+1),\ g'_{j+1}=g_{j\mp1}+1,$$

$$f'_k = f_k (k \neq j),\ f'_j = f_j + 1.$$

因此,不论是情况 1°或情况 2°,总有

$$\sum_{k=1}^{n} g'_k - \sum_{k=1}^{n} f'_k = \sum_{k=1}^{n} g_k - \sum_{k=1}^{n} f_k. \tag{7}$$

于是将 n 与它的右面相邻的那项对调时,差 $\sum_{k=1}^{n} g_k - \sum_{k=1}^{n} f_k$ 的值不变. 经过若干次对调后,即化为情况 1. 但对于情况 1, $\sum_{k=1}^{n} g_k - \sum_{k=1}^{n} f_k = 0$, 所以在情况 2 中也有 $\sum_{k=1}^{n} g_k - \sum_{k=1}^{n} f_k = 0$, 即这时定理成立. 证毕.

3

本节用两种方法来证明 Gupta 原来的定理 A.

第一种证法. 与 § 2 中的证法相同,采用归纳法,并分为两种情况进行讨论:

(1) $a_n = n$

比较(3)、(4)可知

$$g'_k = g_k (1 \leqslant k \leqslant n-1),\ g_n = 0,$$
$$f'_k = f_k (1 \leqslant k \leqslant n-1),\ f_n = 0,$$

因此由归纳假设,

$$\sum_{k=1}^{n} g_k = \sum_{k=1}^{n-1} g'_k = \sum_{k=1}^{n-1} f'_k = \sum_{k=1}^{n} f_k.$$

(2) $a_j = n$, $(1 \leqslant j \leqslant n-1)$

比较(5)、(6)可知

$$g'_k = g_k (k \neq j,\ j+1),\ g'_j = g_{j+1} - 1,\ g'_{j+1} = g_j = 0,$$
$$f'_k = f_k (k \neq j,\ j+1),\ f'_j = f_{j+1},\ f'_{j+1} = f_j - 1,$$

因此

$$\sum_{k=1}^{n} g'_k - \sum_{k=1}^{n} f'_k = (\sum_{k=1}^{n} g_k - 1) - (\sum_{k=1}^{n} f_k - 1) = \sum_{k=1}^{n} g_k - \sum_{k=1}^{n} f_k,$$

按照§2中的推理可知这时(2)成立.

第二种证法. 考虑集合

$$S=\{(i,j)\mid i>j,\ a_i<a_j,\ i,j=1,2,\cdots,n\}$$

的元数. 如果先固定 i,这时 j 的个数即 g_i,因此 S 的元数 $|S|=\sum_{i=1}^{n}g_i$. 如果先固定 j,这时 i 的个数即 f_j,因此

$$|S|=\sum_{j=1}^{n}g_j,$$

从而(2)式成立.

如果 g_k 表示 a_k 左面 $<a_k$ 的数的个数, f_k 表示 a_k 右面 $>a_k$ 的数的个数,则用上述两种方法同样能证得(2)式成立.

值得注意的是,若 g_k 表示 a_k 左面 $>k$ 的数的个数, f_k 表示 a_k 右面 $<k$ 的数的个数,则(2)式并不成立. 例如序列 5, 3, 1, 2, 4 即是一例.

On Composite n for Which $\varphi(n) \mid n-1$

1 Introduction

In 1932, D. H. Lehmer[1] asked if there are any composite integers n for which $\varphi(n) \mid n-1$, $\varphi(n)$ being Euler's function. The answer to this question is still not known.

If S is any set of positive integers, denote by $N(S, x)$ the number of members of S which do not exceed x. Let a be an arbitrary integer,

$$F_{(a)} = \{n : n \equiv a (\bmod \varphi(n))\}$$

and

$$F'_{(a)} = \{n \in F_{(a)},\ n \neq pa \text{ for } p \text{ prime},\ p \nmid a\}.$$

The best result up to now was

$$N(F^{1}_{(a)}, x) = O(x^{\frac{1}{2}} (\log x)^{\frac{3}{4}} (\log\log x)^{-\frac{1}{2}})$$

obtained by C. Pomerance(1).

In this paper, our aim is to prove the following:

Theorem For every integer a,

$$N(F^{1}_{(a)}, x) = O(x^{\frac{1}{2}} (\log x)^{\frac{1}{2}} (\log\log x)^{-\frac{1}{2}})$$

where the implied constant depends on a.

2 Lemmas

Lemma 1. Let $F''_{(a)} = \{n \in F'_{(a)} : n \text{ is square free}\}$, then

$$N(F'_{(a)}, x) \leqslant 4a^2 + \sum_{d \mid a} N\left(F''\left(\frac{a}{d}, \frac{x}{d}\right)\right).$$

Proof. This is Lemma 1. of [4].

Lemma 2. Suppose $n \geqslant 16a^2$, $n \in F''(a)$, $K = \omega(n)$. Let the prime factorization of n be $p_1 p_2 \cdots p_K$ where $p_1 > p_2 > \cdots > p_K$. Then for $1 \leqslant i \leqslant K$, we have

$$p_i < (i+1)\Bigl(1+\prod_{j=i+1}^{K} p_j\Bigr).$$

Proof. This is Theorem 1. of [4].

Lemma 3. Suppose $\delta \geqslant 0$, $a_1 \geqslant a_2 \geqslant \cdots \geqslant a_t = 0$ and

$$a_i \leqslant \delta + \sum_{j=i+1}^{t} a_j \tag{1}$$

for $1 \leqslant i \leqslant t-1$. Then given any y with $0 \leqslant y < \sum_{j=1}^{t} a_j$, there is a subset S of $\{1, 2, \cdots, t\}$ with

$$y - \delta < \sum_{j \in S} a_j \leqslant y. \tag{2}$$

Proof. We use induction on t. The case $t = 1$ is obvious. Suppose the conclusion is true with $t < m$. For $t = m$, since

$$0 = a_m \leqslant y < \sum_{j=1}^{m} a_j,$$

there exists one term in the sequence

$$a_m,\ a_m + a_{m-1},\ \cdots,\ a_m + a_{m-1} + \cdots + a_2,\ a_m + a_{m-1} + \cdots + a_j,$$

which is greater than y and every term before it is not greater than y. Suppose

$$a_i + a_{i+1} + a_{i+2} + \cdots + a_m > y \tag{3}$$

and

$$a_{i+1} + a_{i+2} + \cdots + a_m \leqslant y. \tag{4}$$

If

$$a_{i+1} + a_{i+2} + \cdots + a_m > y - \delta,$$

then we take

$$S=\{i+1,\ i+2,\ \cdots,\ m\}$$

which satisfies (2). Otherwise we have

$$a_{i+1}+a_{i+2}+\cdots+a_m \leqslant y-\delta. \tag{5}$$

From (1),

$$a_i \leqslant a_{i+1}+a_{i+2}+\cdots+a_m+\delta. \tag{6}$$

By (5) and (6),

$$a_i \leqslant y. \tag{7}$$

So from (3) and (7),

$$0 \leqslant y-a_i < a_{i+1}+a_{i+2}+\cdots+a_m. \tag{8}$$

By the hyprothesis of induction, there is a subset S_1 of $\{i+1,\ i+2,\ \cdots,\ m\}$ with

$$y-a_i-\delta < \sum_{j\in S_1} a_j \leqslant y-a_i. \tag{9}$$

Let $S=S_1 \cup \{i\}$, then

$$y-\delta < \sum_{j\in S} a_j \leqslant y.$$

This completes the proof.

3 Proof of Theorem

In view of Lemma 1, it will be sufficient to prove for every a that $N(F''_{(a)},\ x)=O(x^{\frac{1}{2}}(\log x)^{\frac{1}{2}}(\log\log x)^{-\frac{1}{2}})$ where the implied constant depends on a.

Let $a \in F''_{(a)}$, $16a^2 \leqslant n \leqslant x$, $K=\omega(n)$. Let the prime factorization of n be $p_1 p_2 \cdots p_K$ where $p_1 > p_2 > \cdots > p_K$. we may assume $n > x^{\frac{1}{2}}(\log x)^{\frac{1}{2}}(\log\log x)^{-\frac{1}{2}}$ Lemma 2 implies

$$\log p_i < \log(2K)+\sum_{j=i+1}^{K} \log p_i,\ 1 \leqslant i \leqslant K.$$

We apply Lemma 3 with

$$\delta=\log(2K),\ t=K+1,\ a_i=\log p_i\,(1\leqslant i\leqslant K),\ a_t=0.$$

and

$$y=\frac{1}{2}\log x+\frac{1}{2}\log\log x-\frac{1}{2}\log\log\log x.$$

Hence there is an integer m with $m\mid n$ and

$$y-\delta<\log m\leqslant y.$$

Then

$$x^{\frac{1}{2}}(\log x)^{\frac{1}{2}}(\log\log x)^{-\frac{1}{2}}/2K<m\leqslant x^{\frac{1}{2}}(\log x)^{\frac{1}{2}}(\log\log x)^{-\frac{1}{2}}.$$

Since

$$K=\omega(n)\ll\log x/\log\log x.$$

we have

$$cx^{\frac{1}{2}}(\log x)^{-\frac{1}{2}}(\log\log x)^{\frac{1}{2}}\leqslant m\leqslant x^{\frac{1}{2}}(\log x)^{\frac{1}{2}}(\log\log x)^{-\frac{1}{2}},$$

i. e. $m\in[d_1,\ d_2]$ where $d_1=cx^{\frac{1}{2}}(\log x)^{-\frac{1}{2}}(\log\log x)^{\frac{1}{2}}$, $d_2=x^{\frac{1}{2}}(\log x)^{\frac{1}{2}}(\log\log x)^{-\frac{1}{2}}$.

For each integer $m\in[d_1,\ d_2]$, we now count the number of choices for $n\in F''(a)$ with $n<x$ and $m\mid n$. Since $\varphi(m)\mid\varphi(n)$ for such n, we have

$$n\equiv 0\ (\bmod m),\ n\equiv a\,(\bmod \varphi_{(m)})$$

so by the Chinese remainder theorem, there are at most $1+\frac{x}{[m,\ \varphi_{(m)}]}$ choices for such n (here $[\ ,\ ]$ denotes least common multiple). Now $(m,\ \varphi(m))\mid(n,\ \varphi(n))$ and $(n,\ \varphi_{(a)})\mid a$. Hence for each m, there are at most

$$1+\frac{x}{[m,\ \varphi(m)]}=1+\frac{x(m,\ \varphi(m))}{m\varphi(m)}\leqslant 1+\frac{ax}{m\varphi(m)}$$

choices for $n\in F''(a)$ with $n\leqslant x$ and $m\mid n$.

Hence we have

$$N(F''(a),\ x) \leqslant 16a^2 + x^{\frac{1}{2}}(\log x)^{\frac{1}{2}}(\log\log x)^{-\frac{1}{2}} + \sum_{d_1 \leqslant m \leqslant d_2}\left(1 + \frac{ax}{m\varphi(m)}\right)$$

$$= O(x^{\frac{1}{2}}(\log x)^{\frac{1}{2}}(\log\log x)^{-\frac{1}{2}}) + O\left(x\sum_{d_1 \leqslant m \leqslant d_2}\frac{1}{m\varphi(m)}\right).$$

By a theorem of Landau,

$$\sum_{d_1 \leqslant m}\frac{1}{m\varphi(m)} = O\left(\frac{1}{d_1}\right),$$

Hence

$$N(F''(a),\ x) = O(x^{\frac{1}{2}}(\log x)^{\frac{1}{2}}(\log\log x)^{-\frac{1}{2}}).$$

References

[1] Lehmer. D. H. , *Bull. Amer. Math. Soc.* , **38**(1932), 745 - 757.

[2] Erdös. p. , *Publ. Math. Debrecen*, **4**(1956), 201 - 206.

[3] Pomerance. C. , *Acta Arith.* , **28**(1976), 387 - 389.

[4] Pomerance. C. , *Pacific J. of Math.* , **69**(1977), 177 - 186.

A Conditional Result on Goldbach Problem

1 Main Result

In the present note, the following result is proved.

Theorem 1. *Let q and N be two positive integers such that $q \leqslant N/c(\log N)^2$ and N is even if q is even, where c is a constant. Then under the assumption of the (GRH) (generalized Riemann hypothesis), the equation*

$$N = p + p' + hq$$

has always a solution in prime numbers p, p' and integer h satisfying $0 \leqslant h \leqslant c(\log N)^2$, when N is sufficiently large.

The proof of the theorem is based on Selberg's method[1], and it gives an improvement on the corresponding results due to Ju. V. Linnik[2] and Wang Yuan[3] obtained by the circle method of Hardy and Littlewood. In their results, the range of h should be replaced by $0 \leqslant h \leqslant (\log N)^{6+\varepsilon}$ and $0 \leqslant h \leqslant (\log N)^{3+\varepsilon}$ respectively, where ε is any pre-assigned positive number.

Now let us state the (GRH) as follows:

All zeros of all the Dirichlet's L-functions $L(s, \chi)$ $(s = \sigma + it)$ in the strip $0 < \sigma < 1$ lie on the line $\sigma = 1/2$.

We use p, p' to denote prime numbers, q, N, k, l, $\cdots$ positive integers, x, T real numbers $\geqslant 2$, $\Lambda(n)$ the von Mangoldt function, $\varphi(n)$ the Euler function, $\chi_0(n)$ the principal character mod q and $\chi(n)$ a character mod q. We use also

$$E_0(\chi) = \begin{cases} 1, & \text{if } \chi = \chi_0, \\ 0, & \text{if } \chi \neq \chi_0, \end{cases}$$

$$\psi(x, \chi) = \sum_{n \leqslant x} \chi(n)\Lambda(n),$$

$$\psi(x, q, l)=\sum_{\substack{n\leqslant x\\ n\equiv l(\bmod q)}} \Lambda(n),$$

$$\vartheta(x, q, l)=\sum_{\substack{p\leqslant x\\ p\equiv l(\bmod q)}} \log p,$$

and

$$I(f, x, q)=\sum_{\substack{l=1\\ (l, q)=1}}^{q} \int_{x}^{2x} \left| f(y+\tau y, q, l)-f(y, q, l)-\frac{\tau y}{\varphi(q)}\right|^{2} \mathrm{d}y,$$

where $0<\tau\leqslant 1$.

2 Some Lemmas

We assume hereafter the (GRH). To prove the theorem, we shall need

Lemma 1. *Suppose that* $2\leqslant T\leqslant x\log qx$. *Then*

$$\psi(x, \chi)=E_0(\chi)+A(x, \chi, T)+B(x, \chi),$$

where

$$A(x, \chi, T)=-\sum_{|\gamma|\leqslant T}\frac{x^{\rho}}{\rho}+O\left(\frac{x}{T}(\log qx)^{2}\right),$$

and

$$B(x, \chi)=\begin{cases}0, & \text{if } \chi \text{ is a primitive character},\\ -\sum\limits_{\substack{p^m\leqslant x\\ p\mid q,\, p\nmid q^*}} \chi^*(p^m)\log p, & \text{otherwise},\end{cases}$$

in which $\rho=\frac{1}{2}+i\gamma$ *denotes a typical zero of* $L(s, \chi)$ *and* χ^* *is the primitive character* $\bmod q^*$ *that induces* χ.

Cf. K. Prachar[4].

Lemma 2. *We have*

$$I(\psi, x, q)\ll\frac{1}{\varphi(q)}\sum_{\chi}(J_1(x, \chi, T)+J_2(x, \chi))+\frac{x^3}{T^2}(\log qx)^4,$$

where

$$J_1(x, \chi, T)=\int_1^2 \mathrm{d}\lambda \int_{\frac{\lambda x}{2}}^{2\lambda x}\left|\sum_{|\gamma|\leqslant T}\frac{(1+\tau)^{\rho}-1}{\rho}y^{\rho}\right|^2 \mathrm{d}y,$$

$$J_2(x, \chi)=\int_x^{2x}|B(y+\tau y, \chi)-B(y, \chi)|^2 \mathrm{d}y,$$

and $\sum_{\chi}$ *denotes a sum over all characters* $\bmod q$.

Proof. It is well-known that

$$\psi(x, q, l)=\frac{1}{\varphi(q)}\sum_{\chi}\bar{\chi}(l)\psi(x, \chi),$$

and thus by Lemma 1,

$$\psi(y+\tau y, q, l)-\psi(y, q, l)-\frac{\tau y}{\varphi(y)}$$
$$=\frac{1}{\varphi(q)}\sum_{\chi}\bar{\chi}(l)(A(y+\tau y, \chi, T)-A(y, \chi, T)$$
$$+B(y+\tau y, \chi)-B(y, \chi)).$$

Therefore

$$I(\psi, x, q)\ll K_1+K_2,$$

where

$$K_1=\frac{1}{\varphi(q)^2}\sum_{\substack{l=1\\(l, q)=1}}^{q}\int_x^{2x}\left|\sum_{\chi}\bar{\chi}(l)(A(y+\tau y, \chi, T)-A(y, \chi, T))\right|^2 \mathrm{d}y$$

and

$$K_2=\frac{1}{\varphi(q)^2}\sum_{\substack{l=1\\(l, q)=1}}^{q}\int_x^{2x}\left|\sum_{\chi}\bar{\chi}(l)(B(y+\tau y, \chi)-B(y, \chi))\right|^2 \mathrm{d}y.$$

Since

$$K_1\leqslant\frac{1}{\varphi(q)^2}\int_1^2 \mathrm{d}\lambda\int_{\frac{\lambda x}{2}}^{2\lambda x}\sum_{\chi}\sum_{\chi'}\sum_{\substack{l=1\\(l, q)=1}}^{q}\bar{\chi}(l)\chi'(l)(A(y+\tau y, \chi, T)-A(y, \chi, T))$$
$$\times\overline{(A(y+\tau y, \chi', T)-A(y, \chi', T))}\mathrm{d}y$$
$$=\frac{1}{\varphi(q)}\int_1^2 \mathrm{d}\lambda\int_{\frac{\lambda x}{2}}^{2\lambda x}\sum_{\chi}|A(y+\tau y, \chi, T)-A(y, \chi, T)|^2 \mathrm{d}y$$
$$=\frac{1}{\varphi(q)}\sum_{\chi}J_1(x, \chi, T)+O\left(\frac{x^3}{T^2}(\log qx)^4\right)$$

and

$$K_2=\frac{1}{\varphi(q)^2}\int_x^{2x}\sum_{\chi}\sum_{\chi'}\sum_{\substack{l=1\\(l,q)=1}}^{q}\bar{\chi}(l)\chi'(l)(B(y+\tau y,\chi)-B(y,\chi))$$
$$\times\overline{(B(y+\tau y,\chi')-B(y,\chi'))}$$
$$=\frac{1}{\varphi(q)}\sum_{\chi}J_2(x,\chi),$$

the lemma follows.

Lemma 3. $J_1(x,\chi,T)\ll\tau x^2(\log q(\tau^{-1}+2))^2$.

Proof. By mean value theorem, we have

$$\frac{(1+\tau)^{\rho}-1}{\rho}=\int_1^{1+\tau}t^{\rho-1}\mathrm{d}t\ll\min(\tau,|\gamma|^{-1}),$$

and thus by Schwarz inequality,

$$J_1(x,\chi,T)$$
$$=\int_1^2\sum_{|\gamma_1|\leqslant T}\sum_{|\gamma_2|\leqslant T}\frac{(1+\tau)^{\rho_1}-1}{\rho_1}\cdot\frac{(1+\tau)^{\bar{\rho}_2}-1}{\bar{\rho}_2}\cdot\frac{(2\lambda x)^{1+\rho_1+\bar{\rho}_2}-\left(\frac{\lambda x}{2}\right)^{1+\rho_1+\bar{\rho}_2}}{1+\rho_1+\bar{\rho}_2}\mathrm{d}\lambda$$
$$=\sum_{|\gamma_1|\leqslant T}\sum_{|\gamma_2|\leqslant T}\frac{(1+\tau)^{\rho_1}-1}{\rho_1}\cdot\frac{(1+\tau)^{\bar{\rho}_2}-1}{\bar{\rho}_2}$$
$$\times\frac{2^{1+\rho_1+\bar{\rho}_2}-\left(\frac{1}{2}\right)^{1+\rho_1+\bar{\rho}_2}}{1+\rho_1+\bar{\rho}_2}\cdot\frac{2^{2+\rho_1+\bar{\rho}_2}-1}{2+\rho_1+\bar{\rho}_2}x^{1+\rho_1+\bar{\rho}_2}$$
$$\ll x^2\sum_{|\gamma_1|\leqslant T}\sum_{|\gamma_2|\leqslant T}\frac{\min(\tau,|\gamma_1|^{-1})\min(\tau,|\gamma_2|^{-1})}{(1+|\gamma_1-\gamma_2|)^2}$$
$$\ll x^2\sum_{|\gamma_1|\leqslant T}\sum_{|\gamma_2|\leqslant T}\frac{\min(\tau^2,|\gamma_1|^{-2})}{(1+|\gamma_1-\gamma_2|)^2}.$$

Let $N(T,\chi)$ denote the number of zeros of $L(s,\chi)$ on the segment $s=\frac{1}{2}+it$, $|t|\leqslant T$. Then

$$N(T,\chi)\ll T\log qT$$

and

$$N(T+1,\chi)-N(T,\chi)\ll \log qT$$

(cf. K. Prachar[4]). Therefore for given γ_1,

$$\sum_{|\gamma_2|\leqslant T}\frac{1}{(1+|\gamma_1-\gamma_2|)^2}\leqslant\sum_{k=0}^{\infty}\sum_{k\leqslant|\gamma_1-\gamma_2|<k+1}\frac{1}{(k+1)^2}$$

$$\ll\sum_{k=1}^{\infty}\frac{\log(|\gamma_1|+k+2)q}{k^2}\ll\log q(|\gamma_1|+2)$$

and

$$\begin{aligned}J_1(x,\chi,T)&\ll x^2\sum_{|\gamma_1|\leqslant T}\min(\tau^2,|\gamma_1|^{-2})\log q(|\gamma_1|+2)\\&\ll x^2\sum_{|\gamma_1|\leqslant\tau^{-1}}\tau^2\log q(|\gamma_1|+2)\\&\quad+x^2\sum_{|\gamma_1|>\tau^{-1}}|\gamma_1|^{-2}\log q(|\gamma_1|+2)\\&\ll\tau x^2(\log q(\tau^{-1}+2))^2.\end{aligned}$$

The lemma is proved.

Lemma 4. $\frac{1}{\varphi(q)}\sum_{\chi}J_2(x,\chi)\ll\tau x^2\log qx+x(\log qx)^2\log\log qx.$

Proof. Let $\sum_{\chi'}$ denote a sum over all imprimitive characters mod q. Let $y_k=y+kq^*$ for $k=0,1,\cdots,\left[\frac{\tau y}{q^*}\right]$, and $y_{\left[\frac{\tau y}{q^*}\right]+1}=y+\tau y$. Then

$$\begin{aligned}\sum_{\chi}|B(y+\tau y,\chi)-B(y,\chi)|^2&=\sum_{\chi'}\Big|\sum_{\substack{y<p^m\leqslant y+\tau y\\ p\nmid q}}\chi^*(p^m)\log p\Big|^2\\&\leqslant\sum_{q^*|q}\sum_{\chi(\bmod q^*)}\Big|\sum_{\substack{y<p^m\leqslant y+\tau y\\ p\nmid q}}\chi(p^m)\log p\Big|^2\\&=\sum_{kq^*|q}\sum_{\chi(\bmod q^*)}\Big|\sum_{0\leqslant k<\frac{\tau y}{q^*}}\sum_{\substack{y_n<p^m\leqslant y_{k+1}\\ p\nmid q}}\chi(p^m)\log p\Big|^2,\end{aligned}$$

and by Schwarz inequality, it is

$$\leqslant \sum_{q^*|q}\sum_{\chi(\bmod q^*)}\left(\frac{\tau y}{q^*}+1\right)\sum_{0\leqslant k<\frac{\tau y}{q^*}}\left|\sum_{\substack{y_k<p^m\leqslant y_{k+1}\\ p|q}}\chi(p^m)\log p\right|^2$$

$$\leqslant \sum_{q^*|q}\left(\frac{\tau y}{q^*}+1\right)\sum_{0\leqslant k<\frac{\tau y}{q^*}}\sum_{\substack{y_k<p_1^m\leqslant y_{k+1}\\ p_1|q}}\sum_{\substack{y_k<p_2^m\leqslant y_{k+1}\\ p_2|q}}\log p_1\log p_2\sum_{\chi(\bmod q^*)}\chi(p_1^m)\bar{\chi}(p_2^m)$$

$$=\sum_{q^*|q}\left(\frac{\tau y}{q^*}+1\right)\varphi(q^*)\sum_{\substack{y<p^m<y+\tau y\\ p|q}}(\log p)^2$$

$$\leqslant \sum_{q^*|q}\left(\frac{\tau y}{q^*}+1\right)\varphi(q^*)\sum_{p|q}\frac{\log 2y}{\log p}(\log p)^2$$

$$\leqslant \log 2y\cdot\log q\sum_{q^*|q}\left(\frac{\tau y}{q^*}+1\right)\varphi(q^*)\ll \log 2y\cdot\log q(\tau y q^*+q).$$

Hence

$$\frac{1}{\varphi(q)}\sum_{\chi}J_2(x,\chi)\ll\int_x^{2x}\frac{\log 2y\cdot\log q}{\varphi(q)}(\tau y q^{\varepsilon}+q)\mathrm{d}y$$
$$\ll \tau x^2\log qx+x(\log qx)^2\log\log qx.$$

The lemma is proved.

Lemma 5. $I(\psi, x, q)\ll \tau x^2(\log qx)^2$.

Proof. Suppose first that $\tau>x^{-1}(\log qx)^{1/2}$. Set $T=x\log qx$. The lemma follows immediately by Lemmas 2, 3 and 4. Now we proceed to prove the lemma for the case $\tau\leqslant x^{-1}(\log qx)^{1/2}$. Since

$$\sum_{\substack{l=1\\ (l,q)=1}}\left|\psi(y+\tau y,q,l)-\psi(y,q,l)-\frac{\tau y}{\varphi(q)}\right|^2$$
$$\ll\left(\sum_{l=1}^{q}(\psi(y+\tau y,q,l)-\psi(y,q,l))\right)^2+\frac{\tau^2y^2}{\varphi(q)}$$
$$\ll\left(\sum_{y<p^m\leqslant y+\tau y}\log p\right)^2+\tau^2y^2$$

and

$$\sum_{y<p^m\leqslant y+\tau y}\log p\ll\log y\sum_{m\leqslant 2\log y}\frac{1}{m}\sum_{y<p^m\leqslant y+\tau y}1,$$

we have by Schwarz inequality,

$$I(\psi,\ x,\ q) \ll (\log x)^2 \sum_{m \leqslant 4\log x} \int_x^{2x} \Big(\sum_{y < p^m \leqslant y+\tau y} 1 \Big)^2 \mathrm{d}y + \tau^2 x^3.$$

For a fixed m,

$$\begin{aligned}\int_x^{2x} \Big(\sum_{y<p^m \leqslant y+\tau y} 1 \Big)^2 \mathrm{d}y &\leqslant \sum_{x-2\tau x \leqslant p'^m \leqslant x+2\tau x} \int_{p'^m-2\tau x}^{p'^m+2\tau x} \Big(\sum_{y<p^m \leqslant y+\tau y} 1 \Big)^2 \mathrm{d}y \\ &\ll \frac{m x^{\frac{1}{m}}}{\log x} \tau x (1+\tau^2 x^2),\end{aligned}$$

and therefore

$$\begin{aligned}I(\psi,\ x,\ q) &\ll \tau x(1+\tau^2 x^2) \log x \cdot \sum_{m \leqslant 4\log x} m x^{\frac{1}{m}} + \tau^2 x^3 \\ &\ll \tau x (\log qx)^2 \Big(x + x^{\frac{1}{2}} \sum_{m \leqslant 4\log x} m\Big) + \tau^2 x^3 \ll \tau x^2 (\log qx)^2.\end{aligned}$$

The lemma is proved.

Lemma 6. $I(\vartheta,\ x,\ q) \ll \tau x^2 (\log qx)^2$.

Proof. For the case $\tau \leqslant x^{-1}(\log qx)^{1/2}$, the lemma follows by the same argument as in the proof of Lemma 5. Now suppose that $\tau \geqslant x^{-1}(\log qx)^{1/2}$. Since

$$(1+\tau)^{\frac{1}{m}} - 1 = \frac{1}{m} \int_1^{1+\tau} t^{\frac{1}{m}-1} \mathrm{d}t \leqslant \frac{\tau}{m},$$

$$\begin{aligned}\sum_{\substack{y<p^m \leqslant y+\tau y \\ m \geqslant 2}} \log p &\leqslant 2\log y \cdot \sum_{2 \leqslant m \leqslant 2\log p} \frac{1}{m} \sum_{y^{\frac{1}{m}} < p \leqslant (1+\tau)^{\frac{1}{m}} y^{\frac{1}{m}}} 1 \\ &\ll \log y \cdot \sum_{m \leqslant 2\log y} \frac{1}{m} \Big(1 + \frac{\tau}{m} y^{1/2}\Big) \ll \log y \cdot \log\log y + \tau y^{1/2} \log y\end{aligned}$$

and

$$\begin{aligned}&\psi(y+\tau y,\ q,\ l) - \psi(y,\ q,\ l) \\ =& \vartheta(y+\tau y,\ q,\ l) - \vartheta(y,\ q,\ l) + \sum_{\substack{y<p^m \leqslant y+\tau y \\ p^m \equiv l (\mathrm{mod}\, q) \\ m \geqslant 2}} \log p,\end{aligned}$$

we have

$$
\begin{aligned}
I(\vartheta, x, q) &\ll I(\psi, x, q)+\sum_{\substack{l=1\\(l, q)=1}}^{q}\int_{x}^{2x}\Big(\sum_{\substack{y<p^{m}\leqslant y+\tau y\\ p^{m}\equiv l(\bmod q)\\ m\geqslant 2}}\log p\Big)^{2}\mathrm{d}y\\
&\ll I(\psi, x, q)+\int_{x}^{2x}\Big(\sum_{\substack{y<p^{m}\leqslant y+\tau y\\ m\geqslant 2}}\log p\Big)^{2}\mathrm{d}y\\
&\ll I(\psi, x, q)+x(\log x)^{2}(\log\log x)^{2}+\tau^{2}x^{2}(\log x)^{2},
\end{aligned}
$$

the lemma follows.

3 The Proof of Theorem 1

Suppose that $N-kq=p+p'$ has no solution in p, p' for any given k satisfying $0\leqslant k\leqslant h$, where $hq\leqslant N$. Then for any given y satisfying $\frac{1}{2}qh\leqslant y\leqslant N-\frac{1}{2}qh$, it is impossible that the interval $\left[y-\frac{1}{2}qh, y\right]$ contains a prime number$\equiv l\ (\bmod q)$ and the interval $\left[N-y-\frac{1}{2}qh, N-y\right]$ contains a prime number$\equiv N-l\ (\bmod q)$ simultaneously.

Consider a set of intervals

$$
\mathscr{M}:\ \left[\frac{N}{2}+\frac{1}{2}khq, \frac{N}{2}+\frac{1}{2}(k+1)hq\right], -\left[\frac{N}{2hq}\right]-2\leqslant k\leqslant\left[\frac{N}{2hq}\right].
$$

We may rewrite $\mathscr{M}$ as follows:

$$
\left[\frac{N}{2}-\frac{1}{2}(k+2)hq, \frac{N}{2}-\frac{1}{2}(k+1)hq\right], -\left[\frac{N}{2hq}\right]-2\leqslant k\leqslant\left[\frac{N}{2hq}\right].
$$

Hence for any l, if the number of intervals in $\mathscr{M}$ that contain a prime $\equiv l\ (\bmod q)$ is $\geqslant\left[\frac{N}{2hq}\right]+2$, the number of those intervals that contain a prime $\equiv N-l(\bmod q)$ must be $\leqslant\left[\frac{N}{2hq}\right]+1$. Let $F(q)$ denote the number of integers l satisfying $1\leqslant l\leqslant q$ and $(l(N-l), q)=1$. Then

$$F(q)=\sum_{1\leqslant l\leqslant qk}\sum_{|(l(N-l),\, q)}\mu(k)=\sum_{k|q}\mu(k)\sum_{\substack{1\leqslant l\leqslant q\\ l(N-l)\equiv 0(\bmod k)}}1=q\prod_{p|q}\left(1-\frac{\omega(p)}{p}\right),$$

where $\omega(p)=1$ if $p|N$, and $\omega(p)=2$ otherwise. Since q is even if N is even, we have $F(q)\neq 0$. Therefore there are more than $\frac{1}{2}F(q)$ integers $l_1,\ l_2,\ \cdots,\ l_s$, say, such that for $1\leqslant k\leqslant s$, the number of intervals in m that contain no prime $\equiv l_k\ (\bmod\ q)$ is $\geqslant\left[\frac{N}{2hq}\right]+2$, and we denote these intervals by $I_{k_i}=\left[y_i,\ y_i+\frac{hq}{2}\right]$ $(1\leqslant i\leqslant j(k))$. Let $\tau=\frac{hq}{4N}$. Since

$$\frac{N}{2}+\frac{1}{2}(k+1)hq\leqslant\frac{4}{5}N,\ \frac{N}{2}+\frac{1}{2}khq\geqslant\frac{1}{5}N$$

for $-\left[\frac{N}{2hq}\right]-2\leqslant k\leqslant\left[\frac{N}{2hq}\right]$ and N is sufficiently large, we have

$$\left(y_i+\frac{hq}{4}\right)(1+\tau)<y_i+\frac{hq}{2},$$

and

$$\begin{aligned}\mathfrak{S}&=\sum_{k=1}^{s}\sum_{i=1}^{j(k)}\int_{I_{ki}}\left|\vartheta(y+\tau y,\ q,\ l_k)-\vartheta(y,\ q,\ l_k)-\frac{\tau y}{\varphi(q)}\right|^2\mathrm{d}y\\ &\geqslant\sum_{k=1}^{s}\sum_{i=1}^{j(k)}\int_{y_i}^{y_i+\frac{hq}{4}}\left|\vartheta(y+\tau y,\ q,\ l_k)-\vartheta(y,\ q,\ l_k)-\frac{\tau y}{\varphi(q)}\right|^2\mathrm{d}y\\ &\geqslant\frac{F(q)}{2}\left(\left[\frac{N}{2hq}\right]+2\right)\frac{hq}{4}\left(\frac{\tau N}{5\varphi(q)}\right)^2.\end{aligned}$$

On the other hand, it follows by Lemma 6 that

$$\begin{aligned}\mathfrak{S}&\leqslant\sum_{\substack{l=1\\(l,\ q)=1}}^{q}\left(\int_{\frac{1}{5}N}^{\frac{2}{5}N}+\int_{\frac{2}{5}N}^{\frac{4}{5}N}\right)\left|\vartheta(y+\tau y,\ q,\ l)-\vartheta(y,\ q,\ l)-\frac{\tau y}{\varphi(q)}\right|^2\mathrm{d}y\\ &\ll\tau N^2(\log N)^2.\end{aligned}$$

Hence

$$h\ll\frac{\varphi(q)^2}{qF(q)}(\log N)^2.$$

Since

$$\frac{\varphi(q)^2}{qF(q)}=\frac{q^2\prod_{p|q}\left(1-\frac{1}{p}\right)^2}{q^2\prod_{p|q}\left(1-\frac{\omega(p)}{p}\right)}=\prod_{\substack{p|q\\p|N}}\frac{p-1}{p}\prod_{\substack{p|q\\p\nmid N}}\left(1+\frac{1}{p^2-2p}\right),$$

we have

$$0<\frac{\varphi(q)^2}{qF(q)}<\prod_{k=3}^{\infty}\left(1+\frac{1}{k^2-2k}\right)<\infty,$$

and the theorem is proved.

References

[1] Selberg, A., On the Normal Density of Primes in Small Interval and the Difference between Consecutive Primes, *Arch. Math. Naturvid*, 47 (1943), 87-105.

[2] Linnik, Ju. V., Some Conditional Theorems Concerning Binary Problems with Prime Numbers, *Dokl. Akad. Nauk SSSR*, 77(1951), 15-18, and *Iev. Akad. Nauk SSSR*, Ser. Mat. 16(1952), 503-520.

[3] Wang Yuan, On Linnik's Method Concerning Goldbach Number, *Sci. Sin.*, 20 (1977), 16-30.

[4] Prachar, K., *Primzahlverteilung*, Springer-Verlag, 1957.

Шнирельман 常数的估计

1 引言

在[6]、[7]中,Шнирельман 引入密率的概念并证明了每一个大于 1 的整数可以表为至多 K 个素数的和. 1969 年,Климов[2] 证明了 $k\leqslant 6\times 10^9$. 1972 年,Климов 等[3] 又将这一结果改进为 $k\leqslant 115$. 最近, R. C. Vaughan[8] 证明了 $k\leqslant 27$, J. M. Deshouillers[1] 证明了 $k\leqslant 26$, 张明尧与丁平[9] 证明了 $k\leqslant 25$, 即有:

定理 每个正偶数是至多 24 个素数的和.

本文应用 Vaughan[8] 所提出的另一种方法给这个定理以一个简单的证明.

2 若干定理

引理 1 若 $y\geqslant 0$, $x\geqslant 10^9$,

$$R(n,\ x)=\sum_{\substack{y\leqslant p\leqslant x+y\\ n-p=p_1}} 1, \tag{1}$$

其中 p、p_1 均表素数,则有

$$R(n,\ x)<\frac{16Cx}{(\alpha+\log x)\log x}\prod_{\substack{p\mid n\\ p>2}}\frac{p-1}{p-2}, \tag{2}$$

其中

$$C=\prod_{p>2}\frac{p(p-2)}{(p-1)^2}, \tag{3}$$

$\alpha=\alpha(x)$

$$=\frac{\log^2 x+1.666\log x-6.52-16C\left(\frac{1}{16}\log^2 x+0.238\log x\right)\frac{2.61}{x^{1/4}}}{\log x\left(1+2x^{-1/2}\left(\frac{1}{16C}(\log^2 x+1.666\log x-6.52)-\left(\frac{1}{16}\log^2 x+0.2381\log x\right)\frac{2.61}{x^{1/4}}\right)\right)-\log x} \tag{4}$$

是 x 的增函数,并且

$$\alpha \geqslant \begin{cases} 1, & x = 10^9; \\ 1.508, & x = e^{42}; \\ 1.557, & x = e^{60}; \\ 1.575, & x = e^{72}; \\ 1.588, & x = e^{86}; \\ 1.594, & x = e^{91}. \end{cases} \tag{5}$$

引明　基本上即 Vaughan[8] 引理 8.

引理 2　$0.6601 < C < 0.6602.$ (6)

证明　即 Vaughan[8] 引理 3(3.7).

引理 3

$$\pi(x) > \frac{x}{\log x - \frac{1}{2}} \quad (x \geqslant 67), \tag{7}$$

$$\theta(x) < 1.001102x, \tag{8}$$

$$\theta(x) > 0.998684x \quad (x \geqslant 1319007), \tag{9}$$

$$|\theta(x) - x| < 0.0242269 \frac{x}{\log x} \quad (x \geqslant 10^9). \tag{10}$$

证明　见 Rosser 与 Schoenfeld[5].

引理 4　设 $x > k$, 则

$$\pi(x+y, k, l) - \pi(y, k, l) < \frac{2x}{\phi(k)\log^{x/k}}. \tag{11}$$

证明　即 Vaughan[8] 引理 3.

引理 5　若 $10^9 \leqslant a \leqslant x$, $10^9 \leqslant b \leqslant x$, $a + b = x$, 则

$$F(a) = \frac{a}{\log^2 a} + \frac{b}{\log^2 b}$$

的最大值为 $\dfrac{x}{\log^2 \dfrac{x}{2}}$.

证明　不妨设 $a \leqslant \dfrac{x}{2}$, 这时

$$\frac{\mathrm{d}F}{\mathrm{d}a}=\frac{\log a-2}{\log^3 a}-\frac{\log b-2}{\log^3 b}=\frac{\log b-\log a}{\log^3 a\log^3 b}(\log a\log b(\log a+\log b)$$
$$-2(\log^2 a+\log^2 b+\log a\log b))$$
$$\geqslant\frac{\log b-\log a}{\log^3 a\log^3 b}(\log^2 b+\log^2 a-2\log a\log b)\geqslant 0,$$

所以 $F(a)$ 递增，从而

$$F(a)\geqslant\frac{\frac{x}{2}}{\log^2\frac{x}{2}}+\frac{\frac{x}{2}}{\log^2\frac{x}{2}}=\frac{x}{\log^2\frac{x}{2}}.$$

引理 6 令

$$A(x)=\sum_{n=p_1+p_2<x}1,\tag{12}$$

其中 p_1、p_2 表素数，则在

$$A(x)>\frac{x}{24}\tag{13}$$

时定理成立.

证明 这是熟知的结果.

在§3、§4 中我们证(13)成立.

3 $x\leqslant e^{95}$ 时，$A(x)$ 的估计

引理 7 $2\leqslant x\leqslant e^{24}$ 时，(13)成立.

证明 $x\leqslant 70$ 时，(13)显然. 设 $70<x\leqslant e^{24}$，则因为

$$A(x)\geqslant\pi(x-3)\geqslant\pi(x)-2,$$

由引理 3(7)，

$$A(x)>\frac{x}{\log x-\frac{1}{2}}-2>\frac{x}{24}.$$

引理 8 $e^{24}\leqslant x\leqslant e^{95}$ 时，(13)成立.

证明　令 $p_1=3$，$p_2=5$，$p_3=7$，$p_4=11$，$p_5=17$，$p_6=13$，$p_7=19$，p 表示素数，

$$a_i(n)=\begin{cases}1, & n=p_i+p.\\ 0, & \text{其他}.\end{cases}\quad (i=1,2,\cdots,7)$$

则在 $2\leqslant s\leqslant 7$ 时，

$$\begin{aligned}A(x)&\geqslant\sum_{n\leqslant x}\left(1-\prod_{i=1}^{s}(1-a_i(n))\right)\geqslant\sum_{r=1}^{s}\sum_{n\leqslant x}a_i(n)-\sum_{1\leqslant i<j\leqslant s}\sum_{n\leqslant x}a_i(n)a_j(n)\\&\geqslant s(\pi(x)-19)-\sum_{1\leqslant i<j\leqslant s}R(|p_i-p_j|,x).\end{aligned}$$

由引理 1 及引理 3，

$$A(x)\geqslant\frac{sx}{\log x-\frac{1}{2}}-19s-\sum_{1\leqslant i<j\leqslant s}\frac{16Cx}{(\alpha+\det x)\log x},\prod_{\substack{p\,\|\,p_i-p_j\,|\\p>2}}-\frac{p-1}{p-2},$$

考虑函数

$$F(u,s)=su(\alpha+u)-\beta\left(u-\frac{1}{2}\right)-\frac{1}{24}u(\alpha+u)\left(u-\frac{1}{2}\right),$$

其中

$$\beta=16C\sum_{1\leqslant i<j\leqslant s}\prod_{\substack{p\,\|\,p_i-p_j\,|\\p>2}}\frac{p-1}{p-2}=\begin{cases}73.95, & s=4;\\130.30, & s=5;\\296.48, & s=7.\end{cases}$$

$\alpha=\alpha(x)=\alpha(\mathrm{e}^u)$ 由(4)、(5)给出

易知在 $23\leqslant u\leqslant 72$，$36\leqslant u\leqslant 83$，$83\leqslant u\leqslant 95$ 时分别有 $F(u,4)>0$，$F(u,5)>0$，$F(u,7)>0$，因而在 $\mathrm{e}^{24}\leqslant x\leqslant \mathrm{e}^{95}$ 时，$A(x)>\frac{x}{24}$.

4　$x\geqslant \mathrm{e}^{95}$ 时，$A(x)$ 的估计

设 $x\geqslant \mathrm{e}^{91}$，记

$$R_k(n)=\sum_{\substack{\frac{kx}{2}<p_1,\ p_2\leqslant\frac{kx}{2}+x\\ 3<p_1,\ p_2\\ n=p_1+p_2}}1,$$

其中 p_1、p_2 为素数, k 为非负整数.

由引理 1,

$$\sum_n R_k(n)\prod_{\substack{p\mid n\\ p>2}}\frac{p-2}{p-1}$$

$$=\sum_{kx<n\leqslant ks+s}R_k(n)\prod_{\substack{p\mid u\\ p>2}}\frac{p-2}{p-1}+\sum_{kx+n<n\leqslant kx+2s}R_k(n)\prod_{\substack{p\mid u\\ p>2}}\frac{p-2}{p-1}$$

$$\leqslant\sum_{\substack{10^9<n-kx\leqslant x\\ R_{k(u)}>0}}\frac{16C(n-kx)}{\log^2(n-kx)}+\sum_{\substack{10^9<(k+2)x-n\leqslant x\\ R_{k(n)}>0}}\frac{16C((k+2)x-n)}{\log^2((k+2)x-n)}+2\times10^{18},$$

$$\sum_{k=0}^{24}\sum_n R_k(n)\prod_{\substack{p\mid n\\ p>2}}\frac{p-2}{p-1}$$

$$\leqslant\sum_{k=0}^{24}\sum_{\substack{10^9<n-kx\leqslant x\\ R_{k(n)}>0}}\frac{16C(n-kx)}{\log^2(n-kx)}+\sum_{k=0}^{25}\sum_{\substack{10^9<(k+1)x-n\leqslant x\\ R_{k(n)}>0}}\frac{16C((k+1)x-n}{\log^2((k+1)x-n)}+5\times10^{19}$$

$$\leqslant\sum_{k=0}^{25}\sum_{\substack{10^9<n-ks\leqslant x-10^9\\ R_{k(n)}>0}}16C\left(\frac{n-kx}{\log^2(n-kx)}+\frac{(k+1)x-n}{\log^2((k+1)x-n)}\right)$$

$$+\sum_{k=0}^{24}\sum_{\substack{x-10^9\leqslant n-kx<x\\ R_{k(n)}>0}}\frac{16C(n-kx)}{\log^2(n-kx)}+\sum_{k=1}^{25}\sum_{\substack{x-10^9\leqslant(k+1)x-n<x\\ R_{k(n)}>0}}\frac{16C((k+1)x-n)}{\log^2((k+1)x-n)}$$

$$+5\times10^{19},$$

由引理 5 及$\frac{x}{\log^2 x}$的单调性得

$$\sum_{k=0}^{24}\sum_n R_k(n)\prod_{\substack{p\mid n\\ p>2}}\frac{p-2}{p-1}$$

$$\leqslant\sum_{k=0}^{25}\sum_{\substack{10^9<n-kx<x\\ R_{n(k)}>0}}\frac{16Cx}{\log^2\frac{x}{2}}+\sum_{k=0}^{24}\sum_{\substack{x-10^9\leqslant n-kx<x\\ R_{k(n)}>0}}\frac{16Cx}{\log^2\frac{x}{2}}+\sum_{k=1}^{25}\sum_{\substack{x-10^9\leqslant(k+1)x-n<x\\ R_{k(n)}>0}}\frac{16Cx}{\log^2\frac{x}{2}}$$

$$+5\times10^{19}\leqslant\frac{16Cx}{\det^2\frac{x}{2}}\sum_{k=0}^{25}\sum_{\substack{0<k-nx<x\\R_{k(n)}>0}}1+5\times10^{19}\leqslant\frac{16Cx}{\text{and}^2\frac{x}{2}}A(26x)+5\times10^{19}.\tag{14}$$

另一方面，

$$\begin{aligned}\sum_n R_k(n)\prod_{\substack{p\mid n\\p>2}}\frac{p-2}{p-1}&=\sum_n R_k(n)\prod_{\substack{p\mid n\\p>2}}\left(1-\frac{1}{p-1}\right)\\&\geqslant\sum_n R_k(n)-\sum_n R_k(n)\sum_{\substack{p\mid n\\p>2}}\frac{1}{p-1}\\&=\Big(\sum_{\frac{kx}{2}<p_1,\ p_2<\frac{kx}{2}+x}1\Big)-\sum_{p>2}\frac{1}{p-1}\sum_{p\mid n}R_k(n)\\&=(\Sigma_1)^2-\frac{1}{2}\sum_{3\mid n}R_k(n)-\sum_{p>3}\frac{1}{p-1}\sum_{p\mid n}R_k(n),\end{aligned}\tag{15}$$

其中 $\Sigma_1=\sum\limits_{\frac{kx}{2}<p_1\leqslant\frac{kx}{2}+x}1.$

因为

$$\begin{aligned}&\sum_{k\mid n}R_k(n)\\&=\sum_{\substack{k\mid p_1+p_2\\\frac{kx}{2}\leqslant p_1,\ p_2<\frac{kx}{2}+x}}1=2\sum_{\substack{\frac{kx}{2}\leqslant p_1,\ p_2<\frac{kx}{2}+x\\p_1\equiv1,\ p_2\equiv-1(\bmod 3)}}1\\&=2\left(\pi\left(\frac{kx}{2}+x,\ 3,\ 1\right)-\pi\left(\frac{kx}{2},\ 3,\ 1\right)\right)\left(\pi\left(\frac{kx}{2}+x,\ 3,\ -1\right)-\pi\left(\frac{kx}{2},\ 3,\ -1\right)\right)\\&\leqslant\frac{1}{2}\left(\left(\pi\left(\frac{kx}{2}+x,\ 3,\ 1\right)-\pi\left(\frac{kx}{2},\ 3,\ 1\right)\right)+\left(\pi\left(\frac{kx}{2}+x,\ 3,\ -1\right)\right.\right.\\&\qquad\left.\left.-\pi\left(\frac{kx}{2},\ 3,\ -1\right)\right)\right)^2\leqslant\frac{1}{2}(\Sigma_1)^2,\end{aligned}\tag{16}$$

所以由(15)、(16)得

$$\sum_n R_k(n)\prod_{\substack{p\mid n\\p>2}}\frac{p-2}{p-1}\geqslant\frac{3}{4}(\Sigma_1)^2-\sum_{p>3}\frac{1}{p-1}\sum_{p\mid n}R_k(n).\tag{17}$$

记

$$\sum_{p>3}\frac{1}{p-1}\sum_{p\mid n}R_k(n)=\Sigma_2+\Sigma_3,\tag{18}$$

其中

$$\Sigma_2=\sum_{x^{1/3}\leqslant p}\frac{1}{p-1}\sum_{p\mid n}R_k(n)\leqslant\frac{x}{\log x}\Sigma_1\cdot 10^{-11},\tag{19}$$

$$\begin{aligned}\Sigma_3&=\sum_{3<p<x^{1/3}}\frac{1}{p-1}\sum_{p\mid n}R_k(n)=\sum_{3<p<x^{1/3}}\frac{1}{p-1}\sum_{\frac{kx}{2}\leqslant p_1<\frac{kx}{2}+x}\ \sum_{\substack{\frac{kx}{2}\leqslant p_2<\frac{kx}{2}+x\\ p_2\equiv p_1\,(\mathrm{mod}\,p)}}1\\&\leqslant\sum_{3<p<x^{1/3}}\frac{1}{p-1}\sum_{\frac{kx}{2}\leqslant p_1<\frac{kx}{2}+x}\frac{2x}{(p-1)\log\frac{x}{p}}\quad\text{（利用引理 4）}\qquad(20)\\&\leqslant\Sigma_1\cdot\sum_{3<p<x^{1/3}}\frac{2x}{(p-1)^2\log\frac{x}{p}},\end{aligned}$$

记

$$\sum_{3<p<x^{1/3}}\frac{2x}{(p-1)^2\log\frac{x}{p}}=\Sigma_4+\Sigma_5,\tag{21}$$

其中

$$\begin{aligned}\Sigma_4&=\frac{2x}{\log x}\sum_{3<p<149}\frac{1}{(p-1)^2}\left(1+\frac{\log p}{\log x-\log p}\right)\\&\leqslant\frac{2x}{\log x}(0.0625\times1.0180045619+0.02777777\times1.021850879+0.01\\&\quad\times1.027063678+0.006944444\times1.0290037642+0.00390625\\&\quad\times1.0321347011+0.0030864197\times1.0334384216+0.0020661157\\&\quad\times1.0356855613+0.001275510 2\times1.0384251046+0.0011111111\\&\quad\times1.0392159822+0.00077160493\times1.0413200119+0.000625\\&\quad\times1.0425446676+0.00056689342\times1.0431138452+0.00047258979\\&\quad\times1.0441784753+0.00036982248\times1.0456199612+0.00029726516\\&\quad\times1.0469100543+0.00027777777\times1.0473117183+0.00022956841\\&\quad\times1.048443778+0.00020408163\times1.0491447034+0.00019290123\\&\quad\times1.0494808231+0.00016436554\times1.0504377236+0.00014872099\\&\quad\times1.0510369766+0.00012913223\times1.0518849337+0.00010850694\\&\quad\times1.0529325505+0.0001\times1.0534250964+0.000096116878\\&\quad\times1.05366426672+0.000088999644\times1.054129294+0.000085733882\end{aligned}$$

$$\times 1.0543554761+0.000079719387\times 1.0547959276+0.000062988158$$
$$\times 1.0562258879+0.000059171597\times 1.0566061944+0.000054065743$$
$$\times 1.0571558995+0.000052509976\times 1.0573339193)$$
$$<0.126749\cdot\frac{2x}{\log x}. \tag{22}$$

$$\Sigma_5=\sum_{169\leqslant p<x^{1/3}}\frac{2x}{(p-1)^2\log\frac{x}{p}}$$
$$=2x\sum_{149\leqslant p<x^{1/3}}\frac{1}{(p-1)^2}\left(\frac{1}{\log\frac{x}{149}}+\int_{149}^{p}\frac{\mathrm{d}u}{u\log^2\frac{x}{u}}\right)$$
$$\leqslant 2x\left(\frac{1}{\log\frac{x}{149}}\sum_{149\leqslant p}\frac{1}{(p-1)^2}+\sum_{151\leqslant p}\frac{p^{3/4}}{(p-1)^2}\int_{149}^{x1/3}\frac{\mathrm{d}u}{u^{7/6}\log^2\frac{x}{u}}\right). \tag{23}$$

由于

$$\sum_{149\leqslant p}\frac{1}{(p-1)^2}\leqslant 2\left(\frac{149}{148}\right)^2\frac{1}{\log 149}\int_{169}^{\infty}\frac{\theta(u)-\theta(139)}{u^3}\mathrm{d}u$$
$$\leqslant 2\left(\frac{149}{148}\right)^2\frac{1}{\log 149}\left(1.001102\times\frac{1}{149}-\frac{1}{149^2}\times\frac{1}{2}\times 126.64\right)\text{(利用(8))}$$
$$\leqslant 0.0015665, \tag{24}$$

$$\sum_{151\leqslant p}\frac{p^{3/4}}{(p-1)^2}\leqslant\frac{5}{4}\left(\frac{151}{150}\right)^2\int_{151}^{\infty}\frac{\theta(u)-\theta(149)}{u^{9/6}}\mathrm{d}u$$
$$\leqslant\frac{5}{4}\left(\frac{151}{150}\right)^2\left(1.001102\times\frac{4}{151^{1/4}}-\frac{4}{5}\times\frac{131.64}{151^{5/4}}\right) \tag{25}$$
(利用(8))
$$\leqslant 0.2381772.$$

所以

$$\Sigma_5\leqslant\frac{2x}{\log x}\left(0.0015665\left(1+\frac{\log 149}{\log x-\log 149}\right)+0.2381772\right.$$

$$\times \frac{\log x}{149^{3/4}}\left(\frac{4}{3\log^2 \frac{x}{149}}+\frac{32}{9\log^3 \frac{x}{149}}+\frac{128}{9\log^4 \frac{x}{x^{1/3}}}\right)\Bigg)$$

$$\leqslant \frac{2x}{\log x}(0.0016578+0.0000947)$$

$$\leqslant \frac{2x}{\log x}\cdot 0.0017525. \tag{26}$$

由(17)、(18)、(19)、(20)、(21)、(22)、(26)得

$$\sum_n R_k(n)\prod_{\substack{p\mid n\\ p>2}}\frac{p-2}{p-1}\geqslant \frac{3}{4}(\Sigma_1)^2-\Sigma_1\cdot\frac{x}{\log x}\cdot 0.257004$$

$$=\Sigma_1\cdot\left(\frac{3}{4}\Sigma_1-0.257004\frac{x}{\log x}\right). \tag{27}$$

在 $k=0$ 时,由引理 3 的(7)式

$$\Sigma_1=\pi(x)-1>\frac{x}{\log x-\frac{1}{2}}-1>\frac{x}{\log x}. \tag{28}$$

在 $k>0$ 时,由引理 3,

$$\Sigma_1=\pi\left(\frac{kx}{2}+x\right)-\pi\left(\frac{kx}{2}\right)=\frac{\theta\left(\frac{k}{2}x+x\right)}{\log\left(\frac{kx}{2}+x\right)}+\int_{\frac{kx}{2}}^{\frac{kx}{2}+x}\frac{\theta(u)}{u\log^2 u}-\mathrm{d}u-\frac{\theta\left(\frac{kx}{2}\right)}{\log\frac{kx}{2}}$$

$$\geqslant \frac{\left(\frac{k}{2}+1\right)x}{\log\left(\frac{k}{2}+1\right)x}-\frac{\frac{k}{2}x}{\log\frac{k}{2}x}-\frac{0.0242269\left(\frac{k}{2}+1\right)x}{\log^2\left(\frac{k}{2}+1\right)x}-\frac{0.0242299\cdot\frac{k}{2}x}{\log^2\frac{k}{2}x}$$

$$+\int_{\frac{kx}{2}}^{\frac{kx}{2}}\frac{0.998684}{\log^2 u}\mathrm{d}u.$$

由于$\frac{1}{\log^2 u}$凸,所以

$$\int_a^b\frac{\mathrm{d}u}{\log^2 u}\geqslant (b-a)\frac{1}{\log^2\frac{b+a}{2}},$$

$$\Sigma_1 \geqslant \frac{\left(\frac{k}{2}+1\right)x}{\log\left(\frac{k}{2}+1\right)x}-\frac{\frac{k}{2}x}{\log\frac{k}{2}x}-\frac{0.0242269\left(\frac{k}{2}+1\right)x}{\log^2\left(\frac{k}{2}+1\right)}$$

$$-\frac{0.0242269\cdot\frac{k}{2}x}{\log^2\frac{k}{2}x}+\frac{0.998684x}{\log^2\frac{k+1}{2}x}\geqslant -\frac{x}{\log x}\cdot t_k. \tag{29}$$

其中 t_k 的具体数值在下面的(30)中.

由(27)、(28)、(29)得

$$\sum_{k=0}^{24}R_k(n)\prod_{\substack{p\mid n\\ p>2}}\frac{p-2}{p-1}>\sum_{k=0}^{24}\left(\Sigma_1\cdot\left(\frac{3}{4}\Sigma_1-0.257004\frac{x}{\log x}\right)\right)$$

$$\geqslant\sum_{k=0}^{24}t_k\left(\frac{3}{4}t_k-0.257004\right)\frac{x^2}{\log^2 x}\geqslant\frac{x^2}{\log^2 x}(1\times 0.492924202+0.9999519974$$

$$\times 0.492864207+0.9949678885\times 0.486688794+0.9914930371\times 0.4824052$$

$$+0.9887855351\times 0.479080516+0.9865482508\times 0.476341776+0.9846298739$$

$$\times 0.473998998+0.9829423769\times 0.4719421942194+0.9814298926$$

$$\times 0.470103368+0.980054698\times 0.468433527+0.9787961043\times 0.466900672$$

$$+0.9776165274\times 0.465480807+0.9765191678\times 0.464154933$$

$$+0.9754865675\times 0.462908051+0.9745096731\times 0.461731163$$

$$+0.9735812073\times 0.460613269+0.9726952319\times 0.45954737+0.9718468378$$

$$\times 0.458528467+0.9710319226\times 0.45755056+0.9702470181\times 0.456609649$$

$$+0.9694891696\times 0.455710735+0.9687558407\times 0.454824819+0.9680448347$$

$$\times 0.4539749+0.9673542369\times 0.453149978+0.9666823679\times 0.452348055$$

$$\geqslant 11.677\frac{x^2}{\log^2 x}.$$

由(14)、(30)得到 $x\geqslant e^{91}$ 时,

$$A(26x)\geqslant\frac{x}{16C}\left(\frac{\log x/2}{\log x}\right)^2\cdot 11.67,$$

因为 $x\geqslant e^{95}$ 时, $\frac{x}{26}\geqslant e^{91}$, 所以这时

$$A(x) \geqslant \frac{x}{16C \cdot 26}\left(\frac{\log x/52}{\log x/26}\right)^{2} \cdot 11.67 \geqslant \frac{x}{23.72}.$$

故(13)式成立.

References

[1] Deshouillers. J. M., *Séminaire Delange—Pisot—Poiston*, 17e annće: 1975/76. Théorie des numbers: Fasc. 2, Exp. No. G16, 6pp. *Secretariat Math.*, *Pairs*, 1977.

[2] Klimov. N. I., *Volz. Math. Sb. Vpp.*, 7 (1969), 32 - 40.

[3] Klimov. N. I., Piltai. G. Z. and Sheptitskaya. T. A., *Issled. Teor. Chisel*, Saratov 4(1972), 35 - 51.

[4] Montgomery. H. L. and Vanghen. R. C., *Mathematika*, 20(1973), 119 - 134.

[5] Rosser. J. B. and Schoenfeld. L., *Math. Comp.*, 29(1975), 243 - 269.

[6] Schnirelman. L. G., *Izv. Donskowo Politeh.*, Inst., **14** (1930), 3 - 28.

[7] Schnirelman. L. G., *Math. Ann.*, 107(1933), 649 - 690.

[8] Vanghan. R. C., J. Reine. Angew. *Math.*, 290(1977), 93 - 108.

[9] Chang, M. Y. and Ding. P., *J. of China University of Science and Technology*, Math. Issue, 13(1983), 31 - 53.

Hilbert 不等式

1974 年, H. L. Montgomery 与 R. C. Vaughan 在研究大筛法时讨论了一般形式的 Hilbert 不等式,我们改进了他们的结果,简化了他们的证明,建立了以下的两个定理.

定理 1 设 $\Delta_1=1, \Delta_2, \Delta_3, \cdots, \Delta_n, \cdots$ 为正数, $\theta_r=\min(\Delta_r, \Delta_{r+1})$, 则

$$\sum_{r=1}^{\infty}\frac{\theta_r}{(\Delta_1+\Delta_2+\cdots+\Delta_r)^2}\leqslant\frac{\pi^2}{6}.$$

定理 2 设 $\lambda_1, \lambda_2, \cdots, \lambda_n$ 为一组实数, $|\lambda_r-\lambda_s|\geqslant\delta_r>0$ $(r, s=1, 2, \cdots, n, r\neq s)$, 则对任意复数 $u_1, u_2, \cdots, u_n$,有

$$\left|\sum_{r, s}\frac{u_r\bar{u}_s}{\lambda_r-\lambda_s}\right|<4.452\sum_r\delta_r^{-1}\cdot|u_r|^2.$$

在证明定理 1 时用到以下引理

引理 1 若 $\sum_{r=1}^{n}\frac{\theta_r}{(\Delta_1+\Delta_2+\cdots+\Delta_n)^2}$ 在$(\Delta_2^{(0)}, \Delta_3^{(0)}, \cdots, \Delta_n^{(0)})$处取得最大值 M_n,则 $\Delta_n^{(0)}\geqslant\Delta_{n-1}^{(0)}\geqslant\cdots\geqslant\Delta_2^{(0)}\geqslant1$.

引理 2 若有某个 r $(2\leqslant r\leqslant n)$,使 $\Delta_n^{(0)}>\cdots>\Delta_r^{(0)}=\Delta_{r-1}^{(0)}$,则 $\Delta_{r-1}^{(0)}=\Delta_{r-1}^{(0)}=\cdots=\Delta_2^{(0)}=\Delta_1=1$.

引理 3 在 $r<\frac{n}{3}$ 时, $\Delta_{r+1}^{(0)}=\Delta_r^{(0)}=\cdots=\Delta_2^{(0)}=\Delta_1=1$.

在证明定理 2 时用到以下引理

引理 4 $\sum_r\frac{\delta_r}{(\lambda_r-\lambda_s)^2}<\frac{\pi^2}{3}\delta_s^{-1}$,并且$\frac{\pi^2}{3}$ 是最佳常数.

引理 5 $\sum_r\frac{\delta_r}{(\lambda_r-\lambda_s)^4}<\frac{\pi^4}{45}\delta_s^{-3}$,并且$\frac{\pi^4}{45}$ 是最佳常数.

引理 6 $s\neq t$ 时,

$$\sum_r\frac{\delta_r}{(\lambda_r-\lambda_s)^2(\lambda_r-\lambda_s)^2}\leqslant\frac{4}{(\lambda_s-\lambda_t)^2}(\delta_s^{-1}+\delta_t^{-1}).$$

Hilbert 不等式

1 引言

1974 年, H. L. Montgomery 与 R. C. Vaughan[1] 提出并证明了以下命题:

设 $\lambda_1, \lambda_2, \cdots, \lambda_n$ 为一组实数,并且 $r \neq s$ 时, $|\lambda_r - \lambda_s| \geqslant \delta, > 0$, 则对任意复数 $u_1, u_2, \cdots, u_n$,有

$$\left|\sum_{r, s} \frac{u_r \bar{u}_s}{\lambda_r - \lambda_s}\right| < c \sum_r \delta_r^{-1} |u_r|^2, \tag{1}$$

其中 C 为某个正的常数(约定和号中不出现分母为零的项).

在 $\lambda_r = r$ 时,(1)就是普通的 Hilbert 不等式,其中 C 可取最佳值 π.

在一切 δ_r 全相等时, C 也可取最佳值 π,这个结果是 H. L. Montgomery 与 R. C. Vaughan[1] 证明的. 利用这一结果,他们建立了著名的大筛法不等式:

$$\sum_{r=1}^{R} \left| \sum_{n=M+1}^{M+N} a_n e(n\alpha_r) \right|^2 \leqslant (N + \delta^{-1} - 1) \sum_{n=M+1}^{M+N} |a_n|^2$$

对于一般情况, H. L. Montgomery 与 R. C. Vaughan[1] 用相当复杂的方法证明了不等式(1)对于 $C = \frac{3\pi}{2}$ 成立,猜测 C 的最佳值为 π.

1977 年, S. Srinivasan[3] 对于一般情况给出了另一个证明,他的方法仍然很复杂,并且没有给出常数的 C 值.

本文证明了以下两个定理:

定理 I 设 $\Delta_1 = 1, \Delta_2, \cdots, \Delta_n, \cdots$ 为正数,

$\theta_r = \min(\Delta_r, \Delta_{r+1})$, 则

$$\sum_{r=1}^{\infty} \frac{\theta_r}{(\Delta_1 + \Delta_2 + \cdots + \Delta_r)^2} \leqslant \frac{\pi^2}{6}, \tag{2}$$

并且这里的 $\frac{\pi^2}{6}$ 是最佳常数.

定理Ⅱ　设 $\lambda_1, \lambda_2, \cdots, \lambda_n$ 为一组实数，$|\lambda_r-\lambda_s|\geqslant\delta_r>0\ (r, s=1, 2, \cdots, n;\ r\neq s)$，则对任意复数 $u_1, u_2, \cdots, u_n$，有

$$\left|\sum_{r, s}\frac{u_r\bar{u}_s}{\lambda_r-\lambda_s}\right|<4.452\sum_r\delta_r^{-1}|u_r|^2 \tag{3}$$

利用定理Ⅰ，我们改进了 Montgomery 与 Vanghan 对 Σ_1、Σ_2(意义见下文)的估计，并且所得结果是最佳的，从而改进了他们的结果，得到定理Ⅱ；同时本文的引理 3 使 Montgomery 与 Vanghan 原来的证明大为简化.

作者对陆鸣皋老师的鼓励与帮助表示衷心的感谢.

2　定理Ⅱ的证明

本节的目的是证明定理Ⅱ，其中引理 1、2 改进了 Montgomery 与 Vanghan[1]的引理 2、4，引理 3 简化了他们的引理 5.

引理 1　$\displaystyle\sum_r\frac{\delta_r}{(\lambda_r-\lambda_s)^4}<\frac{\pi^2}{3}\delta_s^{-2}$，并且常数$\dfrac{\pi^2}{3}$是最佳的.

证明见下节.

引理 2　$\displaystyle\sum_r\frac{\delta_r}{(\lambda_r-\lambda_s)^4}<\frac{\pi^4}{45}\delta_s^{-2}$，并且常数$\dfrac{\pi^4}{45}$是最佳的.

证明见下节.

引理 3　在 $s\neq t$ 时，

$$\sum_r\frac{\delta_r}{(\lambda_r-\lambda_s)^2(\lambda_r-\lambda_s)^2}\leqslant\frac{4}{(\lambda_s-\lambda_t)^2}(\delta_s^{-1}+\delta_t^{-1}).$$

证明　不妨设 $\lambda_s<\lambda_t$，由于$(x-\lambda_s)^{-2}(x-\lambda_t)^{-2}$ 是凸函数，所以

$$\frac{\delta_r}{(\lambda_r-\lambda_s)^2(\lambda_r-\lambda_s)^2}\leqslant\int_{\lambda_r-\frac{\delta_r}{2}}^{\lambda_r+\frac{\delta_r}{2}}\frac{\mathrm{d}x}{(x-\lambda_s)^2(x-\lambda_t)^2},$$

$$\sum_r\frac{\delta_r}{(\lambda_r-\lambda_s)^2(\lambda_r-\lambda_s)^2}\leqslant\left(\int_{-\infty}^{\lambda_s-\frac{\delta_s}{2}}+\int_{\lambda_s+\frac{\delta_s}{2}}^{\lambda_t-\frac{\delta_t}{2}}+\int_{\lambda_s+\frac{\delta_s}{2}}^{+\infty}\right)\frac{\mathrm{d}x}{(x-\lambda_s)^2(x-\lambda_t)^2}$$

$$=\left(\int_{-\infty}^{\lambda_s-\frac{\delta_s}{2}}+\int_{\lambda_s+\frac{\delta_s}{2}}^{\lambda_s-\frac{\delta_s}{2}}+\int_{\lambda_t+\frac{\delta_t}{2}}^{+\infty}\right)\left[\frac{A}{(x-\lambda_s)^2}+\frac{A}{(x-\lambda_s)^2}-\frac{C}{x-\lambda_s}+\frac{C}{x-\lambda_s}\right]\mathrm{d}x,$$

其中 $$A=\frac{1}{(\lambda_s-\lambda_t)^2},\quad C=\frac{2}{(\lambda_s-\lambda_t)^2}<0.$$

因为

$$\int\left[\frac{A}{(x-\lambda_s)^2}+\frac{A}{(x-\lambda_s)^2}-\frac{C}{x-\lambda_s}+\frac{C}{x-\lambda_s}\right]\mathrm{d}x$$

$$=\frac{-A}{x-\lambda_s}+\frac{-A}{x-\lambda_s}+C\ln\left|\frac{x-\lambda_t}{x-\lambda_s}\right|,$$

所以

$$\left(\int_{-\infty}^{\lambda_s-\frac{\delta_s}{2}}+\int_{\lambda_s+\frac{\delta_s}{2}}^{\lambda_s-\frac{\delta_s}{2}}+\int_{\lambda_s+\frac{\delta_s}{2}}^{+\infty}\right)\left[\frac{A}{(x-\lambda_s)^2}+\frac{A}{(x-\lambda_s)^2}-\frac{C}{x-\lambda_s}+\frac{C}{x-\lambda_s}\right]\mathrm{d}x$$

$$=\frac{2A}{\delta_s}+\frac{2A}{\delta_s}-\frac{A}{\lambda_s-\lambda_s-\frac{\delta_s}{2}}+\frac{A}{\lambda_s-\lambda_s+\frac{\delta_s}{2}}+\frac{A}{\lambda_s-\lambda_s+\frac{\delta_s}{2}}-\frac{A}{\lambda_s-\lambda_s-\frac{\delta_s}{2}}$$

$$+\frac{2A}{\delta_s}+\frac{2A}{\delta_s}+C\ln\frac{\left(\lambda_s-\lambda_s+\frac{\delta_s}{2}\right)\left(\lambda_s-\lambda_s+\frac{\delta_s}{2}\right)}{\left(\lambda_s-\lambda_s-\frac{\delta_s}{2}\right)\left(\lambda_s-\lambda_s-\frac{\delta_s}{2}\right)}$$

$$\leqslant\frac{4A}{\delta_s}+\frac{4A}{\delta_t}.$$

定理Ⅱ的证明. 我们可以设 u_1, u_2, …, u_n 已经“标准化”,即满足条件

$$\begin{cases}\sum_r|u_r|^2=1, & (4)\\ \sum_r\frac{u_r}{\lambda_s-\lambda_r}=i\lambda u_s\delta_s^{-1}. & (5)\end{cases}$$

其中纯虚数 $i\lambda$ 为反对称阵$\left(\frac{\delta_r^{\frac{1}{2}}\delta_s^{\frac{1}{2}}}{\lambda_r-\lambda_s}\right)$的特征根.

由 Canchy-Schwarz 不等式,

$$\left|\sum_{r,s}\frac{u_r\bar{u}_s}{\lambda_r-\lambda_s}\right|^2\leqslant\left(\sum_r|u_r|^2\delta_r^{-1}\right)\left(\sum_r\left|\sum_s\frac{\bar{u}_s}{\lambda_r-\lambda_s}\delta_r^{\frac{1}{2}}\right|^2\right)$$

$$=\sum_r\delta_r\sum_s\sum_t\frac{\bar{u}_s u_t}{(\lambda_r-\lambda_s)(\lambda_r-\lambda_t)}=\Sigma_1+\Sigma_3.$$

其中

$$\Sigma_1=\sum_r\delta_r\sum_s|u_s|^2(\lambda_r-\lambda_s)^{-2}=\sum_s|u_s|^2\sum_r\frac{\delta_r}{(\lambda_r-\lambda_s)^2},$$

$$\Sigma_3=\sum_r\delta_r\sum_{s\neq t}\frac{\bar{u}_su_t}{(\lambda_r-\lambda_s)(\lambda_r-\lambda_t)}$$

$$=\sum_{r,s,t}\frac{\delta_r\bar{u}_su_t}{\lambda_r-\lambda_t}\left(\frac{1}{\lambda_r-\lambda_s}-\frac{1}{\lambda_r-\lambda_t}\right)$$

$$=\sum_r\delta_r\sum_{\substack{s\\t\neq r}}\frac{\bar{u}_su_t}{(\lambda_s-\lambda_t)(\lambda_r-\lambda_s)}-\sum_r\delta_r\sum_{\substack{t\\s\neq r}}\frac{\bar{u}_su_t}{(\lambda_s-\lambda_t)(\lambda_r-\lambda_s)}$$

$$=\sum_{r,s,t}\frac{\delta_r\bar{u}_su_t}{(\lambda_s-\lambda_t)(\lambda_r-\lambda_s)}-\sum_{r,s,t}\frac{\delta_r\bar{u}_su_t}{(\lambda_s-\lambda_t)(\lambda_r-\lambda_s)}$$

$$+\sum_{s,t}\frac{\delta_t\bar{u}_su_t}{(\lambda_s-\lambda_t)^2}+\sum_{s,t}\frac{\delta_s\bar{u}_su_t}{(\lambda_s-\lambda_t)^2}$$

$$=\Sigma_3-\Sigma_4+2\mathrm{Re}\Sigma_5.$$

其中
$$\Sigma_3=\sum_{r,s}\frac{\delta_r\bar{u}_s}{\lambda_r-\lambda_s}\sum_t\frac{u_t}{\lambda_s-\lambda_t}=i\lambda\sum_{r,s}\frac{\delta_r|u_s|^2\delta_t^{-1}}{\lambda_r-\lambda_s}.$$

$$\Sigma_4=\sum_{r,t}\frac{\delta_ru_s}{\lambda_r-\lambda_s}\sum_s\frac{\bar{u}_s}{\lambda_s-\lambda_t}=i\lambda\sum_{r,t}\frac{\delta_r|u_s|^2\delta_t^{-1}}{\lambda_r-\lambda_s}=\Sigma_3.$$

$$\Sigma_5=\sum_{s,t}\frac{\delta_s\bar{u}_su_t}{(\lambda_s-\lambda_t)^2}.$$

所以
$$\Sigma_5=2\mathrm{Re}\Sigma_5\leqslant 2|\Sigma_5|\leqslant 2\Sigma_6.$$

其中
$$\Sigma_6=\sum_{s,t}\frac{\delta_s|u_s|\cdot|u_t|}{(\lambda_s-\lambda_t)^2},$$

$$(\Sigma_6)^2\leqslant\left(\sum_r\delta_r^{-1}|u_r|^2\right)\left[\sum_r\delta_r\left(\sum_s\frac{\delta_s|u_s|}{(\lambda_s-\lambda_r)^2}\right)^2\right]$$

$$=\sum_r\delta_r\sum_{s,t}\frac{\delta_s\delta_t|u_s|\cdot|u_t|}{(\lambda_s-\lambda_r)^2(\lambda_t-\lambda_r)^2}$$

$$=\sum_{s,t}\delta_s\delta_t|u_s|\cdot|u_t|\sum_r\frac{\delta_r}{(\lambda_r-\lambda_s)^2(\lambda_r-\lambda_s)^2}$$

$$=\sum_s\delta_s^2|u_s|^2\sum_r\frac{\delta_r}{(\lambda_r-\lambda_s)^6}+\sum_{s\neq t}\delta_s\delta_t|u_s|\cdot|u_t|\sum_r\frac{\delta_r}{(\lambda_r-\lambda_s)^2(\lambda_r-\lambda_s)^2}$$

$$=\Sigma_7+\Sigma_8.$$

由引理 2，

$$\Sigma_7=\sum_s\delta_s^2\,|u_s|^2\sum_r\frac{\delta_r}{(\lambda_r-\lambda_s)^6}\leqslant\sum_s\delta_s^2\,|u_s|^2\cdot\frac{\pi^4}{45}\delta_s^{-2}$$

$$=\sum_s\delta_s^{-1}\,|u_s|^2\cdot\frac{\pi^4}{45}=\frac{\pi^4}{45}.$$

由引理 3，

$$\Sigma_8=\sum_{s\neq t}\delta_s\delta_t\,|u_s|\cdot|u_t|\sum_r\frac{\delta_r}{(\lambda_r-\lambda_s)^2(\lambda_r-\lambda_s)^2}$$

$$\leqslant\sum_{s\neq t}\delta_s\delta_t\,|u_s|\cdot|u_t|\,\frac{4}{(\lambda_s-\lambda_t)^2}(\delta_s^{-1}+\delta_t^{-1})$$

$$=\sum_{s,t}(\delta_s+\delta_t)\cdot\frac{4}{(\lambda_s-\lambda_t)^2}|u_s|\cdot|u_t|=4\times2\Sigma_6$$

所以

$$(\Sigma_6)^2\leqslant\frac{\pi^4}{45}+8\Sigma_6,$$

$$\Sigma_6\leqslant4+\sqrt{\frac{\pi^4}{45}+16}<0.8372\pi^2.$$

由引理 1，

$$\Sigma_1\leqslant\sum_s|u_s|^2\cdot\frac{\pi^2}{3}\delta_s^{-1}=\frac{\pi^2}{3},$$

所以

$$\left|\sum_{r,s}\frac{u_r\bar{u}_s}{\lambda_r-\lambda_s}\right|\leqslant\sqrt{\frac{\pi^2}{3}+2\times0.8372\pi^2}\leqslant1.4169\pi<4.452.$$

3 定理Ⅰ的证明

本节的目的是证明定理Ⅰ，并由定理Ⅰ导出引理 1、2.

考虑函数 $f(\Delta_s,\cdots,\Delta_n)=\sum_{r=1}^{n}\frac{\theta_r}{(\Delta_1+\Delta_2+\cdots+\Delta_r)^2}$ 在区 $0\leqslant\Delta_2$、Δ_3、…、

$\Delta_n<+\infty$中的最大值 M_n. 我们证明

$$M_n<\frac{\pi^2}{6} \tag{6}$$

由于 M_n 随 n 递增,所以可假定 n 为一个大的整数.

当某个 $\Delta_r\to 0$ 或 $\Delta_r\to+\infty$ $(2\leqslant r\leqslant n)$ 时, $\sum_{r=1}^{n}\frac{\theta_r}{(\Delta_1+\Delta_2+\cdots+\Delta_r)^2}$ 中至少有一项趋于零,此时 f 的值趋于 $\leqslant M_{n-1}<M_n$ 的数,因此最大值 M_n 一定在区域 $0<\Delta_r<+\infty$ $(r=2, \cdots, n)$, 内的某一点取得.

引理 4　若 $f(\Delta_2^{(0)}, \cdots, \Delta_n^{(0)})=M_n$,则 $\Delta_n^{(0)}\geqslant\Delta_{n-1}^{(0)}\geqslant\cdots\geqslant\Delta_2^{(0)}\geqslant\Delta_1=1$.

证明　若 $\Delta_n^{(0)}<\Delta_{n-1}^{(0)}$, 那么在$(\Delta_2^{(0)}, \cdots, \Delta_n^{(0)})$的充分小的邻域内, $\Delta_n<\Delta_{n-1}$,

$$f(\Delta_2, \cdots, \Delta_n)=\varphi(\Delta_3, \cdots, \Delta_{n-1})+\frac{\Delta_n}{(\Delta_1+\cdots+\Delta_{n-1})^2}+\frac{\Delta_n}{(\Delta_1+\cdots+\Delta_{n-1}\Delta_n)^2}$$

$$\frac{\partial f}{\partial\Delta_n}=\frac{1}{(\Delta_1+\Delta_3+\cdots+\Delta_{n-1})^2}+\frac{1}{(\Delta_1+\Delta_3+\cdots+\Delta_n)^2}-\frac{2\Delta_n}{(\Delta_1+\Delta_3+\cdots+\Delta_n)^3}$$

$$=\frac{1}{(\Delta_1+\cdots+\Delta_{n-1})^2}+\frac{\Delta_1+\cdots+\Delta_{n-1}-\Delta_n}{(\Delta_1+\cdots+\Delta_n)^3}>0,$$

这与 $f(\Delta_2^{(0)}, \cdots, \Delta_n^{(0)})=M_n$ 矛盾,所以 $\Delta_n^{(0)}\geqslant\Delta_{n-1}^{(0)}$.

设已有 $\Delta_n^{(0)}\geqslant\Delta_{n-1}^{(0)}\geqslant\cdots\geqslant\Delta_{r+1}^{(0)}\geqslant\Delta_r^{(0)}$. 若 $\Delta_r^{(0)}<\Delta_{r-1}^{(0)}$,固定 $\Delta_i=\Delta_i^{(0)}$ $(i=n, n-1, \cdots, r+1)$, 考虑空间 R^{r-1} 中的点$(\Delta_3, \cdots, \Delta_r)$,在$(\Delta_2^{(0)}, \cdots, \Delta_r^{(0)})$的充分小的邻域 U 内, $\Delta_r<\Delta_{r-1}$. 这时有两种情况:

(1) $\Delta_r^{(0)}<\Delta_{r+1}^{(0)}$.

在$(\Delta_2^{(0)}, \cdots, \Delta_r^{(0)})$的充小的邻域 $U_1\subset U$ 内, $\Delta_r<\Delta_{r+1}^{(0)}$,

$$f(\Delta_2, \cdots, \Delta_n)=\psi(\Delta_2, \cdots, \Delta_{r-1})+\frac{\Delta_r}{(\Delta_1+\cdots+\Delta_{r-1})^2}+\frac{\Delta_r}{(\Delta_1+\cdots+\Delta_r)^2}$$
$$+\frac{\Delta_{r+1}^{(0)}}{(\Delta_1+\cdots+\Delta_{r+1}^{(0)})^2}+\cdots+\frac{\Delta_n^{(0)}}{(\Delta_1+\cdots+\Delta_n^{(0)})^2},$$

$$\frac{\partial f}{\partial\Delta_r}=\frac{1}{(\Delta_1+\cdots+\Delta_{r-1})^2}+\frac{1}{(\Delta_1+\cdots+\Delta_r)^2}-\frac{2\Delta_r}{(\Delta_1+\cdots+\Delta_r)^3}$$
$$-\frac{2\Delta_{r+1}^{(0)}}{(\Delta_1+\cdots+\Delta_{r+1}^{(0)})^3}-\cdots-\frac{2\Delta_n^{(0)}}{(\Delta_1+\cdots+\Delta_n^{(0)})^3}.$$

在 $y \geqslant x \geqslant 0$ 时,

$$\frac{1}{x^2}-\frac{1}{y^2}=\int_y^x\left(\frac{1}{x^2}\right)'\mathrm{d}x=\int_x^y\frac{2}{x^2}\mathrm{d}x\geqslant\frac{2(y-x)}{y^2}.$$

所以

$$\begin{aligned}\frac{\partial f}{\partial \Delta_r}&\geqslant\frac{1}{(\Delta_1+\cdots+\Delta_r)^2}+\frac{2\Delta_r}{(\Delta_1+\cdots+\Delta_r)^3}+\frac{1}{(\Delta_1+\cdots+\Delta_r)^2}\\&\quad-\frac{2\Delta_r}{(\Delta_1+\cdots+\Delta_r)^3}-\frac{2\Delta_{r+1}^{(0)}}{(\Delta_1+\cdots+\Delta_{r+1}^{(0)})^3}-\cdots-\frac{2\Delta_n^{(0)}}{(\Delta_1+\cdots+\Delta_n^{(0)})^3}\\&=\frac{1}{(\Delta_1+\cdots+\Delta_r)^2}+\frac{1}{(\Delta_1+\cdots+\Delta_r)^2}-\frac{2\Delta_{r+1}^{(0)}}{(\Delta_1+\cdots+\Delta_{r+1}^{(0)})^3}\\&\quad-\cdots-\frac{2\Delta_n^{(0)}}{(\Delta_1+\cdots+\Delta_n^{(0)})^3}\\&\geqslant\frac{1}{(\Delta_1+\cdots+\Delta_r)^2}+\frac{1}{(\Delta_1+\cdots+\Delta_{r+1}^{(0)})^2}-\frac{2\Delta_{r+2}^{(0)}}{(\Delta_1+\cdots+\Delta_{r22}^{(0)})^3}\\&\quad-\cdots-\frac{2\Delta_n^{(0)}}{(\Delta_1+\cdots+\Delta_n^{(0)})^3}\\&\geqslant\cdots\\&\geqslant\frac{1}{(\Delta_1+\cdots+\Delta_r)^2}+\frac{1}{(\Delta_1+\cdots+\Delta_n^{(0)})^2}>0.\end{aligned}$$

与 $f(\Delta_2^{(0)},\cdots,\Delta_n^{(0)})=M_n$ 矛盾.

(2) $\Delta_r^{(0)}=\Delta_{r+1}^{(0)}$

考虑邻域 U 中 $\Delta_r\geqslant\Delta_r^{(0)}$ 的部分,因为

$$\begin{aligned}f(\Delta_s,\cdots,\Delta_n)&=\psi(\Delta_s,\cdots,\Delta_{r-1})+\cdots+\frac{\Delta_r}{(\Delta_1+\cdots+\Delta_{r-1})^2}+\frac{\Delta_{r+1}^{(0)}}{(\Delta_1+\cdots+\Delta_r)^2}\\&\quad+\cdots+\frac{\Delta_n^{(0)}}{(\Delta_1+\cdots+\Delta_n^{(0)})^2},\end{aligned}$$

所以偏导数

$$\begin{aligned}\frac{\partial f}{\partial \Delta_r}&=\frac{1}{(\Delta_1+\cdots+\Delta_{r-1})^2}-\frac{2\Delta_{r+1}^{(0)}}{(\Delta_1+\cdots+\Delta_r)^3}-\frac{2\Delta_{r+1}^{(0)}}{(\Delta_1+\cdots+\Delta_{r+1}^{(0)})^3}\\&\quad-\cdots-\frac{2\Delta_n^{(0)}}{(\Delta_1+\cdots+\Delta_n^{(0)})^3}\geqslant\frac{1}{(\Delta_1+\cdots+\Delta_r)^2}+\frac{2\Delta_r}{(\Delta_1+\cdots+\Delta_r)^3}\end{aligned}$$

$$-\frac{2\Delta_{r+1}^{(0)}}{(\Delta_1+\cdots+\Delta_r)^3}-\frac{2\Delta_{r+1}^{(0)}}{(\Delta_1+\cdots+\Delta_{r+1}^{(0)})^3}-\cdots-\frac{2\Delta_n^{(0)}}{(\Delta_1+\cdots+\Delta_n^{(0)})^3}$$

$$\geqslant\frac{1}{(\Delta_1+\cdots+\Delta_r)^2}-\frac{2\Delta_{r+1}^{(0)}}{(\Delta_1+\cdots+\Delta_{r+1}^{(0)})^3}-\cdots-\frac{2\Delta_n^{(0)}}{(\Delta_1+\cdots+\Delta_n^{(0)})^3}$$

$$\geqslant\cdots$$

$$\geqslant\frac{1}{(\Delta_1+\cdots+\Delta_n^{(0)})^2}>0,$$

仍与 $f(\Delta_2^{(0)},\cdots,\Delta_n^{(0)})=M_n$ 矛盾.

因此 $\Delta_r^{(0)}\geqslant\Delta_{r-1}^{(0)}$，从而引理 4 成立.

由引理 4，$M_n=\max\limits_{\Delta_1\leqslant\Delta_2\leqslant\cdots\leqslant\Delta_n}\sum\limits_{r=1}^{n}\frac{\Delta_r}{(\Delta_1+\cdots+\Delta_r)^2}$

记 $h(\Delta_2,\cdots,\Delta_n)=\sum\limits_{r=1}^{n}\frac{\Delta_r}{(\Delta_1+\cdots+\Delta_r)^2}$，$1=\Delta_1\leqslant\Delta_2\leqslant\cdots\leqslant\Delta_n$.

又设 $h(\Delta_2^{(0)},\cdots,\Delta_n^{(0)})=\max h=M_n$.

引理 5　若 $\Delta_{r-1}^{(0)}<\Delta_r^{(0)}\leqslant\Delta_{r+1}^{(0)}$，则左偏导数

$$\left.\frac{\partial h}{\partial\Delta_r}\right|_{(\Delta_2^{(0)},\cdots,\Delta_n^{(0)})}\geqslant 0,\tag{7}$$

若 $\Delta_{r-1}^{(0)}\leqslant\Delta_r^{(0)}<\Delta_{r+1}^{(0)}$，则右偏导数

$$\left.\frac{\partial h}{\partial\Delta_r}\right|_{(\Delta_2^{(0)},\cdots,\Delta_n^{(0)})}\leqslant 0,\tag{8}$$

若 $\Delta_{r-1}^{(0)}<\Delta_r^{(0)}<\Delta_{r+1}^{(0)}$，则偏导数

$$\left.\frac{\partial h}{\partial\Delta_r}\right|_{(\Delta_2^{(0)},\cdots,\Delta_n^{(0)})}=0.\tag{9}$$

证明　若 $\Delta_{r-1}^{(0)}<\Delta_r^{(0)}\leqslant\Delta_{r+1}^{(0)}$，则左偏导数

$$\left.\frac{\partial h}{\partial\Delta_r}\right|_{(\Delta_2^{(0)},\cdots,\Delta_n^{(0)})}$$

$$=\lim_{\Delta_r\to\Delta_r^{(0)-}}\frac{h(\Delta_2^{(0)},\cdots,\Delta_r^{(0)},\cdots,\Delta_n^{(0)})-h(\Delta_2^{(0)},\cdots,\Delta_r,\cdots,\Delta_n^{(0)})}{\Delta_r^{(0)}-\Delta_r}\geqslant 0.$$

同理可证(8)，(9).

引理 6 若有某个 $\gamma(2\leqslant\gamma\leqslant n)$,使 $\Delta_n^{(0)}>\cdots>\Delta_n^{(0)}=\Delta_{r-1}^{(0)}$,

则 $\Delta_{r-1}^{(0)}=\Delta_{r-2}^{(0)}=\cdots=\Delta_2^{(0)}=\Delta_1=1.$

证明 由引理 5 的(8),右偏导数

$$\left.\frac{\partial h}{\partial\Delta_r}\right|_{(\Delta_2^{(0)},\ \cdots,\ \Delta_n^{(0)})}\leqslant 0 \tag{10}$$

(易知在 $r=n$ 时,(10)仍然成立).

即

$$\frac{1}{(\Delta_1+\Delta_2^{(0)}+\cdots+\Delta_r^{(0)})^2}\leqslant\frac{2\Delta_r^{(0)}}{(\Delta_1+\Delta_2^{(0)}+\cdots+\Delta_r^{(0)})^3}$$
$$+\frac{2\Delta_{r+1}^{(0)}}{(\Delta_1+\Delta_2^{(0)}+\cdots+\Delta_{r+1}^{(0)})^3}+\cdots+\frac{2\Delta_n^{(0)}}{(\Delta_1+\Delta_2^{(0)}+\cdots+\Delta_n^{(0)})^3}, \tag{11}$$

记 $S=\Delta_1+\Delta_2^{(0)}+\cdots+\Delta_r^{(0)}$,则有显然的不等式

$$S^2(S-3\Delta_r^{(0)})<S^2(S-3\Delta_r^{(0)})+3\Delta_r^{(0)\,2}S-\Delta_r^{(0)\,3}=(S-\Delta_r^{(0)})^2,$$

即

$$\frac{\Delta_1+\Delta_2^{(0)}+\cdots+\Delta_r^{(0)}-3\Delta_r^{(0)}}{(\Delta_1+\Delta_2^{(0)}+\cdots+\Delta_{r-1}^{(0)})^3}=\frac{S-3\Delta_r^{(0)}}{(S-\Delta_r^{(0)})^3}<\frac{1}{S^2}=\frac{1}{(\Delta_1+\Delta_2^{(0)}+\cdots+\Delta_r^{(0)})^2}, \tag{12}$$

由(11)、(12)得(注意 $\Delta_r^{(0)}=\Delta_{r-1}^{(0)}$):

$$\frac{1}{(\Delta_1+\cdots+\Delta_{r-1}^{(0)})^2}<\frac{2\Delta_{r-1}^{(0)}}{(\Delta_1+\cdots+\Delta_{r-1}^{(0)})^3}+\frac{2\Delta_r^{(0)}}{(\Delta_1+\cdots+\Delta_r^{(0)})^3}$$
$$+\cdots+\frac{2\Delta_n^{(0)}}{(\Delta_1+\cdots+\Delta_n^{(0)})^3}, \tag{13}$$

若 $\Delta_{r-2}^{(0)}<\Delta_{r-1}^{(0)}$,则由引理 5,左偏导数

$$\left.\frac{\partial h}{\partial\Delta_{r-1}}\right|_{(\Delta_2^{(0)},\ \cdots,\ \Delta_n^{(0)})}\geqslant 0,$$

与(13)矛盾,因此 $\Delta_{r-1}^{(0)}=\Delta_{r-2}^{(0)}$.

同样由(13)及 $\Delta_{r-1}^{(0)}=\Delta_{r-2}^{(0)}$ 可得

$$\frac{1}{(\Delta_1+\cdots+\Delta_{r-2}^{(0)})^2}<\frac{2\Delta_{r-2}^{(0)}}{(\Delta_1+\cdots+\Delta_{r-2}^{(0)})^3}+\frac{2\Delta_{r-1}^{(0)}}{(\Delta_1+\cdots+\Delta_{r-1}^{(0)})^3}$$
$$+\cdots+\frac{2\Delta_n^{(0)}}{(\Delta_1+\cdots+\Delta_n^{(0)})^3},\tag{14}$$

于是 $\Delta_{r-2}^{(0)}=\Delta_{r-3}^{(0)}$. 依此类推即得结论.

如果有 $\Delta_n^{(0)}>\Delta_{n-1}^{(0)}>\cdots>\Delta_k^{(0)}$, 那么由引理 5 的(9)式,

$$\left.\frac{\partial h}{\partial\Delta_r}\right|_{(\Delta_2^{(0)},\cdots,\Delta_n^{(0)})}=0\quad(r=n,\ n-1,\ \cdots,\ k+1),\tag{15}$$

由 $\left.\frac{\partial h}{\partial\Delta_n}\right|_{(\Delta_2^{(0)},\cdots,\Delta_n^{(0)})}=0$ 可得

$$\Delta_n^{(0)}=\Delta_1+\cdots+\Delta_{n-1}^{(0)}=a_n(\Delta_1+\cdots+\Delta_{n-1}^{(0)}).$$

其中 $a_n=1$.

假定 $\Delta_j^{(0)}=a_j(\Delta_1+\cdots+\Delta_{j-1}^{(0)})$, $j=n,\ n-1,\ \cdots,\ r\geqslant k+1$.

则由(15)

$$0=\frac{1}{(\Delta_1+\cdots+\Delta_r^{(0)})^2}-\frac{2\Delta_r^{(0)}}{(\Delta_1+\cdots+\Delta_r^{(0)})^3}-\frac{2\Delta_{r+1}^{(0)}}{(\Delta_1+\cdots+\Delta_{r+1}^{(0)})^3}$$
$$-\cdots-\frac{2\Delta_n^{(0)}}{(\Delta_1+\cdots+\Delta_n^{(0)})^3}$$
$$=\frac{1}{(\Delta_1+\cdots+\Delta_r^{(0)})^2}-\frac{2\Delta_r^{(0)}}{(\Delta_1+\cdots+\Delta_r^{(0)})^3}-\frac{2\Delta_{r+1}^{(0)}}{(\Delta_1+\cdots+\Delta_{r+1}^{(0)})^3}$$
$$-\left[\frac{1}{(\Delta_1+\cdots+\Delta_{r+1}^{(0)})^2}-\frac{2\Delta_{r+1}^{(0)}}{(\Delta_1+\cdots+\Delta_{r+1}^{(0)})^3}\right]$$
$$=\frac{1}{(\Delta_1+\cdots+\Delta_r^{(0)})^2}-\frac{2\Delta_r^{(0)}}{(\Delta_1+\cdots+\Delta_r^{(0)})^3}-\frac{1}{(\Delta_1+\cdots+\Delta_{r+1}^{(0)})^2},$$

从而

$$\frac{(\Delta_1+\cdots+\Delta_r^{(0)})-2\Delta_r^{(0)}}{(\Delta_1+\cdots+\Delta_r^{(0)})^3}=\frac{1}{(\Delta_1+\cdots+\Delta_{r+1}^{(0)})^2}=\frac{1}{(a_{r+1}+1)^2(\Delta_1+\cdots+\Delta_r^{(0)})^2},$$
$$(a_{r+1}+1)^2(\Delta_1+\cdots+\Delta_r^{(0)}-2\Delta_r^{(0)})=\Delta_1+\cdots+\Delta_r^{(0)},$$
$$\Delta_r^{(0)}=\frac{(a_{r+1}+1)^2-1}{(a_{r+1}+1)^2+1}(\Delta_1+\cdots+\Delta_{r+1}^{(0)}),$$

因此有递推公式

$$a_r=\frac{(a_{r+1}+1)^2-1}{(a_{r+1}+1)^2+1}$$

(当 $r=n-1$ 时,此公式亦正确,可直接验证).

引理 7 设 $b_1=1,\ b_r=\frac{(1+b_{r-1})^2-1}{(1+b_{r-1})^2+1}\quad(r=2,\ 3,\ \cdots)$,

则

$$0<b_r<\frac{2}{r}.$$

证明 显然 $0<b_1<\frac{2}{1}=2$. 设 $0<b_{r-1}<\frac{2}{r-1}$, 则

$$0<b_r<\frac{\left(1+\frac{2}{r-1}\right)^2-1}{\left(1+\frac{2}{r-1}\right)^2+1},$$

而不难证明

$$\frac{\left(1+\frac{2}{r-1}\right)^2-1}{\left(1+\frac{2}{r-1}\right)^2+1}<\frac{2}{r}.$$

引理 8 当 $r<\frac{n}{3}$ 时, $\Delta_{r+1}^{(0)}=\Delta_r^{(0)}=\cdots=\Delta_2^{(0)}=\Delta_1=1$.

证明 若有 $\Delta_n^{(0)}>\cdots>\Delta_{r+1}^{(0)}>\Delta_r^{(0)}$, 则

$$a_{r+1}(\Delta_1+\cdots+\Delta_r^{(0)})>\Delta_r^{(0)},$$

$$a_{r+1}(\Delta_1+\cdots+a_{r-1}^{(0)})>(1-a_{r+1})\Delta_r^{(0)}>(1-a_{r+1})\frac{\Delta_1+\cdots+\Delta_{r-1}^{(0)}}{r-1},$$

$$a_{r+1}>\frac{1-a_{r+1}}{r-1}.$$

从而

$$a_{r+1}>\frac{1}{r}, \tag{16}$$

但由引理 7(取 $a_r=b_{n-r+1}$),

$$a_{r+1} < \frac{2}{n-r} < \frac{2}{n-\frac{n}{3}} = \frac{3}{n} < \frac{1}{r}, \tag{17}$$

(16)与(17)矛盾！所以必有某个 $k \geqslant \frac{n}{3}$，使得

$$\Delta_{k+1}^{(0)} = \Delta_k^{(0)},$$

由引理 6 即得结论.

定理Ⅰ的证明,对于任意给定的 $\varepsilon > 0$,取定 $n > N(\varepsilon)$，使得

$$\int_{\frac{n}{3}}^{\infty} \frac{\mathrm{d}x}{x^2} < \varepsilon.$$

对这固定 n 的,由引理 8 可知

$$\Delta_r^{(0)} = \Delta_{r-1}^{(0)} = \cdots = \Delta_2^{(0)} = \Delta_1 = 1,\ r = \left[\frac{n}{3}\right],$$

从而

$$M_n \leqslant \sum_{r=1}^{\left[\frac{n}{3}\right]} \frac{\Delta_r^{(0)}}{(\Delta_1 + \cdots + \Delta_r^{(0)})^2} + \int_{\frac{n}{3}}^{\infty} \frac{\mathrm{d}x}{x^2} < \frac{\pi^2}{6} + \varepsilon. \tag{18}$$

由于 M_n 随 n 递增,所以(18)不仅对于 $n > N(\varepsilon)$ 的 n 成立,而且对所有的 n 成立.

由 ε 的任意性,即有 $M_n \leqslant \frac{\pi^2}{6}$.

再由 $M_n < M_{n+1} \leqslant \frac{\pi^2}{6}$,可知一切 $M_n < \frac{\pi^2}{6}$. 定理Ⅰ证毕.

显然定理Ⅰ的$\frac{\pi^2}{6}$是最佳常数.

用完全同样的方法可以证明:

定理Ⅰ′ 设 $\Delta_1 = 1,\ \Delta_2,\ \cdots,\ \Delta_n,\ \cdots$为函数,

$$\theta_r = \min(\Delta_r,\ \Delta_{r+1}),\text{则}$$

$$\sum_{r=1}^{\infty} \frac{\theta_r}{(\Delta_1 + \cdots + \Delta_r)^4} \leqslant \frac{\pi^4}{90}, \tag{19}$$

并且常数$\frac{\pi^4}{90}$是最佳的.

引理 1 的证明,不妨设$\lambda_1<\lambda_2<\cdots<\lambda_s<\cdots<\lambda_n$. 对于固定的$s$,

$$\sum_{r>s}\frac{\delta_r}{(\lambda_r-\lambda_s)^2}=\frac{1}{\lambda_{s+1}-\lambda_s}\sum_{r>s}\frac{\frac{\delta_r}{\lambda_{s+1}-\lambda_s}}{\left(\frac{\lambda_r-\lambda_s}{\lambda_{s+1}-\lambda_s}\right)^2}$$

$$\leqslant\frac{1}{\delta_s}\sum_r\frac{\theta_r}{(\Delta_1+\cdots+\Delta_r)^2},$$

其中$\Delta_1=1$, $\Delta_r=\frac{\lambda_{s+r}-\lambda_{s+r-1}}{\lambda_{s+1}-\lambda_s}$, $\theta_r=\min(\Delta_r, \Delta_{r+1})$.

由定理Ⅰ,
$$\sum_{r>s}\frac{\delta_r}{(\lambda_r-\lambda_s)^2}<\frac{\pi^2}{6}\delta_s^{-1}.$$

同样
$$\sum_{r<s}\frac{\delta_r}{(\lambda_r-\lambda_s)^2}<\frac{\pi^2}{6}\delta_s^{-1}.$$

因此
$$\sum_{r}\frac{\delta_r}{(\lambda_r-\lambda_s)^2}<\frac{\pi^2}{3}\delta_s^{-1}.$$

显然,常数$\frac{\pi^2}{3}$是最佳的.

同样地,由定理Ⅰ′可得引理 2.

参考文献

[1] H. L. Montgomery and R. C. Vapghan, Hilbert's Inequality, *J. London Math. Soc*. (2)8(1974), 73-81.

[2] H. L. Montgomery. The Analytic Principle of the Large Sieve, *Bull. Amer. Soc*., Vol. 84, 4(1978), 547-567.

[3] S. Srinivasan, A Footnote to the Large Sieve, *J. Number Theory* 4(1977), 493-498.

初等数论中的一个猜测（Ⅱ）

柯召、孙琦[1]猜测任意的 $2n-1$ 个数(本文中的数均指整数)中必有 n 个数的和为 n 整除. 单墫[2]证明了这一猜测是正确的.

这里的 $2n-1$ 不能改成更小的数,因为在任意的 $2n-2$ 个数中未必能找到 n 个数,其和为 n 整除. 例如在 $n-1$ 个 0 与 $n-1$ 个 1(或与 n 互素的数 a)组成的 $2n-2$ 个数中,任意 n 个的和不被 n 整除.

有趣的是,如果 $2n-2$ 个数中没有 n 个数的和被 n 整除,那么这 $2n-2$ 个数基本上就是上面所举的反例. 这结论较柯召、孙琦的猜测强,由它可得到柯召、孙琦猜测的又一证明. 确切地说,本文特证明下面的定理及其推论.

定理　如果在 $2n-2$ 个数中($n>1$),任 n 个数的和不被 n 整除,那么这 $2n-2$ 个数可分成两组,每组 $n-1$ 个数,分别同余于模 n 的两个数 a、b,并且 $(a-b, n)=1$.

为了证明这个定理,需要几条引理.

引理 1　(Kemperman 与 Scherk)设有 r 个不同的剩余类 $a_0=0, a_1, a_2, \cdots, a_{r-1} \pmod{n}$ 及 s 个不同的剩余类 $b_0=0, b_1, b_2, \cdots, b_{s-1} \pmod{n}$,并且在 i、j 不同为 0 时 $a_i+b_j \not\equiv 0 \pmod{n}$,则

$$a_i+b_j \quad i=0, 1, \cdots, r-1, j=0, 1, \cdots, s-1.$$

中至少有 $\min(n, r+s-1)$ 个不同的剩余类$\pmod{n}$.

证明　这就是 H. Halberstam 与 K. F. Roth[3]第一章的 Th. 15′.

引理 2　如果 $n-1$ 个数中任意个数的和(包括仅一个数的情况,下同)不被 n 整除,则这些数属于同一剩余类 $a \pmod{n}$,并且 $(a, n)=1$.

证明　设 a_1, a_2 为其中二数,我们断定 $a_1 \equiv a_2 \pmod{n}$,事实上

$$0, a_1, a_1+a_2, a_1+a_2+a_3, \cdots, a_1+a_2+\cdots+a_{n-1},$$

这 n 个数互不同余$\pmod{n}$,否则它们的差同余于 $0 \pmod{n}$ 与已知矛盾. 因此它

们是 $\bmod n$ 的完系,从而 a_2 必与其中之一同余,显然只能是 $a_2 \equiv a_1 (\bmod n)$.

这就证明了 $n-1$ 个数属于同一剩余类 $a(\bmod n)$. 若 $(a, n)=d>1$, $\frac{n}{d} \leqslant n-1$,这 $n-1$ 个数中的 $\frac{n}{d}$ 个数的和为 n 整除矛盾,因此 $(a, n)=1$.

引理 3 任意的 n 个数中一定可选出若干个数,它们的和被 n 整除.

证明 即[2]中引理 2. 由本文引理 2 亦不难推出.

引理 4 将 $n-1$ 个数按照 $\bmod n$ 分类,设每一类至多有 k 个数. 如果这 $n-1$ 个数中有若干个数的和被 n 整除,那么必可从这 $n-1$ 个数中选出 $\leqslant k+1$ 个数,它们的和必被 n 整除.

证明 考虑这 $n-1$ 个数中的数所组成的被 n 整除的和,设其中最短的(即加数个数最少的)长为 k_2(即有 k_2 个加数),我们要证明 $k_2 \leqslant k+1$.

将这 $n-1$ 个数写成

$$\begin{array}{l} x_1 \quad x_2 \quad x_3 \cdots x_{l_1} \\ x_1 \quad x_2 \quad x_3 \cdots x_{l_2} \\ \qquad\cdots \\ x_1 \quad x_2 \quad x_3 \cdots x_{l_k} \end{array}$$

其中每一列表示一个剩余类,并且同一类中的元素(至多 k 个)用同一字母表示. $l_1 \geqslant l_2 \geqslant \cdots \geqslant l_k$. 显然

$$l_1+l_2+\cdots+l_k=n-1$$

若 $x_1, x_2, \cdots, x_{l_1}$ 中有数属于零类,结论显然,故可设 $x_1, x_2, \cdots, x_{l_1}$ 均不属于零类. 令 $x_0=0$,如果有 $x_{i_1}+x_{i_2}+\cdots+x_{i_k}$ 被 n 整除 $(0 \leqslant i_t \leqslant l_t, t=1, 2 \cdots k)$, 并且 $i_1, \cdots i_k$ 不全为 0,那么结论成立,因此可设 $x_{i_1}+x_{i_2}+\cdots+x_{i_k}$ 仅在 $i_1=i_2=\cdots=i_k=0$ 时被 n 整除.

根据引理 1. $x_{i_1}+x_{i_2}(i_1=0, 1, \cdots, l_1, i_2=0, 1, \cdots l_2)$ 中至少有 l_1+l_2+1 个剩余类$(\bmod n)$.

于是 $x_{i_1}+x_{i_2}+x_{i_3}(0 \leqslant i_t \leqslant l_t, t=1, 2, 3)$ 中至少有 $l_1+l_2+l_3+1$ 个剩余类$(\bmod n)$. 依此类推, $x_{i_1}+x_{i_2}+\cdots+x_{i_k}(0 \leqslant i_t \leqslant l_t, t=1, 2 \cdots, k)$ 中有

$$l_1+l_2+\cdots+l_k+1=n$$

个剩余类(mod$n$)，即它们组成完系(mod$n$).

于是$-x_j$必与其中之一同余，设

$$-x_j \equiv x_{i_1}+x_{i_2}+\cdots+x_{i_k}(\bmod n),$$

则上式右边必有l_j个x_j(否则

$$x_{i_1}+\cdots+x_{i_k}+x_j \equiv 0(\bmod n),$$

从而$k_2 \leqslant k+1$，结论成立). 我们把这样的和$x_{i_1}+x_{i_2}+\cdots+x_{i_k}$称为(对$x_j$是)满的.

现在考虑n个数：

$$\begin{gathered}0,\ x_1,\ x_2 \cdots x_{l_1} \\ 2x_1,\ 2x_2 \cdots 2x_{l_2} \\ \cdots \\ kx_1,\ kx_2 \cdots kx_{l_k}\end{gathered} \tag{1}$$

其中若有两类同余：

$$m' x_i \equiv m x_j(\bmod n),$$

而$m' \leqslant m$，那么在与$-x_j$同余的剩余类$x_{i_1}+x_{i_2}+\cdots+x_{i_k}$中可将$m$个$x_j$换为$m'$个$x_i$，化为对$x_j$不满的和，而添上$x_j$被$m$整除，所以

$$k_2 \leqslant k+1 \text{ 成立}.$$

假定(1)中两两不同余，那么(1)构成完系(mod$n$)，这时$x_{j_1}+x_{j_2}(j_1 \neq j_2)$必与(1)中某一类同余，若有

$$x_{j_1}+x_{j_2} \equiv m x_j(\bmod n),$$

而$m \geqslant 2$，则仍可将与$-x_j$同余的，(对x_j是)满的和$x_{i_1}+\cdots+x_{i_k}$换成不满的，导出结论. 假设对所有$j_1 \neq j_2$，均有

$$x_{j_1}+x_{j_2} \equiv x_j(\bmod n),$$

那么考虑和被n整除的k_2个数，这时有三种情况：

(1) 这和对所有x_j都不是满的，则将其中两项之和$x_{j_1}+x_{j_2}$改为x_j，得到更短的和被n整除，与k_2定义矛盾.

(2) 这和仅对某个 x_{j_1} 是满的,若 $k_2 \geqslant k+1$,则存在加数 $x_{j_2} \neq x_{j_1}$,和仍可缩短,矛盾.

(3) 这和对 $x_{j_1}, x_{j_2}, \cdots, x_{j_t}$ 是满的,则应用上面的方法将 $x_{j_1}+x_{j_2}+\cdots+x_{j_t}$ 尽量缩短.如果其中出现 $2x_{j_t}$,则将 x_{j_t} 与另一项相加,即

$$2x_{j_t}+x_{j_s} \equiv x_{j_t}+x'_{j_s} \equiv x'_{j_t} \pmod{n},$$

最后化为一项,或者三项 $x_j+x'_j+x''_j$, x_j, x'_j, x''_j互不同余,并且

$$x_j+x'_j \equiv x''_j \pmod{n},$$

但这时可先加出 $x_j+x''_j$,若又有

$$x_j+x''_j \equiv x'_j \pmod{n},$$

则

$$x_j+x'_j+x''_j \equiv 2x''_j \equiv 2x'_j \pmod{n},$$

矛盾,所以

$$x_j+x''_j \equiv x'''_j \not\equiv x'_j \pmod{n},$$
$$x_j+x''_j+x'_j \equiv x'''_j+x'_j \equiv x''''_i \pmod{n},$$

仍化为一项.总之,和的长 k_2 可以缩短,与 k_2 定义矛盾.引理 4 证毕.

定理的证明:将 $2n-2$ 个数按 n 的剩余类分类,设其中最多的一类含 k_0 个数,不妨设这类为零类(如果这一类为 a,将每一个数减去 a,即化为这种情形)显然 $k_0 \leqslant n-1$.

若 $k_0=n-1$,则剩下的 $n-1$ 个数中,任意个数的和不被 n 整除(否则添上若干零,得 n 个数的和被 n 整除),由引理 2,结论成立.

若 $k_0<n-1$,则剩下的数的个数为

$$2n-2-k_0 \geqslant 2n-2-(n-2)=n,$$

所以其中必有个数$\leqslant n$ 个数,其和为 n 整除(引理 3),设长$\leqslant n$ 的、被 n 整除的和中最长的长为 k_1,则

$$k_1+k_0<n$$

(否则有 n 个数的和为 n 整除)除去这 k_1+k_0 个数后剩下

$$2n-2-(k_1+k_0)\geqslant n-1$$

个数，取其中 $n-1$ 个数，若这 $n-1$ 个数中不存在被 n 整除的和，则由引理 2，它们在同一类，与 $k_0<n-1$ 矛盾. 所以其中存在着被 n 整除的和，由引理 4，这种和的最短长度 $k_2\leqslant k_0+1$，于是

$$k_1+k_2\leqslant k_1+k_0+1\leqslant n.$$

但这时有 k_1+k_2 个数的和被 n 整除，与 k_1 的定义矛盾，证毕.

推论　任意的 $2n-1$ 个数中必有 n 个数的和为 n 整除.

证明　若在其中的 $2n-2$ 个数中，每 n 个数的和不为 n 整除，则由定理，这些数属于两个剩余类 a、b，每个剩余类有 $n-1$ 个数. 将这 $2n-2$ 个数中任一数换为剩下的一个数，如果这 $2n-2$ 个数中有 n 个数的和为 n 整除，则结论成立，否则它们仍属于剩余类 a、b. 于是 a 类和 b 类中有一类含 n 个数，它们的和显然为 n 整除.

参考文献

[1]　柯召、孙琦，初等数论 100 例，上海教育出版社，1979.

[2]　单墫，初等数论中的一个猜测，《数学进展》12 卷，4(1983)，299 - 301.

[3]　H. Halberstam and K. F. Roth, *Sequences*. Vsl. 1. (Oxford, 1966), 50 - 51.

On the Diophantine Equation $\sum_{i=0}^{k} \frac{1}{x_i} = \frac{a}{n}$

Abstract

In this paper, the author proves the following result:

Let $E_{a,k}(N)$ denote the number of natural numbers $n \leqslant N$ for which equation

$$\sum_{0}^{k} \frac{1}{x_i} = \frac{a}{n}$$

is insoluble in positive integers $x_i (i = 0, 1, \cdots, k)$. Then

$$E_{a,k}(N) \ll N \exp\{-C(\log N)^{1-\frac{1}{k+1}}\}$$

where the implied constant depends on a and k.

1 Introduction

The main result in this paper is the following

Theorem 1. *Let $E_{a,k}(N)$ denote the number of natural numbers $n \leqslant N$ for which equation*

$$\sum_{0}^{k} \frac{1}{x_i} = \frac{a}{n} \tag{1}$$

is insoluble in positive integers $x_i (i = 0, 1, \cdots, k)$. Then

$$E_{a,k}(N) \ll N \exp\{-C(\log N)^{1-\frac{1}{k+1}}\}, \tag{2}$$

where the implied constant depends on a and k.

(2) is better than the result of Viola[1], where the index $1 - \dfrac{1}{k+1}$ is

replaced by $1-\frac{1}{k}$. When $k=2$, (2) is just the same early result of Vaughan[2]. So we may assume k is an integer >2.

2 Lemma

The ordered $k+1$-tuple $(y_1, y_2, \cdots, y_{k+1})$ is called an admissible $k+1$-factorization of v whenever $\prod_1^{k+1} y_i = v$, y_i positive integer, and y_{k+1} square free.

Lemma 1. *Let integers* $v_1 \geqslant v \geqslant (k-1)^8$. *Let* $(y_1, y_2, \cdots, y_{k+1})$ *and* $(y'_1, y'_2, \cdots, y'_{k+1})$ *be admissible* $k+1$*-factorization of* v. *If* $y_i \geqslant v^{\frac{3}{4^i}}$, $y'_i \geqslant v^{\frac{3}{4^i}}$ $(i=1, 2, \cdots, k-1)$ $v^{\frac{1}{k\cdot 4^{k+2}}} \leqslant y_k$, $y'_k \leqslant v^{\frac{1}{k\cdot 4^{k+1}}}$ *and there exists a natural number* n *satisfying*

$$\left.\begin{aligned} y_k n &\equiv -(y_1+y_2+\cdots+y_{k-1}) \\ y'_k n &\equiv -(y'_1+y'_2+\cdots+y'_{k-1}) \end{aligned}\right\} \pmod{v_1}, \tag{3}$$

then $$y_i = y'_i (i=1, 2, \cdots, k+1).$$

Proof From (3) we have

$$y_k(y'_1+y'_2+\cdots+y'_{k-1}) \equiv y'_k(y_1+y_2+\cdots+y_{k-1}) \pmod{v_1} \tag{4}$$

Since

$$\begin{aligned} y_k(y'_1+y'_2+\cdots+y'_{k-1}) &\leqslant y_k \cdot y'_1(k-1) \leqslant (k-1)\cdot \frac{v}{y_1 y_2}\cdot \frac{v}{y'_2} \\ &\leqslant (k-1)v^{\frac{1}{16}}v^{1-\frac{3}{16}} = (k-1)v^{\frac{7}{8}} \leqslant v \leqslant v_1, \\ y'_k(y_1+y_2+\cdots+y_{k-1}) &\leqslant v_1, \end{aligned}$$

(4) becomes

$$y_k(y'_1+y'_2+\cdots+y'_{k-1}) = y'_k(y_1+y_2+\cdots+y_{k-1}). \tag{5}$$

If $y_k y'_1 \neq y'_k y_1$, then

$$|y_k y_1' - y_k' y_1| = \left| y_k \cdot \frac{v}{y_2' y_3' \cdots y_{k+1}'} - y_k' \cdot \frac{v}{y_2 y_3 \cdots y_{k+1}} \right|$$

$$\geqslant \frac{v}{y_2' y_3' \cdots y_{k+1}' y_2 y_3 \cdots y_{k+1}} = \frac{y_1 y_1'}{v} \geqslant v^{\frac{1}{2}}.$$

But on the otherhand, from (5)

$$\begin{aligned}|y_k y_1' - y_k' y_1| &= |y_k'(y_2 + y_3 + \cdots + y_{k-1}) - y_k(y_2' + y_3' + \cdots + y_{k-1}')| \\ &< \max(y_k'(y_2 + y_3 + \cdots + y_{k-1}), y_k(y_2' + y_3' + \cdots + y_{k-1}')) \\ &\leqslant \max(y_2 y_3 \cdots y_{k-1} y_k', y_2' y_3' \cdots y_{k-1}' y_k) \\ &\leqslant \max\left(\frac{v y_k'}{y_1}, \frac{v y_k}{y_1'}\right) \leqslant v^{1-\frac{3}{4}+\frac{1}{4}} = v^{\frac{1}{2}}.\end{aligned}$$

It is a contradiction.

Now we assume that $y_k y_1' = y_k' y_1$, $y_k y_2' = y_k' y_2$, $\cdots$, $y_k y_{s-1}' = y_k' y_{s-1}$ $(s < k-1)$. Let $y_k^{1-s} \cdot u = \dfrac{v}{y_1 y_2 \cdots y_{s-1}} = y_s y_{s+1} \cdots y_{k+1}$. Since

$$y_s (y_1 y_2 \cdots y_{s-1})^{\frac{3}{4}} \geqslant v^{\frac{3}{4}\sum_{1}^{s-1}\frac{3}{4^i}+\frac{3}{4^s}} = v^{\frac{3}{4}},$$

we have

$$y_s \geqslant \left(\frac{v}{y_1 y_2 \cdots y_{s-1}}\right)^{\frac{3}{4}} = (u y_k^{1-s})^{3/4}.$$

If $y_k y_s' \neq y_k' y_s$, then

$$|y_k y_s' - y_k' y_s| = \left| \frac{y_k u y_k'^{1-s}}{y_{s+1}' y_{s+2}' \cdots y_{k+1}'} - \frac{y_k' u y_k^{1-s}}{y_{s+1} y_{s+2} \cdots y_{k+1}} \right|$$

$$\geqslant \frac{u y_k'^{1-s} y_k^{1-s}}{y_{s+1}' y_{s+2}' \cdots y_{k+1}' y_{s+1} y_{s+2} \cdots y_{k+1}} = \frac{y_s y_s'}{u}$$

But on the otherhand, from (5)

$$\begin{aligned}|y_k y_s' - y_k' y_s| &= |y_k'(y_{s+1} + y_{s+2} + \cdots + y_{k-1}) - y_k(y_{s+1}' + y_{s+2}' + \cdots + y_{k-1}')| \\ &< \max(y_k'(y_{s+1} + y_{s+2} + \cdots + y_{k-1}), y_k(y_{s+1}' + y_{s+2}' + \cdots + y_{k-1}')) \\ &\leqslant \max(y_{s+1} y_{s+2} \cdots y_{k-1} y_k', y_{s+1}' y_{s+2}' \cdots y_{k-1}' y_k) \\ &\leqslant \max\left(\frac{u y_k'}{y_s} y_k^{1-s}, \frac{u y_k}{y_s'} y_k'^{1-s}\right) \leqslant \frac{y_s y_s'}{u}.\end{aligned}$$

It is a contradiction again. Hence we have

$$y_k y_1' = y_k' y_1,\ y_k y_2' = y_k' y_2,\ \cdots,\ y_k y_{k-1}' = y_k' y_{k-1}.$$

i. e.

$$\frac{y_1}{y_1'} = \frac{y_2}{y_2'} = \cdots = \frac{y_k}{y_k'}. \tag{6}$$

From (6) and

$$y_1 y_2 \cdots y_k y_{k+1} = y_1' y_2' \cdots y_k' y_{k+1}' (= v)$$

we have

$$y_k^k y_{k+1} = y'^k_k y_{k+1}'.$$

Since y_{k+1} and y_{k+1}' are square free, it follows that

$$y_k = y_k',\ y_{k+1} = y_{k+1}'. \tag{7}$$

Thus by (6) and (7)

$$y_1 = y_1',\ y_2 = y_2',\ \cdots,\ y_{k+1} = y_{k+1}'.$$

The proof of Lemma 1 is complete.

Lemma 2 (Brun-Titchmarsh). *If $q \leqslant x^\alpha$, $0 < \alpha < 1$, $(q, l) = 1$, then*

$$\pi(x; q, l) \ll \frac{x}{\varphi(q)\log x}.$$

Lemma 3 (Bombieri). *For any $A > 0$, there is $B > 0$ such that*

$$\sum_{q \leqslant x^{\frac{1}{2}}(\log x)^{-B}} \max_{y \leqslant x} \max_{(q, l)=1} \left| \pi(y; q, l) - \frac{l_i y}{\varphi(q)} \right| \ll x(\log x)^{-A}.$$

Lemma 4[1]. *Let $d_{k(n)} = \sum_{x_1 x_2 \cdots x_k = n} 1$. Then*

$$\sum_{n \leqslant x} \frac{\mathrm{d}_{k(n)}^l}{\varphi(n)} \ll (\log x)^{k^l}.$$

Lemma 5 (Montgomery). *If $\omega(p)$ $(0 < \omega(p) < p)$ residue classes* $(\bmod p)$ *are removed from the first N natural numbers for each prime $p \leqslant$*

$\sqrt{N}$, *then the number Z of natural numbers which remain satisfies*

$$Z \leqslant \frac{4N}{\sum_{m \leqslant \sqrt{N}} \mu^2_{(m)} \prod_{p \mid m} \frac{\omega(p)}{p-\omega(p)}}.$$

3 Theorem 2 and its Proof

Let

$$f_k(v, \xi) = \sum_{\substack{x_1 x_2 \cdots x_{k+1} = v \\ x_i > \xi^{1/4^i} \\ (i=1, 2, \cdots, k)}} |\mu_{(x_{k+1})}|$$

$$\omega_{a, k(p)} = \begin{cases} f_k\left(\frac{p+1}{a}, \frac{p+1}{a}\right), & \text{if } p \equiv -1(\bmod a),\ p > a(k-1)^8. \\ 0, & \text{otherwise.} \end{cases}$$

In this section, we give the average order of the faction $\omega_{a, k(p)}$ i. e. the following

Theorem 2.

$$(\log \xi)^k \ll \sum_{p \leqslant \xi} \frac{\omega_{a, k(p)}}{p} \ll (\log \xi)^k.$$

Proof By partial summations, it suffices to prove that

$$\xi(\log \xi)^{k-1} \ll \sum_{p \leqslant \xi} \omega_{a, k(p)} \ll \xi(\log \xi)^{k-1}. \tag{8}$$

The upper bound. If $p \equiv -1(\bmod a)$, $p \geqslant a(k-1)^8$, then

$$\omega_{a, k(p)} = \sum_{\substack{x_1 x_2 \cdots x_{k+1} = \frac{p+1}{a} \\ x_i \geqslant \left(\frac{p+1}{a}\right)^{1/4^i} \\ (i=1, 2, \cdots, k)}} |\mu_{(x_{k+1})}| \leqslant \sum_{\substack{x_1 x_2 \cdots x_{k+1} = \frac{p+1}{a} \\ x_1 \geqslant \left(\frac{p+1}{a}\right)^{1/4}}} 1$$

$$= \sum_{\substack{r \mid \frac{p+1}{a} \\ r \leqslant \left(\frac{p+1}{a}\right)^{3/4}}} \sum_{x_1 x_2 \cdots x_k = r} 1 = \sum_{\substack{r \mid \frac{p+1}{a} \\ r \leqslant \left(\frac{p+1}{a}\right)^{3/4}}} \mathrm{d}_{k(r)}.$$

Hence

$$\sum_{p\leqslant\xi}\omega_{a,\,k(p)}\leqslant\sum_{\substack{p\leqslant\xi\\ p\equiv-1((\bmod a)}}\ \sum_{\substack{r\,\mid\,\frac{p+1}{a}\\ r\leqslant\left(\frac{p+1}{a}\right)^{3/4}}}\mathrm{d}_{k(r)}\leqslant\sum_{r\leqslant(\xi+1)^{3/4}}\mathrm{d}_{k(r)}\,\pi(\xi;\ a,\ -1).$$

Therefore, by Lemmas 2 and 4

$$\sum_{p\leqslant\xi}\omega_{a,\,k(p)}\ll\frac{\xi}{\log\xi}\sum_{\tau\leqslant(\xi+1)^{1/4}}\frac{\mathrm{d}_{k(r)}}{\varphi(r)}\ll\xi(\log)^{k-1}.$$

The lower bound.

$$\sum_{p\leqslant\xi}\omega_{a,\,k(p)}\geqslant\sum_{\substack{\frac{\xi}{2}<p\leqslant\xi\\ p\equiv-1(\bmod a)}}f_k\left(\frac{p+1}{a},\ \xi\right)=\sum_{\substack{\frac{\xi}{2}<p\leqslant\xi\\ p\equiv-1(\bmod a)}}\ \sum_{\substack{x_1x_2\cdots x_{k+1}=\frac{p+1}{a}\\ x_i\geqslant 3/4^i\\ (i=1,\,2,\,\cdots,\,k)}}|\mu_{(x_{k+1})}|$$

$$=\sum_{\substack{\frac{\xi}{2}<p\leqslant\xi\\ p\equiv-1(\bmod a)}}\ \sum_{\substack{x_1\,\mid\,\frac{p+1}{a}\\ x_1\geqslant\xi^{3/4}}}\ \sum_{\substack{x_1x_2\cdots x_{k+1}=\frac{p+1}{a}\\ x_i\geqslant\xi^{3/4^i}\\ (i=2,\,3,\,\cdots,\,k)}}|\mu_{(x_{k+1})}|$$

$$\geqslant\sum_{\substack{\frac{\xi}{2}<p\leqslant\xi\\ p\equiv-1(\bmod a)}}\ \sum_{\substack{r\,\mid\,\frac{p+1}{a}\\ r\leqslant\frac{\xi^{1/4}}{2a}}}\ \sum_{\substack{x_2x_3\cdots x_{k+1}=r\\ x_i\geqslant\xi^{3/4^i}\\ (i=2,\,3,\,\cdots,\,k)}}|\mu_{(x_{k+1})}|$$

$$=\sum_{r\leqslant\frac{\xi^{1/4}}{2a}}\ \sum_{\substack{x_1x_2\cdots x_k=r\\ x_i\geqslant\xi^{3/4^{i+1}}\\ (i=1,\,2,\,\cdots,\,k-1)}}|\mu_{(x_k)}|\sum_{\substack{\frac{\xi}{2}<p\leqslant\xi\\ p\equiv-1(\bmod ar)}}1$$

$$=\sum_{r\leqslant\frac{\xi^{1/4}}{2a}}f_{k-1}(r,\ \xi^{1/4})\left\{\pi(\xi;ar,\ -1)-\pi\left(\frac{\xi}{2};\ ar,\ -1\right)\right\},$$

hence

$$\sum_{p\leqslant\xi}\omega_{a,\,k(p)}\geqslant\left(l_i\xi-l_i\ \frac{\xi}{2}\right)\sum_{r\leqslant\frac{\xi^{1/4}}{2a}}\frac{f_{k-1}(r,\ \xi^{1/4})}{\varphi(ar)}+R\tag{9}$$

where

$$R=\sum_{r\leqslant\frac{\xi^{1/4}}{2a}}f_{k-1}(r,\ \xi^{1/4})\left\{\left(\pi(\xi;\ ar,\ -1)-\frac{l_i\xi}{\varphi_{(ar)}}\right)-\left(\pi\left(\frac{\xi}{2};\ ar,\ -1\right)-\frac{l_i\ \frac{\xi}{2}}{\varphi_{(ar)}}\right)\right\}.$$

The Cauchy-Schwarz inequality yields

$$|R| \leqslant \Big(\sum_{r \leqslant \frac{\xi^{1/4}}{2a}} f_{k-1}^2(r,\ \xi^{1/4}) / \varphi_{(ar)} \Big)^{1/2} \Bigg(\sum_{r \leqslant \frac{\xi^{1/4}}{2a}} \varphi_{(ar)} \Bigg\{ \Big(\pi(\xi;\ ar,\ -1) - \frac{l_i \xi}{\varphi_{(ar)}} \Big)$$

$$- \left(\pi\Big(\frac{\xi}{2};\ ar,\ -1\Big) - \frac{l_i \frac{\xi}{2}}{\varphi_{(ar)}} \right) \Bigg\}^2 \Bigg)^{1/2}$$

$$\leqslant \Big(\sum_{r \leqslant \xi^{1/4}} \frac{d_k^2(r)}{\varphi(r)} \Big)^{1/2} \Bigg(\sum_{r \leqslant \frac{\xi^{1/4}}{2a}} \varphi_{(ar)} \Bigg\{ \Big(\pi(\xi;\ ar,\ -1) - \frac{l_i \xi}{\varphi_{(ar)}} \Big)$$

$$- \left(\pi\Big(\frac{\xi}{2};\ ar,\ -1\Big) - \frac{l_i \frac{\xi}{2}}{\varphi_{(ar)}} \right) \Bigg\}^2 \Bigg)^{1/2}. \tag{10}$$

By Lemma 4

$$\Big(\sum_{r \leqslant \xi} \frac{d_k^2(r)}{\varphi(r)} \Big)^{1/2} \ll (\log \xi)^{k^i/2} \tag{11}$$

and by Lemma 2 with $\alpha = \frac{1}{4}$

$$\varphi_{(s)} \cdot \left| \pi(\xi;\ s,\ -1) - \frac{l_i \xi}{\varphi_{(s)}} - \left(\pi\Big(\frac{\xi}{2};\ s,\ -1\Big) - \frac{l_i \frac{\xi}{2}}{\varphi_{(s)}} \right) \right| \ll \frac{\xi}{\log \xi}. \tag{12}$$

Lemma 3 gives, for any $A > 0$,

$$\sum_{s \leqslant \xi^{1/4}} \left| \pi(\xi;\ s,\ -1) - \frac{l_i \xi}{\varphi_{(s)}} - \left(\pi\Big(\frac{\xi}{2};\ s,\ -1\Big) - \frac{l_i \frac{\xi}{2}}{\varphi_{(s)}} \right) \right| \ll \xi (\log \xi)^{-A}. \tag{13}$$

From (10), (11), (12) and (13) we obtain

$$|R| \ll \xi (\log \xi)^{-A_1}. \tag{14}$$

For the main term in (9) we have

$$\sum_{r \leqslant \frac{\xi^{1/4}}{2a}} \frac{f_{k-1}(r,\ \xi^{1/4})}{\varphi_{(ar)}} = \sum_{r \leqslant \frac{\xi^{1/4}}{2a}} \sum_{\substack{x_1 x_2 \cdots x_k = r \\ x_i \geqslant \xi^{3/4^{i+1}} \\ (i=1,\ 2,\ \cdots,\ k-1)}} \frac{|\mu_{(x_k)}|}{\varphi_{(ar)}}$$

$$= \sum_{\substack{x_1 x_2 \cdots x_k \leqslant \frac{\xi^{1/4}}{2a} \\ x_i \geqslant \xi^{3/4^{i+1}} \\ (i=1,\ 2,\ \cdots,\ k-1)}} \frac{|\mu_{(x_k)}|}{\varphi_{(a x_1 x_2 \cdots x_k)}} \geqslant \sum_{\substack{x_1 x_2 \cdots x_k \leqslant \frac{\xi^{1/4}}{2a} \\ x_i \geqslant \xi^{3/4^{i+1}} \\ (i=1,\ 2,\ \cdots,\ k-1)}} \frac{|\mu_{(x_k)}|}{\varphi_{(x_k)} a x_1 x_2 \cdots x_{k-1}}$$

$$=\frac{1}{a}\sum_{\substack{x_1x_2\cdots x_{k-1}\leqslant\frac{\xi^{1/4}}{2a}\\ x_i\geqslant\xi^{3/4^{i+1}}}}\frac{1}{x_1x_2\cdots x_{k-1}}\sum_{x_k\leqslant\frac{\xi^{1/4}}{2ax_1x_2\cdots x_{k-1}}}\frac{|\mu_{(x_k)}|}{\varphi_{(x_k)}}.$$

Since

$$\frac{|\mu_{(x_k)}|}{\varphi_{(x_k)}}=|\mu_{(x_k)}|\cdot\prod_{p|x_k}\frac{1}{p-1}=\frac{|\mu_{(x_k)}|}{x_k}\prod_{p|x_k}\sum_{h=0}^{\infty}\frac{1}{p^h},$$

we have

$$\sum_{r\leqslant\frac{\xi^{1/4}}{2a}}\frac{f_{k-1}(r,\ \xi^{1/4})}{\varphi_{(ar)}}\geqslant\frac{1}{a}\sum_{\substack{x_1x_2\cdots x_{k-1}\leqslant\frac{\xi^{1/4}}{2a}\\ x_i\geqslant\xi^{3/4i+1}}}\frac{1}{x_1x_2\cdots x_{k-1}}\sum_{x_k\leqslant\frac{\xi^{1/4}}{2ax_1x_2\cdots x_{k-1}}}\frac{|\mu_{(x_k)}|}{x_k}$$

$$\gg\sum_{x_1x_2\cdots x_{k-1}\leqslant\frac{\xi^{1/4}}{2a}}\frac{1}{x_1x_2\cdots x_{k-1}}\log\frac{\xi^{1/4}}{2ax_1x_2\cdots x_{k-1}}$$

$$\gg\log\xi\sum_{\substack{x_1x_2\cdots x_{k-1}\leqslant\frac{\xi^{1/4-\epsilon}}{2a}\\ x_i\geqslant\xi^{3/4i+1}}}\frac{1}{x_1x_2\cdots x_{k-1}}$$

$$\left(\text{where }\epsilon\text{ is a positive number}<\frac{1}{4^k}\right)$$

$$\gg\log\xi\sum_{\xi^{3/4i+1}\leqslant x_i\leqslant\xi^{3/4i+1+\epsilon_i}}\frac{1}{x_1x_2\cdots x_{k-1}}$$

$$\left(\text{where }\epsilon_i\text{ are positive satisfing }\sum_{i=1}^{k-1}\epsilon_i<\frac{1}{4^k}-\epsilon\right)$$

$$\gg(\log\xi)^k. \tag{15}$$

Since $l_i\xi-l_i\,\frac{\xi}{2}\gg\frac{\xi}{\log\xi}$, we obtain

$$\left(l_i\xi-l_i\,\frac{\xi}{2}\right)\sum_{r\leqslant\frac{\xi^{1/4}}{2a}}\frac{f_{k-1}(r,\ \xi^{1/4})}{\varphi_{(ar)}}\gg\xi(\log\xi)^{k-1}. \tag{16}$$

(9), (14) and (16) give

$$\sum_{p\leqslant\xi}\omega_{a,\ k(r)}\gg\xi(\log\xi)^{k-1},$$

which proves the theorem.

4 The Proof of Theorem 1

Let $p\equiv -1 \pmod a$ and $(y_1, y_2, \cdots, y_{k+1})$ be admissible $k+1$-factorization of $\frac{p+1}{a}$. If a natural number n satisfies

$$y_k n\equiv -(y_1+y_2+\cdots+y_{k-1}) \pmod p,$$

then

$$y_k n+(y_1+y_2+\cdots+y_{k-1})=y_0 p=y_0(ay_1y_2\cdots y_{k+1}-1),$$
$$y_0+y_1+y_2+\cdots+y_{k-1}+y_k n=ay_0y_1\cdots y_{k+1},$$
$$\frac{a}{n}=\frac{y_0+y_1+\cdots+y_{k-1}+y_k n}{ny_0y_1\cdots y_{k+1}}=\frac{1}{x_0}+\frac{1}{x_1}+\cdots+\frac{1}{x_k}$$

with

$$x_0=ny_1y_2\cdots y_{k+1},\ x_1=ny_0y_2\cdots y_{k+1},\ \cdots,$$
$$x_{k-1}=ny_0y_1\cdots y_{k-2}y_ky_{k+1},\ x_k=y_0y_1\cdots y_{k-1}y_{k+1},$$

i.e. the equation

$$\frac{a}{n}=\frac{1}{x_0}+\frac{1}{x_1}+\cdots+\frac{1}{x_k} \tag{1}$$

is soluble. Therefore by Lemma 1, there are at least $\omega_{a,k(p)}$ residue classes $(\bmod p)$ such that, for any n belonging to one of them, equation (1) is soluble. It follows from Lemma 5 that

$$E_{a,k}(N)\leqslant \frac{4N}{\sum\limits_{m\leqslant\sqrt{N}}\mu^2_{(m)}\prod\limits_{p\mid m}\frac{\omega_{a,k(p)}}{q-\omega_{a,k(p)}}}. \tag{17}$$

Let $\omega_{a,k(m)}$ be, for any integer $m>0$, the completely multiplication function generated by $\omega_{a,k(p)}$. Then

$$\sum_{m\leqslant\sqrt{N}}\mu^2_{(m)}\prod_{p\mid m}\frac{\omega_{a,k(p)}}{p-\omega_{a,k(p)}}\geqslant\sum_{m\leqslant\sqrt{N}}\mu^2_{(m)}\prod_{p\mid m}\frac{\omega_{a,k(p)}}{p}=\sum_{m\leqslant\sqrt{N}}\frac{\mu^2_{(m)}\omega_{a,k(m)}}{m}.$$

Hence, in view of (17), Theorem 1 is proved if we show that

$$\sum_{m\leqslant\sqrt{N}}\frac{\mu^2_{(m)}\omega_{a,\,k(m)}}{m}\gg\exp\{C(\log N)^{1-\frac{1}{k+1}}\}. \tag{18}$$

Since $\omega_{a,\,k(m)}\ll m^s$ for any $s>0$, the Dirichlet series

$$F_{a,\,k(s)}=\sum_{m=1}^{\infty}\frac{\mu^2_{(m)}\omega_{a,\,k(m)}}{m^2}$$

converges for Re $s>1$. Also

$$\sum_p\left(\frac{\omega_{a,\,k(p)}}{p}\right)^2<\infty,$$

hence for any $s>0$,

$$F_{a,\,k(1+s)}=\prod_p\left(1+\frac{\omega_{a,\,k(p)}}{p^{1+s}}\right)=\exp\left\{\sum_p\frac{\omega_{a,\,k(p)}}{p^{1+s}}+0_{(1)}\right\}=\exp\sum_p\frac{\omega_{a,\,k(p)}}{p^{1+s}}. \tag{19}$$

It follows from Theorem 2 that

$$\sum_p\frac{\omega_{a,\,k(p)}}{p^{1+s}}\geqslant\sum_{p\leqslant x}\frac{\omega_{a,\,k(p)}}{p^{1+s}}\geqslant x^{-s}\sum_{p\leqslant x}\frac{\omega_{a,\,k(p)}}{p}\gg x^{-s}(\log x)^k;$$

taking $x=\exp\frac{1}{s}$ we obtain

$$\sum_p\frac{\omega_{a,\,k(p)}}{p^{1+s}}\gg s^{-k}. \tag{20}$$

Next

$$\begin{aligned}\sum_p\frac{\omega_{a,\,k(p)}}{p^{1+s}}&=\sum_{n=0}^{\infty}\sum_{2^n\leqslant p\leqslant 2^{n+1}}\frac{\omega_{a,\,k(p)}}{p^{1+s}}\leqslant\sum_{n=0}^{\infty}2^{-sn}\sum_{2^n\leqslant p<2^{n+1}}\frac{\omega_{a,\,k(p)}}{p}\\&=\sum_{n=0}^{\infty}(2^{-sn}-2^{-s(n+1)})\sum_{p<2^{n+1}}\frac{\omega_{a,\,k(p)}}{p}\\&=(1-2^{-s})\sum_{n=0}^{\infty}2^{-sn}\sum_{p<2^{n+1}}\frac{\omega_{a,\,k(p)}}{p};\end{aligned}$$

hence by Theorem 2

$$\sum_p\frac{\omega_{a,\,k(p)}}{p^{1+s}}\ll(1-2^{-s})\sum_{n=0}^{\infty}(n+1)^k2^{-sn}.$$

Sence $$\sum_{n=0}^{\infty}(n+1)^k 2^{-sn} \ll s^{-k-1}$$

we obtain $$\sum_{p}\frac{\omega_{a,\,k(p)}}{p^{1+s}} \ll s^{-k}. \tag{21}$$

Moreover

$$\sum_{m\leqslant\sqrt{N}}\frac{\mu^2_{(m)}\omega_{a,\,k(m)}}{m} \geqslant \sum_{m\leqslant\sqrt{N}}\frac{\mu^2_{(m)}\omega_{a,\,k(m)}}{m^{1+s}} = F_{a,\,k}(1+\varepsilon) - \sum_{m>\sqrt{N}}\frac{\mu^2_{(m)}\omega_{a,\,k(m)}}{m^{1+s}}$$
$$\geqslant F_{a,\,k}(1+\varepsilon) - \sum_{m=1}^{\infty}\left(\frac{m}{\sqrt{N}}\right)^{\varepsilon/2}\frac{\mu^2_{(m)}\omega_{a,\,k(m)}}{m^{1+s}}$$
$$= F_{a,\,k}(1+\varepsilon) - N^{-s/4}F_{a,\,k}\left(1+\frac{\varepsilon}{2}\right)$$

From (19), (20), (21) and (22) we deduce that for any sufficiently small ε,

$$\sum_{m\leqslant\sqrt{N}}\frac{\mu^2_{(m)}\omega_{a,\,k(m)}}{m} \geqslant C_1\exp(C_2\varepsilon^{-k}) - C_3N^{-\frac{s}{4}}\exp(C_4\varepsilon^{-k})$$
$$= C_1\exp(C_2\varepsilon^{-k}) - C_3\exp\left(-\frac{\varepsilon}{4}\log N + C_4\varepsilon^{-k}\right).$$

Putting $\varepsilon = (4C_4)^{\frac{1}{k+1}}(\log N)^{-\frac{1}{k+1}}$ we obtain.

$$-\frac{\varepsilon}{4}\log N - C_4\varepsilon^{-k} = 0.$$

Hence the above choice for ε gives, if N is large,

$$\sum_{m\leqslant\sqrt{N}}\frac{\mu^2_{(m)}\omega_{a,\,k(m)}}{m} \geqslant C_1\exp\{C_2(\log N)^{1-\frac{1}{k+1}}\} - C_3 \gg \exp\{C_2(\log N)^{1-\frac{1}{k+1}}\},$$

and (18) is proved.

References

[1] Viola. C., On the Diophantine Equations $\sum_{i=0}^{k}x_i - \sum_{0}^{k}x_i = n$ and $\sum_{i=0}^{k}\frac{1}{x_i} = \frac{a}{n}$, *Acta Arith.*, **22**(1973), 339 - 352.

[2] Vaughan. R. C., On a Problem of Erdos. Straus and Schinzel, *Mathematika*, **17** (1970), 193 - 198.

一个丢番图不等式

1　引言

设 $\lambda_1, \lambda_2, \cdots, \lambda_5$ 为非零实数,不全同号并且两两的比不全为有理数. Davenport 与 Heilbronn 曾经证明对任意正数 η,一定有不全为 0 的整数 $x_1, x_2, \cdots, x_5$ 使得

$$|\lambda_1 x_1^2 + \lambda_2 x_2^2 + \cdots + \lambda_5 x_5^2| < \eta$$

(见[3]或[5]).

本文的目的是证明更强的结果,即下面的:

定理　设 $\lambda_1, \lambda_2, \cdots, \lambda_5$ 满足上述条件,则对任意实数 η 及正数 ε,一定有无限多组不全为 0 的整数 $x_1, x_2, \cdots, x_5$,使得

$$|\eta + \lambda_1 x_1^2 + \lambda_2 x_2^2 + \cdots + \lambda_5 x_5^2| < (\max_{1\leqslant i\leqslant 5}|x_i|)^{-\frac{1}{3}+\varepsilon}. \tag{1}$$

2　对 I_4 的估计

我们需要一个引理:

引理(Davenport [4])　对任一自然数 n,存在实函数 $K_n(\alpha) \ll \min(1, |\alpha|^{-n-1})$,对所有实数 β,

$$0 \leqslant \int_{-\infty}^{+\infty} e(\alpha\beta) K_n(\alpha)\mathrm{d}\alpha \leqslant 1,$$

并且

$$\int_{-\infty}^{+\infty} e(\alpha\beta) K_n(\alpha)\mathrm{d}\alpha = \begin{cases} 0, & \forall\, |\beta| \geqslant 1. \\ 1, & \forall\, |\beta| \leqslant \dfrac{1}{3}. \end{cases}$$

如[5]所述，不妨设$\frac{\lambda_1}{\lambda_2}$为负无理数，$q$ 是它的一个渐近分数的分母，$N=q^{3/2}$. 记

$$f(\alpha)=\sum_{x=1}^{N}e(\alpha x^{\lambda}), \tag{2}$$

$$f_j(\alpha)=f(\lambda_j\alpha), \tag{3}$$

$$Y=N^{\frac{1}{3}-\varepsilon}, \tag{4}$$

根据上述引理，在 N 充分大时

$$|\eta+\lambda_1x_1^2+\lambda_2x_2^2+\cdots+\lambda_5x_5^2|<Y^{-1} \tag{1'}$$

的满足 $\max\limits_{1\leqslant l\leqslant 5}|x_j|<N$ 的解数

$$K(N)\gg Y^{-1}\int_{-\infty}^{+\infty}e(\eta\alpha)\left(\prod_{j=1}^{5}f_j(\alpha)\right)K_n\left(\frac{\alpha}{Y}\right)\mathrm{d}\alpha=Y^{-1}\cdot I. \tag{5}$$

将$(-\infty,\ +\infty)$分为

$$E_1=\{|\alpha|\leqslant N^{-2+\nu}\}, \tag{6}$$

$$E_2=\{N^{-2+\nu}\leqslant|\alpha|\leqslant 1\}, \tag{7}$$

$$E_3=\{1<|\alpha|\leqslant YN^{\delta}\}, \tag{8}$$

$$E_4=\{|\alpha|>YN^{\delta}\}, \tag{9}$$

其中 $0<\nu<\frac{1}{3}$, $0<\delta<\frac{1}{5}\varepsilon$. 记相应于这些区间的积分为 I_1、I_2、I_3、I_4，则

$$I=I_1+I_2+I_3+I_4. \tag{10}$$

首先估计 I_4，由引理

$$\begin{aligned}I_4&=\int_{E_1}e(\eta\alpha)\left(\prod_{j=1}^{5}f_j(\alpha)\right)K_n\left(\frac{\alpha}{Y}\right)\mathrm{d}\alpha\\&\ll\sum_{X\geqslant YN^{\delta}}\int_{X}^{X+1}\left(\prod_{j=1}^{5}f_j(\alpha)\right)\left(\frac{\alpha}{Y}\right)^{-n-1}\mathrm{d}\alpha\\&\ll Y^{n+1}\sum_{X\geqslant YN^{\delta}}\frac{1}{X^{n+1}}\sum_{j=1}^{5}\int_{0}^{1}|f_j(\alpha)|^5\mathrm{d}\alpha\end{aligned}$$

$$\ll Y^{n+1}\cdot(YN^{\delta})^{-n}\cdot N^{3},$$

取 $n=\left[\frac{1}{3\delta}\right]+1$, 则由上式及(4)得

$$I_4=o(N^3). \tag{11}$$

3　对 I_1 与 I_2 的估计

令

$$\nu_j(\alpha)=\int_0^N \mathrm{e}(\lambda_j\alpha x^2)\mathrm{d}x, \tag{12}$$

则

$$\begin{aligned} I_1&=\int_{E_1}\mathrm{e}(\eta\alpha)\Big(\prod_{j=1}^{5}f_j(\alpha)\Big)K_n\left(\frac{\alpha}{Y}\right)\mathrm{d}\alpha\\ &=\int_{E_1}\mathrm{e}(\eta\alpha)\Big(\prod_{j=1}^{5}f_j(\alpha)-\prod_{j=1}^{5}\nu_j(\alpha)\Big)K_n\left(\frac{\alpha}{Y}\right)\mathrm{d}\alpha\\ &\quad+\int_{E_1}\mathrm{e}(\eta\alpha)\Big(\prod_{j=1}^{5}\nu_j(\alpha)\Big)K_n\left(\frac{\alpha}{Y}\right)\mathrm{d}\alpha, \end{aligned} \tag{13}$$

熟知在 E_1 上(例如参见[5], p. 149 第 10 行),

$$f_j(\alpha)-\nu_j(\alpha)\ll N^{2\nu},$$

于是

$$\begin{aligned} &\int_{E_1}e(\eta\alpha)\Big(\prod_{j=1}^{5}f_j(\alpha)-\prod_{j=1}^{5}\nu_j(\alpha)\Big)K_n\left(\frac{\alpha}{Y}\right)\mathrm{d}\alpha\\ &\qquad\ll\int_{E_1}\sum_j(f_j-\nu_j)\Big(\prod_{i<j}f_j\Big)\Big(\prod_{i>j}\nu_j\Big)\mathrm{d}\alpha\\ &\qquad\ll N^{2\nu+4}\int_{E_1}\mathrm{d}\alpha\\ &\qquad\ll N^{2\nu+4+(-2+\nu)}=o(N^3), \end{aligned} \tag{14}$$

又

$$\nu_j(\alpha)\ll\frac{1}{2\sqrt{|\lambda_j\alpha|}}\int_0^{\lambda_j\alpha N^2}\frac{\mathrm{e}(x)}{\sqrt{x}}\mathrm{d}x\ll\frac{1}{\sqrt{|\alpha|}},$$

所以

$$\int_{N^{-2+\nu}}^{+\infty} \mathrm{e}(\eta\alpha)\left(\prod_{j=1}^{5}\nu_j(\alpha)\right)K_n\left(\frac{\alpha}{Y}\right)\mathrm{d}\alpha$$

$$\ll \int_{N^{-2+\nu}}^{+\infty} |\alpha|^{-5/2}\mathrm{d}\alpha$$

$$\ll (N^{-2+\nu})^{-3/2} = o(N^3), \tag{15}$$

由(13)、(14)、(15)

$$\begin{aligned} I_1 &= \int_{-\infty}^{+\infty} \mathrm{e}(\eta\alpha)\left(\prod_{j=1}^{5}\nu_j(\alpha)\right)K_n\left(\frac{\alpha}{Y}\right)\mathrm{d}\alpha + o(N^3) \\ &= Y\int_0^{N^2}\cdots\int_0^{N^2} 2^{-5/2}(y_1y_2\cdots y_5)^{-1/2}\times\left(\int_{-\infty}^{+\infty}\mathrm{e}((\lambda_1y_1+\lambda_2y_2+\cdots\right. \\ &\quad \left.+\lambda_5y_5+\eta)\alpha Y)K_n(\alpha)\mathrm{d}\alpha\right)\mathrm{d}y_1\cdots\mathrm{d}y_5 + o(N^3) \\ &\gg YN^{-5}\int_{\mathscr{B}}\mathrm{d}y_2\cdots\mathrm{d}y_5\int_{\mathscr{A}(y_2,\,\cdots,\,y_5)}\mathrm{d}y_1 + o(N^3), \end{aligned} \tag{16}$$

其中

$$\mathscr{B} = \{\delta N^2 \leqslant y_2 \leqslant 2\delta N^2,\ \delta^2N^2 \leqslant y_j \leqslant 2\delta^2N^2,\ j=3,4,5\},$$

$$\mathscr{A} = \left\{y_1:\ |\lambda_1y_1+\lambda_2y_2+\cdots+\lambda_5y_5+\eta| < \frac{1}{3Y}\right\},$$

并且 δ 足够小,使

$$2\delta^2N^2 < -(\lambda_2y_2+\cdots+\lambda_5y_5+\eta)\lambda_1^{-1} < \frac{1}{2}N^2$$

(注意 λ_1 与 λ_2 异号),从而 $\mathscr{A}(y_2,\cdots,y_5)$ 中的 y_1 适合

$$\delta^2N^2 < y_1 < N^2.$$

于是

$$I_1 \gg YN^{-5}\cdot(N^2)^4\cdot Y^{-1} = N^3. \tag{17}$$

现在来估计 I_2. 取

$$Q_1 = Q_2 = N^{2-\frac{\nu}{2}}, \tag{18}$$

对于 $\alpha \in E_2$，由 Dirichlet 定理，存在整数 a_j、q_j，$(a_j, q_j)=1$，$0<q_j\leqslant Q_j$，并且

$$\left|\lambda_j\alpha-\frac{a_j}{q_j}\right|<\frac{1}{q_jQ_j}\quad(j=1,\ 2). \tag{19}$$

$a_j\neq 0$，否则由(19)得出

$$|\lambda_j\alpha|<\frac{1}{q_jQ_j}\leqslant N^{-2+\frac{\nu}{2}},$$

从而 $\alpha\in E_1$，矛盾. 于是 $\frac{a_j}{q_j}\ll\lambda_j\alpha\ll\frac{a_j}{q_j}$ $(j=1,\ 2)$.

如果 $q_j\leqslant N^{1/3}(j=1,\ 2)$，那么

$$\begin{aligned}\left|a_2q_1\frac{\lambda_1}{\lambda_2}-a_1q_2\right|&=\left|q_1q_2\left(\frac{a_2/q_2}{\lambda_2\alpha}\left(\lambda_1\alpha-\frac{a_1}{q_1}\right)-\frac{a_1/q_1}{\lambda_2\alpha}\left(\lambda_2\alpha-\frac{a_2}{q_2}\right)\right)\right|\\&\ll q_1q_2\left(\left|\lambda_1\alpha-\frac{a_1}{q_1}\right|+\left|\lambda_2\alpha-\frac{a_2}{q_2}\right|\right)\\&\ll\frac{q_1}{Q_2}+\frac{q_2}{Q_1}\\&\ll N^{-\frac{5}{3}+\frac{\nu}{2}}\ (\text{由}(18))\\&\ll q^{-1}\ (\text{由 } q=N^{2/3}),\end{aligned}$$

而

$$a_2q_1=q_1q_2\times\frac{a_2}{q_2}\ll q_1q_2\ll N^{2/3}=q,$$

这与关于连分数最佳逼近的 Legendre 定理相背. 因此 q_1、q_2 中必有一个大于 $N^{1/3}$. 不妨设

$$q_1>N^{1/3}, \tag{20}$$

由 Weyl 不等式

$$f_1(\alpha)\ll N^{1+\delta}\left(\frac{1}{q_1}+\frac{1}{N}+\frac{q_1}{N^2}\right)^{\frac{1}{2}}=o(N). \tag{21}$$

所以

$$\begin{aligned}I_2 &= \int_{E_2} e(\eta\alpha)\Big(\prod_{j=1}^{5} f_j(\alpha)\Big) K_n\Big(\frac{\alpha}{Y}\Big)\,\mathrm{d}\alpha \\ &\ll \max_{\alpha\in E} |f_1(\alpha)| \cdot \sum_{j=2}^{4}\int_0^1 |f_j(\alpha)|^4\,\mathrm{d}\alpha \\ &\ll o(N)\cdot N^2 = o(N^3).\end{aligned} \tag{22}$$

4 对 I_3 的估计

定义集合

$$M=\{\alpha:\ |f_j(\alpha)|\geqslant N^{2/3+\delta},\ j=1,\ 2\}\cap E_3,$$

则

$$\begin{aligned}&\int_{E\backslash M} e(\eta\alpha)\Big(\prod_{j=1}^{5} f_j(\alpha)\Big) K_n\Big(\frac{\alpha}{Y}\Big)\,\mathrm{d}\alpha \\ &\ll N^{2/3+\delta}\sum_{j=1}^{5}\int_1^{YN^{\delta}} |f_j(\alpha)|^4\,\mathrm{d}\alpha \\ &\ll N^{2/3+\delta}\cdot YN^{\delta}\cdot N^2 \\ &\ll N^{2/3+\delta+1/3-\varepsilon+\delta} = o(N^3).\end{aligned} \tag{23}$$

将集 M 分为若干子集 $M(Z)$,其中

$$N^{2/3+\delta}\leqslant Z\leqslant N,$$

并且在 $M(Z)$ 上

$$Z\leqslant |f_2(\alpha)|\leqslant 2z.$$

显然 $M(Z)$ 的个数 $\ll N^{\delta}$.

令 $Q_1=N^{4/3}$. 对于 $\alpha\in M(Z)$,取整数 a_1、q_1, $(a_1,\ q_1)=1$, $0<q_1\leqslant Q_1$,

$$\left|\lambda_1\alpha-\frac{a_1}{q_1}\right|<\frac{1}{q_1Q_1} \tag{24}$$

和 § 3 相同, $a_1\neq 0$ 并且 $\dfrac{a_1}{q_1}\ll\lambda_1\alpha\ll\dfrac{a_1}{q_1}$.

由 Weyl 不等式,

$$N^{2/3+\delta} \leqslant f_1(\alpha) \ll N^{1+\frac{\delta}{2}}\left(\frac{1}{q_1}+\frac{1}{N}+\frac{q_1}{N^2}\right)^{1/2}, \tag{25}$$

于是

$$q_1 < N^{2/3-\delta}. \tag{26}$$

又取 $Q_2 = Z^2 N^{-\delta}$. 同样有整数 a_2、q_2, $(a_2, q_2)=1$, $0 < q_2 \leqslant Q_2$, $a_2 \neq 0$,

$$\left|\lambda_2\alpha - \frac{a_2}{q_2}\right| < \frac{1}{q_2 Q_2}. \tag{27}$$

由 Weyl 不等式

$$Z \leqslant f_2(\alpha) \ll N^{1+\frac{\delta}{3}}\left(\frac{1}{q_2}+\frac{1}{N}+\frac{q_2}{N^2}\right)^{1/2}, \tag{28}$$

得

$$q_2 \ll \left(\frac{N}{Z}\right)^2 \cdot N^{\delta}. \tag{29}$$

我们有(与§3类似)

$$\begin{aligned}\left|a_2 q_1 \frac{\lambda_1}{\lambda_2} - a_1 q_2\right| &\ll \frac{q_1}{Q_2}+\frac{q_2}{Q_1}\\ &\ll N^{2/3+\delta} Z^{-2}\\ &\ll N^{-2/3} = q^{-1},\end{aligned}$$

所以由[2]中引理9(iii),对于任两组满足上述要求的 a_1, q_1, a_2, q_2 及 a_1', q_1', a'_2, q'_2 有

$$|a_2 q_1' - a_2 q_1| \gg q. \tag{30}$$

再将 $M(Z)$ 分为 $\ll N^{\delta}$ 个子集 $M(Z, p_1, p_2)$,对于 $\alpha \in M(Z, p_1, p_2)$,其相应的 q_j 满足

$$p_j \leqslant q_j \leqslant 2p_j \quad (j=1, 2), \tag{31}$$

其中

$$p_1 \leqslant N^{2/3-\delta}, \tag{32}$$

$$p_2 \leqslant \left(\frac{N}{Z}\right)^2 \cdot N^{\delta}, \tag{33}$$

这时

$$a_2 q_1 = \frac{a_2}{q_2} \times q_1 q_2 \ll YN^{\delta} p_1 p_2. \tag{34}$$

而由(30),在长度为 q 的区间中只有$\ll$一个 $a_2 q_1$,所以 $a_2 q_1$ 的个数

$$\ll \frac{YN^{\delta} p_1 p_2}{q},$$

从而满足(31)的 a_1, a_2, q_1, q_2 的个数

$$\ll \frac{YN^{2\delta} p_1 p_2}{q}. \tag{35}$$

由于每个区间 $\left(\frac{a_2}{q_2} - \frac{1}{q_2 Q_2}, \frac{a_2}{q_2} + \frac{1}{q_2 Q_2}\right)$ 的长度

$$\ll \frac{1}{q_2 Q_2} \ll \frac{N^{\delta}}{p_2 z^2}, \tag{36}$$

所以

$$\begin{aligned}
&\int_{M(z,\, p_1,\, p_2)} |f_1(\alpha) f_2(\alpha)|^4 \mathrm{d}\alpha \\
&\ll Z^4 (N^{1+\frac{\delta}{2}} p_1^{-\frac{1}{2}})^4 \int_{M(z,\, p_1,\, p_2)} \mathrm{d}\alpha \ (\text{由}(25),\ (31)) \\
&\ll Z^4 (N^{1+\frac{\delta}{2}} p_1^{-\frac{1}{2}})^4 \cdot \frac{N^{\delta}}{p_2 z^2} \cdot \frac{p_1 p_2 YN^{2\delta}}{q} (\text{由}(35),\ (36)) \\
&\ll Z^2 p_1^{-1} N^{4+5\delta} Y q^{-1},
\end{aligned}$$

$$\begin{aligned}
&\int_{M(z,\, p_1,\, p_2)} \prod_{j=1}^{5} |f_j| \mathrm{d}\alpha \\
&\ll \left(\int_{M(z,\, p_1,\, p_2)} |f_1(\alpha) f_2(\alpha)|^4 \mathrm{d}\alpha\right)^{1/4} \left(\int_{M(z,\, p_1,\, p_2)} |f_3(\alpha)|^4 \mathrm{d}\alpha\right)^{1/4} \\
&\quad \times \left(\int_1^{YN^{\delta}} |f_4(\alpha)|^4\right)^{1/4} \left(\int_1^{YN^{\delta}} |f_5(\alpha)|^4\right)^{1/4} \\
&\ll (Z^2 p_1^{-1} N^{4+5\delta} Y q^{-1})^{1/4} \times N \times \left(\frac{N^{\delta}}{p_2 z^2} \cdot \frac{p_1 p_2 YN^{2\delta}}{q}\right)^{1/4} \times (YN^{\delta})^{1/2} \times N
\end{aligned}$$

$$\ll Yq^{-1/2}N^{1+\frac{5}{4}\delta+1+\frac{3\delta}{4}+\frac{\delta}{2}+1}$$

$$\ll N^{\frac{1}{3}-\varepsilon-\frac{1}{3}+1+\frac{5}{4}\delta+1+\frac{3\delta}{4}+\frac{\delta}{2}+1}$$

$$\ll N^{3-2\delta},$$

于是

$$\int_M e(\eta\alpha)\left(\prod_{j=1}^{5} f_j(\alpha)\right) K_n\left(\frac{\alpha}{Y}\right) d\alpha = o(N^3), \tag{37}$$

由(23)、(37)得

$$I_3 = o(N^3), \tag{38}$$

综合(11)、(17)、(22)、(38)及(5)、(10)得

$$R(N) \gg Y^{-1}N^3,$$

从而定理成立.

本文是作者1985年在数学所访问时完成的，谨对王元所长和数学所表示衷心的感谢.

参考文献

[1] Baker, R. C., Cubic Diophantine Inequalities, *Mathematika*, **29**(1982), 201-215.

[2] Baker, R. C. & Harman, G., Diophantine Approximation by Prime Numbers, *J. Lond. Math. Soc.*, **25**(1982), 201-215.

[3] Davenport, H. & Heilbronn, H., On Indefinote Quadratic Forms in Five Variables, *J. Lond. Math. Soc.*, **21**(1946), 185-193.

[4] Davenport, H., *Analytic Methods for Diophantine Equations and Diophantine Inequalities*, Ann Arbor Publishers, 1962.

[5] Vaughan, R. C., *The Hardy-Littlewood Method*, Camb. Univ. Press, 1981, 144-151.

[6] Vaughan, R. C., Diophantine Approximation by Prime Numbers I & II, *Proc. Lond math. Soc.*, *Third Series Part* 2 & *Part* 3, **28**(1974), 373-384, 385-401.

A Note on Irrationality of Some Numbers

1 Introduction

Let g, $h \geqslant 2$ be two fixed integers, and let $a_h(g)$ denote the positive real number $O \cdot (g^0)_h (g')_h (g^2)_h \cdots$, where $(g^n)_h$ means the number g^n written in base h. Mahler [1] first proved that $a_h(g)$ is irrational if $h = 10$ by p-adic theory. Recently Bundschuh[2] extended Mahler's result to any base $h \geqslant 2$, but his proof was more complicated. In this paper, we shall use the well-known Kronecker Theorem to give a simple and elementary proof of the above result, i. e., the following:

THEOREM. *Let g, $h \geqslant 2$ be two fixed integers, then $a_h(g)$ is irrational.*

2 Proof

If $a_h(g)$ is rational, then $O \cdot (g^0)_h (g')_h \cdots (g^s)_h \cdots$ is a recurring decimal. Let the length of its recurring period be L digits.

Case 1. $\text{Log}_h g$ is rational. Say, $\log_h g = (n/m)$, n, m are positive integers. In this case, we have

$$g^{Km} = h^{Kn} \tag{1}$$

for any integers $K > 0$. Let K be a sufficiently large integer ($Kn > 2L$), then by (1),

$$(g^{Km})_h = |0\underbrace{0\cdots0}_{Kn>2L}$$

this contradicts the assumption that $O(g^0)_h (g')_h \cdots (g^s)_h \cdots$ is a recurring decimal with recurring period of L digits.

Case 2. $\log_h g$ is irrational.

Let $b = h^K$ where K is any sufficiently large integers ($K > 2L$). By Kronecker Theorem [3], there are infinitely many integers m, n with $n > 0$ such that

$$\left| -m + n\log_h g - \frac{\log_h(b+1) + \log_h b}{2} \right| < \frac{\log_h(b+1) - \log_h(b)}{2}. \qquad (2)$$

(2) is equivalent to

$$b \times h^m < g^n < (b+1) \times h^m \qquad (3)$$

and we can assume $m > 0$. From (3) we have

$$(g^n)_h = |\underbrace{0\ 0\cdots 0}_{K>2L}\underbrace{*\ \ *\cdots *}_{m}$$

This is also a contradiction. Therefore, the proof is complete.

References

[1] K. Mahler, On some irrational decimal fractions, *J. Number Theory* **13**(1981), 268 - 269.

[2] P. Bundschuh, Generalization of a recent irrationality result of Mahler, *J. Number Theory* **19**(1984), 248 - 253.

[3] G. H. HARDY and E. M. Wright, "An Introduction to the Theory of Numbers," 5th ed., pp. 375 - 377.

An Application of Number Theory to Tournaments

Abstract

For a tournament on n vertices v_i, let $w_i = d^+(v_i)$ and $l_i = d^-(v_i)$ denote the outdegree and indegree of v_i, respectively. It is well known that $\sum_{i=1}^{n} w_i^j = \sum_{i=1}^{n} l_i^j$ for $j = 1, 2$. In this paper we use known results in number theory on the Tarry-Escott problem of equal sum of powers to prove that for any given integer $k > 2$, there exists a positive integer n and a tournament on n vertices such that $\sum_{i=1}^{n} w_i^j = \sum_{i=1}^{n} l_i^j$ for all $j = 1, 2, \cdots, k$, and the multisets $\{w_i\}$ and $\{l_i\}$ are distinct.

1 Motivation to the Problem

We consider a round robin tournament with $n \geqslant 2$ players in which every two players play against each other exactly once with no ties possible. If we let w_i and l_i denote the number of games won and lost by the ith player respectively, $i = 1, 2, \cdots, n$, then it is well-known and easy to show [2; p. 3, problem B-2] that $\sum_{i=1}^{n} w_i = \sum_{i=1}^{n} l_i$ and $\sum_{i=1}^{n} w_i^2 = \sum_{i=1}^{n} l_i^2$. These lead us naturally to the question as to whether there exists a positive integer $k \geqslant 3$ such that $\sum_{i=1}^{n} w_i^k = \sum_{i=1}^{n} l_i^k$ for all tournaments. On the other hand, we observe that it is possible that, for a particular tournament, $\sum_{i=1}^{n} w_i^k = \sum_{i=1}^{n} l_i^k$ holds for all positive integers k. This would be the case if, e.g. the multisets $\{w_i \mid i = 1, 2, \cdots, n\}$ and $\{l_i \mid i =$

1, 2, ⋯, $n\}$ are identical. Conversely, if $\sum_{i=1}^{n} w_i^k = \sum_{i=1}^{n} l_i^k$ holds for all k, then clearly the multisets $\{w_i\}$ and $\{l_i\}$ must be identical. What if we replace "for all k" by "for some $k \geqslant 3$"? This leads us to our second question: If $\sum_{i=1}^{n} w_i^k = \sum_{i=1}^{n} l_i^k$ holds for some $k \geqslant 3$, must the two multisets $\{w_i\}$ and $\{l_i\}$ be identical?

2 Statements of Results

Our first question raised in §1 has a relatively easy negative answer.

Theorem 1. Let k denote an arbitrary positive integer. Then $\sum_{i=1}^{n} w_i^k = \sum_{i=1}^{n} l_i^k$ for all tournaments with n players if and only if $k=1$ or 2.

Proof. In view of the known results mentioned earlier, it suffices to find a tournament for which $\sum_{i=1}^{n} w_i^k \neq \sum_{i=1}^{n} l_i^k$ for all $k \neq 1, 2$. We consider a tournament with four players in which one player beats the other three who beat one another in a circular manner. Thus the values of w_i's are 3, 1, 1, 1 and the corresponding values of l_i's are 0, 2, 2, 2. Suppose for some $k \geqslant 3$ we have $\sum_{i=1}^{4} w_i^k = \sum_{i=1}^{4} l_i^k$. Then $3^k + 3 = 3 \times 2^k$ or $3^{k-1} + 1 = 2^k$ which is clearly impossible since

$$3^{k-1} - 2^k = 9 \times 3^{k-3} - 8 \times 2^{k-3} > 3^{k-3} \geqslant 1.$$

We now turn to our second question which turns out to be much harder to answer. Interestingly enough, we shall prove that the answer is in the negative despite the fact that no counterexamples have been found for small values of n. In fact, we shall establish the following stronger result:

Theorem 2. For any $k \geqslant 3$ there exists a positive integer n and a tournament with n players such that $\sum_{i=1}^{n} w_i^j = \sum_{i=1}^{n} l_i^j$ holds for all $j = 1, 2, \cdots, k$ and the two multisets $\{w_i\}$ and $\{l_i\}$ are different.

Our proof, though constructive in a sense will not yield the exact value of n

for which such a tournament exists. Though it is possible to estimate some bounds for n, the actual value is in general quite large.

3 Proof of the Main Theorem

Before proving Theorem 2 we need a few lemmas. Throughout, we denote by N the set of positive integers.

Lemma 1. Given any $k \in N$ there exist $m \in N$ and two distinct multisets $\{a_1, a_2, \cdots, a_m\}$ and $\{b_1, b_2, \cdots, b_m\}$ of integers such that $\sum_{i=1}^{m} a_i^j = \sum_{i=1}^{m} b_i^j$ for all $j = 1, 2, \cdots, k$.

Proof. This is the Tarry-Escott problem on equal sum of powers in number theory. A proof can be found in [1, §18.6, Theorem 6.3].

Lemma 2. Suppose a_i's and b_i's are all in Lemma 1 and that $l \in N$ is fixed. Then $\sum_{i=1}^{m} (l \pm a_i)^j = \sum_{i=1}^{m} (l \pm b_i)^j$ for all $j = 1, 2, \cdots, k$.

Proof. Since $\sum_{i=1}^{m} a_i^j = \sum_{i=1}^{m} b_i^j$ for $j = 1, 2, \cdots, k$, the Binomial Theorem implies

$$
\begin{aligned}
\sum_{i=1}^{m} (l - a_i)^j &= \sum_{i=1}^{m} \sum_{t=0}^{j} (-1)^t \binom{j}{t} l^{j-t} a_i^t \\
&= \sum_{t=0}^{j} (-1)^t \binom{j}{t} l^{j-t} \sum_{i=1}^{m} a_i^t \\
&= \sum_{t=0}^{j} (-1)^t \binom{j}{t} l^{j-t} \sum_{i=1}^{m} b_i^t \\
&= \sum_{i=1}^{m} \sum_{t=0}^{j} (-1)^t \binom{j}{t} l^{j-t} b_i^t \\
&= \sum_{i=1}^{m} (l - b_i)^j.
\end{aligned}
$$

The case when "$-$" is replaced by "$+$" is similar.

To facilitate the terminology in the proof of Theorem 2 we shall now follow the standard approach of identifying a tournament with a complete directed graph on $n \geqslant 2$ vertices with vertex v_i representing the ith player. If player i defeats player j, then there is an (directed) are from v_i to v_j and we say that v_i dominates v_j. The number of vertices that v_i dominates is the outdegree of v_i and is denoted by $d^+(v_i)$. Similarly, the number of vertices that dominate v_i is called the indegree of v_i and is denoted by $d^-(v_i)$. Of course, these numbers are just the w_i's and l_i's mentioned in §1.

Lemma 3. For any $n \in N$ there exists a tournament on $2n+1$ vertices such that $d^+(v_i) = d^-(v_i) = n$ for all vertices v_i.

Proof. We use induction on n. When $n=1$ a positively oriented triangle suffices. Suppose the assertion holds for some $n \geqslant 1$ and T is a tournament on $2n+1$ vertices $v_1, v_2, \cdots, v_{2n+1}$ such that $d^+(v_i) = d^-(v_i) = n$ for all $i = 1, 2, \cdots, 2n+1$. We construct a tournament T' from T as follows: the vertices of T' are obtained from the vertices of T by adjoining two new vertices v_{2n+2} and v_{2n+3}. For each $i = 1, 2, \cdots, n$ and for each $j = n+1, n+2, \cdots, 2n+1$, draw an arc from v_i to v_{2n+2}, one from v_{2n+2} to v_j, one from v_{2n+3} to v_i, and one from v_j to v_{2n+3}. Finally, add an arc from v_{2n+3} to v_{2n+2}. Then it is easy to check that $d^+(v_i) = d^-(v_i) = n+1$ for all $i = 1, 2, \cdots, 2n+3$.

Proof of Theorem 2. By Lemma 1, there exist two different multisets $\{a_1, a_2, \cdots, a_m\}$ and $\{b_1, b_2, \cdots, b_m\}$ of integers such that $\sum_{i=1}^{m} a_i^j = \sum_{i=1}^{m} b_i^j$ for all $j = 1, 2, \cdots, k$.

Without loss of generality, we may assume that $a_1 < b_1$. Furthermore, by Lemma 2 we may also assume that $2m-1 < a_1 \leqslant a_2 \leqslant \cdots \leqslant a_m$ and $0 < b_1 \leqslant b_2 \leqslant \cdots \leqslant b_m$. We construct a tournament T as follows: let the set of vertices be $X \cup Y \cup Z$ where $X = \{x_1, x_2, \cdots, x_m\}$ and $Y = \{y_1, y_2, \cdots, y_m\}$ denote two sets of m points each such that the $2m$ points are all distinct. The set Z will be described later on. We draw a_i arcs from x_i and b_i arcs to y_i for each $i = 1, 2, \cdots, m$ so that $d^+(x_i) = a_i$ and $d^-(y_i) = b_i$. The other end vertices of these arcs are to be specified later on. We draw an arc from x_1 to x_i for all $i > 1$, an arc from x_2 to x_j for all $j>2$, etc. Then among all the arcs which are incident with vertices in X there are $c=\sum_{i=1}^{m}a_i-\frac{1}{2}m(m-1)$ left to be assigned the terminal vertices. In a similar manner, we draw an arc from y_i to y_1 for all $i>1$, an arc from y_j to y_2 for all $j>2$, etc. Then among all the arcs which are incident with vertices in Y there are $\sum_{i=1}^{m}b_i-\frac{1}{2}m(m-1)=c$ left to be assigned the initial vertices. Next, we draw an arc from each x_i to every y_j, $i, j=1, 2, \cdots, m$. This way, m^2 more arcs are added. Thus there will be $d=c-m^2$ arcs with initial vertices in X for which we have to specify the terminal vertices, and d arcs with terminal vertices in Y for which we have to specify the initial vertices. Choose sufficiently large n so that $2n+1>d$. By Lemma 3 there exists a tournament on $2n+1$ vertices with vertex set $Z=\{z_1, z_2, \cdots, z_{2n+1}\}$ such that $d^+(z_i)=d^-(z_i)=n$ for all $i=1, 2, \cdots, 2n+1$. Now we use $z_1, z_2, \cdots, z_d$ as the terminal vertices of the d arcs with initial vertices in X mentioned above, one for each arc. If an arc from some x_q to some z_s is drawn, then we also draw an arc from z_s to x_t for all $t\neq q$. Simultaneously, we use $z_1, z_2, \cdots, z_d$ as the initial vertices of the d arcs with terminal vertices in Y mentioned above, one for each arc. If an arc from some z_s to some y_q is drawn, then we also draw an arc from y_t to z_s for all $t\neq q$. Finally, for each of the remaining $2n+1-d$ vertices z_j in Z, we draw an arc from z_j to x_i and an arc from y_i to z_j for all $i=1, 2, \cdots, m$. This construction yields a tournament T on $2m+2n+1$ vertices v_i. Let $l=m+n$. Then clearly $d^+(x_i)=a_i$, $d^+(y_i)=2l-b_i$ and $d^+(z_j)=l$ for $i=1, 2, \cdots, m$, $j=1, 2, \cdots, 2n+1$. Since

$$\sum_{i=1}^{m}a_i^j+\sum_{i=1}^{m}(2l-b_i)^j+\sum_{i=1}^{2n+1}l^j=\sum_{i=1}^{m}b_i^j+\sum_{i=1}^{m}(2l-a_i)^j+\sum_{i=1}^{2n+1}l^j$$

we have $\sum_{i=1}^{N}w_i^j=\sum_{i=1}^{N}l_i^j$ for all $j=1, 2, \cdots, k$ where $N=2m+2n+1$, $w_i=d^+(v_i)$ and $l_i=d^-(v_i)$. To complete the proof it remains to show that the multisets $\{w_i\}$ and $\{l_i\}$ are indeed different. This is the case since if we choose

n large enough so that $a_1 < \min\{2l - a_m, l^{**}\}$, then it is easy to see that the two multisets $\{2l - a_1, 2l - a_2, \cdots, 2l - a_m, b_1, b_2, \cdots, b_m, l, l, \cdots l\}$ and $\{a_1, a_2, \cdots, a_m, 2l - b_1, 2l - b_2, \cdots, 2l - b_m, l, l, \cdots l\}$ where there are $2n + 1$ l's in each set, are distinct.

References

[1] Loo Ken Hua, *Introduction to Number Theory* (translated from the Chinese by Peter Shiu), Springer-Verlag, Berlin/Heidelberg/New York, 1982.

[2] The William Lowell Putnam Mathematical Competition, *Problems and Solutions: 1965 - 1984*, edited by G. L. Alexanderson *et al*, The Mathematical Association of America, 1985.

含有全部 k 元排列的短数列

设 n、k 都是正整数，$k\leqslant n$. 设函数 $F(n, k)$ 具有下述性质：存在一个长度为 $F(n, k)$ 的数列 $S_{n,k}$，对每一个 i，$1\leqslant i\leqslant k$，它的前 $F(n, i)$ 项以 $1, 2, \cdots, n$ 的全部 i 元排列为其子数列，并且任何长度小于 $F(n, k)$ 的数列不再满足这一条件.

本文证明了下面的定理

定理 设 $1\leqslant k\leqslant n-1$，$F(n, k)$ 的定义如上所述，则

$$F(n, k)\leqslant k(n-1)+1-\left[\frac{k}{6}\right]-\left[\frac{k+2}{6}\right],$$

这里 $[x]$ 表示实数 x 的整数部分.

V. Chvatal, D. A. Klarner 与 D. E. Knuth, M. Newey, L. Adelman, G. Galbitai 与 F. P. Preparata Savage Carla，蔡茂诚(Cai Maocheng)等人[1-6]研究过下面的问题：设 $k\leqslant n$ 为自然数，考虑自 $1, 2, \cdots, n$ 的 n 个数所组成的数列 S.

(i) 如果 S 以 $1, 2, \cdots, n$ 这 n 个数的每一个 k 元排列为其子数列，数列 S 的最小长度 $g(n, k)$ 是多少？猜测

$$g(n, n)=n^2-2n+4\quad(n\geqslant 3),$$

现在业已证明 $g(n, n)\leqslant n^2-2n+4$.

(ii) 对于 $k\leqslant n-1$，求函数 $F(n, k)$，具有性质：存在一个长度为 $F(n, k)$ 的数列 $S_{n,k}$，对每一个 i，$1\leqslant i\leqslant k$，它的前 $F(n, i)$ 项以 $1, 2, \cdots, n$ 的全部 i 元排列为其子数列，并且任何长度小于 $F(n, k)$ 的数列不再满足这一条件.

关于第 2 个问题，以蔡茂诚的结果为最佳. 他证明了：

$$F(n, k)\leqslant k(n-1)+1-\left[\frac{k}{4}\right]\quad(1\leqslant k\leqslant n-1).\tag{1}$$

本文在蔡的基础上，证明下面的定理.

定理　设 $1 \leqslant k \leqslant n-1$, $F(n, k)$ 的定义如上所述,则

$$F(n, k) \leqslant k(n-1)+1-\left[\frac{k}{6}\right]-\left[\frac{k+2}{6}\right], \tag{2}$$

这里$[x]$表示 x 的整数部分.

由于 $\left[\frac{k}{6}\right]+\left[\frac{k+2}{6}\right] \geqslant \left[\frac{k}{4}\right]$,并且在 $k \geqslant 6$ 时有严格的不等式,所以(2)优于(1).

证明　不妨设 $k \geqslant 6$, 设

$$\pi_1 = 2,\ 3,\ \cdots,\ n-2,\ n-1,\ n$$
$$\pi_2 = 1,\ 2,\ \cdots,\ n-3,\ n-2,\ n-1$$
$$\pi_3 = 1,\ 2,\ \cdots,\ n-3,\ n,\ n-2$$
$$\pi_4 = 1,\ 2,\ \cdots,\ n-4,\ n-1,\ n-3$$
$$\pi_5 = 1,\ 2,\ \cdots,\ n-4,\ n-2,\ n,\ n-1$$
$$\pi_6 = 2,\ 3,\ \cdots,\ n-3,\ n-2,\ 1$$

并且在 $t \equiv i(\mathrm{mod} 6)$ 时,令

$$\pi_1 = \pi_i.$$

容易验证数列 $1, \pi_1$ 以所有 1 -排列为子列. $1, \pi_1, \pi_2$ 以所有 2 -排列为子列. $1, \pi_1, \pi_2, \pi_3$ 以所有 3 -排列为其子列. $1, \pi_1, \pi_2, \pi_3, \pi_4$ 以所有 4 -排列为其子列. 由于 π_4 的最后一个元素是 $n-3$,它与 π_5 合在一起时,含有所有的 1 -排列,所以 $1, \pi_1, \pi_2, \pi_3, \pi_4, \pi_5$,以所有 5 -排列为其子列.

对于$\{1, 2, \cdots, n\}$的 6 元排列 $a_1a_2a_3a_4a_5a_6$,如果 $a_6 \neq n$ 或$a_6=n$,而 $a_5 \neq n-1$ 或 $a_6=n$, $a_5=n-1$,而 $a_4 \neq n-3$ 则易知 $1, \pi_1, \pi_2, \pi_3, \pi_4, \pi_5, \pi_6$ 均以 $a_1a_2a_3a_4a_5a_6$ 为其子列. 如果 $a_6=n$, $a_5=n-1$, $a_4=n-3$, 这时又有两种情况:

(i) $a_3 \neq n-2$, 由于 $1, \pi_1, \pi_2, \pi_3$ 以所有 3 -排列为其子列,所以 $1, \pi_1, \pi_2 \pi_3$ 以 $a_1a_2a_3$ 为其子列, $1, \pi_1, \pi_2, \pi_3, \pi_4, \pi_5, \pi_6$ 以 $a_1a_2a_3(n-3)(n-1)n$ 为其子列.

(ii) $a_3=n-2$, 由于 $1, \pi_1, \pi_2$ 以 2 -排列 a_1a_2 为其子列,所以 $1, \pi_1, \pi_2$,

π_3, π_4, π_5, π_6 以 $a_1a_2\cdot(n-2)(n-3)(n-1)n$ 为其子列.

于是我们证明了 1, π_1, π_2, π_3, π_4, π_5, π_6 以所有 6-排列为其子列. 假设对于 $i<k$, 1, π_1, π_2, …, π_i 以所有 i-排列为其子列. 考虑

$$1,\ \pi_1,\ \pi_2,\ \cdots,\ \pi_{k-1},\ \pi_k. \tag{3}$$

我们要证明它以任一个 k 元排列 $a_1a_2\cdots a_k$ 为其子列.

当 $k-1\equiv 0(\mathrm{mod}6)$ 时,由于 $\pi_{k-1}(=\pi_6)$的最后一个元 1 与 $\pi_k(=\pi_1)$含有所有的 1-排列,所以由归纳假设即知结论成立,$k-1\equiv 1, 2, 4(\mathrm{mod}6)$ 的情况与此类似.

当 $k-1\equiv 3(\mathrm{mod}6)$ 时,由于 π_6 的最后一个元素为 1,且 1, π_1, π_2, π_3, π_4 以所有 4-排列为其子列,特别地,以 $a_{k-3}a_{k-2}a_{k-1}a_k$ 为其子列. 于是由于归纳假设 1, π_1, π_2, …, π_{k-4} 以 $a_1a_2\cdots a_{k-4}$ 为其子列,而 1, π_{k-3}, π_{k-2}, π_{k-1}, π_k 以 $a_{k-3}a_{k-2}a_{k-1}a_k$ 为其子列,所以 1, π_1, π_2, …, π_k 以 $a_1a_2\cdots a_k$ 为其子列. 同样可证 $k-1\equiv 5(\mathrm{mod}6)$ 的情况.

注意 $|\pi_1|=|\pi_2|=|\pi_3|=|\pi_5|=n-1$, $|\pi_4|=|\pi_6|=n-2$,所以数列(3)的长度为

$$k(n-1)+1-\left[\frac{k}{6}\right]-\left[\frac{k+2}{6}\right].$$

参考文献

[1] Chvatal V, Klarner D A & Knuth D E, Selected Combinatorial Research Problems, *Computer Science Department Report*, CS - 7L - 292, Stanford University, Stanford, Calif., June 1972.

[2] Newey M, Notes on a Problem Involving Permutations as Subsequences, *Computer Science Department Report*, CS - 73 - 340, Stanford University, Stanford, Calif. March 1973.

[3] Adelman I, *Discrete Math*. **10**, (1974), 197 - 200.

[4] Galbitai G & Preparata, F P, On Permutation Embedding Sequences, *SIAM. J. Appl. Math*., 1976, No. 3, 421 - 423.

[5] Savage Carla, *Discrete Math*., **42**(1982), 281 - 285.

[6] Cai Maocheng, *Discrete Math*., **39**(1982), 329 - 330.

Shortest String Containing All k-Element Permutation

Abstract

Let k, n be two positive integers and $k \leqslant n$. Let $F(n, k)$ be the length of the shortest string $S_{n,k}$ consisting of $\{1, 2, \cdots, n\}$ with the following property: for $1 \leqslant i \leqslant k$ the string consisting of the first $F(n, i)$ symbols of $S_{n,k}$ contains every permutation of each i-element subset of $\{1, 2, \cdots, n\}$ as a subsequence.

Theorem $F(n, k) \leqslant k(n-1)+1-\left[\frac{k}{6}\right]-\left[\frac{k+2}{6}\right]$, $1 \leqslant k \leqslant n-1$ is proved.

自然数集的（a，b，k）型可加划分

本文讨论了自然数集合的一类所谓$(a、b、k)$型可加划分，证明了其存在的条件和计算公式.

1 引言

设 $\mathbf{N}$ 为自然数集. 数列 $U=\{u_m\}$ 由

$$u_1=a,\ u_2=b, \tag{1.1}$$

$$u_m=u_{m-1}+u_{m-2}+k \tag{1.2}$$

给出，其中 a、b 为自然数，k 为非负整数.

集 $M\subset\mathbf{N}$. 如果 M 有一个划分 $M=M_1\cup M_2$，$M_1\cap M_2=\varnothing$，使得 $M_i(i=1, 2)$ 中任意两个不同的数 n_1、n_2 的和 $n_1+n_2\overline{\in}U$，则称这个分划为 M 的一个(a, b, k)型可加划分.

K. Alladi, P. Erdös 与 V. E. Hoggatt[1] 讨论了 $a=1$，$b>1$，$k=0$ 的情况. J. Evans[2]，张观容[3, 4]，朱迎宪[5]，朱平天[6, 7] 也对 a、b、k 的一些取值情况进行了研究. 本文的目的是证明：

定理 当 $a>b+k$，并且 $c=a+b+k$ 为偶数时，$\mathbf{N}$ 的(a, b, k)型可加划分不存在.

当 $a\leqslant b+k$ 或 c 为奇数时，$\mathbf{N}$ 的(a, b, k)型可加划分存在，且有 2^{l-1} 种，其中

$$l=\begin{cases}\dfrac{k+e+1}{2}, & \text{若 } k+1\geqslant d,\\[2ex] \left\lceil\dfrac{d+e-1}{2}\right\rceil, & \text{若 } k+2\leqslant d,\end{cases}$$

e 为 a、b、c 中偶数的个数，$d=(a+k,\ b+k)$，$\lceil x\rceil$为天花板函数，即不小于 x 的最小整数.

2　底集

定义　$B=\{n \mid n \in N,\ n<c\}$ 称为底集.

命题1　自然数集**N**的每一个$(a,\ b,\ k)$型可加划分产生底集B的一个$(a,\ b,\ k)$型可加划分,并且由B的这个划分唯一确定.

证明　只需证明后一断言,设$B=B_1\cup B_2$是一个$(a,\ b,\ k)$型可加划分.我们用归纳法来构造**N**的$(a,\ b,\ k)$型可加划分$N=N_1\cup N_2$,使$N_i\supset B_i(i=1,\ 2)$.

设小于n的数已经确定属于N_1或N_2,这里$u_{m-1}\leqslant n<u_m(m\geqslant 4)$.考虑$n$的归属.由于

$$u_m-n=u_{m-1}+u_{m-2}+k-n\leqslant u_{m-2}+k<u_{m-1}$$

根据归纳假设u_m-n的归属已定.令n在u_m-n所不在的那个集中.

不妨设$n\in N_1$.对任一自然数$x<n$,$u_{m-1}<x+n<2u_m<u_{m+2}$.如果$x+n=u_{m+1}$,则

$$(u_m-x)+(u_m-n)=2u_m-u_{m+1}=u_m-u_{m-1}-k=u_{m-2},$$

根据归纳假设$u_m-x\in N_1$(因$u_m-n\in N_2$),$x\in N_2$,于是N_1中任意一个数与n的和$\notin U$.这样,每个自然数n都有确定的归属,因而产生一个唯一的、N的$(a,\ b,\ k)$型可加划分.

由命题1,我们只需讨论底集B的$(a,\ b,\ k)$型可加划分.

3　基本划分与基本图

命题2　当$a>b+k$,并且$c=a+b+k$为偶数时,底集B的$(a,\ b,\ k)$型可加划分不存在.

证明　$\dfrac{a-(b+k)}{2}$,$\dfrac{a+(b+k)}{2}$,$\dfrac{a+3(b+k)}{2}$都在B中,两两的和为$u_1=a$,$u_3=c$,$u_4=a+2(b+k)$,因此B的$(a,\ b,\ k)$型可加划分不存在.

定义　设$B=B_1\cup B_2$,$B_1\cap B_2=\varnothing$.如果对于同一个B_i $(i=1,\ 2)$中任意两个不同的数n_1、n_2,都有$n_1+n_2\notin\{a,\ b,\ c\}$,则称此划分为$B$的基本

划分.

命题 3 在 $a \leqslant b+k$ 或 c 为奇数时, B 的每一个基本划分都是 (a, b, k) 型可加划分.

证明 设 $B=B_1 \cup B_2$ 是一个基本划分,对于 B_1 中的数 $n_1>n_2$,恒有 $n_1+n_2<2c=u_3+u_3<u_5$,所以,若 $n_1+n_2 \in U$, 则

$$n_1+n_2=u_4=a+2b+2k. \tag{3.1}$$

但 $n_1<c=a+b+k$,所以 $n_2>b+k$. 由(3.1)得

$$(n_1-b-k)+(n_2-b-k)=a, \tag{3.2}$$

$$(n_1-b-k)+n_2=c, \tag{3.3}$$

$$(n_2-b-k)+n_1=c. \tag{3.4}$$

(3.3), (3.4)表明 n_1-b-k, n_2-b-k 同属 B_2, (3.2)表明这两个不同的数的和为 a,与基本划分定义矛盾. 所以(3.1)不成立, $n_1+n_2 \notin U$. 同样 B_2 也是如此. 证毕.

于是,我们只需讨论底集 B 的基本划分的个数. 由于在基本划分的定义中, a, b 完全平等,我们设 $a \leqslant b$.

将 B 中的数用点表示. 如果两个数的和 $\in\{a, b, c\}$,就在相应的两个点之间连一条边,并分别称之为 a 边、b 边或 c 边. 这个图称为基本图.

B 的每个基本划分就是基本图的一个 2-染色:将点染上颜色 B_1 或 B_2,使相邻的点颜色不同.

如果基本图有圈,那么圈上每一点 x 引出的两条边中必有一条 b 边. 不然的话,设 x 引出一条 a 边一条 c 边,则 $x<a$, $c-x>c-a \geqslant b$,从而点 $c-x$ 无法与其他点相连,矛盾. 于是圈上的边数必为偶数(圈上 b 边的 2 倍),基本图是二部分图,可以 2 染色. 即有(不考虑 B_1, B_2 顺序)

命题 4 设 B 的基本图有 l 个连通分支,则 B 有 2^{l-1} 种基本划分.

问题化为计算 l.

4 a = b 与 c ≥ 2b 的情况

在 $a=b$ 时,易知基本图的连通分支仅有如图的两种(后者包括在 c 为偶数

时,退化为点$\frac{c}{2}$的那个分支),没有圈. 而 a 边有 $\left\lceil \frac{a-2}{2} \right\rceil$ 条,所以有

命题 5　在 $a=b$ 时, $l=\left\lceil \frac{c-1}{2} \right\rceil-\left\lceil \frac{a-2}{2} \right\rceil$.

以下设 $a<b$. 在 $c\geqslant 2b$ 时,仅由 c 边组成的分支共 $\left\lceil \frac{c-2b+1}{2} \right\rceil$ 个. 其余的 c 边有且仅有一端与 a 边或 b 边相连,我们称之为“毛刺”,将它去掉(同时去掉不与其他边相连的那个端点),不影响 l. 再考虑 b 边,分三种情况讨论:

若 $b\geqslant 2a$,则有$\left\lceil \frac{b-2a+1}{2} \right\rceil$个仅由 b 边组成的分支. 其余的 b 边又可作为毛刺除去,剩下 $\left\lceil \frac{a-1}{2} \right\rceil$ 个仅由 a 边组成的分支,于是

$$l=\left\lceil \frac{c-2b+1}{2} \right\rceil+\left\lceil \frac{b-2a+1}{2} \right\rceil+\left\lceil \frac{a-1}{2} \right\rceil=\frac{k+1+e}{2}.$$

若 $b=2a-1$, 所有的 b 边均为毛刺,上式仍然成立.

若 $b<2a-1$. 先设 a 为奇数,将毛刺除去后,每一条 b 边的两端均有 a 边. 图中没有圈(否则由于圈上 a、b 边个数相等,导出 $a=b$), 每个分支有两个端点,端点$\leqslant b-a$ (当 b 为偶数时,还有一个退化情况,即点$\frac{b}{2}$成一分支),于是

$$l=\left\lceil \frac{c-2b+1}{2} \right\rceil+\left\lceil \frac{b-a}{2} \right\rceil=\frac{k+1+e}{2}.$$

当 a 为偶数时,无论 $b-a/2\geqslant a$ 或 $b-a/2<a$, 均有

$$l=\left\lceil \frac{c-2b+1}{2} \right\rceil+\left\lceil \frac{b-a+1}{2} \right\rceil=\frac{k+1+e}{2}.$$

于是有

命题 6　在 $c\geqslant 2b$, $b>a$ 时, $l=\frac{k+1+e}{2}$.

5　c < 2b 的情况

这时没有仅由一条 c 边组成的分支,除非 b 为偶数且 $2(a+k)\geqslant b\geqslant 2a$,

有一个例外(如图).

$$\frac{b}{2} \qquad c \qquad c-\frac{b}{2}\quad(\geqslant b)$$

去掉作为毛刺的 c 边(上述例外也仅保留点$\frac{b}{2}$). 剩下的数即 $1, 2, \cdots, b-1$. 将它们按照 $\bmod d$ 分类 $(d=(a+k, b+k))$.

如果基本图有圈,则圈上 b 边与其他边相间,由 x 出发沿圈前进,先过一条 a 边或 c 边,再过一条 b 边,这样继续下去,经过形如

$$x+s(b-a)-t(c-b) \text{ 及 } c-x+s(b-a)-t(c-b)$$

的数. 它们都在 $\bmod d$ 的某两个或某一个剩余类里. 不仅如此,我们有以下命题.

命题 7 剩余类中的数如果有一个在圈上,则必全在圈上.

证明 设 x 在圈上, $x+d<b$. 由于 $d=(c-b, b-a)$,必有正整数 u, v 使 $d=u(b-a)-v(c-b)$,并且 $u\leqslant i, v\leqslant j$,这里自然数 i、j 满足 $i(b-a)=j(c-b)$, 且 i 是此等式成立的最小数.

当 x 沿圈前进时,经过点 $x+s(b-a)-t(c-b)$, s 与 t 中必有一个先达到 u 或 v. 不妨设 s 先达到 u,这点为 $y=x+u(b-a)-t(c-b)$, $(t<v)$. 由于 $y\geqslant x+d+c-b>a$, 所以接下去是 c 边、b 边,如此继续下去, u 不变, t 增加直至 $t=v$,从而 $x+d$ 也在圈上. 类似地,可以证明在 $x>d$ 时, $x-d$ 在圈上.

每个非圈的连通分支必有两个端点(仅前述例外由一点$\frac{b}{2}$组成,这时 $\frac{b}{2}\in[a, a+k]$), 端点为区间 $[a, a+k]$ 中整数及$\frac{c}{2}$(c 为偶数时), $\frac{b}{2}$(b 为偶数时), $\frac{a}{2}$(a 为偶数时),于是共有 $\frac{k+1+e}{2}$ 个非圈的连通分支.

命题 8 若 $k+1\geqslant d$,则 $l=\frac{k+e+1}{2}$.

证明 $a, a+1, \cdots, a+k$ 构成 $\bmod d$ 的完系,因此由命题 7 基本图无圈.

命题 9 若 $k+2\leqslant d$,则 $l=\left\lceil\frac{d+e-1}{2}\right\rceil$.

证明　设 $a=q_1d+r$, $0<r\leqslant d$. 由于 $(a+k, d)=(r+k, d)=d$,所以 $r+k=d$. a, $a+1$, $\cdots$, $a+k$ 跑过剩余类 r, $r+1$, $\cdots$, $r+k=d$. $\frac{c}{2}$, $\frac{b}{2}$, $\frac{a}{2}$ (在为整数时)都是 $2x\equiv r\pmod d$ 的解,当 r 为奇数时,这同余方程在 1, 2, $\cdots$, $r-1$ 中无解,所以在圈上共有 $r-1$ 个剩余类,每个圈上 2 个. 当 r 为偶数时,这同余方程在 1, 2, $\cdots$, $r-1$ 中有一解 $\frac{r}{2}$. 如果 d 为奇数,则 a, b, c 中至少有一个为偶数,不妨设 b 为偶数,则 $\frac{b}{2}\equiv\frac{r}{2}\pmod d$; 如果 d 为偶数,则 a、b、c 全为偶数,设 $b=q_2d+r$, $c=q_3d+r$, 则 q_1、q_2、q_3 不可能全为奇数,否则 $2d\mid(c-a, b-a)=d$. 不妨设 q_2 为偶数,则 $\frac{b}{2}\equiv\frac{r}{2}\pmod d$. 总之,这时圈上共有 $r-2$ 个剩余类,每个圈上两个. 因此有 $\left[\frac{r-2}{2}\right]$ 个圈.

$$l=\frac{k+1+e}{2}+\left\lceil\frac{r-2}{2}\right\rceil=\left\lceil\frac{k+r+e-1}{2}\right\rceil=\left\lceil\frac{d+e-1}{2}\right\rceil.$$

由命题 1—9 即得定理.

参考文献

[1]　Alladi K., Erdös P. and Hoggatt. Jr V., On Additive Partitions of Integers, *Discrete Math*. 22(1978), 201 - 211.

[2]　Evans R., On Additive Partitions of Sets of Positive Integers, ibid. 36(1981), 239 - 245.

[3]　张观容,自然数集合的一种 U 划分,南京师范大学学报(自然科学版),2(1988), 36 - 40.

[4]　张观容,关于集合 N 的一类 U 划分,南京师范大学学报(自然科学版),4(1988), 33 - 38.

[5]　朱迎宪,关于一类正整数集合的 U 划分,南京工学院学报,1(1988), 10 - 17.

[6]　朱平天,自然数集的一类可加划分,待发表.

[7]　朱平天,自然数集的泛可加划分,待发表.

On (a, b, k)-Partitions of Positive Integers*

For given integers a, b and k, with a, $b > 0$ and $k \geqslant 0$, the sequence $\{u_m\}_{m=1}^{\infty}$ is defined by $u_1 = a$, $u_2 = b$ and $u_m = u_{m-1} + u_{m-2} + k$ $(m \geqslant 3)$. We study partitions of the set of positive integers into two parts with the property that the sum of two distinct members of the same part is never in $\{u_m\}$. Such a partition is called an (a, b, k)-partition. This paper gives a necessary and sufficient condition for an (a, b, k)-partition to exist. It also determines the number of (a, b, k)-partitions as a function of a, b and k.

1 Introduction

Let N be the set of positive integers.

For $a \in N$, $b \in N$, $k \in N \cup \{0\}$, let $c = a + b + k$ and $d = (a + k, b + k) = (c - a, c - b)$, the GCD of $a + k$ and $b + k$.

We denote by e the number of even numbers among a, b and c, and by $\lceil x \rceil$ the least integer not less than x.

The sequence $U = \{u_m\}$ of integers is defined recursively by

$$u_1 = a, \ u_2 = b \tag{1.1}$$

$$u_m = u_{m-1} + u_{m-2} + k. \tag{1.2}$$

Let $M \subseteq N$. A pair of sets M_1, M_2 is called an (a, b, k)-*partition* of M if

$$M = M_1 \cup M_2, \ M_1 \cap M_2 = \varnothing$$

and the sum of two distinct members of the same M_i $(i = 1, 2)$ is never in U.

K. Alladi, P. Erdös and V. E. Hoggatt[1] first discussed the $(1, b, 0)$-

* This work is supported by the Natural Science Foundation of Jiangsu Province, China.

partition. The subsequent papers of J. Evans[2], Chang Guan-rong[3], [4], Zhu Ying-xian[5], Zhu Ping-tian[6], [7] dealt with some other special cases.

The purpose of this paper is to prove the following result.

Theorem. *If* $a > b + k$ *and* c *is an even number, then there are no* (a, b, k)*-partitions of* N. *If* $a \leqslant b + k$ *or* c *is an odd number, then there are* 2^{l-1} *different* (a, b, k)*-partitions of* N, *where*

$$l = \begin{cases} \dfrac{k+e+1}{2}, & \text{if } k+1 \geqslant d, \\ \left\lceil \dfrac{d+e-1}{2} \right\rceil, & \text{if } k+2 \leqslant d. \end{cases}$$

2 Base

Definition. The set

$$B = \{n \mid n \in N, n < c\}$$

is called a *base*.

Proposition 1. *Every* (a, b, k)*-partition of the base* B *can be extended uniquely to an* (a, b, k)*-partition of* N *and conversely every* (a, b, k)*-partition* $N = N_1 \cup N_2$ *induces an* (a, b, k)*-partition* $B = (B \cap N_1) \cup (B \cap N_2)$.

Proof. We need only to prove the first claim.

Let $B = B_1 \cup B_2$ be an (a, b, k)-partition of B. By induction, we will construct an (a, b, k)-partition $N_1 \cup N_2$ of N, such that $N_i \supseteq B_i (i = 1, 2)$.

Suppose that $n \in N$ satisfies $u_{m-1} \leqslant n < u_m (m \geqslant 4)$ and for every positive number less than n, we know whether it should belong to N_1 or N_2, such that the sum of any two of these numbers from the same $N_i (i = 1, 2)$ is not in U. Now let us consider whether N_1 or N_2 should n belong to. Since

$$0 < u_m - n = u_{m-1} + u_{m-2} + k - n \leqslant u_{m-2} + k < u_{m-1} \leqslant n,$$

we know which one, N_1 or N_2, $u_m - n$ belongs to by the induction hypothesis.

Let n belong to the set which does not contain $u_m - n$. Without loss of generality, we may assume that $u_m - n \in N_2$ and $n \in N_1$.

For any positive integer $x < n$, we have

$$u_{m-1} < x + n < 2u_m < u_{m+2}.$$

If $x + n = u_{m+1}$, then

$$(u_m - x) + (u_m - n) = 2u_m - u_{m+1} = u_m - u_{m-1} - k = u_{m-2}.$$

By the induction hypothesis, $u_m - x \in N_1$. This implies $x \in N_2$. Thus the sum of n and any member (which is less than n) of N_1 can never be in U.

In this way, we obtain a unique (a, b, k)-partition $N_1 \cup N_2$ of N such that $N_i \supseteq B_i (i = 1, 2)$.

By Proposition 1, it is enough to discuss the (a, b, k)-partition of the base B instead of the (a, b, k)-partition of N.

3 Basic Partition

Proposition 2. *If $a > b + k$ and c is an even number, then there are no (a, b, k)-partition of base B.*

Proof. In any partition $B = B_1 \cup B_2$, one of the sets B_1 and B_2 contains at least two of the three positive integers:

$$\frac{a - (b + k)}{2}, \frac{a + (b + k)}{2}, \frac{a + 3(b + k)}{2}.$$

But the sums of pairs of these numbers are

$$u_1 = a, \ u_3 = c \text{ and } u_4 = a + 2(b + k).$$

Hence there are no (a, b, k)-partition of B.

Definition. Let $B = B_1 \cup B_2$, $B_1 \cap B_2 = \varnothing$. If $n + n' \notin \{a, b, c\}$ for any two distinct members $n, n' \in B_i (i = 1, 2)$, then the partition $B_1 \cup B_2$ is called a *basic partition* of B.

Proposition 3. *If $a \leqslant b + k$ or c is an odd number, then every basic* *partition of B is an (a, b, k)-partition.*

Proof. Let $B=B_1\cup B_2$ be a basic partition and n_1, n_2 be a pair of elements of B_1 with $n_1>n_2$. Since

$$n_1+n_2<2c=u_3+u_3<u_5,$$

if $n_1+n_2\in U$, then the only possibility is that

$$n_1+n_2=u_4=a+2b+2k. \tag{3.1}$$

For $n_1<c=a+b+k$, we have $n_2>b+k$. From (3.1), it follows that

$$(n_1-b-k)+(n_2-b-k)=a, \tag{3.2}$$

$$(n_1-b-k)+n_2=c, \tag{3.3}$$

and

$$(n_2-b-k)+n_1=c. \tag{3.4}$$

By (3.3) and (3.4), we have that $n_1-b-k\in B_2$ and $n_2-b-k\in B_2$. But (3.2) shows that the sum of two distinct members n_1-b-k and n_2-b-k of B_2 is equal to a, this contradicts the definition of basic partition. Hence $n_1+n_2\notin U$ the same conclusion holds for the set B_2. The proof is thus complete.

Thus we need only to determine the number of basic partitions of base B. Since a and b are in equal positions in the definition of basic partition, from now on, we will assume that $a\leqslant b$.

4 Basic Graph

We will use some terminologies in graph theory. The numbers of base B are denoted by points. If the sum of two numbers of B belongs to the set $\{a, b, c\}$, we join the two corresponding points by a line and it is called *a-line*, *b-line* or *c-line* according to the sum a, b, c respectively. The resulting graph is called a *basic graph*.

Every basic partition of base B is equivalent to a 2-colouring of the basic graph, i. e. to assign colour B_1 or B_2 to its points such that no two adjacent points have the same colour.

If the basic graph contains a cycle, then one and only one of the two lines incident with any point x in this cycle is a b-line. For otherwise, supposing the lines incident with x are an a-line and c-line, then $x < a$ and $c - x > c - a \geqslant b$. So the point $c - x$ cannot be adjacent to other points except the point x. This is a contradiction.

From this, it follows that the length of any cycle is even (twice of the number of b-lines in that cycle). In other words, the basic graph is a bipartite graph. It is well known that every bipartite graph is 2-colourable. So it is not difficult to show that the following proposition holds.

Proposition 4. *Suppose the basic graph consists of l connected components. Then the base B has 2^{l-1} basic partitions (without specifying the order of B_1 and B_2).*

Now the proof of our theorem is reduced to the calculation of the number of connected components l.

5 The Cases a = b and c ⩾ 2b

Obviously $c - a = b + k \geqslant b \geqslant a$. If $a = b$, then it is easy to see that there are only two kinds of connected components of the basic graph as shown below:

c a c c

The latter may include a degenerate case, i. e. the isolated point $\frac{c}{2}$ when c is even.

Since there are $\left[\frac{c-1}{2}\right]$ c-lines (for convenience we consider the isolated point $\frac{c}{2}$ as a c-line) and $\left[\frac{a-2}{2}\right]$ a-lines, we have:

Proposition 5. *If* $a=b$, *then* $l=\left\lceil\frac{c-1}{2}\right\rceil-\left\lceil\frac{a-2}{2}\right\rceil$.

In the following, we will assume that $b>a$.

Now suppose $c\geqslant 2b$. There are $\left\lceil\frac{c-2b+1}{2}\right\rceil$ components consisting of only a c-line because their endpoints x and $c-x$ must be greater than or equal to b. Every other c-line is adjacent to one and only one a-line or b-line. We call these *c-burrs*.

Eliminate all c-burrs (at the same time we delete an endpoint of every c-burr which is not adjacent to other lines). Obviously, the number l does not change in this polishing process.

Put the c-lines aside, we turn to the discussion about b-lines and a-lines. There are three possible cases.

Case i) $b\geqslant 2a$. In this case, for any component consisting of only one b-line, any endpoint of the b-line must satisfy the conditions $x\geqslant a$ and $b-x\geqslant a$, i. e. $b-a\geqslant x\geqslant a$. Hence the number of components of this kind is $\left\lceil\frac{b-2a+1}{2}\right\rceil$ (including a degenerate component, i. e. the isolated point $\frac{b}{2}$ when b is even). Every other b-line is adjacent to one and only one a-line since $b-a\geqslant a$. After these b-burrs are eliminated there remain $\left\lceil\frac{a-1}{2}\right\rceil$ components consisting of only one a-line (including the isolated point $\frac{a}{2}$ when a is even). Hence

$$l=\left\lceil\frac{c-2b+1}{2}\right\rceil+\left\lceil\frac{b-2a+1}{2}\right\rceil+\left\lceil\frac{a-1}{2}\right\rceil=\frac{k+1+e}{2}.$$

Case ii) $b=2a-1$. All b-lines are b-burrs. The above formula of l still holds.

Case iii) $b<2a-1$. First we assume that a is odd. There is no component consisting of only one b-line. We eliminate the c-burrs and b-burrs. If a remaining point is incident with a b-line, then it is incident with an a-line also.

We claim that there are no cycles in the graph. If there is a cycle, then the a-lines and b-lines appear alternatively in the cycle. Going around the cycle, the point will plus $m(b-a)$ when it returns to the original position, where m is equal to half of the length of the cycle. This forces that $a=b$, a contradiction.

Hence every component has two endpoints. When b is odd, any endpoint x is incident with an a-line and a b-burr. This implies that $x<a$ and $b-x\geqslant a$, i. e. $x\leqslant b-a$. When b is even, $\frac{b}{2}$ is an endpoint also. Hence

$$l=\left\lceil\frac{c-2b+1}{2}\right\rceil+\left\lceil\frac{b-a}{2}\right\rceil=\frac{k+1+e}{2}.$$

Now suppose a is even. If $b-\frac{a}{2}\geqslant a$, then there is a component consisting of only one b-line. The endpoints of this component are $\frac{a}{2}$ and $b-\frac{a}{2}$. No matter whether b is odd or even, we always have

$$l=\left\lceil\frac{c-2b+1}{2}\right\rceil+\left\lceil\frac{b-a+1}{2}\right\rceil=\frac{k+1+e}{2}.$$

If $b-\frac{a}{2}<a$, then there are no components consisting of only one b-line, and the point $\frac{a}{2}$ is deleted in the polishing process. The point $b-\frac{a}{2}$ is an endpoint of a component. Each of the other endpoints x must satisfy the condition $x\leqslant b-a$, except $x=\frac{b}{2}$ when b is even. Hence the above formula still holds.

Thus we have proved

Proposition 6. *If $c\geqslant 2b$ and $b>a$, then $l=\frac{k+1+e}{2}$.*

6 The Case c < 2b

In this case there are no components consisting of only one c-line except the possible component

$$\frac{b}{2} \qquad\qquad c \qquad\qquad c-\frac{b}{2}(b)$$

$$\bullet\!\!-\!\!-\!\!-\!\!-\!\!-\!\!-\!\!-\!\!-\!\!-\!\!-\!\!\bullet$$

which occurs only when b is even and $2(a+k) \geqslant b \geqslant 2a$.

Eliminate all c-burrs (for the above exceptional component only the endpoint $\frac{b}{2}$ is retained). The remaining points are $1, 2, \cdots, b-1$. They can be classified into d residue classes (mod d).

If there are some cycles in the graph, then it is easy to see that the b-lines are alternated with other lines in any cycle. Let us go along a cycle, starting at a point x and an a-line or c-line which is incident with x, then we follow a b-line, and so on. In this trip, the points we passed through can be written in the form

$$x+s(b-a)-t(c-b) \quad \text{or} \quad c-x+s(b-a)-t(c-b)$$

where s and t are integers. They are all in one class or in two residue classes (mod d). We will now prove the following:

Proposition 7. *If one of the points* $1, 2, \cdots, b-1$ *lies in a cycle, then all other points in the residue class* (mod$d$) *of it lie also in the cycle.*

Proof. Suppose x lies in a cycle and $x+d<b$. Since $d=(c-b, b-a)$, there are two positive integers $u \leqslant i$ and $v \leqslant j$ such that

$$d=u(b-a)-v(c-b), \tag{6.1}$$

where i, j are the least positive integers satisfying the equation

$$i(b-a)=j(c-b). \tag{6.2}$$

Starting at the point x and proceeding along the cycle, we pass through some points of the form $x+s(b-a)-t(c-b)$. After we pass two lines, s or t increases by 1, and at last s and t will satisfy Eq. (6.2). Hence s or t will attain u or v at a certain time. Without loss of generality, we may assume s attains u before t attains v. When $s=u$, the point is

$$y = x + u(b-a) - t(c-b) \quad (t < v).$$

Since $y \geqslant x + d + (c-b) > a$, the next step will go through a c-line, then a b-line, etc. So t increases up to $t = v$ while s remains at the same value u. Hence the point $x + d$ lies also in the cycle.

By the same argument, we can prove that if a point x lies in a cycle and $x > d$, then $x - d$ lies in the same cycle.

The proposition follows immediately.

If a connected component of the graph is not a cycle, it has exactly two endpoints (in the above exception, it is convenient to take the isolated point $\frac{b}{2}$ as two endpoints). When an endpoint x is not equal to $\frac{a}{2}$ or $\frac{c}{2}$, it must satisfy the conditions $x \geqslant a$ and $c - x \geqslant b$, i.e. $x \in [a, a+k]$. Conversely, if $x \in [a, a+k]$, then x must be an endpoint. The points $\frac{a}{2}$, $\frac{b}{2}$, $\frac{c}{2}$ (when a, b, c are even) are also endpoints of some components. Hence there are $k + 1 + e$ endpoints and $\frac{k+1+e}{2}$ connected components which are not cycles.

Proposition 8. *If* $k + 1 \geqslant d$, *then* $l = \frac{k+1+e}{2}$.

Proof. Since a, $a+1$, $\cdots$, $a+k$ constitute a complete system of residues (mod d), there are no cycles in the basic graph by Proposition 7.

Proposition 9. *If* $k + 2 \leqslant d$, *then* $l = \left\lceil \frac{d+e-1}{2} \right\rceil$.

Proof. Let $a = q_1 d + r$, $0 < r \leqslant d$. From $r + k < 2d$ and

$$(a+k, d) = (r+k, d) = d,$$

it follows that $r + k = d$. Furthermore

$$a, a+1, \cdots, a+k$$

pass through the residue classes (mod d)

$$r, r+1, \cdots, r+k = d.$$

If r is odd, the congruence equation

$$2x \equiv r(\bmod d) \tag{6.3}$$

has no solution in $\{1, 2, \cdots, r-1\}$. Therefore x and $c-x$ (or $a-x$) are not in the same residue class (mod d) for $x \in \{1, 2, \cdots, r-1\}$. Hence every cycle has exactly two classes of residues. Since $\frac{a}{2}$, $\frac{b}{2}$ and $\frac{c}{2}$ (when they are integers) are solutions of (6.3), they cannot be congruent to the elements of $\{1, 2, \cdots, r-1\}$. It follows that there are $\left\lceil\frac{r-2}{2}\right\rceil$ cycles.

If r is even, Eq. (6.3) has one solution $\frac{r}{2}$ in $\{1, 2, \cdots, r-1\}$.

There are two possible cases.

Case i) d is odd. Since $d=(b-a, c-a)$, at least one of a, b and c is even. Suppose a is an even number. Then $\frac{a}{2} \equiv \frac{r}{2}(\bmod d)$. Hence $\frac{r}{2}$ is not in any cycle and every cycle has exactly two residue classes. There are $\left\lceil\frac{r-2}{2}\right\rceil$ cycles again.

Case ii) d is even. In this case, a, b, c are all even numbers. Let $b=q_2 d+r$, $c=q_3 d+r$. Then the three umbers q_1, q_2, q_3 cannot be all odd. For otherwise, it implies that $2d \mid (c-a, b-a)=d$, a contradiction. Suppose q_2 is even, then $\frac{r}{2} \equiv \frac{b}{2}(\bmod d)$. Hence $\frac{r}{2}$ is not in any cycle. Since $\frac{a}{2}$, $\frac{b}{2}$ and $\frac{c}{2}$ are solutions of (6.3), they cannot be congruent to the elements of $\{1, 2, \cdots, r-1\} \backslash \left\{\frac{r}{2}\right\}$. Therefore the number of cycles is $\left\lceil\frac{r-2}{2}\right\rceil$.

Since there are $\left\lceil\frac{r-2}{2}\right\rceil$ cycles and $\frac{k+1+e}{2}$ connected components which are not cycles, we have

$$l=\frac{k+1+e}{2}+\left\lceil\frac{r-2}{2}\right\rceil=\left\lceil\frac{k+r+e-1}{2}\right\rceil=\frac{d+e-1}{2}.$$

The theorem now follows from Propositions 1 to 9.

References

[1] K. Alladi, P. Erdös and V. Hoggatt, Jr, On additive partitions of integer, *Discrete Math.* **22**(1978), 201 - 211.

[2] R. Evans, On additive partitions of sets of positive integers, ibid. **36**(1981), 239 - 245.

[3] Zhang Guanrong, A new U-partition of natural number set, *J. Nanjing Normal Univ.* (*Natural Science*) **2** (1988), 36 - 40.

[4] Zhang Guanrong, A U-partition of natural number set, ibid. **4**(1988), 33 - 35.

[5] Zhu Yingxian, On the U-partition of class of positive sets, *J. Nanjing Inst. of Technology* **1**(1988), 26 - 31.

[6] Zhu Pingtian, A class of additive partitions of the set of natural numbers, *J. Nanjing Normal Univ.* (*Natural Science*) **4**(1989), 1 - 6.

[7] Zhu Pingtian, On universal additive partitions of the set of natural numbers, ibid. **2**(1990), 1 - 5.

A Diophantine Inequality (Ⅱ)

Abstract

This paper proves the existence of infinitely many integer solutions to a Diophantine inequality.

1 Introduction

The object of this paper is to prove the following

Theorem. *Let λ_1, λ_2, λ_3, λ_4, λ_5 be non-zero real numbers not all in rational ratics and not all of the same sign. Then, given real number η, the inequality*

$$|\lambda_1 x_1^2+\lambda_2 x_2^2+\lambda_3 x_3^2+\lambda_4 x_4^2+\lambda_5 x_5^2+\eta|<(\max_{1\leqslant i\leqslant 5}|x_i|)^{-1/2+\varepsilon} \tag{1}$$

has infinitely many solutions in positive integers x_1, x_2, x_3, x_4, x_5 for any $\varepsilon>0$.

This is an improvement of the result in [6], where the exponent of the right side of inequality (1) is $-1/3+\varepsilon$.

2 The Estimation of I_1, I_2, I_4

Without loss of generality, we may assume that λ_1/λ_2 is negative and irrational. Suppose that q is the denominator of a convergent fraction of λ_1/λ_2.

Let

$$N=q,\ Y=N^{1/2-\varepsilon},\ 0<v<1/3,\ 0<\delta<\varepsilon/5$$

and write

$$f(\alpha)=\sum_{x=1}^{N} e(\alpha x^2),$$

$$f_j(\alpha)=f(\lambda_j\alpha),\ j=1,\ 2,\ 3,\ 4,\ 5.$$

Similar to [6], we can show that $R(N)$, the number of the solutions of inequality

$$|\lambda_1 x_1^2+\lambda_2 x_2^2+\lambda_3 x_3^2+\lambda_4 x_4^2+\lambda_5 x_5^2+\eta|<Y^{-1}$$

in integers satisfying

$$\max_{1\leqslant i\leqslant 5}|x_i|<N,$$

satisfies

$$R(N)\gg Y^{-1}\int_{-\infty}^{+\infty} e(\eta\alpha)\Big(\prod_{j=1}^{5} f_j(\alpha)\Big)K_n\Big(\frac{\alpha}{Y}\Big)\,\mathrm{d}\alpha=Y^{-1}I=Y^{-1}(I_1+I_2+I_3+I_4),$$

where I_1, I_2, I_3, I_4 are the integrals over the intervals

$$E_1=\{|\alpha|\leqslant N^{-2+v}\},$$

$$E_2=\{N^{-2+v}<|\alpha\leqslant 1|\},$$

$$E_3=\{1<|\alpha|\leqslant YN^{\delta}\},$$

$$E_4=\{|\alpha|>YN^{\delta}\},$$

respectively.

It is almost the same of [6] (we take $v=\left[\frac{1}{2\delta}\right]+1$ now) that

$$I_4=o(N^3), \tag{2}$$

$$I_2=o(N^3) \tag{3}$$

and

$$I_1\gg N^3. \tag{4}$$

3 The Estimation of I_3

Let $M=\{\alpha:|f_j(\alpha)|\geqslant N^{1/2+\delta},\ j=1,\ 2\}\cap E_3$. Then

$$\int_{E_3|M} e(\eta\alpha)\Big(\prod_{j=1}^{5} f_j(\alpha)\Big) K_n\Big(\frac{\alpha}{Y}\Big)\,\mathrm{d}\alpha \ll N^{1/2+\delta}\sum_{j=1}^{5}\int_1^{YN^\delta} |f_j|^4\,\mathrm{d}\alpha$$

$$\ll N^{1/2+\delta}\cdot N^2\cdot YN^\delta \ll N^{1/2+1/2+2+2\delta-\varepsilon} = o(N^3).$$

In order to estimate the integral I_3, we need a lemma below.

Lemma. *Let $Y \geqslant 1$, $A \geqslant N^{1/2+\delta}$, $B \geqslant N^{1/2+\delta}$ and the total length of the set*

$$\mathscr{E} = \{\alpha \mid \leqslant \alpha \leqslant Y,\ A \leqslant |f_1(\alpha)| < 2A,\ B \leqslant |f_2(\alpha)| < 2B\}$$

be $m(\mathscr{E})$. Then there is a positive constant C such that

$$m(\mathscr{E}) \ll YN^{2+\delta}(AB)^{-2}q^{-1},\ if\ AB > 2Cq^{1/2}N$$

and

$$m(\mathscr{E}) \ll YN^{4+\delta}(AB)^{-4} + Y^{1/4}q^{3/4}N^{1/2+\delta}(AB)^{-2},\ if\ AB \leqslant 2Cq^{1/2}N.$$

Proof This is essentially the Lemma 1 of [3], only N and η being replaced by N^2 and q.

Now, we subdivide M into several sets

$$\mathscr{E} = M(A, B) = \{\alpha : A \leqslant |f_1(\alpha)| < 2A,\ B \leqslant |f_2(\alpha)| < 2B\} \cap E_3.$$

Obviously the number of $M(A, B)$ is $\ll (\log N)^2 \ll N^{\delta/2}$.

If $AB > 2Cq^{1/2}N$, then

$$\int_{M(A,B)} e_{(\eta\alpha)}\Big(\prod_{j=1}^{5} f_{j(\alpha)}\Big) K_n\Big(\frac{\alpha}{Y}\Big)\,\mathrm{d}\alpha$$

$$\ll \Big(\int_{M(A,B)} |f_1 f_2|^4\,\mathrm{d}\alpha\Big)^{1/4}\Big(\int_{M(A,B)} |f_3|^4\,\mathrm{d}\alpha\Big)^{1/4}\Big(\int_1^{YN^\delta} |f_4|^4\,\mathrm{d}\alpha\Big)^{1/4}\Big(\int_1^{YN^\delta} |f_5|^4\,\mathrm{d}\alpha\Big)^{1/4}$$

$$\ll (AB(m(\mathscr{E}))^{1/4})((m(\mathscr{E})^{1/4}N)NY^{1/2}N^{\delta/2} \ll AB(m(\mathscr{E}))^{1/2}N^{2+\delta/2}Y^{1/2}$$

$$\ll AB(YN^\delta N^{2+\delta}(AB)^{-2}q^{-1})^{1/2}N^{2+\delta/2}Y^{1/2} \ll YN^{3+3\delta/2-1/2} \ll N^{3-\delta}.$$

If $AB \leqslant 2Cq^{1/2}N$, then

$$\int_{M(A,B)} e(\eta\alpha)\Big(\prod_{j=1}^{5} f_j(\alpha)\Big) K_n\Big(\frac{\alpha}{Y}\Big)\,\mathrm{d}\alpha$$

$$\ll \left(\int_{M(A,\,B)} |f_1 f_2|^4 \mathrm{d}\alpha\right)^{1/4} \left(\int_1^{YN^\delta} |f_3| \mathrm{d}\alpha\right)^{1/4} \left(\int_1^{NY^\delta} |f_4|^4 \mathrm{d}\alpha\right)^{1/4} \left(\int_1^{YN^\delta} |f_5|^4 \mathrm{d}\alpha\right)^{1/4}$$

$$\ll (AB(m(\mathscr{E}))^{1/4})(YN^\delta N^2)^{3/4}$$

$$\ll AB(YN^\delta N^{4+\delta}(AB)^{-4} + Y^{1/4}N^{\delta/4}q^{3/4}N^{1/2+\delta}(AB)^{-2})^{1/4}Y^{3/4}N^{3/2+3\delta/4}$$

$$\ll YN^{5/2+5\delta/4} + (AB)^{1/2}Y^{1/16+3/4}q^{3/16}N^{13/8+5\delta/4}$$

$$\ll N^{3-\delta} + Y^{13/16}q^{7/16}N^{17/8+5\delta/4}$$

$$\ll N^{3-\delta} + N^{05/32+5\delta/4-13\varepsilon/16} \ll N^{3-\delta}.$$

Hence

$$I_3 = o(N^3). \tag{5}$$

From (2), (3), (4), (5) we have

$$R(N) \gg Y^{-1}N^3.$$

The theorem follows immediately.

References

[1] Baker, R. C., Cubic Diophantine Inequalities, *Mathematika*, **29** (1982), 201-215.

[2] Baker, R. C. & Harmann, G., Diophantine Approximation by Prime Numbers, *J. Lond. Math. Soc.*, **25**(1982), 201-215.

[3] Baker, R. C. & Harmann, G., Diophantine Inequalities with Mixed Powers, *J. of Number Theory*, **18**(1984), 69-85.

[4] Davenport, H., & Heibronn, H., On Indefinite Quadratic Forms in Five Variables, *J. Lond. Math. Soc.*, **21**(1946), 185-193.

[5] Davenport, H., *Analytic Methods for Diophantine Equations and Diophantine Inequalities*, Ann Arbor publishers, 1962.

[6] Shan Zun, A Diophantine Inequality, *Acta Math. Sinica*, **55**(1987), 598-604.

[7] Vanghan, R. C., *The Hard-Littlewood Method*, Camb, Univ Press, 1981, 144-151.

Z_n 中的 D. F 集

若对集 $S \subset Zn$ 中每个元素 x，均有 $2x \in S$，则称 S 为 $D.\ F(double\text{-}free)$ 集. 设 $f(n)=\max|S|$. 本文给出了计算 $f(n)$的公式.

0　引言

设 G 为一加法群. 集 $S \subset G$，如果对 S 中每个元素 x，均有 $2x \notin S$，则称 S 为 $D.F$ 集(*double-free set*). 文[1]研究了 G 为自然数集 **N** 时，$N_n=\{1, 2, \cdots, n\}$ 中最大的 D. F 集 S 的元数 $f(n)$，给出了递推公式 $f(n)=\left\lceil \frac{n}{2} \right\rceil + f\left\lfloor \frac{n}{4} \right\rfloor$，其中 $\lceil x \rceil$、$\lfloor x \rfloor$分别为天花板函数与地板函数.

本文讨论 G 是以自然数 n 为模的剩余类加群 Z_n 的情况，研究了它的 $D.F$ 集的最大元数(记为 $f(n)$).

令 $f(n)=f(Z_n)=\max\{|S|: S$ 为 Z_n 中的 $D.F$ 集$\}$.

1　计算 f(n) 的公式

定理 1　若 $n=2^{\alpha}$，α 为自然数，则

$$f(n)=\left\lfloor \frac{2n}{3} \right\rfloor, \tag{1}$$

即在 α 为奇数时，

$$f(n)=2^{\alpha-1}+2^{\alpha-3}+\cdots+1=\frac{1}{3}(2^{\alpha+1}-1), \tag{2}$$

在 α 为偶数时，

$$f(n)=2^{\alpha-1}+2^{\alpha-3}+\cdots+2=\frac{1}{3}(2^{\alpha+1}-2). \tag{3}$$

定理 2 若 $n=2m$, m 为奇数,则

$$f(n)=m. \tag{4}$$

定理 3 若 $n=4m$, 则

$$f(n)=2m+f(m). \tag{5}$$

定理 4 若 $n=2^{2\alpha+1}m$, α 为自然数, m 为奇数,则

$$f(n)=\frac{1}{3}(2^{2\alpha+2}-1)m. \tag{6}$$

定理 5 若 $n=2^{2\alpha}m$, α 为自然数,为奇数,则

$$f(n)=\frac{1}{3}(2^{2\alpha+1}-2)m+f(m). \tag{7}$$

定理 6 若 n 为奇数,则

$$f(n)=\sum_{d/n}\frac{\varphi(d)}{f_d}\left\lfloor\frac{f_d}{2}\right\rfloor, \tag{8}$$

即

$$f(n)=\frac{n}{2}-\frac{1}{2}\sum_{d/n}\frac{\varphi(d)}{f_d}\lambda_d. \tag{9}$$

其中 φ 为欧拉函数, f_d 为 2 在 Z_d^* 中的次数(即使 $2^f=1$ 的最小正数),而

$$\lambda_d=\begin{cases}1, & \text{若 } f_d \text{ 为奇数;}\\ 0, & \text{若 } f_d \text{ 为偶数.}\end{cases} \tag{10}$$

2 定理 1—5 的证明

将 Z_n 中的数用点表示,如果点(数)x 与 y 满足 $y=2x$, 则在 x、y 之间连一条线,这样获得一个由 n 个点组成的图.

定理 1 的证明 这时 2 的幂不会为 1,所得的图是树,最上面一层是奇数 1, 3, … $n-1$;然后是“半偶数”,即奇数的两倍;……;第 j 层是形如 $2^{j-1}(2k+1)$ 的数;……;最后是这树的根 0(即 n). 如下图表示 $n=32$ 的情况:

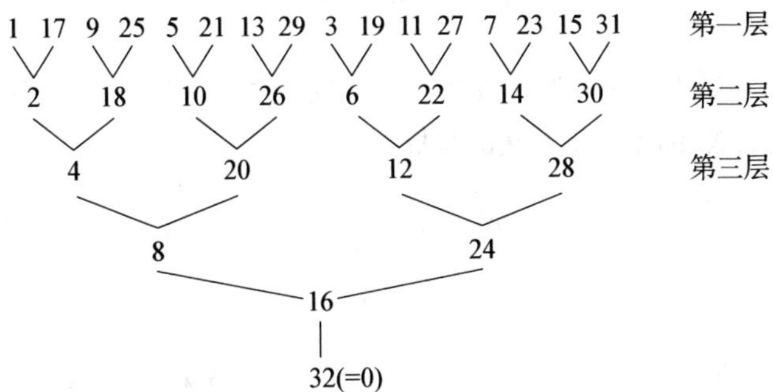

取第1、3、5、…层(注意根是不可以取的,因为$0=2\times0$),这些数显然组成$D.F$集.这集的元数即为(1)((2)或(3))

另一方面,元数最大的$D.F$集总是存在的($D.F$集仅有有限多个).任一$D.F$集,若有第二层的数,均可将它换成两个第一层的、与它相连的数.这时所得的集仍为$D.F$集,但元数增多,因而可以认为这$D.F$集不含第二层的数.类似地,可以处理第4、6、…诸层.所以上面的$D.F$集的元数$\left[\frac{2n}{3}\right]$确为最大.

定理2的证明　全部奇数的集$\{1、3、5、\cdots、2m-1\}$即是Z_n中的$D.F$集(因为$2x$不再是奇数,它的元数为m)

另一方面,若x, y为奇数并且$2x=2y$,则$x=y$或$x=y+m$,由于$y+m$是偶数,后一种情况不可能发生,所以必有$x=y$.即由$\{1、3、5、\cdots、2m-1\}$到Z_n的映射$x\to2x$是单射.形如$(x, 2x)$, x为奇数的数组共m个.每个$D.F$集在每一对$(x, 2x)$中至多取一个,从而$f(n)<m$.综合以上两方面即得定理2.

定理3的证明　这时$\{x, x$为奇数$\}\cup(\{y, 4\mid y\}$中最大的$D.F$集)是Z_n的$D.F$集,它的元数为$2m+f(m)$.

另一方面,最大的$D.F$集一定不含半偶数,含有全部奇数(理由与定理1证明相同,虽然这时的图未必是树),所以(5)式成立.

由定理3及定理1立即得出定理4、5.

3　定理6的证明

现在设n为奇数,n的质因数分解为

$$n = p_1^{\alpha_1} p_2^{\alpha_2} \cdots p_k^{\alpha_k}. \tag{11}$$

其中 p_i 为不同的奇素数，$\alpha_i \in \mathbf{N}$, $i=1, 2, \cdots K$. 用 Z_n^* 表示 Z_n 中与 n 互质的数所成的集(缩系)，则有集的分拆

$$Z_n = \bigcup p_1^{\beta_1} p_2^{\beta_2} \cdots p_k^{\beta_k} Z^*_{p_1^{\alpha_1-\beta_1} p_2^{\alpha_2-\beta_2} \cdots p_k^{\alpha_k-\beta_k}}, \tag{12}$$

其中　$0 \leqslant \beta_i \leqslant \alpha_i$, $i=1, 2, \cdots, k$.

(12)中的各个集 $p_1^{\beta_1} p_2^{\beta_2} \cdots p_k^{\beta_k} Z^*_{p_1^{\alpha_1-\beta_1} p_2^{\alpha_2-\beta_2} \cdots\cdots p_k^{\alpha_k-\beta_k}}$ 互不相交，并且由于 2 与 n 互质，x 与 $2x$ 必属于同一个集 $p_1^{\beta_1} p_2^{\beta_2} \cdots p_k^{\beta_k} Z^* p_1^{\alpha_1-\beta_1} p_2^{\alpha_2-\beta_2} \cdots\cdots p_k^{\alpha_k-\beta_k}$ 中，

显然　$p_1^{\beta_1} p_2^{\beta_2} \cdots\cdots p_k^{\beta_k} Z^*_{p_1^{\alpha_1-\beta_1} p_2^{\alpha_2-\beta_2} \cdots p_k^{\alpha_k-\beta_k}} \cong Z^*_{p_1^{\alpha_1-\beta_1} \cdots p_k^{\alpha_k-\beta_k}}$,

所以　$f(n) = \sum f(Z^*_{p_1^{\alpha_1-\beta_1} p_2^{\alpha_2-\beta_2} \cdots p_k^{\alpha_k-\beta_k}})$. (13)

从而问题化为计算 $f(Z_n^*)$

现在讨论 Z_n^* 中的 D. F 集，仍将 x 与 $2x$ 用线相连构成图.

设 2 在 Z_n^* 中的次数为 f，即 $2^f=1$，并且 f 是使这种等式成立的最小正整数，则这时图由 $e=\dfrac{\varphi(n)}{f}$ 个圈组成(熟知 $f \mid \varphi(n)$)，每个圈的长为 f，在每个圈中可以而且至多可以取 $\left\lfloor \dfrac{f}{2} \right\rfloor$ 个数成 D. F 集，于是

$$f(Z_n^*) = \frac{\varphi(n)}{f}\left\lfloor \frac{f}{2} \right\rfloor = \begin{cases} \dfrac{1}{2}\varphi(n), & \text{若 } f \text{ 为偶数；} \\ \dfrac{1}{2}(\varphi(n)-e), & \text{若 } f \text{ 为奇数.} \end{cases} \tag{14}$$

回到 $f(n)$，由(13)、(14)得

$$f(n) = \sum_{d/n} \frac{\varphi(d)}{f_d} \frac{f_d}{2},$$

其中 f_d 为 2 在 Z_d^* 中的次数

由于 $\sum\limits_{d/n} \varphi(d) = n$，所以(8)可化为

$$f(n) = \sum_{d/n} \frac{\varphi(d)}{f_d}\left(\frac{f_d}{2} - \frac{\lambda_d}{2}\right) = \frac{n}{2} - \frac{1}{2}\sum_{d/n} \frac{\varphi(d)}{f_d}\lambda_d.$$

定理 6 证毕.

注　在利用定理 6(公式(8)或(9))计算 $f(n)$时,需先定出 f_d. 熟知

$$f_{p_1^{r_1}p_2^{r_2}\cdots p_k^{r_k}}=[f_{p_1^{r_1}},\ f_{p_2^{r_2}},\ \cdots f_{p_k^{r_k}}], \tag{15}$$

$$f_{p_1^{r_1+1}}=f_{p_1^{r_1}}\ \text{或}\ p_1 f_{p_1^{r_1}}, \tag{16}$$

所以可以先算出 f_{p_1} 再逐步算出 f_d. 这一过程可利用下面的推理而大大简化.

设 2 对于模 p_1, $p_2\cdots p_s$ 的次数 f_{p_1}、f_{p_2}、$\cdots f_{p_s}$ 为奇数,对于模 p_{s+1}、p_{s+2}、$\cdots p_k$ 的次数 $f_{p_{s+1}}$、$f_{p_{s+2}}$、$\cdots f_{p_k}$ 为偶数,则由于(15)、(16),在 $(d,\ p_{s+1}\cdots p_k)>1$ 时, f_d 为偶数, $\lambda_d=0$,在$(d,\ p_{s+1},\ p_{s+2}\cdots p_k)=1$ 时, f_d 为奇数, $\lambda_d=1$, 从而(9)可化为

$$\begin{aligned} f(n)&=\frac{n-1}{2}-\frac{1}{2}\sum_{\substack{d/n\\ d>1}}\frac{\varphi(d)}{f_d}\lambda_d \\ &=\frac{n-1}{2}-\frac{1}{2}\sum_{\substack{1\leqslant d-p_1^{r_1}p_2^{r_2}\cdots p_s^{r_s}\\ 0\leqslant r_i\leqslant \alpha_i\\ (i=1,\ 2,\ \cdots s)}}\frac{\varphi(d)}{f_d}. \end{aligned} \tag{9'}$$

定出 f_{p_1}、f_{p_2}、$\cdots f_{p_s}$ 后,可定出

$$f_{p_i^2}=\begin{cases} f_{p_i},\ \text{若}\ p_1^2 \mid 2^{f_{p_i}}-1; \\ p_i f_{p_i},\ \text{若}\ p_i^2 \nmid 2^{f_{p_i}}-1. \end{cases} \tag{17}$$

然后利用定理:若 $p_i^i \,/\!/\, 2^{f_{p_1^2}}-1$, 则

$$f_{p_i r_i}=\begin{cases} f_{p_i^2}, & \text{若}\ 2\leqslant r_i\leqslant i; \\ p_i^{r_i-1}f_i^{p_2}, & \text{若}\ r_i>i. \end{cases} \tag{18}$$

便可定出一切 $f_{p_i}r_i$, $(r=1、2、3、\cdots s)$ 及 $f_d=f_{p_1^{r_1}p_2^{r_2}\cdots p_s^{r_s}}$.

参考文献

[1] Edward T. H. Wanh. On Double-Free Sets of Integers. *Discrete Math*. (to appear).

[2] Wallis W. D, Anne Penfold Street. Jennifeer Seberry Wallis, Combinatorics: Room-Squares, Sum-Free Sets, Hadamard Matrices; *Lecture Notes in Mathematics*, Vol. 292, Springr-Verlag, Berlin-Heidelberg-New York (1972).

On Double-free Sets of Zn

Abstract

Let G be a group. A set $S \subset G$ is called double-free (D. F) if $x \in s$ implies $2x \overline{\in} s$. Let $f(n) = f(Zn) = \max\{|s|: S$ is D. F of $Zn\}$ Some formulas for $f(n)$ are given in this paper.

Mutual Multiples in $\mathbb{Z}_n$

The following problem appeared in [1]:

Problem: *Let R be a commutative ring with unit element 1. Prove or disprove: If a, $b \in R$ are multiples of one another, then they are* unit *multiples of one another; that is, there is an invertible element $u \in R$ such that $a = ub$.*

The given statement is false for a general commutative ring R (see [3]). We show here, however, that it is true for $\mathbb{Z}_n$.

In what follows we will use ϕ to denote the Euler phi-function; $\tau(n)$ will denote the number of positive divisors of n. The following definition will be convenient:

Definition. Let $\mathbb{Z}_n$ be the commutative ring of integers modulo n, where $n > 1$ is a given natural number. Two elements a and b of $\mathbb{Z}_n$, not necessarily distinct, are said to form an MM pair (mutual multiple pair) if there exist i, $j \in \mathbb{Z}_n$ such that $a = ib$ and $b = ja$ in $\mathbb{Z}_n$; that is, $a \equiv ib \pmod{n}$ and $b \equiv ja \pmod{n}$. In this case, i and j are called *multipliers*.

Note that 0 cannot form an MM pair with any element of $\mathbb{Z}_n$ but itself. Also, if $n = p$ is a prime, then clearly any two non-zero elements of Z_p form an MM pair, and the multipliers i and j are both unique.

Following is a more meaty example.

Example: In $\mathbb{Z}_6$, the numbers 2 and 4 form an MM pair since $4 = 2 \times 2$ and $2 = 5 \times 4$. Similarly, 1 and 5 form an MM pair since $5 = 5 \times 1$ and $1 = 5 \times 5$. On the other hand, 3 and 4 do not form an MM pair since $3j \equiv 0$ or $3 \pmod{6}$ depending on whether j is even or odd.

Observe that when a and b form an MM pair, the multipliers i and j need not be unique in general even if $a \neq 0$ and $b \neq 0$; e. g., in $\mathbb{Z}_6$ we could also write

$4 = 5 \times 2$ and/or $2 = 2 \times 4$.

Lemma 1. *Let a, $b \in \mathbb{Z}_n$. Then a and b form an MM pair if and only if* $\gcd(a, n) = \gcd(b, n)$.

Proof. Suppose a and b form an MM pair. Then there exist i, $j \in \mathbb{Z}_n$ such that $a = ib$ and $b = ja$. Since $\gcd(a, n) \mid a$ implies $\gcd(a, n) \mid ja$, we have $\gcd(a, n) \mid b$ and so $\gcd(a, n) \mid \gcd(b, n)$. Similarly, $\gcd(b, n) \mid \gcd(a, n)$ and thus $\gcd(a, n) = \gcd(b, n)$.

Conversely, suppose $\gcd(a, n) = \gcd(b, n) = d$. Let $a = da'$ and $b = db'$. Then either $a = b = 0$ or $\gcd(a', n) = \gcd(b', n) = 1$. Since $\gcd(a', n/d) = \gcd(b', n/d) = 1$ there exist i, $j \in \mathbb{Z}_n$ such that $a' \equiv b'i$ and $b' \equiv a'j \pmod{n/d}$. Hence $a \equiv ib$ and $b \equiv ja \pmod{n}$. This completes the proof.

Theorem 1. *Suppose a, $b \in \mathbb{Z}_n$ form an MM pair. Then there exists an invertible element $u \in \mathbb{Z}_n$ such that $a = ub$.*

Proof. As in the proof of the lemma, let $\gcd(a, n) = \gcd(b, n) = d$, $a = da'$, and $b = db'$.

Then there exists $i \in \mathbb{Z}_n$ such that $a' \equiv b'i \pmod{n/d}$. Clearly $\gcd(i, n/d) = 1$ as $\gcd(a', n/d) = 1$. By the celebrated theorem of Dirichlet, there are infinitely many primes in the sequence $\{i + k(n/d)\}_{k=0}^{z}$, and hence, *a fortiori*, there are primes in this sequence that exceed n. Thus there exists $k_0 \in \mathbb{N}$ for which $i + k_0(n/d)$ is such a prime, and so $\gcd(i + k_0(n/d), n) = 1$. If we let u denote the least positive residue of $i + k_0(n/d)$ modulo n, then $n \in \mathbb{Z}_n$ is such that

$$\gcd(u, n) = 1 \text{ and } ub' \equiv ib' \equiv a' \pmod{n/d};$$

it follows that $a \equiv ub \pmod{n}$, which completes the proof.

Remark 1. The key to the preceding proof is the existence of an integer in the sequence $\{i + k(n/d)\}_{k=0}^{z}$ that is coprime with n. This result, which is a consequence of Dirichlet's theorem, appeared in [4, p. 12, Ex. 3] with an elementary proof.

As we explored MM pairs in $\mathbb{Z}_n$, we were led to wonder how many there are. We found the following answer:

Theorem 2. *Let $f(n)$ denote the number of unordered MM pairs in $\mathbb{Z}_n$. Then $f(n)=\frac{1}{2}\left[n+\sum_{d\mid n}\phi(d)^2\right]$; the summation is over all positive divisors d of n.*

Proof. For each divisor d of n and for any $a \in \mathbb{Z}_n$, note that $\gcd(a, n)=d$ if and only if $a=da'$ for some $a' \in \mathbb{Z}_{n/d}$ such that $\gcd(a', n/d)=1$. Hence if we let $Z^*_{n/d}=\{m \in \mathbb{Z}_{n/d} \mid \gcd(m, n/d)=1\}$, then, by Lemma 1, any two elements of $\mathbb{Z}^*_{n/d}$ would form an MM pair and no elements of $\mathbb{Z}^*_{n/d}$ can form an MM pair with elements not in the set. Since $|\mathbb{Z}^*_{n/d}|=\phi(n/d)$, we have

$$f(n)=\sum_{d\mid n}\left[\phi(n/d)+\binom{\phi(n/d)}{2}\right]=\sum_{d\mid n}\left[\phi(d)+\binom{\phi(d)}{2}\right]$$

$$=\frac{1}{2}\sum_{d\mid n}[\phi(d)+\phi(d)^2]=\frac{1}{2}\left[n+\sum_{d\mid n}\phi(d)^2\right],$$

where the last equality holds because $\sum_{d\mid n}\phi(d)=n$ (see, e. g., [2, Thm. 6.7, p. 212]).

Remark 2. Since there is no known closed form expression for $\sum_{d\mid n}\phi(d)^2$, the only way to find the exact value of $f(n)$ is to compute $\phi(d)$ for all divisors d of n. A corollary, however, gives a lower bound for $f(n)$.

Corollary: $f(n)\geqslant\frac{n}{2}\left(\frac{n}{\tau(n)}+1\right)$.

Proof. By the Cauchy-Schwarz inequality,

$$\sum_{d\mid n}\phi(d)^2\sum_{d\mid n}1^2\geqslant\left(\sum_{d\mid n}\phi(d)\right)^2=n^2,$$

so $\sum_{d\mid n}\phi(d)^2\geqslant n^2/\tau(n)$. Substituting this into the formula from Theorem 2 completes the proof.

若干数论问题的注记

本文主要对几个与同余类有关的命题给出新的简化证明.

1

张树生在[1]中证明了:

定理 对一切素数 $p \geqslant 5$,

$$\sum_{k=1}^{p-1} \frac{1}{tp+k} \equiv 0 (\bmod p^2), \tag{1}$$

其中 t 为任一整数.

他认为这个定理推广了华罗庚《数论导引》第二章 § 10 的 Wolstenholme 定理.

张的证明比较复杂. 其实华罗庚书中的证法稍作修改即可得出(1):

命

$$\sum_{k=1}^{p-1} (x-(tp+k)) = x^{p-1} - s_1 x^{p-2} + \cdots + s_{p-1}, \tag{2}$$

因

$$\prod_{k=1}^{p-1} (x-(tp+k)) \equiv x^{p-1} - 1 (\bmod p),$$

所以

$$p \mid (s_1, s_2, \cdots, s_{p-2}). \tag{3}$$

在(2)中令 $x=(2t+1)p$, 则

$$\prod_{k=1}^{p-1} (tp+p-k) = ((2t+1)p)^{p-1} - s_1((2t+1)p)^{p-2} + \cdots + s_{p-1},$$

即

$$0=((2t+1)p)^{p-2}-s_1((2t+1)p)^{p-3}+\cdots-s_{p-2}.$$

结合(3)得

$$s_{p-2}\equiv 0(\bmod p^2),$$

即

$$p^2\left|\left(\prod_{k=1}^{p-1}(tp+k)\cdot\sum_{k=1}^{p-1}\frac{1}{tp+k}\right)\right.,$$

亦即(1)式成立.

2

设 $S_{r(n,d)}$ 表示 $\bmod n$ 的缩系中指数为 d 的元素的 r 次方幂和. 1952 年 R. Moller[2] 将 Gauss 等人的古典结果[3],[4]

$$S_{1(p,p-1)}\equiv\mu(p-1)(\bmod p),$$
$$S_1(p,d)\equiv\mu(d)(\bmod p),$$

推广为

$$S_{r(p,d)}\equiv\frac{\varphi(d)}{\varphi(d_1)}\mu(d_1)(\bmod p),\tag{4}$$

其中 $d_1=\dfrac{d}{(r,d)}$.

1980 年, H. Gupta[5] 给出(4)的一个简化证明. 1987 年方玉光[6] 又将(4)推广为

$$S_r(p^\alpha,d)\equiv\frac{\varphi(d)}{\varphi(l_0)}\mu(l_0)(\bmod p^\alpha),\tag{5}$$

其中 l_0 定义如下:

$$d_1=\frac{d}{(r,d)}=p^m l_0,\ p\nmid l_0.\tag{6}$$

在这里我们给出(5)的一个简短的新证明.

引理　设 g 为 $\bmod p^\alpha$ 的原根,则对 $0\leqslant m<\alpha$,

$$\sum_{k=1}^{p^m} g^{\varphi(p^\alpha)rh/p^m} \equiv p^m (\bmod p^\alpha). \tag{7}$$

证明 对 α 进行归纳.

$\alpha=1$ 时, $m=0$. (7)即 Fermat 小定理.

假设命题对于 $\alpha-1$ ($\geqslant 1$) 成立. 则对 $0\leqslant m<\alpha$,

$$\begin{aligned}\sum_{k=1}^{p^m} g^{\varphi(p^\alpha)rh/p^\alpha} &= \sum_{k=1}^{p^m} g^{\varphi(p^{\alpha-1})rh/p^{m-1}} = \sum_{k=0}^{p-1}\sum_{k=1}^{p^{m-1}} g^{\varphi(p^{\alpha-1})r(h+kp^{m-1})/p^{m-1}} \\ &= \sum_{k=1}^{p-1} g^{\varphi(p^{m-1})rh/p^{m-1}} \cdot \sum_{k=0}^{p-1} g^{\varphi(p^{m-1})rk}.\end{aligned} \tag{8}$$

由归纳假设,(8)的前一个因子为 $p^{m-1}+a\cdot p^{\alpha-1}$. 由 Euler 定理,

$$g^{\varphi(p^{\alpha-1})}=1+bp^{\alpha-1},$$

所以

$$\begin{aligned}\sum_{k=1}^{p^m} g^{\varphi(p^\alpha)rh/p^m} &= (p^{m-1}+ap^{\alpha-1})\cdot\sum_{k=0}^{p-1}(1+bp^{\alpha-1})^{rk} \\ &= (p^{m-1}+ap^{\alpha-1})\sum_{k=0}^{p-1}(1+brkp^{\alpha-1}+\cdots) \\ &= (p^{m-1}+ap^{\alpha-1})\left(p+br\cdot p^{\alpha-1}\cdot\frac{p(p-1)}{2}+\cdots\right) \\ &\equiv p^m (\bmod p^\alpha),\end{aligned}$$

因此引理成立.

设 $r_1=\dfrac{r}{(d,r)}$,由于 $h\equiv h'(\bmod d_1)$ 时,

$$g^{\frac{\varphi(p^\alpha)r_1h}{d_1}} \equiv g^{\frac{\varphi(p^\alpha)r_1h'}{d_1}} (\bmod p^\alpha),$$

所以

$$\begin{aligned}S_r(p\alpha, d) &= \sum_{h(\bmod d)}{}^{*} g^{\frac{\varphi(p^\alpha)rh}{d}} = \sum_{h(\bmod d)}{}^{*} g^{\frac{\varphi(p^\alpha)r_1h}{d_1}} \\ &\equiv \frac{\varphi(d)}{\varphi(d_1)}\sum_{h(\bmod d_1)}{}^{*} g^{\frac{\varphi(p^\alpha)r_1h}{d_1}} (\bmod p^\alpha),\end{aligned} \tag{9}$$

而

$$\sum_{h(\bmod d_1)}{}^{*} g^{\frac{\varphi(p^\alpha)r_1h}{d_1}}=\sum_{h(\bmod d_1)} g^{\frac{\varphi(p^\alpha)r_1h}{d_1}}\sum_{\substack{s|h\\ s|d_1}}\mu(s)=\sum_{s|d_1}\mu(s)\sum_{h\left(\bmod\frac{d_1}{s}\right)} g^{\frac{\varphi(p^\alpha)r_1hs}{d_1}}$$

$$=\sum_{s|p^m}\mu(s)\sum_{t|l_0}\mu(t)\sum_{h\left(\bmod\frac{p^ml_0}{st}\right)} g^{\frac{\varphi(p^\alpha)r_1hst}{p^ml_0}}$$

$$=\sum_{s|m}\mu(s)\sum_{t|l_0}\mu(t)\sum_{h\left(\bmod\frac{p^ml_0}{st}\right)} g^{\frac{\varphi(p^\alpha)r_1hst}{p^ml_0}},$$

其中最里面的和,在 $t\neq l_0$ 时,值为

$$\frac{g^{\varphi(p^\alpha)r_1\cdot\frac{st}{p^ml_0}\cdot\frac{p^ml_0}{st}}-1}{g^{\varphi(p^\alpha)r_1\cdot\frac{st}{p^ml_0}-1}}\equiv 0(\bmod p^\alpha),$$

在 $t=l_0$ 时,值为

$$\sum_{h\left(\bmod\frac{p^m}{s}\right)} g^{\varphi(p^m)r_1hs/p^m}.$$

因此

$$\sum_{h(\bmod d_1)}{}^{*} g^{\varphi(p^\alpha)r_1h/d_1}=\sum_{s|p}\mu(s)\mu(l_0)\sum_{h\left(\bmod\frac{p^m}{s}\right)} g^{\varphi(p^\alpha)r_1hs/p^m}$$

$$=\mu(l_0)\Big(\sum_{h(\bmod p^m)} g^{\varphi(p^\alpha)r_1h/p^m}-\sum_{h(\bmod p^{m-1})} g^{\varphi(p^\alpha)r_1h/p^{m-1}}\Big)$$

$$\equiv\mu(l_0)(p^m-p^{m-1})(\bmod p^\alpha)$$

$$\equiv\mu(l_0)\cdot\frac{\varphi(d_1)}{\varphi(l_0)}(\bmod p^\alpha).\tag{10}$$

由(9)、(10),

$$S_r(p^\alpha,d)\equiv\frac{\varphi(d)}{\varphi(d_1)}\cdot\mu(l_0)\cdot\frac{\varphi(d_1)}{\varphi(l_0)}\equiv\frac{\varphi(d)}{\varphi(l_0)}\mu(l_0)(\bmod p^\alpha).$$

当 $m=0$ 时,由 $d_1=l_0$ 及上面推导过程容易看出(5)式仍然成立.

3

设 $\mathbf{M}$ 为某代数数域中的整理想，$\mathbf{M}\neq 0$、1，$\mathbf{M}\nmid 2$，

$$\alpha_1, \alpha_2, \cdots \alpha_n \tag{11}$$

与

$$\beta_1, \beta_2, \cdots, \beta_n \tag{12}$$

为 $\bmod \mathbf{M}$ 的两个完全剩余系，其中 $n=N(\mathbf{M})$. 旷京华等[7]曾证明积

$$\alpha_1\beta_1, \alpha_2\beta_2, \cdots, \alpha_n\beta_n \tag{13}$$

不是 $\bmod \mathbf{M}$ 的完系. 我们在这里另给一个证明：

若(13)是 $\bmod \mathbf{M}$ 的完系，不妨设前面 $\varphi(\mathbf{M})=t$ 个是 $\bmod \mathbf{M}$ 的缩系，这时 $\alpha_1, \alpha_2, \cdots, \alpha_t$ 与 $\beta_1, \beta_2, \cdots, \beta_t$ 也必然是 $\bmod \mathbf{M}$ 的缩系.

设素理想 $N|\mathbf{M}$，若 $(\alpha_i\beta_i, \mathbf{M})=\mathbf{P}$，则由于 α_i，β_i 均不在 $\bmod \mathbf{M}$ 的缩系中，必有 $\mathbf{P}|\alpha_i$，$\mathbf{P}|\beta_i$ 并且 $\mathbf{P}\parallel \mathbf{M}$，于是可设 $\mathbf{M}=\mathbf{p}_1\mathbf{p}_2\cdots\mathbf{p}_m$，其中 $\mathbf{p}_i\,(1\leqslant i\leqslant m)$ 为不同的素理想.

用归纳法易知当且仅当 $(\alpha_i, \mathbf{M})=(\beta_i, \mathbf{M})=\mathbf{p}_{j_1}\mathbf{p}_{j_2}\cdots\mathbf{p}_{j_m}$ 时，$(\alpha_i\beta_i, \mathbf{M})=\mathbf{p}_{i_1}\mathbf{p}_{i_2}\cdots\mathbf{p}_{i_m}$，并且这样的 i 恰有 $\varphi\left(\dfrac{\mathbf{M}}{\mathbf{p}_{j_1}\mathbf{p}_{j_2}\cdots\mathbf{p}_{j_m}}\right)$ 个.

特别地，设 $\alpha_{i_1}, \alpha_{i_2}, \cdots, \alpha_{i_k}$ 及 $\beta_{i_1}, \beta_{i_2}, \cdots, \beta_{i_k}$ 满足 $(\alpha_{i_1}, \mathbf{M})=\cdots=(\alpha_{i_k}, \mathbf{M})=(\beta_{i_1}, \mathbf{M})=\cdots=(\beta_{i_k}, \mathbf{M})=\mathbf{p}_1\mathbf{p}_2\cdots\mathbf{p}_{m-1}$，其中 $k=\varphi(\mathbf{p}_m)$，则 $(\alpha_{i_1}\beta_{i_1}, \mathbf{M})=\cdots=(\alpha_{i_k}\beta_{i_k}, \mathbf{M})=\mathbf{p}_1\mathbf{p}_2\cdots\mathbf{p}_{m-1}$.

由于 $\alpha_{i_1}, \cdots, \alpha_{i_k}$ 是 $\bmod \mathbf{M}$ 的不同的剩余类，所以 $\alpha_{i_1}, \cdots \alpha_{i_k}$ 是 $\bmod \mathbf{p}_m$ 的不同的剩余类，同样 $\beta_{i_1}, \cdots \beta_{i_k}$ 及 $\alpha_{i_1}\beta_{i_1}, \cdots \alpha_{i_k}\beta_{i_k}$ 也都是 $\bmod \mathbf{p}_m$ 的不同剩余类，于是它们均为 $\bmod \mathbf{p}_m$ 的缩系. 由 Wilson 定理，

$$\alpha_{i_1}\alpha_{i_2}\cdots\alpha_{i_k}\equiv -1(\bmod \mathbf{p}_m),$$
$$\beta_{i_1}\beta_{i_2}\cdots\beta_{i_k}\equiv -1(\bmod \mathbf{p}_m),$$
$$(\alpha_{i_1}\beta_{i_1})(\alpha_{i_2}\beta_{i_2})\cdots(\alpha_{i_k}\beta_{i_k})\equiv -1(\bmod \mathbf{p}_m),$$

前两个式子相乘所得结果与第三个式子矛盾，这表明(13)不是 $\bmod \mathbf{M}$ 的完系.

4

旷京华[7]证明了以下定理.

定理 设 $\mathbf{A}$ 为代数数域 K 中的理想, $\mathbf{B}=(2, \mathbf{A})$,并且 $N(\mathbf{B})=2$. 若 $n=N(\mathbf{A})$,

$$\alpha_1, \alpha_2, \cdots, \alpha_n \tag{14}$$

与

$$\beta_1, \beta_2, \cdots, \beta_n \tag{15}$$

是 mod$\mathbf{A}$ 的两组完全剩余系,则

$$\alpha_1+\beta_1, \alpha_2+\beta_2, \cdots, \alpha_n+\beta_n \tag{16}$$

不是 mod$\mathbf{A}$ 的完全剩余系.

这定理显然是从 $K=Q$ 时相应的命题推广而来. 我们的证明如下:

因为 $N(\mathbf{B})=2$, 所以 $\mathbf{B}$ 是素理想,设

$$\mathbf{A}=\mathbf{B}^a\mathbf{N},$$
$$(2)=\mathbf{B}^b\mathbf{M},$$

其中 $\mathbf{M}$、$\mathbf{N}$、$\mathbf{B}$ 两两互素; $a\geqslant 1$, $b\geqslant 1$ 并且至少有一个为 1.

由于 $-\alpha_1, -\alpha_2, \cdots, -\alpha_n$ 也是 mod$\mathbf{A}$ 的完系,所以

$$\sum(\alpha_i+\beta_i)\equiv\sum(\alpha_i+(-\alpha_i))\equiv 0 \pmod{\mathbf{A}}. \tag{17}$$

另一方面,对 mod$\mathbf{A}$ 的任一完系(14),考虑和 $\sum\alpha_i \pmod{\mathbf{A}}$, 将其中形如 α_i 与 $\alpha'_i\equiv-\alpha_i \pmod{\mathbf{A}}$ 的两项互相抵消,只剩下满足

$$\alpha_i\equiv-\alpha_i \pmod{\mathbf{A}} \tag{18}$$

的那些 α_i.

(18)即 $\mathbf{B}^a\mathbf{N}\mid(2\alpha_i)$,从而 $\mathbf{B}^a\mathbf{N}\mid\mathbf{B}^b\mathbf{M}(\alpha_i)$. 由于 $\mathbf{M}$、$\mathbf{N}$、$\mathbf{B}$ 两两互素,不论 $a=1$ 或 $b=1$, 均有

$$\mathbf{B}^{a-1}\mathbf{N} \mid (\alpha_i).$$

由于 $N(\mathbf{A}) \div N(\mathbf{B}^{a-1}\mathbf{N}) = N(\mathbf{B}) = 2$，所以恰有两个 α_i 满足(18). 其中一个当然是 $\alpha_i \equiv 0(\mathrm{mod}\mathbf{A})$，另一个 $\alpha_i \not\equiv 0(\mathrm{mod}\mathbf{A})$，这样，对 mod$\mathbf{A}$ 的任一完系(14)，

$$\sum \alpha_i \not\equiv 0(\mathrm{mod}\mathbf{A}).$$

因而，由(17)即知结论成立.

参考文献

[1] 张树生. 一个数论定理的推广. 数学的实践与认识，1989；1：86－91.

[2] Moller. R. Sums of powers of number having a Given Exponent Modular a Prime. *Amer Monthly*. 1952；59：180－182.

[3] Gauss C. *Disquisitions Arithmeticae arts*. 80－81.

[4] Stern M. Bernerkungen über Hohere Arithmetrik *Journal für Mathmetick*. 1830；6：180－185.

[5] Gupta H. *Selected Topics in Namber Theory*. ABACUS Press，1980：56－57.

[6] 方玉光. 再论某一类指数幂和的同余问题. 曲阜师范大学学报，1987；72－75.

[7] 旷京华，万大庆. 关于覆盖剩余类的注记. 数论研究与评论，1984；4:1.

关于哥德巴赫猜想、孪生素数猜想和余新河猜想的若干新结果

本文给出了我们所得到的关于哥德巴赫猜想、孪生素数猜想和余新河猜想的若干结果. 详细证明将另文发表.

1　大偶数表为 1 个素数与 1 个殆素数之和：素数属于某个等差数列

设 N 为充分大的偶数，p 为素数，P_r 为至多含 r 个素因子(重复者依重数计)的殆素数. 置

$$C_N=\prod_{p>2}(1-(p-1)^{-2})\prod_{2<p\mid N}(p-1)(p-2)^{-1}.$$

1973 年,陈景润证明了关于哥德巴赫猜想的著名定理[1]：

$$\#\{p:p<N,\ p+P_2=N\}>0.67C_N N\ln^{-2}N. \tag{1}$$

1990 年我们证明了[2~3]

$$\begin{aligned}&\#\{p:p<N,N-p=p_1\cdots p_{r-1}\text{ 或 }p_1\cdots p_r,\ p_1<\cdots<p_r\}\\&>\frac{0.77}{(r-2)!}C_N N\ln^{-2}N(\ln\ln N)^{r-2},\ \forall r\geqslant 3.\end{aligned} \tag{2}$$

由式(2)可得

$$\#\{p:p<N,\ p+P_r=N\}>\frac{0.77}{(r-2)!}C_N N\ln^{-2}N(\ln\ln N)^{r-2},\ \forall r\geqslant 3. \tag{3}$$

这里研究在式(1)～(3)中素数 p 取自某个等差数列的情况. 设 a 为自然数，b 为整数，φ 表示 Euler 函数. 对关于偶数的哥德巴赫猜想,我们得到下述结果.

定理 1　设 δ 满足 $0<\delta<1$，$\forall r\geqslant 3$，有

$$\#\{p: p < N,\ p \equiv b(a),\ (a,b)=1,\ (N-b,a)=1,\ N-p=p_1\cdots p_{r-1}$$
$$\text{或 } p_1\cdots p_r,\ p_r > \cdots > p_1 > \exp(\ln^4 N)\}$$
$$> \frac{0.77(1-\delta)^{r-2}}{(r-2)!\ \varphi(a)} \prod_{p\mid a,\ p\nmid N} \frac{p-1}{p-2} C_N N \ln^{-2} N (\ln\ln N)^{r-2}.$$

定理 2 （本定理是陈氏定理的深化或精细化）

$$\#\{p: p < N,\ p \equiv b(a),\ (a,b)=1,\ (N-b,a)=1,$$
$$N-p=p_1 \text{ 或 } p_1p_2,\ p_2 > p_1\} > \frac{0.77}{\varphi(a)} \prod_{p\mid a,\ p\nmid N} \frac{p-1}{p-2} C_N N \ln^{-2} N.$$

定理 3 $\forall r \geqslant 2$，有

$$\#\{p: p < N,\ p \equiv b(a),\ (a,b)=1,\ (N-b,a)=1,\ N-p=p_1\cdots p_{r-1} \text{ 或}$$
$$p_1\cdots p_r,\ p_r > \cdots > p_1\} > \frac{0.77}{(r-2)!\ \varphi(a)} \prod_{p\mid a,\ p\nmid N} \frac{p-1}{p-2} C_N N \ln^{-2} N (\ln\ln N)^{r-2}.$$

定理 4 $\forall r \geqslant 2$，有

$$\#\{p: p < N,\ p \equiv b(a),\ (a,b)=1,\ (N-b,a)=1,\ p+P_r=N\}$$
$$> \frac{0.77}{(r-2)!\ \varphi(a)} \prod_{p\mid a,\ p\nmid N} \frac{p-1}{p-2} C_N N \ln^{-2} N (\ln\ln N)^{r-2}.$$

关于孪生素数猜想，我们的方法可导致类似的结果. 设 x 为充分大的正数，h 为偶数，

$$C_1 = \prod_{p>2} (1-(p-1)^{-2}) \prod_{2<p\mid h} (p-1)(p-2)^{-1}.$$

定理 1. 设 δ 满足 $0 < \delta < 1$，$\forall r \geqslant 3$，有

$$\#\{p: p < x,\ p \equiv b(a),\ (a,b)=1,\ (b+h,a)=1,\ p+h=p_1\cdots p_{r-1}$$
$$\text{或 } p_1\cdots p_r,\ p_r > \cdots > p_1 > \exp(\ln^{\delta} x)\}$$
$$> \frac{0.77(1-\delta)^{r-2}}{(r-2)!\ \varphi(a)} \prod_{p\mid a,\ p\nmid h} \frac{p-1}{p-2} C_1 x \ln^{-2} x (\ln\ln x)^{r-2}$$

定理 2. （本定理为“1+2”型定理的加强或精细化）

$$\#\{p: p < x,\ p \equiv b(a),\ (a,b)=1,\ (b+h,a)=1,\ p+h=p_1 \text{ 或}$$
$$p_1p_2,\ p_2 > p_1\} > \frac{0.77}{\varphi(a)} \prod_{p\mid a,\ p\nmid h} \frac{p-1}{p-2} C_1 x \ln^{-2} x.$$

定理 3. $\forall r \geqslant 2$, 有

$$\#\{p: p < x,\ p \equiv b(a),\ (a,\ b)=1,\ (b+h,\ a)=1,\ p+h = p_1 \cdots p_{r-1} \text{ 或 } p_1 \cdots p_r,\ p_r > \cdots > p_1\}$$

$$> \frac{0.77}{(r-2)!\ \varphi(a)} \prod_{p|a,\ p\nmid h} \frac{p-1}{p-2} C_1 x \ln^{-2} x (\ln \ln x)^{r-2}.$$

定理 4. $\forall r \geqslant 2$, 有

$$\#\{p: p < x,\ p \equiv b(a),\ (a,\ b)=1,\ (b+h,\ a)=1,\ p+h = P_r$$

$$> \frac{0.77}{(r-2)!\ \varphi(a)} \prod_{p|a,\ p\nmid h} \frac{p-1}{p-2} C_1 x \ln^{-2} x (\ln \ln x)^{r-2}.$$

2　余新河猜想对几乎所有的自然数成立

1993 年余新河提出了一个与偶数哥德巴赫猜想密切相关的猜想[4]. 该猜想可重新叙述如下：令 $i \in \{1,\ 7,\ 11,\ 13,\ 17,\ 19,\ 23,\ 29\}$, $A_i = \{n:\ 30n - i$ 为素数$\}$, 则 $A_i + A_j = N - \{1\}$, 即每个大于 1 的自然数均可表为一个 A_i 中的数与一个 A_j 中的数之和.

更一般地, 设 a 为自然数, b_1 和 b_2 为整数, $(a,\ b_1) = (a,\ b_2) = 1$, 则由 Dirichlet 定理, 存在无穷多个形如 $an + b$, $(t = 1,\ 2)$ 的素数. 我们猜测：所有使得 $an + b_1 + b_2$ 为偶数的自然数 n, 均可写成 $n = n_1 + n_2$, 而 $an_1 + b_1$, $an_2 + b_2$ 为素数；或等价地说, 所有满足 $m \equiv b_1 + b_2 (\bmod a)$ 的正偶数 m, 均可表为 2 个分别属于等差数列 $\{an + b_1\}$ 和 $\{an + b_2\}$ 的素数之和. 我们证明了这一猜测对几乎所有满足上述要求的数成立. 确切地说, 有下述定理.

定理 5　设 a 为自然数, b_1 和 b_2 为整数,

$$(a,\ b_1) = (a,\ b_2) = 1,$$

$\mathscr{P}_i = \{p: p$ 为素数且 $p \equiv b_i(a)\}(i = 1,\ 2)$, x 为大的正实数, m 为正偶数, $m \equiv b_1 + b_2(a)$.

$$E(x) = \{m: m \leqslant x \text{ 且 } m \notin \mathscr{P}_1 + \mathscr{P}_2\}.$$

则对任意正数 A,

$$E(x) \ll x\ln^{-A}x,$$

并且当 $m \notin E(x)$ 时，$m \leqslant x$ 可表成 $p_1 + p_2$，$p_i \in p_{\mathscr{P}_i}$ 的方法数为

$$\sim \frac{[a, m]}{\varphi(a)\varphi([a, m])} \prod_{p \nmid am} (1-(p-1)^{-2}) \frac{m}{\ln^2 m}.$$

注 (1) 当 $a=1, 2$ 时，上述猜测即为哥德巴赫猜想，而当 $a=30$ 时，即为余新河猜想.

(2) 由第1和第2节的定理可以看出，对于余新河猜想[实际上对于更普遍(即素数或殆素数限制在等差数列中)的猜测]，我们已经得到了与偶数哥德巴赫猜想的目前结果一样好的结论.

3 三素数定理的加强

1937年 I. M. Vinogradov 证明了关于奇数哥德巴赫猜想的著名结果，即

三素数定理[5] 每个充分大的奇数均可表为三个素数之和.

如果这三个素数被限制于等差数列中，结论是否仍然成立？对这一问题的回答是肯定的. 我们证明了下述定理.

定理6 每个充分大的奇数 $m \equiv b_1 + b_2 + b_3 (a)$，可以表成三个素数 $p_i (i=1, 2, 3)$ 之和，这三个素数分别取自等差数列 $\{an + b_i\} (i=1, 2, 3)$，其中 a 和 b_i 为整数，$a > 1$，且 $(a, b_i) = 1 (i=1, 2, 3)$. 关于表示法的个数有下述渐近公式：

$$T(m) = \frac{1}{2} S_a(m) \frac{m^2}{\ln^3 m} + O\left(\frac{m^2}{\ln^4 m}\right)$$

式中，

$$S_a(m) = \frac{a}{\varphi^3(a)} \prod_{p \nmid am} (1+(p-1)^{-3}) \prod_{p \nmid n、p \mid m} (1-(p-1)^{-2}).$$

参考文献

[1] 陈景润. 大偶数表为一个素数及一个不超过两个素数的乘积之和. 中国科学, 1973, 16(2): 111 - 128.

[2] 阚家海. 关于序列 $p+h$ 与方程 $N-p=P_r$ 的解数. 科学通报, 1990, 35(7): 558.

[3] 阚家海. On the Number of Solutions of $N-p=P_r$, *Jour reine angew Math*. 1991, 414: 117-130.

[4] 余新河. 余新河数学题. 人民日报, 1993 年 4 月 6 日, 第 8 版.

[5] Vinogradov I M. Representation of an Odd Number as a Sum of Three Primes. *C R Acad Sci USSR*, 1937, 15(6-7): 291-294.

Some New Results about the Goldbach Conjecture, the Twin Prime Conjecture and the Yu Xinhe Conjecture

Abstract Some new results about the Goldbach conjecture, the twin prime conjecture and the Yu Xinhe conjecture are given.

Key words Goldbach's conjecture, the twin prime conjecture, Yu's conjecture, arithmetic progression, Chen's theorem, the three prime theorem

哥德巴赫问题的一个推广

设整数 a、b_1、b_2 满足 $a>1$, $(a, b_1)=(a, b_2)=1$. 本文证明了几乎所有适合条件 $m \equiv b_1+b_2(\bmod a)$ 的偶数 m 可以表示为 p_1+p_2 的形式,其中 p_i 为素数,并且 $p_i \equiv b_i(\bmod a)$, $i=1, 2$.

1 引言及记号

设 a 为正整数, b_1, b_2 为整数, $(a, b_1)=(a, b_2)=1$. 则由 Dirichlet 定理知,有无限多个形如 $an+b_i(i=1, 2)$ 的素数. 哥德巴赫问题的一个推广是(猜测):每个适合条件 $m \equiv b_1+b_2(\bmod a)$ 的正偶数 m 可以写成 p_1+p_2 的形式,其中 p_i 为素数并且 $p_i \equiv b_i(\bmod a)$, $i=1, 2$. 我们注意,在 $a=1$, 2 时,这一猜测即哥德巴赫猜测;在 $a=30$ 时,则化为余新河猜测[1].

本文证明上述猜测对几乎所有适合上述条件的 m 成立,确切地说,我们有下面的

定理 设 a、b_1、b_2 满足上述条件,记

$$P_i=\{p \mid p \text{ 为素数且 } p \equiv b_i(\bmod a),\ i=1, 2\},$$

$$E(x)=\{m \mid m \leqslant x,\ m \equiv b_1+b_2(\bmod a),\ 2 \mid m \text{ 且 } m \notin P_1+P_2\},$$

则对任意正数 A,

$$|E(x)| \ll x/\log^A x.$$

并且在 $m \notin E(x)$ 时,将 m 表成 p_1+p_2, $p_i \in P_i(i=1, 2)$ 的表法个数

$$\sim \frac{[a, m]}{\varphi(a)\varphi([a, m])} \prod_{p \nmid am}\left(1-\frac{1}{(p-1)^2}\right) \frac{m}{\log^2 m}.$$

我们采用圆法证明定理(参见[2]中第十一章). 为此,我们取 $Q=\log^{3A+9} x$, $\tau=xQ^{-1}$, 令

$$E_1=\bigcup_{1\leqslant q\leqslant Q}\bigcup_{\substack{0\leqslant h\leqslant q-1\\(h,q)=1}}\left[\frac{h}{q}-\frac{1}{\tau},\frac{h}{q}+\frac{1}{\tau}\right], E_2=\left[-\frac{1}{\tau},1-\frac{1}{\tau}\right]\backslash E_1,$$

对 $i=1,2$ 记

$$S_i(\alpha,x)=\sum_{ak+b_i=p\leqslant x}e(\alpha(ak+b_i)),$$

$$D_i(m,x)=\int_{E_i}S_1(\alpha,x)S_2(\alpha,x)e(-\alpha m)\mathrm{d}\alpha.$$

2　$D_2(m, x)$ 的估计

引理 1　设

$$\alpha=\frac{h}{q}+z,\ (q,h)=1,\ q\geqslant 1,\ |z|\leqslant\frac{1}{q^2}. \tag{1}$$

若 $q\leqslant x$，$H=\exp\left(\frac{1}{2}\sqrt{\log x}\right)$，则

$$S_i(\alpha,x)\ll x\log^2 x\cdot(\sqrt{qx^{-1}+q^{-1}}+H^{-1}).$$

证明　类似于[2]中第五章§1.

引理 2　设 M 为满足 $|D_2(m,x)|>xQ^{-1/3}$ 的 m 的个数，则

$$M\ll xQ^{-1/3}\log^3 x.$$

证明　由引理 1 及[2]中第十一章引理 1 的证法即知.

3　$D_1(m, x)$ 的估计

引理 3　当 $\alpha\in E_1$ 时，

$$S_i(\alpha,x)=A(q,b_i)\sum_{k=2}^{[x]-1}\frac{e(kz)}{\log k}+O(xe^{-c\sqrt{\log x}}).$$

其中 c 是正常数，z 由(1)给出，而

$$A(q,b_i)=\frac{\varphi((a,q))}{\varphi(a)\varphi(q)}\sideset{}{'}\sum_{\substack{l=1\\l\equiv b_i(\bmod (a,q))}}^{q}e\left(\frac{h}{q}l\right)\ll\log\log q. \tag{2}$$

证明 类似于[2]中第六章引理3的证明.

引理4

$$\sum_{q\leqslant Q}\left|\sum_{h=1}^{q-1}{}'(q,\ b_1)A(q,\ b_2)e\left(-\frac{h}{q}m\right)\right|\ll\log\log m. \tag{3}$$

证明 设 $q=q_1q_2$, $(q_1,\ a)=1$, $p\mid q_2\Rightarrow p\mid a$,则 $d=(a,\ q)\mid q_2$.所有适合 $(l,\ q)=1$, $l\equiv b_i(\bmod d)$ 的 l 可写成 $l=q_2l'+q_1b'_i$,其中 l' 过 $\bmod q_1$ 的缩系,条件 $l\equiv b_i(\bmod d)$ 导出 $(b'_i,\ d)=(b_i,\ d)=1$,即 $(b'_i,\ q_2)=1$. 我们有

$$\sum_{\substack{l=1\\ l\equiv b_i(\bmod d)}}^{q}{}' e\left(\frac{h}{q}l\right)=\sum_{l'=1}^{q_1}\left(\frac{h}{q_1}l'\right)\sum_{q_1b'_i\equiv b_i(\bmod d)}e\left(\frac{h}{q_2}b'_i\right), \tag{4}$$

其中 $b'_i=b_iq_1^{-1}+kd$, k 过 $\bmod\frac{q_2}{d}$的完系.

若 $d\neq q_2$, 则易见(4)右端的内和为零;当 $q_2=d$ 时,(4)的右边是 $\mu(q_1)e\left(\frac{h}{q_2}b'_i\right)$. 因此,在后一条件下,

$$A(q,\ b_1)A(q,\ b_2)=\frac{\mu^2(q_1)}{\varphi^2(q_1)\varphi^2(a)}e\left(\frac{h}{q}q_1(b'_1+b'_2)\right). \tag{5}$$

熟知 Ramanujan 和

$$C_q(-m)=\sum_{h=1}^{q}{}' e\left(-\frac{h}{q}m\right)$$

是 q 的积性函数,故由(5)知(记 $m=b_1+b_2+an$)

$$\begin{aligned}
&\sum_{h=1}^{q-1}{}' A(q,\ b_1)A(q,\ b_2)e\left(-\frac{h}{q}m\right)\\
=&\frac{\mu^2(q_1)}{\varphi^2(q_1)\varphi^2(a)}C_q(-m+q_1b'_1+q_1b'_2)\\
=&\frac{\mu^2(q_1)}{\varphi^2(q_1)\varphi^2(a)}C_{q_1}(-m)C_{q_2}(-an+q_1b'_1-b_1+q_1b'_2-b_2)\\
=&\frac{\mu^2(q_1)}{\varphi^2(q_1)\varphi^2(a)}C_{q_1}(-m)\varphi(q_2),
\end{aligned}$$

注意: $q_2=d$,并且在 $q\mid m$ 时, $C_q(-m)=\varphi(q)$.

于是,(3)的左边

$$\leqslant \sum_{\substack{q_1 \leqslant Q \\ (q_1, a)=1}} \sum_{q_2 \mid a} \left| \sum_{h=1}^{q_1 q_2 - 1}{}' A(q_1 q_2, b_1) A(q_1 q_2, b_2) e\left(-\frac{h}{q_1 q_2} m\right) \right|$$

$$= \sum_{\substack{q_1 \leqslant Q \\ (q_1, a)=1}} \sum_{q_2 \mid a} \frac{\mu^2(q_1)}{\varphi^2(q_1)\varphi^2(a)} \varphi(q_2) \left| C_{q_1}(-m) \right|$$

$$= \sum_{\substack{q_1 \leqslant Q \\ (q_1, a)=1}} \frac{\mu^2(q_1) a}{\varphi^2(q_1)\varphi^2(a)} \left| C_{q_1}(-m) \right|$$

$$\leqslant \frac{a}{\varphi^2(a)} \sum_{q_1 \leqslant Q} \frac{\mu^2(q_1)}{\varphi^2(q_1)} \varphi((m, q_1))$$

$$\ll \sum_{d \mid m} \frac{\mu^2(d)}{\varphi(d)} = \frac{m}{\varphi(m)} \ll \log \log m.$$

引理 5 设 $A(q, b_i)$由(2)给出,则

$$\sum_{q > Q} \left| \sum_{h=1}^{q-1}{}' A(q, b_1) A(q, b_2) e\left(-\frac{h}{q} m\right) \right| \ll d(m) Q^{-1} (\log \log Q)^2 \log \log m.$$

证明 类似于引理 4 及[2]中第十一章引理 3 的证明.

引理 6 级数

$$\mathfrak{S} = \sum_{q=1}^{\infty} \sum_{h=1}^{q-1}{}' A(q, b_1) A(q, b_2) e\left(-\frac{h}{q} m\right)$$

绝对收敛,并且

$$\mathfrak{S} = \frac{[a, m]}{\varphi(a)\varphi([a, m])} \prod_{p \nmid am} \left(1 - \frac{1}{(p-1)^2}\right) > 0. \qquad (6)$$

证明 由引理 5 即知 $\mathfrak{S}(m)$绝对收敛. 又由引理 4 的证明,我们有

$$\mathfrak{S}(m) = \frac{a}{\varphi^2(a)} \sum_{(q_1, a)=1} \frac{\mu^2(q_1)}{\varphi^2(q_1)} C_{q_1}(-m)$$

$$= \frac{a}{\varphi^2(a)} \prod_{p \nmid a} \left(1 + \frac{C_p(-m)}{(p-1)^2}\right),$$

由此易导出(6).

引理 7 设 $\frac{1}{2}x < m \leqslant x$，则

$$D_1(m, x) = \mathfrak{S}(m)\frac{m}{\log^2 m} + O\left(\frac{x}{\log^2 x}d(m)Q^{-1}(\log\log x)^3\right) + O\left(\frac{x(\log\log x)^2}{\log^2 x}\right).$$

证明 这可由引理 3～6 及熟知的方法导出来(参见[2]中第十一章引理 5 的证明).

最后，我们由引理 2，引理 11 及熟知的方法(见[2]中第十一章 §1)便推出了定理.

注：(1) 对公差 a 在范围 $a \ll \log^2 x$ 中变化时，上面的证明仍然有效.

(2) 用更精确的方法，$E(x)$的估计可进一步改进.

参考文献

[1] 余新河. 哥德巴赫猜想的新尝试. 福建师范大学学报，1993，9:1-8.
[2] 潘承洞，潘承彪. 哥德巴赫猜想. 北京：科学出版社，1981.

A Generalization of Goldbach's Problem

Abstract Let a, b_1, b_2 be integers, $a > 1$, $(a, b_1) = (a, b_2) = 1$. prove that almost all even numbers m satisfying $m \equiv b_1 + b_2 \pmod a$ can be represented as $p_1 + p_2$, where p_i is a prime and $p_i \equiv b_i \pmod a$, $i = 1, 2$.

ON THE DIVISOR FUNCTION $d(n)$

1 Introduction

In 1984 Heath-Brown [5] proved the following conjecture of Erdös and Mirsky [2] (which seemed at one time as hard as the twin prime problem):

"There exist infinitely many integers n for which $d(n)=d(n+1)$."

Erdös also conjectured that every positive real number is a limit point of the sequence $\{d(n+1)/d(n)\}$. In other words, let E_1 be the set of the limit points of the sequence $\{d(n+1)/d(n)\}$, then $E_1=\mathbb{R}_{\geqslant 0}$ (the set of all nonnegative real numbers). This conjecture has been studied by some authors, *e.g.*, by Erdös-Pomerance-Sarközy [3] and by Hildebrand [6].

From the result of Heath-Brown, it follows that $1\in E_1$. In this paper we shall show that, for any positive real number α, at least one of the two numbers α, $\alpha/2$ (or α, 2α) must belong to E_1.

More generally, let b be a given non-zero integer and E_b the set of the limit points of the sequence $\{d(n+b)/d(n)\}$. The main results of the present paper are as follows.

Theorem 1. *For all $\alpha>0$, at least one of the two numbers α, $\alpha/2$ (or α, 2α) belongs to E_b.*

Theorem 2. *For all $x>0$,*

$$|(0,x)\cap E_b|\geqslant x/3,$$

where $|\cdot|$ denotes the Lebesgue measure.

2 Lemmas

Lemma 1. *Let a, b be integers satisfying*

$$ab \neq 0, \ \gcd(a, b) = 1, \ 2 \mid ab.$$

Then, for sufficiently large x,

$$\#\{p: p < x, ap + b = p_1 \text{ or } p_1 p_2, \ p_1 < p_2\}$$
$$> C \prod_{p>2} (1 - (p-1)^{-2}) \prod_{2<p \mid ab} \frac{p-1}{p-2} x \ln^{-2} x.$$

Proof. This may be obtained by the lower bound sieve with Chen's switching-procedure (*cf*. [1] or [4, Ch. 11]).

Lemma 2. *Let a, b be integers satisfying*

$$ab \neq 0, \ \gcd(a, b) = 1, \ 2 \mid ab.$$

Then, for any $r \geqslant 2$ and for all sufficiently large x,

$$\#\{p: p < x, ap + b = p_1 \cdots p_{r-1} \text{ or } p_1 \cdots p_r, \ p_1 < p_2 < \cdots < p_r\}$$
$$> \frac{C}{(r-2)!} \prod_{p>2} (1 - (p-1)^{-2}) \prod_{2<p \mid ab} \frac{p-1}{p-2} x \ln^{-2} x (\ln \ln x)^{r-2}.$$

Proof. From Lemma 1, this may be proved by the general unified method given in [7] (*cf*. [7, p281, Remark 3] especially).

Note. The relationship between Lemma 1 and Lemma 2 is similar to the relationship between Chen's theorem (*cf*. [1] or [4, Ch 11]) and [8, Theorem 2].

Lemma 3. *For any positive real number α and any small $\varepsilon > 0$, there exist infinitely many pairs of positive integers s, t, such that*

$$\left| \frac{2^t}{s} - \alpha \right| < \varepsilon.$$

Proof. Choose t large, so that the greatest integer $s \leqslant 2^t/\alpha$ is also large. Then $s > 2^t/\alpha - 1$, leading to $\alpha \leqslant 2^t/s < \alpha + (\alpha/s)$, and the conclusion follows.

3 Proof of Theorem 1

In Lemma 2 let

$$a=\begin{cases}2^{s-1}, & \text{if } b \text{ is odd}\\ p_0^{s-1}, & \text{if } b \text{ is even}\end{cases}$$

where p_0 denotes an odd prime, $p_0 \nmid b$. Then, for $p \nmid a$,

$$d(ap)=2s.$$

While the value of the divisor function at

$$ap+b=p_1 \cdots p_{r-1} \text{ or } p_1 \cdots p_r$$

(with $p_1 < p_2 < \cdots < p_r$) is

$$d(ap+b)=2^{r-1} \text{ or } 2^r.$$

Hence

$$d(ap+b)/d(ap)=2^t/s \tag{1}$$

where $t=r-2$ or $r-1$. Combining (1) with Lemma 3, the proof of Theorem 1 is completed.

4 Proof of Theorem 2

From the definition of E_b it follows that E_b is closed and hence for each (u, v), $(u, v)\backslash E_b$ is open. Thus this latter set is a countable union of disjoint open intervals and so is measurable. Thus so is $(u, v) \cap E_b$. Let

$$\mathscr{A}=(x/4, x/2) \cap E_b, \quad \mathscr{B}=\{\alpha: x/4<\alpha<x/2, 2\alpha \in E_b\}.$$

Then

$$\mathscr{A} \cup 2\mathscr{B}=((x/4, x/2) \cup (x/2, x)) \cap E_b,$$
$$\mathscr{A} \cap 2\mathscr{B}=\varnothing,$$

$$\mathscr{A} \cup \mathscr{B} = (x/4,\ x/2).$$

Hence

$$|(x/4,\ x) \cap E_b| = |((x/4,\ x/2) \cup (x/2,\ x)) \cap E_b| = |\mathscr{A}| + |2\mathscr{B}|$$

and

$$|\mathscr{A}| + |2\mathscr{B}| = |\mathscr{A}| + 2|\mathscr{B}| \geqslant |\mathscr{A}| + |\mathscr{B}| \geqslant |\mathscr{A} \cup \mathscr{B}| = x/4.$$

Therefore

$$|(x/4,\ x) \cap E_b| \geqslant x/4$$

and it follows easily that

$$|(0,\ x) \cap E_b| \geqslant x/3.$$

Acknowledgements. The first author is grateful to Professors R. C. Vaughan, D. Masser and the referee(s). Their precious advice greatly improves the quality of this paper. This research was supported by the National Natural Science Foundation of China.

References

[1] J. Chen. On the Representation of a Large Even Integer as the Sum of a Prime and the Product of at most Two Primes. *Sai. Sinica*, 16(1973), 157 - 176.

[2] P. Erdös and L. Mirsky. The Distribution of the Values of the Divisor Function $d(n)$. *Proc. LMS* (3), 2(1952), 257 - 271.

[3] P. Erdös, C. Pomerance and A. Sárközy. On Locally Repeated Values of Certain Arithmetic Functions. II. *Acta Math. Hung.*, 49(1987), 251 - 259.

[4] H. Halberstam and H.-E. Richert. *Sieve Methods* (Academic Press, London, 1974).

[5] D. R. Heath-Brown. The Divisor Function at Consecutive Integers. *Mathematika*, 31(1984), 141 - 149.

[6] A. Hildebrand. The Divisor Function at Consecutive Integers. *Pacif. J. Math.*, 129(1987), 307 - 319.

[7] J. Kan. On the Lower Bound Sieve. *Mathematika*, 37(1990), 273 - 286.

[8] J. Kan. On the Number of Solutions of $N - p = P_r$. *J. reine und angew. Math.*, 414(1991), 117 - 130.

On the divisor function $d(n)$: II

1 Introduction

In 1984 Heath-Brown[2] proved the conjecture of Erdös and Mirsky[1], and obtained the following result.

Theorem 1. *There are infinitely many integers n for which $d(n+1)=d(n)$. Indeed, for large x, the number of such $n \leqslant x$ is at least of order $x \ln^{-7} x$.*

In [3], Hildebrand improved the order to $x(\ln \ln x)^{-3}$.

Heath-Brown's proof depends on his "Key Lemma" and uses the N-dimensional sieve. However the method seems incapable of dealing with $d(n+1)=2d(n)$, for example.

In this paper, we shall deal with the more general case and prove the following theorem.

Theorem 2. *For any given positive integer l, there are infinitely many integers n for which $d(n+1)=ld(n)$, and for large x, the number of such $n \leqslant x$ is at least of order $x(\ln \ln x)^{-3}$.*

2 Definitions and Theorem 1

Let p, q denote primes, $v_p(a)$ be defined by $p^{v_p(a)} \| a$, and (a, b), $[a, b]$ be the greatest common divisor and the least common multiple of a, b, respectively.

Let $a_1, a_2, \cdots, a_n$ be n distinct positive integers. The prime factors p of a_i of the first class $P_i^{(1)}$ are those for which $p \mid a_i \Rightarrow p \mid a_j (j \neq i)$. If there is a set $P_i^{(2)}$ of $|P_i^{(1)}|$ prime factors of a_i, not in $P_i^{(1)}$, and a bijection $q(p)$ from $P_i^{(1)}$ to

$P_i^{(2)}$ with the property that $v_{q(p)}(a_i)=v_p(a_i)$ (in the sequel we say that such p and q are paired with each other), then we say that the elements of $P_i^{(2)}$ are the prime factors of a_i of the second class. Let $U_i P_i^{(1)}=P^{(1)}$, and $U_i P_i^{(2)}=P^{(2)}$. We call a_i's remaining prime factors (if they exist) its prime factors of the third class, and define similarly $P_i^{(3)}$, and $U_i P_i^{(3)}=P^{(3)}$.

We now state the following theorem which plays a crucial role in the proof of Theorem 2.

Theorem 1. *For any fixed positive integers n and l, there exist n positive integers* $a_1<a_2<\cdots<a_n$, *with the following properties.*

(1) $a_j-a_i=(a_i,a_j)\quad(1\leqslant i<j\leqslant n)$.

(2) $\dfrac{d(a_i)}{d\left(\dfrac{a_i}{(a_i,a_j)}\right)}\Big/\dfrac{d(a_j)}{d\left(\dfrac{a_j}{(a_i,a_j)}\right)}=l\quad(1\leqslant i<j\leqslant n)$.

(3) *Every* a_i *has prime factors of the first, the second, and the third classes. There is at least one* $q_i\in P_i^{(3)}$ *with* $v_{q_i}(a_i)=l$, *and* $2\in P^{(1)}\cup P^{(2)}$.

3 Proof of Theorem 1

Fix five distinct primes a, b, c, d, p. Since $(abp,cd)=1$, there exist positive integers u_0, v_0, with $u_0 d^{l^2+l-1}b^l p^{l^2-1}-v_0 c^l d^l=1$. By the Chinese Remainder Theorem, the simultaneous congruences

$$u_0+tc^l d^l\equiv 1(\operatorname{mod} pab),\quad v_0+ta^{l^2+l-1}b^l p^{l^2-1}\equiv 1(\operatorname{mod} cd)$$

have a common solution $t_0>0$. Let $u=u_0+t_0c^l d^l$, $v=v_0+t_0a^{l^2+l-1}b^l p^{l^2-1}$. Set $a_2=ua^{l^2+l-1}b^l p^{l^2+l-1}$ and $a_1=vc^l d^l p^l$, and let $a=2$. Then it is easy to verify that (1), (2) and (3) are all satisfied for $n=2$, *i.e.*, the theorem holds for $n=2$.

Now we assume that the theorem holds for $n\geqslant 2$, and go on to prove that it also holds for $n+1$.

Note that (2) is just (for $1\leqslant i<j\leqslant n$)

$$\prod_p\frac{(v_p(a_i)+1)(v_p(a_j)-v_p((a_i,a_j))+1)}{(v_p(a_j)+1)(v_p(a_i)-v_p((a_i,a_j))+1)}=l,\tag{4}$$

where p runs over all the prime factors of a_i and a_j.

If p divides only one of a_i, a_j, then its contribution to the left side of (4) is 1. So we need only consider the prime factors of (a_i, a_j). Moreover, if $v_p(a_i)=v_p(a_j)$, then the contribution of such p to the left side of (4) is also 1, and we may neglect such p.

Assume that $a_1<a_2<\cdots<a_n$ satisfy all the requirements of the theorem. Set

$$M=[a_1, a_2, \cdots, a_n]\prod_{p\in P^{(1)}} p, \tag{5}$$

and define inductively $n+1$ numbers

$$KM,\ KM+a_1,\ KM+a_2,\ \cdots,\ KM+a_n, \tag{$*$}$$

where the positive integer K will be determined in the sequel.

Obviously we have

$$\begin{aligned}(KM+a_i, KM+a_j)&=(KM+a_i, a_j-a_i)=(KM+a_i, (a_i, a_j))\\&=(a_i, a_j)=a_j-a_i=(KM+a_j)-(KM+a_i),\end{aligned}$$

and

$$(KM+a_i, KM)=(a_i, KM)=a_i=(KM+a_i)-KM.$$

Thus (1) holds for ($*$), while (2) (*i.e.*, (4)) becomes

$$\prod_{p\mid(a_i, a_j)}\frac{(v_p(KM+a_i)+1)(v_p(KM+a_j)-v_p((a_i, a_j))+1)}{(v_p(KM+a_j)+1)(v_p(KM+a_i)-v_p((a_i, a_j))+1)}=l, \tag{6}$$

and

$$\prod_{p\mid a_i}\frac{(v_p(KM)+1)(v_p(KM+a_i)-v_p(a_i)+1)}{(v_p(KM+a_i)+1)(v_p(KM)-v_p(a_i)+1)}=l. \tag{7}$$

From (5), for $p\mid(a_i, a_j)$,

$$v_p(KM+a_i)=v_p(a_i),\quad v_p(KM+a_j)=v_p(a_j),$$

so (6) holds by the inductive hypothesis.

Let $M/a_i=M_i$. Let K satisfy the following congruence conditions (such a

K exists, see the next paragraph). First,

$$K \equiv 1(\bmod p), \quad p \in P^{(1)}. \tag{8}$$

$$KM_i + 1 \equiv q^{v_p(M)-v_p(a_i)}(\bmod q^{v_p(M)-v_p(a_i)+1}), \tag{9}$$

where $q \in P_i^{(2)}$, and $p \in P_i^{(1)}$ is the prime factor pairing with q. Next,

$$KM_i + 1 \equiv q_i^{l^2-1}(\bmod q_i^{l^2}), \tag{10}$$

where $q_i \in P_i^{(3)}$ and $v_{q_i}(a_i) = l$ (take only one such prime factor as q_i). Finally,

$$KM_i + 1 \equiv 2(\bmod q), \tag{11}$$

where $q \in P_i^{(3)}$ and $q \neq q_i$.

Since $(M_i, q) = 1$ for $q \in P_i^{(2)} \cup P_i^{(3)}$, (9), (10) and (11) are solvable. Moreover, because the moduli in (8), (9), (10) and (11) are mutually coprime, by the Chinese Remainder Theorem the simultaneous congruences (8)-(11) have common solutions. Let K be such a common solution; then from (8)-(11), $(K, M) = 1$. Since $2 \notin P^{(3)}$ (by the inductive hypothesis (3)), by (11), for prime factor $q \in P_i^{(3)}$ and $q \neq q_i$ (for q_i, see (10)), we have

$$v_q(KM + a_i) = v_q(a_i) + v_q(KM_i + 1) = v_q(a_i) = v_q(KM),$$

and such prime factors may be neglected in (7). Therefore, by (8), (9) and (10), the left side of (7) becomes

$$\prod_{p \in P_i^{(1)}} \frac{v_p(M)+1}{(v_p(a_i)+1)(v_p(M)-v_p(a_i)+1)}$$

$$\times \prod_{q \in P_i^{(2)}} \frac{(v_q(a_i)+1)(v_p(M)-v_p(a_i)+1)}{v_p(M)-v_p(a_i)+v_q(a_i)+1} \times \frac{(v_{q_i}(a_i)+1)(l^2-1+1)}{v_{q_i}(a_i)+(l^2-1)+1}$$

$$= \frac{(l+1)l^2}{l+l^2} = l.$$

So (7), and thus (2) also, hold for (*).

Note that any prime factor of M belongs to $P^{(1)}(n+1)$, where $P^{(1)}(n+1)$ denotes the set of prime factors of the first class for (*). It is easy to see that

$2 \in P^{(1)}(n+1)$. (This is obvious if 2 divides some a_i— by (5) $2 \mid KM$ and $2 \mid KM+a_i$. However, for $n=2$ we have set $a=2$, and hence $2 \mid a_2$, and by the inductive hypothesis, $2 \in P^{(1)}(n+1)$ for any $n \geqslant 2$.) On the other hand, since the greatest common divisor of any two numbers in ($*$) is a_i or (a_i, a_j), $p \in P^{(1)}(n+1) \Rightarrow p \mid M$. Further, $v_p(KM)=v_p(M)$ for any $p \in P^{(1)}(n+1)$. For some prime factor of M, with K determined as above, we have

(i) $p \in P_i^{(1)} \Rightarrow v_p(KM+a_i)=v_p(a_i)$;

(ii) $q \in P_i^{(2)} \Rightarrow v_q(KM+a_i)=v_q(a_i)+v_q(KM_i+1)=v_P(M)$, where p is the prime factor pairing with q;

(iii) $q_i \in P_i^{(3)}$ and $v_{q_i}(a_i)=l \Rightarrow v_{q_i}(KM+a_i)=v_{q_i}(a_i)+v_{q_i}(KM_i+1)=l^2+l-1$;

(iv) $q \in P_i^{(3)}$, $q \neq q_i \Rightarrow v_q(KM+a_i)=v_q(a_i)$.

Hence we see that, in the prime factorization of some number in ($*$), the exponent of any prime factor in $P^{(1)}(n+1)$ is independent of K, provided that K satisfies (8)-(11).

Take sufficiently many other primes q, q', each greater than the largest prime factor of M, and let

$$KM+a_i \equiv q^{m_q} \left(\bmod q^{m_q+1}\right), \tag{12}$$

$$KM \equiv q'^{m_{q'}} \left(\bmod q'^{m_{q'}+1}\right), \tag{13}$$

where m_q, $m_{q'}$ are suitably chosen positive integers. By the Chinese Remainder Theorem, the simultaneous congruences (8)-(11), with (12) and (13) added, still have a common solution K. Further, when q, q' are sufficiently large, and m_q, $m_{q'}$ are suitably chosen, (3) will hold for $n+1$ numbers KM, $KM+a_1$, $\cdots$, $KM+a_n$. This completes the proof of Theorem 1.

4 Proof of Theorem 2

The proof is similar to §4 of [2], so we only sketch it. With the same argument and notation we will have, for $a_i < a_j$,

$$\frac{a_j}{(a_i, a_j)} r_i F_i(x) = 1 + \frac{a_i}{(a_i, a_j)} r_j F_j(x),$$

$$d\left(\frac{a_j}{(a_i, a_j)} r_i F_i(x)\right) = d\left(\frac{a_j}{(a_i, a_j)}\right) d(a_i) d(F_i(x)),$$

$$d\left(\frac{a_i}{(a_i, a_j)} r_j F_j(x)\right) = d\left(\frac{a_i}{(a_i, a_j)}\right) d(a_j) d(F_j(x)).$$

Hence, by Theorem 1, we have $d(u+1) = ld(u)$ with

$$u = \frac{a_i}{(a_i, a_j)} r_j F_j(x),$$

whenever

$$d(F_i(x)) = d(F_j(x)) \quad (i \neq j).$$

The remaining proof is just the same with that of [2], and by the method of [3] Theorem 2 follows.

References

[1] P. Erdös and L. Mirsky. The Distribution of the Values of the Divisor Function $d(n)$. *Proc. London Math. Soc.* (3), 2(1952), 257 - 271.

[2] D. R. Heath-Brown. The Divisor Function at Consecutive Integers. *Mathematika*, 31(1984), 141 - 149.

[3] A. Hildebrand. The Divisor Function at Consecutive Integers. *Pacific J. Math.*, 129(1987), 307 - 309.

[4] J, Kan and Z. Shan. On the Divisor Function $d(n)$. *Mathematika*, 43(1996), 320 - 322.

组合数论中的一个猜测

摘要 证明如下定理:设 $a^{(i)} = (a_1^{(i)}, a_2^{(i)})$ 为整向量, $i = 1, 2, \cdots, m$, 则对任意自然数 n,当 $m > 2(n-1)$ 时,必有非空集合 $I \subseteq \{1, 2, \cdots, m\}$, 使 $\sum\limits_{i\in I} a^{(i)} \equiv 0(\bmod n)$.

设 $a^{(i)}=(a_1^{(i)}, a_2^{(i)}, \cdots, a_k^{(i)})$, $(i=1, 2, \cdots, m)$ 为 k 维整向量(坐标为整数的向量). Alon 等[1]猜测:对任意自然数 n,当 $m > k(n-1)$ 时,必有非空集合 $I \subseteq \{1, ,2, \cdots, m\}$,使 $\sum\limits_{i\in I} a^{(i)} \equiv 0(\bmod n)$. Olson[2]证明对于 $n=p^d$ (p 为素数),这一猜测成立. 本文作者证明对于 $k=2$, 这一猜测成立,即有

定理 设 $a^{(i)}=(a_1^{(i)}, a_2^{(i)})$ 为整向量, $i=1, 2, \cdots, m$, 则对任意自然数 n, 当 $m>2(n-1)$ 时,必有非空集合 $I \subseteq \{1, 2, \cdots, m\}$,使 $\sum\limits_{i\in I} a^{(i)} \equiv 0(\bmod n)$.

作为定理的特例,当第 1 分量全为 1 时, $a^{(i)}=(1, a_i)$, $i=1, 2, \cdots, m$, 我们得到 Erdös 等人的著名结果:

推论 对任意自然数 n 及一组整数 $a_1, a_2, \cdots, a_m$,当 $m=2n-1$ 时,必有 n 个 a_i 的和恒等于 $0(\bmod n)$.

定理的证明需要以下几个引理.

引理 1 对任意 $2p-1$ 个二维整向量 $a^{(1)}(1 \leqslant i \leqslant 2p-1)$,必有一个非空集合 $S \subseteq \{1, 2, \cdots, 2p-1\}$,使 $p \mid \sum\limits_{i\in S} a^{(i)}$.

证明 由 Olson[2]的结果 $(d=1)$ 立即得出.

引理 2 对任意 $3p-2$ 个二维整向量 $a^{(i)}(1 \leqslant i \leqslant 3p-2)$,必有一个非空集合 $S \subseteq \{1, 2, \cdots, 3p-2\}$,使 $p \mid \sum\limits_{i\in S} a^{(i)}$,并且 $|S| \leqslant p$.

证明 考虑三维向量 $b^{(i)}=(a^{(i)}, 1)=(a_1^{(i)}, a_2^{(i)}, 1)$, 由三维向量的 Olson 的结果[因为 $3(p-1)+1=3p-2$] ,有非空集合 $S_1 \subseteq \{1, 2, \cdots, 3p-2\}$,使 $p \mid \sum\limits_{i\in S_1} b^{(i)}$. 从而 $p \mid \sum\limits_{i\in S_1} a^{(i)}$,并且 $|S_1|=p$ 或 $2p$. 若 $|S_1|=p$, 则引理已经成立. 若 $|S_1|=2p$,不妨设 $S_1=\{1, 2, \cdots, 2p\}$, 则由引理 1,有非空集合 $S_2 \subseteq \{1,$

2, …, $2p-1\}$,使 $p\left|\sum_{i\in S_2}a^{(i)}\right.$,所以 $p\left|\sum_{i\in S_1\backslash S_2}a^{(i)}\right.$. S_2 与 $S_1\backslash S_2$ 中必有一个元数$\leqslant p$,所以引理成立.

引理 3 对任意 $3p-1$ 个二维整向量 $a^{(i)}(1\leqslant i\leqslant 3p-1)$,必有非空集合 S_1, $S_2\subseteq\{1, 2, \cdots, 3p-1\}$,互不相交,使得 $p\left|\sum_{i\in S_j}a^{(i)}\right.$, $j=1, 2$.

证明 由引理 2,可取 S_1 满足 $|S_1|\leqslant p$ 及 $p\left|\sum_{i\in S_1}a^{(i)}\right.$. 再对剩下的 $\geqslant 2p-1$ 元所成的集应用引理 2,得 S_2 即可.

引理 4 对任意 $hp-1$ $(h\geqslant 2)$ 个二维整向量 $a^{(i)}(1\leqslant i\leqslant hp-1)$, 必有互不相交的非空集合 S_1, S_2, …, S_{h-1} 满足 $p\left|\sum_{i\in S_j}a^{(i)}\right.$, $j=1, 2, \cdots, h-1$.

证明 应用归纳法,仿照引理 3 的证法即得.

现在证明定理. 对于素数 p 结论已真. 设对于$<n$ 的数结论成立,考虑 n,设 n 至少有两个素因数,p 是其中之一,由引理 4,对 $2n-1=\left(2\cdot\frac{n}{p}\right)p-1$ 个整向量 $a^{(i)}$,有 $2\cdot\frac{n}{p}-1$ 个互不相交的非空集合 S_1, S_2, …, $S_{2\cdot\frac{n}{p}-1}$,满足

$$p\left|\sum_{i\in S_j}a^{(i)}\right., \ j=1, 2, \cdots, 2\cdot\frac{n}{p}-1.$$

对于 $2\cdot\frac{n}{p}-1$ 个二维整向量,

$$b_j=\frac{1}{p}\sum_{i\in S_j}a^{(i)}, \ j=1, 2, \cdots, 2\cdot\frac{n}{p}-1.$$

由归纳假设,存在非空集合 $T\subseteq\left\{1, 2, \cdots, 2\cdot\frac{n}{p}-1\right\}$,满足$\frac{n}{p}\left|\sum_{j\in T}b_j\right.$. 于是,令 $S=\bigcup_{j\in T}S_j$, 则有

$$\sum_{i\in S}a^{(i)}=\sum_{j\in T}\sum_{i\in S_j}a^{(i)}=\sum_{j\in T}pb_j,$$

被 $\frac{n}{p}\cdot p=n$ 整除,定理证毕.

参考文献

[1] Alon N, Frvedland S, Kalai G. Regular Subgraphs of Almost Regular Graphs. *J*.

Combin Theory, Ser B, 1984, 37: 79 - 91.
[2] Olson J E. Some Recent Advances in Number Theory. *J. Number Theory*, 1969, 1: 8 -10.

A Conjecture in Number Theory

Abstract This paper proves the following theorem: Let $a^{(i)}=(a_1^{(i)}, a_2^{(i)})$, $i=1, 2, \cdots, m$, be m integer vectors. If $m>2(n-1)$, then there exists a non-empty set $I\subseteq\{1, 2, \cdots, m\}$, such that

$$\sum_{i\in I}a^{(i)}\equiv 0(\bmod n).$$

A Simple Proof of A Curious Congruence by SUN

Abstract. In this note, we give a simple and elementary proof of the following curious congruence which was established by Zhi-Wei Sun:

$$\sum_{k=1}^{(p-1)/2} \frac{1}{k \cdot 2^k} \equiv \sum_{k=1}^{[3p/4]} \frac{(-1)^{k-1}}{k} (\bmod p).$$

In [4], the following curious congruence for odd prime p was established by Zhi-Wei Sun:

$$\sum_{k=1}^{(p-1)/2} \frac{1}{k \cdot 2^k} \equiv \sum_{k=1}^{[3p/4]} \frac{(-1)^{k-1}}{k} (\bmod p). \tag{1}$$

The author's proof, using Pell sequences, is fairly complicated. In fact, a recent article [3] on congruence modulo p ends in the remark that "It seems unlikely that (1) can be proved with the simple approach that we have used here." In the present note, we give a simple and elementary proof of (1). Throughout, p denotes an odd prime.

First of all, it is well known (e. g. [1], [2]) that for $k = 0, 1, 2, \cdots, p-1$,

$$\binom{p-1}{k} \equiv (-1)^k (\bmod p). \tag{2}$$

From (2) we get

$$\begin{aligned} \frac{2^{p-1}-1}{2} &= \frac{(1+1)^p - 2}{2p} = \frac{1}{2p} \sum_{k=1}^{p-1} \binom{p}{k} = \frac{1}{2} \sum_{k=1}^{p-1} \frac{1}{k} \binom{p-1}{k-1} \\ &\equiv \frac{1}{2} \sum_{k=1}^{p-1} \frac{(-1)^{k-1}}{k} (\bmod p). \end{aligned} \tag{3}$$

Let $\varepsilon = e^{\pi i/4}$. Then

$$(1+\varepsilon)^p+(1-\varepsilon)^p=2+2\sum_{\substack{1\leqslant k\leqslant p\\ k\text{ even}}}\binom{p}{k}\varepsilon^k$$

$$=2+2p\sum_{\substack{1\leqslant k\leqslant p\\ k\text{ even}}}\frac{1}{k}\binom{p-1}{k-1}\varepsilon^k$$

$$\equiv 2-2p\sum_{\substack{1\leqslant k\leqslant p\\ k\text{ even}}}\frac{\varepsilon^k}{k}(\operatorname{mod} p^2)$$

$$=2-2p\left(\sum_{k=1}^{\left[\frac{p-1}{4}\right]}\frac{(-1)^k}{4k}+i\sum_{k=1}^{\left[\frac{p+1}{4}\right]}\frac{(-1)^{k-1}}{4k-2}\right)$$

$$=2-\frac{p}{2}\sum_{k=1}^{\left[\frac{p-1}{4}\right]}\frac{(-1)^k}{k}+ip\sum_{k=1}^{\left[\frac{p+1}{4}\right]}\frac{(-1)^k}{2k-1}$$

$$=2-\frac{p}{2}A+ipB, \tag{4}$$

where

$$A=\sum_{k=1}^{\left[\frac{p-1}{4}\right]}\frac{(-1)^k}{k}\quad\text{and}\quad B=\sum_{k=1}^{\left[\frac{p-1}{4}\right]}\frac{(-1)^k}{2k-1}.$$

Since $\bar{\varepsilon}=\varepsilon^{-1}$, taking modulus of both sides of (4) yields

$$4-2pA\equiv\left(2-\frac{p}{a}A\right)^2+p^2B^2\equiv 4-2pA$$

$$\equiv((1+\varepsilon)^p+(1-\varepsilon)^p)((1+\varepsilon^{-1})^p+(1-\varepsilon^{-1})^p)$$

$$=(2+\varepsilon+\varepsilon^{-1})^p+(2-\varepsilon-\varepsilon^{-1})^p=(2+\sqrt{2})^p+(2-\sqrt{2})^p$$

$$=2^{p+1}+2\sum_{\substack{1\leqslant k\leqslant p\\ k\text{ even}}}\binom{p}{k}2^{p-k}(\sqrt{2})^k=2^{p+1}+2^{p+1}\sum_{k=1}^{(p-1)/2}\binom{p}{2k}\frac{1}{2^k}$$

$$=2^{p+1}+2^p p\sum_{k=1}^{(p-1)/2}\frac{1}{k\cdot 2^k}\binom{p-1}{2k-1}$$

$$\equiv 2^{p+1}-2^p p\sum_{k=1}^{(p-1)/2}\frac{1}{k\cdot 2^k}(\operatorname{mod} p^2). \tag{5}$$

From (5) and (3) we obtain, since $2^{p-1}\equiv 1(\operatorname{mod} p)$,

$$A \equiv -\frac{2^p-2}{p}+2^{p-1}\sum_{k=1}^{(p-1)/2}\frac{1}{k\cdot 2^k}$$
$$\equiv \sum_{k=1}^{p-1}\frac{(-1)^k}{k}+\sum_{k=1}^{(p-1)/2}\frac{1}{k\cdot 2^k}\pmod p,$$

and so

$$\sum_{k=1}^{(p-1)/2}\frac{1}{k\cdot 2^k} \equiv -\sum_{k=1}^{p-1}\frac{(-1)^k}{k}+A=\sum_{k=1}^{p-1}\frac{(-1)^{k-1}}{k}+\sum_{k=1}^{\left[\frac{p-1}{4}\right]}\frac{(-1)^k}{k}$$
$$=\sum_{k=1}^{p-1}\frac{(-1)^{k-1}}{k}+\sum_{k=p-\left[\frac{p-1}{4}\right]}^{p-1}\frac{(-1)^{p-k}}{p-k}$$
$$\equiv \sum_{k=1}^{p-1}\frac{(-1)^{k-1}}{k}-\sum_{k=p-\left[\frac{p-1}{4}\right]}^{p-1}\frac{(-1)^{k-1}}{k}\pmod p$$
$$=\sum_{k=1}^{\left[\frac{3p}{4}\right]}\frac{(-1)^{k-1}}{k},$$

and (1) is proved.

References

[1] Louis Comet, *Advanced Combinatorics*, D. Reidel Publishing Company, 1974.

[2] G. H. Hardy and E. M. Wright, *An Introduction to the Theory of Numbers*, Fourth Edition, Clarendon Press, Oxford, 1960.

[3] Winfried Kohnen, A Simple Congruence Modulo p, *Amer. Math. Monthly* **104** (1997), 444 - 445. MR **98e**: 11004.

[4] Zhi-Wei Sun, A Congruence for Primes, *Proc. Amer. Math. Soc.* **123**(1995), 1341 - 1346. MR **95f**: 11003.

模一个理想的 d 阶元的个数

摘要：设 A 为一个代数数域的理想，d 为正整数，给出 $\bmod A$ 的 d 阶元的个数的计算公式.

对于正整数 m，有一些文章讨论过 Z_m 的 d 阶元的个数.

本文讨论更一般的问题. 设在一个代数数域中，A 是一个理想，$R(A)$是 $\bmod A$ 的缩系，d 为正整数，$\psi(d)$为 $R(A)$的 d 阶元的个数，即 $\psi(d)$是 $R(A)$中满足下列条件的 x 的个数：

$$x^d \equiv 1(\bmod A) \tag{1}$$

且 d 为使(1)成立的最小的自然数. 我们的目的是给出 $\psi(d)$的公式.

由于 $x^d \equiv x^{(d,\varphi(A))}$，$\varphi$ 是欧拉函数，所以在 $d \nmid \varphi(A)$时，$\psi(d)=0$，以下假定 $d \mid \varphi(A)$.

引理 1　设 d_1，d_2 互素，则

$$\psi(d_1 d_2)=\psi(d_1)\psi(d_2).$$

证明　设 x_1，…，x_s 满足

$$x_i^{d_1} \equiv 1(\bmod A),\ (1 \leqslant i \leqslant s) \text{ 并且 } d_1 \text{ 最小，} s=\psi(d_1),$$

又设 y_1，…，y_t 满足

$$y_j^{d_2} \equiv 1(\bmod A),\ (1 \leqslant j \leqslant t) \text{ 并且 } d_2 \text{ 最小，} t=\psi(d_2).$$

显然 $(x_i y_j)^{d_1 d_2} = x_i^{d_1 d_2} y_j^{d_1 d_2} \equiv 1(\bmod A)$.

并且设 $(x_i y_j)^l \equiv 1(\bmod A)$，

则

$$(x_i y_j)^{l \cdot d_1} \equiv 1(\bmod A).$$

从而

$$y_j^{l\cdot d_1} \equiv 1(\bmod A).$$

又由 d_2 的最小性,熟知 $d_2 \mid ld_1$,因为 d_1, d_2 互素,所以 $d_2 \mid l$,同理 $d_1 \mid l$, 即 $d_1 d_2$ 是使

$$(x_i y_j)^l \equiv 1(\bmod A)$$

成立的最小的自然数.

最后,如果

$$x_i y_j \equiv x_{i'} y_{j'}(\bmod A),$$

那么,两边 d_1 次方得

$$y_j^{d_1} \equiv y_{j'}^{d_1}(\bmod A),$$

由于 d_1、d_2 互素,存在自然数 m, n 使

$$md_1 - nd_2 = 1,$$

从而

$$y_j^{nd_2+1} \equiv y_{j'}^{nd_2+1}(\bmod A),$$

即

$$y_j \equiv y_{j'}(\bmod A),$$

同理

$$x_i \equiv x_{i'}(\bmod A).$$

因此 $x_i y_j$ $(1 \leqslant i \leqslant s,\ 1 \leqslant j \leqslant t)$ 是

$$x^{d_1 d_2} \equiv 1(\bmod A)$$

的 st 个互不同余的解,引理 1 成立.

由引理 1,设 d 的素因数分解式为

$$d = \prod_{q \mid d} q^{\beta},$$

则 $\psi(d)=\prod_{q|d}\psi(q^\beta)$,

设 $f(d)$为 $R(A)$中满足

$$x^d\equiv 1(\bmod A)$$

的 x 的个数,则

$$\psi(q^\beta)=f(q^\beta)-f(g^{\beta-1}).$$

因此问题在于求出 $f(q^\beta)$.

设理想 A 的素理想分解式为

$$A=\prod_{P|A}P^\alpha,$$

对于素理想 P,设它的范数 $N(P)=p^f$, 分歧指数为 e,令 $h(P^\alpha, q^\beta)$为

$$x^{q^\beta}\equiv 1(\bmod P^\alpha)$$

的解数,则由中国剩余定理,

$$f(q^\beta)=\prod_{P|A}h(P^\alpha, q^\beta).$$

引理 2　如果 $q\neq p$, 那么

$$h(P^\alpha, q^\beta)=(q^\beta, p^f-1).$$

证明　由于 $R(P)$是 $\varphi(P)$阶循环群,所以方程

$$x^{q^\beta}\equiv 1(\bmod P) \tag{2}$$

的解数是 $(q^\beta, \varphi(P)))=(q^\beta, p^f-1)$,假设对于 $k<\alpha$, 方程

$$x^{g^\beta}\equiv 1(\bmod P^k) \tag{3}$$

的解数是 (q^β, p^f-1), 则对(3)的每一个解 x,取 $\pi\in P\backslash P^2$,

由 $(x+\pi^k y)^{q^\beta}\equiv 1(\bmod P^{k+1})$　(4)

得

$$(x^{q^\beta}-1)+\pi^k q^\beta y\equiv 0(\bmod P^{k+1}).$$

从而有唯一确定的

$$y\equiv -\frac{x^{q^\beta}-1}{\pi^k}\cdot\frac{1}{q^\beta}(\bmod P).$$

于是

$$x^{q^\beta} \equiv 1(\bmod P^{k+1})$$

的解数也是 $(q^\beta, p^f - 1)$. 由归纳法,引理 2 成立.

引理 3 $h(p^\alpha, p^\beta) = (p^{f(\alpha-1)}, p^{\beta e f})$.

证明 因为 $(p^\beta, p^f - 1) = 1$, 所以方程

$$x^{p^\beta} \equiv 1(\bmod P)$$

只有一个解 $x \equiv 1(\bmod P)$.

如果 $\beta e \geqslant \alpha - 1$, 那么对于 P 中的任一个数 $\pi y(\pi \in P \backslash P^2)$, 均有

$$(1 + \pi y)^{p^\beta} = 1 + p^\beta \pi y + \cdots$$

其中未写出的项形状为 $\binom{p^\beta}{k} \pi^k y^k = \frac{p^\beta \pi^k}{k} \cdot \binom{p^\beta - 1}{k - 1} y^k (2 \leqslant k \leqslant p^\beta)$, 次数

$\operatorname{Ord}_P \frac{p^\beta \pi^k}{k} \geqslant \beta e + k - \operatorname{ord}_P k \geqslant \beta e + 1$, 所以

$$(1 + \pi y)^{p^\beta} \equiv 1(\bmod P^\alpha)$$

因此这时

$$x^{p^\beta} \equiv 1(\bmod P^\alpha) \tag{5}$$

的解数

$$h(P^\alpha, p^\beta) = \frac{\varphi(P^\alpha)}{\varphi(P)} = p^{f(\alpha-1)}.$$

如果 $\beta e < \alpha - 1$, 那么与上面类似,

$(1 + \pi^j y) p^\beta = 1 + p^\beta \pi^j y +$ 次数更高的项,所以当且仅当 $j \geqslant \alpha - \beta e$ 时, $1 + \pi^j y$ 是

$$x^{p^\beta} \equiv 1(\bmod P^\alpha) \tag{6}$$

的解,于是(5)的解数

$$h(P^\alpha, p^\beta) = \frac{\varphi(P^\alpha)}{\varphi(P^{\alpha-\beta e})} = p^{f\beta e}.$$

综合以上所说,得到如下定理:

定理 设理想 A 的素理想分解式为

$$A=\prod P^{\alpha},$$

P 的范数为 f,分歧指数为 e.

设 d 的素因数分解式为

$$d=\prod q^{\beta},$$

则 mod$A$ 的 d 阶元的个数

$$\psi(d)=\prod_{q|d}\left(\prod_{P|A}h(P^{\alpha},\ q^{\beta})-\prod_{P|A}h(P^{\alpha},\ q^{\beta-1})\right),$$

其中

$$h(P^{\alpha},\ q^{\beta})=\begin{cases}(q^{\beta},\ p^{f}-1), & 若\ q\neq p\\ (p^{\beta ef},\ p^{f(\alpha-1)}), & 若\ q=p\end{cases}$$

特别地,在 $d\nmid\varphi(A)$时, $\psi(d)=0$.

推论 $x^{d}\equiv 1(\bmod A)$ 的解数是 $\prod_{q|d}\prod_{P|A}h(P^{\alpha},\ q^{\beta})$.

参考文献

[1] Hecke E. Lectures on the Theory of Algebraic Number [M]. Springer-Verlag, New York Inc, 1981.

[2] Ireland K. Rosen M. A Classical Introduction to Mordem Number Theory [M]. Springer-Verlay, New York Inc, 1982.

[3] 刘澄清,倪谷炎. 关于模 m 次数为 d 的整数个数的计算[J]. 江西师范大学学报(自然科学版),1997, 1(21): 40-42.

The Number of the Elements of Order d (mod A)

Abstract Let A be an ideal and d be a positive integer. We give a formula for the number of the solutions of the congruence equation $x^{d}\equiv 1(\bmod A)$ and analogue formula for the number of the elements of order d (mod A).

A remarkable class of congruences

Abstract In this paper, we deduce from one common source a class of congruences modulo odd primes p most of which are new, e. g.

$$\sum_{k=1}^{[p/6]}\frac{(-1)^k}{k}\equiv\sum_{k=1}^{(p-1)/2}\frac{3^k}{k}\pmod p,$$

$$\sum_{k=1}^{[p/8]}\frac{1}{k}\equiv\frac{5}{2}\sum_{k=1}^{(p-1)/2}\frac{1}{k}+\sum_{k=1}^{(p-1)/2}\frac{1}{k\cdot 2^k}\pmod p.$$

1 Introduction

The topic to which we contribute here came up quite recently when Zhi-Wei Sun [5] proved the rather curious congruence

$$\sum_{k=1}^{[3p/4]}\frac{(-1)^{k-1}}{k}\equiv\sum_{k=1}^{(p-1)/2}\frac{1}{k\cdot 2^k}\pmod p \tag{1}$$

where p denotes, as *always in the present paper*, an odd prime. His proof, using Pell sequences, was relatively complicated, and W. Kohnen's note [1] on a similar subject ended with the following remark. "It seems unlikely that (1) can be proved with the simple approach we have used here." What Kohnen [1] showed, in fact, was the congruence

$$\sum_{k=1}^{(p-1)/2}\frac{(-1)^{k-1}}{k}\equiv\sum_{k=1}^{p-1}\frac{1}{k\cdot 2^k}\pmod p. \tag{2}$$

Then, only very recently, there appeared two short articles containing really elementary proofs of (1), one by Zun Shan and E. T. H. Wang [4], and another one by Kohnen[2]. Whereas the procedure in [4] is based on certain manipulations with eight's roots of unity, in [2] the approach of [4] is generalized, by working with arbitrary 2-power roots of unity.

The purpose of our note is to state and prove the following result containing (1) as very special case.

Theorem. *Let p be an odd prime and n a positive integer. Then*

$$\sum_{k=1}^{[p/2n]}\frac{1}{k}\equiv\frac{n+1}{2}\sum_{k=1}^{(p-1)/2}\frac{1}{k}+\sum_{j=1}^{[(n-1)/2]}\sum_{k=1}^{(p-1)/2}\frac{\left(\cos\frac{\pi j}{n}\right)^{2k}}{k}\pmod{p},\tag{3}$$

where $[x]$ denotes the greatest integer not exceeding the real number x.

Note that the sum $\sum_{1\leqslant j<n/2}\left(\cos\frac{\pi j}{n}\right)^{2k}$ is rational for each positive integer $k<p/2$.

Since sums of type $\sum_{k=1}^{(p-1)/2}\cdots$ as in (1), (2), (3) occur very often in the sequel, we shall write them $\sum^{*}$ for short if the summation index is obvious.

In the subsequent Corollary we list some applications of the Theorem.

Corollary. *Modulo an odd prime p the following congruences hold*

$$\sum_{k=1}^{[p/3]}\frac{1}{k}\equiv\sum{}^{*}\frac{1}{k}+\frac{1}{2}\sum_{k=1}^{p-1}\frac{3^k}{k}-\frac{1}{2}\sum{}^{*}\frac{(-3)^k}{k},\tag{4}$$

$$\sum_{k=1}^{[p/4]}\frac{(-1)^k}{k}\equiv\sum{}^{*}\frac{1}{k}+\sum{}^{*}\frac{1}{k\cdot 2^k},\tag{5}$$

$$\sum_{k=1}^{[p/6]}\frac{1}{k}\equiv2\sum{}^{*}\frac{1}{k}+\sum{}^{*}\frac{1}{k\cdot 4^k}\equiv\frac{3}{2}\sum{}^{*}\frac{1}{k}+\frac{1}{2}\sum_{k=1}^{p-1}\frac{3^k}{k},\tag{6}$$

$$\sum_{k=1}^{[p/6]}\frac{(-1)^k}{k}\equiv\sum{}^{*}\frac{3^k}{k},\tag{7}$$

$$\sum_{k=1}^{[p/8]}\frac{1}{k}\equiv\frac{5}{2}\sum{}^{*}\frac{1}{k}+\sum{}^{*}\frac{1}{k\cdot 2^k},\tag{8}$$

$$\sum_{k=1}^{[p/12]}\frac{1}{k}\equiv\frac{7}{2}\sum{}^{*}\frac{1}{k}+\sum{}^{*}\frac{1}{k\cdot 4^k}+\sum{}^{*}\frac{3^k}{k\cdot 4^k}.\tag{9}$$

By Lemma 3(i) below, the equivalence of the two congruences (1) and (5) becomes obvious. It should be remarked also that our formula (7) was annonced at the end of [5], and proved in [6].

Furthermore, it is well known that $B_{p-1}-B_{p-1}(t)\equiv\sum_{k=1}^{t-1}1/k \pmod p$ holds for all $t\in\{1,\cdots,p\}$. Here B_{p-1} and $B_{p-1}(.)$ are the $(p-1)$-th Bernoulli number and polynomial, respectively. Therefore, we can obtain a lot of new congruences for the values of Bernoulli polynomials, too.

2 Some auxiliary lemmas

The first lemma is so easy that we omit any proof.

Lemma 1. *Modulo p^2 we have the following congruences*

(i) $(-1)^{k-1}\binom{p}{k}\equiv\frac{p}{k}$ for $0<k\leqslant p$,

(ii) $\binom{p}{2k}\equiv-\frac{p}{2k}$ for $0<k<\frac{p}{2}$,

(iii) $(-1)^{k-1}\binom{2p}{k}\equiv(-1)^{[(k-1)/p]}\frac{2p}{k}$ for $0<k\leqslant 2p$.

Lemma 2. *With $\zeta_n:=e^{2\pi i/n}$ and the rational integers $S_n:=\sum_{j=1}^{n}(1+\zeta_n^j)^p$ we have*

$$\sum_{k=1}^{[p/n]}\frac{(-1)^{nk}}{k}\equiv\frac{n-S_n}{p}\pmod p.$$

Proof. From the definition of S_n we get

$$\begin{aligned}S_n&=\sum_{j=1}^{n}\sum_{k=0}^{p}\binom{p}{k}\zeta_n^{jk}=n+\sum_{k=1}^{p}\binom{p}{k}\sum_{j=1}^{n}\zeta_n^{jk}=n+n\sum_{k=1}^{[p/n]}\binom{p}{nk}\\&\equiv n-p\sum_{k=1}^{[p/n]}\frac{(-1)^{nk}}{k}\pmod{p^2},\end{aligned}$$

using Lemma 1(i), and thus the asserted result.

Lemma 3. *The subsequent congruences hold modulo p*

(i) $\sum_{k=1}^{[p/2]}\frac{1}{k}\equiv\frac{2-2^p}{p}\equiv2\sum{}^{*}\frac{(-1)^k}{k}$,

$$\text{(ii)} \sum_{k=1}^{[p/3]} \frac{(-1)^k}{k} \equiv \delta_{3,p} + \sum{}^* \frac{1}{k} \equiv \frac{1}{2}\Big(\sum{}^* \frac{1}{k} + \sum{}^* \frac{(-3)^k}{k}\Big),$$

$$\text{(iii)} \sum_{k=1}^{[p/4]} \frac{1}{k} \equiv \frac{3}{2}\sum{}^* \frac{1}{k}.$$

In (ii), $\delta_{3,p}$ *denotes Kronecker's symbol.*

Proof. Applying Lemma 2 with $n=2$ we get the first part of (i); the second one comes from

$$2-2^p = -\sum_{k=1}^{p-1}\binom{p}{k} \equiv p\sum_{k=1}^{p-1}\frac{(-1)^k}{k} \equiv 2p\sum{}^* \frac{(-1)^k}{k} \pmod{p^2}$$

where we used Lemma 1(i). Since (ii) is obvious for $p=3$, we prove it only for $p>3$. Then, from the definition of S_3, we get

$$S_3 = 2^p + 2\mathrm{Re}(1+e^{2\pi i/3})^p = 2^p + 2\mathrm{Re}e^{\pi i p/3} = 2^p + 1.$$

Lemma 2 as well as (i) of the present lemma lead to the first congruence in (ii), the second one coming from

$$\begin{aligned} 2^{p-1} &= 2^p\mathrm{Re}e^{\pi i/3} = 2^p\mathrm{Re}(1+e^{2\pi i/3})^p = \mathrm{Re}(1+\sqrt{-3})^p \\ &= 1+\sum{}^*\binom{p}{2k}(-3)^k \equiv 1-\frac{p}{2}\sum{}^*\frac{(-3)^k}{k} \end{aligned}$$

modulo p^2, whence $\sum{}^*(-3)^k/k \equiv (2-2^p)/p \pmod{p}$ for $p>3$. To prove finally (iii) we combine

$$\begin{aligned} S_4 &= 2^p + 2\mathrm{Re}(1+i)^p = 2^p + 2\sum_{k=0}^{[p/2]}(-1)^k\binom{p}{2k} \\ &\equiv 2^p + 2 - p\sum{}^*\frac{(-1)^k}{k} \pmod{p^2}, \end{aligned}$$

using Lemma 1(ii), with Lemma 2 and (i) above.

Our last lemma is concerned with the Chebyshev polynomials T_k of the first kind, defined by $T_k(x)=\cos k\Theta$, $x=\cos\Theta$, for $k=0, 1, \cdots$

Lemma 4. (i) *The polynomials* T_k *satisfy* $T_0(x)=1$ *and*, *for* $k=1, 2, \cdots$

$$T_k(x)=\sum_{l=0}^{[k/2]}a_{k,\,k-2l}x^{k-2l} \quad with \quad a_{k,\,k-2l}:=(-1)^l\frac{k}{k-l}\binom{k-l}{l}2^{k-1-2l}.$$

(ii) *For any* $\lambda\in\{1,\cdots,(p-1)/2\}$ *the following congruence holds*

$$\sum_{\kappa=\lambda}^{(p-1)/2}\frac{1}{\kappa}a_{2\kappa,\,2\lambda}\equiv\frac{1}{2\lambda}(\bmod p).$$

Proof. For (1) we refer to [3], p. 54. Using (i) and $\binom{n}{m}=\frac{n}{n-m}\binom{n-1}{m}$ for any integer m, n with $n>m\geqslant 0$ we find

$$\begin{aligned}2\sum_{\kappa=\lambda}^{(p-1)/2}\frac{1}{\kappa}a_{2\kappa,\,2\lambda}&=2^{1+2\lambda}\sum_{\kappa=\lambda}^{(p-1)/2}(-1)^{\kappa-\lambda}\frac{1}{\kappa+\lambda}\binom{\kappa+\lambda}{\kappa-\lambda}\\&=\frac{2^{2\lambda}}{\lambda}\sum_{\kappa=\lambda}^{(p-1)/2}(-1)^{\kappa-\lambda}\binom{k+\lambda-1}{\kappa-\lambda}=\frac{2^{2\lambda}}{\lambda}\sum_{\mu=0}^{\frac{p-1}{2}-\lambda}(-1)^{\mu}\binom{2\lambda+\mu-1}{\mu}\\&=\frac{2^{2\lambda}}{\lambda}\sum_{\mu=0}^{\frac{p-1}{2}-\lambda}\binom{-2\lambda}{\mu}=\frac{2^{2\lambda}}{\lambda}\sum_{\mu=0}^{\nu}\binom{2\nu+1-p}{\mu}\end{aligned}$$

where $\nu:=\frac{p-1}{2}-\lambda\in\left\{0,\cdots,\frac{p-3}{2}\right\}$. Here the last sum is

$$\equiv\sum_{\mu=0}^{\nu}\binom{2\nu+1}{\mu}=\frac{1}{2}\sum_{\mu=0}^{2\nu+1}\binom{2\nu+1}{\mu}=2^{2\nu}=2^{p-1-2\lambda}\equiv 2^{-2\lambda}(\bmod p),$$

and thus (ii) is proved, too.

3 Proof of the main results

Proof of the Theorem. First, Lemma 2 implies

$$\sum_{k=1}^{[p/2n]}\frac{1}{k}\equiv\frac{2n-S_{2n}}{p}(\bmod p), \tag{10}$$

and we are looking for an appropriate representation of S_{2n}. From its above definition we get

$$S_{2n}=\sum_{j=1}^{2n}(1+e^{\pi ij/n})^{p}=2^{p}+2\mathrm{Re}\sum_{j=1}^{n-1}(1+e^{\pi ij/n})^{p}$$
$$=2^{p}+2(n-1)+2\sum_{j=1}^{n-1}\sum_{k=1}^{p}\binom{p}{k}\cos\frac{\pi jk}{n}. \tag{11}$$

Since $\cos\frac{\pi jk}{n}=T_k\left(\cos\frac{\pi j}{n}\right)$, the last double sum in (11) equals, by Lemma 4 (1),

$$\sum_{j=1}^{n-1}\sum_{k=1}^{p}\binom{p}{k}\sum_{\ell=0}^{[k/2]}a_{k,\,k-2\ell}\left(\cos\frac{\pi j}{n}\right)^{k-2\ell}=\sum_{k=1}^{p}\sum_{\ell=0}^{[k/2]}\binom{p}{k}a_{k,\,k-2\ell}\sum_{j=1}^{n-1}\left(\cos\frac{\pi j}{n}\right)^{k-2\ell}$$

where the last inner sum vanishes for odd k. Therefore the last triple sum equals

$$\sum_{\kappa=1}^{(p-1)/2}\sum_{\ell=0}^{\kappa}\binom{p}{2\kappa}a_{2\kappa,\,2(\kappa-\ell)}\sum_{j=1}^{n-1}\left(\cos\frac{\pi j}{n}\right)^{2(\kappa-\ell)}$$
$$=(n-1)\sum_{\kappa=1}^{(p-1)/2}(-1)^{\kappa}\binom{p}{2\kappa}+2\sum_{\kappa=1}^{(p-1)/2}\sum_{\lambda=1}^{\kappa}\binom{p}{2\kappa}a_{2\kappa,\,2\lambda}\sum_{j=1}^{[(n-1)/2]}\left(\cos\frac{\pi j}{n}\right)^{2\lambda}$$

where we used $a_{2\kappa,\,0}=(-1)^{\kappa}$, compare Lemma 4(i). Using Lemma 1(ii), our last calculations lead from (11) to

$$2n-S_{2n}$$
$$\equiv 2-2^{p}+p(n-1)\sum{}^{*}\frac{(-1)^{\kappa}}{\kappa}+2p\sum_{j=1}^{[(n-1)/2]}\sum_{\lambda=1}^{(p-1)/2}\left(\cos\frac{\pi j}{n}\right)^{2\lambda}\sum_{\kappa=\lambda}^{(p-1)/2}\frac{1}{\kappa}a_{2\kappa,\,2\lambda}$$

modulo p^2. Dividing this congruence by p and using (10), Lemma 3(i), and Lemma 4(ii) we deduce immediately the asserted congruence (3).

Proof of the Corollary. Applying (3) with $n=4$ gives immediately formula (8). Using (8) and Lemma 3(iii) we get (5) taking $m=4$ in

$$\sum_{k\leqslant p/m}(-1)^{k}/k=\sum_{k\leqslant p/2m}1/k-\sum_{k\leqslant p/m}1/k. \tag{12}$$

The first congruence in (6) again is a direct consequence of (3) with $n=3$; the second one can be seen from the following congruences modulo p^2

$$p\sum_{k<p}\frac{3^k}{k}\equiv-\sum_{k<p}\binom{p}{k}(-3)^k=-(1-3)^p+1-3^p$$
$$=(2-2^p)-((2+1)^p+(2-1)^p-2\cdot 2^p)$$
$$\equiv(2-2^p)-2\sum{}^*\binom{p}{2k}2^{p-2k}\equiv p\sum{}^*\frac{1}{k}+2p\sum{}^*\frac{1}{k\cdot 4^k}.$$

Here we used Lemma 1(i), (ii) and Lemma 3(i). In the same manner, (4) follows from (6) and Lemma 3(ii) if we apply (12) with $m=3$. Taking $n=6$ in (3) we find (9).

From (6), (9) and (12) we conclude

$$\sum_{k=1}^{[p/6]}\frac{(-1)^k}{k}\equiv\frac{3}{2}\sum{}^*\frac{1}{k}+\sum{}^*\frac{3^k}{k\cdot 4^k}\pmod p, \tag{13}$$

and for (7) we have to show that the right-hand side of (13) is $\equiv \sum^* 3^k/k \pmod p$. To do so, we apply all three items of Lemma 1 to deduce the following congruence modulo p^2

$$-\frac{p}{2}\sum{}^*\frac{3^k}{k\cdot 4^k}$$
$$=\sum{}^*\binom{p}{2k}\left(\frac{\sqrt{3}}{2}\right)^{2k}=\frac{1}{2}\left(\left(1+\frac{\sqrt{3}}{2}\right)^p+\left(1-\frac{\sqrt{3}}{2}\right)^p\right)-2$$
$$=\frac{1}{2\cdot 4^p}((1+\sqrt{3})^{2p}+(1-\sqrt{3})^{2p}-2\cdot 4^p)=\frac{1}{4^p}\sum_{k=1}^{p-1}\binom{2p}{2k}3^k+\frac{3^p+1-4^p}{4^p}$$
$$=-\frac{p}{4}\sum{}^*\frac{3^k}{k}+\frac{p}{4}\sum_{k=(p+1)/2}^{p-1}\frac{3^k}{k}-\frac{1}{4}\sum_{k=1}^{p-1}\binom{p}{k}3^k$$
$$=-\frac{p}{2}\sum{}^*\frac{3^k}{k}+\frac{p}{4}\sum_{k=1}^{p-1}\frac{3^k}{k}+\frac{p}{4}\sum_{k=1}^{p-1}\frac{(-3)^k}{k}.$$

From this chain of congruences we find, modulo p,

$$\sum{}^*\frac{3^k}{k\cdot 4^k}\equiv\sum{}^*\frac{3^k}{k}-\frac{1}{2}\sum{}^*\frac{3^{2k}}{k}\equiv\sum{}^*\frac{3^k}{k}+\frac{1}{p}\sum{}^*\binom{p}{2k}3^{2k}$$
$$\equiv\sum{}^*\frac{3^k}{k}+\frac{4^p-2^p-2}{2p}=\sum{}^*\frac{3^k}{k}-\frac{2^p+1}{2}\cdot\frac{2-2^p}{p}$$

$$\equiv \sum{}^{*} \frac{3^k}{k} - \frac{3}{2}\sum{}^{*} \frac{1}{k}$$

using once more Lemma 1 (ii) and Lemma 3 (i). Thus (7) is proved, too.

Let us finally remark that Kohnen's congruence (2) is an easy by-product of our considerations. Namely, using Lemma 1(i) and Lemma 3(i) we have

$$\sum_{k=1}^{p-1} \frac{1}{k \cdot 2^k} \equiv \sum_{k=1}^{p-1} \frac{2^{p-1-k}}{k} \equiv -\frac{1}{2p}\sum_{k=1}^{p-1}\binom{p}{k}(-1)^k 2^{p-k} = -\frac{1}{2p}((2-1)^p - 2^p + 1)$$

$$= -\frac{2-2^p}{2p} \equiv \sum{}^{*} \frac{(-1)^{k-1}}{k} (\bmod p).$$

References

[1] W. KOHNEN, A Simple Congruence Modulo p, *Amer. Math. Monthly*, **104** (1997), 444 - 445.

[2] W. KOHNEN, Some Congruences Modulo Primes, *Monatsh. Math.*, **127** (1999), 321 - 324.

[3] J. RIORDAN, *Combinatorial Identities*, John Wiley, New York et al., 1968.

[4] ZUN SHAN and E. T. H. WANG, A Simple Proof of a Curious Congruence by Sun, *Proc. Amer. Math. Soc.*, **127**(1999), 1289 - 1291.

[5] Zhi-Wei Sun, A Congruence for Primes, *Proc. Amer. Math. Soc.*, **123**(1995), 1341 - 1346.

[6] ZHI-WEI SUN, On the Combinatorial Sum $\sum_{k=0,\ k\equiv r(\bmod 12)}^{n}\binom{n}{k}$ and its Number-Theoretical Applications, to appear.

第二章　　数学竞赛

数学奥林匹克与奥林匹克数学　曲阜师范大学学报(自然科学版),1987(2)

互补的数列与一道国际数学竞赛题　数学教学,1979(2)

漫谈第二十一届国际数学竞赛题(与常庚哲、程龙合作) 数学通报,1980(7)

美国(1975,1976 年)数学竞赛题和解答(与常庚哲,程龙合作)　中学数学教学,1980(1)

第二十二届国际数学竞赛试题及解答(与常庚哲,程龙合作) 数学通报,1982(2)

第十届美国数学竞赛试题及解答　中学数学教学,1981(4)

1980 年芬兰等四国国际数学竞赛试题及解答　教学与研究(中学数学),1982(2)

第 13 届加拿大数学竞赛试题及解答(与程龙合作)　数学通讯,1982(6)

1978 年罗马尼亚数学竞赛决赛试题解答(与程龙合作) 数学通讯,1983(4)

1978 年罗马尼亚数学竞赛决赛试题解答(续)(与程龙合作) 数学通讯,1983(5)

第十一届美国数学竞赛试题及解答(与程龙合作)　中等数学,1983(3)

第二十五届国际数学竞赛　数学通讯,1985(3)

保加利亚数学竞赛试题及解答(与沈骅合作)　数学教师,1986(3)

第二十七届国际数学奥林匹克题解(与严镇军,杜锡录合作)　中学数学杂志,1986(9)

从批改一道竞赛题所想到的　中学数学教学,1987(1)

IMO 中的几何问题(与沈骅合作)　中等数学,1987(2)

奥林匹克集训班选拔考试题及解答　中学数学杂志,1987(4)

IMO 中的数论问题(上)(与余红兵合作)　中等数学,1988(2)

IMO 中的数论问题(下)(与余红兵合作)　中等数学,1988(3)

一道美国竞赛题的推广　中等数学,1989(4)

第 30 届 IMO 预选题解答(与刘亚强合作)　中等数学,1989(5)

1989 年亚洲太平洋地区数学竞赛题及解答(与曹鸿德合作)　数学通报,1989(12)

1989 年澳大利亚数学竞赛试题及解答　数学通报,1990(4)

第 31 届国际数学奥林匹克试题与解答(与杜锡录、刘鸿坤合作)　数学通报,1990(8)

我们的训练工作　数学通报,1990(10)

第 6 届巴尔干数学竞赛试题及解答(与刘亚强合作)　中等数学,1990(5)

1990 年加拿大数学竞赛(与刘亚强合作)　中等数学,1991(1)

1990 年亚太地区数学竞赛(与刘亚强合作)　中等数学,1991(1)

1992 年中国数学奥林匹克(与胡大同、孙瑞清合作)　数学通报,1992(4)

第 33 届国际数学奥林匹克　数学通报,1992(10)

评《首届全国数学奥林匹克命题比赛精选》　中等数学,1993(6)

1993 年德国数学奥林匹克(与浦敏亚合作)　中等数学,1994(2)

谈谈 1995 年全国高中联赛试题　数学通讯,1996(3)

谈谈 1996 年全国高中联赛　数学通讯,1997(3)

近两年的美国数学奥林匹克　中等数学,1998(5)

评 2001 年高中联赛　数学通讯,2002(1)

第 31 届(2002 年)美国数学奥林匹克　中等数学,2003(1)

评 2002 年全国高中数学联赛　数学通讯,2003(1)

评 2003 年高中联赛　数学通讯,2004(9)

谈 2004 年高中联赛　数学通讯,2005(9)

评 2005 年全国高中数学联赛　中学数学研究,2005(11)

2005 年中国数学奥林匹克的不等式题　中等数学,2005(6)

评 2006 年全国高中数学联赛　中学数学研究,2006(12)

怎样解今年(2007 年)国际数学奥林匹克试题　中学数学研究,2007(9)

评 2007 年安徽省高中数学竞赛　中学数学研究,2008(1)

不等式竞赛题的解法探究　中学教研(数学),2009(6)

评 2009 年全国高中数学联赛试题　中等数学,2009(12)

2009 年国家集训队几道试题的另解　中等数学,2010(8)

评 2010 年全国高中数学联赛试题　中等数学,2010(12)

评 2011 年全国高中数学联赛试题　中等数学,2011(12)

评 2012 年全国高中数学联赛试题　中等数学,2012(12)

数学奥林匹克与奥林匹克数学

本文介绍了数学奥林匹克的历史以及我国参加数学奥林匹克竞赛的情况. 文中阐述了奥林匹克数学的特点及其作用,指出这是介于中学数学与大学数学之间的"中间数学",它起着桥梁作用,提出在师范院校开设奥林匹克数学的课程与研究是有意义的.

1 数学奥林匹克

现代的数学奥林匹克主要是指中学生的数学竞赛,其目的在于发现与培养数学人才.

开竞赛先河的是东欧的一些国家. 早在 1894 年,匈牙利数学物理协会就通过了在全国举办中学数学竞赛的决议. 在匈牙利数学竞赛的优胜者中,有被誉为匈牙利现代数学之父的 Fejer,航天动力学的奠基人 Von Karman,组合数学家 Konig,哈尔测度与哈尔积分的提出者 Haar,对泛函分析有重大贡献的 Riesz,分析学家 Szego(他与 Polya 合著的《数学分析的定理与问题》"造就了两代数学家")及在函数论、微分几何、组合等领域都有建树的"多面手"Rado. 仅仅举出这些名字就足以表明数学竞赛确实是发现人才的重要途径.

1959 年,在罗马尼亚首都布加勒斯特举行了首届国际数学奥林匹克(IMO). 开始只有东欧及苏联参加. 后来西欧、美洲、亚洲及非洲的一些国家也陆续参加,IMO 发展成真正全球性的中学数学竞赛.

IMO 每年举行一次,每次的东道国经协商产生. 除 1979 年因主办国波兰政局不稳中断外,至 1986 年是第 27 届. 1987 年的第 28 届已确定在古巴举行. 希望能在不远的将来(例如 1990 年)在我国举办一届①.

每届竞赛在 7 月份举行. 试题共 6 题,分为两试,第一天上午 $4\frac{1}{2}$ 小时做一

① 1990 年 IMO 确由中国主办.

试的 3 道题,第二天上午 $4\frac{1}{2}$小时再做二试的 3 道题,通常是每道题 7 分,满分为 42 分.

试题由各参加国提供,在 4 月底交东道国的命题委员会集中.经过初次筛选后,剩下的题目交由各国的领队参加的领队会讨论,用民主协商的方式确定出 6 道题,译成各国文字.选手的解答是用本国文字写的,由本国的领队、副领队根据拟定的标准答案批改评分.

我国 1985 年首次参加比赛.由于时间仓促,来不及选拔,临时拟定北京王锋与上海吴思皓两人参加,集训也只有 10 天.这届试题难度甚大,我国选手吴思皓获三等奖,两人的总分 28 分,成绩是中等偏下(东道国芬兰 6 名选手共 25 分,远不及我国).第一次参加,是为了了解情况、取得经验,应当说这一目的是达到了.

1986 年我国派 6 名选手参加.为了准备这届比赛,在 1985 年省市联赛的基础上,选出 70 名选手,于 1986 年 1 月举办冬令营,再选出 21 名选手.这 21 名选手于 1986 年 3 月下旬在北京集中,参加集训.集训班在北京 101 中学,由胡大同、张宏才老师担任班主任,学生除语文、外语、体育等科(均由 101 中学的老师授课)外,主要学习数学,由北大、中科大、复旦、南开等学校派人讲课.整个工作由中国数学会普及工作委员会裘宗沪同志负责安排,他也是这次代表队的副领队.5 月初,经过选拔,采取民主的方式,确定正式参加比赛的 6 名队员.

由于准备比较充分,队员发挥较好,这届比赛中,方为民(河南郑州)、张浩(上海)、李平立(天津)三人获得一等奖,成为获得一等奖最多的两个国家之一(另一获得三个一等奖的是美国).我国女选手荆秦(陕西西安)获得二等奖,高二学生林强(湖北黄冈)获三等奖,江苏泰县沈健略差几分未能得奖.我国总分居第四位,这是一个出乎意料的好成绩.

成绩属于这 6 名选手,当然也属于参加训练工作的许多同志及广大从事或关心中学数学教育的同志.但是,我们必须有冷静的头脑.1986 年的试题比 1985 年容易,而且第二道题是我国提供的,虽然事前学生不知道试题,但他们熟悉用复数解几何题的方法,因而每人都稳拿了 7 分.第三道题(见下文),我国学生的理解与原意有出入,评分偏高.这届的第一名(美、苏并立)二百多分,第三名一百九十多分,比我国高出不少,而第五、六名与我们都是一百七十多分,相差甚微. 因此 1987 年的二十八届 IMO,如果我们能保持在前十名之列,就应当说不错了.

2　奥林匹克数学

随着数学竞赛的开展,一个新的名词——奥林匹克数学应运而生. 它不是大学数学,因为它的内容并不超出中学或中学生所能接受的范围. 它也不是中学数学,因为它有很多高等数学的背景,采用了很多大学数学中的方法. 这是一种"中间数学",起着桥梁作用,联系着中学数学与大学数学. 很多新的内容、方法通过这一座桥,源源不断地输入中学,对于中学数学的改革,从内容、思想到教师的素质,都起着重要的作用. 国外 20 世纪六十至七十年代的"新数学"运动效果不太理想,其原因之一是缺少一个渐变阶段. 如果将新的内容先在数学竞赛中进行试验,在培养尖子学生的奥林匹克学校或数学物理学校(匈、罗等国都有这样的学校,我国也有一些省市已经或正在开设)讲授,再逐步渗入中学,是一种很好的方法,也是目前的一种潮流. 由此可见,在师范院校开设奥林匹克数学的课程与研究,也是十分有意义的事.

陈省身教授曾经说过,"一个好的数学家与一个蹩脚的数学家,差别在于前者手中有很多具体的例子,后者则只有抽象的理论. "奥林匹克数学给大学数学提供了很多具体的例子. 因此,学习它可以加深对一般理论的理解,真正将知识学"活"了,"化"为己有.

下面举两个例子,从这两个例子可以窥见奥林匹克数学的一斑.

例 1　$2n+1$ 个正整数 $a_1, a_2, \cdots, a_{2n+1}$ 具有这样的性质:从中任取一个数,剩下的 $2n$ 个数总可以分为两组,每组 n 个数,并且两组的和相等. 求证:这 $2n+1$ 个数相等.

证明　我们把上述性质称为性质 P,显然,在 $a_1, a_2, \cdots, a_{2n+1}$ 具有性质 P 时, $a_1+1, a_2+1, \cdots, a_{2n+1}+1$ 与 $a_1/2, a_2/2, \cdots, a_{2n+1}/2$ 仍具有性质 P.

在取出 a_i 后,剩下的 $2n$ 个数的和是偶数(等于和相等的两组数的和的两倍),因此 a_i 的奇偶性与这 $2n+1$ 个数的和的奇偶性相同,从而 $a_1, a_2, \cdots, a_{2n+1}$ 的奇偶性一定相同.

对 $a_1, a_2, \cdots, a_{2n+1}$ 施行变换 S:如果它们都是偶数,将每个数都除以 2;

如果它们都是奇数,将每个数加上 1 再除以 2. 经过变换 S,得到的 $2n+1$ 个数仍然具有性质 P,但每个数比原来的数小,仅在原来的数为 1 时,所得的数也是 1.

继续施行变换 S,如果 $2n+1$ 个数中至少有一个不是 1. 由于严格单调递减的自然数的数列只能有有限多项,所以上述变换进行有限多次后就应当停止. 这时 $2n+1$ 个数都是 1. 它们是相等的,因而原来的 $2n+1$ 个数也必定是相等的.

这个问题并未用到超出中学范围的知识,但其中"变换""不变性(变换后仍具有性质 P)""单调递减""有限"等都是大学数学中常见的思想方法,并且又是(如果通过具体的例子)中学生(至少是优秀的中学生)可以接受的.

这道题对于大学生,也是颇有好处的. 获得上面的解答后,一个优秀的学生(更不必说优秀的教师)应当考虑结论能否推广? 这种推广可以说是从事研究工作的开端.

首先,不难看出条件"自然数"可以改为"整数",只要将每个整数同加上一个足够大的正数就可以化为自然数的情况. 同样地,"整数"可进一步推广为"有理数",因为每一个有理数乘以它们分母的公倍数即可化为整数.

结论能不能推广到"实数"? 这有一点困难. 不过,如果结论对实数成立,那么结论对复数也一定成立. 因为每个复数可写成 $a+bi$, 将实部与虚部分开,即化为实数的情形. 这正好启发我们找出从有理数推广到实数的一个证明.

事实上,上面是说复数是实数域上的二维向量空间. 实数域是有理数域上的无限维空间,但这里并不需要全体实数,只需要所给的 $2n+1$ 个实数在有理数域上生成的向量空间,它当然是有限维的,设其维数为 m(显然 $m \leqslant 2n+1$),考虑这 $2n+1$ 个数在一组基上的分解,问题便(和复数的情况类似)化为有理数的情况了.

熟悉线性代数的同学还可以用线性方程组的理论来处理(不必分为自然数、整数、……逐步推广,也不必利用域论的知识). 取出 a_1 后得到方程

$$\pm a_2 \pm a_3 \pm \cdots \pm a_{2n+1} = 0,$$

其中有一半取"+"号,另一半则取"−"号.

类似地,有

$$\pm a_1 \pm a_2 \pm \cdots \pm a_{i-1} \pm a_{i+1} \pm \cdots \pm a_{2n+1} = 0,\ (i = 1, 2, \cdots, 2n+1). \quad (1)$$

方程组(1)显然有解 $a_1 = a_2 = \cdots = a_{2n+1} = K.$ (2)

我们要证明它的解一定为(2),即它的解空间是一维的.

事实上,$2n$ 阶行列式

$$D = \begin{vmatrix} \pm 1 & \pm 1 & \pm 1 & \cdots & \pm 1 & \pm 1 \\ 0 & \pm 1 & \pm 1 & \cdots & \pm 1 & \pm 1 \\ \pm 1 & 0 & \pm 1 & \cdots & \pm 1 & \pm 1 \\ & & \cdots & & & \\ \pm 1 & \pm 1 & \pm 1 & \cdots & 0 & \pm 1 \end{vmatrix} \neq 0 \quad (3)$$

也就是(1)的系数矩阵的秩为 $2n$(它不是满秩的,因为有非零解(2)),所以解空间是一维的.

(3) 式 $D \neq 0$ 的证明可用除以 2 的余数来考虑(更确切些说是在有限域 F_2 上来考虑),这时

$$D \equiv \begin{vmatrix} 1 & 1 & 1 & \cdots & 1 & 1 \\ 0 & 1 & 1 & \cdots & 1 & 1 \\ 1 & 0 & 1 & \cdots & 1 & 1 \\ & & \cdots & & & \\ 1 & 1 & 1 & \cdots & 0 & 1 \end{vmatrix} (\mathrm{mod} 2).$$

易知 $D \equiv 1 \not\equiv 0 (\mathrm{mod} 2)$,因而 $D \neq 0$.

这个例子大概已足够说明,指导奥林匹克数学,必须有更高一点的水平,才能高屋建瓴,居高临下,所以师范院校的同学学习高等数学是非常必要的.

例 2　在正五边形的每个顶点各有一个整数,记为 x、y、z、u、v,已知 $x + y + z + u + v > 0$. 如果这五个数中有某个数,例如 $y < 0$,则施行如下的变换:将 x、y、z、u、v 变为 $x + y$, $-y$, $y + z$, u, v. 证明:经过有限多步变换后,这五个数全变为非负的.

这就是 27 届国际数学竞赛的第 3 题,题目有两种不同的理解. 第一种是在作变换时没有选择,见到负数便施行变换,而不是从几个负数中挑选一个(比如说绝对值最大的)来施行变换. 第二种是可以选择. 前一种较难而且它的解答当

然也适用于第二种,应当以前一种理解为准.

证明 解答的要点是建立一个函数 $f(x, y, z, u, v)$,取非负整数值,而且在施行变换时,数值严格递减(于是在有限步后,不再能施行变换,也就是五个数都是非负的),这样的函数可取 $|x|+|y|+|z|+|u|+|v|+|x+y|+|y+z|+|z+u|+|u+v|+|v+x|+|x+y+z|+|y+z+u|+|z+u+v|+|u+v+x|+|v+x+y|+|x+y+z+u|+|y+z+u+v|+|z+u+v+x|+|u+v+x+y|+|v+x+y+z|$ 或者

$$(x-z)^2+(y-u)^2+(z-v)^2+(u-x)^2+(v-y)^2.$$

本题中的"正"字是多余的.五边形可推广到 n 边形,没有实质性的困难.整数也可以改为实数,只要用一点极限知识(但并不容易),这是一道很好的分析习题.

奥林匹克数学的侧重点不在知识,而在能力,尤其有助于使学生"在数学上成熟起来".这是很好的活动,也很值得我们去研究.

Mathematics Olympic and Olympic Mathematics

Abstract

In this paper, an introduction to the history of Mathematics Olympic and the situation that China takes part in it are given. The author expounded the character and usefulness of Olympic Mathematics; and pointed out that it is a "secondary mathematics" lies between the school mathematics and the mathematics, playing a role as bridge; and also illustrated that it is significant to offer a course of Olympic Mathematics and study it in normal colleges.

互补的数列与一道国际数学竞赛题

1

数列互补的定义及定理：

定义　如果两个递增的正整数数列$\{f(n)\}$、$\{g(m)\}$满足下面两个条件：

(i) 这两个数列没有相同的项，即对任意的正整数m、n，有$f(n)\neq g(m)$；

(ii) 每一个正整数k，都必定在数列$\{f(n)\}$或$\{g(m)\}$中出现，即总可以找到正整数n或m，使得$k=f(n)$或$k=g(m)$.

那么$\{f(n)\}$与$\{g(m)\}$就叫做**互补的数列**.

例如，奇数数列1，3，5，…与偶数数列2，4，6，…就是互补的数列.

下面介绍两个关于互补数列的定理.

定理1　如果α、β是正的无理数，并且$\frac{1}{\alpha}+\frac{1}{\beta}=1$，那么数列$\{[n\alpha]\}$与数列$\{[m\beta]\}$是互补的数列，其中$[x]$表示不超过$x$的最大整数.

证明　因为α、β是正的无理数，所以

$$\frac{1}{\alpha}<\frac{1}{\alpha}+\frac{1}{\beta}=1,\ \alpha>1,$$

从而

$$[(n+1)\alpha]=[n\alpha+\alpha]\geqslant[n\alpha+1]=[n\alpha]+1>[n\alpha],$$

即$\{[n\alpha]\}$是递增的. 同理，$\{[m\beta]\}$也是递增的.

对于任意一个正整数k，取

$$n=\left[\frac{k}{\alpha}\right]+1,\ m=\left[\frac{k}{\beta}\right]+1,$$

又令

$$\xi=n-\frac{k}{\alpha}=\left[\frac{k}{\alpha}\right]+1-\frac{k}{\alpha},$$

$$\eta = m - \frac{k}{\beta} = \left[\frac{k}{\beta}\right] + 1 - \frac{k}{\beta}.$$

因为 α 是无理数，

$$\left[\frac{k}{\alpha}\right] < \frac{k}{\alpha} < \left[\frac{k}{\alpha}\right] + 1,$$

所以 $0 < \xi < 1$. 同理 $0 < \eta < 1$. 所以

$$0 < \xi + \eta < 2.$$

又 $\xi + \eta = n - \frac{k}{\alpha} + m - \frac{k}{\beta} = n + m - k$ 是一个整数,所以只能是 $\xi + \eta = 1$, 即

$$\frac{\alpha\xi}{\alpha} + \frac{\beta\eta}{\beta} = 1.$$

显然 $\alpha\xi = \alpha n - k$ 及 $\beta\eta = \beta m - k$ 都是无理数,不可能等于 1. 如果 $\alpha\xi$ 与 $\beta\eta$ 都大于 1,那么

$$\frac{\alpha\xi}{\alpha} + \frac{\beta\eta}{\beta} > \frac{1}{\alpha} + \frac{1}{\beta} = 1,$$

与上式矛盾. 如果 $\alpha\xi$ 与 $\beta\eta$ 都小于 1,那么

$$\frac{\alpha\xi}{\alpha} + \frac{\beta\eta}{\beta} < \frac{1}{\alpha} + \frac{1}{\beta} = 1,$$

也与上式矛盾. 所以 $\alpha\xi$ 与 $\beta\eta$ 中有一个大于 1,另一个小于 1. 不妨假定 $\alpha\xi > 1$, 而 $\beta\eta < 1$. 于是 $[\alpha\xi] \geqslant 1$, $[\beta\eta] = 0$.

因为 $\beta\eta = m\beta - k$, 所以

$$[m\beta] = [k + \beta\eta] = k + [\beta\eta] = k,$$

这就证明了 $\{[n\alpha]\}$ 与 $\{[m\beta]\}$ 满足条件(ii).

因为 $\alpha\xi = n\alpha - k$, 所以

$$[n\alpha] = [k + \alpha\xi] = k + [\alpha\xi] \geqslant k + 1 > k,$$

又 $n - 1 = \left[\frac{k}{\alpha}\right] < \frac{k}{\alpha}$, 所以

$$[(n-1)\alpha]<(n-1)\alpha<k.$$

因此在数列$\{[m\beta]\}$中出现的k不会在数列$\{[n\alpha]\}$中出现. 这也就证明了$\{[n\alpha]\}$与$\{[m\beta]\}$满足条件(i).

这就证明了$\{[n\alpha]\}$与$\{[m\beta]\}$是互补的数列,定理1证毕.

定理2 如果γ是正的无理数,

$$g(m)=[m\gamma]+m,\ f(n)=\left[\frac{n}{\gamma}\right]+n,$$

那么$\{g(m)\}$与$\{f(n)\}$是互补的数列.

证明 令$\alpha=\gamma+1$, $\beta=\dfrac{1}{\gamma}+1$, 那么$\alpha$、$\beta$都是正的无理数,并且

$$\frac{1}{\alpha}+\frac{1}{\beta}=\frac{1}{\gamma+1}+\frac{1}{\frac{1}{\gamma}+1}=1,$$

利用定理1就得到定理2. 当然,反过来,由定理2也可以推出定理1.

2

第二十届国际中学生数学竞赛的第3题,是一个有关互补数列的问题. 原题如下:

设f、g: $\mathbf{Z}^+\to\mathbf{Z}^+$为递增函数,且

$$f(\mathbf{Z}^+)\cup g(\mathbf{Z}^+)=\mathbf{Z}^+,\ f(\mathbf{Z}^+)\cap g(\mathbf{Z}^+)=\varnothing,$$
$$g(n)=f(f(n))+1,$$

求$f(n)$,其中$\mathbf{Z}^+$是正整数集合,$\varnothing$是空集.

利用我们的术语,上题就是“求出满足

$$g(n)=f(f(n))+1 \tag{1}$$

的互补数列$\{f(n)\}$与$\{g(m)\}$.”

解 令

$$f(n)=\left[\frac{\sqrt{5}-1}{2}n\right]+n=\left[\frac{\sqrt{5}+1}{2}n\right],$$

$$g(n)=f(f(n))+1=\left[\frac{\sqrt{5}-1}{2}f(n)\right]+f(n)+1$$
$$=\left[\frac{\sqrt{5}-1}{2}\left[\frac{\sqrt{5}+1}{2}n\right]\right]+f(n)+1.$$

因为

$$\frac{\sqrt{5}-1}{2}\left[\frac{\sqrt{5}+1}{2}n\right]<\frac{\sqrt{5}-1}{2}\cdot\frac{\sqrt{5}+1}{2}n=n,$$

且

$$\frac{\sqrt{5}-1}{2}\left[\frac{\sqrt{5}+1}{2}n\right]>\frac{\sqrt{5}-1}{2}\left(\frac{\sqrt{5}+1}{2}n-1\right)=n-\frac{\sqrt{5}-1}{2}>n-1,$$

所以

$$\left[\frac{\sqrt{5}-1}{2}\left[\frac{\sqrt{5}+1}{2}n\right]\right]=n-1.$$

从而

$$g(n)=n+f(n), \tag{2}$$

即 $$g(n)=n+\left[\frac{\sqrt{5}+1}{2}n\right]=\left[\frac{\sqrt{5}+3}{2}n\right].$$

由于 $\left(\frac{\sqrt{5}+1}{2}\right)^{-1}=\frac{\sqrt{5}-1}{2}$，根据定理 2，数列$\{f(n)\}$与$\{g(m)\}$是互补数列.

所以原竞赛题的解是：

$$f(n)=\left[\frac{\sqrt{5}+1}{2}n\right],\ g(n)=\left[\frac{\sqrt{5}+3}{2}n\right].$$

3

可以证明，该竞赛题的解是唯一的.

设 $f_1(n)$、$g_1(n)$及 $f_2(n)$、$g_2(n)$都是问题的解. 因为$g(n)=f(f(n))+1>1$，所以 1 不在数列$\{g_1(n)\}$及$\{g_2(n)\}$中出现，只能是 $f_1(1)=f_2(1)=1$. 下

面用数学归纳法来讨论.

假定有 $f_1(k)=f_2(k)$, $k\leqslant n$,我们要证明 $f_1(n+1)=f_2(n+1)$.

1° 如果有 $n=f_1(u)$, 那么由于 $f(1)$, $f(2)$, $\cdots$, $f(n)$, $\cdots$是递增的正整数数列, $f_1(n)\geqslant n$,所以 $u\leqslant n$. 从而

$$f_2(u)=f_1(u)=n,\ g_1(u)=f_1(n)+1,$$
$$g_2(u)=f_2(n)+1.$$

$f_1(n)+2$ 必在 $f_1(1)$, $f_1(2)$, $\cdots$, $f_1(n)$, $f_1(n+1)$, $\cdots$中出现,不然的话, $f_1(n)+2=g_1(v)$,

即
$$g_1(u)+1=f_1(f_1(v))+1,$$

从而有 $g_1(u)=f_1(f_1(v))$, 此与互补数列的定义矛盾.

因此 $f_1(n+1)=f_1(n)+2$,同样 $f_2(n+1)=f_2(n)+2$, 所以

$$f_1(n+1)=f_2(n+1).$$

2° 如果有 $n=f_2(v)$, 同样可得

$$f_1(n+1)=f_2(n+1).$$

3° 如果 $n\neq f_1(u)$ 并且 $n\neq f_2(v)$, 那么

$$f_1(n)+1\neq g_1(u),$$

从而 $f_1(n)+1$ 一定在数列 $f_1(1)$, $f_1(2)$, $\cdots$, $f_1(n)$, $f_1(n+1)$, $\cdots$中出现, 因此 $f_1(n+1)=f_1(n)+1$, 同样 $f_2(n+1)=f_2(n)+1$, 所以

$$f_1(n+1)=f_2(n+1).$$

于是 $f_1(n)\equiv f_2(n)$,从而 $g_1(n)\equiv g_2(n)$. 所以 $f(n)=\left[\frac{\sqrt{5}+1}{2}n\right]$ 是竞赛题的唯一解.

4

对于互补的数列$\{f(n)\}$与$\{g(n)\}$,前面的(1)、(2)两式是等价的,即由(1)式可以推出(2)式,由(2)式也可以推出(1)式.

证明 如果(1)式成立,那么不超过 $g(n)$ 的正整数 $1, 2, 3, \cdots, g(n)$(共有 $g(n)$ 个)中,有 n 个在数列 $\{g(n)\}$ 中,即 $g(1), g(2), \cdots, g(n)$. 又有 $f(n)$ 个在数列 $\{f(n)\}$ 中,即 $f(1), f(2), \cdots, f(f(n))$,所以 $g(n)=n+f(n)$,即证得了(2)式成立.

反过来,如果(2)式成立,在 $1, 2, 3, \cdots, g(n)$ 中除去 n 个数 $g(1), g(2), \cdots, g(n)$,剩下的 $g(n)-n=f(n)$ 个数都在数列 $\{f(n)\}$ 中,即 $f(1), f(2), \cdots, f(f(n))$ 都不超过 $g(n)$,但其中不超过 $g(n-1)$ 的只有 $f(n-1)$ 个,即 $f(1), f(2), \cdots, f(f(n-1))$,所以 $g(n-1)<f(f(n))<g(n)$,因此,$g(n)=f(f(n))+1$. 即证得了(1)式成立.

所以,竞赛题也就是求出满足(2)式的互补数列 $\{f(n)\}$ 与 $\{g(n)\}$. 这当然可以按照上面的步骤,用定理 1 或 2 来解决.

5

有了定理 1 或 2,我们也就可以知道解是怎样得到的,或者说题目是怎样造出来的.

我们希望使互补的数列 $g(m)=[m\gamma]+m$ 与 $f(n)=\left[\dfrac{n}{\gamma}\right]+n$ 满足(1)式或(2)式,即希望对任何自然数 n,有

$$[n\gamma]=\left[\frac{n}{\gamma}\right]+n, \tag{3}$$

从而
$$n\gamma-\varepsilon_1=\frac{n}{\gamma}-\varepsilon_2+n,$$

其中 $\varepsilon_1=n\gamma-[n\gamma]$ 和 $\varepsilon_2=\dfrac{n}{\gamma}-\left[\dfrac{n}{\gamma}\right]$ 都是小于 1 的正数. 由此 $\gamma=\dfrac{1}{\gamma}+1+\dfrac{\varepsilon_1-\varepsilon_2}{n}$,令 $n\to\infty$,得

$$\gamma=\frac{1}{\gamma}+1. \tag{4}$$

由(4)式立即得到 $\gamma=\dfrac{\sqrt{5}+1}{2}$.

6

问题显然可以推广为：求满足 $g(n)=f(f(n)+kn)+1$ 的互补数列 $\{f(n)\}$、$\{g(m)\}$，其中 k 为非负整数.

答案为 $f(n)=\left[\frac{n}{\gamma}\right]+n$，$g(m)=[m\gamma]+m$，其中 $\gamma=\frac{k+1+\sqrt{(k+1)^2+4}}{2}$.

7

定理 2 其实是一个更一般的定理的特殊情况. 我们先引入伴随数列的定义：

如果 $\{F(n)\}$ 为不降的非负整数数列，令 $F^*(m)$ 为满足 $F(n)<m$ 的 n 的个数，则 $\{F^*(m)\}$ 称为 $\{F(n)\}$ 的**伴随数列**.

例 1　$\{F(n)\}$：1, 3, 5, 7, …，则 $\{F^*(m)\}$：0, 1, 1, 2, ….

例 2　$\{F(n)\}$：0, 1, 2, 3, …，则 $\{F^*(m)\}$：1, 2, 3, 4, ….

例 3　设 γ 为正无理数，$F(n)=[n\gamma]$，则 $F^*(m)=\left[\frac{m}{\gamma}\right]$.

例 4　设 $F(n)=n^2$，则 $F^*(m)=[\sqrt{m-1}]$.

例 5　一般地，设 $\psi(x)\geqslant 0$ 为递增函数，$F(n)=[\psi(n)]$，则 $F^*(m)$ 为小于 $\psi^{-1}(m)$ 的自然数的个数，如果 $\psi^{-1}(m)$ 不是整数，$F^*(m)=[\psi^{-1}(m)]$，如果 $\psi^{-1}(m)$ 是整数，$F^*(m)=\psi^{-1}(m)-1$，其中 ψ^{-1} 是 ψ 的反函数.

从 $F^*(m)$ 的定义，立即得出 $\{F^*(m)\}$ 是不降的非负整数数列，并且在 $m\leqslant F(n)$ 时，如果 $F(k)<m$，那么 $k<n$，因此满足 $F(k)<m$ 的 k 的个数 $<n$，即 $F^*(m)<n$. 反过来，在 $F^*(m)<n$ 时，满足 $F(k)<m$ 的 k 的个数 $<n$，因此 $F(n)\not< m$，即 $m\leqslant F(n)$.

于是满足 $F^*(m)<n$ 的 m 恰好有 $F(n)$ 个，所以 $(F^*)^*(n)=F(n)$，我们可以说 $\{F(n)\}$ 与 $\{F^*(m)\}$ 互为伴随数列，简称为互伴数列.

从上面的推导还可以看出,如果 $F(n)$ 与 $G(m)$ 是互伴数列,那么 $F(n)<m$ 与 $G(m)<n$ 有且仅有一个成立.

8

下面的定理指出了互补数列与互伴数列的关系.

定理 3 令 $f(n)=F(n)+n$, $g(m)=G(m)+m$, 则当且仅当 $\{F(n)\}$ 与 $\{G(m)\}$ 是互伴数列时, $\{f(n)\}$ 与 $\{g(m)\}$ 是互补数列.

证明 如果 $\{F(n)\}$ 与 $\{G(m)\}$ 是互伴数列,那么 $F(n)<m$ 与 $G(m)<n$ 有且仅有一个成立,因此 $f(n)<m+n$ 与 $g(m)<m+n$ 有且仅有一个成立.假如有 $f(n)=g(m)$, 那么 $f(n)$ 与 $g(m)$ 同时小于 $m+n$ 或者同时不小于 $m+n$, 这就导出矛盾,所以 $f(n)$ 与 $g(m)$ 决不相等.

对于任一正整数 k,考虑不超过 $f(k)$ 的正整数 $1, 2, \cdots, f(k)$. 其中属于数列 $\{f(n)\}$ 的有 k 个,即 $f(1), f(2), \cdots, f(k)$. 而在 $m \leqslant F(k)$ 时, $G(m)<k$, 所以

$$g(m)=m+G(m)<F(k)+k=f(k),$$

因此有 $F(k)$ 个属于 $\{g(m)\}$ 的数,即 $g(1), g(2), \cdots, g(F(k))$ 不超过 $f(k)$. 因为 $f(k)=k+F(k)$, 所以 $f(k)$ 个数 $1, 2, \cdots, f(k)$ 必定在 $\{f(n)\}$ 或 $\{g(m)\}$ 中出现,当然 k 也在 $\{f(n)\}$ 或 $\{g(m)\}$ 中出现.

又由于 $\{F(n)\}$ 与 $\{G(m)\}$ 是不降的,所以 $\{f(n)\}$ 与 $\{g(m)\}$ 是递增的,这就证得了 $\{f(n)\}$ 与 $\{g(m)\}$ 是互补数列.

反过来,如果 $\{f(n)\}$ 与 $\{g(m)\}$ 是互补数列,那么 $\{f(n)\}$ 递增,所以

$$f(n)-f(n-1) \geqslant 1,$$

从而

$$\begin{aligned} F(n)-F(n-1) &= (f(n)-n)-(f(n-1)-(n-1)) \\ &\geqslant f(n)-f(n-1)-1 \geqslant 0, \end{aligned}$$

即 $\{F(n)\}$ 不降,同样 $\{G(m)\}$ 也是不降的.

令 $h(m)=F^{*}(m)+m$, 由于 $\{F^{*}(m)\}$ 与 $\{F(n)\}$ 互伴,所以 $\{h(m)\}$ 与

$\{f(n)\}$互补，从而 $h(m)=g(m)$，即 $F^*(m)+m=G(m)+m$，所以 $G(m)=F^*(m)$，即$\{G(m)\}$与$\{F(n)\}$是互伴数列.

至此，定理 3 完全得证.

由定理 3 及例 3，立即得到定理 2.

从定理 3 出发，我们还可以构造出许多关于互补数列的例子和问题. 例如：求满足 $g(n)=nf(n^2+1)-n^3$ 的互补数列$\{f(n)\}$与$\{g(m)\}$.

答案是 $f(n)=[\sqrt{n-1}]+n$，$g(m)=[m^2]+m=m(m+1)$.

漫谈第二十一届国际数学竞赛题

最近,我们的一位正在美国学习的友人寄来一份第二十一届国际数学竞赛题,其中未附解答,也没有这届竞赛的情况介绍.在这里,把我们的解答奉献给读者,并对每个题目都作了一些评注,即指明本题是否有推广的可能,举出与本题内容相近或解法上有共同之处的若干例题,目的是着重于解题方法的探讨.虽然我们认真地研究过这些解法,但它们很可能不是最佳的,期望读者批评指正.

1. 设 p 与 q 为自然数,使得

$$\frac{p}{q}=1-\frac{1}{2}+\frac{1}{3}-\cdots-\frac{1}{1318}+\frac{1}{1319}.$$

证明:p 可被 1979 整除. (命题国:西德)

证明

$$\begin{aligned}\frac{p}{q}&=1-\frac{1}{2}+\frac{1}{3}-\cdots-\frac{1}{1318}+\frac{1}{1319}\\&=\left(1+\frac{1}{2}+\cdots+\frac{1}{1319}\right)-2\left(\frac{1}{2}+\frac{1}{4}+\cdots+\frac{1}{1318}\right)\\&=\left(1+\frac{1}{2}+\cdots+\frac{1}{1319}\right)-\left(1+\frac{1}{2}+\cdots+\frac{1}{659}\right)\\&=\frac{1}{660}+\frac{1}{661}+\cdots+\frac{1}{1319}\\&=\left(\frac{1}{660}+\frac{1}{1319}\right)+\left(\frac{1}{661}+\frac{1}{1318}\right)+\cdots+\left(\frac{1}{989}+\frac{1}{990}\right)\\&=1979\times\left(\frac{1}{660\times1319}+\frac{1}{661\times1318}+\cdots+\frac{1}{989\times990}\right),\end{aligned}$$

所以 $1319!\times\frac{p}{q}$ 可被 1979 整除,从而 $1319!\times p$ 可被 1979 整除.

但 1979 是一个素数(不难验证 2、3、5、7、11、13、17、19、23、29、31、

37、41、43 均不能整除 1979)，1319！不能被 1979 整除，所以 p 可被 1979 整除.

评注　(1) 本题的 1979 可以换成任一形如 $3k+2$ 的素数，相应地将 1319 改为 $2k+1$(k 为自然数).

(2) 数的整除问题属于初等数论的研究内容. 数论是一个历史悠久、十分有趣的数学分枝. 初等数论的问题常常在数学竞赛中出现. 下面我们再举一个例子.

例 1－1　证明 $\frac{1}{2}+\frac{1}{3}+\cdots+\frac{1}{n}$ 不是整数，这里 n 是大于 1 的自然数.

证明　将各个分母分解为素因数的乘积，设其中出现的 2 的最高次幂为 2^m，则这一项恰为 $\frac{1}{2^m}$.

如果 $\frac{1}{2}+\frac{1}{3}+\cdots+\frac{1}{n}$ 是整数，那么乘以 M 后仍为整数，这里 M 为去掉 $\frac{1}{2^m}$ 这一项外、各项分母的最小公倍数.

但 $M\left(\frac{1}{2}+\frac{1}{3}+\cdots+\frac{1}{n}\right)=$ 整数 $+\frac{M}{2^m}$，而 M 不能被 2^m 整除，即 $\frac{M}{2^m}$ 不是整数，这就引起矛盾，所以 $\frac{1}{2}+\frac{1}{3}+\cdots+\frac{1}{n}$ 不是整数.

下面的几个问题可供读者考虑：

问题 1－1　证明 $\left(1+\frac{1}{2}+\cdots+\frac{1}{p-1}\right)(p-1)!$ 可被 p 整除，这里 p 是大于 3 的素数(可以进一步证明它被 p^2 整除，参看华罗庚《数论导引》37 页).

问题 1－2　p 是大于 3 的素数，a、b 是整数，且 $\frac{a}{b}=1+\frac{1}{2^2}+\frac{1}{3^2}+\cdots+\frac{1}{(p-1)^2}$，证明 a 可被 p 整除.

问题 1－3　证明 k 为正的奇数时，$1^k+2^k+\cdots+n^k$ 可被 $1+2+\cdots+n$ 整除.

问题 1－4　m、n 都是自然数，证明

$$\frac{1}{3}+\frac{1}{5}+\cdots+\frac{1}{2n+1}, \tag{1}$$

$$\frac{1}{n}+\frac{1}{n+1}+\cdots+\frac{1}{n+m} \tag{2}$$

都不是整数.

2. 一棱柱以五边形 $A_1A_2A_3A_4A_5$ 与 $B_1B_2B_3B_4B_5$ 为上、下底,这两个多边形的每一条边及每一条线段 $A_iB_j(i, j=1, 2, \cdots, 5)$ 均涂上红色或绿色. 每一个以棱柱顶点为顶点的、以已涂色的线段为边的三角形均有两条边颜色不同. 证明:上、下底的 10 条边颜色一定相同. (命题国:保加利亚)

证明 首先我们证明上底的五条边颜色完全相同.

如果上底的五条边颜色不完全相同,那么必有两条相邻的边颜色不同,不妨设 A_1A_2 是绿的,A_1A_5 是红的.

根据抽屉原则(请参看常庚哲著《抽屉原则及其他》,上海教育出版社 1979 年出版),自 A_1 引出的五条线段 A_1B_1、A_1B_2、A_1B_3、A_1B_4、A_1B_5 中至少有三条有相同的颜色,这三条线段的端点 B_i 中必有两个相邻,不妨设 A_1B_1、A_1B_2 均为绿色,那么 B_1B_2 必须为红色,但由$\triangle B_1A_1A_2$ 推出 A_2B_1 为红色,由$\triangle B_2A_1A_2$ 推出 A_2B_2 也为红色,从而$\triangle A_2B_1B_2$ 的三条边都是红色的,与已知矛盾. 这就证明了上底的五条边颜色必须相同.

同样,下底的五条边颜色也必须相同.

现在再来证明,上、下底的颜色必须是同样的.

如果上、下底的颜色不同,不妨设上底的五条边全为绿色,下底的五条边全为红色.

前面已经说过自 A_1 点引出的五条线段 A_1B_1、A_1B_2、A_1B_3、A_1B_4、A_1B_5 中一定有两条相邻的线段是同一种颜色,不妨假定 A_1B_1 与 A_1B_2 颜色相同. 由于 B_1B_2 是红的,A_1B_1、A_1B_2 都必须是绿的,从而和前面的证明完全一样,$\triangle B_1A_1A_2$、$\triangle B_2A_1A_2$、$\triangle A_2B_1B_2$ 中必有一个三条边是同一种颜色的三角形,这与已知矛盾.

所以棱柱上、下底的 10 条边颜色一定相同.

评注 (1) 本题的五边形可改为 $2n+1$ 边形($n\geqslant 2$),证明完全相同,但不能改为 $2n$ 边形($n\geqslant 2$),例如我们可以将上底与下底的边任意涂色;每一条线段 A_iB_j,当 $i+j$ 为偶数时涂红色,当 $i+j$ 为奇数时涂绿色;则每一个所述的三角形都有颜色不同的边.

(2) 这类涂色问题属于图论的研究范围. 图论中有许多饶有趣味的问题,不需要很多预备知识,只要有一定的推理能力就可以解决,因而常常被选作数学竞赛的题目,我们也举几个例子,以见一斑.

例 2-1 证明任意六个人中一定有三个人互相认识或者三个人互不相识(这里甲认识乙就意谓着乙也认识甲).

证明 用六个点表示六个人,每两个点之间连一条线(称之为边),如果两个人互相认识,就将这条边涂上红色,否则涂上蓝色. 要证明的结论就是一定有一个以这些点为顶点的三角形,它的三条边颜色相同.

因为从任一点 V_1 可引出五条边,而颜色只有红、蓝两种,所以必有三条(或更多条)边颜色相同,不妨设 V_1V_2、V_1V_3、V_1V_4 是红的.

(1) 如果$\triangle V_2V_3V_4$ 的三条边都是蓝的,那么这个三角形就是所要求的三角形.

(2) 如果$\triangle V_2V_3V_4$ 的三条边不全是蓝的,比如说 V_2V_3 是红的,那么$\triangle V_1V_2V_3$ 就是所要求的三角形.

例 2-1 及下面的例 2-2 都是图论中 Ramsey 定理的特例.

例 2-2 九个点,每两个点之间连一条线,每条线涂上红色或蓝色,证明一定可以找到三个点,这三个点之间的连线全是红的,或者可以找到四个点,这四个点之间的连线(六条)全是蓝的.

证明 我们分两种情况来证明.

(1) 有一个点引出四条或更多条红线.

设这时 V_1V_2、V_1V_3、V_1V_4、V_1V_5 为红色. 如果 V_2、V_3、V_4、V_5 之间的连线有一条是红的,比如说 V_2V_3 是红的,那么 V_1、V_2、V_3 就是所求的三个点. 如果 V_2、V_3、V_4、V_5 之间的连线全是蓝的,那么 V_2、V_3、V_4、V_5 就是所求的四个点.

(2) 每个点至多引出三条红线.

这时每个点至少引出五条蓝线. 如果每个点都恰好引出五条蓝线,那么九个点共引出 $\frac{5\times 9}{2}$ 条蓝线,而$\frac{5\times 9}{2}$ 不是整数,所以这是不可能的,因此必有一个点引出 6 条或更多条蓝线.

设 V_1V_2、V_1V_3、V_1V_4、V_1V_5、V_1V_6、V_1V_7 都是蓝线,根据例 2-1,在 V_2、

V_3、V_4、V_5、V_6、V_7 这六个点中必有三个点组成一个三条边颜色相同的三角形. 不妨设$\triangle V_2V_3V_4$ 的三条边颜色相同,如果同为红色,那么 V_2、V_3、V_4 这三个点之间的连线全为红色. 如果$\triangle V_2V_3V_4$ 的三条边全为蓝色,那么 V_1、V_2、V_3、V_4 这四个点之间的连线全为蓝色,证毕.

下面的问题供读者思考.

问题 2-1 17 个科学家中每一个和所有其他的人都通信. 在他们的通信中仅仅讨论三个题目,并且任两个科学家仅仅讨论一个题目. 证明:其中至少有三个科学家,他们互相通信中讨论的是同一个题目(第六届国际数学竞赛题).

问题 2-2 九位数学家在一次国际会议上相遇,他们之中的任意三个人中,至少有两人会说同一种语言,如果每一位数学家最多只能说三种语言,试证明:至少有三位数学家能用同一种语言交谈(第七届美国数学竞赛题).

3. 平面上两圆相交, A 为一个交点,两点同时自 A 出发,以常速度分别在各自的圆周上依相同方向绕行,旋转一周后两点同时回到原出发点. 证明:在这平面上有一定点 P,使得在任何时刻从 P 到两动点的距离相等.

(命题国:苏联)

证法一(复数证法) 不妨设这两圆在复平面上,其方程为

$$|Z|=1 \quad \text{及} \quad |Z-a|=\rho,$$

其中 $a>0$, 即第一个圆为单位圆,第二个圆的圆心在实轴上.

点 A 有两种表示方法(见图 1):

$$A=\mathrm{e}^{\mathrm{i}\theta}=a+\rho\mathrm{e}^{\mathrm{i}\varphi},$$

其中 $\mathrm{e}^{\mathrm{i}\theta}=\cos\theta+\mathrm{i}\sin\theta$, 等等.

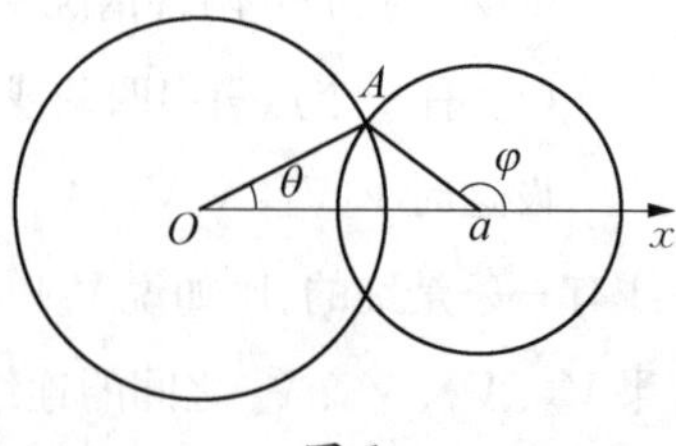

图 1

我们应当求出一个复常数 p,使对一切实数 t 有

$$|P-\mathrm{e}^{\mathrm{i}\theta}\cdot\mathrm{e}^{\mathrm{i}t}|=|P-(a+\rho\mathrm{e}^{\mathrm{i}\varphi}\cdot\mathrm{e}^{\mathrm{i}t})|,$$

在上式左边用 $\mathrm{e}^{\mathrm{i}\theta}=a+\rho\mathrm{e}^{\mathrm{i}\varphi}$ 代入,得

$$|P-a\mathrm{e}^{\mathrm{i}t}-\rho\mathrm{e}^{\mathrm{i}\varphi}\mathrm{e}^{\mathrm{i}t}|=|P-\rho\mathrm{e}^{\mathrm{i}\varphi}\cdot\mathrm{e}^{\mathrm{i}t}-a|,$$

再将上式左边的数变为它的共轭复数,提取因子 $\mathrm{e}^{-\mathrm{i}t}$,得

$$|\bar{P}e^{it}-\rho e^{-i\varphi}-a|=|P-\rho e^{i\varphi}e^{it}-a|.$$

若能使 $\bar{P}e^{it}-\rho e^{-i\varphi}=P-\rho e^{i\varphi}e^{it}$，则上式自然成立. 而这也就是

$$P+\rho e^{-i\varphi}=e^{it}(\bar{P}+\rho e^{i\varphi}).$$

由此显见，若取 $P=-\rho e^{-i\varphi}$，则上式对一切 t 两边均为零，所以 $-\rho e^{-i\varphi}$ 就是所求的点.

由复数的表示法可知，$-\rho e^{-i\varphi}$ 正是点 A 关于两圆圆心连线的中垂线的对称点，这就提示了如下的纯几何证法.

证法二(纯几何证法)　如图 2，设$\odot O_1$与$\odot O_2$为所述的两个圆，作 O_1O_2 的中垂线 l，设 A 关于 l 的对称点为 A'，我们来证明 A'即为所求.

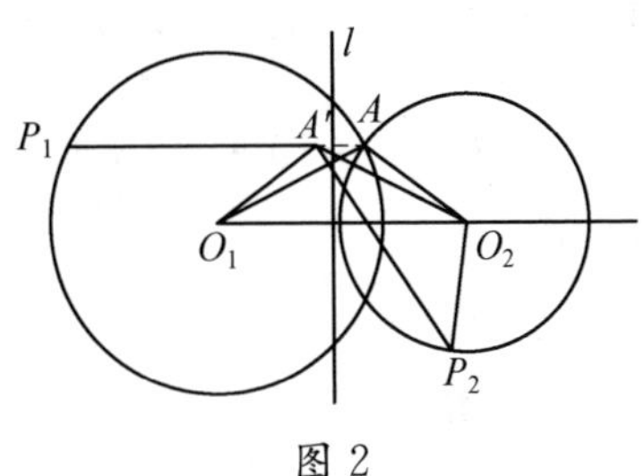

图 2

设一对点 P_1 与 P_2 分别在$\odot O_1$与$\odot O_2$上，使得

$$\angle AO_1P_1=\angle AO_2P_2.$$

由于对称性

$$O_1A'=O_2A=O_2P_2,$$
$$O_1P_1=O_1A=O_2A',$$
$$\angle A'O_1P_1=\angle AO_1P_2-\angle AO_1A'=\angle AO_2P_2-\angle AO_2A'=\angle A'O_2P_2,$$

所以

$$\triangle A'O_1P_1\cong\triangle P_2O_2A'.$$

从而有

$$A'P_1=A'P_2.$$

证完.

评注　从本题的证法一可以看出：用复数来描写点的圆周运动往往是非常方便的. 为了进一步说明这一事实，再看一个例子.

例 3-1　设 Z_1 与 Z_2 两点分别在两个圆周上以相同的角速度按相同的方向旋转. 求证：以 Z_1 与 Z_2 的连线为一边所作成的正三角形的另一顶点 p 也一

定在一个圆周上运动.

证明 设此二圆分别为单位圆 $|Z|=1$ 及 $|Z-a|=\rho$，其中 a 为一复数，而 $\rho>0$

不妨设

$$Z_1=\mathrm{e}^{\mathrm{i}\theta},$$

$$Z_2=a+\rho\mathrm{e}^{\mathrm{i}\theta}\cdot\mathrm{e}^{\mathrm{i}\varphi},$$

其中 φ 为一实常数，它表征着第二个圆周上的点 Z_2 的初始位置. 于是点 P 可以表为 $P=Z_1+(Z_2-Z_1)\mathrm{e}^{\mathrm{i}\frac{\pi}{3}}$（当然也可以表为 $P=Z_2+(Z_1-Z_2)\mathrm{e}^{\mathrm{i}\frac{\pi}{3}}$），所以

$$P=\mathrm{e}^{\mathrm{i}\theta}+[a+\mathrm{e}^{\mathrm{i}\theta}(\rho\mathrm{e}^{\mathrm{i}\varphi}-1)]\mathrm{e}^{\mathrm{i}\frac{\pi}{3}}=a\mathrm{e}^{\mathrm{i}\frac{\pi}{3}}+\mathrm{e}^{\mathrm{i}\theta}[1+(\rho\mathrm{e}^{\mathrm{i}\varphi}-1)\mathrm{e}^{\mathrm{i}\frac{\pi}{3}}],$$

所以

$$|P-a\mathrm{e}^{\mathrm{i}\frac{\pi}{3}}|=|1+(\rho\mathrm{e}^{\mathrm{i}\varphi}-1)\mathrm{e}^{\mathrm{i}\frac{\pi}{3}}|,$$

由此可见 P 只能在以 $a\mathrm{e}^{\mathrm{i}\frac{\pi}{3}}$ 为中心. 以 $|1+(\rho\mathrm{e}^{\mathrm{i}\theta}-1)\mathrm{e}^{\mathrm{i}\frac{\pi}{3}}|$ 为半径的一个圆周上运动.

4. 已知平面 π，π 上一点 P 及 π 外一点 Q，在 π 上求出点 R，使得 $\dfrac{QP+PR}{QR}$ 为最大.

（命题国：美国）

解 如图 3，设 R 点为所求，记 $\angle QPR=\alpha$，$\angle PQR=\beta$，$\angle QRP=\gamma$，则由正弦定理，得

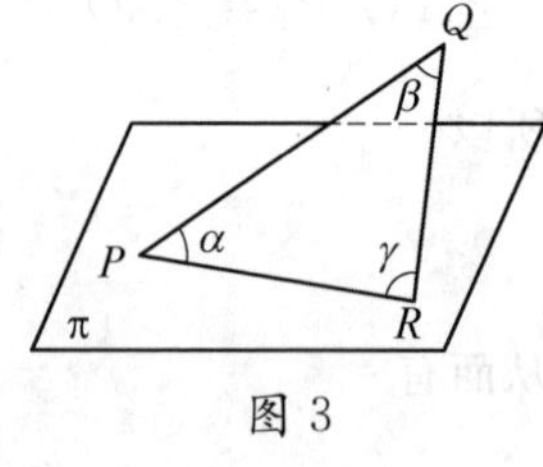

图 3

$$PR=PQ\cdot\frac{\sin\beta}{\sin\gamma},$$

$$QR=PQ\cdot\frac{\sin\alpha}{\sin\gamma},$$

所以

$$\frac{QP+PQ}{QR}=\frac{QP+QP\cdot\dfrac{\sin\beta}{\sin\gamma}}{QP\cdot\dfrac{\sin\alpha}{\sin\gamma}}=\frac{\sin\gamma+\sin\beta}{\sin\alpha}=\frac{2\sin\dfrac{\beta+\gamma}{2}\cos\dfrac{\beta-\gamma}{2}}{\sin\alpha}=\frac{\cos\dfrac{\beta-\gamma}{2}}{\sin\dfrac{\alpha}{2}}.$$

先假定 α 为定值，显然当 $\cos\frac{\beta-\gamma}{2}=1$，也就是 $\beta=\gamma$，$PR=PQ$ 时，比 $\frac{QP+PR}{QR}$ 取得最大值，最大值为$\frac{1}{\sin\frac{\alpha}{2}}$.

由于 $\sin\frac{\alpha}{2}(0<\alpha<\pi)$ 随着 α 的增加而增加，因此当 α 最小时，$\frac{1}{\sin\frac{\alpha}{2}}$最大. 显然当 PR 为 PQ 的射影时 α 最小，因此在 PQ 的射影上取一点 R 使 $PR=PQ$，则 R 就是使$\frac{QP+PR}{QR}$为最大的点(图 4).

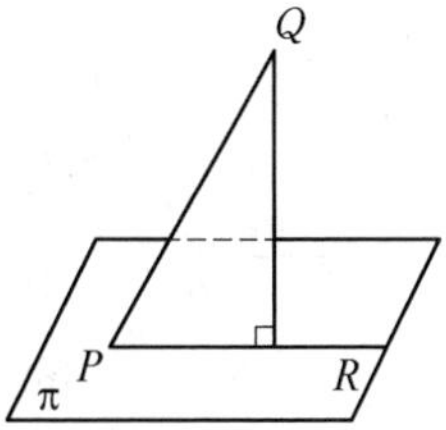

图 4

评注　本题使用的方法可以称为"局部调整法". 在研究多变量的极值问题时有时采用这种方法，它的要点是先假定一些变量是固定的，调整其余(个数较少的)变量，看看在什么情况下，函数达到极值，然后再综合起来考虑. 由于每次考虑的变量个数比原来的少，这就把问题简化了.

本题也可以先固定 PR 的长，然后考虑 α 取什么值时，所述的比最大.

下面再举一个例题.

例 4－1　在直线 l 的同侧有 P、Q 两点，试求一点 R，使 R 到 l 的距离与 R 到 P、Q 的距离的和为最小.

解　如图 5，设 R 到 l 的距离为 RM，先假定 M 为固定点，我们考虑一下，R 在什么位置时，和 $RP+PQ+RM$ 为最小. 这正是熟知的 Fermat 问题(参看《几何不等式》，上海教育出版社 1980 年出版)，如果 R 在 $\triangle PQM$ 内，R 点应当是 $\triangle PQM$ 的 Fermat 点，即 R 点具有性质：

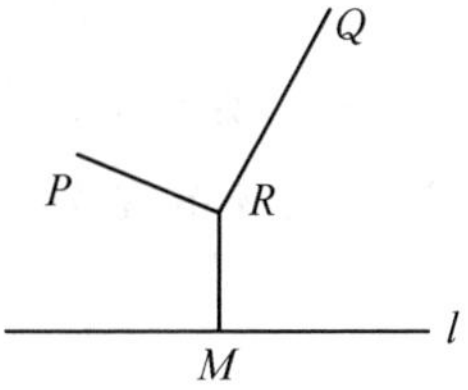

图 5

$$\angle PRQ=\angle QRM=\angle MRP=120^{\circ}.$$

又因为 $RM\perp l$，所以如果过 P、Q 分别作 PQ、RQ 的垂线，那么它们与 l 构成一个正三角形.

于是可以用下面的方法来确定 R：

过 P、Q 各作一条直线与 l 构成一个正三角形 ABC，再作 $PR\perp AB$，$QR\perp AC$，PR 与 QR 相交于 R，如果 R 在 $\triangle ABC$ 内，那么 R 就是所求的点(图 6)

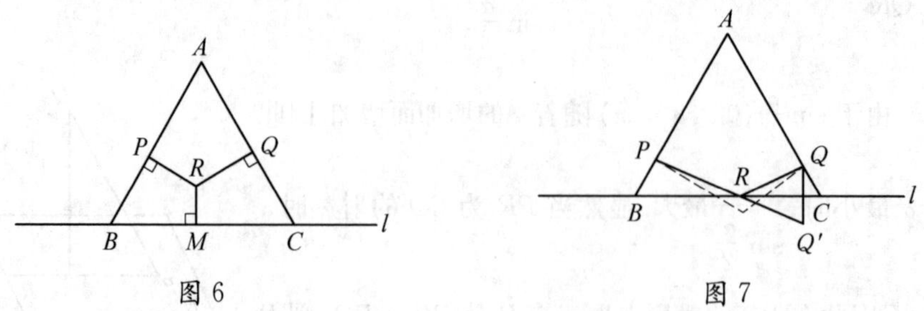

图 6　　图 7

如果上述的两条垂线在 $\triangle ABC$ 外相交，那么应当改用下面的作法(图 7)：作 Q 关于 l 的对称点 Q'，连 PQ'，则 PQ' 与 l 的交点就是所求的点 R.

证明请读者自己补足.

下面的问题留给读者考虑.

问题 4-1　给定两条相交的射线 AB 和 CD，试在射线 AB 上找一点 M，在 CD 上找一点 N，使得 $AM=CN$，并且线段 MN 为最短.

5. 求出所有的实数 a，使得有非负实数 x_1，x_2，x_3，x_4，x_5 适合

$$\sum_{k=1}^{5} kx_k=a,\ \sum_{k=1}^{5} k^3x_k=a^2,\ \sum_{k=1}^{5} k^5x_k=a^3.$$

(命题国：以色列)

解　如果引用 Canchy-Schwarz 不等式，本题可以很顺利地解决. Canchy-Schwarz 不等式即：

设 a_1，a_2，…，a_n 与 b_1，b_2，…，b_n 为两组实数，则

$$\begin{aligned}&(a_1b_1+a_2b_2+\cdots+a_nb_n)^2\\ \leqslant\ &(a_1^2+a_2^2+\cdots+a_n^2)(b_1^2+b_2^2+\cdots+b_n^2),\end{aligned}$$

其中等号成立的充要条件是

$$\frac{a_1}{b_1}=\frac{a_2}{b_2}=\cdots=\frac{a_n}{b_n}.$$

(约定 $b_t=0$ 时，$a_t=0$) 即等号当且仅当 a_1，a_2，…，a_n 与 b_1，b_2，…，b_n 这两

组数成比例时成立,在解某些问题时,这个条件是十分重要的.

假定有非负实数 x_1, x_2, x_3, x_4, x_5 适合题目要求,则

$$a^4=(\Sigma k^3x_k)^2=[\Sigma(k^{\frac{1}{2}}x_k^{\frac{1}{2}})(k^{\frac{5}{2}}x_k^{\frac{1}{2}})]^2\leqslant(\Sigma kx_k)(\Sigma k^5x_k)=a\cdot a^3=a^4,$$

所以式中只能出现等号,于是

$$\frac{k^{\frac{5}{2}}x_k^{\frac{1}{2}}}{k^{\frac{1}{2}}x_k^{\frac{1}{2}}}=\frac{k^2x_k^{\frac{1}{2}}}{x_k^{\frac{1}{2}}}=\text{常数}(k=1,\ 2,\ 3,\ 4,\ 5).\qquad(*)$$

如果 $x_1=x_2=\cdots=x_5=0$,则 $a=0$. 现在设 x_1、x_2、x_5 不全为零. 如果 x_1, x_2, $\cdots$, x_5 中有两个不为零,则($*$)决不能成立,所以其中只能有一个 $x_j\neq0$,这时

$$jx_j=a,\ j^3x_j=a^2,\ j^5x_j=a^3,$$

由前两式得出 $a=\frac{a^2}{a}=\frac{(j^3x_j)}{(jx_j)}=j^2$, 这样 $x_j=\frac{a}{j}=j$ 其余各数均为零.

这就是说,只有当 $a=0,\ 1,\ 4,\ 9,\ 16,\ 25$ 时,才能有非负实数 x_1, x_2, x_3, x_4, x_5 适合题目要求.

评注 (1) 本题的 5 可以改为任意的自然数 n,这时 a 只能是以下各值之一:

$$0,\ 1,\ 2^2,\ 3^2,\ \cdots,\ n^2.$$

(2) 若把题中条件改为

$$\sum\frac{1}{k}x_k=a,\ \sum\frac{1}{k^3}x_k=a^2,\ \sum\frac{1}{k^5}x_k=a^3,$$

也可以推出类似的结论.

(3) 近年来,国外的数学竞赛题中有不少题目是可以用 Canchy-Schwarz 不等式来解的,兹举数例如下.

例 5-1 已知 a_1, a_2, $\cdots$, a_k, $\cdots$为两两各不相同的正整数. 求证对任意正整数 n,下面的不等式

$$\sum_{k=1}^{n}\frac{a_k}{k^2}\geqslant\sum_{k=1}^{n}\frac{1}{k}$$

成立(第二十届国际数学竞赛题).

证明 利用 Canchy-Schwarz 不等式可得

$$\left(\sum_{k=1}^{n}\frac{1}{k}\right)^2=\left(\sum_{k=1}^{n}\frac{\sqrt{a_k}}{k}\cdot\frac{1}{\sqrt{a_k}}\right)^2\leqslant\left(\sum_{k=1}^{n}\frac{a_k}{k^2}\right)\left(\sum_{k=1}^{n}\frac{1}{a_k}\right),$$

所以

$$\sum_{k=1}^{n}\frac{a_k}{k^2}\geqslant\left(\sum_{k=1}^{n}\frac{1}{k}\right)\cdot\left(\frac{\sum_{k=1}^{n}\frac{1}{k}}{\sum_{k=1}^{n}\frac{1}{a_k}}\right).$$

因为 $a_1, a_2, \cdots, a_n$ 是互不相同的正整数,故其中最小者不会小于 1,次小者不会少于 2, ……, 最大者不会小于 n,由此可知,上式右边的第二个因子决不会小于 1,于是得出欲证之不等式.

例 5-2 给定实数 a, b, c, d, e 适合

$$a+b+c+d+e=8,$$
$$a^2+b^2+c^2+d^2+e^2=16,$$

试确定 e 的最大值(第七届美国数学竞赛题).

解 由于

$$\begin{aligned}(8-e)^2=(a+b+c+d)^2&\leqslant(1^2+1^2+1^2+1^2)(a^2+b^2+c^2+d^2)\\&=4(16-e^2)=64-4e^2,\end{aligned}$$

所以

$$e(16-5e)\geqslant 0,$$

解得

$$0\leqslant e\leqslant\frac{16}{5}.$$

这表明 $\frac{16}{5}$ 是 e 的一个上界.

当 $a=b=c=d=\frac{6}{5}$ 时,e 正好取得 $\frac{16}{5}$,所以 $\frac{16}{5}$ 就是 e 的最大值.

6. A、E 为正八边形的相对顶点，一只青蛙从 A 点开始跳跃，如果青蛙在任一个不是 E 的顶点，那么它可以跳向两个相邻顶点中的任一点，当它跳到 E 点时就停在那里. 设 e_n 为经过 n 步到达 E 的不同的路的个数，证明

$$e_{2n-1}=0,\ e_{2n}=\frac{1}{\sqrt{2}}(x^{n-1}-y^{n-1}),\ n=1,\ 2,\ 3,\ \cdots,$$

其中 $x=2+\sqrt{2}$，$y=2-\sqrt{2}$.

（命题国：西德）

原注：一个 n 步的路是指顶点的一个序列$(P_0,\ P_1,\ \cdots,\ P_n)$，满足

(i) $P_0=A$，$P_n=E$.

(ii) 对每一 i，$0\leqslant i\leqslant n-1$，$P_i$ 与 E 不同.

(iii) 对每一 i，$0\leqslant i\leqslant n-1$，$P_i$ 与 P_{i+1} 是相邻的顶点.

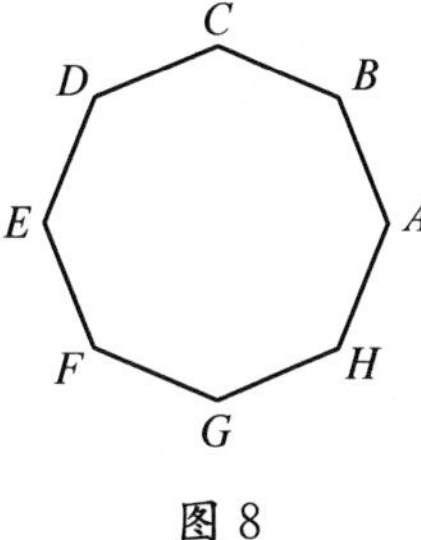

图 8

证明　如图 8，设正八边形为 $ABCDEFGH$，从 A 出发经过 n 步到达 B、C、D、A 的路（意义见原注，只需把 E 分别改为 B、C、D、A）的个数分别记为 b_n、c_n、d_n、a_n. 由于对称性，由 A 出发经过 n 步到达 H、G、F 的路的个数也分别是 b_n，c_n，d_n. 因此有

$$e_n=2d_{n-1},\tag{1}$$

$$c_n=b_{n-1}+d_{n-1},\tag{2}$$

$$b_n=c_{n-1}+a_{n-1},\tag{3}$$

$$a_n=2b_{n-1},\tag{4}$$

由于青蛙跳到 E 点后就停止不动了，所以

$$d_n=c_{n-1}.\tag{5}$$

依据关系式(1)—(5)，得

$$\begin{aligned}e_n&=2d_{n-1}=2c_{n-2}=2[b_{n-3}+d_{n-3}]\\&=2[c_{n-4}+a_{n-4}+d_{n-3}]=2[d_{n-3}+2b_{n-5}+d_{n-3}]\\&=2[d_{n-3}+2(d_{n-3}-b_{n-5})+d_{n-3}]=8d_{n-3}-4d_{n-5},\end{aligned}$$

从而

$$e_n = 4e_{n-2} - 2e_{n-4}. \tag{6}$$

由于 $e_1 = e_3 = 0$,

$$e_2 = 0 = \frac{1}{\sqrt{2}}(x^{1-1} - y^{1-1}),$$

$$e_4 = 2 = \frac{1}{\sqrt{2}}(x^{2-1} - y^{2-1}) = \frac{1}{\sqrt{2}}[(2+\sqrt{2}) - (2-\sqrt{2})],$$

所以命题对 $n=1, 2$ 是成立的,假定命题对一切 $n \leqslant k$ 已经成立,那么

$$e_{2k+1} = 4e_{2k-1} - 2e_{2k-3} = 0,$$

$$\begin{aligned} e_{2k+2} &= 4e_{2k} - 2e_{2k-2} = \frac{4}{\sqrt{2}}(x^{k-1} - y^{k-1}) - \frac{2}{\sqrt{2}}(x^{k-2} - y^{k-2}) \\ &= \frac{1}{\sqrt{2}}[(x+y)(x^{k-1} - y^{k-1}) - xy(x^{k-2} - y^{k-2})] \\ &= \frac{1}{\sqrt{2}}[x^k - y^k + x^{k-1}y - xy^{k-1} - x^{k-1}y + xy^{k-1}] \\ &= \frac{1}{\sqrt{2}}(x^k - y^k), \end{aligned}$$

即命题对于 $n=k+1$ 也是成立的,因此命题对一切自然数 n 皆成立.

评注 (1) 如果对于数列 $\{u_n\}$,有 k 个数 $\alpha_1, \alpha_2, \cdots, \alpha_k$,使得关系式

$$u_{n+k} = \alpha_1 u_{n+k-1} + \alpha_2 u_{n+k-2} + \cdots + \alpha_k u_n \tag{7}$$

对所有的自然数 n 都成立,那么 $\{u_n\}$ 称为 k 阶循环数列. 本题的 $\{e_{2n}\}$ 就是一个二阶循环数列,它满足(8)式

$$e_{2n} = 4e_{2(n-1)} - 2e_{2(n-2)}. \tag{8}$$

(2) 循环数列的通项公式可以利用它的特征方程

$$t^k = \alpha_1 t^{k-1} + \alpha_2 t^{k-2} + \cdots + \alpha_k$$

来求,以本题为例,从(8)式可知它的特征方程是

$$t^2 = 4t - 2.$$

解出两个根为

$$x=2+\sqrt{2},\ y=2-\sqrt{2}.$$

根据循环数列的一般理论可知

$$e_{2n}=ax^{n-1}+by^{n-1},$$

其中常数 a、b 可用初始条件 $e_2=0$, $e_4=2$ 来确定：

$$\begin{cases}0=a+b,\\2=ax+by.\end{cases}$$

所以

$$a=\frac{1}{\sqrt{2}},\ b=-\frac{1}{\sqrt{2}},$$

$$e_{2n}=\frac{1}{\sqrt{2}}(x^{n-1}-y^{n-1}),$$

而 $e_{2n-1}=0$ 是很容易看出来的.

关于循环数列的详细讨论请参看华罗庚《从杨辉三角谈起》(人民教育出版社).

(3) 如果用 S_n 表示从 A 点出发,至多经过 n 步到达 E 点的路的个数,那么 $s_n=\sum\limits_{k=1}^{n}e_k$,不难求出 $s_{2n+1}=s_{2n}=x^{n-1}+y^{n-1}-2$.

一般地,循环数列的和也可以通过循环关系(7)来求得.

(4) 值得注意的是 e_{2n}(以及 s_n)是一个整数,却利用两个无理数 x、y 来表示. 由于 $2-\sqrt{2}$ 是小于 1 的正数,因此整数 $\frac{1}{\sqrt{2}}[(2+\sqrt{2})^{n-1}-(2-\sqrt{2})^{n-1}]$ 就等于 $\frac{1}{\sqrt{2}}(2+\sqrt{2})^{n-1}$ 的整数部分,这种性质也有很多的应用,请读者参看下面的问题.

问题 6－1 (Fibonacci 问题)假定每对大兔子每月能生一对小兔子,而每对小兔子过一个月就能长成大兔子. 设 u_n 表示过了 n 个月后大兔的对数,证明

$$u_n=\frac{1}{\sqrt{5}}(x^{n+1}-y^{n+1})\quad n=1,\ 2,\ \cdots,$$

其中 $x=\frac{(1+\sqrt{5})}{2}$，$y=\frac{(1-\sqrt{5})}{2}$.

问题 6-2 设$\{x_n\}$、$\{y_n\}$是如下的两个数列：

$$x_0=1,\ x_1=1,\ x_{n+1}=x_n+2x_{n-1}\quad(n=1,\ 2,\ \cdots),$$

$$y_0=1,\ y_1=7,\ y_{n+1}=y_n+2y_{n-1}\quad(n=1,\ 2,\ \cdots),$$

于是这两个数列的项为

$$\{x_n\}:\ 1,\ 1,\ 3,\ 5,\ 11,\ 21,\ \cdots;$$

$$\{y_n\}:\ 1,\ 7,\ 17,\ 55,\ 161,\ 487,\ \cdots.$$

证明：除“1”外，这两个数列中没有相同的项.（第二届美国数学竞赛题）

问题 6-3 证明：当 n 为正整数时，

(1) $(2+\sqrt{3})^n$ 的整数部分是一个奇数；

(2) $(7+4\sqrt{3})^n$ 的小数部分是以 n 个 9 为开头的.

美国（1975，1976年）数学竞赛题和解答

第四届（1975年）

1.（Ⅰ）证明

$$[5x]+[5y]\geqslant[3x+y]+[3y+x],$$

这里 x，$y\geqslant 0$，其中$[u]$表示不大于 u 的最大整数(例如 $[\sqrt{2}]=1$).

（Ⅱ）利用(Ⅰ)或不利用(Ⅰ)，证明

$$\frac{(5m)!\ (5n)!}{m!\ n!\ (3m+n)!\ (3n+m)!},$$

对任何正整数 m，n 均为整数.

证明　（Ⅰ）实际上我们可以证明

$$[5x]+[5y]\geqslant[3x+y]+[3y+x]+[x]+[y]. \tag{1}$$

记 $x'=x-[x]$，$y'=y-[y]$，那么

$$0\leqslant x'<1,\ 0\leqslant y'<1.$$

$$\begin{aligned}[5x]+[5y]&=[5[x]+5x']+[5[y]+5y']\\&=5[x]+[5x']+5[y]+[5y'],\end{aligned}$$

$$\begin{aligned}&[3x+y]+[3y+x]+[x]+[y]\\=&3[x]+[y]+[3x'+y']+3[y]+[x]+[3y'+x']+[x]+[y]\\=&5[x]+5[y]+[3x'+y']+[3y'+x'].\end{aligned}$$

因此(1)式就等价于

$$[5x']+[5y']\geqslant[3x'+y']+[3y'+x']. \tag{1'}$$

令
$$\frac{i}{5}\leqslant x'<\frac{i+1}{5},$$

$$\frac{j}{5}<y'<\frac{j+1}{5}\ (i,\ j=0,\ 1,\ 2,\ 3,\ 4),$$

即
$$[5x']+[5y']=i+j,$$

$$3x'+y'<\frac{3(i+1)}{5}+\frac{j+1}{5}=\frac{3i+j+4}{5},$$

$$3y'+x'<\frac{3j+i+4}{5}.$$

若 $i+j>2$(即 $i+j\geqslant 3$),

$$[3x'+y']+[3y'+x']<\frac{4i+4j+8}{5}\leqslant\frac{5(i+j)+5}{5}=i+j+1,$$

所以

$$[3x'+y']+[3y'+x']\leqslant[5x]+[5y].$$

若
$$i+j=2,$$

则

$$3x'+y'<\frac{3i+j+4}{5}<\frac{6+4}{5}=2,$$

所以 $[3x'+y']\leqslant 1$. 同样 $[3y'+x']\leqslant 1$, $(1')$成立.

若 $i+j\leqslant 1$, 则 i、j 中必有一个为 0,不妨设 $i=0$, 则

$$3x'+y'<\frac{3i+j+4}{5}=\frac{j+4}{5}\leqslant 1,$$

所以 $[3x'+y']=0$,

而

$$3y'+x'<\frac{3j+i+4}{5}=\frac{3j+4}{5}\leqslant j+1,$$

所以 $[3y'+x']\leqslant j$. $(1')$成立.

(Ⅱ) 在 $(5m)!$、$(5n)!$、$m!$、$n!$、$(3m+n)!$、$(3n+m)!$ 里,出现任一素因数 p 的方次分别是

$$\left[\frac{5m}{p}\right]+\left[\frac{5m}{p^2}\right]+\cdots,$$

$$\left[\frac{5n}{p}\right]+\left[\frac{5n}{p^2}\right]+\cdots,$$

$$\left[\frac{m}{p}\right]+\left[\frac{m}{p^2}\right]+\cdots,$$

$$\left[\frac{n}{p}\right]+\left[\frac{n}{p^2}\right]+\cdots,$$

$$\left[\frac{3m+n}{p}\right]+\left[\frac{3m+n}{p^2}\right]+\cdots,$$

$$\left[\frac{3n+m}{p}\right]+\left[\frac{3n+m}{p^2}\right]+\cdots.$$

根据(1)式,对于每一个自然数 k,都有

$$\left[\frac{5m}{p^k}\right]+\left[\frac{5n}{p^k}\right]\geqslant\left[\frac{m}{p^k}\right]+\left[\frac{n}{p^k}\right]+\left[\frac{3m+n}{p^k}\right]+\left[\frac{3n+m}{p^k}\right],$$

这就是说,分式

$$\frac{(5m)!\ (5n)!}{m!\ n!\ (3m+n)!\ (3n+m)!}$$

分母里任一素因子的方次都不大于分子里同一素因子的方次. 因此,对任意正整数 m、n,上式是整数.

2. 如果 A、B、C、D 表示空间中四点, AB 表示 A、B 之间的距离,等等. 证明:

$$AC^2+BD^2+AD^2+BC^2\geqslant AB^2+CD^2.$$

证明　线段 AB 与 CD 的中点分别记为 E 与 F,在$\triangle ABC$ 中,有

$$AC^2+BC^2=2AE^2+2CE^2.$$

在$\triangle ABD$ 中,有

$$AD^2+BD^2=2AE^2+2DE^2.$$

将以上二式两边分别相加,得到

$$AC^2+BD^2+AD^2+BC^2=2CE^2+2DE^2+4AE^2=2CE^2+2DE^2+AB^2.$$

但是,在$\triangle CED$ 中,又有

$$2CE^2+2DE^2=CD^2+4EF^2,$$

所以

$$AC^2+BD^2+AD^2+BC^2=AB^2+CD^2+4EF^2.$$

由于 $4EF^2\geqslant 0$, 最后得到

$$AC^2+BD^2+AD^2+BC^2\geqslant AB^2+CD^2,$$

显然式中等号当且只当 $E=F$, 即 AB 的中点与 CD 的中点重合时成立,此时 $ACBD$ 是一个平行四边形.

3. 如果 $P(x)$表示 n 次多项式,且对 $k=0,1,2,\cdots,n$ 有$P(k)=\frac{k}{k+1}$,试确定 $P(n+1)$.

解 考察 $n+1$ 次多项式

$$Q(x)=(x+1)P(x)-x.$$

根据已知条件,我们有

$$Q(0)=Q(1)=Q(2)=\cdots=Q(n)=0,$$

这表明 $0,1,2,\cdots,n$ 是多项式$Q(x)$的 $n+1$ 个根,所以可以把 $Q(x)$写成

$$Q(x)=Ax(x-1)(x-2)\cdots(x-n),$$

其中 A 是一个待定的常数,它可以用 $Q(-1)=1$ 这一条件定出:

$$A=\frac{(-1)^{n+1}}{(n+1)!},$$

从而

$$Q(x)=\frac{(-1)^{n+1}}{(n+1)!}x(x-1)(x-2)\cdots(x-n).$$

由函数 $Q(x)$的定义得到

$$F(x)=\frac{Q(x)+x}{x+1},$$

令 $x=n+1$ 代入,便得

$$P(n+1)=\frac{n+1+(-1)^{n+1}}{n+2}=\begin{cases}1, & \text{当 } n \text{ 为奇数;} \\ \dfrac{n}{n+2}, & \text{当 } n \text{ 为偶数.}\end{cases}$$

4. 两给定圆相交于 P 和 Q 两点，作出一条过 P 且交两圆于 A、B 的线段 AB，使 $AP \cdot PB$ 为最大.

解　如图 1，设两定圆 O_1、O_2 的半径分别是 r_1、r_2. 两交点为 P 和 Q，连 O_1P、O_2P. 则 $\angle O_1PO_2$ 为定角，记 $\angle O_1PO_2=\alpha$，延长 O_1P 至 R，则 $\angle O_2PR=180^\circ-\alpha$.

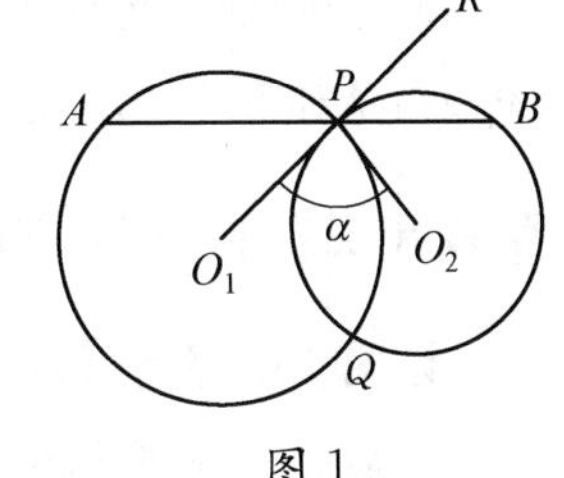

图 1

(1) 若过 P 点交两圆的弦 AB 在 $\angle O_2PR$ 及其对顶角的内部变动(如图 1)，这时

$$AP=2r_1\cos\angle APO_1,$$
$$BP=2r_2\cos\angle BPO_2.$$

$$\begin{aligned}AP\cdot BP&=4r_1r_2\cos\angle APO_1\cdot\cos\angle BPO_2\\&=2r_1r_2[\cos(\angle APO_1+\angle BPO_2)+\cos(\angle APO_1-\angle BPO_2)]\\&=2r_1r_2[\cos(180^\circ-\alpha)+\cos(\angle APO_1-\angle BPO_2)].\end{aligned}$$

这里 r_1、r_2、$180^\circ-\alpha$ 均为定值，因而当 $\angle APO_1-\angle BPO_2=0$ 时，$AP\cdot BP$ 达到最大值. 即当 AB 是$\angle O_2PR$ 的平分线时，$AP\cdot BP$ 达到最大值，这时最大值等于 $2r_1r_2[\cos(180^\circ-\alpha)+1]$.

(2) 若过 P 点交两圆的弦 AB 在$\angle O_1PO_2$ 内部变动时(如图 2)，这时

$$\begin{aligned}AP\cdot BP&=4r_1r_2\cos\angle O_1PA\cdot\cos\angle O_2PB\\&=2r_1r_2[\cos\alpha+\cos(\angle O_1PA-\angle O_2PB)].\end{aligned}$$

所以当 $\angle O_1PA-\angle O_2PB=0$ 时，$AP\cdot BP$ 达到最大值. 即 PA 是$\angle O_1PO_2$ 的平分线时，$AP\cdot BP$ 最大，最大值是 $2r_1r_2(\cos\alpha+1)$.

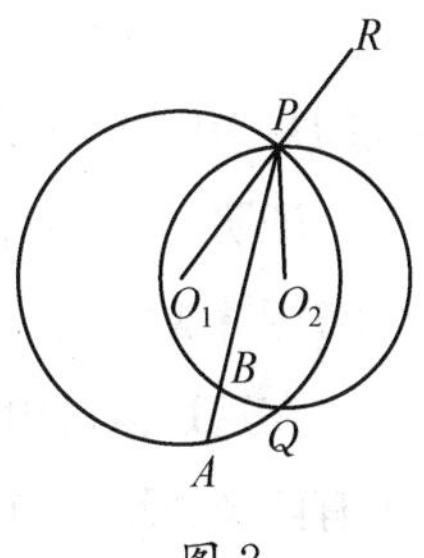

图 2

由于当 $\alpha<90^\circ$ 时，$\cos\alpha>\cos(180^\circ-\alpha)$；

当 $90^\circ<\alpha<180^\circ$ 时，$\cos\alpha<\cos(180^\circ-\alpha)$；

当 $\alpha=90^\circ$ 时，$\cos\alpha=\cos(180^\circ-\alpha)$.

则本题的解答应是：

当 $\angle O_1PO_2 < 90^\circ$ 时，是$\angle O_1PO_2$ 的平分线；

当 $\angle O_1PO_2 > 90^\circ$ 时，是$\angle O_1PO_2$ 的外角的平分线；

当 $\angle O_1PO_2 = 90^\circ$ 时，则题目有两解. $\angle O_1PO_2$ 及其外角的两平分线都满足要求.

5. 一副纸牌共有 N 张，其中有三张 A，现随机地洗牌(假设牌的所有可能的分布是等可能地出现的)，然后从顶上开始一张接一张地翻牌，直至翻到第二张 A 出现为止. 证明：翻过的纸牌数的期望(平均)值是 $\dfrac{N+1}{2}$.

证明 记翻到第 K 张出现第二张 A 的事件为 E_K. 因为 E_{N+1-K} 也可以看成是从最后一张往前翻，翻到第 K 张出现第二张 A 的事件，所以事件 E_{N+1-K} 与 E_K 出现的概率相同，而 K 与 $N+1-K$ 的平均值为 $\dfrac{N+1}{2}$ $(K=1, 2, \cdots, N)$，因此总的平均值也是 $\dfrac{N+1}{2}$.

第五届(1976 年)

1. (Ⅰ) 假设一个 4×7 的国际象棋棋盘(如图 3 所示)的每一个方块用黑色或白色着了色，证明对任何一种着色方式，在棋盘中必定包含一个由棋盘的水平线和垂直线所构成的矩形(例如图中所画出的一个那样)，它的四个不同角上的方块是同样颜色的.

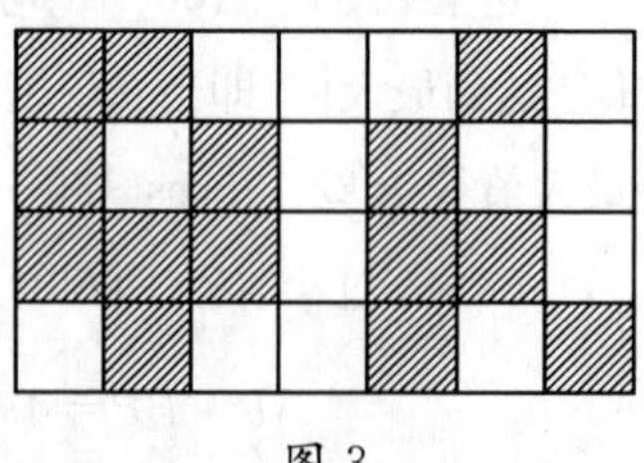

图 3

(Ⅱ) 试给出 4×6 的棋盘上的一种黑白着色，使得在棋盘内每一个如上所述的矩形中，四个角上的方块不是同样颜色的.

证明 (Ⅰ) 我们可以证明更强一些的结果，即：对于 3×7 的棋盘，结论(Ⅰ)能够成立.

在这种棋盘中，每一列有三个方块，由于只能着两种颜色，所以每一列中至少有两个方块颜色相同. 如果棋盘中有两列着色的方式相同，那么四角上的方块同色的矩形显然就已经找到了!

现在来考察棋盘的 7 个列的着色方式互不相同的情况. 由于三个方块、两种颜色、按顺序和颜色不同计有八种涂法. 因此,我们先来考察如下图 4 中 3×8 的棋盘及其着色情况(其中 0 代表白色,1 代表黑色):

0	0	0	1	0	1	1	1
0	0	1	0	1	0	1	1
0	1	0	0	1	1	0	1

图 4

任何一个 3×7 的棋盘,当它的七个列的着色方式互不相同时,在不计较行的次序和列的次序的场合,都可以看成是从上表中去掉一列而得到的. 但是,只去掉一个列时,无论怎样也不会把上图中那个全由 0 组成的正方形与全由 1 组成的正方形同时破坏. 由于行(列)的次序的调换不会影响结论成立与否,这样就证明了在七个列着色方式互不相同时,也有四角由同色方块组成的矩形存在.

(Ⅱ) 如果在每一列的四个方块中,两个方块着白色,另外两个方块着黑色,一共有 $C_4^2=6$ 种不同的着色方式. 我们把这六种方式在一个 4×6 的棋盘中实现,例如图 5:

0	0	0	1	1	1
0	1	1	0	0	1
1	0	1	0	1	0
1	1	0	1	0	0

图 5

这样就可以保证在某两行与某两列交叉位置上的四个方块中不可能全是 0,也不可能全是 1.

2. 如果 A、B 是一个给定圆上的固定点, XY 是这个圆中的一条变动的直径,试确定(并证明)直线 AX 与 BY 交点的轨迹(你可以假设 AB 不是一条直径.)

证明　如图 6,设 AX 与 BY 相交于 P, P 是所求轨迹上的点;又设 AY 与

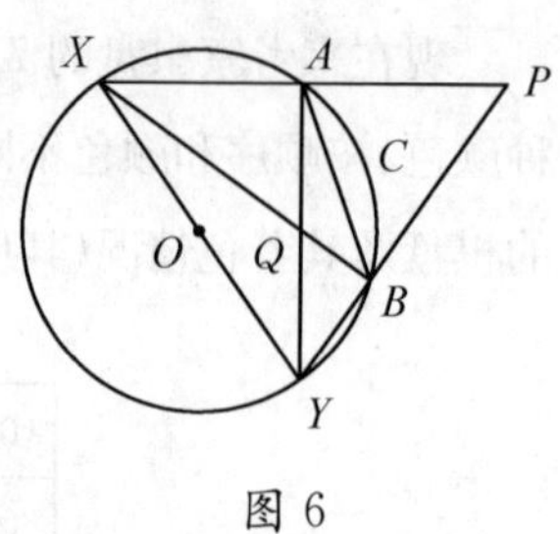

图 6

BX 相交于 Q,注意当 XY 变动到 YX 的位置时,P 点变动到 Q,可知 Q 也是所求轨迹的点.

由于 AB 是定弦,所以 $\overset{\frown}{ACB}$ 的度数一定,从而 $\angle AQB=\angle AYB+\angle XBY$ 也一定(其中 $\angle AYB$ 用 $\frac{1}{2}\overset{\frown}{ACB}$ 的度数来量,而 $\angle XBY=90°$). 同样 $\angle APB=\angle XAY-\angle AYB$ 也一定(其中 $\angle XAY=90°$). 所以所求轨迹是以 AB 为弦,弓形角分别以 $90°+\frac{1}{2}\overset{\frown}{ACB}$ 及 $90°-\frac{1}{2}\overset{\frown}{ACB}$ 来度量的两个弓形弧,而 A、B 为其极限点(图中 XY 与 AB 不相交,若 XY 变动到与 AB 相交的位置,可以看到$\angle APB$、$\angle AQB$ 都以 $90°-\frac{1}{2}\overset{\frown}{ACB}$ 来度量). 进一步看到,这两个弓形弧正好合成一个圆,这个圆的圆心是圆 O 在 A 点的切线与 AB 的垂直平分线的交点. 反过来,容易证明轨迹上的点都满足条件.

3. 试确定(并证明) $a^2+b^2+c^2=a^2b^2$ 的所有整数解.

解　设有整数 a、b、c 适合 $a^2+b^2+c^2=a^2b^2$, 由此可得

$$c^2+1=(a-1)(a+1)(b-1)(b+1). \tag{1}$$

假设 c 是奇数: $c=2k+1$, 于是

$$c^2+1=(2k+1)^2+1=4k(k+1)+2,$$

由此可见此时(1)的左边被 8 除余数为 2. 这时(1)的右边的四个因子中至少有一个是偶数,不妨设 $a-1=2l$, 故

$$\begin{aligned}&(a-1)(a+1)(b-1)(b+1)\\=&2l\cdot 2(l+1)(b-1)(b+1)\\=&4l(l+1)(b-1)(b+1),\end{aligned}$$

这个数正好被 8 整除. 所以(1)式不可能成立. 这就表明 c 不能是奇数.

再设 c 为偶数. 首先考察 $c=0$,这时 $c^2+1=1$. 由(1)可知,右边的四个因子只能是-1 与 1. 如果 $a-1=-1$,则 $a=0$;若 $a-1=1$,则 $a+1=3$,这不可能. 所以说只能是 $a=b=c=0$; 显然,这也正是方程的一个解.

其次再设 c 是非零的偶数,仍由(1)推知,(1)的右边的四个因子应均为奇

数,所以 a、b 应同时为偶数. 用 m 来记 a、b、c 的最大公因数,显然 m 为偶数. 设 $a=ma_1$, $b=mb_1$, $c=mc_1$, 于是可以得出

$$a_1^2+b_1^2+c_1^2=m^2a_1^2b_1^2, \tag{2}$$

a_1、b_1、c_1 中至少有一个是奇数,否则 m 就不是 a、b、c 的最大公因数了. 如果 a_1、b_1、c_1 全是奇数或恰有一个是奇数,那么(2)的左边将是一奇数,而(2)的右边总可以被 4 整除,这是矛盾. 所以最后只须讨论 a_1、b_1、c_1 中恰有两个奇数的情况,这时(2)的左边被 4 除余 2,(2)式仍不能成立!

所以 $a=b=c=0$ 是方程的唯一整数解.

4. 如果一个三直角的四面体 $PABC$(即 $\angle APB=\angle BPC=\angle CPA=90^\circ$)的六条棱的长度之和为 s,试确定该四面体的最大体积,须证明你的结论.

解 如图 7,设 $PA=a$, $PB=b$, $PC=c$, 于是

$$s=a+b+c+\sqrt{a^2+b^2}+\sqrt{b^2+c^2}+\sqrt{c^2+a^2},$$

并且该四面体的体积是 $V=\frac{1}{6}abc$.

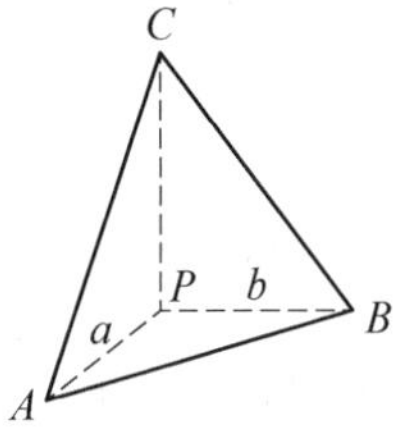

图 7

由关于几何平均—算术平均不等式可知

$$a+b+c\geqslant 3\sqrt[3]{abc},$$

$$\sqrt{a^2+b^2}+\sqrt{b^2+c^2}+\sqrt{c^2+a^2}\geqslant\sqrt{2ab}+\sqrt{2bc}+\sqrt{2ca}\geqslant 3\sqrt{2}\sqrt[3]{abc},$$

上述不等式中的等号当且仅当 $a=b=c$ 时成立. 这样一来

$$s\geqslant 3(1+\sqrt{2})(abc)^{\frac{1}{3}}=3(1+\sqrt{2})(6V)^{\frac{1}{3}},$$

即

$$\frac{1}{6}\left[\frac{s}{3(1+\sqrt{2})}\right]^3\geqslant V,$$

式中等号当且仅当 $a=b=c$ 时成立. 由于 s 为一常数,所以当且仅当 $a=b=c$ 时体积 V 达到它的最大值

$$V_{\max}=\frac{s^3}{162(1+\sqrt{2})^3}=\frac{s^3(\sqrt{2}-1)^3}{162}=\frac{5\sqrt{2}-7}{162}s^3.$$

5. 如果 $P(x)$、$Q(x)$、$R(x)$ 及 $S(x)$ 是多项式,且使得

$$P(x^5)+xQ(x^5)+x^2R(x^5)=(x^4+x^3+x^2+x+1)S(x).$$

证明: $x-1$ 是 $P(x)$ 的一个因子.

证明 令 $\omega=\cos\frac{2\pi}{5}+\mathrm{i}\sin\frac{2\pi}{5}$, 即设 ω 是一个五次单位根,于是 $\omega^5=1$. 依次将 $x=\omega$、ω^2、ω^3 代入题目所给的方程,得到

$$P(1)+\omega Q(1)+\omega^2R(1)=0, \tag{1}$$

$$P(1)+\omega^2Q(1)+\omega^4R(1)=0, \tag{2}$$

$$P(1)+\omega^3Q(1)+\omega^6R(1)=0. \tag{3}$$

(2)-(1)得

$$\omega(\omega-1)Q(1)+\omega^2(\omega^2-1)R(1)=0,$$

即

$$Q(1)+\omega(\omega+1)R(1)=0. \tag{4}$$

(3)-(2)得

$$\omega^2(\omega-1)Q(1)+\omega^4(\omega^2-1)R(1)=0,$$

即

$$Q(1)+\omega^2(\omega+1)R(1)=0. \tag{5}$$

(5)-(4)得

$$\omega(\omega+1)(\omega-1)R(1)=0,$$

所以 $R(1)=0$.

代入(4)得 $Q(1)=0$.

再代入(1)得 $P(1)=0$.

由余式定理可知 $x-1$ 是 $P(1)$ 的一个因子.

第二十二届国际数学竞赛试题及解答

第二十二届国际中学生数学竞赛于1980年7月在卢森堡举行.我们拟了一份解答,不妥之处期望读者批评指正.

1. 设 Q 是全体有理数的集合,求所有适合下列条件的从 Q 到 Q 的函数 f:

(i) $f(1)=2$;

(ii) 对 Q 中所有的 x、y,

$$f(xy)=f(x)f(y)-f(x+y)+1.$$

(6分,英国供题)

解　令 $y=1$,则由(ii)得

$$f(x)=f(x)\cdot f(1)-f(x+1)+1, \tag{1}$$

将 $f(1)=2$ 代入(1)式并化简得

$$f(x+1)=f(x)+1.$$

从而在 n 为正整数时,

$$f(x+n)=f(x+n-1)+1=f(x+n-2)+2=\cdots\cdots=f(x)+n;$$

在 n 为负整数时,

$$f(x+n)=f(x+n+1)-1=\cdots\cdots=f(x)-(-n)=f(x)+n.$$

因此在 n 为整数时,

$$f(x+n)=f(x)+n. \tag{2}$$

在(2)式中,令 $x=1$ 即得

$$f(n+1)=n+2. \tag{3}$$

对于任意的有理数 $\frac{n}{m}$,其中 n、m 为整数,在(ii)中,令 $x=m$, $y=\frac{n}{m}$ 得

$$f(n)=f(m)\cdot f\left(\frac{n}{m}\right)-f\left(m+\frac{n}{m}\right)+1. \tag{4}$$

由于(2)、(3),上式即

$$n+1=(m+1)f\left(\frac{n}{m}\right)-f\left(\frac{n}{m}\right)-m+1,$$

化简得

$$mf\left(\frac{n}{m}\right)=n+m,$$

即

$$f\left(\frac{n}{m}\right)=\frac{n}{m}+1.$$

因此所求的函数是唯一的,即

$$f(x)=x+1,\ x\in Q,$$

它显然满足条件(i)、(ii).

附注 这种类型的题目,过去在国外数学竞赛中也出现过,下面就是其中的一道:

问题 从 Q 到 Q 的函数 $f(x)$ 具有下述性质,试确定函数 $f(x)$:

(i) $f(x+y)=f(x)f(y)$

(ii) $f(x)\not\equiv 0$

我们把它留给读者练习.

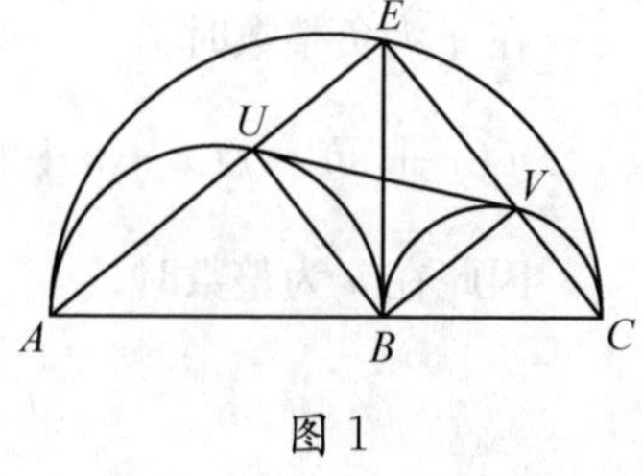

图 1

2. 如图 1,设 A、B、C 三点共线,并且 B 在 AC 之间. 在 AC 同侧,分别以 BA、BC 与 AC 为直径画三个半圆. 前两个半圆在 B 点公切线与第三个半圆相交于 E, UV 是前两个半圆的另一条公切线, U、V 是切点. 将比值

$$\frac{\triangle EUV\text{ 的面积}}{\triangle EAC\text{ 的面积}}$$

表为 $r_1=\frac{1}{2}AB$ 与 $r_2=\frac{1}{2}BC$ 的函数.　　　　　　　(7 分,卢森堡供题)

解　连 AE、CE 分别交半圆弧 $\overset{\frown}{AB}$、$\overset{\frown}{BC}$ 于 U', V'. 连 $U'V'$、$U'B$、$V'B$.

由于直径上的圆周角为直角,所以

$$\angle AU'B=\angle AEC=\angle CV'B=90^\circ.$$

于是,四边形 $EU'BV'$ 是矩形,从而

$$\angle BU'V'=\angle BEC=90^\circ-\angle ACE=\angle EAC.$$

因而 $U'V'$ 与以 AB 为直径的半圆相切. 同理亦与以 BC 为直径的半圆相切,即 $U'V'$ 是两个半圆的公切线. 这就是说, U', V' 分别与 U, V 重合. 显然有

$$\begin{aligned}\frac{\triangle EUV\text{ 的面积}}{\triangle EAC\text{ 的面积}}&=\frac{EU\times EV}{EA\times EC}=\frac{BV\times BU}{EA\times EC}=\frac{BV}{EA}\times\frac{BU}{EC}\\&=\frac{CB}{CA}\times\frac{AB}{CA}=\frac{2r_2\times 2r_1}{4(r_1+r_2)^2}=\frac{r_1r_2}{(r_1+r_2)^2}.\end{aligned}$$

附注　不难看出,上述比值的最大值是 $\frac{1}{4}$,仅当 B 点是 AC 中点时达到.

3. p 为系数, n 是正整数. 证明下面的两个陈述(i)和(ii)等价:

(i) 对 $k=0, 1, \cdots, n$, 二项式系数 $\binom{n}{k}$ 中没有一个能被 p 整除;

(ii) n 可以表成 $n=p^sq-1$, 其中 s、q 为整数, $s\geqslant 0$, $0<q<p$.

(7 分,南斯拉夫供题)

证明　首先来证明(ii)⇒(i). 因为

$$\binom{n}{k}=\frac{n(n-1)\cdots(n-k+1)}{k!}=\prod_{l=1}^{k}\frac{n+1-l}{l},\tag{1}$$

如果 $(n+1-l)$ 是 p 的整数倍,于是有 $n+1-l=qp^s-l=rp^t$, 其中 r、t 都是正整数, $r\nmid p$, $t\leqslant s$ 由此即得

$$l=qp^s-rp^t=p^t(qp^{s-t}-r).$$

因此(1)式中分子里的 p 的幂也都在分母中出现,即 $\binom{n}{k}$ 不能被 p 整除.

现在将 $n+1$ 用 p 进制表示:

$$n+1=q_s p^s+q_{s-1} p^{s-1}+\cdots+q_0,$$

其中 $0 \leqslant q_i<p,\ i=0,\ 1,\ \cdots,\ s-1;\ 0<q_s<p$.

这时我们来证明:如果(ii)不成立,也就是说有某个 $i \leqslant s-1$,使得 $q_0=q_1=\cdots=q_{i-1}=0,\ q_i \neq 0$,那么(i)也不成立.事实上,取 $k=q_t p^t$,那么在 $l<k$ 时,如果

$$l=r p^t,\ t<i,\ r \nmid p \text{ 或 } t=i,\ r<q_t,$$

那么

$$n+1-l=p^t\left(q_s p^{s-t}+\cdots+q_i p^{t-i}-r\right),$$

因此(1)中 $\frac{n+1-l}{l}$ 的分子、分母中 p 的幂是相同的.

但

$$\frac{n+1-k}{k}=\frac{q_s p^s+\cdots+q_{i+1} p^{i+1}}{q_t p^t}$$

中,分子里含 p 的幂次比分母高,因此 $\binom{n}{k}$ 可被 p 整除.

综上,这就证明了(i)与(ii)是等价的.

附注 И. М. 维诺格拉陀夫著《数论基础》第一章问题 12(题目见下面)是本题的一个特例,读者可以考虑如何解这道题(不一定用上面的方法).

问题 设 n 是整数, $n>0$. 证明:牛顿二项式 $(a+b)^n$ 的展开式的所有系数,在而且只在 n 有形状 2^k-1 时,才都是奇数.

4. 两圆相切(外切或内切)于 P 点,一直线与其中一圆相切于 A 点,与另一圆相交于 B、C 两点. 证明: PA 为 $\angle BPC$ 的角(或外角)平分线.

(6 分,比利时供题)

证明 先证明两圆内切时的情况. 设 PB、PC 与 A 点所在的圆分别交于 E、F 两点. 联 AE、AF. 并过 P 点作两圆的外公切线 PQ,如图 2.

由 $\angle AFP$ 是 $\triangle AFC$ 的外角,因此有

$$\angle AFP=\angle FAC+\angle C.$$

另一方面,由弦切角定理,有

$$\angle QPA=\angle AFP,\ \angle FAC=\angle APC,$$

即得

$$\angle QPA=\angle APC+\angle C,$$

又
$$\angle QPB=\angle C,$$

故得
$$\angle APB=\angle APC,$$

这就是说, AP 是$\angle BPC$ 的平分线.

当两圆外切时,辅助线添法同前,如图 3 所示.

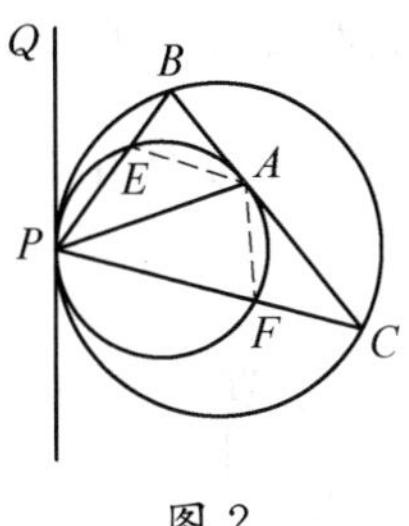

图 2

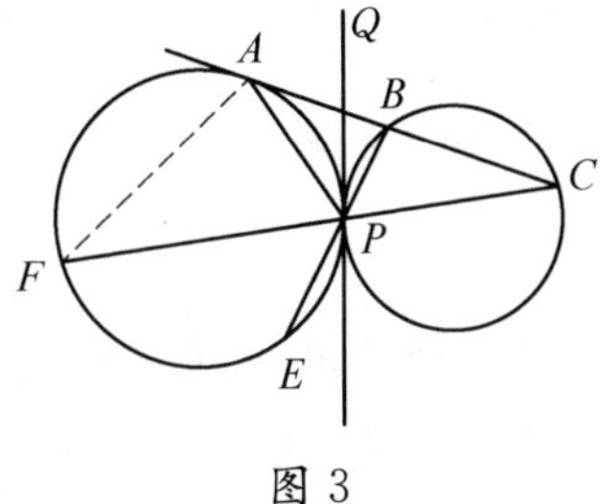

图 3

这时, $\angle APF$ 是$\triangle PAC$ 的外角,于是有

$$\angle APF=\angle PAC+\angle C,$$

另一方面,由于

$$\angle CAP=\angle QPA,\ \angle QPB=\angle C,$$

即得

$$\angle BPA=\angle APF.$$

这就是说, AP 是$\angle BPC$ 的外角平分线.

附注　这一道试题并不难,我国一般水平的初中毕业生都能解决它! 当然,这里我们所关心的是它有没有推广的可能. 例如题目里直线 BC 与另一圆不是相切而是相割,能够得到什么结果? 又例如题目里两圆相切变成两圆相交,又能得到什么结果? 或者更一般地,上面两种情况同时发生,又能得到什么结果？下面的三个问题就是对此作出回答.它们的证明留给读者.

问题 1 两圆内切(或外切)于 P 点,一直线和一圆交于 A_1、A_2 两点,且与另一圆交于 B、C 两点.则有 $\angle A_1PB=\angle A_2PC$ (或外角).

问题 2 两圆相交于 P_1、P_2 两点,一直线和一圆切于 A 点且和另一圆交于 B、C 两点.若切点 A 在另一圆内(或另一圆外),则有

$$\angle BP_1A=\angle CP_2A(\text{或外角}).$$

问题 3 两圆相交于 P_1、P_2 两点,一直线和一圆交于 A_1、A_2 两点,且和另一圆交于 B、C 两点,则有 $\angle BP_1A_1=\angle CP_2A_2$(或外角).

不难看出,竞赛题是这三个问题的特例,而问题 1、问题 2 又都是问题 3 的特例.

5. 十个赌徒在一起赌钱,开始时各人赌本相同.他们轮流掷五枚骰子.每个掷的人付给九个对手的钱等于该对手当时钱数的 $\frac{1}{n}$,这里 n 是五个骰子掷出的点数的和.

他们一个接一个地轮流掷下去,每次都付清所有款项.已知第十次掷出的点数是 12.并且付清款项后,每个赌徒所有的钱数恰好与其原来的赌本相同.试确定其他九次出现的点数. (7 分,荷兰供题)

解 设开始时各个赌徒的赌本为 A.第 i 次掷出的点数为 n_i,又记

$$K=\left(1+\frac{1}{n_1}\right)\left(1+\frac{1}{n_2}\right)\cdots\left(1+\frac{1}{n_{10}}\right).$$

第 i 个赌徒,在掷第 i 次之前的钱数为 $A\left(1+\frac{1}{n_1}\right)\cdots\left(1+\frac{1}{n_{i-1}}\right)$, $(i\leqslant 10)$,这时其他九个赌徒钱数的和为

$$10A-A\left(1+\frac{1}{n_1}\right)\cdots\left(1+\frac{1}{n_{i-1}}\right), \tag{1}$$

掷第 i 次以后,第 i 个赌徒余下钱数为

$$\begin{aligned}&A\left(1+\frac{1}{n_1}\right)\cdots\left(1+\frac{1}{n_{i-1}}\right)-\frac{1}{n_i}\left[10A-A\left(1+\frac{1}{n_1}\right)\cdots\left(1+\frac{1}{n_{i-1}}\right)\right]\\&=A\left[\left(1+\frac{1}{n_1}\right)\cdots\left(1+\frac{1}{n_{i-1}}\right)\left(1+\frac{1}{n_i}\right)-\frac{10}{n_i}\right].\end{aligned}$$

掷到最后,即掷出十次后,他的钱为

$$A\left[\left(1+\frac{1}{n_1}\right)\cdots\left(1+\frac{1}{n_i}\right)-\frac{10}{n_i}\right]\left(1+\frac{1}{n_{i+1}}\right)\cdots\left(1+\frac{1}{n_{10}}\right)=AK-\frac{10}{n_i}A\left(1+\frac{1}{n_{i+1}}\right)\cdots\left(1+\frac{1}{n_{10}}\right).$$

同样可得,第 $i+1$ 个赌徒在掷完十次后,他的钱数为

$$AK-\frac{10}{n_{i+1}}A\left(1+\frac{1}{n_{i+2}}\right)\cdots\left(1+\frac{1}{n_{10}}\right).$$

根据题意,有

$$AK-\frac{10}{n_i}A\left(1+\frac{1}{n_{i+1}}\right)\cdots\left(1+\frac{1}{n_{10}}\right)=AK-\frac{10}{n_{i+1}}A\left(1+\frac{1}{n_{i+2}}\right)\cdots\left(1+\frac{1}{n_{10}}\right)=A, \tag{2}$$

从而

$$n_i=\left(1+\frac{1}{n_{i+1}}\right)\cdot n_{i+1}=n_{i+1}+1,$$

由于 $n_{10}=12$,所以 $n_9=13$, $n_8=14$, $\cdots$, $n_2=20$, $n_1=21$.

这些数可使(2)成立,并且(1)为正. 所以确实是问题的解.

附注　完全类似地,可把试题推广到一般情况. 即把“十个赌徒”改为“K 个赌徒”,把“已知第十次掷出的点数是 12”改为“第 K 次掷出的点数是 t”. 那么问题的答案是:依次掷出的点数是: $n_1=K+t-1$, $n_2=K+t-2$, $\cdots$, $n_{K-1}=t+1$.

6. 求适合方程

$$x^3+x^2y+xy^2+y^3=8(x^2+xy+y^2+1)$$

的所有整数对(x, y).　　　　（7 分,荷兰供题）

解　原式即

$$(x+y)(x^2+y^2)=8(x^2+y^2+xy+1), \tag{1}$$

易见 x、y 的奇偶性必须相同. 而

$$4(x^2+y^2)<8(x^2+y^2+xy+1)\leqslant 12(x^2+y^2)+8, \tag{2}$$

所以

$$4(x^2+y^2)<(x+y)(x^2+y^2)\leqslant 12(x^2+y^2)+8. \tag{3}$$

如果 $x+y>12$, 那么

$$(x+y)(x^2+y^2)-12(x^2+y^2)\geqslant x^2+y^2>\frac{(x+y)^2}{2}>8,$$

与(3)矛盾,所以

$$4<x+y\leqslant 12,$$

从而 $x+y=6, 8, 10$ 或 12.

若 $x+y=12$, 则由(1)得

$$x^2+y^2=2xy+2,$$
$$(x-y)^2=2,$$

这是不可能的. 同样, $x+y=6$ 或 8 也是不可能的.

若 $x+y=10$, 则由(1)得

$$\begin{cases}x=8,\\ y=2\end{cases} \text{或} \begin{cases}x=2,\\ y=8,\end{cases}$$

这就是原方程的所有整数解.

第十届美国数学竞赛试题及解答

美国第十届数学竞赛于 1981 年五月举行. 下面是竞赛题和解答.

1. 已知一角大小为$\frac{180^\circ}{n}$,其中 n 为不能被 3 整除的正整数. 证明:这个角可以用欧几里得的作图工具(圆规与直尺)三等分.

证明　因为 n 是不能被 3 整除的正整数,所以 $n=3k\pm1$.

如果 $n=3k+1$, 由于

$$\frac{180^\circ}{3}-k\times\frac{180^\circ}{n}=\frac{180^\circ}{3n}(n-3k)=\frac{180^\circ}{3n},$$

且$\frac{180^\circ}{n}$为已知角,所以 $k\times\frac{180^\circ}{n}$ 可用圆规与直尺作出,显然 $\frac{180^\circ}{3}=60^\circ$ 可用圆规直尺作出,所以$\frac{180^\circ}{3n}$可作. 也就是说,这时$\frac{180^\circ}{n}$可以用圆规直尺三等分.

如果 $n=3k-1$, 那么由于

$$k\times\frac{180^\circ}{n}-\frac{180^\circ}{3}=\frac{180^\circ}{3n}(3k-n)=\frac{180^\circ}{3n},$$

同上理,这时$\frac{180^\circ}{n}$也可以用圆规直尺三等分.

2. 某个县下辖某些区. 已知每两个区都恰好由以下三种交通方式:汽车、火车、飞机中的一种直接联系. 在全县里三种交通方式全有,但没有一个区三种方式全有;并且没有任何三个区中两两联系的方式全相同. 试问这个县至多有几个区?

解　将每一个区用一个点表示(我们称它为顶点),根据两个区之间的交通方式:汽车、火车、飞机,将相应的两个顶点之间的连线(我们称它为边)涂上红色、蓝色、白色. 这样就得到一个图. 根据题意,这个图具有以下性质:

① 这个图的每两个顶点之间有一条边连结,每条边都涂上红、蓝、白三种

颜色中的一种.

② 整个图中三种颜色的边全有.

③ 每一个顶点引出的边中,至多只有两种颜色.

④ 在这个图中不存在同色三角形.即没有一个三角形,它的三条边的颜色全相同.

首先我们来证明:由图中的任何一个顶点引出的边中,没有三条颜色相同.

假设由点 A 引出的边中,有三条边颜色相同.如图 1,设边 AB、AC、AD 同色.不失一般性,设为红色.那么根据④可知,B、C、D 三条连线只能是蓝、白两色.但若这三边中,蓝、白两色都有,这将出现 B、C、D 三点中,有某一点引出的边有三种颜色,这与③矛盾.但若只有蓝、白两色中的一种,则$\triangle BCD$ 是同色三角形,这又与④矛盾.因此,图中任何一点绝不能引出三条颜色相同的边.

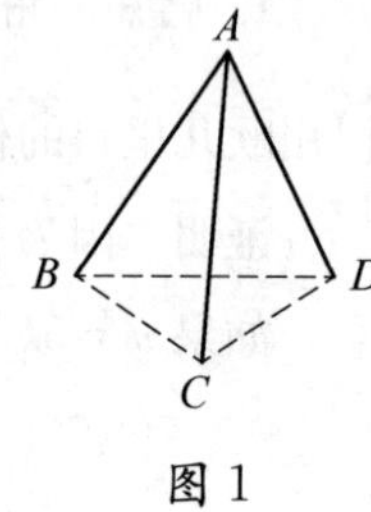

图 1

我们再来证明,图中顶点个数 $n\leqslant 4$,也就是说,这个县至多有四个区.

假设图中顶点个数 $n\geqslant 6$,那么根据①图中每一顶点引出的边数不小于 5 条.由于③,根据抽屉原则可知,其中至少有 3 条边同色,但这与上述的结论矛盾.所以顶点个数 $n<6$.

但 $n=5$ 也不行.假设 $n=5$,那么由图中任一点要引出四条边,根据上述讨论可知,这四条边必然具有两种颜色,且每种颜色有两条.如图 2,不失一般性,假设由 A 点引出的四条边中,AB、AC 同为红色.那么 BC 边一定不是红色.由于每一点引出的四条边中,必有两条同色,所以 BD、BE 两边以及 CD、CE 两边之中,必然各有一条是红色的,这就是说,在整个图中同一种颜色的边至少有四条.由于图中三种颜色都有,所以图中至少应有 $4\times 3=12$ 条边.

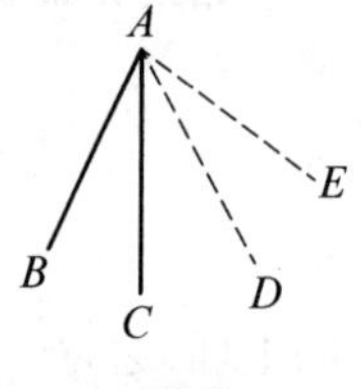

图 2

但另一方面,图中 5 个顶点至多只有 $C_5^2=10$ 条边.矛盾.

于是,我们证得图中的顶点数 $n\leqslant 4$,也就是说,这个县至多有四个区.

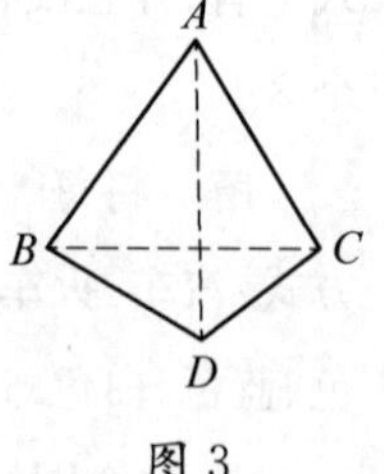

图 3

$n=4$ 的情况是可能的.图 3 所示是满足条件的许多方

案之一. 其中以实线、点线、间断线分别代表三种颜色.

3. 若 A、B、C 是三角形的三个角,证明 $-2\leqslant\sin3A+\sin3B+\sin3C\leqslant\frac{3}{2}\sqrt{3}$. 并确定等号何时成立?

解　不失一般性,设 A 为 $\triangle ABC$ 的最大角, C 为最小角,则

$A\geqslant60^\circ$, $C\leqslant60^\circ$, $B+C\leqslant180^\circ-60^\circ=120^\circ$, $3C\leqslant180^\circ$, $\sin3C\geqslant0$.

于是 $\sin3A+\sin3B+\sin3C\geqslant\sin3A+\sin3B\geqslant-2$.

要使 $\sin3A+\sin3B+\sin3C=-2$. 则必须同时有:

$$\sin3A=-1,\ \sin3(B+C)=-1,\ \sin3C=0,$$

从而得 $A=B=90^\circ$, $C=0^\circ$. 这与已知 A、B、C 为三角形的三个内角相矛盾. 因此 $\sin3A+\sin3B+\sin3C>-2$.

再证不等式的另一边.

$$\sin3B+\sin3C=2\sin\frac{3}{2}(B+C)\cos\frac{3}{2}(B-C)\leqslant2\sin\frac{3}{2}(B+C).$$

记 $\alpha=\frac{3}{2}(B+C)$, 则

$$\begin{aligned}\sin3A+\sin3B+\sin3C&\leqslant\sin(3\times180^\circ-2\alpha)+2\sin\alpha=\sin2\alpha+2\sin\alpha\\&=2\sin\alpha(1+\cos\alpha)=2\sqrt{(1-\cos\alpha)(1+\cos\alpha)^3}.\end{aligned}$$

由于 $(1+\cos\alpha)+(1-\cos\alpha)=2$ 是定值,所以当 $3(1-\cos\alpha)=1+\cos\alpha$,即 $\cos\alpha=\frac{1}{2}$ 时,三角式 $2\sqrt{(1-\cos\alpha)(1+\cos\alpha)^3}$ 取得最大值. 最大值是

$$2\sqrt{\left(1-\frac{1}{2}\right)\left(1+\frac{1}{2}\right)^3}=\frac{3}{2}\sqrt{3}.$$

因此,
$$\sin3A+\sin3B+\sin3C\leqslant\frac{3}{2}\sqrt{3},$$

并且等号在且仅在 $\cos\alpha=\frac{1}{2}$, $B=C$,即 $B=C=20^\circ$, $A=140^\circ$ 时取得.

4. 已知一凸多面角的所有面角之和等于它所有二面角之和. 证明:这个多面角为三面角.

原注:凸多面角是由一个凸多边形所在平面外一点到凸多边形的所有顶点引射线所得出的图形.

证明 在立体几何中,有如下两个定理(两个定理的证明略):

① 凸多面角的所有面角之和 $S<2\pi$.

② 凸 n 面角的所有二面角之和 S',有 $(n-2)\pi<S'<n\pi$.

由于题目里的凸多面角具有 $S=S'$,由上述两个定理,即得 $(n-2)\pi<2\pi$, $n<4$ 因此,只有 $n=3$. 即满足题设的多面角是三面角.

5. 若 x 为正实数,n 为正整数. 证明

$$[nx]\geqslant\frac{[x]}{1}+\frac{[2x]}{2}+\frac{[3x]}{3}+\cdots+\frac{[nx]}{n},$$

其中$[t]$表示小于或等于 t 的最大整数. 例如: $[\pi]=3$, $[\sqrt{2}]=1$.

证明 用数学归纳法.

当 $n=1, 2$ 时,命题显然成立. 假设命题对 $n<k-1$ 时均成立. 现在来证明 $n=k$ 时命题也成立.

记 $x_i=\sum\limits_{j=1}^{i}\frac{[jx]}{j}$, 则有

$$kx_n=kx_{j-1}+[kx]=(k-1)x_{j-1}+x_{j-1}+[kx],$$
$$(k-1)x_{k-1}=(k-2)x_{k-2}+x_{k-2}+[(k-1)x],$$
$$\cdots\cdots$$
$$2x_2=x_1+x_1+[2x].$$

于是: $kx_k=x_{k-1}+x_{k-2}+\cdots+x_1+x_1+[kx]+[(k-1)x]+\cdots+[2x]$.

根据归纳假设,上式可写为

$$kx_k\leqslant[kx]+2([(k-1)x]+[(k-2)x]+\cdots+[x]),$$

由于有 $[(k-m)x]+[mx]\leqslant[(k-m)x+mx]$,其中 $m<k$. 所以

$$kx_k\leqslant[kx]+([(k-1)x]+[x])+([(k-2)x]+[2x])+\cdots+([x]+[(k-1)x])\leqslant k[kx],$$

即 $$x_k\leqslant[kx].$$

1980 年芬兰等四国国际数学竞赛试题及解答

1980 年 7 月 1 日至 2 日在芬兰的马里安哈米纳举行了一次由芬兰、英国、匈牙利和瑞典等四国共 32 名选手参加的小型国际数学竞赛. 比赛结果,得 20 分(40 分制)以上的有 3 人,最高分 28 分. 全体平均成绩只有 11.06 分,其中以第 2 题最难,32 名选手总共得 11 分,平均每人只有 0.34 分. 这次比赛匈牙利的成绩较好.

下面是比赛试题,以及我们所拟的一份解答. 不妥之处祈请读者批评指正.

1. 在三角形 ABC 中,边 AB 和 AC 的垂直平分线分别交 BC(如果存在的话)于 X 和 Y. 证明:使 $BC=XY$ 的一个充分条件是 $\tan B\cdot\tan C=3$. 再证明这个条件不是必要的,并且找出使 $BC=XY$ 成立的充分必要条件.

(6 分,英国供题)

解　不难看出,若$\triangle ABC$ 的角 B、C 中有一个为直角,那么将有一条垂直平分线与 BC 不相交. 现不妨假定 B、C 都不是直角(图 1).

约定下面出现的线段为有向线段,即 $CB=-BC$.

记$\triangle ABC$ 的三边长为 a、b、c,外接圆半径为 R,显然有(包括符号在内)

$$BX=\frac{\frac{1}{2}c}{\cos B},\ CY=-\frac{\frac{1}{2}b}{\cos C}.$$

原题中的条件“$BC=XY$”,在上述约定下,即是 $|BC|=|XY|$,它等价于

$$BC\pm XY=0.$$

将 $XY=XB+BC+CY$ 代入上式,得

$$BC\pm(XB+BC+CY)=0.$$

这就是说,$|BC|=|XY|$ 等价于:$2BC+XB+CY=0$ 或 $XB+CY=0$.

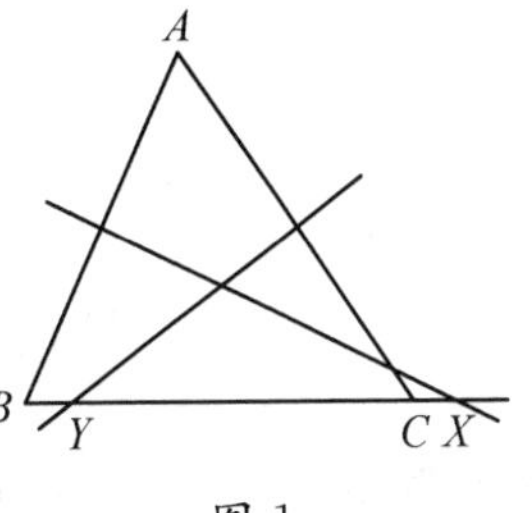

图 1

(1) 由于 $2BC+XB+CY$

$$=2a-\frac{\frac{1}{2}c}{\cos B}-\frac{\frac{1}{2}b}{\cos C}$$

$$=R\left(4\sin A-\frac{\sin C}{\cos B}-\frac{\sin B}{\cos C}\right)$$

$$=R\left[4(\sin B\cos C+\sin C\cos B)-\frac{\sin C}{\cos B}-\frac{\sin B}{\cos C}\right]$$

$$=R\cos B\cos C[4\tan B+4\tan C-\tan C(1+\tan^2 B)-\tan B(1+\tan^2 C)]$$

$$=R\cos B\cos C[3(\tan B+\tan C)-\tan B\tan C(\tan B+\tan C)]$$

$$=R\cos B\cos C(\tan B+\tan C)(3-\tan B\tan C).$$

因 B、C 均不是直角,所以 $\cos B\neq 0$, $\cos C\neq 0$. 又因 $B\neq\pi-C$, 所以 $\tan B+\tan C\neq 0$. 因此, $3-\tan B\tan C=0$,即 $\tan B\tan C=3$ 与 $2BC+XB+CY=0$ 等价,而它是 $|BC|=|XY|$ 的一个充分条件.

(2) 由于 $XB+CY=-\dfrac{\frac{1}{2}c}{\cos B}-\dfrac{\frac{1}{2}b}{\cos C}=-R\left(\dfrac{\sin C}{\cos B}+\dfrac{\sin B}{\cos C}\right)$

$$=-R\cos B\cos C[\tan C(1+\tan^2 B)+\tan B(1+\tan^2 C)]$$

$$=-R\cos B\cos C(\tan B+\tan C)(\tan B\tan C+1).$$

同上,因 $\cos B\neq 0$, $\cos C\neq 0$, $\tan B+\tan C\neq 0$,所以 $\tan B\tan C+1=0$ 即 $\tan B\cdot\tan C=-1$ 与 $XB+CY=0$ 等价.

综上可知,使 $|BC|=|XY|$ 的充要条件是 $\tan B\cdot\tan C=3$ 或 $\tan B\cdot\tan C=-1$.

不难看出,后者在 $B=90^\circ+C$(或 $C=90^\circ+B$)时成立. 这时$\triangle ABC$ 是钝角三角形.

2. 数列 a_0, a_1, …, a_n 是用下式定义的:

$$\begin{cases}a_0=\dfrac{1}{2},\\ a_{K+1}=a_K+\dfrac{1}{n}a_K^2,\text{其中 } K=0, 1, \cdots, n-1.\end{cases}$$

证明: $1-\dfrac{1}{n}<a_n<1$. (7 分,瑞典供题)

证明　首先证明 $a_n<1$. 为此,证明在 $K\leqslant n$ 时,有 $a_K<\dfrac{n}{2n-K}$.　　(1)

用数学归纳法. 显然有, $a_1=\left(\dfrac{1}{2}+\dfrac{1}{4n}\right)=\dfrac{2n+1}{4n}<\dfrac{n}{2n-1}$.

假定(1)对于 $K-1<n$ 是成立的,那么

$$a_K=\left(a_{K-1}+\frac{1}{n}a_{K-1}^2\right)=a_{K-1}\left(1+\frac{a_{K-1}}{n}\right)<\frac{n}{2n-(K-1)}\left(1+\frac{1}{2n-(K-1)}\right)$$

$$=\frac{n[2n-(K-1)+1]}{[2n-(K-1)]^2}<\frac{n}{2n-(K-1)-1}=\frac{n}{2n-K}.$$

因此(1)对于 $K\leqslant n$ 成立. 于是 $a_n<\dfrac{n}{2n-n}=1$.

再证 $a_n>1-\dfrac{1}{n}$. 为此,证明在 $K\leqslant n$ 时,有 $a_K>\dfrac{n+1}{2n-K+2}$.　　(2)

用数学归纳法. 当 $K=1$ 时,有 $a_1=\dfrac{2n+1}{4n}>\dfrac{n+1}{2n+1}$.

假设(2)式对于 $K-1<n$ 成立. 那么

$$a_K=a_{K-1}+\frac{1}{n}a_{K-1}^2>\frac{n+1}{2n-K+3}+\frac{(n+1)^2}{n[2n-K+3]^2}$$

$$=\frac{n+1}{2n-K+2}-\frac{n+1}{(2n-K+3)(2n-K+2)}+\frac{(n+1)^2}{n[2n-K+3]^2}$$

$$=\frac{n+1}{2n-K+2}+\frac{n+1}{n(2n-K+3)}\left[\frac{n+1}{2n-K+3}-\frac{n}{2n-K+2}\right].$$

因为当 $K\leqslant n$ 时,有$\dfrac{n+1}{2n-K+3}>\dfrac{n}{2n-K+2}$,所以对 $K\leqslant n$ 时有 $a_K>\dfrac{n+1}{2n-K+2}$.

于是 $a_n>\dfrac{n+1}{2n-n+2}=\dfrac{n+1}{n+2}>\dfrac{n-1}{n}=1-\dfrac{1}{n}$.

3. 考虑方程 $x^n+1=y^{n+1}$,其中 n 是自然数,且 $n\geqslant 2$. 证明:上面的方程当 x、y 是自然数,且 x 与 $n+1$ 无公因子时无解.

(7 分,匈牙利供题)

证明　原方程可写为 $x^n=y^{n+1}-1$.　　(1)

显然 $y=1$ 不满足方程(1)(因 $x\neq 0$). 同时 $y=2$ 也不满足(1). 因若 $y=2$,

(1)的右边为 $2^{n+1}-1=2^n+2^{n-1}+\cdots+2+1$. 但由于

$$\begin{aligned}2^n&<2^n+2^{n-1}+\cdots+2+1\\&<2^n+C_n^1 2^{n-1}+C_n^2 2^{n-2}+\cdots+C_n^{n-1}\cdot 2+1\\&=(2+1)^n=3^n.\end{aligned}$$

这就是说,不存在自然数 x,满足 $x^n=2^{n+1}-1$.

现在设 $y>2$, 我们用反证法,即假设有自然数 x 及 $y>2$ 满足

$$x^n=y^{n+1}-1, \text{并且}(x, n+1)=1.$$

由于
$$x^n=y^{n+1}-1=(y-1)(y^n+y^{n-1}+\cdots+1), \tag{2}$$
所以,对于 $(y-1)$ 的任一个素因子 p,有 $p\mid x$.

又因 $(x, n+1)=1$, 所以 $p\nmid(n+1)$.

由 $y\equiv 1(\bmod p)$,得 $y^n+y^{n-1}+\cdots+1\equiv n+1(\bmod p)$.

所以 $p\nmid(y^n+y^{n-1}+\cdots+1)$,因而$(y-1, y^n+y^{n-1}+\cdots+1)=1$.

根据(2)式,可令 $y-1=r^n$, $y^n+y^{n-1}+\cdots+1=s^n$. 其中 $rs=x$.

由于 $s^n>y^n$. 所以 $s\geqslant y+1$. 即 $s^n\geqslant(y+1)^n>y^n+y^{n-1}+\cdots+1$.
矛盾,于是命题得证.

4. 一个凸 $2n$ 边形内接于一圆,其中 $(n-1)$ 对对边互相平行(即设此多边形为 $A_1A_2\cdots A_nA_{n+1}\cdots A_{2n}$, $A_1A_2 \parallel A_{n+1}A_{n+2}$, $A_2A_3 \parallel A_{n+2}A_{n+3}$, $\cdots$, $A_{n-1}A_n \parallel A_{2n-1}A_{2n}$), 问:当 n 为何值时,另一对对边(即 A_nA_{n+1} 与 $A_{2n}A_1$)是互相平行的?

(6 分,匈牙利供题)

解　如图 2,记圆心角 $\angle A_iOA_{i+1}$ 为 α_i ($i=1, 2, \cdots, 2n$, $A_{2n+1}\equiv A_1$).

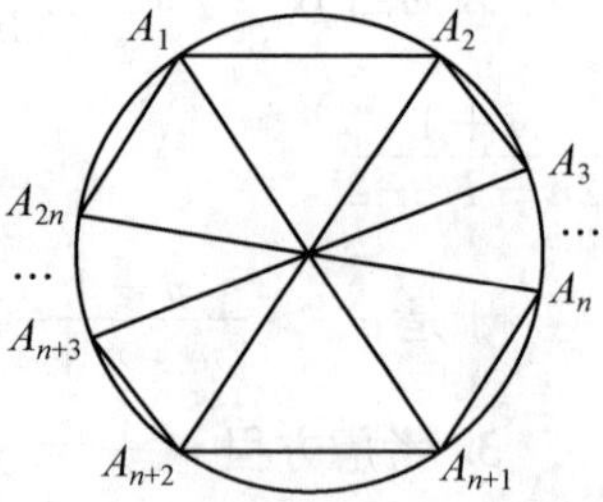

因为 $A_1A_2 \parallel A_{n+1}A_{n+2}$, $A_2A_3 \parallel A_{n+2}A_{n+3}$,

所以 $\angle A_1A_2A_3=\angle A_{n+1}A_{n+2}A_{n+3}$.

得 $\alpha_1+\alpha_2=2\pi-2\angle A_1A_2A_3$

$=2\pi-2\angle A_{n+1}A_{n+2}A_{n+3}=\alpha_{n+1}+\alpha_{n+2}$.

同理可得: $\alpha_3+\alpha_4=\alpha_{n+3}+\alpha_{n+4}$, $\cdots$.

图 2

(1) 若 n 为奇数,则有 $\alpha_{n-2}+\alpha_{n-1}=\alpha_{2n-2}+\alpha_{2n-1}$, 从而有

$$(\alpha_1+\alpha_2)+(\alpha_3+\alpha_4)+\cdots+(\alpha_{n-2}+\alpha_{n-1})$$

$$=(\alpha_{n+1}+\alpha_{n+2})+(\alpha_{n+3}+\alpha_{n+4})+\cdots+(\alpha_{2n-2}+\alpha_{2n-1}).$$

于是有 $\overset{\frown}{A_1A_n}=\overset{\frown}{A_{n+1}A_{2n}}$，则 $A_{2n}A /\!/ A_nA_{n+1}$.

因此，当 n 为奇数时，恒有 A_nA_{n+1} 与 $A_{2n}A_1$ 平行.

(2) 若 n 为偶数，只能得到 $\alpha_{n-3}+\alpha_{n-2}=\alpha_{2n-3}+\alpha_{2n-2}$，即只有

$$(\alpha_1+\alpha_2)+(\alpha_3+\alpha_4)+\cdots+(\alpha_{n-3}+\alpha_{n-2})$$
$$=(\alpha_{n+1}+\alpha_{n+2})+(\alpha_{n+3}+\alpha_{n+4})+\cdots+(\alpha_{2n-3}+\alpha_{2n-2}).$$

上式只能说明有 $\overset{\frown}{A_1A_{n-1}}=\overset{\frown}{A_{n+1}A_{2n-1}}$，即 $A_{n-1}A_{n+1} /\!/ A_{2n-1}A_1$.

图 3 就表明了，虽然这时 $A_{n-1}A_n /\!/ A_{2n-1}A_{2n}$，但也有可能出现 A_nA_{n+1} 与 $A_{2n}A_1$ 不平行(这时却有 $A_nA_{n+1}=A_{2n}A_1$). 一个最简单的反例是，如图 4 的圆内接梯形.

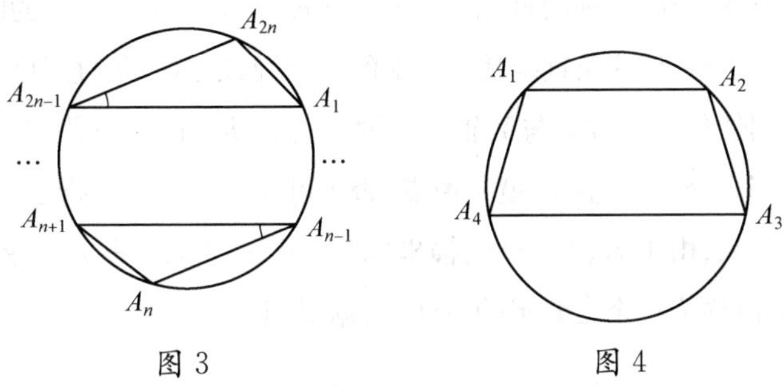

图 3　　图 4

事实上可以证明：当 n 是偶数时，使命题成立的充要条件是：A_1A_{n+1} 是圆的直径.

若 A_1A_{n+1} 是直径，由 $A_1A_2 /\!/ A_{n+1}A_{n+2}$，易见 A_2A_{n+2} 也是直径. 同理可以推得 A_3A_{n+3}，…，A_nA_{2n} 也是直径. 因两条直径 A_1A_{n+1}，A_nA_{2n} 的端点构成的是矩形，即有 $A_nA_{n+1} /\!/ A_{2n}A_1$.

另一方面，如果 $A_nA_{n+1} /\!/ A_{2n}A_1$，那么有

$$\alpha_1+\alpha_2=\alpha_{n+1}+\alpha_{n+2},\ \alpha_3+\alpha_4=\alpha_{n+3}+\alpha_{n+4},\ \cdots,\ \alpha_{n-1}+\alpha_n=\alpha_{2n-1}+\alpha_{2n}.$$

从而 $\alpha_1+\alpha_2+\cdots+\alpha_{n-1}+\alpha_n=\alpha_{n+1}+\alpha_{n+2}+\cdots+\alpha_{2n-1}+\alpha_{2n}=\dfrac{2\pi}{2}=\pi$.

这就是说，A_1A_{n+1} 是圆的直径.

综上讨论知：当 n 是奇数时，命题成立；当 n 是偶数时，当且仅当 A_1A_{n+1} 是

圆的直径时,命题成立.

5. 一条平行于 X 轴的直线,如果它与函数 $y=x^4+px^3+qx^2+\gamma x+s$ 的图像相交于互异的四点 A、B、C、D. 而线段 AB、AC 与 AD 可以构成某个三角形的三条边,那么就称此直线是"三角形"的. 证明:平行 X 轴而与上述函数的图像相交于互异四点的直线中,要么全都是三角形的,要么没有一条是三角形的. (7 分,芬兰供题)

证明 首先我们提请读者注意一个引理:

引理 方程 $x^3+px^2+qx+r=0$ 的三个实根 a、b、c 能够作为三角形的三边的充要条件是 $p<0$, $q>0$, $r<0$, $p^3>4pq-8r$.

这个引理是 1965 年全国高考数学试题中的附加题. 它的证明可参看有关资料,这里从略.

设有一条平行于 X 轴的直线与 $y=x^4+px^3+qx^2+\gamma x+s$ 的图像相交于四点 A、B、C、D,且这条直线是三角形的. 不妨假定这一条直线就是 X 轴(否则可将函数图像作一个方向与 Y 轴平行的平行移动),还可以假定 A 为图像与 X 轴最左的那个交点,并且 A 就是原点(否则可作一个方向平行于 X 轴的平行移动). 这样一来,由于 B、C、D 的横坐标为方程 $x^3+px^2+qx+\gamma=0$ 的三个实根,并且可以作为一个三角形的三边. 根据引理,

$$p<0,\ q>0,\ \gamma<0,\ p^3>4pq-8\gamma.$$

对于任一条与 $y=x^4+px^3+qx^2+\gamma x$ 的图像相交于四点的直线 $y=y_0$,记最左边的一个交点横坐标为 x_0,其余三点的横坐标为 x_1, x_2, x_3. 于是,正数 $a=x_1-x_0$, $b=x_2-x_0$, $c=x_3-x_0$ 及 0 满足 $y_0=(x+x_0)^4+p(x+x_0)^3+q(x+x_0)^2+\gamma(x+x_0)$. 也就是说,正数 a、b、c 满足方程

$$x^3+(4x_0+p)x^2+(6x_0^2+3px_0+q)x+(4x_0^3+3px_0^2+2qx_0+\gamma)=0.$$

根据韦达定理,有 $4x_0+p=-(a+b+c)<0$, $6x_0^2+3px_0+q=ab+bc+ca>0$, $4x_0^3+3px_0^2+2qx_0+\gamma=-abc<0$.

另外,因为 $(4x_0+p)^3-4(4x_0+p)(6x_0^2+3px_0+q)+8(4x_0^3+3px_0^2+2qx_0+\gamma)=p^3-4pq+8\gamma>0$.

于是,根据引理 a、b、c 可以构成一个三角形.

这就是说,平行于 X 轴而与函数 $y=x^4+px^3+qx^2+\gamma x+s$ 的图像相交

于不同的四点的直线中，只要有一条是三角形的，那么全都是三角形的. 否则将没有一条是三角形的. 命题证毕.

评注　证明中我们假设四点是如图 5 的顺序配置的. 事实上，不论四点顺序如何配置，我们都可以认为 A 点是左起第一个点. 因为：

(i) 若四点配置如图 6，由 AB、AC、AD 成三角形，显然 BA、BC、BD 亦成三角形.

图 5 　　　　图 6

(ii) 若四点配置如图 7，则对函数作 $x=-t$ 的代换，即化为(i).

(iii) 若四点配置如图 8，则同样可作代换 $x=-t$，即化成图 5 的配置.

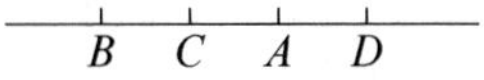

B　C　A　D　　　　B　C　D　A

图 7 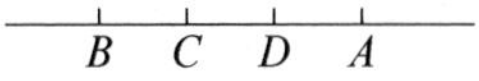　　　　图 8

6. 找出数 $(\sqrt{2}+\sqrt{3})^{1980}$ 的十进小数表达式中紧靠着小数点的右面第一位数字(即第一位小数)和左面一位数字(即个位数). 并证明你的结论.

(7 分，英国供题)

解　显然，数 $A=(\sqrt{3}+\sqrt{2})^{1980}+(\sqrt{3}-\sqrt{2})^{1980}$ 是一个正整数. 而 $(\sqrt{3}-\sqrt{2})^{1980}=\left[\frac{1}{\sqrt{3}+\sqrt{2}}\right]^{1980}<1$. 所以 $(\sqrt{3}+\sqrt{2})^{1980}$ 个位数等于 A 的个位数减去 1. 而

$$\begin{aligned}A&=(5+2\sqrt{6})^{990}+(5-2\sqrt{6})^{990}\\&=2\times[5^{990}+C_{990}^{2}5^{988}(2\sqrt{6})^{2}+\cdots+C_{990}^{988}5^{2}(2\sqrt{6})^{988}+(2\sqrt{6})^{990}]\\&\equiv 2\times 2^{990}\times 6^{495}\pmod{10}.\end{aligned}$$

因为　$2\times 2^{990}\times 6^{495}\equiv 2\times 2^{990}\equiv 2\times(-1)\equiv 3\pmod{5}$.

所以　$A\equiv 2\times 2^{990}\times 6^{495}\equiv 8\pmod{10}$.

因此，$(\sqrt{3}+\sqrt{2})^{1980}$ 的个位数为 7.

另一方面，由于 $\sqrt{3}-\sqrt{2}=0.32\cdots$，所以 $(\sqrt{3}-\sqrt{2})^{1980}$ 小数点后的第一位是 0，因此，$(\sqrt{3}+\sqrt{2})^{1980}$ 的小数点后第一位数字是 9.

第 13 届加拿大数学竞赛试题及解答

第13届加拿大数学竞赛于1981年5月举行,题目颇为灵活,下面是我们所拟的解答,谨供读者参考.

1. 对任意实数 t, $[t]$表示小于或等于 t 的最大整数. 如 $[8]=8$, $[\pi]=3$, $\left[-\frac{5}{2}\right]=-3$. 证明方程

$$[x]+[2x]+[4x]+[6x]+[8x]+[16x]+[32x]=12345$$

无实数解.

证明 设 $x=k+x'$,其中 k 为整数, $x'\in[0,1)$, 则有

$$[nx]=[n(k+x')]=nk+[nx'], \ (n\text{ 为整数})$$

于是原方程变为

$$63k+[x']+[2x']+[4x']+[8x']+[16x']+[32x']=12345,$$

如果这一方程有解,则 $63k+63x'>12345>63k$,而 $12345=63\times195+60$,所以 $k=195$,并且

$$[x']+[2x']+[4x']+[8x']+[16x']+[32x']=60$$

因为 $x\in[0,1)$, 于是

$$\begin{aligned}&[x']+[2x']+[4x']+[8x']+[16x']+[32x']\\&\leqslant 0+1+3+7+15+31=57<70.\end{aligned}$$

所以原方程无实数解.

2. 已知一半径为 r 的圆及过这圆上一已知点 P 的切线 l. 自圆上动点 R 引 l 的垂线 RQ. Q 在 l 上,确定$\triangle PQR$ 的面积的最大值.

解 如图 1,以圆心 O 为圆点,直线 OP 为 x 轴建立坐标系. 于是得圆的参数方程为

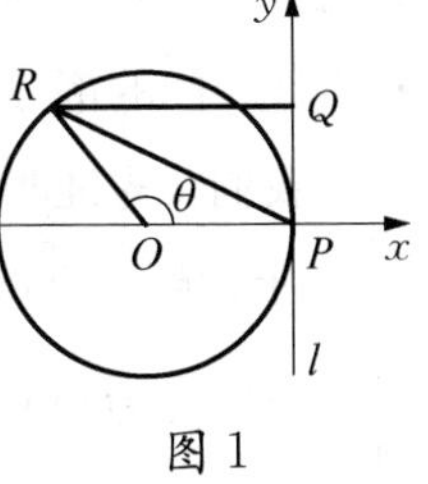

图 1

$$\begin{cases} x = r\cos\theta, \\ y = r\sin\theta \end{cases} \quad (\theta \text{ 为参数}, 0 \leqslant \theta < 2\pi),$$

切线 l 的方程为 $x = r$.

若 R 点的坐标为 $(r\cos\theta, r\sin\theta)$, 则 Q 点的坐标为 $(r, r\sin\theta)$.

于是$\triangle PQR$ 的面积

$$S = \frac{1}{2}RQ \cdot PQ = \frac{1}{2}r(1-\cos\theta) \cdot r|\sin\theta|$$

$$= \frac{1}{2}r^2\sqrt{(1-\cos\theta)^2\sin^2\theta} = \frac{1}{2}r^2\sqrt{(1-\cos\theta)^3(1+\cos\theta)}.$$

因

$$(1-\cos\theta)+(1-\cos\theta)+(1-\cos\theta)+3(1+\cos\theta)=6$$

是定值,所以,当 $1-\cos\theta=3(1+\cos\theta)$,即 $\cos\theta=-\frac{1}{2}$,也就是当 $\theta=120°$ 或 $\theta=240°$ 时, $\triangle PQR$ 面积达到最大值,且

$$S_{\max} = \frac{1}{2}r^2\sqrt{\left(\frac{3}{2}\right)^2 \cdot \frac{1}{2}} = \frac{3}{8}\sqrt{3}r^2.$$

如果利用“圆内接三角形中以正三角形面积为最大”这一结论,还可以得到一种简单的解法,即作 $RR' /\!/ PQ$,交已知圆于 R'(图 2),易知$\triangle PRR'$的面积是$\triangle PRQ$ 的两倍,而$\triangle PRR'$的面积的最大值是$\frac{3}{4}\sqrt{3}r^2$(这时$\triangle PRR'$为圆内接正三角形),因此$\triangle PRQ$ 的面积的最大值为$\frac{3}{8}\sqrt{3}r^2$.

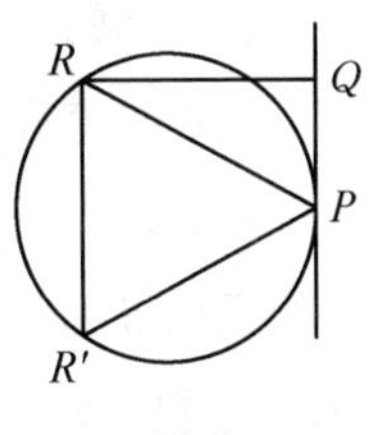

图 2

3. 已知平面 P 上有有限条直线,证明:可以在平面 P 上作一个任意大的圆,这圆与这些直线无公共点. 另一方面,证明可以在平面 P 上作可数条直线(即作无限条直线,并且可以顺序地数出第一条直线,第二条直线,第三条直线……),使得平面 P 上任一个圆至少与其中一条直线相交(一个点不认为是一个圆).

证明 先证命题的前一部分. 设平面 P 上有 n 条直线 $l_1, l_2, \cdots, l_n$. 对于

任意大的正数 r,我们作 n 个宽为 $2r$ 的"带形"b_1, b_2, $\cdots$, b_n. 带形 b_i 由两条平行直线之间的部分组成,这两条平行直线与直线 l_i 的距离为 r.

我们首先证明这 n 个带形不能把整个平面 P 完全盖住. 换句话说可以在平面 P 上找到一个点不在这 n 个带形内. 为此,作一个半径为 R 的圆 K,取 $R>\frac{4nr}{\pi}$,则圆的面积 $S=\pi R^2>R\times 4nr$.

每个带形 b_i 与圆 K 的公共部分的面积 A_i 小于这样一个矩形的面积,这个矩形的宽为 $2r$. 而长不超过圆 K 的直径 $2R$,因此 $A_i\leqslant 2r\times 2R=4Rr$. 从而有

$$A_1+A_2+\cdots+A_n\leqslant n\times 4Rr<S,$$

这就是说在圆 K 内可以找到一个点 O 不在这 n 个带形内. 换句话说,O 到直线 $l_i(i=1, 2, \cdots, n)$ 的距离大于 r. 因此,以 O 为圆心,以 r 为半径的圆与每一条直线 $l_i(i=1, 2, \cdots, n)$ 均不相交.

再看定理的后一部分. 在平面 P 上建立一直角坐标系 XOY. 作无数条与 Y 轴平行的直线 $x=\alpha$, 其中 α 为任一有理数. 由于有理数是可数的,所以这无数条直线可数. 注意到任一个圆内一定有点的横坐标为有理数,因而一定与这些直线的至少某一条相交.

本题还可用如下的简单通俗的解法:

若已知的 n 条直线都互相平行,结论显然成立.

若这些直线不全互相平行,它们的交点的个数是有限的(不超过 C_n^2 个),在平面上任取一点 O,设那些交点中与 O 的距离最远的点之一是 A. 以 O 为圆心, OA 为半径作圆 O,则这圆的外部再没有已知直线的交点了. 显然,在 A 分已知直线中的若干条所成的射线中,至少有两条射线构成的劣角内不再有已知直线上的点,如图 3 中的 α 角. 在 α 角的平分线上取一点 B,使 $|AB|>\dfrac{\gamma}{\sin\frac{\alpha}{2}}$ (γ 是事先给定的任意大的正数). 则以 B 为圆心, γ 为半径的圆外在 α 的内部(因 B 到 α 两边的距离 $|AB|\sin\frac{\alpha}{2}>\gamma$), 从而与已知的 n 条直线没有公共点.

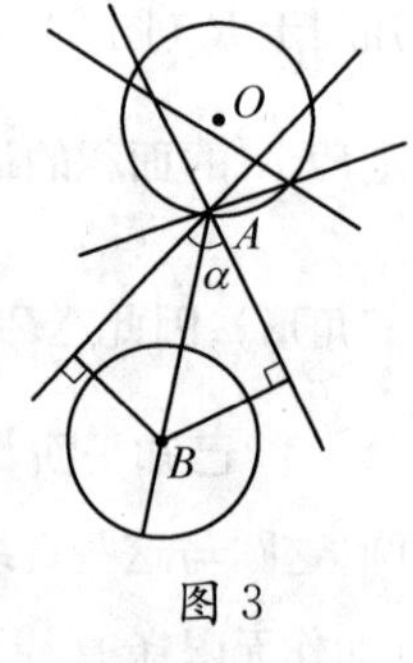

图 3

4. 已知 $P(x)$、$Q(x)$为两个实系数多项式，且对所有实数 x 满足恒等式 $P(Q(x)\equiv Q(P(x))$. 若方程 $P(x)=Q(x)$ 无实数解，证明方程

$$P(P(x))=Q(Q(x))$$

也无实数解.

证明　方程 $P(x)-Q(x)=0$ 没有实数根，它的根都是虚数. 而实系数多项式的虚数根是共轭出现的. 因此，

$$\begin{aligned}P(x)-Q(x)&=A[x-(a_1+b_1\mathrm{i})]\cdot[x-(a_1-b_1\mathrm{i})]\cdots\cdot\\&\quad[x-(a_n+b_n\mathrm{i})][x-(a_n-b_n\mathrm{i})]\\&=A[x^2-2a_1x+a_1^2+b_1^2]\cdots[x^2-2a_nx+a_n^2+b_n^2]\\&=A[(x+a_1)^2+b_1^2]\cdots[(x-a_n)^2+b_n^2].\end{aligned}$$

其中 A 为实常数. 不妨假定 $A>0$，那么对于任一实数 x，有 $P(x)-Q(x)>0$. 因此，对于任一实数 x，由于 $P(x)$、$Q(x)$也都是实数.

$$\begin{aligned}&P[P(x)]-Q[Q(x)]\\=&\{P[P(x)]-Q[P(x)]\}+\{P[Q(x)]-Q[Q(x)]\}>0,\end{aligned}$$

这就是说，方程 $P[P(x)]=Q[Q(x)]$ 无实数解.

5. 有十一个剧团参加一联欢节演出. 每一天有一些剧团参加演出，其余的剧团观看演出，联欢节结束时，每个剧团在这期间至少看过其他任一个剧团一场演出，问这联欢节至少持续多少天?

解　我们先证明两个更一般的结论:

(1) 在 n 天内可以安排 $C_n^{\left[\frac{n}{2}\right]}$ 个剧团演出，使得每个剧团可以看到其他的任一剧团的演出.

证明:由 $1,\ 2,\ \cdots,\ n$ 个取$\left[\dfrac{n}{2}\right]$个数的组合共有 $C_n^{\left[\frac{n}{2}\right]}$ 个，对应于每一个组合$(i_1,\ i_2,\ \cdots\cdots,\ i_{\left[\frac{n}{2}\right]})$，我们安排一个剧团，它在第 i_1 天，第 i_2 天，……，第 $i_{\left[\frac{n}{2}\right]}$ 天演出. 这样就安排了 $C_n^{\left[\frac{n}{2}\right]}$ 个剧团. 对于一个剧团 A，它与其他的任一剧团 B 对应于不同的组合，因而一定有一天 B 演出而 A 不演出，所以 A 能看到其他任一个剧团的演出.

(2) 在 n 天内至多安排 $C_n^{\left[\frac{n}{2}\right]}$ 个剧团,使得每一个剧团至少看到其他任一个剧团的一场演出.

证明:如果一个剧团恰在第 i_1 天,第 i_2 天,……,第 i_m 天演出,$1 \leqslant i_1 < i_2 < \cdots < i_m \leqslant n$,我们就将这剧团与集合$\{i_1, i_2, \cdots, i_m\}$对应,如果每一个剧团至少看到其他任一个剧团一场演出,那么相对应的各个集合满足:(i)是互不相同的;(ii)没有一个包含另一个.只要证明这些满足(i)、(ii)的集合的个数不超过 $C_n^{\left[\frac{n}{2}\right]}$.

设演出 m 天的剧团有 f_m 个,对于与$\{i_1, i_2, \cdots, i_m\}$对应的剧团,它可产生 $m!\,(n-m)!$ 个 $1, 2, \cdots, n$ 的全排列,前 m 个数为 $i_1, i_2, \cdots, i_m$ 的全排列,后 $n-m$ 个数为 $1, 2, \cdots, n$ 中其余数的全排列.由于(i)、(ii),这些全排列互不相同,所以 $\sum f_m \cdot m!\,(n-m)! \leqslant n!$ 而 $C_n^m = \dfrac{n!}{m!\,(n-m)!} \leqslant C_n^{\left[\frac{n}{2}\right]}$,所以

$$1 \geqslant \sum f_m \frac{m!\,(n-m)!}{n!} \geqslant \sum f_m \cdot \frac{1}{C_n^{\left[\frac{n}{2}\right]}},\ \sum f_m \leqslant C_n^{\left[\frac{n}{2}\right]}.$$

1978年罗马尼亚数学竞赛决赛试题解答

在第二十届IMO中,罗马尼亚以237分的好成绩获得团体总分第一名.有关专家认为:其重要因素之一是他们为挑选和训练选手而精心拟定的43道题,每一道题都是由罗马尼亚专家花了一定功夫编出的,这些题目灵活性大,有一定的难度,广泛地涉及集合、矩阵、微积分、数论以及初等数学的许多基础知识.

下面是决赛试题和我们所拟的解答,它可供我们了解罗马尼亚进行智力开发和选拔人才的一些做法,或许能给我们一些有益的启示.

解答中的错误与不妥之处,祈请读者批评指正.

九年级

1. 对于 $a \in \mathbf{R}$,确定 $\sqrt{a^2+a+1}-\sqrt{a^2-a+1}$ 的所有可能的值.

解　记 $y=\sqrt{a^2+a+1}-\sqrt{a^2-a+1}$, (1)

先假定 $a \geqslant 0$,这时 $y \geqslant 0$. 由(1)两边平方得

$$y^2=2a^2+2-2\sqrt{a^4+a^2+1}, \tag{2}$$

即

$$2a^2+2-y^2=2\sqrt{a^4+a^2+1}, \tag{3}$$

再两边平方并整理得

$$y^4-4y^2-4(y^2-1)a^2=0, \tag{4}$$

从而

$$a^2=\frac{y^2(y^2-4)}{4(y^2-1)} \geqslant 0. \tag{5}$$

由(2)得 $y^2<2a^2+2-2\sqrt{a^4}=2$; 再由(5)知 $y^2 \leqslant 1$, 从而

$$0 \leqslant y<1.$$

反过来,对于$[0, 1)$中每一个 y 值,由(5)可以定出 a,并且这时 $2a^2+2-y^2>0$,故可由(5)逆推出(2)及(1).因而在 $a\geqslant 0$ 时,$\sqrt{a^2+a+1}-\sqrt{a^2-a+1}$ 的值域为$[0, 1)$.

同样,在 $a<0$ 时,$\sqrt{a^2+a+1}-\sqrt{a^2-a+1}$ 的值域为$(-1, 0)$.

综合起来知 $\sqrt{a^2+a+1}-\sqrt{a^2-a+1}$ 的值域为$(-1, 1)$.

2. 设 $ABCD$ 为任意凸四边形,M 为对角线 AC 上的点,过 M 平行于 AB 的直线交 BC 于 P,过 M 平行于 DC 的直线交 AD 于 Q.

(Ⅰ) 证明:$MP^2+MQ^2\geqslant\dfrac{AB^2\cdot DC^2}{AB^2+DC^2}$,并指出等号何时成立;

(Ⅱ) 当 M 在 AC 上移动时,求 PQ 中点的轨迹.

解 (Ⅰ) 易知 $\dfrac{MP^2}{AB^2}=\dfrac{MC^2}{AC^2}$,$\dfrac{MQ^2}{DC^2}=\dfrac{AM^2}{AC^2}$,故

$$MP^2+MQ^2=\frac{AB^2\times MC^2+DC^2\times AM^2}{AC^2}, \tag{1}$$

但 $2AM\times MC=2\cdot\dfrac{AM\times DC}{AB}\cdot\dfrac{MC\times AB}{DC}\leqslant\dfrac{AM^2\times DC^2}{AB^2}+\dfrac{MC^2\times AB^2}{DC^2}$,

$$\begin{aligned}AC^2&=(AM+MC)^2=AM^2+MC^2+2AM\cdot MC\\&\leqslant AM^2+MC^2+\frac{MC^2\cdot AB^2}{DC^2}+\frac{AM^2\cdot DC^2}{AB^2}\\&=\frac{MC^2(AB^2+DC^2)}{DC^2}+\frac{AM^2(AB^2+DC^2)}{AB^2},\end{aligned}$$

由此得

$$\frac{AB^2\cdot MC^2+DC^2\cdot AM^2}{AC^2}\geqslant\frac{AB^2\cdot DC^2}{AB^2+DC^2},$$

由(1)知,上式即欲证之不等式.其中等号当且仅当 $\dfrac{MC^2\cdot AB^2}{DC^2}=\dfrac{AM^2\cdot DC^2}{AB^2}$,即 $\dfrac{MC}{AM}=\dfrac{DC^2}{AB^2}$ 时成立.

(Ⅱ) 我们首先考虑点 M 与 A 重合的极端性形(图 1),这时 P 与 B 重合,Q 与 A 重合,R 为 AB 的中点 E.同样,在点 M 与 C 重合时,R 为 CD 的中点

F.因此我们可以猜测 R 的轨迹就是线段 EF.

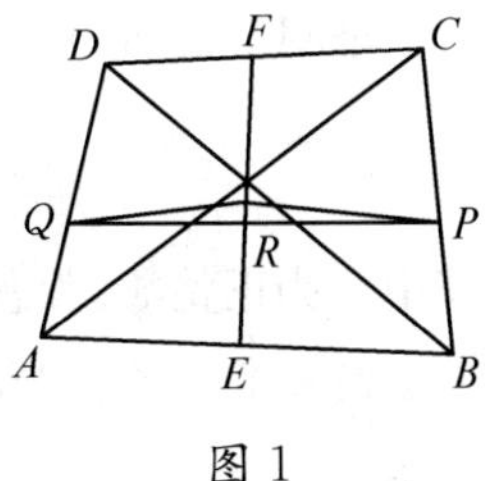

图 1

事实上,记顶点 A、B、C、D 的坐标为(x_i, y_i),$(i=1, 2, 3, 4)$.则当$\frac{AM}{AC}=\frac{BP}{BC}=\frac{AQ}{AD}=\lambda\geqslant 0$时,由定比分点公式得点 P 的坐标为

$$\begin{cases}x_P=\lambda x_3+(1-\lambda)x_2,\\ y_P=\lambda y_3+(1-\lambda)y_2.\end{cases}\tag{2}$$

同样,点 Q 的坐标为

$$\begin{cases}x_Q=\lambda x_4+(1-\lambda)x_1,\\ y_Q=\lambda y_4+(1-\lambda)y_1.\end{cases}\tag{3}$$

于是点 R 的坐标为

$$\begin{cases}x_R=\lambda\cdot\frac{x_2+x_4}{2}+(1-\lambda)\cdot\frac{x_1+x_2}{2},\\ y_R=\lambda\cdot\frac{y_3+y_4}{2}+(1-\lambda)\cdot\frac{y_1+y_2}{2}.\end{cases}\tag{4}$$

但$\left(\frac{x_1+x_2}{2}, \frac{y_1+y_2}{2}\right)$与$\left(\frac{x_3+x_4}{2}, \frac{y_3+y_4}{2}\right)$恰好是 F、E 的坐标,因此点 R 在线段 EF 上.

反之,设点 R 在线段 EF 上,那么(4)成立;由(2)、(3)及$\frac{AM}{AC}=\lambda$,即可确定点 P、Q、M,则从$\frac{AM}{AC}=\frac{BP}{BC}=\lambda$ 及$\frac{AM}{AC}=\frac{AQ}{AD}=\lambda$ 知$MP\parallel AB$, $MQ\parallel CD$.又由(2)、(3)、(4)知点 R 为线段 PQ 的中点,故线段 EF 上的点均合乎题述条件.

综上所述,知线段 EF 为所求轨迹.

评注 如果点 M 在直线AC 上移动,那么轨迹为直线 EF.这时,上面的证明仍然有效,只是$\frac{AM}{AC}$可取负值.

3. 设 m、n 为整数,且 $m\geqslant 1$、$n\geqslant 1$ 使$\sqrt{7}-\frac{m}{n}>0$,证明$\sqrt{7}-\frac{m}{n}>\frac{1}{mn}$.

证明 易知 $7n^2>m^2$, 从而

$$7n^2=m^2+a, \tag{1}$$

其中 a 为正整数,考虑(1)两边除以 8 所得的余数. 因为一个奇数的平方除以 8 的余数为 1,一个偶数的平方除以 8 的余数为 0 或 4,因此(1)的左边除以 8 的余数为 0, 7 或 4($7\times4=28=3\times8+4$);而在 $a=1$ 或 2 时, (1)式右边除以 8 的余数为 1, 2, 3, 5 或 6,因为 $a=1$ 或 2 时, (1)不可能成立,从而 $a\geqslant3$. 这时

$$7m^2n^2=(m^2+a)m^2=m^4+am^2\geqslant m^4+2m^2+m^2\geqslant(m^2+1)^2, \tag{2}$$

其中等号当且仅当 $a=3$, $m=1$ 时成立,这时(1)变为 $7n^2=1^2+3=4$, 这是不可能的,所以恒有

$$7m^2n^2>(m^2+1)^2,$$

即

$$\sqrt{7}-\frac{m}{n}>\frac{1}{mn}.$$

4. 在实数范围内解方程

$$\sqrt{x}+\sqrt{y-1}+\sqrt{z-2}=\frac{1}{2}(x+y+z).$$

解 原方程即

$$(\sqrt{x}-1)^2+(\sqrt{y-1}-1)^2+(\sqrt{z-2}-1)^2=0.$$

所以 $\sqrt{x}=1$, $\sqrt{y-1}=1$, $\sqrt{z-1}=1$, 从而

$$x=1,\ y=2,\ z=3,$$

经检验知,这确实是原方程的解.

十年级

1. 设 $A=\{a_1, a_2, \cdots, a_n\}$ 是一个实数的集合,且 $\varphi: A\to A$ 是一个可逆映射,假如 $a_1<a_2<\cdots<a_n$,并且 $a_1+\varphi(a_1)<a_2+\varphi(a_2)<\cdots<a_n+\varphi(a_n)$.

证明 φ 是集合 A 的恒等映射. 如果把 A 换成全体实数的集合,这个结果还对吗?

解 我们断定必有 $\varphi(a_1)=a_1$. 不然的话,设

$$\varphi(a_1)=a_i,\ i>1. \tag{1}$$

由于 φ 是可逆映射,这时

$$\varphi(a_1),\ \varphi(a_2),\ \cdots,\ \varphi(a_{i-1}) \tag{2}$$

均不等于 a_1,从而均大于 a_1,并且(2)中 $i-1$ 个数不可能均在 $i-2$ 个元素的集合$\{a_2,\ a_3,\ \cdots,\ a_{n-1}\}$中. 于是(2)中必有数$>a_{i-1}$,因而$\geqslant a_j$,设

$$\varphi(a_j)\geqslant a_i,\ j<i, \tag{3}$$

则由(3)、(1)得

$$a_j+\varphi(a_j)\geqslant a_j+a_i\geqslant a_1+a_i=a_i+\varphi(a_j), \tag{4}$$

这与题设矛盾,因而 $a_1=\varphi(a_1)$.

同法可证 $a_2=\varphi(a_2),\ a_3=\varphi(a_3),\ \cdots,\ a_n=\varphi(a_n)$.

当 A 为全体整数集时,结论不成立. 例如 $\varphi(n)=n+1$ 不是恒等映射,但它满足全部已知条件.

2. 确定整数 $k\geqslant 1$, 使得表达式

$$\sin kx\cdot\sin^k x+\cos kx\cdot\cos^k x-\cos^k 2x \tag{1}$$

不依赖于 x.

解 当 $x=0$ 时, (1)的值为 0,因此即求整数 k,使

$$\sin kx\cdot\sin^k x+\cos kx\cdot\cos^k x-\cos^k 2x\equiv 0. \tag{2}$$

取 $x=\frac{\pi}{k}$, 由(2)得

$$-\cos^k\frac{\pi}{k}=\cos^k\frac{2\pi}{k}.$$

因此 k 为奇数,且

$$-\cos\frac{\pi}{k}=\cos\frac{2\pi}{k},$$

即

$$\cos\frac{2\pi}{k}=\cos\left(\pi-\frac{\pi}{k}\right), \tag{3}$$

由(3)得

$$\pi-\frac{\pi}{k}=2n\pi\pm\frac{2\pi}{k}, \tag{4}$$

由(4)及 $k\geqslant 1$，得奇数 k 的值为 1，3. 但代入(2)知，只有 $k=3$ 是本题的解.

3. 已知凸多面体的面数 $n\geqslant 5$. 其中每个顶点恰好都引出三条棱. 两个人做下面的游戏：两人交替地在没有被签上名的面上签名，以先在具有公共顶点的三个面上签上名者为胜. 证明：在任何情况下，都是先开始签名的人获胜.

证明 首先证明这个凸多面体至少有一个面的边数≥4. 不然的话，每一面均为三角形，从而 $3n=2E$ (其中 E 为多面体的棱数)；但由欧拉公式 $V+n-E=2$(其中 V 为多面体的顶点数)；又由于每一个顶点恰好引出三条棱，所以 $3V=2E$. 由这三等式消去 V、E 得 $n=4$，这与已知 $n\geqslant 5$ 矛盾.

第一个人先在边数≥4 的这个面 S_1 上签名，然后在与面 S_1 相邻的面 S_2, S_3, …, $S_k(k\geqslant 5)$ 中选择一个面签名. 选择的方法是这样的(图 2)：如果第二个人未在 S_2, …, S_k 上签名，则可在 S_2, …, S_k 中任取一个；如果第二人在 S_2, …, S_k 中某一个，不妨设为 S_2 上签名，那么 S_3, …, S_k 中必有与 S_2 不相邻者，第一个人就在这种面中的一个上签名. 设第一个人在 S_4 上签了第二次名，那么他一定可以在与 S_4 相邻的面 S_3、S_6 中的一个上再签一次名，从而取得了胜利.

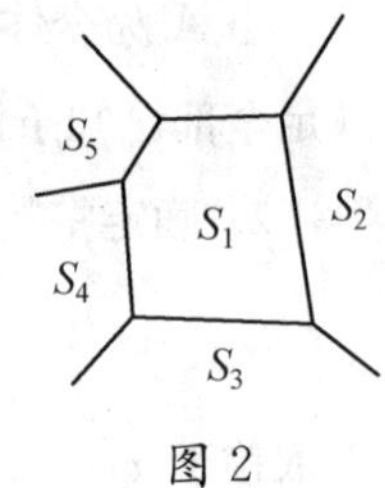

图 2

4. 在一次国际象棋比赛中有 n 个选手，每人要与其他所有参加者比赛一局(每人每天最多比赛一局)，进行全部比赛最少要用多少天？

解 显然，比赛的总局数为 C_n^2；每天至多进行 $\left[\frac{n}{2}\right]$ 局比赛(记录 $[x]$ 表示不超过 x 的最大整数). 因此，天数至少为

$$\frac{C_n^2}{\left[\frac{n}{2}\right]}=\frac{n(n-1)}{2\left[\frac{n}{2}\right]}=\begin{cases}n-1\ (n\text{ 为偶数时}),\\ n,\ (n\text{ 为奇数时}).\end{cases}$$

困难在于要证明在上述的时间内确实能安排完所有的棋赛.

首先考虑 n 为偶数的情形. 为此,用数字 1, 2, …, $n-1$ 表示一个正 $n-1$ 边形的顶点,用 n 表示它的中心(图 3). 它们分别代表 n 个选手. 第一天的比赛用实线表示;把这些实线绕中心 n 顺时针旋转 $\frac{2\pi}{n-1}$ 弧度(注意,点 1, 2, 3, …, $n-1$, n 的位置不动),得虚线构成的图形,它们就表示第二天比赛的安排;以后各天的安排方案依此类推. 这样就得到了用 $n-1$ 天完成全部比赛的安排.

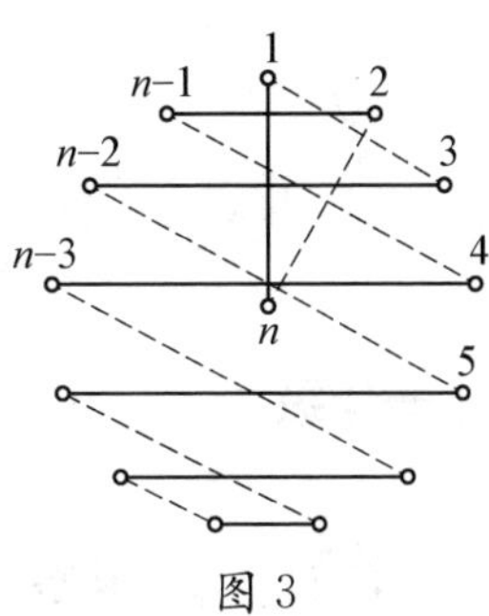

图 3

当 n 为奇数时,我们添上一名"假定选手",这时有选手 $n+1$ 名,是个偶数. 根据前面的证明,最少可用 $(n+1)-1=n$ 天进行比赛. 在实际比赛时,只须将前述方案中每天与"假定选手"比赛的那名选手轮空即可. 这就证明了,当 n 为奇数时,用 n 天确实可以进行全部比赛.

(待续)

1978 年罗马尼亚数学竞赛决赛试题解答（续）

十一年级

1. 设 $f: \mathbf{R} \to \mathbf{R}$ 是一实函数，它定义为：若 x 是无理数，则 $f(x)=0$；若 p、q 为整数，$q>0$ 且 $\frac{p}{q}$ 不可约，则

$$f\left(\frac{p}{q}\right)=\frac{1}{q^3},$$

证明：f 在每一个无理点 $x_0=\sqrt{k}$（k 为自然数）处有导数.

证明 我们证明在无理点 $\sqrt{k}$ 处，$f(x)$ 以 0 为它的导数值.

显然，当 x 的值取无理数而趋近 $\sqrt{k}$ 时，

$$\frac{f(x)-f(\sqrt{k})}{x-\sqrt{k}}=0.$$

因此只要证明当 x 值取有理数 $\frac{p}{q}$ 而趋近于 $\sqrt{k}$ 时，

$$\frac{f\left(\frac{p}{q}\right)-f(\sqrt{k})}{\frac{p}{q}-\sqrt{k}} \to 0,$$

即

$$\frac{1}{q^3\left(\sqrt{k}-\frac{p}{q}\right)} \to 0. \tag{1}$$

易知对每个 $Q>0$，在 $\sqrt{k}$ 的邻域 $(0, \sqrt{k}+1)$ 中只有有限多个分母 $\leqslant Q$ 的有理数 $\frac{p}{q}$，因此，在 $\sqrt{k}$ 的充分小的邻域中，一切有理数 $\frac{p}{q}$ 的分母均大于 Q，换句话说，当 $\frac{p}{q} \to \sqrt{k}$ 时，$q \to +\infty$.

设 $F(x)=x^2$，由拉格朗日中值定理，

$$\left|F\left(\frac{p}{q}\right)-F(\sqrt{k})\right|=|F'(\xi)|\cdot\left|\sqrt{k}-\frac{p}{q}\right|,$$

即

$$\left|\frac{p^2}{q^2}-k\right|=2\xi\left|\sqrt{k}-\frac{p}{q}\right|, \tag{2}$$

其中 $\frac{p}{q}<\xi<\sqrt{k}$ 或 $\sqrt{k}<\xi<\frac{p}{q}$. 由于 $\frac{p}{q}\to\sqrt{k}$，故不妨设

$$0<\xi<\sqrt{k}+1, \tag{3}$$

由(2)、(3)得

$$q^3\left|\sqrt{k}-\frac{p}{q}\right|=\frac{q^3\left|\frac{p^2}{q^2}-k\right|}{2\xi}>\frac{q\ |\ p^2-kq^2\ |}{2(\sqrt{k}+1)}\geqslant\frac{q}{2(\sqrt{k}+1)} \tag{4}$$

(因为 $\sqrt{k}$ 为无理数，所以 $p^2-kq^2\neq 0$，从而 $|p^2-kq^2|\geqslant 1$). 由于 $q\to+\infty$，故由(4)得知(1)成立.

2. $\{C\}$ 里的任何 C_n 是大于或等于 n 的最小整数幂 4^k，其中 n 是自然数，记 $b_n=C_1+C_2+\cdots+C_n-\frac{1}{5}$.

(Ⅰ) 以 $\frac{C_n}{n}$ 表达 $\frac{b_n}{n^2}$；

(Ⅱ) 证明：对于任何实数 $d\in\left[\frac{4}{5},\frac{5}{4}\right]$，有一个自然数序列 $n_k\to\infty$，使得 $\lim\limits_{k\to\infty}\frac{bn_k}{(n_k)^2}=d$.

解 (Ⅰ) 设 $4^{k-1}<C_n\leqslant 4^k$，则 $C_n=4^k$，从而

$$\begin{aligned}b_n&=1+(4-1)\times 4+(4^2-4)\times 4^2+\cdots+(4^{k-1}-4^{k-2})\times 4^{k-1}+\\&\quad(n-4^{k-1})\times 4^k-\frac{1}{5}\\&=1-4+4^2-4^3+4^4+\cdots-4^{2k-1}+n\times 4-\frac{1}{5}\\&=\frac{1-4^{2k}}{5}+n\times 4^k-\frac{1}{5}=n\times 4^k-\frac{4^{2k}}{5}=nC_n-\frac{(C_n)^2}{5},\end{aligned}$$

所以

$$\frac{b_n}{n^2}=\frac{C_n}{n}-\frac{1}{5}\left(\frac{C_n}{n}\right)^2. \tag{1}$$

（Ⅱ）记 $y_n=\frac{C_n}{n}$ 易知二次函数

$$y-\frac{1}{5}y^2,\ (1\leqslant y\leqslant 4)$$

的值域为$\left[\frac{4}{5},\ \frac{5}{4}\right]$. 因此,对于 $d\in\left[\frac{4}{5},\ \frac{5}{4}\right]$,总可以找到 y_0,使到 $y_0-\frac{1}{5}y_0^2=d$ $(1\leqslant y_0\leqslant 4)$.

由于(1),要证明有一个自然数列 $n_k\to\infty$,使

$$\lim_{k\to\infty}\frac{b_{n_k}}{(n_k)^2}=d, \tag{2}$$

只要证明存在 $n_k\to\infty$,使得 $y_{n_k}\to y_0$.

因为

$$4=\frac{4_k}{4^{k-1}+0},\ \frac{4^k}{4^{k-1}+1},\ \cdots,\ \frac{4^k}{4^{k-1}+4^{k-1}}=1 \tag{3}$$

中每相邻两项的距离 $\leqslant\frac{4^k}{(4^{k-1}+0)(4^{k-1}+1)}<\frac{1}{4^{k-2}}$,所以在(3)中必有一项与 y_0 的距离$<\frac{1}{4^{k-2}}$,取这项作为 y_{n_k},即 $yn_k=\frac{4^k}{4^{k-1}+a}$,其中 a 为 0, 1, …, 4^{k-1} 中一数,它具有所述的性质,而 $n_k=4^{k-1}+a$,则 $n_k\to+\infty$,并且 $y_{n_k}\to y_0$,从而(2)成立.

3. A 和 B 两人用轮流确定一个实数作为一个元素的方法构成一个 3×3 矩阵,证明:无论是哪个人先开始, A 总能做到使最终的矩阵为奇异矩阵(即行列式为零的矩阵).

证明 轮到 A 填时, A 总是填零,由于行列式是否为零不受行(列)调动的影响,故可假定在 B 先填时,B 填的第一个数 B_1 位置为图 4(1)所示(A、B 第 i 次填的数分别记为 A_i、B_i),则 A 可如图 4(1)所示填上 A_1,由对称性,可假定 B_2 位于图 4(1)中某×处, A 再填上A_2,这时 B_2 或 B_3 必在第 2 行第 1 列,否则

在这里填上 A_3,行列式为 0. 不妨设 B_2 在第 2 行第 1 列,划去第 2 行及第 1 列,得到一个 2 阶子行列式,如果这个子行列式为 0,则原行列式为 0. 不妨设在这个子行列式中 B_3 的位置如图 4(2), A 可如图填上 A_3. 那么无论 B_4 填在什么位置, A 总可在图 4(2)的两个记"×"的位置中的适当的一个填写 A_4,这样这个子式为 0,从而原行列式为 0.

$$\begin{vmatrix} B_1 & & \\ \times & A_1 & A_2 \\ \times & \times & \times \end{vmatrix} \qquad \begin{vmatrix} A_3 & \times \\ \times & B_3 \end{vmatrix}$$

(1)　　(2)

图 4

在 A 先填时,仍按上法,显然结论仍成立.

4. 设 f: **R**→**R** 是由

$$f(x)=x|x-a_1|+|x-a_2|+\cdots+|x-a_n|$$

所定义的函数,这里 a_1, a_2, …, a_n 是定实数. 试给出 f 在整个 **R** 上可微的条件.

解　$|x-a_i|$ $(i=2, 3, \cdots, n)$ 在 x_j 处不可微,因此,如果有某个 $a_i \neq a_1 (i \geqslant 2)$,那么在 $x=a_i$ 处

$$f(x)=(x|x-a_1|+|x-a_{k_1}|+\cdots+|x-a_{k_m}|)+j|x-a_i|$$

(其中 a_{k_1}, a_{k_2}, …, a_{k_m} 是a_2, a_3, …, a_n 中不等于a_j 的数, j 是a_2, a_3, …, a_n 中等于a_j 的数的个数),当 $x=a_1$ 时,上式的前一个括号中各项可微,而 $j|x-a_i|$ 不可微,于是 f 不可微. 这样, f 在整个 **R** 上可微时,必有 $a_j=a_2=\cdots=a_n$, 并且

$$f(x)=(x+n-1)|x-a_1|, \tag{1}$$

由(1)易知 $f(x)$在 $x=a_1$ 处可微的条件, $a_1=1-n$, 从而

$$a_1=a_2=\cdots=a_n=1-n,$$

这是 f 在整个 **R** 上可微的充分必要条件.

十二年级

1. 设 $f: \mathbf{R} \to \mathbf{R}$ 是连续函数,且对于任何 $x \in \mathbf{R}$, 记

$$g(x)=f(x)\int_0^x f(t)\mathrm{d}t,$$

证明:如果 g 是递减的,则在 $\mathbf{R}$ 上 $f \equiv 0$.

证明 先考虑 $x \geqslant 0$, 分几步进行.

(i) 至少有一个 $x_0>0$,使 $f(x_0)=0$.

若对所有 $x>0$, $f(x) \neq 0$, 又 $f(x)$连续,则 $x>0$ 时 $f(x)$保持定号,无论它大于零或小于零,都易知

$$g(x)=f(x)\int_0^x f(t)\mathrm{d}t>0 \ (x>0), \tag{1}$$

但 $g(0)=0$, 又 $g(x)$递减,(1)与此矛盾,故至少有一个 $x_0>0$,使 $f(x_0)=0$.

(ii) 若 $f(x_0)=0(x_0>0)$,则 $f(x) \equiv 0 \ (0<x \leqslant x_0)$.

因为 $f(x_0)=0$, 从而

$$g(x_0)=f(x_0)\int_0^{x_0} f(t)\mathrm{d}t=0, \tag{2}$$

但 $g(0)=0$, 又 g 递减,故

$$g(x) \equiv 0 \ (0 \leqslant x \leqslant x_0). \tag{3}$$

对于任一点 $a \ (0 \leqslant a<x_0)$,若 $f(a) \neq 0$, 那么在 a 的一个右邻域 $(a, a+\delta)$ 中, $f(x)$保持定号,从而

$$\int_a^{a+\delta} f(t)\mathrm{d}t \neq 0, \tag{4}$$

但由(2)

$$f(a)\int_0^a f(t)\mathrm{d}t=0,$$

$$f(a+\delta)\int_0^{a+\delta} f(t)\mathrm{d}t=0,$$

由此得

$$\int_{a}^{a+\delta} f(t)\mathrm{d}t=0,$$

与(4)矛盾,所以

$$f(x)\equiv 0\ (0<x\leqslant x_0). \tag{5}$$

(iii) 若 $x_1>x_2>0$,且 $f(x_1)=f(x_2)=0$,则 $f(x)\equiv 0\ (x_1\leqslant x\leqslant x_2)$.

仿(ii)证之.

(iv) 对于任何 $x_0>0$,总存在 $x_1>x_0$,使 $f(x)\equiv 0\ (0<x\leqslant x_1)$.

仿(i)可知,必有 $x_1>x_0$,使 $f(x_1)=0$;再由(ii)知 $f(x)\equiv 0\ (0<x\leqslant x_1)$.

(v) $f(x)\equiv 0\ (0\leqslant x<+\infty)$.

由(iv)知 $f(x)\equiv 0\ (0<x<+\infty)$;由 $f(x)$的连续性即得 $f(0)=0$,从而 $f(x)\equiv 0\ (0\leqslant x<+\infty)$.

同样可证,$x<0$ 时,$f\equiv 0$.

2. 设 P 和 Q 为两个复系数的非零多项式. 证明:P 和 Q 有相同的根(相同根的重数也相同),当且仅当 $f(z)=|P(z)|-|Q(z)|$ 所定义的函数 f: $\mathbf{C}\to\mathbf{R}$ 在整个 $\mathbf{C}$ 上有固定的符号,也可能是零.

证明 必要性是显然的. 设 $|P(z)|-|Q(z)|$ 在整个 $\mathbf{C}$ 上有固定符号,不妨设

$$|P(z)|\geqslant|Q(z)|, \tag{1}$$

设 α 为$P(z)$的根,则 $P(\alpha)=0$, 由(1)得 $Q(\alpha)=0$, 即 α 也是$Q(z)$的根. 设 α 分别为 $P(z)$、$Q(z)$的 m 重根与 n 重根,我们断言 $m\leqslant n$. 不然的话,设 $m>n$,

$$|P(z)|-|Q(z)|=|z-\alpha|^{n}(|z-\alpha|^{m-n}\times|P_1(z)|-|Q_1(z)|), \tag{2}$$

其中 $P_1(z)$、$Q_1(z)$都是多项式,并且 $Q_1(\alpha)\neq 0$.

由于 $m>n$,在 $z=\alpha$ 时

$$|z-\alpha|^{m-n}\cdot|P_1(z)|-|Q_1(z)|=-|Q_1(\alpha)|<0, \tag{3}$$

由 $|z-\alpha|^{m-n}\cdot|P_1(z)|-|Q_1(z)|$的连续性,存 α 的邻域$|z-\alpha|<\delta$, 在此

邻域中

$$|z-\alpha|^{m-n}\cdot|P_1(z)|-|Q_1(z)|<0, \tag{4}$$

于是在 $0<|z-\alpha|<\delta$ 时,由(2)、(4)

$$|P(z)|-|Q(z)|<0, \tag{5}$$

(5)与(3)矛盾.因此 $m\leqslant n$.

又由(1), $P(z)$的次数$\geqslant Q(z)$的次数(否则当$|z|$充分大时,(1)不成立),因此 $Q(z)$的根也必须是 $P(z)$的根,并且对两个多项式,相同的重数都相等.

3. (Ⅰ)设适合$\int_0^1 f(x)\mathrm{d}x=1$ 的所有连续函数 f: $[0,1]\to[0,2]$的集合为 F,当 f 遍历 F 时,$\int_0^1 xf(x)\mathrm{d}x$ 的所有可能的值是什么?

(Ⅱ)证明:若 f, f^2 属于 F,则 f 是一个常量.

解 (Ⅰ)令 $g_1(x)=\begin{cases}2\ \left(0\leqslant x<\frac{1}{2}\right),\\ 0\ \left(\frac{1}{2}\leqslant x\leqslant 1\right),\end{cases}$ 则

$$\int_0^1 xg_1(x)\mathrm{d}x=\int_0^{\frac{1}{2}}2x\,\mathrm{d}x=\frac{1}{4}, \tag{1}$$

并且

$$\begin{aligned}&\int_0^1 xf(x)\mathrm{d}x-\int_0^1 xg_1(x)\mathrm{d}x\\ =&\int_0^{\frac{1}{2}}x(f(x)-2)\mathrm{d}x+\int_{\frac{1}{2}}^1 f(x)\mathrm{d}x\geqslant\frac{1}{2}\int_0^{\frac{1}{2}}(f(x)-2)\mathrm{d}x+\frac{1}{2}\int_{\frac{1}{2}}^1 f(x)\mathrm{d}x\\ =&\frac{1}{2}\int_0^1 f(x)\mathrm{d}x-\int_0^{\frac{1}{2}}\mathrm{d}x\\ =&\frac{1}{2}-\frac{1}{2}=0,\end{aligned} \tag{2}$$

所以

$$\int_0^1 xf(x)\mathrm{d}x\geqslant\int_0^1 xg_1(x)\mathrm{d}x=\frac{1}{4}, \tag{3}$$

并且(3)中等号只能在 $f(x)=g_1(x)$ 时成立,由于 f 连续,所以 $f(x)\neq g_1(x)$,即(3)中不等式应是严格的不等式.

同样,考虑

$$g_2(x)=\begin{cases}0\ \left(0\leqslant x<\dfrac{1}{2}\right),\\ 2\ \left(\dfrac{1}{2}\leqslant x\leqslant 1\right),\end{cases}$$

可得

$$\int_0^1 xf(x)\mathrm{d}x<\frac{3}{4}, \tag{4}$$

因此

$$\frac{1}{4}<\int_0^1 xf(x)\mathrm{d}x<\frac{3}{4}.$$

现在我们证明 $\left(\dfrac{1}{4},\ \dfrac{3}{4}\right)$ 中的值均可为 $\int_0^1 xf(x)\mathrm{d}x$ 取得. 为此考虑矩形 $\{(x,\ y)\mid 0\leqslant x\leqslant 1,\ 0\leqslant y\leqslant 2\}$,设点 B 在矩形的周界上移动,并且点 B 的横坐标$<\dfrac{1}{2}$,纵坐标$\geqslant 1$,过 B 及 $C\left(\dfrac{1}{2},\ 1\right)$作直线交矩形周界于 D,则在图 5(1)中考虑折线 $ABCDE$ 所表示的函数 $f(x)$;在图 5(2)中考虑直线段 BD 所表示的函数 $f(x)$. 可知 $\int_0^1 xf(x)\mathrm{d}x$ 可取$\left(\dfrac{1}{4},\ \dfrac{1}{2}\right]$中的一切值. 类似地,当 B 点横坐标$>\dfrac{1}{2}$,纵坐标$\geqslant 1$ 时可知 $\int_0^1 xf(x)\mathrm{d}x$ 可取$\left[\dfrac{1}{2},\ \dfrac{3}{4}\right)$中的一切值(在图 5(1)中用 F 取代 A,用 O 取代 E).

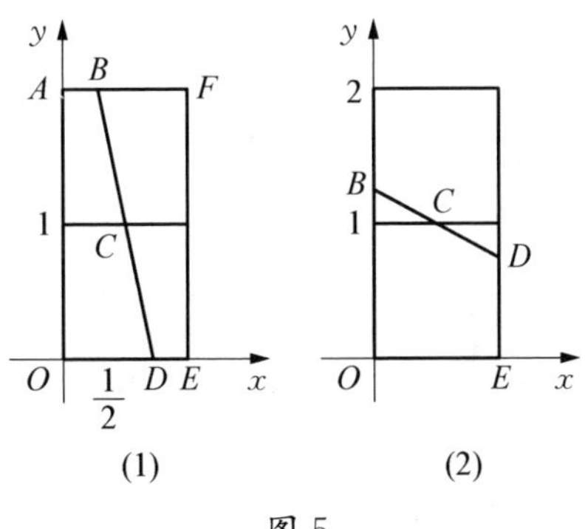

图 5

（Ⅱ）由柯西不等式，得

$$1=\int_0^1 f(x)\mathrm{d}x \leqslant \left(\int_0^1 \mathrm{d}x\int_0^1 f^2(x)\mathrm{d}x\right)^{\frac{1}{2}}=1,$$

因此上式中间等号成立，从而 $f(x)\equiv 0$.

4. 设 $P(z)$ 为复系数多项式，证明：$(P*P*\cdots*P)(z)=z$ 可被 $P(z)-z$ 整除，这里"复合"运算 $*$ 被取 n 次.

证明 记 $(P*P*\cdots*P)(z)=P_n(z)$.

用数学归纳法，显然 $P_1(z)-z=P(z)-z$ 能被 $P(z)-z$ 整除.

假定 $P_{n-1}(z)-z$ 能被 $P(z)-z$ 整除，则

$$P(z)-z=[P(z)-P_{n-1}(z)]+(P_{n-1}(z)-z), \tag{1}$$

而设 $$P(z)-z=(z-\alpha_1)(z-\alpha_2)\cdots(z-\alpha_n), \tag{2}$$

则 $$\begin{aligned}P_1(z)-P_{n-1}(z)&=P[P_{n-1}(z)]-P_{n-1}(z)\\&=(P_{n-1}(z)-\alpha_1)(P_{n-1}(z)-\alpha_2)\times\cdots\times(P_{n-1}(z)-\alpha_n).\end{aligned} \tag{3}$$

由(2)得

$$\alpha_j=P(\alpha_j)=P(P(\alpha_j))=P_2(\alpha_j)=\cdots=P_{n-1}(\alpha_j),\ j=1,\ 2,\ \cdots,\ m. \tag{4}$$

所以由(3)、(4)得

$$P_n(z)-P_{n-1}(z)=(P_{n-1}(z)-P_{n-1}(\alpha_1))\times\cdots\times(P_{n-1}(z)-P_{n-1}(\alpha)), \tag{5}$$

显然 $z-\alpha$ 整除 $P_{n-1}(z)-P_{n-1}(\alpha)$（即 α_j 是这个多项式的根），因此由(2)、(5)得 $P(z)-z$ 整除 $P_n(z)-P_{n-1}(z)$，再由(1)及归纳假设，$P(z)-z$ 整除 $P_n(z)-z$.

第十一届美国数学竞赛试题及解答

第十一届美国数学竞赛于 1982 年 5 月 4 日举行. 下面是这届竞赛的试题及我们所拟的一份解答. 限于水平, 不妥之处, 祈请读者指正.

1. 一次集会有 1982 个人参加, 其中任意的四个人中至少有一个人认识其余的三个人. 问在这次集会上, 认识全体到会者的人至少有多少位?

解　我们证明认识全体到会者的人至少有 1979 名. 换句话说, 至多有三个人不认识全体到会者.

采用反证法. 假设至少有四个人不认识全体到会者. A 为其中之一, A 不认识 B. 这时, 除 A、B 两人外, 还有 C 不认识全体到会者. 假如 C 不认识 D, 并且 D 不是 A、B, 那么在 A、B、C、D 这四个人中, 每一个都不全认识其余的三个人, 与已知矛盾. 因此 C 不认识的人一定是 A 或 B. 这时, 除 A、B、C 三个人外, 还有 D 不认识全体到会者. 根据刚才对 C 的同样的推理, D 所不认识的人一定是 A、B 或 C. 这时在 A、B、C、D 四个人中, 每个人都不全认识其余的三个人, 仍与已知矛盾.

恰好有 1979 个人认识全体到会者是可能的, 例如在 1982 个人中, A 不认识 B 和 C. 此外每两个人都互相认识, 则在任何四个人中至少有一个人认识其余的三个人, 并且恰好有 1979 个人认识全体到会者.

附注　(1) 本题实质是一个简单的图论问题, 用图论的语言来叙述和证明本题可参阅《趣味的图论问题》一书第 4 页的例一(上海教育出版社, 1980 年出版).

(2) 初等图论问题经常在数学竞赛试题中出现, 下面的问题可供读者练习.

问题一　一次集会有 1982 人参加, 且知每一位到会者的熟人不少于 991 人. 证明: 总可以在到会者中找出四个人来, 让这四个人围坐在圆桌周围, 且使每一个人的邻座都是自己的熟人.

问题二　九位数学家在一次会议上相遇, 发现他们中的任意三个人中, 至少有两个人可以用同一种语言对话. 如果每一位数学家至多可说三种语言, 证明至少有三个数学家可以用同一种语言对话.

问题三　十七位学者,每一位都给其余的人写一封信,信的内容是讨论三个论文题目中的任一个,而且两个人互相通信所讨论的是同一个题目.证明至少有三位学者,他们互相之间通信所讨论的是同一个论文题目.

2. 设 $S_n = x^n + y^n + z^n$,其中 x、y、z 为实数.已知在 $S_1 = 0$ 时,对 $(m, n) = (2, 3), (3, 2), (2, 5)$ 或 $(5, 2)$,有

$$\frac{S_{m+n}}{m+n} = \frac{S_m}{m} \cdot \frac{S_n}{n}. \quad (*)$$

试确定所有的其他适合($*$)式的正整数组(m, n)(如果有这样的数组存在的话).

解　令 $x = k+1$, $y = -k$, $z = -1$,则它们适合 $S_1 = 0$.如果($*$)式成立,则以 x、y、z 的上述值代入后,($*$)式成立 k 的恒等式.

在 m 为偶数时,$S_m = (k+1)^m + k^m + 1 = 2k^m + k$ 的次数低于 m 的项;

在 m 为奇数时,$S_m = (k+1)^m - k^m - 1 = mk^{m-1} + k$ 的次数低于 $m-1$ 的项.

考虑以下的三种情况:

(i) m、n 都是奇数,这时 $m+n$ 是偶数.如果($*$)式成立,那么左边 k 的最高次项为 $\frac{2}{m+n}k^{m+n}$,而右边 k 的最高次项为$\frac{k^{m-1}}{m} \cdot \frac{k^{n-1}}{n} = \frac{k^{m+n-2}}{mn}$.因此,在这种情况中($*$)式不可能成立.

(ii) m、n 都是偶数,这时 $m+n$ 也是偶数.如果($*$)式成立,那么比较两边的最高次 ($m+n$ 次)项的系数便得

$$\frac{2}{m+n} = \frac{2}{m} \cdot \frac{2}{n}, \ \frac{1}{\frac{m}{2}} + \frac{1}{\frac{n}{2}} = 1.$$

由于 m、n 是正偶数,所以$\frac{m}{2}$, $\frac{n}{2}$都是正整数,由上式易知 $\frac{m}{2} = \frac{n}{2} = 2$,即 $m = n = 4$.

但在 $m = n = 4$ 时,取 $k = 1$,则 $x = 2$, $y = -1$, $z = -1$, $S_4 = 2^4 + 1 + 1 = 18$, $S_8 = 2^8 + 1 + 1 = 258$,因此 $\frac{S_8}{8} \neq \frac{S_4}{4} \cdot \frac{S_4}{4}$,即在这种情况中($*$)式也不可能

成立.

(iii) m、n 一奇一偶. 不妨设 m 为奇数, n 为偶数,这时 $m+n$ 为奇数. 比较 (*)式两边 k 的最高次 ($m+n-1$ 次)项的系数得 $1=1\cdot\frac{2}{n}$. 从而 $n=2$.

显然 $m=1$ 时,(*)式右边为零,因此这时(*)式不可能成立.

假设 $m>3$, 这时(*)式左边为

$$\frac{S_{m+2}}{m+2}=\frac{(k+1)^{m+2}-k^{m+2}-1}{m+2}=k^{m+1}+\frac{m+1}{2}k^{m}+\frac{(m+1)m}{6}k^{m-1}+\cdots,$$

右边为

$$\begin{aligned}\frac{S_m}{m}\cdot\frac{S_2}{2}&=\frac{(k+1)^m-k^m-1}{m}\cdot\frac{(k+1)^2+k^2+1}{2}\\&=\left(k^{m-1}+\frac{m-1}{2}k^{m-2}+\frac{(m-1)(m-2)}{6}k^{m-3}+\cdots\right)(k^2+k+1)\\&=k^{m+1}+\frac{m+1}{2}k^{m}+\left(\frac{(m-1)(m-2)}{6}+\frac{m-1}{2}+1\right)k^{m-1}+\cdots.\end{aligned}$$

因此,如果(*)式成立,则

$$\frac{(m-1)(m-2)}{6}+\frac{m-1}{2}+1=\frac{(m+1)m}{6}.$$

从而 $m=5$.

于是 $(m, n)=(2, 3), (3, 2), (2, 5), (5, 2)$就是满足(*)式的全部正整数组.

附注　(1) 要证明 $(m, n)=(2, 3)$ 或$(2, 5)$满足(*)式可以采用下面的方法:

记 $xyz=\sigma_3$, $xy+yz+zx=\sigma_2$, 则由根与系数的关系可知 x、y、z 是方程

$$\mu^3+\sigma_2\mu-\sigma_3=0 \tag{1}$$

的三个根.

将(1)式两边同乘 μ^{m-3},得

$$\mu^m+\sigma_2\mu^{m-2}-\sigma_3\mu^{m-3}=0.$$

令 $\mu=x$、y、z, 然后将所得的三个等式相加得

$$S_m+\sigma_2 S_{m-2}-\sigma_3 S_{m-3}=0,\ (m=3,\ 4,\ \cdots).$$

由于

$$S_0=x^0+y^0+z^0=3,\ S_1=0,\ S_2=x^2+y^2+z^2=-2\sigma_2,$$

于是

$$S_3=-\sigma_2 S_1+\sigma_3 S_0=3\sigma_3,$$
$$S_4=-\sigma_2 S_2+\sigma_3 S_1=2\sigma_2^2,$$
$$S_5=-\sigma_2 S_3+\sigma_3 S_2=-3\sigma_2\sigma_3-2\sigma_2\sigma_3=-5\sigma_2\sigma_3.$$

于是,

$$\frac{S_5}{5}=-\sigma_2\sigma_3=\frac{S_3}{3}\cdot\frac{S_2}{2}.$$

同样,

$$S_7=-\sigma_2 S_5+\sigma_3 S_4=5\sigma_2^2\sigma_3+2\sigma_2^2\sigma_3=7\sigma_2^2\sigma_3.$$

因此,
$$\frac{S_7}{7}=\sigma_2^2\sigma_3=\frac{S_5}{5}\cdot\frac{S_2}{2}.$$

(2) 对于其他的正整数组$(m,\ n)$,$(\ *\ )$式虽不成立,但可以建立类似的等式,如

$$\frac{S_2}{2}\cdot\frac{S_2}{2}=\frac{S_4}{2},$$
$$\frac{S_7}{7}=\frac{S_3}{3}\cdot\frac{S_4}{2},$$

等等.

(3) 对这类问题的另一种处理方法可参见《一类恒等式的证明》一文(载华东师范大学数学系编《数学教学》1982 年第 6 期).

3. 若点 A_1 在等边三角形 ABC 的内部,点 A_2 在三角形 A_1BC 的内部.证明

$$I.Q.\ (A_1BC)>I.Q.\ (A_2BC).$$

其中一个图形 F 的等周商定义为

$$I.Q.=\frac{F\text{ 的面积}}{(F\text{ 的周长})^2}.$$

证明 首先指出，对于任意三角形 ABC 其等周商为：

$$I.Q.(ABC)=\frac{1}{4}\tan\frac{A}{2}\tan\frac{B}{2}\tan\frac{C}{2}. \tag{1}$$

这是因为

$$\begin{aligned}I.Q.(ABC)&=\frac{\frac{1}{2}ab\sin C}{(a+b+c)^2}=\frac{2R^2\sin A\sin B\sin C}{4R^2(\sin A+\sin B+\sin C)^2}\\&=\frac{\sin A\sin B\sin C}{2\left(4\cos\frac{A}{2}\cos\frac{B}{2}\cos\frac{C}{2}\right)^2}=\frac{\sin\frac{A}{2}\sin\frac{B}{2}\sin\frac{C}{2}}{4\cos\frac{A}{2}\cos\frac{B}{2}\cos\frac{C}{2}}\\&=\frac{1}{4}\tan\frac{A}{2}\tan\frac{B}{2}\tan\frac{C}{2}.\end{aligned}$$

如图 1，延长 BA_2 交 CA_1 于 A'. 现在来考察$\triangle A_1BC$ 与$\triangle A'BC$ 的等周商的大小.

为方便，记 $\angle BA_1C=A_1$，$\angle BA_2C=A_2$，$\angle BA'C=A'$，$\angle A_1BC=B_1$，$\angle A_2BC=B_2$，$\angle A_1CB=C_1$，$\angle A_2CB=C_2$. 于是

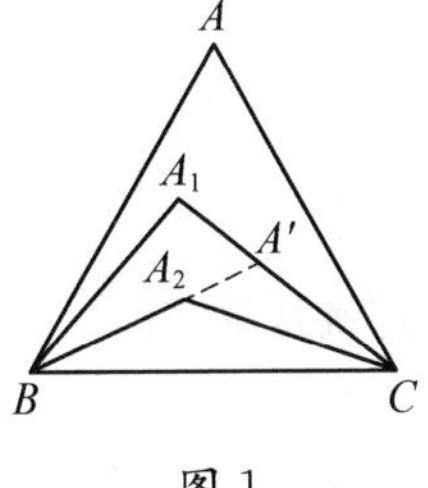

图 1

$$I.Q.(A_1BC)=\frac{1}{4}\tan\frac{A_1}{2}\tan\frac{B_1}{2}\tan\frac{C_1}{2},$$

$$I.Q.(A'BC)=\frac{1}{4}\tan\frac{A'}{2}\tan\frac{B_2}{2}\tan\frac{C_1}{2}.$$

因此，要比较 $I.Q.(A_1BC)$与 $I.Q.(A'BC)$的大小，只需比较$\tan\frac{A_1}{2}\tan\frac{B_1}{2}$ 与 $\tan\frac{A'}{2}\tan\frac{B_2}{2}$的大小. 由于

$$\tan\frac{A_1}{2}\tan\frac{B_1}{2}=\frac{\sin\frac{A_1}{2}\sin\frac{B_1}{2}}{\cos\frac{A_1}{2}\cos\frac{B_1}{2}}=\frac{\cos\frac{A_1-B_1}{2}-\cos\frac{A_1+B_1}{2}}{\cos\frac{A_1-B_1}{2}+\cos\frac{A_1+B_1}{2}},$$

因 $\frac{1}{2}(A_1+B_1)=90^\circ-\frac{C_1}{2}$，所以

$$\tan\frac{A_1}{2}\tan\frac{B_1}{2}=1-\frac{2\cos\frac{C_1}{2}}{\cos\frac{A_1-B_1}{2}+\cos\frac{C_1}{2}}.$$

同理可得

$$\tan\frac{A'}{2}\tan\frac{B_2}{2}=1-\frac{2\cos\frac{C_1}{2}}{\cos\frac{A'-B_2}{2}+\cos\frac{C_1}{2}}.$$

由于 $A_1>60^\circ>B_1$，$A'>60^\circ>B_2$ 及 $A'>A_1$，所以有 $0^\circ<\frac{A_1-B_1}{2}<\frac{A'-B_2}{2}<90^\circ$，从而 $\cos\frac{A_1-B_1}{2}>\cos\frac{A'-B_2}{2}$.

又因 $0<\frac{C_1}{2}<90^\circ$，即 $\cos\frac{C_1}{2}>0$，于是得

$$\tan\frac{A_1}{2}\tan\frac{B_1}{2}>\tan\frac{A'}{2}\tan\frac{B_2}{2}.$$

这就是说，有

$$I.Q.\ (A_1BC)>I.Q.\ (A'BC).$$

同理可证

$$I.Q.\ (A'BC)>I.Q.\ (A_2BC).$$

所以有

$$I.Q.\ (A_1BC)>I.Q.\ (A_2BC).$$

附注 一般地，由于周长一定的 n 边形以正 n 边形的面积最大，所以 n 边形等周商的最大值是在正 n 边形时达到，其最大值为 $\frac{1}{4n}\cot\frac{\pi}{n}$. 更进一步，由于周长一定的平面闭图形以圆的面积最大，因此平面图形 F 的等周商的最大值在 F 为圆时达到，最大值为 $\frac{1}{4\pi}$.

4. 证明:存在一个正整数 k,使得 $k \cdot 2^n + 1$ 对每一个正整数 n,均为合数.

证明　首先证明每一个正整数 n 至少适合下列同余式的一个同余式(这样的一组同余式称为覆盖同余式):

$$n \equiv 1, \quad (\text{mod} 2) \tag{1}$$

$$n \equiv 1, \quad (\text{mod} 3) \tag{2}$$

$$n \equiv 2, \quad (\text{mod} 4) \tag{3}$$

$$n \equiv 4, \quad (\text{mod} 8) \tag{4}$$

$$n \equiv 0, \quad (\text{mod} 12) \tag{5}$$

$$n \equiv 8, \quad (\text{mod} 24) \tag{6}$$

事实上,如果 n 为奇数,那么它适合(1);如果 n 为偶数但不是 4 的倍数,那么它适合(3);如果 n 为 4 的倍数,但不是 8 的倍数,那么它适合(4). 如果 n 为 8 的倍数,即 $n=8m$, 那么在 m 为 3 的倍数时,n 适合(5);在 m 除以 3 余 1 时, n 适合(6);在 m 除以 3 余 2 时, n 适合(2). 因此,n 至少适合同余式(1)—(6)中的一个同余式. 由于 $2^2 \equiv 1(\text{mod} 3)$, $2^3 \equiv 1(\text{mod} 7)$, $2^4 \equiv 1(\text{mod} 5)$, $2^8 \equiv 1(\text{mod} 17)$, $2^{12} \equiv 1(\text{mod} 13)$, $2^{24} \equiv 1(\text{mod} 241)$, 因此,在 $n \equiv 1(\text{mod} 2)$ 时,记 $n=2m+1$, 则

$$k \cdot 2^n + 1 = k \cdot 2^{2m+1} + 1 = 2k \cdot 2^{2m} + 1 \equiv 2k + 1. \ (\text{mod} 3)$$

同样在 n 适合(2)、(3)、(4)、(5)、(6)时,分别有

$$k \cdot 2^n + 1 \equiv 2k + 1, \quad (\text{mod} 7)$$
$$k \cdot 2^n + 1 \equiv 4k + 1, \quad (\text{mod} 5)$$
$$k \cdot 2^n + 1 \equiv 16k + 1, \quad (\text{mod} 17)$$
$$k \cdot 2^n + 1 \equiv k + 1, \quad (\text{mod} 13)$$
$$k \cdot 2^n + 1 \equiv 256k + 1, \quad (\text{mod} 241)$$

因此,只要 k 适合下面的同余方程组:

$$2k + 1 \equiv 0, (\text{mod} 3)$$
$$2k + 1 \equiv 0, (\text{mod} 7)$$

$$4k+1\equiv 0, \pmod{5}$$
$$16k+1\equiv 0, \pmod{17}$$
$$k+1\equiv 0, \pmod{13}$$
$$256k+1\equiv 0, \pmod{241}$$

则 $k\cdot 2^n+1$ 至少被 3、7、5、17、13、241 中某一个整除,从而 $k\cdot 2^n+1$ 为合数.

但上述同余方程组等价于

$$k\equiv 1, \pmod{3}$$
$$k\equiv 3, \pmod{7}$$
$$k\equiv 1, \pmod{5}$$
$$k\equiv 1, \pmod{17}$$
$$k\equiv -1, \pmod{13}$$
$$k\equiv 16, \pmod{241}$$

由于 3、7、5、17、13、241 都是素数,根据著名的中国剩余定理(即孙子定理),上述方程组一定有解,因而一定存在正整数 k,使 $k\cdot 2^n+1$ 对每一个 n 都是合数(具体地可以算出 $k=1\,207\,426+5\,592\,405m$,其中 m 为非负整数).

附注 (1) 将 $k\cdot 2^n+1$ 改为 $k\cdot a^n+1$,其中 a 为大于 2 的整数,则证明容易得多.

(2) 在 1980 年国际数论会议上,著名数学家 P. Erdös 曾提出这样的问题:能否不用覆盖同余式来找出一个整数 k,使得 $k\cdot 2^n+1$ 对每一个自然数 n 均为合数?更确切些说,是否存在一个整数 k,使得 $k\cdot 2^n+1$ 对每一个自然数 n 均为合数,并且对于任意一组(有限个)素数 p_1、p_2、…、p_n,都存在一个 n,使得 $k\cdot 2^n+1$ 与 $p_1p_2\cdots p_n$ 互素?这是一个目前尚未解决的问题.

5. 已知点 A、B、C 为球 S 内三点,且 AB、AC 垂直于 S 的过 A 点的直径. 过 A、B、C 可作两个球均与 S 相切. 证明它们的半径之和等于 S 的半径.

证明 设所作的两个球的球心为 S_1、S_2. 因为 S_1 过 A、B、C 三点,所以 S_1 在过 $\triangle ABC$ 的外心 O 并且与平面 ABC 垂直的直线 OD 上,同样 S_2 也在 OD 上.

因为 $SA\perp$ 平面 ABC,所以 $SA/\!/OD$. 过 SA 与 OD 可作一个平面 M. 平面 M 与球 S、S_1、S_2 相截得$\odot S$、$\odot S_1$、$\odot S_2$,并且$\odot S_1$、$\odot S_2$ 都与$\odot S$ 相切,点

A 在$\odot S_1$ 与$\odot S_2$ 上(图 2(b)).

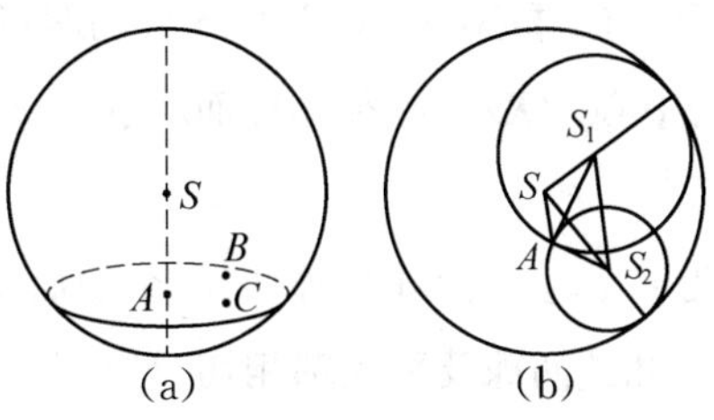

图 2

现在的问题已经化为平面几何中的问题:设$\odot S$、$\odot S_1$、$\odot S_2$ 的半径分别为 r、r_1、r_2,要证明 $r=r_1+r_2$.

考虑梯形 SS_1S_2A(图 3). 由于$\odot S_1$ 过 A,所以 $S_1A=r_1$. 由于$\odot S_1$ 与$\odot S$ 相切,所以 $SS_1=r-r_1$. 同样, $S_2A=r_2$, $SS_2=r-r_2$. 因而$\triangle S_1SA$ 与$\triangle S_2SA$ 不但有一条公共边 $SA=a$, 而且有相等的周长 $2p=a+r$.

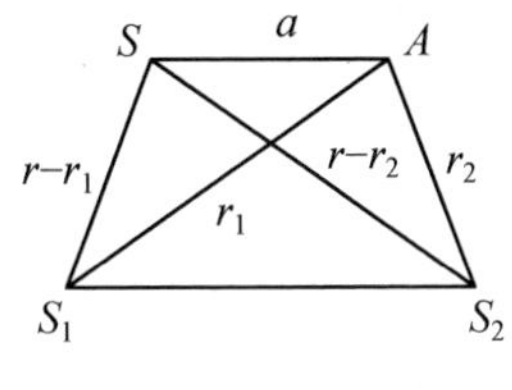

图 3

因为 $S_1S_2/\!/SA$,所以$\triangle S_1SA$ 与$\triangle S_2SA$ 的面积相等. 由面积公式得

$$p(p-a)(p-r_1)[p-(r-r_1)]=p(p-a)(p-r_2)[p-(r-r_2)],$$

即
$$(p-r_1)[p-(r-r_1)]=(p-r_2)[p-(r-r_2)].$$

展开得

$$p^2-rp+r_1(r-r_1)=p^2-rp+r_2(r-r_2).$$

从而

$$r_1(r-r_1)=r_2(r-r_2),$$
$$r_1=r_2 \text{ 或 } r=r_1+r_2.$$

但在 $r_1=r_2$ 时, $\triangle AS_1S_2$ 与$\triangle OS_1S_2$ 都是以 S_1S_2 为底的等腰三角形,并且底 S_1S_2 上的高是相等的,因此 O 与 A 重合,从而 $r-r_1=r_1=r_2$, $r_1=r_2=\frac{r}{2}$. 这时仍有 $r_1+r_2=r$.

附注 本题的解答思路是,恰当地作一截面,从而把较复杂的立体几何问题转化为我们较熟悉的平面几何问题来解. 很多立体几何问题的解决都可以采用这一方法. 下面的问题供读者练习.

问题一 设 O 为球 S 的球心, P、Q 为球 S 外的两点, 以 P 为中心, FO 为半径作球面, 又以 Q 为中心, QO 为半径作球面. 试证: 这两个球面位于 S 内部分的面积相等.

问题二 设半径为 r_1、r_2 的两球相外切, 一个平面与此两球都相切, 切点分别为 A_1、A_2. 若球 S 和这两球及平面都相切、且与平面的切点在直线 A_1A_2 上, 试求 S 的半径.

第二十五届国际数学竞赛

第二十五届国际数学竞赛于1984年6月29至7月10日在捷克首都布拉格举行.由于我国将在近年内参加国际数学竞赛,笔者谨将这次竞赛的情况与试题作一些介绍,并拟就一份解答,供有关方面的同志及广大数学爱好者参考.

这次比赛有34个国家参加,比创纪录的二十四届又多了两个.按照惯例,每个国家可派6名选手,但卢森堡、挪威只派了一名,阿尔及利亚派了4名,因此共有192名选手.

比赛仍分两试,每试$4\frac{1}{2}$小时,三道题,每道题7分,满分为42分.这次获满分的选手共有8名,其中苏联3名,罗马尼亚、东德、保加利亚、美国、越南各1名.东德得满分的是一名女选手.

各国得分情况如下表(按6名选手计算,满分为$6\times42=252$分):

1. 苏　联　　235分
2. 保加利亚　203分
3. 罗马尼亚　199分
4. 匈牙利　　195分
5. 美　国　　195分
6. 英　国　　169分
7. 越　南　　162分
8. 东　德　　161分
9. 西　德　　150分
10. 蒙　古　　146分
11. 波　兰　　140分
12. 法　国　　126分
13. 捷　克　　125分
14. 南斯拉夫　105分

15. 澳大利亚　　103 分
16. 奥地利　　97 分
17. 荷　兰　　93 分
18. 巴　西　　92 分
19. 希　腊　　88 分
20. 加拿大　　83 分
21. 哥伦比亚　　80 分
22. 古　巴　　67 分
23. 比利时　　56 分
24. 墨西哥　　56 分
25. 瑞　典　　53 分
26. 塞浦路斯　　47 分
27. 西班牙　　43 分
28. 阿尔及利亚　　36 分
29. 芬　兰　　31 分
30. 突尼斯　　29 分
31. 挪　威　　24 分
32. 卢森堡　　22 分
33. 科威特　　9 分
34. 意大利　　0 分

这次比赛有一个突出的缺点:各国提供给命题组选择的试题太少.1983 年有 72 个题目提供命题组选择,而今年仅有 15 道题,其中有 4 道题是众所周知的陈题,有 5 道数论,5 道几何或组合几何的问题.命题组只好临时又征集了 5 道题目,但其中一道是已知的、两道太容易,加拿大提供的两题很好,但加队领队已给该国选手作为练习,所以命题组挑选的余地很小,最后几何(组合几何)占 3 题,数论 2 题,不等式 1 题.

26 届竞赛(1985 年)将在芬兰举行,27 届(1986 年)将在波兰举行.

这届竞赛的试题如下:

第一试（7月4日）

1. 证明

$$0 \leqslant yz + zx + xy - 2xyz \leqslant \frac{7}{27},$$

其中 x、y、z 为非负实数，满足 $x + y + z = 1$.

2. 求一对正整数 a、b，满足

(i) $ab(a+b)$ 不被 7 整除；

(ii) $(a+b)^7 - a^7 - b^7$ 被 7^7 整除.

验证你的答案.

3. 平面上已给两个不同的点 O、A. 对这平面上的每一个不同于 O 的点 X，将从 OA 依反时针转到 OX 时所转过的角的大小（用弧度制表示）记为 $a(X)(0 \leqslant a(X) < 2\pi)$. 令 $C(X)$ 为以 O 为圆心，长 $OX + \frac{a(X)}{OX}$ 为半径的圆. 已知这平面上每一个点都被涂上颜色，而颜色的种数是有限的. 证明一定存在一点 Y，$a(Y) > 0$，并且点 Y 的颜色在圆 $C(Y)$ 的圆周上出现.

第二试（7月5日）

4. 设在凸四边形 $ABCD$ 中，直线 CD 与以 AB 为直径的圆相切. 证明当且仅当直线 $BC \parallel AD$ 时，直线 AB 与以 CD 为直径的圆相切.

5. 设一个有 $n(n > 3)$ 个顶点的平面凸多边形的所有对角线的长度之和为 d，周长为 p，证明

$$n - 3 < \frac{2d}{p} < \left[\frac{n}{2}\right]\left[\frac{n+1}{2}\right] - 2,$$

其中 $[x]$ 表示不超过 x 的最大整数.

6. 设 a、b、c、d 为奇整数，$0 < a < b < c < d$，并且 $ad = bc$，证明：如果

$$a + d = 2^k,\ b + c = 2^m,$$

k、m 为整数,那么 $a=1$.

下面是笔者所拟的解答,不妥之处敬请同志们指正.

1. 不妨设 $x\geqslant y\geqslant z$. 由于 $x+y+z=1$, 所以

$$3z\leqslant x+y+z=1,\ z\leqslant\frac{1}{3}. \tag{1}$$

从而(因为 x、y、z 均非负)

$$2xyz\leqslant\frac{2}{3}xy\leqslant xy.$$

于是

$$0\leqslant yz+zx+xy-2xyz. \tag{2}$$

现在来证明题中右边的不等式. 我们有

$$2y\leqslant x+y\leqslant x+y+z=1,$$

所以
$$y\leqslant\frac{1}{2}.$$

又有
$$3x\geqslant x+y+z=1,$$

所以 $x\geqslant\frac{1}{3}$, 故

$$\begin{aligned}
yz+zx+xy-2xyz&=y(z+x)+zx(1-2y)\\
&\leqslant y(z+x)+zx(1-2y)+\left(x-\frac{1}{3}\right)\left(\frac{1}{3}-z\right)(1-2y)\\
&=y(z+x)+\frac{1}{3}\left(x+z-\frac{1}{3}\right)(1-2y)\\
&=y\left(w+\frac{1}{3}\right)+\frac{1}{3}w(1-2y)\\
&=\frac{1}{3}yw+\frac{1}{3}(y+w).\ \left(\text{令 } w=x+z-\frac{1}{3}\right)
\end{aligned}$$

显然
$$y+w=y+x+z-\frac{1}{3}=\frac{2}{3},$$

所以
$$yw\leqslant\frac{1}{4}(y+w)^2=\frac{1}{9}.$$

从而 $$yz+zx+xy-2xyz\leqslant\frac{1}{3}\cdot\frac{1}{9}+\frac{1}{3}\cdot\frac{2}{3}=\frac{7}{27}.\tag{3}$$

评注　在证右边不等式时，我们采用的是“调整法”. 先将 x、z 调整为 $\frac{1}{3}$、$w=x+z-\frac{1}{3}$，这时和 $x+z$（从而 y）不变，而 $yz+zx+xy-2xyz$ 的值增大. 然后再将 y、w 均调整为 $\frac{1}{3}$. 显然这一不等式在（且仅在）$x=y=z=\frac{1}{3}$ 时，变为等式.

如果用 Cauchy-Schwarz 不等式，易得

$$\frac{1}{x}+\frac{1}{y}+\frac{1}{z}=\left(\frac{1}{x}+\frac{1}{y}+\frac{1}{z}\right)(x+y+z)\geqslant 9,\tag{4}$$

去分母得

$$yz+zx+xy\geqslant 9xyz,$$

即 $$yz+zx+xy-9xyz\geqslant 0.\tag{5}$$

这比题中左边的不等式强.

2. $b=1$，$a=18$ 就是满足要求的一组数，事实上，显然有 $7\nmid ab(a+b)$（我们用 $c\mid d$ 表示 c 整除 d，$c\nmid d$ 表示 c 不整除 d），并且

$$18^2+18+1=324+18+1=343=7^3,$$

所以

$$(a+b)^7-a^7-b^7=7ab(a+b)(a^2+ab+b^2)^2\tag{6}$$

被 7^7 整除.

这组解是如何求出的呢？我们可设 $b=1$. 由(6)可知，当且仅当

$$7^3\mid a^2+a+1\tag{7}$$

时，

$$7^7\mid(a+b)^7-a^7-b^7.$$

设 $a=7k+a_1$，k 为整数，$a_1\in\{1,\pm 2,\pm 3\}$，容易知道在 $a_1=-3$ 时，

$7 \mid a^2+a+1$，并且

$a^2+a+1=(7k-3)^2+(7k-3)+1=7^2k^2-5\times 7k+7=7(7k^2-5k+1)$,

于是(7)成为

$$7^2 \mid 7k^2-5k+1. \tag{8}$$

设 $k=7h+a_2$, h 为整数, $a_2\in\{0, \pm 1, \pm 2, \pm 3\}$,容易知道在 $a_2=3$ 时, $7 \mid 7k^2-5k+1$, 并且

$$\begin{aligned}7k^2-5k+1&=7(7h+3)^2-5(7h+3)+1=7(7h+3)^2-7\times 5h-14\\&=7((7h+3)^2-5h-2),\end{aligned}$$

于是(8)成为

$$7 \mid (7h+3)^2-5h-2=7^2h^2+6\times 7h-5h+7, \tag{9}$$

于是 $7|h$. 令 $h=7l$ (l 为非负整数),则

$$k=7^2l+3,\ a=7^3l+7\times 3-3=7^3l+18.$$

特别地,取 $l=0$ 得 $a=18$.

评注 用上面的方法不难得出正整数 $a=7^3l+18b$ 或 $a=7^3l+324b$ (l 为整数)与 b, $7+b$, 是满足题中要求的全部整数.

3. 另取一个坐标平面,但我们只考虑 $0\leqslant x$, $0\leqslant y<2\pi$ 的那一部分. 使原点与 O 点颜色相同(不妨仍记作 O),点(x_0, y_0)的颜色与 $OX=x_0$, $a(X)=y_0$ 的点 X 的颜色相同. 圆 $C(X)$变成了直线 $x=x_0+\dfrac{y_0}{x_0}$(的一段:$0\leqslant y<2\pi$), 其中(x_0, y_0)是与 X 相对应的点. 要证明的结论就是:在所讨论的平面区域内一定有一个点(x_0, y_0),它与直线 $x=x_0+\dfrac{y_0}{x_0}$ 上某一点的颜色相同,并且 $y_0\neq 0$.

问题还可以换一个提法:证明存在一个数 $a\in(0, \sqrt{8\pi}\,]$,在直线 $x=a$ 与抛物线 $y=x(a-x)\left(\text{也就是 } a=x+\dfrac{y}{x}\right)$ 上各有一个点(a, b)与 $(x_0, x_0(a-x_0))$, 它们的颜色相同,其中 $0<b<2\pi$, $0<x_0<a$.

为了证明这个结论,在所述区域里任取一抛物线 $y=x(a_0-x)$, 比如说 $y=x(\sqrt{8\pi}-x)=-(x-\sqrt{2\pi})^2+2\pi$. 这抛物线上有无限多个点,而只有有

限种颜色，其中必有无限多个点是同一种颜色，不妨认为是第 1 种颜色，并且这些点的横坐标为 $a_1 > a_2 > a_3 > \cdots$(图 1).

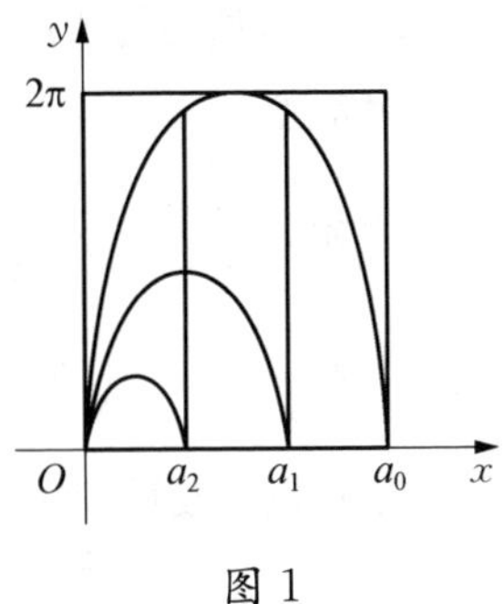

图 1

在抛物线 $y=x(a_1-x)$ 上，考虑横坐标为 a_2，a_3，…的点，如果有一点的颜色为第 1 种，那么它与直线 $x=a_1$ 上的点$(a_1, a_1(a_0-a_1))$的颜色相同，结论已经成立. 假如这无限多个点都不是第一种颜色. 那么，由于颜色种数有限，其中必有无限多个点是同一种颜色. 为不使记号繁复起见，不妨设抛物线 $y=x(a_1-x)$ 上横坐标为 $a_2 > a_3 > a_4 > \cdots$ 的点都是第 2 种颜色.

依此类推，如果颜色的种数为 n，那么进行到第 n 步，在直线 $x=a_n$ 上将有 n 个点，颜色分别为第 1, 2, …, n 种，这些点是这直线与抛物线 $y=x(a_i-x)$ 的交点 $(i=0, 1, \cdots, n-1)$. 因此在抛物线 $y=x(a_n-x)$ 上任一点必与上述 n 个点中的某一点颜色相同.

评注　完全类似地(而且更简单)，如果平面上的点都涂上颜色，且颜色的种类有限，那么一定能找到一对点(a, b)与(c, a)，它们的颜色相同，且 $c<a$.

4. 如图 2，设 E、F 分别为 AB、CD 的中点. 如果 $BC \parallel AD$，那么 $EF \parallel AD$，从而

$$S_{\triangle AEF}=S_{\triangle DEF}, \tag{10}$$

即

$$E\text{ 到 }DF\text{ 的距离}\times DF=F\text{ 到 }AE\text{ 的距离}\times AE, \tag{11}$$

由于直线 CD 与以 AB 为直径的圆相切，所以

$$E\text{ 到 }DF\text{ 的距离}=AE, \tag{12}$$

图 2

于是由(11)

$$DF=F\text{ 到 }AE\text{ 的距离}, \tag{13}$$

即直线 AB 与以 CD 为直径的圆相切.

反之，若直线 AB 与以 CD 为直径的圆相切，则(13)成立，由(12)及(13)，导出(11)即(10)，于是 $EF \parallel AD$，同样 $EF \parallel BC$，所以 $BC \parallel AD$.

5. 设多边形的顶点为 $A_1, A_2, \cdots, A_n$. 并约定 A_{i+n} 即 A_i.

对角线

$$A_iA_{i+j} < A_iA_{i+1} + A_{i+1}A_{i+2} + \cdots + A_{i+j-1}A_{i+j} = \sum_{k=1}^{j} A_{j+k-1}A_{j+k}.$$

$$\left(i=1, 2, \cdots, n;\ j=2, 3, \cdots, \left[\frac{n}{2}\right]\right)$$

在 n 为奇数时,将以上各式加起来,便得

$$d < \sum_{i=1}^{n} \sum_{j=2}^{\left[\frac{n}{2}\right]} \sum_{k=1}^{j} A_{i+k-1}A_{i+k} = \sum_{j=2}^{\left[\frac{n}{2}\right]} \sum_{k=1}^{j} \sum_{i=1}^{n} A_{i+k-1}A_{i+k}$$

$$= \sum_{j=2}^{\left[\frac{n}{2}\right]} \sum_{k=1}^{j} p = \sum_{j=2}^{\left[\frac{n}{2}\right]} jp = p\left(\frac{\left[\frac{n}{2}\right]\left(\left[\frac{n}{2}\right]+1\right)}{2} - 1\right),$$

即 $$\frac{2d}{p} < \left[\frac{n}{2}\right]\left(\left[\frac{n}{2}\right]+1\right) - 2 = \left[\frac{n}{2}\right] \cdot \left[\frac{n+1}{2}\right] - 2.$$

在 n 为偶数时,需要注意对角线 $A_iA_{i+\frac{n}{2}}$ 在总和 d 中只能计算一次,因此

$$d < \sum_{i=1}^{n} \sum_{j=2}^{\left[\frac{n}{2}\right]-1} \sum_{k=1}^{j} A_{i+k-1}A_{i+k} + \frac{1}{2} \sum_{i=1}^{n} \sum_{k=1}^{\left[\frac{n}{2}\right]} A_{i+k-1}A_{ik}$$

$$= \sum_{i=1}^{n} \sum_{j=2}^{\left[\frac{n}{2}\right]} \sum_{k=1}^{j} A_{i+k-1}A_{i+k} - \frac{1}{2} \sum_{i=1}^{n} \sum_{k=1}^{\left[\frac{n}{2}\right]} A_{i+k-1}A_{k}$$

$$= p\left(\frac{\frac{n}{2}\left(\frac{n}{2}+1\right)}{2} - 1\right) - \frac{1}{2}p \cdot \frac{n}{2} = \frac{1}{2}p\left(\left(\frac{n}{2}\right)^2 - 2\right),$$

即 $$\frac{2d}{p} < \left[\frac{n}{2}\right] \cdot \left[\frac{n+1}{2}\right] - 2.$$

现在证明 $\dfrac{2d}{p} > n-3$.

设 A_iA_{i+k} 与 $A_{i+1}A_{i+1+k}$ 相交于 O,我们有

$$A_iA_{i+k} + A_{i+1}A_{i+1+k} = A_iO + OA_{i+1} + OA_{i+k} + OA_{i+1+k} > A_iA_{i+1} + A_{i+k}A_{i+k+1}$$

$$(i=1, 2, \cdots, n,\ k=2, 3, \cdots, n-2)$$

将以上各式相加便得 $2d > (n-3)p$. 证毕.

6. 由 $a+d=2^k$, $b+c=2^m$, 得 $d=2^k-a$, $c=2^m-b$.

代入 $ad=bc$ 中得

$$a(2^k-a)=b(2^m-b),$$

即
$$b\cdot 2^m-a\cdot 2^k=b^2-a^2. \tag{14}$$

又

$$(a+d)^2=4ad+(d-a)^2=4bc+(d-a)^2>4bc+(c-b)^2=(b+c)^2,$$

所以
$$2^k>2^m,\ k>m.$$

由(14)得

$$2^m(b-a\cdot 2^{k-m})=(b+a)(b-a), \tag{15}$$

因为 a、b 都是奇数,所以 $b+a$, $b-a$ 都是 2 的倍数. 又

$$(b+a)-(b-a)=2a$$

是奇数的 2 倍,所以 $b+a$, $b-a$ 中必有一个不是 4 的倍数,这样从(15)得到

$$\begin{cases}b+a=2^{m-1}e,\\ b-a=2f\end{cases} \quad 或 \begin{cases}b-a=2^{m-1}e,\\ b+a=2f,\end{cases}$$

其中 e、f 为正整数,并且 $ef=b-a\cdot 2^{k-m}$.

由于 $k>m$,所以

$$ef\leqslant b-2a<b-a\leqslant 2f,$$

于是 $e=1$, $f=b-a\cdot 2^{k-m}$,

若 $\begin{cases}b+a=2^{m-1},\\ b-a=2f=2(b-a\cdot 2^{k-m}),\end{cases}$ 则由第二式得

$$b+a=2^{k+1-m}\cdot a,$$

从而
$$2^{m-1}=2^{k+1-m}\cdot a,$$

于是奇数 $a=1$.

同理,若 $\begin{cases}b-a=2^{m-1}e,\\ b+a=2f,\end{cases}$ 也导出 $a=1$.

保加利亚数学竞赛试题及解答

下面是1985年举行的第三十四届保加利亚数学竞赛(选拔赛)的试题与解答.

1. k 与 n 为正整数,证明

$$(n^4-1)(n^3-n^2+n-1)^k+(n+1)\cdot n^{4k-1} \tag{1}$$

被 n^5+1 整除.

证明 这类问题的解法很多,我们对 k 施行归纳. 当 $k=1$ 时,(1)成为

$$\begin{aligned}&(n^4-1)(n^3-n^2+n-1)+(n+1)\cdot n^3\\ =&n^7-n^6+n^5+n^2-n+1=(n^5+1)(n^2-n+1),\end{aligned}$$

被 n^5+1 整除. 假设(1)式被 n^5+1 整除,考虑

$$(n^4-1)(n^3-n^2+n-1)^{k+1}+(n+1)\cdot n^{4(k+1)-1}, \tag{2}$$

显然(2)可化为

$$\begin{aligned}&[(n^4-1)(n^3-n^2+n-1)^k+(n+1)\cdot n^{4k-1}]\\ &\cdot(n^3-n^2+n-1)+(n^5+1)\cdot n^{4k-1},\end{aligned}$$

于是(2)被 n^5+1 整除. 命题成立.

2. 确定参数 a 的范围,使方程

$$\log 2x\cdot\log 3x=a \tag{3}$$

有两个不同的解,求这两个解的积.

解 令 $y=\log x$,则由(3)易得

$$y^2+(\log 6)\cdot y+\log 2\cdot\log 3-a=0, \tag{4}$$

其判别式

$$(\log 6)^2-4\cdot(\log 2\cdot\log 3-a)>0,$$

所以 $$(\log 3-\log 2)^2>-4a.$$

于是 $$a>-\frac{\log^2\frac{3}{2}}{4}.$$

而 $$\log x_1+\log x_2=y_1+y_2=-\log 6,$$

所以 $$x_1x_2=\frac{1}{6}.$$

3. 一个四面体 $ABCD$ 的内切球的中心 I 与棱 AB、CD 的中点共线,证明这四面体的外接球的中心也在这条直线上.

证明　如图 1,设 E、F 分别为 AB、CD 的中点,由于 I 到面 ACD、BCD 的距离相等,I 在 EF 上,所以 E 到面 ACD、BCD 的距离相等,从而 A 到面 BCD 的距离与 B 到面 ACD 的距离相等(都等于 E 到这两个面距离的两倍).

由于四面体的体积等于底面积乘高的 $\frac{1}{3}$,所以由以上的两个高相等得 $S_{\triangle BCD}=S_{\triangle ACD}$. $\triangle BCD$ 与 $\triangle ACD$ 有公共边 CD,所以从上式又推出 A、B 到 CD 的距离相等. 同理, C、D 到 AB 的距离相等.

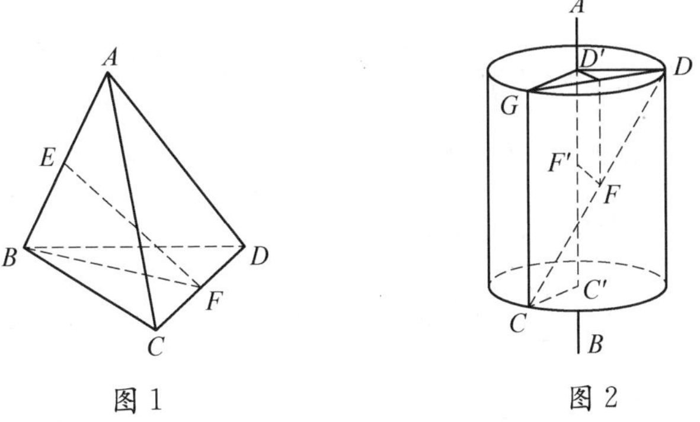

图 1　　　　　　图 2

现在我们来证明 EF 为 AB、CD 的公垂线. 如图 2,设 C、D 在 AB 上的射影分别为 C'、D',由于 $CC'=DD'$,所以, D、C 是在以 AB 为轴的圆柱面上. 过 $C'D'$ 中点 F' 作平面与轴 AB 垂直,则这平面平分 CD,因而过 F 点,设 CD 在圆柱的上底面的射影为 GD,则 F 在上底面的射影为 GD 中点 H. 因为 $D'H\perp GD$, $F'F\parallel D'H$,所以 $F'F\perp GD$,从而 $FF'\perp$ 平面 GCD, $FF'\perp CD$. 即 $F'F$

是 AB、CD 的公垂线. 同理,过 E 作 CD 的垂线 EE'(E'在 CD 上),则 EE' 为 AB、CD 的公垂线,因而 EE' 与 $F'F$ 重合.

在图 1 中,作线段 BC 的中垂面 M,它与 EF 一定相交于一点 O(如果平面 $M \parallel EF$,那么 $BC \perp EF$,过 AB、CD 分别作平面与 EF 垂直,这两个平面 M_1、M_2 互相平行. B 在平面 M_1 内,并且 $BC \perp EF$,所以 BC 在平面 M_1 内,C 是 M_1、M_2 的公共点,与平面 $M_1 \parallel M_2$ 矛盾),$OB = OC$. 又由于 OE 垂直平分 AB、CD,所以 $OA = OB$,$OC = OD$. 从而 O 到 A、B、C、D 距离相等,O 是四面体 $ABCD$ 的外心. 证毕.

4. 设正整数 a_n 与 b_n 满足关系

$$a_n + b_n\sqrt{2} = (2+\sqrt{2})^n,\ (n=1,\ 2,\ 3,\ \cdots), \tag{5}$$

证明: $\lim\limits_{n\to\infty}\dfrac{a_n}{b_n}$ 存在,并求出这个极限.

证明 略.

5. 面积为 S 的 $\triangle ABC$ 内接于半径为 1 的圆 K,内心 I 在边 BC、CA、AB 上的正射影分别为 A_1、B_1、C_1,S_1 为 $\triangle A_1B_1C_1$ 的面积. 如果直线 AI 交圆 K 于 A_2,证明:

$$4S_1 = AI \cdot A_2B \cdot S.$$

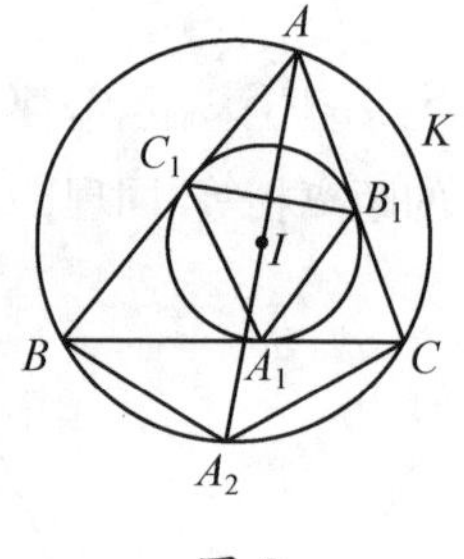

图 3

证明 设内切圆半径为 r,则

$$\begin{aligned}S_1 &= S_{\triangle IA_1B_1} + S_{\triangle IB_1C_1} + S_{\triangle IC_1A_1}\\ &= \frac{1}{2}r^2\sin A + \frac{1}{2}r^2\sin B + \frac{1}{2}r^2\sin C\\ &= \frac{1}{2}r^2 \times \frac{a}{2} + \frac{1}{2}r^2 \times \frac{b}{2} + \frac{1}{2}r^2 \times \frac{c}{2} = \frac{1}{2}rS.\end{aligned}$$

另一方面,由 $\triangle IAC_1$ 及 $\triangle A_2BC$ 得

$$AI = \frac{r}{\sin\dfrac{A}{2}},\ A_2B = \frac{\dfrac{a}{2}}{\cos\dfrac{A}{2}},$$

所以

$$4S_1=2rS=\frac{a}{\sin A}\cdot r\cdot S=AI\cdot A_2B\cdot S.$$

6. 平面上的五个已知点具有下列性质:任意四点中有三个构成一个正三角形.

(Ⅰ)证明这五点中有四点构成一个有一个角等于60°的菱形;

(Ⅱ)求以这些点为顶点的正三角形的个数.

解　设其中A、B、C三点构成正三角形.如果B、C、D或B、C、E三点构成正三角形,那么A、B、D、C或A、B、E、C构成一个菱形,$\angle CAB=60°$.如果B、C、D或B、C、E都不是正三角形的三个顶点,那么B、C、D、E四点中,$\triangle BDE$或$\triangle CDE$是正三角形.如果两者都是,那么B、D、C、E构成一个有一个角等于60°的菱形.如果只有一个是正三角形,不妨设$\triangle BDE$是正三角形,$\triangle CDE$不是,这时,在A、D、C、E四点中必有一个正三角形:如果$\triangle ADE$是正三角形,那么A、D、E、B构成一个所要的菱形;如果$\triangle ACE$或$\triangle ACD$是正三角形,那么菱形$ABCE$或菱形$ABCD$即为所求.总之(Ⅰ)成立.

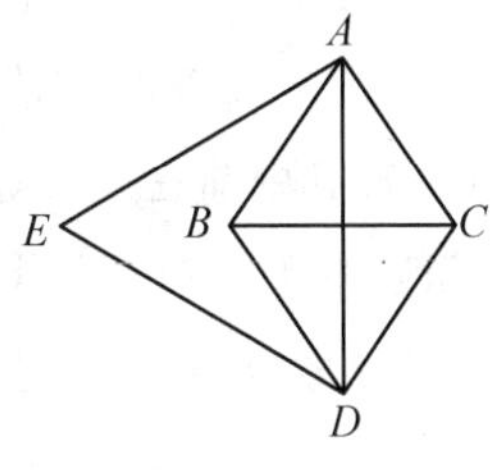

图4

以这些点为顶点的正三角形应有三个.理由如下:

设A、B、D、C构成(Ⅰ)中所说的菱形,这时E、A、D、C中有三点构成正三角形.如果$\triangle EAC$是正三角形,那么E、A、B、D中任三点不构成三角形,与已知矛盾.同理$\triangle ECD$不是正三角形.所以$\triangle EAD$是正三角形(如图4).易知这时只有三个三角形,即$\triangle ABC$、$\triangle BCD$、$\triangle ADE$是正三角形.

第二十七届国际数学奥林匹克题解

编者按:第二十七届国际数学奥林匹克竞赛 1986 年 7 月在波兰华沙举行,我国首次派出代表队参加这次竞赛,并取得了辉煌成绩. 我国派出的六名选手是河南郑州方为民、上海张浩、天津李平立、陕西西安荆秦(女)、湖北黄冈林强(高二)、江苏泰县沈建. 他们的得分依次为(满分为 42 分)41, 39, 37, 26, 19, 15 分. 其中前三人获得一等奖,荆秦获二等奖,林强获三等奖. 总分为 177 分,居第四位(美国与苏联代表队总分均为 203 分,并列第一). 消息传来,全国人民无不为之欢欣. 这一成绩的取得除了队员及其中学老师的努力之外,与集训队的指导老师努力也是分不开的. 为满足广大读者的渴望,我们特邀请部分导师对此试题给出解答. 另外,这套竞赛试题的第二题系中国科学技术大学常庚哲教授命题,被竞赛委员会选中,亦反映我国竞赛命题之水平.

第一天试题(7月9日)

1. 设正整数 d 不等于 2、5、13. 证明在集合 $\{2, 5, 13, d\}$ 中可以找到两个不同元素 a、b,使 $ab-1$ 不是完全平方数.

证明 在本题中,所有字母都表示正整数. 为了证明本题的结论,只需证 $2d-1$, $5d-1$, $13d-1$ 中至少有一个不是平方数. 由于平方数 n^2 只可能是 $4l$ 或 $4l+1$(n、l 为正整数)两种形式的数,我们只要证明 $2d-1$、$5d-1$、$13d-1$ 中至少有一个不是这两种形式的数即可,下面分一些情况讨论:

(1) 若 d 为偶数,令 $d=2m$ (m 为整数),则

$$2d-1=4m-1=4(m-1)+3,$$

故 $2d-1$ 不是平方数.

(2) 若 d 为奇数,令 $d=2m+1$, 则

$$5d-1=10m+4,$$
$$13d-1=26m+12.$$

这时又分两种情况：

(i) 若 m 为奇数，令 $m=2k+1$（k 为正整数），则

$$5d-1=10m+4=20k+14=4(5k+3)+2,$$

故 $5d-1$ 不是平方数.

(ii) 若 m 为偶数，令 $m=2k$，则

$$13d-1=26m+12=52k+12=4(13k+3),$$
$$5d-1=10m+4=20k+4=4(5k+1),$$

由于 4 是平方数，所以为了证明这两个数中至少有一个不是平方数，只要证明 $13k+3$ 及 $5k+1$ 中至少有一个非平方数即可.

设 $13k+3$ 是平方数. 若 $13k+3=4h$（h 为正整数），则 $5k+1=4h-8k-2=4(h-2k-1)+2$，不是平方数；若 $13k+3=4h+1$，则 $5k+1=4h+1-8k-2=4(h-2k-1)+3$，不是平方数. 同样可证，$5k+1$ 是平方数，则 $13k+3$ 不是平方数. 这就证得 $5k+1$ 及 $13k+3$ 中至少有一个不是平方数.

2. 平面上给定$\triangle A_1A_2A_3$ 及点 P_0，定义 $A_s=A_{s-3}$，$s\geqslant 4$. 造点列 P_0，P_1，P_2，…使得 P_{k+1} 为绕中心 A_{k+1} 顺时针旋转 120° 时 P_k 所达到的位置，$k=0$，1，2，…，若 $P_{1986}=P_0$，证明$\triangle A_1A_2A_3$ 为等边三角形.

证明　用复数方法求解，并且约定点 A 的复数表示亦记作 A. 记 $\alpha=\dfrac{2}{3}\pi\mathrm{i}$，由题设，有

$$P_k=A_k+\mathrm{e}^{-\alpha}(P_{k-1}-A_k),$$

即
$$P_k-\mathrm{e}^{-\alpha}P_{k-1}=A_k(1-\mathrm{e}^{-\alpha}).$$

在上式中，令 $k=1986$，1985，…，2，1，并将所得各式分别乘以 1，$\mathrm{e}^{-\alpha}$，…，$\mathrm{e}^{-1984\alpha}$，$\mathrm{e}^{-1985\alpha}$，再将所得各式相加，得

$$P_{1986}=\sum_{k=1}^{1985}A_k(1-\mathrm{e}^{-\alpha})\mathrm{e}^{-\alpha(1986-k)}+P_0$$

因为
$$\mathrm{e}^{-1986\alpha}=1,\ P_{1986}=P_0,$$

所以
$$\sum_{k=1}^{1986}A_k\mathrm{e}^{k\alpha}=0,$$

即 $$\sum_{j=0}^{661}A_{3j+1}\mathrm{e}^{(3j+1)\alpha}+\sum_{j=0}^{661}A_{3j+2}\mathrm{e}^{(3j+2)\alpha}+\sum_{j=0}^{661}A_{3j+3}\mathrm{e}^{(3j+3)\alpha}=0.$$

由题设 $$A_{3j+1}=A_1,\ A_{3j+2}=A_2,\ A_{3j+3}=A_3,$$

又 $$\mathrm{e}^{(3j+1)\alpha}=\mathrm{e}^{\alpha},\ \mathrm{e}^{(3j+2)\alpha}=\mathrm{e}^{2\alpha},\ \mathrm{e}^{(3j+3)\alpha}=1,$$

所以 $$A_1\mathrm{e}^{2\pi i/3}+A_2\mathrm{e}^{4\pi i/3}+A_3=0.$$

即 $A_3-A_1=-A_1(1+\mathrm{e}^{2\pi i/3})-A_2\mathrm{e}^{4\pi i/3}=-A_1\mathrm{e}^{\pi i/3}+A_2\mathrm{e}^{\pi i/3}=\mathrm{e}^{\pi i/3}(A_2-A_1)$，所以$\triangle A_1A_2A_3$ 为等边三角形.

3. 正五边形的每个顶点对应一个整数，使得这五个整数的和为正. 若其中三个相连顶点相应的整数依次为 x、y、z，而中间的 $y<0$，则要进行如下的操作：整数 x、y、z 分别换为 $x+y$、$-y$、$z+y$，只要所得的五个整数中至少还有一个为负时，这种操作继续进行，问：是否这种操作进行有限次后必定终止？

解 设五个顶点对应的整数依次为 x、y、z、u、v. 定义值为整数的函数

$$f(x,y,z,u,v)=\sum|x|+\sum|x+y|+\sum|x+y+z|+\sum|x+y+z+u|,$$

$$\begin{aligned}&f(x,y,z,u,v)\\=&|x|+|y|+|z|+|u|+|v|+|x+y|+|y+z|+|z+u|+|u+v|+\\&|v+x|+|x+y+z|+|y+z+u|+|z+u+v|+|u+v+x|+\\&|v+x+y|+|x+y+z+u|+|y+z+u+v|+|z+u+v+x|+\\&|u+v+x+y|+|v+x+y+z|\end{aligned}$$

当 $y<0$ 时，按题设要求作一次操作，这时上述函数成为

$$\begin{aligned}&f(x+y,-y,y+z,u,v)\\=&|x+y|+|y|+|y+z|+|u|+|v|+|x|+|z|+|y+z+u|+\\&|u+v|+|v+x+y|+|x+y+z|+|z+u|+|y+z+u+v|+\\&|u+v+x+y|+|v+x|+|x+y+z+u|+|z+u+v|+\\&|x+2y+z+u+v|+|u+v+x|+|v+x+y+z|,\end{aligned}$$

所以 $$\begin{aligned}&f(x+y,-y,y+z,u,v)-f(x,y,z,u,v)\\&=|x+2y+z+u+v|-|x+z+u+v|\end{aligned}$$

$$=\begin{cases}x+2y+z+u+v-(x+z+u+v)=2y\leqslant -2<0,\\-(x+2y+z+u+v)-(x+z+u+v)\end{cases}$$

$$=-2(x+y+z+u+v)\leqslant -2<0,$$

这就是说,每经一次操作,f 的值严格减少,且减少的值不小于 2. 由于开始时,f 的值是有限的,所以经有限次操作必停止.

第二天试题(7月10日)

4. 以点 O 为心的正 n 边形($n\geqslant 3$)的两个相邻顶点记为 A、B. $\triangle XYZ$ 与 $\triangle OAB$ 全等,最初令 $\triangle XYZ$ 重叠于 $\triangle OAB$. 然后在平面上移动 $\triangle XYZ$ 使点 Y 和 Z 都沿着多边形周界移动一周,而点 X 保持在多边形内移动,求 X 的轨迹.

解　设正 n 边形 $ABC\cdots G$ 的外接圆半径为 1, O 为外接圆圆心. $\triangle XYZ$ 的位置如图 1,由题设

$$XZ=XY=1,$$

$$\angle YXZ=\angle AOB,$$

$$\angle YXZ+\angle YBZ=\angle AOB+\angle ABC=\frac{n-2}{n}\pi+\frac{2}{n}\pi=\pi,$$

从而 X、Y、B、Z 四点共圆,故

$$\angle XBY=\angle XZY=\angle OBY,$$

所以,点 X 在直线 BO 上.

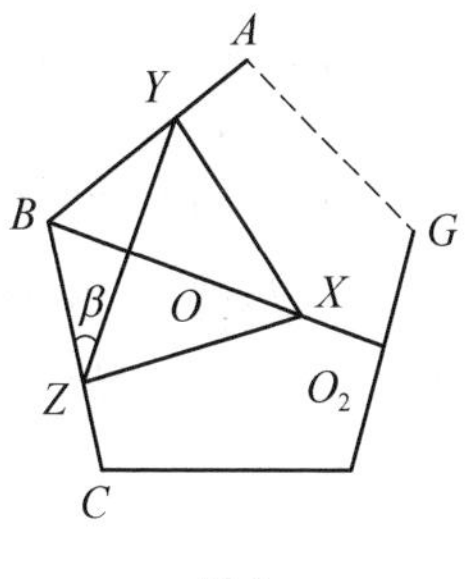

图 1

记 $\angle BZY=\beta$. 对 $\triangle BZX$ 应用正弦定理,得

$$\frac{BX}{XZ}=\frac{BX}{1}=\frac{\sin\angle BZX}{\sin\angle XBZ}$$

$$=\frac{\sin\left(\frac{n-2}{2n}\pi+\beta\right)}{\sin\frac{n-2}{2n}\pi}=\frac{\cos\left(\frac{\pi}{n}-\beta\right)}{\cos\frac{\pi}{n}},$$

即

$$BX=\frac{\cos\left(\frac{\pi}{n}-\beta\right)}{\cos\left(\frac{\pi}{n}\right)},$$

所以
$$1 \leqslant BX \leqslant \frac{1}{\cos\left(\frac{\pi}{n}\right)},$$

于是当 Z 从 B 变到 BC 上的 K 点时(这时 $\beta = \frac{\pi}{n}$), X 在直线 BO 上由 O 移动到 $O_2\left(OO_2 = \frac{1}{\cos\left(\frac{\pi}{n}\right)} - 1\right)$；当 Z 从 K 移动到 C 时, X 在直 BO 上从 O_2 回到 O, 这样点 X 的轨迹是分别在直线 AO, BO, …, GO 上的 n 条线段 OO_j ($y=1$, 2, …, n), 且在每条线段上通过两次.

5. f 为定义于非负实数上的且取非负实数值的函数,求所有满足下列条件的 f:

(1) $f(xf(y))f(y) = f(x+y)$;

(2) $f(2) = 0$;

(3) $f(x) \neq 0$, $0 \leqslant x < 2$.

解 分两种情况讨论:

(i) 当 $x > 2$ 时,令 $x = t + 2$, $t > 0$. 于是由题条件,得
$$f(x) = f(t+2) = f(tf(2))f(2) = 0.$$

(ii) 当 $x < 2$ 时,先令 $t = 2 - x > 0$,则 $x + t = 2$.

$\therefore$ $0 = f(2) = f(t+x) = f(tf(x))f(t)$.

$\because$ $f(t) \neq 0$, $\therefore$ $f(tf(x)) = 0$.

$\therefore$ $tf(x) \geqslant 2$.

$\therefore$ $f(x) \geqslant \frac{2}{t} = \frac{2}{2-x}$.

再取 $t < 2 - x$,则 $x + t < 2$,

$\therefore$ $f(tf(x))f(x) = f(t+x) \neq 0$.

$\therefore$ $f(tf(x)) \neq 0$, $\therefore$ $tf(x) < 2$.

$\therefore$ $f(x) < \frac{2}{t}$.

令 $t \to 2 - x$, 取极限,得

$$f(x)\leqslant\frac{2}{2-x}. \tag{2}$$

由(1)、(2)得

$$f(x)=\frac{2}{2-x}.$$

综上讨论知

$$f(x)=\begin{cases}0 & (x\geqslant 2),\\ \dfrac{2}{2-x} & (x<2).\end{cases}$$

不难验证这 $f(x)$满足题中要求.

6. 平面上有限个点构成一个集合，其中每个点的坐标为整数，可不可以把此集合中某些点染红色，而其余的点染白色，使得与纵横坐标轴平行的每一条直线 l 上所包含的红、白点的个数至多相差一个？

解　我们用数学归纳法证明题中的结论可以实现. 设有 n 个点. 当 $n=1$ 时，命题显然成立. 设 $n\leqslant k$ 时，命题成立. 对 $k+1$ 个点的情形分几种情况考虑：

(1) 若存在一点 $P(i,j)$，在直线 l_1：$x=i$ 及 l_2：$y=j$ 上均有集合的奇数个点. 去掉 P 点. 由归纳假设，对其余 k 个点可以完成题设要求的染色. 而且由于 l_1 及 l_2 上均有偶数个点，故 l_1 及 l_2 上的红点与白点数相同(如不相同，其差是偶数$\geqslant 2$). 加进 P 点后，除 l_1、l_2 外其他纵线和横线上的点不受影响. 所以，这时把 P 点染成任何一种颜色都可以.

(2) 如果存在一点 $P(i,j)$，在直线 l_1：$x=i$ 及 l_2：$y=j$ 上的点数为一奇一偶. 不妨设 l_1 上有奇数个点；l_2 上有偶数个点. 这时去掉 P 点后，余下的 k 个点可以完成所要求的染色. 由于 l_2 上有偶数个点，故红点与白点数相同. 而 l_1 上有奇数个点，红点与白点数相差为 1，若红点多 1 个，把 P 点染成白色；若白点多 1 个，把 P 点染成红色，即可完成所要求的染色.

(3) 若对任一点 $P(i,j)$，直线 l_1：$x=i$ 及直线 l_2：$y=j$ 上都有偶数个点. 现在去掉某条纵线 l：$x=i_0$ 上的所有点，所余的$\leqslant k$ 个点可以完成染色. 且对每一条纵线上红点与白点的个数相同，从而全体红点与全体白点的个数相同.

由于去掉了纵线 l 上的点，横线上的点可能有变化：

(i) 对不受影响的横线，点数仍为偶数，红点与白点的个数相同.

(ii) 对受影响的横线，每条横线上的点数为奇数，红点与白点个数相差为 1. 由于 l 上点数即这类横线的总数为偶数，而全体红点数与白点数相同. 所以红点多 1 个的横线与白点多 1 个的横线数相同. 这样只要把 l 上的点这样染色：如该点所在横线上红点多 1 个，此点染白色；如白点多 1 个，此点染红色. 就能完成所要求的染色.

从批改一道竞赛题所想到的

1986 年全国高中数学联赛的最后一题并不属于“难题”之列，然而考生的成绩却不甚理想. 应当说这是一道好题，它揭露了目前中学数学教学中存在的一些问题，值得我们大家深思.

首先，相当多的同学不能正确地理解题意. 原题是：

“平面直角坐标系中，纵、横坐标都是整点的点称为整点，请设计一种方法将所有的整点染色，每一个整点染成白色、红色或黑色中的一种颜色，使得

(1) 每一种颜色的点出现在无穷多条平行于横轴的直线上；

(2) 对于任意白点 A、红点 B 及黑点 C，总可以找到一个红点 D，使 $ABCD$ 为一平行四边形.

证明你设计的方法符合上述要求.

很明确，它要求同学做两件事：一、设计一种染色的方法；二、证明你设计的方法满足要求(1)与(2).

可是不少同学不知道要自己去设计染色的方法，而误认为整点已经预先染好了色. 有的同学将要求(1)、(2)割裂开，各作为一道题去做. 更离奇的是有的同学通篇没有一个字提到染色，有的又增加了第四种颜色——黄色，甚至还有的人花了很长篇幅去计算点的个数，真是“下笔千言，离题万里”. 这表明同学的阅读、理解能力亟需加强.

其次，从试卷中还可以看出很多学生的表达能力相当差. 他们无法说清楚自己设计的染色法(即使他们已经想出了正确的染色方法). 叙述欠条理，欠通顺，逻辑混乱，颠三倒四(他们中有些人喜欢采用一种不正规的“逆证法”). 不少人说了半天，还没有接触到问题的实质. 例如，有人这样写：

“先过 x 轴上坐标为整数的点作平行于 y 轴的直线，再过 y 轴上坐标为整数的点作平行于 x 轴的直线，这些线的交点就是坐标为整数的点，也就是问题中所说的整点.”

这些话只不过是解释整点的定义，对这道题来说，尽是废话.

叙述不清也反映了这部分同学的思路是不清晰的. 题目中有些关键的字,如"所有的""任意的""总可以找到"是数学中特有的,尤其需要特别注意,弄清其含义. 一些同学匆匆忙忙,没有仔细审题,但更多的人根本不懂这些字的意义,有的仅将一部分整点染上颜色,有的选择三个特殊的点 A、B、C,有的先将 A、B、C、D 四点取定然后涂色(完全颠倒了条件与结论).

除了语文水平不高的原因外,平时很少接触课本外的知识,眼界过窄也是一个严重的问题.

试卷还反映出学生的基本功不够扎实,在平时学习中没有很好地消化、总结(缺少华罗庚教授所说的从厚到薄的过程). 一个好的教师应当引导学生去完成这一过程.

基本功的重要性不必多说. 基本问题做好了,复杂的问题也就不难解决. 比如下面的两个问题都是常见的基本问题.

问题 1 已知四边形 $ABCD$ 的顶点为 $A(3, 5)$, $B(7, 1)$, $C(9, -3)$, $D(5,1)$,求证 $ABCD$ 是平行四边形.

问题 2 已知$\square ABCD$ 的顶点为 $A(3, 5)$, $B(7, 1)$, $C(9, -3)$,求 D 点坐标.

这样的问题,课本或任何一本(包含直角坐标的)习题集中都可以找到. 第一个问题有多种解法,如用斜率去证明 $AB \parallel CD$, $AC \parallel BD$;用距离公式计算边长证明 $AB = CD$, $AC = BD$;或将前两者结合起来证明 $AB \underline{\parallel} CD$. 但是最简单的方法却是证明对角线互相平分,即线段 AC 与 BD 的中点重合,也就是证明

$$x_A + x_C = x_B + x_D,\ y_A + y_C = y_B + y_D \tag{$*$}$$

(这里 x_A、y_A 分别表示点 A 的横坐标与纵坐标,其他类同).

同样地,第二个问题也应当利用($*$)式.

一个优秀的同学不仅要知道一个问题有几种解法,而且应当知道用最简单的方法去解决问题.

最后,谈谈如何去解上面的竞赛题.

第一步是仔细审题,搞清题目的要求. 我们应当做两件事:设计与证明. 这两件事当然是密切联系的,必须有正确的设计,满足要求(1)、(2),否则证明无法进行.

设计要满足(1),可见每种点都有无限多个. 如果在一条不与 x 轴平行的直线上有无穷多个白点(黑点或红点),那么白点(黑点或红点)就出现在无穷多条

平行于横轴的直线上.

设计要满足(2),即"对于任意……,总可以找到一个红点 D,……",可见如果红点的个数越多(或更确切地说密度越大),那么证明起来比较方便.我们应当将尽可能多的点染成红色点.

于是,我们的设计方法是(当然不是唯一的一种):将 y 轴上的整点染成白色或黑色,每种各有无穷多个(比如说上半轴为白色点,下半轴为黑色点),其余的整点全染成红色.

现在证明这样的染色确实符合要求(1)、(2).

证明并不困难.白色的点有无穷多个,而且有无穷多个纵坐标互不相同,因而白点出现在无穷多条平行于横轴的直线上.同样,黑点、红点也是如此.

对于任意的白点 A、红点 B、黑点 C,由(*)式,可以算出点 D 的坐标

$$x_D=x_A+x_C-x_B=-x_B(x_A=x_C=0),\ y_D=y_A+y_C-y_B.$$

x_A, y_A, x_B, y_B, x_C, y_C 都是整数,所以 x_D、y_D 也是整数,D 是整点.由于红点 B 的横坐标 $x_B\neq 0$,所以 $x_D\neq 0$,因而 D 是红点.并且 $x_D\neq x_B$,所以 D 不与 B 重合.这样的四个点组成平行四边形 $ABCD$,因而要求(2)满足.

关于解答还想多说几句话:

(1) 要注意设计时切勿使黑点、白点、红点共线.有人认为白点 A、红点 B、黑点 C 共线可以导出"退化的平行四边形",但原题所说的平行四边形并不包括"退化的"情况在内,不应当自己增添解释.

(2) 为了避免出现 A、B、C 共线的情况,将所有白点与黑点设计在一条直线上,在这直线外的点染红色,是最方便的一种设计方法.这条直线可以是 y 轴,也可以是任一条不与横轴平行的直线,例如直线 $y=x$.如果将一条直线上的整点染白,另一条直线上的整点染黑就难以避免 A、B、C 共线的情况.

(3) 如果计算斜率或长度,都非常麻烦,甚至连 D 为整点都难以证明.由此可见,平时练好基本功,对每个问题掌握最简单的解法是何等重要.

(4) 有人建议将要求(1)改为"每一种颜色的点既出现在无穷多条平行于横轴的直线上、又出现在无穷多条平行于纵轴的直线上".这时上面的设计与证明只需略加修改,即用直线 $y=x$ 来代替 y 轴,注意到由(*)或导出的 D 点坐标满足 $y_D-x_D=x_B-y_B\neq 0$,就不难导出结论.

(5) 本题利用奇偶性也可以解决,但这里不拟介绍了.

IMO 中的几何问题

在国际数学竞赛中，传统的(古典的)几何题是相当多的. 在最近几届中，每年都有两道题. 因此，为了参加 IMO，我们的高中数学竞赛中，这部分内容也应有适当的地位.

不少几何问题，采用几何变换(旋转、对称、平移、相似等)的观点来处理，较为明快. 本文的目的正是通过一些 IMO 的赛题来说明这一观点. 在推导中，为了突出主要的线索，我们故意省略了若干细节. 事实上，IMO 的标准答案也是简单明了，不沉溺于细节之中的.

例 1 (IMO-24)已知 A 为平面上两半径不等的圆 O_1 和圆 O_2 的一个交点，外公切线 P_1P_2 的切点为 P_1、P_2，另一外公切线的切点为 Q_1、Q_2，M_1、M_2 分别为 P_1Q_1、P_2Q_2 的中点. 求证 $\angle O_1AO_2=\angle M_1AM_2$.

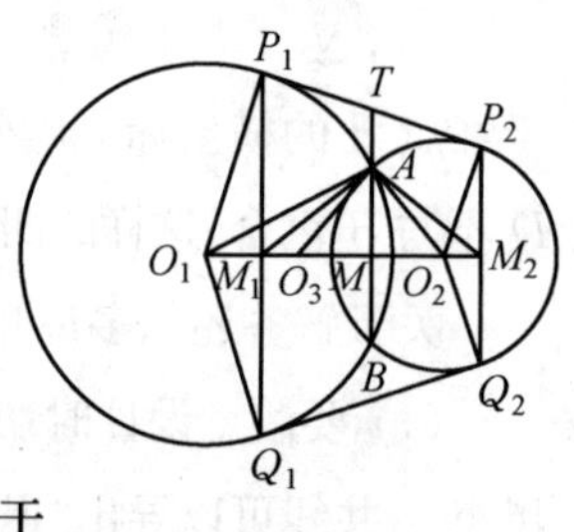

图 1

证明 如图 1，延长公共弦 AB，交 P_1P_2 于 T. 由于

$$TP_1^2=TA\times TB=TP_2^2,$$

所以 T 是 P_1P_2 的中点.

由对称性(整个图形关于 O_1O_2 对称)，M_1、M_2 都在连心线 O_1O_2 上，并且 P_1Q_1、AB、P_2Q_2 都与 O_1O_2 垂直.

由于 T 是 P_1P_2 的中点，所以 AB 平分 M_1M_2. 以 AB 为对称轴，将 $\triangle AO_2M_2$ 翻转，得到 $\triangle AO_3M_1$，这时

$$\frac{O_1A}{O_3A}=\frac{O_1A}{O_2A},\ \frac{O_1M_1}{M_1O_3}=\frac{O_1M_1}{O_2M_2}.$$

由于 $\triangle O_1P_1Q_1\backsim\triangle O_2P_2Q_2$，所以

$$\frac{O_1M_1}{O_2M_2}=\frac{O_1P_1}{O_2P_2}=\frac{O_1A}{O_2A}.$$

从而

$$\frac{O_1A}{O_3A}=\frac{O_1M_1}{M_1O_3},$$

AM_1 是$\angle O_1AO_3$ 的平分线,

$$\angle O_1AM_1=\angle M_1AO_3=\angle O_2AM_2,$$

$$\angle O_1AO_2=\angle O_1AM_1+\angle M_1AO_2=\angle O_2AM_2+\angle M_1AO_2=\angle M_1AM_2.$$

在例 1 中,通过轴对称,使图形“集中”,原来分开的两个三角形($\triangle O_2AM_2$ 与$\triangle O_1AM_1$)变为相邻的两个三角形,便于利用有关定理(如分角线定理)去证明. 这种手法值得注意.

例 2　(IMO-19-1)在已知正方形 $ABCD$ 内,作等边三角形 ABK, CDM, BCL, DAN. 试证 KL、LM、MN 和 NK 四条线段的中点及 AK、BK、BL、CL、CM、DM、DN、AN 八条线段的中点是一个正十二边形的 12 个顶点.

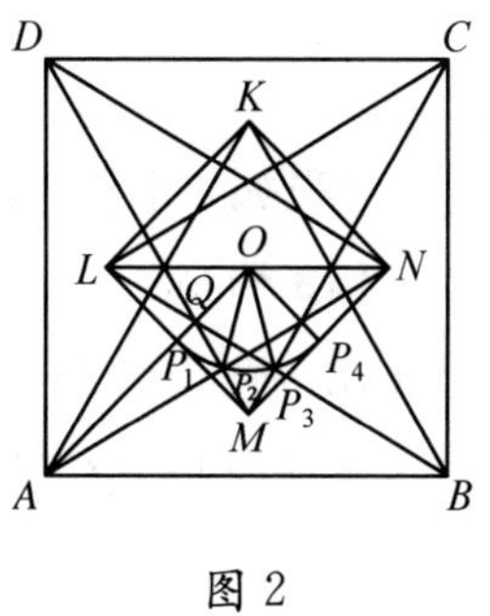

图 2

证明　如图 2,设正方形的中心为 O,当 A 绕 O 点旋转到 B、C、D、A 时, K 旋转到 L、M、N、K,所以四边形 $KLMN$ 是正方形,以 O 为中心,并且 $DM\perp AN$.

设 DM 与 AN 交于 P_2,则由于 DM 是正三角形 DAN 的高,所以 P_2 是 AN 的中点. 同理 BL 与 CM 的交点 P_3 是 BL 的中点.

容易看出,直线 OA 是整个图形的对称轴, OA 与 LM 的交点 P_1 是 LM 的中点. 只要证明

$$P_1P_2=P_2P_3,\ \angle P_1P_2P_3=150^\circ$$

就可以了.

关于 OA 对称的直线 DM、BL 交于轴 OA 上的一点 Q. 在直角三角形 QMP_3 中,

$$\angle QMP_3=60^\circ,\ \angle MQP_3=30^\circ.$$

又易知 OM 也是整个图形的对称轴, P_2、P_3 关于 OM 对称,所以

$$MP_2 = MP_3 = P_2P_3 = \frac{1}{2}MQ.$$

而在直角三角形 QMP_1 中，斜边中线

$$P_1P_2 = \frac{1}{2}QM.$$

故

$$P_1P_2 = P_2P_3.$$

易知

$$\begin{aligned}\angle P_1P_2P_3 &= \angle P_1P_2Q + 120^\circ = 2\angle P_1MP_2 + 120^\circ \\ &= \angle P_1MN - \angle P_2MP_3 + 120^\circ = 30^\circ + 120^\circ = 150^\circ.\end{aligned}$$

本题如果不利用旋转与对称，很容易陷入繁琐的论证.

例 3 (IMO-17-3)在任意$\triangle ABC$ 的三边上向外作$\triangle BPC$、$\triangle CQA$ 和$\triangle ARB$，使$\angle PBC = \angle CAQ = 45^\circ$, $\angle BCP = \angle QCA = 30^\circ$, $\angle ABR = \angle BAR = 15^\circ$，求证：(1) $\angle QRP = 90^\circ$；(2) $QR = RP$.

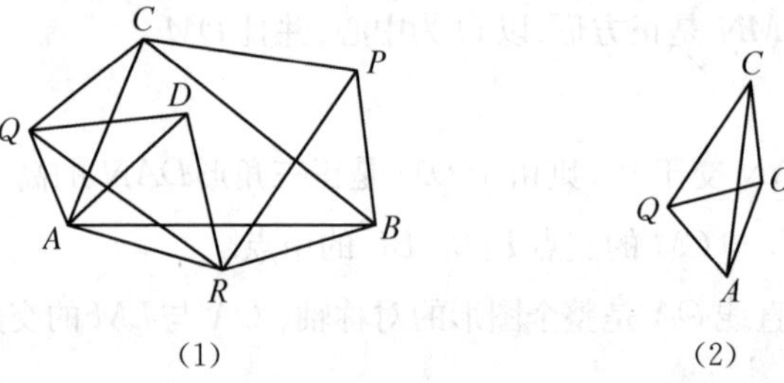

图 3

证明 如图 3(1)，将 B 点绕 R 旋转 90°至 D(我们希望证明 $\triangle RBP \cong \triangle RDQ$，即 P 绕 R 旋转 90°变为 Q). 显然，$\angle ARD = \angle ARB - 90^\circ = (180^\circ - 2\times 15^\circ) - 90^\circ = 60^\circ$, $RD = RA$. 所以，$\triangle ARD$ 是正三角形. $\angle DAB = 60^\circ - 15^\circ = 45^\circ$.

如果把$\triangle ABC$ 绕 A 旋转 45°，边 AB、AC 分别落到射线 AD、AQ 上，我们断言

$$\triangle ABC \cong \triangle ADQ. \quad (*)$$

假如(*)已经成立,那么

$$\frac{DQ}{BC}=\frac{AQ}{AC}, \angle ADQ=\angle ABC.$$

而显然 $$\triangle AQC \cong \triangle BPC, \frac{AQ}{AC}=\frac{BP}{BC}.$$

所以, $$\frac{DQ}{BC}=\frac{BP}{BC}, DQ=BP.$$

又 $$\angle RDQ=60^{\circ}+\angle ABC=15^{\circ}+\angle ABC+45^{\circ}=\angle RBP,$$

所以

$$\triangle RBP \cong \triangle RDQ.$$

从而(1)、(2)均成立.

剩下的工作是证明(*)式成立,也就是

$$\frac{AR}{AB}=\frac{AQ}{AC}. \quad (**)$$

如果利用三角,上式不难证明.下面介绍一个纯几何的证法.

设 O 为$\triangle AQC$ 的外心(见图 3(2)),

$$\angle QOA=2\angle QCA=2\times 30^{\circ}=60^{\circ}.$$

又 $OQ=OA$,所以$\triangle OQA$ 是正三角形,

$$QA=OA, \angle OAC=60^{\circ}-\angle CAQ=60^{\circ}-45^{\circ}=15^{\circ}.$$

又

$$\angle OCA=\angle OAC=15^{\circ},$$

从而

$$\triangle OAC \backsim \triangle RAB, \frac{AQ}{AC}=\frac{OA}{AC}=\frac{RA}{AB}.$$

即(* *)成立.这就完成了我们的证明.

例4 (IMO-1-5)在线段 AB 上取一点 M,在 AB 同侧分别以 AM、MB 为边作正方形 $AMCD$ 和 $MBEF$. 这两个正方形的外接圆为⊙P、⊙Q,它们相交于 M 及另一点 N. 直线 AF 与 BC 交于点 N'. (1)证明 N 与 N'重合;(2)证明不论 M 怎样选取, MN 总通过一个定点 L;(3)当 M 在 A、B 间变动时,求线段 PQ 中点的轨迹.

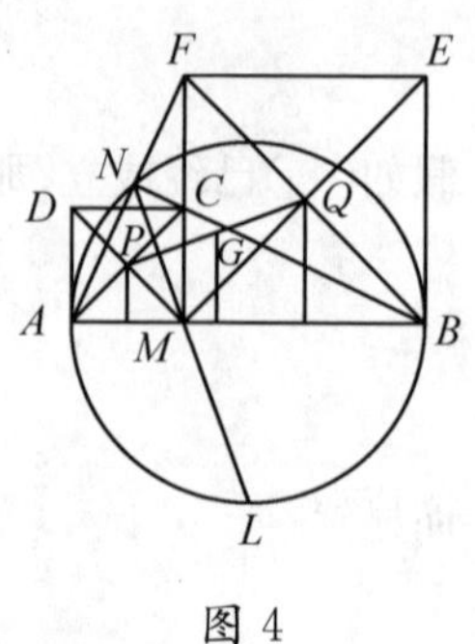

图 4

证明 如图 4,如果将 $\triangle AMF$ 绕 M 点旋转 90°,则 A 变为 C, F 变为 B,从而 AF 变为 BC,这就表明 BC 与 AF 垂直,从而 $\angle AN'C=90^\circ$,因而 N'在以 AC 为直径的圆上,即 N'在⊙P 上. 同样 N'在⊙Q 上,所以 N'与 N 重合.

由于 $\angle ANB=90^\circ$,所以 N 在以 AB 为直径的圆上. 又 $\angle ANM=\angle ACM=45^\circ=\frac{1}{2}\angle ANC$,所以 NM 是$\angle ANB$ 的平分线,从而 NM 必过定点 L, L 是以 AB 为直径的、位于 N 异侧的半圆弧 $\overset{\frown}{AB}$ 的中点.

设 G 为 PQ 的中点,则 G 到 AB 的距离为 P、Q 到 AB 的距离的平均值. 而 P 到 AB 的距离显然是$\frac{1}{2}AM$, Q 到 AB 距离是$\frac{1}{2}MB$. 因此 G 到 AB 的距离为

$$\frac{1}{2}\left(\frac{1}{2}AM+\frac{1}{2}MB\right)=\frac{1}{4}AB.$$

作与 AB 平行并且距离为$\frac{1}{4}AB$ 的直线,交 AC 于 G_1、交 BF 于 G_2. 则由于 AC、BF 都是确定的射线(均与 AB 成 45°),与 M 点位置无关,所以 G_1、G_2 都是确定的点. P、Q 分别在 AC、BF 上,因而 G 一定在这两条射线所夹的角的内部,这就是说 G 一定在线段 G_1G_2 上.

另一方面,我们可以证明线段 G_1G_2 上每一点 G 都是轨迹中的点(纯粹性). 为此,如图 5,设 G 在 AB 上的投影为 G',由

$$MB-MA=2(G'B-G'A),\ MB+MA=AB,$$

可定出 AB 上的一点 M,然后作出正方形 $AMCD$ 与

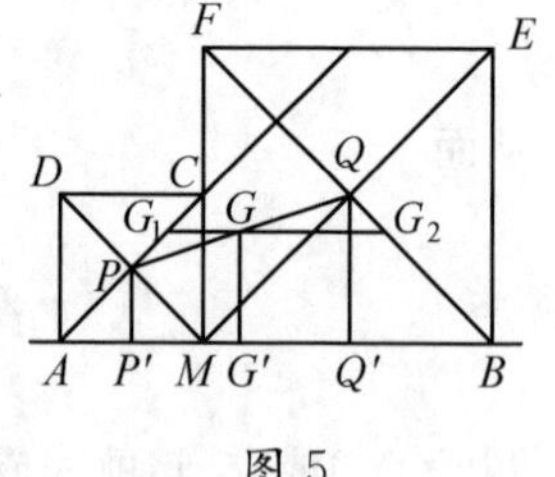

图 5

$MBEF$,它们的中心分别为 P、Q. 设 P, Q 在 AB 上投影分别为 P'、Q',又设 PQ 中点为 G'',则由前面的论证, G''在线段 G_1G_2 上. 而

$$G'P'=G'A-P'A=G'A-\frac{1}{2}MA=G'B-\frac{1}{2}MB=G'B-Q'B=G'Q',$$

所以 G' 是 $P'Q'$ 的中点,从而 $G'G''\ /\!/\ P'P$. G'既是G、也是G''在 AB 上的投影, G''与 G 又同在 G_1G_2 上,所以 G 与 G''重合,即 G 为 PQ 的中点.

这样,我们完成了完备性与纯粹性的证明, G_1G_2 确实为所求的轨迹.

这类纯粹性的证明,通常要作出符合要求的图形,并采用同一法导出某两个点重合. 例 4 就是一个典型的例子.

例 5　(IMO－22－5)三个全等的圆有一个公共点 O,并且都在一个已知三角形 ABC 内,每一个圆与 $\triangle ABC$ 的两条边相切. 证明这三角形的内心,外心和 O 点共线.

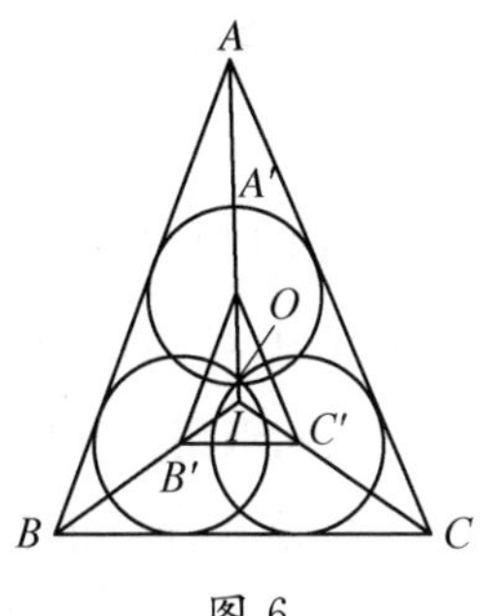

图 6

证明　如图 6,设三个圆的圆心分别为 A'、B'、C'. 因为 AB 是等圆$\odot A'$、$\odot B'$的公切线, A'、B'到 AB 的距离相等,都等于半径,因而 $A'B'\ /\!/\ AB$. 同理, $B'C'\ /\!/\ BC$, $C'A'\ /\!/\ CA$.

又 AB、AC 是$\odot A'$的切线,所以 AA'是$\angle BAC$ 的平分线. 同样 CC'、BB'也是$\triangle ABC$ 的角平分线,因而它们相交于$\triangle ABC$ 的内心 I.

由于 O 是等圆$\odot A'$、$\odot B'$、$\odot C'$的公共点,所以 $OA'=OB'=OC'$, O 是$\triangle A'B'C'$的外心.

以 I 为相似中心,$\frac{IA'}{IA}$为相似比作相似变换,则$\triangle A'B'C'$变为$\triangle ABC$. 因而$\triangle A'B'C'$的外心 O 变为$\triangle ABC$ 的外心 K,这就表明 O、K 与相似中心 I 在一条直线上.

相似也是证题中常用的手法. 一个三角形的三边的中点组成一个三角形,与原三角形相似,相似中心为重心 G,相似比为$\frac{1}{2}$. 由此可导出许多重要的性质,如原三角形的外心 O、垂心 H 与G 共线,并且 $OG:GH=1:2$(因为 O 是新三角形的垂心),等等. 有兴趣的读者不妨自己去研究一番.

奥林匹克集训班选拔考试题及解答

第一天（1987年5月2日上午，4小时30分）

1. 对于任意正整数 k，试求最小的正整数 $f(k)$，使得存在 5 个集合 S_1，S_2，…，S_5，满足

(i) $|S_i|=k$，$i=1, 2, \cdots, 5$；

(ii) $S_1 \cap S_{i+1}=\varnothing$，$i=1, 2, \cdots, 5$，$S_6=S_1$；

(iii) $|\bigcup\limits_{i=1}^{5} S_i|=f(k)$.

其中 $|Q|$ 表示集合 Q 中数的个数.

又问当集合个数是一般的不小于 3 的正整数 n 时，有何结果？

2. 在平面直角坐标系中给定一个 100 边形 P，满足

(i) P 的顶点坐标都是整数，

(ii) P 的边都与坐标轴平行，

(iii) P 的边长都是奇数.

试证 P 的面积是奇数.

3. 已知数列 $\{r_n\}$ 满足

$$r_1=2,\ r_n=r_1 r_2 \cdots r_{n-1}+1,\ n \geqslant 2,$$

自然数 $a_1, a_2, \cdots, a_n$ 满足

$$\frac{1}{a_1}+\frac{1}{a_2}+\cdots+\frac{1}{a_n}<1,$$

试证

$$\frac{1}{a_1}+\frac{1}{a_2}+\cdots+\frac{1}{a_n} \leqslant \frac{1}{r_1}+\frac{1}{r_2}+\cdots+\frac{1}{r_n}.$$

第二天（1987年5月3日上午，4小时30分）

4. 设 S 是直角坐标平面上关于两坐标轴都对称的任意凸图形. 在 S 中作

一个四边都平行于坐标轴的矩形A,使其面积最大.把矩形A按相似比$1:\lambda$放大为矩形A',使A'完全盖住S.试求对任意平面凸图形S都适用的最小的λ.(注:若平面图形中任意两点所连线段包含在这个图形之中,则称这个图形为凸图形)

5. 试求所有正整数n,使方程

$$x^3+y^3+z^3=nx^2y^2z^2$$

有正整数解.

6. 空间$2n\ (n\geqslant 2)$个点,其中任意四点都不共面,试证连接这些点的任何n^2+1条直线段必可构成两个有公共边的三角形.

解　答

1. 在集合S_i的个数n为偶数$2m$时,$f(k)=2k$.事实上,k元集合$S_1=S_3=\cdots=S_{2m-1}$与$S_2=S_4=\cdots=S_{2m}$.在$S_1\cap S_2=\varnothing$时满足(i)、(ii),并且$\left|\bigcup\limits_{i=1}^{2m}S_i\right|=2k$.由要求(i)、(ii),这显然是最小的值.

在n为奇数$2m+1$时,

$$f(k)=2k+\left[\frac{k+m-1}{m}\right],\tag{1}$$

其中$[\alpha]$表示不超过α的最大整数.证明如下:考虑表

	x_1　x_2　$\cdots$
S_1	
S_2	
$\vdots$	
S_{2m+1}	

如果元素$x_i\in S_i$,则在第i行第j列的交叉点处记上1,否则记上0.由于(i),表中每行有k个1,所以$2m+1$行的总和为$k(2m+1)$.

另一方面,由于(ii),每列至多有m个1,因此列数$\geqslant\dfrac{k(2m+1)}{m}=2k+\dfrac{k}{m}$,

即
$$f(k) \geqslant 2k + \left[\frac{k+m-1}{m}\right].$$

为了证明上面的等号成立，取

$$2k + \left[\frac{k+m-1}{m}\right]$$

个列，并将 $k(2m+1)$ 个 1 填入表中. 填法是从第一列开始填，行数则依然 1, 3, 5, …, $2m+1$, 2, 4, …, $2m$, 1, 3, …的顺序，每填 m 个 1 就转入下一列. 这种填法是均匀的，所以(i)满足，(ii)也显然满足，所以

$$f(k) \leqslant 2k + \left[\frac{k+m-1}{m}\right].$$

从而(1)式成立. 特别地，

$$n=3 \text{ 时}, \ f(k)=3k;$$
$$n=5 \text{ 时}, \ f(k)=2k+\left[\frac{k+1}{2}\right].$$

下面是 $n=5$ 时填成的表：

	x_1	x_2	x_3	x_4	x_5	x_6	x_7	x_8
S_1	1		1			1		1
S_2		1			1		1	
S_3	1			1		1		
S_4			1		1			1
S_5		1		1			1	

这种表示元素与集合关系的表是常常采用的.

2. 设多边形的顶点为 $P_i(x_i, y_i)$ $(i=1, 2, \cdots, 100)$，解析几何中有一个求多边形面积的公式：

$$S=\frac{1}{2}\sum_{i=1}^{100}\begin{vmatrix} x_i & y_i \\ x_{i+1} & y_{i+1}\end{vmatrix} \ (x_{101}=x_1, \ y_{101}=y_1).$$

不妨假定第一条边 P_1P_2 是平行于 y 轴的，则

$$x_{2i-1}=x_{2i},\ y_{2i}=y_{2i+1},\ x_{2i+1}-x_{2i}\equiv 1,\ y_{2i+2}-y_{2i+1}\equiv 1(\mathrm{mod}2),$$
$$i=1,2,\cdots,50.$$

所以

$$\begin{aligned}S&=\frac{1}{2}(x_1y_2+x_2y_2+x_3y_4+x_4y_5+\cdots\\&\quad-x_2y_1-x_3y_2-\cdots-x_1y_{100})\\&=(x_1y_2+x_3y_4+\cdots+x_{40}y_{100}-x_2y_2-x_3y_4-\cdots-x_1y_{100})\\&=(x_1-x_3)y_2+(x_3-x_5)y_4+\cdots+(x_{99}-x_1)y_{100}\\&\equiv y_2+y_4+\cdots+y_{100}\equiv 50y_2+25\equiv 1(\mathrm{mod}2).\end{aligned}$$

3. 容易看出(例如用归纳法证明)

$$1-\frac{1}{r_1}-\frac{1}{r_2}-\cdots-\frac{1}{r_n}=\frac{1}{r_1r_2\cdots r_n}. \tag{1}$$

假定对于对 $n<k$,

$$\frac{1}{a_1}+\frac{1}{a_2}+\cdots+\frac{1}{a_n}\leqslant\frac{1}{r_1}+\frac{1}{r_2}+\cdots+\frac{1}{r_n} \tag{2}$$

对所有满足

$$\frac{1}{a_1}+\frac{1}{a_2}+\cdots+\frac{1}{a_n}<1 \tag{3}$$

的自然数 a_1, a_2, $\cdots$, a_n 成立. 而在 $n=k$ 时,对于某组满足(3)的自然数 a_1, a_2, $\cdots$, a_n 有

$$\frac{1}{a_1}+\frac{1}{a_2}+\cdots+\frac{1}{a_n}>\frac{1}{r_1}+\frac{1}{r_2}+\cdots+\frac{1}{r_n}. \tag{4}$$

不妨设 $a_1\leqslant a_2\leqslant\cdots\leqslant a_n$, 由(归纳)假设,对于 $n\leqslant k$,

$$\frac{1}{a_1}\leqslant\frac{1}{r_1},$$

$$\frac{1}{a_1}+\frac{1}{a_2}\leqslant\frac{1}{r_1}+\frac{1}{r_2},$$

$$\cdots$$

$$\frac{1}{a_1}+\frac{1}{a_2}+\cdots+\frac{1}{a_{n-1}}\leqslant\frac{1}{r_1}+\frac{1}{r_2}+\cdots+\frac{1}{r_{n-1}},$$

将以上各式分别乘以 a_1-a_2, a_2-a_3, $\cdots$, $a_{n-1}-a_n$, 将(4)乘以 a_n,然后相加得

$$n>\frac{a_1}{r_1}+\frac{a_2}{r_2}+\cdots+\frac{a_n}{r_n}.$$

由算术—几何平均不等式得

$$1>\frac{1}{n}\left(\frac{a_1}{r_1}+\frac{a_2}{r_2}+\cdots+\frac{a_n}{r_n}\right)\geqslant\sqrt[n]{\frac{a_1a_2\cdots a_n}{r_1r_2\cdots r_n}},$$

从而
$$r_1r_2\cdots r_n>a_1a_2\cdots a_n. \tag{5}$$

但正数

$$1-\left(\frac{1}{a_1}+\frac{1}{a_2}+\cdots+\frac{1}{a_n}\right)\geqslant\frac{1}{a_1a_2\cdots a_n},$$

故由(1)、(4)得

$$\frac{1}{a_1a_2\cdots a_n}\leqslant\frac{1}{r_1r_2\cdots r_n},$$

即
$$r_1r_2\cdots r_n\leqslant a_1a_2\cdots a_n,$$

与(5)矛盾,所以在 $n=k$ 时,对于每一组满足(8)的自然数 a_1, a_2, $\cdots$, a_n,(2)均成立.

又显然在 $\frac{1}{a_1}<1$ 时,自然数 $a_1\geqslant 2$, 所以本题的结论成立.

可以进一步证明(2)式仅在 a_1, a_2, $\cdots$, a_n 是 r_1, r_2, $\cdots$, r_n 的一个排列时,等号成立.

4. 如图 1,设点 $P(a, 0)$, $Q(0, b)$分别为凸形 S 的边界与 x、y 轴正向的交点.

S 一定被矩形

$$A'=\{(x, y)\mid |x|\leqslant a, |y|\leqslant b\}$$

覆盖,事实上,若点 $B\in$ 凸形 S,不妨假定 B 在第一象限,若 $B\notin$ 矩形 A',则有(图 2)

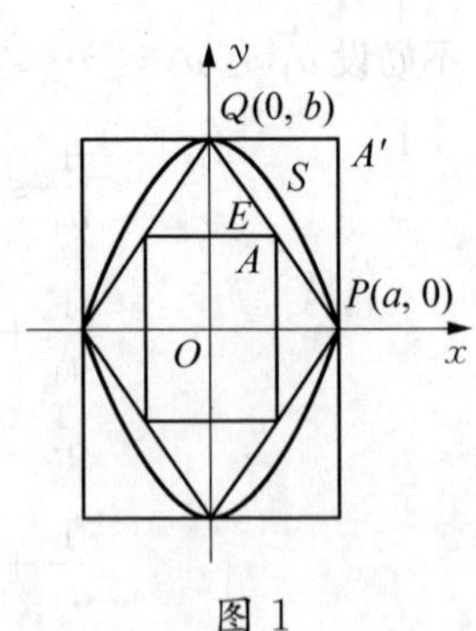

图 1

$$\angle BPO > 90^\circ \text{ 或 } \angle BQO > 90^\circ.$$

不妨设前者成立. 点 B 关于 x 轴的对称点 $B' \in S$,并且

$$\angle BPB' = 2\angle BPO > 180^\circ, \tag{1}$$

但凸形 S 的点应当在过 P 点的一条直线的同侧,因此(1)不可能成立. 这就说明凸形 S 的每一点 $B \in$ 矩形 A'. 即矩形 A'覆盖 S.

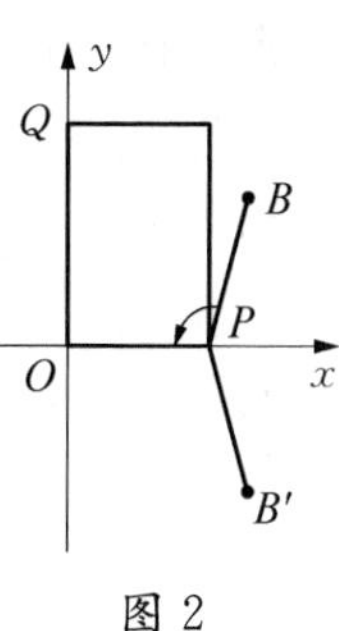

图 2

连 PQ,线段 PQ 的中点E 为$\left(\frac{a}{2}, \frac{b}{2}\right)$, E 在凸形 S 中,从而矩形

$$A = \left\{(x, y) \,\middle|\, |x| \leqslant \frac{a}{2}, |y| \leqslant \frac{b}{2}\right\}$$

在凸形 S 中(图 1).

A 与 A'的相似比为 $1:2$,因此 $\lambda \leqslant 2$.

另一方面,考虑菱形

$$S_1 = \left\{(x, y) \,\middle|\, \left|\frac{x}{a}\right| + \left|\frac{y}{b}\right| \leqslant 1\right\},$$

S_1 是凸形,并且覆盖 S_1 的最小的、四边与坐标轴平行的矩形是上面所说的 A'.

由于 $xy = ab \cdot \frac{x}{a} \cdot \frac{y}{b}$,在$\frac{x}{a} + \frac{y}{b} = 1$时, xy 的最大值在$\frac{x}{a} = \frac{y}{b} = \frac{1}{2}$,即$x = \frac{a}{2}$, $y = \frac{b}{2}$ 时达到,因此 S_1 中的四边与坐标轴平行的矩形以上为所说的 A 面积最大,从而 $\lambda \geqslant 2$,于是 $\lambda = 2$.

5. 若 $(x, y) = d > 1$,则 $d \mid nx^2y^2z^2$,从而 $d \mid (x^3 + y^3 + z^3)$, $d \mid z^3$. 可令

$$x = x_1 d,\ y = y_1 d,\ z = z_1 d,\ n_1 = nd^2,$$

原方程化为

$$x_1^3 + y_1^3 + z_1^3 = n_1 x_1^2 y_1^2 z_1^2,$$

所以我们设 x、y、z 两两互质.

若 $x = y$,则 $x = y = 1$(因为 x、y 互质),从而 $2 + z^3 = nz^2$,于是 $z \mid 2$, $z =$

1 或 2. 前者导出 $n=3$，后者无解.

若 x、y、z 互不相等，不妨设 $x<y<z$，由于

$$\frac{x^3}{z^2y^2}+\frac{y}{z^2}+\frac{z}{y^2}=nx^2, \tag{1}$$

如果 $z>y^2$，则

$$\frac{x^3}{z^2y^2}+\frac{y}{z^2}<\frac{2y}{z^2}<\frac{2}{y^3}\leqslant\frac{1}{y^2},$$

从而(1)式左边不为整数(因为$\frac{z}{y^2}$的分数部分不为 0，与 1 至少相差$\frac{1}{y^2}$)，矛盾. 所以 $z<y^2$.

从而(1)式左边

$$\frac{x^3}{z^2y^2}+\frac{y}{z^2}+\frac{z}{y^2}<\frac{1}{z}+\frac{1}{z}+1<\frac{1}{2}+\frac{1}{2}+1=2,$$

所以(1)式两边均为 1，$n=x=1$.

本题的答案是 $n=3$ 或 1.

如果需进一步求出 x、y、z 的值. 注意在 $n=1$ 时 $x=1$，从而

$$1+y^3+z^3=y^2z^2, \tag{2}$$

注意上面的讨论中已有 $y<z<y^2$，令

$$y^2=z+k,\ k \text{ 为正整数}.$$

若 $k\geqslant 2$，则

$$\begin{aligned}
&y^2z^2-z^3-y^3-1\\
&=z^2\cdot k-y^3-1=z^2\cdot k-y(z+k)-1\\
&\geqslant z^2\cdot k-z(z+k)\geqslant z(kz-z-k)\\
&=z((k-1)(z-1)-1)>0,
\end{aligned}$$

矛盾！所以 $k=1$，从而由(2)得

$$1+y(z+1)=z^2,$$

即
$$(1+y)+yz=z^2,$$

从而 $z \mid 1+y$，即 $z=1+y$，从而由 $y^2=z+1$ 得

$$y^2-1=z=y+1,$$

所以 $y=2$，$z=3$.

即本题的解只有两组(假定 $x\leqslant y\leqslant z$)，

$$\begin{cases}n=3,\\x=1,\\y=1,\\z=1,\end{cases}\quad\begin{cases}n=1,\\x=1,\\y=2,\\z=3.\end{cases}$$

6. $n=2$ 时结论显然成立. 假定命题对于 n 成立，考虑 $2(n+1)$ 个点，有 $(n+1)^2+1$ 条线段连结这些点.

在其中取点 A、B，A、B 之间有线段相连. 用 $d(A)$、$d(B)$分别表示以 A、B 为端点的线段的条数.

(i) 若 $d(A)+d(B)\leqslant 2n+2$，则除去 A、B 两点及以它们为端点的线段后，还剩下 $\geqslant (n+1)^2+1-(2n+1)\geqslant n^2+1$ 条线段，因而由归纳假设，结论成立.

(ii) 若 $d(A)+d(B)\geqslant 2n+4$，在其余的 $2n$ 个点中至少有两个点C、D 同时与 A、B 相连，否则

$$d(A)+d(B)\leqslant 2n+1+2=2n+3.$$

(iii) 最后考虑 $d(A)+d(B)=2n+3$ 的情况，可以假定有且只有一个点 C 同时与 A、B 相连(否则与上面讨论的情况相同)，这时其他的点均恰与 A、B 中一点相连. 假如其中有一点 D 与C 相连，则 A、B、C、D 之间有 5 条线段相连，故有两个有公共边的三角形 ($n=2$ 的情况)，假如每一点均不与 C 相连，则 $d(C)=2$，而 $d(A)$、$d(B)$ 中较小的(比如说)$d(B)$ 满足 $d(B)\leqslant n+1$，$d(B)+d(C)\leqslant n+3$，这又化为情况(1).

至此命题得证.

IMO中的数论问题(上)

初等数论(算术)问题在IMO中经常出现.本文试图就这类问题的解法作些粗浅探讨.在解答问题过程中,顺便介绍一些与问题有关的初等数论基础知识,因这些内容可在任何一本数论教本中找到,故略去其证明.虽然,熟悉这些知识是解题的先决条件,但最重要的仍然是灵活、巧妙地应用这些知识.

首先,我们谈谈初等数论中的两个基本问题.

1 带余除法

若a、b是两个整数,$b>0$,则存在两个整数q、r,使得

$$a=bq+r,\ 0\leqslant r<b,$$

且q、r是唯一的.

特别,如果$r=0$,则$a=bq$.这时,a被b整除,我们记作$b|a$,也说b是a的约数.

2 最大公约数

设a_1, $\cdots$, a_n是n个整数,如果正整数d是它们每一个的约数,则d称为a_1, $\cdots$, a_n的公约数, a_1, $\cdots$, a_n的公约数中最大的一个称为它们的最大公约数,记作$(a_1, \cdots, a_n)$.

如果$(a_1, \cdots, a_n)=1$,我们说a_1, $\cdots$, a_n互素(也称互质).如果a_1, $\cdots$, a_n中每两个都互素,就说它们两两互素.应当注意,$(a_1, \cdots, a_n)=1$不能保证a_1, $\cdots$, a_n两两互素.

在IMO问题中,更多的是涉及$n=2$的情况,下面的裴蜀(Be′zout)定理是非常有用的.

定理　设 a、b 是整数,则 $(a,\ b)=d$ 的充要条件是存在整数 x、y,使得

$$ax+by=d.$$

特别,a、b 互质(也就是 $(a,\ b)=1$) 的充要条件是,存在整数 x、y,使得

$$ax+by=1.$$

在 a、b 都是正整数时,可以改述成如下的形式:

正整数 a、b 互质(互素)的充要条件是存在正整数 x、y,使得

$$ax-by=1.$$

例 1　(IMO - 1 - 1)证明:对任何自然数 n,分数 $\dfrac{21n+4}{14n+3}$ 不可约简.

证明　要证所说的分数不可约简,实际上就是要证分子和分母互质. 注意

$$3(14n+3)-2(21n+4)=1.$$

在裴蜀定理中取

$$a=14n+3,\ b=21n+4,\ x=3,\ y=-2$$

即得结果.

我们也可以不直接用定理,而采用一个变形的证法. 设

$$d=(21n+4,\ 14n+3).\ (\text{我们要证 } d=1)$$

因为

$$d\mid 21n+4,$$

故

$$d\mid 2(21n+4).$$

同样

$$d\mid 3(14n+3).$$

于是,我们有
$$d\mid 3(14n+3)-2(21n+4),$$

即 $d\mid 1$,从而 $d=1$.

评注　上面的解答中,裴蜀恒等式

$$3(14n+3)-2(21n+4)=1$$

起了决定性的作用,这个恒等式可以通过观察"凑"出来,也可以"正规"一点,采

用待定系数法来求,求法如下:

设 x、y 是整数,满足

$$x(14n+3)+y(21n+4)=1.$$

整理得

$$(14x+21y)n+(3x+4y)=1.$$

令(等式两边 n 的同次幂的系数相等)

$$14x+21y=0,\ 3x+4y=1.$$

解得
$$x=3,\ y=-2.$$

例 2 m 个盒子,每个盒子中有一些球. 设 $n<m$ 为一已知自然数,施行下面的操作:从这些盒子中取 n 个,在取定的盒子中各放一个球. 证明:如果 $(m, n)=1$, 则可以施行有限次操作,使所有的盒子中有相等的球.

证明 由于 m、n 互素,根据裴蜀恒等式,存在自然数 x、y 使

$$xn-ym=1.$$

此即
$$xn=ym+1=y(m-1)+(y+1).$$

即可以将 xn 个球放入盒中,使 $m-1$ 个盒子各增加 y 个,一个盒子增加 $y+1$ 个,这就相当于将一个盒子增加一个,其余的盒子不增加. 因此,重复这一操作至足够多次,就可使盒中的球数全相等.

例 3 (IMO-27-1)集 $M=\{1, 2, \cdots, n-1\}$, $n\geqslant 3$, M 中每个数染上红或黑两种颜色中的一种,使得

(i) 对每个 $i\in M$, i 与 $n-i$ 同色;

(ii) 对每个 $i\in M$, $i\neq k$,有 i 与 $|k-i|$ 同色,这里 k 是 M 中的一个固定的数,它和 n 互素. 证明:M 中所有的数同色.

证明 为了下面叙述的方便,将 0、n 这两个数补充到 M 中,并规定它们与 k 的颜色相同. 容易验证性质(i)、(ii)仍然保持.

由裴蜀恒等式,有正整数 x、y,使

$$xk-yn=1$$

(凡是问题中给出两个互质的数——这里是 k 与 n,我们都可以立即写出这样的等式用以解题). 于是,对于 M 中的任意一个正数 d,有正整数 u、v 使得

$$uk - vn = d \tag{1}$$

(取 $u = dx$, $= dy$ 即可). 这表明可以将 u 个 k 逐个相加,每超过 n 时便从和中减去一个 n,最后 u 个 k 加完后所得的结果就是 d.

下面来证明在这个过程中,所得的(M 中的)数是同一种颜色,从而 d 与 k 是同一种颜色.

事实上,对于 $l \in M$ 与 $l+k \in M$,根据(ii) (取 $i = l+k$),有 $l = (l+k) - k$ 与 $l+k$ 是同一种颜色. 因此,每次加 k 时,只要不超过 n,所得的数是同一种颜色.

如果 $0 \leqslant l \leqslant k$,而 $l+k > n$,根据(i), $n-l$ 与 l 同色;再根据(ii)(取 $i = n-l$),

$$l + k - n = k - (n - l)$$

与 $n-l$ 同色. 从而, l 与 $l+k-n$ 同色. 即每次加 k 时,每超过 n 便从和中减去一个 n,所得的数是同色的.

由于 d 是 M 中任意一数,所以 M 中所有的数都(与 k)同一种颜色.

以下的知识也极有用.

3 素(质)数及其特性

如果 $p > 1$ 是整数,其约数只有 1 和本身,则称 p 为素数(质数). 按定义, 2 是素数,这也是唯一的偶素数. 我们约定:1 不是素数,且把不是 1、又不是素数的正整数称为合数.

素数有如下一个重要性质:

设 a、b 是整数, p 是素数. 若 $p \mid ab$,则 $p \mid a$ 或者 $p \mid b$ 至少有一个成立(当然也可能两者都成立). 特别地,如果 $p \nmid a$(即 p 不能除尽 a),则必有 $p \mid b$.

4 唯一分解定理

任一大于 1 的整数 n 能够唯一地表示成

$$n = p_1^{\alpha_1} \cdot \cdots \cdot p_k^{\alpha_k} (\alpha_i > 0,\ i = 1,\ \cdots,\ k),$$

其中 $p_i \neq p_j (i \neq j,\ i,\ j = 1,\ \cdots,\ k)$ 都是素数.

唯一分解定理是初等数论中最重要的定理之一，它表明，素数在正整数集合中所占的重要地位.

例 4 (IMO-4-1)求下列性质的最小自然数 n，其十进制表示法中以 6 结尾，当删去最后一位的 6 并把它写在余下的数字的第一位时，成为 n 的四倍.

解 n 的十进制表示中以 6 结尾，意味着它被 10 除后余数为 6. 按带余除法，我们有

$$n=10m+6,$$

其中 m 是正整数，我们设它是 l 位数，则由题设

$$4(10m+6)=6\times 10^{l}+m.$$

于是，

$$m=\frac{2(10^{l}-4)}{13}.$$

要使 m 是整数，则需 $13\mid 2(10^{l}-4)$. 但 13 是素数，$13\nmid 2$，故应有 $13\mid 10^{l}-4$(这儿用到素数的特性). 再者，欲使 n 最小，由上式可知，这等价于要使 m 最小. 这样，我们的问题就转化成求最小的正整数 l，使 $13\mid 10^{l}-1$.

取 $l=1, 2, \cdots$，逐一验证，得知适合条件的最小的 $l=5$(我们用了“最小”这一限制)，此时 $m=15384$，而所求的最小的 $n=153846$.

评注 本题的数不大，因而验证起来并不困难. 这种“试验法”对于大数的计算，借助计算机也容易获得解决. 但涉及某些理论问题，计算机仍然无能为力.

下面的问题与例 4 有密切联系，提供读者思考：

问题 证明：存在无穷多个满足题中条件的自然数 n.

例 5 (IMO-9-3)若 k, m, n 是正整数，$m+k+1$ 是一个比 $n+1$ 大的素数，记 $C_s=s(s+1)$，证明乘积

$$(C_{m+1}-C_k)(C_{m+2}-C_k)\cdots(C_{m+n}-C_k)$$

能被 $C_1C_2\cdots C_n$ 整除.

证明 易见 $C_1\cdots C_n=(n+1)!\ n!$，而 $n!$ 必然整除连续 n 个整数的积，从而

$$n!\mid[(m+1-k)(m+2-k)\cdot\cdots\cdot(m+n-k)].$$

同样

$$(n+1)! \mid [(m+k+1)(m+k+2)\cdot\cdots\cdot(m+k+n+1)].$$

但 $m+k+1$ 为大于 $n+1$ 的素数，从而 $(n+1)!$ $n!$ 实际上不能整除 $m+k+1$(素数的性质). 故有

$$\begin{aligned}&C_1\cdots C_k \mid (m+k+2)\cdots(m+k+n+1)\cdot(m+1-k)(m+2-k)\cdot\cdots\\&\quad\cdot(m+n-k)\\&=[(m+1-k)(m+1+k+1)]\cdot[(m+2-k)(m+2+k+1)]\cdot\cdots\\&\quad\cdot[(m+n-k)(m+n+k+1)]\\&=(C_{m+1}-C_k)(C_{m+2}-C_k)\cdots(C_{m+n}-C_k).\end{aligned}$$

评注　(1) 本题并不困难，解题的主要手法是作恒等变形. 但我们在关键之处用到了素数的特性.

(2) 解答中，我们还用到了这样一个结果：n 个连续整数的积被 $n!$ 整除. 读者可以利用组合数 C_m^n 是整数来证明之(不妨设连续 n 个数都是正的. 如果其中有正，有负，则必然有零，而此时结论明显).

例 6　设 p 是素数，$1\leqslant k<p$，则 C_p^k 被 p 整除.

证明　由组合数的意义可见 C_p^k 是整数.

我们有 $kC_p^k=pC_{p-1}^{k-1}$，C_{p-1}^{k-1} 也是整数，所以 $p\mid pC_{p-1}^{k-1}$，即 $p\mid kC_p^k$.

但 p 为素数，$1\leqslant k<p$，$p\nmid k$，所以 $p\mid C_p^k$.

下面我们看一个比较有技巧的问题.

例 7　(IMO－21－1)设 p、q 都是自然数，使得

$$\frac{p}{q}=1-\frac{1}{2}+\frac{1}{3}-\cdots-\frac{1}{1318}+\frac{1}{1319}.$$

证明 p 可被 1979 整除.

证明　首先，我们作恒等变形

$$\begin{aligned}\frac{p}{q}&=1-\frac{1}{2}+\frac{1}{3}-\cdots-\frac{1}{1318}+\frac{1}{1319}\\&=\left(1+\frac{1}{2}+\cdots+\frac{1}{1319}\right)-2\left(\frac{1}{2}+\frac{1}{4}+\cdots+\frac{1}{1318}\right)\\&=\left(1+\frac{1}{2}+\cdots+\frac{1}{1319}\right)-\left(1+\frac{1}{2}+\cdots+\frac{1}{659}\right)=\frac{1}{660}+\frac{1}{661}+\cdots+\frac{1}{1319}\end{aligned}$$

$$=\left(\frac{1}{660}+\frac{1}{1319}\right)+\left(\frac{1}{661}+\frac{1}{1318}\right)+\cdots+\left(\frac{1}{989}+\frac{1}{990}\right)$$

$$=1979\left(\frac{1}{660\times 1319}+\frac{1}{661\times 1318}+\cdots+\frac{1}{989\times 990}\right).$$

所以，$1319!\times\frac{p}{q}$ 可被 1979 整除. 从而 $1319\times p$ 可被 1979 整除.

但 1979 是一个素数(为什么？请见下面的评注)，从而 1319！不能被 1979 整除，故 p 被 1979 整除(利用了素数的特性).

评注 (1) 上面解答的关键之处与前两题类似，只是问题的核心比较隐蔽，需通过适当的变形才能揭示出来.

(2) 解答中我们用到了 1979 是一个素数这一事实，这并非是一件显然的事. 要判断一个数是否为素数，我们有下面的定理.

定理 设 $n>1$ 是正整数. 如果不大于$\sqrt{n}$的所有素数都不能整除n，则 n 是素数.

实际上，设 n 不是素数，则它必有一个最小的素约数 p，从而

$$n=pn'\geqslant p\times p=p^2,$$

即 $p\leqslant\sqrt{n}$. 与假设矛盾，故 n 是素数.

根据这定理并注意到 $\sqrt{1979}<45$. 而不超过 45 的素数 2, 3, 5, 7, 11, 13, 17, 19, 23, 29, 31, 37, 41, 43 都不能整除 1979，故它必是素数.

(3) 以上几题都属于初等数论中的整除问题. 下面提出两个问题供读者考虑.

问题 1 证明 $\left(1+\frac{1}{2}+\cdots+\frac{1}{p-1}\right)\times(p-1)!$ 可被 p 整除. 这里 $p>3$ 是素数.

问题 2 对怎样的自然数 n，有 $n\mid(n-1)!$.

我们再举一例说明如何灵活运用素数的性质解题.

例 8 (IMO-12-4)找出具有下列 性质的所有正整数 n：集合 $\{n, n+1, n+2, n+3, n+4, n+5\}$ 可以划分成两个无公共元素的非空子集，使得一个子集中所有元素的积与另一子集中所有元素的积相等.

解 本题无解，证明如下.

假定 n 具有所述性质,则 n, $n+1$, $n+2$, $n+3$, $n+4$, $n+5$ 中任一个素约数 p 必然整除每个子集的元素的乘积,从而整除这六个数中的至少两个数(素数的性质).

如果 p 整除两个整数,则必然整除它们的差.但上面六个数两两之差的素约数只有 2, 3 或 5,故 p 只能是 2, 3, 5.

再考虑 $n+1$, $n+2$, $n+3$, $n+4$. 它们中任一素约数不能为 5. 否则,因此时 $5\nmid n$, $5\nmid n+5$,而这四个数中显然不能有两个数都被 5 整除.这样,六个数中只有一个被 5 整除,这和上面所得结论矛盾.

于是,这四个数的素约数只能是 2 或 3,但它们中恰有两者为奇数(n 为奇数时, $n+2$, $n+4$ 为奇数;n 为偶数时, $n+1$, $n+3$ 为奇数),而这两个奇数的素约数只有 3,因而必须是 3^r 和 3^s(r、s 为正整数)的形式.这两奇数的差为 2,而显然 3^r-3^s 是 3 的倍数,决不能是 2,矛盾.故所说的 n 不存在.

例 9 (IMO-11-1)证明存在无限多个自然数 a 有如下性质:对任何自然数 n, $z=n^4+a$ 都不是素数.

证明 题目的意思是:给定 n,一定可找到无限个 a,使 n^4+a 非素数.由素数的定义可知,这等价于说,有无限个 a,使 n^4+a 能分解成两个真约数的积(真约数指不是 1,也不是这个数自身的其他约数).于是问题转化为一个因式分解的问题.

注意到对 $m>1$, m、$n\in\mathbf{N}$,

$$n^4+4m^4=(n^2+2m^2)^2-4m^2n^2=(n^2+2mn+2m^2)\cdot(n^2-2mn+2m^2).$$

我们已经得到了 n^4+4m^4 的分解式.再有

$$n^2+2mn+2m^2>n^2-2mn+2m^2=(n-m)^2+m^2\geqslant m^2>1.$$

所以,这两个约数都是 n^4+4m^4 的真约数.从而,取 $a=4m^4$ 时($m=2, 3, \cdots$), n^4+a 都不是素数.

用和上面类似的想法,可以解决下面的两个问题.

问题 1 如果 n 是正整数, $n>1$, 2^n-1 是素数,则 n 必是素数.

问题 2 设 $N_m=8^m+9m^2$, m 为正奇数.证明:存在无限个奇数 m,使 N_m 不是素数.

作为本文的结束,我们介绍一个重要的概念——同余.

5 同余

同余是初等数论的一个基本概念,它的引入使数论中许多问题得到简化.

给定一个正整数 m,如果 a、b 是整数,$a-b$ 被 m 整除(在 a、b 为正整数时这等于说 a、b 被 m 除后的余数相同),此时,我们记 $a\equiv b(\bmod m)$ (读做 a 和 b 关于模 m 同余).

下面是同余的基本性质(与等式的性质相同或相近):

(1) $a\equiv a(\bmod m)$;

(2) 如果 $a\equiv b(\bmod m)$,则 $b\equiv a(\bmod m)$;

(3) 如果 $a\equiv b(\bmod m)$, $b\equiv c(\bmod m)$,则 $a\equiv c(\bmod m)$;

(4) 如果 $a\equiv b(\bmod m)$, $c\equiv d(\bmod m)$,则 $ax+cy\equiv bx+dy(\bmod m)$ (x、y 是任意整数)及 $ac\equiv bd(\bmod m)$.

特别,我们推出,对 $n\in\mathbf{N}$,

$$a^n\equiv b^n(\bmod m).$$

以上性质,都不难由同余的定义得出.

例 10 (IMO-6-1) (Ⅰ) 求所有能使 2^n-1 被 7 整除的正整数 n.

(Ⅱ) 证明:没有正整数 n,使得 2^n+1 被 7 整除.

解 (Ⅰ) 首先注意,对 $k\geqslant 0$, k 是整数, $2^{3k}-1$ 能被 7 整除. 证明如下:

因 $2^3\equiv 1(\bmod 7)$, 由同余性质立得

$$2^{3k}\equiv 1(\bmod 7),\ 7\mid 2^{3k}-1$$

(或者,利用 $2^{3k}-1=(2^3)^k-1=(2^3-1)\cdot(2^{3k-3}+\cdots+1)$ 也可推得)

于是, $2^{3k+1}-1=2(2^{3k}-1)+1$, $2^{3k+2}-1=4(2^{3k}-1)+3$ 都不能被 7 整除.

故当且仅当 $3|n$ 时,有 $7\mid 2^n-1$.

(Ⅱ) 由于 $2^{3k}\equiv 1(\bmod 7)$,有 $2^{3k}+1\equiv 2(\bmod 7)$, $2^{3k+1}+1\equiv 3(\bmod 7)$, $2^{3k+2}+1\equiv 5(\bmod 7)$. 故对任何 $n\in\mathbf{N}$,总有 $7\nmid 2^n+1$.

评注 (1) 上面的解答中,关键在于将自然数按 mod3 分类(即将任一自然数按被 3 除后的余数分类). 这是初等数论中一个基本而又十分灵活的手法. 难点之一在于究竟应怎样分,这当然要根据具体情况而定. 本题我们是从 $2^{3k}\equiv$

$1 \pmod 7$ 而联想到了上面的分法. 大家知道,为了计算 i^n ($i=\sqrt{-1}$, $n\in\mathbf{N}$),而想到 $i^4=1$, 这就暗示我们应将自然数按 mod4 分类.

(2) 本题也可以用二项式定理来解:分别考虑 $n=3k$, $3k+1$, $3k+2$($k\geqslant 0$). 注意到 $2^{3k}+1=(1+7)^k+1$, 然后用二项式定理展开.

例 11 (IMO-17-4)设 A 是十进制数 4444^{4444} 的各位数码的和, B 是 A 的各位数码的和,求 B 的各位数码的和.(所有讨论的数都是在十进制数系中)

解 用 $S(n)$表示正整数 n 在十进制中各位数码的和. 本题的关键是反复利用同余式

$$S(n)\equiv n \pmod 9. \tag{1}$$

(1) 可以这样证明:设

$$n=a_k\times 10^k+a_{k-1}10^{k-1}+\cdots+a_1\times 10+a_0.$$

这里, $0\leqslant a_i\leqslant 9$, $a_k\neq 0$ ($0\leqslant i\leqslant k$), 则

$$S(n)=a_0+\cdots+a_k.$$

$$n-S(n)=a_k(10^k-1)+a_{k-1}(10^{k-1}-1)+\cdots+a_1(10-1).$$

因 $10\equiv 1 \pmod 9$,故 $10^i\equiv 1^i\equiv 1 \pmod 9$, $1\leqslant i\leqslant k$. 从而上式右端和式中每一项都被 9 整除,故 $n-S(n)$ 被 9 整除,即(1)成立.

下面解本题. 易见 4444^{4444} 的位数不超过 $4\times 4444=17\,776$. 因而其各位数码的和$\leqslant 177\,760$,即 A 至多为首位数字是 1 的 6 位数,从而 $B=S(A)\leqslant 1+5\times 9=46$. 而 $S(B)\leqslant 4+9=13$.

另一方面,

$$S(B)\equiv B=S(A)\equiv A=S(4444^{4444})\equiv 4444^{4444}\equiv 7^{4444}\equiv 7^4$$

(因为 $7^3\equiv 1 \pmod 9$) $\equiv 7 \pmod 9$. 结合上面得到的 $1\leqslant S(B)\leqslant 13$, 可见 $S(B)=7$.

例 12 (IMO-20-1)数 1978^n 与 1978^m 的最后三位数相等,试求出正整数 n 和 m,使得 $n+m$ 取最小值,这里 $n>m\geqslant 1$.

解 由题设,我们有

$$\begin{aligned}1978^n-1978^m&=1978^m(1978^{n-m}-1)\\&=2^m\times 989^m\times(1978^{m}-1)\equiv 0 \pmod{10^3}.\end{aligned} \tag{1}$$

由于 $$10^3=2^3\times5^3,$$

而 989^m，1978^m-1 都是奇数，所以 $m\geqslant3$. 又 5 与 2^m，989^m 互质，所以

$$1978^{n-m}-1\equiv0(\bmod 5^3).$$

即 $$1978^{n-m}\equiv1(\bmod 125).\tag{2}$$

从而，$$3^{n-m}\equiv1978^{n-m}\equiv1(\bmod 25).\tag{3}$$

而用例 10 中的方法(现在按 mod4 来分类)，可从 $3^{n-m}\equiv1(\bmod 5)$ 得 (4)

$$n-m=4k.$$

从而，$$3^{4k}=81^k\equiv1(\bmod 25).$$

即 $$(80+1)^k=1+k\times80+C_k^2\times80^2\equiv1(\bmod 25).$$

于是(由于 $25\mid80^2$)易得 $5\mid k$. 再由(2)得

$$1\equiv1978^{4k}=(3+1975)^{4k}\equiv3^{4k}+4k\times3^{k-1}\times1975\equiv3^{4k}(\bmod 125),$$

$$1\equiv3^{4k}=81^k\equiv1+k\times80+\frac{k(k-1)}{2}\times80^2\equiv1+80k(\bmod 125)$$

(请注意 $5\mid k$). 于是 $125\mid80k$，$25\mid k$. 这样 k 至少为 25，$n-m$ 至少为 $4\times25=100$.

$$n+m=(n-m)+2m\geqslant100+2\times3=106.$$

故 $n-m$ 的最小值为 106(上面的推导可以表明这时(1)成立).

评注 在本例中考虑了几个不同的模，特别是先后以 5，$5^2=25$，$5^3=125$ 为模. 这种 5—adic(5 进制)的方法非常重要.

IMO 中的数论问题（下）

不定方程是数学竞赛中经常出现的问题.所谓不定方程,简言之,就是未知数的个数多于方程的个数,但它们的解受某种限制(如正整数解,有理数解等等).我们将通过具体的例子来说明处理这些问题的一些常用方法.读者将会看到,这些方法与初等数论的基础知识有关,而且非常灵活和多样化.

例 1　(IMO-27-1)设正整数 d 不等于 2、5、13.证明在集合 $\{2, 5, 13, d\}$ 中可找到两个不同的元素 a、b,使得 $ab-1$ 不是完全平方.

证明　用反证法.假设结论不对,则

$$2d-1=x^2, \tag{1}$$

$$5d-1=y^2, \tag{2}$$

$$13d-1=z^2, \tag{3}$$

x、y、z 都是整数.

显然,由(1)知 x 是奇数,从而有

$$2d=x^2+1\equiv 2 \pmod 4,$$

(我们使用了同余式,这在上一篇文章中已作过介绍)即 d 是奇数.从(2)、(3)可知 y、z 都是偶数.

设 $y=2y_1$, $z=2z_1$, (3)－(2), 有

$$z^2-y^2=8d,$$

即

$$(z_1-y_1)(z_1+y_1)=2d. \tag{4}$$

注意到 z_1-y_1 和 z_1+y_1 同为奇数或同为偶数(因它们的和是偶数),但(4)式表明 $(z_1-y_1)(z_1+y_1)$ 是偶数,故 z_1-y_1 和 z_1+y_1 都是偶数.再由(4)式知 d 是偶数,和上面得到的 d 是奇数矛盾.

例 2　证明: $x^2+y^2+z^2=7k^{2n}$ 无正整数解,这里 k 是已知的奇数.

证明 这里有四个未知量 n, x, y, z. 我们仍用反证法.

首先,注意到一个简单却重要的事实:

$$x^2 \equiv 0, 1, 4 \pmod 8.$$

因为,如果 x 是奇数,设 $x=2l+1$, $x^2=4l(l+1)+1$,但 $l(l+1)$ 是偶数,故 $8 \mid 4l(l+1)$ 即 $x^2 \equiv 1 \pmod 8$; 如果 x 是偶数,设 $x=2l$,如 l 也是偶数,则显然 $x^2 \equiv 0 \pmod 8$; 如 l 是奇数,则按上面已得到的,有 $x^2=4l^2 \equiv 4 \pmod 8$,故总有 $x^2 \equiv 0, 1, 4 \pmod 8$.

这样由于 k 是奇数,故 $k^2 \equiv 1 \pmod 8$. 从而(我们用同余式的性质)

$$k^{2n}=(k^2)^n \equiv 1 \pmod 8,$$

即
$$7k^{2n} \equiv 7 \pmod 8,$$

这就是说右边$\equiv 7 \pmod 8$.

那么左边怎样呢? (当然也考虑模 8,即 mod8)因 x^2, y^2, $z^2 \equiv 0, 1, 4 \pmod 8$,列举一下可能的组合可知:

$$x^2+y^2+z^2 \text{ 只能} \equiv 0, 1, 2, 3, 4, 5, 6 \pmod 8,$$

和右边$\equiv 7 \pmod 8$相矛盾.

评注 上面两例本质上是利用同余式,这是处理不定方程的基本方法. 但这方法也非常灵活,主要的难点在于确定所取的模(上面我们分别取模 4 和模 8),这都应当根据问题的特点来确定.

问题 1 设 $n \equiv 4 \pmod 9$,证明 $x^3+y^3+z^3=n$ 无正整数解.

不定方程既然是方程,解方程的某些技巧当然能使用. 一些不定方程通过恒等变形(变量代换,分解,等等)往往就易于处理,见下面的例子.

例 3 证明 $x^2+y^2=z+z^5$ 有无穷组正整数解,满足条件 $(x, y)=1$.

证明 我们的方法是直接找出所要求的解(注意不必是全部解). 有一个关键的恒等式读者应当熟记,它在许多地方都有用:

$$(a^2+b^2)(c^2+d^2)=(ad+bc)^2+(ac-bd)^2. \tag{1}$$

我们把原方程变形为

$$x^2+y^2=z(1^2+(z^2)^2)$$

就可看出如何应用(1)式了.

在(1)中取 $z=a^2+b^2$, $c=z^2$, $d=1$, 则

$$x=(a^2+b^2)^2+ab,$$
$$y=a(a^2+b^2)^2-b.$$

这样得到的 x、y、z(a、b 都是正整数)当然是方程的解. 为了使 $(x, y)=1$,我们取 $a=1$, 则

$$x=(b^2+1)^2+b,\ y=(b^2+1)^2-b.$$

现在 x、y 都是奇数(读者自证).

下面来证 $(x, y)=1$.

假设 $(x, y)>1$, 则有素数 p,使

$$p \mid (x, y).$$

故 $p \mid x$, $p \mid y$,从而 $p \mid x-y$,即 $p \mid 2b$. 但 $p \neq 2$(因 x、y 都是奇数),所以 $p \mid b$. 再结合 $p \mid x$,知 $p \mid 1$,矛盾,从而$(x, y)=1$.

最后,取 $x=(b^2+1)^2+b$, $y=(b^2+1)^2-b$, $z=b^2+1$, 则给出无穷组满足条件的正整数解(这里 b 是任意正整数).

问题 2 求出 $x^3+y^3+z^3=3xyz$ 的全部整数解.

例 4 (IMO-25-6)设 a, b, c, d 都是奇数, $0<a<b<c<d$,且 $ad=bc$. 证明:如果

$$a+d=2^k,\ b+c=2^m,$$

k、m 为整数,则 $a=1$.

证明 由 $a+d=2^k$, $b+c=2^m$ 得

$$d=2^k-a,\ c=2^m-b.$$

代入 $ad=bc$ 中,有

$$a(2^k-a)=b(2^m-b), \tag{1}$$

即

$$b\cdot 2^m-a\cdot 2^k=b^2-a^2.$$

又,我们有

$$\begin{aligned}(a+d)^2&=4ad+(d-a)^2\\&=4bc+(d-a)^2>4bc+(c-b)^2\\&=(b+c)^2.\text{(注意 } d-a>c-b>0)\end{aligned}$$

所以,$2^k>2^m$,即 $k>m$.(这一点很重要)

由(1)可知

$$2^m(b-a\cdot 2^{k-m})=(b+a)(b-a). \tag{2}$$

已知 a、b 都是奇数,所以 $a+b$, $b-a$ 都是偶数. 又$(b+a)-(b-a)=2a$ 是奇数的 2 倍,故 $b+a$, $b-a$ 中必有一个不是 4 的倍数(否则其差应是 4 的倍数). 注意到这一点是解题的关键.

这样,从(2)知(请读者思考理由)

$$\begin{cases}b+a=2^{m-1}e,\\b-a=2f\end{cases}\text{或}\begin{cases}b-a=2^{m-1}e,\\b+a=2f.\end{cases}$$

其中 e、f 为正整数,且 $ef=b-a\cdot 2^{k-m}$ (是奇数).

由于 $k>m$, 故

$$ef\leqslant b-2a<b-a\leqslant 2f$$

(如果 $b-a=2f$ 则取等号,如 $b+a=2f$ 则应为严格不等式).

从而有 $$e=1,\ f=b-a\cdot 2^{k-m}.$$

若 $$\begin{cases}b+a=2^{m-1},\\b-a=2f=2(b-a\cdot 2^{k-m}),\end{cases}$$

由第二个式子得

$$b+a=2^{k+1-m}a,$$

故 $$2^{m-1}=2^{k+1-m}a.$$

所以奇数 $$a=1.$$

同理,若

$$\begin{cases} b-a=2^{m-1}e, \\ b+a=2f, \end{cases}$$

也必有

$$a=1.$$

评注　上面的解答中有两个技巧：第一是利用"放缩法"证明了 $k>m$；第二是把(2)分解，读者应能看出，数论知识在分解中所起的重要作用.

勾股数也经常在数学竞赛中出现. 如果正整数 x、y、z 满足

$$x^2+y^2=z^2, \tag{1}$$

则称(x, y, z)是一组勾股数，即 x、y、z 构成边长为整数的直角三角形.

显然，可设 $(x, y)=1$（否则，如 $d=(x, y)$，则 $d^2 \mid z^2$，所以 $d \mid z$，两边可约去 d）. 此外，x、y 不能同为奇数，否则(1)两边取模 4，有 $z^2 \equiv 2 \pmod 4$，这不可能，故 x、y 一奇一偶.

下面是一个重要定理.

定理　不定方程(1)满足

$$(x, y)=1, x>0, y>0, z>0,$$

$2|x$ 的全部解可表示成

$$x=2ab, y=a^2-b^2, z=a^2+b^2. \tag{2}$$

其中，$a>b>0$，a，b 一奇一偶，且$(a, b)=1$.

评注　如果不要求 $(x, y)=1$，我们设$(x, y)=d$，则方程(1)的所有正整数解可表示成

$$x=2abd, y=d(a^2-b^2), z=d(a^2+b^2).$$

其中 a、b 的限制和定理中所说的一致.

下面举几例说明上面定理的应用.

例 5　(IMO－17－5)证明：在单位圆上可放置 1975 个点，使任两点之间的直线距离都是有理数.

证明　我们的方法是利用勾股数把所说的点构造出来(即直接找出来).

取

$$\theta_n=\arctan\frac{n^2-1}{2n}, (1\leqslant n\leqslant 1975),$$

则 $$\sin\theta_n=\frac{n^2-1}{n^2+1},\ \cos\theta_n=\frac{2n}{n^2+1}$$

都是有理数(注意关键之处是利用 n^2-1, $2n$, n^2+1 是一组勾股数).

又 $2\theta_n$ 互不相同,我们在单位圆上找出相应于幅角 $2\theta_1$, …, $2\theta_{1975}$ 的点 P_1, …, P_{1975}.

易知, $|P_iP_j|=2|\sin\theta_i\cos\theta_j-\cos\theta_i\sin\theta_j|$ $(1\leqslant i<j\leqslant 1975)$. 显然, $|P_iP_j|$ 是有理数. 从而,上面找出的 1975 个点符合要求.

问题 3 证明:单位圆 $x^2+y^2=1$ 上有无穷个有理点(即横、纵坐标都是有理数的点).

例 6 (IMO-27 预选题)设 A、B、C 是一个圆形水池边上的三点, B 在 C 的正西方,且 ABC 构成边长为 86 米的正三角形. 一游泳者从 A 径直游向 B. 在游了 x 米后,抵达 E 点,然后他转向,往正西方向游去,游过 y 米后,抵达岸边 D 点. 如果 x、y 都是整数,试求 y.

解 如图 1,由于△AEF 是正三角形,故

$$AF=AE=x.$$

由对称性, $FG=DE=y$. 又因

$$AE\cdot EB=DE\cdot EG,$$

从而有

$$x(86-x)=y(x+y). \qquad (1)$$

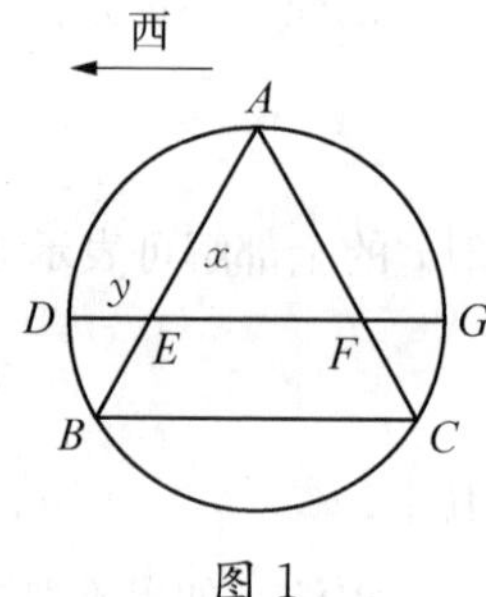

图 1

注意到,如果 x 是奇数,则 $x(86-z)$ 也是奇数,而 $y(x+y)$ 此时总是偶数,导出矛盾(因为,如 y 是偶数,显然 $y(x+y)$ 为偶;如 y 是奇数,则 $x+y$ 为偶数, $y(x+y)$ 仍为偶数).

因此, x 一定是偶数. 由(1)知 $y(x+y)$ 为偶数,从而 y 必是偶数.

上面的方法实际上是对(1)式取模 2,确定了 x、y 的奇偶性.

下面的步骤是关键的. 把(1)配方成

$$\left(x+\frac{y}{2}-43\right)^2+y^2=\left(43-\frac{y}{2}\right)^2.$$

这表明 $\left|x+\frac{y}{2}-43\right|$，$y$，$43-\frac{y}{2}$ 组成勾股数.

设 d 为 $\left|x+\frac{y}{2}-43\right|$ 与 y 的最大公约数，则按上述定理后面的评注知：

有正整数 a、b，$a>b$，使得

$$y=2abd(\text{因 } y \text{ 是偶数}),$$

$$43-\frac{y}{2}=(a^2+b^2)d.$$

由这两式可得

$$(a^2+b^2+ab)d=43.\ (a>b>0)$$

因 43 是素数，故 $d=1$.

最后，我们用“放缩法”来解方程

$$a^2+ab+b^2=43.$$

由 $a>b>0$，可得 $3b^2<43$，$b^2<\frac{43}{3}$，$y^2\leqslant 14$，故 $b\leqslant 3$. 易见，只有 $b=1$ 及 $a=6$ 为正整数解答.

于是，立知 $y=12$. 进而从 $\left|x+\frac{y}{2}-43\right|=a^2-b^2$ 求出 $x=2$ 或者 $x=72$.

评注　不定方程(1)也可以不用勾股数定理，而直接求解.

例 7　(IMO - 24 - 3)设 a、b、c 为三个正整数，其中任意两个都没有大于 1 的公因数，证明：

$$2abc-bc-ca-ab$$

是不能表成 $xbc+yca+zab$ 的最大整数(x、y、z 都是非负整数).

证明　题目的意思是要我们做两件事：

首先证明 $2abc-bc-ac-ab$ 不能表成所说的形式；然后再证明大于 $2abc-bc-ac-ab$ 的整数都可以表成所说的形式.

我们先证第一件事.

用反证法. 如果有非负整数 x、y、z 使

$$2abc-bc-ac-ab=xbc+yca+2ab,$$

则 $$2abc=(x+1)bc+(y+1)ac+(z+1)ab. \tag{1}$$

(1)式左边是 a 的倍数,所以右边也是 a 的倍数,即应有 $a \mid (x+1)bc$. 但据假设 $(a, b)=(a, c)=1$,从而 $(a, bc)=1$,即必有 $a \mid x+1$.

另一方面,由(1)式,注意 $x+1$, $y+1$, $z+1$ 都是正数,故 $2abc>(x+1)bc$,即 $2a>x+1$.

这样,只能有 $$x+1=a.$$

同理 $$y+1=b,\ z+1=c.$$

这样一来,(1)式就成为 $2abc=3abc$,矛盾. 故 $2abc-bc-ca-ab$ 不能表成所说的形式.

下面来证第二件事.

我们要引用初等数论中的一条定理:

如果正整数 A、B、C 满足 $(A, B, C)=1$, 则存在整数 α、β、γ,使得

$$\alpha A+\beta B+\gamma C=1.$$

这一定理是我们论证的出发点.

因 ab、bc、ca 的最大公约数为 1,故有整数 x_1、y_1、z_1,使得

$$abx_1+bcy_1+caz_1=1.$$

从而对任一整数 u,有

$$abux_1+bcuy_1+cauz_1=u.$$

这就是说,一定可把 u 表示成 $abx+bcy+caz$ 的形式,其中 x、y、z 是整数(自然,不一定都非负).

另一方面,对任意整数 m、n,有恒等式

$$xbc+acy+zbc=(x+na)bc+(y+mb)ca+(z-nc-mc)ab.$$

这样,我们可设 $a>x\geqslant 0$, $b>y\geqslant 0$(否则,总可以选择适当的 m、n 使

$$a>x+na\geqslant 0,\ b>y+mb\geqslant 0.$$

令 $$x'=x+na,\ y'=y+mb,\ z'=z-nc-mc,$$

即得 $$u=x'bc+y'ca+z'ab.$$

而
$$a>x'\geqslant 0,\ b>y'\geqslant 0,$$
这时,如果 $z<0$, 则
$$u=xbc+yca+zab\leqslant(a-1)bc+(b-1)ca-ab=2abc-bc-ca-ab.$$
故只要
$$u>2abc-bc-ca-ab,$$
必然有 $z\geqslant 0$.

这样,当整数 $>2abc-bc-ca-ab$ 时,一定可写成 $xbc+yca+zab$ 的形式,其中 x、y、z 都是非负整数.

用和上面类似的方法读者可证明下面的定理:

定理　设 a、b 是正整数,且 $(a,b)=1$,则 $ab-a-b$ 是不能表示成 $ax+by$ 形式的最大整数(x、y 是非负整数).

最后,我们举几个应用上面定理的例子来结束本文.

例 8　证明:仅用 2 分和 5 分的币值可兑换大于 3 分的任意币值.

证明　本题可以用归纳法来证.但应用上述定理将更加简明.

在定理中取 $a=2$, $b=5$,则 $(a,b)=1$. 据上述定理可知:大于 $2\times5-2-5=3$ 的正整数可表示成
$$2x+5y,$$
x、y 是非负整数.

因此,只要用 x 个 2 分及 y 个 5 分就可以兑换大于 3 分的任意币值.

例 9　证明大于 36 的任意整数可表示成两个有平方因子数的和.

证明　所谓有平方因子的正整数,是指它可被一个大于 1 的平方数除尽. 当然,完全平方是有平方因子数,但反之不真. 所谓无平方因子数,就是它不能被大于 1 的平方数除尽.

如 4、12、27 是有平方因子数,而 2、6、31 是无平方因子数.

现在回到本题,我们的方法是直接把所说的表示法写出来.

注意 4、9 是互素的最小一对有平方因子数. 按照定理,大于 $4\times9-4-9$ 的整数可写成
$$4x+9y,$$

x、y 是非负整数.

但我们有一点麻烦,就是 x 或 y 可以是 0,所以这样的表达式并不完全适合要求. 其实,这点困难只要把定理稍稍变形即可克服.

根据定理知,大于 4×9 的整数可表成

$$4x+9y+4+9=4(x+1)+9(y+1),$$

x、y 是非负整数.

此时, $x+1$, $y+1$ 都是正整数,而 $4(x+1)$, $9(y+1)$ 都是有平方因子数(分别被 2^2、3^2 除尽). 这样就证明了结论.

评注 上面的变形实际上是一种"平移". 这个技巧在下例中也是有效的.

例 10 证明:大于 11 的整数可表示成两个合数的和.

证明 在定理中取 $a=2$, $b=3$,显然, $(a, b)=1$. 由定理,大于 $2\times 3-2-3$ 的整数可表成

$$2x+3y,$$

x、y 是非负整数.

麻烦在于:第一, x 和 y 可能为 0;第二,即使 x、y 都不为 0,但它们可以为 1,从而 $2x$、$3y$ 不一定是合数.

我们使用平移技巧,由定理可知,凡是大于 $2\times 3+2+3$ 的整数,可表示成

$$2x+3y+2\times 2+2\times 3=2(x+2)+3(y+3),$$

x、y 为非负整数.

即大于 11 的整数可表成

$$2x'+3y',$$

这里 $$x'\geqslant 2,\ y'\geqslant 2.$$

这时 $2x'$和 $3y'$都是合数.

一道美国竞赛题的推广

第 17 届(1988 年)美国数学奥林匹克试题(见《中等数学》1988 年第 6 期)，与以往各届相比，有一个显著的特点：没有一道难题. 稍难一点的只有第 5 题，原题如下："多项式 $(1-z)^{b_1}(1-z^2)^{b_2}\cdots(1-z^{32})^{b_{32}}$，其中 b_1 为正整数，具有如下性质：将它乘开后，如果忽略 z 的高于 32 次的那些项，留下的是 $1-2z$. 试求 b_{32}.(答案可以表为 2 的两个方幂之差)"

这道题原来的解法需要一点特殊的技巧与略多的计算(见《中等数学》1988 年第 6 期)，所求得的结果是表达式中的一个特殊的幂指数 b_{32}. 有没有办法求出表达式中所有的幂指数 b_j $(j=1, 2, \cdots, 32)$ 呢？原来的解法显然无法做到这一点.

本文的目的就是介绍如何求出 b_j. 为此，我们要利用一下对数函数的展开式(log 表示自然对数)

$$\log(1-x)=-\left(x+\frac{x^2}{2}+\frac{x^3}{3}+\cdots\right)=-\sum_{n=1}^{\infty}\frac{x^n}{n}. \tag{1}$$

在(1)的两边求导数就得到中学里常见的无穷级数

$$\frac{1}{1-x}=1+x+x^2+\cdots.$$

已知条件可以写成

$$(1-z)^{b_1}(1-z^2)^{b_2}\cdots(1-z^{32})^{b_{32}}\equiv 1-2z(\bmod z^{33}), \tag{2}$$

这里 $\bmod z^{33}$ 即表示忽略次数高于 32 的项.

在(2)的两边取对数，并利用(1)展开，左边成为

$$\sum_{j=1}^{32}b_j\log(1-z^j)=-\sum_{j=1}^{32}b_j\sum_{n=1}^{\infty}\frac{z^{nj}}{n}. \tag{3}$$

而右边(忽略掉一些高于 32 次的项)成为

$$\log(1-2z) \equiv -\sum_{k=1}^{32} \frac{2^k}{k} z^k \pmod{z^{33}}. \tag{4}$$

将(3)中高于 32 次的项略去,并将 nj 改记为 k,得到

$$-\sum_{k=1}^{32}\left(\sum_{nj=k} \frac{b_j}{n}\right) z^k = -\sum_{k=1}^{32}\left(\sum_{j|k} \frac{jb_j}{k}\right) z^k, \tag{3'}$$

这里 $j|k$ 表示 j 是 k 的因数,$\sum_{j|k}$ 表示对 k 的所有因数 j 求和.

比较(3′)与(4)中 z^k 的系数得

$$2^k = \sum_{j|k} jb_j, (k=1, 2, \cdots, 32). \tag{5}$$

由方程组(5)就可定出 b_j. 如果熟悉莫比乌斯(Möbius)函数 $\mu(n)$及莫比乌斯反转公式(请参见任何一本数论书),那么由(5)立即可得

$$jb_j = \sum_{d|j} \mu(d) \cdot 2^{\frac{j}{d}},$$

$$b_j = \frac{1}{j} \sum_{d|j} \mu(d) 2^{\frac{j}{d}}, (j=1, 2, \cdots, 32). \tag{6}$$

这里莫比乌斯函数的定义是

$$\mu(n) = \begin{cases} 0, & \text{若 } n \text{ 被一质数的平方整除;} \\ +1, & \text{若 } n=1; \\ (-1)^m, & \text{若 } n \text{ 是 } m \text{ 个不同质数的积.} \end{cases}$$

特别地,从(6)可以得到

$$b_{32} = \frac{1}{32}(\mu(1) \cdot 2^{32} + \mu(2) \cdot 2^{16}) = \frac{1}{32}(2^{32} - 2^{16}) = 2^{27} - 2^{11},$$

$$b_{31} = \frac{1}{31}(\mu(1) \cdot 2^{31} + \mu(31) \cdot 2) = \frac{1}{31}(2^{31} - 2),$$

等等.

所得的结果还可以推广,例如在忽略掉高于 32 次的项后留下的是 $1-gz$,那么结果就是

$$b_j = \frac{1}{j} \sum_{d|j} \mu(d) g^{\frac{j}{d}}. \tag{6'}$$

此外, 32 也可以换为任意一个自然数.

第 30 届 IMO 预选题解答

1　说明

各国向第 30 届 IMO 共提供了 107 道试题(我国提交的试题由于某种原因,组织委员会未能收到),选题委员会从中初选出 32 道题供主试委员会决定.这些题可以列成一个简表:

编号	提供国	难度	知识范畴
1*△	澳大利亚	B	平面几何、不等式
2	澳大利亚	C_-	数论
3	澳大利亚	B	数论
4	保加利亚	B_-	多项式、数论
5	哥伦比亚	B	多项式(不等式)
6	捷克斯洛伐克	B	平面几何、不等式
7	芬　　兰	B	平面几何
8*	法　　国	A	组合几何、不变量
9	法　　国	A	数论、多项式
10	希　　腊	C_-	复数、函数方程
11	匈 牙 利	A	数论、组合
12*	匈 牙 利	B_-	组合
13*△	冰　　岛	C_-	平面几何
14	印　　度	A_+	平面几何
15	爱 尔 兰	C_-	数论
16	以 色 列	B_+	数列、不等式
17	蒙　　古	C	组合几何

(续表)

编号	提供国	难度	知识范畴
18	蒙　古	C	平面几何
19*	蒙　古	C_+	组合
20*△	荷　兰	A	组合几何、不等式
21	荷　兰	B	立体几何
22△	菲律宾	C_{--}	组合
23△	波　兰	A_{++}	组合
24*	波　兰	B_+	立体几何
25*	韩　国	A	数论
26	韩　国	C	不等式
27	罗马尼亚	B	数论
28	罗马尼亚	B_+	平面几何、不等式
29	罗马尼亚	A	组合
30*△	瑞　典	B_-	数论
31	瑞　典	A	数论、不等式
32	美　国	C	平面几何

其中 * 表示选题委员会看中的题,△表示主试委员会最后选定的 6 道试题. 在正式试卷上,这 6 道题的顺序是:1.(22 题),2.(1 题),3.(20 题),4.(13 题),5.(30 题),6.(23 题).

A 表示最难的题,B 表示中等,C 表示容易. 但这只是选题委员们的主观看法,未必准确. 例如第 22 题,需要构造出合乎要求的子集(命题委员会把其中的 17 改为 117 以增加难度),虽不算难题,但也未见得十分容易. 第 23 题,并非过难,很多学生的解答都远比原先提供的解答简单.

2　预选题

下面是 32 道问题(其中被选作正式试题的 6 道题不再列入).

2. 地毯商人阿里巴巴有一块长方形的地毯,其大小未知. 不幸他的量尺坏了,而他又没有其他测量工具.但他发现将这地毯平铺在他两间店房的每一间中,地毯的每一个角恰好与房间的不同的墙相遇.如果两间房间的尺寸为38呎×55呎与50呎×55呎,求地毯的尺寸.

3. 地毯商人阿里巴巴有一块长方形的地毯,尺寸未知.不幸他的量尺坏了,又没有其他测量工具.但他发现如果将地毯平铺在他两间店房的任一间,地毯的每一个角恰好与房间的不同的墙相遇.他知道地毯长的呎数为整数,两间房子的一边有相同的长(不知多长),另一边分别为38呎与55呎,求地毯的尺寸.

4. 证明:对每一整数 $n>1$, 方程

$$\frac{x^n}{n!}+\frac{x^{n-1}}{(n-1)!}+\cdots+\frac{x^2}{2!}+\frac{x}{1!}+1=0$$

无有理根.

5. 考虑多项式

$$p(x)=x^n+nx^{n-1}+a_2x^{n-2}+\cdots+a_n.$$

若 $|r_1|^{16}+|r_2|^{16}+\cdots+|r_n|^{16}=n$, 这里 r_1 为 $p(x)$的全部根,求这些根(译注:原题$|r_1|$误作 r_1,与解答不符).

6. $\triangle ABC$ 的外接圆 K 的半径为R,内角平分线分别交圆 K 于 A'、B'、C'.证明不等式:

$$16Q^3\geqslant 27R^4P.$$

其中 Q、P 分别为$\triangle A'B'C'$与$\triangle ABC$ 的面积.

7. 证明在正 n 边形 E 中,任两点可以用两条在 E 内的圆弧连结起来(译注:意指每条圆弧以已给两点为端点),这两条弧的交角至少为 $\left(1-\frac{2}{n}\right)\pi$.

8. R 为一长方形,它是若干个长方形 R_i, $1\leqslant i\leqslant n$ 的并集,满足

(1) R_i 的边与R 的边平行,

(2) R_i 互不重叠,

(3) 每个 R_i 都至少有一条边长为整数.

证明:R 至少有一条边长为整数.

9. n 为非负整数,将 $(1+4\sqrt[3]{2}-4\sqrt[3]{4})^n$ 写成$(1+4\sqrt[3]{2}-4\sqrt[3]{4})^n=a_n+b_n\sqrt[3]{2}+$

$c_n\sqrt[3]{4}$，其中 a_n、b_n、c_n 为整数. 证明：若 $c_n=0$，则 $n=0$.

10. $g: C\to C$, $\omega\in C$, $a\in C$, $\omega^3=1$, $\omega\neq 1$. 证明有且仅有一个函数 f: $C\to C$，满足

$$f(z)+f(\omega z+a)=g(z),\ z\in C,$$

求出 f.

11. 由 $\sum\limits_{d\mid n}a_d=2^n$ 定义 a_n，证明 $n\mid a_n$.

12. 圆形跑道上 n 个不同点处有 n 辆汽车正准备出发. 每辆车 1 小时跑一圈. 听到信号后，它们各选一个方向立即出发. 如果两辆汽车相遇，则同时改变方向以原速前进. 证明必有一时刻，每一辆车在原出发点.

14. 双心四边形是指既有内切圆又有外接圆的四边形，证明对这样的四边形，两个心与对角线交点共线.

15. 设 a, b, c, d, m, n 为正整数，$a^2+b^2+c^2+d^2=1989$, $a+b+c+d=m^2$，并且 a, b, c, d 中最大的为 n^2，确定(并证明) m、n 的值.

16. 实数集 $\{a_0, a_1, \cdots, a_n\}$ 满足以下条件：

(1) $a_0=a_n=0$,

(2) 对 $1\leqslant k\leqslant n-1$,

$$a_k=c+\sum_{1=k}^{n-1}a_{1-k}(a_i+a_{i+1}).$$

证明：$c\leqslant\dfrac{1}{4n}$.

17. 平面上已给 7 个点，用一些线段连结它们，使得

(1) 每三点中至少有两点相连，

(2) 线段的条数最少.

问有多少条线段？给出一个这样的图.

18. 平面上有凸多边形 $A_1A_2\cdots A_n$，面积为 S，又有一点 M, M 绕 A_1 旋转 α 角后得点 M_i $(i=1, 2, \cdots, n)$. 求多边形 $M_1M_2\cdots M_n$ 的面积.

19. 一个 $m\times n$ 的长方形表中填写了自然数. 可以将相邻方格中的两个数同时加上一个整数 k，使所得的数为非负整数(有一条公共边的两个方格称为相邻的). 试确定充分必要条件，使可以经过有限多次这种运算后，表中各数为 0.

21. 证明平面与一个正四面体的交可以是一个钝角三角形. 并且在任一个这样的钝角三角形中,钝角总小于 $120°$.

24. 点 A_1, …, A_5 在半径为 1 的球上, $\min\limits_{1\leqslant i,\ j\leqslant 5} A_iA_j$ 的最大值是多少? 确定所有取得最大值的情况.

25. 设 a、b 为整数,不是完全平方. 证明:若 $x^2-ay^2-bz^2+abw^2=0$ 有非平凡的整数解(即不全为 0 的整数解),则 $x^2-ay^2-bz^2=0$ 也有非平凡的整数解.

26. n 为正整数, a、b 为给定实数, x_0, x_1, …, x_n 为实变数,

$$\sum_{i=0}^{n} x_i=a,\ \sum_{i=0}^{n} x_i^2=b.$$

确定 x_0 的变化范围.

27. 设 m 为大于 2 的正奇数,求使 2^{1989} 整除 m^n-1 的最小的自然数 n.

28. 考虑同一平面的点 O, A_1, A_2, A_3, A_4,满足 $S_{\triangle OA_iA_j}\geqslant 1(i,\ j=1,\ 2,\ 3,\ 4;\ i\neq j)$. 证明至少有一对 i_0, $j_0\in\{1,\ 2,\ 3,\ 4\}$ 满足 $S_{\triangle OA_{i_0}A_{j_0}}\geqslant\sqrt{2}$.

29. 155 只鸟停在一圆 C 上. 如果 $\overset{\frown}{P_iP_j}\leqslant 10°$, 称鸟 P_j 与 P_i 是互相可见的. 求互相可见的鸟对的最小个数(可以假定一个位置同时有几只鸟).

31. 设 $a_1\geqslant a_2\geqslant a_3$ 为已知正整数, $N(a_1,\ a_2,\ a_3)$为方程

$$\frac{a_1}{x_1}+\frac{a_2}{x_2}+\frac{a_3}{x_3}=1$$

的解$(x_1,\ x_2,\ x_3)$的个数,这里 x_1、x_2、x_3 为正整数. 证明:

$$N(a_1,\ a_2,\ a_3)\leqslant 6a_1a_2(3+\ln(2a_1)).$$

32. 锐角三角形 ABC 的顶点 A 到外心 O 与垂心 H 的距离相等,求$\angle A$ 的所有可能的值.

3　解答

以下解答,大部分由各国提供的解答译出,有些作了局部的变更,也有些换用了我们自己做的解答.

2. 如图 1,设地毯的边长为 x、y. 易知

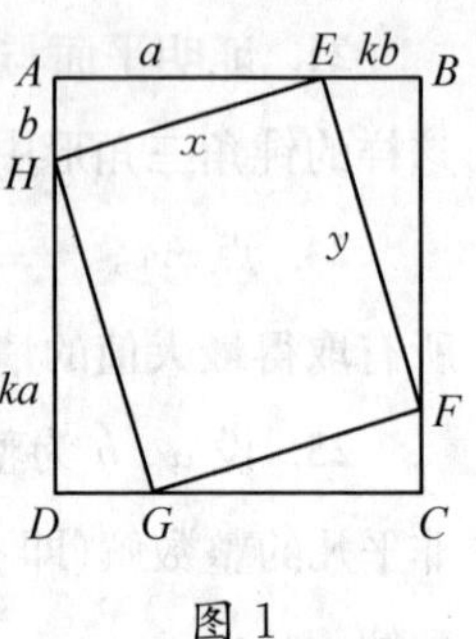

图 1

$$\triangle AEH \cong \triangle CGF \backsim \triangle BFE \cong \triangle DHG.$$

设 $\frac{y}{x}=k$, 则 k 是上述两组三角形的相似比. 设 $AE=a$, $AH=b$. 由相似, $BE=kb$, $DH=ka$. 于是,

$$a+kb=50,\ ka+b=55.$$

从而,

$$a=\frac{55k-50}{k^2-1},\ b=\frac{50k-55}{k^2-1}.$$

$$\begin{aligned} x^2 &= \left(\frac{55k-50}{k^2-1}\right)^2+\left(\frac{50k-55}{k^2-1}\right)^2 \\ &= \frac{3025k^2-5500k+2500+2500k^2}{(k^2-1)^2}-\frac{5500k-3025}{(k^2-1)^2}. \end{aligned}$$

类似地,由另一间房子得

$$x^2=\left(\frac{55k-38}{k^2-1}\right)^2+\left(\frac{38k-55}{k^2-1}\right)^2.$$

化简为

$$\begin{aligned} x^2(k^2-1)^2 &= 3025k^2-4180k+1444+1444k^2-4180k+3025 \\ &= 4469k^2-8360k+4469. \end{aligned}$$

比较以上两式得

$$5525k^2-11000k+5525=4469k^2-8360k+4469,$$

即 $\qquad 1056k^2-2640k+1056=0,\ 2k^2-5k+2=0,\ k=2$ 或 $\frac{1}{2}$,

也就是一条边长是另一条的两倍. 不失一般性,设 $y=2x$,有 $a+2b=50$, $2a+b=55$. 于是 $a=20$, $b=15$. $x^2=a^2+b^2=20^2+15^2=400+225=625$, $x=25$, $y=50$.

地毯的尺寸为 25 呎×50 呎.

3. 设房间未知的一边长为 q 呎. 同上题可得

$$(kq-50)^2+(50k-q)^2=(kq-38)^2+(38k-q)^2,$$

即

$$\begin{aligned}&k^2q^2-100kq+2500+2500k^2-100kq+q^2\\=&k^2q^2-76kq+1444+1444k^2-76kq+q^2.\end{aligned}$$

于是
$$1056k^2-48kq+1056=0,$$

即
$$22k^2-kq+22=0,\ kq=22(k^2+1),\ q=22\left(k+\frac{1}{k}\right).$$

由于 $k=\frac{y}{x}$ 是有理数,令 $k=\frac{c}{d}$,其中 c、d 是正整数,且 $(c,\ d)=1$. 则

$$q=22\left(\frac{c}{d}+\frac{d}{c}\right).$$

于是 $22\left(c+\frac{d^2}{c}\right)=dq$ 是整数. 从而 c 是 22 的约数,同样 d 也是 22 的约数,因而

$$c,\ d\in\{1,\ 2,\ 11,\ 22\}.$$

由于 $(c,\ d)=1$,所以(不妨设 $c>d$)

$$k=1,\ 2,\ 11,\ 22,\ \frac{11}{2}.$$

相应地,

$$q=44,\ 55,\ 244,\ 485,\ 125.$$

与上题类似,

$$x^2(k^2-1)^2=(k^2+1)(q^2-1900). \tag{3.1}$$

$k=1$ 导致 $0=72$ 矛盾. $k=2$, $q=55$, 得出上题的解.

若 $k=11$,则 $61\mid(k^2+1)$, $61^2\nmid(k^2+1)$. $61\mid q^2$, 61 不整除 1900. 所以 $61\mid$(3.1)的右边, 61^2 不整除(3.1)的右边. 这与(3.1)左边为平方数矛盾.

若 $k=22$, 同样可得 $97\mid$(3.1)右边,而 97^2 不整除(3.1)右边,仍导致矛盾.

若 $k=\frac{11}{2}$, $k^2+1=\frac{125}{4}$, $k^2-1=\frac{117}{4}$, $q=125$, 则

$$x^2(117)^2=4(125)(125^2-1900)=50^2\times 5\times 549.$$

5^5 整除上式右边,而 5^6 不能,矛盾.

所以 $k=2$, $q=55$ 呎, $x=25$ 呎, $y=50$ 呎.

评注 第2题是本题的一种变形.

4. 首先证明对每个整数 $k>0$ 及每个素数 p, $p^k\nmid k!$.

设 $s\geqslant 0$ 为整数,满足 $p^s\leqslant k<p^{s+1}$, 则满足 $p^r\mid k!$ 的最大整数为

$$r=\left[\frac{k}{p}\right]+\left[\frac{k}{p^2}\right]+\cdots+\left[\frac{k}{p^s}\right]\leqslant\frac{k}{p}+\frac{k}{p^2}+\cdots+\frac{k}{p^s}=k\cdot\frac{1-\frac{1}{p^s}}{p-1}<k.$$

所以 $p^k\nmid k$.

设有理数 α 为所给方程的根,则

$$\alpha^n+n\alpha^{n-1}+\cdots+\frac{n!}{k!}\alpha^k+\cdots+\frac{n!}{2!}\alpha^2+\frac{n!}{1!}\alpha+n!\ =0.$$

由此易知, α 为整数(设 $\alpha=\frac{c}{d}$, c、d 为互质整数,则 $c^n+nc^{n-1}d+\cdots+\frac{n!}{1!}cd^{n-1}+n!\cdot d^n=0$, 从而 $d\mid c$, d 必须为1).

设 p 为 n 的素因数,则由上面的方程, $p\mid\alpha^n$,从而 $p\mid\alpha$. 设 r 为满足 $p^r\mid n!$ 的最大整数,由于 $p^k\mid\alpha^k$, $p^k\nmid k!$,所以 $p^{r+1}\left|\frac{n!}{k!}\cdot\alpha^k\right.$, $k=1, 2, \cdots, n$. 从而由上面的方程得 $p^{r+1}\mid n!$. 矛盾.

5. 设 $a_1, a_2, \cdots, a_n, b_1, b_2, \cdots, b_n$ 为复数,则有柯西不等式

$$\left|\sum_{i=1}^n a_ib_i\right|^2\leqslant\sum_{i=1}^n|a_i|^2\cdot\sum_{i=1}^n|b_i|^2,$$

当且仅当有常数 $k\in C$,使 $a_i=kb_i(i=1, 2, \cdots, n)$ 时,上式等号成立.

应用该不等式得

$$n^2=|r_1+r_2+\cdots+r_n|^2\leqslant n(|r_1|^2+|r_2|^2+\cdots+|r_n|^2).\quad(5.1)$$

$$\begin{aligned}n^4&=|r_1+r_2+\cdots+r_n|^4\\&\leqslant n^2(|r_1|^2+|r_2|^2+\cdots+|r_n|^2)^2\\&\leqslant n^3(|r_1|^4+|r_2|^4+\cdots+|r_n|^4),\end{aligned}\quad(5.2)$$

$$\begin{aligned} n^8 &= |r_1 + r_2 + \cdots + r_n|^8 \\ &\leqslant n^6(|r_1|^4 + |r_2|^4 + \cdots + |r_n|^4)^2 \\ &\leqslant n^7(|r_1|^8 + |r_2|^8 + \cdots + |r_n|^8), \end{aligned} \tag{5.3}$$

$$\begin{aligned} n^{16} &= |r_1 + r_2 + \cdots + r_n|^{16} \\ &\leqslant n^{14}(|r_1|^8 + |r_2|^8 + \cdots + |r_n|^8)^2 \\ &\leqslant n^{15}(|r_1|^{16} + |r_2|^{16} + \cdots + |r_n|^{16}). \end{aligned} \tag{5.4}$$

但 $|r_1|^{16} + |r_2|^{16} + \cdots + |r_n|^{16} = n$，所以在(5.4)中等号成立，从而

$$|r_1|^8 + |r_2|^8 + \cdots + |r_n|^8 = n.$$

再由(5.3)同样导出

$$|r_1|^4 + |r_2|^4 + \cdots + |r_n|^4 = n.$$

由(5.2)得

$$|r_1|^2 + |r_2|^2 + \cdots + |r_n|^2 = n.$$

最后由(5.1)中等号成立得

$$r_1 = r_2 = \cdots = r_n.$$

但由韦达定理，$$r_1 + r_2 + \cdots + r_n = -n,$$

所以 $$r_1 = r_2 = \cdots = r_n = -1,\ p(x) = (x+1)^n.$$

6. 设$\triangle ABC$的内角为α、β、γ，则

$$P = \frac{1}{2}R^2(\sin 2\alpha + \sin 2\beta + \sin 2\gamma).$$

由于$\triangle A'B'C'$的内角为$\frac{\beta+\gamma}{2}$，$\frac{\alpha+\gamma}{2}$，$\frac{\alpha+\beta}{2}$，所以

$$Q = \frac{1}{2}R^2(\sin(\beta+\gamma) + \sin(\alpha+\gamma) + \sin(\alpha+\beta)).$$

所以，由算术—几何平均不等式，

$$\begin{aligned}16Q^3&=2R^6(\sin(\beta+\gamma)+\sin(\alpha+\gamma)+\sin(\alpha+\beta))^3\\&\geqslant 2R^6\cdot 27\sin(\beta+\gamma)\sin(\alpha+\gamma)\sin(\alpha+\beta)\\&=27R^6(\cos(\alpha-\beta)-\cos(\alpha+\beta+2\gamma)\cdot\sin(\alpha+\beta)\\&=27R^6(\cos(\alpha-\beta)+\cos\gamma)\sin(\alpha+\beta)\\&=\frac{27}{2}R^6(\sin(\alpha+\beta+\gamma)+\sin(\alpha+\beta-\gamma)+\sin 2\alpha+\sin 2\beta)\\&=\frac{27}{2}R^6(\sin 2\alpha+\sin 2\beta+\sin 2\gamma)=27R^4P.\end{aligned}$$

评注 由于 $Q\leqslant\frac{3\sqrt{3}}{4}R^2$(在半径为 R 的圆中,内接三角形的面积以正三角形为最大),所以

$$27R^4P\leqslant 16Q^3\leqslant 16Q\cdot\frac{27}{16}R^4.$$

从而 $P\leqslant Q$. 这是加拿大给 29 届 IMO 提供的问题(第 4 题).

7. 设 P_1、P_2 为多边形 E 中的点. 直线 P_1P_2 交 E 的周界于 Q_1、Q_2. 如果 Q_1、Q_2 可以用两条合乎要求的圆弧 α_1'、α_2' 连结,那么作两个相似变换便可将 α_1'、α_2' 变为连结 P_1、P_2 的弧 α_1、α_2,它们在 P_1、P_2 处的交角与 α_1'、α_2' 在 Q_1、Q_2 处的交角相同. 于是,只需讨论所给两点在周界上的情况.

设 Q_1、Q_2 是 E 的两个端点. 对 E 的内切圆作两次相似或平移,便可得到两条在 E 内,连结 Q_1、Q_2 的弧(当 Q_1、Q_2 为 E 的一条边的端点时,一条是弧,一条是 E 的边 Q_1Q_2),它们与端点为 Q_1、Q_2 的 E 的边相切.

若 Q_1 为顶点, Q_2 在边 V_2V_3 上. 作两个(通常是不同的)以 Q_1 为位似中心的相似变换,将连结 Q_1、V_2 及连结 Q_1、V_3 的弧变为连结 Q_1 与 Q_2 的弧,这些弧仍与 E 的过 Q_1 的边相切,它们的交角仍为 $\left(1-\frac{2}{n}\right)\pi$.

若 Q_1、Q_2 分别在边 V_1V_2、V_iV_{i+1} 上,用一对弧连结 V_1、Q_2,一对弧连结 V_2、Q_2,每对弧的交角为 $\left(1-\frac{2}{n}\right)\pi$. 其中有两条弧与直线 Q_1Q_2 的另一个交点在 E 外,这两条弧在 Q_2 的交角 $\geqslant\left(1-\frac{2}{n}\right)\pi$. 以 Q_2 为心作两个相似变换将这两条弧变为另一交点为 Q_1 的两条弧,这两条弧具有同样的交角.

8. 以 A 为原点，AB 为 x 轴，AD 为 y 轴建立直角坐标系. 只要证明 B、C、D 中至少有一个是整点即可.

每一个小矩形 R_i 至少有一条边长为整数，并且边与坐标轴平行，所以它的顶点中整点的个数为 0、2 或 4. 因而所有 R_i 的顶点中，整点的个数是偶数. 这里一个顶点如果同时是 k 个 R_i 的顶点，则计算 k 次.

A 是整点，并且只被计算 1 次. 矩形 $ABCD$ 内部的整点 E 可能被计算 4 次，也可能被计算 2 次. 于是内部的顶点中，整点的个数是偶数(按出现的重数计算).

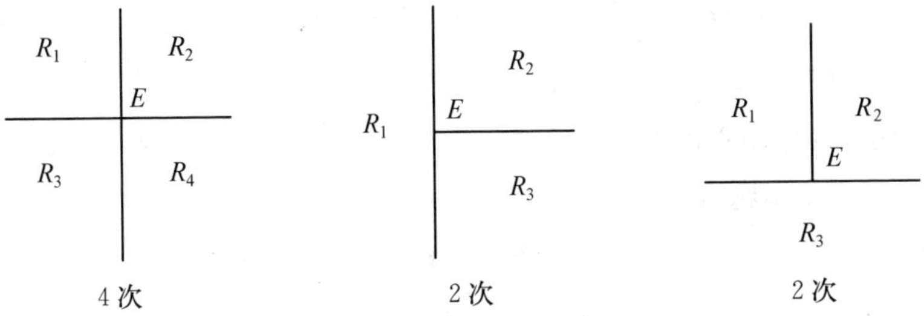

R_1　R_2　E　R_3　R_4

4 次

R_1　E　R_2　R_3

2 次

R_1　R_2　E　R_3

2 次

图 2

综合上面的讨论，B、C、D 中至少有一个是整点.

9. 由 $(1+4\sqrt[3]{2}-4\sqrt[3]{4})^{n+1}=(a_n+b_n\sqrt[3]{2}+c_n\sqrt[3]{4})(1+4\sqrt[3]{2}-4\sqrt[3]{4})$

得

$$a_{n+1}=a_n-8b_n+8c_n.$$

因为 $a_0=1$，所以 a_n 全是奇数.

每一非零整数 k 可写成 $k=2^p\cdot k'$，其中 k' 是奇数，$p\geqslant 0$. 定义 $v(k)=p$. 令

$$\beta_n=v(b_n),\ \gamma_n=v(c_n).$$

(1) 对非负整数 t，$\beta_{2^t}=\gamma_{2^t}=t+2$.

证明：当 $t=0$ 时，$b_1=4$，$c_1=-4$. 上述断言成立. 设 $\beta_{2^t}=\gamma_{2^t}=t+2$，则由于

$$\begin{aligned}&(a+2^{t+2}(b\sqrt[3]{2}+c\sqrt[3]{4}))^2\\=&a^2+2^{t+3}a(b\sqrt[3]{2}+c\sqrt[3]{4})+2^{2t+4}\cdot(b\sqrt[3]{2}+c\sqrt[3]{4})^2\\=&A+2^{t+3}(B\sqrt[3]{2}+C\sqrt[3]{4}).\end{aligned}$$

在 a、b、c 为奇数时，A、B、C 为奇数，于是 $\beta_{2^{t+1}}=\gamma_{2^{t+1}}=t+3$，从而所述断言对一切非负整数 t 成立.

(2) 若 n、m 为整数，b_n，c_n，b_m，c_m 非零，$\beta_n=\gamma_n=\lambda$，$\beta_m=\gamma_m=\mu$，$\mu<\lambda$，则 b_{n+m}，c_{n+m} 非零，并且 $\beta_{m+n}=\gamma_{m+n}=\mu$.

证明：由 $(a'+2^{\lambda}(b'\sqrt[3]{2}+c'\sqrt[3]{4}))a''+2^{\mu}(b''\sqrt[3]{2}+c''\sqrt[3]{4}))=A+2^{\mu}(B\sqrt[3]{2}+C\sqrt[3]{4})$，在 a'，a''，b'，b''，c'，c'' 为奇数时，A、B、C 均为奇数.

(3) 对任一整数 $n\geqslant 1$，设

$$n=2^{t_r}+2^{t_{r-1}}+\cdots+2^{t_n},\ 0\leqslant t_0<\cdots<t_r,$$

则 c_n 非零，并且 $r_n=t_0+2$.

证明：由(1)、(2)即得.

由(3)即知本题结论成立.

评注 b_{1989} 与 c_{1989} 被 4 整除，不被 8 整除.

10. 在函数方程中用 $\omega z+a$ 代替 z 得

$$f(\omega z+a)+f(\omega^2 z+\omega a+a)=g(\omega z+a). \tag{10.1}$$

重复这一作法得 $$f(z)+f(\omega^2 z+\omega a+a)=g(\omega^2 z+\omega a+a). \tag{10.2}$$

解由原方程(10.1)，(10.2)组成的线性方程组得

$$f(z)=\frac{1}{2}(g(z)+g(\omega^2 z+\omega a+a)-g(\omega z+a)).$$

11. 一个长为 n 的 0、1 序列 s，如果对某个 $d\mid n$，s 可以分为 d 个恒等的块，那么 s 便称为循环的. 显然，每一个长为 n 的 0，1 序列可以由它的唯一的、最长的第一个不循环的块重复若干次而得到. 由于长为 n 的 0、1 序列共 2^n 个，因此设 b_n 为长为 n 的不循环的 0，1 序列的个数，则有

$$\sum_{d\mid n}b_d=2^n. \tag{11.1}$$

(11.1)对任一自然数 n 成立. 特别地，$b_1=2$. 于是 $b_1=a_1$. 由已知条件 $\sum_{d\mid n}a_d=2^n$ 易知，对一切 n，$b_n=a_n$，即 a_n 为长为 n 的不循环的 0、1 序列的个数.

这样，由于从每个不循环的长为 n 的 0、1 序列，将第 1 项移到最后可逐步产生 n 个不同的(不循环的)0、1 序列，$n\mid a_n$.

12. 设两辆车在相会时交换其编号，这样我们看到的是每一辆车，例如 1 号

车,依同样速度和方向一圈一圈地绕圆周运动. 所以在一个小时后(经过若干次交换编号),每一个出发点被一辆与原先号码相同的车占据,并且它准备运行的方向与一个小时前在那里出发的车完全相同.

恢复每辆车原先的号码,就回到实际发生的状态. 由于汽车的顺序不会改变,仅有的可能是将原先出发时的状态作了一个旋转(也可以就是原先的状态). 于是对某个 $d\mid n$,经过 d 小时后,每辆车回到了出发点.

14. 如图 3,设四边形 $ABCD$ 为双心四边形,其外接圆圆心为 O,内切圆圆心为 I,对角线交点为 K.

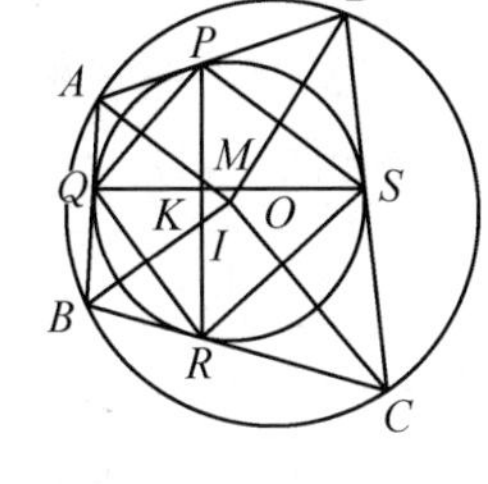

图 3

引理 1 对圆外切四边形 $ABCD$,设切点为 P, Q, R, S,则 PR、QS 的交点就是对角线 AC、BD 的交点 K.

这是一个熟知的结论,证明略去.

引理 2 若 K 为$\odot I$ 内一定点,则对 K 点张直角的弦 EF 的中点的轨迹是一个圆,圆心为 IK 的中点 M.

引理 3 在引理 1 中,如果 $ABCD$ 有外接圆,则 $PR\perp QS$.

引理 2、3 的证明均不难.

由引理 3, PQ, QR, RS, SP 对 K 点张直角,因而它们的中点 A', B', C', D'均在以 IK 的中点 M 为圆心的一个圆上.

由于 IA 与 PQ 相交于 A',所以 A'就是以 I 为反演中心, $\odot I$ 为反演圆时, A 经反演所得的象,同样 B', C', D'分别为 B, C, D 的象. 因此$\odot O$ 经过反演成为 $A'B'C'D'$的外接圆,从而 O 与这圆的圆心 M、反演中心 I 共线. 于是, O 在直线 IM 上. 因此 O, M, K 共线.

15. 由柯西不等式 $a+b+c+d\leqslant 2\sqrt{1989}<90$,由于 $a^2+b^2+c^2+d^2$ 为奇数,所以 $a+b+c+d$ 也是奇数, $m^2\in\{1, 9, 25, 49, 81\}$.

由 $(a+b+c+d)^2>a^2+b^2+c^2+d^2$ 推出 $m^2=49$ 或 81.

不妨设 $a\leqslant b\leqslant c\leqslant d=n^2$.

若 $m^2=49$, 则

$$(49-d)^2=(a+b+c)^2>a^2+b^2+c^2=1989-d^2.$$

从而,

$$d^2-49d+206>0.$$

得 $d>44$ 或 $d\leqslant 4$. 但

$$45^2>1989>d^2\Rightarrow d<45,$$
$$4d^2\geqslant a^2+b^2+c^2+d^2=1989\Rightarrow d>22.$$

所以, $m^2\neq 49$. 从而, $m^2=81$, $m=9$, 并且

$$d=n^2\in\{25,\ 36\}.$$

若 $d=n^2=25$,令 $a=25-p$, $b=25-q$, $c=25-r$, p, q, $r\geqslant 0$,则由已知条件导出

$$p+q+r=19,\ p^2+q^2+r^2=439,$$

与 $(p+q+r)^2>p^2+q^2+r^2$ 矛盾. 所以 $n^2=36$, $n=6$.

评注 不难推出 $a=12$, $b=15$, $c=18$, $d=36$.

16. 定义 $S_k=\sum\limits_{i=0}^{k}a_i(k=0,\ 1,\ \cdots,\ n)$,则

$$\begin{aligned}S_n&=\sum_{k=0}^{n}a_k=\sum_{k=0}^{n-1}a_k=nc+\sum_{k=0}^{n-1}\sum_{i=k}^{n-1}a_{i-k}(a_i+a_{i+1})\\&=nc+\sum_{i=0}^{n-1}\sum_{k=0}^{i}a_{i-k}(a_i+a_{i+1})=nc+\sum_{i=0}^{n-1}(a_i+a_{i+1})\sum_{k=0}^{i}a_{i-k}\\&=nc+\sum_{i=0}^{n-1}(a_i+a_{i+1})\sum_{t=0}^{i}a_t(\text{令 } t=i-k)=nc+\sum_{i=0}^{n-1}(a_i+a_{i+1})S_1\\&=nc+\{S_1S_0+(S_2-S_0)S_1+(S_3-S_1)S_2+\cdots+(S_n-S_{n-2})S_{n-1}\}\\&=nc+S_n^2(\text{由于 } S_{n-1}=S_n),\end{aligned}$$

即

$$S_n^2-S_n+nc=0.$$

但 S_n 为实数,所以 $1\geqslant 4nc$.

17. 图 4 表明 9 条线段已经足够了.

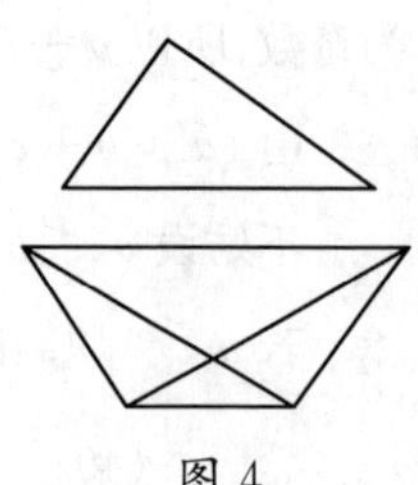

图 4

现在证明至少需要 9 条线段.

如果点 A 不作为线段的端点,则由于条件(1),其他 6 点至少要连 $C_6^2>9$ 条线段.

如果点 A 只作为 1 条线段的端点,则由于条件(1),不与 A 相连的 5 点之间至少要连 $C_5^2>9$ 条线段.

设每一点至少作为两条线段的端点. 则若点 A 只作为两条线段 AB、AC 的端点,由条件(1),不与 A 相连的 4 点之间至少要连 $C_4^2=6$ 条线段. 自 B 点至少还要引出一条线段,所以这时至少有 $2+6+1=9$ 条线段.

若每一点至少作为 3 条线段端点,则至少有多于 $\left[\frac{3\times 7}{2}\right]=10$ 条线段.

18. 由于 $R_{A_1}^{\alpha}(M)=M_1$, 这里 $R_{A_1}^{\alpha}$ 表示绕 A_1 旋转 α 角的变换,所以以 M 为位似中心,相似比为 $2\sin\frac{\alpha}{2}$,作一位似变换 $H_{M^{2\sin(\alpha/2)}}$,然后再绕 M 旋转 $\frac{\alpha}{2}-90°R_{M^{\alpha/2-90°}}$, 便将 A_i 变为 M_i,即

$$R_{M^{\alpha/2-90°}}\cdot H_{M^{2\sin(\alpha/2)}}(A_1)=M_1, i=1,\ 2,\ \cdots,\ n.$$

从而,

$$R_{M^{\alpha/2-90°}}\cdot H_{M^{2\sin(\alpha/2)}}(A_1A_2\cdots A_n)=M_1M_2\cdots M_n,$$

$$S_{M_1M_2\cdots M_n}=\left(4\sin^2\frac{\alpha}{2}\right)S.$$

19. 将 $m\times n$ 的表中相邻的方格涂上两种不同的颜色 $*$ 与□,两种方格中的数的和分别记为 S_*, $S_\square$,令 $S=S_*-S_\square$. 由于每次运算 S 均保持不变,所以 $S=0$ 是经过若干次运算后,表中各数为 0 的必要条件.

现在证明 $S=0$ 也是充分条件,从表中的第一列开始,设第一列第一行的数为 a,第一列第二行的数为 b,第一列第三行的数为 c.

如果 $a>b$,将 b、c 同时加上 $a-b$,然后再将 a 与 $b+(a-b)=a$ 同时加上 $-a$.

如果 $a\leqslant b$, 将 a、b 同时加上 $-a$.

这样进行下去,直至表成为:

<table><tr><td>0
0
⋮
0
g</td><td>

r
h</td></tr></table>

或

<table><tr><td>0
⋮
0
g
0</td><td>

r
h
</td></tr></table>

如果 $g \leqslant h$，则将 g 与 h 同时加上 $-g$. 如果 $g > h$，则将 r、h 同时加上 $g-h$，然后将 g 与 $h+(g-h)$ 同时加上 $-g$. 总之，我们可以使第一列的数全变为 0. 如此继续下去，可以使表中只有第 n 列的一个数可能非零，其余各数都变成 0.

由于 $S=0$，所以这时每一个数都是零.

21. 设平面截正四面体 $ABCD$ 的棱 AB 于 P，AC 于 Q，AD 于 R. 设 $AP=x$，$AQ=y$，$AR=z$. $\angle QPR=\alpha$，则 PQ^2、PR^2、QR^2 分别为

$$x^2+y^2-xy,\ x^2+z^2-xz,\ y^2+z^2-yz,$$

并且

$$\begin{aligned}&2\sqrt{(x^2+y^2-xy)(x^2+z^2-xz)}\cos\alpha\\&=(x^2+y^2-xy)+(x^2+z^2-xz)-(y^2+z^2-yz)\qquad(21.1)\\&=2x^2-xy-xz+yz.\end{aligned}$$

当 z 很小，而 $y>2x$ 时，上式右边为负，这时 α 为钝角.

现在设 α 为钝角，则

$$2x^2-xy-xz+yz<0. \qquad (21.2)$$

我们要证明

$$-\sqrt{(x^2+y^2-xy)(x^2+z^2-xz)}<2x^2-xy-xz+yz, \qquad (21.3)$$

即

$$(x^2+y^2-xy)(x^2+z^2-xz)>(2x^2-xy-xz+yz)^2.$$

不妨设 $x=1$. 上式即

$$\begin{aligned}&1+y^2+z^2+y^2z^2-z-y^2z+yz-y-yz^2\\&>4+y^2+z^2+y^2z^2-4y-4z+4yz+2yz-2y^2z-2yz^2,\end{aligned}$$

等价于

$$y^2z+yz^2+3y+3z>3+5yz. \qquad (21.4)$$

而由(21.2)，

$$3(y+z)>3(2+yz).$$

因此，(21.4)成立，从而(21.3)也成立，$\alpha<120^\circ$.

24. 我们可以用$\angle A_iOA_j$ 的大小来代替距离A_iA_j，这里 O 为球心.

存在一组点 A_1，A_2，…，A_5，使

$$\underset{1\leqslant i,\,j\leqslant 5}{n}\angle A_iOA_j \leqslant \frac{\pi}{2}. \tag{24.1}$$

例如，取一个正八面体的 5 个顶点.

现在设 A_1，A_2，…，A_5 为任一组使(24.1)成立的点.我们断言其中必有两个点为对经点.

首先设 A_5 为南极.这时，A_1，A_2，A_3，A_4 均在北半球(包括赤道).若无对经点，则北极没有点.由赤道的$\frac{1}{4}$及两条互相垂直的经线所围成的、北半球的象限内如果含有两个 A_1，则由于(24.1)，其中一个必在“角落”(即赤道与一条经线的交点)，另一个必在它所对的经线上.于是 A_1，…，A_4 中每两个的经度之差不小于$\frac{\pi}{2}$.从而它们在将北半球分为四个象限的四条经线上.最后，如果有一个点不在赤道上，则它的两个“邻居”必须在赤道上(各在一个象限的“角落”)，这两个点是对经点.

由于总有两个对经点 A_1、A_2，而第三个点 A_3^n 不能使 $\angle A_1OA_3 > \frac{\pi}{2}$，$\angle A_2OA_3 > \frac{\pi}{2}$ 同时成立，所以(24.1)中严格的不等号不可能成立，即 $\min\angle A_iOA_j$ 的最大值为$\frac{\pi}{2}$.在等号成立时，设 A_1、A_2 为两极，则 A_3、A_4、A_5 均在赤道上，并且 OA_3、OA_4、OA_5 两两夹角不小于$\frac{\pi}{2}$.

$\min A_iA_j$ 的最大值为$\sqrt{2}$，在上述的情况中达到.

25. a、b 不可能均为负数，不失一般性，可设 $a>0$(a、b 非完全平方数，当然都不为 0).设 $(x_0, y_0, z_0, w_0)\neq(0, 0, 0, 0)$ 为

$$x^2-ay^2-bz^2+abw^2=0$$

的解，则

$$x_0^2-ay_0^2-b(z_0^2-aw_0^2)=0. \tag{25.1}$$

将上式乘以 $(z_0^2-aw_0^2)$ 得

$$(x_0^2-ay_0^2)(z_0^2-aw_0^2)-b(z_0^2-aw_0^2)^2=0.$$

而

$$\begin{aligned}(x_0^2-ay_0^2)(z_0^2-aw_0^2)&=x_0^2z_0^2-ay_0^2z_0^2-ax_0^2w_0^2+a^2y_0^2w_0^2\\&=x_0^2z_0^2-2ax_0y_0z_0w_0+a^2y_0^2w_0^2\\&\quad -a(y_0^2z_0^2-2x_0y_0z_0w_0+x_0^2w_0^2)\\&=(x_0z_0-ay_0w_0)^2-a(y_0z_0-x_0w_0)^2,\end{aligned}$$

所以

$$(x_0z_0+ay_0w_0)^2-a(y_0z_0-x_0w_0)^2-b(z_0^2-aw_0^2)^2=0.$$

令 $x_1=x_0z_0-ay_0w_0,\ y_1=y_0z_0-x_0w_0,\ z_1=z_0^2-aw_0^2$,

则$(x_1,\ y_1,\ z_1)$适合方程

$$x^2-ay^2-bz^2=0.$$

若 $z_1=0$,则 $z_0^2=aw_0^2$. 由于 a 不是完全平方,所以 $z_0=w_0=0$. 再由(25.1)得 $x_0^2=ay_0^2$. 根据同样的理由, $x_0=y_0=0$. 这与$(x_0,\ y_0,\ z_0,\ w_0)\neq(0,\ 0,\ 0,\ 0)$矛盾. 所以 $z_1\neq0$, $(x_1,\ y_1,\ z_1)\neq(0,\ 0,\ 0)$.

26. 由柯西不等式,

$$\left(\sum_{i=1}^n x_i\right)^2\leqslant n\sum_{i=1}^n x_i^2.$$

因此,
$$(a-x_0)^2\leqslant n(b-x_0^2),$$
即
$$(n+1)x_0^2-2ax_0+a^2-nb\leqslant0.$$

这个二次三项式的判别式

$$D=4n(n+1)\left(b-\frac{a^2}{n+1}\right).$$

(1) 若 $b<\dfrac{a^2}{n+1}$,则 $D<0$, x_0 不存在.

(2) 若 $b=\dfrac{a^2}{n+1}$,则 $D=0$, $x_0=\dfrac{a}{n+1}$.

(3) 若 $b>\dfrac{a^2}{n+1}$, 则

$$\frac{a-\sqrt{\frac{D}{4}}}{n+1} \leqslant x_0 \leqslant \frac{a+\sqrt{\frac{D}{4}}}{n+1}.$$

27. 表 n 为 $n=2^s q$，其中 q 为奇数，则

$$\begin{aligned} m^n-1 &= m^{2^s \cdot q}-1 \\ &= (m^{2^s})q-1 \\ &= (m^{2^s}-1)((m^{2^s})^{q-1}+(m^{2^s})^{q-2}+\cdots+m^{2^s}+1) \\ &= (m^{2^s}-1)A. \end{aligned}$$

其中 $A\equiv 1(\bmod 2)$. 于是，

$$2^{1989} \mid m^n-1 \Leftrightarrow 2^{1989} \mid m^{2^s}-1.$$

因此，可设 $n=2^s$. 这时有两种情况：

(1) 若 $m\equiv 1(\bmod 4)$，则 m 的二进制表示为 $m=1\cdots\underbrace{100\cdots01}_{k\text{个数字}}$，即 k 是使 $m\equiv 1(\bmod 2^k)$ 的最大整数. 于是，

$$m^2-1=(m+1)(m-1)$$

被 2^{k+1} 整除而不被 2^{k+2} 整除. 设 $m^{2^s}-1$ 被 2^{k+t} 整除而不被 2^{k+t+1} 整除，则

$$m^{2^{t+1}}-1=(m^{2^t}+1)(m^{2^t}-1)$$

被 2^{k+t+1} 整除，而不被 2^{k+t+2} 整除. 所以，对所有自然数 s，

$$2^{s+k} \mid m^{2^s}-1,\ 2^{s+k+1} \nmid m^{2^s}-1.$$

(2) 若 $m\equiv 3(\bmod 4)$，则 m 的二进制表示为

$$m=1\cdots 0\underbrace{11\cdots 1}_{k\text{个数字}},$$

即 k 为使 $m\equiv -1(\bmod 2^k)$ 成立的最大整数.

同样可证对所有自然数 s，$2^{s+k} \mid m^{2^s}-1$，$2^{s+k+1} \nmid m^{2^s}-1$.

于是，由 $2^{1989} \mid m^{2^s}-1 \Rightarrow 1989 \leqslant s+k$（$k$ 的定义见上面）$\Rightarrow s \geqslant 1989-k$.

因此，在 $k\leqslant 1989$ 时，最小的指数 $n=2^{1989-k}$，在 $k>1989$ 时，$n=2^0=1$.

28. 如图 5，记 $OA_1=a$，$OA_2=b$，$OA_3=c$，$OA_4=d$，则

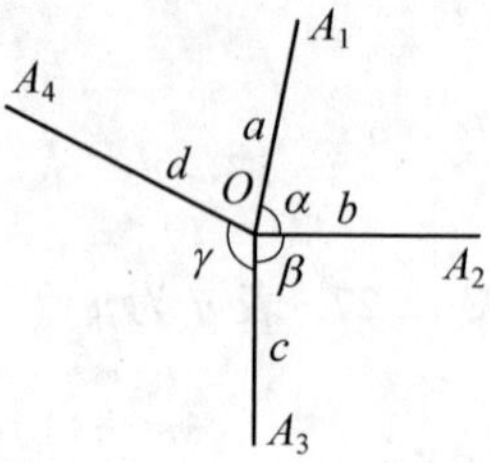

$$S_1 = S_{OA_1A_2} = \frac{1}{2}ab\,|\sin\alpha|,$$

$$S_2 = S_{OA_1A_3} = \frac{1}{2}ac\,|\sin(\alpha+\beta)|,$$

$$S_3 = S_{OA_1A_4} = \frac{1}{2}ad\,|\sin(\alpha+\beta+\gamma)|,$$

$$S_4 = S_{OA_2A_3} = \frac{1}{2}bc\,|\sin\beta|,$$

$$S_5 = S_{OA_2A_4} = \frac{1}{2}bd\,|\sin(\beta+\gamma)|,$$

$$S_6 = S_{OA_3A_4} = \frac{1}{2}cd\,|\sin\gamma|.$$

图 5

由于

$$\sin(\alpha+\beta+\gamma)\sin\beta + \sin\alpha\sin\gamma = \sin(\alpha+\beta)\sin(\beta+\gamma),$$

所以选择适当的＋号与－号后有

$$S_3S_4 \pm S_1S_6 \pm S_2S_5 = 0.$$

比如说

$$S_3S_4 = S_1S_6 + S_2S_5,$$

则 $$(\max S_1)^2 \geqslant S_3S_4 = S_1S_6 + S_2S_5 \geqslant 1+1=2,$$

即 $$\max S_1 \geqslant \sqrt{2}.$$

29. 设 A 为圆 C 上一点，鸟 p_1 停在 A，从 A 可以看到在 B 处 $(B \neq A)$ 的鸟 p_j. 设 k 为从 B 可以看到而从 A 看不到的鸟的个数. l 为从 A 可以看到而从 B 看不到的鸟的个数，不妨设 $k \geqslant l$.

如果所有在 B 处的鸟都飞往 A 处，那么对其中的每一只来说，减少了 k 个可见对，同时增加了 l 个可见对，因此互相可见的对数不会增加.

经过上述运算，停鸟的位置减少 1. 重复若干次可能的运算(至多 154 次)可以使得每两只鸟只有在同一位置时才是互相可见的，这时有鸟的位置至多 35 个(若有 36 个位置，则至少有一段弧$\leqslant 10°$. 从而弧的两端的鸟还可以并到一处).

于是问题化为求

$$\min\left\{\left|\sum_{j=1}^{35} C_{x_j}^2\right| x_1+x_2+\cdots+x_{35}=155,\ x_j\in N\cup\{0\}\right\}$$

(当 $x_j<2$ 时，$C_{x_j}^2=0$).

设 $x_1=\min x_j$，$x_2=\max x_j$. 若 $x_2-x_1\geqslant 2$,则令 $x'_1=x_1+1$，$x'_2=x_2-1$,这时 $\sum x_j$ 仍为 155,而 $C_{x_1}^2+C_{x_2}^2-C_{x'_1}^2-C_{x'_2}^2=\frac{x_1(x_1-1)}{2}+\frac{x_2(x_2-1)}{2}-\frac{(x_1+1)x_1}{2}-\frac{(x_2-1)(x_2-2)}{2}=-x_1+(x_2-1)\geqslant 1$,即 $\sum C_{x_j}^2$ 较原来为小.

因此,可以 $x_2-x_1\leqslant 1$. 从而 x_j 中有 20 个为 4，15 个为 5. 所求的最小值为

$$20C_4^2+15C_5^2=270.$$

在圆 C 上取 35 个互相不可见的位置,各位置上停 4 或 5 只鸟,则可以达到上述最小值.

31. 设$(i,\ j,\ k)$为$(1,\ 2,\ 3)$的排列. $N_{i,j,k}$ 为满足 $\frac{a_1}{x_1}\geqslant\frac{a_j}{x_j}\geqslant\frac{a_k}{x_k}$ 的方程

$$\frac{a_1}{x_1}+\frac{a_2}{x_2}+\frac{a_3}{x_3}=1 \tag{31.1}$$

的正整数解的个数. 于是

$$N_{(a_1,a_2,a_3)}\leqslant\sum_{(i,j,k)}N_{i,j,k}\leqslant 6\max N_{i,j,k}.$$

这里 $\sum$ 是对$(1,\ 2,\ 3)$的所有排列$(i,\ j,\ k)$求和,max 是对$(1,\ 2,\ 3)$的所有排列求最大值.

设 $N_{p,q,r}=\max N_{i,j,k}$，并且$(x_1,\ x_2,\ x_3)$是方程(31.1)的解,满足 $\frac{a_p}{x_p}\geqslant\frac{a_q}{x_q}\leqslant\frac{a_r}{x_r}>0$. 则

$$\frac{3a_p}{x_p}\geqslant\frac{a_p}{x_p}+\frac{a_q}{x_q}+\frac{a_r}{x_r}=1\Rightarrow x_p\leqslant 3a_p,$$

$$\frac{a_p}{x_p}=1-\frac{a_q}{x_q}-\frac{a_r}{x_r}<1\Rightarrow x_p\geqslant a_p+1.$$

类似地,利用 $1-\frac{a_p}{x_p}>0$ 得

$$\frac{2a_q}{x_q}\geqslant\frac{a_q}{x_q}+\frac{a_r}{x_r}=1-\frac{a_p}{x_p}\Rightarrow x_q\leqslant\frac{2a_q}{1-\frac{a_p}{x_p}},$$

及

$$\frac{a_q}{x_q}=1-\frac{a_p}{x_p}-\frac{a_r}{x_r}<1-\frac{a_p}{x_p}\Rightarrow x>\frac{a_q}{1-\frac{a_p}{x_p}}.$$

于是,对给定的 x_p,至多有 $\frac{a_q}{1-\frac{a_p}{x_p}}$ 个 x_q 的值. 给定 x_p、x_q,则 x_r 唯一确定,

所以

$$N_{p,q,r}\leqslant\sum_{k=ap+1}^{3a_p}\frac{a_q}{1-\frac{a_p}{k}}=a_q\sum_{k=ap+1}^{3a_p}\frac{k}{k-a_p}$$

$$=a_q\sum_{h=1}^{2a_p}\frac{h+a_p}{h}=a_pa_q\left(2+\sum_{h=1}^{2a_p}\frac{1}{h}\right)$$

$$\leqslant a_pa_q\left(3+\sum_{h=2}^{2a_p}\int_{h-1}^{h}\frac{\mathrm{d}x}{x}\right)=a_pa_q\left(3+\sum_{h=2}^{2a_p}\ln\frac{h}{h-1}\right)$$

$$=a_pa_q(3+\ln(2a_p))\leqslant a_1a_2(3+\ln(2a_1)).$$

从而,

$$N_{(a_1,a_2,a_3)}\leqslant 6a_1a_2(3+\ln(2a_1)),$$

32. 如图 6,设 CC' 为高. 由于 $AH=R$(外接圆半径),所以

$$AC'=R\cdot\sin\angle AHC'=R\sin B.$$

从而, $$CC'=R\sin B\tan A.$$

又 $$CC'=BC\sin B=2R\sin A\sin B,$$

比较以上三式得 $2\sin A=\tan A$.

从而, $A=60^\circ$.

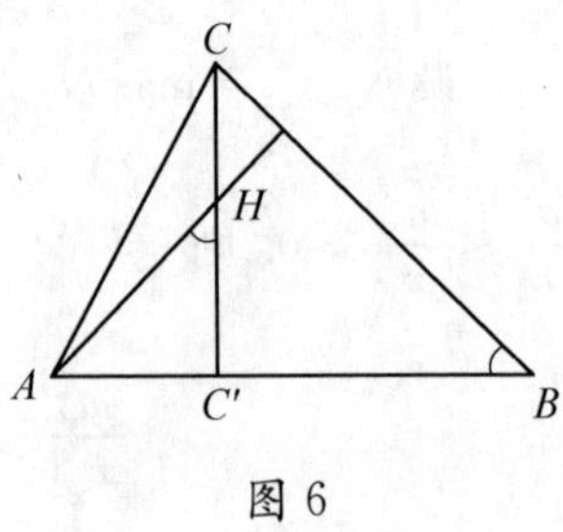

图 6

1989年亚洲太平洋地区数学竞赛题及解答

首届亚洲太平洋地区数学竞赛于1989年3月14日举行.这一竞赛的目的是为了发现、鼓励亚太地区的有数学才能的学生,促进这一地区师生的国际交往与合作.竞赛的时间定在每年三月的第二个星期内.虽然有统一的考题,但不需要将选手集中.在各自的地区分开考试,可以节省许多经费.

试题共5道,时间为4小时,每道题7分.组织、选题、发奖等大致与国际数学奥林匹克相同.目前参加这一比赛的有澳大利亚、加拿大、新加坡和中国香港地区.预计1990年会有更多的国家和地区参加,其中包括中国大陆.

下面是这一届的试题与解答.

1. 设 $x_1, x_2, \cdots, x_n$ 为正实数,令

$$s=x_1+x_2+\cdots+x_n,$$

证明:

$$(1+x_1)(1+x_2)\cdots(1+x_n)\leqslant 1+s+\frac{s^2}{2!}+\frac{s^3}{3!}+\cdots+\frac{s^n}{n!}.$$

证明 由算术—几何平均不等式

$$(1+x_1)(1+x_2)\cdots(1+x_n)\leqslant\left(\frac{n+x_1+\cdots+x_n}{n}\right)^n$$

$$=\left(1+\frac{s}{n}\right)^n=1+s+\cdots+\frac{\mathrm{C}_n^k}{n^k}s^k+\cdots+\frac{s^n}{n^n}.$$

而对于 $k=1, 2, \cdots, n$,

$$\frac{\mathrm{C}_n^k}{n^k}\div\frac{1}{k!}=\frac{n(n-1)\cdots(n-k+1)}{n^k}\leqslant 1,$$

所以原式成立.

2. 证明方程 $6(6a^2+3b^2+c^2)=5n^2$ 除去 $a=b=c=n=0$ 外无正整数解.

证明 不妨设非零解 a, b, c, n 的最大公约数 $(a, b, c, n)=1$.

方程
$$6(6a^2+3b^2+c^2)=5n^2 \tag{1}$$

表明 $6|n^2$,从而 $6|n$, $6^2 \mid 6(6a^2+3b^2+c^2)$, $3 \mid c$.

令 $n=6m$, $c=3d$, m、d 为整数. 则

$$2a^2+b^2+3d^2=10m^2. \tag{2}$$

由于平方数 x^2 除以 8 时余数为 1(如果 x 是奇数),或 0 与 4(如果 x 是偶数). 所以(2)式右边除以 8 的余数为 2(如果 m 是奇数)或 0(如果 m 是偶数). (2)式左边 b、d 的奇偶性必定相同(如果 b、d 中仅有一个奇数,则左边为奇数).

在 b、d 均为奇数时, (2)式左边除以 8 后余数为 2. $1+1+3=6$ 或 $0+1+3=4$, 均与右边不同. 所以 b、d 均为偶数. 这时 a 必为奇数(否则 2 是 a、b、c、n 的公约数,与 $(a, b, c, n)=1$ 矛盾). 从而(2)式左边除以 4 余 2,右边的 m 也必为奇数.

令 $b=2b_1$, $d=2d_1$, 则

$$a^2+2b_1^2+6d_1^2=5m^2, \tag{3}$$

$5m^2-a^2$ 除以 8 时余数为 $5-1=4$,所以 $\frac{5m^2-a^2}{2}$ 除以 4 时余数为 2. 这样由(3)得 $b_1^2+3d_1^2$ 除以 4 时余数为 2. 但平方数除以 4 时余 0 或 1. 所以 $b_1^2+3d_1^2$ 除以 4 时余数应为 0 或 1 或 3. 矛盾. 这表明原方程仅有一组整数解 $a=b=c=n=0$.

3. A_1、A_2、A_3 为平面上三点. 为方便起见,约定 $A_4=A_1$, $A_5=A_2$. 对于 $n=1, 2, 3$, 令 B_n 为 A_nA_{n+1} 的中点, C_n 为 A_nB_n 的中点, A_nC_{n+1} 与 B_nA_{n+2} 交于 D_n. A_nB_{n+1} 与 C_nA_{n+2} 交于 E_n. 求 $\triangle D_1D_2D_3$ 与 $\triangle E_1E_2E_3$ 的面积之比.

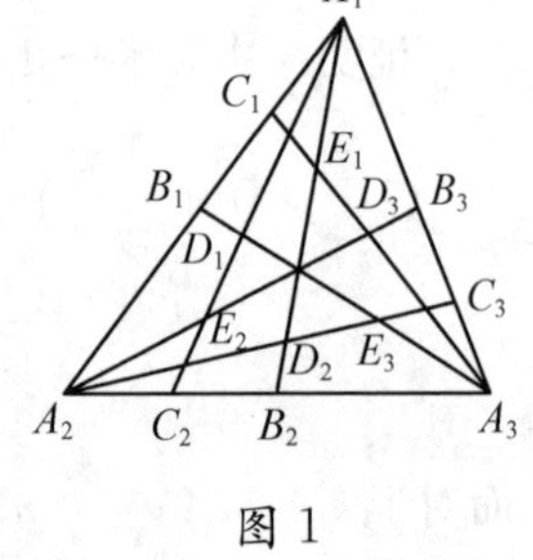

图 1

解　如图 1,设 A_n 的坐标为 (x_n, y_n), $n=1, 2, 3$. 如果在 A_1、A_2 处各放一个重为 3 的小球,在 A_3 放一个重为 1 的小球,那么这个系统的重心应当在 A_3B_1 上(B_1 是 A_1、A_2 的重心),也应当在 A_1C_2 上(C_2 是 A_2、A_3)的重心). 因而 D_1 就是它们的重心,即 D_1 的坐标为 $\left(\frac{3x_1+3x_2+x_3}{7}, \frac{3y_1+3y_2+y_3}{7}\right)$.

同样可得 E_1 的坐标为 $\left(\frac{3x_1+x_2+x_3}{5}, \frac{3y_1+y_2+y_3}{5}\right)$.

将下标 1, 2, 3 轮换便产生其他点的坐标.

于是由面积公式

$$\frac{S_{\triangle D_1D_2D_3}}{S_{\triangle E_1E_2E_3}}=\frac{5^2}{7^2}\cdot\frac{\begin{vmatrix}1 & 3x_1+3x_2+x_3 & 3y_1+3y_2+y_3\\ 1 & 3x_2+3x_3+x_1 & 3y_2+3y_3+y_1\\ 1 & 3x_3+3x_1+x_2 & 3y_3+3y_1+y_2\end{vmatrix}}{\begin{vmatrix}1 & 3x_1+x_2+x_3 & 3y_1+y_2+y_3\\ 1 & 3x_2+x_3+x_1 & 3y_2+y_3+y_1\\ 1 & 3x_3+x_1+x_2 & 3y_3+y_1+y_2\end{vmatrix}}$$

$$=\frac{5^2}{7^2}\cdot\frac{\begin{vmatrix}2x_3-2x_1 & 2y_3-2y_1\\ 2x_1-2x_2 & 2y_1-2y_2\end{vmatrix}}{\begin{vmatrix}2x_2-2x_1 & 2y_2-2y_1\\ 2x_3-2x_1 & 2y_3-2y_1\end{vmatrix}}=\frac{25}{49}.$$

4. S 为 m 个正整数对 $(a, b)(1\leqslant a<b\leqslant n)$ 所成的集. 证明至少有 $4m\cdot\dfrac{m-\dfrac{n^2}{4}}{3n}$ 个三元数组(a, b, c)使得(a, b), (a, c)与(b, c)都属于 S(译注:(a, b)与(b, a)被认为是相同的).

证明　考虑 n 个点 $1, 2, \cdots, n$. 如果 $(i, j)\in S$, 则在 i 与 j 之间连一条线. 我们来求这个图中的三角形的个数(孔就是具有所述性质的三元组(a, b, c)的个数)T.

设 $(i, j)\in S$, 自 i 引出的线有 $d_{(i)}$ 条,则以(i, j)为边的三角形至少有 $d_{(i)}+d_{(j)}-n$ 个. 由于每个三角形有三条边,所以 S 中至少有

$$\frac{1}{3}\sum_{(i, j)\in S}(d_{(i)}+d_{(j)}-n) \tag{1}$$

个三角形.

$$\sum_{(i, j)\in S}n=n\cdot\sum_{(i, j)\in S}1=nm. \tag{2}$$

对于每个固定的 i,恰有 $d_{(i)}$个 j 使 $(i, j)\in S$, 所以在(1)中的 $d_{(i)}$ 出现了 $d_{(i)}$次. 注意(i, j)既可作为自 i 引出的边,又可作为自 j 引出的边,被计算了 2 次. 因此

$$\sum_{(i,j)\in S}(d_{(i)}+d_{(j)})^2\sum_{(i,j)\in S}d_{(i)}=\sum_{i=1}^{n}d_{(i)}^2.$$

由 Cauchy 不等式,

$$\sum_{i=1}^{n}d_{(i)}^2\geqslant\frac{1}{n}\left(\sum_{i=1}^{n}d_{(i)}\right)^2=\frac{1}{n}(2m)^2=\frac{4m^2}{n}.$$

(1)、(2)及上式导出

$$T\geqslant\frac{1}{3}\left(\frac{4m^2}{n}-nm\right)=4m\cdot\frac{m-\frac{n^2}{4}}{3n}.$$

5. 求所有从实数集到实数集的、满足下列条件的函数 f:

(1) $f(x)$严格增.

(2) 对所有实数 x, $f(x)+g(x)=2x$. 这里 $g(x)$是 $f(x)$的反函数.

解 我们证明 $f(x)-x=$常数 $c=f(0)$. 令 $S_c=\{x: f(x)-x=c\}$, 则 $0\in S_c$, S_c 非空.

如果 $a\in S_c$,那么 $f(a)=a+c$,从而 $g(a+c)=a$, $f(a+c)=2(a+c)-a=(a+c)+c$,即 $a+c\in S_c$.

设 $c'<c$, 可证 $S_{c'}$一定是空集. 取 $a\in S_c$,若 b 满足 $a\leqslant b<a+(c-c')$, 则 $b+c'<a+c=f(a)\leqslant f(b)$,所以 $b\notin S_{c'}$.

如果 b 满足 $a+(k-1)(c-c')\leqslant b<a+k(c-c')$ 时, $b\overline{\in} S_{c'}$. 那么在 $a+k(c-c')\leqslant b<a+(k+1)(c-c')$ 时, $(a+c)+(k-1)(c-c')\leqslant b+c'<(a+c)+k(c-c')$. 由归纳假设(用 $a+c\in S_c$ 代替a), $b+c'\notin S_{c'}$,从而 $b\notin S_{c'}$($b\in S_{c'}\Rightarrow b+c'\in S_{c'}$). 于是一切$\geqslant a$ 的实数均$\notin S_{c'}$.

如果有实数 $b\in S_{c'}$,那么$b+c'$, $b+2c'$, $\cdots$ 均 $\in S_{c'}$. 但这串数中必有$\geqslant a$ 的,与上面所证矛盾. 因此 $S_{c'}$是空集.

如果 $c'>c$ 并且$S_{c'}$不是空集,那么上面的证明表明 S_c 将是空集,矛盾. 因此对于一切 $c'\neq c$, $S_{c'}$是空集. 换言之, S_c 包含所有实数, $f(x)=x+c$.

1989 年澳大利亚数学竞赛试题及解答

随着经济实力的增长,澳大利亚在各个领域里都崭露头角.它主办了第二十九届国际数学奥林匹克,又发起了亚洲太平洋地区数学竞赛.堪培拉高等教育学院的 P. J. O'Halloran 教授尤为活跃,担任了世界各国数学竞赛联合会的主席与亚太地区数学竞赛委员会的主席,出版了一本杂志《数学竞赛》,专门报道各国数学竞赛的动态.

下面是 1989 年澳大利亚数学竞赛的试题及我们所拟的解答.

第一试(2 月 14 日,4 小时,每题 7 分)

1. 设正整数 n 的不同因数的个数为 $N_{(n)}$.例如 24 有因数 1, 2, 3, 4, 6, 8, 12, 24.所以 $N_{(24)}=8$. 试确定和

$$N_{(1)}+N_{(2)}+\cdots+N_{(1998)} \tag{1}$$

是奇数还是偶数.

解　由于 d 是 n 的因数时, $\frac{n}{d}$ 也是 n 的因数,并且只有在 n 为平方数, $d=\sqrt{n}$ 时 $d=\frac{n}{d}$. 所以 n 的因数是成对(d 与 $\frac{n}{d}$ 是一对)出现的,只有在 n 为平方数,有一个因数 $\sqrt{n}$ 自身配对.这就表明当且仅当 n 为平方数时, $N_{(n)}$ 为奇数.

由于 $45^2>1989>44^2$, 所以 1, 2, …, 1989 中有 44 个平方数,即和(1)中有 44 个奇数,从而这和是偶数.

2. 设 BP、CQ 是 $\triangle ABC$ 的内角平分线. AH、AK 分别为 A 至 BP、CQ 的垂线.证明 $KH \parallel BC$.

证明　延长 AH 交 BC 于 H',易知 $\triangle ABH$ 与 $\triangle H'BH$ 全等, $AH=H'H$, 同样延长 AK 交 BC 于 K',则 $AK=K'K$. 因此 $KH \parallel BC$.

3. 正数 $u_1, u_2, u_3, \cdots$满足条件 $u_1=1$,

$$u_n=\frac{1}{u_1+\cdots+u_{n-1}},\ n=2,\ 3,\ \cdots,$$

证明存在正整数 N,使 $u_1+u_2+\cdots+u_N>1989$.

证明 假设结论不存在,则恒有

$$u_1+u_2+\cdots+u_k\leqslant 1989,\ k=1,\ 2,\ \cdots,$$

故由递推公式

$$u_{k+1}\geqslant\frac{1}{1989},\ k=1,\ 2,\ \cdots,$$

从而取 $N=1989^2+1$, 则

$$u_1+u_2+\cdots+u_N>1989^2\times\frac{1}{1989}=1989,$$

矛盾!

4. 设 n 为偶数. 从整数 $1, 2, \cdots, n$ 中选出四个不同的数 a、b、c、d,满足 $a+c=b+d$. 证明选取的方法(译注:不考虑 a、b、c、d 的顺序)共 $\frac{n(n-2)(2n-5)}{24}$ 种.

证明 不妨设 $a>b>d$,则 $d>c$. 考虑从 $1, 2, \cdots, n$ 中选出三个数 $a>b>c$,满足 $a+c-b\neq b$ 的选法:选三个数 $a>b>c$ 有 C_n^3 种,其中满足 $a+c=2b$ 的就是 a 与 c 同奇偶的,共有 $\frac{n\left(\frac{n}{2}-1\right)}{2}$ 种(先从 $1, 2, \cdots, n$ 中取定一个有 n 种方法,与它同奇偶的数有 $\frac{n}{2}-1$ 个,有 $\frac{n}{2}-1$ 种选法. 由于不计顺序,共 $\frac{1}{2}n\left(\frac{n}{2}-1\right)$ 种选法,因此三元数组 $\{(a, b, c), n\geqslant a>b>c\geqslant 1, a+c-b\neq b\}$ 共有

$$C_n^3-\frac{n\left(\frac{n}{2}-1\right)}{2}=\frac{1}{12}n(n-2)(2n-5)$$

种. 上述每个三元数组确定一个符合题述要求的四元数组,但每个四元数组$(a,$

b, c, d)出现两次(a, b, c)及(a, d, c)产生相同的四元数组,所以问题的答案是$\frac{1}{24}n(n-2)(2n-5)$.

第二试(2月15日,4小时,每题7分)

5. n为非负整数,$d_0, d_1, \cdots, d_n$为0,1或2. $d_0+3d_1+\cdots+3^k d_k+\cdots+3^n d_n$是正整数的平方.证明在$0\leqslant i\leqslant n$中至少有一个$i$,使$d_i=1$.

证明 易知平方数除以3余数为0或1.如果$d_0\neq 1$,则$d_0=0$.从而

$$3(d_1+3d_2+\cdots+3^{n-1}d_n)$$

是平方数,括号中的数应被3整除,即$d_1=0$.于是$d_2+3d_3+\cdots+3^{n-2}d_n$是正整数的平方.

仿此进行,如果d_i均不为1,那么逐步得出$d_2, d_3, \cdots, d_n$均必须为0(如果愿意,可以用归纳法来论证),这与$d_0+3d_1+\cdots+3^n d_n$为正整数矛盾.

6. 四根木棒AB、BC、CD、DA在点A、B、C、D处自由地连结,可以在平面上移动使四边形$ABCD$的形状变更. P、Q、R分别为AB、BC、CD的中点.若木棒在某一位置时,$\angle PQR$为锐角,证明无论$ABCD$的形状如何变更,这角始终是锐角.

证明 连对角线AC、BD,相交于O,则易知$\angle AOD=\angle PQR$.

由于在某一四边形$ABCD$中,$\angle PQR$为锐角,所以这时$\angle AOD=\angle BOC$为锐角,$\angle AOB=\angle COD$为钝角,

$$AO^2+OD^2>AD^2,\ BO^2+OC^2>BC^2, \tag{1}$$

$$AO^2+BO^2<AB^2,\ CO^2+OD^2<CD^2, \tag{2}$$

从而

$$AB^2+CD^2>AD^2+BC^2. \tag{3}$$

若$\angle PQR$变为钝角或直角,则(1)、(2)中不等式均变为不等号方向相反的不等式(或等式),从而与(3)矛盾.所以$\angle PQR$始终为锐角.

7. $f(n)$定义在正整数集上,并且

(i) 对任一正整数n, $f(f(n))=4n+9$;

(ii) 对任一非负整数 k, $f(2^k)=2^{k+1}+3$. 试确定 $f(1789)$.

解 $$1789=9+4\times 9+4^2\times 9+4^3\times 9+4^4\times 2^2,$$

而 $$f(4n+9)=f(f(f(n)))=4f(n)+9.$$

所以

$$\begin{aligned}f(1789)&=9+4f(9+4\times 9+4^2\times 9+4^3\times 2^2)\\&=9+4\times 9+4^2f(9+4\times 9+4^2\times 2^2)\\&=9+4\times 9+4^2\times 9+4^3f(9+4\times 2^2)\\&=9+4\times 9+4^2\times 9+4^3\times 9+4^4f(2^2)\\&=9+4\times 9+4^2\times 9+4^3\times 9+4^4\times(2^3+3)\\&=1789+1792=3581.\end{aligned}$$

评注 可以证明 $f(n)=2n+3$.

8. 点 X、Y、Z 分别在$\triangle ABC$ 的边 BC、CA 与 AB 上,并且$\triangle ABC\backsim\triangle XYZ$,在 X、Y、Z 处的角分别等于在 A、B、C 处的角. 求 X、Y、Z 使得$\triangle XYZ$ 的面积为最小.

解 由于$\angle ZAY=\angle ZXY$, 所以$\triangle AZY$ 与$\triangle XYZ$ 的外接圆相等. 同样, $\triangle BXZ$、$\triangle CXY$ 的外接圆也都与它们相等.

设$\triangle ABC$ 的外接圆半径为 1, $\triangle XYZ$ 的外接圆半径为 e. 则 e 就是$\triangle XYZ$ 与$\triangle ABC$ 的相似比.

如图 1,连 AX, β、γ 分别为三角形 AZX、AYX 的外角,所以不难得知 $\beta+\gamma=2\alpha$. 令 $\beta=\alpha+x$, 则 $\gamma=\alpha-x$.

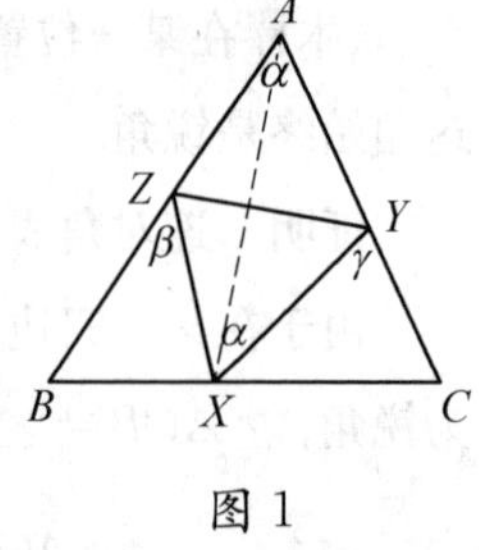

图 1

由正弦定理及 $BC=BX+XC$ 得

$$2\sin\alpha=2e(\sin(\alpha+x)+\sin(\alpha-x)).$$

即 $$2e\cos x=1.$$

所以 $$e=\frac{1}{2\cos x}\geqslant\frac{1}{2}.$$

即$\triangle XYZ$ 的面积最小时,相似比为 $\frac{1}{2}$. 不难推出这时 X、Y、Z 为$\triangle ABC$ 各边的中点.

第 31 届国际数学奥林匹克试题与解答

第 31 届国际数学奥林匹克于 1990 年 7 月在北京举行，我国选手成绩如下：

选手 \ 分数 \ 题号	1（平面几何）	2（组合数学）	3（数论）	4（函数方程）	5（博弈论）	6（组合几何）	总分
汪建华	7	7	7	7	7	7	42
周　彤	7	7	7	7	7	7	42
库　超	7	7	6	7	5	1	33
王　崧	7	7	6	7	7	7	41
余嘉联	7	7	2	7	7	6	36
张朝晖	7	7	7	7	7	1	36
	42	42	35	42	40	29	230

（每题满分为 7 分）下面是这次竞赛的试题与解答.

第一天(北京，1990 年 7 月 12 日 9：00—13：30）

1. 在一圆中，两条弦 AB、CD 相交于 E 点. M 为弦 AB 上严格在 E、B 之间的点，过 D、E、M 的圆在 E 点的切线分别交直线 BC、AC 于 F、G. 已知 $\frac{AM}{AB}=t$. 求 $\frac{GE}{EF}$（用 t 表示）.

2. 设 $n\geqslant 3$. 考虑在同一圆周上的 $2n-1$ 个互不相同的点所成的集合 E. 将 E 中一部分点染成黑色，其余的点不染颜色. 如果至少有一对黑点，以它们为端点的两条弧中有一条的内部（不包含端点）恰含 E 中 n 个点，则称这种染色方式为好的. 如果将 E 中 k 个点染黑的每一种染色方式都是好的，求 k 的最小值.

3. 求出（并予以证明）所有大于 1 的整数 n，使 $\frac{2^n+1}{n^2}$ 为整数.

第二天（北京，1990年7月13日9：00—13：30）

4. $\mathbf{Q}^+$ 是全体正有理数所成的集合. 试作一个函数 $f: \mathbf{Q}^+ \to \mathbf{Q}^+$，使得对任意的 x、$y \in \mathbf{Q}^+$，均有 $f(xf(y)) = \dfrac{f(x)}{y}$.

5. 给定一个初始整数 $n_0 > 1$ 后，两名选手 A、B 按以下规则轮流取整数 $n_1, n_2, n_3, \cdots$：

在已知 n_{2k} 时，选手 A 可以取任一整数 n_{2k+1}，使得 $n_{2k} \leqslant n_{2k+1} \leqslant n_{2k}^2$；

在已知 n_{2k+1} 时，选手 B 可以取任一整数 n_{2k+2}，使得 $\dfrac{n_{2k+1}}{n_{2k+2}}$ 是一个素数的正整数幂.

若 A 取到 1990，则 A 胜. 若 B 取到 1，则 B 胜.

对怎样的初始值 n_0，(1) A 有必胜策略，(2) B 有必胜策略，(3) 双方均无必胜策略？

6. 证明存在一个凸 1990 边形，同时具有下面的性质(i)与(ii)；

(i) 所有的内角均相等；

(ii) 1990 条边的长度是 $1^2, 2^2, \cdots, 1989^2, 1990^2$ 的一个排列.

解答

1. $\angle AEG = \angle FEM = \angle EDM$,

$\angle GEC = \angle AEG + \angle AEC = \angle EDM + \angle DEM = \angle DMB$,

$\angle GCE = \angle DBM$.

所以

$$\triangle GCE \backsim \triangle DBM, \ \frac{GE}{CE} = \frac{DM}{BM}. \tag{1}$$

同样

$$\triangle AMD \backsim \triangle CEF, \ \frac{AM}{MD} = \frac{CE}{EF}. \tag{2}$$

由(1)、(2)得

$$\frac{GE}{EF}=\frac{AM}{BM}=\frac{t}{1-t}.$$

2. 将 E 中的点依次记为 $1, 2, 3, \cdots, 2n-1$,并将点 i 与 $i+(n+1)$ 用一条边相连(我们约定 $j+(2n-1)k$, $k\in\mathbf{Z}$, 表示同一个点 j). 这样得到一个图 G. G 的每个点的次数均为 2(即与两个点相连),并且相差为 3 的两个点与同一个点相连.

由于 G 的每个点的次数为 2, G 由一个或几个圈组成.

(i) 在 $3\nmid 2n-1$ 时, $1, 2, \cdots, 2n-1$ 中每一个点 j 都可以表示成 $3k$ 的形式(即方程

$$3x\equiv j\,(\bmod\,(2n-1))$$

有解),因此图 G 是一个长为 $2n-1$ 的圈. 在这圈上可以取出 $n-1$ 个互不相邻的点,而且至多可以取出 $n-1$ 个互不相邻的点.

(ii) 在 $3\mid(2n-1)$ 时,图 G 由三个长为$\frac{2n-1}{3}$的圈组成,每个圈的顶点集合为

$$\left\{1+3k,\ k=0,\ 1,\ \cdots,\ \frac{2n-4}{3}\right\},\ \left\{2+3k,\ k=0,\ 1,\ \cdots,\ \frac{2n-4}{3}\right\},$$

$$\left\{3k,\ k=1,\ \cdots,\ \frac{2n-1}{3}\right\},$$

每个圈上至多可以取出 $\dfrac{\frac{2n-1}{3}-1}{2}=\dfrac{n-2}{3}$ 个点,两两互不相邻,总共可以取出 $n-2$ 个点互不相邻.

综上所述,在 $3\nmid 2n-1$ 时, $\min k=n$,在 $3\mid(2n-1)$ 时, $\min k=n-1$.

3. $n=3$.

显然 n 为奇数,设 p 为 n 的最小素因数,则

$$2^n\equiv-1(\bmod\,p). \tag{1}$$

设 a 为使

$$2^a \equiv -1 (\bmod p) \tag{2}$$

的最小正整数，

$$n = ka + r,\ 0 \leqslant r < a, \tag{3}$$

则

$$2^n = 2^{ka+r} \equiv (-1)^k \cdot 2^r (\bmod p). \tag{4}$$

若 k 为偶数，则由(1)、(4)

$$2^r \equiv -1 (\bmod p),$$

由 a 的最小性，$r = 0$.

若 k 为奇数，则由(1)、(4)，

$$2^r \equiv 1 (\bmod p),$$

从而

$$2^{a-r} \equiv -1 (\bmod p).$$

而 $0 < a - r \leqslant a$，由 a 的最小性，仍有 $r = 0$. 总之，$a \mid n$. 由费尔马小定理

$$2^{p-1} \equiv 1 (\bmod p),$$

若 $a > p - 1$，则 $a - (p-1) < a$，并且

$$2^{a-(p-1)} \equiv -1 (\bmod p),$$

与 a 的最小性矛盾. 因此 $a < p$.

由 p 的最小性及 $a \mid n$ 得 $a = 1$，即

$$2^1 \equiv -1 (\bmod p).$$

从而 $p = 3$.

令 $n = 3^k \cdot d$，$(d, 3) = 1$. 若 $k \geqslant 2$，则

$$2^n + 1 = (3-1)^n + 1 = 3n - \sum_{h=2}^{n} (-1)^h \mathrm{C}_n^h 3^h. \tag{5}$$

由于 $h!$ 中 3 的幂指数为

$$\left[\frac{h}{3}\right]+\left[\frac{h}{3^2}\right]+\cdots<\frac{h}{3}+\frac{h}{3^2}+\cdots=\frac{h}{2},$$

所以

$$C_n^h=\frac{n(n-1)\cdots(n-h+1)}{h!}$$

中 3 的幂指数 $>k-\frac{h}{2}$, (5)式右边和号中,各项都被 $3^{k-\frac{h}{2}+h+1}=3^{k+\frac{h}{2}+1}$ 整除 $(h\geqslant 2)$,而 $3n$ 恰被 3^{k+1} 整除,从而 $2^n+1=3^{k+1}d$, $(3, d)=1$. 由于 $n^2=3^{2k}d^2$, $n^2\nmid 2^n+1$, 矛盾!

因此 $n=3d$, $(3, d)=1$.

若 $d>1$, 令 q 为 d 的最小素因数,则

$$2^n\equiv -1(\bmod q),$$

令 j 为使

$$2^j\equiv -1(\bmod q)$$

的最小正整数,与前面类似,可得 $j\mid n$,再由 q 的最小性及 $j\leqslant q-1$,得 $j=1, 3$, 从而 $q\mid 3$ 或 3^2 均与 $(3, d)=1$ 矛盾. 因而 $d=1$, $n=3$.

4. 将质数依大小排列为 $p_1, p_2, p_3, \cdots$.

设 $a\in\mathbf{Q}^+$, 并且

$$a=p_1^{\alpha_1}p_2^{\alpha_2}p_3^{\alpha_3}\cdots p_{2k}^{\alpha_{2k}},\ \alpha_i\in\mathbf{Z},\ i=1, 2, \cdots, 2k.$$

则令

$$f(a)=p_2^{-\alpha_1}p_1^{\alpha_2}p_4^{-\alpha_3}p_3^{\alpha_4}\cdots p_{2k-1}^{\alpha_{2k}}.$$

对于 $$a=p_1^{\alpha_1}p_2^{\alpha_2}\cdots p_{2k}^{\alpha_{2k}},\ b=p_1^{\beta_1}p_2^{\beta_2}\cdots p_{2k}^{\beta_{2k}},$$

我们有

$$\begin{aligned}f(a\cdot b)&=f(p_1^{\alpha_1+\beta_1}p_2^{\alpha_2+\beta_2}\cdots p_{2k}^{\alpha_{2k}+\beta_{2k}})\\&=p_2^{-(\alpha_1+\beta_1)}p_1^{\alpha_2+\beta_2}\cdots p_{2k}^{-(\alpha_{2k-1}+\beta_{2k-1})}p_{2k-1}^{\alpha_{2k}+\beta_{2k}}=f(a)\cdot f(b),\end{aligned}$$

又

$$f(f(a))=f(p_2^{-\alpha_1}p_1^{\alpha_2}\cdots p_{2k-1}^{\alpha_{2k}})=p_1^{-\alpha_1}p_2^{-\alpha_2}\cdots p_{2k}^{-\alpha_{2k}}=a^{-1},$$

从而

$$f(xf(y))=f(x)\cdot f(f(y))=\frac{f(x)}{y}.$$

5. $n_0=2$ 时, A 只能取 2、3、4, B 可取 1, B 胜.

$n_0=3$ 时, A 只能取 3 至 9,均为形如 p^k 或 $2p^k$ 的数(p 为素数),从而 B 可取 1 或 2, B 胜.

$n_0=4$ 时, A 只能取 4 至 16,均为形如 p^k、$2p^k$ 或 $3p^k$ 的数,从而 B 可取 1、2 或 3, B 胜.

$n_0=5$ 时, A 只能取 5 至 25,均为形如 p^k、$2p^k$、$3p^k$ 或 $4p^k$ 的数, B 可取 1、2、3、4, B 胜.

若 $45\leqslant n_0\leqslant 1990$, 则 A 可取 1990, A 胜.

若 $21\leqslant n_0\leqslant 44$,则 A 可取 $420=2^2\times 3\times 5\times 7$, B 所取的数在 45 与 1990 之间. 由上一种情况, A 胜.

若 $13\leqslant n_0\leqslant 20$,则 A 取 $n_1=2^3\times 3\times 7=168$ B 所取的数在 21 与 1990 之间,因而 A 胜.

若 $11\leqslant n_0\leqslant 12$,则 A 取 $n_1=3\times 5\times 7=105$ B 所取的数在 15 与 1990 之间,因而 A 胜.

若 $8\leqslant n_0\leqslant 10$, 取 $n_1=2^2\times 3\times 5=60$, B 所取的数在 12 与 1990 之间, A 胜.

若 $n_0>1990$,取 $n_1=2^{r+1}\times 3^2$ 满足

$$2^r\times 3^2<n_0\leqslant 2^{r+1}\times 3^2<n_0^2,$$

则 B 所取的数 n_2,满足 $8\leqslant n_2<n_0$. 用 n_2 代替 n_0,继续采取上面的方法,经有限多步后得到

$$8\leqslant n_{2k}\leqslant 1990,$$

于是 A 胜.

最后,在 $n_0=6$ 或 7 时, A 取 $n_1=30$, 则 B 只能取 6(B 取 10 或 15,均 A 胜), A 再取 30,这样 A 立于不败之地.

另一方面,在≤49的数中,A只有取30与42才能保证$n_2 \geqslant 6$,这时B总可以取$n_2=6$. 因此,B可立于不败之地. 即在n_0为6或7时,两人均无必胜策略.

6. 问题等价于存在$1^2, 2^2, \cdots, 1990^2$的一个排列$a_1, a_2, \cdots, a_{1990}$使

$$\sum_{k=1}^{1990} a_k e^{ik\theta}=0, \tag{1}$$

其中
$$\theta=\frac{2\pi}{1990}=\frac{\pi}{995}.$$

令$\{(a_{2k-1}, a_{2k-1+995}), k=1, 2, \cdots, 995\}=\{((2n-1)^2, (2n)^2), n=1, 2, \cdots, 995\}$

(约定$a_{j+1990}=a_j$, $j=1, 2, \cdots$), 则(1)等价于

$$\sum_{k=1}^{995} b_k e^{2k\theta i}=0, \tag{2}$$

其中b_k是

$$(2n)^2-(2n-1)^2=2n-1, (n=1, 2, \cdots, 995)$$

的一个排列.

令

$$S_r=\sum_{i=0}^{4} b_{199t+r} e^{2(199t+r)\theta i}, (r=1, 2, \cdots, 199).$$

并取

$$b_{199t+r}=2(5r+t)-1, \tag{3}$$

则由于

$$\sum_{i=0}^{4} e^{2\times 199t\theta i}=0,\ S_r=2e^{2r\theta i}\sum_{i=0}^{4} t e^{2\times 199t\theta i}=2e^{2r\theta i}\cdot s,$$

其中s与r无关.

$$\sum_{k=1}^{995} b_k e^{2k\theta i}=2s\sum_{r=1}^{199} e^{2r\theta i}=0,$$

即(2)成立.

我们的训练工作*

1 成绩

第 31 届 IMO 中,我国选手取得五块金牌一块银牌(其中库超同学距金牌线只差一分).尽管试题比上一届难,6 名选手的总分仍高达 230 分,是唯一总分超过 200 分大关的队(居第二位的苏联队,总分为 197).好的成绩,首先应归功于我们的 6 位选手,他们是数十万高中学生的代表,聪明、勤奋、热爱祖国.

在这些学生的背后是广大的数学工作者,尤其是广大的中小学数学教师及其他从事普及工作的同志.他们辛勤地耕耘、播种,发行中学数学的刊物,出版数学奥林匹克的小册子,组织课外活动小组,举办各种水平的讲习班、报告会(数量,是一项世界之最).即使从 1981 年第一次全国高中数学联赛算起,也已经历时十年,现在(指 1990 年)开花结果,喜获丰收,是十分自然的,是一个集体努力的结晶.

2 做数学

学数学的最好办法是"做数学".IMO 是解题的竞赛,"解题是一种实践性的技能".因此,要在数学竞赛中取得好成绩,必须做大量练习.

在《数学竞赛史话》(广西教育出版社 1990 年出版)这本书中,我曾说过去年我国的选手,"每个人都至少做了 1000 道难题,每一本笔记本都是难题的海洋".第 31 届 IMO 的中国队队员也是如此.蒋步星(上届队员,金牌获得者)的一本笔记就在本届队员手中流传,全国各地有不少学生继承了这样的"秘本",他们做过的题比我们多,解题的能力与速度比我们强.作为教练,常常感到捉襟见肘,难以应付.

---

* 本文作者时任第 31 届 IMO 中国队的领队兼总教练.

教练比学生高明的地方,大概在对数学的理解.通过教练对问题的分析,学生逐步了解了数学中的思想、方法,提高了观点.这是一个潜移默化的过程.

对于能力强的学生,题的质量比数量更重要.我们的选手能够解各种类型的问题:平面几何、组合数学、数论、函数方程、组合几何……要求教练也精通每一个方面是困难的.但我们可以发挥集体的力量,每一位教练讲一两个专题,同一个专题也可以请几位不同的教练来讲,百花齐放,切忌单一、僵化,一两位教练"包打天下".

虽然今年有很多数学家参加选题工作,不能给集训队上课,但我们仍尽量邀请其他同志以弥补这一损失.今年参加训练工作的,除刘鸿坤副领队与我外,还有严镇军、杜锡禄、黄宣国、李大元、康士凯、曹鸿德、胡大同、马明、冯惠愚、李克正、周春荔等十多位名流学者.讲授风格不尽相同,更能活跃学生的思维.

教练的一个重要作用是引导学生讨论.让各地优秀的学生较早地集中在一起,切磋琢磨,对他们的成长有好处.汪建华、张朝晖、余嘉联在去年就被清华大学与清华附中合办的数学试验班录取(严镇军、孔令颐等老师给这个班上了一年数学),湖北黄冈中学的库超与王崧更是"难兄难弟",从小就在一起研究难题.

与1986年相比,现在大部分训练工作已由基层完成.国家集训队的重点已经由数学方面的训练转为如何选准6名队员与加强心理训练.

3 训练的第一阶段

训练的第一阶段在南京师范大学进行,由葛军担任班主任,南京师大的领导对这一工作极为重视,江苏教育出版社提供了赞助.

这一阶段是强化训练,"大运动量":每天上、下午都上课,还布置习题晚上完成.每周至少一次测试,每次测试3个小时、4道题,难度不亚于IMO,时间则少得多.集训的24名选手中只有四分之一能被选上,因而有一定的压力.在这样紧张的训练中仍能脱颖而出,当然是得过硬的,不仅数学好,而且有坚韧不拔的毅力,十分顽强.

需要指出,被淘汰的18名学生也绝非弱者.或许由于偶然的失误,或许由于他们不是"竞赛型"的人物.要求在短时间内解答难题的能力,对于研究工作是没有必要的. 年轻时受到一些挫折可能还是好事,因为“天将降大任于斯人也”.

4 训练的第二阶段

第二阶段的训练原定在北大附中,后因该校修理房屋而移至清华大学. 由高珍光担任班主任. 清华大学领导,清华附中郑增仪副校长,北大附中孙承彪副校长对训练均很关心.

“真正的胜利,并不是你能用武器争取到的,那一定要用你的信心.”

去年我们首次取得总分第一,因此本届选手大受鼓舞,有信心再拿第一.

中国数学会理事长王元、中国数学奥林匹克委员会主席王寿仁、中国数学会普及工作委员会主任裘宗沪都没有要我们“立军令状”. 他们很清楚我们会全力以赴,争取好的成绩. 这样的信任比什么都重要.

我们并未感到有沉重的压力,也从未对学生施加压力. 因为我们有信心,相信这些学生:他们都是出类拔萃的,没有必要反复唠叨“戒骄戒躁”“为国争光”之类的套话.

这一阶段,由于 6 名队员已经选定,我们的训练已逐步由紧到松,每天只上半天课. 平时不测验,只做了两次模拟考试,每次两天,每天在 4 或 $4\frac{1}{2}$ 小时内完成 3 道题,与 IMO 接近或相同. 最后一次的成绩如下表:

题目 / 得分 / 人	1 (平面几何)	2 (组合数学)	3 (代数与数论)	4 (组合数学)	5 (组合数学)	6 (组合几何)	Σ
汪建华	6	7	7	7	0	7	34
周 彤	6	6	0	7	7	7	33
库 超	7	7	7	7	6	7	41
王 崧	7	7	7	7	7	7	42
余嘉联	6	7	0	7	3	7	30
张朝晖	6	7	7	7	7	7	41
Σ	38	41	28	42	30	42	221

从这张表可以预测每名选手都至少能得银牌，如果运气好一些，就可以夺得金牌. 这样的模拟考试，增强了选手的信心.

这一阶段不再做大量的难题，更多的时间用于讨论与总结. 通过讨论，将一道或几道题吃透，总结出最好的解法. 训练过多有一个副作用：一些学生将简单的问题想复杂了，走了不少弯路(尤其是做了太多的难题后，以为每一道题都是难题). 我们则希望他们“返璞归真”，找出最简单、最自然的解法，也就是要求他们单刀直入，立刻剖析问题核心，而不是在外围大兜圈子.

我们还强调了表达，因为表达的清晰反映了思维的清晰. 这方面，刘鸿坤老师做了极细致的工作，对学生的作业逐字逐句地批改. 看来，表达与书写在一个相当长的时期内仍然是严重的问题. 目前中学语文课作文很少，恐怕是造成这一问题的主要原因.

5 考前休整

考试前的半个月，学生在北戴河“休整”，彻底放松：游泳、打牌、看足球赛(世界杯)与武侠小说. 王崧买了金庸的《鹿鼎记》《天龙八部》《倚天屠龙记》等 14 本书. 他说：“我的父母要我一心一意学习数学，万万想不到在这里成天看小说. ”学生能够彻底放松，正表明他们满怀信心.

6 难度

本届试题比上一届难. 我们认为这样的难度恰到好处，与平时训练的路子也大致相近，其中前 5 道题估计问题不大，第 6 道题则有赖于选手的临场发挥.

出乎意料的是其他强国：苏、美、罗、德……考得不很好，如第一道平面几何，相当容易，苏联的成绩却很差. 我们的选手则发挥正常，1、2、4 三题都拿了满分. 当然，张朝晖与库超对自己的表现还不够满意.

将赛前赛后成绩表对照一下，就会发现正像王崧所说的“考试无常”. 我们只能保证在总体上不出现大的差错，而每一位选手的成绩有起有伏则是正常现象.

张朝晖，解难题的能力最强；库超，全面发展，“具有拿金牌的实力”，他们很可能是由于要好心切,反而影响了发挥."怪才"王崧,有点懵懵懂懂,对考分置之度外,成绩最为稳定,第二天每道题他只用了半个小时.文静的周彤,在集训中一直名列前茅,模拟考试却首次出现一道题失误.看来,赛前的小挫折倒是值得庆贺的,因为它减少了在比赛中出现失误的机会.被誉为"一号种子"的汪建华,去年已参加过集训,在南京师大时,心理压力较大,"并不是每次都考第一",经过在清华、北戴河的调整,恢复了自信.他与周彤都取得了满分(另两名获得满分的选手是法国的维森特·莱夫格、苏联的叶甫盖尼娅·玛林列科娃).兴趣广泛的余嘉联"担心自己考不好",第一天有一道题失误,但第二天,他十分顽强,背水一战终于完成了最难的第6道题.

沉着、顽强,对于IMO这样大的比赛是十分重要的.很多人喜欢思维敏捷的学生.但对于难度较高的问题,敏捷与否并不起决定作用.

7 教练

我们的学生经过严格的选拔与训练,基础扎实、全面,近几年内中国队将稳定在十强或六强的行列里,不会出现滑坡.谁也不能保证每次都拿总分第一.因为我们的队员是年年更换的,而且考试总有偶然性.更重要的,IMO是为了促进数学科学的发展,增进友谊.过分强调总分的多少,反倒背离了这个目标.

迫切的问题可能在教练方面.目前中国数学奥林匹克委员会教练组的成员都是超过或接近五十岁的人.必须年轻化!今年,我们曾请一些平均年龄不到30的同志担任教练,效果很好,深受学生欢迎.他们是熊斌、余红兵、陈计,吴伟朝、刘亚强、葛军、高珍光.希望年轻人能在今后的训练中发挥更大的作用.

苏联、美国、罗马尼亚等国都有很多数学家热心数学竞赛,将自己的研究成果编为问题,转入竞赛之中.在这方面,我们有较大的差距.我们的题目引进的多,出口的少.科研与数学竞赛联系不多,研究命题的人不多.而且,所有的教练都是业余的,绝大部分时间需用于本职工作.要保持我国在IMO中领先的地位,这些都是亟待解决的问题.

第 6 届巴尔干数学竞赛试题及解答

1. 设 $d_1, d_2, \cdots, d_k$ 为正整数 n 的全部因数，$1=d_1<d_2<d_3<\cdots<d_k=n$. 求出使 $k\geqslant 4$ 并且 $d_1^2+d_2^2+d_3^2+d_4^2=n$ 的所有 n.

解　若 n 为奇数，则 $d_1, d_2, \cdots$ 都是奇数，但四个奇数的平方和为偶数，不可能等于 n. 矛盾.

于是 n 为偶数，$d_2=2$.

若 n 是 4 的倍数，则 $4\in\{d_3, d_4\}$，前 4 个因数的平方中已有两个(2^2 与 4^2)为 4 的倍数，1 个为 1. 另一个因数为奇数时，它的平方除以 4 余 1；为偶数时，它的平方被 4 整除. 因此这 4 个因数的平方和不是 4 的倍数，仍得矛盾.

因此，$n=2m$，m 为奇数. d_3 是 m 的最小质因数. 由于 $1^2+2^2+d_3^2+d_4^2=n$ 为偶数，所以 d_4 为偶数，从而 $d_4=2d_3$.

$n=1+2^2+d_3^2+4d_3^2\Rightarrow 5$ 是 n 的因数. 由于 $d_4\neq 4$，所以 $d_4\geqslant 6$，从而 $d_3=5$，$d_4=10$，$n=1+2^2+5^2+10^2=130$.

2. 设 $\overline{a_na_{n-1}\cdots a_1a_0}=10^na_n+10^{n-1}a_{n-1}+\cdots+10a_1+a_0$ 为一个质数的十进表示. $n>1$，$a_n>1$. 证明：多项式 $P(x)=a_nx^n+a_{n-1}x^{n-1}+\cdots+a_1x+a_0$ 不可约，即不能分解为两个次数为正的整系数多项式的积.

本题甚难. 关键是要知道一个引理.

引理　如果 $a_0, a_1, \cdots, a_n\in\{0, 1, 2, \cdots, 9\}$，$n\geqslant 1$，$a_n\geqslant 1$，则多项式

$$P(x)=a_nx^n+a_{n-1}x^{n-1}+\cdots+a_1x+a_0$$

的根的实部小于 4.

证明　设复数 z 的实部 $R(z)\geqslant 4$，则

$$R\left(\frac{1}{z}\right)\geqslant 0,\ |z|\geqslant 4,$$

因此

$$\left|\frac{P(z)}{z^n}\right| \geqslant \left|a_n + \frac{a_{n-1}}{z}\right| - \frac{a_{n-2}}{|z|^2} - \frac{a_{n-3}}{|z|^3} - \cdots - \frac{a_0}{|z|^n} >$$

$$R\left(a_n + \frac{a_{n-1}}{z}\right) - \frac{9}{|z|^2} - \frac{9}{|z|^3} - \cdots \geqslant 1 - \frac{9}{|z|^2 - |z|} > 0.$$

所以 $P(x)$的根的实部必小于 4.

现在回到原来的问题.

证明 如果 $P(x)=g(x)h(x)$, 其中 g、h 都是次数为正的整系数多项式,那么设 $\alpha_1, \alpha_2, \cdots, \alpha_k$ 为$g(x)$的根,则$R(\alpha_j)<4$, $(1\leqslant j\leqslant k)$. 于是$g(4)\neq 0$,从而$|g(4)|\geqslant 1$. 同时$|10-\alpha_j|>|4-\alpha_j|$, $j=1, 2, \cdots, k$ (利用复数的几何表示易知),所以

$$|g(10)|>|g(4)|\geqslant 1.$$

同理

$$|h(10)|>|h(4)|\geqslant 1.$$

于是,质数 $P(10)$分解为两个大于 1 的整数$|g(10)|$与$|h(10)|$的积,矛盾. 因此, $P(x)$不可约.

3. 直线 l 交$\triangle ABC$ 的AB 边于B_1,交AC 于C_1, $\triangle ABC$ 的重心G 与A 点在 l 的同一侧. 证明:

$$S_{BB_1GC_1} + S_{CC_1GB_1} \geqslant \frac{4}{9} S_{\triangle ABC}.$$

等号何时成立?

解 如图 1,设 D 为BC 中点. 连 DB_1、DC_1,则

$$\begin{aligned} S_{BB_1GC_1} + S_{CC_1GB_1} &= 2S_{GB_1C_1} + S_{BC_1B_1} + S_{CB_1C_1} \\ &= 2S_{GB_1C_1} + 2S_{BB_1C_1C} - S_{BC_1C} - S_{B_1BC_1} \\ &= 2(S_{GB_1C_1} + S_{BB_1C_1C} - S_{DC_1C} - S_{B_1BD}) \\ &= 2S_{GB_1DC_1} \\ &= 2(S_{GB_1D} + S_{GDC_1}) \\ &= \frac{2}{3}(S_{AB_1D} + S_{ADC_1}). \end{aligned}$$

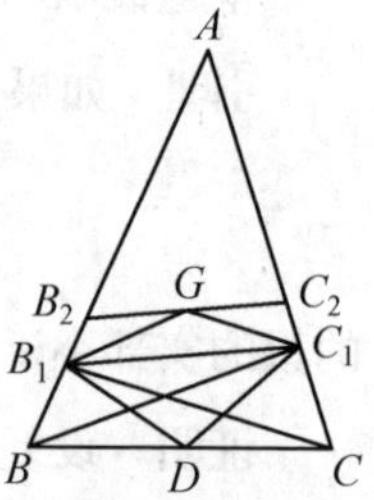

图 1

过 G 作直线平行于 l，分别交 AB、AC 于 B_2、C_2. 由于 G 与 A 在 l 的同侧，所以 B_2 在线段 AB_1 内，C_2 在 AC_1 内. 设

$$\frac{AB_2}{AB}=\lambda,\ \frac{AC_2}{AC}=\mu,$$

则
$$\frac{S_{AB_2C_2}}{S_{ABC}}=\lambda\mu.$$

但另一方面，

$$\frac{S_{AB_2C_2}}{S_{ABD}}=\frac{S_{AB_2G}}{S_{ABC}}+\frac{S_{AGC_2}}{S_{ABC}}=\frac{1}{2}\left(\frac{S_{AB_2G}}{S_{ABD}}+\frac{S_{AGC_2}}{S_{ADC}}\right)=\frac{1}{2}\left(\frac{2\lambda}{3}+\frac{2\mu}{3}\right)=\frac{1}{3}(\lambda+\mu),$$

所以，
$$\lambda\mu=\frac{1}{3}(\lambda+\mu).$$

即
$$\frac{1}{\lambda}+\frac{1}{\mu}=3,$$

从而，

$$\lambda+\mu=\frac{1}{3}\left(\frac{1}{\lambda}+\frac{1}{\mu}\right)(\lambda+\mu)\geqslant\frac{4}{3}.$$

于是，

$$S_{BB_1GC_1}+S_{CC_1GB_1}\geqslant\frac{2}{3}(S_{AB_2D}+S_{ADC_2})$$
$$=\frac{2}{3}(\lambda S_{ABD}+\mu S_{ADC})=\frac{1}{3}(\lambda+\mu)S_{ABC}\geqslant\frac{4}{9}S_{ABC}.$$

4. F 为 $\{1,2,\cdots,n\}$ 的一个子集族，满足

(1) 若 $A\in F$，则 $|A|=3$；

(2) 若 $A\in F$, $B\in F$, $A\neq B$，则 $|A\cap B|\leqslant 1$.

设 $f(n)=\max|F|$，证明：在 $n\geqslant 3$ 时，

$$\frac{1}{6}(n^2-4n)\leqslant f(n)\leqslant\frac{1}{6}(n^2-n).$$

（$|S|$ 表示集 S 的元素个数）

证明 F 中每个集 $A=\{a, b, c\}$ 有 3 个二元子集 $\{a, b\}$, $\{b, c\}$, $\{c, a\}$. 由(2), F 中每两个集的二元子集互不相同,这样的二元子集共有 $3|F|$ 个. 而 $\{1, 2, \cdots, n\}$ 的二元子集共 C_n^2 个,因此,

$$3|F| \leqslant C_n^2.$$

从而,

$$|F| \leqslant \frac{1}{3}C_n^2 = \frac{1}{6}(n^2-n),$$

$$f(n) \leqslant \frac{1}{6}(n^2-n).$$

另一方面,考虑三元集 $A=\{a, b, c\}$, 其中

$$c=\begin{cases} n-(a+b), & (a+b<n), \\ 2n-(a+b), & (a+b \geqslant n). \end{cases} \qquad (*)$$

c 被 a、b 唯一确定. b 也被 a、c 唯一确定:

$$b=\begin{cases} n-(a+c), & (a+c<n), \\ 2n-(a+c), & (a+c \geqslant n). \end{cases}$$

同样有 a. 因此满足($*$)的三元集两两的交至多 1 个元素. 这种三元集的族 F 满足题设要求(1)、(2).

a 有 n 种选择, $b \neq a$, $n-2a$, $2n-2a$, $\frac{n-a}{2}$, $\frac{2n-a}{2}$(即不使 $n-(a+b)$或 $2n-(a+b)$ 等于 a 或 b),但 b 等于 $n-2a$ 与 $2n-2a$ 不可能同时发生(前者发生时 $a<\frac{n}{2}$,后者 $2n-2a \leqslant n$ 即 $a \geqslant \frac{n}{2}$), 所以 b 至少有 $n-4$ 种选择, c 由 a、b 唯一确定. 所以族 F 的元素 $\{a, b, c\}$ 的个数 $\geqslant \frac{1}{6}n(n-4)$. 因而,

$$f(n) \geqslant \frac{1}{6}n(n-4).$$

1990 年加拿大数学竞赛

1　试题

1. 一次竞赛有 $n \geqslant 2$ 名选手参加,历时 k 天.每天选手的得分为 $1, 2, \cdots, n$,每两名选手的得分互不相同.在第 k 天末,每名选手的总分均为 26 分.求出使这成为可能的所有的数对 (n, k).

2. $\frac{1}{2}n(n+1)$ 个不同的数随机地排成三角阵(图 1),设 M_k 为第 k 行(自上往下数)的最大数,求 $M_1 < M_2 < M_3 < \cdots < M_n$ 成立的概率.

```
         *
       *   *
     *   *   *
       ……
       ……
*  *            *  *
```

图 1

3. 设 $ABCD$ 为圆内接四边形,对角线 AC 与 BD 相交于 X. X 到边 AB, BC, CD, DA 的垂线的垂足分别为 A', B', C', D'.证明:$A'B' + C'D' = A'D' + B'C'$.

4. 一质点可以每秒 2 米的速度沿 x 轴前进,以每秒 1 米的速度在平面上其他地方前进,质点从原点出发,求它可以达到的区域.

5. 设函数 f 定义在正整数上,满足 $f(1)=1$, $f(2)=2$,

$$f(n+2)=f[n+2-f(n+1)]+f[n+1-f(n)]\ (n \geqslant 1).$$

(Ⅰ) 证明:

(i) $0 \leqslant f(n+1)-f(n) \leqslant 1$;

(ii) 若 $f(n)$为奇数,则 $f(n+1)=f(n)+1$.

(Ⅱ) 确定(并证明)使 $f(n)=2^{10}+1$ 的所有 n.

2　解答

1. 易知 k 天的分数之总和为

$$k \times \frac{1}{2} n(n+1) = 26n.$$

所以，
$$k(n+1) = 52.$$

从而，$(n, k) = (51, 1), (25, 2), (12, 4), (3, 13)$. 只有$(51, 1)$不可能实现.

对$(25, 2)$有

$$(26, 26, \cdots, 26) = (1, 2, \cdots, 24, 25) + (25, 24, \cdots, 2, 1).$$

对$(12, 4)$有

$$(26, 26, \cdots, 26) = 2(1, 2, \cdots, 11, 12) + 2(12, 11, \cdots, 2, 1).$$

对$(3, 13)$有

$$\begin{aligned}(26, 26, 26) = &(1, 2, 3) + 2(2, 3, 1) + 2(3, 1, 2) \\ &+ 3(1, 3, 2) + 2(3, 2, 1) + 3(2, 1, 3).\end{aligned}$$

2. 设所求概率为 p_n，显然，$p_1 = 1$, $p_2 = \frac{2}{3}$.

由于最大的数必须在最后一行出现，才能有题述的不等式. 这种情况发生的概率为
$$\frac{n}{\frac{1}{2} n(n+1)} = \frac{2}{n+1}.$$

所以，
$$p_n = \frac{2}{n+1} p_{n-1} = \cdots = \frac{2^n}{(n+1)!} \quad (n \geqslant 2).$$

3. 由正弦定理，

$$A'B' = XB \sin\angle ABC,\ C'D' = XD \sin\angle ADC.$$

由于 $\angle ABC + \angle ADC = \pi$，所以，

$$A'B' + C'D' = (XB + XD)\sin\angle ABC = BD \sin\angle ABC.$$

同理，
$$A'D' + B'C' = AC \sin\angle BAD,$$

而
$$\frac{BD}{\sin\angle BAD} = \frac{AC}{\sin\angle ABC} = \text{圆 } ABC \text{ 的直径},$$

所以，
$$A'B' + C'D' = A'D' + B'C'.$$

4. 由于对称性,我们只考虑第一象限.设质点沿 x 轴运动至 $(t, 0)$,然后再到点 (x, y),则

$$(x-t)^2+y^2 \leqslant \left(\frac{2-t}{2}\right)^2,$$

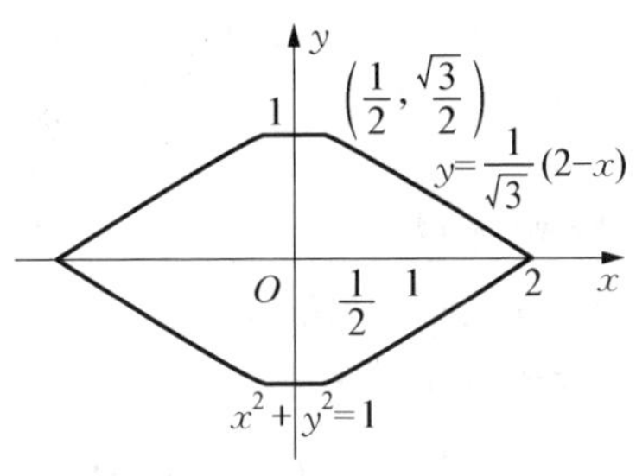

图 2

即

$$y^2 \leqslant \frac{3}{4}\left(\frac{4}{9}(2-x)^2-\left(t-\frac{2(2x-1)}{3}\right)^2\right).$$

当 $\frac{1}{2} \leqslant x \leqslant 2$ 时, y 在 $t=\frac{2}{3}(2x-1)$ 时取得最大值 $\frac{1}{\sqrt{3}}(2-x)$.

当 $0 \leqslant x \leqslant \frac{1}{2}$ 时, y 在 $t=0$ 时取得最大值 $\sqrt{1-x^2}$.

于是所求区域如图 2 所示.

5. (Ⅰ) (i) 设对 $n=1, 2, \cdots, m-1$, 有

$$f(n+1)-f(n)=0 \text{ 或 } 1. \tag{1}$$

则对每个 $n \in \{1, 2, \cdots, m-1\}$,

$$[(n+2)-f(n+1)]-[(n+1)-f(n)]=1-[f(n+1)-f(n)] \in \{0, 1\}.$$

因此,由归纳假设

$$f[n+2-f(n+1)]-f[n+1-f(n)] \in \{0, 1\}. \tag{*}$$

情况 1: $f(m)=f(m-1)+1$. 这时,

$$\begin{aligned} f(m+1)-f(m) &= f[m+1-f(m)]-f[m-1-f(m-2)] \\ &= f[m-f(m-1)]-f[m-1-f(m-2)] \in \{0, 1\}. \end{aligned}$$

(最后一步根据(*))

情况 2: $f(m)=f(m-1)$. 这时由已知的递推式

$$f[m-f(m-1)]=f[m-2-f(m-3)].$$

根据(*),它们均等于 $f[m-1-f(m-2)]$. 从而,

$$f(m+1)-f(m)=f[m+1-f(m)]-f[m-1-f(m-2)]$$
$$=f[m+1-f(m)]-f[m-f(m-1)]\in\{0,1\}.$$

(根据(∗))

无论哪种情况，(∗)均对 $n=m$ 成立. 从而(∗)对所有自然数成立,即(1)对所有自然数成立.

(ii) 设结论在 $n<m$ 时成立. 若 $f(m)$为奇数,则 $f(m-1)$ 必为偶数(否则,由归纳假设, $f(m)=f(m-1)+1$, 与 $f(m)$为奇数矛盾). 于是,

$$f(m)=f(m-1)+1,$$
$$f(m+1)=f[m+1-f(m)]+f[m-f(m-1)]$$
$$=2f[m-f(m-1)].$$

即 $f(m+1)$ 为偶数,因而由(i)

$$f(m+1)=f(m)+1.$$

(Ⅱ) 我们用归纳法来证明:对任意整数 $k>1$, $n=2^k$ 是方程 $f(n)=2^{k-1}+1$ 的唯一解. 从而 $k=2^{11}$ 是 $f(n)=2^{10}+1$ 的唯一解.

$k=2$ 时,结论显然. 设 $n=2^m$ 是 $f(n)=2^{m-1}+1$ 的唯一解.

由于 $f(n)$的值每次增加 0 或 1,并且从已知的递推式可以看出 $f(n)\to\infty$ (否则从某一时刻起, $f(n)$将为(正的)常数值,而对于足够大的 n, $f(n+2)$ 却为此值的两倍,矛盾),所以必有整数 u,使 $f(u)=2^m+1$.

这时, $f(u-1)$ 必为偶数,并且 $f(u-1)=2^m$.

由于

$$f[u-f(u-1)]+f[u-1-f(u-2)]=f(u)=2^m+1,$$

并且左端两项之差为 0 或 1,所以

$$f[u-f(u-1)]=f[u-1-f(u-2)]+1=2^{m-1}+1.$$

由归纳假设 $$u-f(u-1)=2^m,$$

从而, $$u=2^m+f(u-1)=2^{m+1}.$$

1990年亚太地区数学竞赛

1　试题

1. 在$\triangle ABC$中，D、E、F分别为BC、CA、AB的中点，G为重心，对$\angle BAC$的每个值，有多少互不相似的$\triangle ABC$，使$AEGF$为圆内接四边形？

2. a_1，a_2，…，a_n为正实数，S_k为从a_1，a_2，…，a_n中每次取k个所得的乘积的和. 证明：

$$S_k S_{n-1} \geqslant (C_n^k)^2 a_1 a_2 \cdots a_n, \quad (k=1, 2, \cdots, n-1).$$

3. 考虑所有底AB固定，C点引出的高为常数h的$\triangle ABC$. 什么时候它的三条高的乘积为最大？

4. 1990个人分为若干互不相交的子集，使得

(a) 每个子集中没有人认识这子集中所有人.

(b) 每个子集中，任意三个人中至少有两个人互不相识.

(c) 每个子集中，对任意两个不相识的人，这子集中恰有一个人认识这两个人.

(Ⅰ) 证明在每个子集中，各个人认识的人数相等.

(Ⅱ) 求上述子集最多能有几个？

注：约定A认识B，则B认识A，每个人认识他自己.

5. 证明对每个整数$n \geqslant 6$，存在一凸六边形，它可以分为n个全等的三角形.

2　解答

1. 如图1，由A，E，G，F共圆，得

$$\angle CGE = \angle BAC = \angle CED. \tag{1}$$

设 CG 交 DE 于 M，则由(1)易导出

$$CM \times CG = CE^2,$$

即
$$\frac{1}{2}m_c \times \frac{2}{3}m_c = \left(\frac{1}{2}b\right)^2. \tag{2}$$

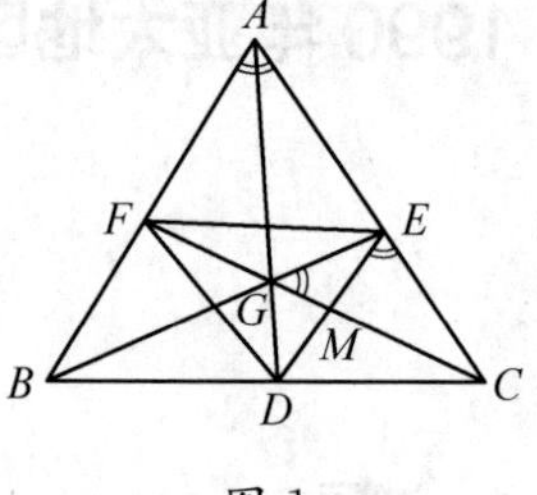

图 1

这里中线

$$m_c^2 = \frac{1}{2}a^2 + \frac{1}{2}b^2 - \frac{1}{4}c^2. \tag{3}$$

由(2)、(3)得

$$2a^2 = b^2 + c^2. \tag{4}$$

从而，由余弦定理及(4)有

$$b^2 + c^2 = 4bc\cos\angle BAC. \tag{5}$$

令 $\frac{b}{c} = \lambda$，则

$$\lambda^2 + 1 = 4\lambda\cos\angle BAC. \tag{6}$$

(6)式中当且仅当 $\angle BAC \leqslant 60^\circ$ 时有实数解，所得的解 λ_1、λ_2 互为倒数，因而产生相似的三角形. 于是，对 $\angle BAC \leqslant 60^\circ$，满足条件的互不相似的 $\triangle ABC$ 只有一个. 在 $\angle BAC > 60^\circ$ 时，无满足条件的 $\triangle ABC$.

2. 由算术—几何平均值不等式

$$\begin{aligned} S_k S_{n-k} &\geqslant \mathrm{C}_n^k \cdot \mathrm{C}_n^{n-k}\left(\prod a_{j_1 j_2 \cdots j_k}\right)^{\frac{1}{\mathrm{C}_n^k}} \cdot \left(\prod a_{j_1 j_2 \cdots j_{n-k}}\right)^{\frac{1}{\mathrm{C}_n^{n-k}}} \\ &= (\mathrm{C}_n^k)^2 (a_1 a_2 \cdots a_n)^{\mathrm{C}_{n-1}^{k-1}/\mathrm{C}_n^k} \cdot (a_1 a_2 \cdots a_n)^{\mathrm{C}_{n-1}^{n-k-1}/\mathrm{C}_n^k} \\ &= (\mathrm{C}_n^k)^2 (a_1 a_2 \cdots a_n) \frac{(\mathrm{C}_{n-1}^{k-1} + \mathrm{C}_{n-1}^k)}{\mathrm{C}_n^k} = (\mathrm{C}_n^k)^2 a_1 a_2 \cdots a_n. \end{aligned}$$

3. 由于 $h_a \cdot a = h_b \cdot b = h \cdot c$ 为定值，所以，

$$h_a h_b h = \frac{h^3 \cdot c^2}{ab},$$

只需使 ab 为最小.

当 $a=b$ 时，$\triangle ABC$ 为等腰三角形. 设这时 C 为 C_0，外接圆 $\odot O$ 在 C_0 点的切线为 l，则 $l /\!/ AB$ 并且与 AB 的距离为 h.

$\triangle ABC$ 的顶点 C 必在 l 上，由于 l 为切线，C 在 $\odot O$ 外，$\angle ACB < \angle AC_0B$，从而

$$CA \times CB = \frac{ch}{\sin\angle ACB} > \frac{ch}{\sin\angle AC_0B} = C_0A \times C_0B.$$

即当 C 与 C_0 重合时，ab 为最小.

4. (Ⅰ) 只考虑一个小组. 设 y_1 与 y_2 在同一组中互不相识，由(c)，存在 x 与 y_1、y_2 均相识. 设除去 x 与自身外，y_1 认识的人为 z_{11}，z_{12}，…，z_{1h}. y_2 认识的人为 z_{21}，z_{22}，…，z_{2k}.

由(b)，x 与 z_{ij} 均不相识. 由(c)，z_{11}，z_{12}，…，z_{1h} 与 y_2 均不相识. 所以，z_{11}，z_{12}，…，z_{1h}，z_{21}，z_{22}，…，z_{2k} 互不相同，并且与 x、y_1、y_2 也互不相同.

由(c)，z_{11} 与 y_2 有一公共的熟人，不妨设为 z_{21}，z_{12} 与 y_2 也有一公共的熟人，此人决非 z_{21}（否则，y_1 与 z_{21} 有两个公共熟人 z_{11}、z_{12}，与(c)矛盾). 设 z_{12} 与 y_2 的公共熟人为 z_{22}. 如此继续下去，可知 $h \leqslant k$ (z_{11}，z_{12}，…，z_{1h} 在 z_{21}，z_{22}，…，z_{2k} 中各有一个熟人，并且这些熟人互不相同). 由对称性，亦有 $k \leqslant h$. 所以，$k=h$，即互不相识的两个人，熟人的个数相同.

对于 y_1 的熟人 x，x 的熟人 y_2 与 y_1 互不相识. 由上面的论证，我们知道 x 与 z_{11} 的熟人个数相同，z_{11} 与 y_2 的熟人个数相同. 从而，这一组中，每个人的熟人个数均为 $h+1$.

(Ⅱ) 在一组中，x 有一不认识的人 y，x、y 有一公共熟人 z. z 有一不认识的人 u. u 不可能与 x、y 都认识(否则与(c)矛盾). 设 u 不认识 x，则 u、x 有一公共熟人 v，v 不同于 y、z. 于是每一组至少 5 个人. 5 个人的组是可以存在的，只需 x 认识 y，y 认识 z，z 认识 u，u 认识 v，v 认识 x，其余的每两对人互不相识.

于是小组的个数至多为 $\frac{1990}{5}=398$.

5. 考虑▱$ABCD$，其中 $\angle BAD$ 为钝角. 将这平行四边形用与 AD 平行的直线分为 n 个全等的平行四边形，每个平行四边形又用对角线分为两个三角形，

从而得到 $2n$ 个全等的三角形.

将图 2 中$\triangle AED$ 关于 AD 对称得$\triangle AE'D$. 再将$\triangle BFC$ 关于 BC 对称得$\triangle BF'C$. 则凸六边形 $F'BAE'DC$ 可分为 $2n+2$ 个全等的三角形.

若将$\triangle GFC$ 关于 FC 对称得$\triangle FG'C$, 则凸六边形 $G'FAE'DC$可分为 $2n+1$ 个全等的三角形,这里 n 为任一自然数.

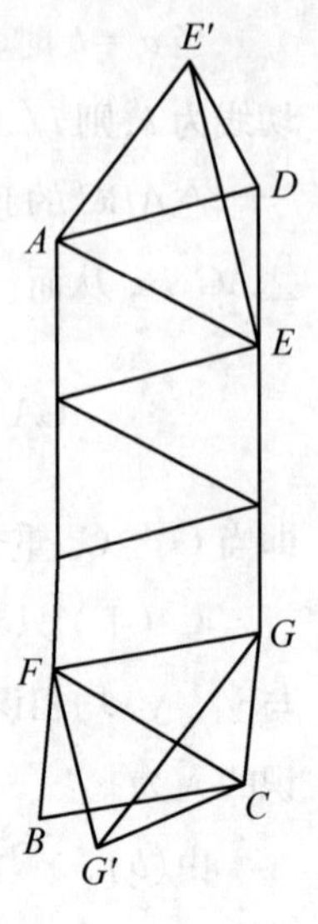

图 2

1992 年中国数学奥林匹克

1992 年中国数学奥林匹克(第七届冬令营)于 1 月 11 日在北京举行. 本届人数(包括自费生)近 100 名,可谓盛况空前.

按照惯例,考试分两天举行. 第二天的试题似过于容易,第 4 题实即 26 届 IMO 第 5 题,大轴的第 6 题与 1991 年全国联赛二试第 3 题惊人地接近(都是三阶递推数列),难度还降低了不少. 第一天试题恰恰相反,第 3 题与第 1 题不少选手翻船,大概是一开始就判断错了(第 3 题的答案是否定的,需要构造一个反例,第 1 题不能断定 $a_{n-1}=a_0=1$). 在解题过程中当然需要判断,需要假设,但不可武断,不可顽固坚持"一条道走到黑". 看来,选手的"识"(判断是非的能力与感觉)及"灵活性"(修正错误的速度)尚有待提高. 因而第 3 题这样需要判断(可能需要及时修正)的问题值得提倡(如改为:"证明……"就索然无味了).

下面是试题及我们所拟的解答,不妥之处敬请指正.

1 试题

第一天(1992 年 1 月 12 日)

1. 设方程 $x^n+a_{n-1}x^{n-1}+\cdots+a_1x+a_0=0$ 的系数都是实数,且适合条件 $0<a_0\leqslant a_1\leqslant\cdots\leqslant a_{n-1}\leqslant 1$,已知 λ 为此方程的复数根,且适合条件 $|\lambda|\geqslant 1$. 试证: $\lambda^{n+1}=1$.

2. 设 $x_1, \cdots, x_n$ 为非负实数,记 $x_{n+1}=x_1$, $a=\min\{x_1, \cdots, x_n\}$.

求证: $\sum\limits_{j=1}^{n}\frac{1+x_j}{1+x_{j+1}}\leqslant n+\frac{1}{(1+a)^2}\sum\limits_{j=1}^{m}(x_j-a)^2$. 且等式成立,当且仅当 $x_1=x_2=\cdots=x_n$.

3. 在平面上画出一个 9×9 的方格,在这 81 个小方格中任意填入 $+1$ 或 -1,下面一种改变填入之数字的方式称为作了一次变动:对任意一个小方格,凡与此小方格有一条边相邻(不包含此小方格本身)之所有小方格中的数作连乘积,于是每取一格,就算出一个数,在所有小方格都取遍后,再将这些算出之数放入相应所取之小方格中.试问:是否总可以经过有限次变动,使得所有小方格中的数都变为1?

第二天(1992年1月13日)

4. 凸四边形 $ABCD$ 内接于圆 O,对角线 AC 与 BD 相交于 P. $\triangle ABP$、$\triangle CDP$ 的外接圆相交于 P 和另外一点 Q,且 Q、P、Q 三点两两不重合.试证:$\angle OQP=90°$.

5. 在有8个顶点的简单图中,没有四边形的图的边数的最大值是多少?(简单图是指此图中任一点与自己没有边相连,而且任何两点之间,如果有边相连,就只有一条边相连).

6. 已知整数序列$\{a_0, a_1, a_2, \cdots\}$适合条件

(i) $a_{n+1}=3a_n-3a_{n-1}+a_{n-2}$, $n=2, 3, \cdots$;

(ii) $2a_1=a_0+a_2-2$;

(iii) 对任意自然数 m,在序列$\{a_0, a_1, a_2, \cdots\}$中必有相继的 m 项 a_k, a_{k+1}, $\cdots$, a_{k+m-1} 都是完全平方数.

试证:序列$\{a_0, a_1, a_2, \cdots\}$的所有项都是完全平方数.

2 解答

1.
$$\lambda^n+a_{n-1}\lambda^{n-1}+\cdots+a_1\lambda+a_0=0, \tag{1}$$
$$\lambda^{n+1}+a_{n-1}\lambda^n+a_{n-2}\lambda^{n-1}+\cdots a_0\lambda=0, \tag{2}$$

两式相减得

$$\lambda^{n+1}=(1-a_{n-1})\lambda^n+(a_{n-1}-a_{n-2})\lambda^{n-1}+\cdots+(a_1-a_0)\lambda+a_0 \tag{3}$$

取模并利用 $|\lambda|\geqslant 1$ 及 $0<a_0\leqslant a_1\leqslant\cdots\leqslant a_{n-1}\leqslant 1$,得

$$\begin{aligned}|\lambda|^{n+1}&\leqslant(1-a_{n-1})|\lambda|^n+(a_{n-1}-a_{n-2})|\lambda|^{n-1}+\cdots+(a_1-a_0)|\lambda|+a_0\\&\leqslant|\lambda|^n(1-a_{n-1})+(a_{n-1}-a_{n-2}+\cdots+a_0)=|\lambda|^n.\end{aligned} \tag{4}$$

从而

$$|\lambda|=1, \tag{5}$$

并且(4)是等式，$(1-a_{n-1})\lambda^n$，$(a_{n-1}-a_{n-2})\lambda^{n-1}$，…，$a_0$ 均为非负实数,进而由(3)得出 λ^{n+1} 为非负实数. 结合(5)得 $\lambda^{n+1}=1$.

评注　从本题的条件并不能导出 $a_{n-1}=\cdots=a_1=a_0=1$. 如

$$(x^2+x+1)\left(x^3+\frac{1}{2}\right)=x^5+x^4+x^3+\frac{1}{2}x^2+\frac{1}{2}x+\frac{1}{2}.$$

2. 设 $y_j=x_j-a$,则 y_i 为非负实数($j=1, 2, \cdots, n$),

$$\begin{aligned}\sum_{j=1}^{n}\frac{1+x_j}{1+x_{j+1}}&=\sum_{j=1}^{n}\left(1+\frac{x_j-x_{j+1}}{1+x_{j+1}}\right)\\&=n+\sum_{j=1}^{n}\frac{y_j-y_{j+1}}{1+a+y_{j+1}}\\&=n+\sum_{j=1}^{n}\left(\frac{y_j}{1+a+y_{j+1}}-\frac{y_{j+1}}{1+a+y_{j+1}}\right)\\&=n+\sum_{j=1}^{n}y_j\left(\frac{1}{1+a+y_{j+1}}-\frac{1}{1+a+y_j}\right)\\&=n+\sum_{j=1}^{n}\frac{y_j(y_j-y_{j+1})}{(1+a+y_{j+1})(1+a+y_j)}\\&\leqslant n+\frac{1}{(1+a)^2}\sum_{j=1}^{n}y_j^2\\&=n+\frac{1}{(1+a)^2}\sum_{j=1}^{n}(x_j-a)^2,\end{aligned}$$

显然,当且仅当 $y_1=y_2=\cdots=y_n=0$,即 $x_1=x_2=\cdots=x_n=a$ 时等式成立.

3. 否. 解法 1:如图 1 仅有 4 个-1,经过 6 次变动后又恢复原状,因此无法经过有限次变动使所有的数都变成 1. 图 2 方格中的数 j ($j=1, 2, 3, 4, 5, 6$),表示在第 j 次变动后-1 所在位置.

解法 2:如图 3(有 12 个-1)经 1 次变动后得图 4,再变动又恢复为图 3,反复无穷.

				−1				
		−1				−1		
				−1				

图 1

	5	4	3	4，2	3	4	5	
5		5		1，3		5		5
4	5		5，1	4，6	5，1		5	4
3		5，1		5，1		5，1		3
4，2	1，3	4，6	5，1		5，1	4，6	1，3	4，2
3		5，1		5，1		5，1		3
4	5		5，1	4，6	5，1		5	4
5		5		1，3		5		5
	5	4	3	4，2	3	4	5	

图 2

−1								−1
−1		−1				−1		−1
−1		−1				−1		−1
−1								−1

图 3

	-1						-1	
		-1				-1		
			-1		-1			
-1		-1				-1		-1
-1		-1				-1		-1
			-1		-1			
		-1				-1		
	-1						-1	

图 4

4. 设$\odot ABP$、$\odot CDP$ 的圆心分别为 O_1、O_2. 由于 $\angle O_1PB=\frac{\pi}{2}-\angle BAP=\frac{\pi}{2}-\angle BDC$，所以 $O_1P\perp CD$.

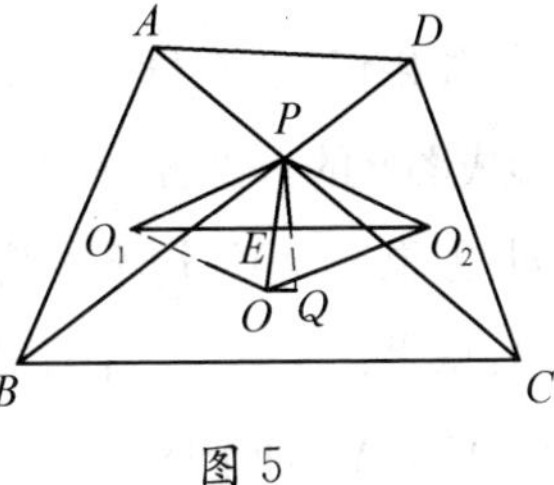

图 5

又$\odot O$、$\odot O_2$ 的公共弦为 CD，所以 $OO_2\perp CD$，从而 $OO_2\,/\!/\,O_1P$.

同理 $OO_1\,/\!/\,O_2P$，所以四边形 O_1OO_2P 是平行四边形，O_1O_2 过 OP 中点 E，又 O_1O_2 平分$\odot O_1$、$\odot O_2$ 的公共弦 PQ，所以 $O_1O_2\,/\!/\,OQ$. 而 $O_1O_2\perp PQ$，于是 $OQ\perp PQ$，即 $\angle OQP=90^\circ$.

5. 记 n 个顶点的、边数最多而不含四边形的简单图为 M_n，它的边数记为 e_n. 图 6 表明 $e_8\geqslant 11$.

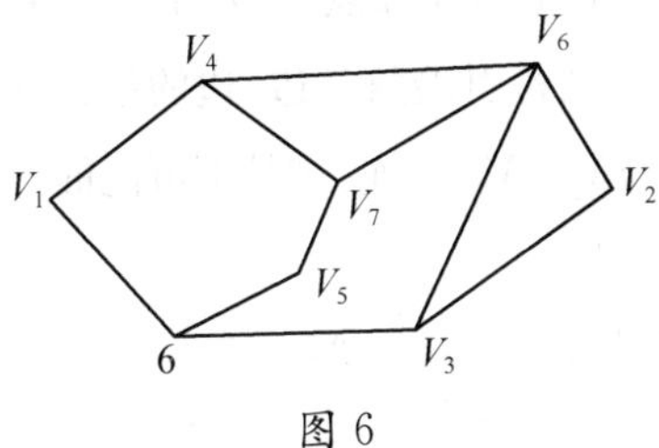

图 6

我们证明 $e_4=4$，$e_5=6$，$e_6=7$，$e_7=9$，$e_8=11$.

首先 M_n 必有圈(否则 $e_n\leqslant n-1$). M_4 中最小圈为三角形，所以 M_4 必为图 7.

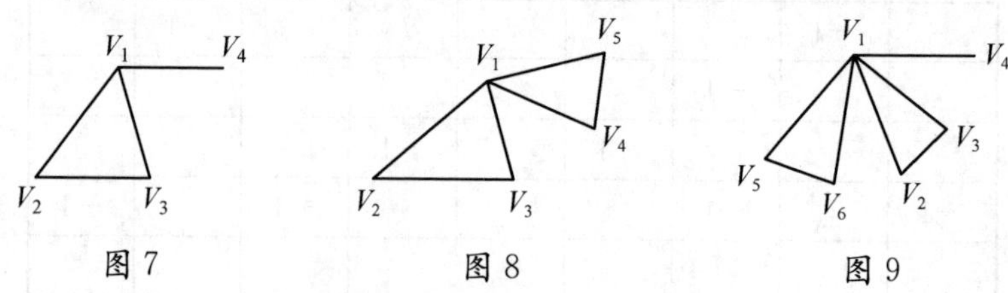

图 7　　图 8　　图 9

M_5 中的圈必为三角形(五边形只有五条边),并有 $e_5 \geqslant 6$ 时,有两个圈(去掉一条边后仍有圈),这两个三角形必须有公共点,所以 M_5 必为图 8.

图 6 的前 6 个顶点构成的图表明 $e_6 \geqslant 7$. M_6 有两个圈. 如果最小圈为五或六边形,导出边数$\leqslant 6$,所以,最小圈必为$\triangle V_1V_5V_6$. 其他顶点至多与 V_1、V_5、V_6 中一个相连. 如果 V_1、V_5、V_6 均各与一点相连,则只有这 6 条边,与 $e_6 \geqslant 7$ 矛盾. 如果仅 V_6 不与其他点相连,设有边 V_1V_2、V_5V_4,在 V_3 不与 V_1、V_5 相连时,形成图 6 的一部分. 在 V_3 与 V_1 相连时, V_3 只能与 V_2 相连,形成的图与 9 相同. 如果 V_6, V_5 均不与其他点相连,那么 V_1、V_2、V_3、V_4 至多连四条边 ($e_4 = 4$),总之, $e_6 = 7$, 并且 M_6 只有两种,即图 6 的前 6 点所构成的图与图 9 ($\triangle V_1V_5V_6$ 加上图 7).

同样, $e_7 \geqslant 9$ 并且 M_7 中最小圈为$\triangle V_1V_5V_6$,其他顶点至多与 V_1、V_5、V_6 中一个相连. 如果 V_6 不与其他顶点相连,那么去掉 V_6 后 M_7 只有 7 条边 ($e_6 = 7$), 如果 V_1、V_5、V_6 分别与 V_2、V_4、V_7 相连,那么 V_2、V_4、V_7 均只能与 V_3 相连,去掉 V_2、V_4、V_7 中任一个 M_7 只剩下 7 条边. 总之 $e_7 = 9$, 并且 M_7 有一个点只引出两条边,去掉这点后,M_7 变为 M_6.

M_8 中最小圈为$\triangle V_1V_2V_3$. 若每点至少引出 3 条边,那么应有边 V_1V_4, V_2V_5, V_3V_6, V_4、V_5、V_6 又各引出两条边,边的端点不同于 V_1、V_2、V_3、V_4、V_5、V_6,于是 V_7、V_8 均与 V_4、V_5 相连,这是不可能的. 因此, M_8 有一个点只引出两条边, $e_8 = e_7 + 2 = 11$.

6. $\{a_n\}$的特征方程为 $\lambda^3 - 3\lambda^2 + 3\lambda - 1 = (\lambda - 1)^3 = 0$ 有三个重根,所以

$$a_n = c_1n^2 + c_2n + c_3. \tag{1}$$

由(1)得

$$a_0 = c_3,$$

$$a_1 = c_1 + c_2 + a_0,$$

$$a_2 = 4c_1 + 2c_2 + a_0.$$

由后两式及已知 $2a_1=a_0+a_2-2$ 得 $c_1=1$, $c_2=a_1-1-a_0\in\mathbf{Z}$,
于是

$$a_n=n^2-c_2n+a_0, \tag{2}$$

$$a_{n+1}-a_n=(n+1)^2-n^2+c_2=2n+1+c_2,$$

所以取 $M>-c_2$,则在 $n>M$ 时, a_n 严格递增.

取 $m>M+1$, 则$\{a_n\}$中的 m 个连续平方数里有 a_n、a_{n+1} 的下标均超过 M,

$$a_n=n^2+c_2n+a_0=(n+b)^2, \tag{3}$$

$$a_{n+1}=(n+1)^2+c_2(n+1)+a_0=(n+d)^2, \tag{4}$$

相减得

$$2n+1+c_2=2(d-b)n+d^2-b^2, \tag{5}$$

$d-b\geqslant 1$, n 可任意地大(因为 m 可任意地大),所以由(5)式导出 $d-b=1$, $c_2+1=d^2-b^2=d+b=2b+1$,即 $c_2=2b$, 代入(3)得 $a_0=b^2$, 再代入(2)得 $a_n=(n+b)^2$.

第33届国际数学奥林匹克

第33届IMO于1992年7月在俄罗斯首都莫斯科举行.我国代表队在苏淳、严镇军两位领队的率领下,取得6枚金牌,总分第一.中华台北队首次参赛,也取得3银2铜,总分124,在60多个代表队中居17位.前6名成绩如下:

名次	国家(地区)	总分	题1	题2	题3	题4	题5	题6	金	银	铜
1	中国	240	41	42	42	38	35	42	6		
2	美国	181	37	32	32	36	11	33	3	2	1
3	罗马尼亚	177	38	42	29	38	9	21	2	2	2
4	独联体	176	41	30	35	26	15	29	2	3	
5	英国	168	42	24	23	26	21	32	2	2	2
6	俄罗斯	158	35	34	35	22	8	24	2	2	2

我国6名选手的成绩如下:

	题1	题2	题3	题4	题5	题6	总分	学校
沈凯	7	7	7	7	7	7	42	南京师大附中
杨保中	7	7	7	7	7	7	42	郑州一中
罗炜	7	7	7	7	7	7	42	哈尔滨师大附中
章寅	7	7	7	6	7	7	41	成都七中
何斯迈	7	7	7	5	7	7	40	安庆一中
周宏	6	7	7	6	0	7	33	北大附中

其中前4名曾在国家教委的理科试验班学习,后2名为高二学生.

下面是这一届的试题与所拟的解答.

1　试题

第一天(7 月 15 日)

1. 试求出所有的正整数 a、b、c,其中 $1<a<b<c$,且使得 $(a-1)(b-1)(c-1)$ 是 $abc-1$ 的约数.(新西兰供题)

2. 设 **R** 是全体实数的集合.试求出所有的函数 f: **R→R**,使得对于 **R** 中的一切 x 和 y,都有

$$f(x^2+f(y))=y+(f(x))^2.$$

(印度供题)

3. 给定空间中的 9 个点,其中任何 4 点都不共面.在每一对点之间都连有一条线段.试求出最小的 n 值,使得将其中任意 n 条线段任意地染为红色或蓝色,在这 n 条线段的集合中都包含有一个各边同色的三角形.(中国供题)

第二天(7 月 16 日)

4. 在一个平面中,C 为一个圆周,直线 L 是圆周的一条切线,M 为 L 上一点.试求出具有如下性质的所有点 P 的集合:在直线 L 上存在两个点 Q 和 R,使得 M 是线段 QR 的中点,且 C 是三角形 PQR 的内切圆.(法国供题)

5. 设 O-xyz 是空间直角坐标系,S 是空间中的一个由有限个点所形成的集合,S_x、S_y、S_z 分别是 S 中所有的点在 Oyz 平面、Ozx 平面、Oxy 平面上的正交投影所成的集合.证明

$$|S|^2 \leqslant |S_x| \cdot |S_y| \cdot |S_z|,$$

其中 $|A|$ 表示有限集合 A 中的元素数目.(注:所谓一个点在一个平面上的正交投影是指由点向平面所作垂线的垂足.)(意大利供题)

6. 对于每个正整数 n,以 $S(n)$ 表示满足如下条件的最大正整数:对于每个正整数 $k \leqslant S(n)$,n^2 都可以表示成 k 个正整数的平方之和.

(Ⅰ)证明,对于每个正整数 $n \geqslant 4$,都有 $S(n) \leqslant n^2-14$;

(Ⅱ)试找出一个正整数 n,使得 $S(n)=n^2-14$;

(Ⅲ)证明,存在无限多个正整数 n,使得 $S(n)=n^2-14$.

(英国供题)

2 解答

1. 令 $x=a-1$, $y=b-1$, $z=c-1$, 则 x、y、z 为正整数,并且 $z>y>x$, xyz 是

$$(x+1)(y+1)(z+1)-1=xyz+xy+yz+zx+x+y+z$$

的约数. 因此 xyz 是 $xy+yz+zx+x+y+z$ 的约数. 若 $x\geqslant 3$, 则

$$\begin{aligned}\frac{xy+yz+zx+x+y+z}{xyz}&=\frac{1}{x}+\frac{1}{y}+\frac{1}{z}+\frac{1}{xy}+\frac{1}{yz}+\frac{1}{zx}\\&\leqslant\frac{1}{3}+\frac{1}{4}+\frac{1}{5}+\frac{1}{3\times 4}+\frac{1}{4\times 5}+\frac{1}{5\times 3}=\frac{59}{60}<1\end{aligned}$$

不是整数,所以 $x\leqslant 2$.

在 $x=1$ 时, yz 是 $2(y+z)+1$ 的约数,因而 y 是奇数.

因为 $yz\leqslant 2(y+z)+1\leqslant 4z-1$,所以 $y<4$. 从而 $y=3$. 由 $3z$ 是 $2z+7$ 的约数得 $z=7$,从而$(a, b, c)=(2, 4, 8)$.

在 $x=2$ 时,由 $2\mid xy+yz+zx+x+y+z$ 得y、z 都是偶数. 从而$y\geqslant 4$. 由

$$\begin{aligned}xyz=2yz&\leqslant xy+yz+zx+x+y+z\\&=yz+3y+3z+2\leqslant yz+6z-4<yz+6z,\end{aligned}$$

得 $y<6$,所以 $y=4$. 由 $8z\mid 7z+14$ 得 $z=14$. 从而$(a, b, c)=(3, 5, 15)$.

2. 在已知等式 $$f(x^2+f(y))=y+f^2(x) \tag{1}$$

中令 $x=0$ 得 $$f(f(y))=y+f^2(0). \tag{2}$$

由(1)、(2)得

$$f(y+f^2(x))=f(f(x^2+f(y)))=x^2+f(y)+f^2(0). \tag{3}$$

在(3)中令 $y=f(z)$ 并利用(2)得

$$f(f(z)+f^2(x))=x^2+f(f(z))+f^2(0)=x^2+z+2f^2(0). \tag{4}$$

另一方面,由(1)、(2)得

$$f(f(z)+f^2(x))=z+f^2(f(x))=z+(x+f^2(0))^2. \tag{5}$$

比较(4)、(5)得　$$2xf^2(0)+f^4(0)=2f^2(0).$$

对所有 x 都成立,从而 $f(0)=0$, (2)成为

$$f(f(y))=y. \tag{6}$$

因此,对任一实数 a,由(6)、(1)得

$$f(x^2+a)=f(x^2+f(f(a)))=f(a)+f^2(x)\geqslant f(a),$$

从而 f 为增函数.

若有 x 使 $f(x)>x$,则 $f(f(x))\geqslant f(x)>x$ 与(6)矛盾,所以 $f(x)\leqslant x$,同理 $f(x)\geqslant x$,于是 $f(x)=x$.

显然 $f(x)=x$ 满足题中条件.

另解:由(1)及 y 可取一切实数得 f 为满映射.

若 $f(y_1)=f(y_2)$, 则由(1)得

$$y_1+f^2(x)=f(x^2+f(y_1))=f(x^2+f(y_2))=y_2+f^2(x),$$

从而 $y_1=y_2$, 因此 f 为单映射.

由(1) $y+f^2(x)=f(x^2+f(y))=f((-x)^2+f(y))=y+f^2(-x)$. 因此 $f^2(x)=f^2(-x)$. 当 $x\neq 0$ 时, $x\neq -x$,由于 f 为单映射 $f(x)\neq f(-x)$,所以 $f(x)=-f(-x)\neq 0$. 由于 f 为满映射,必有 $f(0)=0$. 从而(6)成立. 以下解法与上面相同.

3. 设染色的线段至少有 33 条,则由于线段共 $C_9^2=36$ 条,不染色的线段至多 3 条.

若点 A_1 引出不染色的线段,去掉 A_1 及所引出的线段. 若剩下的图中,还有点 A_2 引出不染色的线段,去掉 A_2 及所引出的线段. 依此进行,由于不染色线段至多 3 条,所以至多去掉 3 个顶点(及从它们引出的线段),即有 6 个点,每两点之间的连线染上红色(实线)或蓝色(虚线).

评注　熟知这时存在一个同色三角形(参见《趣味的图论问题》,单墫著,上海教育出版社 1980 年出版).

图 1 表明染色的边少于 33 条时,未必有同色三角形(不染色的边 19, 28,

37, 46 没有画出).

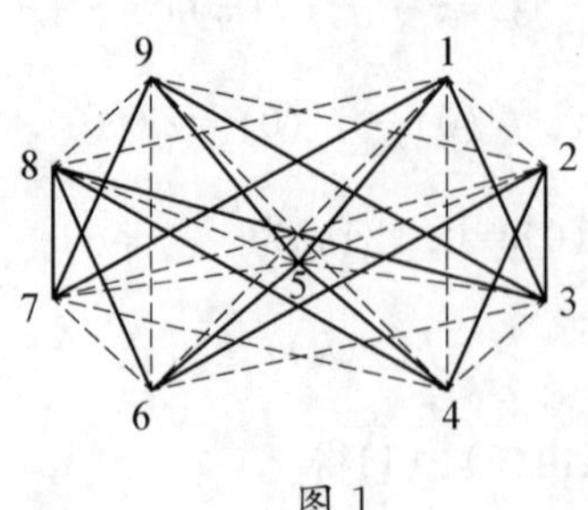

图 1

因此 n 的最小值为 33.

4. 如图 2,设圆 C 与 L 切于 T 点, TS 为直径. 过 S 作直线平行于 RQ,分别交 PR 于 R'、交 PQ 于 Q'. 显然$\triangle PQ'R'$与$\triangle PQR$ 位似, P 为位似中心. 由于圆 C 是$\triangle PQ'R'$的旁切圆,所以在上述位似变换下,圆 C 变为$\triangle PQR$ 的傍切圆, PS 与直线 L 的交点 N 即傍切圆与 QR 的切点.

设 $PQ=c$, $PR=b$, $QR=a$, $s=\dfrac{a+b+c}{2}$,则易知 $QN=RT=s-c$,所以 $NM=MT$.

于是在直线上取点 N,使 $NM=MT$, 过 N、S 作直线,则 P 在直线 NS 上. 从而 P 点轨迹为直线 NS 的一部分:以 S 为端点不含点 N 的射线(S 点除外).

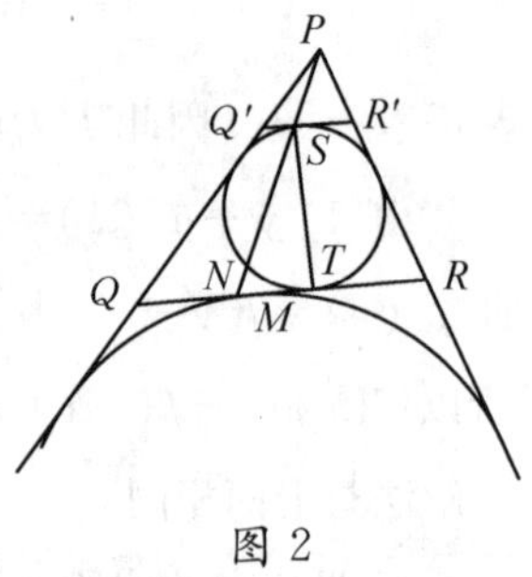

图 2

评注 纯粹性的证明是简单的:设 P 在上述射线 l 上. 过 P 作圆 C 的切线交直线 L 于 R,在 L 上取 Q,使 $QM=MR$. 过 Q 作圆 C 的切线交 PR 于 P',则由上面的证明 P'在射线 l 上. 但 l 与 PR 只有一个公共点,所以 P'与 P 重合, P 满足条件(在 IMO 中,这一部分的证明通常略去).

5. 本题是得分最低的一道题. 1990 年曾作为预选题提出,参见北京大学出版社"数学奥林匹克系列"图书《31 届 IMO 预选题》第 43 题. 这里的解法稍有不同. 实际上,如果弄清符号的意义,本题是非常简单的.

S 的点在$|S_z|$条平行于 z 轴的直线上. 设各直线上 s 的点数分别为 t_{ij} ($(i, j, 0)\in s_z$),则

$$
\begin{aligned}
|S|^2 &= \Big(\sum_{(i,j,0)\in S_z} t_{ij}\Big)^2 \\
&\leqslant \sum_{(i,j,0)\in S_z} 1^2 \cdot \sum_{(i,j,0)\in S_z} t_{ij}^2 \quad (\text{Cauchy 不等式}) \\
&= |S_z| \sum_{(i,j,0)\in S_z} t_{ij}^2.
\end{aligned}
$$

设 $u_i = |\{(i, 0, k) \in S_y\}|$, $v_j = |\{(0, j, k) \in S_x\}|((i, j, 0) \in S_z)$, 则 $u_i \geqslant t_{ij}$, $v_j \geqslant t_{ij}$,

$$
|S_x| = \sum_i u_i, \quad |S_y| = \sum_j v_j,
$$

$$
|S_x| \cdot |S_y| = \sum_{(i,j,0)\in S_z} u_i v_j \geqslant \sum_{(i,j,0)\in B_z} t_{ij}^2. \tag{2}
$$

由(1)、(2)即得

$$
|S|^2 \leqslant |S_x| \cdot |S_y| \cdot |S_z|.
$$

6. (Ⅰ) 若 $n^2 = a_1^2 + a_2^2 + \cdots + a_k^2$, $k = n^2 - 13$,并且 $a_1 \leqslant a_2 \leqslant \cdots \leqslant a_k$ 为正整数,则由

$$
\underbrace{1^2 + 1^2 + \cdots + 1^2}_{k-1\text{个}} + 4^2 = n^2 + 2 > n^2,
$$

所以 $a_k \leqslant 3$.

设 $a_1, a_2, \cdots, a_k$ 中有 s 个 2, t 个 3,其余为 1(s、t 为非负整数并且 $s + t \leqslant k$), 则

$$
a_1^2 + a_2^2 + \cdots + a_k^2 = k + 3s + 8t,
$$

从而

$$
3s + 8t = 13.
$$

易知 t 只能为 0 或 1,但相应的 s 均不为整数. 因此 $k = n^2 - 13$ 时, $n^2 = a_1^2 + a_2^2 + \cdots + a_k^2$ 不可能成立. 即 $S(n) \leqslant n^2 - 14$.

(Ⅱ) 每一个大于 13 的正整数 l 可以表为 $3s + 8t$, 其中 s、t 为非负整数. 事实上,若 $t = 3s_1 + 1$,则 $s_1 \geqslant 5$, $l = 3(s_1 - 5) + 2 \times 8$. 若 $l = 3s_1 + 2$,则 $s_2 \geqslant 4$, $l = 3(s_1 - 2) + 8$.

于是对满足 $k^2 - 14 \geqslant k \geqslant s + t$ 的 k,取 s、t 使 $3s + 8t = n^2 - k$, 则

$$\underbrace{1^2+\cdots+1^2}_{(k-s-t)\text{个}}+\underbrace{2^2+2^2+\cdots+2^2}_{s\text{个}}+\underbrace{3^2+3^2+\cdots+3^2}_{t\text{个}}=n^2.$$

即 n^2 可表为 k 个平方数的和. 当 $k\geqslant\frac{1}{4}n^2$ 时,

$$s+t\leqslant\frac{1}{3}(n^2-k)\leqslant\frac{1}{4}n^2\leqslant k.$$

因此,每个 $n\geqslant 4$ 可表为 n^2-14, n^2-15, $\cdots$, $\left[\frac{1}{4}n^2\right]+1$ 个平方数的和.

令 $n=13$, 则

$$n^2=12^2+5^2=12^2+4^2+3^2=8^2+8^2+5^2+4^2.$$

因为 8^2 可表为 4 个 4^2 的和, 4^2 可表为 4 个 2^2 的和, 2^2 可表为 4 个 1^2 的和,所以 $n^2=8^2+8^2+5^2+4^2$ 可表为 4, 7, 10, $\cdots$, 43 个平方数的和. 又因为 $5^2=4^2+3^2$, n^2 可表为 5, 8, 11, $\cdots$, 44 个平方数的和.

因为 12^2 可表为 4 个 6^2 的和, 6^2 可表为 4 个 3^2 的和,4^2 可表为 4 个 2^2 的和,2^2 可表为 4 个 1^2 的和,所以 $n^2=12^2+4^2+3^2$ 可表为 3, 6, 9, $\cdots$, 33 个平方数的和.

又 $n^2=\underbrace{3^2+3^2+\cdots+3^2}_{17\text{个}}+4^2$,而 $3^2=2^2+2^2+1$,所以 n^2 可表为 $18+2\times 9=36$, $18+2\times 12=42$ 个平方数的和,再由 4^2 为 4 个 2^2 的和, n^2 也可表为 39 个平方数的和. 因此 n^2 可表为 1, 2, $\cdots$, 44 个平方数的和. 由于 $\left[\frac{13^2}{4}\right]+1=43$,根据前面所说, $n=13$ 满足要求.

(Ⅲ) 令 $n=2^m\times 13$, 则

$$n^2=4^h\times(2^{m-h}\times 13)^2\quad(0\leqslant h\leqslant m).$$

因为 13^2 可表为 1, 2, $\cdots$, 155 个平方数的和,所以 n^2 可表为 1, 2, $\cdots$, 155×4^m 个平方数的和.

因为 $155\times 4^m>\frac{1}{4}(2^m\times 13)^2=\frac{1}{4}n^2$,所以根据(Ⅱ)中的结论, n^2 可表为 k 个平方数的和,其中 $k\leqslant n^2-14$,即 $S(2^m\times 13)=(2^m\times 13)^2-14$.

评《首届全国数学奥林匹克命题比赛精选》

1

1988 年 10 月,《中等数学》杂志编辑部等单位举办全国数学竞赛命题有奖比赛,引起强烈反响,在半年时间内,近千人踊跃参加,提供 1200 多道题. 这一创举推动了国内数学竞赛的命题工作,对于数学竞赛、数学教学、初等数学研究都有重要的意义.

我认为这次比赛有以下特点:第一,参加人员多,其中有大、中学校的教师、研究生及大、中学生,还有不少数学爱好者;第二,题目涉及范围广,数论、几何、代数(包括不等式、函数等)、组合数学一应俱全;第三,有不少创造性的新题目,包括个人研究中的一些发现,例如江苏海安中学高二学生江焕新的题.

比赛是首次举行,因而也有不足之处:(1)高层次的数学工作者少,西方与独联体均有相当多的著名数学家参与命题,近年我国的数学家对于竞赛甚少问津(华罗庚先生曾亲自领导竞赛活动),原因值得探讨;(2)传统的题目(尤其平面几何)及不等式比例较大,组合与数论方面的好题不多;(3)从现代数学研究中产生的问题较少. 固然,初等数学研究中的问题应占竞赛问题的主要位置,但仅限于初等数学的研究. 就难以使竞赛"现代化".

最近,《中等数学》编辑部经过认真挑选,将其中 73 道题目及解答整理成《首届全国数学奥林匹克命题比赛精选》(以下简称《精选》),这是对首届命题竞赛的一个很好的总结,是一件极有益的工作.

2

笔者位列命题比赛的评审委员会,但由于工作忙,未能尽到应尽的责任. 借《精选》出版,写这篇评论,作为弥补.

古人曾说对书应当采取"师、友、敌"三种态度,即学习、讨论、争辩. 争辩时不免“评头论足”“吹毛求疵”，但目的是希望推动、促进命题研究工作，使下一次命题竞赛搞得更好.

首先，73 道题太少了，其余的题中可能还有“珍珠”，最好能将题全附在书后(为节省篇幅解答从略).

其次，还有一些创造性的新题可以给奖，如高书宽(11 页)，宿晓阳(65 页)，熊军华(122 页)，万金寿(134 页)，盛宏礼(159 页)等. 鉴于组合问题较少，应多加鼓励，陈育祥(148 页)，李忠旺(151 页)等也可以给奖.

再次，有几道题是陈题，可能提供者与前人暗合(如 20 页、22 页、186 页). 有些问题太接近熟题(如 84 页、98 页、169 页). 例如 98 页题目“不存在这样的四边形 $PQRS$，满足 $PQRS$ 内接于$\triangle ABC$(如图 1)，且 $S_1=S_2=S_3=S_4$”.

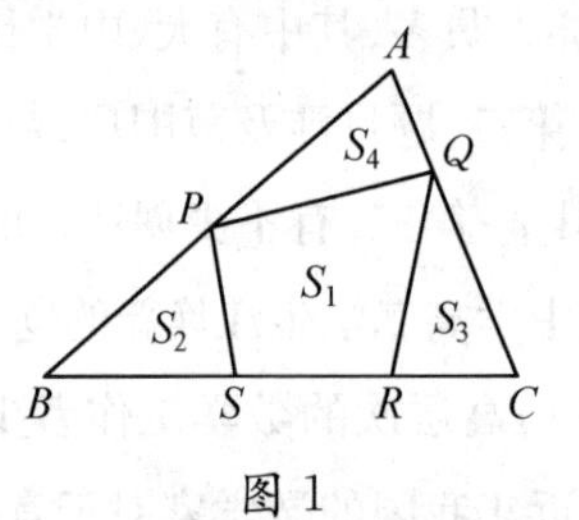

图 1

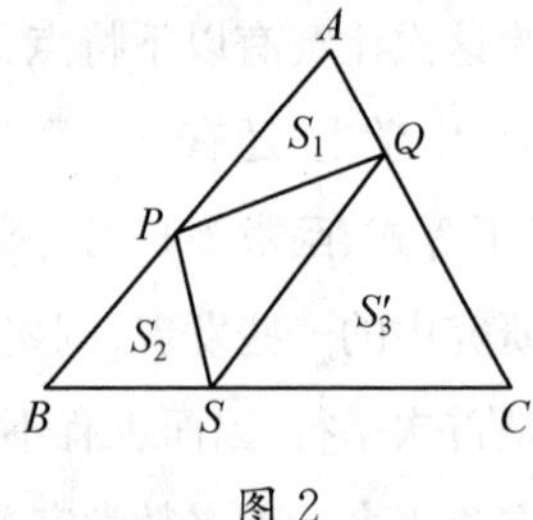

图 2

熟知“如图 2，P、Q、S 分别在$\triangle ABC$ 的边 AB、AC、BC 上时，图中面积 S_1、S_2、S_3' 满足 $S_1S_2S_3'\leqslant\left(\frac{1}{4}\right)^3$”.

如果图 1 中 $S_1=S_2=S_3=S_4=\frac{1}{4}$，那么 $S_1S_2S_3'>S_1S_2S_3=\left(\frac{1}{4}\right)^3$，矛盾.

所以，98 页题目只是熟题的微小变动，没有本质的差异.

还有些题太容易，如 175 页、182 页题目(容斥原理的直接应用).

3

下面主要谈一些问题的解法，因为一个问题的优劣往往通过它的解法才能看出来. 同时，解题本身也有独立的趣味与意义.

首先《精选》中有几道题的解法有错或有疑.

24 页题目. 判断并证明：是否存在正整数 n 和 m，使 $n+m$ 最小且满足 1989^n 与 1989^m 的差是 3375 的倍数（$n>m\geqslant 1$）.

原解认为 $n+m$ 的最小值不存在. 实际上，解答存在. 不难得出 m 的最小值为 2，$n-m$ 的最小值为 2×5^2，于是 $n+m$ 的最小值为 54.

26 页题目. 设 x、y、z 为自然数，$1<x<y<z$. 问两质数 p 与 q 为何值时，方程

$$\frac{1}{x}-\frac{p}{y}+\frac{q}{z}=\frac{1\times 88+2\times 89}{1989} \tag{1}$$

有解？并求出此解.

原解由 $1989=3\times 3\times 13\times 17$ 且 p、q 为质数，得出 x、y、z 只能取 1989 的因数. 这一论断是错误的. 实际上解不只 $p=2$，$q=3$，$x=9$，$y=13$，$z=17$ 这一组. 例如 $p=7$，$q=5$，$x=2$，$y=18$，$z=13\times 17$ 也是解（其中 x、y 均非 1989 的因数）.

22 页题目. 设 a、b、c 为正整数，且满足 $\frac{1}{a}+\frac{1}{b}=\frac{1}{c}$. 试证当 $(a,b,c)=1$ 时，$a+b$ 为平方数.

本题可用如下解法：由已知易得 $\frac{a-c}{c}=\frac{c}{b-c}$.

设以上比值为 $\frac{p}{q}$，其中 p、q 为互质的正整数. 则 $\frac{a-c}{c}=\frac{p}{q}$，$\frac{a}{c}=\frac{p+q}{q}$.

于是，可设 $c=qd_1$，$a=(p+q)d_1$，d_1 为自然数.

同样可设 $c=pd_2$，$b=(p+q)d_2$，d_2 为自然数.

由 $(a,b,c)=1$，可得 $(d_1,d_2)=1$. 于是，由 $qd_1=pd_2$ 得 $d_1=p$，$d_2=q$，从而 $a+b=(p+q)^2$.

原解稍长，在得出 $\frac{a}{p(p+q)}=\frac{b}{q(p+q)}=\frac{c}{pq}$ 后，由 $(p,q)=(p,p+q)=(q,p+q)=1$ 及 $(a,b,c)=1$ 导出 $a=p(p+q)$，$b=q(p+q)$，$c=pq$. 似少了一些必要的推理.

99 页题目. 本题与高桥进一的猜测有关，即构造一条空间闭曲线，任意五等分点组不共球. 我怀疑这里的球（面）包括极限情况平面（半径为∞的球面）在

内.因此,题目中的曲线不合要求(否则,$\overset{\frown}{BnC}$ 不必取不必折即已构成所需的反例).

其次,《精选》中某几道题有简单的解法.

32 页题目. 已知 $k_1x_1+k_2x_2+\cdots+k_nx_n=a$,且 $\sum\limits_{i=1}^{n}k_1=S_0$. 其中 $k_i>0$ $(i=1, 2, \cdots, n)$, a 与 S_0 都是常数.求 $y=k_1x_1^2+k_2x_2^2+\cdots+k_nx_n^2$ 的最小值.

这只需利用熟知的 Cauchy 不等式,由

$$\sum k_i\cdot\sum k_ix_i^2\geqslant\left(\sum k_ix_i\right)^2=a^2$$

即得 $y\geqslant\dfrac{a^2}{S_0}$,且在 $x_1=x_2=\cdots=x_n=\dfrac{a}{S_0}$ 时, y 取最小值$\dfrac{a^2}{S_0}$,不必用原解答所说的待定系数法.

113 页题目. 如图 3,已知凹四边形 $ABCD$, $\angle A=\angle B=\angle D=45°$. 求证 $AC=BD$.

只需延长 BC 交 AD 于 E. 由 $\angle A=\angle B=45°$ 立得 $\angle E=90°$, $AE=BE$. 再由 $\angle D=45°$, $\angle E=90°$ 得 $CE=DE$. 因此,两直角三角形 AEC, BED 全等, $AC=BD$. 原解答过于复杂.

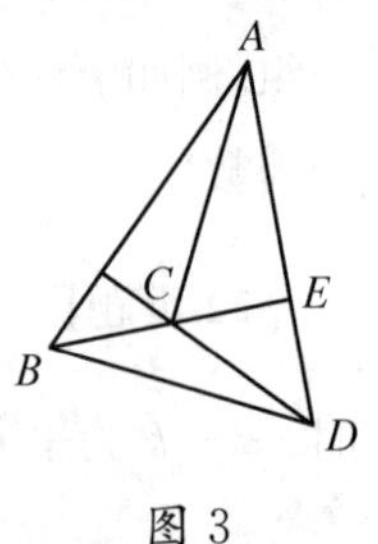

图 3

137 页题目. 实际上即两个小题组成:“在 $S=\pi\cdot(1-h^2)$ 时,求 $V_{锥}=\dfrac{1}{3}Sh$ 的最大值”,“在棱锥的高 h 与底面面积 S 固定时,求内接棱柱的体积的最大值”.

利用“若干正数和一定,在它们相等时积最大”,易知

$$V_{柱}=S\times(h-h_1)\times\left(\frac{h_1}{h}\right)^2=\frac{1}{2}S\times(2h-2h_1)\times\left(\frac{h_1}{h}\right)^2$$

$$\leqslant\frac{1}{2}S\times\left(\frac{2}{3}\right)^3\times h=\frac{4}{9}V_{锥}(h-h_1\text{ 为棱柱的高}),$$

$$V_{锥}=\frac{1}{3}Sh=\frac{\pi}{3}(1-h^2)h$$

$$=\frac{\pi}{3\sqrt{2}}\sqrt{2h^2(1-h^2)(1-h^2)}\leqslant\frac{\pi}{3\sqrt{2}}\left(\frac{2}{3}\right)^{\frac{3}{2}}=\frac{2\pi}{9\sqrt{3}}.$$

不必利用祖暅原理,不必引进三角函数.

111 页题目. 已知 O 是$\triangle ABC$ 的外心, AO 或 AO 的延长线交 BC 于 M. 求证:

$$\frac{BM}{MC}=\frac{\sin 2C}{\sin 2B}. \tag{2}$$

证明可由正弦定理得

$$\frac{BM}{\sin\angle BOM}=\frac{OM}{\sin\angle OBM}=\frac{OM}{\sin\angle OCM}=\frac{CM}{\sin\angle COM}.$$

再利用 $\angle BOM=\pi-\angle BOA=\pi-2C$, $\angle COM=\pi-2B$ 即导出(2). 不需要像原解答那样作辅助线,利用重心的性质与外心的关系,也不需要太多的几何、三角、代数的知识与特别的机智.

96 页题目. 凸四边形 $ABCD$ 的两组对边互不平行,线段 P_1P_2 位于四边形内部. 如果 P_1、P_2 两点分别到四边距离之和都等于 m,那么 P_1P_2 上任意一点到四边距离之和也等于 m.

设 P 为 P_1P_2 上任一点, $P_1P:PP_2=\lambda:\mu\ (\lambda+\mu=1)$.

设 P_1、P_2 到边 AB 等的距离分别为 d_i 与 $d_i'(i=1,2,3,4)$,则 P 到四边距离之和为

$$\sum_{i=1}^{4}(\lambda d_i'+\mu d_i)=\lambda\sum d_i'+\mu\sum d_i=\lambda m+\mu m=m.$$

这道题是定比分点公式的简单应用.

4

下面再谈几道题的解法.

63 页题目. 已知正数 $m_i\in\mathbf{R}^+\ (i=1,2,\cdots,n)$, $r\geqslant 2$ 且 $r\in\mathbf{N}$ 满足

$$\frac{1}{1+m_1^r}+\frac{1}{1+m_2^r}+\cdots+\frac{1}{1+m_n^r}=1. \tag{3}$$

求证: $m_1m_2\cdots m_n\geqslant(n-1)^{\frac{n}{r}}$. $\quad(4)$

此题原解用了三角代换. 但实际上,对这道题用三角代换并不自然(人工色

彩很浓),题中的条件也显得多余,问题可改成稍一般的形式:

已知正数 $x_i(i=1, 2, \cdots, n)$, 且

$$\frac{1}{1+x_1}+\frac{1}{1+x_2}+\cdots+\frac{1}{1+x_n}=1. \tag{5}$$

求证: $x_1x_2\cdots x_n \geqslant (n-1)^n$. (6)

证明也很简单. 由(5)得

$$\sum_{\substack{i=1\\ i\neq j}}^{n}\frac{1}{1+x_i}=1-\frac{1}{1+x_j}=\frac{x_j}{1+x_j}.$$

从而由平均不等式

$$\frac{x_j}{1+x_j}\geqslant (n-1)\left(\prod_{i\neq j}\frac{1}{1+x_i}\right)^{\frac{1}{n-1}}\ (j=1, 2, \cdots, n),$$

相乘得

$$\prod_j \frac{x_j}{1+x_j}\geqslant (n-1)^n\prod_j\left(\prod_{i\neq j}\frac{1}{1+x_i}\right)^{\frac{1}{n-1}}=(n-1)^n\prod_i\frac{1}{1+x_i}.$$

约去 $\prod\limits_i \frac{1}{1+x_i}\left(=\prod\limits_j\frac{1}{1+x_j}\right)$ 即得(6).

47 页题目. 设实数 λ_i 满足 $0<\lambda_i<1\ (i=1, 2, 3)$. 定义 $X=\lambda_1(1-\lambda_3)$, $Y=\lambda_2(1-\lambda_1)$, $Z=\lambda_3(1-\lambda_2)$. 试证 X、Y、Z 中必定有一个不大于 $1-(X+Y+Z)$.

这道题从几何中来,原解将它化为几何问题去解,相当复杂. 事实上,可以从其他途径单刀直入,或许不如原解那样富于技巧,然而却更为实用(因为这是解题者实际的解法,而不是命题者理想的解法).

这道题可以用代数方法直接解决:

若 X、Y、Z 均不大于$\frac{1}{4}$,则 $1-(X+Y+Z)\geqslant\frac{1}{4}$, 结论显然.

若 X、Y、Z 中至少有一个大于$\frac{1}{4}$,不妨设 $X>\frac{1}{4}$, 则

$$\begin{aligned}1-(X+Y+Z)&=(1-\lambda_1)(1-\lambda_2)(1-\lambda_3)+\lambda_1\lambda_2\lambda_3\\&\geqslant 2\sqrt{(1-\lambda_1)(1-\lambda_2)(1-\lambda_3)\lambda_1\lambda_2\lambda_3}\\&=2\sqrt{XYZ}>\sqrt{YZ}\geqslant \min(Y, Z).\end{aligned}$$

于是,代数帮了几何的忙.我们立即可以得到48页的几何命题(这是一道熟题):

设 D、E、F 分别在 $\triangle ABC$ 的边 AB、BC、CA 上,则 $\triangle ADF$、$\triangle BDE$、$\triangle CEF$ 中必有一个面积不大于 $\triangle DEF$ 的面积.

115 页题目. AD、BE、CF 是正三角形 ABC 的高,在 $\triangle ABC$ 内任取一点 P. 试证 $\triangle PAD$、$\triangle PBE$、$\triangle PCF$ 中,最大一个的面积等于其余两个的面积之和.

这是一个颇为有趣的问题.设 O 为 $\triangle ABC$ 中心,不妨设 P 在 $\triangle OBF$ 内,取 OC 中点 M,易知 $S_{\triangle POK}+S_{\triangle PMO}=S_{\triangle PME}-S_{\triangle OME}$,即 P 到 BE、CF 的距离之和等于 P 到 AD 的距离,从而结论成立.

上面的证法表明这道题与一道著名的习题:"点 P 到正三角形 OME 三边距离的代数和等于 $\triangle OME$ 的高"密切相关.

熟练的解题者往往不需要一步一步地推导,一些已经证过的结论(如上面所说的"三个距离的代数和")可以作为"定式"应用.因此,解题的过程不是由多而乱的晶体管电阻电容组成,而是整齐排列的集成块.命题也应当如此.

基本知识,例如定比分点公式有不少应用,《精选》中还可找到两个例子.

119 页题目. P 是 $\triangle ABC$ 内一点,自 P 向三边作垂线 PL、PM、PN,L、M、N 为垂足.若 PL、PM、PN 可以组成三角形,求 P 点所在区域.

设三条角平分线为 AA'、BB'、CC',则 $\triangle A'B'C'$ 的内部就是 P 点所在区域.证明的关键是当 P 在 $B'C'$ 上时,$PL=PM+PN$.

如图4,设 $C'P:PB=\lambda:\mu(\lambda+\mu=1)$. 则

$$PL=\mu\cdot C'F+\lambda\cdot B'D=\mu\cdot CF'+\lambda\cdot B'D'=PM+PN.$$

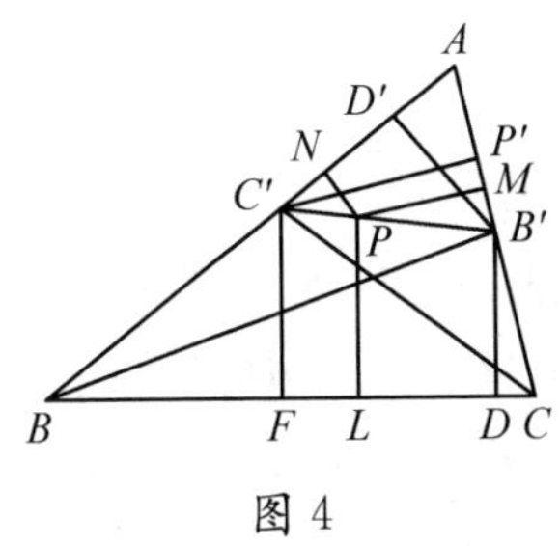

图4

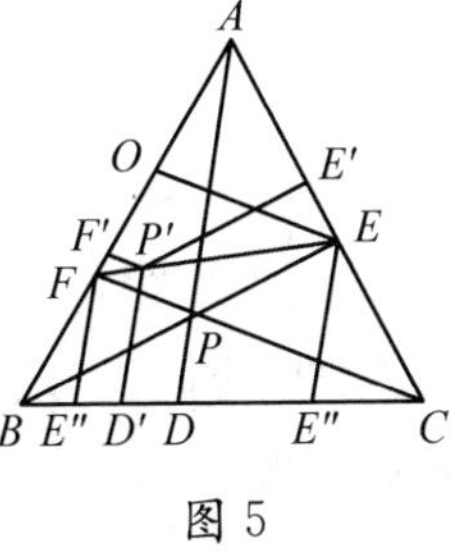

图5

129 页题目. 设 P 为 $\triangle ABC$ 内任一点,顶点 A、B、C 与 P 的连线分别与

BC、CA、AB 交于 D、E、F. P'为$\triangle DEF$ 的周界上任一点,过 P'引 PD、PE、PF 的平行线分别与 BC、CA、AB 交于 D'、E'、F'. 证明比值$\frac{P'D'}{PD}\cdot\frac{P'E'}{PE}\cdot\frac{P'F'}{PF}$中必有一个等于另两个的和.

如图 5,设 $FP':P'E=\lambda:\mu$. 过 E、F 作 AD 的平行线分别交 BC 于 E''、F'',则 $P'D'=\mu\cdot FF''+\lambda\cdot EE''$. 再作 $EG\parallel CF$ 交 AB 于 G,则$\frac{\lambda\cdot EE''}{PD}=\frac{\lambda\cdot BE}{BP}=\frac{\lambda\cdot EG}{PF}=\frac{P'F'}{PF}$. 类似地,有$\frac{\mu\cdot FF''}{PD}=\frac{P'E'}{PE}$. 于是结论成立.

命题者希望用这道题考查利用面积关系证明几何题的技能技巧,但利用面积的证法较为麻烦(参见原解答),学生未必只走这一条路. 我认为还是不必限定或坚持一条预定的道路,顺其自然为好.

说到自然的解法,不由想起被称为“宇宙流”的围棋高手武宫正树,他自称为“自然流”,即“棋该怎么下,我就怎么下”,自然的解法也就是“题该怎么解,就怎么解”,不故弄玄虚,不花里胡哨.

9 页题目. 已知 $f_1(x)=x-2$, $f_2(x)=2x^2-5x+2$, $f_n(x)=\frac{f_{n-1}^2(x)+f_1(x)f_{n-2}(x)}{f_{n-2}(x)+f_1(x)}$ $(n\geqslant 3)$. 试求方程 $f_n(x)+f_1(x)+4=0$ 的根.

由 $f_1(x)+x=2(x-1)$, $f_2(x)+x=2(x-1)^2$ 等可推想 $f_n(x)+x=2(x-1)^n$. 这似乎比原解答更自然些(然后再用归纳法证明).

还有一些题的解法可以简化,这里不一一列举了.

5

某些问题可略作修改,使它的难度提高或降低.

151 页题目. 在直角坐标系中,把每个整点都染上给定的 $n(\geqslant 2)$ 种颜色之一. 证明:必能找到四个同色的整点,它们恰好是某个矩形的四个顶点,且该矩形的边的斜率分别为$+1$ 与-1.

最后一句改为“该矩形的边与坐标轴平行”,便稍容易一些.

解题时可由较容易的情况入手,发现规律,然后再考虑一般情况.

159 页题目. 试证对任意的 1990 个连续整数：

(1) 两端 994 个整数之积的差，必能被中间两个整数之和的 $(1990^2-4)\cdot 2^{495}$ 倍整除.

(2) 两端 994 个整数之积的和，能表示成中间两个整数之积一半的 497 次整系数多项式.

题中 994 用更一般的 $4k+2$ 代替后，可采用归纳法. 在证明中需同时利用(1)、(2)的归纳假设. 因此，这道题如果藏起一半，只证明(1)或只证明(2)，难度反而提高.

《精选》中有一些"有背景"的题. 如 132 页的题与单形有关，80 页的题与"顺相似心"有关.

114 页题目. AB 是$\odot O$ 的弦，不是直径. 过 AB 中点 P 作两弦 A_1B_1、A_2B_2，过 A_1、B_1 分别作$\odot O$ 的切线得交点 C_1，过 A_2、B_2 分别作$\odot O$ 的切线得交点 C_2. 求证：$C_1C_2 \parallel AB$.

本题与极线有关，C_1 的极线过 P，C_2 的极线也过 P，所以 P 的极线为 C_1C_2. 因此，$C_1C_2 \perp OP$，即 $C_1C_2 \parallel AB$.

54 页题目. 已知 $\varphi_i > 0\ (i=1, 2, \cdots, 1989)$，且 $\sum_{i=1}^{1989}\varphi_i = 2\pi$. 则

$$\sum_{i=1}^{1989}\sin\varphi_i \leqslant 1989\sin\frac{2\pi}{1989}. \tag{7}$$

当且仅当 $\varphi_1=\varphi_2=\cdots=\varphi_{1989}$ 时等号成立.

本题与凸函数有关. $y=\sin x$ 在$[0, \pi]$内上凸，但在$[0, 2\pi]$内并不完全上凸，为了克服这一困难，我们令

$$\varphi_i' = \begin{cases}\varphi_i, \text{若 } \varphi_i \in [0, \pi], \\ 0, \text{若 } \varphi_i \notin [0, \pi].\end{cases}$$

则
$$\sum_{i=1}^{1989}\sin\varphi_i \leqslant \sum_{i=1}^{1989}\sin\varphi_i' \leqslant 1989\sin\frac{\sum\varphi_i'}{1989} \leqslant 1989\sin\frac{2\pi}{1989}.$$

当然，凸函数的知识超出中学数学大纲，不宜作为全国联赛的内容. 但对于数学爱好者，有一点凸函数的知识(可以借助直观)还是有益的. 它比逐步调整简单.

数学竞赛重在参与.希望吸引更多的人了解、喜爱数学.因此,应有较多富有现代数学背景的问题.

本文已经很长了.最后,应当指出,《精选》中好题不少,如176页的翻硬币,178页斐波那契数的应用,等等,建议读者买一本书自已去读.

1993 年德国数学奥林匹克

1993 年德国数学奥林匹克的两轮比赛的试题，我们曾经组织国家教委 93 年数学试验班的同学进行讨论，大家提供了不少好的解法，整理如下.

第一轮

1. 每一个大于 2 的自然数 n 都可以表成若干两两不等的自然数的和. 设个数的最大值为 $A(n)$，求 $A(n)$（用 n 表示）.

2. 平面有限点集 M 具有性质：对 M 中任意两点 A、B，必存在第三点 $C \in M$，使 $\triangle ABC$ 为正三角形. 求 $|M|$ 的最大值.

3. 证明有无穷多对自然数 a、b，满足：

(i) a、b（均用十进制表示）位数相同；

(ii) a、b 都是平方数；

(iii) 将 b 写在 a 后面，产生一个平方数.

（例如 $a = 16$，$b = 81$，$1681 = 41^2$）

4. $\triangle ABC$ 的边长 $AB = c$，$BC = a$，$CA = b$. 延长 AB 到 A''，使 $BA'' = a$，再反向延长到 B'，使 $AB' = b$. 类似地得到 A'，C'，B''，C''，如图 1. 证明：$\dfrac{S_{A'B''B'C''C'A''}}{S_{ABC}} \geqslant 13$.

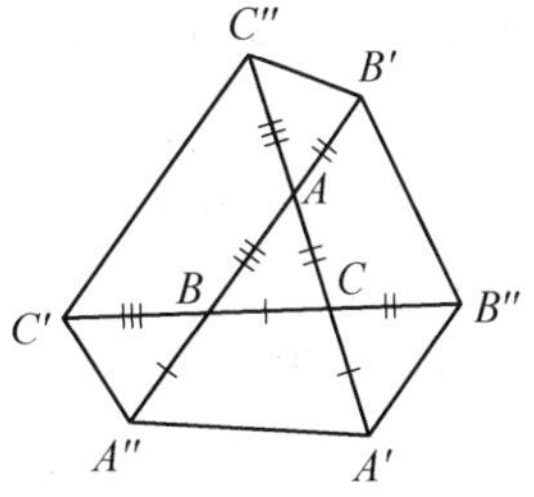

图 1

第二轮

1. 一正九边形各顶点分别染红、绿两色. 任三顶点确定一个三角形，若三顶点同色，则称之为绿（红）三角形. 求证：必存在两个同色三角形，颜色相同且全等.

2. 已知实数 a 满足：有且仅有一正方形，其四顶点均在曲线 $y = x^3 + ax$

上.试求正方形的边长.

3. 在$\triangle ABC$中，$\angle A$平分线交AB边中垂线于A'，$\angle B$平分线交BC中垂线于B'，$\angle C$平分线交CA边中垂线于C'.求证：

(1) A'与B'重合,则$\triangle ABC$为等边三角形;

(2) 若A'、B'、C'互异,则

$$\angle B'A'C'=90^\circ-\frac{1}{2}\angle BAC.$$

4. 是否存在自然数n,使$n!$的前面四位数为1993?

参考答案

第一轮

1. 设$n=a_1+a_2+\cdots+a_{A(n)}$，$a_1<a_2<\cdots<a_{A(n)}$均为自然数.

显然，$a_1\geqslant 1$，$a_2\geqslant 2$，…，$a_{A(n)}\geqslant A_{(n)}$.因此，

$$n\geqslant 1+2+\cdots+A(n)=\frac{1}{2}A(n)(A(n)+1). \qquad (*)$$

从而,自然数$A(n)\leqslant\left[\frac{\sqrt{8n+1}-1}{2}\right]$.

又$n=1+2+\cdots+\left(\left[\frac{\sqrt{8n+1}-1}{2}\right]-1\right)+\left(\left[\frac{\sqrt{8n+1}-1}{2}\right]+a\right)$.

其中a为整数.

由$(*)$可知$a\geqslant 0$.所以

$$A(n)\geqslant\left[\frac{\sqrt{8n+1}-1}{2}\right].$$

综上所述，$A(n)=\left[\frac{\sqrt{8n+1}-1}{2}\right]$.

2. 设A、$B\in M$且AB最长,由已知存在$C\in M$,使$\triangle ABC$为正三角形.显然,点集M完全在分别以A、B、C为圆心，AB为半径的三个扇形CAB、ABC、BCA的并集内,如图2.

若有第4个点 $D \in M$，则另有点 $D' \in M$，D'、A、D 成正三角形，$\angle DAD' = 60°$，所以必有一点在 $\triangle ABC$ 外. 因此，不妨设 D 在弓形 AC 内，BD 与弦 AC 相交于 K. 又不妨设 D' 在弓形 BC 内. 将 $\triangle ABD$ 绕 B 旋转 $60°$ 与 $\triangle CBD'$ 重合，则 K 与 BD' 上一点 K' 重合，$\angle BCK' = \angle BAK = 60° > 30°$. 从而，$K'$ 不在弓形 BC 内，D' 更不在弓形 BC 内，矛盾. 因此，M 中不可能含有第4点.

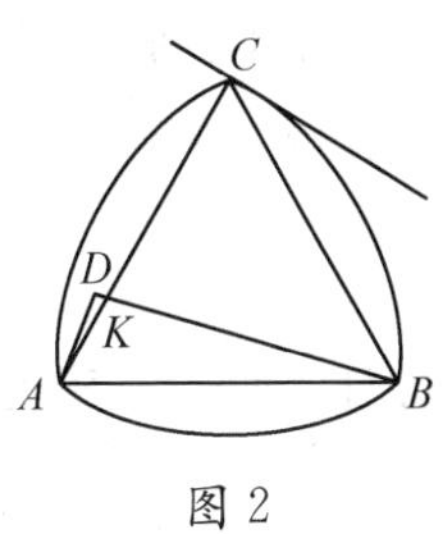

图 2

因此，M 恰含三个点，它们组成正三角形.

3. 令 $a = (5 \times 10^{n-1} - 1)^2$，$b = (10^n - 1)^2$，则 a、b 都是 $2n$ 位数，并且

$$(5 \times 10^{n-1} - 1)^2 \times 10^{2n} + (10^n - 1)^2 = (5 \times 10^{2n-1})^2 - (10^n - 1) \times 10^{2n} + (10^n - 1)^2 = (5 \times 10^{2n-1} - (10^n - 1))^2.$$

所以对任意 $n \in N$，a、b 都是符合要求的数对.

4. 显然，$\triangle AB'C'' \cong \triangle C'BA'' \cong \triangle B''CA' \cong \triangle ABC$.

$$\frac{S_{\triangle AA''A'}}{S_{\triangle ABC}} = \frac{(a+b)(a+c)}{bc} = 1 + \frac{a(a+b+c)}{bc}.$$

所以，$\dfrac{S_{\triangle AA''A'} + S_{\triangle BB''B'} + S_{\triangle CC''C'}}{S_{\triangle ABC}} = 3 + \sum \dfrac{a(a+b+c)}{bc}$.

于是，$\dfrac{S_{A'B''B'C''C'A''}}{S_{\triangle ABC}} = 3 + \sum \dfrac{a(a+b+c)}{bc} - 2 + 3 = 4 + (a+b+c)\sum \dfrac{a}{bc}$

$$\geqslant 4 + 3\sqrt[3]{abc} \cdot 3\sqrt[3]{\frac{a}{bc} \cdot \frac{b}{ca} \cdot \frac{c}{ab}} = 13.$$

第二轮

1. 正九边形外接圆圆周被顶点分为九段，设每段弧长为1. 不妨设红点个数 $\geqslant 5$. 对每三个红点，考虑所分成的三段弧长，其中较小的两段只有7种可能，即(1, 1), (1, 2), (1, 3), (1, 4), (2, 2), (2, 3), (3, 3).

而五个红点可构成 $C_5^2 = 10$ 个同色三角形. $10 > 7$. 因此，必有两个红三角形所分的三段弧长完全相同，即相应的两个三角形全等.

注：本题使用枚举法也不困难.

2. 由于曲线 $y = x^3 + ax$ 关于原点(0, 0)对称，所以该正方形的中心必为原点(否则将这正方形关于原点作对称又得一个顶点均在曲线上的正方形).

设正方形的一顶点为$A(m, n)$,则$B(n, -m)$, $C(-m, -n)$, $D(-n, m)$是其他三个顶点. 令OA与OB的斜率分别为k_1、k_2,则

$$k_1k_2=-1. \tag{1}$$

并且(由于A、B在曲线上).

$$k_1=\frac{n}{m}=m^2+a, \tag{2}$$

$$k_2=-\frac{m}{n}=n^2+a. \tag{3}$$

由(2)、(3),

$$\begin{aligned} k_1^2+k_2^2&=(m^2+a)^2+(n^2+a)^2\\ &=(m^2+a)m^2+(n^2+a)n^2+(m^2+a)a+(n^2+a)a\\ &=mn-mn+ak_1+ak_2=a(k_1+k_2). \end{aligned} \tag{4}$$

结合(1)得

$$(k_1+k_2)^2-a(k_1+k_2)+2=0. \tag{5}$$

方程

$$z^2-az+2=0 \tag{6}$$

有解$z=k_1+k_2$. 并且,对(6)的任一解k_1+k_2,结合(1)可求出k_1、k_2(均为实数,因为$k_1k_2=-1$),再由(2)定出m、n,它们满足(3)的第一个等式. 由(5)可知,

$$\begin{aligned} 0&=k_1^2+k_2^2-ak_1-ak_2=(m^2+a)^2-a(m^2+a)+k_2^2-ak_2\\ &=(m^2+a)m^2+k_2^2-ak_2\\ &=mn+\frac{m^2}{n^2}+\frac{am}{n}=\frac{m}{n}\left(n^2+a+\frac{m}{n}\right). \end{aligned}$$

从而,(3)的第二个等式也成立. 于是,由m、n确定的四点A、B、C、D构成曲线上的正方形.

已知曲线上只有一个正方形,所以(6)只有两个相同的解. 由$\Delta=a^2-2\times4=0$得

$$a=\pm2\sqrt{2}.$$

由(2)、(3)有

$$m^2+n^2=k_1+k_2-2a=\frac{a}{2}-2a=-\frac{3}{2}a.$$

可知 $a=-2\sqrt{2}$，$m^2+n^2=3\sqrt{2}$.

正方形的边长

$$AB=\sqrt{2}\cdot OA=\sqrt{2}\cdot\sqrt{3\sqrt{2}}=\sqrt[4]{72}.$$

3. (1) A'与B'重合时，$\triangle ABC$ 的内心与外心重合,因而是正三角形.

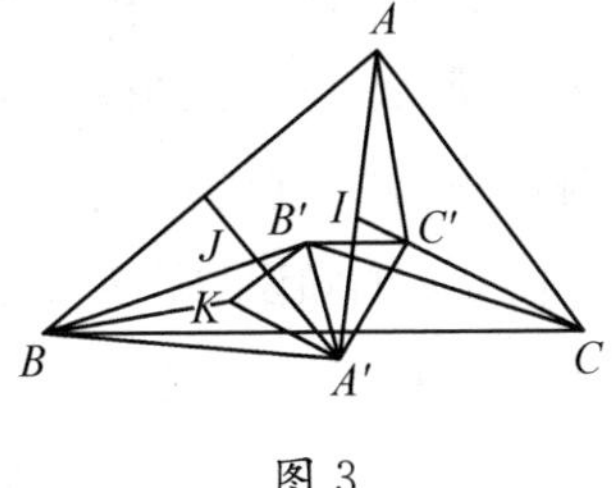

图 3

(2) 将$\triangle A'AC'$绕A'旋转,使 A 与 B 重合.设这时C'转到K,则$BK=AC'=CC'$，$\angle A'BK=\angle A'AC'=\frac{1}{2}(\angle BAC-\angle ACB)$. 又$BB'=B'C$,

$$\angle B'BK=\frac{1}{2}\angle BAC-\frac{1}{2}\angle ABC-\angle A'BK$$

$$=\frac{1}{2}(\angle BAC-\angle ABC-\angle BAC+\angle ACB)=\angle B'CC',$$

所以 $\triangle B'BK\cong\triangle B'CC'$，$B'K=B'C'$.

从而，$\triangle B'A'K\cong\triangle B'A'C$,

$$\angle B'A'C'=\angle B'A'K=\frac{1}{2}\angle C'A'K=\frac{1}{2}\angle AA'B=90^\circ-\frac{1}{2}\angle BAC.$$

注:设 I 为内心，AB 的垂直平分线交 BB'于 J,则可以证明 $\triangle A'C'I\backsim\triangle A'B'J$,从而导出结论,但需要稍多的计算.

4. 存在. 设 $m=1\,000\,100\,000$. 当 $k<99\,999$ 时,若$(m+k)!=\overline{abcd\cdots}$,则$(m+k+1)!=(m+k)!\times(m+k+1)=\overline{abcd\cdots}\times10001\cdots=\overline{abcx\cdots}$,其中$x=d$ 或 $d+1$. 于是,设 $m!=\overline{abcd\cdots}$,则$(m+1)!$，$(m+2)!$，…，$(m+99\,999)!$ 中每一个的前四位数与前一个的相等或增加 1. 而且(由于左起第五位数字增加 a),至多经过 10 个数,前四位数就需增加 1. 这样 100 000 个数 $m!$，$(m+1)!$，…，$(m+99\,999)!$ 的前四位数跑遍 10 000 个值,其中必有 1993 出现.

谈谈 1995 年全国高中联赛试题

1995 年全国高中数学联赛的命题工作由广西数学会负责主持,编拟出了不少好题.如一试选择第 3 题,这题可用横坐标代表身高,纵坐标代表体重,每个人用坐标平面第一象限中的一个点表示.只要在直线 $x+y=k$ (常数)上取 100 个点,那么每一点比它左方的点横坐标大,比它右方的点纵坐标大,因此都代表"棒小伙子"(直线 $x+y=k$ 也可以换成任一减函数的图像).这种题新颖、简明,不需太多的知识与计算,最能考查学生的机敏及对概念(减函数)的领悟.

再如一试选择第 6 题,需要学生根据一些特例判断.作为选择题,可以很快得出答案:如果过 O、P 作平面,那么$\frac{1}{PQ}$的分母为零,严格地说最好在题目中指明平面不过 P,但这样做也就会泄漏底牌,即当平面与 PA 的交点 Q 非常接近 P 时$\frac{1}{PQ}$非常大,因此无最大值,从而只有选(B).这道题还可细细讨论一下,先考虑平面上类似的问题:过 $\triangle PAB$ 的重心 O 作直线与 PA、PB 或其延长线相交于 Q、R,$\frac{PA}{PQ}+\frac{PB}{PR}$ 的取值如何?当 Q、R 分别在线段 PA、PB 内部时,$\frac{PA}{PQ}+\frac{PB}{PR}=3$.最简单的证法是利用斜坐标(如图 1),以 P 为原点,直线 PA、PB 为 x、y 轴,并分别以线段 PA、PB 的长作为两轴的单位,则重心 O 的坐标是$\left(\frac{1}{3},\frac{1}{3}\right)$(过 O 作 x 轴平行线交 PB 于 N,设 PA 中点为 D,则 $\frac{PN}{PB}=\frac{DO}{DB}=\frac{1}{3}$.即 O 的纵坐标 $PN=\frac{1}{3}$).

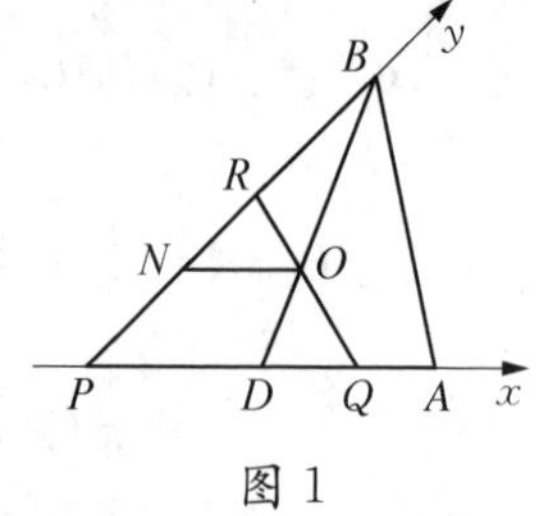

图 1

由截距式,直线 QR 的方程是 $\frac{x}{PQ}+\frac{y}{PR}=1$.因为 O 在这直线上,所以

$$\frac{\frac{1}{3}}{PQ}+\frac{\frac{1}{3}}{PR}=1,\ \frac{PA}{PQ}+\frac{PB}{PR}=3. \tag{1}$$

当 Q 在 AP 的延长线上时,如果约定 PQ 表示负值(即 $PQ=-|PQ|$),那么(1)仍保持成立,如果 PQ 永远表示线段的长度,不取负值,那么(1)不成立.这时

$$\frac{PA}{PQ}+\frac{PB}{PR}=\frac{PB}{PR}-\frac{PA}{PQ}+2\cdot\frac{PA}{PQ}=3+2\cdot\frac{PA}{PQ}$$

可从 3 增至无穷. R 在 BP 延长线上时,情况类似.

现在回到立体情况.如果以 PA、PB、PC 为轴, $PA=PB=PC$ 为长度单位,那么同样有动平面的截距式

$$\frac{x}{PQ}+\frac{y}{PR}+\frac{z}{PS}=1.$$

用 O 点坐标 $\left(\frac{1}{3},\ \frac{1}{3},\ \frac{1}{3}\right)$ 代入上式得

$$\frac{1}{PQ}+\frac{1}{PR}+\frac{1}{PS}=3. \tag{2}$$

如果 PQ、PR、PS 作为有向线段,值可正可负,那么(2)永远成立.否则 $\frac{1}{PQ}+\frac{1}{PR}+\frac{1}{PS}$ 可由 3 增至无穷.这一题的条件只需 $PA=PB=PC$,不必限定三棱锥是正三棱锥(中心 O 需改为重心 O).

一试的填空第 6 题也是一道好题,需将 $\{1,\ 2,\ \cdots,\ 1995\}$ 分为三个子集 $\{134,\ 135,\ \cdots,\ 1995\}$, $\left\{9,\ \cdots,\ 133=\frac{1995}{15}\right\}$, $\left\{1,\ 2,\ \cdots,\ 8=\left[\frac{133}{15}\right]\right\}$,一、三两个子集的并就是元数最多时的 A.作为填空题,写出答案为 $1995-133+8=1870$ 就可以了.严格证明也不难:9 与 9×15, 10 与 10×15, $\cdots$, 133 与 133×15,这 $133-8$ 对中,每一对只能有一个数在 A 中,所以 $|A|\leqslant1995-(133-8)=1870$.类似的题在过去的国内外竞赛中曾出现过.但所谓“陈题”只要是好题,也可以再出.虽然见多识广的教师可能熟悉它,对于大多数学生,却还是新鲜的.

一、二试均有一些题接近高考复习资料上的测试题,计算量较大.有一种说法:这可以吸收更多学生参加.现在看来,或许准备高考的学生仍觉得过难,而喜爱竞赛的选手又可能在这种常规题中不慎“落马”.不如将一、二试并作五或

六道大题的一次考试,人数少些,时间短些为好.

二试的四道题都较好,第四道尤为精彩.可惜的是学生在前三道题耗费时间精力太多,做到这题绝大多数的人已经力不从心了.前三题应当降低一点难度,减少一些计算量,或者改为两道.数学竞赛本不应设立太多的"规矩".规矩既是人定的,也可以改一改嘛.下面谈一谈二试的四道题.

第一题当然是先求曲线与直线的交点,一个交点是原点,另一个的横坐标 $x=\frac{8\sin\theta+\cos\theta+1}{2\sin\theta-\cos\theta+3}$,纵坐标 $y=2x$.弦长 $=\sqrt{x^2+y^2}=\sqrt{5}|x|$,只需求 $|x|$ 的最大值.先估算,在 $\sin\theta=1$、-1 时,x 的值分别为 $\frac{9}{5}$、-7.易知 $\frac{8\sin\theta+\cos\theta+1}{2\sin\theta-\cos\theta+3}<7$(去分母后有 $20>-6\sin\theta+8\cos\theta$).再考虑 $x<0$ 时,$|x|$ 的最大值,由 $-|x|=\frac{8\sin\theta+\cos\theta+1}{2\sin\theta-\cos\theta+3}$ 得 $(8+2|x|)\sin\theta+(1-|x|)\cos\theta+1+3|x|=0$,从而 $1+3|x|\leqslant\sqrt{(8+2|x|)^2+(1-|x|)^2}$,解得 $|x|\leqslant 8$.在 $\theta=\pi-\arcsin\frac{24}{25}$ 时等号成立.因此 $|x|$ 的最大值是 8,弦长的最大值是 $8\sqrt{5}$.

这类与 $A\sin\theta+B\cos\theta$ 有关的极值问题似已出现过几次.

第二题应看出 1 是方程的根,从而化为方程

$$5x^2-5px+66p-1=0 \tag{3}$$

的两个根 x_1、x_2 均为自然数.于是 $p=x_1+x_2$ 也是自然数,(3)虽是 x 的方程,这时却应将 x 作为已知自然数**解出** $p=\frac{5x^2-1}{5x-66}=x+\frac{66x-1}{5x-66}$,再乘以 5 得

$$5p=5x+\frac{66\times5x-5}{5x-66}=5x+66+\frac{66^2-5}{5x-66}.$$

$66^2-5=4351=19\times229$,$5x-66$ 是它的因数,只可能为 ±1,±19,±229,±4351.不难得出只有 $p=76$,-844 合乎要求,相应的 $x_1=17$,$x_2=59$ 及 $x_1=13$,$x_2=-857$.

第三题是一道几何证明,也许将这种问题作为初中竞赛更好.证法甚多,我觉得第一步应是看出 $MQ\,/\!/\,MP \Leftrightarrow \angle AMQ=\angle CPN \Leftarrow \triangle AMQ\backsim\triangle CPN \Leftarrow \frac{AM}{AQ}=\frac{CP}{CN} \Leftarrow AM\times CN=AQ\times CP$.于是只需证明 $AM\times CN$ 是与切线 MN 位置

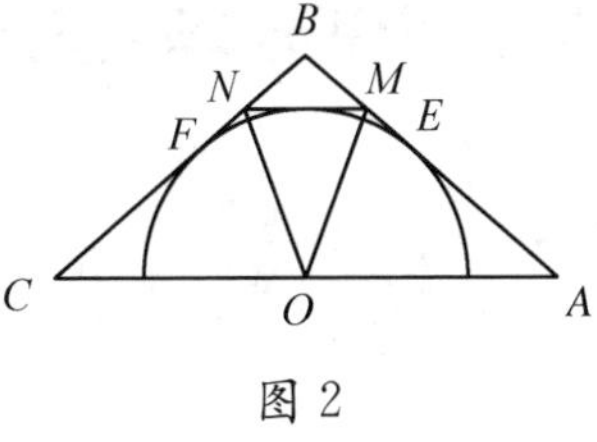

图 2

无关的常量(同理 $CP \times AQ$ 也等于这一常量). 问题化成只需考虑半个图(图 2).

易证 $2\angle CNO + 2\angle AMO + \angle C + \angle A = 360^\circ$, 所以 $\angle CON = 180^\circ - \angle CNO - \angle C = \angle AMO$, 从而 $\triangle AMO \backsim \triangle CON$, $AM \times CN = OC^2$.

第四题的证法也不止一种. 多数学生是先证对任意正数 d, 有两个同色点距离为 d, 再以这两点为直径两端作圆, 它们与圆上另外四点构成正六边形, 易知六点中有三点同色并且构成一个锐角为 30° 与 60° 的直角三角形. 将 d 改为 $\frac{d}{1995}$ 就得到另一个同色三角形, 两者的相似比为 1995.

本题若加上限制"两个三角形都有两条边分别与两个坐标轴平行", 那么上面的证法便不能奏效, 难度也增加了. 这时可用如下证法:

先在 x 轴上取两个同色点 A、B, 设它们的横坐标为 a、b, $C(a, 1)$, $D(b, 1)$ 中或者有一点与 A、B 构成同色三角形, 或者 C、D 均与 A、B 异色, 对后一种情况, 再考虑点 $E(b, 2)$. $\triangle EAB$ 或 $\triangle ECD$ 中必有一个是同色三角形. 于是我们有一个同色的直角三角形 ABC ($\angle B = 90^\circ$), 将它补全为矩形 $ABCD$, 然后用平行线 B_1C_1, B_2C_2, …, $B_{1994}C_{1994}$ 将 AB、CD 等分为 1995 份, B_i、C_i 为分点. 不妨设 A、B、C 为红. 若 B_1、C_1 中有一个为红, 我们得到一边 $=\frac{AB}{1995}$, 一边为 BC 的同色三角形, 否则 B_1、C_1 同蓝, 将 B、C 改为 B_1、C_1 进行同样讨论. 如果没有一边等于 $\frac{AB}{1995}$、一边等于 BC 的直角三角形, 那么 B_1, B_2, …, A 交错地为蓝色与红色, 与 A 为红色矛盾. 所以必有一个同色三角形一边等于 $\frac{AB}{1995}$、一边等于 BC, 并且它的这两条边分别与 AB、BC 平行. 用这三角形代替 $\triangle ABC$ 进行同样的讨论便可得到结论.

笔者提供给命题组的原始问题就是"平面上的点染成红蓝两色, 如果有一个直角三角形顶点同色, 两条直角边的长均为奇数(比如说 1995), 那么必有一个同色的直角三角形, 两条直角边的长度为 1". 命题组将它改成现在的形式, 改得好, 但如果添上两个三角形的边互相平行的要求, 那就更好了.

谈谈 1996 年全国高中联赛

1996 年全国高中数学联赛，试题比较平稳，没有特别新奇的难题.

先说二试.前两题较为常规，解法大概也只有依照标准答案，不会有太大差异.

第三道平面几何题，如图 1，A 可以看成两个圆$\odot O_1$、$\odot O_2$ 的内公切线的交点，因而是它们的位似中心，所以$\frac{AG}{AH}$，$\frac{AO_1}{AO_2}$都等于两圆半径之比，如果结论成立，那么$\frac{ED}{DF}=\frac{AO_1}{AO_2}$. 反过来，如果$\frac{ED}{DF}=\frac{AO_1}{AO_2}$，那么结论成立. 要计算$\frac{ED}{DF}$，只需注意 PE 与$\triangle CAD$ 三边的延长线相截，由梅氏定理

$$\frac{PD}{PA}\cdot\frac{GA}{GC}\cdot\frac{EC}{ED}=1,$$

而 $EC=GC$，所以 $ED=\frac{PD}{PA}\cdot GA$，同样 $DF=\frac{PD}{PA}\cdot AH$. 所以

$$\frac{ED}{DF}=\frac{AG}{AH}=\frac{AO_1}{AO_2}.$$

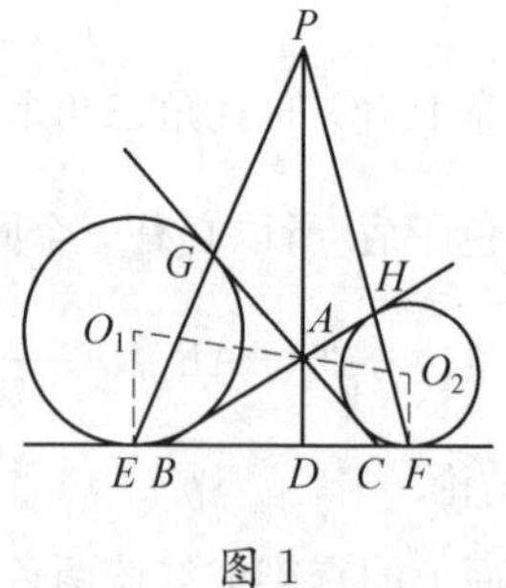

图 1

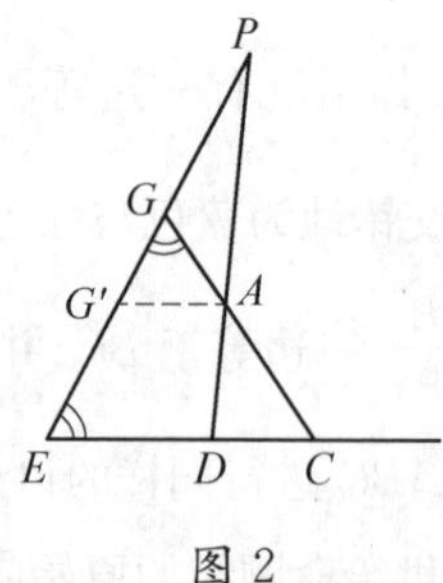

图 2

求 ED 也可不用梅氏定理. 如图 2，$\triangle CGE$ 中，$CG=CE$，那么对任一条截线 PD，均有$\frac{ED}{AG}=\frac{PD}{PA}$. 这只需过 A 作 $AG' \parallel DE$ (G'在GE 上)，则由等腰三角

形的性质，$\angle AG'G=\angle GEC=\angle EGC$，$AG'=AG$. 所以

$$\frac{DE}{AG}=\frac{DE}{AG'}=\frac{PD}{PA}.$$

第四题是图论问题. 将人用点表示，两人相识就在相应的两点间连一条线. 要证明所得的图中有(以这些点为顶点的)三角形.

条件(2)是需要的. 否则，取$\left[\frac{n}{2}\right]$个点互不相连组成集 X，$\left[\frac{n+1}{2}\right]$个点互不相连组成集 Y，将集 X 的每一点与集 Y 的每一点相连. 所得的图，每一点至少引出$\left[\frac{n}{2}\right]$条线，但图中没有三角形.

我们的证明可以从条件(2)出发，分情况讨论. 设$\left[\frac{n}{2}\right]=k$.

若每 k 个点中至少有两点相连，则任取一点 A，与 A 相连的 k 个点中有 B，C 相连，$\triangle ABC$ 即为所求.

以下设有 k 个点互相不连，并称它们为前 k 个点(类似于上面的集 X).

在 $n=2k$ 时，前 k 个点中的任一点 A 与其余的 k 个点均相连，并且这与 A 相连的 k 个点中必有两点 B、C 相连，结论成立.

在 $n=2k+1$ 时，若有 $k+1$ 个点互相不连，则其余的 k 个点中必有两点 B、C 相连. 前 $k+1$ 个点中的任一点 A 与 B、C 构成三角形.

最后，可设 $n=2k+1$ 并且至多有 k 个点互相不连. 这时其余 $k+1$ 个点中必有两点 B、C 相连. 前 k 个点中每一点仅与 B、C 之一相连(否则结论已经成立)，从而与其他的 $k-1(=k+1-2)$ 个点均相连，若 B 与这 $k-1$ 个点中某点 D 相连，则由于至多有 k 个点互不相连，B 必与前 k 个点中某点 A 相连，$\triangle ABD$ 即为所求. 设 B 与这 $k-1$ 个点均不相连，C 也如此(否则结论成立)，则 B，C 各向前 k 个点至少引 $k-1$ 条线. 由于

$$(k-1)+(k-1)>k\ (k\geqslant 3),$$

所以前 k 个点中必有一点 A 与 B、C 均相连，组成$\triangle ABC$.

一试题目并不容易. 高考中，各种题型(选择、填空、综合题)的题都有难有易，由易到难，并不是综合题都难，选择、填空都容易. 现在的高中竞赛似乎也走这条路，但难度当然超过高考，尤其选择的第 6 题，填空的第 5、6 题都很不容

易. 填空的第 6 题要用到勾股数的知识,已超过中学数学大纲,但学生很可能猜到答案,因此结论是否正确不一定反映水平的高低,这种题如果一定要用,或许作为综合题更好. 填空的第 5 题难而且好,这道题不大可能猜出答案.

选择第 1 题很好,但选项可加(或换成)恰有两边相等的三角形. 选择 2 亦好,高考题曾出现过 n 项和最大的情况,要积最大,一应为正,二应在为正的积中值最大,从而只需比较 π_9, π_{12}($a_{13} < |a_{13}| = \frac{1536}{2^{11}} < 1$,所以不必考虑 $n > 12$ 的 π_n) 由 $a_{16} = -3$, $a_{10}a_{11}a_{12} = 3^3 \times \left(\frac{1}{2}\right)^3 > 1$ 即知 π_{12} 最大. 选择 3 是个机敏的好题,题目的文字有点拗口,但也没有更好的表述方式. 让学生读读这种数学中特有的句子也有好处. 如果用集合,可以说成:"使集 $\{n: n \in Z$ 且 $\sqrt{p+n} + \sqrt{n} \in Z\}$ 不是空集的质数 p".

填空第 1 题注意到 $\frac{1}{x} < 1$,化对数为指数便得 $x \geqslant 10 > x^{\frac{1}{2}}$,即 $10 \leqslant x < 100$, 要求$\{10, 11, \cdots, 99\}$的真子集的个数可从元素 a 是否属于这子集考虑. 由于每一元素均有两种选择(属于或不属于),所以共有 2^{90} 个子集, $2^{90}-1$ 个真子集. 填空第 3 题并不需要知道曲线 C 是心脏线,只需知道 A 与曲线上的点的距离可由 0 连续增至$\sqrt{\frac{16}{3}}$,也是一道好题. 但用到极坐标,中学需在高三才学习,竞赛中也很少出现过.

试题中,极值问题似太多. 如果面能更宽广一些,当然更好.

数学竞赛原本起着普及数学的作用,但由于不少地方的奥校停办,今年的题目虽不算难,平均水平恐怕有所下降.

近两年的美国数学奥林匹克

因为南京大学出版社要出新版的《理科(数、理、化、信息)奥林匹克题典》，我翻阅了一些国内杂志，未发现近两年(1996、1997)的美国数学奥林匹克，题目载在加拿大的杂志 *Crux* 上，但没有解答. 解题过程中有些想法，写出来供读者参考.

近年来，美国数学奥林匹克已与 IMO“接轨”，由原来的 5 题改为 6 题. 其中有些难题，也有些题比较容易；有不少新题，也有一些“陈题”.

题 25.1：证明数 $n\sin n^\circ(n=2, 4, 6, \cdots, 180)$的平均值是 $\cot 1^\circ$.

就是极常见的. 将和 $\sum\limits_{k=1}^{90} 2k\sin 2k^\circ$ 乘以 $\sin 1^\circ$再积化和差变成 $\sum\limits_{k=1}^{90}\cos(2k-1)^\circ-90\cos 181^\circ$. 前面的和是熟知的，再用 $2\sin 1^\circ$乘并积化和差，即化为 0. 所以，

$$原式=\frac{90\cos 1^\circ}{\sin 1^\circ}=90\cot 1^\circ.$$

题 26.1：令 $p_1, p_2, p_3, \cdots$为依递增次序排列的全体质数. 实数 x_0 在 0、1 之间，对正整数 k，定义

$$x_k=\begin{cases}0, & 若\ x_{k-1}=0,\\ \left\{\dfrac{p_k}{x_{k-1}}\right\}, & 若\ x_{k-1}\neq 0.\end{cases}$$

其中 $\{x\}=x-[x]$ 表示 x 的分数部分.

求出所有适合 $0<x_0<1$，并使序列 $x_0, x_1, x_2, \cdots$最终成为 0 的 x_0，并予以证明.

就比较有意思. 答案是 x_0 为有理数.

因为对于真分数 $x_{k-1}=\dfrac{n}{m}$，代入 $x_k=\left\{\dfrac{p_k}{x_{k-1}}\right\}$ 后所得的值仍为真分数，而分母为 n 的约数，严格小于 m，所以，经有限次迭代后成为 0.

对无理数 x_0，一切 x_k 都是无理数，其分数部分永不为 0.

这道题不难,却很好地刻画了有理数与无理数的不同特性. 国内竞赛中,这样的题目似乎不多.

题 26.1 还有一个特点,即答案需要自己去找.

题 25.3: 已知$\triangle ABC$. 证明:存在一条直线 l(在$\triangle ABC$ 所在平面内),使得$\triangle ABC$ 关于 l 的对称图形$\triangle A'B'C'$与$\triangle ABC$ 的公共部分面积大于$\triangle ABC$ 面积的$\frac{2}{3}$.

题 25.6: 确定(并证明)是否有整数集的子集 X 具有下面的性质:对任意整数 n,恰有一组 a、$b \in X$ 使 $a+2b=n$.

这两题也是如此. 题 25.3 中 l 究竟在哪里? 得先探索一下. 初看以为与重心有关(因为出现了$\frac{2}{3}$),有趣的是,如果过重心作一边的平行线,用它作为对称轴 l,翻折后公共部分的面积恰好为$\triangle ABC$ 面积的$\frac{2}{3}$,可惜不是大于$\frac{2}{3}$. 有没有更合适的对称轴? 当然有,这就是三角形的角平分线.

如设 $AB=c \geqslant AC=b \geqslant BC=a$, 则以角平分线 AD 作为对称轴,易得公共部分面积为

$$\left(1-\frac{c-b}{c} \times \frac{c}{c+b}\right) S_{\triangle ABC}=\frac{2b}{c+b} S_{\triangle ABC}>\frac{2}{3} S_{\triangle ABC}.$$

题 25.6 的判定则比较困难. 究竟是有还是没有,恐怕一下子难以确定. 判定"没有",当然要用反证法来证明,但又证不出,因而(经过一番失败后)应转而断定"有",并采用构造法(即设法将 X 构造出来). 此题是 25 届的大轴,当然比较困难. 需采取归纳构造的方法,即先作较小的集,保证对集中任二元素 a、b(a、b 可以为同一元素), $a+2b$ 互不相同(唯一性),然后再逐步增加元素,直至每个整数 n 都能表成 $a+2b$.

具体做法是,令 $X_2=\{0, 1\}$. 若已有

$$X_k=\{a_1, a_2, \cdots, a_k\}$$

使得 $a_i+2a_j(1 \leqslant i, j \leqslant k)$ 互不相同. 对任一与这些 a_i+2a_j 都不相同的整数 n,取正整数 a_{k+1},再令 $a_{k+2}=-2a_{k+1}+n$. 只要 a_{k+1} 充分大,对于 $1 \leqslant i, j, s, t \leqslant k$, 有

$$3a_{k+1} > a_s + 2a_{k+1} > a_{k+1} + 2a_t > n > a_{k+2} + 2a_j$$
$$> a_{k+1} + 2a_{k+2} > a_i + 2a_{k+2} > 3a_{k+2},$$

而且 $a_{k+1} + 2a_t$ 大于$\{a_i + 2a_j \mid 1 \leqslant i, j \leqslant k\}$中一切正数，$a_{k+2} + 2a_j$ 则小于其中一切负数. 于是，

$$X_{k+2} = \{a_{k+1}, a_{k+2}\} \cup X_k$$

具有上面所说的唯一性.

令 $X = X_2 \cup X_4 \cup \cdots \cup X_{2k} \cup X_{2k+2} \cup \cdots$，则 X 满足要求.

此题的关键在于对无穷的理解：

第一，只要 a_{k+1} 足够大，就可以使唯一性仍然保持.

第二，任一整数 n 可以表出是逐步实现的，每一次扩大至少解决前面一个遗漏的 n 从而在扩大的过程中实现“无遗漏”.

第三，X 是经无穷多次增添元素而得到的.

以上这些在初等数学中极少出现，在高等数学中却屡见不鲜. 所以这类问题的确是“初等的面貌，高等的背景”.

纯粹的不等式在这两届中仅出现一题(相比之下，国内各种竞赛中，不等式的比重较大).

题 26.5：证明对所有正实数 a、b、c，有 $(a^3 + b^3 + abc)^{-1} + (b^3 + c^3 + abc)^{-1} + (c^3 + a^3 + abc)^{-1} \leqslant (abc)^{-1}$.

此题无法用柯西不等式，因为不等号是“$\leqslant$”，而不是“$\geqslant$”. 对这类不等式，其实不必挖空心思去想巧解，老老实实去分母并化简(虽然似乎繁一点)，原式等价于

$$a^6(b^3 + c^3) + b^6(c^3 + a^3) + c^6(a^3 + b^3) \geqslant 2a^2b^2c^2(a^3 + b^3 + c^3).$$

易知右边 $\leqslant (a^4 + b^4)c^5 + (b^4 + c^4)a^5 + (c^4 + a^4)b^5$.

不妨设 $a \geqslant b \geqslant c$，则有

$$\begin{aligned} & a^6(b^3 + c^3) + b^6(c^3 + a^3) + c^6(a^3 + b^3) \\ & \quad - (a^4 + b^4)c^5 - (b^4 + c^4)a^5 - (c^4 + a^4)b^5 \\ = & (a - b)a^3b^3(a^2 - b^2) + (a - c)a^3c^3(a^2 - c^2) + (b - c)b^3c^3(b^2 - c^2) \\ \geqslant & 0. \end{aligned}$$

另两道与不等式有关的问题是

题 26.6:设非负整数 $a_1, a_2, \cdots, a_{1997}$ 满足 $a_i+a_j \leqslant a_{i+j} \leqslant a_i+a_j+1$ $(1 \leqslant i, j, i+j \leqslant 1997)$. 证明:存在实数 x,对所有 $1 \leqslant n \leqslant 1997$,满足 $a_n=[nx]$.

题 25.2:对任意非空实数集 S,令 $\sigma(s)$ 为 S 的元素之和. 已知 n 个正整数的集 A,考虑 S 跑遍 A 的非空子集时,所有不同和 $\sigma(s)$ 的集. 证明:这些和可分为 n 类,每一类中最大的和与最小的和的比不超过 2.

题 26.6 的结论表明,对 $1 \leqslant n \leqslant 1997$,有

$$\frac{a_n+1}{n}>x \geqslant \frac{a_n}{n}.$$

如取定 $x=\max \frac{a_n}{n}$,只需证明对一切 $1 \leqslant m \leqslant 1997$, $\frac{a_m+1}{m}>x$,即 $\frac{a_m+1}{m}>\frac{a_n}{n}$,或

$$n a_m+n>m a_n. \tag{1}$$

式(1)可用归纳法证明. 奠基显然. 设 m、n 均小于 k 时(1)成立. 当 m、n 中较大的为 k 时,有两种情况:

(i) 当 $n=k$ 时,设 $n=qm+r$, q 为自然数, $0 \leqslant r<m$. 由已知,有

$$a_n \leqslant a_{qm}+a_r+1 \leqslant a_{(q-1)m}+a_m+a_r+2 \leqslant \cdots \leqslant q a_m+a_r+q.$$

再由归纳假设 $r a_m+r>m a_r$ 得

$$m a_n \leqslant m q a_m+m a_r+m q<n a_m+n.$$

(ii) 当 $m=k$ 时,设 $m=qn+r$, q 为自然数, $0 \leqslant r<n$. 类似(i),有

$$a_m \geqslant q a_n+a_r,$$
$$n a_m+n \geqslant n q a_n+n a_r+n=m a_n+n a_r+n-r a_n>m a_n.$$

题 25.2 似比题 26.6 还难一些. 题中 A 可为任一正实数集. 此题我很想用归纳法,若命题对 $n-1$ 成立,则对 $A=\{a_1, a_2, \cdots, a_n\}$,集 $\{a_1, a_2, \cdots, a_{n-1}\}$ 的子集和已分为 $n-1$ 类,合乎要求,且最大元为 $a_1+a_2+\cdots+a_{n-1}$. 若 $a_n \geqslant a_1+a_2+\cdots+a_{n-1}$,则所有含 a_n 的子集和可作为一类,最大和 $a_1+a_2+\cdots+a_n \leqslant$ 最小和 a_n 的 2 倍. 但若 $a_n<a_1+a_2+\cdots+a_{n-1}$,则 a_n 不能属于新添的一类,反

倒是 $a_1+a_2+\cdots+a_{n-1}$ 可归于这一类. 这时 a_n 应归到哪一类,似不易处理. 因此,归纳法不能奏效. 不过上面的考虑已给出了分类的办法,即第 n 类的最小数应为 $\max\{a_n, a_1+a_2+\cdots+a_{n-1}\}$(最大数当然是 $a_1+a_2+\cdots+a_n$).

一般地,令

$$f_j=a_1+a_2+\cdots+a_j, e_j=\max\{a_j, f_{j-1}\},$$

则

$$f_j=f_{j-1}+a_j\leqslant 2e_j.$$

只要证明每个和 $a_{i_1}+a_{i_2}+\cdots+a_{i_t}$, $i_1<i_2<\cdots<i_t$,必在某个区间 $[e_j, f_j]$ 中.

显然,这种和必在某个区间 $(f_{j-1}, f_j]$ 中,由

$$a_{i_1}+a_{i_2}+\cdots+a_{i_t}>f_{j-1}=a_1+a_2+\cdots+a_{j-1}$$

得 $i_t\geqslant j$.

从而, $a_{i_1}+a_{i_2}+\cdots+a_{i_t}\geqslant a_j$.

于是, $a_{i_1}+a_{i_2}+\cdots+a_{i_t}\in[e_j, f_j]$.

除题 25.3 外,还有三道平面几何题.

题 25.5:$\triangle ABC$ 具有下面性质:存在一个内部的点 P 使 $\angle PAB=10°$, $\angle PBA=20°$, $\angle PCA=30°$, $\angle PAC=40°$. 证明:$\triangle ABC$ 是等腰三角形.

似为一道陈题的改造.

如图 1,作高 BD,又作 AQ,使 $\angle QAD=30°$,且 AQ 交 BD 于 Q. 连 PQ,易知 P 是$\triangle ABQ$ 的内心(AP、PB 是角平分线), $\angle PQA=60°$. 设 PQ 交 AC 于 C',则

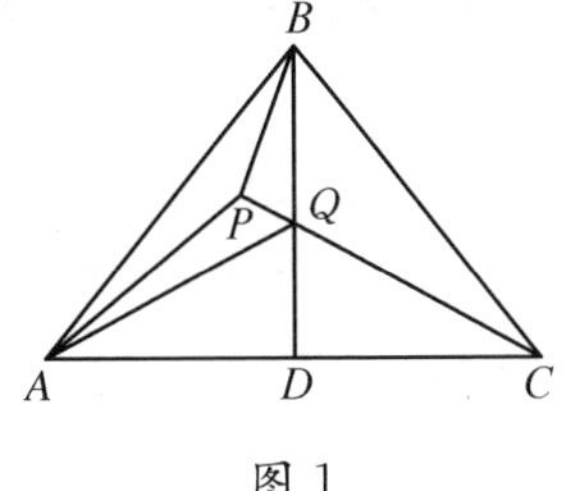

图 1

$$\angle PC'A=\angle PQA-\angle QAC=60°-30°=30°=\angle PCA.$$

故 C' 与 C 重合. 从而, $QA=QC$, $BA=BC$.

题 26.2:分别以$\triangle ABC$ 的边 BC、CA、AB 为底向外作等腰$\triangle BCD$、$\triangle CAE$、$\triangle ABF$. 分别过 A、B、C 作 EF、FD、DE 的垂线,证明:这三条垂线共点.

此题也不难. 分别以 D、E、F 为圆心, DB、EC、FA 为半径作圆,则这三个圆两两的公共弦就是题中的三条垂线. 而熟知这三条弦所在直线交于一点

(三圆的根心).

题 26.4:切一个凸 n 边形时选出一对相邻的边 AB、BC,切去$\triangle MBN$ 得到一个凸 $n+1$ 边形,其中 M、N 分别为 AB、BC 的中点. 一个正六边形 P_6,面积为 1,切成七边形 P_7;再将 P_7(用七种可能的切法之一)切成八边形 P_8. 如此继续下去. 证明:不论怎么切,对所有 $n \geqslant 6$, P_n 的面积大于$\frac{1}{3}$.

这是一道有趣的题目,新颖却不需要用很多知识.

一个凸 n 边形,任一顶点 A 有两个相邻顶点 B、C,称线段 BC 为小对角线. 设 P_6 为正六边形 $ABCDEF$,它的小对角线围成一个六边形 S,易知 S 为正六边形,且面积为$\frac{1}{3}$.

P_6 严格含有 S 且小对角线不经过 S 的内部. 假设 $P_n(n \geqslant 6)$ 严格含有 S 且小对角线不经过 S 的内部. 切一刀切去$\triangle LM'N'$,其中 M'、N'分别为 $P_n = KMLNH\cdots$ 的边 LM、LN 的中点. 这时 P_{n+1} 新增加的小对角线 KM、$M'N'$、$N'H$ 都不过 S 的内部,于是, P_{n+1} 严格含有 S 且小对角线不过 S 的内部. 从而结论成立.

题 25.4:n 项的 0、1 序列$(x_1, x_2, \cdots, x_n)$称为长为 n 的二元序列. a_n 为无连续三项 0, 1, 0 的、长为 n 的二元序列的个数. b_n 为无连续四项成 0, 0, 1, 1 或 1, 1, 0, 0 的、长为 n 的二元序列的个数. 证明:对每一个正整数 n, $b_{n+1} = 2a_n$.

这是一道组合题. 只要注意 0、1 序列 $y=(y_1, y_2, \cdots, y_{n+1})$ 可与 $x=(x_1, x_2, \cdots, x_n)$ 对应,其中

$$x_i \equiv y_i + y_{i+1} (\operatorname{mod} 2).$$

由此易得 $b_{n+1} = 2a_n$. 不用对应似相当棘手.

题 26.3:证明:对任意整数 n,存在一个唯一的多项式 Q,系数 $\in \{0, 1, \cdots, 9\}$, $Q(-2) = Q(-5) = n$.

这可能是最难的一题. 唯一性易证:

若 $Q(x)$、$Q_1(x)$ 均合乎要求,则令 $f(x) = Q(x) - Q_1(x) = b_k x^k + b_{k-1} x^{k-1} + \cdots + b_1 x + b_0$,其中 $b_i \in \{0, \pm 1, \cdots, \pm 9\}$, $0 \leqslant i \leqslant k$.

由 $f(-2) = 0$ 得 $2 \mid b_0$.

由 $f(-5)=0$ 得 $5|b_0$.

所以 $10|b_0$. 但 $|b_0|<10$,所以 $b_0=0$.

再由 $\frac{1}{2}f(-2)=\frac{1}{5}f(-5)=0$ 得 $10|b_1$, $b_1=0$. 依此类推即有 $f(x)=0$.

由唯一性的证明,易想到合乎要求的 $Q(x)$,其系数 a_0, a_1, …,应满足

$$a_0\equiv n(\bmod 2),\ a_0\equiv n(\bmod 5)$$

及
$$a_k(-2)^k+a_{k-1}(-2)^{k-1}+\cdots+a_1(-2)+a_0\equiv n(\bmod 2^{k+1}),$$
$$a_k(-5)^k+a_{k-1}(-5)^{k-1}+\cdots+a_1(-5)+a_0\equiv n(\bmod 5^{k+1}),$$

其中 a_k 为 x^k 的系数.

由这些方程组及中国剩余定理确实能逐步定出 a_0, a_1, …, a_k, …$\in\{0, 1, 2, \cdots, 9\}$. 但如何证明只有有限多个系数非 0? 仅用初等方法,不涉及 p-adic 收敛概念似难以解决. 因此改用下面的解法.

如果多项式 $f(x)$ 与 $g(x)$ 在 $x=-2$ 与 $x=-5$ 时值均相等,就记成 $f(x)=g(x)$. 如 $x^2+7x+10=0$.

在 $n\in\{0, 1, 2, \cdots, 9\}$ 时,常数 n 就是满足要求的多项式 $Q(x)$. 在 $n=10$ 时, $Q(x)=x^3+6x^2+3x$ 满足要求. 将它简记为 $(0, 3, 6, 1)$. 一般地, $Q(x)=a_kx^k+a_{k-1}x^{k-1}+\cdots+a_0$ 简记为 $(a_0, a_1, \cdots, a_k)$.

设 $Q(x)=(a_0, a_1, \cdots, a_k)$ 的系数 $\in\{0, 1, 2, \cdots, 9\}$. 我们证明存在多项式 $P(x)=Q(x)+1$, $p(x)$ 的系数 $\in\{0, 1, 2, \cdots, 9\}$ 且 $P(x)$ 的系数和等于 $Q(x)$ 的系数和 $+1$. 为此,对 $Q(x)$ 的系数和 $a_0+a_1+\cdots+a_k$ 进行归纳. 奠基显然,设对系数和较小的多项式结论成立.

若 $a_0<9$,结论显然. 若 $a_0=9$, 则

$$(a_0, a_1, \cdots, a_k)+1=(0, a_1, \cdots, a_k)+(0, 3, 6, 1).$$

(i) 若 $3+a_1\leqslant 9$, 则

$$(0, a_1, \cdots, a_k)+(0, 3, 6, 1)=(0, a_1+3, a_2, \cdots, a_k)+(0, 0, 6, 1).$$

对多项式 $(a_2, a_3, \cdots, a_k)$ 用归纳假设,得多项式 $(a_2', a_3', \cdots, a_k')=(a_2, a_3, \cdots, a_k)+1$. 继续对所得多项式用归纳假设,直至得到

$$(a_2^{(6)}, \cdots, a_t^{(6)})=(a_2, a_3, \cdots, a_k)+6.$$

再由归纳假设得

$$(a_2^{(6)}, \cdots, a_t^{(6)}) + (0, 1) = (a_2^{(7)}, a_3^{(7)}, \cdots, a_r^{(7)}).$$

多项式 $(0, a_1+3, a_2^{(7)}, a_3^{(7)}, \cdots, a_r^{(7)})$ 即为所求的 $P(x)$.

(ii) 若 $3+a_1 \geqslant 10$,则令 $a_1' = a_1 - 7$,

$$(0, a_1, \cdots, a_k) + (0, 3, 6, 1) = (0, a_1', a_2, \cdots, a_k) + (0, 0, 9, 7, 1).$$

当 $a_2=0$ 时,上式即 $(0, a_1', 9, a_3, \cdots, a_k)+(0, 0, 0, 7, 1)$,情况与(i)类似. 当 $a_2 \geqslant 1$ 时,令 $a'_2 = a_2 - 1$,则

$$\text{上式} = (0, a_1', a_2', a_3, \cdots, a_k) + (0, 0, 10, 7, 1) = (0, a_1', a_2', a_3, \cdots, a_k).$$

同样,可以证明存在多项式 $R(x)$,系数 $\in \{0, 1, 2, \cdots, 9\}$,且 $R(x) = Q(x) - 1$(这只要注意 $Q(x) - 1 = Q(x) + (9, 7, 1)$,再多次利用上面关于 $Q(x)+1$ 的结果即得).

因此,对一切整数 n,均有合乎要求的多项式 $Q(x)$ 存在.

本题解法类似于十进制的加法. 但十进制中逢 10 进 1,因而和的位数至多比大的加数多 1. 这里 10 却变成(0361),不能保证和的位数只多 1 位. 所以需要对数字和进行归纳,以保证和也是多项式(即只有有限多项).

上面各题的解法未必是最好的,一得之见而已.

评 2001 年高中联赛

2001 年高中联赛,接近高中教材与高考.试题大多中正平和,解法也简明扼要.

选择题第 6 题是作图题的讨论,应选(D),不小心就会做错.第 6 题买花是不等式的应用.设玫瑰与康乃馨的单价分别为每枝 x 元与 y 元,建立不等式组

$$\begin{cases}6x+3y>24, & (1)\\ 4x+5y<22. & (2)\end{cases}$$

不等式组与方程组的解法类似,(1)÷3(如果除以负数,不等号方向要改变)得

$$2x+y>8. \tag{3}$$

(2)−2×(3)消去 x(不等号同向的两个不等式可以相加;不等号异向的两个不等式可以相减,所得差的不等号与用作被减式的不等号同向)得

$$3y<6,\ y<2.$$

同样可消去 y(或将 $y<2$ 代入(3))得 $x>3$.

所以
$$2x>3y.$$

填空题第 8 题可令 $u=3z_1$, $v=2z_2$,这时

$$u-v=\frac{3}{2}-\mathrm{i}, \tag{4}$$

$$|u|=|v|=6, \tag{5}$$

本题有两个关键,一是由(4)取共轭得

$$\bar{u}-\bar{v}=\frac{3}{2}+\mathrm{i}. \tag{6}$$

二是注意 $\bar{u}=\frac{|u|^2}{u}$，由(5)、(6)得

$$\frac{6^2}{u}-\frac{6^2}{v}=\frac{-6^2(u-v)}{uv}=\frac{3}{2}+\mathrm{i}. \tag{7}$$

最后由(4)、(7)得

$$uv=\frac{-6^2\left(\frac{3}{2}-\mathrm{i}\right)}{\frac{3}{2}+\mathrm{i}}=\frac{-6(30-72\mathrm{i})}{13}.$$

即
$$z_1z_2=-\frac{30}{13}+\frac{72}{13}\mathrm{i}.$$

填空题第 12 题的做法很多. A 有 4 种栽法，A 栽好后 B 有 3 种栽法，共 $4\times3=12$ 种栽法. 不妨固定 A 栽第 1 种，B 栽第 2 种.

可以考虑更一般的情况：A、B 后面有 n 个区域 C_1, C_2, $\cdots$, C_n，C_1 与 B 相邻，C_{i+1} 与 C_i 相邻 $(i=1, 2, \cdots, n-1)$，C_n 与 A 相邻. 设有 a_n 种栽法，则显然 $a_1=2$，并且

$$a_{n-1}+a_n=3^n. \tag{8}$$

理由是如果 C_n 不与 A 相邻，那么 C_1, C_2, $\cdots$, C_n 各有 3 种栽法，共 3^n 种栽法. 其中 C_n 栽第 1 种的，即相当于将 C_n 与 A 合并，应当有 a_{n-1} 种栽法，所以(8)成立.

由(8)得

$$a_n=3^n-3^{n-1}+3^{n-2}+\cdots+(-1)^{n-1}\cdot3+(-1)^n.$$

特别地 $a_4=3^4-3^3+3^2-3+1=61$.

本题答案是 $12\times61=732$ 种.

解答题第 14 题虽然是解析几何的问题，但主要内容却是方程的根的讨论，在由方程组

$$\begin{cases}\frac{x^2}{a^2}+y^2=1, & (9)\\ y^2=2(x+m), & (10)\end{cases}$$

消去 y 得

$$x^2+2a^2x+2a^2m-a^2=0 \tag{11}$$

后,可直接配方得

$$(x+a^2)^2=a^2(a^2-2m+1). \tag{12}$$

方程有解,所以(12)式右边 $\geqslant 0$, $m\leqslant\dfrac{a^2+1}{2}$.

(i) 若 $m=\dfrac{a^2+1}{2}$,则 $x=-a^2$, 代入(9)得 $y=\pm\sqrt{1-a^2}$.

此时当然有 $a<1$(否则两条曲线在 x 轴上方没有交点).

(ii) 若 $m<\dfrac{a^2+1}{2}$, 则

$$x=-a^2\pm a\sqrt{a^2-2m+1},$$

代入(9)得

$$y^2=1-(a\pm\sqrt{a^2-2m+1})^2. \tag{13}$$

因为两条曲线在 x 轴上方仅有一个公共点,所以必有

$$a+\sqrt{a^2-2m+1}\geqslant 1, \tag{14}$$

$$\left|a-\sqrt{a^2-2m+1}\right|<1. \tag{15}$$

由(14)得 $a\geqslant 1$;或者 $m\leqslant a<1$. 由(15)得 $a<1$, $m>-a$;或者 $a\geqslant 1$, $a>m>-a$.

于是 m 的取值范围是: $a\geqslant 1$ 时, $a>m>-a$; $a<1$ 时, $-a<m\leqslant a$ 或$m=\dfrac{a^2+1}{2}$.

至于△OAP 的面积,在 P 点纵坐标 y 最大时最大. 由于 $a<\dfrac{1}{2}$,所以 $m=\dfrac{a^2+1}{2}$ 或 $m\leqslant a$.

(i) $m=\dfrac{a^2+1}{2}$ 时, $y=\sqrt{1-a^2}$.

(ii) $m \leqslant a$ 时，$m < \frac{1}{2}$，$x = -a^2 + a\sqrt{a^2 - 2m + 1} > -a^2 + a \cdot a = 0$. 并且在 $m = a$ 时，x 取最小值 $a(1-2a)$，而 y 取得最大值

$$\sqrt{1-\frac{x^2}{a^2}} = \sqrt{1-(1-2a)^2} = 2\sqrt{a(1-a)}.$$

比较 $\sqrt{1-a^2}$ 与 $2\sqrt{a(1-a)}$ 的大小得:

(i) 在 $a \leqslant \frac{1}{3}$ 时，$2\sqrt{a(1-a)} \leqslant \sqrt{1-a^2}$，从而面积的最大值是 $\frac{1}{2}a\sqrt{1-a^2}$.

(ii) 在 $\frac{1}{3} < a < \frac{1}{2}$ 时，$2\sqrt{a(1-a)} > \sqrt{1-a^2}$，从而面积的最大值是 $a\sqrt{a(1-a)}$.

解答题第 15 题是个很好的题,既涉及一个重要的数学概念——序,又有明显的实际应用.

复杂的情况应分解为简单的情况. 简单的情况处理好了,复杂的情况也就迎刃而解. n 个电阻全部并联或全部串联,总电阻值显然与各个电阻的位置无关. 只需考虑既有串联又有并联的情况. 最简单的是 3 个电阻,情况有两种:

(i) 由 R_1、R_2 并联后与 R_3 串联，容易知道总电阻最小时，$R_1 > R_3$，$R_2 > R_3$.

(ii) 由 R_1、R_2 串联后与 R_3 并联，容易知道总电阻最小时，$R_1 > R_3$，$R_2 > R_3$.

对于本题,由(i)，R_1，$R_2 > R_3$. 由(ii)(将 R_1，R_2 的组合看成 一个电阻)，$R_3 > R_4$，再由(i)将 R_1、R_2、R_3 当作一个电阻，$R_4 > R_5$. 最后由(ii)，$R_5 > R_6$. 即 $R_i = a_i$，$i = 3, 4, 5, 6$，R_1、R_2 可以是 a_1、a_2，也可以是 a_2、a_1(如下图).

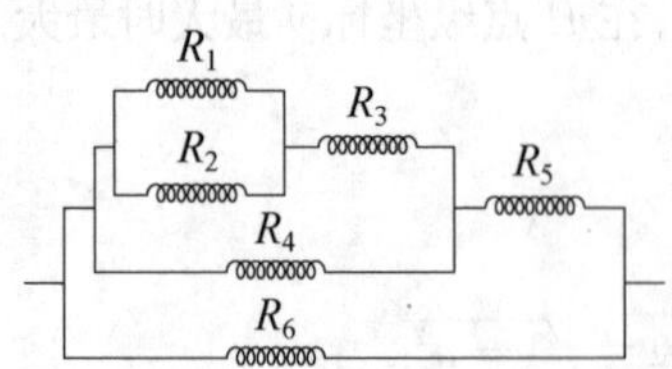

加试题第一题是一个有背景的几何题：如果有两个三角形 $A_1B_1C_1$、$A_2B_2C_2$，自顶点 A_1、B_1、C_1，分别向边 B_2C_2、C_2A_2、A_2B_2 所作的垂线交于一点 P，那么我们就称$\triangle A_1B_1C_1$ 与$\triangle A_2B_2C_2$ 正交.

正交具有对称性，即如果$\triangle A_1B_1C_1$ 与$\triangle A_2B_2C_2$ 正交，那么$\triangle A_2B_2C_2$ 也与$\triangle A_1B_1C_1$ 正交. 证明如下：设 B_2、C_2 分别向 C_1A_1、A_1B_1 所作的垂线交于 Q，则有

$$PB_2^2 - PC_2^2 = A_1B_2^2 - A_1C_2^2, \tag{16}$$

$$PC_2^2 - PA_2^2 = B_1C_2^2 - B_1A_2^2, \tag{17}$$

$$PA_2^2 - PB_2^2 = C_1A_2^2 - C_1B_2^2, \tag{18}$$

$$QC_1^2 - QA_1^2 = B_2C_1^2 - B_2A_1^2, \tag{19}$$

$$QA_1^2 - QB_1^2 = C_2A_1^2 - C_2B_1^2, \tag{20}$$

以上五式相加得

$$QC_1^2 - QB_1^2 = C_1A_2^2 - B_1A_2^2. \tag{21}$$

(21) 表明 A_2 向 B_1C_1 所作的垂线也过 Q，即$\triangle A_2B_2C_2$ 也与$\triangle A_1B_1C_1$ 正交.

加试题第一题是这个命题的特殊情况. 其中的三角形 DMN 与 HBC 正交，所以$\triangle HBC$ 也与$\triangle DMN$ 正交，即 $OH \perp MN$. 如果直接证，方法也完全一样，只是其中的对称美不易体现(不如上面由(16)—(20)推出(21)那样整齐清晰).

用(21)这样的、关于平方差的等式来证明两线垂直学生不太熟悉，其实是一种常用方法.

本题还有更深刻的背景：$\triangle DEF$ 与$\triangle ABC$ 的对应顶点的连线 AD、BE、CF 交于一点，这样的一对三角形称为透视的. 由著名的笛沙格定理，透视三角形的对应边的交点共线，这条直线称为透视轴. 与正交性结合，有下面的定理：

如果两个三角形是透视的，又互相正交，那么它们的透视中心(对应顶点连线的交点)、正交中心(前面所述的三条垂线的交点 P、Q 称为正交中心)三点共线，而且这条直线垂直于透视轴.

在第一题的图中，OH 就是过$\triangle DEF$ 与$\triangle ABC$ 的正交中心、透视中心的直线，它垂直于透视轴 MN.

第一题还有许多证法.熟悉根轴与九点圆的,立即看出 MN 是$\triangle ABC$ 的外接圆与九点圆 DEF 的根轴.事实上,由 $\angle ADB=\angle AEB=90^\circ$, A、B、D、E 四点共圆,所以

$$MB\times MA=MD\times ME,$$

即 M 到$\odot O$($\triangle ABC$ 的外接圆)的幂等于 M 到圆 DEF 的幂,所以 M 在$\odot O$ 与圆 DEF 的根轴上.这里圆 DEF 是$\triangle DEF$ 的外接圆,也就是$\triangle ABC$ 的九点圆.

同理, N 也在上述根轴上.因此 MN 就是$\triangle ABC$ 的外接圆与九点圆的根轴.

熟知九点圆的圆心 K 在直线 OH 上(K 正好是线段 OH 的中点.以 K 为圆心,外接圆半径的一半为半径的圆,过九个特殊的点,即 D, E, F, AH、BH、CH 的中点,边 BC、CA、AB 的中点,称为九点圆).于是 $OH\perp MN$.

关于根轴与九点圆可参看《近代欧氏几何学》(约翰逊著,单墫译,上海教育出版社出版).

加试题第二题是求条件极值,在条件

$$\sum_{i=1}^{n}x_i^2+2\sum_{1\leqslant k<j\leqslant n}\sqrt{\frac{k}{j}}x_kx_j=1 \tag{22}$$

下,求 $\sum\limits_{i=1}^{n}x_i$ 的最大值(最小值为 1 易求).

通常的条件是平方和

$$u_1^2+u_2^2+\cdots+u_n^2=1, \tag{23}$$

而函数是线性的: $\sum\limits_{i=1}^{n}a_iu_i$. 由柯西不等式

$$\left(\sum_{i=1}^{n}a_iu_i\right)^2\leqslant\sum_{i=1}^{n}a_i^2\cdot\sum_{i=1}^{n}u_i^2=\sum_{i=1}^{n}a_i^2,$$

易得最大值为 $\sqrt{\sum\limits_{i=1}^{n}a_i^2}$.

现在条件是一般的二次形.应当先用拉格朗日配方法,将它化为平方和.由于系数$\sqrt{\frac{k}{j}}$有根号,并且根号下有分母,为运算简便起见,可先令 $x_j=\sqrt{j}\,y_j$(1

$\leqslant j \leqslant n$)，化二次形为整系数的二次形

$$y_1^2+2y_1(y_2+y_3+\cdots+y_n)+2y_2^2+2y_2(2y_3+2y_4+\cdots+2y_n)+ \\ 3y_3^2+2y_3(3y_4+3y_5+\cdots+3y_n)+\cdots+ny_n^2.$$

用标准的配方法，上式

$$\begin{aligned}
&=(y_1+y_2+\cdots+y_n)^2+y_2^2+2y_2(y_3+\cdots+y_n)+2y_3^2+\cdots+(n-1)y_n^2\\
&=(y_1+y_2+\cdots+y_n)^2+(y_2+y_3+\cdots+y_n)^2+y_3^2\\
&\quad+2y_3(y_4+\cdots+y_n)+\cdots+(n-2)y_n^2\\
&=\cdots\\
&=(y_1+y_2+\cdots+y_n)^2+(y_2+y_3+\cdots+y_n)^2\\
&\quad+\cdots+(y_{n-1}+y_n)^2+y_n^2.
\end{aligned}$$

令 $u_1=y_1+y_2+\cdots+y_n$, $u_2=y_2+y_3+\cdots+y_n$, $\cdots$, $u_n=y_n$，就化为上面所说的标准的、在平方和已知(等于1)的条件下，求线性函数极值的问题.

最大值为 $\sqrt{\sum_{k=1}^{n}(\sqrt{k}-\sqrt{k-1})^2}$.

加试题第三题在感觉上，首先应分出一个边长尽可能大，即等于 n 的正方形. 照这样做下去，在 $m=qn+r$ $(0\leqslant r<n)$ 时，可化为 $n\times r$ 的长方形的问题($r=0$ 甚易，只讨论 $r>0$). 继续做下去，形成 m、n 的辗转相除，可猜出答案应为 $m+n-(m, n)$.

猜到答案后，用归纳法，假定长边$<m$ 时，结论成立(奠基 $m=1$ 显然). 考虑 $m\times n(n\leqslant m)$ 的长方形. 这时先去掉一个 $n\times n$ 的正方形，再用归纳假设可得

$$f(m, n)\leqslant n+(m-n)+n-(m-n, n)=m+n-(m, n). \tag{24}$$

感觉上，先去掉一个 $n\times n$ 的正方形是当然的、最佳做法，所以(24)应该是等式. 但感觉不能代表严格的证明. 证明虽不复杂，却并不容易想到：假如每一个正方形的边长都小于 n，那么每一条与长方形的长边 AB 垂直的直线都至少与两个正方形的下底相交. 从而这些正方形的下底射影到 AB 上后，至少将 AB 覆盖两次. 所以这些正方形的边长的和 $\geqslant 2m>m+n-(m, n)$，于是(24)中等号成立，结论对一切 m 正确.

第31届（2002）美国数学奥林匹克

师：2002年的美国竞赛题，难度与IMO相当．涉及的知识范围，如极限、多项式性质等，比我国的CMO稍广．但几乎没有平面几何，这是他们的弱点．

下面是第1道试题，你做做看．

1．设S是2002元集，N为整数，满足$0 \leqslant N \leqslant 2^{2002}$．证明：可将$S$的所有子集染上黑色或白色，使得下列条件成立：

(a) 任两个白色子集的并集是白的；

(b) 任两个黑色子集的并集是黑的；

(c) 恰好存在N个白色的子集．

生：我从简单的情况做起，设$S=\{1, 2, \cdots, 2002\}$．在$N=1$时，可将空集$\varnothing$或任一个一元子集，例如$\{1\}$，染黑（其余子集染白），这时(a)、(b)、(c)显然满足．

$N=2$时，将$\varnothing$与$\{1\}$染黑．

$N=3$时，将$\varnothing$、$\{1\}$、$\{1, 2\}$染黑．

$N=4$时，将$\varnothing$、$\{1\}$、$\{1, 2\}$、$\{2\}$染黑．

显然，在上述几种情况，(a)、(b)、(c)均成立．

师：接下去怎么办呢？

生：当然不能一直这样做下去．我想采用数学归纳法．

师：很好．归纳假设是什么？

生（想了一想）：我假设对于n元集S及$N \leqslant 2^n$，可以实现所说的染色．考虑$n+1$元集$S \cup \{n+1\}$的染色．不过，不知道怎样利用这个假设将$S \cup \{n+1\}$的2^n+k $(k=1, 2, \cdots, 2^n)$个子集染黑．

师：还是回到简单具体的情况．你已经解决了$S=\{1, 2\}$及$N \leqslant 2^2$的染色情况．考虑将集$\{1, 2, 3\}$的2^2+1个子集染黑．

生：这只要在已有的基础上再增加一个黑的子集就可以了．$\varnothing$、$\{1\}$、$\{2\}$、$\{1, 2\}$都是黑的，再增加谁呢？$\{3\}$不行（因为$\{2, 3\}$是白的），$\{1, 2, 3\}$也不行

(因为 $\{1, 2, 3\}=\{1, 3\}\cup\{2, 3\}$). 只剩下$\{1, 3\}$与$\{2, 3\}$. 奇怪,它们也不行.

师: 所以,你必须抛弃原有的"基础",重新开始. 不要先将$\{1, 2\}$的 4 个子集染黑. 相反地,应当将$\{1, 2, 3\}$的另 4 个子集,也就是含有 3 的$\{3\}$、$\{1, 3\}$、$\{2, 3\}$、$\{1, 2, 3\}$染黑. 然后在$\{1, 2\}$的 4 个子集中再增加一个黑的.

生: 我明白了. 2^2+2、2^2+3、2^2+4 的情况也是如此.

师: 一般情况呢?

生: 我先将含 $n+1$ 的 2^n 个子集染黑. 然后运用归纳假设,将 $S=\{1, 2, \cdots, n\}$ 的 k 个子集染黑,使得它们满足(a)、(b). 这样,对 2^n+k $(k=1, 2, \cdots, 2^n)$,有满足要求的染色.

师: 应当验证一下这样的染色确实符合要求.

生: 这并不困难.

师: 那我们就看第 2 道题吧.

2. 设$\triangle ABC$ 满足

$$\left(\cot\frac{A}{2}\right)^2+\left(2\cot\frac{B}{2}\right)^2+\left(3\cot\frac{C}{2}\right)^2=\left(\frac{6s}{7r}\right)^2,$$

其中 $s=\frac{1}{2}(a+b+c)$, r 为内切圆半径. 证明:$\triangle ABC$ 与一个三角形 T 相似, T 的边长均为整数,并且三边的最大公约数为 1,确定 T 的边长.

生: 我知道

$$\cot\frac{A}{2}=\frac{s-a}{r},\ \cot\frac{B}{2}=\frac{s-b}{r},\ \cot\frac{C}{2}=\frac{s-c}{r}.$$

所以,原来的条件即

$$(s-a)^2+[2(s-b)]^2+[3(s-c)]^2=\left(\frac{6s}{7}\right)^2. \qquad ①$$

接下去如何进行却看不太清楚.

师: 应当利用①求出 $a:b:c$.

生: 要求比 $a:b:c$,可假定 $c=1$,实际上有两个未知数 a、b. 但只有①一个等式,通常只能定出一个未知数. 除非……

师: 除非什么?

生：除非这个等式是一些实数平方的和等于 0. 但现在是 3 个平方的和等于一个非零的平方，要作恒等变形有点烦.

师：不必作恒等变形. 因为有一个著名的不等式可以利用.

生：我知道您说的是柯西不等式，①的左边是三个平方的和，应当再乘上三个平方的和. 这三个平方的和，应当是右边的分母 7^2. 但这三个平方分别是什么？我要凑一下.

师：$s-a$、$2(s-b)$、$3(s-c)$ 应当分别乘上不同的数，使得所得结果中 s 的系数相等. 这才便于与右边的分子比较.

生：1、2、3 的公倍数是 6，所以 $s-a$、$2(s-b)$、$3(s-c)$ 应当分别乘上 6、3、2. 而

$$6^2+3^2+2^2=36+9+4=49=7^2.$$

则

$$\begin{aligned}&7^2\{(s-a)^2+[2(s-b)]^2+[3(s-c)]^2\}\\&=(6^2+3^2+2^2)\{(s-a)^2+[2(s-b)]^2+[3(s-c)]^2\}\\&\geqslant[6(s-a)+6(s-b)+6(s-c)]^2=(6s)^2.\end{aligned}$$

其中等号成立. 所以，

$$\frac{s-a}{6}=\frac{2(s-b)}{3}=\frac{3(s-c)}{2}.$$

由等比定理，这比就是

$$\frac{6(s-a)+6(s-b)+6(s-c)}{6^2+3^2+2^2}=\frac{6s}{49}.$$

从而，

$$\frac{a}{49-6^2}=\frac{b}{49-3^2}=\frac{c}{49-2^2},$$

即

$$a:b:c=13:40:45.$$

三角形 T 的三边是 13、40、45.

师：再看第 3 道题.

3. 证明：任一 n 次实系数的首一(首项系数为 1)多项式是两个 n 次的、有 n 个实根的首一多项式的平均.

师：这是一个关于多项式的问题. 你应当知道一个关于根的定理：

如果实系数多项式 $f(x)$在区间$[c, d]$上变号,即 $f(c)f(d)<0$,那么,$f(x)$在$[c, d]$内必有一个根.

生: 听说过这个定理.

师: 那么,你就可以做这道题了.

生: 设 $f(x)$是已知的多项式. 又设 $g(x)$是另一个 n 次的首一多项式,则有

$$f(x)=\frac{1}{2}(2f(x)-g(x)+g(x)).$$

只要设法选择 $g(x)$,使得 $g(x)$与 $2f(x)-g(x)$ 都有 n 个实数根.

师: 你先将 $f(x)$表成两个 n 次首一多项式的平均,然后再设法满足其他要求,这种想法很好.

生: 怎么选择 $g(x)$比较困难.

师: 先假定 n 是偶数. 这时,在 x 的绝对值很大时, $g(x)>0$.

然后任取 n 个值,例如 $1, 2, \cdots, n$,再取一个正数 M,使得

$$g(1)=g(3)=\cdots=g(n-1)=-M, \tag{①}$$

$$g(2)=g(4)=\cdots=g(n)=M. \tag{②}$$

根据前面所说的定理, $g(x)$在$(-\infty, 1)$, $(1, 2)$, $\cdots$, $(n-2, n-1)$, $(n-1, n)$ 上各有一个根,即 $g(x)$有 n 个实根.

生: 有一个问题,满足①、②的 n 次、首一多项式 $g(x)$存在吗?

师: 我忘记说了,这要用到拉格朗日插值定理.

生: 我知道这个定理. 它是说对任意的两组实数 $a_1<a_2<\cdots<a_k$ 及 $b_1, b_2, \cdots, b_k$,有一个 $k-1$ 次多项式 $h(x)$存在,满足 $h(a_i)=b_i(1\leqslant i\leqslant k)$. 具体的表达式是

$$h(x)=\sum_{i=1}^{k}\prod_{j\neq i}\frac{x-a_j}{a_i-a_j}\cdot b_i.$$

但我不知道现在如何应用这个定理. 而且, $h(x)$并不是首一的.

师: 我们只定了 n 个点的函数值,所以你说的 $h(x)$是 $n-1$ 次多项式. 再加上一个首一的 n 次多项式 $(x-1)(x-2)\cdots(x-n)$ 就得到合乎要求的 $g(x)$了.

生: 原来这么简单！不过，怎么能保证另一个多项式 $2f(x)-g(x)$ 也有 n 个实根呢?

师: 上面的 M 可以由我们自由地选择. 现在希望 $2f(x)-g(x)$ 在 $x=1$, 3, …, $n-1$ 处的值大于 0，在 $x=2, 4, \cdots, n$ 的值小于 0.

生: 这只要取

$$M > \max\{2|f(1)|, 2|f(2)|, \cdots, 2|f(n)|\}$$

就可以了.

师: n 是奇数的情况证明只需稍作修改.

4. 设 $\mathbf{R}$ 为实数集，确定所有满足下列条件的函数 $f: \mathbf{R} \to \mathbf{R}$,

$$f(x^2-y^2)=xf(x)-yf(y), \ \forall x、y \in \mathbf{R}.$$

师: 你先猜猜看，$f(x)$是什么函数?

生: 我猜想 $f(x)$是正比例函数 kx. $f(x)=kx$ 确实符合要求. 但要证明只有 $f(x)=kx$，似乎不太容易.

师: 试试看.

生: 令 $x=y=0$ 得 $f(0)=0$. 令 $y=0$ 得

$$f(x^2)=xf(x). \quad ①$$

令 $x=0$ 得

$$f(-y^2)=-yf(y)=-f(y^2). \quad ②$$

所以 $f(x)$是奇函数，只需在$(0, +\infty)$上讨论.

由于①,

$$f(x^2-y^2)=f(x^2)-f(y^2).$$

将 x^2、y^2 改写为 x、$y(x、y>0)$，则

$$f(x-y)=f(x)-f(y),$$

即(将 x 改记为 $x+y$, $x-y$ 改记为 x)

$$f(x+y)=f(x)+f(y). \quad ③$$

由③，运用熟知的柯西方法可知，对 $x \in \mathbf{Q}$,

$$f(x)=kx,\ k=f(1).\qquad ④$$

如何证明④对于 $x\in\mathbf{R}$ 成立,好像很难.

师: 由③只能得出④对 $x\in\mathbf{Q}$ 成立. 要证明④对 $x\in\mathbf{R}$ 成立,通常要利用连续性. 本题未给出这一条件,但除③外,还有一个重要的式①,利用它可以得出所欲的结果.

你可以考虑一下 $f((x+1)^2)$.

生: 一方面

$$f((x+1)^2)=(x+1)f(x+1)=(x+1)(f(x)+k).$$

另一方面,

$$f((x+1)^2)=f(x^2+2x+1)=xf(x)+2f(x)+k.$$

比较以上两个方面即得

$$f(x)=kx.$$

师: 这是典型的"算两次",是一种很有用的方法.

5. 设 a、b 为大于 2 的整数. 证明:存在一个正整数 k 及正整数的有限序列 $n_1,n_2,\cdots,n_k$,满足 $n_1=a$, $n_k=b$ 且对所有 i $(1\leqslant i\leqslant k)$, n_in_{i+1} 被 n_i+n_{i+1} 整除.

生: 这道题,我还是从简单的例子做起. 首先,3 可以用一步变为 6,记为 3→6. 一般地,

$$a\rightarrow a(a-1).\qquad ①$$

师: 很好! ①可以算做第 1 条引理. 你再做几个简单的例子,如将 4、6、9 变为 3.

生: 因为 3→6,所以反过来便有 6→3.

师: 这也可以算做一条引理,即引理 2:

如果 a 经若干步可以变成 b,那么 b 也可以经过若干步变为 a.

生: $4\rightarrow4\times3\rightarrow2\times6\times5\rightarrow6\times5\rightarrow6\rightarrow3$

$9\rightarrow3\times3\times2\rightarrow3\times6\times5\rightarrow2\times6\times5\rightarrow\cdots\rightarrow3$

我还可以得出第 3 条引理:

如果 a 能经过若干步变成 b，那么 ac 能经过若干步变为 bc.

师：一般情况，你准备怎么做?

生：我想用归纳法，证明一切大于 2 的整数 a 都可以变为 3. 这样，由引理 2，a 可以先变为 3，再变为任一大于 2 的整数 b.

师：很好的想法，就这样做下去.

生：假设 $a>3$，并且小于 a、大于 2 的整数都可以经若干步变为 3，考虑 a.

如果 a 是合数，设 $a=mn$，m、n 是两个大于 1 的整数，$m\geqslant n$.

若 $m=2$，则 $n=2$，$a=4$. 若 $m=3$，则 $n=3$ 或 2，$a=9$ 或 6. 这些情况前面已经做过，以下设 $m>3$.

由归纳假设，m 可以经若干步变为 3. 由引理 3，a 可经若干步变为 $3n$. 因为 $3n<a$，再用归纳假设便得出 a 可经若干步变为 3.

如果 a 是质数，我还没有想好.

师：如果 a 是质数，那么 $a+1$ 是合数，设 $a+1=mn$，m、n 是两个大于 1 的整数，它们能相等吗?

生：如果 $m=n$，那么，

$$a=m^2-1=(m+1)(m-1)$$

不是质数(除非 $m=2$，$a=3$. 但前面假设 $a>3$).

师：设 $m>n$，由前面的引理 1、3，

$$\begin{aligned}&a\to a(a-1)\to a(a-1)(a-2)\to\cdots\to a(a-1)\cdots m\cdots n\\&=(a+1)a(a-1)\cdots(m+1)(m-1)(m-2)\cdots(n+1).\end{aligned}$$

生：接下去我知道怎么做了. 由引理 1、2、3，

$$\begin{aligned}\text{上式}&\to(a+1)a\cdots(m+2)(m-1)(m-2)\cdots(n+1)\\&\to\cdots\to(a+1)a(m-1)(m-2)\cdots(n+1)\\&\to(a+1)(m-1)(m-2)\cdots(n+1)\\&=m(m-1)(m-2)\cdots(n+1)n\\&\to m(m-1)(m-2)\cdots(n+1)\\&\to\cdots\\&\to m.\end{aligned}$$

最后再用归纳假设就证完了.

6. 有一版 $n \times n$ 的邮票. 要从中分出由 3 张在同一行或同一列相连的邮票组成的块(只能沿着针孔线分,而且每个块就是分下的一片纸). 分下一些块以后,不可能再分下更多的块了. 令这时的块数的最小值为 $b(n)$. 证明:存在实常数 c、d,对所有 $n > 0$, 都有

$$\frac{1}{7}n^2 - cn \leqslant b(n) \leqslant \frac{1}{5}n^2 + dn.$$

生: 这道题看上去很复杂,又是大轴题,恐怕很不容易.

师: 那倒不见得. 解题得有信心,不要先折了自己的锐气. 你先考虑右边的上界.

生: 上界即 $5b(n) \leqslant n^2 + 5dn$. 意味每一块在整版上的影响大致为 5. 我从第一行左端开始,每隔 2 张邮票剪下一块,如图 1 所示:

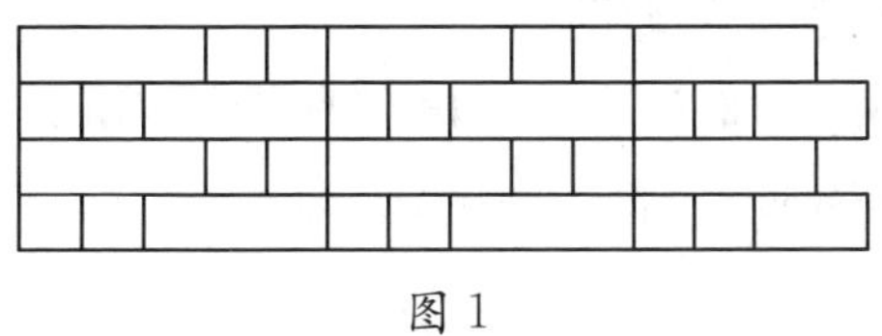

图 1

其中黑色表示剪下的块,每块由 3 张邮票组成. 这样剪下去,第一行剪到最后,可能出现以下两种情况:

(1) 恰好剪完或剩下 1 张或剩下 2 张邮票;

(2) 剩下 3 张或 4 张邮票. 这时再剪下 1 块.

其他行与第一行相同或类似.

设这时剪下的块数为 $c(n)$.

一般情况,第一、三、……行(第二、四、……行)每一块"控制"自身的 3 张邮票及随后(前面)的 2 张邮票. 对于最后的一、两块,在 $n \times n$ 的这版邮票右边再添 4 列($4n$ 张邮票)作为被它们控制的,那么每个块都至少可以"控制"5 张邮票. 因此

$$5c(n) \leqslant n^2 + 4n.$$

从而 $$b(n) \leqslant c(n) \leqslant \frac{1}{5}n^2 + \frac{4}{5}n.$$

师： 边缘的(或添加的)几列,邮票张数是 n 的一次式,对结果(有 n^2 出现,它是主项)没有什么影响(只改变常数 d 或 c 的值).

生： 下界要难一些. 每个块,无论横竖,右面与下面有 5 张相邻的邮票,如图 2(包括与这块仅有一个公共顶点的邮票).

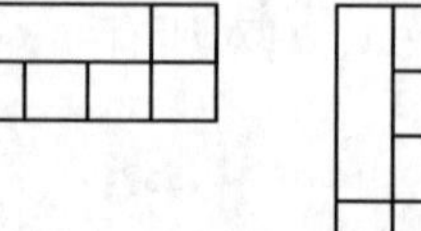

图 2

如果已经不能再剪下更多的块,那么下方或右方的 4 张邮票中至少有 1 张属于其他被剪下的块. 规定每一块控制如图 2 所示的 8 张邮票中的 7 张(去掉属于其他块的 1 张). 如果这版邮票的每张邮票都被控制了,那么就有

$$7b(n) \geqslant n^2. \quad ①$$

问题是如何证明每张邮票都被控制了.

师： 最上面的两行与最左面的两列可能有未被控制的.

生： 这些邮票不超过 $4n$ 张,只需将①改为

$$7b(n) \geqslant n^2 - 4n. \quad ②$$

师： 如果其他地方有 1 张邮票 A 未被控制,那么 A 的上方 B、左方 C 及左上方 D 都不属于任何一块(图 3).

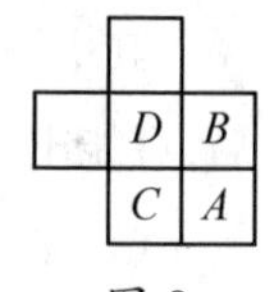

图 3

这时 D 的上方必属于一块,而 D 的左方又属于另一块. D 受到双重控制,将 D 改为只受一块控制, A 改为受另一块控制.

这样处置后,除去第一、二行与第一、二列,每张邮票都受到控制. 所以②成立,即

$$\frac{1}{7}n^2 - \frac{4}{7}n \leqslant b(n).$$

评 2002 年全国高中数学联赛

2002 年全国高中数学联赛的试题，普遍反映难度不大，与高考题接近，受到师生的欢迎. 其中与不等式有关的内容(包括函数的单调性、极值、参数的取值范围等)似乎偏多一些.

标准答案大多简明，有几道题的解法可以稍有不同，如**选择题第 6 题**，原题如下：

直线 $\frac{x}{4}+\frac{y}{3}=1$ 与椭圆 $\frac{x^2}{16}+\frac{y^2}{9}=1$ 相交于 A、B 两点，该椭圆上点 P，使得 $\triangle PAB$ 面积等于 3，这样的 P 点共有　（　　）

(A) 1 个　　(B) 2 个　　(C) 3 个　　(D) 4 个

如果采用直线的法线式，椭圆上一点 $P(4\cos\alpha,\ 3\sin\alpha)$ 与原点 O 在 AB 异侧时，它到 AB 的距离为

$$\frac{3(4\cos\alpha)+4(3\sin\alpha)-12}{5}$$
$$=\frac{12}{5}(\cos\alpha+\sin\alpha-1)\leqslant\frac{12}{5}(\sqrt{2}-1)<\frac{12}{5}\times\frac{1}{2}=\frac{6}{5}.$$

而 $AB=5$，所以 $\triangle PAB$ 的面积 $<\frac{1}{2}\times5\times\frac{6}{5}=3$.

因此，在 $\triangle PAB$ 面积为 3 时，P 与 O 在 AB 同侧，这样的点有 2 个，所以选 (B).

第 10 题为填空题，原题如下：

已知 $f(x)$ 是定义在 $\mathbf{R}$ 上的函数，$f(1)=1$，且对任意的 $x\in\mathbf{R}$ 都有

$$f(x+5)\geqslant f(x)+5,\ f(x+1)\leqslant f(x)+1.$$

若 $g(x)=f(x)+1-x$，则 $g(2002)=$________.

本题以先定出 $f(2002)$ 为好，因为

$$f(x)+5\leqslant f(x+5)\leqslant f(x+4)+1\leqslant f(x+3)+2$$
$$\leqslant f(x+2)+3\leqslant f(x+1)+4\leqslant f(x)+5,$$

所以其中等号均成立 $f(x+1)=f(x)+1$.

再由 $f(1)=1$,逐步得出 $f(2)=2$, $f(3)=3$, $\cdots$, $f(2002)=2002$,从而 $g(2002)=1$.

第 12 题原题如下:

使不等式 $\sin^2 x+a\cos x+a^2\geqslant 1+\cos x$ 对一切 $x\in\mathbf{R}$ 恒成立的负值 a 的取值范围是________.

本题"对一切 $x\in\mathbf{R}$"不等式成立,所以可取 $x=0$,得 $a+a^2\geqslant 2$,从而 $a\leqslant -2$.

反过来,在 $a\leqslant -2$ 时,

$$a^2+a\cos x\geqslant a^2+a\geqslant 2\geqslant \cos^2 x+\cos x=1+\cos x-\sin^2 x.$$

于是 a 的取值范围是 $a\leqslant -2$.

2002 年的加试题比 2001 年容易,当然也有一定的难度,由于要通过加试来选拔,所以题目应当有一定的难度与较好的区分度.如果题目中有很好的数学思想,而不是过偏过繁,那就更为理想了.

加试第一题原题如下:

如图 1,在$\triangle ABC$ 中,$\angle A=60^\circ$, $AB>AC$,点 O 是外心,两条高 BE、CF 交于 H 点.点 M、N 分别在线段 BH、HF 上,且满足 $BM=CN$.求$\dfrac{MH+NH}{OH}$的值.

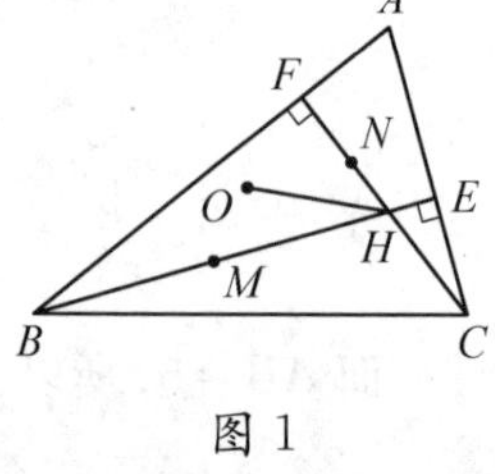

图 1

图中 N 在线段 CH 的延长线上,而 M 在线段 BH 内.其实本题 N 也可以与 H 重合,或者 N 在线段 CH 内,M 也可以在 BH 的延长线上.

先看一个特殊的情况(也是最简单的情况),即 N 与 H 重合,此时 $NH=0$, $BM=CH$, M 在线段 BH 内(严格说来,这一点是要证明的.当然并不难证:由 $\angle BCH=90^\circ-\angle ABC>90^\circ-\angle ACB=\angle HBC$,得 $BH>CH$,所以 M 在线段 BH 内).

因为 $\angle BOC=2\angle BAC=120^\circ=180^\circ-\angle BAC=\angle BHC$,所以 B、O、H、

C 共圆,从而易知

$$\triangle OBM \cong \triangle OCH,\ OM = OH,\ \angle MOH = \angle BOC = 120^\circ.$$

在$\triangle MOH$ 中,$\dfrac{MH}{OH} = \sqrt{3}$.

如果 N 在 CH 的延长线上,那么 M 将从上面所说的位置 $M_0(BM_0 = CH)$ 上移,这时有两种可能:

(1) M 在线段 M_0H 内.

由于 $M_0M = BM - BM_0 = CN - CH = HN$,所以 $MH + HN = MH + M_0M = M_0H$,根据上面所证

$$\frac{MH + HN}{OH} + \frac{M_0H}{OH} = \sqrt{3}.$$

(2) M 在线段 M_0H 的延长线上.

由于 $M_0M = HN$,所以

$$HN - HM = M_0M - HM = M_0H.$$

约定 MH 为有向线段,即 $MH = -HM$,那么仍有

$$\frac{MH + HN}{OH} = \frac{M_0H}{OH} = \sqrt{3}.$$

如果 N 在 CH 内,那么有向线段 $HN = -NH = M_0M$,仍有

$$\frac{MH + HN}{OH} = \frac{M_0H}{OH} = \sqrt{3}.$$

于是,约定 MH、HN 为有向线段(向上为正,向下为负),恒有

$$\frac{MH + HN}{OH} = \sqrt{3}.$$

加试第二题与三次方程有关,这是教材中没有、学生普遍生疏的内容,因而做对的人不太多. 原题如下:

实数 a、b、c 和正数 λ 使得 $f(x) = x^3 + ax^2 + bx + c$ 有三个实根 x_1、x_2、x_3,且满足

(1) $x_2 - x_1 = \lambda$;

(2) $x_3>\frac{1}{2}(x_1+x_2)$.

求 $\frac{2a^3+27c-9ab}{\lambda^3}$ 的最大值.

本题的解法很多,这里提供两种.

解法 1 首先搞清 $2a^3+27c-9ab$ 的意义.如果看不清楚,可先提出 27,将它写成

$$27\left(\frac{2}{27}a^3-\frac{1}{3}ab+c\right),$$

从 $-\frac{1}{3}ab$ 与 c 这两项可以猜出

$$f\left(-\frac{1}{3}a\right)=\frac{2}{27}a^3-\frac{1}{3}ab+c,$$

而这立即可以验证是成立的,于是问题化为在条件(1)、(2)下,求

$$\begin{aligned}S&=\frac{27f\left(-\frac{1}{3}a\right)}{\lambda^3}\\&=\frac{27\left(-\frac{1}{3}a-x_1\right)\left(-\frac{1}{3}a-x_2\right)\left(-\frac{1}{3}a-x_3\right)}{(x_2-x_1)^3}\\&=\frac{-27\left(x_1+\frac{1}{3}a\right)\left(x_2+\frac{1}{3}a\right)\left(x_3+\frac{1}{3}a\right)}{(x_2-x_1)^3}\end{aligned}$$

的最大值.

为简便起见,令 $u_i=x_i+\frac{a}{3}(i=1,2,3)$,这时 $u_2-u_1=x_2-x_1=\lambda_1$, $u_3>\frac{1}{2}(u_1+u_2)$,并且 $u_1+u_2+u_3=x_1+x_2+x_3+a=0$.

所以 u_1、u_2、u_3 仍满足相应的条件,而

$$S=\frac{-27u_1u_2u_3}{(u_2-u_1)^3}.$$

由 $u_3=-(u_1+u_2)>\frac{u_1+u_2}{2}$ 得 $-(u_1+u_2)>0$. 从而 u_1、u_2 中至少有一个小于 0. 设 $u_1<0$, 如果 $u_2<0$, 那么 $S<0$, 不可能取最大值. 于是可设 $u_2>0$.

问题可进一步简化,即令

$$v_1=\frac{-u_1}{u_2-u_1},\ v=v_2=\frac{u_2}{u_2-u_1},$$

则 $v_1+v_2=1$, v_1、v_2 均正,

$$v_1-v_2=\frac{u_3}{u_2-u_1}>0,\ S=27v_1v_2(v_1-v_2)=27v(1-v)(1-2v).$$

现在 S 的最大值已很容易求出:

$$\begin{aligned}S&=27\sqrt{(v-v^2)^2(1-2v)^2}\\&=27\sqrt{2(v-v^2)(v-v^2)\left(\frac{1}{2}-2v+2v^2\right)}\\&\leqslant 27\sqrt{2\times\left(\frac{1}{6}\right)^3}=\frac{3}{2}\sqrt{3}.\end{aligned}$$

在 $v=\frac{1}{2}\left(1-\frac{\sqrt{3}}{3}\right)$ 时等号成立,相应的三次方程为

$$x^3-\frac{1}{2}x+\frac{1}{6}\sqrt{\frac{1}{3}}=0,$$

$$\lambda=1\left(x_1=-\frac{1}{2}\left(1+\sqrt{\frac{1}{3}}\right),\ x_2=\frac{1}{2}\left(1-\sqrt{\frac{1}{3}}\right),\ x_3=\sqrt{\frac{1}{3}}\right).$$

解法 2　先考虑 $a=0$ 的特殊情况,这时问题即求

$$S=\frac{2a^3+27c-9ab}{\lambda_3}=\frac{27c}{\lambda^3}=\frac{-27x_1x_2x_3}{(x_2-x_1)^3}$$

的最大值,可参见上面的解答.

对于一般情况,由于

$$x^3+ax^2+bx+c=\left(x+\frac{a}{3}\right)^3+p\left(x+\frac{a}{3}\right)+q$$

(其中 p、q 无须具体定出),令 $u=x+\frac{a}{3}$,原方程化为 u 的三次方程

$$u^3+pu+q=0.$$

而

$$\begin{aligned}q&=-u_1u_2u_3=-\left(\frac{a}{3}+x_1\right)\left(\frac{a}{3}-x_2\right)\left(\frac{a}{3}+x_3\right)\\&=-\left[\left(\frac{a}{3}\right)^3+\left(\frac{a}{3}\right)^2(x_1+x_2+x_3)\right.\\&\qquad\left.+\left(\frac{a}{3}\right)(x_1x_2+x_1x_3+x_2x_3)+x_1x_2x_3\right]\\&=-\frac{a^3}{27}+\frac{a^2}{9}\cdot a-\frac{a}{3}b+c\\&=\frac{1}{27}(2a^3+27c-9ab),\end{aligned}$$

$$u_2-u_1=\lambda,\ u_3>\frac{1}{2}(u_1+u_2),$$

所以 $S=\frac{27q}{(u_2-u_1)^3}$. 问题化归为上面 $a=0$ 的特殊情况.

在将 S 用 x_1、x_2、x_3 表出后,也可以将它看成 x_3 的函数,用导数来求极值. 有的学生就是这样做的.

评 2003 年高中联赛

与往年的试题相比，2003 年的试题计算量较小，而思维的程度有所增加，更有利于培养人才.

选择题第 3 题是过抛物线 $y^2=8(x+2)$ 的焦点 F 作倾斜角为 $60°$ 的弦 AB，AB 的中垂线交 x 轴于 P，求 PF. 本题焦点 F 为原点，直线 AB 方程为 $y=\sqrt{3}x$，所以 A、B 横坐标适合方程

$$3x^2-8x-16=0.$$

由韦达定理，AB 中点 E 的横坐标为 $\frac{1}{2}\times\frac{8}{3}=\frac{4}{3}$.

由于 AB 倾斜角为 $60°$，所以 $FE=2\times\frac{4}{3}$，$PF=2FE=4\times\frac{4}{3}=\frac{16}{3}$.

选择题第 4 题是 $x\in\left[-\frac{5\pi}{12},-\frac{\pi}{3}\right]$，求 $y=\tan\left(x+\frac{2\pi}{3}\right)-\tan\left(x+\frac{\pi}{6}\right)+\cos\left(x+\frac{\pi}{6}\right)$ 的最大值.

本题可先化 y 为同角的三角函数的代数和.

$$\begin{aligned}y&=-\cot\left(x+\frac{\pi}{6}\right)-\tan\left(x+\frac{\pi}{6}\right)+\cos\left(x+\frac{\pi}{6}\right)\\&=\cot z+\tan z+\cos z,\ z\in\left[\frac{\pi}{6},\frac{\pi}{4}\right].\end{aligned}$$

由于 $\cot z\cdot\tan z=1$，所以在 $\cot z$ 与 $\tan z$ 的差越大时，$\cot z+\tan z$ 越大，而

$$\cos z\leqslant\cos\frac{\pi}{6}=\frac{\sqrt{3}}{2},$$

所以 y 在 $z=\frac{\pi}{6}$ 时取最大值

$$\sqrt{3}+\frac{\sqrt{3}}{3}+\frac{\sqrt{3}}{2}=\frac{11}{6}\sqrt{3}.$$

选择题第 6 题是在 $AB=1$, $CD=\sqrt{3}$, AB 与 CD 距离为 2,夹角为 $\frac{\pi}{3}$ 时,求四面体 $ABCD$ 的体积.

本题由简单向量知识可知由 $\overrightarrow{AB}$、$\overrightarrow{CD}$ 所形成的平行四边形面积为

$$1\times\sqrt{3}\times\sin\frac{\pi}{3}=\frac{3}{2},$$

所求体积为

$$\frac{1}{6}\times\frac{3}{2}\times 2=\frac{1}{2}.$$

填空题第 10 题:已知 a, b, c, d 均为正整数,且 $\log_a b=\frac{3}{2}$, $\log_c d=\frac{5}{4}$,若 $a-c=9$,求 $b-d$.

本题由已知得 $a=b^{\frac{2}{3}}$, $c=d^{\frac{4}{5}}$,所以 $b=s^3$, $d=t^5$, s、t 为正整数,并且 $s^2-t^4=9$,从而 $s=5$, $t=2$, $b-d=5^3-2^5=93$.

解答题第 14 题:设 A, B, C 是复数 $z_0=a\mathrm{i}$, $z_1=\frac{1}{2}+b\mathrm{i}$, $z_2=1+c\mathrm{i}$ 对应的不共线的三点(a, b, c 都是实数),证明曲线

$$z=z_0\cos^4 t+2z_1\cos^2 t\sin^2 t+z_2\sin^4 t\,(t\in\mathbf{R})$$

与$\triangle ABC$ 中平行于 AC 的中位线只有一个公共点,并求出此点.

本题大多数人都是解方程组求出曲线与中位线的公共点.如果具有一些重心坐标的知识,那么由 $\cos^4 t$, $2\cos^2 t\sin^2 t$, $\sin^4 t$ 均为正数,并且和为 1 可以知道曲线上的点到 AC 的距离$=AC$ 上的高 $h\times 2\cos^2 t\sin^2 t$,当且仅当 $t=\frac{\pi}{4}$(由周期性,我们只在一个周期 $\left[0,\frac{\pi}{2}\right)$ 内考察 $2\cos^2 t\sin^2 t$)时,这值为 $\frac{h}{2}$,即曲线与平行于 AC 的中位线恰有一个公共点,这点为

$$(z_0+2z_1+z_2)\times\left(\frac{\sqrt{2}}{2}\right)^4=\frac{1}{2}+\frac{a+c+2b}{4}\mathrm{i}.$$

解答题第 15 题是纸上画有半径为 R 的圆 O 内有一定点 A, $OA=a$,折叠纸片,使圆周上某点 A' 与 A 重合.当 A' 跑遍圆周时,求折痕所在线上点的集合.

本题当然以 O 为原点，OA 为 x 轴建立坐标轴，如图 1，设折痕与 OA' 的交点为 E，则

$$EO+EA=EO+EA'=R.$$

所以 E 点在以 O、A 为焦点，R 为长轴的椭圆

$$\frac{\left(x-\frac{a}{2}\right)^2}{\left(\frac{R}{2}\right)^2}+\frac{y^2}{\left(\frac{R}{2}\right)^2-\left(\frac{a}{2}\right)^2}=1$$

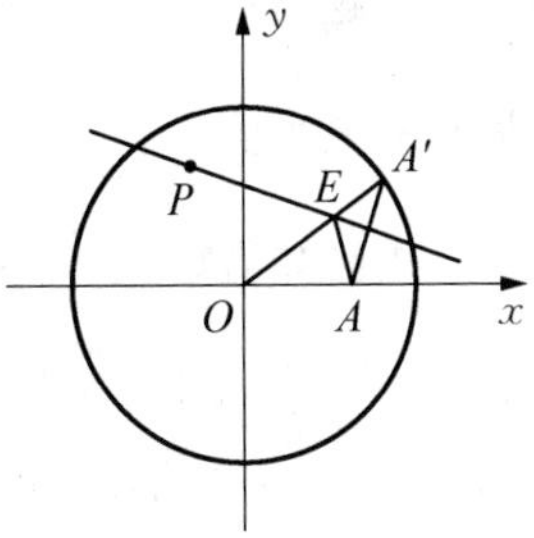

上，而对折痕上其他点 P，$PO+PA=PO+PA'>R$. 因此折痕上的点在上述椭圆上或椭圆外.

图 1

反之，对椭圆上或椭圆外任一点 P，可过 P 作椭圆的切线，设切点为 E，延长 OE 交圆于 A'，则 $EA'=R-EO=EA$，且由椭圆切线的性质，$\angle AEP=\angle A'EP$，从而 A、A' 关于 PE 对称，即 PE 是所说的折痕.

综上所述所求集合为椭圆上及椭圆外的点的全体.

加试题第 1 题仍是平面几何，原题为：

过圆外一点 P 作图的两条切线和一条割线，切点为 A、B，割线交圆于 C、D，C 在 P、D 之间，在弦 CD 上取一点 Q，使 $\angle DAQ=\angle PBC$，求证 $\angle DBQ=\angle PAC$.

本题证法很多，这里介绍一种. 如图 2，首先易知

$$\frac{AC}{AD}=\frac{PA}{PD}=\frac{PB}{PD}=\frac{BC}{BD}. \qquad (1)$$

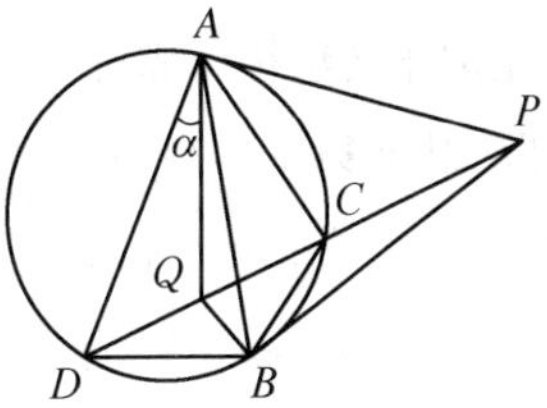

又 $\angle BAC=\angle PBC=\alpha$，$\angle ABC=\angle ADC$，

所以 $\triangle ADQ\backsim\triangle ABC$，$DQ=\dfrac{AD\times BC}{AB}$.

同样，$\triangle ADB\backsim\triangle AQC$，

$$QC=\frac{AC\times DB}{AB}.$$

图 2

结合(1)得 $DQ=QC$，即在 $\angle PBC=\alpha$ 时，Q 是 DC 的中点，由于中点的唯一性. 反过来，在 Q 是 DC 的中点时，$\angle PBC=\alpha$，这也就得出，由于 Q 是 DC 的中点，所以 $\angle PAC=\angle DBQ$.

加试题第 2 题与数论有关,但我们希望解法用到的知识较少(否则会造成学生"恶补"的现象),下面的解法只用到同余式.

原题为:设三角形的三边长是整数 $l > m > n$. $\left|\frac{3^l}{10^4}\right| = \left|\frac{3^m}{10^4}\right| = \left|\frac{3^n}{10^4}\right|$,求周长的最小值.

由已知得 $\frac{3^l - 3^m}{10^4} = \frac{3^l}{10^4} - \frac{3^m}{10^4} =$ 整数,所以 $3^l - 3^m$ 被 10 整除,即 $3^l \equiv 3^m (\mathrm{mod} 10^4)$. 因为 3 与 10 互质,所以 $3^{l-m} \equiv 1(\mathrm{mod} 10^4)$. 同理 $3^{m-n} \equiv 1(\mathrm{mod} 10^4)$, $3^{l-n} \equiv 1(\mathrm{mod} 10^4)$.

我们来寻求 $3^h \equiv 1(\mathrm{mod} 10^4)$ 时,正整数 h 满足的条件.

首先 $3^2 = 9 \equiv -1$, $3^3 \equiv -3$, $3^4 \equiv 1(\mathrm{mod} 10)$,现在 $3^h \equiv 1(\mathrm{mod} 10)$,所以 $h = 4k$, k 为正整数.

$$l \equiv 3^h = 3^{4k} = (10-1)^{2k}$$
$$\equiv 1 - 2k \times 10 + k(2k-1) \times 10^2 - \frac{k(2k-1)(2k-2)}{3} \times 10^3 (\mathrm{mod} 10^4),$$

所以

$$2k - k(2k-1) \times 10 + \frac{2k(2k-1)(k-1)}{3} \times 10^2 \equiv 0(\mathrm{mod} 5^3),$$

左边后两项被 5 整除,所以 $2k$ 被 5 整除, $k = 5t$, t 为正整数,并且

$$2t - 10t(2k-1) \equiv 0(\mathrm{mod} 5^2).$$

左边第 2 项被 5 整除,所以第 1 项 $2t$ 被 5 整除,这样一来,左边第 2 项被 5^2 整除,从而第 1 项也被 5^2 整除,即 t 被 5^2 整除,于是 h 被 $4 \times 125 = 500$ 整除.

$l-m$, $m-n$, $l-n$ 都是 500 的倍数,所以 $m-n \geqslant 500$, $l-n \geqslant 2 \times 500$, $n > l-m \geqslant 500$. 周长 $l+m+n \geqslant 3n + 3 \times 500 \geqslant 3 \times 501 + 3 \times 500 = 3003$.

另一方面,在 $n = 501$, $m = 501 + 500$, $l = 501 + 2 \times 500$ 时,周长为 3003,并且由

$$3^{500} = (10-1)^{250}$$
$$\equiv 1 - 2500 + \frac{250 \times 249}{2} \times 100 \equiv 1 - 2500 + 250 \times 50$$
$$= 1 - 250 \times 40 \equiv 1(\mathrm{mod} 10000)$$

知 l, m, n 均满足条件. 因此,周长的最小值为 3003.

加试题第 3 题研究一个 n 点 l 条线的图中何时有一个四边形,这方面有一个著名的结果,在 $l > \left[\frac{1}{4}n \cdot (1+\sqrt{4n-3})\right]$ 时,图中必有一个四边形,证法是考虑一个点 u 及这点引出的两条边 uv_1、uv_2 所组成的"角". 如果图中没有四边形,每个"角"对应的点对(v_1, v_2)是互不相同的,所以点对的总数 $C_n^2 \geqslant$ 角的总数 $\sum_{i=1}^{n} C_{x_i}^2$, 其中 x_1, x_2, $\cdots$, x_n 是各点的次数.

由 Cauchy 不等式及 $\sum_{i=1}^{n} x_i = 2l$ 可得

$$n(n-1) \geqslant \sum_{i=1}^{n} x_i(x_i - 1) \geqslant \frac{4}{n}l^2 - 2l,$$

从而

$$l \leqslant \frac{n}{4}(1+\sqrt{4n-3}).$$

现在的题目给出 $n = q^2 + q + 1$, $l \geqslant \frac{1}{2}q(q+1)^2$, $q \in \mathbf{N}$, $q \geqslant 2$,并且有一个点次数 $\geqslant q+2$, 要证明有一个四边形,直接援引上面的结果是不成立的. 必须去掉与次数 $x_1 \geqslant q+2$ 的点 A_1 相连的 x_1 个点,剩下的点对有 $C_{n-x_1}^2$ 个. 与上面的类似,在无四边形时有

$$C_{n-x_1}^2 \geqslant \sum_{i=2}^{n} C_{x_i-1}^2.$$

同样用 Cauchy 不等式及 $\sum_{i=2}^{n}(x_i - 1) = 2l - n + 1 - x_1$ 得

$$(n-1)(n-x_1)(n-x_1-1) \geqslant (nq-q+2-x_1)(nq-q-n+3-x_1),$$

这与 $(q+1)(n-x_1) < nq-q+2-x_1$ 及 $q(n-x_1-1) \leqslant nq-q-n+3-x_1$ 矛盾,所以图中必有四边形.

二、三两题的解法似嫌单一,第三题尤其过于细致,更像研究的问题,而不很适合竞赛. 因为在考场上很难(除非事先做过)从这些不太简明的条件看出问题的本质. 能够完整地解出这种题目的学生恐怕不会很多.

谈 2004 年高中联赛

2004年的高中联赛，计算相当多，我国著名数学家华罗庚先生曾经说过："不要怕算"，可是目前我国中学生，运算能力大多不很强，或不够熟练，或不够简捷，或忙中出错，或不会检查. 因此，增加一些计算对提高学生的运算能力是有益的. 当然，计算过多，也有些"矫枉过正"，或许命题者认为"不过正则不能矫枉"吧.

2004年联赛，增多概率、向量等方面内容，这是值得赞赏的.

1 试题

先谈联赛的 1～15 题.

大多数试题与标准解答均很好，不必多说，本文只讨论几道有其他解法的问题.

第4题涉及向量，这是过去较少出现的内容.

4. 设 O 点在 $\triangle ABC$ 内部，且有

$$\overrightarrow{OA}+2\overrightarrow{OB}+3\overrightarrow{OC}=\vec{0},$$

则 $\triangle ABC$ 的面积与 $\triangle AOC$ 的面积的比为 (　　)

(A) 2　　(B) $\dfrac{3}{2}$　　(C) 3　　(D) $\dfrac{5}{3}$

这道题的图1画得太准，有60%以上的学生用尺去量 BO 及 OD（D 为 BO 延长后与 AC 的交点)，发现 $BO=2OD$，从而选(C). 所以这题最好不要画图，让学生自己画. 如果画，也不要画得太准.

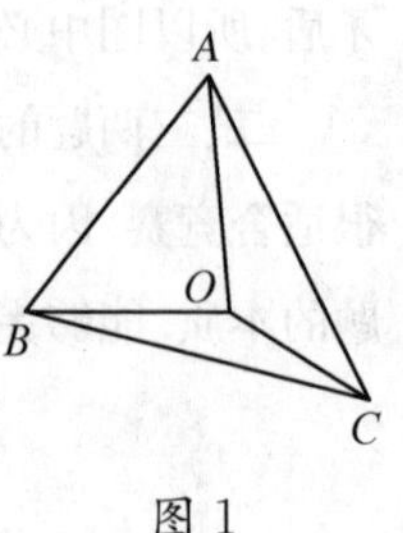

图 1

不靠量，可作出以 OA 及 $3OC$ 为边的平行四边形 $OAEF$（图2). 这时 $OE=2BO$，并且 B, O, E 共线. 设 P 为

平行四边形的中心，则 $OP=BO$，由 $\triangle ACF$ 的中位线 $PQ \parallel CD$，可得 $OD=DP=\frac{1}{2}OP=\frac{1}{2}BO$. 从而选(C).

更一般地，设 l, m, n 为正实数，

$$l\cdot\overrightarrow{OA}+m\cdot\overrightarrow{OB}+n\cdot\overrightarrow{OC}=\vec{0},$$

求$\triangle BOC$ 的面积与$\triangle ABC$ 的面积的比.

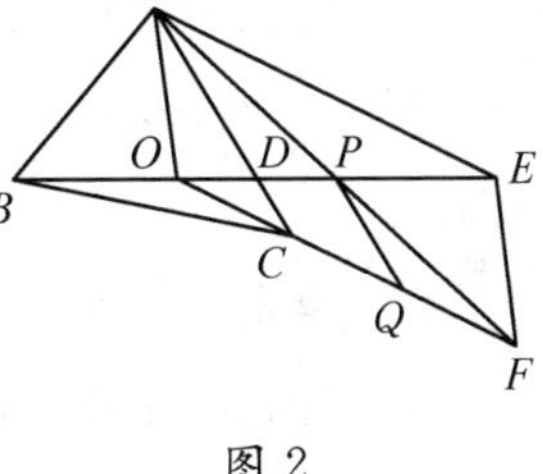

图 2

我们知道 $\frac{m\cdot\overrightarrow{OB}+n\cdot\overrightarrow{OC}}{m+n}=\overrightarrow{OD}$，$D$ 是 BC 上的一点，分 BC 为 $n:m$，由已知 $\overrightarrow{OD}=-\frac{l\cdot\overrightarrow{OA}}{m+n}$，所以 D 即是 AO 延长线与 BC 的交点，而且

$$\overrightarrow{AD}=\overrightarrow{AO}+\overrightarrow{OD}=\frac{l+m+n}{m+n}\cdot\overrightarrow{AO}.$$

于是

$$\frac{S_{\triangle BOC}}{S_{\triangle ABC}}=\frac{OD}{AD}=\frac{l}{l+m+n}.$$

(其实 l, m, n 也可取 0 或负数，但不能全都为 0)

应用这一结果，立即得出第 4 题的答案是

$$\frac{1+2+3}{2}=3.$$

第 5 题需要列举种种情况.

5. 设三位数 $n=\overline{abc}$，若以 a, b, c 为三条边的长可以构成等腰三角形，则 n 有　　　(　　)

(A) 45 个　　(B) 81 个　　(C) 165 个　　(D) 216 个

显然 $a=b=c$ 的有 9 种，其他的等腰三角形均不等边，由对称性，不妨设 $b=c\neq a$，再将算得的结果乘以 3. 这时有两种情况：

1°　$b=c>a$.

有 $(b, c, a)=(2, 2, 1), (3, 3, 1), (3, 3, 2), \cdots, (9, 9, 1), \cdots, (9, 9, 8)$，共 $1+2+\cdots+8=\frac{8\times 9}{2}=36$(种).

$2°\quad a > b = c > \frac{a}{2}$.

有 $(a, b, c)=(3, 2, 2), (4, 3, 3), (5, 3, 3), (5, 4, 4), (6, 4, 4), (6, 5, 5), (7, 4, 4), (7, 5, 5), (7, 6, 6), (8, 5, 5), (8, 6, 6), (8, 7, 7), (9, 5, 5), (9, 6, 6), (9, 7, 7), (9, 8, 8)$,共 16 种.

所以 n 的个数是 $9+3\times(36+16)=165$. 这是不被 9 整除的数,应当选(C).

填空题的 8 实际上是一个函数方程.

8. 设函数 f: $\mathbf{R}\to\mathbf{R}$,满足 $f(0)=1$,且对 $\forall x$、$y\in\mathbf{R}$,都有 $f(xy+1)=f(x)f(y)-f(y)-x+2$,则 $f(x)=$________.

这种问题,当然是取一些特殊的 x 或 y 代入函数方程,看看能得什么结果.由于已知的函数值只有 $f(0)$,所以取 $y=0$ 代入得

$$f(1)=f(x)-x+1,$$

即

$$f(x)=x+f(1)-1. \tag{1}$$

(1)已经是 $f(x)$的表达式了,只是常数 $f(1)$还不知道,再在(1)中令 $x=0$,得 $f(1)=2$,所以 $f(x)=x+1$.

第 9 题是求二面角的度数.

9. 在正方体 $ABCD-A_1B_1C_1D_1$ 中,二面角 $A-BD_1-A_1$ 的度数是________.

本题可利用 $A_1D\perp$平面 AD_1B, $AB_1\perp$平面 A_1D_1B,法线 A_1D、AB_1 所成的角就是两个平面的夹角.由于$\triangle AB_1C$ 是正三角形,所以$\angle AB_1C=60°$,即 A_1D、AB_1 所成的角是 $60°$,二面角 $A-BD_1-A_1$ 的度数也是 $60°$(平面 $ABA_1\perp$平面 A_1D_1B),所以二面角 $A-BD_1-A_1$ 的度数小于 $90°$).

第 10 题是数论题.

10. 设 p 是给定的奇质数,正整数 k 使 $\sqrt{k^2-pk}$ 也是一个正整数,则 $k=$________.

我们当然设 $k^2-pk=n^2(n\in\mathbf{N})$. 标准答案是将它看作二次方程. 另一种做法是分解:

$$pk = k^2 - n^2 = (k+n)(k-n).$$

由于 $k+n>k$,所以 $p \nmid k-n$, $p \mid k+n$, 设

$$k+n = mp \ (m \in \mathbf{N}), \tag{1}$$

则

$$m(k-n) = k, \tag{2}$$

多出了两个字母 m、n,但可以消去 n,即(1)$\times m+$(2)得

$$(2m-1)k = m^2 p. \tag{3}$$

因为 $(2m-1, m)=1$,所以 $2m-1 \mid p$, 从而

$$2m-1=1 \text{ 或 } 2m-1=p.$$

前者导出 $k=p$,与 k^2-pk 为正整数不合,后者得出

$$k = m^2 = \left(\frac{p+1}{2}\right)^2.$$

12. 在平面直角坐标系 xOy 中,给定点 $M(-1, 2)$和 $N(1, 4)$. 点 P 在 x 轴上移动,当$\angle MPN$ 最大时,点 P 的横坐标是________.

熟知圆 MPN 与 x 轴相切时, $\angle MPN$ 最大. 先求出 MN 的方程为 $x=y-3$,它交 x 轴于点 Q, Q 的横坐标为 -3. $[x_P-(-3)]^2 = QM \times QN = 2\sqrt{2} \times 4\sqrt{2} = 4^2$,所以 $x_P=1$ 或 -7,由于 MN 与 x 轴正方向成锐角,应取在 Q 右边的 P 点,即 $x_P=1$.

解答题中第 13 题是概率,这是过去未有的内容.

13. 在第 n 关要掷一颗骰子 n 次,如果 n 次的点数和大于 2^n,则算过关:问

(1) 最多能过几关?

(2) 连过前 3 关的概率是多少?

由于每次最多掷出 6 点, $6>2$, $6\times 2>2^2$, $6\times 3>2^3$, $6\times 4>2^4$,所以可过 4 关. 但 $6\times 5<2^5$,所以最多能过 4 关.

另一方面,过第一关需掷出的点数>2,过关概率为$\frac{4}{6}$. 过第二关需两次点数之和>4,而 $1+1$, $1+2$, $2+1$, $1+3$, $3+1$, $2+2$ 均不超过 4,所以过关概

率为 $\frac{36-6}{36}=\frac{30}{36}$，过第3关需3次掷出的点数之和$>8$，3次点数之和不大于8的情况有：

$1+1+1$, $1+1+2$, $1+1+3$, $1+1+4$, $1+1+5$, $1+1+6$, $1+2+2$, $1+2+3$, $1+2+4$, $1+2+5$, $1+3+3$, $1+3+4$, $2+2+2$, $2+2+3$, $2+2+4$, $2+3+3$.

其中三次点数全相同的(如$1+1+1$, $2+2+2$)各1种，三次中有二次相同的各3种，三次点数全不同的各6种，所以共有$2\times1+10\times3+4\times6=56$种.点数之和大于8的有$6^3-56=160$种.过第三关的概率$\frac{160}{216}$.

连过前3关的概率为

$$\frac{4}{6}\times\frac{30}{36}\times\frac{160}{216}=\frac{100}{243}.$$

第14、15两题原题如下：

14. 在平面直角坐标系xOy中，给定三点$A\left(0,\ \frac{4}{3}\right)$, $B(-1,\ 0)$, $C(1,\ 0)$.点P到直线BC的距离是该点到直线AB、AC距离的等比中项.

(1) 求点P的轨迹方程；

(2) 若直线L经过$\triangle ABC$的内心(设为D)，且与点P的轨迹恰好有3个公共点.求L的斜率k的取值范围.

15. 已知α、β是方程$4x^2-4tx-1=0(t\in\mathbf{R})$的两个不等实根.函数$f(x)=\frac{2x-t}{x^2+1}$的定义域为$[\alpha,\ \beta]$.

(1) 求$g(t)=\max f(x)-\min f(x)$.

(2) 证明：对于$u_i\in\left(0,\ \frac{\pi}{2}\right)(i=1,\ 2,\ 3)$，若$\sin u_1+\sin u_2+\sin u_3=1$，则

$$\frac{1}{g(\tan u_1)}+\frac{1}{g(\tan u_2)}+\frac{1}{g(\tan u_3)}<\frac{3}{4}\sqrt{6}.$$

这两题似乎是“华山一条路”，没有什么实质不同的解法.解法也都不难想到.只需要有耐心与时间去做，如果在前面已经花去很多时间，做这两题就很紧张了.第14题如果改为只求轨迹(或加上求内心)，也许学生的时间会更宽裕些，水平能更正常发挥，第15题也可只要第(1)问.现在题量似乎过大一些，除

非平时做过类似问题,成竹在胸,否则在考场上恐怕大多同学难以完成,更无法检查一遍.而从培养学生兴趣,重在参与这一点考虑,似乎减少一点题量会吸引更多学生的参加.

此外,第14题中,切点只算一个公共点,而不算作两个(重合的)公共点,这最好能够说明,以免歧义.第15题(Ⅱ)中,$\sum_{i=1}^{3}\frac{1}{\tan u_i}$的最大值是$\frac{9\sqrt{2}}{7}$,比题中上界$\frac{3}{4}\sqrt{6}$稍小一些.有人认为第15题可以改作加试题(换去加试的第二题).目前试题的难度似乎偏大,在100分钟内完成,要求似偏高.希望别再升高了.

2　加试题

加试题第一题仍是平面几何题,但变成了一道计算题.

1. 在锐角三角形ABC中,AB上的高CE与AC上的高BD相交于点H,以DE为直径的圆分别交AB、AC于F、G两点,FG与AH相交于点K.已知$BC=25$,$BD=20$,$BE=7$,求AK的长.

如图3,本题有三个三角形相似:$\triangle ABC\backsim\triangle ADE\backsim\triangle AFG$.题中$B$,$C$,$D$,$E$共圆,$\angle ADE=\angle ABC$.这种情况,我们说$DE$与$BC$(关于$AB$,$AC$)逆平行.同样,$FG$与$DE$(关于$AB$、$AC$)逆平行.两次逆平行导出$FG$与$BC$平行.从而与$BC$垂直的直线$AH$也与$FG$平行.$\triangle AFG$的高$AK$与$\triangle ABC$对应高的比等于相似比$\frac{AF}{AB}$.

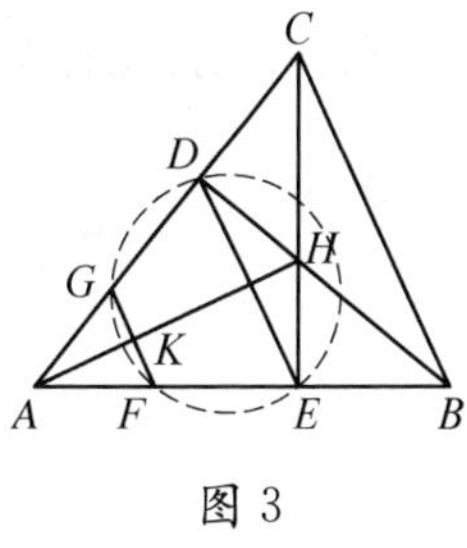

图3

在$\triangle ABC$中,由已知条件可算出$AE=18$.算法很多,除标准答案外,如用三角法.

由$\angle ACE=90^\circ-\angle CAB=\angle ABC+\angle ACB-90^\circ$得

$$AE=CE\tan\angle ACE=-CE\cot(\angle ABC+\angle ACB)=24\times\frac{\frac{24}{7}\times\frac{4}{3}-1}{\frac{24}{7}+\frac{4}{3}}=18.$$

从而 $AB=25=BC$，$\triangle ABC$，$\triangle ADE$，$\triangle AFG$ 均为等腰三角形，$\triangle ABC$ 的 BC 边上的高 $=CE=24$，$AF=FE=\frac{1}{2}AE=9$. $AK=24\times\frac{9}{25}=\frac{216}{25}$.

2. 在平面直角坐标系 xOy 中，y 轴正半轴上的点列 $\{A_n\}$ 与曲线 $y=\sqrt{2}x$ $(x\geqslant 0)$ 上的点列 $\{B_n\}$ 满足 $|OA_n|=|OB_n|=\frac{1}{n}$. 直线 A_nB_n 在 x 轴上截距为 a_n，点 B_n 的横坐标为 b_n，$n\in\mathbf{N}$.

(1) 证明 $a_n>a_{n+1}>4$.

(2) 证明有 $n_0\in\mathbf{N}$，使得 $\forall n>n_0$，都有

$$\frac{b_2}{b_1}+\frac{b_3}{b_2}+\cdots+\frac{b_{n+1}}{b_n}<n-2004.$$

这道题也有相当多的计算. 首先，不难求出：

$$b_n=\sqrt{\frac{1}{n^2}+1}-1=\frac{\sqrt{n^2+1}-n}{n}=\frac{1}{n(\sqrt{n^2+1}+n)},$$

$$a_n=\frac{b_n}{1-n\sqrt{2b_n}}.$$

要证(1)，当然是将 b_n 的表达式代入，证明 a_n 递减.

$$\begin{aligned}
a_n&=\frac{b_n}{1-n\sqrt{2b_n}}\\
&=\left[n(n+\sqrt{n^2+1})\cdot\left(1-\sqrt{\frac{2n}{\sqrt{n^2+1}+n}}\right)\right]^{-1}\\
&=\left[n(n+\sqrt{n^2+1}-\sqrt{2n(n+\sqrt{n^2+1})}\right]^{-1}\\
&=\left[\frac{n}{n+\sqrt{n^2+1}+\sqrt{2n(n+\sqrt{n^2+1})}}\right]^{-1}\\
&=1+\sqrt{1+\frac{1}{n^2}}+\sqrt{2+2\sqrt{1+\frac{1}{n^2}}},
\end{aligned}$$

显示随 n 增加而减少，并且

$$a_n>1+\sqrt{1}+\sqrt{2+2\sqrt{1}}=4.$$

(在 $n \to +\infty$ 时, $a_n \to 4$)

要证(2),先估计 $\frac{b_{n+1}}{b_n}$(要敢于去掉碍手碍脚的因子):

$$\frac{b_{n+1}}{b_n}=\frac{n}{n+1}\times\frac{\sqrt{n^2+1}+n}{\sqrt{(n+1)^2+1}+n+1}<\frac{n}{n+1}.$$

所以

$$\frac{b_2}{b_1}+\frac{b_3}{b_2}+\cdots+\frac{b_{n+1}}{b_n}<\frac{1}{2}+\frac{2}{3}+\frac{3}{4}+\cdots+\frac{n}{n+1}=n-\left(\frac{1}{2}+\frac{1}{3}+\cdots+\frac{1}{n+1}\right).$$

括号里是著名的"调和级数".学过一点高等数学的人都知道趋于无穷,因此,存在一个 n_0,当 $n>n_0$ 时,

$$\frac{1}{2}+\frac{1}{3}+\cdots+\frac{1}{n+1}<2004,$$

即(2)成立.具体一点,可用

$$\frac{1}{2}+\left(\frac{1}{3}+\frac{1}{4}\right)+\left(\frac{1}{5}+\frac{1}{6}+\frac{1}{7}+\frac{1}{8}\right)+\cdots+\left(\frac{1}{2^k+1}+\cdots+\frac{1}{2^{k+1}}\right)$$
$$>\frac{1}{2}+\frac{3}{4}+\frac{4}{8}+\cdots+\frac{1}{2^{k+1}}=\frac{k+1}{2}>2004.$$

得出 $n_0=2^{4008}$.

3. 对于整数 $n \geqslant 4$,求出最小的整数 $f(n)$,使得对于任何正整数 m,集合 $\{m, m+1, \cdots, m+n-1\}$ 的任一个 $f(n)$元子集中,均有至少 3 个两两互素的元素.

这是一道与整式有关的组合问题.

首先要弄懂题意,其中"任何""任一个""均有""至少""最小"等词汇是组合中常见的.

可以从简单情况做起.

$n=4$ 时,4 元集$\{2, 3, 4, 5\}$的三元子集$\{2, 3, 4\}$中的三个数并不两两互素,所以 $f(4)>3$. 另一方面,4 个连续自然数中,必有 3 个连续自然数,第 1 个是奇数,这 3 个自然数两两互素,所以 $f(4)=4$.

$n=5$ 时,5 元集$\{2, 3, 4, 5, 6\}$的 4 元子集$\{2, 3, 4, 6\}$中的三个数并不两两互素(三个中有两个偶数),所以 $f(5)>4$. 另一方面,5 个连续自然数中,必

有 3 个连续自然数,第一个是奇数,这 3 个自然数两两互素,所以 $f(5)=5$.

$n=4, 5$ 时, $f(n)=n$, 是不是对所有 n,均有 $f(n)=n$ 呢? 当然不能这样武断,应再往下看.

$n=6$ 时, 6 元集 $\{2, 3, 4, 5, 6, 7\}$ 的 4 元子集 $\{2, 3, 4, 6\}$ 中没有 3 个数两两互素,另一个方面对任 6 个连续自然数.

$m, m+1, m+2, m+3, m+4, m+5$, 在 m 为偶数时 5 元子集 $\{m, m+1, m+2, m+4, m+5\}$ 中, $m+1$ 与 $m+5$ 是互素的奇数,在 $3\nmid m+2$ 时, $m+2$ 与它们都互素,在 $3\mid m+2$ 时, $m+4$ 与 $m+1$, $m+5$ 都互素. 其他的 5 元子集中, $m+1, m+2, m+3$ 两两互素或 $m+3, m+4, m+5$ 两两互素. 在 m 为奇数时,情况类似(将 $m, m+1, m+2, \cdots, m+5$ 分别换成 $m+5, m+4, m+3, \cdots, m$). 因此 $f(6)=5$.

进一步,可考虑 $n=6k$ (6 是 2 与 3 的最小公倍数). 一方面在 $\{2, 3, \cdots, 6k+1\}$ 中, $4k$ 元子集

$$\{2, 4, 6, \cdots, 6k, 3, 9, \cdots, 6k-3\}$$

中没有 3 个两两互素的数(3 个数中或有 2 个偶数,或有 2 个被 3 整除)所以 $f(6k)\geqslant 4k+1$.

另一方面, $6k$ 个连续自然数可分为 k 组,每组 6 个连续的自然数,任一 $4k+1$ 元子集必与 k 组至少一组中有 5 个公共元,根据 $f(6)=5$, 可知这 5 个数中有 3 个两两互素,所以 $f(6k)=4k+1$. 类似地,可以得出 $f(6k+1)=4k+2$, $f(6k+2)=4k+3$, $f(6k+3)=4k+4$, $f(6k+4)=4k+4$, $f(6k+5)=4k+5$,以 $f(6k+4)$ 为例. 一方面 $\{2, 3, \cdots, 6k+5\}$ 的 $4k+3$ 元子集 $\{2, 4, 6, \cdots, 6k+4, 3, 9, \cdots, 6k+3\}$ 中没有 3 个两两互素的数,另一方面, $6k+4$ 个连续自然数可分为 $k+1$ 组,前 k 组,每组 6 个连续自然数,最后一组是 4 个连续自然数,任一 $4k+4$ 元子集或者与前 k 组中某一组有 5 个公共元或者包含最后的 4 元组,从而由 $f(6)=5$ 及 $f(4)=4$ 可知有 3 个数两两互素.

不难验证,上面的结果与标准答案 $f(n)=\left[\frac{n+1}{2}\right]+\left[\frac{n+1}{3}\right]-\left[\frac{n+1}{6}\right]+1$ 是一致的,而且 $\left[\frac{n+1}{2}\right]+\left[\frac{n+1}{3}\right]-\left[\frac{n+1}{6}\right]$ 正是 $\{2, 3, \cdots, n+1\}$ 中 2 或 3 的倍数的个数.

评 2005 年全国高中数学联赛

全国高中数学联赛,年复一年地举行. 2005 年总的印象是所出题目比较平稳,有一些好题,但能激发学习兴趣的题不多. 少数题过偏过繁(如加试第三题)或过难(如加试第二题). 一试宜尽可能接近高考,以利调动更多学生的积极性. 最好不要出现超过大纲或课程标准的内容,有些选择题(如第 5 题)或填空题(如第 11、12 题)可改为解答题,而有些解答题(如第 13、14 题)可放入加试,加试第二题可用于全国中学生数学冬令营考试.

具体意见如下.

选择题中,真正的选择题只有第 3 题,第 6 题也还可以算作选择题,其他 4 道题实际上都是填空题或解答题,需要花一番力气算出结果.

第 3 题当然用最特殊的三角形,即正三角形来猜结果. $A=60^\circ$, AA_1 是直径 2. 答案显然为 2(即选(A)). 标准答案仍然是认真地按一般情形去做,似乎命题者也忘了"选择题"的特点是"选择"一个(唯一的)正确答案,忘了选择题与填空题、解答题的区别.

第 6 题与七进制有关. 将 M 中每个数乘以 7^4,七进制的小数便化为整数. 形如 $a_1\times 7^3+a_2\times 7^2+a_3\times 7+a_4$ 的数中从大到小的第 2005 个数应是 7^4-2005. 这个数除以 7 余 4,所以选 $a_4=4$ 的(C)为答案. 原解答在十进制、七进制之间变来变去,本无必要.

第 1 题应知道 $(a+b)^2\leqslant 2(a^2+b^2)$,即平方和一定,两个正数在相等时,和最大,所以 $k\leqslant\sqrt{x-3}+\sqrt{6-x}\leqslant 2\sqrt{\dfrac{(x-3)+(6-x)}{2}}=\sqrt{6}$. 应选(D).

第 2 题中,四个数 3, 7, 11, 9 之间有关系: $11^2-9^2=7^2-3^2$,即 $CD^2-DA^2=BC^2-AB^2$. 这关系表明 D 在 CA 上的射影与 B 在 CA 上的射影是同一个点 E,如图 1. CA 与平面 EBD 中两条相交直线 EB、ED 垂直,所以 $CA\perp BD$,即 $\overrightarrow{AC}\cdot\overrightarrow{BD}=0$. 选(A).

顺便说一下,虽然勾股定理常叙述成 $a^2+b^2=c^2$,但平方差的形式 $c^2-a^2=$

b^2 在应用中更为方便.

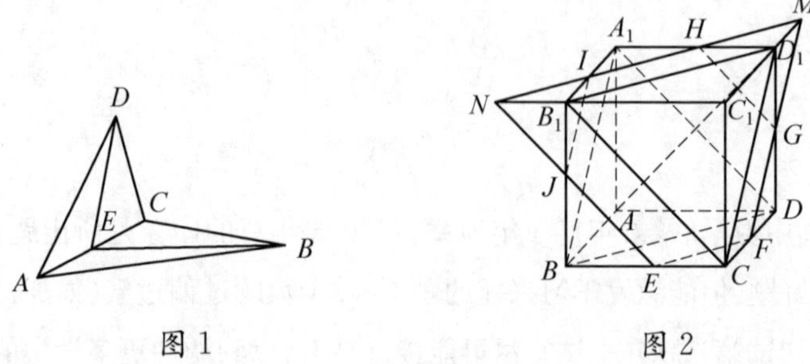

图 1　　　　图 2

第 4 题应先看看 α 的位置. α 与各面的交线均与 AC' 垂直. 而面的对角线 B_1C、CD_1、D_1B_1、A_1B、BD、DA_1 与 AC' 垂直,所以 α 与各面的交线应分别与这些面的对角线平行,即 α 与面 B_1CD_1、A_1BD 平行. 而且 α 在这两个平行平面之间时,才与正方体的各个面都相交.

设 α 与棱 A_1B_1 的交点为 I,则在面 $A_1B_1C_1$ 内作 $IH \parallel B_1D_1$,交 A_1D_1 于 H,交 C_1B_1、C_1D_1 的延长线于 N、M. 再在面 B_1BC 内作 $ME \parallel B_1C$,交 B_1B 于 J,交 BC 于 E;在面 C_1CD 内作 $MF \parallel CD_1$,交 D_1D 于 G,交 CD 于 F(如图 2). 六边形 $EFGHIJ$ 就是所得的截面.

截面的面积似乎随着 α 由平面 B_1CD_1 开始向平面 A_1BD 平移,先逐渐变大,再逐渐变小(当 α 通过 AC_1 中点时,截面面积最大). 但周长有无变化却不易看清.

我们可设 $C_1N=\lambda C_1B_1$, $\lambda \geqslant 1$. 这时 $NB_1=(\lambda-1)B_1C_1$, $NI=(\lambda-1)B_1D_1$. 所以,

$$S_{EFGHIJ}=[\lambda^2-3(\lambda-1)^2]S_{\triangle B_1CD_1}$$

随 α 位置的变化而变化,而截面的周长

$$L=(IH+IJ)\times 3=NH\times 3=B_1D_1\times 3,$$

即与 $\triangle B_1CD_1$ 的周长相等. 因而选(B).

第 5 题答案为(C). 应证明

$$0<\sin\sqrt{2}-\sin\sqrt{3}<\cos\sqrt{2}-\cos\sqrt{3}, \tag{1}$$

即 $$0<-\sin\alpha\cos\beta<\sin\alpha\sin\beta. \tag{2}$$

其中 $\alpha=\frac{\sqrt{3}-\sqrt{2}}{2}$，满足 $0<\alpha<\frac{\pi}{2}$；$\beta=\frac{\sqrt{3}+\sqrt{2}}{2}=1.732\cdots+1.414\cdots=3.146\cdots$，大于 π，而小于 1.5π，所以 $\cos\beta<0$，(2)左边的不等式成立. 并且

$$-\cos\beta=\sin\left(\beta-\frac{\pi}{2}\right)<\sin(\pi-\beta)=\sin\beta,$$

所以(2)右边的不等式也成立.

以上六道选择题，我们详细写来，说明"小题不小"，要做好得花一定的时间.

填空题似乎(平均说来)比选择题还容易一些.

第 7 题中 $g(\gamma)$的系数和 $a_0+a_1+\cdots+a_{20}=g(1)=f(5)=1-5+5^2-\cdots+5^{20}=\frac{1}{6}(5^{21}+1)$.

第 9 题是一道好题. 可使学生认识"任意 $x\in\mathbf{R}$"的作用. 令 $x=-\alpha$、$-\beta$、$-\gamma$(或其他值，如 $x=0$，$x=\frac{\pi}{2}-\beta$ 等)可得

$$\beta=\frac{\alpha+\gamma}{2},\ \gamma-\alpha=\frac{4\pi}{3}. \tag{3}$$

反过来，在(3)成立时，

$$\begin{aligned}&\cos(x+\alpha)+\cos(x+\beta)+\cos(x+\gamma)\\=&\cos(x+\beta)\cos\frac{\gamma-\alpha}{2}+\cos(x+\beta)\\=&\cos(x+\beta)\left(1+2\cos\frac{4\pi}{3}\right)=0.\end{aligned}$$

第 12 题中，较小的吉祥数可以直接计数. 四位以下的吉祥数，百位有 0～7 这 8 种. 相应地，十位有 8～1 种(个位随之确定). 因此，共有 $1+2+\cdots+8=36$ 个. 1000～1999 之间的吉祥数，百、十、个三位的数字和为 6，与上面类似，这样的吉祥数共有 $1+2+\cdots+7=28$ 个. 于是 $n=36+28+1=65$，$5n=325$.

位数不超过 5 的吉祥数，个数是不定方程

$$x_1+x_2+x_3+x_4+x_5=7 \tag{4}$$

的非负整数解的个数.由允许重复的组合公式(稍稍超出中学课程内容),(4)的解数为

$$C_{7+5-1}^{7}=C_{11}^{7}=330>325.$$

五位吉祥数中最大的六个是70000,61000,60100,60010,60001,52000.所以$a_{325}=52000$.

解答题中第13、14题的解法基本上只有一种.

第13题中,由$a_{n+1}=\dfrac{7a_n+\sqrt{45a_n^2-36}}{2}$可得

$$a_{n+1}^2-7a_{n+1}a_n+a_n^2+9=0. \tag{5}$$

将n换为$n-1$可得

$$a_{n-1}^2-7a_{n-1}a_n+a_n^2+9=0. \tag{6}$$

于是a_{n-1}、a_{n+1}都是方程

$$x^2-7a_nx+a_n^2+9=0 \tag{7}$$

的根.而且由已知,$\{a_n\}$严格递增(这一点,标准答案似缺论证.应先用归纳法得$a_n>0$,再得出$a_{n+1}\geqslant 3.5a_n>a_n$),所以$a_{n-1}$、$a_{n+1}$是两个不同的根.由韦达定理

$$a_{n-1}+a_{n+1}=7a_n. \tag{8}$$

(本题显然可增加一小问,即证明$a_{n-1}a_{n+1}=a_n^2+9$)

由(8)及$a_0=1$,$a_1=\dfrac{7a_0+\sqrt{45a_0^2-36}}{2}=5$可得$a_n$为正整数.再由(5),

$$a_{n+1}a_n-1=\left(\frac{a_{n+1}+a_n}{3}\right)^2, \tag{9}$$

即整数$a_{n+1}a_n-1$是一个有理数的平方,因而也是整数的平方.((9)不必再用归纳法证明)

第14题如考虑n个球(编号为$1\sim n$),答案是$\dfrac{2^{n-2}}{(n-1)!}$.这概率很小,随n

的增大快速趋向于 0.

第 15 题是一道好题,可用解析法,也可兼用解析法与几何意义,以兼用为好.

首先用解析法,切线 AB 方程为 $y=2x-1$. B 为 $(0, -1)$, D 为 $(0.5, 0)$,所以 D 为 AB 的中点(其实过这抛物线上任一点 P 作切线,设切线交 y 轴于 Q,则线段 PQ 必被 x 轴平分).

然后,问题化为一道纯粹的平面几何题:

如图 3, $\triangle ABC$ 中, CD 为中线, E、F 分别在 CA、CB 上,并且 $\dfrac{AE}{EC}=\lambda_1$, $\dfrac{BF}{FC}=\lambda_2$, $\lambda_1+\lambda_2=1$. EF 交 CD 于 P. 求证 P 为 $\triangle ABC$ 的重心.

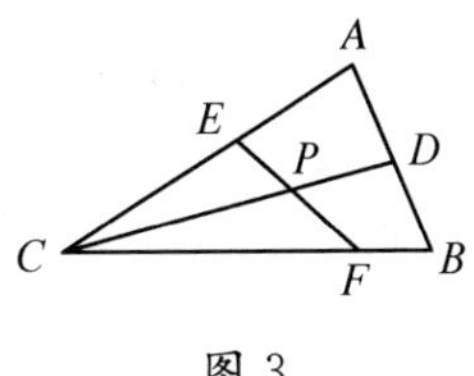

图 3

这题可照标准答案,利用面积来证(证法很好). 也可以利用斜坐标(与第 15 题原来的坐标毫无关系),以 C 为原点, CB、CA 为 x、y 轴,又两个轴上分别以 CF、CE 为单位长,则 EF 方程为

$$x+y=1. \tag{10}$$

A、B 的坐标分别为 $(0, \lambda_1+1)$, $(\lambda_2+1, 0)$. D 的坐标为 $\left(\dfrac{\lambda_2+1}{2}, \dfrac{\lambda_1+1}{2}\right)$, CD 方程为

$$\frac{x}{y}=\frac{\lambda_2+1}{\lambda_1+1}. \tag{11}$$

由(10)、(11)及比例的性质得

$$\frac{x}{1}=\frac{\lambda_2+1}{3} \Rightarrow \frac{x}{0.5(\lambda_2+1)}=\frac{2}{3},$$

上式表明 $CP : CD=2 : 3$, 所以 P 为重心.

上面的结论在 λ_1、λ_2 不全为正(E 或 F 不在线段 CA 或 CB 内)时仍然成立.

最后,以 D 为位似中心,相似比为 $\dfrac{1}{3}$ 的位似变换中, C 的像是 P. 所以 P 的

轨迹是与抛物线 $y=x^2$ 位似的抛物线. 设 P 的坐标为 (x, y), C 的坐标为 (x_0, y_0),则 $3\left(x-\frac{1}{2}\right)=x_0-\frac{1}{2}$, $3y=y_0$. 消去 x_0, $y_0(=x_0^2)$ 便得 P 的轨迹方程 $y=\frac{1}{3}(3x-1)^2$.

加试的第一题是一道好题. 条件与结论都很简明,用到的知识不多,证法多种多样. 如果"吹毛求疵",那就是稍嫌容易:

(1) 作 $\angle BAC$ 的平分线,交 DE 于 I. 只需证明 CI 平分 $\angle ACB$. 易知 $\triangle ADI \cong \triangle ACI$,所以

$$\angle ACI=\angle ADI=\angle AED=\frac{180^\circ-\angle DAE}{2}.$$

又由弦切角的性质, $\angle EAC=\angle ABC$. 所以

$$\angle ACI=\frac{1}{2}(180^\circ-\angle BAC-\angle ABC)=\frac{1}{2}\angle ACB.$$

从而 I 为 $\triangle ABC$ 的内心.

这一证明没有用到四点共圆.

(2)的证明与(1)类似,本可省略. 但也可以作为"同理"或"同法可证"的一个案例练习写在这里:

作 $\angle BAC$ 的平分线,交直线 DF 于 I_1,易得 $\triangle ADI_1 \cong \triangle ACI_1$,所以 $180^\circ-\angle ACI_1=180^\circ-\angle ADI_1=\angle ADF=\angle AFD=\frac{180^\circ-\angle DAF}{2}=\frac{180^\circ-\angle ACB}{2}$,即 CI_1 是 $\angle ACB$ 的外角平分线, I_1 为旁心.

加试的第二题中有六个字母 a, b, c, x, y, z. 前三者通常表示常数,最好说明它们也是变数以免歧义. 数学表述要求简明,但在易混淆、不明确时,宁愿多说些话,增加些字(本题只需在"正数"两字中加一个"变"字),以保障"明".

由 x, y, z 的方程 $cy+bz=a$ 等易解得 x, y, z. 比如用行列式,

$$x=\begin{vmatrix} a & c & b \\ b & 0 & a \\ c & a & 0 \end{vmatrix} \div \begin{vmatrix} 0 & c & b \\ c & 0 & a \\ b & a & 0 \end{vmatrix}=\frac{b^2+c^2-a^2}{2bc}.$$

于是看出 x, y, z 是一个锐角三角形 ABC 的三个角的余弦 $\cos A$, $\cos B$,

$\cos C$.

做到这里都很容易. 接下去求极值却有困难. 标准答案所作代换及以后的"组合拳"有点匪夷所思. 想不到(至少我是想不到)! 它有点像武侠小说中的独门功夫或秘制兵器.

有无其他方法呢? 特别是更普遍的方法呢?

很自然地想到凸函数. 函数 $\varphi(x)=\dfrac{x^2}{1+x}$ 是(下)凸的. 但函数 $\psi(\theta)=\dfrac{\cos^2\theta}{1+\cos\theta}$ 却不是 $\left[0, \dfrac{\pi}{2}\right]$ 上的凸函数: 因为 $\psi(\theta)=\cos\theta-1+\dfrac{1}{1+\cos\theta}$,

$$\psi'(\theta)=\left[1-\frac{1}{(1+\cos\theta)^2}\right]\cdot(-\sin\theta),$$

$$\begin{aligned}\psi''(\theta)&=\frac{2}{(1+\cos\theta)^3}\sin^2\theta-\left[1-\frac{1}{(1+\cos\theta)^2}\right]\cos\theta\\&=\frac{1}{(1+\cos\theta)^2}[2(1-\cos\theta)-(2+\cos\theta)\cos^2\theta]\\&=\frac{1}{(1+\cos\theta)^2}(2-2\cos\theta-2\cos^2\theta-\cos^3\theta).\end{aligned}$$

$\psi''(\theta)>0$ 在 $\left[0, \dfrac{\pi}{2}\right]$ 上不恒成立, 所以 $\psi(\theta)$ 不是 $\left[0, \dfrac{\pi}{2}\right]$ 上的凸函数.

事实上, 本题的最小值 $\dfrac{1}{2}$, 不仅在 $A=B=C=\dfrac{\pi}{3}$ (即 $x=y=z=\dfrac{1}{2}$) 时取得, 而且也在 $A=0, B=C=\dfrac{\pi}{2}$ (即 $x=1, y=z=0$) 时取得. 因此, 简单地使用凸函数是不能完成的.

但上面所作努力不会白费. 注意到 $2-2\cos\theta-2\cos^2\theta-\cos^3\theta$ 是 θ 的增函数, 所以它只有一个零点 θ_0. 函数 $\psi(\theta)$ 在 $[0, \theta_0]$ 上凹, 在 $\left[\theta_0, \dfrac{\pi}{2}\right]$ 上是凸函数.

设 $A\leqslant B\leqslant C\left(\dfrac{\pi}{2}\right)$. 如果 $A\geqslant\theta_0$, 那么由琴生不等式立即得出结论. 如果 $A<\theta_0\leqslant B$, 那么令 $\beta=\dfrac{B+C}{2}$, 由琴生不等式

$$\frac{\cos^2 B}{1+\cos B}+\frac{\cos^2 C}{1+\cos C}\geqslant\frac{2\cos^2\beta}{1+\cos\beta},\tag{12}$$

于是只需证

$$\frac{\cos^2 A}{1+\cos A}+\frac{2\cos^2\beta}{1+\cos\beta}\geqslant\frac{1}{2}. \tag{13}$$

如果 $B<\theta_0$，那么设 $\theta_0-B=\delta>0$. 因为在 $[0,\theta_0]$ 上 $\psi(\theta)$ 凹 $(\psi''(\theta)<0)$，所以 $\psi'(\theta)\downarrow$.

$$(\psi(x+\delta)-\psi(x))'=\psi'(x+\delta)-\psi'(x)<0,$$

所以 $\psi(x+\delta)-\psi(x)$ 递减. 于是有

$$\psi(A)-\psi(A-\delta)>\psi(B+\delta)-\psi(B),$$

即
$$\psi(A)+\psi(B)>\psi(A-\delta)+\psi(B+\delta).$$

因此，总可假定 $B\geqslant\theta_0$（否则将 A、B 改为 $A-\delta$ 与 $B+\delta=\theta_0$），从而只需证明(13).

(13)左边其实是一个一元函数，因为 $\beta=\frac{1}{2}(180^\circ-A)$. 如果改记 $A=2\alpha$，那么 $\beta=90^\circ-\alpha$，(13)变为

$$\frac{\cos^2 2\alpha}{1+\cos 2\alpha}+\frac{\sin^2\alpha}{1+\sin\alpha}\geqslant\frac{1}{2}. \tag{14}$$

(14)又等价于

$$\cos^2 2\alpha(1+\sin\alpha)+4\sin^2\alpha\cos^2\alpha\geqslant\cos^2\alpha(1+\sin\alpha).$$

剩下的是一些化简工作. 上式

$$\begin{aligned}
&\Leftrightarrow\quad 1+\sin\alpha\cos^2 2\alpha\geqslant\cos^2\alpha+\sin\alpha\cos^2\alpha\\
&\Leftrightarrow\quad \sin\alpha\geqslant\cos^2\alpha-\cos^2 2\alpha\\
&\Leftrightarrow\quad 2\sin\alpha\geqslant\cos 2\alpha-\cos 4\alpha\\
&\Leftrightarrow\quad 2\sin\alpha\geqslant 2\sin\alpha\sin 3\alpha.
\end{aligned}$$

最后的不等式显然成立.

上面的解法有一定的普遍性（如 2003 年全国中学生数学冬令营试题第 3 题，2004 年中国西部数学奥林匹克第 8 题均可使用此法）. 其要点是对先凹后凸的函数，关于 n 个函数值的极值或不等式，在不能运用琴生不等式时，有可能对

$n-1$ 个函数值运用琴生不等式,从而将问题化为一个一元函数的极值或不等式,再设法解决.

过去竞赛中,只用“初等方法”,不用微积分. 其实微积分也很初等,它已经进入我国中学教材. 在越南、罗马尼亚等国的竞赛题中都出现微积分的内容. 应当提倡使用微积分,因为用微积分处理一些问题比用判别式等初等方法更具有普遍性. 与其花很大功夫学一些今后没有多少用处的奇技淫巧,恶补各种证不等式的特殊方法,不如学一点微积分. 或许有一天会在国际竞赛中出现微积分的题,应当未雨绸缪,早作准备.

加试的第三题并不难,却有点繁. 在得到 $\sum_{i=1}^{2k}\left[\frac{2k}{i}\right]$ 后应想一想它的几何意义. 它表示双曲线 $xy=2k$ 与坐标轴之间的整点的个数,即

$$\sum_{i=1}^{2k}\left[\frac{2k}{i}\right]=\sum_{xy\leqslant 2k}1=\sum_{h=1}^{2k}\sum_{xy=h}1=\sum_{h=1}^{2k}d(h).$$

其中 $d(h)$ 表示 h 的(正)因数的个数(也就是“标准答案”中的 $T(h)$,但 $d(h)$ 或 $\tau(h)$ 是数论中的标准记号). 这样问题就与数论中著名的除数(即因数)问题挂上了钩. 但 $\sum_{h=1}^{2k}d(h)$ 迄今没有获得简单公式. 由和号变换,

$$\begin{aligned}\sum_{k=1}^{n}\sum_{h=1}^{2k}d(h)&=\sum_{h=1}^{2n}d(h)\sum_{\frac{h}{2}\leqslant k\leqslant n}1=\sum_{h=1}^{2h}d(h)\left(n-\left[\frac{h-1}{2}\right]\right)\\&=n\sum_{h=1}^{2n}d(h)-\sum_{h=1}^{2n}\left[\frac{h-1}{2}\right]d(h).\end{aligned}$$

剩下的就只有将 $n=15$ 代入计算,幸好 n 还不太大!

加试第三题虽有一点数论背景,基本上还是不等式. 第二题也是不等式. 这就显得不等式的分量畸重. 希望今后能出现一道饶有趣味的数论题或组合题,少一点不等式,少一点繁琐的计算. 谢天谢地!

2005 年中国数学奥林匹克的不等式题

2005 年中国数学奥林匹克的试题比 2004 年容易些,只有第 4 题较困难. 如下:

已知数列 $\{a_n\}$ 满足条件 $a_1=\frac{21}{16}$,

$$2a_n-3a_{n-1}=\frac{3}{2^{n+1}},\ n\geqslant 2. \tag{1}$$

设 m 为正整数, $m\geqslant 2$. 证明:当 $n\leqslant m$ 时,有

$$\left(a_n+\frac{3}{2^{n+3}}\right)^{\frac{1}{m}}\left[m-\left(\frac{2}{3}\right)^{\frac{n(m-1)}{m}}\right]<\frac{m^2-1}{m-n+1}. \tag{2}$$

解这道题当然是先求出 $\{a_n\}$ 的通项,即由式(1)得

$$2\left(a_n+\frac{3}{2^{n+3}}\right)=3\left(a_{n-1}+\frac{3}{2^{n+2}}\right),$$

从而,有

$$a_n+\frac{3}{2^{n+3}}=\left(\frac{3}{2}\right)^{n-1}\left(a_1+\frac{3}{16}\right)=\left(\frac{3}{2}\right)^{n}.$$

于是,式(2)化为

$$\left(\frac{3}{2}\right)^{\frac{n}{m}}\left[m-\left(\frac{2}{3}\right)^{\frac{n(m-1)}{m}}\right]<\frac{m^2-1}{m-n+1}. \tag{3}$$

本题的难点是式(3)的证明. 命题者给出了一个很好的解答,本文将提供另一种解答.

首先,指出在 $n\geqslant 2$ 时,式(3)可加强为

$$\left(\frac{3}{2}\right)^{\frac{n}{m}}\cdot m\leqslant\frac{m^2-1}{m-n+1}. \tag{4}$$

式(4)的证明如下：设 $f(n)=\left(\frac{3}{2}\right)^{\frac{n}{m}}(m-n+1)$，当 $n<m$ 时，有

$$\frac{f(n)}{f(n+1)}=\left(\frac{2}{3}\right)^{\frac{1}{m}}\cdot\frac{m-n+1}{m-n}.$$

因为 $\left(1+\frac{1}{m-n}\right)^{m}>\left(1+\frac{1}{m}\right)^{m}>1+m\cdot\frac{1}{m}=2>\frac{3}{2}$，

所以，$f(n)>f(n+1)$.

由于 $f(n)$ 递减，要证式(4)只须证 $f(2)\leqslant\frac{m^{2}-1}{m}$，即

$$\left(\frac{3}{2}\right)^{\frac{2}{m}}\leqslant\frac{m+1}{m}. \tag{5}$$

因为

$$\left(\frac{m+1}{m}\right)^{m}\geqslant 1+m\cdot\frac{1}{m}+\frac{m(m-1)}{2}\cdot\frac{1}{m^{2}}=\frac{5}{2}-\frac{1}{2m}\geqslant\frac{9}{4},$$

所以，式(5)成立，式(4)也随之成立.

易看出，当且仅当 $m=n=2$ 时，式(4)中等号成立.

于是，剩下的问题是证明 $n=1$ 时，式(3)成立，即

$$\left(\frac{3}{2}\right)^{\frac{1}{m}}\left[m-\left(\frac{2}{3}\right)^{\frac{m-1}{m}}\right]<\frac{m^{2}-1}{m}. \tag{6}$$

为了证明式(6)，考虑关于 t 的二次函数 $y=mt-\frac{2}{3}t^{2}$.

由于该函数在 $t\leqslant\frac{3}{4}m$ 时递增，而

$$\left(\frac{3}{2}\right)^{\frac{1}{m}}<1+\frac{1}{2m}\left(\left(1+\frac{1}{2m}\right)^{m}>1+m\cdot\frac{1}{2m}=\frac{3}{2}\right),$$

$$1+\frac{1}{2m}<1+\frac{1}{2}\leqslant\frac{3}{4}m,$$

所以，$m\left(\frac{3}{2}\right)^{\frac{1}{m}}-\frac{2}{3}\left(\frac{3}{2}\right)^{\frac{2}{m}}<m\left(1+\frac{1}{2m}\right)-\frac{2}{3}\left(1+\frac{1}{2m}\right)^{2}$

$$<m+\frac{1}{2}-\frac{2}{3}-\frac{2}{3m}$$

$$=m-\frac{1}{m}+\frac{1}{3m}-\frac{1}{6}\leqslant\frac{m^2-1}{m}.$$

故式(6)成立.

在上面的证明中，n 为正整数，这与原来的题意相符. 其实，不等式(3)对一切满足 $1\leqslant n\leqslant m$ 的实数 n 均成立(m 为大于或等于 2 的实数). 原解答仍可用于 n 为实数的情形，而上面的解答则需要略加修改.

关于 $f(n)$ 的单调性，应当利用导数：

$$f'(n)=-\left(\frac{3}{2}\right)^{\frac{n}{m}}+(m-n+1)\left(\frac{3}{2}\right)^{\frac{n}{m}}\cdot\frac{1}{m}\ln\frac{3}{2}$$

$$=\left(\frac{3}{2}\right)^{\frac{n}{m}}\left(\frac{m-n+1}{m}\ln\frac{3}{2}-1\right)<\left(\frac{3}{2}\right)^{\frac{n}{m}}\left(\ln\frac{3}{2}-1\right)<0.$$

而在 $1\leqslant n\leqslant 2$ 时，应当先证明

$$g(n)=(m+1-n)\left(\frac{3}{2}\right)^{\frac{n}{m}}\left[m-\left(\frac{2}{3}\right)^{\frac{n(m-1)}{m}}\right]$$

是单调递减的(从而，问题化为式(6)的证明). 这也要利用导数.

因为

$$g'(n)=-\left[\left(\frac{3}{2}\right)^{\frac{n}{m}}m-\left(\frac{2}{3}\right)^{\frac{n(m-2)}{m}}\right]+(m-n+1)$$

$$\cdot\left[\left(\frac{3}{2}\right)^{\frac{n}{m}}+\left(\frac{2}{3}\right)^{\frac{n(m-2)}{m}}\cdot\frac{m-2}{m}\right]\ln\frac{3}{2},$$

于是，有

$$g'(n)<0$$

$$\Leftarrow m\left(\frac{3}{2}\right)^{\frac{n}{m}}\left(1-\ln\frac{3}{2}\right)>(m-2)\left(\frac{2}{3}\right)^{\frac{n(m-2)}{m}}\ln\frac{3}{2}+\left(\frac{2}{3}\right)^{\frac{n(m-2)}{m}}$$

$$\Leftarrow m\left(1-\ln\frac{3}{2}\right)>(m-2)\left(\frac{2}{3}\right)^{\frac{n(m-1)}{m}}\ln\frac{3}{2}+\left(\frac{2}{3}\right)^{\frac{n(m-1)}{m}}$$

$$\Leftarrow m\left(1-\ln\frac{3}{2}\right)>(m-2)\ln\frac{3}{2}+1\Leftarrow m\left(1-2\ln\frac{3}{2}\right)>1-2\ln\frac{3}{2}.$$

而 $1-2\ln\frac{3}{2}>0\left(e=2.71828\cdots>\left(\frac{3}{2}\right)^2=2.25\right)$，

所以,上式成立.从而,$g(n)$递减.

当指数由整数变为任意实数时,不可避免地要利用导数(原解答用到贝努利不等式,而实指数的贝努利不等式必须用导数来证明).冬令营的选手中有一些也试图利用导数,但却未能得出正确结果.这表明导数的学习应当加强,因为它是很重要的.

评 2006 年全国高中数学联赛

2006 年全国高中数学联赛重视基础. 加试题出乎许多人的意料，没有数论、组合问题，也没有技巧很高的不等式证明. 似乎"返朴归真"，返回到三十多年前的数学竞赛，加强了代数的基本技巧. 这在目前普遍忽视基本训练，存在一味钻偏题怪题的情况时，来一次"矫枉过正"，倒的确是有好处的，至少可以引发"竞赛题应当如何出"的议论，缓和竞赛题越来越难的趋势.

选择题 1 是一道向量题，颇有新意. 用数量积可以用平方取代"绝对值"(向量的模)，即由已知得

$$c^2+t^2a^2-2tac\cos B\geqslant b^2, \tag{1}$$

其中 a、b、c、A、B、C 是$\triangle ABC$ 的边长与角(我们用 c^2，不用 $|\overrightarrow{BA}|^2$，因为前者比后者简便). 于是

$$(ta-c\cos B)^2+c^2\sin^2 B\geqslant b^2. \tag{2}$$

取 $t=\dfrac{c\cos B}{a}$，由(2)得

$$b\leqslant c\sin B=b\sin C(\text{正弦定理}), \tag{3}$$

所以

$$\sin C=1,\ C=90°. \tag{4}$$

填空题 8 是一道复数题，可先表 z 为

$$z=a(1+2\mathrm{i})+(\cos\theta-\mathrm{i}\sin\theta). \tag{5}$$

$\cos\theta-\mathrm{i}\sin\theta$ 在单位圆上，模为 1, $a(1+2\mathrm{i})$ 的模为$\sqrt{5}\,|a|$，所以两者的和，最长为$\sqrt{5}\,|a|+1$. 由

$$\sqrt{5}\,|a|+1\leqslant 2, \tag{6}$$

得

$$|a|\leqslant\frac{1}{\sqrt{5}}, \tag{7}$$

即 $a \in \left[-\frac{1}{\sqrt{5}}, \frac{1}{\sqrt{5}}\right]$.

第 11 题方程左边的项均为偶数幂，显然为正，所以右边也为正，$x > 0$. 由算术—几何平均不等式得

$$\begin{aligned}&(x^{2006}+1)(1+x^2+x^4+\cdots+x^{2004})\\ \geqslant\ & 2x^{1003}\cdot 1003x^{(2+4+\cdots+2004)\div 1003}=2006x^{2005},\end{aligned}\tag{8}$$

等号成立，所以 $x=1$.

第 12 题，袋内球数始终为 10（题目中"放回 1 个白球"宜改为"放进 1 个白球"）. 第 4 次取红球，前面还有 1 次也取红球，这 2 次的概率为 0.2×0.1. 另两次取白球，可以是第一、二次，第一、三次或第二、三次. 概率分别为 0.8^2，0.8×0.9，0.9^2. 所求概率为 $(8^2+8\times 9+9^2)\times 2\times 1\div 10^4=0.0434$.

解答题 13 实际上是一道数列题. M_0 在直线 $y=x$ 上，显然 $x_0=y_0$，代入抛物线方程得 $x_0^2=nx_0-1$，即 $x_0+x_0^{-1}=n$. 因为 n 是正整数，所以 x_0 是正数. 不必求出 x_0. 只需证明 $x_0^m+x_0^{-m}$ 是正整数 $\geqslant 2$. 显然 $x_0^m+x_0^{-m}\geqslant 2\sqrt{x_0^m\cdot x_0^{-m}}=2$. 为了证明 $a_m=x_0^m+x_0^{-m}$ 是整数，可用二阶递推数列

$$a_m=\alpha a_{m-1}+\beta a_{m-2},\tag{9}$$

其中 $\alpha=x_0+x_0^{-1}=n$ 是整数，$\beta=x_0\cdot x_0^{-1}=1$ 是整数，$a_0=2$，$a_1=n$ 是整数，所以一切 a_m 是整数.

第 14 题可令 $A=\sum\limits_{i=1}^{5}x_i^2$，$B=\sum\limits_{1\leqslant i<j\leqslant 5}(x_i-x_j)^2$，利用恒等式

$$(x_1+x_2+\cdots+x_5)^2-A=2S,\tag{10}$$

及

$$B+(x_1+x_2+\cdots+x_5)^2=5A.\tag{11}$$

(10)表明 S 在 A 最小(大)时取最大(小)值. (11)表明 A 在 B 最小(大)时取最小(大)值.

不妨设 $x_1\geqslant\cdots\geqslant x_5$，因 5 不整除 2006，故 $x_1>x_5$. 若 $x_1-x_5\geqslant 2$，则 $(x_1-x_5)^2\geqslant 4$，$B\geqslant 4$. 故 B 的最小值为 4. 此值在 $x_1-x_5=1$ 即 $x_5=(2006-1)\div 5=401=x_4=x_3=x_2$，$x_1=402$ 时取得. 这时 S 取最大值 $\frac{1}{2}\Big(2006^2-$

$\left.\dfrac{4+2006^2}{5}\right)=\dfrac{2}{5}(2006^2-1)=1\,609\,614.$

在限制条件 $x_1-x_5\leqslant 2$ 下,要使 B 尽可能大,应有 $x_1=x_5+2$,故 $x_1=402$, $x_5=400$,其余各数只有两种可能:$x_2=402$, $x_3=x_4=401$; $x_2=x_3=402$, $x_4=400$. 后者使 B 取最大值 $6\times 4=24$. 这时 S 取最小值 $\dfrac{1}{2}\Big(2006^2-$

$\left.\dfrac{24+2006^2}{5}\right)=\dfrac{2}{5}(2006^2-6)=1\,609\,612.$

条件 $|x_1-x_5|\leqslant 2$ 限制得太死,似可取消或修改.

第 15 题,如不给出 $M=\left[-2,\dfrac{1}{4}\right]$,改为"求 M"似更难一点,也更有趣(增加探索的过程).

在 $a\leqslant 0$ 时,由 $|f(0)|=|a|\leqslant 2$ 立得 $a\geqslant -2$. 同时,在 $0\geqslant a\geqslant -2$ 时,可以用数学归纳法证明 $|f^n(0)|\leqslant |a|\leqslant 2$.

在 $a>0$ 时,记 $a_n=f^n(0)$,则 $a_2=a^2+a>a=a_1$,由数学归纳法易知 $a_{n+1}=a_n^2+a>a_{n-1}^2+a=a_n$. $|a_n|$ 单调递增且有上界,所以有极限 c, $c=c^2+a$,从而 $c=\dfrac{1}{2}(1\pm\sqrt{1-4a})$,可见必须 $a\leqslant\dfrac{1}{4}$. 同时,在 $0<a\leqslant 1/4$ 时,由数学归纳法可知 $a_n<\dfrac{1}{2}$. 于是极限 $c=\dfrac{1}{2}(1-\sqrt{1-4a})$ 而不是 $\dfrac{1}{2}(1+\sqrt{1-4a})$,除非 $a=\dfrac{1}{4}$.

通常加试第一题是一道平面几何题. 2006 年也不例外,只不过增加了一些解析几何的外包装,其实除了

$$C_1B_1+C_1B_0=C_0B_1+C_0B_0 \tag{12}$$

这一性质外,与解析几何完全无关.

第(1)问不难证. 第(2)问只需根据三角形的外角等于不相邻的两个内角的和(不必利用切线):

$$\begin{aligned}\angle P_0Q_0P_1&=\angle P_0Q_0B_0+\angle B_0Q_0P_1\\&=\frac{1}{2}(\angle Q_0B_0A+\angle B_1C_1B_0)\\&=\frac{1}{2}(\angle B_0AB_1+\angle B_0C_1A+\angle B_1C_1B_0)\end{aligned}$$

$$=\frac{\pi}{2}+\frac{1}{2}\angle B_0AB_1. \tag{13}$$

同理
$$\angle P_0Q_1P_1=\frac{\pi}{2}+\frac{1}{2}\angle B_0AB_1. \tag{14}$$

故 $\angle P_0Q_0P_1=\angle P_0Q_1P_1$，$P_0$、$Q_0$、$Q_1$、$P_1$ 四点共圆.

加试第二题的递推公式
$$a_{n+1}=\frac{a_na_{n-1}+1}{a_n+a_{n-1}} \tag{15}$$

使人想到余切的和角公式

$$\cot(\alpha+\beta)=\frac{\cot\alpha\cot\beta-1}{\cot\alpha+\cot\beta}, \tag{16}$$

不过还差一个符号. 若熟悉双曲函数，可知双曲余切的和角公式

$$\operatorname{cth}(\alpha+\beta)=\frac{\operatorname{cth}\alpha\operatorname{cth}\beta+1}{\operatorname{cth}\alpha+\operatorname{cth}\beta}. \tag{17}$$

于是，若设
$$a_0=x=\operatorname{cth}u,\ a_1=y=\operatorname{cth}v, \tag{18}$$

则
$$a_2=\operatorname{cth}(u+v),\ a_3=\operatorname{cth}(u+2v),\ \cdots,$$
$$a_n=\operatorname{cth}(F_{n-2}u+F_{n-1}v), \tag{19}$$

即
$$a_n=\frac{e^{F_{n-2}u+F_{n-1}v}+e^{-(F_{n-2}u+F_{n-1}v)}}{e^{F_{n-2}u+F_{n-1}v}-e^{-(F_{n-2}u+F_{n-1}v)}}. \tag{20}$$

因

$$x=\operatorname{cth}u=\frac{e^u+e^{-u}}{e^u-e^{-u}},\ \frac{x+1}{x-1}=e^{2u},\ \frac{y+1}{y-1}=e^{2v},$$

故

$$\frac{a_n+1}{a_n-1}=e^{2(F_{n-2}u+F_{n-1}v)}=\left(\frac{x+1}{x-1}\right)^{F_{n-2}}\left(\frac{y+1}{y-1}\right)^{F_{n-1}}, \tag{21}$$

$$a_n=\frac{(x+1)^{F_{n-2}}(y+1)^{F_{n-1}}+(x-1)^{F_{n-2}}(y-1)^{F_{n-1}}}{(x+1)^{F_{n-2}}(y+1)^{F_{n-1}}-(x-1)^{F_{n-2}}(y-1)^{F_{n-1}}}. \tag{22}$$

当然对一切 $u\in\mathbf{R}$，$\operatorname{cth}u>1$，所以(18)仅在 $x>1$，$y>1$ 时才成立. 但不难验证所得到的公式(22)(或(21))却在 $x\neq\pm1$，$y\neq\pm1$，$x+y\neq0$ 时，适合

递推关系(15)及初始条件 $a_0=x$, $a_1=y$. 而由初始条件及(15)唯一确定$\{a_n\}$. 因此(22)就是所求的递推公式(并不仅限于 $x>1$, $y>1$).

由(22)或(21)可以看出, $a_{n+1}=a_n$ 导出

$$\left(\frac{x+1}{x-1}\right)^{F_{n-2}}\left(\frac{y+1}{y-1}\right)^{F_{n-1}}=\left(\frac{x+1}{x-1}\right)^{F_{n-1}}\left(\frac{y+1}{y-1}\right)^{F_n}, \tag{23}$$

从而
$$\left(\frac{x+1}{x-1}\right)^{F_{n-3}}\left(\frac{y+1}{y-1}\right)^{F_{n-2}}=1, \tag{24}$$

即
$$\frac{a_{n-1}+1}{a_{n-1}-1}=1. \tag{25}$$

但(25)是不可能成立的. 所以在 $x\neq\pm1$, $y\neq\pm1$, $x+y\neq0$ 时,不会出现 $a_{n+1}=a_n(n\geqslant1)$ 的情况.

另一方面,不难验证在 $y=\pm1$ 时, $a_n=\pm1(n=1, 2, \cdots)$. 在 $x=\pm1$ 时, $a_n=\pm1(n=2, 3, \cdots)$. 当然 $x+y\neq0$ 总是必要的.

加试第三题是解方程组的问题. 在过去(20 世纪 20 年代以来),很著名的代数书,如 Hall 与 Knight 的《大代数》或上野清的《大代数讲义》中常有类似的问题. 如果 y 与 w 的幂,系数全改为$+1$ 或者改为$-y$, y^2, $-y^3$, y^4 与$-w$, w^2, $-w^3$, w^4,那么可用一个以 x、y、z、w 为根或以 x、$-y$、z、$-w$ 为根的四次方程来处理. 现在似难以办到. 只有分别造两个二次方程.

$$u^2-pu+q=0, \tag{26}$$

$$v^2-sv+t=0. \tag{27}$$

前者以 x、z 为根,后者以 y、w 为根,即

$$p=x+z,\ q=xz; \tag{28}$$

$$s=y+w,\ t=yw. \tag{29}$$

以①, ②, ③, ④表原方程组的 4 个方程,由①及(28)、(29)得

$$p=s+2. \tag{30}$$

又由②及(28)、(29)得
$$p^2-2q=6+s^2-2t. \tag{31}$$

用(30)代入得
$$2(2s+2)-2q=6-2t,$$

即 $$q=t+2s-1. \tag{32}$$

(32)的另一种推导方法是②$-p\times$①得

$$6-2p+y^2-py+q+w^2-pw+q=0, \tag{33}$$

即 $$6-2p-2y+q-t-2w+q-t=0.$$

所以 $$q=p-3+(y+w)+t=p-3+s+t=t+2s-1. \tag{34}$$

类似地,将③$-p\times$②$+q\times$①得

$$\begin{aligned}&20-6p+2q+y(y^2-py+q)+w(w^2-pw+q)\\ =&20-6p+2q+y(-2y+q-t)+w(-2w+q-t)\\ =&20-6p+2q+y(-2y+2s-1)+w(-2w+2s-1)(\text{利用 }34))\\ =&20-6p+2q-2(y^2-sy+t)-y+2t-2(w^2-sw+t)-w+2t\\ =&20-6p+2q+4t-s=0.\end{aligned} \tag{35}$$

再结合(30)、(34),可将(35)化为

$$s=2t+2. \tag{36}$$

同理,由④$-p\times$③$+q\times$②可得

$$\begin{aligned}&66-20p+6q+y^2(y^2-py+q)+w^2(w^2-pw+q)\\ =&66-20p+6q+y(-y+2t)+w(-10+2t)\\ =&66-20p+6q+(2ty-sy+t)+(2tw-sw+t)\\ =&66-20p+6q+(2t-s)s+2t\\ =&66-20(s+2)+6(t+2s-1)-2s+2t\\ =&20-10s+8t=0.\end{aligned} \tag{37}$$

由(36)、(37)得 $$t=0,\ s=2. \tag{38}$$

从而 $$p=4,\ q=3. \tag{39}$$

于是 x、z 为方程 $u^2-4u+3=0$ 的根, y、w 为 $v^2-2v=0$ 的根,所以

$(x,y,z,w)=(3,2,1,0),(3,0,1,2),(1,2,3,0),(1,0,3,2).$

怎样解今年(2007年)国际数学奥林匹克试题

甲：您写了一本《我怎样解题》. 那么，可不可以谈谈今年(2007年)的国际数学奥林匹克试题?

师：当然可以. 我们一同讨论吧!

乙：第1题比较容易就不必谈了. 谈第2题吧.

师：今年出现了两道几何题. 做几何题得先画一个草图，熟悉题目中的条件与结论.

甲：我已经画了一个草图.

乙：画得太差了，得不到一点启发. 应当重画一个好些的.

师：虽然是草图，不要求十分准确，但也应当"神似"，其中线段或角的大小，直线的平行或垂直等不应与实际完全不符. 现在 E 是 $\triangle CFG$ 的外心，所以应先画 $\odot E$ 及它的弦 CG. 如果 l 平分 $\angle DAB$，那么 $\angle DAF=\angle BAF=\angle DFA=\angle CFG=\angle CGF$，所以 $CG=CF$，$AF=AD$. 在 $\odot E$ 中应弦 $CF=CG$，然后在 GF 的延长线上任取一点 A，完成图中其余部分(易知这图中，B、D、C、E 共圆).

甲：从现在的图中可以看出 $\triangle DEF$ 与 $\triangle BEC$ 应当全等. 事实上，由 B、C、E、D 共圆，$\angle EDF=\angle EBC$. 又 $EF=EC$. 只是还差一个条件.

乙：从 $AD\parallel BC$ 可得

$$\frac{DF}{FC}=\frac{AD}{GC}=\frac{BC}{GC}. \tag{1}$$

可是这个条件不能用于证明 $\triangle DEF\cong\triangle BEC$.

甲：相等的角也找不出. 连 $\triangle DEF\backsim\triangle BEC$，也无法证明.

师：不必拘泥于 $\triangle DEF$ 与 $\triangle BEC$. 我们可以造出一对相似三角形，它们虽不是 $\triangle DEF$ 与 $\triangle BEC$，却与这两个三角形密切相关.

乙：是不是作 $EK\perp CD$，$EC\perp CG$ (垂足分别为 K、L)?

甲: 显然 $\mathrm{Rt}\triangle EKD \backsim \mathrm{Rt}\triangle ELB$. 但能否得出 $\triangle DEF \backsim \triangle BEC$ 呢?

师: 由(1)得 $\dfrac{DF}{2FK}=\dfrac{DF}{FC}=\dfrac{BC}{GC}=\dfrac{BC}{2CL}$. 这表示在相似三角形 EKD 与 ELB 中, F 与 C 是对应点,所以 $\triangle DEF \backsim \triangle BEC$.

乙: 从而由 $EF=EC$, $\triangle DEF \cong \triangle BEC$, $FD=BC=AD$, $\angle BAF=\angle AFD=\angle FAD$.

甲: 第 4 题不难,图中有两个直角三角形 CKP 与 CLQ,它们是相似的. 而 $\triangle RPK$、$\triangle RQL$ 各与一个直角三角形密切相关.

乙: 容易得出 $2\angle KPR=2\angle LQR=180^\circ+\angle ABC$. 所以只需证明

$$KP\times PR=LQ\times QR. \tag{2}$$

甲: 由 $\triangle CKP \backsim \triangle CLQ$, 有

$$KP\times CQ=LQ\times CP. \tag{3}$$

要证(2)只需证明

$$CQ=PR,\ CP=QR. \tag{4}$$

乙: 显然 AC、BC 的中垂线 LQ、KP 相交于外心 O. 由 $\angle KPR=\angle LQR$ 得 $OQ=OP$. 故 QP 的中垂线过 O,因而也是 CK 的中垂线. 从而(4)成立.

师: 证得很好,说明国际数学奥林匹克试题也并不可怕. 只要扎扎实实,一步一步向前进,你们也都能够解决.

甲: 我们再看看第 5 题.

乙: 这一题, a、b 不对称.

师: 作一点恒等变形,可以使 a、b 地位对称.

甲: 怎么样?

乙:
$$\begin{aligned}(4a^2-1)^2-(4ab-1)^2&=(4a^2-4ab)(4a^2+4ab-2)\\&=8a(a-b)(2a^2+2ab-1)\\&=8a(a-b)(2a^2-2ab+4ab-1)\\&=16a^2(a-b)^2+8a(a-b)(4ab-1)\end{aligned}$$

被 $4ab-1$ 整除,所以 $4ab-1$ 整除 $16a^2(a-b)^2$,而 $4a$ 与 $4ab-1$ 互质,所以 $4ab-1$ 整除$(a-b)^2$.

甲: 现在 a、b 对称,可设 $a\geqslant b$.

乙：还可设 $(a-b)^2=k(4ab-1)$，$k\in\mathbf{Z}^*$. (5)

甲：也就是

$$a^2-2b(2k+1)a+b^2+k=0. \tag{6}$$

师：(6)是所谓马尔柯夫型的不定方程. 可以用无穷递降法证明它没有 $a>b$ 的(正整数)解.

乙：如果(a, b)是(5)的一个解，$a>b$，那么由韦达定理 $a_1=\dfrac{b^2+k}{a}=2b(2k+1)-a$，也是 $x^2-2b(2k+1)x+b^2+k=0$ 的解. (7)

甲：而且 $\dfrac{b^2+k}{a}$ 是正数，$2b(2k+1)-a$ 是整数，所以 a_1 是正整数. (b, a_1)也是(6)的一个解.

乙：由(5)，$a^2-ab>k(4ab-1)$，$a>b+4kb-k>b+k$，所以

$$a_1=\frac{b^2+k}{a}<b.$$

甲：从(6)的一组解(a, b)，$a>b$，产生一组新的解(b, a_1)，$a_1<b$. 这样继续下去，形成一个正整数的严格递减数列

$$a>b>a_1>\cdots,$$

每两个相邻项组成(6)的解. 但这样的无穷数列不可能存在. 所以必须有 $a=b$.

乙：第 3 题比第 5 题难吧？

师：也不见得. 可以设最大的团有 $2m$ 个人. 将他们分配在教室 A，其余的人分配到教室 B. 然后再作调整.

甲：如果 B 中最大团的人数也是 $2m$，结论已经成立. 否则，我将 A 中的人一个一个地调入 B 中，每调走 1 个，A 中最大的团，人数减少 1. 而 B 中最大团的人数不变或增加 1. 这样做下去，总会出现 A 中最大数(“最大团的人数”简称为“最大数”)$\leqslant B$ 中最大数. 如果是等号，结论成立.

乙：A 中最大数第一次$\leqslant B$ 中最大数时，由于上一次还是 A 中最大数多，所以这时 A 中最大数比 B 中最大数少 1(如果相等，不必再证).

甲：再往下怎么办呢？

师：可以把 B 中的人往 A 中调. 设 B 中最大的团 K 有 s 个人，而 A 中剩下

$s-1$ 个人组成团 H. 如果有一个从 A 中调过去的 a 不在这个团 K 中，那么将 a 调回 A，A 的最大数就增加为 s. 因此，可设 B 中每一个最大的团 K 都包括所有 A 中调过去的人，即 $K \supseteqq N$，N 为 A 中调过去的人所成的集.

由于最初 A 中的 $2m$ 个元素均在 K 及 H 中，而 $H \cup K$ 的元素个数为奇数 $2s-1$. 所以 K 中必有一个元素 c 不是最初 A 中的 $2m$ 个元素. 将 c 调到 A 中，K 变为 $s-1$ 元的团. 如果 B 中还有元数为 s 的团 K'，也照此处理，调一个元 c' 到 A 中，直至 B 中最大数为 $s-1$.

乙： 这时 A 中最大数会不会超过 $s-1$？

师： 设 M 是 A 中最大的团. M 中的人或者是 A 中原来的人，或者是从 K、K' 等中调过来的人. 这些人都认识 N 中的每一个人. 所以 $M \cup N$ 是团，人数 $\leqslant 2m$，从而 M 中的人数不超过 H 中的人数. 即 M 的人数 $\leqslant s-1$.

甲： 当然 M 的人数也不会少于 $s-1$，所以 A 中最大数仍为 $s-1$.

师： 因此，经过调整总可使 A、B 中的最大数相等.

乙： 听说第 6 题极难，做对的人很少.

甲： 我觉得这道题应当从较简单的二维问题做起. 似乎也不很难. 在平面上给定 n^2+1 个格点的集

$$I=\{x,\ y):x,\ y=0,\ 1,\ \cdots,\ n\}.$$

我们显然可以作 $2n$ 条线 $x=k(k=1,\ 2,\ \cdots,\ n)$，$y=h(h=1,\ 2,\ \cdots,\ n)$ 将 I 中 n^2 个点覆盖，但不覆盖点 $(0,\ 0)$.

乙： 困难在于证明 $2n$ 条线最少.

师： 可以用直线的方程 $ax+by+c=0$ 来表示直线. 设覆盖 $I-\{(0,\ 0)\}$ 而不覆盖 $\{(0,\ 0)\}$ 所需直线为 $a_ix+b_iy+c_i=0$，$i=1,\ 2,\ \cdots,\ m$，则多项式 $f(x,\ y)=\prod\limits_{i=1}^{m}(a_ix+b_iy+c_i)$ 满足 $f(k,\ h)=0(k,\ h=0,\ 1,\ 2,\ \cdots,\ n$ 但 $k,\ h$ 不同时为 0) 及 $f(0,\ 0)\neq 0$.

然后证明 f 的次数 $m\geqslant 2n$.

甲： 这怎么证呢？

师： 可以按照 y 的降幂排列写出 $f(x,\ y)$，而且可以用

$$y(y-1)(y-2)\cdots(y-n)=0,$$

也就是

$$y^{n+1}=(1+2+\cdots+n)y^{n}+\cdots+(-1)^{n-1}n!\ y,$$

将 $f(x, y)$ 中高于 n 次的 y 的幂化为不超过 n 次的幂,得到多项式

$$g(x, y)=a_{n}(x)y^{n}+a_{n-1}(x)y^{n-1}+\cdots+a_{0}(x),$$

满足 $g(k, h)=0$ ($k, h=0, 1, 2, \cdots, n$, 但 k, h 不同时为 0)及 $g(0, 0)\neq 0$. $m\geqslant n+a_{n}(x)$ 的次数.

乙: 所以只需要证明 $a_{n}(x)$ 的次数 $\geqslant n$.

师: 多项式 $a_{n}(1)y^{n}+a_{n-1}(1)y^{n-1}+\cdots+a_{0}(1)$ 有 $n+1$ 个根 $y=0, 1, \cdots, n$,所以它必是零多项式, $a_{n}(1)=0$. 同样 $a_{n}(2)=\cdots=a_{n}(n)=0$.

又 $a_{n}(0)y^{n}+a_{n-1}(0)y^{n-1}+\cdots+a_{0}(0)$ 不是零多项式(因为 $a_{n}(0)=g(0, 0)\neq 0$),有 n 个根 $y=1, 2, \cdots n$,所以 $a_{n}(0)y^{n}+a_{n-1}(0)y^{n-1}+\cdots+a_{0}(0)$ 是 n 次多项式, $a_{n}(0)\neq 0$.

$a_{n}(x)$ 有 n 个根 $x=1, 2, \cdots, n$, 不是零多项式,所以 $a_{n}(x)$ 次数 $\geqslant n$.

甲: 三维情况与这类似.

乙: 那就不必多说了,我们回去自己补吧.

评 2007 年安徽省高中数学竞赛

安徽，是“文化大革命”后最早举行高中数学竞赛的省份之一. 当时中国科学技术大学已迁至合肥，有不少数学名家，如龚昇、曾肯成、陆洪文、常庚哲、史济怀、严镇军、杜锡禄等先生参与命题，所以安徽省数学竞赛的试题有很多好题，不少已经成为经典. 2007 年，安徽省数学会秘书长、中国科学技术大学数学系陈发来主任代表安徽省数学会委托吴康先生命题. 这一届试题很有特色，有不少自编的新题，题目不过难，着重基本运算与基本技巧. 如第 6 题：“设

$$A=\sqrt{1+\cos 3^{\circ}}+\sqrt{1+\cos 7^{\circ}}+\sqrt{1+\cos 11^{\circ}}+\cdots+\sqrt{1+\cos 87^{\circ}},$$

$$B=\sqrt{1-\cos 3^{\circ}}+\sqrt{1-\cos 7^{\circ}}+\sqrt{1-\cos 11^{\circ}}+\cdots+\sqrt{1-\cos 87^{\circ}},$$

则 $A:B=$________.”

其中就有 $\sqrt{1+\cos\alpha}=\sqrt{2}\cos\frac{\alpha}{2}\ (0<\alpha<90^{\circ})$，$2\sin\theta[\cos\alpha+(\cos\alpha+2\theta)+\cos(\alpha+4\theta)+\cdots+\cos(\alpha+2k\theta)]=\sin[\alpha+(2k+1)\theta]-\sin(\alpha-\theta)=2\cos(\alpha+k\theta)\sin(k+1)\theta$，$\dfrac{\cos\frac{\alpha}{2}}{\sin\frac{\alpha}{2}}=\dfrac{1+\cos\alpha}{\sin\alpha}$ 等三角运算的技巧.

这次竞赛的得分比较低，这正反映了我国数学教育的现状与存在问题. 尤其是：(1)近年来忽视基本运算与基本技巧的倾向日益严重，以致中上等水平的学生连第 6 题这样的题也难以完成；(2)一些教师与学生用对付高考的办法来对付竞赛，即总结若干题型，考试时生搬硬套，这种做法扼杀了创造性，只能促成头脑僵化.

出现这些问题，当然也与现在数学竞赛的命题人员较为固定、思路不够开阔有关. 因此，命题应力求推陈出新，人员也应常有变化.

从试题方面也可以找出一些原因. 新颖，是好的，即使学生不适应，也应坚持. 但这次试题也有几点似可改进：

1. 没有特别容易的、所谓送分的题. 就连第 1 题,也不是轻而易举就能拿下. 这样,可能增加参赛选手的心理压力. 其实,可以出几道送分的题,而且要真正送到手上. 这不仅减少参赛选手的心理压力,也有助于调动师生参加竞赛活动的积极性. 题目还是由易到难,逐步升高为好,不要上来就一棍子将人打矇.

2. 运算量稍大. 几乎每题都有运算,都有技巧. 题题是好题,但时间恐怕不够,不少学生难以做到最后几道.

3. 选择题基本上是填空题,看不出哪道题可用排除法立即得出正确的选项.

或许,难度稍降一点更好,这并不难做到,如第 1 题,可改为:"非空集合 A_1、B_1,若同时满足是 $A_1 \cup B_1=\{2, 3, 4\}$, $A \cap B=\varnothing$, 就称有序集对 (A, B) 为'好的'. 好的有序集对一共有________个."还可再进一步改为:"集合 $\{2, 3, 4\}$ 有多少个非空的真子集."这就容易得多,但也离原题越来越远了.

我们再就试题与解答,作些具体的讨论,顺便说说自己解这些题的体会.

第 3 题"求 $A=100\,101\,102\,103\cdots499\,500$ 除以 126 的余数."这题颇有新鲜感,没有看到过去出现过 126 这样的除数. 显然 $126=2\times9\times7$. 被 9 整除的判定法是众所周知的. A 被 9 除,余数与 $B=100+101+\cdots+500=(100+500)\times401\div2=300\times401$ 被 9 除的余数相同,即为 6. 因此 $A-6$ 被 9 整除. 偶数 $A-6$ 也被 2 整除,所以 $A-6$ 被 2×9 整除.

本题的关键是被 7 整除的判定法. 注意

$$999\,999=999\times1001=999\times7\times11\times13$$

被 7 整除. 所以

$$10^6\equiv1(\bmod 7).$$

可将多位数 A 自右到左,每 6 位一节,各节的和被 7 除,余数与 A 被 7 除的余数相同. 即

$$A\equiv100+101\,102+103\,104+\cdots+499\,500(\bmod 7).$$

而又由 $1001=7\times11\times13$ 被 7 整除,有 $1000\equiv-1(\bmod 7)$, 所以

$$\begin{aligned}A&\equiv100+(102-101)+(104-103)+\cdots+(500-499)\\&=100+200=300\equiv6(\bmod 7).\end{aligned}$$

于是 $A-6$ 被 $2\times7\times9=126$ 整除,即 A 除以 126 余 6.

第 5 题是“从 1, 3, 5, 7, …中删去与 55 不互质的数后得到数列 $\{a_n\}$,求 a_{2007}.”

这题也有新意,通常的问题从 1, 2, 3, 4, …中删去与 $2\times3\times5$ 不互质的数,而现将自然数数列改为正奇数数列.

本题可以不用任何公式(甚至不需要知道容斥原理),直接从简单情况做起,即在前 55 个奇数 1, 3, 5, …, 109 中应删去 15 个(即 5, 15, 25, …, 105 及 11, 33, 77, 99),剩下 40 个.

这 55 是周期,即后续 55 个奇数(前 55 个奇数分别加上 110)中,同样有 40 个数剩下,依此类推. 由 $2007=40\times50+7$, $50\times55\times2=5500$,以及 $a_7=19$,可知 $a_{2007}=5500+19=5519$.

第 7 题可设三边为 $20-d$, 20, $20+d$. 由

$$(20+d)^2>(20-d)^2+20^2$$

及

$$(20-d)+20>20+d,$$

得

$$10>d>5.$$

所以 d 可取 6, 7, 8, 9 这 4 个值. 所求三角形有 4 种.

第 10 题用向量解非常简单(即原解答). 这是一道体现向量作用的好题.

第 11 题亦颇新颖. 线性函数的迭代永远是线性函数,可以设 $f^{(9)}(x)=ax+b$, $g^6(x)=cx+d$, a, b, c, d 待定. 由方程组得

$$\begin{cases}ax+b=cx+d, & (1)\\ ay+b=cz+d, & (2)\\ az+b=cx+d. & (3)\end{cases}$$

从而

$$a(x-y)=c(y-z), \tag{4}$$

$$a(y-z)=c(z-x), \tag{5}$$

$$x-y=\frac{c}{a}(y-z)=\frac{c^2}{a^2}(z-x). \tag{6}$$

同理

$$y-z=\frac{c}{a}(z-x)=\frac{c^2}{a^2}(x-y). \tag{7}$$

若 $x>y$，则由(6)得 $z>x$，而由(7)得 $y>z$. 产生矛盾. 同理 $x<y$ 也产生矛盾，故必有 $x=y=z=\dfrac{d-b}{a-c}$.

而 a, b, c, d 并不难求：

$$f^{(3)}(x)=2(2(2x-3)-3)-3=8x-21,$$

$$f^{(9)}(x)=8(8(8x-21)-21-21=8^3x-21\times(8^2+8+1)=8^3x-21\times 73,$$

$$g^{(2)}(x)=3(3x+2)+2=9x+8,$$

$$g^{(6)}(x)=9(9(9x+8)+8)+8=9^3x+8\times(9^2+9+1)=9^3x+8\times 91,$$

故
$$x=\frac{8\times 91+21\times 73}{8^3-9^3}=\frac{8\times 13+3\times 73}{-31}=-\frac{323}{31}.$$

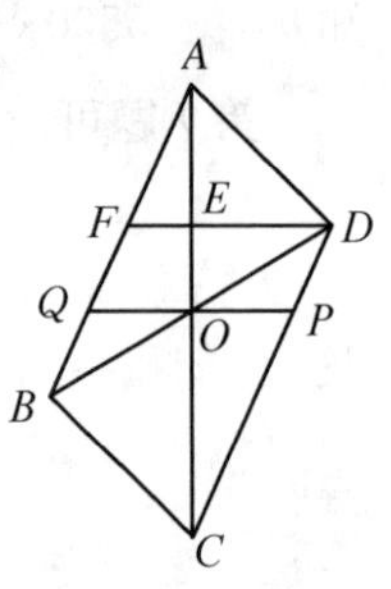

图 1

第 **12** 题求旋转球的体积. 首先要确定这个旋转体(在以往的竞赛中亦不多见). 为此，如图 1，我们将轴 AC 画在竖直位置上，过 D 作 AC 的垂线，交 AC 于 E，交 AB 于 F. 过 O 作 AC 的垂线，分别交 CD、AB 于 P、Q. 由于 $CD=4>AD$, $\angle CAD>\angle ACD=\angle BAC$. 所以$\triangle ADF$ 绕AC 旋转所得立体即$\triangle AED$(绕 AC 旋转所得圆锥，体积为 $\dfrac{1}{3}\pi\times AE\times DE^2$. 易知 $OP=OQ$，又 $DE>EF$，故四边形 $DFQP$ 绕 AC 旋转所得立体，即梯形 $DEOP$ 绕 AC 旋转所得圆台，体积为 $\dfrac{1}{3}\pi(DE^2\times CE-OP^2\times CO)$. 整个旋转体体积$V=\dfrac{2}{3}\pi(AE\times DE^2+DE^2\times CE-OP^2\times CO)$ $=\dfrac{1}{3}\pi(2DE^2\times AC-AC\times OP^2)=\dfrac{2\pi}{3}\times OA\times DE^2\times\left[2-\left(\dfrac{OC}{CE}\right)^2\right]$. 由 $AD^2+BD^2=4+12=AB^2$，得 $\angle ADB=90°$. $OA\times OE=AD\times DO=2\sqrt{3}$. 又 $AC^2=2(AB^2+AD^2)-BD^2=28$, $AC=2\sqrt{7}$, $OA=\sqrt{7}$, $AE=\dfrac{AD^2}{OA}=\dfrac{4}{\sqrt{7}}$, $CE=2\sqrt{7}-\dfrac{4}{\sqrt{7}}=\dfrac{10}{\sqrt{7}}$, $\left(\dfrac{OC}{CE}\right)^2=\dfrac{49}{100}$. 于是 $V=\dfrac{2\pi}{3}\times 2\sqrt{3}\times\dfrac{2\sqrt{3}}{\sqrt{7}}\times\left(2-\dfrac{49}{100}\right)=\dfrac{302\pi}{25\sqrt{7}}$.

解析几何题目有一些技巧.

第 13 题自点 $Q(4,\ 0)$向椭圆 Γ：$\frac{x^2}{4}+\frac{y^2}{3}=1(1)$ 作直线 l 交 Γ 于 A、B. 当 A、B 重合时(l 为切线)，$\angle AOB=0°$. 当 l 绕 Q 逆时针转动至 x 轴时，$\angle AOB$ 由 $0°$增加至 $180°$. 因此若设 l 的斜率为 $-k$ 时，$\angle AOB=90°$，则当 l 的斜率在区间$(-k,\ k)$中，$\angle AOB$ 为钝角或平角. 第(Ⅰ)小题即求$-k$.

可采用截距式,设 l 的方程为

$$\frac{x}{4}+\frac{y}{b}=1. \tag{2}$$

方程

$$\frac{x^2}{4}+\frac{y^2}{3}=\left(\frac{x}{4}+\frac{y}{b}\right)^2 \tag{3}$$

表示过 O 点的两条直线 OA、OB,它们的斜率 k_1、k_2 的乘积 k_1k_2 即(由(3)整理而得的)方程

$$y^2\left(\frac{1}{3}-\frac{1}{b^2}\right)+\cdots+\frac{3}{16}x^2=0 \tag{4}$$

的系数比

$$\frac{3}{16}:\left(\frac{1}{3}-\frac{1}{b^2}\right)=\frac{9b^2}{16(b^2-3)}. \tag{5}$$

在 $\angle AOB=90°$时，$k_1k_2=-1$，即(5)右$=-1$,解得 $b=\frac{4}{5}\sqrt{3}$,斜率$-k=-\frac{1}{4}b$ $=-\frac{\sqrt{3}}{5}$.

第(Ⅱ)(Ⅲ)小题仍可用上述方法. 只是原点 O 改为焦点 $F(1,\ 0)$,即以 F 为原点. 这时 Γ 的方程为 $\frac{(x+1)^2}{4}+\frac{y^2}{3}=1$；$l$ 的方程为$\frac{x}{q-1}+\frac{y}{b}=1$. 而 A_1、F、B 三点共线,即 FA 的斜率k_1 与 FB 的斜率k_2 的和 $k_1+k_2=0$. FA、FB 的方程是

$$\frac{1}{4}\left(x+\frac{x}{q-1}+\frac{y}{b}\right)^2+\frac{1}{3}y^2=\left(\frac{x}{q-1}+\frac{y}{b}\right)^2,$$

其中 xy 的系数易算得为 $\frac{q-4}{2b(q-1)}$. 当且仅当 $q=4$ 时，A_1、F、B 三点共线.

第 14 题基本上是一道陈题. 可设 $y_n=\frac{1}{x_n}$，则

$$y_1=1,\ y_2=3,\ y_{n+1}=y_n+\frac{2}{y_n}. \tag{1}$$

显然 y_n 为正,而且递增、估计的关键即在将(1)平方得

$$y_{n+1}^2=y_n^2+\frac{4}{y_n^2}+4. \tag{2}$$

于是一方面
$$y_{n+1}^2>4n; \tag{3}$$

另一方面 $y_{n+1}^2<y_n^2+\frac{40}{9}<\cdots<y_2^2+\frac{40}{9}(n-1)=\frac{1}{9}(40n+41)$, (4)

从而
$$y_{2007}>\sqrt{4\times 2006}>10, \tag{5}$$

$$y_{2007}<\sqrt{\frac{1}{9}(40\times 2005+41)}<100. \tag{6}$$

故
$$1<|\lg x_{2007}|=\lg y_{2007}<2,\ [\lg x_{2007}]=-2.$$

这道题的估计不必过细,过细反倒是自寻烦恼.

第 15 题如何着手,是一大困难. 我认为应先用特殊值猜出结果. 这特殊值即 $x=y=z$,从而都等于$\frac{P}{a+b+c}$. 代入得 $S=\frac{P^2}{a+b+c}$. 然后证明

$$(a+b+c)S\leqslant P^2. \tag{1}$$

$$\begin{aligned}(1)\Leftrightarrow &\sum[(b+c)^2-a^2]x^2\\ &\leqslant\sum\frac{a^2y^2z^2}{x^2}+2\sum bcx^2 \qquad (2)\\ &\Leftrightarrow\sum(b^2+c^2-a^2)x^2\leqslant\sum\frac{a^2y^2z^2}{x^2}. \qquad (3)\end{aligned}$$

而
$$(3)\text{ 左}\leqslant\sum(b^2+c^2-a^2)x^2\cdot\frac{1}{2}\left(\frac{y^2}{z^2}+\frac{z^2}{y^2}\right) \tag{4}$$

$$=\sum\frac{x^2y^2}{z^2}\left(\frac{b^2+c^2-a^2}{2}+\frac{c^2+a^2-b^2}{2}\right) \tag{5}$$

$$=\sum\frac{c^2x^2y^2}{z^2}=(3)\text{ 右}.$$

其中(1)～(3),等价变形容易想到.(4)是关键,也不难想到(将x^2改造为$\frac{x^2y^2}{z^2}$,$\frac{x^2z^2}{y^2}$).(4)变(5)是恒等变形,只是重新集项(按$\frac{x^2y^2}{z^2}$,$\frac{y^2z^2}{x^2}$,$\frac{z^2x^2}{y^2}$集项)而已.锐角三角形的条件,保证了(4)中每一项为正.

参考文献

[1] 吴康,黄宗明.2007年全国高中数学联合竞赛安徽赛区预赛试题解答[J].中学数学研究(广州),2007(10).

附注:

第1题也许穷举法更简单:(A, B)只有$(\{1, 2\}, \{1, 3, 4\})$, $(\{1, 3\}, \{1, 2, 4\})$, $(\{1, 4\}, \{1, 2, 3\})$,或A、B互换,共6种情形.

第6题中国科学技术大学数学系王建伟先生给出简解:

注意到

$$\begin{aligned}&\sqrt{1+\cos\theta}+\sqrt{1+\cos(90^\circ-\theta)}\\=&\sqrt{2}\left(\cos\frac{\theta}{2}+\cos\frac{90^\circ-\theta}{2}\right)\\=&2\sqrt{2}\cos\frac{45^\circ}{2}\cos\frac{45^\circ-\theta}{2},\end{aligned}$$

$$\begin{aligned}&\sqrt{1-\cos\theta}+\sqrt{1-\cos(90^\circ-\theta)}\\=&\sqrt{2}\left(\sin\frac{\theta}{2}+\sin\frac{90^\circ-\theta}{2}\right)=2\sqrt{2}\sin\frac{45^\circ}{2}\cos\frac{45^\circ-\theta}{2},\end{aligned}$$

因此

$$\begin{aligned}A:B&=\sum_{i=0}^{10}\cos\frac{45^\circ}{2}\cos(21-2i)^\circ\div\sum_{i=0}^{10}\sin\frac{45^\circ}{2}\cos(21-2i)^\circ\\&=\cos\frac{45^\circ}{2}\div\sin\frac{45^\circ}{2}=(1+\cos45^\circ)\div\sin45^\circ=\sqrt{2}+1.\end{aligned}$$

不等式竞赛题的解法探究

这里介绍最近见到的 8 道不等式,以及它们的解法. 我们更希望通过这些不等式及其解法提倡 2 件事.

1. 提倡努力寻找好的解法

一道问题可能有多种解法,有的解法很优雅,很好,值得学习,值得提倡;有的解法丑陋,不好,应当改进,不值得提倡. 好的解法,简单自然,抓住问题的本质,直剖核心. 而差的解法,往往大兜圈子,没有抓住本质,矫揉造作,十分冗长.

2. 提倡努力寻找解题的方法

一道题,怎样才能找到解法? 这是一个很难说清楚的事情. 但如果解完题以后,回顾一下,或许可以总结出一些规律性的东西,这些东西有助于我们今后找到解题的方法. 琢磨已有的解法,也能学到一些解题的方法. 所谓"金针线迹分明在,但把鸳鸯仔细看",能学到一般的解题方法显然比只学一道题的解法高明.

例 1 已知 $(a+c)(a+b+c)<0$, 证明:

$$(b-2c)^2>4\sqrt{2}a(a+b+c).$$

证明 题设条件表明 $a+c$ 与 $a+b+c$ 异号,可分 2 种情况进行讨论.

(1) 当 $a+c<0$, $a+b+c>0$ 时, $b>-(a+c)>0$, 而 a、c 的正负仍需讨论,又可分为 2 种情况:

① 当 $a<0$ 时, $(b-2c)^2\geqslant 0>4\sqrt{2}a(a+b+c)$;

② 当 $a\geqslant 0$ 时, $-c>a\geqslant 0$, $(b-2c)^2>(b+2a)^2\geqslant 8ab>4\sqrt{2}ab>4\sqrt{2}a(a+b+c)$.

(2) 当 $a+c>0$, $a+b+c<0$ 时,可令 $a'=-a$, $b'=-b$, $c'=-c$, 条件化为

$$a'+c'<0,\ a'+b'+c'>0,$$

因此由情况(1)可得

$$(b'-2c')^2>4\sqrt{2}a'(a'+b'+c'),$$

即不等式得证.

分情况讨论(枚举法)是数学中常用的方法之一. 每一种情况均增加了一些条件:譬如情况(1)中的 $a+c<0$, $a+b+c>0$, 情况①中的 $a<0$, 情况②中的 $a\geqslant 0$. 正是由于增加了条件,使原来难以奏效的工作变得相当容易.

情况(2)可以化为情况(1). 将新的情况化为已解决的情况或较为简单的情况,这种手法称为化归,也是数学中常用的方法.

本题不少人喜欢用判别式来求解,但不及上面的解法好.

这道题是由笔者改编的,其中所证不等式右边的系数故意用一个无理数,而且故意不用最佳的系数 $8(>4\sqrt{2})$. 在证不等式时,常常有人会证最紧的(最佳的)结果,却不会证稍宽的结果,或许这也是一种定势思维.

例 2　在锐角三角形 ABC 中,已知

$$\begin{aligned}&\left(\sqrt{\frac{\sin B}{\sin A}}+\sqrt{\frac{\sin C}{\sin A}}\right)\cos A+\left(\sqrt{\frac{\sin C}{\sin B}}+\sqrt{\frac{\sin A}{\sin B}}\right)\cos B\\&\quad+\left(\sqrt{\frac{\sin A}{\sin C}}+\sqrt{\frac{\sin B}{\sin C}}\right)\cos C\\&=2(\cos A+\cos B+\cos C),\end{aligned}\tag{1}$$

判断$\triangle ABC$ 的形状,并给出证明.

解　当$\triangle ABC$ 是正三角形时,式(1)显然成立. 问题是它的逆命题,即当式(1)成立时, $\triangle ABC$ 一定是正三角形. 为此,我们证明:对任意的锐角三角形 ABC,有

$$\sum\left(\sqrt{\frac{\sin B}{\sin A}}+\sqrt{\frac{\sin C}{\sin A}}\right)\cos A\geqslant 2\sum\cos A,\tag{2}$$

仅当$\triangle ABC$ 是正三角形时,式(2)中的等号成立.

式(2)的两边都有 3 个角: A, B, C. 将它们进行“分离”,确切地说,证明仅含 A、B 的不等式

$$\sqrt{\frac{\sin B}{\sin A}}\cos A+\sqrt{\frac{\sin A}{\sin B}}\cos B\geqslant\cos A+\cos B,\tag{3}$$

类似地,一个仅含 B、C 的不等式与一个仅含 C、A 的不等式. 不妨设 $A \geqslant B$,这时

$$\cos A \leqslant \cos B,\ \sin A \geqslant \sin B.$$

又
$$\sqrt{\frac{\sin B}{\sin A}}+\sqrt{\frac{\sin A}{\sin B}} \geqslant 2\sqrt{\sqrt{\frac{\sin A}{\sin B}}\sqrt{\frac{\sin A}{\sin B}}}=2,$$

所以式(3)的左边减去右边

$$\begin{aligned}
&=\left(\sqrt{\frac{\sin A}{\sin B}}-1\right)\cos B+\left(\sqrt{\frac{\sin B}{\sin A}}-1\right)\cos A \\
&\geqslant\left(\sqrt{\frac{\sin B}{\sin A}}-1\right)\cos A+\left(\sqrt{\frac{\sin B}{\sin A}}-1\right)\cos A \\
&=\left(\sqrt{\frac{\sin A}{\sin B}}+\sqrt{\frac{\sin B}{\sin A}}-2\right)\cos A \geqslant 0,
\end{aligned}$$

即式(3)成立,当且仅当 $A=B$ 时,等号成立. 从而式(2)成立,当且仅当 $A=B=C$ 时,等号成立.

"分离"是关键. 含 3 个变量的不等式,如果能分成 3 个不等式,每个各含 2 个变量甚至各含一个变量,那么问题的难度就大大减少了(当然并不是所有含 3 个变量的不等式都能分离的).

例 3 $\{a_n\}$是由正数组成的等比数列.

(1) 证明: $\dfrac{\lg S_n+\lg S_{n+2}}{2}<\lg S_{n+1}$.

(2) 是否存在常数 $C>0$, 使得

$$\frac{\lg(S_n-C)+\lg(S_{n+2}-C)}{2}<\lg(S_{n+1}-C)? \tag{4}$$

分析 本题是一道高考题. 等比数列有 2 个重要的参数:首项 a_1 及公比 q, 将 S_n, S_{n+2}, S_{n+1} 都用 a_1 及 q 表示便可以证明第(1)小题,但需要注意分 $q=1$ 与 $q\neq 1$ 两种情况,不可漏去 $q=1$ 这种情况. 以下是更为简单的证法. 不必将 S_n, S_{n+2}, S_{n+1} 用 a_1 与 q 表示:

$$S_nS_{n+2}=S_{n+1}S_{n+2}-a_{n+1}S_{n+2}<S_{n+1}S_{n+2}-a_{n+1}\cdot qS_{n+1}=S_{n+1}^2,$$

两边取对数即得所证不等式.

对于第(2)小题,要注意式(4)与

$$(S_n-C)(S_{n+2}-C)=(S_{n+1}-C)^2 \tag{5}$$

并不等价. 式(5)与 $S_n>C(>0)$ 合在一起才与式(4)等价,不可漏去式(4)!

将 S_n 等用 a_1 与 q 表示也可解第(2)小题,但更繁(也要当心别漏掉 $q=1$ 的情况),简单的解法如下: $(S_n-C)(S_{n+2}-C)-(S_{n+1}-C)^2$ 是 C 的一次函数或常数. 当 $C=0$ 时,由第(1)小题知,函数的值为负;当 $C=S_n$ 时,值也为负,因此在区间 $[0, S_n]$ 上,函数值恒为负. 式(5)不可能成立,即满足要求的 C 不存在.

例 4　D, E, F 分别在 $\triangle ABC$ 的边 BC, CA, AB 的内部,记 $\triangle AFE$, $\triangle BDF$, $\triangle CED$, $\triangle DEF$ 的面积为 α, β, γ, δ,证明: $\frac{3}{\delta^2}\leqslant\frac{1}{\alpha\beta}+\frac{1}{\beta\gamma}+\frac{1}{\gamma\delta}$.

这是一道关于几何的不等式问题,与之相关的还有

$$\min(\alpha, \beta, \gamma)\leqslant\frac{1}{4}S_{\triangle ABC};$$

$$\min(\alpha, \beta, \gamma)\leqslant\delta.$$

证明　设 $S_{\triangle ABC}$ 为 1, $\frac{BD}{BC}=\lambda$, $\frac{CE}{CA}=\mu$, $\frac{AF}{AB}=v$, 则

$$\alpha=v(1-\mu),\ \beta=\lambda(1-v),\ \gamma=\mu(1-\lambda),$$

$$\delta=1-\alpha-\beta-\gamma=\lambda\mu v+(1-\lambda)(1-\mu)(1-v), \tag{6}$$

因此　$\alpha\beta\gamma=\lambda(1-\lambda)\mu(1-\mu)v(1-v)\leqslant\left(\frac{1}{2}\right)^2\times\left(\frac{1}{2}\right)^2\times\left(\frac{1}{2}\right)^2$.

于是
$$\min(\alpha, \beta, \gamma)\leqslant\left(\frac{1}{2}\right)^2,$$

即
$$\min(\alpha, \beta, \gamma)\leqslant\frac{1}{4}S_{\triangle ABC}.$$

由式(6)及常见不等式 $(a+b)^2\geqslant 4ab$, 得

$$\delta^2\geqslant 4\lambda\mu v(1-\lambda)(1-\mu)(1-\lambda)=4\alpha\beta\gamma. \tag{7}$$

不妨设 $\alpha\geqslant\beta\geqslant\gamma$. 如果 $\alpha\leqslant\frac{1}{4}$, 那么

$$\delta=1-\alpha-\beta-\gamma\geqslant 1-\frac{1}{4}\times 3=\frac{1}{4}\geqslant\alpha;$$

如果 $\alpha>\frac{1}{4}$，那么由式(7)，得

$$\delta^2\geqslant\beta\gamma,\ \delta\geqslant\sqrt{\beta\gamma}\geqslant\gamma,$$

因此 $$\min(\alpha,\beta,\gamma)\leqslant\delta.$$

于是，所证不等式可转化为

$$\delta^2(\alpha+\beta+\gamma)\geqslant 3\alpha\beta\gamma.$$

当 $\delta\geqslant\frac{1}{4}$ 时，令 $t=1-\delta=\alpha+\beta+\gamma$，则 $t\leqslant\frac{3}{4}$，于是

$$\begin{aligned}\delta^2(\alpha+\beta+\gamma)-3\alpha\beta\gamma&=(1-t)^2t-3\alpha\beta\gamma\geqslant(1-t)^2t-\frac{1}{9}t^3\\&=\frac{t}{9}\left(\frac{3}{2}-t\right)\left(\frac{3}{4}-t\right)\geqslant 0,\end{aligned}$$

即所证不等式成立.

当 $\delta<\frac{1}{4}$ 时，由式(7)得

$$\begin{aligned}\delta^2(\alpha+\beta+\gamma)-3\alpha\beta\gamma&=\delta^2(1-\delta)-3\alpha\beta\gamma\geqslant\delta^2(1-\delta)-\frac{3}{4}\delta^2\\&=\delta^2\left(\frac{1}{4}-\delta\right)>0,\end{aligned}$$

即所证不等式成立.

例 5 已知 $\frac{1}{2}\leqslant a,\ b,\ c,\ d\leqslant 2$，并且 $abcd=1$，求 $\left(a+\frac{1}{b}\right)\left(b+\frac{1}{c}\right)\left(c+\frac{1}{d}\right)\left(d+\frac{1}{a}\right)$ 的最大值.

解 由于 $abcd=1$，因此

$$\left(a+\frac{1}{b}\right)\left(c+\frac{1}{d}\right)=2ac+\frac{a}{d}+\frac{c}{b}=\left(\sqrt{\frac{a}{d}}+\sqrt{\frac{c}{b}}\right)^2,$$

$$\left(b+\frac{1}{c}\right)\left(d+\frac{1}{a}\right)=\left(\sqrt{\frac{b}{a}}+\sqrt{\frac{d}{c}}\right)^2.$$

从而
$$\left(a+\frac{1}{b}\right)\left(b+\frac{1}{c}\right)\left(c+\frac{1}{d}\right)\left(d+\frac{1}{a}\right)$$
$$=\left(\sqrt{\frac{a}{d}}+\sqrt{\frac{c}{b}}\right)^2\left(\sqrt{\frac{b}{a}}+\sqrt{\frac{d}{c}}\right)^2$$
$$=\left(\sqrt{\frac{a}{c}}+\sqrt{\frac{c}{a}}+\sqrt{\frac{b}{d}}+\sqrt{\frac{d}{b}}\right)^2\leqslant\left(2+\frac{1}{2}+2+\frac{1}{2}\right)^2=25,$$

其中利用了函数 $y=x+\frac{1}{x}$ 在 $\left[\frac{1}{2}, 2\right]$ 上的最大值是 $2+\frac{1}{2}$（将 $\sqrt{\frac{a}{c}}$ 或 $\sqrt{\frac{d}{b}}$ 当作 x）.

此解法是在浙江奥数网 2008 年的夏令营中一位学生给出的，比常用方法要简单得多.

例 6　设 m、n 都是大于 1 的整数，$a_{ij}(i=1, 2, \cdots, m; j=1, 2, \cdots, n)$ 是不全为 0 的 mn 个非负实数，求

$$f=\frac{m\sum_{i=1}^{m}\left(\sum_{j=1}^{n}a_{ij}\right)^2+n\sum_{j=1}^{n}\left(\sum_{i=1}^{m}a_{ij}\right)^2}{\left(\sum_{i=1}^{m}\sum_{j=1}^{n}a_{ij}\right)^2+mn\sum_{i=1}^{m}\sum_{j=1}^{m}a_{ij}^2}$$

的最大值和最小值.

解　用一些特殊的 a_{ij} 代入，猜一猜 f 的最大值和最小值. 设 $m\geqslant n$，当所有 $a_{ij}=1$ 时，$f=1$；当 $a_{11}=a_{22}=\cdots=a_{nn}=1$，$a_{ij}=0(i\neq j)$ 时，

$$f=\frac{mn+n^2}{n^2+mn^2}=\frac{m+n}{n(m+1)};$$

当 $a_{11}=1$，其余 $a_{ij}=0$ 时，$f=\frac{m+n}{1+mn}$.

比较这 3 个值，可以猜想最大值为 1，最小值为 $\frac{m+n}{n(m+1)}$. 由拉格朗日恒等式，可得

$$\left(\sum_{i=1}^{m}\sum_{j=1}^{n}a_{ij}\right)^2+mn\sum_{i=1}^{m}\sum_{j=1}^{n}a_{ij}^2-m\sum_{i=1}^{m}\left(\sum_{j=1}^{n}a_{ij}\right)^2-n\sum_{j=1}^{n}\left(\sum_{i=1}^{m}a_{ij}\right)^2$$

$$=m\sum_{i=1}^{m}\left[n\sum_{j=1}^{n}a_{ij}^2-\left(\sum_{j=1}^{n}a_{ij}\right)^2\right]-\left\{n\sum_{j=1}^{n}\left(\sum_{i=1}^{m}a_{ij}\right)^2-\left[\sum_{j=1}^{n}\left(\sum_{i=1}^{m}a_{ij}\right)\right]^2\right\}$$

$$=m\sum_{i=1}^{m}\sum_{j<k}(a_{ij}-a_{ik})^2-\sum_{j<k}\left(\sum_{i=1}^{m}a_{ij}-\sum_{i=1}^{m}a_{ik}\right)^2$$

$$=\sum_{j<k}\left\{m\sum_{i=1}^{m}(a_{ij}-a_{ik})^2-\left[\sum_{i=1}^{m}(a_{ij}-a_{ik})\right]^2\right\}$$

$$=\sum_{j<k}\sum_{i<h}(a_{ij}-a_{ik}-a_{hj}+a_{hk})^2\geqslant 0,$$

因此 $f\leqslant 1$, 即 1 为 f 的最大值.

另一方面,设法将 f 的分子、分母中的 $\sum$ 都变成相同的,然后约去,即

$$f\geqslant\frac{m\sum_{i=1}^{m}\left(\sum_{j=1}^{n}a_{ij}\right)^2+n\sum_{j=1}^{n}\left(\sum_{i=1}^{m}a_{ij}\right)^2}{n\cdot\sum_{j=1}^{n}\left(\sum_{i=1}^{m}a_{ij}\right)^2+mn\sum_{i=1}^{m}\sum_{j=1}^{n}a_{ij}^2}\geqslant\frac{m\sum_{i=1}^{m}\sum_{j=1}^{n}a_{ij}^2+n\sum_{j=1}^{n}\left(\sum_{i=1}^{m}a_{ij}\right)^2}{n\sum_{j=1}^{n}\left(\sum_{i=1}^{m}a_{ij}\right)^2+mn\sum_{i=1}^{m}\sum_{j=1}^{n}a_{ij}^2}$$

$$=1-\frac{mn\sum_{i=1}^{m}\sum_{j=1}^{n}a_{ij}^2-m\sum_{i=1}^{m}\sum_{j=1}^{n}a_{ij}^2}{n\sum_{j=1}^{n}\left(\sum_{i=1}^{m}a_{ij}\right)^2+mn\sum_{i=1}^{m}\sum_{j=1}^{n}a_{ij}^2}\geqslant 1-\frac{mn\sum_{i=1}^{m}\sum_{j=1}^{n}a_{ij}^2-m\sum_{i=1}^{m}\sum_{j=1}^{n}a_{ij}^2}{n\sum_{j=1}^{n}\sum_{i=1}^{m}a_{ij}^2+mn\sum_{i=1}^{m}\sum_{j=1}^{n}a_{ij}^2}$$

$$=1-\frac{mn-m}{n+mn}=\frac{m+n}{n(m+1)},$$

因此 f 的最小值为 $\dfrac{m+n}{n(m+1)}$.

例 7 已知正数 a, b, c 满足 $ab+bc+ca=1$. 求证:

$$\sqrt{a^3+a}+\sqrt{b^3+b}+\sqrt{c^3+c}\geqslant 2\sqrt{a+b+c}. \tag{8}$$

证明 利用已知条件可将根号内的项都写成 3 次的,即式(8)等价于

$$\sum\sqrt{a(a+b)(a+c)}\geqslant 2\sqrt{(a+b+c)(ab+bc+ca)},$$

平方后整理得

$$\sum a^3-3abc\geqslant 3(a+b)(b+c)(c+a)-2\sum\sqrt{ab(a+b)^2(a+c)(b+c)}.$$

由于上式右边

$$=(a+b)(b+c)(c+a)\left[3-2\sum\sqrt{\frac{ab}{(c+a)(c+b)}}\right]$$

$$=(a+b)(b+c)(c+a)\sum\left(\sqrt{\frac{a}{c+a}}-\sqrt{\frac{b}{c+b}}\right)^2$$

$$=\sum(a+b)[\sqrt{a(b+c)}-\sqrt{b(c+a)}]^2 \tag{9}$$

$$\leqslant\sum(a+b)[\sqrt{a(b+c)}-\sqrt{b(c+a)}]^2\cdot\frac{[\sqrt{a(b+c)}+\sqrt{b(c+a)}]^2}{ac+bc}$$

$$=\sum c(a-b)^2, \tag{10}$$

因此只需证明

$$\sum a^3-3abc\geqslant\sum c(a-b)^2,$$

而这就是舒尔(Schur)不等式

$$\sum a^3-\sum a^2(b+c)+3abc\geqslant 0.$$

本题的解法完全用代数式的变形，这似乎显得笨拙，甚至“野蛮”，却是普遍适用的一种方法. 变形中又以恒等变形为主，这可以检验学生对基本运算的熟练程度. 其中不是恒等的变形(放缩)只有一步，即由式(9)到式(10)，这一步正是本题的关键. 它不仅消除了“无理性”，而且将不等式化归为熟悉的舒尔不等式.

例 8　$n\geqslant 3$，n 元实数组$(a_1, a_2, \cdots, a_n)$，$(b_1, b_2, \cdots, b_n)$，$(c_1, c_2, \cdots, c_n)$满足

$$a_1^2+a_2^2+\cdots+a_n^2=b_1^2+b_2^2+\cdots+b_n^2=c_1^2+c_2^2+\cdots+c_n^2=1, \tag{11}$$

$$b_1c_1+b_2c_2+\cdots+b_nc_n=0. \tag{12}$$

求证：$(a_1b_1+a_2b_2+\cdots+a_nb_n)^2+(a_1c_1+a_2c_2+\cdots+a_nc_n)^2\leqslant 1.$ (13)

证明 n 元实数组可以作为 n 维向量，与三元向量相同. n 维向量 $\alpha=(a_1, a_2, \cdots, a_n)$, $\beta=(b_1, b_2, \cdots, b_n)$ 可以相加减，即 $\alpha\pm\beta=(a_1\pm b_1, a_2\pm b_2, \cdots, a_n\pm b_n)$；也可以与实数 k 相乘，即 $k\alpha=(k\alpha_1, k\alpha_2, \cdots, k\alpha_n)$；向量 α、β 还可以作数量积，即 $\alpha\cdot\beta=\sum_{j=1}^{n}a_jb_j$. 当 $\alpha\cdot\beta=0$ 时，我们说向量 α、β 互相垂直或正交，式(12)表明 β 与 $\gamma=(c_1, c_2, \cdots, c_n)$ 正交. $\alpha\cdot a=a_1^2+a_2^2+\cdots+a_n^2\geqslant 0$, $\sqrt{\alpha\cdot a}$ 被称为 α 的长，式(11)表示 α, β, γ 的长都是 1.

向量的数量积与加(减)法适合分配律，诸如此类的性质均与三维向量相同. 令 $\alpha'=\alpha-(\alpha\cdot\beta)\beta-(\alpha\cdot\gamma)\gamma$, 则由式(11)、(12)得

$$\alpha'\cdot\beta=\alpha\cdot\beta-(\alpha\cdot\beta)=0, \tag{14}$$

$$\alpha'\cdot\gamma=\alpha\cdot\gamma-(\alpha\cdot\gamma)=0, \tag{15}$$

即 α' 与 β, γ 均正交，于是

$$\begin{aligned}\alpha'\cdot\alpha'&=\alpha'\cdot[\alpha-(\alpha\cdot\beta)\beta-(\alpha\cdot\gamma)\gamma]=\alpha'\cdot\alpha\\&=[\alpha-(\alpha\cdot\beta)\beta-(\alpha\cdot\gamma)\gamma]\cdot\alpha\\&=1-(\alpha\cdot\beta)^2-(\alpha\cdot\gamma)^2.\end{aligned} \tag{16}$$

因为 $\alpha'\cdot\alpha'\geqslant 0$, 所以由式(16)得

$$(\alpha\cdot\beta)^2+(\alpha\cdot\gamma)^2\leqslant 1,$$

这也就是式(13).

本题是由笔者编制的，它的“背景”是线性代数中将向量组正规正交化的施密特(Schmidt)方法. 正规即长度为 1(式(11))，正交即不但有式(12)(β 与 γ 正交)，而且将 α 改为 α' 后，有式(14)、式(15)，亦即 α', β, γ 两两正交. 此时式(16)表示 α' 的长度的平方.

评 2009 年全国高中数学联赛试题

2009 年全国高中数学联赛试题，首次将加试增加为四道题，引起众多的关注.

此套试题中，填空题第 7 题、解答题第 11 题、加试的第一题和第三题都是好题.

题目及标准答案本文不再重复.

1　一试

1. 填空题第 7 题.

可以倒过来考虑(如图 1). 其中，

$$a=b+c=(d+e)+(e+f)=d+C_2^1e+f=\cdots.$$

$$\begin{array}{ccccc} d & & e & & f \\ & b & & c & \\ & & a & & \end{array}$$

图 1

倒推上去可见，在第 n 行时，第一行的数依次乘以 C_{n-1}^0，C_{n-1}^1，…，C_{n-1}^{n-1}，然后相加便得到第 n 行的 a. 从而，

$$\begin{aligned} a &= 1\times C_{n-1}^0+2C_{n-1}^1+\cdots+nC_{n-1}^{n-1} \\ &= \frac{1}{2}[(1+n)C_{n-1}^0+(2+n-1)C_{n-1}^1+\cdots+(n+1)C_{n-1}^{n-1}] \\ &= \frac{n+1}{2}(C_{n-1}^0+C_{n-1}^1+\cdots+C_{n-1}^{n-1}) \\ &= (n+1)2^{n-2}. \end{aligned}$$

故 $a_{100}=101\times 2^{98}$.

2. 填空题第 8 题不难,但计算可以巧一些:

$$10\times\frac{1}{2}+30\times\frac{1}{3}+\frac{1}{6}\times\left(50\times\frac{1}{6}+70\times\frac{1}{2}+90\times\frac{1}{3}\right)$$

$$=5+10+\frac{1}{6}\times\left(50+20\times\frac{1}{2}+40\times\frac{1}{3}\right)$$

$$=15+\frac{1}{6}\times\left(60+40\times\frac{1}{3}\right)=25+\frac{20}{9}\approx 27.$$

这里利用了 $\frac{1}{6}+\frac{1}{3}+\frac{1}{2}=1$. 这些简单的等式应当烂熟于胸,解题时才能得心应手.

3. 解答题第 10 题.

此题以归纳法最为简单.

当 $\alpha\neq\beta$ 时,

$$a_1=p=\alpha+\beta=\frac{\alpha^2-\beta^2}{\alpha-\beta},$$

$$a_2=p^2-q=(\alpha+\beta)^2-\alpha\beta=\frac{\alpha^3-\beta^3}{\alpha-\beta}.$$

接下去便可假设

$$a_{n-1}=\frac{\alpha^n-\beta^n}{\alpha-\beta},\ a_{n-2}=\frac{\alpha^{n-1}-\beta^{n-1}}{\alpha-\beta}.$$

故
$$a_n=(\alpha+\beta)\frac{\alpha^n-\beta^n}{\alpha-\beta}-\alpha\beta\cdot\frac{\alpha^{n-1}-\beta^{n-1}}{\alpha-\beta}$$

$$=\frac{1}{\alpha-\beta}(\alpha^{n+1}-\beta^{n+1}-\alpha\beta^n+\alpha^n\beta-\alpha^n\beta+\alpha\beta^n)=\frac{\alpha^{n+1}-\beta^{n+1}}{\alpha-\beta}.$$

$\alpha=\beta$ 的情况类似(更为简单),也可作为 $\beta\to\alpha$ 的极限.

有些学生学过递推数列,可直接利用 $a_n=A_1\alpha^n+A_2\beta^n$ 来解. 严格地说,应当给出这一公式的证明(恐怕很多学生给不出). 不过,从标准答案解法 2 看,似乎默认这是一个可用的定理.

当用这一方法定系数时,可先倒推出 $a_0=\frac{1}{q}(pa_1-a_2)=1$, 再解方程组

$$\begin{cases}A_1+A_2=1,\\ A_1\alpha+A_2\beta=\alpha+\beta,\end{cases}$$

较为简单.

4. 解答题第 11 题.

此题是求函数的极值. 应用导数方法求

$$2y'=\frac{1}{\sqrt{x+27}}+\frac{1}{\sqrt{x}}-\frac{1}{\sqrt{13-x}}$$

的极值点,即使上式为 0 的 x,它应满足方程

$$\frac{1}{\sqrt{27+x}}+\frac{1}{\sqrt{x}}=\frac{1}{\sqrt{13-x}}.$$

应当立即看到当 $x=9$ 时,各根号下的数都是平方数,而且方程变为

$$\frac{1}{6}+\frac{1}{3}=\frac{1}{2}.$$

所以, $x=9$ 是方程的根.

当 $x<9$ 时,方程左边大;当 $x>9$ 时,方程右边大. 因此,方程只有 $x=9$ 这唯一的根. 另外,导数由正而 0 而负,函数先增后减.

当 $x=9$ 时, $y=11$. 这是函数唯一的极值,既是极大值,也是最大值. 最小值则在定义区间的端点取得.

因为当 $x=0$ 与 $x=13$ 时的函数值分别为 $3\sqrt{3}+\sqrt{13}$ 与 $\sqrt{40}+\sqrt{13}$,所以,最小值为 $3\sqrt{3}+\sqrt{13}$.

标准答案是用柯西不等式求最大值. 柯西不等式众所周知,困难在于如何发现函数在 $\sqrt{x}$ 、$\sqrt{x+27}$ 、$\sqrt{13-x}$ 与 3、6、2 成比例,即 $x=9$ 时最大.

2　加试

1. 第一题是一道很漂亮的平面几何题. 之所以说漂亮是指无需用太多的计算(如用三角或解析几何),直接用纯几何方法即可解决. 但作为试题,笔者认为第(1)问就已经足够了,第(2)问可以用作考后的研讨问题.

2. 第二题是一道标准的数学分析习题,但作为中学竞赛题似较偏.

如果熟悉不等式

$$\frac{1}{n+1}<\ln\left(1+\frac{1}{n}\right)<\frac{1}{n}$$

$$\left(\text{或}\left(1+\frac{1}{n}\right)^{n}<e<\left(1+\frac{1}{n}\right)^{n+1}\right),$$

求和便得

$$\ln n=\sum_{k=1}^{n-1}\ln\left(1+\frac{1}{k}\right)>\sum_{k=1}^{n-1}\frac{1}{k+1}=\sum_{k=2}^{n}\frac{1}{k}>\sum_{k=2}^{n}\frac{k}{k^2+1}=\sum_{k=1}^{n}\frac{k}{k^2+1}-\frac{1}{2},$$

及
$$\ln n=\sum_{k=1}^{n-1}\ln\left(1+\frac{1}{k}\right)<\sum_{k=1}^{n-1}\frac{1}{k}$$

$$=1+\sum_{k=1}^{n-2}\frac{1}{k+1}\leqslant 1+\sum_{k=1}^{n-2}\frac{k}{k^2+1}<1+\sum_{k=1}^{n}\frac{k}{k^2+1}.$$

3. 第三题这道数论题很好,有一定的难度,但经过努力可以做出. 由于

$$C_m^k=\frac{m(m-1)\cdots(m-k+1)}{k!},$$

只要分子 $m(m-1)\cdots(m-k+1)$ 与 l 互质即可. 当 $k<l$ 时,这可以做到. 例如,当 $k=3$, $l=7$ 时,可以取 $m=l^n-1$ $(n\in\mathbf{N})$.

而在一般情况下, $m+1-j$ $(1\leqslant j\leqslant k)$ 可能不与 l 互质,需要与分母中的因数相约后才与 l 互质. 为此,仍取 $m=l^n-1$.

如果 $m+1-j=l^n-j$ 与 l 有公共的质因数 p,则 p 也是 j 的质因数. 设 $p^t\,|\,j$. 则当 $n>t$ 时, $p^t\,|\,(l^n-j)$.

于是,只要取 n 足够大(大于 $k!$ 中一切质因数的幂指数),则 C_m^k 与 l 互质. 这样的 n、m 显然有无穷多个.

4. 第四题题目太长,内涵却并不复杂.

先要搞清题意,特别是"三列具有性质(O)"的意义,它是指对于表中任一列 t(这列可以是所说三列中的一列)中,必有一个元素是这三列及 t 中同行元素的最小值.

于是,取 t 为这三列中的任一列,便得出这一列有一个同行的最小元素. 因

此,问题(1)成立.

不妨设 $u_i=x_{ii}(i=1, 2, 3)$, $x_{31}>x_{32}$.

如果有 $k\neq 1, 2$, 使第1、2、k 列具有性质(O),由 $x_{21}>x_{22}$, $x_{31}>x_{32}$,则必有

$$x_{11}=\min\{x_{11}, x_{12}, x_{1k}\}.$$

由性质(O), x_{22}、x_{32} 中恰有一个是同行的最小元素.

(1) 若 $x_{22}<x_{2k}$, 即 x_{22} 是同行的最小元素,则由第1、2、k 列具有性质(O)及 $x_{13}>x_{11}$, $x_{23}>x_{22}$,得 $x_{33}\leqslant x_{3k}$. 而由第1、2、3列具有性质(O)及 $x_{1k}>x_{11}$, $x_{2k}>x_{22}$,得 $x_{3k}\leqslant x_{33}$.

因此, $x_{33}=x_{3k}\Rightarrow k=3$.

(2) 若 $x_{22}>x_{2k}$, 即 x_{2k} 是同行的最小元素, x_{32} 也是同行的最小元素,此时,必有 $x_{3k}>x_{32}$, $x_{1k}>x_{11}$, 满足

$$x_{3j}>x_{32}, x_{1j}>x_{11} \quad ①$$

的 j 是存在的. 例如, $j=8$.

对于这样的 j,由性质(O)有 $x_{2j}\leqslant x_{2k}$. 因此, k 是满足式①的那些 j 中使 x_{2j} 最大的 j 从而,与1、2列合在一起具有性质(O)的列 k ($k\neq 1, 2, 3$) 至多一个(唯一性).

另一方面,如果 k 是式①中使 x_{2j} 最大的 j,则对第 t ($1\leqslant t\leqslant 9$) 列,或者 $x_{1j}\leqslant x_{11}$,或者 $x_{3t}\leqslant x_{32}$,或者 $x_{1t}>x_{11}$, $x_{3t}>x_{32}$. 从而,由 k 的最大性得 $x_{2k}\geqslant x_{2t}$.

因此,这样的 k 使1、2、k 这三列具有性质(O).

2009年国家集训队几道试题的另解

每年华东师范大学出版社都要出一本《走向IMO·数学奥林匹克试题集锦》,其中收集了当年国家集训队的测试题与选拔考试题.虽然这些题较难,但解法值得讨论.

本文从文献[1]中选出几题加以讨论.

题1 求所有的复系数多项式 $P(x)$,使得对任意和不为0的三个整数 a、b、c,均有

$$\frac{P(a)+P(b)+P(c)}{a+b+c} \tag{1}$$

为整数.

分析与解 题目给出的"对任意和不为0的三个整数 a、b、c 均有代数式(1)为整数"是一个极为有用的条件,要充分利用.

首先,取整数 $a \neq 0$, $b=c=0$,得

$$a \mid (P(a)+2P(0)). \tag{2}$$

再取整数 $a=b=c \neq 0$,得

$$a \mid P(a). \tag{3}$$

由式(2)、(3)得

$$a \mid 2P(0)$$

对任意非零整数 a 均成立.从而,

$$P(0)=0. \tag{4}$$

取整数 a、b、c 的和不为0,则 a、$b+c$、0的和也不为0.因此(利用式(4)),

$$[a+(b+c)] \mid (P(a)+P(b+c)). \tag{5}$$

又
$$(a+b+c)\mid(P(a)+P(b)+P(c)),\tag{6}$$

则由式(5)、(6)得

$$(a+b+c)\mid(P(b+c)-P(b)-P(c)).\tag{7}$$

对于固定的 b、c，a 可任意选取(只要不等于 $-(b+c)$)，从而，式(7)导出对任意整数 b、c，有

$$P(b+c)=P(b)+P(c).$$

又取 $a=b=c=1$，得 $P(1)$为整数.

设 $P(1)=k$. 则对正整数 n 有

$$P(n)=P(1)+P(n-1)=2P(1)+P(n-2)=\cdots=nP(1)=nk.$$

由多项式恒等定理得

$$P(x)=kx\ (k\in\mathbf{Z}).$$

评注　原解答上来就用有关整值多项式的定理求解，比较麻烦.

事实上，解竞赛题时，应尽量利用题目的条件，少用中学生不熟悉的大定理. 与其用复杂的方法解决简单的问题，不及用简单的方法解决复杂问题.

题 2　求证：对于任意的质数 p，满足 $p\mid(n!+1)$ 的正整数 n 的个数不超过 $cp^{\frac{2}{3}}$，其中，c 是一个与 p 无关的常数.

分析与解　显然，当 $n\geqslant p$ 时，$n!+1$ 不被 p 整除，所以，满足 $p\mid(n!+1)$ 的 $n\in\{1, 2, \cdots, p-1\}$. 将这些 n 表示在数轴上，设它们共有 k 个，即图 1 中的 $a_1<a_2<\cdots<a_k$.

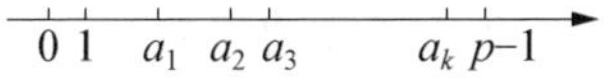

0 1 a_1 a_2 a_3 a_k $p-1$

图 1

考虑这 k 个点构成区间 $[1, p-1]$ 的 $k-1$ 个子区间：

$$[a_1, a_2], [a_2, a_3], \cdots, [a_{k-1}, a_k].$$

首先，长度不小于 $p^{\frac{1}{3}}$ 的区间个数不超过

$$p \div p^{\frac{1}{3}} = p^{\frac{2}{3}}.$$

其次，长度 h 小于 $p^{\frac{1}{3}}$ 的区间 $[a, a+h]$，由于端点满足

$$a! + 1 \equiv 0 \pmod p,$$
$$(a+h)! + 1 \equiv 0 \pmod p.$$

故
$$(a+1)(a+2)\cdots(a+h) \equiv 1 \pmod p. \tag{8}$$

式(8)是 a 的 h 次同余方程. 由拉格朗日定理(它与普通的 h 次方程至多 h 个根相当)知，式(8)至多有 h 个根.

因此，长度小于 $p^{\frac{1}{3}}$ 的区间个数不超过

$$\sum_{h=1}^{[p^{\frac{1}{3}}]} h = 1 + 2 + \cdots + [p^{\frac{1}{3}}] \leqslant p^{\frac{2}{3}} - 1.$$

综上，
$$k - 1 \leqslant p^{\frac{2}{3}} + (p^{\frac{2}{3}} - 1) = 2p^{\frac{2}{3}} - 1.$$

从而，满足要求的 n 的个数 $k \leqslant 2p^{\frac{2}{3}}$.

评注 原解答考虑各区间长度之和，比较麻烦，得到的常数 c 也远比 2 大(虽然 c 的大小并不重要). 其实，与 n 的个数密切相关的是区间的个数，而不是区间的长度.

题 3 设 n 是合数. 证明：存在正整数 m，满足 $m \mid n$ $(m \leqslant \sqrt{n})$，且 $d(n) \leqslant d^3(m)$，其中，$d(k)$表示正整数 k 的正约数的个数.

分析与解 可以先考虑一些特殊情况.

(1) $n = p^{\alpha}$，p 为质数，α 为大于 1 的整数.

此时，取 $m = p^{\left[\frac{\alpha}{2}\right]}$. 则 $m \mid n$ $(m \leqslant \sqrt{n})$，并且

$$\begin{aligned} d^3(m) &= \left(1 + \left[\frac{\alpha}{2}\right]\right)^3 \\ &\geqslant \left(1 + \left[\frac{\alpha}{2}\right]\right)^2 \left(1 + \frac{\alpha - 1}{2}\right) \\ &\geqslant 4\left(1 + \frac{\alpha - 1}{2}\right) = 2(\alpha + 1) > \alpha + 1 = d(n). \end{aligned}$$

(2) $n=p_1p_2$, $p_1<p_2$ 均为质数.

此时,取 $m=p_1$. 则

$$d^3(m)=2^3=8>4=d(n).$$

(3) $n=p_1p_2\cdots p_s$, $p_1<p_2<\cdots<p_s$ 均为质数, $s\geqslant 3$.

此时,取 $m=p_1p_2\cdots p_{\left[\frac{s}{2}\right]}$. 则

$$d^3(m)=2^{3\left[\frac{s}{2}\right]}\geqslant 2^{\frac{3(s-1)}{2}}\geqslant 2^{\frac{3s-s}{2}}=2^s=d(n).$$

对于一般情况,

$$n=p_1p_2\cdots p_sq_1^{\alpha_1}q_2^{\alpha_2}\cdots q_k^{\alpha_k},$$

其中, $p_1<p_2<\cdots<p_s$, $q_1<q_2<\cdots<q_k$ 都是质数,且互不相同, α_1, α_2, $\cdots$, α_k 都是大于 1 的整数,并且 $k\geqslant 1$.

取 $m=p_1p_2\cdots p_{\left[\frac{s}{2}\right]}q_1^{\left[\frac{\alpha_1}{2}\right]}q_2^{\left[\frac{\alpha_2}{2}\right]}\cdots q_k^{\left[\frac{\alpha_k}{2}\right]}$. 则显然,

$$m\mid n\ (m\leqslant\sqrt{n}).$$

而

$$\begin{aligned}&d(n)\leqslant d^3(m)\\&\quad\Leftrightarrow 2^s(\alpha_1+1)(\alpha_2+1)\cdots(\alpha_k+1)\\&\quad\leqslant 2^{3\left[\frac{s}{2}\right]}\left(\left[\frac{\alpha_1}{2}\right]+1\right)^3\cdots\left(\left[\frac{\alpha_k}{2}\right]+1\right)^3.\end{aligned}$$

由(1)中的证明知,只需证明

$$2^s\leqslant 2^{3\left[\frac{s}{2}\right]}\times 2^k.$$

最后一个不等式在 $s=1$、2 及 $s\geqslant 3$ 时均成立.

评注　我们给出了 m 的具体表达式(当然要借助 n 的标准分解式).

本题从特殊到一般,是解题的常用方法. 前面特殊的情况可以启迪同学们获得一般情况的解法. 也表明一般情况的解法并非凭空从天而降,而是由实际的、具体的例子逐步一般化而得到的.

题 4　设 $a>b>1$, b 为奇数, n 为正整数. 若 $b^n\mid(a^n-1)$,求证: $a^b>\dfrac{3^n}{n}$.

分析与解 因为 b 为大于 1 的奇数,所以, b 必有奇质因数 p. 不妨设 b 就是 p,即设

$$p^n \mid (a^n - 1). \tag{9}$$

往证:
$$a^p > \frac{p^n}{n}.$$

$$\text{式}(9) \Leftrightarrow a^n \equiv 1 (\bmod p^n).$$

因此,
$$(a,\ p) = 1.$$

而由费马小定理得

$$a^{p-1} \equiv 1 (\bmod p).$$

设正整数 d 满足

$$a^d \equiv 1 (\bmod p),$$

且最小. 则熟知
$$d \mid n,\ d \mid (p-1).$$

设 $n = md\ (m \in \mathbf{N}_+)$,

$$a^d = lp^t + 1\ (t \in \mathbf{N}_+,\ p \nmid l).$$

用 $v_p(m)$表示 m 中 p 的次数(即 $p^{v_p(m)} \parallel m$).

如果能够证明

$$v_p(a^n - 1) = t + v_p(m), \tag{10}$$

那么,由于 $p^n \mid (a^n - 1)$, 则

$$t + v_p(m) \geqslant n.$$

从而, $a^p > a^d > p^t \geqslant p^{n - v_p(m)} \geqslant p^{n - v_p(n)} \geqslant \dfrac{p^n}{n}$.

因此,只需证明式(10).

事实上,

$$\begin{aligned} a^n - 1 &= a^{md} - 1 = (lt^t + 1)^m - 1 \\ &= lmp^t + \frac{lm(m-1)}{2} p^{2t} + \cdots + l \cdot \frac{m}{k} \mathrm{C}_{m-1}^{k-1} p^{kt} + \cdots. \end{aligned}$$

在 $k \geqslant 3$ 时,

$$v_p\left(l \cdot \frac{m}{k} \mathrm{C}_{m-1}^{k-1} p^{kt}\right) \geqslant kt + v_p(m) - v_p(k)$$
$$\geqslant t + v_p(m) + k - 1 - v_p(k) > t + v_p(m).$$

(当 $v_p(k)=0$ 时显然;当 $v_p(k)>0$ 时, $k \geqslant p^{v_p(k)} > 1 + v_p(k)$)

又 $$v_p\left(\frac{lm(m-1)}{2} p^{2t}\right) \geqslant v_p(m) + 2t > t + v_p(m),$$

因此, $$v_p(a^n - 1) = v_p(lmp^t) = t + v_p(m).$$

评注　计算 p 在 $(1+lp^t)^m$ 中的次数,上面的方法是常用的,其中,组合数 C_m^k 写成$\frac{m}{k}\mathrm{C}_{m-1}^{k-1}$ 有利于估算 p 的次数.

题 5　设非负实数 a_1, a_2, a_3, a_4,满足 $a_1+a_2+a_3+a_4=1$. 求证:

$$\max\left\{\sum_{i=1}^{4}\sqrt{a_i^2 + a_i a_{i-1} + a_{i-1}^2 + a_{i-1}a_{i-2}},\right.$$
$$\left.\sum_{i=1}^{4}\sqrt{a_i^2 + a_i a_{i+1} + a_{i+1}^2 + a_{i+1}a_{i+2}}\right\} \geqslant 2,$$

其中, $a_{i+4}=a_i$ 对所有整数 i 成立.

分析与解　不喜欢用太多的下标,改记 a_1, a_2, a_3, a_4 为 a, b, c, d. 问题即为:

设非负实数 a, b, c, d,满足

$$a+b+c+d=1.$$

往证: $\max\{A, B\} \geqslant 2$,

其中, $A=\sum\sqrt{a^2+ad+d^2+dc}$, $B=\sum\sqrt{a^2+ab+b^2+bc}$.

事实上,有

$$A-2=\sum\left[\sqrt{a^2+ad+d^2+dc}-(a+d)\right]$$
$$=\sum\frac{d(c-a)}{\sqrt{a^2+ad+d^2+dc}+(a+d)},$$

$$B-2=\sum\left[\sqrt{a^2+ab+b^2+bc}-(a+b)\right]$$
$$=\sum\frac{b(c-a)}{\sqrt{a^2+ab+b^2+bc}+(a+b)}$$
$$=\sum\frac{d(a-c)}{\sqrt{c^2+cd+d^2+da}+(c+d)}.$$

故

$$(A-2)+(B-2)=\sum\left\{d(a-c)\left[\frac{1}{\sqrt{c^2+cd+d^2+da}+(c+d)}-\frac{1}{\sqrt{a^2+ad+d^2+dc}+(a+d)}\right]\right\}.$$

显然，$a-c$ 与后面的因式同号，且上式轮换对称，故每一项都非负.

因此，$(A-2)+(B-2)\geqslant 0$.

故 $A\geqslant 2$ 或 $B\geqslant 2$ 至少有一个成立.

评注 本题的证明只用了“分子有理化”，并无特别的技巧.

题 6 设 m 是大于 1 的整数，n 是一个奇数，且 $3\leqslant n<2m$. 数 $a_{i,j}$(i、$j\in\mathbf{N}$, $1\leqslant i\leqslant m$, $1\leqslant j\leqslant n$) 满足：

(1) 对于任意的 $1\leqslant j\leqslant n$, $a_{1,j}$, $a_{2,j}$, $\cdots$, $a_{m,j}$ 是 1, 2, $\cdots$, m 的一个排列.

(2) 对于任意的 $1\leqslant i\leqslant m$, $1\leqslant j\leqslant n-1$，有 $|a_{i,j}-a_{i,j+1}|\leqslant 1$.

求 $M=\max\limits_{1\leqslant i\leqslant m}\sum\limits_{j=1}^{n}a_{i,j}$ 的最小值.

分析与解 设 $n=2k+1$ $(k\geqslant 1)$. 不妨设第 $k+1$ 列的 m 在第 1 行.

如果 $a_{1,k}=m$，则由条件(1)有

$$a_{1,k-1}\geqslant m-1,\ a_{1,k-2}\geqslant m-2,\ \cdots\cdots$$
$$a_{1,1}\geqslant m-(k-1),$$
$$a_{1,k+2}\geqslant m-1,\ a_{1,k+3}\geqslant m-2,\ \cdots\cdots$$
$$a_{1,n}\geqslant m-k,$$

$$\sum_{j=1}^{n}a_{1,j}\geqslant mn-[1+2+\cdots+(k-1)+1+2+\cdots+k]$$
$$=mn-[1+3+5+\cdots+(2k-1)]$$
$$=mn-k^2. \tag{11}$$

如果 $a_{1,k+2}=m$，同样得式(11).

如果 $a_{1,k}$、$a_{1,k+2}$ 都不是 m,那么,设第 k、$k+2$ 列的 m 分别在第 i、t 行.此时,

$$a_{i,k+1}=m-1,\ a_{t,k+1}=m-1.$$

但第 $k+1$ 列只有一个 $m-1$,从而，$i=t$.

与上面推理相同,则有

$$a_{i,k-1}\geqslant m-1,\ a_{i,k-2}\geqslant m-2,\ \cdots\cdots$$
$$a_{i,1}\geqslant m-(k-1),$$
$$a_{i,k+3}\geqslant m-1,\ a_{i,k+4}\geqslant m-2,\ \cdots\cdots$$
$$a_{i,n}\geqslant m-k+1.$$

与式(11)相比较,显然,

$$\sum_{j=1}^{m}a_{i,j}>mn-k^2.$$

因此,得到

$$M\geqslant mn-k^2. \tag{12}$$

困难在于构造一个实例使式(12)中等号成立.

下面用 $b_{i,j}=m-a_{i,j}$ 代替 $a_{i,j}$.

此时，$m\times n$ 的数表 $(b_{ij})_{1\leqslant i\leqslant m,\ 1\leqslant j\leqslant n}$ 中,每一列都是 $0,\ 1,\ \cdots,\ m-1$ 的一个排列,每一行相邻两个数的差不超过 1.

接下来要证明:行和的最小值可以是 k^2.

$k=1$ 的情况不难(见图 2 的数表).

$m-1$	$m-1$	$m-1$
	……	
3	3	3
2	2	2
1	0	0
0	1	1

图 2

$k=2$ 的情况也不难(见图 3 的数表).

$m-1$	$m-1$	$m-1$	$m-1$	$m-1$
		……		
4	4	4	4	4
3	3	3	3	3
2	1	0	0	1
0	0	1	2	2
1	2	2	1	0

图 3

于是,希望构造一个 $k+1$ 行、$2k+1$ 列的数表,每列是 $0, 1, \cdots, k$ 的排列,每行相邻两数的差不超过 1,而且行和的最小值是 k^2. 再在该数表上面再加 $m-(k+1)$ 行,各行的数相同,分别是 $k+1, k+2, \cdots, m-1$,就得到 $m\times n$ 的数表.

下面就构造所需的 $(k+1)\times(2k+1)$ 的数表.

先构造一个 $(k+1)\times(2k+2)$ 的数表,其中,第 1 行依次为 $k, k-1, \cdots, 0, 0, 1, \cdots, k$,以后逐行将第 i $(i=1, 2, \cdots, k)$行、第 j $(j=3, 4, \cdots, 2k+2)$ 列的数放在第 $i+1$ 行、第 $j-2$ 列上,第 1、2 列的数放在下一行第 $2k+1$、$2k+2$ 列上. 完成图 4 的数表(以 $k=5$ 为例).

5	4	3	2	1	0	0	1	2	3	4	5
3	2	1	0	0	1	2	3	4	5	5	4
1	0	0	1	2	3	4	5	5	4	3	2
0	1	2	3	4	5	5	4	3	2	1	0
2	3	4	5	5	4	3	2	1	0	0	1
4	5	5	4	3	2	1	0	0	1	2	3

图 4

再将图 4 的数表去掉最后一行,得到所需的 $(k+1)\times(2k+1)$ 的数表.

显然,图 4 的数表满足性质(1)(2),且数表中各行的和均为 k^2+k. 从而,所构造的数表中各行的和互不相同,组成集合 $\{k^2, k^2+1, \cdots, k^2+k\}$,其中,第1行的和为 k^2.

评注　用 $b_{i,j}$ 代替 $a_{i,j}$ 只是为叙述方便,并非实质性的改进. 但将行数 m 减为 $k+1$,加强了结果,而且省去讨论 m、n 大小的麻烦.

构造法,光看书往往看不清构造的理由,应当从简单的情形做起,发现规律,推广至一般. 自己动手做最为重要.

当然,还有一些试题,请参见文献[1].

文献[1]中解法虽未必尽善尽美,有些解法值得商榷,也有些印刷错误(如将⌈　⌉印成[　]),但它提供了一份原始资料,这是最宝贵的. 既使是上述不足之处,也可以锻炼同学们的阅读能力,看看能否发现错误或不妥之处,能否改进原来的解法,甚至找出新的解法.

总之,它可以启迪我们思考,努力地独立思考.

参考文献

[1]　2009年IMO中国国家集训队教练组. 走向IMO·数学奥林匹克试题集锦(2009)[M]. 上海:华东师范大学出版社,2009.

[2]　邹明. 2009年中国国家集训队几道试题另解[J]. 中等数学,2010(4).

评 2010 年全国高中数学联赛试题

1　一试

2010 年全国高中数学联赛一试整体看来还是相当平稳的，若要选一道最佳题目，第 11 题可以算是新颖（以前联赛中未出现过）的了，解答也不是很难.

1. 填空题第 8 题.

可求方程 $x+y+z=6n$（n 为自然数）的满足 $x\leqslant y\leqslant z$ 的正整数解 (x, y, z) 的个数.

此时，$x=y=z$ 的解为 $x=y=z=2n$. x、y、z 中恰有两个相等的解有 $3(3n-2)$ 个.

设 x、y、z 互不相等的解有 $6k$ 个. 则

$$1+3(3n-2)+6k=\mathrm{C}_{6n-1}^{2}.$$

故满足 $x\leqslant y\leqslant z$ 的正整数解的个数为

$$1+(3n-2)+k=\frac{\mathrm{C}_{6n-1}^{2}+5+3(3n-2)}{6}=3n^{2}.$$

当 $n=\frac{2010}{6}=335$ 时，答案为

$$3\times 335^{2}=336\,675.$$

由此可见，将 2010 换成 $6n$，不但得到更一般的结论，而且避免了繁琐的数的计算.

将 $6n$ 换成 $6n+1$, $6n+2$, $6n+3$, $6n+4$, $6n+5$, 相应的结论为

$$3n^{2}+n,\ 3n^{2}+2n,\ 3n^{2}+3n+1,\ 3n^{2}+4n+1,\ 3n^{2}+5n+1.$$

今年与函数有关的题目比较多，如第 2 题、第 5 题、第 9 题.

2. 解答题第 9 题.

此题与契比雪夫三角多项式有关.

设 $\cos\theta=x$,则 $\cos n\theta$ 是 x 的多项式,记 $\cos n\theta=T_n(x)$.

$T_n(x)$称为契比雪夫多项式.

例如, $\cos 2\theta=2\cos^2\theta-1$, $T_2(x)=2x^2-1$.

命题 1 设 $P(x)$是首项系数为 1 的 n 次多项式. 则在区间$[-1, 1]$上, $|P(x)|$的最大值大于或等于$\frac{1}{2^{n-1}}$,并且在等号成立时,

$$P(x)=\frac{1}{2^{n-1}}T_n(x).$$

作代换 $t=2x-1$ 就可将区间$[-1, 1]$换成区间$[0, 1]$. 此时,命题 1 成为

命题 2 设 $P(x)$是首项系数为 2^n 的 n 次多项式. 则在区间 $[0, 1]$ 上, $|P(x)|$的最大值大于或等于 $\frac{1}{2^{n-1}}$,并且在等号成立时, $P(x)=\frac{1}{2^{n-1}}T_n(2x-1)$.

由命题 2,如果多项式 $P(x)$首项系数为 A,并且在区间 $[0, 1]$ 上, $|P(x)|\leqslant 1$,则$1\geqslant\frac{A}{2^n}\times\frac{1}{2^{n-1}}$. 所以, A 最大为 2^{2n-1},而且在$P(x)=T_n(2x-1)$时 A 取最大值.

特别地,在第 9 题中, $f'(x)=3ax^2+\cdots$, 所以, a 的最大值为$\frac{8}{3}$. 此时,

$$f'(x)=2(2x-1)^2-1=8x^2-8x+1,$$

$$f(x)=\frac{8}{3}x^3-4x^2+x+m,$$

其中, m 为常数.

其他题目与标准答案出入不大,在此不再赘述.

2 二试

1. 第一题.

此题可以分为两个部分,更好的表述是"……当且仅当 A、B、C、D 四点共圆时, $OK\perp MN$". 前一部分已知四点共圆,证明两线垂直,很多书上出现

过.如华东师范大学出版社出版的《高中数学竞赛多功能题典》题 8.2.7,《走向 IMO 数学奥林匹克试题集锦(2009)》中国国家队选拔考试第一天第一题中也包含这一部分.事实上,它是一个定理,可见《近代欧氏几何学》第 143 页.

熟悉射影几何的人知道, MN 是点 K 关于$\odot O$ 的极线,而圆心与极点的连线垂直于极线是一个熟知的定理(如用解析几何,点 $K(x_0, y_0)$关于圆 $x^2+y^2=1$ 的极线为 $x_0x+y_0y=1$, 它显然与 OK 垂直).

反过来,四点不共圆时,两线不垂直.首先,应当知道如果 $BC \parallel MN$, 则 K 为 BC 的中点.但已知 K 不是 BC 的中点,则 BC 与 MN 不平行.设 BC 与 MN 交于点 R.利用标准答案所绘的图及相关记号,对$\triangle AMN$ 与$\triangle ECB$ 用笛沙格定理立即得出 PQ 与 MN 交于点 R(可见,这是一道射影几何的问题).而由前一部分可导出 $OK\perp PQ$, 所以, OK 与 MN 不垂直.

2. 第二题.

可从简单情况开始.

$$f(r)=k(k+1)+\frac{k+1}{2}.$$

在 k 为奇数时, $m=1$.

在 k 为偶数 $2h$ 时, $f(r)=A+\frac{1}{2}$, 其中,

$$A=2h(k+1)+h=h(2k+3).$$

于是, $f(f(r))=A(A+1)+\frac{A+1}{2}$.

在 A 为奇数,即 h 为奇数时, $m=2$.

接下去,当然用数学归纳法了.

设 $k=2^n h$ (n 为非负整数, h 为奇数).假设在 $n=t$ 时, $m=t+1$.则在 $n=t+1$ 时,有

$$f(r)=k(k+1)+\frac{k+1}{2}=2^{t+1}h(k+1)+2^t h+\frac{1}{2}=2^t h(2k+3)+\frac{1}{2},$$

其中, $h(2k+3)$ 是奇数.

因此,由归纳假设得

$$f^{(t+2)}(r)=f^{(t+1)}(f(r))=f^{(t+1)}\left(2^t h(2k+3)+\frac{1}{2}\right)$$

为整数,即 $m=t+2$.

从而,对所有的 n,有 $m=n+1$.

3. 第三题.

这是一道好题.好在不用各种著名不等式,不依赖知识,而要依靠良好的感觉,直接去比较、估计.

首先,可将 $\sum_{i=1}^{n}a_k$ 中的每一项换成相等的 A_n,然后,再估计 $|A_n-A_k|$.

当 $A_k>A_n$ 时,

$$\begin{aligned}|A_n-A_k|=A_k-A_n&\leqslant\frac{a_1+a_2+\cdots+a_k}{k}-\frac{a_1+a_2+\cdots+a_k}{n}\\&=(a_1+a_2+\cdots+a_k)\left(\frac{1}{k}-\frac{1}{n}\right)\leqslant\frac{n-k}{n}.\end{aligned}$$

当 $A_n\geqslant A_k$ 时,

$$\begin{aligned}|A_n-A_k|=A_n-A_k&=\frac{a_1+a_2+\cdots+a_n}{n}-\frac{a_1+a_2+\cdots+a_k}{k}\\&\leqslant\frac{a_{k+1}+a_{k+2}+\cdots+a_n}{n}\leqslant\frac{n-k}{n}.\end{aligned}$$

故

$$\begin{aligned}&\left|\sum_{i=1}^{n}a_k-\sum_{i=1}^{n}A_k\right|=\left|\sum_{k}(A_n-A_k)\right|\\\leqslant&\sum_{i=1}^{n}|A_n-A_k|\leqslant\sum_{k=1}^{n}\frac{n-k}{n}=\sum_{h=0}^{n-1}\frac{h}{n}=\frac{n-1}{2}.\end{aligned}$$

此题不宜用归纳法(因 n 出现在分母,而且由 $a_i\leqslant1$, $a_j\leqslant1$,不能得出 $a_i+a_j\leqslant1$).

4. 第四题.

可以用递推数列.

首先重新标数.原来标 0 并染红色的点标 0,原来标 0 并染蓝色的点标 1,原来标 1 并染红色的点标 2,原来标 1 并染蓝色的点标 3.这样,相邻的点标的数和不为 3.

解法 1 去掉边 A_nA_1. 将 $A_i\ (1\leqslant i\leqslant n)$ 标上 0、1、2、3,使得相邻的点标的数和不为 3,并且 A_1 标 0. 设此时 A_n 标 0、1、3 的标法分别有 b_n、c_n、d_n 种. 则 A_n 标 2 的标法也是 c_n 种.

易知

$$b_2=1,\ b_3=3,\ b_4=7;$$

$$c_2=1,\ c_3=2,\ c_4=7;$$

$$d_2=0,\ d_3=2,\ d_4=6.$$

$$b_n=b_{n-1}+2c_{n-1},$$

$$c_n=b_{n-1}+c_{n-1}+d_{n-1},$$

$$d_n=2c_{n-1}+d_{n-1}.$$

消去 b_{n-1}、b_n 得

$$c_{n+1}-2c_n-c_{n-1}=d_n-d_{n-1}.$$

再消去 c_{n+1}、c_n、c_{n-1} 得

$$d_{n+2}=3d_{n+1}+d_n-3d_{n-1}.$$

于是, $4d_n=3^{n-1}-2-(-1)^n$.

设本题答案为 a_n. 则

$$a_n=4\times 3^{n-1}-4d_n=3^n+2+(-1)^n.$$

解法 2 去掉边 A_nA_1. 将 $A_i(1\leqslant i\leqslant n)$ 标上 0、1、2、3,使得相邻的点标的数不同,并且 A_n 与 A_1 所标的数不同.

设此时的标法为 x_n 种. 则

$$x_2=4\times 3,$$

且

$$x_n=4\times 3^{n-1}-x_{n-1}.$$

故 $x_n-3^n=-(x_{n-1}-3^{n-1})=\cdots=(-1)^n(x_2-3^2)=(-1)^n\times 3$.

因此,在 n 为偶数时, $x_n=3^n+3$. 而且此时恢复边 A_nA_1,并将下标为偶数的点标的数 a 改为 $3-a$, 就得到合乎本题要求的标法. 反之亦然. 故在 n 为偶

数时，答案是 3^n+3.

去掉边 A_nA_1. 将 A_i $(1\leqslant i\leqslant n)$ 标上 0、1、2、3，使得相邻的点标的数不同，并且 A_n 与 A_1 所标的数和不为 3.

设此时的标法为 y_n 种. 则

$$y_3=4\times(3+2+2)=28=3^3+1,$$

且
$$y_n=4\times 3^{n-1}-y_{n-1}.$$

同样可得 n 为奇数时，$y_n=3^n+1$. 而且此时恢复边 A_nA_1，并将下标为偶数的点标的数 a 改为 $3-a$，就得到合乎本题要求的标法. 反之亦然. 故在 n 为奇数时，答案是 3^n+1.

评 2011 年全国高中数学联赛试题

2011 年全国高中数学联赛试题,无论是一试还是加试都比较难.

近年来,我国实行新的课程标准,中学生的数学水平似乎下降,尤其是解题能力比以前差了.一方面,初中对代数式的运算、恒等变形都没有足够的练习,以至于到了高中,稍有技巧的运算便无法进行,不等式更难以处理.最糟糕的是很多学生初中毕业竟不知道什么是数学证明.另一方面,高中数学竞赛试题似乎越来越难.

2011 年的题,恐怕参加的学生得分普遍较低.这就会影响学生的积极性.

1 一试

每道填空题都有一点技巧或困难.其实可以出两三道容易的题,"送"点分给学生,即开始的几道题尽量贴近课本,类似高考题,这样,大多数同学都能做出来.

1. 第 2 题.

求函数 $f(x)=\dfrac{\sqrt{x^2+1}}{x-1}$ 的值域可以不用三角函数.

令 $t=x-1$, 则原式化为

$$g(t)=\frac{\sqrt{t^2+2t+2}}{t}.$$

在 $t>0$ 时,

$$g(t)=\sqrt{1+\frac{2}{t}+\frac{2}{t^2}}\in(1,+\infty);$$

在 $t<0$ 时,

$$g(t)=-\sqrt{1+\frac{2}{t}+\frac{2}{t^2}}=-\sqrt{2\left(\frac{1}{t}+\frac{1}{2}\right)^2+\frac{1}{2}}\in\left(-\infty,-\frac{\sqrt{2}}{2}\right].$$

2. 第 3 题.

此题得出 $ab=1$ 后,立即有

$$\log_a b=\log_a a^{-1}=-1.$$

不必求出 a、b.

3. 第 4 题.

本题只需注意 x^3、x^5 都是增函数. 所以,当且仅当 $\cos\theta<\sin\theta$ 时,

$$\cos^5\theta-\sin^5\theta<0<7(\sin^3\theta-\cos^3\theta),$$

即
$$\theta\in\left(\frac{\pi}{4},\frac{5\pi}{4}\right).$$

4. 第 6 题.

本题△ABD 是正三角形. 于是,设球心 O 在它上面的射影为 N(如图 1). 则

$$DN=3\times\frac{2}{3}\times\frac{\sqrt{3}}{2}=\sqrt{3}$$

是△ABD 的外接圆半径,直径 $DH=2\sqrt{3}$.

设点 C 在 DA、平面 ABD 上的射影分别为 E、F(如图 2). 则 DF 平分∠ADB,

$$DE=1,\ DF=\frac{2}{\sqrt{3}}.$$

因 $CD^2=4=DF\cdot DH$,所以,点 C 在以 DH 为直径的外接球上,即四面体 $ABCD$ 的外接球直径为 DH(如图 3),半径长为$\sqrt{3}$(图 1 中的点 N 即球心 O).

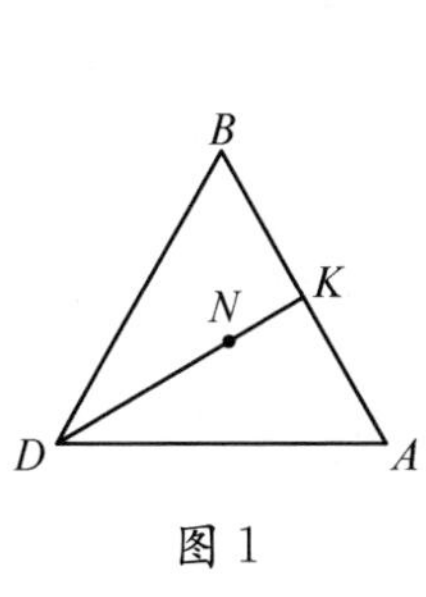

图 1

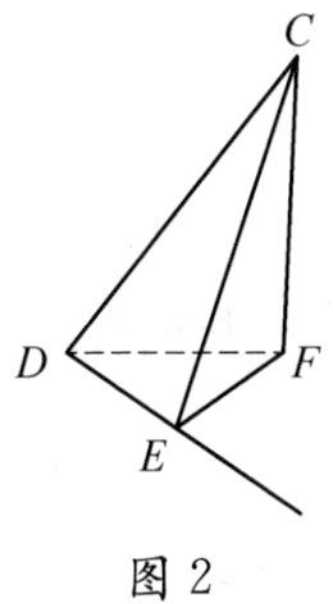

图 2

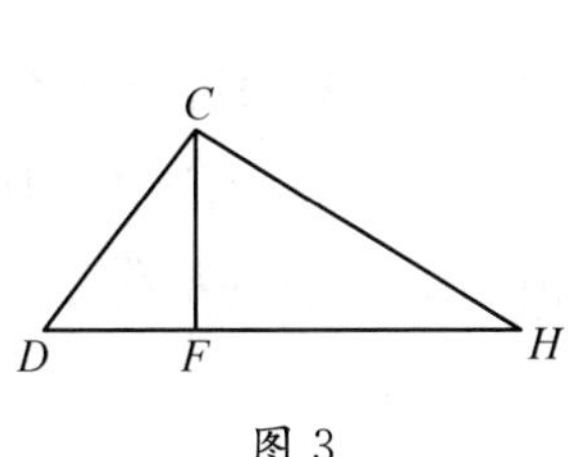

图 3

5. 第 7 题.

此题点 C 的坐标以用 $(x,\ y)$ 表示为好，用参数坐标 $(t^2,\ 2t)$ 反而不整齐.

6. 第 10 题.

此题设 $c_n=\dfrac{2(t^n-1)}{a_n+1}$，$c_1=1$ 为好.

易得 $c_{n+1}=1+c_n$，$c_n=n$.

7. 第 11 题.

此题直接计算较为麻烦. 先将原点移到点 P 较为简单. 此时，椭圆方程为

$$\frac{(x-3\sqrt{2})^2}{36}+\frac{(y-\sqrt{2})^2}{4}=1,$$

即
$$\frac{x^2}{36}+\frac{y^2}{4}-\frac{\sqrt{2}}{6}x-\frac{\sqrt{2}}{2}y=0. \tag{1}$$

直线 l 的方程为

$$y-\sqrt{2}=\frac{1}{3}(x-3\sqrt{2})+m,$$

即
$$\frac{1}{m}\left(y-\frac{1}{3}x\right)=1. \tag{2}$$

由式(1)、(2)得

$$\frac{x^2}{36}+\frac{y^2}{4}-\left(\frac{\sqrt{2}}{6}x+\frac{\sqrt{2}}{2}y\right)\times\frac{1}{m}\left(y-\frac{1}{3}x\right)=0. \tag{3}$$

因为式(3)的左边是 x、y 的二次齐次式，所以，式(3)表示两条过新原点 P 的直线.

由于点 A 坐标同时适合式(1)、(2)，故它也适合式(3). 同样，点 B 坐标也适合式(3).

故式(3)表示的两条直线就是 PA、PB.

由韦达定理知，这两条直线斜率的和是式(3)中 xy 的系数除以 y^2 的系数的相反数. 但 xy 的系数为

$$-\frac{1}{m}\left(\frac{\sqrt{2}}{6}-\frac{\sqrt{2}}{2}\times\frac{1}{3}\right)=0,$$

因此，PA、PB 与新 y 轴的夹角相等，即直线 $x=3\sqrt{2}$ 是$\angle APB$ 的平分线.

故$\triangle APB$ 的内心在直线 $x=3\sqrt{2}$ 上.

当 $\angle APB=60^\circ$ 时，在以 P 为原点的新坐标系中 PA、PB 的方程为

$$y=\pm\sqrt{3}x. \tag{4}$$

将式(4)代入式(1)得 A、B 的横坐标 x_1、x_2 为

$$x_{1,2}=\left(\frac{\sqrt{2}}{6}\pm\frac{\sqrt{2}}{2}\times\sqrt{3}\right)\div\left(\frac{1}{36}+\frac{3}{4}\right)=\frac{9}{7}\left(\frac{\sqrt{2}}{6}\pm\frac{\sqrt{6}}{2}\right).$$

则　　$PA=\sqrt{x_1^2+y_1^2}=\sqrt{x_1^2+3x_1^2}=2|x_1|$，$PB=2|x_2|$.

故

$$\begin{aligned}S_{\triangle APB}&=\frac{1}{2}PA\cdot PB\sin 60^\circ=\sqrt{3}\,|x_1,x_2|\\&=\sqrt{3}\left(\frac{9}{7}\right)^2\left[\left(\frac{\sqrt{6}}{2}\right)^2-\left(\frac{\sqrt{2}}{6}\right)^2\right]=\frac{117\sqrt{3}}{49}.\end{aligned}$$

评注　在直线 l 成为斜率为$\frac{1}{3}$的切线时，点 A、B 成为切点 M，$\angle APB$ 的平分线成为 PM. 因此，可以猜出内心在直线 PM 上. 而 OM 是与直线 l 共轭的方向，其斜率 k 满足

$$k\times\frac{1}{3}=-\frac{4}{36}\Rightarrow k=-\frac{1}{3}.$$

因此，点 M 与点 P 关于 x 轴对称.

于是，PM 的方程为 $x=3\sqrt{2}$.

总之，一试运算量稍大，时间就不够了，另外，今年一试缺少概率题.

2　加试

1. 第一题.

此题大概是旧题改造的.

在笔者给《中等数学》2011 年第二届陈省身杯数学奥林匹克夏令营准备的

教材“平面几何(二)”就已讲过,题目是:

在圆内接四边形 $ABCD$ 中,$\dfrac{AB}{AD}=\dfrac{BC}{CD}$,$E$ 为边 AC 中点. 证明:$\angle BEA=\angle AED$.

显然,与第一题密切相关.

本题的关键是利用对称找出相等的角,进而找出相似三角形.

在图 4 中,圆是对称的,添上弦 AC 后,仍是对称的,对称轴是过中点 P 的直线 OP. 延长 DP 与圆交于点 E.

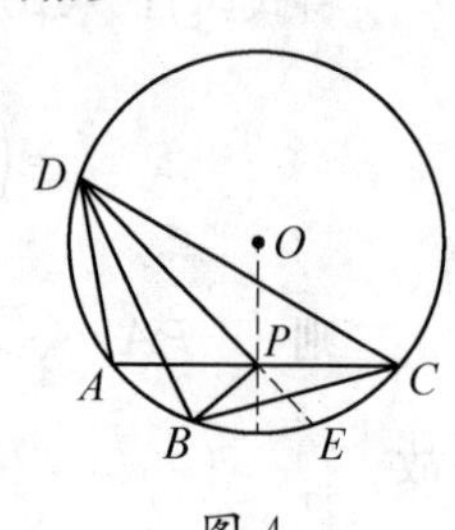

图 4

因为 $\angle BPA=\angle DPA=\angle EPC$,所以,点 B、E 关于 OP 也对称.

于是,$\angle DPC$ 的度数是弧 $\overset{\frown}{DC}$ 与 $\overset{\frown}{AE}$ 的和的一半. 因而,也是弧 $\overset{\frown}{DC}$ 与 $\overset{\frown}{CB}$ 的和的一半.

从而,$\angle DPC=\angle DAB$ (做过上面笔者提供的那道题,就知道这两个角相等,而且应当证明它们相等).

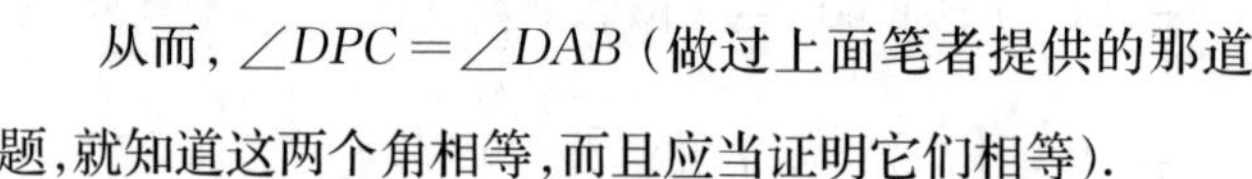

由此易得 $\triangle DPC \backsim \triangle DAB$,则

$$AB\cdot CD=PC\cdot BD.$$

进而,$\triangle DBC \backsim \triangle DAP$,故

$$AD\cdot BC=PA\cdot DB=AB\cdot CD.$$

这就与笔者提供的那道题接上了榫.

2. 第二题.

要对任意整数 n $(n\geqslant 4)$ 构造一个首项系数为 1 的整系数多项式 $f(x)$,使得对任意正整数 m 及 k 个正整数 r_1, r_2, $\cdots$, r_k,均有

$$f(m)\neq f(r_1)f(r_2)\cdots f(r_k), \tag{5}$$

可以先考虑 $n=4$ 的情形.

因为任意四个连续整数的积被 4 整除,所以,$m(m+1)(m+2)(m+3)$ 被 4 整除.

故 $f(m)=m(m+1)(m+2)(m+3)+2$ 是偶数,但不被 4 整除.

式(5)的右边每个 $f(r_i)$ 也都如此,但它们(至少两个)的积一定被 4 整除.

从而，式(5)的两边不等.

一般情形也是如此.

只需令

$$f(x)=x(x+1)(x+2)(x+3)g(x)+2,$$

其中，$g(x)$是任一个首项系数为 1 的整系数的 $k-4$ 次多项式(如 $g(x)=x^{k-4}$).

3. 第三题.

给定正实数 $a_1<a_2<\cdots<a_n(n\geqslant 4)$. 对正实数 r，满足$\dfrac{a_j-a_i}{a_k-a_j}=r$ $(1\leqslant i<j<k\leqslant n)$ 的三元数组(i, j, k)的个数记为 $f(r)$. 证明：$f(r)<\dfrac{n^2}{4}$.

题中的点 a_j 分区间$[a_i, a_k]$为定比 $r:1$.

其实这是一道老题：

问：$a_1<a_2<\cdots<a_n(n\geqslant 4)$ 中至多有多少对，它们的平均数仍在这个数列中？

换句话说，数列中有多少个数可以充当(以这些数为端点的)区间的中点(重数计算在内)？

现在的问题只不过将 $1:1$ 换成 $r:1$.

原来的问题，在 $j\leqslant\dfrac{n}{2}$ 时，以 a_j 为中点的区间有 $j-1$ 个；在 $j>\dfrac{n}{2}$ 时，则有 $n-j$ 个. 对 j 求和即得结果.

本题的解法完全一样.

4. 第四题.

此题答案是坏方格至多 25 个.

举出 25 个坏方格的例子不难，难在证明至少有两个好方格(即不是坏方格).

有两个老的题目：

(i) 对任意 9 个整数组成的数列 $a_1, a_2, \cdots, a_9$，若和 $\sum\limits_{k=1}^{j}a_k(1\leqslant i\leqslant j\leqslant 9)$ 都不被 10 整除，则 $0, a_1, a_1+a_2, \cdots, a_1+a_2+\cdots+a_9$ 是模 10 的完全剩余系.

(ii) 设 $a_1, a_2, \cdots, a_{10}$ 与 $b_1, b_2, \cdots, b_{10}$ 都是模 10 的完全剩余系. 则 $a_1 + b_1, a_2 + b_2, \cdots, a_n + b_n$ 一定不是模 10 的完全剩余系.

本题基本上是由这两个题目组成.

设好方格至多一个.

不妨设第一行没有好方格,且第一行的数依次为 $a_1, a_2, \cdots, a_9$. 则

$$0, a_1, a_1 + a_2, \cdots, a_1 + a_2 + \cdots + a_9$$

是模 10 的完全剩余系.

设第二行、第三行同在第 i 列的两个数的和是 $b_i (1 \leqslant i \leqslant 9)$. 则

$$0, b_1, b_1 + b_2, \cdots, b_1 + b_2 + \cdots + b_9$$

是模 10 的完全剩余系.

$$0, a_1 + b_1, a_1 + b_1 + a_2 + b_2, \cdots, a_1 + b_1 + a_2 + b_2 + \cdots + a_9 + b_9$$

也是模 10 的完全剩余系.

但由题(ii)知

$$0, a_1 + b_1, a_1 + a_2 + b_1 + b_2, \cdots, a_1 + a_2 + \cdots + a_9 + b_1 + b_2 + \cdots + b_9$$

不是模 10 的完全剩余系,矛盾.

这表明好的方格至少 2 个.

综上,2011 年加试的题目,每一道都是不错的题目. 有些题虽由旧题改造得来,但花样翻新,翻得很好. 不过有一点值得注意,就是改造后的题往往比原来的题难. 一道难也还不要紧,而每道题都比原来难,那解题者就吃不消了. 例如加试第四题,留着中国数学奥林匹克用或许更好.

最后,笔者衷心希望今后命题的难度能够降下来.

评 2012 年全国高中数学联赛试题

2012 年全国高中数学联赛的一试平稳，接近高考题，这当然是正确的，不过，如果苛求一些，似乎竞赛的味道少了一点儿，尤其缺乏新颖的、令人眼前一亮的竞赛题. 与俄罗斯的赛题相比，趣味性更有待提高.

1　一试

1. 第 1 题.

需要求点 P 到直线 $y-x=0$ 的距离，当然以用直线的法线式 $\frac{y-x}{\sqrt{2}}=0$ 为好. 在做向量 $\overrightarrow{PA}$ 与 $\overrightarrow{PB}$ 的数量积时，需知道 $\angle BPA$. 这可由图形立即看出，或者利用四边形 $PAOB$ 的内角和为 $360°$ 得出，不必用四点共圆.

2. 第 2 题.

用正弦定理更好些. 将 $a=2R\sin A$，$b=2R\sin B$，$c=2R\sin C$ 代入已知条件

$$a\cos B-b\cos A=\frac{3}{5}c,$$

消去 $2R$(或设 $2R=1$) 得

$$\begin{aligned}5(\sin A\cdot\cos B-\sin B\cdot\cos A)&=3\sin C=3\sin(A+B)\\&=3(\sin A\cdot\cos B+\cos A\cdot\sin B),\end{aligned}$$

即
$$\sin A\cdot\cos B=4\cos A\cdot\sin B.$$

两边同除以 $\cos A\cdot\cos B$，即得

$$\tan A=4\tan B.$$

3. 第 3 题.

由 $M=\sqrt{|x-y|}+\sqrt{|y-z|}+\sqrt{|z-x|}$ 是 x、y、z 的对称式，则可

设 $x \leqslant y \leqslant z$.

于是，$$M=\sqrt{y-x}+\sqrt{z-y}+\sqrt{z-x}.$$

当 y 固定时，M 显然在 $x=0$，$z=1$ 时最大，即

$$M \leqslant \sqrt{y}+\sqrt{1-y}+1.$$

而对非负数 a、b，由熟知的

$$\sqrt{a}+\sqrt{b} \leqslant \sqrt{2(a+b)},$$

得 $$M \leqslant \sqrt{2}+1.$$

4. 第 4 题.

由抛物线的定义与梯形中位线定理得到 $\frac{MN}{AB}=\frac{AF+BF}{2AB}$ 后，问题转化为：

在 $\triangle AFB$ 中，$\angle AFB=60^{\circ}$，$AF+BF$ 为定值，求 AB 的最小值.

除了标准答案的解法，也可以考虑用几何方法：

在射线 FA、FB 上分别取点 A_1、B_1，使

$$FA_1=FB_1=\frac{1}{2}(FA+FB)=a\text{(定长)}.$$

不妨设 $FA>FA_1$，$FB<FB_1$.

由 $\triangle A_1AB_1$ 与 $\triangle BB_1A$ 易得 $AB>A_1B_1=a$.

故所求最大值为 1.

5. 第 7 题.

最好用计算器. 因为

$$\arcsin\frac{1}{4}=14.47\cdots^{\circ},\ \arcsin\frac{1}{3}=19.42\cdots^{\circ},$$

而 $$\frac{180}{9}=20>19.42\cdots>\frac{180}{10}=18,$$

$$\frac{180}{12}=15>14.47\cdots>\frac{180}{13}=13.84\cdots,$$

所以，$$n=10,\ 11,\ 12.$$

有好的计算工具却不使用,好像是自废武功,与时代潮流也不相符.应当提倡学生使用各种计算工具,而不要加以限制.这件事,数学竞赛赛事大可开先河.

6. 第11题.

此题的第(2)小题,满足 $OA \cdot OC=20$、将点 A 变为点 C 的变换是以 O 为反演中心的反演.熟悉反演的人都知道,过反演中心的圆反演后变为直线.现在 A 所在的圆过反演中心,但 A 只在不包括反演中心的半圆上运动,所以,反演后的像是一条线段,半圆端点 $(2,\ 2)$, $(2,\ -2)$ 变为线段端点 $(5,\ 5)$, $(5,\ -5)$.

此题应当是一个使用极坐标的问题.半圆是 $\rho=4\cos\theta\left(-\frac{\pi}{4}\leqslant\theta\leqslant\frac{\pi}{4}\right)$,经过反演 $\rho\rho'=20$ 后变为 $\rho'\cos\theta=5\left(-\frac{\pi}{4}\leqslant\theta\leqslant\frac{\pi}{4}\right)$. 这条线段是直线 $x=5$ 的一部分,端点的纵坐标的绝对值与横坐标相等(也是5),即端点是

$$(5,\ 5),\ (5,\ -5).$$

总而言之,一试每题都有难度,坡度似乎不大.其实前两题最好再容易些,"送分"给学生也无不可.后几题难度可稍大一些,以便拉开差距.

2 加试

加试题颇有新意,竞赛性较强.如果说一试在这方面有点欠缺,那么,加试就是很好的弥补.

1. 第一题.

此题是真正的平面几何题,是一道好题.证明过程没有繁琐的计算,也比较容易.学生如果知道下面的引理,这道题就迎刃而解.

引理 若 O 为 $\triangle ABC$ 的外心,则

$$\angle OAB=90^\circ-\angle B.$$

由此引理及已知条件得

$$\angle O_2AN=90^\circ-\angle AMN,$$

$$\begin{aligned}\angle O_2AC&=90^\circ-\angle AMN+\angle CAN=90^\circ-\angle AMN+\angle BAM\\&=90^\circ-\angle B=\angle O_1AC.\end{aligned}$$

所以，O_1、O_2、A 三点共线.

2. 第二题.

此题解法很多. 关键是 b 与 $b+1$ 互质.

在第(1)题中，

$$2a=2^k(k\geqslant 2),\ b<b+1<2a,$$

所以，b 与 $b+1$ 均不被 $2a$ 整除. 而 b 与 $b+1$ 中有一个是奇数，与 $2a$ 互质，因此，它与另一个的积也不被 $2a$ 整除.

在第(2)题中，

$$2a=2^k m\ (k\in\mathbf{N}_+,\ m \text{ 为大于 1 的奇数}).$$

又 m 与 2^k 互质，则 2^k 个数 $m\times 1$, $m\times 2$, $\cdots$, $m\times 2^k$ 模 2^k 互不同余，其中必有一个与 -1 同余，它当然不是 $m\times 2^k(\equiv 0(\bmod 2^k))$，即有 $mh+1$ 被 2^k 整除，而

$$h\in\{1,\ 2,\ \cdots,\ 2^k-1\}.$$

令 $b=mh$，则 $b(b+1)$ 被 $2a$ 整除，且

$$b+1=mh+1\leqslant m(2^k-1)+1<2a.$$

3. 第三题.

这是一道好题，但难度较大，恐怕能解出的学生寥寥无几. 此题用于冬令营似乎更好些. 标准答案做得很好，本文不再多说.

4. 第四题.

亦颇有趣. 设想在正半数轴上有许多“坑”，即区间

$$(a,\ b),\ (1+a,\ 1+b),\ (2+a,\ 2+b),\ \cdots$$

一个机器人的足迹是 S_1, S_2, $\cdots$. 一方面，这个机器人不断前进(S_n 递增)，而且没有任何目标不能被它超越(S_n 无界)；另一方面，机器人的步伐越来越小 $\left(S_{n+1}-S_n=\dfrac{1}{n+1}\text{ 趋向于零}\right)$，直至小于固定不变的坑长 $b-a$. 这以后，机器人将无穷次地跌进坑里. 若模 1，则这无数个坑就是同一个坑 $(a,\ b)$. 而机器人的足迹就是 $S_n-[S_n]$. 可怜的机器人无穷次地跌入同一个坑中，真悲惨啊！

第三章　数学教育

大纲、教材及其他(载《面向 21 世纪的中国数学教育——数学家谈数学教育》)　江苏教育出版社,1994

数学是思维的科学　数学通报,2001(6)

访谈笔录(载《与数学家同行》)　南京师范大学出版社,2015

苏格拉底的启发式(与韩龙淑合作)　中学数学研究,2007(5);中学数学教与学,2007(11)

单墫谈数学教育改革　数理天地(初中版),2018(7)

《平面几何的知识与问题》前言　上海辞书出版社,2014

《奥数教程》前言　华东师范大学出版社,2005

《单墫数学科普著作集》序　华东师范大学出版社,2011

《趣味的图论问题》再版前言　中国科学技术大学出版社,2011

《几何不等式》再版前言　中国科学技术大学出版社,2011

《近代欧氏几何学》译者赘言　上海教育出版社,1999

《近代欧氏几何学》译本再版,再说几句　哈尔滨工业大学出版社,2012

《初等数学复习及研究(平面几何)习题解答》序　哈尔滨工业大学出版社,2009

《数学文化与文化数学》序　上海教育出版社,2012

《数学探究学习论》序　高等教育出版社,2008

《数学教学生成论》序　高等教育出版社,2008

《数学课堂学习动力论》序　南京师范大学出版社,2010

《义务教育数学课程与教学论》序　中国矿业大学出版社,2009

《中学数学有效课堂教学法》序　陕西师范大学出版总社,2012

祝《中国初等数学研究》诞生(载《中国初等数学研究(第 1 辑)》)　哈尔滨工业大学出版社,2009

评《三角与几何》　中学数学研究,2007(6)

读《名著的疏漏》　中学数学研究,2006(3)

推荐一本好书——《数学小辞典》　中学教研(数学),1984(4)

一本不可不读的好书　中学生数理化,1987(3)

一本好书——《解题·成长·快乐:陶哲轩教你学数学》 北京大学出版社,2009
《奥博丛书》序 上海教育出版社,2007
《数学奥林匹克命题人讲座》序 上海科技教育出版社,2009
《强基计划深度学习丛书》序 湖南科学技术出版社,2020
谈谈研究式的学习 中学生数学,2002(12)
数学课应当讲数学 中学数学研究,2006(11)
一堂数学课 安徽教育,1981(7)
凤头,还是猪头? 中学数学研究,2007(7)
何不使用计算器 中学数学研究,2007(8)
别扼杀了天才 中学数学研究,2005(10)
没事,别来烦我 中学数学研究,2008(2)
化简,不要“化繁” 中学数学研究,2006(5)
“画蛇添足”与“画龙点睛” 中学数学研究,2006(2)
韦达定理能用吗 中学数学研究,2007(11)
不要滥用反证法 中等数学,2011(1)
学好数学先得学好语文 小学生语文学习(中高年级),2003(10)
新的竖式乘法 小学数学教师,2009(9)
一道高考数学试题解的讨论 安徽教育,1980(12)
一道高考题的求解思路 高中数学教与学,2005(3)
2009 年江苏高考数学试题评析 高中数学教与学,2009(8)
关于 2010 年江苏数学高考题 高中数学教与学,2010(8)
评 2011 年江苏数学高考题 高中数学教与学,2011(9)
关于 2012 年江苏高考数学题 高中数学教与学,2012(17)
抽象与具体 中等数学,2002(3)
直观、慧眼 时代数学学习,2005(1)
解题中的形象思维(与李渺合作) 数学通报,2006(5)
解题漫谈 数学通报,1994(2)
解题面对面(一) 中等数学,2008(7)
解题面对面(二) 中等数学,2008(8)
解题面对面(三) 中等数学,2008(9)
从简单的做起 中等数学,2010(9)

大纲、教材及其他

1

我们正走向21世纪.

在21世纪,中国应成为现代化的国家,为实现这一目标,需要许多的变革与努力,教育是一个非常重要的方面.

四十多年的教育是计划经济下的教育(或许教育是计划经济的最后几块阵地之一),固然也取得很多成绩,但难以适应新世纪的需要.

教育必须改革.

改革不是面壁虚构,应当从现实出发.改革不是另起炉灶,应当利用原有的基础.改革不是全盘否定,应当扬长去短.改革不是追赶时髦,应当保持自己的特色.改革不是鹦鹉学舌,应当百家争鸣,改革不是夸夸其谈,应当勤勤恳恳地办好.

非常乐意参加中国数学会教育工作委员会与江苏教育出版社组织的"面向21世纪的中国数学教育——数学家谈数学教育"的讨论.我虽从事数学教育多年,但并未研究过数学教育,只能就中小学数学教育的"大纲"与"教材"谈一点粗浅的想法.

2

需不需要一个统一的大纲?这个问题,我想,是值得讨论的.

中国幅员辽阔,各个地区的发展极不平衡,教育水准也必然参差不齐.

如果有统一的大纲,它应当具有相当的弹性,允许因时因地制宜.

大纲,可以规定最低的标准,促进落后地区赶上去,但不必限制先进地区的发展.

大纲是纲,不必写得过细.有一版大纲规定十字相乘法中"二次项系数与常数项的积的绝对值不大于 60”，这就太琐细了.

大纲，主要规定知识的范围，教法应有授课教师自己的特点，不必写在大纲里(上述大纲有一节“教学中应该注意的几个问题”).

大纲，应根据数学的发展适时修订，以达到现代化的标准. 例如上述大纲中有“会查表求平方根和立方根”“会查正弦和余弦表、正切和余切表”. 随着计算器的普遍使用，这些内容应当删去.

数学教育应使学生掌握基础知识、发展数学能力. 数学教育不是职业教育. 因此，没有必要强调过于具体专门的应用(如炒股票). 数学的应用日益增多，不可能全在数学课中讲授. 相反地，一些专门的应用可以从数学中分裂出去，自立门户. 如制图、统计、计算机等. 经济数学也可单设一门课程，这样，既可使学生对有关问题有较为透彻的了解，又保持了数学的“纯洁性”. 大纲强调理论联系实际，但从四十多年的教学实践看，往往顾此失彼. 近来，“大众数学”的呼声又高起来. 我想，应当吸取过去的教训.

科学是没有国界的. 开放的中国是世界的一员，我们应有开阔的胸襟，一视同仁地介绍世界各国的成就，其中也包括本国的成就. 不应当搞狭隘的民族主义，更不能学阿 Q:“我的祖上比你阔.”

应当介绍古今优秀的数学思想，大纲强调辩证唯物主义，但很多数学家并不是唯物主义者，很多数学思想也难以归入辩证唯物主义的范围，不能将他们都关在门外.

除了界定知识范围外，数学大纲如果要强调什么原则，我以为应当指出数学教育必须集中发展数学能力.

3

现在已经出现“一纲多本”，打破全国统一课本的局面，这是很大的进步.

四十多年，数学课本基本沿袭苏联吉西略夫的教材. 我国自编的教材，有很好的，如华罗庚、丁尔陞等先生编审的 20 世纪 60 年代的初一教材，体现了“数形结合”的思想；但较多的只是修修补补，有些方面还不如原来吉氏的书.

新的课本，应当有利于发展学生的数学能力.

小学低年级，应强调数的四则运算，培养运算能力，禁止使用计算器.

小学三四年级,通过文字(应用)题,培养分析推理能力,应当加强应用题的算术解法.算术解法是我国数学教育的一大特色,算术解法能提高学生的学习兴趣,在训练思维方面极为重要,这种思维方式与代数解法(列方程)各有千秋,不应当急于用后者取代前者.

初中阶段,培养学生对代数式进行四则运算与恒等变形的能力.因式分解可以适当加强,因为它有利于培养学生的能力,如判断(能否分解)、分析(怎样分解)与创造(往往不是依样画葫芦)的能力.

平面几何的知识十分重要,但至少同样重要的是几何学的推理方式.现在平面几何大大削弱,甚至限定推理步骤不超过三步,这是不妥当的.一般说来中国学生在数学学习上超过欧美学生,没有必要降低水准.如果课时不够,可以只在某一章,例如全等三角形,花较多时间进行推理方面的训练,而其他部分则以介绍知识(不必详细论证)为主.50年代,初中不学相似形,但初中生已经具备了几何推理的能力,这样的经验值得利用.

平面几何削弱后,用什么取代它以培养逻辑推理的能力?长期以来没有解决,方案不只一个(例如莫绍揆先生提出用数理逻辑来代替),但均不易实行.现在看来,可以用组合数学.

实践(特别是各种竞赛活动)表明,组合数学中很多内容,如抽屉原理,奇偶分析,分类等,学生喜爱易于接受(小学生都懂得"从三只白袜子、五只黑袜子中随意抽出四只必有两只同色").组合数学灵活,有众多的应用.它的推理方式比起平面几何更加接近现代数学.因此可以将一些组合数学的内容分布于各个年级,取代平面几何.当然,平面几何不能完全取消,它不仅介绍"形"的知识,而且几何推理也是一种独特的推理.

公理化的思想也很重要.吉西略夫的书,特别是立体几何,脉络分明,论证严谨,仍值得借鉴.为了节省课时,可以将一些定理充作公理,但应慎重选择,形成比较合理、学生易于接受的体系.

课本可适当采用"圆周式",略有重复.有些内容,如应用题,可将代数解法与算术解法进行比较,使学生有更好的领悟.

各种能力的培养都必须通过练习.因此,教材应包含大量的习题,有纯粹为巩固知识的简单练习,有应用所学知识来解决的问题,也有创造性较高的难题.

按波利亚的说法,数学教育的目的是教学生"思考",而"思考"即"解题

(Problem Solving)". 至少,数学教育中应包含解题的教育,而有的课本却忽视这一点,只有少量例、习题,教师学生均感不便(吉西略夫的书有大量习题).

4

数学(数学教育)是什么? 众说纷纭. 有的说数学是算学,应强调计算. 有的说数学是思维的体操,应培养逻辑推理. 小平邦彦则说数学是一种感觉. 高斯推崇数学为自然科学的皇后. 亦有人认为数学只不过是一种工具,是科学的奴仆. 有人认为"数学是模式的科学". 有人认为"应用数学是丑陋的数学". 国内常常引用"数学是研究现实世界的空间形式与数量关系的科学". 而在西方,也曾流行"数学是符号的游戏","数学是这样的,研究它的人也不知道自己在做什么". 近年有人提出"大众数学",强调"数学教育应为就业服务". 也有人指出"数学是一种文化"……

或许数学就是数学. 不应对它有过奢的希望,也不应当忽视它自己的特点. 一门钢琴课应以学会弹琴为主要目的. 至于能否赚钱,为谁服务等,大概不是钢琴课本身的任务.

想起一个著名的故事"瞎子摸象". 这个故事并非嘲笑残疾人,而是说站在不同的位置,每个人都有可能发现部分的真理. 但只有综合起来,才能得到完整的认识.

本文作者只是一名摸象的瞎子,不知所说的有几分接近真正的象.

数学是思维的科学

1

数学是思维的科学.这句话,大概不会有什么反对的意见.谁都知道,数学能够启迪、培养、发展人的思维.虽然也有其他学科或其他方式可以培养人的思维,但在深度、广度、系统性等方面,是无法与数学相比的.

然而,在实际运作时,却有一些人忽视这一点,他们只看重数学是一门实用性的科学.提到式的恒等变形,他们会问:这有什么用?提到不等式的证明,他们更摇头表示怀疑:没有用的东西,学它干什么?

在这些人看来,小学的四则运算日常生活少不得,当然是有用的,要学.目前初中的内容约有二分之一还有些用处(其中几何证明都是绝对无用的).高中内容,大部分是为了应试,都应当取消,只有一小部分可以保留.

这种观点,由来已久.早在20世纪60年代,即已出现轻理论、重实用,过分强调理论必须联系实际的思潮.在文化大革命中,更发展到顶峰.当时有的地方,中学数学课已经被取消,少得可怜的一点数学内容被纳入一门叫做"工业基础知识"的课里面.

仅将数学当作实用科学就是不懂得培养思维能力正是数学的一大功用,即使只谈实用性,也决不可忽略思维能力的培养.

明朝的徐光启先生(1562—1633),见解就很高明.他在万历三十五年(公元1607年)与利玛窦合译了欧几里得的《几何原本》.在译本卷首的《几何原本杂议》中,徐先生指出:"人具上资而意理疏莽,即上资无用;人具中材而心思缜密,即中材有用;能通几何之学,缜密甚矣,故率天下之人而归于实用者,是或其所由之道也."

最近我见到一篇文章《数学与文学》,作者是在人文科学方面卓有成就的朱正先生(著有《鲁迅传略》(1956年)、《鲁迅回忆正谈》(1979年)、《小书生大时代》(1999年)、《辫子、小脚及其他》(1999年)等书).朱先生对数学的作用认识非常深刻,他说:“我在学术研究方面所做的工作,凭仗的也就是当年数学‘体操’所训练出来的思维能力,我的一本《1957年的夏季:从百家争鸣到两家争鸣》,程千帆先生看了,许我为汉学家,说那本书深得段戴钱王之妙,却不知道其实是得益于数学的.”(朱正著《字纸篓》,120—121页,广东人民出版社,2000年出版).

即使一个人“从事的几乎是同数学没有什么关系的职业,原来学的代数几何三角中的定理定律几乎全忘记了”(朱正先生语,同上120页),然而数学对思维的训练还是有用的,这才是数学的最广泛的“实用性”,这才是我们要学数学的主要原因.

2

我国古代曾有四大发明,在数学方面也有很多成就,并出现了《九章算术》《周髀算经》等重要著作,但后来我国的自然科学却停滞了,远远落后于西方.这当然有很多的原因(特别是政府的腐败),但其中有一点是很重要的,即过于强调实用,而缺乏理性的思维.

希腊人比古代的中国、埃及、巴比伦前进了一大步,他们“具有重理知的特性,概括并简化各种科学原则,希望由此求出这些科学的道理”,“柏拉图坚持研究几何学,并不是为了几何学的实际用途,而是想发展思想的抽象力,并训练心智使之能正确而活泼地思考.柏拉图把思想的抽象力和正确的思考能力应用在伦理与政治上,结果奠定了西方社会哲学的基础;亚里士多德把它们应用在研究具体事物的真实性上,结果奠定了特质科学的基础.”

“自然科学能发展到目前的阶段,首先归功于希腊人对大自然的观念以及对有系统的智力训练的爱好,中间经过文艺复兴、宗教革命、法国革命,后来又受到工业革命的大刺激.工业革命使工具的技术逐渐改进.西欧在自然科学的后期发展中,从未忽视科学的实际用途.不断的发现和发明更进一步刺激了科学研究.理论科学和应用科学齐头并进,相辅相成.”

应当承认我国在理论思维方面不及希腊与西欧.数学方面,这样的例子很多,我们古代很早就知道了勾3股4弦5,但没有证明一般的勾股定理(即毕达哥拉斯定理),也没有找出勾股数(满足 $a^2+b^2=c^2$ 的整数组 (a, b, c))的一般规律.这些都是由希腊人完成的.我国古代很早就知道 $2^{p-1}-1$ 是奇质数 p 的倍数,但建立起费马小定理的却是法国人费马(Fermat, 1601—1665).

曾在北京大学担任过十多年校长的蒋梦麟先生(1886—1964),在他的名著《西潮》中早就说到这一点,他说:

“在中国,发明常止于直接的实际用途.我们不像希腊人那样在原理原则上探讨,也不像现代欧洲人那样设法从个别的发现中归纳出普遍的定律.现代欧洲人的这种习惯是从古希腊继承而来的,不过较希腊时代更进步而已.中国人一旦达到一件新的发明的实用目的,就会马上止步不前.因此中国科学的发展是孤立无援的,也没有科学思想作为导向明灯.科学发展在中国停滞不进,就因为我们太重实际.

他又说:“我们中国人最感兴趣的是实用东西.……如果有人拿东西给美国人看,他们多半会说:‘这很有趣呀!’碰到同样情形时,中国人的反应却多半是:‘这有什么用处?’……我们中国对一种东西的用途,比对这种东西的本身更感兴趣.”(《西潮》第七部:现代世界中的中国.本节的引文均出自该处,不一一列举)

时至今日,情况当然与蒋梦麟先生的时代有了很大的不同.但忽视理论、太重实际的倾向仍然值得注意.因此,我们在考虑中学数学教材、大纲或是课程标准时,不能仅考虑实用性,不能简单地罗列数学知识,而更应当考虑需要培养哪些思维品质,如何去进行思维的训练,充分发挥数学是思维的科学的特点.

3

数学有众多的分支,在中小学阶段涉及到的有算术(数论)、代数、几何、三角、解析几何、函数论、组合数学、概率统计等.各个分支在数学中都有一定的地位和作用,它们的思维方式各有特点,不尽相同,彼此之间并无高下之分,而是相辅相成,组成一个整体.不宜过分强调其中的某一个,而忽略其他,对各种思维方式在什么时候引入最为适宜也应当深入研究.这里对几个问题谈谈我们的想法.

3.1　算术与代数

在小学阶段(即九年义务教育的前五、六年),用算术方法解应用题是我国

数学教育的一个传统内容,解题方法多种多样,极富巧思,有利于培养学生的学习兴趣,发展学生的思维.例如"和差问题":

"大、小两数的和是18,差是4,求大数与小数各是多少?"

算术的方法可以先将小数加上4,使小数变成与大数相等,从而

$$大数=(18+4)\div 2=11.$$

也可以先将大数减去4,使大数变成与小数相等,从而

$$小数=(18-4)\div 2=7.$$

甚至还可以先求平均数:

$$18\div 2=9,$$

再加上(或减去) $2(=4\div 2)$,便得大数(或小数).

用代数的方法,通过设未知数、列方程(组)解应用题,方法统一简单,其优点是显然的,但能否就肯定代数方法高于算术方法,甚至取消算术解法而统统代之以代数方法呢?恐怕不能.就思维的品质来说,统一性与多样性,各有千秋,不分轩轾.统一、简单固然好,"百花齐放"也不坏.从教育的观点来看,理解统一方法的优点需要一定的基础,低年级尚难做到这一点,而算术解法多变,易培养他们的兴趣,比冷冰冰地"设 x,列方程"有"人情味",有美学价值,这是极为重要的.因此过早地在小学引入方程,效果可能是西望长安不见家(佳),反倒容易使思维简单化,甚至僵化.而且要引入方程,就不能不讲方程的解法,从而需要了解方程的性质,式的变形,更要引入负数的概念,容易破坏系统性,自乱章法.

3.2　平面几何的地位与所占比重

平面几何原来在中学数学中占有相当大的比重,它的地位是历史上形成的.几何学有很完整的公理体系,又有优美的图形,推理较有规律可以遵循,对培养学生的思维能力是有好处的.完全取消几何,"打倒欧几里得"当然不对,但随着历史的发展,几何学已经不能在中小学独占一大块地盘.应当减少几何的课时,已经成为共识.但怎样缩减方为合理?用什么来代替几何?

文革前,初中几何主要讲全等形与平行线,高一讲相似形与比例线段,当时不升高中的人,虽然少学了一些几何内容,但已经基本上掌握了几何的推理方

法. 因此,可以采取类似的做法,即对几何内容采用两种要求,一部分内容,如三角形全等,需要学生很好掌握,能解决有关的习题,包括较难的习题;其余内容,则只需了解,不花或少花功夫去做题. 了解的内容不应太少. 现在连"旁心"是什么,学生都不知道,这是不恰当的. 旁心这一名称介绍给学生并不增加负担,反而可以体现几何学的优美. 再如三角形的内角平分线的性质也应当介绍,不应删去.

几何课时减少后,用什么来代替几何培养学生的推理能力呢? 首先,应当指出代数、三角或者算术,都有培养推理能力的作用,并不比几何逊色,需要进一步开发. 其次,可以考虑增加其他内容. 20 世纪 60 年代莫绍揆先生曾提议用数理逻辑来代替几何,这是一种好想法. 或许更可行的是用组合数学来代替几何,因为组合数学的很多内容有趣味,易为学生接受,而且灵活多变化,对培养思维能力极为有益,比如"抽屉原理",在小学低年级就可以引入. 像"三只袜子,两种颜色,其中必有两只同色",小朋友都能理解. 还可以采用"圆周式"的讲授,在小学、初中、高中都有抽屉原理,但内容逐步加深,再如"奇偶分析""图论初步",不但有趣,而且也很有用(无论在实用方面,还是在思维方面).

3.3 "函数为纲"与离散数学

函数为纲,是一个不太明确的口号,在众多的数学内容中,突出函数既无必要,也没有太好的借口. 相反地,随着计算机等的发展. 离散数学的内容反倒应当增加,国外已有人认为大学(尤其一年级)没有必要非学微积分,或许大学一年级学离散数学更合适一些. 中小学也应引入这方面的内容,包括上面所说的组合数学、初等数论等. 初等数论(算术)讲数的性质,可以结合代数进行,如

$$(n+1)^2-n^2=2n+1.$$

这个简单的代数式,初中学生人人知道,然而将它理解为"每一个奇数可以表示成平方差"的恐怕寥寥无几. 再如

$$(a^2-b^2)^2+(2ab)^2=(a^2+b^2)^2.$$

即表明勾股数有无穷多组. 这些,都可以提高学生的兴趣与思维能力.

3.4 引入新内容要考虑能否提高学生思维能力

有人主张增加一些统计、建模等内容,并以美国小学生调查冰激凌的口味,测定自己脉搏等为例,我们看不出这些做法对培养思维能力有什么好处. 有些

内容仅仅是罗列一些名词与公式(法则),不如统统去掉以节省课时. 因为思维能力是要花很多时间、花大力气去培养的,而那些套公式的事,待用到时再学也为时不晚.

我们并不反对让学生动手,但动手能力主要依靠理化生等实验科学来培养,而数学应侧重培养"动脑",培养智慧.

"磨刀不误砍柴工". 数学能使人变得聪明,学知识就好像砍柴,刀磨好了,砍柴不难.

再如微积分,有人主张不用极限理论,我们觉得取消了极限理论,等于取消了微积分的核心,剩下部分味同鸡肋,对思维的培养没有多大意思. 讲微积分就应当讲极限思想,只不过应当采用学生易于接受的方式,不一定非用 $\varepsilon-\delta$ 语言. 比如可以先讲无穷小量的求和(一个典型的命题是抛物线 $y=x^2$、x 轴、直线 $x=1$ 所围的面积,从中引出 $\frac{1}{n^3}$, $\frac{n}{n^3}$, $\frac{n^2}{n^3}$, $\frac{n^3}{n^3}$,在 n 无限增加时,极限分别为 0, 0, 0, 1). 先讲积分再讲微分,也是一种可以考虑的方案.

不一定追求形式上的新,原有的内容也可以用新的观点去考察,特别应当进一步挖掘它们在培养思维方面的作用,比如前面所说的和差问题,可以结合计算机,采用尝试法,编好程序,经过几次尝试,调整得出结果.

帕斯卡(B. Pascal, 1623—1662)说得好:"人是一根会思想的芦苇","人的全部尊严全在于思想","人的伟大在于他有思想".

数学是思维的科学,应当在培养"会思想"方面起更大的作用.

访谈笔录

访谈者: 单老师,您好！我们正在编写《与数学家同行》这本书,主要是想请您聊聊您的学习和研究经历,还想请您谈谈对于一些社会热点问题的看法,和中小学生分享您的一些宝贵经验. 是不是可以先请您讲讲您是如何喜欢上数学的？您是从小就花很多功夫在数学学习上吗？

单墫: 说实话,我小时候对数学并不是太上心,我其实更喜欢文史一些,喜欢看小说,看看历史方面的东西. 那时候觉得数学比较容易. 我小时候喜欢玩,喜欢动脑筋的游戏、下棋,比如象棋、军棋. 那时候围棋还不普及,还有打扑克牌、玩牌九. 我还没有读小学的时候,“加法”就算得很好了,因为玩牌九那些都需要算. 其实我小的时候就是喜欢玩,数学也没花什么功夫,那时候也没有像现在,要学这么多奥数. 我们在小学就是觉得数学容易,数学本身也不难学,当然当时的同学中也有学不好的.

我在小学的时候做算术应用题有个特点,那时候解应用题老师都要求分开写几个式子,但是我基本上从头到尾就写一个式子. 因为那时候数学学得不错,脑子转得很快. 有一次考试我印象很深,老师题目出错了,其实也不能算他出错了题,实际他本想出“除以”,但是题目上写的是“除”. 我当时按照除做对了,他看到我做的结果,就怀疑我错了. 因为他以为题目是“除以”,我交卷时,他就说你再看看,我就说我没错. 他发现这个题目实际上是我做对了. 我交卷出教室就听他大声说大家要注意这个题目是“除”不是“除以”.

当时小学数学很容易,初中也学得很少. 像平面几何,我们初中都不学相似,相似是在高中才学的,用的是苏联教材,教材体系比较严谨. 那时候初中对平面几何的表达要求比较严格. 我们做平面几何都打草稿,我语文写作文都不打草稿,但是做平面几何是要打一个草稿,然后再写上去. 我们几何老师要求非常严格,他打的叉都是打对角线那种的大叉. 他个子高,嗓门也大,一叫大家都吓一跳,很认真、很严肃. 那时的初中数学不是很难,我们都觉得很有意思. 我印象很深的是那时候初中数学的一个难题:一个圆内接四边形的两条对角线互相垂直，则过其交点作任意一条边的垂线，必平分其对边. 有个同学来问我，我当时做出来了，这个题目在现在来看是比较简单的题，但当时按我们的水平就是难题了. 初中数学大概就是这样的水平，但是好就好在不是很难，都是因为兴趣喜欢，思维要求也不超过我们那时候学生的水平，学生只要努力就能达到了，要是太难的内容学生就没有兴趣了.

到了高一稍微花了一点功夫，平面几何书刚发下来老师还没讲我就把题目都做了一遍. 那个教材习题比较多，难度适当. 我那时就是喜欢自己看看书，觉得也很容易，也不难，主要还是因为有兴趣. 我们那时候书看得也不多，平面几何好不容易有一本李俨编译的《近世几何学初编》，不过排版的错误非常多. 后来有一个同学买了一本邱丕荣翻译的《近世几何学》，大家当成宝贝，那本书相当好，但也就只有平面几何而已. 现在来看眼界非常浅，那时没看到过像现在这么多东西. 我们还组织了一个数学小组，叫 FSTY，四个同学一起做《数学通报》上的题目. FSTY 就是每个人的姓氏首字母. 我们那时候就是觉得做数学题很好玩，也没有说特别懂什么的，都是自己感觉有兴趣. 但水平也就如此而已，那时候书也少，偶尔看几本书或者小册子，也没有看得太懂. 那时候就喜欢玩儿，看书也当作玩.

上高中那会儿，我哥哥比我高两个年级，他是南京师范大学数学系的，所以有时候我就看看他的书. 他到大学就学微积分了，我也拿他的书来看看，就是看看微积分，比较浅的看了一些. 可能看得最花工夫的一本书是高扬芝的《极限浅说》，她是南京师大非常早的系主任，一位女老师. 说老实话，那本《极限浅说》写得一点都不浅，书中什么都要证明，而且印刷错误非常多，真的看得头疼得要死，但是不管怎么说总想要看下来，确实觉得我看书的能力还是有的. 坚持下来以后，后来看书不觉得吃力，就是说自学的能力还是有提高的.

高扬芝　生于 1906 年，南京师范学院（南京师范大学旧称）数学科主任，高扬芝是中国数学会创始时的少数女性前辈之一. 20 世纪 60 年代，她被选为江苏省数学会副理事长. 她著有《极限浅说》《行列式》等科普读物多部.

后来接触到的微积分教材比较严谨，上大学以前我也看过. 自学过一些东西，乱看看，也没有什么计划，我哥哥买了一些书看，我跟着他看着玩，学校的功课也没怎么用过功.

我上课净是在胡思乱想. 有时候自编武侠小说，很少听课，要不就是上课做作业，上语文课做数学作业，上数学课做语文作业，反正在学校把作业都做完，就是偷偷地做，放学就做完了，也不复习什么的. 那时候应该说课程比较轻松，相对来说比较容易，跟现在比较起来那时作业也比较少.

现在中学学的内容太多，数学、英语学得都多. 我上初中的时候没有学过英语，高中才学英语，学得很少. 我那时候到学校比较早，早上用 5 分钟的时间在操场把单词背一下，所以当时考得还很好. 我为什么到操场呢？因为我哥哥跑长跑，他们喜欢跑长跑，天没亮就去跑了，大家都在操场上跑. 我不跑，我早上跟他一块儿上学，到学校没事儿干就看他们长跑，然后花 5 分钟时间背背英语单词. 数学好，其他科的小测验有时候好有时候不好，总归就是不太用功，那时候不懂得用功.

访谈者：听您说了这些事情，我们可以看出虽然您说您不用功，但还是因为出于对数学的兴趣，其实也学了不少的东西，看来这个兴趣真的是非常重要. 您后来在数论方面也做了很好的研究工作，特别是现在很多人一谈到解题都特别佩服您，都说经过单老师认可的话就是权威的认可. 在解题方面您有哪些经验可以分享给孩子们呢？

单墫：解题能力应该说也是要锻炼的. 我当时在中国科学技术大学做研究生，同宿舍的还有肖刚和李克正. 他们都比我小，李克正比我小 5 岁，肖刚实际上比我小 8 岁，那时候我以为小 10 岁，就常讲“他比我小 10 岁，但是能力比我大 10 倍”，肖刚是一个天才. 他们的能力都很强. 因为我“文革”前两年就教书了，他们那时候都还是学生，所以中学这块知识我觉得我还是很强的，就是做题能力很强，但是跟他们一比却不是. 他们有很好的想法，而且思路很活，我跟他们学到不少的东西. 刚开始大家一起做题，做美国的大学生竞赛题，确实收获很大，他们思维很开阔. 这两个人给我影响很大，使我解题方面的能力提高好多. 你跟一个高手在一起切磋还是很有进步的，就跟下棋一样，你跟一个下得好的一起下，很容易提高.

肖刚　代数几何学家，曾任华东师范大学数学系教授，现定居法国，从事计算机的教研工作. 肖刚是 20 世纪 80 年代代数几何界的先锋人物，被公认为是少数的天才型数学家之一. 主要从事代数曲面的研究工作. 在代数曲面的纤维化、曲面自同构群等方面有着杰出的贡献.

李克正　首都师范大学教授,中国数学会教育委员会委员,美国数学会会员.为数学天元基金的发起人之一,曾任数学天元基金学术领导小组成员8年,并负责人才培养项目.《中国科学》等杂志编委,《中学生教学》杂志主编.

那个时候我就开始写一些东西,写普及的东西.我算是第一批博士,但可能和他们很多人不一样,他们都是雄心壮志想搞研究,我就不大一样.我拿到博士学位都40岁了,应该说一个人做研究的黄金时代已经过去了,做研究可以做一点,但是要做得很好是很难的.所以我当时就花工夫在普及上,同时研究也做一点.我做得最好的就是解决厄尔迪什关于除数函数的一个问题.

保罗·厄尔迪什(1913—1996)　匈牙利数学家.1984年以色列政府颁发的10万美元的"沃尔夫奖金"(Wolf Prize)就是由他和华裔美籍的陈省身教授平分的.厄尔迪什是当代发表数学论文最多的数学家,也是全世界和各种各样不同国籍的数学家合作发表论文最多的人.他发表了近1000多篇的论文,平均一年要写和回答1500多封有关数学问题的信.

我读研究生的时候是带工资去的,我记得是51块多.当时有一个小孩,上面还有老人,分居两地,吃食堂.回家乘火车花十块钱都感觉很多,经济比较困难,必须靠写稿子赚点钱吃饭.如果那时候全力以赴去搞科研,说得不好听一点儿,就会活不下去,所以必须要做点普及工作.再者我是做教师出身,也比较适合搞普及工作,中学教师本身搞教育,加之我过去也比较喜欢文史,文笔上面也还可以,也比较严谨,算是一个优势.另外我还在高校,见识面也广一点,如果仅仅在中学,很多材料也看不到,就比较局限.所以应该说搞数学普及工作我当时还是很有优势的.一方面相对站得高一点,看书多一点,另一方面对中学又比较了解,表达方法也适用于普及工作,所以比较有优势,就搞了普及工作,后来就一直搞下去了.当时我的目标就是一年写一篇数学学术文章,再写一本科普的小册子,大概基本上按照这个目标做下去.一年里,写一篇数学学术论文大概花的时间不需要太多,差不多一个月时间,看一些文章,然后找一个题目做一下,花的时间不算多.普及工作花的时间比较多,你写一个小册子,那时候还没有电脑,你就是抄也得花一点儿时间,所以比较辛苦.这样基本上一年写一个小册子、写一篇学术论文.这么多年,从1978年算起有三十几年了(指到2016年),

小册子和论文也都在三十之上,还写了不计其数的普及文章.

我在解题上有个想法,就是总希望把题目解得更好一点儿,那就需要在做完以后好好总结一下.我觉得解题这方面波利亚写的几本书是相当好的,如《怎样解题》,他确实是有真正的看法.我一直要求自己写完以后要回顾,要注意看能不能写得更好一点,尤其是我们搞普及工作的,不是说一个问题你绕了5个圈子解决了,你还要带着人家再绕5个圈子,那就没有意思了.应该是即使你自己绕了5个圈子,你将来总结的时候,也要让人家少绕圈子,这才是好的,所以要注意这个问题,每次都注意做得更好一点.

波利亚(1887—1985) 美籍匈牙利数学家,法国科学院、美国科学院与匈牙利科学院院士,1963年获美国数学家功勋奖.曾著有《怎样解题》《数学的发现》《数学猜想》等,《怎样解题》围绕探索法这一主题,采用明晰动人的散文笔法,阐述了求得一个证明或解出一个数的数学方法,可以有助于解决任何"推理性"问题.

访谈者: 您刚才说到解题的一些心得,现在的中学生实际上做的题非常多,比您那时候可能要多很多.但是学生做了这么多题目以后,也不一定对他们的解题有特别大的帮助,您怎么看这个问题?

单墫: 第一,现在的学生功利性太强,他不一定是因为喜欢数学而去做题的.第二,是有些人把太难的题目给小孩做了,这个不好.包括教材里面也有很多很难的题目,以前教材题目的水平比较平稳,偶尔有难的题目放在最后.现在你不知道忽然哪个难题就"猫"在里面,这个很不好.教学时可以补充一点课外题,但是基本上来说不要太难,不要过分的难,我觉得现在有些题目是过分难了.一个题目是你自己想出来的,还是看了解答之后想出来的,这当然是很不一样的.我们许多人碰到难题,看了解答觉得不难,就拿给学生做,其实这样不好.所以题目难度也要注意.第三,现在还有一个问题,就是初中教材的要求太低,而课外的题目太难,这个差距明显太大,也不好.以前课内外难度差别不大,课外难一点,但不是难很多,现在这两个根本不能比.

这也许跟现在的课标制定中的一些问题有关系.初中课标把几何证明取消了,把几何的表达要求基本上取消了,把三次方公式、十字相乘法这些内容基本取消了,等于初中的代数、几何都没有学,所以初中教材的内容就很少.在学校里面没有让学生学,学生跑到外面学,这样就很花时间很花精力. 你在学校给他学点儿就好了,他也不累,现在反而造成他负担很重. 如果仅仅学在学校教的内容,又太少,确实也不行,这是一个大问题.

访谈者: 您以前也在《数学通报》上发表过文章,说数学是思维的科学. 但现在很多都在强调应用数学,您认为两者的关系是怎么样的?

单墫: 数学本来就是分好多种,应用数学当然也算数学,但是具体的应用应该放到物理里面或者化学里面,就是哪一行用哪一个. 数学可以为每一行都提供所需要的基本的东西,而那一行专业性的东西不能够全提供. 你比如说统计,数学当然可以提供统计中用到的数学,但是具体的问题应该统计自己去搞. 我觉得数学的本质还是培养思维能力,它自己有一个理论体系. 所以怎么应用,它各有各的特点,比如说经济上怎么应用. 它有很多的应用都需各个学科自己搞,数学不能为所有人服务. 比如说物理,物理包括很多物理知识,数学不可能把物理知识补足了,否则变成物理课了,所以数学课还是只讲数学,你要怎么应用你可以自己拿去应用. 比如说数学里面有群,物理怎么用群,这个主要还是物理. 就是那些"应用"是各个学科以及有关工程师、医师他们自己研究的专业内容.

现在的问题就是给数学的负担太重,刚才讲初中教材内容少而空,高中教材的内容却是多而杂. 比如说算法的东西不应该算在数学里面,应该是计算机的内容,应该把它切掉. 把这些切掉以后剩下的就少很多. 现在属于负担太重,你要数学为大家都服务那是不可能的,它自己还有很多事情要做,你应该看到数学首先是一个学科,首先要把自己学科的事做好,然后你拿数学的知识去应用那是你的事. 但是你要叫数学完全完成这个事情,那就要求太高了,不可能这样子. 现在学生学得确实比较多、比较杂. 作为数学本身应该学习的最基本的东西反而不存在了. 就像我们学生在学校里面需要把身体锻炼好,你不能让学生既要会踢足球,又要会打篮球,他可以自己去搞各个项目,我们教他把身体基本素质锻炼好,提高基本的反应灵敏度,学一些基本的东西. 数学只要把自己的体系搞好. 现在好多人不承认数学有一个体系,要把体系打掉,非要过度强调应用. 数学主要是培养人的思维能力,当然学好数学需要的思维能力也是很强的.

访谈者: 现在有不少的中学生会选择出国留学,大学生出国的人更是越来越多,留学生低龄化的现象越来越显著,对于这种现象您怎么看待?

单墫: 这可能在一定程度上说明我们的教育出现了比较大的问题. 你说现

在学生负担重不重？我想绝对是重的.现在看小学生的书包你就知道负担重不重了.今天说要学书法，就上书法课；明天要唱戏，就上戏曲课；后天要学习踢足球，就要去踢球，那学生怎么吃得消？教育就应该有一个标准.一遇坏事，大家找原因就会说第一应试教育不好，第二奥数不好，这种说法就很有问题.其实教育就应有一个标准，你有一个标准就好说话了，当然这个标准定得好不好那是另外一回事.但是我定了一个标准，大家朝这个标准努力去做就行了，很简单很明确，进而教学就是有目的地在进行.现在就是没有目标，你今年说钢琴重要就去学钢琴，明年说可能要考美术就去学美术，后年还是考奥数就来学奥数，再一年又说学英语.这就是目标不明确，朝令夕改，政策多变.其实原来应试教育就包括素质在内，一个人的基本素质，要有数学，要有语文.学好这些就是素质教育，就包括了基本的素质在内，比如说数学就包括了运算和推理这些基本能力.我觉得在众多素质当中提取出最主要的几个素质再加以考核就行了，这就是考试.

什么是考试？考试以前就是科举，科举是一个进步.因为以前靠推荐、门第都是不行的，科举是“公平竞争”的.第一，考试本身是对的，而且从科举至今一直如此考着，不但中国，其他许多国家也跟中国学了，英国、美国都跟中国学考试.那么就是说这个考试没有那么大错，有些东西是不要改的，改反而会把它改坏了.第二，考试内容基本是对的，考试在众多素质当中已经选出了最主要的素质.当然你说有人是美术天才，天才是另外一回事，那是少数人，对大多数人来说这套考试是对的，不能乱动.现在有些人还没有真正搞清楚，好的东西也要把它改掉，那怎么行呢？好的东西是不能改的.比如说一天吃三顿就蛮好的，你不能改成一天吃五顿或者一天吃八顿，有些是不需要改的，如果早上喝牛奶，你说不喝牛奶改喝豆浆，这种改也没有多大意思.有人喜欢喝牛奶就喝牛奶，有人喜欢喝豆浆就喝豆浆，不要乱动.学习最好是要有兴趣，没有兴趣是不行的.陈省身先生说数学好玩，你要是让学生觉得不好玩就麻烦了.所以应该说学得好的、有兴趣的不会觉得数学很难.

访谈者：现在数学学得好的学生也非常多，也有人在国内外的奥赛中取得很好的成绩.以前有些人说奥赛不好，但其实现在很多一流的数学家也都是曾经在奥赛中获过奖的，比如陶哲轩.您觉得我们应该怎样看待奥赛？如何权衡它和义务教育的课堂学习呢？

单墫：有很多现在在代数、函数方向上搞得不错的数学家都是奥赛出身，只是大家没有好好调查. 有一批人确实搞得不错，但也有一些搞奥赛的并没有继续从事数学相关的研究. 关于美国大学生竞赛有一个调查，调查参加竞赛的这些人，将来如果是工程师、讲师，那么就是人才. 调查结果显示人人都可以算作人才. 我想我们搞奥赛的大概也很多是讲师、工程师，虽然没有人调查过，但我想没有当上讲师或没当上工程师的应该很少. 那些得了奥赛金牌的学生基本都被清华、北大选走了. 如果按照这个标准，那么奥赛获得奖牌的人就没有不是人才的了. 这其实还是一个标准怎么样的问题. 如果说获得诺贝尔奖、菲尔兹奖的才叫人才，没获奖的就不叫人才，这个标准当然不可能，全世界也没有几个能得诺贝尔奖的，这个标准本身就不对，这个标准不能太高. 我们现在很多人都把人才的标准定得太高.

还有一个关于对比的问题，就是参加奥赛和不参加奥赛的学生有什么差距，也没有人做过研究. 你光看到奥赛里有一个人不怎么样，并不能说明问题，因为不参加奥赛的人可能更多的不怎么样，这个事情也不好说. 我只能从感觉上讲，参加奥赛的学生绝大多数是不错的，据我所知，有搞数学研究的，也有不搞数学研究去搞软件的，或搞其他行业的，都还不错.

还有，现在是不是把很多事情搞得失衡了？很多培训机构做培训，其实他也不是在做竞赛，只是打着竞赛的旗号，还有些学校将竞赛跟升学挂钩. 本身奥赛是有兴趣的学生学一学，现在变成功利性很强的事情，一些学生也许对数学没有兴趣，看到其他人都去学，他就也去学，这是没有必要的. 奥赛不应该是人人都参加的，但是现在这个问题在于我们升学的标准不明确. 大学更相信参加过竞赛并且拿奖的人，特别是重点大学. 所有的大学都认为这样的学生好，才这样选. 所以还是说明我们高考本身的标准有一点问题. 现在的高考题目也是越来越容易，虽然总分上去了，但是大学越来越不放心，他觉得你学得不一定好，觉得你不是我要的人才，他宁愿相信参加竞赛的人，这就是高考和高校录取中的一个问题. 同样，中考的题目更容易，这个更容易导致的结果就是重点中学也不相信中考成绩. 学校的信誉就是你的招牌，没有信誉就很麻烦，他宁愿选择其他的方式，预先招一些人来反而更好. 有考试才有进步，一方面学生有求知欲，另一方面他也有强烈的“求胜”欲望，这是你要利用的. 当然我们不是说考得不好就打击，这个是不对的，但是学生有竞赛的欲望，这是需要利用的. 比如说跑步，他想跑前面这个很正常，这种欲望如果不加以利用，所有学生都一起走，实际上也很糟糕.

访谈者：现在电脑和网络在生活中使用非常普遍，人们会经常看一些碎片化的东西，有人说这样是不好的，尤其是对正在学习知识的中学生来说，也有人提出来支持碎片化，只需要看一个两三分钟的东西就能够大致了解一件事情或者一个概念. 您认为看碎片化的信息和认认真真看一本书，两者的利弊如何？

单墫：这两个很不一样，差得很远. 一个是系统地看一本书，那是很了不得的，那是真正的精华. "碎片化"当然也有好处，短平快，但是只能作为一个辅助手段，不能作为主要手段. 现在电脑包括手机在内消息来得很快，信息量也大，这个是有好处的，但是确实不够深入，有时也不够准确、不够真实. 真正过硬的东西，比如历史的，那就需要考证，不能道听途说，必须要有大量的论证，这个还是要靠看书的. 你要真正深入是要读书的，包括人的思想、思维的锻炼当然是要看书的，你不能光看一句话. 现在很多人在卖嘴皮，学了一两句话就在那儿卖弄，觉得说点只言片语就了不得了，这个很糟糕很不好.

但是学生能静下心来读读名著的确实少了，可能还是因为学生太忙. 学生为什么太忙？譬如我们说全面发展，实际是要求全优，这就有毛病了，这就很麻烦了. 每门都要学生达到一个 A，或者每门都达到 90 分，这是很难的. 其实全面发展，每门都能保持及格、良好就很好了. 但是现在的标准是每门都要达到 A，这个要达到是很难的.

一个人如果是天才，我想一定是大部分科目一般，有一两门突出，这样才是这方面的天才. 每门都好的学生也有，但是这个很费劲，不应该普遍这样要求. 每门都优秀，每年都评为五好、三好学生，那就不是为自己活了，而是为老师活了. 但现在的要求就是这样，所以这个很麻烦.

访谈者：您刚才也讲到，学生的负担越来越重，可能去做点儿自己感兴趣的事情的时间都很少有，他们也缺少读一本好书的时间.

单墫：就像我一个朋友的女儿读中学，每天都要搞到晚上 11 点钟，他女儿算是学得比较好的都要搞到 11 点钟. 我觉得一个校长应该严格把学生的作业量控制在晚上 9 点钟以前结束，如果 9 点钟这个作业不能做完就有问题，就必须要减少作业量，现在每科都有作业，而且现在瞎搞，有各种各样的考试. 还有一个很简单的方法，把学生的书包打开看看，你要减轻学生负担，去看看书包里有哪些书,哪些书不需要的统统扔掉.教育部门一定要下决定,就是考的东西要少、作业要少,要严格地控制作业量.

应该把时间腾出来给学生自己去玩,自己去看书,我看的书不算多,我一个中学同学,因为他跟图书馆老师认识,就整天在学校图书馆看书,他看的书很多,一本接一本地看,尤其外国小说他看得甚多.现在的中学生有时间的话也应该要看些,尤其要看原著、名著.这些书之所以能成为名著是有它的道理,很多好的思想在里面.要不然这么多年也不会留传下去,留传的时间比较长了,给大家的启示也比较深刻.经过时间淘汰掉一些,留下的就是金子.不像有些流行的东西,很快就会消失了.

我们现在实际是很浮躁的,对孩子的教育有点儿灌输式,而不是让他们自己去思考,然后学习到对自己有用的.这样做会把学习的积极性扼杀了,学生不喜欢学习,当然也就学不好了.现在虽然小学也会考到四大名著,学生都知道有这几本书,但根本没看过原著.现在对学生最重要的是要学习什么,怎么才能真的学到东西,要培养孩子们独立思考的能力,这对学习才有意义.

访谈后记 单墫老师是国内首批十八位博士之一,在这之前他做过十几年的中学教师,深知教学一线的问题和需要,因此他更想将学术与普及紧密地联系在一起.单老师从三十多年前就开始坚持一手抓学术、一手搞普及,直至今日,只要提到"数学解题"和"奥林匹克数学竞赛",大家都会不约而同地想到单老师.也正因如此,单老师在谈到中小学的数学教育、教学问题时,总是开诚布公地和我们分享他的看法.

单老师直爽的性格很有感染力,和他一起谈论教育热点问题时他毫不避讳,有问题就直说,从不拐弯抹角.他的语言也很"接地气",经常使用贴近生活的实例.比如他说到教育改革出现的问题时,说:"一天吃三顿就蛮好的,你不能改成一天吃五顿或者一天吃八顿,有些是不需要改的.如果早上喝牛奶,你说不喝牛奶改喝豆浆,这种改也没有多大意思.有人喜欢喝牛奶就喝牛奶,有人喜欢喝豆浆就喝豆浆,不要乱动."他也经常列举身边人的真实情况来分析热点问题,比如他的孙子和朋友的孩子都曾在访谈中提及.我们因此也能感受到单老师的实在,他很少高谈阔论,他更愿意在生活中发现问题、揭露问题.对单老师的采访,让我们真正地在思考,关于教育的各种问题,我们是否真的既能"仰望星空"亦可"脚踏实地".

苏格拉底的启发式*

Heuristic,通常译作启发式.西方认为是苏格拉底(前470—399)发明的.不过,苏格拉底并没有著作留下,他的言论基本上都出现在他的学生柏拉图(前427—347)的《对话集》中.(虽然稍后的第欧根尼·拉尔修认为柏拉图写了许多苏格拉底并没有说的话,并援引苏格拉底的话:"天哪,这个年轻人把多少我没说过的话说成我说的!")

1　体现苏格拉底启发式的数学例子

在柏拉图写的大多数对话中,苏格拉底谈的都是伦理方面的问题.但是《枚农篇》里,有一个(也是仅有的一个)数学例子,似乎知道的人并不多,因此先用一些篇幅介绍如下,以便分析.

苏格拉底:你告诉我,小厮(指枚农的一个随从,是家生子,即奴隶生的小奴隶),你知道这个四四方方的正方形吗?

小厮:知道.

苏格拉底:正方形的四条边是相等的吗?

小厮:是相等的.

苏格拉底:正方形能大些或小些吗?

小厮:能.

苏格拉底:如果这条边是二尺长,那条边也是二尺长,整个面积一共有几平方尺?

小厮:是四平方尺,苏格拉底.

苏格拉底:是不是可以有另外一个正方形,它的面积是这个正方形的二倍?

---

* 启发式,一般认为源自我国古代教育家孔子(前551—479).在《论语》"述而第七"中,子曰:"不愤不启,不悱不发.举一隅不以三隅反,则不复也."启、发两个同义词合在一起,就成了"启发".

小斯:是的.

苏格拉底:这另外一个正方形的面积是几平方尺?

小斯:八平方尺.

苏格拉底:好!你试着告诉我,这正方形的每一条边是几尺长.前一个正方形的四条边是二尺长,面积是它的二倍的正方形,四条边是几尺长呢?

小斯:很明显,苏格拉底,是二倍长.(笔者注:小斯自以为自己知道.)

苏格拉底:你告诉我,你是说由二倍长的边构成了面积是二倍的大正方形吗?

小斯:我认为是这样.

苏格拉底:那我们就给你画一个这样的正方形.这是不是你所谓的八平方尺面积的正方形?

小斯:就是.

苏格拉底:这个正方形里不是有四个与原正方形相等的面积吗?

小斯:是的.

苏格拉底:那它有多大?不是四倍大吗?

小斯:怎么不是?

苏格拉底:这四倍和二倍一样大吗?

小斯:绝不是,宙斯在上.(笔者注:小斯陷入自相矛盾的窘境)

苏格拉底:那么,由二倍长的边构成的并不是二倍大面积的正方形,而是四倍大面积的正方形.

小斯:你说的对.

苏格拉底:四的四倍是十六,不是吗?

小斯:是啊!

苏格拉底:那么八平方尺大的正方形是由多长的边构成的呢?它不是该由一条大于原来的边,小于所作边的边构成的吗?对不对?

小斯:我是这样想的.

苏格拉底:很好!你要永远照自己想的回答问题.告诉我,这条边是不是二尺,那条边是不是四尺?

小斯:是的.

苏格拉底:那么,那八平方尺大正方形的边就该大于二尺,小于四尺.

小厮:是啊.

苏格拉底:那你就试着说,你认为它有多长.

小厮:三尺长.

苏格拉底:三尺乘三尺是几平方尺呢?

小厮:九平方尺.

苏格拉底:那么,用三尺长的边也不能构成那八平方尺的正方形哦.

小厮:当然不.(笔者注:小厮再次陷入自相矛盾的窘境)

苏格拉底:那你给我们确切地说出是由哪样的边构成的.如果你不愿意说出数据,只要指出由什么边构成就行.

小厮:宙斯在上,苏格拉底啊,这个我并不知道.(笔者注:小厮认识到自己并不知,由此产生困惑.)

苏格拉底:枚农啊,你仔细瞧瞧,他在回忆中前进了多远,一开头,他完全不知道八平方尺正方形是怎么样的,正如他现在仍然不知道一样.只是那时他认为自己知道,并且冒冒失失地作出回答.好像胸有成竹似的,不认为自己遇到困难.现在他认为自己遇到了困难,他对此无知,也认为自己对此无知.

枚农:你说的对.

苏格拉底:在他所不知道的那件事情上,他现在不是处在较好的状态吗?

枚农:我也这样想.

苏格拉底:我们使他陷于困惑,使他像遇到电鳗时那样发呆,有没有伤害他?

枚农:我想没有.

苏格拉底:我们倒是预先给他作了一些启发,使他能够去发现这一方面的真理,因为现在他很乐意寻求它,以弥补自己的无知.当初他却认为自己可以毫无困难地高谈阔论,说二倍面积的正方形必须由二倍长的边构成.

枚农:看来是这样.

苏格拉底:你有没有想到,他曾经努力去寻求或学习他以为知道而并不知道的东西,后来才被启发到有所怀疑,承认自己无知,因而力求认知?

枚农:我没有想到,苏格拉底.

苏格拉底:使他困惑对他有用吗?

枚农:我想有用.

苏格拉底:现在你再看看,他如何从这种困惑出发,跟我一同寻求,并有所发现.虽然我只是询问他,并没有给他传授什么东西.你仔细观察,看看你是不是在什么地方发觉我在教他,暗示他.总之,我们做的不过是询问他心里想的.

苏格拉底:你告诉我,这不是我们那个四平方尺的正方形吗?你懂吗?

小厮:我懂.

苏格拉底:我们能加上一个与它面积相等的正方形,再加上一个与这两个正方形一样大的面积,然后填上欠缺的那个角,这不是四个一样大的正方形吗?

小厮:是啊!

苏格拉底:整个面积是原来正方形的几倍?

小厮:四倍.

苏格拉底:我们要得到的却是二倍的面积.你不记得吗?

小厮:正是.

苏格拉底:从一个角到另一个角的那条线(即正方形的对角线)不是将这四个正方形的每一个分成两个相等的部分吗?

小厮:是啊.

苏格拉底:这四条相等的线不是围成一个正方形的吗?

小厮:是这样.

苏格拉底:你看这个正方形的面积有多大?

小厮:我不了解.

苏格拉底:在这四个面积相等的正方形中,每一条线不是把每个面积分成两半吗?对不对?

小厮:是啊.

苏格拉底:在围成的那个正方形里,一共有几个相等的面积?

小厮:四个.

苏格拉底:在原来的正方形里有几个?

小厮:两个.

苏格拉底:四与二的比是什么?

小厮:是二倍.

苏格拉底:这样看来,围成的正方形面积是几平方尺?

小厮:八平方尺.

苏格拉底:由什么线构成的?

小斯:由这些线构成的.

苏格拉底:这些线智者们称之为对角线.枚农的小厮啊,你就由这原来正方形的对角线画出了一个二倍大的正方形.

小斯:那当然,苏格拉底.(笔者注:小厮从不知到知)

苏格拉底:你认为如何,枚农?他是不是提出了一个原来不属于他的答案?

2　苏格拉底启发式的基本步骤及思考

从上面的例子可以看出,苏格拉底的启发式有以下步骤.

(1) 提出一个有意义的问题作为对话的主题

这个问题应当是重要的,有深度的,不是浮浅的,无聊的.苏格拉底选择的问题都是意义重大的,在伦理方面是这样,如关于美德和正义等问题.在数学方面也是如此,这次与小厮讨论的问题是:如何作一个面积为已知正方形面积二倍的正方形?如果运用勾股定理,可以知道所求正方形的边长是已知正方形边长的$\sqrt{2}$倍.而$\sqrt{2}$正是毕达哥拉斯学派所发现的无理数.发现无理数或无公度的线段导致数学史上著名的第一次数学危机(数学大发展的机会),这是那个时代最有深度的研究课题.另一方面,如果将问题从平面推广到空间,那么相应的问题就是立方倍积,也是著名的几何作图三大难题之一.

讨论的主题一般应由教师(启发者)提出,因为只有教师(不是学生)才能对所学学科的内容有整体的把握.苏格拉底的启发式利用学生已有的知识和经验作简单、自然的过渡,目标明确地进入主题,将一个有意义的数学问题提出来讨论.他没有故弄玄虚地先搞一大段"情境设计",更没有牵强附会地"联系实际".这表明他是真正懂得数学的,真正能引导数学讨论的.(所以他的弟子柏拉图才会在学园门口挂出"不懂几何者不得入内 "的牌子)

(2) 用一系列的反诘,使学生(被启发者)陷入矛盾的窘境.由原来的自以为是逐渐认识、承认自己的无知,进入困惑的状态

这是关键的一步,也是苏格拉底启发式的主要特征.因为"偏见比无知离真理更远",如果学生不虚心,自以为是,坚持成见,那么他就不可能主动获得任何

新鲜的知识和深刻的思想. 所以,必须首先清除被启发者头脑中的那些陋见、偏见以及似是而非的东西.

困惑对于学生是极为有用的,它能引发学生积极的思维活动. 同时承认自己的无知,才能力求认知. 这种“不知为不知”的状态也就是孔夫子所说的“愤”和“悱”. 我们常常看到一些公开课上,教师的问题刚一说出,学生不需深入思考便能很快回答,完全没有困惑的阶段. 这不是启发式,至少不是苏格拉底的启发式.

(3) 在困惑的基础上,经过启发使学生(被启发者)找到正确的答案,从而学生由承认自己的无知到知其所知

苏格拉底根据前面小厮的回答,作出一个边长为四尺的正方形. 小厮已经知道它应由四个(而不是两个)已知的正方形组成. 苏格拉底只是在这四个正方形中各添一条线(对角线),小厮由此知道这四条线围成的图形是正方形,而且面积是已知正方形的两倍.

苏格拉底的启发简洁明了,紧扣主题. 虽然从原来的问题中,可以产生出许多新问题,例如存在性问题(所要作的正方形是否存在),计算问题(所作正方形的边长是多少),有无公度问题(所作正方形的边长是否是无理数)等. 但苏格拉底抓住作正方形这个主要问题,决不节外生枝. 如果面对许多学生,当然更应不枝不蔓,不能超越多数学生的潜在发展水平. 首先,针对一般学生解决主要问题. 同时,可将有关问题留给学有余力的学生作进一步思考和研究.

苏格拉底启发学生的方法主要是对话,并且主要以问答的方式与学生进行对话,以至于后人容易误将启发式和问答法画上等号. 在当前班级授课制的教学组织形式下,不仅应注重对话的外在形式,更要注重对话的精神和实质. 这里的对话已不限于师生双方言语上简单的你来我往,更是思想和精神上的深层次沟通和平等交流. 这样才能促使学生积极主动地进行思考,提高启发式教学的有效性.

单墫谈数学教育改革

2011 年 3 月,《时代学习报》编辑就目前大家都关注的数学课程改革问题,对单墫教授做了一次专访. 单墫教授对课程改革思索良多,既指出了问题,也提出了许多中肯的建议. 下面的访谈录,希望对当前的教育教学有所启迪和帮助.

访谈者: 近年来基础教育做了很多改革,包括课程标准、教学模式等,请您谈谈对目前中学数学教育的看法.

单墫: 我有时会想,教育是不是一定要改? 要经常改? 好比我们的路是否一定要经常修,年年修呢? 现在用的新课标,说实话,我是不太赞成. 其实我们的教育本来还是挺不错的,文革后基本走上正轨,从 1977 年恢复高考到 20 世纪 90 年代前后我国的教育都是不错的,这样下去很好. 事实上教育应有稳定性和长期性.

访谈者: 经常听到对中国学生的评价,比如学得比较死,动手能力不足,您对这个问题怎么看?

单墫: 这种说法不对. 说中国学生动手能力差? 谁能证明? 就说开车,20 世纪 80 年代因为国内没有车,大多数人都不会开车,而现在的年轻人很多都会开,不会的学学,很快也能掌握. 玩电脑也是,很多孩子家里有电脑,所以他很快能掌握,能玩得很好. 要是没有电脑的地方,再天才的孩子也不会玩啊! 你怎么能说这个没电脑的孩子动手能力就差呢? 中国很多做实验的人都做得很不错,比如吴健雄、李远哲,并不差. 现在有人还相信中国孩子就是动手能力差,只会做题,还以这个来改革,这是不合理的.

访谈者: 初中的教材教辅有意识地增加动手操作,突显动手能力,您如何评价这一现象?

单墫: 各个学科有自己的特点,物理化学培养动手,而数学应当培养思维,不能抹杀特点. 小学升到初中后,孩子的心理和生理都会有很大变化,必须培养孩子从感性到理性的思维,初中学习平面几何就是要告诉孩子眼见不一定属实,要否定一些感性的东西,强调理性的东西. 而课改把平面几何删掉很多,我认为是不对的！平面几何很重要！爱因斯坦提到人类发展两大因素：物理实验科学和欧几里得平面几何体系.前者是动手，后者是思维体系，它的伟大之处是公理体系，不是定理罗列.

访谈者：当前的数学教学比较重视动手、体验、合情推理，有些教材到八年级后面的章节才正式地讲解演绎推理，您认为这样设置有问题吗？

单墫：这么迟才教"证明"，显然是晚了.其实也不能怪教材，和课标有关系.数学是门严谨的学问，尤其平面几何，是有一个体系的，前几个知识点告诉你"是什么"，下面就要告诉你"为什么".现在你只讲了"是什么"，"为什么"要到一年以后再讲，这个体系就被切断了，思维探究的精神就弱了.如果初中不学好平面几何，高中立体几何上来了，更复杂的图形出来了，可学生还没有建立起正确的思维，理性概括能力、抽象能力、科学精神都不足.基本的训练错过了，高中根本补不起来.

初中生数学水平有所下降，一是推理能力，二是运算能力.乘法公式由7个变成3个，外国人也是学7个啊，我们以前也是7个，也能行啊！只剩3个，其实这样学生更难.本来有很大的地方培养思维，现在呢？就好像剩个小圆桌要你翻跟斗，结果一翻就掉下去了，没办法施展，教材这样处理对培养思维是不利的.

访谈者：感觉当前数学教学中有些地方思维跨度比较大，有跳跃，包括一些内容的安排.

单墫：以前编的教材体系严谨，习题配套紧密.现在的课本里面，有的习题忽然很难，偷偷把奥数题目塞入习题里面，叫做培养思维能力.习题要以基本题为主.这又是课标的问题，课标里面的思维有些互相矛盾.一个是低要求的"人人学数学"，另一个是高标准的"培养学生发现能力和创造能力"，这两点本身就矛盾.前者要求不高，而后者要求的具有发现和创造能力的人能有百分之一就不错了.学校是培养一般人才的地方，教材也应该照顾多数.天才是自己产生的.这两点本身就矛盾，又是低要求又是高标准，这两个目标混在一起放在教材里，教材肯定要高一脚低一脚了.好的学生可以去培养，但是所谓培养"创造能力"并不是学校的普遍任务.

我看到一本教材是这样培养发现能力和创造能力的：让学生去填负指数幂的值.学生能这么有创造能力？这个我当学生时候也发现不了，也没兴趣去发现,这就是把学生当天才.

访谈者: 现在的习题除基础题外,也出现较多开放题、创新题、探究题,您如何评价这些题目形式?

单墫: 开放题、探究题很多就是原来的奥数题.放一些在平时的练习中也行,但要选择一些好的、有思想的题目.不是光有难度,有的题目孩子可能一下子没做出来,但如果看了答案后也能恍然大悟,觉得并非高不可攀,其实我也能想到,或许下次他就能做出来,这就很好!而有的超纲的题本身就不是正道的题,不要去做!还有的题命题本身就有问题,很简单的题答案很繁琐,算术题非要用方程做,简单数学题非要数形结合或者用函数做,这样挫伤了孩子积极性,还不如不做.

访谈者: 您对做题方面颇有心得,就做题方面请再谈谈您的建议.

单墫: 学数学还是要做题,这不是因为我是搞数学的,应该说这是数学学科的特点.学生做习题也是一种发现能力的培养.当然做题更讲究质量,有的老师布置很多题给学生但收效甚微,其实好好分析一张认真做过的试卷很重要,能事半功倍.大家在选择辅导书和教材的时候也要注意,现在有的书写得不理想,我最近看过一本书,里面题目解法很差很怪,绕很多路求解.比如用纯代数就能解的不等式,那书里非要构造几何模型去解,太麻烦,误人子弟!关于一题多解也是这样,学生可以多找些解法以拓宽思路,而教师应该帮助学生找出最简便的途径,告诉大家为什么要这么做,而不能对着答案照本宣科就完事.

访谈者: 新课改中常有优质课展示,许多示范课与以往传统的课堂模式有很大区别.单老师肯定也熟悉现在的课堂模式,请谈谈您的评价.

单墫: 我看过一些优质课的录像,我觉得有的名师的观摩课缺乏数学内容.有的所谓名师课上用太多时间介绍新的文化,新的观点理念等,许多东西在数学上看都是"空"的.我见过一节课,是介绍路程、速度、时间公式的.一开始,老师"嗖"地一下从讲台窜到门口,大家没闹清怎么回事,老师问还有没有比自己快的?大家说还有运动员.再讨论还有更快的吗?大家七嘴八舌说还有鸵鸟、猎豹……一节课聊了很多跟数学无关的东西,浪费了大量时间,这对学生是"犯罪".我觉得数学课首先要掌握好数学课本的知识,不要为了创设所谓情境扯太多课外的东西来哗众取宠,这样的模式化之风让人很担忧!更滑稽的是,有一次居然听到一个老师说出这样一句"三个点就能搞定一个平面".怎么能说

“搞定”？“不在同一直线上的三点确定一个平面”，这存在了千百年的公理哪能随便改动？估计老师自己还觉得挺风趣的，其实这种语言很不合适！数学是不能胡来的，课堂可以轻松点，但是老师的语言、板演必须严谨.

访谈者： 请谈谈对年轻教师的建议和期望.

单墫： 作为一位教师，你的一言一行直接影响着每一位学生，所以严格要求自己是最基本，也是最重要的. 教师要不断学习，要多看书，提高自己的修养. 数学教师要做题，你看深圳中学引进教师就是做两套考卷，一份奥数卷，一份高考模拟卷，看看水平如何. 一个数学教师要是一般数学题不能做还怎么跟学生讲？教师备课就是备题啊，给学生做的题首先自己要会做，做得好！数学教师不做题就会退化.

访谈者： 对数学竞赛的开展，您认为怎样才合理？如何找到平衡点？

单墫： “禁”很荒唐，这本身不是违法的东西，强迫人人参与或者阻止大家参与都是错误的，这应该是在自愿的基础上组织起来的. 学校愿不愿意组织那是学校的事，就好像体育竞赛或者游戏一样，哪怕班级之间也可以比一比，赛一赛，玩一玩. 奥数原没有罪，在国外，奥数存在了几十年，并没有遭遇像我们这样密集的“炮轰”. 这是升学造成的，如果小升初没有这些奥数的内容，仅仅考课本，还有这些禁考禁组织的事吗？好学校招生还是觉得考查数学比较靠得住嘛！但是把奥数跟升学挂钩又是有些部门自己搞起来的，现在禁奥数是自己否定自己. 我认为不要禁止或者提倡，应该让它自由发展……张景中院士不是说嘛，该禁的东西有很多，禁奥数实在没道理.

在国外，这是公众基本接受的事实. 因此，家长对于自己的孩子要有比较准确的定位. 但在我们这里，尤其在城市里，几乎 100％的家长都希望让自己孩子成为前 5％～10％，并按照前 5％～10％的标准来要求孩子，奥数就是这样走火入魔的. 不应该无视孩子有无兴趣、有无能力，都逼着孩子学奥数.

《平面几何的知识与问题》前言

平面几何真美.

平面几何美. 它有美丽的图形:稳定的三角形,端方的正方形,滴溜滚圆的圆,黄金分割生成的五角星等. 图形之间有着紧密的联系:点的共线,线的共点,三角形的全等、相似,直线与圆的相切,圆与圆的相切等. 这些图形还可以动起来,可以作种种变换:平移,旋转,对称,放大与缩小等. 令人目不暇接,美不胜收!

平面几何真. 它有许多重要、深刻的结论,揭示了图形的本质与内在联系. 这些结论是颠扑不破的真理,因为它们经过了严格的数学证明. 在这些结论的基础上,欧几里得建立了一个几何体系,它是人类的伟大成就,是近代科学发展的两大基础之一(参见《爱因斯坦文集》第一卷,商务印书馆,1977 年,574 页).

徐光启说:"几何使人缜密."认真学习平面几何就能学会推理,学会证明,学会理性的思维,学会严谨的表述.

青少年朋友很有必要学习平面几何.

现在有些课本几何内容贫乏,缺少系统. 而很多奥数书中的几何题又往往偏难,偏深. 本书将提供充分的几何知识,并形成一个简明的体系. 书中问题分为三个层次. 第一层次注重基础. 第二层次略有难度,培养能力. 第三层次才以竞赛问题为主. 读者可以根据自己的需要进行选择.

这本书是我为初中同学写的三种读物[①]的第一种. 希望青少年朋友喜欢它.

---

① 另两种是《代数的魅力与技巧》《数的故事与灵活策略》.

《奥数教程》前言

据说在很多国家,特别是美国,孩子们害怕数学,把数学作为“不受欢迎的学科”. 但在中国,情况很不相同,很多少年儿童喜爱数学,数学成绩也都很好. 的确,数学是中国人擅长的学科,如果在美国的中小学,你见到几个中国学生,那么全班数学的前几名就非他们莫属.

在数(shǔ)数(shù)阶段,中国儿童就显出优势.

中国人能用一只手表示 1～10,而很多国家的人非用两只手不可.

中国人早就有位数的概念,而且采用最方便的十进制(不少国家至今还有 12 进制、60 进制的残余).

中国文字都是单音节,易于背诵,例如乘法表,学生很快就能掌握,再“傻”的人也都知道“不管三七二十一”. 但外国人,一学乘法,头就大了. 不信,请你用英语背一下乘法表,真是佶屈聱牙,难以成诵.

圆周率 $\pi=3.14159\cdots$. 背到小数点后五位,中国人花一两分钟就够了. 可是俄国人为了背这几个数字,专门写了一首诗,第一句三个单词,第二句一个单词……要背 π 先背诗,我们看来简直自找麻烦,可他们还作为记忆的妙法.

四则运算应用题及其算术解法,也是中国数学的一大特色. 从很古的时候开始,中国人就编了很多应用题,或联系实际,或饶有兴趣,解法简洁优雅,机敏而又多种多样,有助于提高学生学习兴趣,启迪学生智慧. 例如:

“一百个和尚一百个馒头,大和尚一个人吃三个,小和尚三个人吃一个,问有几个大和尚,几个小和尚?”

外国人多半只会列方程解. 中国却有多种算术解法,如将每个大和尚“变”成 9 个小和尚,100 个馒头表明小和尚是 300 个,多出 200 个和尚,是由于每个大和尚变成小和尚,多变出 8 个,从而 $200\div8=25$ 即是大和尚人数,小和尚自然是 75 人. 或将一个大和尚与 3 个小和尚编成一组,平均每人吃一个馒头. 恰好与总体的平均数相等,所以大和尚与小和尚这样编组后不多不少,即大和尚是 $100\div(3+1)=25$ 人,小和尚自然也是 75 人.

中国人善于计算，尤其善于心算. 古代还有人会用手指计算（所谓“掐指一算”）. 同时，中国很早就有计算的器械，如算筹、算盘. 后者可以说是计算机的雏形.

在数学的入门阶段——算术的学习中，我国的优势显然，所以数学往往是我国聪明的孩子喜爱的学科.

几何推理，在我国古代并不发达（但关于几何图形的计算，我国有不少论著），比希腊人稍逊一筹. 但是，中国人善于向别人学习. 目前我国中学生的几何水平，在世界上遥遥领先. 曾有一个外国教育代表团来到我国一个初中班，他们认为所教的几何内容太深，学生不可能接受，但听课之后，不得不承认这些内容中国的学生不但能够理解，而且掌握得很好.

我国数学教育成绩显著. 在国际数学竞赛中，我国选手获得众多奖牌就是最有力的证明. 从 1986 年我国正式派队参加国际数学奥林匹克以来，中国队已经获得了 10 余次团体冠军. 成绩骄人. 当代著名数学家陈省身先生曾对此特别赞赏. 他说“今年一件值得庆祝的事，是中国在国际数学竞赛中获得第一……去年也是第一名. ”（陈省身 1990 年 10 月在台湾成功大学的讲演“怎样把中国建为数学大国”）

陈省身先生还预言：“中国将在 21 世纪成为数学大国. ”

成为数学大国，当然不是一件容易的事，不可能一蹴而就，它需要坚持不懈的努力. 我们编写这套丛书，目的就是：(1) 进一步普及数学知识，使数学为更多的青少年喜爱，帮助他们取得好的成绩；(2) 使喜爱数学的同学得到更好的发展，通过这套丛书，学到更多的知识和方法.

“天下大事，必作于细. ”我们希望，而且相信，这套丛书的出版，在使我国成为数学大国的努力中，能起到一点作用. 本丛书初版于 2000 年，2003 年修订过一次，现根据课程改革的要求对各册再作不同程度的修订.

著名数学家、中国科学院院士、原中国数学奥林匹克委员会主席王元先生担任本丛书顾问，并为青少年数学爱好者题词，我们表示衷心的感谢. 还要感谢华东师范大学出版社及倪明先生，没有他们，这套丛书不会是现在这个样子.

《单墫数学科普著作集》序

科学昌明,既需要科学家筚路蓝缕、披荆斩棘,也需要普及工作者耕耘播种、热心培育.

普及工作很重要.如果将科学研究比作金字塔的塔尖,那么普及工作就是金字塔的底.底宽,塔高.

科学研究不容易.从事研究,需要才能、努力与机遇.能够从事研究的人不多,好像阳春白雪,曲高和寡.他们的成果需要普及工作者通俗化、趣味化,才能广为人知,才能使更多的人关心、了解、理解,才能引起公众的兴趣,吸引更多的新人一同参加研究.Fermat 大定理就是一个典型的例子.虽然只有 Wiles 一个人给出了证明,看懂证明的不过十几个人或几十个人,但对大定理感兴趣的人成千上万.他们都是普及读物的读者.

普及工作使千万人受益,我就是其中之一.

在学生时代,我读过不少数学普及读物.如刘薰宇的《数学园地》,孙泽瀛的《数学方法趣引》,许莼舫的《几何定理和证题》,Casey 的《近世几何学初编》,日本数学家林鹤一的《初等几何作图不能问题》,上野清的《大代数讲义》,苏联数学家写的数学丛书中的《摆线》《双曲线函数》等,以及稍后中国数学家华罗庚等写的数学丛书中的《从杨辉三角谈起》等.还在《科学画报》上看到谈祥柏先生写的妙趣横生的文章《奇妙的联系》等.这些数学读物不仅使我学到许多数学知识、方法和思想,眼界大开,而且使我对数学产生了浓厚的兴趣,甚至立志要当一名数学家.

但当数学家的梦想却难以实现,因为那时政治运动频仍.读书,被认为"走白专道路",会横遭批判.四人帮倒台后,我才有幸到中国科学技术大学做研究生,在 1983 年成为首批 18 名国产博士之一.但这一年我已年届不惑,从事数学研究的黄金时期业已过去.我觉得与其花费时间凑一些垃圾论文,不如做普及工作对社会更有贡献.

对普及工作,我有浓厚的兴趣,也有一定的基础:

一是由于做过一些研究工作,能够了解较新的材料,能够较为准确地把握数学及有关史料.

二是由于当过多年教师，文字也还通顺，能够注意趣味性与深入浅出. 1977年恢复高考后，一度出现读书的热潮. 这时常庚哲先生带头写了《抽屉原则及其他》，受到普遍的好评. 稍后，上海教育出版社的王文才、赵斌两位编辑邀我写稿，我就写了《几何不等式》《趣味的图论问题》，在1980年出版. 以后又陆陆续续写了《覆盖》《组合数学的问题与方法》《趣味数论》《棋盘上的数学》《解析几何的技巧》《算两次》《集合及其子集》《组合几何》《对应》《国际数学竞赛解题方法》（与葛军合作）、《不定方程》（与余红兵合作）、《巧解应用题》《因式分解》《平面几何中的小花》《数学竞赛史话》《解题思路训练》《十个有趣的数学问题》《概率与期望》《小学数学趣题巧解》《快乐的数学》《数列与数学归纳法》《解题研究》《数学竞赛研究教程》等.

文革后，大家渴望读书. 而此前的书大多毁于文革劫火. 因此新出的书颇受欢迎，其中也包括了我写的小册子.

冯克勤先生说："不要小看了这些小册子，它们将数学的美带给大众."（冯克勤《评审意见》）

杨世明、杨学枝先生说："直到1980年，大家才盼来单墫的《几何不等式》一书……不仅普及了基础知识、基本思想方法，而且激发了研究兴趣. 今天初等不等式研究中的许多骨干，都曾从该书获益. 单墫的《几何不等式》一书，无疑是这一阶段的标志性的著作."（杨世明、杨学枝《初等不等式在中国》，中学数学研究（广州），2007第1期）.

还有一些数学教师见到我客气地说："我们都是读您的书长大的."

这些评论当然是过奖的溢美之词，但也说明普及工作是一件有意义的、值得去做的事情.

近年来，急功近利的风气在学校蔓延. 要根治这种歪风，还得提倡读书. 要使广大青少年"热爱知识，渴求学问"（卡耐基《林肯传》，人民文学出版社，16页）.

首先得多出一些好书，供大家阅读.

读书是天下第一件好事. 读好书是人生第一件乐事. 好读书，读好书，进步就迅速. 有些学生学数学，只做题，从不看书. 这种做法是难以进步的. 感谢华东师范大学出版社出版我的科普著作集. 这7种小册子修订后，重新出版，希望能有较多的读者，特别是青少年读者. 希望它们能给爱好数学的朋友们带来乐趣.

《趣味的图论问题》再版前言

这本小册子,是31年前(指1980年)写的.

其时"文化大革命"刚刚过去,广大青少年迫切需要学习科学文化. 我的几本小册子就是作雪中送炭之用.

图论,当时国内很少有人研究. 中学界,更是乏人问津. 中国人写的系统介绍图论的普及读物,这本《趣味的图论问题》,或许是第一本. 我写的时候,缺少借鉴,甚至很多名词术语的中译,都得自己杜撰.

很快,图论的研究就在我国迅猛地发展起来. 数学竞赛中,图论的问题也频繁出现. 这本小册子在普及、传播数学方面起了一点作用,对于参加数学竞赛活动的师生也提供了一点帮助.

这次再版,表明这样的书仍有读者需要. 特别是,目前的中学师生,他们中很多人以前没有看过这本书.

这次再版,我曾考虑要不要作较大的更动. 仔细想了想,觉得还是保持原貌,基本不动为好. 因为本书仍然有两个方面的作用:

第一,普及、传播数学. 这本书对于图论作了较为系统(当然也是初步的)的介绍. 当时就有一位复旦大学学图论的青年教师(现在是复旦大学的教授)告诉我:他们的老师推荐他们读这本书. 目前市场上有不少的数学书,实际上只是习题集. 有些人以为这种书可以立竿见影,收到奇效. 其实,学习需要循序渐进,按部就班,"欲速则不达". 做习题,不能代替系统的学习. 只有对数学的理解比其他人高出一筹的学生,才有可能在前进的道路上走得更好.

第二,有利于数学竞赛. 书中例习题较多,有各种难度,并有解答. 适合各种层次的学生选用. 而且对问题的背景与蕴含的思想,着重介绍. 目前我国的数学竞赛,题目有愈来愈难的趋势. 一味追求难,而忽视其中的数学思想,不是一种好的趋势.

所以,这次再版,仅作了很少的改动. 例如,术语"偶图"改成现在通用的"两部分图".

感谢中国科学技术大学出版社重新出版这本小书.

《几何不等式》再版前言

这本小册子,是 31 年前(指 1980 年)写的.

其时"文化大革命"刚刚过去,广大青少年迫切需要学习科学文化.我的几本小册子就是作雪中送炭之用.

几何不等式,当时国内很少有人研究.中学界,更是乏人问津.中国人写的普及读物,这本《几何不等式》,或许是第一本.当时颇有些影响,例如杨世明、杨学枝先生说:

"直到 1980 年,大家才盼来单墫的《几何不等式》一书……不仅普及了基础知识、基本思想方法,而且激发了研究兴趣.今天初等不等式研究中的许多骨干,都曾从该书获益.单墫的《几何不等式》一书,无疑是这一阶段的标志性的著作."(杨世明、杨学枝《初等不等式在中国》,中学数学研究(广州),2007 年第 1 期).

很快,几何不等式的研究就在我国迅猛地发展起来.数学竞赛中,几何不等式的问题也频繁出现.这本小册子在普及、传播数学方面起了一点作用,对于参加数学竞赛活动的师生也提供了一点帮助.

这次再版,表明这样的书仍有读者需要.特别是,目前的中学师生,他们中很多人以前没有看过这本书.

这次再版,我曾考虑要不要作较大的更动.仔细想了想,觉得还是保持原貌,基本不动为好.理由是:

第一,这本书,对于老的读者,已成为历史,不应多改.

第二,对于新的读者,这本书的内容,也还值得一读.

第三,这本书的主要精神是激发读者的研究兴趣.书中有些地方,可以改进或推广.留给读者去发现,去探索,比作者改写更好.

所以,这次再版,仅作了很少的改动.

感谢中国科学技术大学出版社重新出版这本小书.

《近代欧氏几何学》译者赘言

几何学历史悠久,自欧几里得算起,也已经有两千多年.平面欧几里得几何,既有优美的图形,令人赏心悦目;又有众多的问题,供大家思考探索.它的论证严谨而优雅,命题美丽而精致.入门不难,魅力无限.因此吸引了大批业余的数学家与数学爱好者,在这里大显身手.

平面欧几里得几何学是一座丰富的宝藏.经过两千多年的采掘,大部分菁华已经落入人类手中.然而,在19世纪后半叶,又发现了一个宝库,得出不少新的结果,当时称为近世几何学.约翰逊(R. A. Johnson)的这本书,就是对这一部分内容的一个很好的介绍,问世以来深受欢迎,被欧美不少大学选作教材.直到20世纪60年代,还有新的版本出现.

四九年(指1949年)以前,本书曾由邱丕荣先生翻译,在国内颇有影响.我在中学读书时,就曾与数学小组的朋友们一起学习过.但邱译本是文言意译,每每有与原文不尽符合的地方,也不太适合现在使用白话文的读者.同时,邱译本早已绝版,经过文革劫火,更是很难见到.因此,应当重译此书以适应各方面的需要.

欧几里得几何能否又一次再现辉煌?未来的事难以逆料.或许如著名数学家杨(J. W. Young)在介绍本书时所说的,数学像服装一样,往往重复过去的时尚.至少,几何的训练对于人,是非常需要的.在1998年美国科学年会上,学者们一致认为21世纪的教育应把几何学放在头等重要的位置.硅谷的马克斯韦尔等人甚至喊出"几何学万岁"的口号.由此可见,重译此书确实很有必要.

应上海教育出版社叶中豪先生之邀,承乏重译这本名著.每日课余,"爬"稿纸十页,终于完成任务.但这书篇幅不小,我又老眼昏花,恐怕难免有误译或不妥之处,敬请读者指正.

《近代欧氏几何学》译本再版，再说几句

写近代欧氏几何学，约翰逊这本书最佳，内容丰富，体系严谨，叙述简明，引人入胜.

我们的译本问世后，颇受欢迎，很快销售一空. 不少人询问怎么能买到这本书. 但上海教育出版社负责这本书的编辑叶中豪先生已经离职，该社近期无人关心这书重印的事. 哈尔滨工业大学出版社愿意重新出版这本书，满足了很多人的希望与要求，做了一件大好事.

借这次重新出版的机会，我们将全书仔细校勘一遍. 十分欣慰的是，书中没有严重的错误. 发现的问题极少，基本上都是印刷错误（而且大都属于原著）. 曾有位认真的读者找出一些“严重”错误，但我们研究后发现其实都不是错误，而是这位读者没有读懂. 例如 § 298 的 f，这位读者认为结论 $\overline{OO_1}+\overline{OO_2}+\overline{OO_3}=R+r$ 只对锐角三角形成立. 其实其中 $\overline{OO_1}$ 等都是有向线段，结论对于一切三角形均是对的. 由此我们想到读书的态度问题. 读书时，首先要虚心，向书本学习. 对于经典著作，更应当有一点敬畏之心. 当然不必盲目崇拜，但绝不可过分自以为是. 发现书中与自己想法不合的地方，应当先更多地考虑是否自己没有正确理解，而不要断然认为书上错了. 这次校勘，我们也特别注意绝不随意“纠正”作者. 我们随意“纠正”作者，谁来纠正我们？所以，我们尊重原著，绝不乱改. 同时，我们希望读者多加批评.

感谢对本书提出宝贵意见的王曦、李毅等读者.

《初等数学复习及研究(平面几何)习题解答》序

平面几何,既有优美的图形,又有众多的问题,可以培养想象、思维、论证以及表达的能力,是很多人青睐的数学分支.当下的数学竞赛中,无论是国内还是国际,平面几何的"幽灵"都不断地出现.

平面几何的著作,中文的,以梁绍鸿先生的《初等数学复习及研究(平面几何)》为最佳.这本书,内容丰富,结构谨严.不仅有大量的例题,而且作者独具慧眼,侧重于如何解题.书中既介绍推证通法,又分专题(相等,和差倍分与代数证法,不等,垂直与平行,共线点,共点线,共圆点,共点圆,线段计算)讲述证题术.此书一出,洛阳纸贵.当时国内师范院校数学系均以这本书作为教材或主要参考书.时至今日,对于中学教师及准备参加数学竞赛的中学生,这本书仍然是应当仔细研读的经典著作.

梁先生的书有大量的习题,有的难度甚大.谁能将习题全部做出,平面几何的功力当然不凡.但做这些习题,需要花费很多时间,克服很多困难.若非嗜好平面几何的人,多数读者或望而生畏,逡巡不前;或浅尝即止,半途而废;或心有余而力不足;或深感"我面前缺少一个指引的人".

尚强先生早在17岁(1979年)时,即以惊人的毅力将梁书的习题解之过半.后来又将解答不断完善,整理成书,并于1985年由中国展望出版社出版,但印数不多,很快售罄.这次重新出版,非常符合社会的需求.数学教师可以借助这本题解,随时回答学生所提出的几何问题,免去许多演习的时间,以及临时无法解出的尴尬.勤勉的学生得到这本解答,可以与自己的解法对照,获取印证的快乐.遇到解不出的题,也可以得到提示,有所启发(当然,不要急着看完整的解答),不致陷入困境,无法解脱.

撰写、出版这本习题解答是一件大有功德的事情,造福千万学子.因此,对于尚强先生的辛勤劳动与付出的心血,我们应当说一声:"Thank you very much."

《数学文化与文化数学》序

数学是文化.这个论题,在这部大作中已经有深入的讨论、详细的阐述,不需要我来饶舌.我只想强调一点(本书也已说到):因为数学是文化,而且是人类文化中不可或缺的部分,所以每个人都应当学习数学.

数学中有什么东西是非得学习、无法替代的呢?数学是求真的,这与其他学科相同.但数学还给出了真的判定方法,那就是证明,即由公理出发,经过严密的逻辑推理,得出命题的真实性.

证明就是数学中最重要的、无法替代的东西.证明的大前提,不是一两个人(哪怕是“专家”或“大人物”)的私见,也不是某党某派的偏见,而是公理.证明的过程中依靠的不是耳闻目见的“事实”,也不是实验或归纳的结果,而是严密的逻辑推理.因此数学中的定理(证明了的结论)是确凿可靠的真理.这是人所共知的.人们形容一个结论可信,常说“就像 1+1=2 那样”.

波普尔(Karl Popper, 1902—1994)认为归纳法得出的结论不是放之四海而皆准的,这个看法当然是对的.但数学证明采用的是演绎法,数学中的结论即使由归纳法产生,也都必须经过证明才被承认.所以波普尔学派对数学的一些看法,如数学也是可误的等,乃是不正确的误解.这种数学可误的观点在我国一度甚嚣尘上,表明我国有一些人善于跟在大人物后面鼓噪,却没有真正理解什么是数学的证明,不知道证明才是判定是非的方法.

本书的作者尚强兄,年富力强,从事数学教育与数学研究,对平面几何尤有嗜好.17 岁时,就曾以惊人的毅力将梁绍鸿先生《初等数学复习及研究(平面几何)》一书的习题全部做完(《初等数学复习及研究(平面几何)习题解答》,尚强著,哈尔滨工业大学出版社 2009 年出版).他对于几何证明的理解显然高人一等.而以欧几里得平面几何为代表的逻辑体系正是被西方人称为“世界文明三大支柱”的第一根支柱.

另一位作者胡炳生先生更是数学教育界的元老,执教近五十年,著作等身.胡先生还是一位诗翁,擅长古典诗词,发表了大量词藻优美的诗作.

两位作者,珠联璧合,写成一部好书,先睹为快.

《数学探究学习论》序

探究学习是一个重要的、值得研究而又引起热烈讨论的论题.

宁连华先生在他的著作《数学探究学习论》(以下简称“宁著”)中,对有关文献进行了认真的梳理,接着就数学探究学习的基本要素、数学探究学习的过程要素等中心内容展开了深入的讨论,系统地阐述了自己的理论. 宁著论证缜密,分析细致,是一本严谨的学术著作.

宁著是一本奠基性的著作. 它总结了以往的,特别是宁连华先生自己的研究成果. 但它并没有终结这个论题的研究. 恰恰相反,它为进一步研究开辟了广阔的天地. 书中专用一章,提出了若干思考和建议. 这些思考与建议,值得我们大家去思考,去探究.

一本书的最大价值,就在于它能引起人们的思考.

现在,数学教育的文章与著作越来越多. 据我个人浅见,一本好的著作或一篇好的文章,应当满足以下几个基本要求.

(一) 理论必须联系实际

有不少教育方面的论著,动辄数十万言,或旁征博引,无所不包;或天马行空,自成一统,但与实际教学却毫无关系.

宁著显然不同于这类著作. 书中有大量的实证工作与案例,源于教学实际,又可以指导实际教学. 书中的附录三,就是一个真实的数学探究学习范例,殊值一读.

(二) 应当与数学有关,体现学科特色

近来,有一批搞数学教育的人士在“去数学化”. 他们奢谈思想、文化、情境、理念,而数学本身,则被鄙夷为“低水平”,不屑一提. 过去常说:“给学生一杯水,教师自己要有一桶水.”现在却有人认为:“教师有一杯水已经足够.”更有甚者,有人认为:“教师连一杯水也不必有,一同与学生去找水喝,岂不更好?”这类高超的理论,像我这样的普通人的确难以理解.

宁著中数学内容不少,如第三章的第五节、第六节,都有很多很好的数学探究问题. 我以为可以专辟一章,列举一些可供师生参考的数学中的探究问题.

(三) 深入浅出

把抽象的理论具体化,把复杂的东西讲得简单,深入浅出,是真功夫. 很多著名的学者,如梁启超、胡适,他们的著作都是通俗易懂的. 我最近见到的《李慎之文集》,马幼垣的《水浒论衡》等,也都十分好读.

在数学教育方面,傅种孙先生、华罗庚先生的文章及著作,都是深入浅出的典范.

老实说,有些文章我是很怕读的. 这些文章,文字诘屈聱牙,未加定义的概念、不知所云的术语频频出现,令人头晕目眩. 作者似乎存心让读者受苦受难,以此彰显他的学术水平高不可及.

宁著的文字朴实无华,明白流畅. 再增加一些文采,就更锦上添花了.

(四) 有自己的东西

这最后一点,或许也是最要紧的一点. 论著应当有自己独到的见解,与众不同. 即西方人所说的:

Everyone is supposed to specialize his own line.

宁著有很多自己的东西. 这一点不必我在此废话,打开他的书读一读就很清楚了.

《数学教学生成论》序

上课前,需要备课,很多教师会写一份教案.新教师或实习教师往往写得相当详细,每个问题的回答,甚至每一句话,都预先设定好了.然而,实际上课时,情况却可能与预设的不完全相同,甚或大不相同.对这类意外情况,有经验的教师可以驾驭,而经验不足的新人就难以应付,多数只能坚持原来的计划,勉强把课上完.

怎样认识上述预设教案与实际不能相符的现象?怎样由这种认识来改进我们的教与学?

李祎先生在他的著作《数学教学生成论》中,用他自己的理论回答了这些问题.他指出:

“教学是生成的,动态的”,“学生是重要的生成者”.

“对教师而言,教学即为教学生学”,“教应当是生成性的教”,“不能为教而教,要为学而教”.“教师应具有引导生成的教学能力,捕捉生成性资源的意识,应对生成的教学机智和智慧”.

学生的学也应当是“生成的”,“因学而成”.“在探‘究’中生成”,即“深入思考以自得,善于追问以自省,崇尚感悟以自化”.

李著并未否定备课.恰恰相反,书中提出二十四字的教学策略,强调要“精立内容,大作功夫”.

关于“教学生成”的讨论,应当说还只是开始.李先生的著作是奠基性的.李著中充满了新颖独特的见解.这些见解,发人深思,有的值得进一步商榷、研究、讨论.这正是这本书的价值所在.

一种新的理论,往往并不完善,它特别需要实践的检验.李著中已有一些案例,我认为还不够.还应进一步搜集自己或他人的实际案例,更好地、更全面地阐释教学生成的理论.

关于学生个人的数学知识,书中有一调查表,但调查的结果不够具体.据我个人浅见,在学这方面,学生的个人知识没有也不可能有很大的差别(不可能有很多的个人数学知识)，尤其是使用同一教材，又由同一位教师任教的同一班级．个人数学知识的差别，主要体现在“习”的方面．“习”是数学教育的一大特点．学生的个人数学知识主要是在习的过程中生成或获得的．李著对学生的学习，也提出“大作功夫”，这“作”与“习”，似乎并不矛盾，或许就是同一件事．不知我的理解，李先生是否同意?

《数学课堂学习动力论》序

宋晓平博士的论著即将付梓,希望我作一篇序.给这样的论著作序,我以为,既要对国内外的教育有全面的了解,能够高瞻远瞩,发出一些纲领性、方向性的意见;又要对论著讨论的专题有深入的研究,能够指引读者,使读者尽快地理解、掌握全书的要旨和精神.也就是说既要博大,又要精深.而这二者,我均不具备.幸而这套丛书,已有杨启亮先生的总序,又有作者自己的导言,足以担当上述任务.我这里只需要说一句话:请读者先仔仔细细地读一读总序与导言.

序,可以作一些切中肯綮的评论.这样的评论也要有很高的水平才能写出.我这里只能谈点粗浅的看法.

"真正的知识必须是人们进行有意识的、系统的、有目的的观察和实验所得出的结果;因此,它就不能以信仰和教条为准,而必须以经验的事实为准"(何兆武《文化漫谈——思想的近代化及其他》).现在有些教育方面的论著,动辄数十万言,或旁征博引,无所不包;或天马行空,自成一统.但与实际教学却毫无关系.宋著是研究课堂教学的,并且在导言中强调:"研究教师外部驱动学生有效的学习与学生自主学习动力的生成成为本研究考察的重点."

可见这本论著是与实际的课堂教学密切相关的.由对课堂教学的有意识的、系统的、有目的的实际观察和实验导出自己的理论,再用所得的理论指导实践,同时接受实践的检验.因此,这本书对于实际的教学应当是有指导意义的.

我以为宋著有一大特点:关心人,即关心教师和学生以及他们在课堂上的活动.现在有一种强调物的风气,甚至认为课标与教材决定一切,教师必须无条件地服从它们.其实物是死的,而人是活的.物是被使用的工具,人则是物的使用者.物只是奴仆,人才是主人,是上帝.课堂上的动力产生于教师和学生.教师可以而且应该根据实际的需要,对教材进行更改和补充,产生更大的驱动力.学生的学习也不应当囿于课标所规定的内容,而应当有更大的目标,自主地生成更大的学习动力.

这是我的一点想法,不知是否符合宋博士论著的精神.

《义务教育数学课程与教学论》序

潘小明先生年富力强,博学深思,长期在教育一线工作,既有丰富的实践经验,又勤于著述,已在《数学教育学报》《数学通报》《课程·教材·教法》《教育探索》《电化教育研究》《中小学信息技术》《教育导刊》《上海师范大学学报》《小学数学教师》等有影响的学术刊物发表论文近百篇. 这本《义务教育数学课程与教学论》就是他最近撰写的一本关于数学教育的专著.

潘著有很多创新之处.

长期以来,我国的小学、初中是分开的. 因此,有专门研究小学数学课程与教学的,也有专门研究初中数学课程与教学的. 但似未见到将这两者作为一个整体进行研究. 现在实施九年义务教育,潘小明先生审时度势,将原先的六年小学与三年初中合在一起综合研究,这是非常合理的,也是十分必要的. 在国内,这也许是首创之作.

以往的研究多数偏重于教材教法. 潘著跳出了这种模式,从追寻数学、课程和教学的本质出发,分析了数学的意义、义务教育阶段数学课程与教学的基本性质及主要任务. 居高临下,体现了前瞻性与学术性.

潘著不仅对《全日制义务教育数学课程标准(实验稿)》的有关理念和要求进行了解读,而且非常及时地以新修订的课程标准为依据,引领广大师范生及在职数学教师走进数学课程与教学的研究,呈现了时代的特色.

潘著有坚实的实践基础.

潘小明先生长期主持泰州师范高等专科学校"数学课程与教学论"课程建设,做了大量深入细致的工作和调查研究. 这一课程已被评为省级精品课程,并在申报国家精品课程. 潘著正是对上述课程建设的理论总结.

潘著极有针对性,对现实的数学教育有指导意义.

潘著特别强调指出"数学课程实施的关键在数学教师,最终看教学,尤其是数学课堂教学"(第二章第四节). 对数学教师的培养与课堂教学,书中都辟了专章作比较详细的论述(第七章、第六章).

潘著的很多观点,如对教师及课堂教学的重视,我非常赞成. 但有些论题,如数学的意义等,似有意犹未尽的感觉. 我想可能有以下原因:(1) 这些论题过大,每个论题都需要一本甚或多本专著详细讨论. 囿于篇幅,很多讨论只能适可而止. (2) 本书有一份详细的参考文献,有兴趣的读者可以按图索骥,自己作进一步的阅读. (3) 更可能的是作者故意留下一些空间给读者自己去思考,提出自己的看法,以作进一步的切磋,这是一种极高明的做法.

潘小明先生的大作问世,邀我写一篇序. 不揣浅陋,写了上面的读后感. 可能郢书燕说,不一定符合作者的原意.

《中学数学有效课堂教学法》序

课堂教学是中学教学的重要部分. 如何上好课是每位教师都关心的问题. 什么是好课?

我们看到现在有一些示范的公开课,花样层出不穷,五光十色,令人眼花缭乱,但实际效果却很小,有点像《水浒传》中王进教头说的:"都是花棒,只好看,上阵无用."

教学必须有效. 怎样才能实现有效的课堂教学? 这既需要理论上认真研究,更需要反复实践,不断总结.

李志敏先生的大作《中学数学有效课堂教学法》,就是系统研究中学数学有效课堂教学的一本专著.

这本书的内容丰富,讨论全面、深入. 书中提出了中学数学有效课堂教学的理念、原则与评价,介绍了当代国内外典型的课堂教学模式,指出如何进行中学数学有效课堂教学(包括概念课、原理课、习题课、复习课等)的设计,而且还给出了许多中学数学有效课堂教学设计的案例. 其中,第五章"中学数学有效课堂教学的方法与艺术"尤为精彩,有很多作者精辟、独到的见解. 具体例子这里就不用举了,请读者直接阅读原书.

希望改进教学的教师,有志从事教学研究的人士,都可以从这本书中得到启迪和帮助.

有人说教学不是科学,而是艺术,这话颇有道理. 一堂课,可以有种种不同的上法. 不像 1+1=2 那样,只有唯一正确的答案. 所以书中很多内容都有进一步讨论的余地. 所说的一些原则,在真正上课时,应当根据实际情况灵活运用. 事实上,作者在书中也反复强调注意根据教材的不同特点、学生的水平差异、教师的自身素养等众多因素因"材"施教.

李志敏先生年富力强,大有作为. 他长期在教育一线工作,经验丰富,成绩卓著而又善于总结,36 岁时就已经获得特级教师的称号. 我们期望并且相信他将有更多著作问世.

祝《中国初等数学研究》诞生

《中国初等数学研究》诞生了,值得庆贺!

初等数学有大量的问题,需要研究.

中国研究初等数学的人很多,成千上万!

中国研究初等数学的成绩斐然. 在平面几何、不等式、组合、初等数论等方面,都有很好的结果. 大量初等数学的书籍、文章,就是有力的明证.

有人以为研究只能属于高等的数学,只能是前沿的、深邃的问题. 其实研究本身并无高等、初等的分别. 得到高深的结论是新发现,解决初等的问题同样是新发现,都是人类向未知领域的迈进. 而且很多人们耳熟能详的大问题,如费马大定理,如歌德巴赫问题,论起它们的"出身",无不属于初等数学.

初等数学研究好比"下里巴人","和者众",吸引了大量的数学爱好者,为数学研究输送、储备了人才.

数学研究不仅需要"阳春白雪",同样需要"下里巴人". 如果只有少数"精英"研究数学,如果研究者过于专门,"know more and more about less and less",那么数学也就不能获得大众的认同,数学本身也就难以发展.

初等数学研究并不容易,它也有深刻的内涵与精巧的方法. 陆家羲先生的大集定理就是一个经典的例子. 再如在国际双微会议上曾经讨论过如下的几何不等式:

在三角形 ABC 的边上取三个点 D, E, F,使得沿着三角形的边界从 D 到 E,从 E 到 F,从 F 到 D,所走的长都相等(即都等于三角形 ABC 的周长 $a+b+c$ 的三分之一). 求证:

$$\text{三角形 } DEF \text{ 的周长} \geqslant \frac{1}{2}(a+b+c).$$

这道题难倒了众多与会的数学家,在会议期间没有能够解决. 后来福建的杨学枝先生给出了一个简洁优雅的证明.

对于初等数学,我是很有兴趣的,也颇想做些研究,但却没有多少成绩.我做得最多的是数学的普及与传播,或许这也可以说是初等数学研究的一个部分.

普及与提高,二者不可缺一.普及可以使更多的人了解数学,喜爱数学,甚至参与研究数学.但没有提高,就只能停滞在低水平上,没有长进.反过来,一味强调提高,轻视普及,参与的人将越来越少.大量"凡夫俗子"被摒弃在数学大门之外,虽然他们非常喜爱数学.如果出现这样的情况,提高也就难以为继了.

其实,普及中也有提高.一些理论的更好的阐述,某个问题的更通俗的解释,不少解法的简化或推广,往往更接近事情的实质,未尝不是一种提高.

从事数学的普及与传播,不仅于人有益,对自己也大有好处.从事这种工作,可以培养学习与研究数学的兴趣;可以进行研究的尝试,小试身手.在条件可能的情况下,可以进一步走向研究的道路.甚至进入更高深的研究.当年华罗庚先生就是这样步步登高的.

初等数学十分有趣.但年轻人切不可沉溺于初等问题.年轻人应当志存高远,应当尽快占领尽可能高的数学高地.如果一开始就陷入一两个初等问题中不能自拔,那么就会与真正的数学研究渐行渐远了.初等数学研究可能更适合数学教师,有固定职业的业余数学家或年岁较大的人.

最后,祝中国初等数学研究取得更多的成就.

评《三角与几何》

《三角与几何》是华东师范大学出版社 2005 年出版的数学奥林匹克小丛书高中卷中的一本，著者是田廷彦先生.

这本书的特点是很有原创性. 其中有大量的新材料(包括数学同仁之间交流的内容，自编的问题). 对一些老问题，也有自己的新观点、新解法.

例如第 87 页第 5 章例 14:"设$\odot O_1$与$\odot O_2$交于P、Q两点，过点P任作两条直线APB和CPD，其中点A、C在$\odot O_1$上，点B、D在O_2上. M、N分别是AD、BC中点，O为O_1O_2中点，$\angle APC=\theta$为锐角. 设h为点O至MN的距离，K为PQ中点. 求证：$h=OK\cdot\cos\theta$." 这道题就是叶中豪先生发现的新命题. 书中给出了一种运用三角的证法. 后来叶先生也找到了几何证法.

再如第 35 页第 3 章例 6:"如图 1，已知$\triangle ABC$，$\angle B=90^{\circ}$，内切圆分别切BC、CA、AB于点D、E、F，又AD交内切圆于另一点P，$PF\perp PC$. 求$\triangle ABC$三边之比."书中算出三边之比为$3:4:5$，$\triangle ABC$也就是我们最熟悉的"勾三股四弦五"的直角三角形. 可将这道题与 2005 年 IMO 中国国家队培训的一道题作比较(参见华东师范大学出版社的《数学奥林匹克试题集锦(2005)》130 页):"在直角三角形ABC中，$\angle B=90^{\circ}$，它的内切圆分别与边BC、CA、AB相切于点D、E、F，连接AD，与内切圆相交于另一点P，连接PC、PE、PF. 已知$PC\perp PF$，求证：$PE\parallel BC$." 显然，得出$\triangle ABC$三边之比为$3:4:5$的结论，比"$PE\parallel BC$"更好、更深刻，因为它确定了$\triangle ABC$的形状.

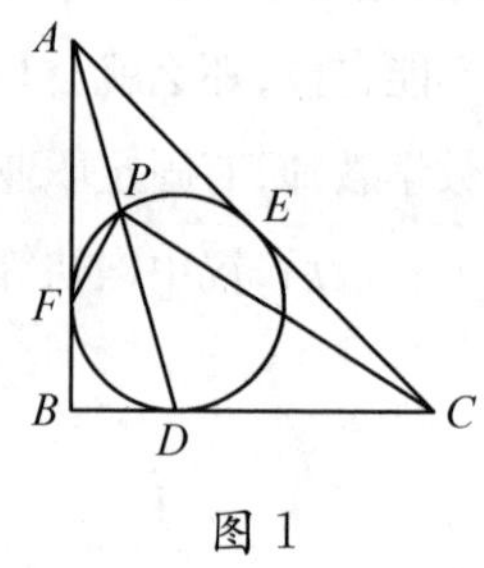

图 1

还可以再与 2006 年中国数学奥林匹克(全国中学生数学冬令营)的第四题比较:"如图 2，在直角三角形ABC中，$\angle ACB=90^{\circ}$，$\triangle ABC$的内切圆O分别与边BC、CA、AB相切于点D、E、F，连接AD，与内切圆O相交于点P，连接BP、CP，若$\angle BPC=90^{\circ}$，求证：$AE+AP=PD$." 其中"$EP\perp BP$"改成了"CP

$\perp BP$”，其他条件完全相同. 可见田先生这本书上的新材料已被很多人关注.

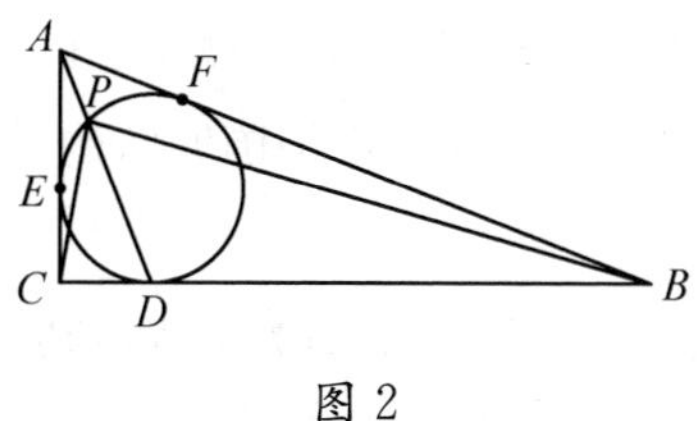

图 2

这样的例子可以举出很多. 不过，我们不必多举，建议读者自己去读一读，以免有人抱怨我们剥夺了他们读书的乐趣.

田先生这本书当然也有缺点，缺点就是其中的解法并非十分完善. 这种缺点是任何一本原创性强的书难以避免的. 换一个角度来看，这种不完善，毋宁是一种优点. 因为读者可以看到著者思考的轨迹. 有些大作者，最著名的是高斯(Gauss)，文章写得非常完善，但思考的过程完全没有，以致有人形容他“像狐狸走过雪地一样，用尾巴把痕迹全扫光了”. 这样的文章，要读出著者的想法非常困难，有时甚至会使读者产生自卑感，误以为自己与著者相差太远，丧失了学习的信心. 田先生的这本书恰恰相反，其中很多解法大有探讨的余地，给读者留下了很多思考的空间. 我在读这本书时，就在书上写了很多的注记. 相信其他的读者也会有许多可以写到书上的心得.

顺便说一下，现在“奥数”的书很多，篇幅越来越大，动辄几十万字，但原创的东西往往只有一点点. 有的甚至将别人书上的文字一字不改地放入自己书中，一口气“吃掉”. 这种行径接近剽窃！对照一下田先生的书，也应当有点脸红或惭愧.

田先生这本书只有 18 万字，内容却很丰富.

这本书还有一个缺点，就是书名限制了书中的解法. 有些题用三角来解未必最佳，这一点，我相信田先生自己也清楚. 但由于用了这样的书名，所有题都一律用三角来解，有点“作茧自缚”. 不过，正如前面已经说过的，这也留给读者寻求其他解法的自由空间. 比如说，书中第 1 页第 1 章例 1：“已知等腰$\triangle ABC$，$AB=AC$，一半圆以 BC 的中点为圆心，且与两腰 AB、AC 分别相切于点 D、G，EF 与半圆相切，交 AB 于点 E，交 AC 于点 F. 过 E 作 AB 的垂线，过 F 作

AC 的垂线,两垂线相交于 P,作 $PQ\perp BC$, Q 为垂足. 求证:$PQ=\frac{EF}{2\sin\theta}$,此处 $\theta=\angle B$."这道题,除书上的解法外,在与学生的讨论中,至少发现了 6 种解法,其中大多不用三角(本题也被用作 IMO 中国国家队的培训题,参见上面所说的《试题集锦》129 页).

当然,专用一种方法也未尝不可. 有时为了练习某一专长,还必须作一些限制. 过去的平几题,常限定只用纯几何的方法. 除非万不得已,不用解几与三角来算,甚至面积证法也在排除之列. 欧几里得的《原本》就尽量延迟平行线的出现,比例与面积更是很晚才用,使得不用平行公设的所谓绝对几何尽可能范围大一些. 现在流行"以算代证",几何味道又似乎太少了. 不知田先生或其他先生有无兴趣写一本《不用三角的几何》,多展现一些几何方法的力量?

读《名著的疏漏》

《中学数学研究》2006 年第 1 期上钱昌本先生的《名著的疏漏》是一篇好文章，影响很大，乃至数学界的一些同行吃饭时都谈起它.

原书的图 1 实际是说“一个正三角形可以分成四份，这四份可以拼成一个正方形(反过来，一个正方形可以分成四份，这四份可以拼成一个正三角形)”. 而原书图 2 则是解释这四份如何分.

原书图 2 肯定是错的. 这一点钱先生的文章已经说得很清楚了.

原书图 1 对不对呢？也就是说“正三角形可以分成四份，这四份可以拼成一个正方形”，这个结论成立不成立呢？如果还成立，那么应当怎样分呢？

我们讨论了这个问题. 认定原书图 1 仍是正确的，即正三角形可以分成四份，这四份可以拼成一个正方形. 原书图 2 则需要改动. 请看这里的图 3. 图 3 中$\triangle ABC$是正三角形. 不妨设其边长为 2. D、E分别为AB、AC中点，F、G在BC上，并且$DF \parallel EG$. 与原书图 2 不同，这里$\angle EGC=\alpha$不是直角，所以$CG \neq BF$. 设$CG=x$.

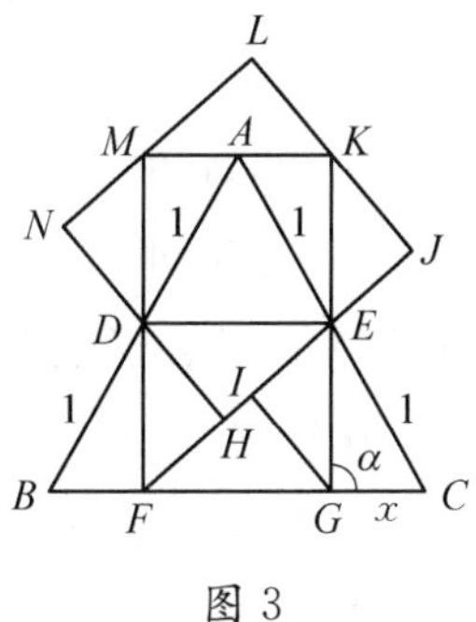

图 3

作DH、GI与EF垂直，H、I为垂足. 又过A作FG的平行线，与FD、GE的延长线相交于M、K. 过M、K分别作DH、HE的垂线，形成矩形$HJLN$.

容易证得$\triangle LMK$、四边形$AEJK$、$AMND$分别与$\triangle IFG$、四边形$CEIG$、$BFHD$全等. 所以$\triangle ABC$可分成四块，重拼成矩形$HJLN$. 只需选择x或α，使得矩形$HJLN$成为正方形.

易知$FE=FH+HE=EJ+HE=HJ$，即正方形的边长. $\triangle ABC$的面积为$\sqrt{3}$，所以正方形的边长应为$\sqrt[4]{3}$. $FG=DE=1$，在$\triangle CEF$中，由余弦定理得

$$(1+x)^2+1-(1+x)=\sqrt{3}. \tag{1}$$

易解得(注意 $x>0$)

$$x=\frac{1}{2}(\sqrt{4\sqrt{3}-3}-1). \tag{2}$$

这时

$$EG=\sqrt{x^2-x+1}=\sqrt{1+\sqrt{3}-\sqrt{4\sqrt{3}-3}}.$$

由正弦定理,得

$$\sin\alpha=\frac{\sin 60^\circ}{EG}=\frac{3}{2\sqrt{3+3\sqrt{3}-3\sqrt{4\sqrt{3}-3}}}. \tag{3}$$

经计算 $\alpha\approx 89.40357868\cdots^\circ$,即 $89^\circ 53.6'$,与直角相差甚微.而 x 与 $\frac{1}{2}$ 也仅相差 $0.009015231\cdots$.所以仅凭肉眼,原书图 2 的错是看不出来的.就实际操作而言,原书图 2 那样做是可以允许的.但从数学的严格性来要求,它当然是错的.

Steinhaus 的这本《数学万花镜》风行全球,经过多次修订,仅中文就有三个译本.分别由开明书店、上海教育出版社、湖南教育出版社出版(均由裘光明先生翻译).何以仍会出现这个疏漏,而史(施)氏竟未察觉呢?

比较三种中译本.前两种均无原图 2,因而并未出现上述所说疏漏.第三种根据 1972 年后的版本翻译的,其时史(施)氏已经去世.所以这疏漏很可能并非原著者的错误,而是后来的人所加的注解出了漏洞.

“疏”,在汉字中有多种解释.可解释为疏忽,也可解释为注解.这名著的疏漏,实际上是注疏出了漏洞.

推荐一本好书——《数学小辞典》

王文才、施桂芬编，科学技术文献出版社1983年出版的《数学小辞典》是中学数学教师的益友，是中学生的良师.

这本小辞典内容丰富，条目约有两千，中学数学范围里的名词术语应有尽有. 如果遇到一个数学词语不知道（例如驴桥定理，完全四边形），或不知其详（例如三大作图问题，等分圆周，等周问题，悖论），或已经忘记了（例如多面角，二项式定理，九点圆），那么，查这本辞典！它可以解答你的问题. 例如在九点圆这个条目下，就有详细的解释：

（九点圆）在一个三角形中，三边的中点、三条高的垂足、垂心与三个顶点所连线段的中点，这九个点在同一个圆周上，这个圆叫做这个三角形的九点圆.

可以证明，三角形的九点圆的半径，等于这个三角形外接圆半径的一半.

关于九点圆的结论，是1820—1821年由日尔刚（J D. Gergcnne, 1771—1859）与庞西里（J. V. Poncelet, 1788—1867）首先发表的. 一位高中教师费尔巴哈（Karl Wilhelm Feuerbach）在1822年也发现了九点圆，并指出九点圆和这个三角形的内切圆、三个旁切圆都相切. 通常就把九点圆叫做费尔巴哈圆.

也有些书刊上把九点圆叫做“欧拉圆”.

电视剧《华罗庚》中的那道数学问题，也可以在下面的条目中查到一般的结论.

（孙子定理）我国古代算书《孙子算经》的下卷记有一些“物不知数”问题，例如：“今有物不知其数，三三数之剩二，五五数之剩三，七七数之剩二，问物几何？答曰：二十三.”从书中记述的具体求解过程可以概括得出：如果正整数m_1，m_2，…，m_k两两互质，那么同余方程组

$$a \equiv a_i (\mathrm{mod} m_i), \quad i=1, 2, \cdots, k$$

有无穷多解，且这些解关于模$M=m_1m_2\cdots m_k$同余，可表成

$$x \equiv a_1M'_1M_1+a_2M'_2M_2+\cdots+a_kM'_kM_k(\mathrm{mod} M),$$

其中 $M_i = \dfrac{M}{m_i}$，而 M_i' 是满足 $M_i'M_i \equiv 1(\bmod m_i)$ 的正整数.

这一标法后来传入西方，被称为中国剩余定理.

这本辞典内容新颖，其中有不少近代数学（例如数理逻辑、抽象代数、数学分析、数论等方面）的知识，有许多著名的问题（例如哥德巴赫问题、华林问题、费马问题、连续统假设、罗素悖论、理发师悖论、三大作图问题）. 阅读这本小辞典可以开拓你的眼界，增长你的知识，提高你的数学修养. 编者注意到知识的更新，不仅使你看到中学数学的昨天和今天，而且展示了中学数学的明天！

这本辞典的解释准确、深入. 编者介绍了很多历史知识，有助你了解有关概念或问题的演变. 不仅如此，编者采用近代数学的观点，居高临下，说得十分透彻（请看复数、反证法、非欧几里得几何、分配律等条目）. 有些问题局限在初等数学的圈子里不容易搞清楚，正如苏东坡所说："不识庐山真面目，只缘身在此山中."现在随着编者的指引，读者好似乘坐飞机俯瞰庐山，看到了整体，看到了全貌！

这本小辞典还有一个鲜明的特点，就是小. 由于条目的筛选严格，所以既是应有尽有，又没有一个条目可有可无，而且文字洗练，要言不繁，因此它没有被编成一个笨重的大部头的厚书，总共只有 300 多页，又是平装，体积小，重量轻，携带方便，乘车坐船都可以带在身边随意浏览.

我相信这本十分有用的好书，将会伴随你度过很多愉快的时光.

一本不可不读的好书

电视剧《华罗庚》中有一个片段:少年华罗庚要借一本好书,未能借到,后来连续在图书馆扫了几天地,方才如愿以偿.

这本好书就是 Hall 与 Knight 的《大代数》,该书上册于 1983 年由科学普及出版社翻译出版. 不久前,中译本的下册(席小云译,何宗莲校)也已问世.

这本书有大量的例题,讲了很多解题的方法与技巧,一般的教科书难以望其项背. 每章还有相当数量的习题供读者练习,有助于提高解题能力. 下册更有综合习题三百道,如能解决其中的一半,就表明读者已经掌握了中学代数. 译者从别处译出了全部解答,与原书合在一起发行,给中学生、自学青年和教师提供了方便. 将它用作第二课堂或课外活动的材料,也是极为合适的.

下面择出两道习题,以见一斑.

190. 如果 $\frac{1}{a}+\frac{1}{c}+\frac{1}{a-b}+\frac{1}{c-b}=0$,证明除了 $b=a+c$ 的情况外, a, b, c 的倒数成等差数列.

185. 如果 $(6\sqrt{6}+14)^{2n+1}=N$, F 为 N 的小数部分,证明 $N\cdot F=20^{2n+1}$.

这两道题的解答如下(笔者将问题与解答的译文作了些更动):

190. 解:由原式得 $\frac{c-b+a}{a(c-b)}+\frac{a-b+c}{c(a-b)}=0$,即 $b=a+c$, 或

$$\frac{1}{a(c-b)}=-\frac{1}{c(a-b)}.$$

后一等式即 $a(c-b)+c(a-b)=0$,从而 $2ac=ab+bc$.

两边同除以 abc,得 $\frac{2}{b}=\frac{1}{a}+\frac{1}{c}$.

即 a, b, c 的倒数成等差数列.

185. 解:容易知道 $14<6\sqrt{6}<15$,故 $0<(6\sqrt{6}-14)^{2n+1}<1$.

另一方面, $(6\sqrt{6}+14)^{2n+1}-(6\sqrt{6}-14)^{2n+1}$ 是一个正整数,所以 N 的小数

部分为 $F=(6\sqrt{6}-14)^{2n+1}$. 从而

$$N\cdot F=(6\sqrt{6}+14)^{2n+1}\cdot(6\sqrt{6}-14)^{2n+1}=(216-196)^{2n+1}=20^{2n+1}.$$

当代大数学家陈省身教授，在他的自传《我的成功之路》中特别指出，他年轻时读过这本《大代数》，终生获益. 想要在数学上有长足进步的青年人，这本书不可不读.

一本好书——《解题·成长·快乐：陶哲轩教你学数学》

书籍，就像是朋友，不一定很多，但要精挑细选.

中学数学的书很多，这本《解题·成长·快乐——陶哲轩教你学数学》（于青林译，潘承彪校，北京大学出版社 2009 年出版），无疑是最值得挑选出的一位"朋友". 正如我国著名数学家潘承彪先生所说："是我从未见到的谈中学生如何学数学的一本好书."

陶哲轩（Terence Tao），华裔澳大利亚数学家，被誉为"数学界的莫扎特". 1986 年，他 10 岁时第一次参加国际数学奥林匹克（IMO），获得铜牌. 次年获得银牌. 1988 年，第 29 届 IMO 在澳大利亚举行，他获得金牌，成为 IMO 最年轻的金牌得主. 后来他成为职业数学家，在 2006 年获得数学界的最高荣誉 Fields 奖.

这本书是他 15 岁时写成出版的. 原名《Solving Mathematical Problems A Personal Perspective》，即他自己解题时的体会.

现在我国市场上见到的中学数学读物，很多都是习题集. 虽然有解答，但解答从何而来，如何寻找解答，却避而不谈. 仿佛是"鸳鸯绣好凭君看，不把金针度于人".

陶哲轩的这本书，与此不同，他介绍了自己解题时的真实历程，指明"刚接触一个数学问题时，我们应该如何去处理它，如何通过努力从不同角度尝试一些想法和排除另一些想法，以及如何通过有计划地处理，最终得到一个满意的解答".

因此，阅读这本书，不仅能学到一些问题的解法，而且能学到一般的解题方法. 后者当然远比前者重要.

当下有不少学习奥数（建议改为，所谓"奥数"）的中学生，喜欢听讲座，不喜欢读书. 以为听讲座，见效快，能够速成. 其实只有自己读书，自己想问题，自己寻找解答才有真正的收获，才能享受到学习的乐趣. 听得多，做得少，读书更少，容易形成眼高手低，缺乏深刻的理解，不能增长智慧与创造力.

还有一些同学喜欢搬用一些偏僻的定理作为“独门武器”,或者抬出某个大定理作为“原子弹”. 这也不是学习数学的正道. 陶哲轩 15 岁时就读了不少好书,这从本书的参考文献就可看出. 这本书中所用的知识和方法都是最普通的、最常见的,没有任何怪招,也不需要“恶补”. 陶哲轩将解题比喻成“为寻找金矿而上的一堂‘捉迷藏’课:你要去寻找一块金子,你知道这块金子是什么样的,它就在附近的某个地方,要到达那个地方不是太困难,是在你的能力所及的范围内,同时适当地给了你去挖掘它的合适工具(例如已知条件)”.

他还指出“无论多么复杂和深奥的结果往往都是由非常简单,甚至是常识性的原理导出”.

所以,这本书也给有志于数学的青少年展示了一个如何正确学习数学的典范.

陶哲轩 15 岁就能写出这样的好书,真令人钦佩!

当然,任何一本书都不是完美无缺的. 陶哲轩自己也说“书中有些问题可以用更便捷、更简洁的‘先进’数学工具来解决”. 例如第四章欧几里得几何中,很多问题用纯几何的方法处理或许更好. 不过,陶哲轩说“我不想太多地改变它们,因为年轻时的我比现在的我更能融入高中生的解题世界”. 他这样做完全正确. 既保留了自己年轻时的记录,又给现在的年轻人提供了一些展现身手改进解法或结论的机会.

译者是数学家,所以译文不仅清楚流畅,而且数学方面准确无讹. 只有数字(digit)译为位数,似不太符合国内中学生的习惯.

潘先生的注释弥补了书中的少数疏漏,并且介绍了读者进一步学习的书籍.

中译本的书名起得很好. 的确阅读这本书的年轻人可以通过陶哲轩的这本书学到很多东西,在解题中成长,并且享受到解题的快乐.

《奥博丛书》序

一本武功秘籍！

找到它，勤加练习，就能成为武林高手.

这是金庸等人常写的故事.

这套奥博丛书，其中就有若干本或许可以称为解题秘籍. 当然，得到它之后，要成为解题高手，还得注意：

一、勤加练习. 因为解题是实践性的技能，只能通过模仿和实践来学到它.

二、循序渐进. 孔子说："欲速则不达."不能操之过急. 一个问题或一种方法，彻底弄清楚了，再往下看. 切忌囫囵吞枣，食而不化.

三、不要迷信书本. "尽信书，则不如无书."作者也有可能出错. "乾坤大挪移"第七层心法的一十九句就是"单凭空想而想错了的". 其实要成为真正的高手，不能依赖秘籍，而要自创新招.

这套奥博丛书，不只是解题的秘籍. 它的作者阵营庞大，视角不尽相同，写法各有特点. 或综述，或专题；或讲思想，或谈策略；或提供翔实材料，或介绍背景知识；……

据我了解，奥博丛书原本并不是一套丛书. 它既没有预先设定的宏伟的出书规划，也不能保证其中的每一本都同样精彩. 时间，才是考验它们的唯一准则. 它不像其他丛书那样，追求在同一时间出齐；而是细水长流，渐渐汇聚成河. 除已出的、即出的十余种外，想必还会继续推出新的品种.

开卷有益. 相信这套丛书能很好地普及数学知识，增加读者对数学的理解，提高数学的品味(taste)，也就是鉴赏能力. 祝愿这套丛书能够伴随读者度过一段愉快的时光.

《数学奥林匹克命题人讲座》序

读书,是天下第一件好事.

书,是老师.他循循善诱,传授许多新鲜知识,使你的眼界与思路大开.

书,是朋友.他与你切磋琢磨,研讨问题,交流心得,使你的见识与能力大增.

书的作用太大了!

这里举一个例子:常庚哲先生的《抽屉原则及其他》(上海教育出版社,1980年)问世后,很快地,连小学生都了解了什么是抽屉原则.而在此以前,几乎无人知道这一名词.

读书,当然要读好书.

常常有人问我:哪些奥数书好?希望我能推荐几本.

我看过的书不多.最熟悉的是上海教育出版社出过的几十本小册子.可惜现在已经成为珍本,很难见到.幸而上海科技教育出版社即将推出一套命题人讲座丛书,帮我回答了这个问题.

这套丛书的书名与作者初定如下:

陆洪文:《解析几何》

施咸亮:《代数函数与多项式》

熊斌:《函数迭代与函数方程》

陈计:《代数不等式》

叶中豪:《平面几何》

冯志刚:《初等数论》

单墫:《集合与对应》、《数列与数学归纳法》

刘培杰:《组合中的著名定理》

任韩、田廷彦:《图论与组合几何》

唐立华:《向量与立体几何》

邵嘉林:《复数与三角函数》

显然,作者队伍非常之强.老辈如陆洪文先生、施咸亮先生都是博士生导师.他们不仅在代数数论、函数逼近等领域的研究取得卓越的成绩,而且关心数学竞赛.中年如陈计先生于不等式,叶中豪先生于平面几何,都是国内公认的首屈一指的专家.其他各位也都是当下国内数学奥林匹克的领军人物.如熊斌、冯志刚是2008年IMO中国国家队的正副领队,中国数学奥林匹克委员会委员.他们为我国数学奥林匹克作出了重大的贡献,培养了很多的人才.如邵嘉林先生指导的张成同学,就获得了第49届IMO的金牌.

这些作者有一个共同的特点:他们都为数学竞赛命过题.如:

设数 a 具有以下性质:对于任意四个实数 x_1, x_2, x_3, x_4,总可以取整数 k_1, k_2, k_3, k_4,使得

$$\sum_{1\leqslant i<j\leqslant 4}((x_i-k_i)-(x_j-k_j))^2\leqslant a,$$

求这样的 a 的最小值.

这是施咸亮先生供给我国国家集训队选拔的试题.又如:

设 $S=\{1, 2, \cdots, 2005\}$.若 S 中任意 n 个两两互质的数组成的集合中都至少有一个质数,试求 n 的最小值.

这是唐立华先生供给西部数学奥林匹克的试题.

叶、熊、冯等几位先生供给竞赛的题举不胜举,这里就不罗列了.

命题人讲座丛书,是田廷彦先生的创意.

命题人写书,富于原创性.有许多新的构想,新的问题,新的解法,新的探讨.

新,是这套书的一大亮点.

读者一定会从这套书中学到很多新的知识,产生很多新的想法.

新,会不会造成深、难呢?

这套书当然会有一定的深度,一定的难度.但作者是命题人,充分了解问题的背景(如刘培杰先生就曾专门研究过一些问题的背景),写来能够深入浅出,"百炼钢化为绕指柔".另一方面,倘若一本书,十分浮浅,一点难度没有,那也就失去了阅读的价值.

读书,难免遇到困难.遇到困难,不能放弃.要顶得住,坚持下去,锲而不舍.这样,您不但读懂了一本好书,而且也学会了读书,享受到读书的乐趣.

书的作者,当然要努力将书写好.但任何事情都难以做到完美无缺.经典著作尚且偶有疏漏,富于原创的书更难免有考虑不足的地方.在某种意义上说,这种不足毋宁是一种优点:它给读者留下了思考、想象、驰骋的空间.

如果您在阅读中,能够想到一些新的问题或新的解法,能够发现书中的不足或改进书中的结果,那就是古人所说的"读书得间",值得祝贺!

我们欢迎各位读者对这套书提出建议与批评.

感谢上海科技教育出版社,特别是编辑卢源先生,策划组织编写了这套书.卢编辑认真把关,使书中的错误减至最少,又在书中设置了一些栏目,使这套书增色很多.

《强基计划深度学习丛书》序

国际数学奥林匹克竞赛(IMO),开始于 1959 年,到现在已经过了整整一个甲子了.

我国 1978 年首次举行全国高中数学联赛. 1985 年首次派出 2 名队员参加 IMO. 以后,每年派出满员(6 人)的代表队,成绩一直名列前茅.

数学竞赛,影响巨大,成绩显著. 很多参赛者后来投身各个领域,尤其是数学研究领域,成为其中的佼佼者. 近年获得数学最高奖菲尔兹奖 (四年一评)的就有约科、高尔斯、博彻兹、拉佛格、佩雷尔曼、陶哲轩、吴宝珠、斯米尔诺夫、阿维拉、莫兹坎尼(女)、舒尔茨等多人. 我国年轻的数学家恽之玮、张伟、袁新意、许晨阳、颜华菲、王崧等,也都是通过数学竞赛崭露头角的.

数学竞赛很重要.

数学竞赛的内容,可以称为竞赛数学或奥林匹克数学,也就是人们常说的"奥数".

但时至今日,虽然大家常谈奥数,仍有很多人不明白什么是奥数.

我看到一个帖子说:

"奥数不是数学,奥数是脑筋急转弯."

这位发帖的先生正好说反了. 正确的说法是:

"奥数是数学,不是脑筋急转弯."

奥数的问题仍是初等的,但却有较为深刻的、高等的思想.

"初等的内容,高等的思想",这就是奥数. 因此,奥数是(初等)数学中最好的一部分,最有趣的一部分. 我们可以说:

奥妙的数学,简称为奥数.

正因为如此,奥数是数学竞赛、大学自主招生,以及基础教育的强基计划所重视的数学.

学习竞赛数学(奥数),当然最好有一本好书,一本好的读本.

1985 年,我们刚准备参加 IMO 时,条件非常之差. 训练地点是中科院计算所的一间空房,旁边正在施工,搅拌机声如雷鸣.训练的材料也没有现成的,完全靠从有关期刊上收集.现在的书太多了,这套《强基计划深度学习丛书》就是其中的一套新书.

这套书至少有四个特点.

(1) 全面

数论、组合、代数、几何、数列、不等式等,凡是竞赛、自主招生、强基计划所需要的内容无不齐备.

(2) 系统

这套书特别重视竞赛数学与高考大轴之间的融合与汇通,讲述问题由浅入深、循序渐进,进可以攻(参与竞赛),退可以守(把握高考),适合培训使用,也适合自学.

(3) 深入

全书有不少专题的讨论,讨论有一定的深度.

(4) 新颖

全书编辑思想比较新颖,反映在立意、布局、选材等各个方面,有不少新材料,如 2019 年第 60 届 IMO 的试题解答也已收入.

参加这套书编写的十几位老师,是一支老、中、青结合的优秀团队.其中有久负盛名的奥坛宿将,有亲自带出 IMO 金牌选手的一线教师.他们学有专长,教有经验,熟悉竞赛,保证了这套书的质量.

单墫

2020 年 8 月

谈谈研究式的学习

一位学生,应当学习课本上的知识,这一过程主要在课堂进行. 但对一位优秀的学生,仅仅依靠课本来学习是远远不够的. 即使就知识方面来说,中学阶段所能学的内容是很多的,而课本上的充其量仅占十分之一,更何况要培养各种能力. 要取得长足的进步,最好是进行研究式的学习.

1　自己看书

应当学会看课外书,这样眼界就大为开阔,能力就大为增强,真正成为学习的主人. 课外书不容易读,数学书更不可能像小说、散文那样有趣味. 但只要读下去,有所领悟,有所收获,也会有无穷的乐趣.

初看书时,不要选择大部头的书,最好从读小册子开始. 好的小册子出得少,由于价格不高,往往很快售罄,而出版社因为利润不高,往往不再重印,所以市场上不易见到(在一些学校的图书馆有可能借到). 因此我在这里稍作介绍.

前苏联曾组织数学家写过许多小册子,20 世纪中期有不少被译为中文. 如《直圆柱》《最短线》《不等式》《数学归纳法》《几何学中的归纳法》《简易的极大与极小》《无穷小量的求和》《循环级数》《斐波那契数》《奇妙的曲线》《双曲线函数》《对数与面积》《摆线》《图形的大小相等和组成相等》《什么是非欧几何》《什么是微分学》《力学在几何中的应用》《奇妙的正方形》《直尺作图》《圆规作图》《正定理和逆定理》.

我国数学家也写了很多小册子. 尤其是华罗庚先生,他先后写了《从杨辉三角形谈起》《从祖冲之的圆周率谈起》《从孙子的神奇妙算谈起》等. 后来又汇集为《华罗庚科普著作选集》. 这时期出的小册子还有《对称》《一笔画与邮递路线》《平均》《复数与几何》《欧拉定理与多面体的拓朴分类》《从刘徽的割圆术谈起》《格点与面积》等.

文革后,小册子更多. 尤其上海教育出版社的王文才、赵斌、冯贤、叶中豪等编辑大力组织,推出了数十种.如《抽屉原则及其他》《单位根》《计数》《组合恒等式》《从正五边形谈起》《从反面看问题》《差分方程》《棋盘上的组合问题》《不定方程》等.

这些小册子深入浅出,大多出自数学名家之手.它们的出版对于我国中学界产生了很大的影响.我也写过不少册子,如《几何不等式》《趣味的图论问题》《覆盖》《棋盘上的数学》《组合数学的问题与方法》《趣味数论》《组合几何》《十个有趣的数学问题》《平面几何中的小花》《集合及其子集》《解析几何的技巧》《算两次》《国际数学竞赛解题方法》等.不能自夸这些书都写得很好,但也确实用了一些力气,费了一番斟酌,可供同学们参考.

读者还可以看一些数学的期刊杂志.例如《中学生数学》,上面经常有一些短小精悍、非常有趣的文章.读一读也大有好处.

读课外书,可以采取"牛吃草的方式".先浏览一遍,囫囵吞下去,大致知道有哪些内容,不求甚解.然后,像牛反刍一样,再对那些有兴趣的、用得上的章节细细咀嚼.

读不懂,怎么办?要有耐心,反复地读,反复思考,反复领会."书读十遍,其义自见".一定要顶得住,不可轻易放弃.遇到困难,可以请教老师、同学,但主要靠自己,硬着头皮往下看.坚持下去,不但学到了很多知识,也培养了读书的能力,学会了如何读书.

"尽信书,则不如无书".看书时要多思考,书中也可能有些欠妥的地方,甚至会有错误.这可以考查你的识别能力.如果你发现书中的疏漏或不足,并写出自己的心得,那就是古人所说的"读书得间"了.

2 想问题

"心之官则思".数学是思维的科学,学数学的最好方法是做数学.只有经常想问题,才能真正掌握数学的思想、方法,才能活用所学的知识.

问题可以是书本上的,也可以是生活中的."处处留心皆学问",随时可以发现好的问题.例如 2002 年高考有一道试题如下:

(1) 给出两块面积相同的正三角形纸片(如图 1、图 2),要求用其中一块剪成一个正三棱锥模型,另一块剪成一个正三棱柱模型,使它们的全面积都与原

三角形的面积相等. 请设计一种剪拼方法,分别用虚线标示在图 1、图 2 中,并作简要说明;

(2) 试比较你剪拼的正三棱锥与正三棱柱的体积的大小;

(3) 如果给出的是一块任意三角形的纸片(如图 3),要求剪拼成一个直三棱柱模型,使它的全面积与给出的三角形的面积相等,请设计一种剪拼方法,分别用虚线标示在图 3 中,并作简要说明.

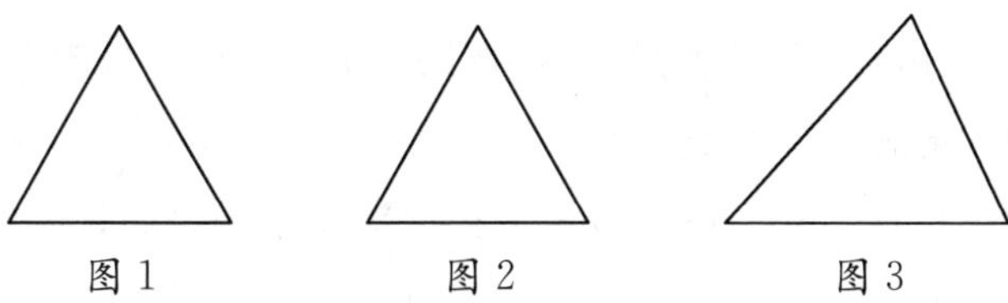

图 1　　图 2　　图 3

从这道高考题出发,可以引出下面的新问题:

一块任意三角形的纸片,是否一定能沿着它的中位线折成一个三棱锥? 如果能,它的体积如何计算? 试与用同样的纸片剪拼成的、有相同底面积的直三棱柱比较,谁的面积大?

这样的问题,可以促使我们带着它去看有关书籍,查找所需的资料,或与同学讨论,向老师请教. 最后,将自己的解答与体会写成文章,这就是研究式的学习.

孔夫子教导我们说:"学而不思则罔,思而不学则殆."读书、想问题应当很好的结合起来,不可偏废.

数学课应当讲数学

题目似乎有点可笑. 数学课当然得讲数学. 不讲数学的课，能叫数学课吗？不讲数学，讲什么呢？

然而，数学课讲数学，这种普通人的常识却并非人人具备，眼下的一些数学课，花在数学上的时间却越来越少.

先回顾一点历史(年过花甲的人往往爱回头看，我也是如此).

在四十年前，中国大陆发生了一场史无前例的浩劫，持续十年. 开始几年，大中小学生停课，书本全部烧毁. 后来才开始"复课闹革命"，由工农兵与一些"改造好的知识分子"编写支离破碎、不成系统的"新教材".

"新教材"有两大特点. 一曰突出政治，二曰理论联系实际. 限于篇幅，这里不能详述，只举一个例子以见一斑.

比如要讲梯形的面积公式，那就得先说上一大段话：

"在伟大领袖毛主席'工业学大庆，农业学大寨，全国学习解放军'的伟大号召指引下，向阳公社前进大队发扬愚公移山的精神，一不怕苦，二不怕死，决心修筑一条红旗渠引水上山，造福子孙万代，真是'为有牺牲多壮志，敢教日月换新天'. 这条红旗渠的横截面是一个梯形，上底长 1.5 米，下底长 1 米，深 2 米. 问横截面积是多少平方米？"

三分之二的话与数学无关，学生的注意力早被分散了.

这种语言现在看来十分可笑，所谓的联系实际也是假的. 那个向阳公社根本是乌有乡，不存在，所有数据都是编造的.

但是这种"穿靴戴帽"的现象现在竟然又频频出现，而且有蔓延的趋势.

"考试是指挥棒"，考试对于各种风气的形成有重大的影响. 我们就举几道考题作例子.

2006 年某市中考的一道题：

今年 5 月 18 日，英美科学家公布了人类第一号染色体的基因测序图，这个染色体是人类"生命之书"中最长也是最后被破解的一章. 据报道，第一号染色

体中共有2.23亿个碱基对. 2.23亿这个数用科学记数法可表示为　　(　　)

A. 2.23×10^5;　　　　B. 2.23×10^6;

C. 2.23×10^7;　　　　D. 2.23×10^8.

这道题不是最长,但废话也不少,其实只有最后一句才有用,前面统统应当删去.(参见《中学数学研究》2006年第10期第28页)

再如2006年某省的一道高考题:

右图中有一个信号源和五个接收器,接收器与信号源在同一个串联线路中时,就能接收到信号,否则就不能接收到信号.若将图中左端的六个接线点随机地平均分成三组、将右端的六个接线点也随机地平均分成三组,再把所有六组中每组的两个接线点用导线连接,则这五个接收器能同时接收到信号的概率是　　(　　)

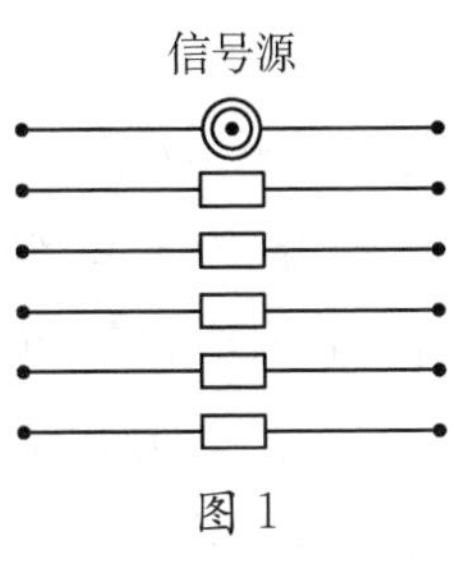

图 1

(A) $\frac{4}{45}$;　　(B) $\frac{1}{36}$;　　(C) $\frac{4}{15}$;　　(D) $\frac{8}{15}$.

题目十分冗长、晦涩,多数考生反映无法看懂.幸而是一道选择题,得分率还不太低(胡选一个,有$\frac{1}{4}$的概率答对).

或许命题者会振振有词地说,这道题可以培养学生解决实际问题的能力.

其实这道题并非真的"实际问题".原为另一个省的高考模拟题.原题是6个电阻,现改为一个信号源与五个接收器,更多人为编造的痕迹,更加艰涩难解.

数学课的主要(或首要)任务是教数学、学数学,解数学问题,而不是解决实际问题.将实际问题化为数学问题,这并不是数学的主要任务.这种能力的培养需要各门学科(如理、化、生、通用技术、信息技术等)的综合,不能全由数学课来承担.否则就成为大家吃饭,数学一家买单.解决实际问题的能力,更需要在今后从事实际工作时提高,不应把将来要吃的饭,现在就全由数学来买单.在数学课时已经不多的时候,数学课更应当坚持讲数学,少讲其他无关的内容.

前面的中考题,像一道生物题.再如某市2006年的另一道数学中考题.

司机在驾驶汽车时,发现紧急情况到踩下刹车需要一段时间,这段时间叫反应时间,之后还会继续行驶一段距离,我们把司机从发现紧急情况到汽车停止所行驶的这段距离叫做“刹车距离”.

已知汽车的刹车距离 s(单位:m)与车速 v(单位:m/s)之间有如下关系:$s=tv+kv^2$,其中 t 为司机的反应时间(单位:s),k 为制动系数.某机构为测试司机饮酒后刹车距离的变化,对某种型号的汽车进行了“醉汉”驾车测试.已知该型汽车的制动系数 $k=0.08$,并测得志愿者在未饮酒时的反应时间 $t=0.7$ s.

(1) 若志愿者未饮酒,且车速为 11 m/s,则该汽车的刹车距离为________ m(精确到 0.1 m).

(2) 当志愿者在喝下一瓶啤酒半小时后,以 17 m/s 的速度驾车行驶,测得刹车距离为 46 m,假如该志愿者当初是以 11 m/s 的车速行驶,则刹车距离将比未饮酒时增加多少(精确到 0.1 m)?

(3) 假如你以后驾驶该型号的汽车以 11 m/s 至 17 m/s 的速度行驶,且与前方车辆的车距保持在 40 m 至 50 m 之间.若发现前方车辆突然停止,为防止“追尾”,则你的反应时间应不超过多少秒(精确到 0.01 s)?

这道题可能是真的联系实际了.但它更像是一道物理题.

数学有很多的应用,很多学科需要运用数学.但数学课应以数学为本、为主体.如果数学课花很多时间去讲应用,那就“侵入”其他学科了,吃力而不讨好.可谓体用倒置,舍本逐末.

数学课应当讲数学!

一堂数学课

"同学们已经学过了一元二次方程的解法，今天我们上一堂练习课."

郝老师转过身，拿起粉笔在黑板上写了十几个方程：

(1) $x^2+5x-6=0$;

(2) $3x^2-4x+1=0$;

(3) $5x^2+12x-17=0$;

(4) $x^2-5x+4=0$;

……

"我们开展竞赛，看谁解得最快，能不能在一分钟之内全部解出来."

同学们都笑了：十几个二次方程怎么能在一分钟之内全部解出来呢？他们抓紧时间做，过了一会，已经有同学举手表示全做好了. 郝老师请他把答案写在黑板上.

各个方程的根分别为

(1) 1, -6;

(2) 1, $\frac{1}{3}$;

(3) 1, $-\frac{17}{5}$;

(4) 1, 4;

……

"大家注意，这些方程都有一个根为 1，为什么呢?"

"因为这些方程左边的系数的和都是 0"，一个同学回答说，"所以将 $x=1$ 代入方程后，左边等于 0，也就是说 $x=1$ 是方程的根."

"说得很好！根据系数和为 0 这个特点，我们可以立即求出一个根 $x=1$，另一个根怎么求呢？我们有一个定理：

设 $ax^2+bx+c=0$ 的两个根为 x_1、x_2，那么

$$x_1+x_2=-\frac{b}{a},$$

$$x_1x_2=\frac{c}{a},$$

因此如果有一个根是 1，那么另一个根就是$\frac{c}{a}$. 根据这个定理，我们就可以很快地把上述方程的根写出来. 这个定理的证明并不难，我们下堂课讲. 现在我们再解几个方程.”

郝老师又在黑板上写了一系列方程：

(5) $25x^2+25x-6=0$;

(6) $x^2+15x-54=0$;

(7) $12x^2-8x+1=0$;

(8) $3x^2-8x+4=0$;

……

“我也参加比赛，看谁解得快，我的答案写在黑板上，你们先不要看.”

郝老师很快地在黑板上写好了答案，然后用纸盖住. 不一会同学们也解完了，大多数同学的结果与老师相同.

“你们解得很好.”

“老师怎么解得这么快呢?”

“这并没有什么稀奇. 比如说，我把(1)中的 x 换为 $5x$ 就得(5)，因此(5)的根恰好是(1)的根的$\frac{1}{5}$，即(5)的根是$\frac{1}{5}$，$-\frac{6}{5}$. 把(1)中的 x 换为$\frac{x}{3}$，再去分母就得到(6)，因此(6)的根是(1)的根的三倍，即 3 与-18. 其余的依此类推.”

“哦！原来是这样的. 我们也会编题目了.”

“对，今天的作业就是每人编十个方程给同桌的同学做.”

凤头，还是猪头?

文章的开头要写好，有人比喻为“凤头”：笔墨不多，却很俏丽，引人入胜.

当然一篇文章，更重要的部分还在后面. 有人比喻为“猪肚”：内容充实，有阅读与思考的价值.

一堂课，也应当用一种好的方法引入，激发学生对新知识的渴望与追求. 引入新课，所花的时间不应太多，否则就不像凤头了.

近来的中学数学教学，似乎出现了一种风尚：花大量的时间在“情境设计”上，并且有人以此作为评课的重要标准.

举一个典型的例子. “多边形的内角和”这节课，如果从三角形的内角和为 180° 引起，说到矩形的内角和为 360°，再到一般的四边形可以用对角线分为两个三角形，内角和也为 360°. 很快就可切入主题：将 n 边形用对角线分为 $n-2$ 个三角形，得出内角和为 $(n-2)\times 180^\circ$.

可是一个试验小组的几位老师却苦心孤诣地构建了一个情境如下：

“银都花园城要设计一个正多边形喷水池. 每个顶点处，各安置一个可旋转的喷头，每个至多旋转 120°，要能恰好覆盖相邻两个水池边所夹角度，有几种设计方案?”(原文很长，这里作了精简. 其中“相邻两个水池边”“角度”等含义不很明确. 我们保留原文未作更动.)

这个“设计”远比这节课的内容难. 学生尚未知道 n 边形的内角和是多少. 当然更无从知道正 n 边形的每个内角是多少，而这个设计甚至还要学生进一步去解不等式 $\frac{n-2}{n}\times 180<120$，真是强人所难. 将这个“实际问题”(其实是试验组闭门造车构建的问题)化为数学问题，又是上述困难之前的一大困难. 结果是试验组承认“对学生来说，这一步很困难”，“花了 15 分钟还有些小组不能理解”，“不得不由教师帮助一些小组将问题化为数学问题”. 而又由于“归纳正多边形公式时出现困难，最终教师不得不放弃让学生探究的设计初衷，而采用了讲授式”.

本来并没有那么多的困难. 由于构建的情境脱离学生的实际,增加了教学的困难. 试验小组可谓“自寻烦恼”.

其实,“情境设计”就是为了引入新课. 如果这一部分过分地长、过分地难,让许多不需要当下讨论的内容(特别是非数学的内容)占据太多的时间,成为一堂数学课的中心,那就成了喧宾夺主,本末倒置. 凤头变成猪头,也就不俏了.

请设计情境的朋友们多想一想,您设计的情境,是凤头,还是猪头?

何不使用计算器

《数学通讯》2006 年第 23 期上，有两种观点的争鸣，争鸣的问题如下：

“在$\triangle ABC$中，已知

$$\cos(2A+C)=-\frac{3}{5}, \tag{1}$$

$$\sin B=\frac{3}{5}, \tag{2}$$

且$A<B<C$，求$\cos A$.”

一种观点认为$B=\pi-(A+C)$，故由(2)可得$\sin(A+C)=\frac{3}{5}$. 又$A<B<C$，所以$0<B<\frac{\pi}{2}$，于是$\frac{\pi}{2}<A+C<\pi$，故$\cos(A+C)=-\frac{4}{5}$. 因$0<2A+C<A+B+C=\pi$，故$\cos(A+C)=-\frac{4}{5}$. 于是$\cos A=\cos[(2A+C)-(A+C)]=\left(-\frac{4}{5}\right)\times\left(-\frac{3}{5}\right)+\left(\frac{3}{5}\right)\times\left(\frac{4}{5}\right)=\frac{24}{25}$.

另一种观点$\sin(A+C)$，$\cos(A+C)$的求法同上. 由$\cos(2A+C)=\cos[(A+C)+A]=\cos(A+C)\cos A-\sin(A+C)\sin A$，可得$4\cos A+3\sin A=3$. 又$\cos^2 A+\sin^2 A=1$，$0<A<\frac{\pi}{3}$，解之得$\cos A=\frac{24}{25}$，$\sin A=-\frac{7}{25}$. 这与$\sin A$应大于零矛盾，故本题应无解.

上面的争论，胜方显然是第二种观点. 因为第一种观点未发现矛盾，并不能说明题目正确，只是尚未发现题目的不正确而已. 第二种观点则已经发现题目中的矛盾，指出题目不正确了.

不过，令人感到奇怪的是，不论哪一种观点都未采用计算器来算一算，而题目中的一些角是可以直接算出的：

由(1)及计算器立即得到

$$2A + C = 180^\circ - 53.13\cdots^\circ. \tag{3}$$

由(2)及计算器得到($B < C$,所以 B 是锐角)

$$B = 36.86\cdots^\circ. \tag{4}$$

(3)+(4)得

$$2A + B + C < 180^\circ, \tag{5}$$

这与 $A + B + C = 180^\circ$ 显然矛盾.

科学技术在发展,我们应当充分利用一切可用的技术. 电脑配上数学软件(如 Maple)不仅可作通常的运算,而且可求导数、不定积分、定积分、矩阵的逆、线性方程组的解……,可进行方差分析、回归分析……. 几何问题可用几何画板显示,数论或组合问题可以编制简易程序进行检验. 杨路先生的软件 Bottema 在几分钟内就能证明一个复杂的不等式. 计算器更是最简单的工具,为何弃而不用呢?

据说,过早使用计算器,会降低学生的运算能力. 那么,在小学高年级,计算能力已经养成后,为何不能用计算器呢?

还有一种理由是计算器可以成为作弊的工具. 我就亲眼目睹广州大学监督监考的人员没收学生的计算器(而在概率统计的计算中,计算器是不可缺少的). 这种说法与做法,对于学生乃至教师极度的不信任,与尊重学生、尊重教师的民主教育理念完全不符.

在西方国家,即使规定低年级学生不许用计算器,仍有许多学生用(这的确导致一些学生运算能力下降). 但在我们这里,情况似乎相反,学生们,甚至不少教师也自己"画地为牢",有计算器,却"自觉地"不用它.

写这篇短文,就是希望大家尽量地"物尽其用". 能用计算器的问题,不要不用计算器.

别扼杀了天才

大家都知道高斯(Gauss)计算

$$1+2+3+\cdots+100 \tag{1}$$

的故事.现在的小学生,只要学过“奥数”,没有一个不会计算(1)的,没有一个不知道得数是5050的.

但是,他们都是天才吗?他们胜过高斯吗?

恐怕不能说他们都是胜过高斯的天才.因为绝大多数小朋友都是老师或家长教的,并不是自己想出来的.

奥数活动中的一种不良倾向就是在短时间给学生“恶补”很多难以消化的内容.这种“拔苗助长”的做法往往适得其反,甚至造成学生“厌学”,扼杀了可以成为高斯的天才.

散文作家梁遇春说过:“只有自己发现出的美景对着我们才会有贴心的亲切感觉,才会感动了整个心灵.”

数学也是这样,只有学生自己发现出“美景”,才会热爱它.

今年在某地,我给初一新生做一道华罗庚金杯赛的题目:

“甲、乙、丙三人跑1000米,同时从起跑线出发,甲到终点时,乙离终点还有50米.乙到终点时,丙离终点还有100米.问甲到终点时,丙离终点多少米?”

出乎我的意料,几个班的学生们竟不约而同地给出下面的解法:

甲、乙速度的比是$1000:950=20:19$.

乙、丙速度的比是$1000:900=10:9$.

所以甲、乙、丙速度的比是

$$200:190:171(171=19\times 9).$$

即甲到终点(跑1000米)时,丙跑

$$(1000\div 200)\times 171=855\text{ 米},$$

距终点还有

$$1000-855=145\text{ 米}.$$

这个解答成立.我不能不表扬他们,但心里却感到莫大的悲哀.这样统一的解法不会是他们自己想出来的,而是“教育”的结果.

算术中的连比,在上个世纪六十年代之前就已经退出数学教材.不知道为什么这种“过时的”内容又被人奉为至宝,灌输给学生.如果学生只会老师教的方法,而不是自己想解法(像高斯那样),久而久之,他们的“灵气”就被销磨殆尽,一点创造性也没有了.

其实这道题,甲、丙速度之比并不需要去求.乙到终点时,丙还差 100 米,所以乙从终点后退 50 米时,丙应退 $\frac{1000-100}{1000}\times 50=45$ 米.即甲到终点时,丙距终点 $100+45=145$ 米.

这种解法比上面那种简单得多.

本是一个可以心算,三言两语就能解决的问题,上面那种解法却绕个大圈,先去求甲、丙速度之比.这不是优雅的解法,而是丑陋的解法.如果学生见到的都是这种解法,他们将会认为数学是繁琐的、丑陋的.千万不要给学生产生这样的印象.

当然,更重要的是让学生自己去想,不要用统一的解法去束缚他们的思想.千万别扼杀天才!

没事，别来烦我

纲，是提网的总绳，比喻事物的最主要部分. 这个词，一度与政治密切相关，如“以粮为纲”“以钢为纲”“三项指标为纲”. 老一辈的人知道，那时听得最多的是“以阶级斗争为纲”.

在我国数学教育界，不知是谁，也不知在什么时候，提出了一个口号：“以函数为纲”. 这个口号把函数先生抬高到 No. 1 的地位，当然也增加了他的负担，许许多多不太相干的事情都扯到他的头上. 请看下面的例子(摘自《超级数学专题题典 · 数列》，37～38 页，上海世界图书出版公司，2007 年第 1 版)：

“**例 9** 等差数列 $\{a_n\}$ 中，已知 $a_3=12$，且有 $S_{12}>0$，$S_{13}<0$，试求公差 d 的取值范围，指出 S_1，S_2，…，S_{12} 中，哪一个值最大，并说明理由.

分析： 由公式 $S_n=\frac{d}{2}n^2+\left(a_1-\frac{d}{2}\right)n$ 可知，只要讨论该函数的零点即可.

解： $\because \dfrac{a_1-\frac{d}{2}}{-\frac{d}{2}}=\dfrac{d-2a_1}{d}=\dfrac{d-2(a_3-2d)}{d}=\dfrac{5d-2a_3}{d}=5-\dfrac{2a_3}{d}$，

$\therefore S_n=\frac{d}{2}n^2+\left(a_1-\frac{d}{2}\right)n$ 的零点分别是 0 和 $5-\frac{2a_3}{d}$.

根据条件可以得到 $5-\frac{2a_3}{d}\in(12,\ 13)$，$\therefore d\in\left(-\frac{24}{7},\ -3\right)$.

$\because a_3=12$，$\therefore a_6=a_3+3d=12+3d\in\left(\frac{12}{7},\ 3\right)$，$a_7=a_3+4d=12+4d\in\left(-\frac{12}{7},\ 0\right)$.

所以 $a_6>0$，$a_7<0$，也就是说 S_6 达到最大.

点评： 将等差数列的前 n 项和化为二次函数解决，更为直观和方便. 如果数列的项变号，数列的前 n 项和达到极值.”

这本书的作者(BSK 高考命题研究组)将函数先生拉出来折腾了一番. 其实

事情完全不需要麻烦他:

因为 $S_{13}=a_1+a_2+\cdots+a_{13}=13a_7<0$,所以 $a_7<0$. 而 $a_7=a_3+4d=12+4d$,所以 $12+4d<0$, $d<-3$. 同样,由 $S_{12}=a_1+a_2+\cdots+a_{12}=6(a_6+a_7)>0$,得 $a_6+a_7>0$,即 $(12+3d)+(12+4d)>0$, $d>-\frac{24}{7}$. 所以 d 的取值范围是 $\left(-\frac{24}{7},\ -3\right)$. 反过来,在这个范围中的 d 使 $a_7<0$, $a_6+a_7>0$,即 $S_{13}<0$, $S_{12}>0$.

又 $a_7<0$, $a_6+a_7>0$,所以 $a_6>-a_7>0$, 即 a_7 是等差数列中的第一个负项. 因此,和 S_6 最大.

函数先生说:把我拉来帮忙,越帮越忙,并不是“更为直观和方便”. 请各位今后没事,别来烦我.

化简，不要"化繁"

化简,学过初中数学的,人人都懂.

一个分式,可以约分的应当约分,一个整式,可以合并同类项的应当合并.这就是式的化简.

在解方程或解(证明)不等式时,也常常化简.通常是利用平方等方法,将无理式化为有理式;利用去分母的方法,将分式化为整式.最后,可以约简或合并时应当约简或合并.

道理虽然简单,做起来却常有人忘记.

偶然翻阅江苏省某名校几位名师编的一本书,就发现很多不符合化简的解法.请看下面的一例.

例　已知 $x, y \in \mathbf{R}$,且 $x \neq y$,求证

$$\left|\frac{1}{1+x^2}-\frac{1}{1+y^2}\right|<|x-y|. \tag{1}$$

原书的解法是:

设 $1+x^2=u \geqslant 1$, $1+y^2=v \geqslant 1$, 则原不等式为

$$\left|\frac{1}{u}-\frac{1}{v}\right|<\left|\pm\sqrt{u-1} \pm \sqrt{v-1}\right|, \tag{2}$$

即

$$\left|\frac{(\sqrt{u-1})^2-(\sqrt{v-1})^2}{uv}\right|<\left|\pm\sqrt{u-1} \pm \sqrt{v-1}\right|. \tag{3}$$

由于 $x \neq y$,故 $\sqrt{u-1} \neq \sqrt{v-1}$. 由此,上式可化简为

$$\left|\sqrt{u-1} \pm \sqrt{v-1}\right| \leqslant uv. \tag{4}$$

因 $uv>1$, 欲证不等式(4),只要证比(4)更强的不等式

$$\left|\sqrt{u-1} \pm \sqrt{v-1}\right| \leqslant \sqrt{uv}, \tag{5}$$

即
$$\left|\frac{1}{\sqrt{v}}\sqrt{1-\frac{1}{u}}\pm\frac{1}{\sqrt{u}}\sqrt{1-\frac{1}{v}}\right|\leqslant 1. \tag{6}$$

再设 $\frac{1}{\sqrt{v}}=\cos\alpha$, $\frac{1}{\sqrt{u}}=\cos\beta$, α, $\beta\in\left[0,\ \frac{\pi}{2}\right)$, 则(6)化为

$$|\sin(\alpha\pm\beta)|\leqslant 1. \tag{7}$$

(7)显然成立.

上面的解法,虽然也出现“化简”的字样,实际做法却与化简的精神背道而驰.原本好端端的有理式,却硬化成无理式.解法中使用了很多“技巧”,如将(2)化为(3),如作代换 $1+x^2=u$, $1+y^2=v$;再作代换 $\frac{1}{\sqrt{v}}=\cos\alpha$, $\frac{1}{\sqrt{u}}=\cos\beta$ 等.令人眼花缭乱.实际上并无多大用处,徒然使学生望而生畏,以为数学真难啊!

所以,上面的解法,虽然是正确的解法,却不是合理的解法.

合理的解法,应当是简单、自然的,它逐步地化繁为简.它不故弄玄虚,而是朴实无华;不炫耀技巧,而是采用最基本的方法.因而每一步都是易于理解,便于操作的.每一步的目标都是清楚的,而且距离总的目标应当越走越近,而解决问题的希望也越来越大,解题者的信心也越来越强.

以上面的不等式证明为例.合理的解法就是逐步化简.先去分母,化(1)为等价的不等式

$$|(1-y^2)-(1-x^2)|<|x-y|(1+x^2)(1+y^2). \tag{8}$$

(8)左即 $|(x-y)(x+y)|$. 因 $x\neq y$, 故可在(8)两边约去 $|x-y|$ 得等价的不等式(约,也是化简)

$$|x+y|<(1+x^2)(1+y^2). \tag{9}$$

可以证明比(9)更强的不等式

$$|x|+|y|<(1+x^2)(1+y^2), \tag{10}$$

也就是不妨设 $x\geqslant 0$, $y\geqslant 0$, 证明

$$x+y<(1+x^2)(1+y^2). \tag{11}$$

(原来 x、y 的正负有多种情况,现在只需证明一种)

x，y 中有一个较大. 不妨设 $x > y$. 这时

$$(1+x^2)(1+y^2) \geqslant 1+x^2 \geqslant 2x > x+y,$$

即(11)成立. 证毕.

比较一下就知道,现在的解法并无特别的技巧,却同样达到目标.

市场上数学书(尤其是高考与竞赛的书)很多,但很遗憾,其中相当多的并不能正确地指导学生学习与解题. 它们有点像那种蹩脚的导游,徒然带着游客漫无目的地乱逛,游客疲倦不堪,却欣赏不到美景.

我们希望导游熟悉旅游的环境,这是作为导游的起码条件. 我们更希望写书的名师熟悉自己所写的内容,这样才能正确地指导学生. 否则"以己之昏昏"是不会"令人昭昭"的. 当然,读者也应当有自己的思考,不要一味被作者"牵着鼻子走". 应当相信,有基本的知识与基本的方法(如化简),就能解决很多问题,不要迷信书本,不要轻信那些所谓的"技巧". 应当相信,越是基本的东西越有用,而很多"花枪",虽然好看,实际上用处却不大. 化简就是一个最基本的方法,应当牢固地掌握.

“画蛇添足”与“画龙点睛”

“画蛇添足”与“画龙点睛”都是脍炙人口的故事.

现代的教育强调学生自主学习,但教师的指点仍是不可缺少的.

指点不应当是“画蛇添足”. 例如学生一口说出 $997\times 3=2991$, 教师却非要求他写出

$$997\times 3=(1000-3)\times 3=3000-9=2991$$

不可. 这就有“蛇足”之嫌了. 这种例子恐怕不是少数,但本文不拟再举.

不久前看到天津某高中的一堂讨论课. 讨论的题目基本是(本文稍作简化):

函数 $f(x)=ax^2+bx+c$, $a>b>c$,满足 $f(1)=0$, $a^2+[f(m_1)+f(m_2)]a+f(m_1)f(m_2)=0$. 求证:$f(m_1+3)$ 与 $f(m_2+3)$ 中至少有一个为正数.

这堂课讨论得很热烈. 师生共同探讨解法. 因

$$f(1)=a+b+c=0, \tag{1}$$

故

$$b=-(a+c). \tag{2}$$

由 $a>b>c$ 及(1)可知 $a>0$. 又由已知可得 $a=-f(m_1)$ 或 $-f(m_2)$. 不妨设

$$a=-f(m_1). \tag{3}$$

即 $-a=am_1^2+bm_1+c$, 从而

$$m_1=\frac{-b\pm\sqrt{b^2-4a(a+c)}}{2a}, \tag{4}$$

$$f(m_1+3)=f(m_1)+6am_1+9a+3b=6am_1+8a+3b.$$

再用(4)代入并结合(2),最后得出 $f(m_1+3)>0$(过程较繁,从略).

一位学生提出,函数 $f(x)$ 的两零点之差是

$$\frac{\sqrt{b^2-4ac}}{a}=\frac{\sqrt{(a+c)^2-4ac}}{a}=\frac{a-c}{a}.$$

因 $3a-(a-c)=2a+c>a+b+c=0$. 故$\frac{a-c}{a}<3$, 即两零点之差小于 3.

现在

$$f(m_1)=-a<0, \tag{5}$$

而函数 $y=ax^2+bx+c$ 是开口向上的抛物线,所以 m_1 在两零点之间. 因两零点的距离小于 3,故 m_1+3 不在两零点之间(在较大的零点之右). 从而 $f(m_1+3)>0$.

这种解法显然简单得多. 老师对此也作了肯定. 但却没有指出这种解法用较弱的条件

$$f(m_1)<0 \tag{5}$$

代替了

$$f(m_1)=-a, \tag{3}$$

因而得出的结论更一般(而且由于较弱的条件是一个不等式,运用起来更加自由,比拘泥于一个等式方便许多). 其实,这里体现了很重要的数学思想. 它说明一个更普遍的命题往往更反映问题的本质,可能比一个特殊的命题更易处理(在特殊的命题中,本质常被特殊的性质所掩盖). 这正是需要"点睛"的地方,却轻易地放过去,十分可惜.

以上一例,说明教师的学养与水平极为重要. 在授课与讨论中,特别要注意"画龙点睛". 这样教学效率就可以大大提高.

韦达定理能用吗?

一位高中生来信,摘录如下:"我写信的目的是有一个问题想请教您. 我和我的数学老师讨论过,但见解不统一,所以特来问您. '设关于 x 的方程

$$x^2-(\tan\theta+\mathrm{i})x-(2+\mathrm{i})=0$$

有实数根,求锐角 θ 和实数根.'老师给出的解法是设实数根 $x=m$ $(m\in\mathbf{R})$, 代入原方程,根据复数相等的条件得出答案 $m=-1$, $\tan\theta=1$, $\theta=45^\circ$. 我是利用韦达定理,即由 $x_1x_2=-2-\mathrm{i}$ 得出,如果方程存在实数根,则一定是一个实数根,一个虚数根. 设这两个根为 $x_1=m$ $(m\in\mathbf{R})$, $x_2=a+b\mathrm{i}$ $(a、b\in\mathbf{R},\ b\neq 0)$. 利用韦达定理得 $\begin{cases}m(a+b\mathrm{i})=-2-\mathrm{i},\\ m+a+b\mathrm{i}=\tan\theta+\mathrm{i},\end{cases}$ 根据复数相等条件求出 $b=1$, $a=2$, $m=-1$, $\tan\theta=1$, $\theta=45^\circ$. 得出的答案与老师的完全相同. 但我向老师提出此方法时,老师则一口否决了. 他认为此方程非实系数方程,不能用韦达定理,韦达定理的条件是实系数方程,并且必须存在实数根."

信中所说的问题不难. 老师给出的解法是正确的,也是简单的(只是设 $x=m$ $(m\in\mathbf{R})$ 并无必要,直接可以看出实根 $x=-1$).

但是,韦达定理能用吗?

这是一个值得讨论的问题,因为认为韦达定理不能用的高中数学老师竟然是存在的.

他(或他们)为什么认为"韦达定理的条件是实系数方程,并且必须存在实数根"呢?

可能是由于推导韦达定理时,先认定一元二次方程 $ax^2+bx+c=0$ 是实系数,有两个实数根,且 $x_1=\dfrac{-b+\sqrt{b^2-4ac}}{2a}$, $x_2=\dfrac{-b-\sqrt{b^2-4ac}}{2a}$; 然后再计算得出 $x_1+x_2=-\dfrac{b}{a}$, $x_1x_2=\dfrac{c}{a}$.

在未学复数之前,当然只能加上这些限制条件. 但是,学过复数,情况就不

同了. 首先,一元二次方程 $ax^2+bx+c=0\ (a\neq 0)$ 必有两个根(一般地,一元 n 次方程必有 n 个根),不需要系数为实数或判别式$\geqslant 0$ 之类条件.

其次,韦达定理可以通过同样计算证明,但更简单的办法是由恒等式 $ax^2+bx+c=a(x-x_1)(x-x_2)$,将右边乘开得 $ax^2-a(x_1+x_2)x+ax_1x_2$ 比较系数即得 $x_1+x_2=-\frac{b}{a}$, $x_1x_2=\frac{c}{a}$. 所以韦达定理总是成立的(甚至可推广至一元 n 次方程),不需要加上"实系数,有实根"之类条件.

随着年级的升高,学生所学的知识应当不断更新. 例如在没有负数时,不能计算 $3-5$,而学过后就有 $3-5=-2$. 在没有分数、小数时,不能计算 $3\div 5$,而学过后就有 $3\div 5=\frac{3}{5}=0.6$. 学习复数以后,关于一元二次方程的知识也应随之更新. 作为教师,理应指导学生学习,帮助学生更新知识. 遗憾的是,并非所有教师都能够指导学生这样做. 更遗憾的是有些教师自己也未将有关知识搞清楚.

认为韦达定理需要上述条件的教师集合,不知有多大. 但它不是空集,这就值得严肃对待了. 当前,不少人热衷于谈思想,说理念,而数学却丢在一边. 其实,一位数学教师,最要紧的是自身的数学素养. 如果数学甚差,滥竽充数,岂不误人子弟?

不要滥用反证法

反证法是一种重要的证明方法.在证题过程中,当直接证法难以奏效时,可采用间接证法.就像打仗一样,正面攻击不能奏效,迂回到侧后或许是一种好的策略.但是,并非任何问题都得用反证法,笔者建议能够直接证明的还是以直接证明为好.

例1 实数 $x_1, x_2, \cdots, x_n$ 满足

$$x_1+x_2+\cdots+x_n=0,\ x_1^2+x_2^2+\cdots+x_n^2=1.$$

证明:$x_1, x_2, \cdots, x_n$ 中至少有两个数的积小于或等于 $-\frac{1}{n}$.

很多同学选择反证法.

证明 假设结论不成立,即

$$x_ix_j>-\frac{1}{n}\ (1\leqslant i<j\leqslant n).$$

由已知得

$$0=(x_1+x_2+\cdots+x_n)^2=\sum x_i^2+2\sum x_ix_j=1+2\sum x_ix_j.$$

因此,
$$\sum x_ix_j=-\frac{1}{2}. \tag{1}$$

另一方面,由反证法的假设有

$$\sum x_ix_j>-\frac{1}{n}\mathrm{C}_n^2=-\frac{n-1}{2}. \tag{2}$$

式(1)、(2)矛盾.所以,结论成立.

事实上,式(1)、(2)并不矛盾,可见,证明是错的.

出现这种错误的同学不在少数.笔者在某地两个竞赛班试验.第一天下午是一个悲惨的下午,因为上黑板的三名同学全做错了.没想到第二天更为悲惨,

自告奋勇上来的四名同学竟也没有一个做对. 更为悲剧的是有几名同学坚持认为从式(1)与反证法的假设可以导出矛盾. 其实这已经是一条死胡同了. 式(1)与反证法的假设并不矛盾. 要找矛盾还得回到已知的两个条件.

正确的证法是直接证明.

证明 不妨设 $x_1, x_2, \cdots, x_n$ 中, x_1 最大, x_n 最小. 则

$$(x_1 - x_i)(x_i - x_n) \geqslant 0 \ (i=1, 2, \cdots, n). \tag{3}$$

将式(3)的左边展开,对 i 求和得

$$x_1 \sum x_i - \sum x_i^2 - n x_1 x_n + x_n \sum x_i \geqslant 0,$$

即 $x_1 x_n \leqslant -\dfrac{1}{n}$.

这道题的证法很多. 下面再介绍一种.

另证 设 $x_1, x_2, \cdots, x_n$ 中,正数为 $y_1 \geqslant y_2 \geqslant \cdots \geqslant y_s$,其余的为 $z_1 \leqslant z_2 \leqslant \cdots \leqslant z_t$, $s+t=n$.

则 $1 = y_1^2 + y_2^2 + \cdots + y_s^2 + z_1^2 + z_2^2 + \cdots + z_t^2 \leqslant y_1(y_1 + y_2 + \cdots + y_s) + z_1(z_1 + z_2 + \cdots + z_t) = -y_1(z_1 + z_2 + \cdots + z_t) - z_1(y_1 + y_2 + \cdots + y_s) \leqslant -t y_1 z_1 - s z_1 y_1 = -n y_1 z_1$.

因此, $y_1 z_1 \leqslant -\dfrac{1}{n}$.

这种证法可以改成反证法,但毫无必要. 不过已知条件中的 1 及(要证明的)结论中的 $-\dfrac{1}{n}$,启示解题者应取 n 个 $x_i x_j$ 的和,而不是 C_n^2 个. 该证法正是这样做的.

例 2 100 个互不相同的实数写在一个圆周上. 证明:一定可以找到四个相邻的数,两端的两个数的和大于中间两个数的和.

此题很多同学毫不犹豫地采用反证法.

证明 假设结论不成立. 设圆周上 100 个实数依次为 $a_1, a_2, \cdots, a_{100}$. 则

$$a_i + a_{i+3} \leqslant a_{i+1} + a_{i+2},$$

其中, $i=1, 2, \cdots, 100$,约定 $a_{i+100} = a_i$, 即

$$a_{i+3}-a_{i+2}\leqslant a_{i+1}-a_i. \tag{4}$$

将式(4)对 i 求和,知式(4)全取等号,即

$$a_{2i}-a_{2i-1}=c\ (c\ 为常数). \tag{5}$$

同理,

$$a_{2i+1}-a_{2i}=d\ (d\ 为常数). \tag{6}$$

在式(5)、(6)中令 $i=1,2,\cdots,50$,然后相加得

$$0=50c+50d.$$

从而,$c=-d$,

$$a_3-a_2=d=-c=a_1-a_2.$$

这就导出 $a_3=a_1$,与已知矛盾.

证明是正确的.但很繁琐.

另证 设 a 为100个数中最小的. a 的左邻为 b,右邻为 c.不妨设 $b>c$.又设 c 的右邻为 d,则 $d>a$.因此,$b+d>c+a$.

看来一些同学误以为存在性的问题一定要用反证法,其实,存在性的证明用构造法或极端性原理(即取最大的或最小的元素)为好.证明"不存在"才用到反证法.一般说来,结论中有"不"或隐含"不"的问题才用反证法.大多数的问题还是应当用直接证法.

此题是俄罗斯的一道竞赛题.可能有一些同学知道标准答案.但这个标准答案并不高明.看别人的解答也要有判别好坏的能力,不要被别人牵着鼻子走.

例3 整数数列$\{a_n\}$中,

$$a_1=1,\ a_{n+1}=a_n+S(a_n)\ (n=1,2,\cdots),$$

$S(a)$是 a 的数字和.问:12345是不是这个数列的项?

此题可以用反证法.

证明 假设12345是这个数列的项 a_{n+1}.

因为 $a_{n+1}=a_n+S(a_n)\equiv 2a_n(\bmod 3)$,而

$$12345\equiv 0(\bmod 3),$$

所以,$2a_n\equiv 0(\bmod 3)$,$a_n\equiv 0(\bmod 3)$.

同理,$a_{n-1}\equiv 0(\bmod 3)$,$\cdots a_1\equiv 0(\bmod 3)$,这与 $a_1=1$ 矛盾.

这道题也可以直接证明(不用反证法).

另证　因

$$a_{n+1}=a_n+S(a_n)\equiv 2a_n(\mathrm{mod}3)\equiv -a_n(\mathrm{mod}3),$$

而 $a_1=1$，所以，

$$a_n\equiv(-1)^{n-1}(\mathrm{mod}3). \tag{7}$$

因为 $12345\equiv 0(\mathrm{mod}3)$，所以，12345 不是数列 $\{a_n\}$ 的项.

两种证法都能解决问题. 不用反证法的证法稍好一点. 因为反证法虽然解决了这个问题，却没有留下任何"正面的"结果. 不用反证法的证明，得出一个结果(即式(7)). 如果要求证明本题的数列 $\{a_n\}$ 中，相邻两项的和被 3 整除，那么，式(7)就发挥了作用.

例 4　已知正整数 m、n 和质数 p 满足：对任意正整数 k，都有

$$(pk-1,\ m)=(pk-1,\ n).$$

证明：存在某个整数 t，使得 $m=p^t n$.

本题的结论表明，对任意质数 $q\neq p$，q 在 m、n 中的次数都是相等的. mn 的质因数可能很多，而解题者对它们一无所知，因此，直接证明它们在 m、n 中的次数相等相当困难. 这道题以用反证法为宜.

证明　设 q 为 mn 的质因数，$q\neq p$，且 $q^\alpha\parallel m$，$q^\beta\parallel n$，$\alpha\neq\beta$. 不妨设 $\alpha>\beta$.

因为 $q\neq p$，所以，$(q,\ p)=1$，k 的同余方程

$$pk-1\equiv 0\ (\mathrm{mod}q^\alpha)$$

有解.

对这个 k，$q^\alpha\parallel(pk-1,\ m)$，但

$$q^\beta\parallel(pk-1,\ n),\ \alpha\neq\beta,$$

与已知 $(pk-1,\ m)=(pk-1,\ n)$ 矛盾.

大致说，在结论涉及的对象很多(任意的，每一个等等)时，可以考虑采用反证法. 如果结论的反面涉及很多对象，则不宜采用反证法.

总之，能够用直接证法解决的问题，应当采用直接证法. 反证法不可滥用.

学好数学先得学好语文

常常有小朋友问我:"怎样学好数学?"

我的回答是:"学好数学得先学好语文."

小学数学的难点是应用题.语文不好,就没法看懂题目.例如下面的一道题:

两只猴子偷水果,一只偷番石榴,一只偷芭蕉.正要吃,发现园主走过来.它们一算,园主要两分半钟才能走到.第一只猴子1分钟可吃10个番石榴,只要用$\frac{2}{3}$的时间就可以吃完,然后去帮第二只猴子吃芭蕉,正好在全部吃完后逃走.如果芭蕉数比番石榴多3倍,第一只猴子吃芭蕉比吃番石榴快1倍.问第二只猴子1分钟吃几个芭蕉?

这样的问题,必须仔细读题,理解题意,分析其中的数量关系,记住条件和要求.而且应当抓紧时间,尽快完成这一过程.要达到这一要求,必须加强语文学习,多读书,多读文章.力求做到古人所说的"一目十行,过目不忘"(外语学习中,也有类似的阅读训练).

题目想好了,还要将你的解法用简洁的语言表达出来,使大家了解你的想法,达到交流的目的.上面这道题的解法需要比较多的文字(应用题也常常称为文字题):

第一只猴子吃芭蕉比吃番石榴快1倍,吃芭蕉的时间是吃番石榴的一半,所以它吃的芭蕉数与番石榴数相等.芭蕉数比番石榴多3倍,多出的3倍就是第二只猴子吃的芭蕉数.第二只吃的芭蕉是第一只吃的芭蕉的3倍,时间也是第一只的3倍,所以它们吃芭蕉的速度相等,都是1分钟吃20个.

解答写好后,应当像作文那样仔细推敲,反复修改.既要正确无误,也要明白易懂.

字,也要写清楚,不要潦草.

以上所说都是语文的基本功.可见要学好数学得先学好语文.

新的竖式乘法

竖式乘法,特别是两位数乘一位数、两位数乘两位数、三位数乘两位数,现在仍然是小学数学教育的重要内容.

例如,$67\times 8=536$ 的竖式通常写成:

$$\begin{array}{r} 6\ 7 \\ \times\quad\ \ 8 \\ \hline 5\ 3\ 6 \end{array}$$

具体做法是:先用 8 乘个位的 7 得 56,“写 6 记 5”,即写下 6 作为积的个位数字,将 5 记在心中;再用 8 乘十位的 6 得 48,将 8 与记在心中的 5 相加得 13,写下 3 作为积的十位数字,而 4 与 1 相加得 5 作为积的百位数字.

其中 $8+5$, $4+1$ 都用心算完成,固然有助于心算能力的提高,但也容易导致错误.很多从事小学教育的人认为这是一个难点.

有人采用在竖式的十位上写一个小小的 5,再与 8 相加后去掉.这减轻了记在心中的负担,但不太美观.而且在相加也发生进位时,百位的数字就不能先写下 4,还得记在心中.

因此.我们建议不如索性多写一行,即将 8×7, 8×6(即 8×60)的结果各写一行:

$$\begin{array}{r} 6\ 7 \\ \times\quad\ \ 8 \\ \hline 5\ 6 \\ 4\ 8\ \ \ \\ \hline 5\ 3\ 6 \end{array}$$

即先写下 8×7 的结果 56,再在下一行写下 8×6 的结果 48.由于十位的 6 表示 60,所以 48 实际上是 480,即 8 应表示 80,写在十位上(也就是 5 的正下方). 这样就完全不必"记在心中",最后相加得到 536.

类似地,两位数乘两位数也是如此. 如 $67 \times 58 = 3886$ 的竖式写成:

$$
\begin{array}{r}
6\ 7 \\
\times \quad 5\ 8 \\
\hline
5\ 6 \\
4\ 8\ \ \\
3\ 5\ \ \\
3\ 0\ \ \ \ \\
\hline
3\ 8\ 8\ 6
\end{array}
$$

其中第五行 $(5 \times 7 = 35)$ 的 5 应写在十位,与通常的竖式乘法相同;第六行 $(5 \times 6 = 30)$ 比第五行再前移一位,0 写在百位上.

这样做的好处,除了上面已说过的减轻心算负担、减少学生的错误之外,还有以下两点.

1. 减少了加法运算的次数. 通常的竖式乘法一边乘一边加. 比如,原先在 67×58 中, 8×67 要进行两次加法, 5×67 又要进行一次加法,最后还要将两次乘法所得的积相加. 新的竖式乘法只需在最后作加法. 而且,现在的加法可能还简单一些:如十位上 $5 + 8 + 5$, 可以先将 5 与 5 相加再加 8.

2. 更清晰地体现了乘法分配律,即两位数 $10 \times a + b$ 与 $10 \times c + d$ (a、b、c、d 都是数字)相乘时,

$$
\begin{aligned}
& (10 \times a + b) \times (10 \times c + d) \\
= & 100 \times a \times c + 10 \times a \times d + 10 \times b \times c + b \times d.
\end{aligned} \tag{*}
$$

其中, $b \times d$, $b \times c$, $a \times d$, $a \times c$ 正好是新竖式运算中的四行.

新的竖式乘法还可以解释一些速算的法则.

例如 36×34, 其中个位数字 6 与 4 互为补数(即和为 10),十位数字都是 3. 这时乘积的前两位是 $3 \times (3 + 1) = 12$,后两位是 $6 \times 4 = 24$, 乘积是 1224. 采用上面的竖式不难解释:

$$
\begin{array}{rrrr}
 & & 3 & 6 \\
\times & & 3 & 4 \\
\hline
 & & 2 & 4 \\
 & 1 & 2 & \\
 & 1 & 8 & \\
 & 9 & & \\
\hline
1 & 2 & 2 & 4
\end{array}
$$

积的后两位就是第三行的 24;第四、五两行相加得 30,也就是 $3\times(4+6)$,第六行的 9 与上两行的和 3 相加得 12,即 $3\times(1+3)$. 将 4、6 换成其他的一对补数,将 3 换成其他数字,同样如此. 一般地,在(*)中,当 $b+d=10$, $a=c$ 时,

$$
\begin{aligned}
&(10\times a+b)\times(10\times a+d)\\
=&100\times a\times a+10\times a\times d+10\times a\times b+b\times d\\
=&100\times a\times a+10\times a\times 10+b\times d\\
=&100\times a\times(a+1)+b\times d.
\end{aligned}
$$

三位数乘两位数,同样可用新的竖式乘法,如:

$$
\begin{array}{rrrrr}
 & & 2 & 6 & 7 \\
\times & & & 5 & 8 \\
\hline
 & & & 5 & 6 \\
 & & 4 & 8 & \\
 & 1 & 6 & & \\
 & & 3 & 5 & \\
 & 3 & 0 & & \\
1 & 0 & & & \\
\hline
1 & 5 & 4 & 8 & 6
\end{array}
$$

有兴趣的读者不妨在小学生中试验一下,看看效果如何.

一道高考数学试题解的讨论

今年(1980 年)高考数学试题的第六题的后一问难度较大,不易得出完全正确的解答,很有讨论之必要.

原题如下:"设三角函数

$$f(x)=\sin\left(\frac{kx}{5}+\frac{\pi}{3}\right),\text{其中 } k\neq 0.$$

(1) 写出 $f(x)$的极大值 M、极小值 m 与最小正周期 T.

(2) 试求最小的正整数 k,使得自变量 x 在任意两个整数间(包括整数本身)变化时,函数 $f(x)$至少有一个值是 M 与一个值是 m."

显然(1)的解是 $M=1,\ m=-1,\ T=\frac{10\pi}{|k|}$.

第(2)问,很多人是这样解的:

"当周期 $T\leqslant 1$ 时,每两个整数之间的距离$\geqslant T$,因而 $f(x)$至少有一个值是 M 与一个值是 m. 而由 $T\leqslant 1$ 得 $\frac{10\pi}{|k|}\leqslant 1$,所以 $10\pi\leqslant |k|$,从而所求的最小的正整数 $k=32$."

这里的结论虽然是正确的,但是这里的根据是有疑问的,因为上面的推理仅仅说明 $T\leqslant 1$ 是在每两个整数间 $f(x)$至少有一个值是 M 与一个值是 m 的充分条件,并没有证明它也是必要条件,即不能排除 $T>1$, 从而最小的正整数 $k<32$ 的可能性. 如果将原来的问题稍为变更一下,那么依照上面的推理便可能导致错误的结论. 例如将 $f(x)$改为

$$f(x)=\sin\left(\frac{k\pi}{5}x+\frac{\pi}{2}\right),$$

这时 $T=\frac{10}{|k|}$,按照上面的推理将得到 $k=10$,但正确的结论却是 $k=5$.

因此,还必须证明在上述高考试题中 $T\leqslant 1$ 也是必要的. 一种较为简单的

证法如下:

因为 x =整数时, $f(x)=\sin\left(\frac{k}{5}x+\frac{\pi}{3}\right)\neq 1$, 即 $f(x)$的最大值 $M(=1)$只能在每个区间$[m, m+1]$的内部取得(m 为正整数),如果每个这样的区间内部 $f(x)$至少有一个值为 M,现在$[0, n]$内(n 为正整数), $f(x)$至少有 n 个值为 M. 每两个使 $f(x)$取最大值的点至少相距 T,所以$[0, n]$的长 $n\geqslant(n-1)T$, $T\leqslant\frac{n}{n-1}$,令 $n\to\infty$,即得 $T\leqslant 1$.

将 $T\leqslant 1$ 的必要性与前面引用的解法结合起来,便得到完整的解答.

如果将 $f(x)$改为

$$f(x)=\sin\left(\frac{k\pi}{5}x+\frac{\pi}{2}\right),$$

那么 $f(x)$有可能在区间 $[m, m+1]$ (m 为正整数)的端点取得极值,因此不能断言 $T\leqslant 1$, 但用类似的推理,可知在半开区间 $[m, m+2]$ 上, $f(x)$至少有一个值为 M,因而在$[0, 2n]$上 $f(x)$至少有 n 个值为 M(n 为正整数),从而 $(n-1)T\leqslant 2n$, $T\leqslant\frac{2n}{n-1}$,令 $n\to\infty$,得 $T\leqslant 2$,从而$\frac{10}{|k|}\leqslant 2$, $|k|\geqslant 5$, 不难验证(例如利用函数 $f(x)$的图像), $k=5$ 是所求的最小的正整数.

一道高考题的求解思路

2004 年高考江苏卷的最后一题比较难,多数考生不熟悉这样的问题,不知从何入手.所公布的参考答案也比较迂回,不易看清其思路,本文将介绍另一种解法.

原题如下:

22. 已知函数 $f(x)$ $(x \in \mathbf{R})$ 满足下列条件:对任意的实数 x_1, x_2 都有

$$\lambda(x_1 - x_2)^2 \leqslant (x_1 - x_2)[f(x_1) - f(x_2)]$$

和

$$|f(x_1) - f(x_2)| \leqslant |x_1 - x_2|,$$

其中 λ 是大于 0 的常数.

设实数 a_0, a, b 满足

$$f(a_0) = 0 \text{ 和 } b = a - \lambda f(a).$$

(1) 证明 $\lambda \leqslant 1$,并且不存在 $b_0 \neq a_0$,使得 $f(b_0) = 0$;

(2) 证明 $(b - a_0)^2 \leqslant (1 - \lambda^2)(a - a_0)^2$;

(3) 证明 $[f(b)]^2 \leqslant (1 - \lambda^2)[f(a)]^2$.

题目比较长,需要仔细看,弄清其中字母 λ, a_0, a, b 的意义.问题涉及自变量的差 $(x_2 - x_1$, $a - a_0$, $b - a_0)$ 与函数值的差 $(f(x_2) - f(x_1)$, $f(a) - f(a_0)$, $f(b) - f(a))$. 这肯定是一位高校教师从数学分析中挖掘出来(或自己研究中产生)的一道题目,需要利用已知条件(其中 x_2, x_1 可取任意值),作适当的估计.

由已知条件,在 $x_1 \neq x_2$ 时,

$$0 < \lambda \leqslant \frac{f(x_1) - f(x_2)}{(x_1 - x_2)^2}(x_1 - x_2) = \frac{f(x_1) - f(x_2)}{x_1 - x_2}.$$

所以 $f(x_1) - f(x_2)$ 与 $x_1 - x_2$ 同号(同正或同负),即 $f(x)$在 $\mathbf{R}$ 上是(严

格的)增函数(即在 $x_1>x_2$ 时, $f(x_1)>f(x_2)$). 当然不会有两个零点,也就是在 $b_0\neq a_0$ 时, $f(b_0)\neq 0$. 并且由已知 $|f(x_1)-f(x_2)|\leqslant|x_1-x_2|$,

$$\lambda\leqslant\frac{f(x_1)-f(x_2)}{x_1-x_2}=\frac{|f(x_1)-f(x_2)|}{|x_1-x_2|}\leqslant 1. \quad ①$$

这就完成了(1)的证明,而且还证得更多些,得出了 $f(x)$ 在 **R** 上是(严格)的增函数.

现在来证(2),注意

$$f(a_0)=0,\ b=a-\lambda f(a),$$

所以

$$b-a_0=a-a_0-\lambda[f(a)-f(a_0)]. \quad ②$$

如果 $a=a_0$,那么 $b=a_0-0=a_0$, (2)式当然成立.

以下设 $a\neq a_0$, 令

$$t=\frac{f(a)-f(a_0)}{a-a_0}, \quad ③$$

(为了将函数值的差变为自变量的差)

则
$$f(a)-f(a_0)=t(a-a_0). \quad ④$$

由于①(取 $x_1=a$, $x_2=a_0$),

$$\lambda\leqslant t\leqslant 1. \quad ⑤$$

将④代入②得

$$b-a_0=(a-a_0)(1-t\lambda),$$

所以

$$(b-a_0)^2=(a-a_0)^2(1-t\lambda)^2.$$

剩下的问题是证明

$$(1-t\lambda)^2\leqslant 1-\lambda^2, \quad ⑥$$

也就是

$$\lambda(1+t^2)\leqslant 2t.$$

由于 $0\leqslant\lambda\leqslant t\leqslant 1$, 所以

$$\lambda(1+t^2)\leqslant 2\lambda\leqslant 2t,$$

即⑥成立,(2)证毕.

最后证(3),类似于上面的证明,在 $b\neq a$ 时,设

$$f(b)-f(a)=t(b-a), \qquad ⑦$$

则⑤成立(在①中取 $x_1=b$, $x_2=a$), 所以⑥成立.

将⑦及 $b=a-\lambda f(a)$ 代入(消去 b),得

$$\begin{aligned}[f(b)]^2&=[f(a)+t(b-a)]^2\\&=[f(a)(1-t\lambda)]^2\\&=[f(a)]^2(1-t\lambda)^2\leqslant(1-\lambda^2)[f(a)]^2.\text{(根据 ⑥)}\end{aligned}$$

在 $b=a$ 时, $\lambda f(a)=0$,根据(1), $a=a_0$,所以 $f(b)=f(a)=f(a_0)=0$.

于是总有 $[f(b)]^2\leqslant(1-\lambda^2)[f(a)]^2$.

③式证毕.

引入一个参数 t,是为了将函数值的差变为自变量的差(在证(2)时,将 $f(a)-f(a_0)$ 变为 $t(a-a_0)$, 在证(3)时,将 $f(b)-f(a)$ 变为 $t(b-a)$),以便比较. 虽然多出一个参数 t,但它满足⑤,范围为已知. 这时(2)与(3)都可以用同一个不等式⑥来证明.

证(2)分 $a=a_0$ 与 $a\neq a_0$ 两种情况. 其实不分也可以. 对于 $a=a_0$, 不论 t 为什么数,④当然成立,这时我们也可取 t 满足⑤,所以两种情况可以合在一起,同样,证(3)也可以不分两种情况.

上面证明的另一个关键之处是消去 b,将它用基本的量 a_0, a 及 $f(a)$, t, λ 的表达式代替后,问题就解决了.

评注 其实函数(二次函数) $y=(1-t\lambda)^2$, $\lambda\leqslant t\leqslant 1$, 在区间端点 $t=\lambda$ 处取最大值 $(1-\lambda^2)^2$, 所以本题(2)(3)中的系数 $1-\lambda^2$ 都可改进为(更小的) $(1-\lambda^2)^2$.

2009 年江苏高考数学试题评析

2009 年江苏数学高考题，总体上较为平稳，没有太难的题. 其中，个别立体几何题有与往年雷同的嫌疑.

填充题大多是基本题，侧重检查学生的基本功，不偏不怪，不蓄意为难学生，这是非常正确的. 其中只有第 13 题与第 14 题稍有难度.

第 13 题 如图 1，在平面直角坐标系 xOy 中，A_1，A_2，B_1，B_2 为椭圆 $\frac{x^2}{a^2}+\frac{y^2}{b^2}=1\ (a>b>0)$ 的 4 个顶点，F 为其右焦点，直线 A_1B_2 与直线 B_1F 相交于点 T，线段 OT 与椭圆的交点 M 恰为线段 OT 的中点，则该椭圆的离心率为________.

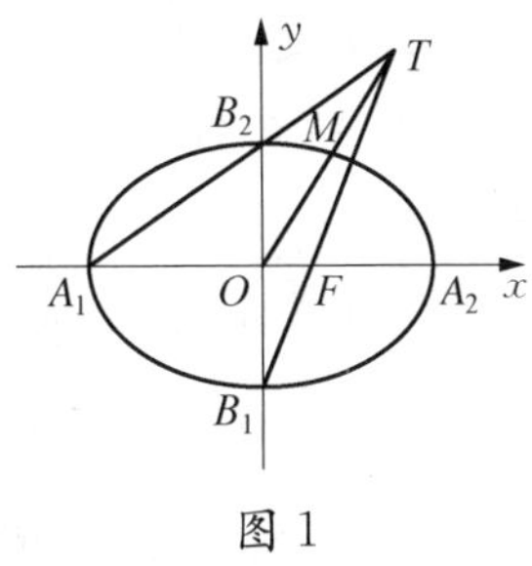

图 1

本题应求出 A_1B_2 方程

$$\frac{x}{-a}+\frac{y}{b}=1,$$

再求出 B_1F 方程

$$\frac{x}{c}+\frac{y}{-b}=1.$$

两式相加即可解出 T 的坐标

$$x_T=\frac{2ac}{a-c},\ y_T=\frac{(a+c)b}{a-c}.$$

OT 中点 M 的坐标是

$$x_M=\frac{1}{2}x_T=\frac{ac}{a-c},$$

$$y_M=\frac{1}{2}y_T=\frac{(a+c)b}{2(a-c)}.$$

代入椭圆方程得

$$\frac{c^2}{(a-c)^2}+\frac{(a+c)^2}{4(a-c)^2}=1,$$

化简得

$$c^2+100ac-3a^2=0.$$

所以离心率(只取正值)

$$e=\frac{c}{a}=\frac{-10+\sqrt{10^2+4\times 3}}{2}=2\sqrt{7}-5.$$

第 14 题　设$\{a_n\}$是公比为q的等比数列，$|q|>1$，令$b_n=a_n+1\ (n=1,2,\cdots)$，若数列$\{b_n\}$有连续 4 项在集合$\{-53,-23,19,37,82\}$中，则$6q=$________.

本题数列$\{a_n\}$有连续 4 项在集合$\{-54,-24,18,36,81\}$中，因而数列中既有正项也有负项，并且正负项交错，公比$q<0$. 连续 4 项中，两正两负，所以-54，-24都是数列的项. 因为$|q|>1$，所以$q^2=\frac{-54}{-24}=\frac{9}{4}$，$q=-\frac{3}{2}$(因为$q<0$)，$6q=-9$.

今年的数列题不难. 没有放在最后作为大轴题，出乎许多人的预料.

第 17 题　设$\{a_n\}$是公差不为零的等差数列，S_n为其前n项和，满足$a_2^2+a_3^2=a_4^2+a_5^2$，$S_7=7$.

(1) 求数列$\{a_n\}$的通项公式及前n项和S_n；

(2) 试求所有的正整数m，使得$\frac{a_m a_{m+1}}{a_{m+2}}$为数列$\{a_n\}$中的项.

可以说一说的是，等差数列中，连续奇数项的和等于中间一项乘以项数，据此立即可以求出$a_4=\frac{1}{7}S_7=1$，再由

$$(1-2d)^2+(1-d)^2=1+(1+d)^2$$

及$d\neq 0$易得$d=2$.

应用题是不等式的应用. 但过分冗长的题目，恐怕会影响考生的心情，影响水平的发挥.

第 19 题　按照某学者的理论，假设一个人生产某产品的单件成本为 a 元，如果他卖出该产品的单价为 m 元，则他的满意度为 $\frac{m}{m+a}$；如果他买进该产品的单价为 n 元，则他的满意度为 $\frac{a}{n+a}$. 如果一个人对两种交易(卖出或买进)的满意度分别为 h_1 和 h_2，则他对这两种交易的综合满意度为$\sqrt{h_1h_2}$.

现假设甲生产 A、B 两种产品的单件成本分别为 12 元和 5 元，乙生产 A、B 两种产品的单件成本分别为 3 元和 20 元. 设产品 A、B 的单价分别为 m_A 元和 m_B 元，甲买进 A 与卖出 B 的综合满意度为 $h_{甲}$，乙卖出 A 与买进 B 的综合满意度为 $h_{乙}$.

(1) 求 $h_{甲}$ 和 $h_{乙}$ 关于 m_A、m_B 的表达式；当 $m_A=\frac{3}{5}m_B$ 时，求证：$h_{甲}=h_{乙}$；

(2) 设 $m_A=\frac{3}{5}m_B$，当 m_A、m_B 分别为多少时，甲、乙两人的综合满意度均最大？最大的综合满意度为多少？

(3) 记(2)中最大的综合满意度为 h_0，试问能否适当选取 m_A、m_B 的值，使得 $h_{甲}\geqslant h_0$ 和 $h_{乙}\geqslant h_0$ 同时成立，但等号不同时成立？试说明理由.

不难求出 $h_{甲}=\sqrt{\frac{12m_B}{(m_A+12)(m_B+5)}}$，$h_{乙}=\sqrt{\frac{20m_A}{(m_A+3)(m_B+20)}}$.

于是 $h_{甲}=h_{乙}$ 经过化简等价于

$$(3m_B-5m_A)(m_Am_B+3m_B+5m_A+60)=0,$$

即 $m_A=\frac{3}{5}m_B$ 是 $h_{甲}=h_{乙}$ 的充分必要条件. 这也说明了 $m_A=\frac{3}{5}m_B$ 从何而来.

当然，题目已经给出条件 $m_A=\frac{3}{5}m_B$ 时，不难验证 $h_{甲}=h_{乙}$. 比较好的(可以避免繁分式的)办法是令

$$m_B=5t,$$

从而

$$m_A=3t.$$

代入得 $h_{甲}=\sqrt{\dfrac{4t}{(t+4)(t+1)}}=h_{乙}$,

并且 $h_{甲}=2\left(t+\dfrac{4}{t}+5\right)^{\frac{1}{2}}$ 在 $t=\dfrac{4}{t}$,即 $t=2$ 时,取得最大值$\dfrac{2}{3}$.

第20题 设 a 为实数,函数 $f(x)=2x^2+(x-a)|x-a|$.

(1) 若 $f(0)\geqslant 1$,求 a 的取值范围;

(2) 求 $f(x)$的最小值;

(3) 设函数 $h(x)=f(x)$, $x\in(a,+\infty)$,直接写出(不需给出演算步骤)不等式 $h(x)\geqslant 1$ 的解集.

本题作为大轴题(不算附加题),实在是不合适的. 解二次不等式的问题,已经在高考试卷中屡次出现,往往繁琐而缺乏数学的美感. 其中第(1)小题,与(2)(3)毫无关系,硬放在一起,易产生混淆,误以为"$f(0)\geqslant 1$"或由此得出的 a 的取值范围"$a\in(-\infty,-1)$"是(2)(3)中的已知条件. 这一小题完全可以而且应当删去. 即使(2)(3)两小题,也嫌得重复,保留一个岂不更好(可能是为了充当"大题",故意叠床架屋)?

解法并不难,只需细心去做. 以(3)为例,解法如下:这时 $x\geqslant a$,

$$h(x)=2x^2+(x-a)^2=3x^2-2ax+a^2.$$

考虑不等式

$$3x^2-2ax+a^2\geqslant 1$$

在 $3-2a^2<0\left(即\ a<-\dfrac{\sqrt{6}}{2}\ 或\ a>\dfrac{\sqrt{6}}{2}\right)$ 时,上式恒成立.

在 $3-2a^2\geqslant 0$ 时,上式的解是

$$x\leqslant\frac{a-\sqrt{3-2a^2}}{3}\ 或\ x\geqslant\frac{a+\sqrt{3-2a^2}}{3}.$$

再比较 $\dfrac{a\pm\sqrt{3-2a^2}}{3}$ 与 a 的大小.

$a<\dfrac{a-\sqrt{3-2a^2}}{3}$ 的解是 $a<-\dfrac{\sqrt{2}}{2}$;

$a<\dfrac{a+\sqrt{3-2a^2}}{3}$ 的解是 $a<\dfrac{\sqrt{2}}{2}$.

因此,在 $a \in \left(-\infty, -\frac{\sqrt{6}}{2}\right] \cup \left[\frac{\sqrt{2}}{2}, +\infty\right)$ 时,解集为$(a, +\infty)$;在 $a \in \left[-\frac{\sqrt{2}}{2}, \frac{\sqrt{2}}{2}\right)$ 时,解集为$\left[\frac{a+\sqrt{3-2a^2}}{3}, +\infty\right)$;在$a \in \left(-\frac{\sqrt{6}}{2}, -\frac{\sqrt{2}}{2}\right)$时,解集为$\left(a, \frac{a-\sqrt{3-2a^2}}{3}\right] \cup \left[\frac{a+\sqrt{3-2a^2}}{3}, +\infty\right)$.

再看附加题.

第 21 题(A) 如图 2,在四边形 $ABCD$ 中,$\triangle ABC \cong \triangle BAD$. 求证:$AB \parallel CD$.

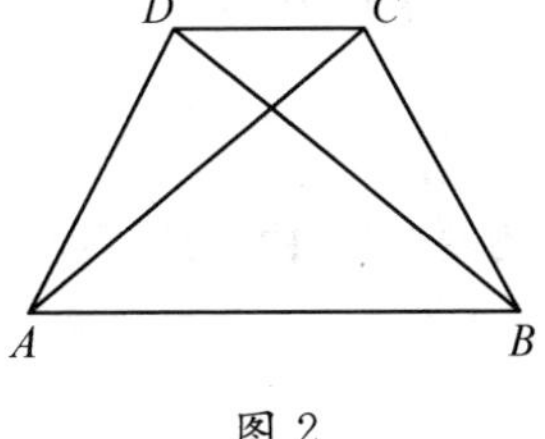

图 2

这题证明并不需要圆的知识,由已知的三角形全等可得

$$\angle CBA = \angle DAB.$$

又可得 $AC = BD$, $AD = BC$,结合 $CD = CD$ 产生 $\triangle ACD \cong \triangle BDC$. 从而

$$\angle BCD = \angle ADC.$$

与上式相加产生

$$\begin{aligned}\angle CBA + \angle BCD &= \angle DAB + \angle ADC \\ &= \frac{1}{2}(\angle CBA + \angle ADC + \angle DAB + \angle BCD) = 180^\circ.\end{aligned}$$

于是 $$AB \parallel CD.$$

第 22 题　在平面直角坐标系 xOy 中,抛物线 C 的顶点在原点,经过点 $A(2, 2)$,其焦点 F 在 x 轴上.

(1) 求抛物线 C 的标准方程;

(2) 求过点 F,且与直线 OA 垂直的直线方程;

(3) 设过点 $M(m, 0)(m>0)$ 的直线交抛物线 C 于 D, E 两点, $ME = 2DM$, 记 D 和 E 两点间的距离为 $f(m)$,求 $f(m)$关于 m 的表达式.

分析　(1) 甚易,抛物线方程为 $y^2 = 2x$.

(2) OA 方程是 $y = x$,与它垂直的方程是 $x + y = a$. 将 F 点坐标$\left(\frac{1}{2}, 0\right)$代

入得 $x+y=\frac{1}{2}$.

(3) 采用直线的参数方程

$$\begin{cases}x=m+\rho\cos\theta,\\ y=\rho\sin\theta,\end{cases}$$

其中 θ 是 EM 的倾斜角, $ME=\rho_1$, $DM=-\rho_2$, $\rho_1=-2\rho_2\cdot\rho_1$, ρ_2 是(消去 x、y 的) ρ 的方程

$$\rho^2\sin^2\theta=2(m+\rho\cos\theta) \quad (*)$$

的两个根. 由韦达定理,有

$$\rho_1\rho_2=\frac{-2m}{\sin^2\theta},$$

所以

$$\rho_2^2\sin^2\theta=m.$$

代入(*)得 $2\rho_2\cos\theta=-m$, 再平方得

$$4\rho_2^2(1-\sin^2\theta)=m^2,$$

即

$$4\rho_2^2=m^2+4m.$$

$$f(m)=\rho_1-\rho_2=-3\rho_2=\frac{3}{2}\sqrt{m^2+4m}.$$

第 23 题 对于正整数 $n\geqslant 2$, 用 T_n 表示关于 x 的一元二次方程 $x^2+2ax+b=0$ 有实数根的有序数组 (a, b) 的组数,其中 $a, b\in\{1, 2, \cdots, n\}$ (a 和 b 可以相等);对于随机选取的 $a, b\in\{1, 2, \cdots, n\}$ (a 和 b 可以相等),记 P_n 为关于 x 的一元二次方程 $x^2+2ax+b=0$ 有实数根的概率.

(1) 求 T_{n^2} 及 P_{n^2};

(2) 求证:对任意正整数 $n\geqslant 2$, 有

$$P_n>1-\frac{1}{\sqrt{n}}.$$

本题有很好的几何解释.

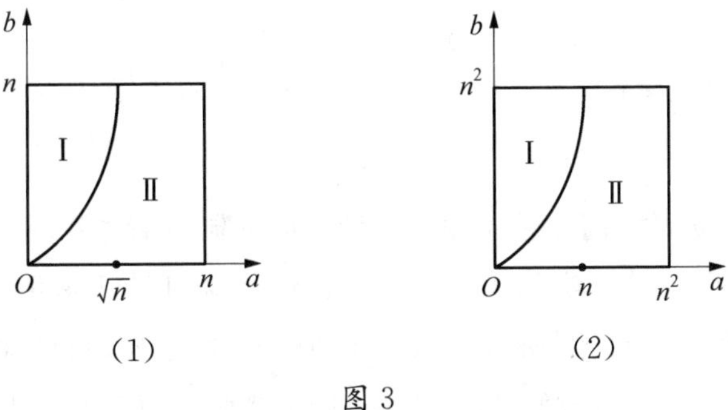

图 3

分析　(1) $x^2+2ax+b=0$ 有实根的充要条件是

$$b\leqslant a^2.$$

因此 T_n 就是区域Ⅱ中整点的个数(图 3(1)),不包括 a 轴上整点,包括其它边界上的整点.

因此,由图 3(2)(将图 3(1)的 n 换成 n^2).

$$T_{n^2}=1^2+2^2+\cdots+n^2+n^2(n^2-n).$$

$$=\frac{n(n+1)(2n+1)}{6}+n^3(n-1)$$

$$=\frac{n}{6}(6n^3-4n^2+3n+1).$$

$$P_{n^2}=\frac{T_n^2}{n^4}=\frac{6n^3-4n^2+3n+1}{6n^3}.$$

(2) 图 3(1)的区域Ⅰ中(不包括 b 轴及曲线边界)整点的个数显然$<n\sqrt{n}$.所以

$$\bar{P}_n<\frac{n\sqrt{n}}{n^2}=\frac{1}{\sqrt{n}},$$

$$P_n=1-\bar{P}_n>1-\frac{1}{\sqrt{n}}.$$

关于 2010 年江苏数学高考题

2010 年江苏数学高考题，普遍反映较去年为难. 细看考卷，不少题都有点小技巧. 技巧多了，区分度较大. 其实无论题目难易，对于广大考生，都是公平的，只要正常发挥，考出自己的水平就可以了. 唯一可虞之处是题目难，总分低. 数学学科比其他学科的分数低，似乎数学老师的教学业绩不如其他学科，虽然实际上，不同学科的成绩并没有可比性.

下面说一说具体的题目与解答.

填空题中，第 6 题右焦点为 $F(4,\ 0)$，M 的纵坐标的平方 $y^2=12\times\left(\frac{3^2}{4}-1\right)=15$，$|MF|=\sqrt{1+15}=4$. 不必求出 $y=\pm\sqrt{15}$.

第 7 题每次得出的 $S=1+2+2^2+\cdots+2^n=2^{n+1}-1$，直至 $S\geqslant 33$，所以最后输出的 $S=64-1=63$.

第 8 题切线方程是 $y-a_k^2=2a_k(x-a_k)$，所以切线与 x 轴交点的横坐标是 $x=\frac{a_k}{2}$，即切点横坐标的一半(最好平时通过练习记住这一结论). $a_{k+1}=\frac{a_k}{2}$，所以 $a_1+a_3+a_5=16+4+1=21$.

第 9 题圆心$(0,\ 0)$到 $\frac{12x-5y+c}{13}=0$ 的距离必须小于 1(半径 2 减去 1)，所以 $|c|<13$.

第 10 题即求 $\sin x$，而 $6\cos x=5\tan x$，$x\in\left(0,\ \frac{\pi}{2}\right)$，所以 $6\sin^2 x+5\sin x-6=0$，$\sin x=\frac{2}{3}\left(-\frac{3}{2}\text{ 舍去}\right)$.

第 11 题 $f(x)$是增函数，而且在 $x\geqslant 0$ 时严格递增，在 $x<0$ 时为常数，所以由 $f(1-x^2)>f(2x)$ 得 $1-x^2>2x$ 并且 $1-x^2>0$. 从而$\sqrt{2}-1>x>-1$.

第 12 题 $\frac{x^4}{y^2}\leqslant 9^2$，$3\leqslant xy^2$，相乘得$\frac{x^3}{y^4}\leqslant 27$. 在 $x=3$，$y=1$ 时，$\frac{x^3}{y^4}$取最大

值 27.

第 13 题应将正切化为弦并利用正弦、余弦定理.

$$\begin{aligned}\frac{\tan C}{\tan A}+\frac{\tan C}{\tan B}&=\left(\frac{\cos A}{\sin A}+\frac{\cos B}{\sin B}\right)\frac{\sin C}{\cos C}=\frac{\cos A\sin B+\sin A\cos B}{\sin A\sin B}\cdot\frac{\sin C}{\cos C}\\&=\frac{\sin^2 C}{\sin A\sin B\cos C}=\frac{c^2}{ab\cos C}=\frac{a^2+b^2-2ab\cos C}{ab\cos C}\\&=\left(\frac{b}{a}+\frac{a}{b}\right)\frac{1}{\cos C}-2=4.\end{aligned}$$

第 14 题可以算一个大题,设剪成的三角形边长(梯形的上底)为 x,则梯形周长为 $3-x$,面积为$\frac{\sqrt{3}}{4}(1-x^2)$. 用导数可以知道在$0<x<1$时, $\frac{(3-x)^2}{1-x^2}$ 的最小值为 8. 更简单的办法是用 $(3-x)^2-8(1-x^2)=9x^2-6x+1=(3x-1)^2\geqslant 0$,在 $x=\frac{1}{3}$ 时, S 取得最小值 $8\div\frac{\sqrt{3}}{4}=\frac{32}{\sqrt{3}}$.

解答题中第 16 题(1)用三垂线定理立即得到. 现在的课程标准去掉三垂线定理,理由是减轻学生负担,其实反倒是增加了学生的负担.

另外,以直线 DP 为 z 轴,D 为原点,则平面 $ABCD$ 正好是坐标平面 xOy. 直线 DC 可作为 y 轴,直线 CB 与 x 轴平行,当然垂直于坐标平面 yOz 及这平面上的直线 PC(如图 1).

第(2)小题如注意到 AB 的中点 A_1 与 B、C、D、P 等点可以是单位立方体的 5 个顶点(如图 1), A_1 到平面 PBC 的距离即面对角线 A_1E 的一半,A 到平面 PBC 的距离即面对角线 A_1E 的长$\sqrt{2}$.

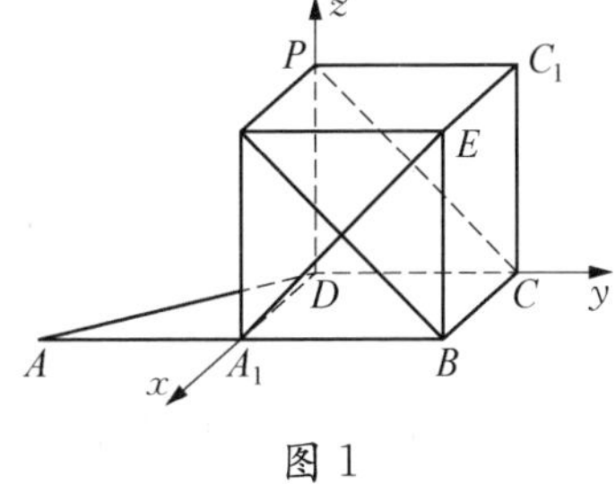

图 1

如果采用向量(利用上面所说的坐标系),不难看出平面 PBC 的法向量是$(0, 1, 1)$(即 A_1E 的方向),而 $\overrightarrow{AB}=(0, 2, 0)$ 在这法向量上的射影,长为$\sqrt{2}$,即所求距离.

今年的应用题第 17 题源于书本上的例题,远胜去年面壁虚构,十分冗长的那道“应用题”.

第 18 题(3)如果用二次曲线束非常简单:过 A、M、N、B 四点的二次曲线

束是

$$\lambda_1[(t+3)y-m(x+3)][(t-3)y-m(x-3)]+\lambda_2\left(\frac{x^2}{9}+\frac{y^2}{5}-1\right)=0. \tag{①}$$

这条曲线束表示直线 $AB(y=0)$ 与 MN 时,形状为

$$y(kx+hy+n)=0. \tag{②}$$

所以①中不含 y 的项应当为 0,其余的项在约去 y 后,再令 $y=0$ 就得到 MN 与 x 轴的交点的横坐标. 于是令(1)中 y 一次项的系数为 0,

$$m(x+3)(t-3)+m(x-3)(t+3)=0. \tag{③}$$

由③得

$$x=\frac{9}{t}. \tag{④}$$

即 MN 过 x 轴上定点 $\left(\frac{9}{t},\ 0\right)$(在 $t=9$ 时,这点是 $(1,\ 0)$).

第 19 题(1)应先将 $n\leqslant 3$ 的情况搞清楚,然后再考虑一般的 n. 可设(为了避免根式)

$$\sqrt{S_1}=b,\ \sqrt{S_2}=b+d,\ \sqrt{S_3}=b+2d,$$

于是

$$a_1=b^2,\ a_1+a_2=(b+d)^2,$$
$$a_1+a_2+a_3=(b+2d)^2,$$
$$a_2=(b+d)^2-b^2=d(2b+d),$$
$$a_3=(b+2d)^2-(b+d)^2=d(2b+3d).$$

由已知(已知条件要用!)

$$2a_2=a_1+a_3,$$

得

$$2d(2b+d)=b^2+d(2b+3d),$$

于是

$$d^2-2bd+b^2=0.$$

从而 $b=d$, $\sqrt{S_1}=d$, $\sqrt{S_2}=2d$, $\sqrt{S_3}=3d$.

当然
$$\sqrt{S_n}=nd,$$
$$a_n=(\sqrt{S_n})^2-(\sqrt{S_{n-1}})^2=(nd)^2-((n-1)d)^2=(2^n-1)d.$$

第(2)小题，显然 c 可以取 $\frac{9}{2}$，因为

$$\frac{S_m+S_n}{S_k}=\frac{9(m^2+n^2)}{(m+n)^2}>\frac{9}{2}.$$

在 $m=n$ 时，上面的式子是等式. 虽然题目中已限制 $m\neq n$，但这是无关紧要的，当 m 很大时，令 $n=m+1$(或者 $m+2$ 也无不可)，n 与 m 的差，相对于 m 来说是微不足道的，即

$$c<\frac{S_m+S_n}{S_k}=\frac{9(m^2+n^2)}{(m+n)^2}\rightarrow\frac{9}{2}.$$

所以 c 的最大值为 $\frac{9}{2}$($\frac{m+n}{3}=\frac{2m+1}{3}$ 是整数不难做到，取 m 除以 3 余 1 即可)

第 20 题(2)中 $g'(x)>0$，$g(x)$递增.

熟悉分点公式的人知道，如果 A、B 的坐标分别为 x_1、x_2，那么满足 $\frac{AC}{AB}=m$ 的点 C 坐标为 $mx_2+(1-m)x_1=\beta$；满足 $\frac{DB}{AB}=m$ 的点 D 坐标为 $mx_1+(1-m)x_2=\alpha$. 在 $m<0$ 与 $m>1$ 时，C、D 都是外分点，即在区间$[A, B]$之外，分别为图 2 的左图与右图，这时由递增性，分别有

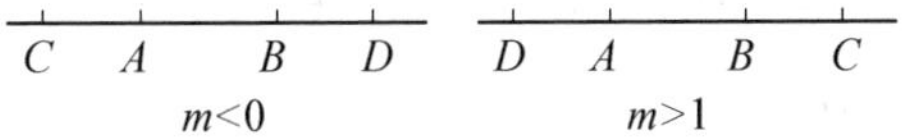

C　A　B　D　　　D　A　B　C

$m<0$　　　　$m>1$

图 2

$$g(\beta)<g(x_1)<g(x_2)<g(\alpha),$$

或
$$g(\alpha)<g(x_1)<g(x_2)<g(\beta),$$

从而

$$|g(\alpha)-g(\beta)|>|g(x_2)-g(x_1)|.$$

在 $0<m<1$ 时，C、D 都是内分点，即在区间(A, B)内，所以由递增性，有

$$g(x_1) < g(\alpha) < g(x_2),$$
$$g(x_1) < g(\beta) < g(x_2),$$

从而

$$|g(\alpha) - g(\beta)| < |g(x_2) - g(x_1)|.$$

在 $m=0$ 或 $m=1$ 时，$\{\alpha, \beta\} = \{x_1, x_2\}$,所以

$$|g(\alpha) - g(\beta)| = |g(x_1) - g(x_2)|.$$

于是 $$0 < m < 1.$$

定比分点的概念及公式非常重要,有了它,对于很多问题的理解与解决,均方便许多. 因此,每一本解析几何学的书都少不了它. 现在的课标将它去掉,不仅增加了学生的负担,而且使知识的系统遭到破坏,殊为不智.

随意删减数学内容,破坏知识体系,往往构成今后学习中的隐患. 从培养人才的角度来看,实在是极不妥当的愚蠢行为.

附加题的反映是不难.

几何题如图 3,可设 $\angle DAB = \alpha$,则由 $DA = DC$ 得

$$\angle DCA = \alpha.$$

又 CD 为切线,所以

$$\angle BDC = \angle DAB = \alpha = \angle DCA.$$
$$BD = BC,$$
$$\angle ABD = \angle BDC + \angle DCA = 2\alpha.$$

因为 AB 为直径,所以

$$\angle ADB = 90°, \ \alpha = 30°,$$
$$AB = 2BD = 2BC.$$

不等式的证明可与第 19 题一样,少用根式:

$$\begin{aligned} & 2(a^3 + b^3) \\ = & (a+b)(2a^2 + 2b^2 - 2ab) \\ \geqslant & (a+b)(a^2 + b^2) \\ \geqslant & 2\sqrt{ab}(a^2 + b^2), \end{aligned}$$

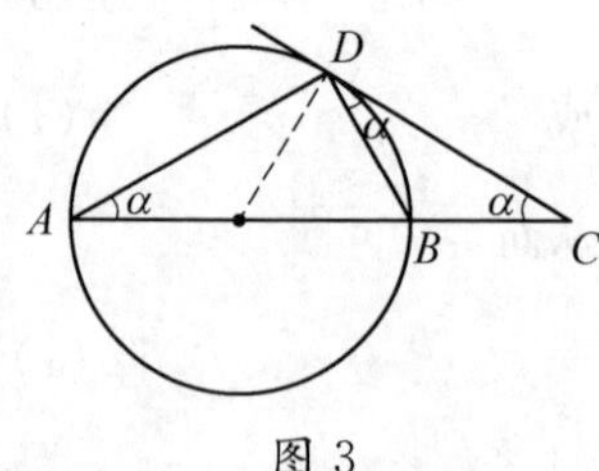

图 3

所以 $$a^3+b^3\geqslant\sqrt{ab}(a^2+b^2).$$

附加题远比第 19 题、第 20 题简单. 所以今后考生应当尽力将附加题做出来,时间不够时,宁肯放弃解答题的最后两题.

总的说来,今年的试题出得很不错. 第 18 题如不用曲线束,计算稍繁. 然而计算能力或许正是需要考查的. 这道题还有深刻的背景:自一点 $T(t, m)$向一圆锥曲线(例如 $\frac{x^2}{9}+\frac{y^2}{5}=1$) 任作两条直线,分别交圆锥曲线于 P、Q、R、S,则 PS 与 QR 的交点,一定在一条直线上,这条直线称为 T 点关于圆锥曲线的极线,方程与切线方程的形状类似. 对本题的曲线,极线方程即 $\frac{tx}{9}+\frac{my}{5}=1$. 它与 x 轴的交点的横坐标即 $x=\frac{9}{t}$, 与 m 无关.

第 19 题的$\{\sqrt{S_n}\}$成等差,$\{a_n\}$也成等差(至少已知前 3 项成等差). 很容易想到 1, 3, 5, …,这样的例子(和 $S_n=n$). 本题大概也是由这个实例引申而得的. 平时学习应当记住这些典型的例子. 使等号成立的值也很有用. 第(2)小题,虽然 $m\neq n$, 但它们可以“差得很少”,从而导出 c 不能超过$\frac{9}{2}$. 这实际上是一种极为重要的思想.

第 20 题的题意可能有人看不懂. 而理解与表达能力正是应当考查与加强的,值得在教与学中注意. 其实这道题只利用函数的单调性,并不是特别困难的题目.

评 2011 年江苏数学高考题

2011 年江苏数学高考题,“没得说的”.

去年高考题,开始的题目不是很容易,引起了不少学生、家长的意见. 今年吸取了去年的经验教训,开始的题目非常容易,后面的题目也大多平易,颇得好评. 当然,区分度可能会降低.

题目的解答,大多数只有一种,就是标准解答. 他们也没得说的.

第 13 题 $q^3=a_7\geqslant a_6=a_4+1=a_2+2\geqslant a_1+2=3$,所以 $q\geqslant\sqrt[3]{3}$. q 的最小值是$\sqrt[3]{3}$(这时数列是 $1,1,\sqrt[3]{3},2,\sqrt[3]{3^2},3,3$).

第 14 题应注意圆 $(x-2)^2+y^2=m^2$ 的斜率为 -1 的切线是 $x+y=2\pm m\sqrt{2}$. 当 $m\leqslant 0$ 时, $2m+1<2+m\sqrt{2}$, $A\cap B$ 为空集. 所以 $m>0$. 这时,由 $m^2\geqslant\frac{m}{2}$ 得 $m\geqslant\frac{1}{2}$. 又由 $2+m\sqrt{2}\geqslant 2m$ 及 $2m+1\geqslant 2-m\sqrt{2}$ 得 $2+\sqrt{2}\geqslant m\geqslant\frac{2-\sqrt{2}}{2}$. 所以 $\frac{1}{2}\leqslant m\leqslant 2+\sqrt{2}$.

第 17 题(1),需求函数 $x(30-x)$ 的最大值. 这种函数的最大值可利用“两个(或几个)正数的和一定时,它们的积在各个数相等时最大”. 由 $x=30-x$ 得出 $x=15$. 同样,对(2)中函数 $x^2(30-x)=4\cdot\frac{x}{2}\cdot\frac{x}{2}(30-x)$,由 $\frac{x}{2}=\frac{x}{2}=30-x$ 得 $x=20$. 没有必要用导数.

第 18 题(3),标准解答提供了两种解法. 显然第二种好. 解答同时说明在 $n=2m$ 时,相应的结论对于椭圆 $mx^2+ny^2=1$ 成立.

第 20 题是压轴题,有一定难度,尤其是(2). 这题符号较多,比较抽象;我们尽量将它具体化. 由已知,

$$S_{n+3}+S_{n-3}=2(S_n+S_3)\ (n\geqslant 4).\qquad ①$$

上式是关于数列和的等式. 为了得到关于数列项的等式,常用的方法是将

上式中的 n 换成 $n+1$,得 $S_{n+4}+S_{n-2}=2(S_{n+1}+S_3)$,再与上式相减,得 $a_{n+4}+a_{n-2}=2a_{n+1}(n\geqslant 4)$,即 $a_{n+4}-a_{n+1}=a_{n+1}-a_{n-2}(n\geqslant 4)$.

从而 $a_2, a_5, a_8, a_{11}, a_{14}, \cdots$ ②

是等差数列. 设它的公差为 $3d_1$(我们用 $3d_1$,而不是 d_1,因为前者更为方便). 同样

$$a_3, a_6, a_9, a_{12}, a_{15}, a_{18}; \quad ③$$

与

$$a_4, a_7, a_{10}, a_{13}, a_{16}, a_{19}, a_{22}, \quad ④$$

也都是等差数列. 设它们的公差分别为 $3d_2$ 与 $3d_3$.

又由 $k=4$ 得

$$S_{n+4}+S_{n-4}=2(S_n+S_4), \quad ⑤$$

同样有

$$a_2, a_6, a_{10}, a_{14}, a_{18}, a_{22}, \cdots \quad ⑥$$

成等差数列,等等. 设它的公差为 $4d_4$.

因为 a_2、a_{14} 在②中分别是第 1 项与第 5 项,在⑥中分别是第 1 项与第 4 项,所以

$$4\times 3d_2=3\times 4d_4,$$

从而

$$d_2=d_4.$$

同样,由 a_6 与 a_{18}, a_{10} 与 a_{22} 可得

$$d_2=d_4d_3=d_4.$$

于是 $a_2, a_3, a_4, \cdots$是等差数列,公差为 d.

最后,回到①,

$$\begin{aligned}2S_3&=(S_{n+4}-S_{n+1})-(S_{n+1}-S_{n-2})\\&=a_{n+4}+a_{n+3}+a_{n+2}-(a_{n+1}+a_n+a_{n-1})=3\times 3d=9d.\end{aligned}$$

同样,由⑤得 $2S_4=16d$,

两式相减得

$$a_4=\frac{7}{2}d.$$

于是 $a_3=\frac{5}{2}d$, $a_2=\frac{3}{2}d$.

$$1=a_1=S_3-a_3-a_2=\frac{9}{2}d-\frac{5}{2}d-\frac{3}{2}d=\frac{1}{2}d.$$

所以 $d=2$. 数列 $\{a_n\}$ 即 1, 3, 5, 7, …, 通项公式为 $a_n=2n-1$.

附加题 23 的(2)稍难. 我们还是尽量具体化. 先看最容易的情形: $n=3k$ 是 3 的倍数.

对 $b=1$, a 可取 4, 7, …, $3k-2$ 共有 $k-1$ 个值. 类似地, 可得下表:

b 的值	1	2	3	4	5	6	…	$3k-5$	$3k-4$	$3k-3$
a 的个数	$k-1$	$k-1$	$k-1$	$k-2$	$k-2$	$k-2$	…	1	1	1

于是点 P 的个数是

$$3(1+2+\cdots+(k-1))=\frac{3k(k-1)}{2}=\frac{n(n-3)}{6}.$$

$n=3k+1$ 时, 上表改成

b 的值	1	2	3	4	5	…	$3k-5$	$3k-4$	$3k-3$	$3k-2$
a 的个数	k	$k-1$	$k-1$	$k-1$	$k-2$	…	2	1	1	1

于是点 P 的个数是

$$3(1+2+\cdots+(k-1))+k=\frac{3k(k-1)}{2}+k=\frac{k(3k-1)}{2}=\frac{(n-1)(n-2)}{6}.$$

$n=3k+2$ 时, 上表改成

b 的值	1	2	3	4	5	…	$3k-3$	$3k-2$	$3k-1$
a 的个数	k	$k-1$	$k-1$	$k-1$	$k-1$	…	1	1	1

于是点 P 的个数是

$$3(1+2+\cdots+(k-1))+2k=\frac{3k(k-1)}{2}+2k=\frac{k(3k+1)}{2}=\frac{(n-1)(n-2)}{6}.$$

不少同学对于最后一题往往采取放弃, 其实最后一题未必很难.

关于 2012 年江苏高考数学题

高考的填空题通常比较容易，所以标准答案往往只给出结果. 但其中也可能有几道稍难的，有些题的解法更有巧拙的不同. 如果能够提供这种题的解答，对于中学的教学有指导意义，对于以后的考生更有莫大的帮助(至少节省了考生的时间).

第 4 题中，二次式 k^2-5k+4 在 $k=1$ 时为 0. 另一个为 0 的值是 $k=4$. 在 $1<k<4$ 时，二次式为负. 所以将 1, 2, 3, 4, 5 逐个代入，第一个使二次式大于 0 的整数值，也就是输出的值是 5. 算法问题，其实应当归入计算机科学，不要硬塞在数学中. 这样数学课就可以多一些时间讲数学.

第 10 题，首先应注意 $f(-1)=f(1)$，所以 $-a+1=\dfrac{b+2}{2}$，从而 $b=-2a$. 再由条件 $f\left(\dfrac{1}{2}\right)=f\left(\dfrac{3}{2}\right)$，得出 $3a+2b=-2$. 解出 $a=2$, $a+3b=-5a=-10$.

第 11 题，可设 $\beta=\alpha+\dfrac{\pi}{6}$，它是锐角(因为它的余弦是正的). 这时

$$
\begin{aligned}
\sin\left(2\alpha+\frac{\pi}{12}\right)&=\sin\left(2\beta-\frac{\pi}{4}\right)=\frac{\sqrt{2}}{2}(\sin 2\beta-\cos 2\beta)\\
&=\frac{\sqrt{2}}{2}(2\sin\beta\cos\beta-2\cos^2\beta+1)\\
&=\frac{\sqrt{2}}{2}\left(2\times\frac{3}{5}\times\frac{4}{5}-2\times\frac{4}{5}\times\frac{4}{5}+1\right)=\frac{17\sqrt{2}}{50}.
\end{aligned}
$$

第 12 题，圆 C 的方程是 $(x-4)^2+y^2=1$，圆心为 $(4, 0)$，半径为 1. 直线 $y=kx-2$ 即 $\dfrac{kx-y-2}{\sqrt{k^2+1}}=0$. 圆心到此直线的距离应当 $\leqslant 2$(为保证所作的圆与已知圆有公共点)，所以 k 最大时，此距离为 2，即 $\dfrac{(4k-2)^2}{k^2+1}=2^2$. 解得 $k=\dfrac{4}{3}$.

第 13 题，首先注意 x^2+ax+b 的最小值为 0，即 x^2+ax+b 有重根 0，所

以 $a^2=4b$. 其次, $x^2+ax+b-c$ 的两根之差为 $(m+6)-m=6$,判别式为 $a^2-4(b-c)=4c$. 于是 $4c=6^2$, $c=9$.

第14题,可令 $x=\frac{b}{a}$, $y=\frac{c}{a}$. 第一个不等式即 $5y-3\leqslant x\leqslant 4y-1$. 由 $5y-3\leqslant 4y-1$ 得 $y\leqslant 2$,所以 $x\leqslant 4y-1\leqslant 7$. 第二个不等式即 $y\ln x\geqslant 1+y\ln y$, $\ln x\geqslant \frac{1}{y}+\ln y$. 利用导数,由 $\left(\frac{1}{y}+\ln y\right)'=-\frac{1}{y^2}+\frac{1}{y}=0$,得 $\frac{1}{y}+\ln y$ 在 $y=1$ 时取最小值 1,所以 $x\geqslant \mathrm{e}$. 于是, $\frac{b}{a}=x$ 的取值范围是 $[\mathrm{e}, 7]$.

解答题应充分说理.

第15题(1),由正弦定理即可得到欲证的等式 $\tan B=3\tan A$. 而由此等式可知 $\tan B$, $\tan A$ 同号,从而 B, A 都是锐角(不可能都是钝角). 在(2)中,由和的正切公式,易得 $\tan A=1$, $-\frac{1}{3}$. 因为 A 是锐角,所以 $\tan A=1$, $A=\frac{\pi}{4}$. 公布的标准解答中,说到 $\cos A>0$, 但理由似未说清.

第18题, (1) $f'(x)=3x^2+2ax+b$ 有两个根 1 和 -1,所以 $f'(x)=3(x^2-1)$,比较系数得 $a=0$, $b=-3$. 不必解方程组 $f'(1)=0$, $f'(-1)=0$.

第(3)问零点个数易引起争议:重根是按重数算,还是只算 1 个? 最好说明一下. 从答案看,重根只算 1 个.

本题函数 $f(x)=x^3-3x$ 的图像是N形,在 $x=-1$ 时极大, $x=1$ 时极小. $y=f(x)+d$ 也是这样(d 为常数).

在 $c=-2$ 时,由 $f(x)-c=x^3-3x+2=(x-1)(x^2+x-2)=(x-1)^2(x+2)$, 得 $h(x)$的零点是 $f(x)-1$ 与 $f(x)+2$ 的零点. 上面已说 $f(x)-1$ 的图像是N形,它在 $x=-1$ 时,值为正;在 $x=1$ 时,值为负. 因此有三个零点,分别在区间 $(-\infty, -1)$, $(-1, 1)$, $(1, +\infty)$内. $f(x)+2$ 有两个零点,即 1, -2. 因此 $h(x)$有 5 个零点.

同样, $c=2$ 时, $h(x)$有 5 个零点,其中三个零点分别在区间 $(-\infty, -1)$, $(-1, 1)$, $(1, +\infty)$内,另两个是 $f(x)-c=x^3-3x-2=(x+1)(x^2-x-2)=(x+1)^2(x-2)$ 的零点 -1 与 2.

在 $-2<c<2$ 时,函数 x^3-3x-c 在 $x=-1$ 时仍为正,在 $x=1$ 时仍为负,而在 $x=-2$ 时为负,在 $x=2$ 时为正,因此有三个零点,分别在区间$(-2,$

$-1)$, $(-1, 1)$, $(1, 2)$内. $h(x)$的零点是$f(x)-c_1$, $f(x)-c_2$, $f(x)-c_3$的零点(c_1, c_2, c_3分别在上述三个区间中). 每一个都有三个零点,而且没有公共零点,所以$h(x)$有9个零点.

第19题,易得$a=\sqrt{2}$, $b=c=1$, $e=\frac{1}{\sqrt{2}}$. 第(2)问可设$\angle AF_1O=\alpha$. 利用AF_1等于A到左准线的距离乘离心率,得

$$AF_1=e\left(\frac{a}{e}-c+AF_1\cos\alpha\right),$$

从而
$$AF_1=\frac{1}{\sqrt{2}(1-e\cos\alpha)},\quad ①$$

同样
$$BF_2=\frac{1}{\sqrt{2}(1+e\cos\alpha)}.\quad ②$$

熟悉极坐标的考生可以立即得出上述结果. 于是在(i)中,由$AF_1-BF_2=\frac{\sqrt{6}}{2}$,得

$$\frac{2e\cos\alpha}{1-e^2\cos^2\alpha}=\sqrt{2}\times\frac{\sqrt{6}}{2},$$

从而
$$\cos^2\alpha+2\sqrt{\frac{2}{3}}\cos\alpha-2=\left(\cos\alpha+\sqrt{\frac{2}{3}}\right)^2-\frac{8}{3}=0,$$

$$\cos\alpha=\sqrt{\frac{8}{3}}-\sqrt{\frac{2}{3}}=\sqrt{\frac{2}{3}}.$$

(因为$AF_1-BF_2=\frac{\sqrt{6}}{2}$,所以①的分母小于②的分母, $\cos\alpha>0$)

所求斜率为

$$\tan\alpha=\sqrt{\frac{3}{2}-1}=\sqrt{\frac{1}{2}}=\frac{\sqrt{2}}{2}.$$

(ii) 也可用①,②来解. 首先有

$$\frac{1}{AF_1}+\frac{1}{BF_2}=2\sqrt{2}.\quad ③$$

(这个结果完全可以作为考题)

显然$\triangle PAF_1 \backsim \triangle PF_2B$,记相似比$\dfrac{BF_2}{AF_1}=\lambda$,则

$$PF_1+PF_2=\frac{1}{\lambda}PB+\lambda PA=\frac{\frac{1}{\lambda}}{1+\frac{1}{\lambda}}BF_1+\frac{\lambda}{1+\lambda}AF_2$$

$$=\frac{1}{1+\lambda}BF_1+\frac{\lambda}{1+\lambda}AF_2$$

$$=\frac{1}{1+\lambda}(2a-BF_2)+\frac{\lambda}{1+\lambda}(2a-AF_1)$$

$$=2a-\frac{2}{1+\lambda}BF_2=2a-2\div\left(\frac{1}{BF_2}+\frac{1}{AF_1}\right)$$

$$=2a-2\div 2\sqrt{2}=2\sqrt{2}-\frac{\sqrt{2}}{2}=\frac{3\sqrt{2}}{2}.$$

第 20 题是一道很好的题.(2) 主要利用一个简单而且大家都熟悉的事实:各项为正的等比数列在公比大于 1 时,通项趋于正无穷;在公比小于 1(大于零)时,通项趋于 0.

首先,$\{a_n\}$是各项为正的等比数列.但$2(a_n^2+b_n^2)\geqslant(a_n+b_n)^2>a_n^2+b_n^2$,所以$1<a_{n+1}\leqslant\sqrt{2}$.通项既不趋于无穷,也不趋于零,因此必有公比$=1$.从而$\{a_n\}$各项相等,设均为$a$,则$a^2(a^2+b_n^2)=(a+b_n)^2$,即$b_n$是方程

$$(a^2-1)x^2-2ax+a^2(a^2-1)=0$$

的根.

其次,$b_{n+1}=\dfrac{\sqrt{2}}{a}b_n$,所以$\{b_n\}$是公比为$\dfrac{\sqrt{2}}{a}$的等比数列,各项为正,而且适合上述方程,即至多存在两个不同的值.因而通项既不趋于无穷,也不趋于零,必有公比$=1$,从而$a=\sqrt{2}$.上述方程化为$x^2-2ax+a^2=0$, $x=a$.即$\{b_n\}$与$\{a_n\}$都是各项为$\sqrt{2}$的常数数列.

附加题不难.第 21 题 B,也可先求出逆矩阵A^{-1}的特征根为-1、$\dfrac{1}{4}$(特征

方程为 $\lambda^2+\frac{3}{4}\lambda-\frac{1}{4}=0$). 从而它们的倒数 -1、4 就是 A 的特征根.

第 22 题要取两条棱. 由对称性，第一条不妨先取定，这时第二条棱有 11 种取法. 与第一条相邻的有 4 种，与第一条平行而且相距为 $\sqrt{2}$ 的 1 种，其余 6 种，$\xi=1$. 所以分布律为

ξ	0	1	$\sqrt{2}$
$P(\xi)$	$\frac{4}{11}$	$\frac{6}{11}$	$\frac{1}{11}$

这题的参考解答太繁了.

本文表明解题的讨论非常重要，它是提高高考成绩的最直接的方法. 要想提高高考成绩，就得在解题讨论方面多下功夫.

抽象与具体

著名数学家柯尔莫哥洛夫在《数学,它的内容、方法和意义》一书中指出数学有三大特点,即抽象性、严谨性、应用的广泛性.

数学是抽象的,但又是具体的. 如果只谈抽象,不谈具体的例子,抽象就会变成无源之水,无本之木,抽象就难于理解,难于感觉. 因此,我们不但要学会如何从具体到抽象,还要学会将抽象的定理、问题具体化.

1 具体的例子可以帮助我们猜出结论

很多命题是从具体的例子抽象出来的,猜就是抽象.

例 1 $X=\{1, 2, 3, \cdots, 2n+1\}$. A 是 X 的子集,具有性质:A 中任意两个数的和不在 A 中,求 $\max|A|$.

我们先考虑一个具体的 A. 题中 A 是具有某种性质的数的集合,而 A 中任两个数的和却不具有这种性质.

举例子应从身边开始,尽量举熟悉的例子. 关于整数,最常见的性质便是奇偶性,而且任两个奇数的和不是奇数. 所以,

$$A=\{1, 3, 5, \cdots, 2n+1\}$$

就是一个合乎要求的例子. 这时

$$|A|=n+1.$$

A 显然不能再添加其他的数,添加任一个偶数就不再合乎要求. 而且去掉一些奇数后,似乎也不能增添更多的偶数(这一点将在下面严格证明). 所以我们大胆猜测

$$\max|A|=n+1.$$

例 2 $X=\{1, 2, 3, \cdots, n\}$, A、B、C 是 X 的分拆,即 $A\cup B\cup C=X$,

并且 A、B、C 两两的交都是空集. 如果从 A、B、C 中各取一个元，那么每两个的和都不等于第三个. 求

$$\max\min(|A|, |B|, |C|).$$

此题比例 1 更难一些，仍应当先举具体的例子. 还是考虑奇偶性. 如果 A 由 X 中的奇数组成，$B\cup C$ 由 X 中的偶数组成，那么它们合乎题述要求. 这时

$$\min(|A|, |B|, |C|)=\left[\frac{n}{4}\right].$$

由此我们可以提出猜测

$$\max\min(|A|, |B|, |C|)=\left[\frac{n}{4}\right],$$

也就是恒有

$$\min(|A|, |B|, |C|)\leqslant\frac{n}{4}. \tag{1}$$

2　具体的例子可以引出命题的证明

命题的证明不是从天上掉下来的，具体的例子往往给我们很多的启发.

在例 1 中，奇偶性给我们很大的帮助. 对一般的、满足要求的 A，可设其中有 k $(k\leqslant n+1)$ 个奇数

$$a_1>a_2>\cdots>a_k.$$

再看一看 A 中至多能有多少个偶数. 显然，偶数

$$a_1-a_2<a_1-a_3<\cdots<a_1-a_k$$

都不能在 A 中(因为 a_1-a_i 与 a_i 之和 a_1 在 A 中)，所以，A 中至多有

$$n-(k-1)=n+1-k$$

个偶数. 从而，

$$|A|\leqslant k+(n+1-k)=n+1.$$

例 2 的证明更复杂一些,但去掉一些比较容易处理的情况,剩下的也就是我们所举的例子. 详细的推导可见《集合及其子集》(单墫著,上海教育出版社 2001 年出版).

3 解题应当从具体出发

前面已经说过,具体的例子会给我们很多启发,所以解题往往从具体开始.

例 3 100 个质量分别为 1、2、…、100 克的砝码放在天平两边,正好达到平衡. 证明:一定可以从每边各取去 2 个砝码,天平仍保持平衡.

从具体出发,不妨设 1(即 1 克的砝码)在左边,进一步设 1, 2, …, k 在左边,而 $k+1$ 在右边. 这时有两种情况:

(i) $k+2$ 在右边.

如果 $k+3$ 在左边,那么,从左边取下 $k+3$ 与 k,从右边取下 $k+2$ 与 $k+1$,天平仍然平衡.

如果 $k+3$ 在右边,而 $k+4$ 在左边,同样,可从左边取下 $k+4$ 与 k,从右边取下 $k+3$ 与 $k+1$, 天平仍然平衡.

依此类推,只要有 $t(t\geqslant k+2)$ 在右边,而 $t+1$ 在左边,那么,就可在左边取下 $t+1$ 与 k,在右边取下 $k+1$ 与 t,天平仍然平衡.

于是,只剩下一种可能不合要求的情况,即 $1\sim k$ 在左边,而 $k+1\sim 100$ 在右边. 但这时应有 $1+2+\cdots+k=\frac{1}{2}(1+2+\cdots+100)$, 即 $k(k+1)=5050=101\times 50$. 而这个 k 的方程无整数解,所以这种情况不会发生.

(ii) $k+2$ 在左边.

如果有 $t(t>k+2)$ 在左边,而 $t+1$ 在右边,那么,可在左边取下 t 与 $k+2$,在右边取下 $t+1$ 与 $k+1$. 如果有 $t(t>k+2)$ 在右边,而 $t+1$ 在左边,那么可在左边取下 $t+1$ 与 k,在右边取下 $k+1$ 与 t. 因此只剩下一种可能不合要求的情况,即 $1\sim k$ 与 $k+2$ 在左边,而 $k+1$, $k+3\sim 100$ 在右边. 同样,由不定方程无整数解可知这种情况不会发生.

这道题,有些同学不从具体出发,解起来就困难重重.

还可以举一个更复杂的例子,用以说明从具体出发是解题的一条重要途径，这就是本文最后的例 6.

4　具体例子中隐含问题的本质

有人认为抽象反映了问题的本质，而具体例子只呈现问题的表象. 其实问题的本质即寓于具体例子之中.

例 4　图 1 中 8 个顶点，每个顶点处各有一个实数，每个顶点处的实数正好等于 3 个相邻顶点（有线段相连的两个顶点称为相邻顶点）处的数的平均数. 求

$$a+b+c+d-(e+f+g+h).$$

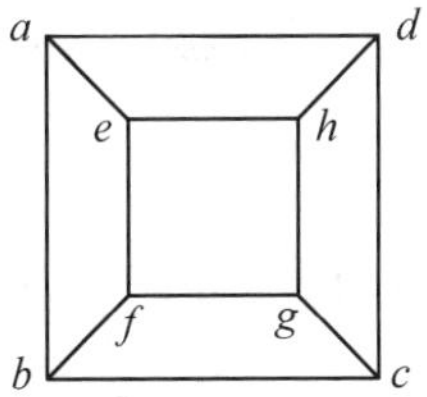

图 1

下面的解法曾在期刊上出现：

由已知

$$a=\frac{b+e+d}{3},\ b=\frac{a+f+c}{3},\ c=\frac{b+g+d}{3},\ d=\frac{c+h+a}{3}.$$

四式相加得

$$a+b+c+d=\frac{1}{3}(2a+2b+2c+2d+e+f+g+h),$$

即

$$a+b+c+d-(e+f+g+h)=0.$$

上述解法，巧诚巧矣！但有两个缺点. 一是太凑巧了. 四式相加可以得出 $a+b+c+d$，这是预料之中的事，但整理后正好得出 $a+b+c+d-(e+f+g+h)=0$ 却是偶然的. 如果题目改为求 $2a+b+c+d-(2e+f+g+h)$，那么上面的解法就不再奏效了. 二是上述解法未能揭示问题的实质，即 8 个数 a、b、c、d、e、f、g、h 之间的关系.

解本题应当先举一个具体的例子. 要使每一个顶点处的数都等于相邻顶点处的数的平均数，最简单的例子是取

$$a=b=c=d=e=f=g=h.$$

而实际上这也就是仅有的可能. 因为不失一般性，可设 a 是 8 个数中最大的，由于 a 是 b、e、d 的平均数，所以 b、e、d 也必须是最大的，即与 a 相等. 同理可知

8 个数全相等. 这样,无论 $a+b+c+d-(e+f+g+h)$ 还是 $2a+b+c+d-(2e+f+g+h)$ 都显然是 0.

所以, (1)(2)等式子只是表面的东西,8 个数相等才是真正的本质.

5 一个具体例子可能包含很多内容，可从不同角度加以抽象

一个极简单也极典型的例子就是高斯(Gauss, 1777—1855)对

$$1+2+\cdots+100$$

的计算. 通常,由这个例子可以抽象出一般的等差数列的求和公式

$$S_n=\frac{(a_1+a_n)n}{2}. \tag{5}$$

但高斯当时的原始想法究竟如何已经无法知道,我们可以有多种的揣想,例如,可以认为高斯知道常数数列的求和(这是每个知道乘法的小朋友都能掌握的),他很可能想将 $1+2+\cdots+100$ 化为常数数列的求和. 为此,他采用“均贫富”的方法,“损有余以补不足”,即将最小的 1 与最大的 100 平均,变成 2 个 $\frac{101}{2}$;再将次小的 2 与次大的 99 平均,也变成 2 个 $\frac{101}{2}$;依此类推,恰好这些平均数都相等,都是全体 100 个数的平均数,所以总和就是 $\frac{101}{2}\times 100$. 因此,可以说高斯的故事中给我们两个重要的思想,即平均的思想与化归(为常数数列)的思想.

例 5 求下列方阵中所有数的和

$$\begin{matrix} 1901 & 1902 & \cdots & 1949 & 1950 \\ 1902 & 1903 & \cdots & 1950 & 1951 \\ \cdots & & & & \\ 1950 & 1951 & \cdots & 1998 & 1999 \end{matrix}$$

很多人先用公式(5)求出第一行的和

$$S_1=\frac{1901+1950}{2}\times 50=96\,275;$$

再求出第 2、3、…、50 行的和；最后将这些和相加. 如果注意到第二行的每个数比上一行的数大 1，那么，各行的和是

$$S_2 = 96\,275 + 50,$$
$$S_3 = 96\,275 + 50 \times 2,$$
$$\cdots$$
$$S_{50} = 96\,275 + 50 \times 49.$$

这样，可以再用(5)求出 $S_1 + S_2 + \cdots + S_{100}$.

但更简单的做法还是应用平均的思想，将每个数与位置关于它中心对称的数平均，每个平均数都是 $\frac{1901+1999}{2} \geqslant 1950$.

于是总和为

$$1950 \times 50 \times 50 = 4\,875\,000.$$

可见，即使是很简单、很常见的例子，也包含很多内容，并非只经过一次抽象，就将它的精、气、神全都抽光了.

6　将抽象问题具体化，是数学素养的一个重要方面

最近在一处讲课，同学问我一个问题：

例 6　集 $S=\{x \mid x$ 是十进制中的 9 位数，数字为 $1, 2, 3\}$. 映射 $f: S \to \{1, 2, 3\}$，且对 S 中任意一对同数位上数字均不相同的 x、y，$f(x) \neq f(y)$. 求 f 及其个数.

首先，应搞清题意，S 可以看作是 9 元有序组 $(x_1, x_2, \cdots, x_9)$ 的集合，其中 $x_i \in \{1, 2, 3\}$，$1 \leqslant i \leqslant 9$. 令 S_i 为 S 中映成 i 的元所成的集 $(i=1, 2, 3)$，则 S_1、S_2、S_3 是 S 的分拆，且 S_i 中任意二个元 x、y，至少有一个“分量”相同. 问题即求这种分拆及其个数.

分拆比映射 f 较为具体，所以对题意的理解往往是具体化的过程.

再进一步，我们举一个符合要求的具体例子. 这也不难，令 $S_i = \{$以 i 为首位的 9 元数组$\}(i=1, 2, 3)$. 显然，这样的 S_i 合乎条件.

首位可改为任何一位，S_1、S_2、S_3 可以任意排列顺序. 因此，我们已经举

出 $9\times 3!=54$ 个合乎要求的例子.

感觉再也没有其他例子了.于是,大胆猜测:只有上述那些情况.证明如下:

不妨设 $(1, 1, \cdots, 1)\in S_1$, $(2, 2, \cdots, 2)\in S_2$, $(3, 3, \cdots, 3)\in S_3$.

设有 $x\in S_1$,并且 x 的首位为 1,记 x 为 $1+y$,其中 y 是 x 的后 8 位所成向量(数组).

若 $2+y\in S_2$,我们往证一切以 2 为首位的数组均在 S_2 中.

事实上,设这个数组为 $2+u$,则有一个 8 元数组 v,每一位与 y 不同,与 u 也不同,所以 $3+v\notin S_1$, $3+v\notin S_2$, $3+v\in S_3$.从而,$2+u\notin S_3$.又 $1+v\notin S_2$.而且有一个 8 元数组 w,每一位与 y 不同,也与 v 不同,显然 $3+w\in S_3$,从而 $1+v\in S_1$, $2+u\notin S_1$.所以,必有 $2+u\in S_2$.

易知这时 S_2 中也只有以 2 为首位的数组,并且,这时 $122\cdots 2\in S_1$, $322\cdots 2\in S_3$. S_1、S_3 分别由 1、3 为首位的数组组成.

因此,若 $21\cdots 1\in S_2$,则结论已经成立.设 $21\cdots 1\in S_1$,同理若 $221\cdots 1\in S_2$,则由上面的推理(将首位换作第二位),S_i 由第二位为 i 的数组组成.设 $221\cdots 1\in S_1$.依此类推,直至 $22\cdots 21\in S_1$.但此时 $22\cdots 22\in S_2$,所以 S_i 由末位为 i 的数组组成.

这样一道题,为什么一些(准备参加冬令营的)同学解不出来呢? 很可能是由于他们已经过分习惯于"抽象",而不先去寻求具体的例子.这一现象值得我们注意,不要难题做得过多,反而忘记了最基本的方法:即应当将抽象问题具体化,从最简单的做起.本文的目的之一也就是提醒师生注意这方面的问题,这应当是最基本的数学素养.

当代数学大师陈省身先生曾经说过一段话:"一位好的数学家与一个蹩脚的数学家之间的差别,在于前者手中有很多具体的例子,而后者只有抽象的理论."所以,我们应当掌握更多的具体的例子,使抽象的东西变为能够感觉的,易于把握的具体的例子.

直观、慧眼

南京师大附中江宁分校高中新生入学，邀我去讲课. 我讲了下面的一道题：

用两个较小的圆纸片，能不能盖住一个较大的圆纸片？请说明理由（圆纸片以下简称圆，通常的圆改称圆周）.

“不能.”学生齐声回答.

师：“很好！请说明理由.”

甲：“显然，无论怎么盖，大圆纸片总有地方没有被两个小圆盖到.”

师：“什么地方未被盖到？”

甲：“直观地看，总有地方未盖到.”

师：“如果不能说明什么地方未被盖到，为什么未被盖到，那么你的理由就不是充分的. 我们需要直观，也需要说理，应当将你‘看到的’说清楚.”

乙：“设两个小圆相交于 A、B，则以公共弦 AB（如图 1）为直径的圆是两个小圆所能盖住的最大的圆. 而公共弦小于小圆的直径，更小于大圆的直径，因此两个小圆不能盖住大圆.”

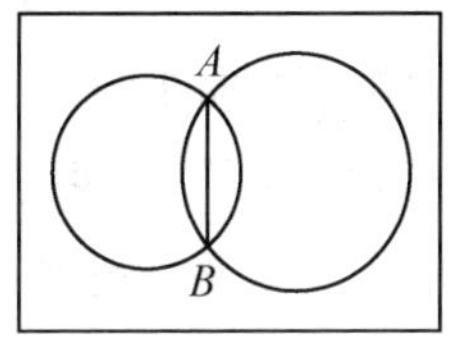

图 1

师：“后一半说理很清楚. 但前一半，为什么以公共弦为直径的圆是两个小圆所能盖住的最大的圆呢？”

乙：“直观地看，应当是这样.”

师：“我倒认为，直观地看，不应当是这样. 比如说，小圆 O_1 可以盖住它自身，这就是一个比以公共弦为直径的圆还大的圆. 由于前一半的说理不正确，所以整个推导就失败了. 不过，谢谢你提供了一个很好的例子，说明直观有时候会误导我们. 那么，这道题怎样做呢？还是要借助直观的力量. 大家看，图 2 中一个小圆盖住了大圆的一部分，但大圆还有一些地方未被盖住. 一般地，大圆的哪些部分未被盖住呢？”

丙：“我看大圆有一半没有被小圆盖住.”

师：“为什么？”

丙:"看上去是这样.不过,也许我看错了."

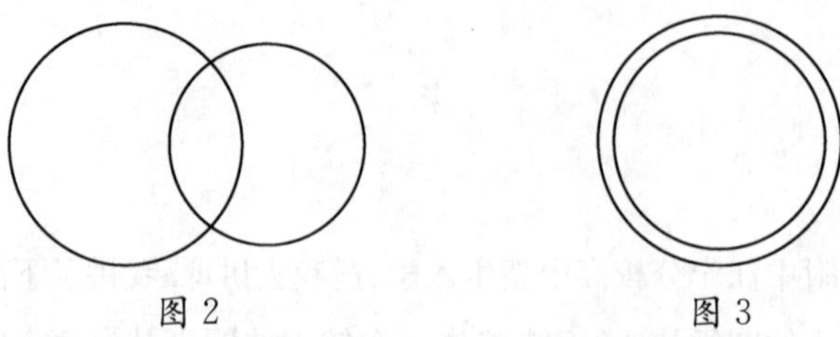

图 2　　　　　　图 3

师:"你是不是说大圆的面积有一半未被小圆盖住?"

丙:"是的."

师:"那么,很遗憾,你看错了.在图 3 中,小圆盖住了大圆的大部分.如果小圆只比大圆小一点点,那么小圆差不多盖住了整个大圆,只差一点点.当然这是就面积而言.不过,即使小圆盖住了大圆的大部分,大圆总还有未被盖到的部分.这部分是……"

丙:"这部分是大圆的圆周.一个小圆盖不住大圆的半个圆周."

师:"你说对了.请说说理由."

丙:"如果小圆盖住大圆的半个圆周,那么它就盖住大圆的一条直径(的两个端点).而小圆中任两点的距离都不超过小圆的直径,所以小于大圆的直径.这就产生矛盾,表明小圆不可能盖住大圆的半个圆周.于是,两个小圆不可能盖住大圆的整个圆周,当然也就不能盖住大圆."

师:"说得很好.这道题目表明直观在解决问题时是有用的,但也要防止它误导我们.要多画些图,多想些情况(尤其是一些极端的情况),避免片面性、局限性.不光用眼睛看,还要用心(脑)想.不能只看到表面,要深入.在这道题中,不仅要看到小圆不能盖住大圆,更要发现大圆的什么地方未被盖住,为什么未被盖住.这不是普通的观察,看一看就过去,而是一种洞察力,它是眼与心的结合,是将直观与理智结合起来的更高一个层次的直观.学习数学就要注意培养这种洞察本质的直观能力.

"刚才我们是先考虑用一个小圆去盖大圆,有什么地方未被盖住.还有一种解法是同时考虑两个小圆.假设两个小圆可以盖住大圆.设大圆半径为 R,小圆半径为 r(如果两个小圆半径不一样大,可以将较小的一个放大,使得两个小圆半径相等).作两个小圆的两条外公切线(图 4).这两条外公切线是平行的,这样

就形成了一个宽为 $2r$ 的带形. 因为两个小圆盖住了大圆,所以这个带形更盖住了大圆. 但是,这个带形能够盖住大圆吗?"

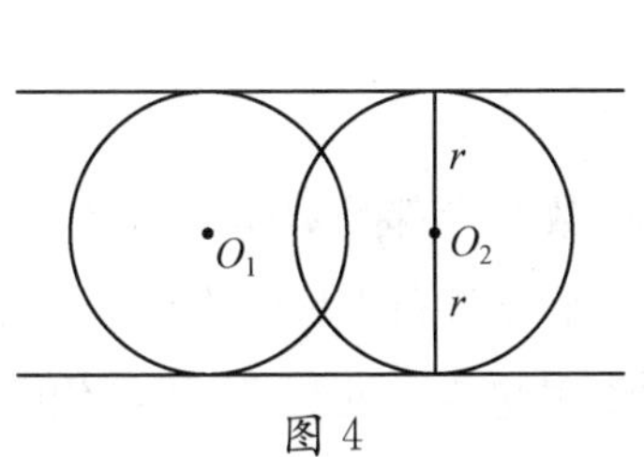

图 4

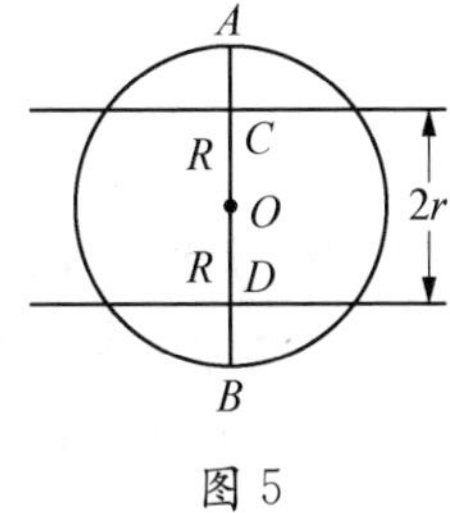

图 5

丁:"不能. 图 5 中,大圆总有一部分露在带形外面."

师:"很直观. 不过能说清楚哪一部分露在外面吗?"

丁:"过大圆圆心 O 作一条直径 AB 与带形的边垂直(图 5). 设直径 AB 与带形的两边相交于点 C、D,则 CD 就是带形的宽. 因为 $AB > CD = 2r$, 所以 A、B 两点至少有一个必在线段 CD 外,因而也就在带形外面."

师:"如果我们把一个图形中任意两点的距离的最大值称为这个图形的直径,那么圆的直径就是原来的直径,而三角形的直径是它最长的边,正方形的直径是它的对角线. 一个直径为 $\sqrt{2}$ 的正方形,能不能被一个宽为 1 的带形覆盖?"

戊:"能. 只要将这个带形移动,使得它的两条边与正方形的两条对边(所在直线)重合就可以盖住正方形."

师:"宽为 1 的带形可以盖住一些直径大于 1 的图形(如正方形),却不能盖住任一个直径大于 1 的圆. 这是因为圆对于每一个方向都具有同样性质,例如在每一个方向上都有一条直径. 上面的证明就利用了圆的这个非常直观的性质."

"再啰嗦几句:直观是重要的,但我们更需要与理智结合后的直观. 这就不是用普通的'肉眼'看,而是用心灵与肉眼结合而成的'慧眼'去看. 有人说:'借我一双慧眼吧!'其实慧眼是每个人都有的. 学习数学就可以帮你提高观察能力与思维能力. 只要你努力去看、去学、去问、去想,你自身具备的'慧眼'就会打开,就可以把一切纷扰看个清清楚楚,明明白白."

附注　本文所讲的题目是 1977 年第一届中国科学技术大学少年班的入学试题,解法很多,可参见拙著《平面几何中的小花》与《覆盖》(均为上海教育出版社出版).

解题中的形象思维

数学需要抽象思维，也需要形象思维. 数与形都是数学中的具体形象(当然，它们又是从实际事物中抽象出来的). 人们常常利用具体的数与形进行形象思维. 在几何中尤其是这样. 几何图形常常引起我们的想象，给我们很多的启迪. 本文试图通过几个解题的实例说明我们怎样利用几何图形进行形象思维. 为了说明形象思维的过程，所举例题稍有难度与层次.

例 1 有两个等腰三角形，一个顶角为 α，腰为 a，底为 b；另一个底角为 α，腰为 b，底为 a. 求 α 及 $\frac{a}{b}$.

题目中没有给出图形，我们应该先将两个等腰三角形画出来以便进行形象思维. 如果 $a=b$，两个三角形都是正三角形，$\alpha=60°$. 现在设 $a>b$. 由于 α 与 $\frac{a}{b}$ 的大小均为未知，所以我们画的图只是一个草图(图 1)，未必准确.

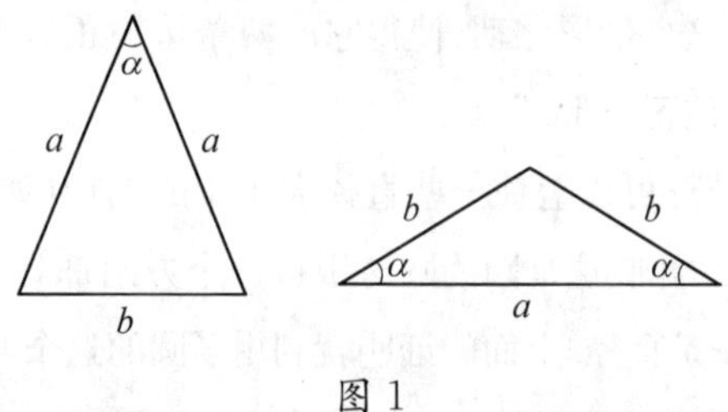

图 1

利用代数与三角分别处理图 1 中的两个三角形也能得到有用的关系，但比较麻烦. 更好的办法是将这两个三角形联系起来(本来这两个三角形就不是孤立的!). 给学生讲解时，可以让同学用硬纸剪下两个三角形，再将它们拼合起来. 拼合时当然将相等的边拼在一起(有相等的边就是两个三角形之间的一种联系)，可以得出图 2 或图 3.

图 2 中不易看到有用的信息，所以我们摒开不论. 图 3 是一个等腰梯形(由 $\angle BAC=\alpha=\angle ACD$，得 $AB \parallel CD$，且 $AB \neq CD$)，所以 $\angle ABC=\angle BAD=2\alpha$，

$\angle ACB=\angle ABC=2\alpha$,从而 $5\alpha=180^{\circ}$, $\alpha=36^{\circ}$.

为了求 $\frac{a}{b}$,可以将图 3 的位置放得更“正”一些(旋转一下),得到图 4.

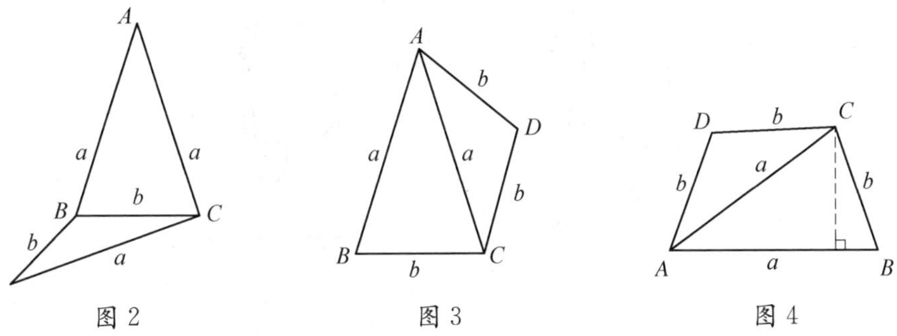

图 2　　图 3　　图 4

从图 4 不难看出, $a^2-\left(\frac{a+b}{2}\right)^2=b^2-\left(\frac{a-b}{2}\right)^2$,从而推出 $\frac{a}{b}=\frac{\sqrt{5}+1}{2}$.

即使不知道 $a>b$, 我们还是照图 3 那样拼合,得出四边形 $ABCD$ 是等腰梯形,它的两条对角线都等于 a. 易知两条对角线的和大于两腰的和,所以一定有 $a>b$.

在例 1 中,虽然图形只是一个草图,并不一定准确,但这种“形象”已经有助于我们思考. 平面几何正是这样一门科学,它可以利用未必准确的图形推出准确的结果.

这未必准确的图形其实不是完全错误的图形,它像漫画、速写,或者中国的写意画. 虽然不是照片,却也能反映事物的本质,“得其神髓”. 培养学生画这种草图的能力也是很重要的.

在例 1 中,我们不仅要看到两个等腰三角形,更重要的,还要看到这些形象之间的联系. 甚至使它们活动起来(如果利用电脑作图,这一过程就很鲜明生动). 上面的拼合,就是数学中所说的运动,或者叫做全等变换(我们知道全等变换包括平移、旋转,还有轴对称). 这些本来是小孩子就知道的方法. 如果有具体的形象在,学生是不难明白的.

图形的内在联系,有时是不易发现的,这就需要思维的力量,甚至需要形而上的抽象思维. 思维,等于一双慧眼,可以透过现象(表象)看到本质,把一切都看得清清楚楚,明明白白.

例 2　如图 5,平行四边形 $ABCD$ 中, E 为 AD 上任一点,过 E 作 EF 与

AB 的延长线交于 F,连 CE、CF,设$\triangle CDE$ 的外心为 O_1, $\triangle EAF$ 的外心为 O_2, $\triangle CBF$ 的外接圆半径为 R,求证: $O_1O_2=R$.

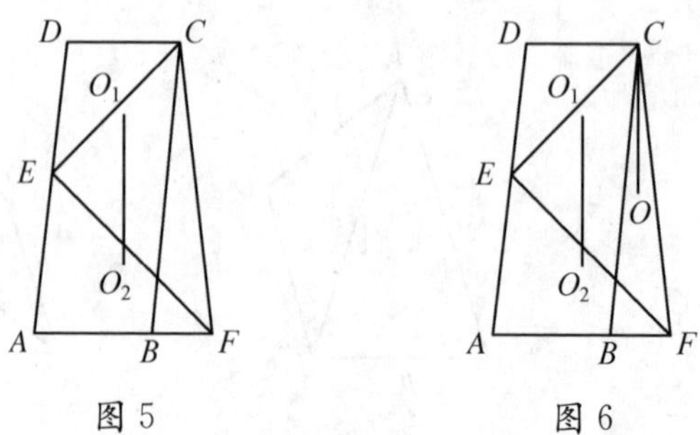

图 5　　　　图 6

原题有图,但这个图并不“完全”,其中半径 R 并未画出,$\triangle CBF$ 的外接圆及$\odot O_1$、$\odot O_2$ 也都没有画出.

解题的第一步就是将一些隐藏的形象显现出来,以便进行形象思维.

首先定出$\triangle CBF$ 的外接圆的圆心 O(图 6,当然也只是在草图上凭眼睛大致定出,不一定十分准确). OB、OF、OC 都是 R. 但这三条半径不必全画,有一条就足够了. 应当画 OC,因为看上去 OC 与 O_1O_2 平行. 而 OB、OF 与 O_1O_2 没有这种关系. $OC\parallel O_1O_2$,这只是一个猜想,并未证明. 值得注意:正是形象思维使我们产生这样的猜想.

$\odot O_1$、$\odot O_2$ 是相交的圆, E 是它们的一个公共点. 两个相交圆的公共弦在解题中极为有用(有很多性质可以利用,例如公共弦与连心线 O_1O_2 垂直). 因此凡遇到相交圆均需找出它们的公共弦,现在应当找出第二个公共点.

再画一个草图(图 7), $\odot O_1$ 与$\odot O_2$ 的第二个交点 G 似乎(差不多)在 CF 上.

形象是这样. 想来也应如此(这是一种正常的感觉). 交点 G 不会与圆上已有的内容毫无关系. 如果 G 不在 CF 上,而是“悬在半空”,解题将难以措手,也与图形的和谐优美相悖.

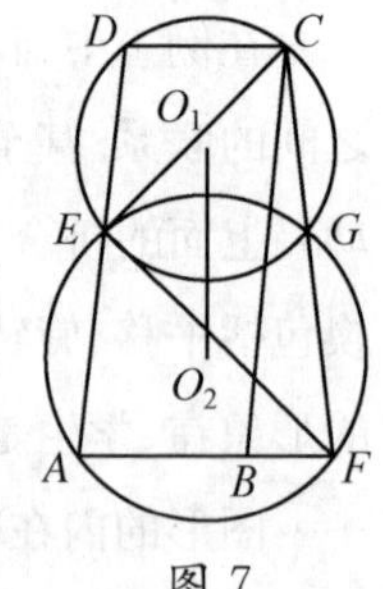

图 7

所以我们大胆断言 G 在 CF 上. 当然这一点有待证明. 但证明并不难(所谓“只怕想不到,不怕证不出”): 设$\odot O_1$ 与 CF 除 C 外还有一个交点 G,则 $\angle EGC=180^\circ-\angle D=\angle A$. 所以 G 也在$\odot O_2$ 上,即$\odot O_1$ 与$\odot O_2$ 的第二个交点 G 在 CF 上.

进一步,再证明前面的猜测: $O_1O_2 /\!/ OC$ (这也是由形象导出的猜测).

由于 $O_1O_2 \perp EG$,所以只需证 $OC \perp EG$. 这也是不难证明的:

图 8 中, $\angle EGC = \angle A = \angle CBF$.

图 9 中, $\angle CBF = \frac{1}{2}\angle COF = \frac{1}{2}(180^\circ - 2\angle OCF) = 90^\circ - \angle OCF$,所以 $OC \perp EG$.

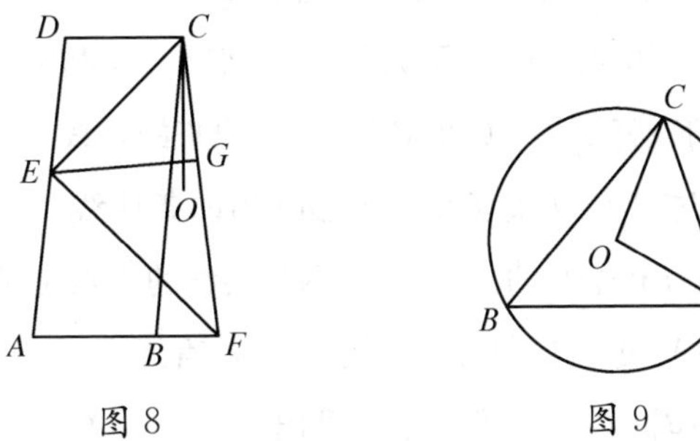

图 8　　　　图 9

很多线画在同一个图上,形象就不清楚,不鲜明了. 所以图应有分有合,例如上面我们单独画一个图 9,以突出 $\angle CBF$ 与 $\angle OCF$ 的关系.

图 6 还暗示我们四边形 O_1O_2OC 是平行四边形,即也应当有 $OO_2 /\!/ CO_1$. 这可以用与 $O_1O_2 /\!/ OC$ 同样的方法证明: $\odot O_2$ 与 $\odot O$ 的第二个交点 H 在 CE 上, $\angle FHC = \angle A = 180^\circ - \angle D = 180^\circ - \frac{1}{2}\angle EO_1C = 90^\circ + \angle O_1CE$, $FH \perp O_1C$. 于是平行四边形 O_1O_2OC 的对边 $O_1O_2 = CO = R$.

在上面的证明中,各种形象有隐有显,有分有合. 它们帮助我们提出一些猜测,帮助我们找到证明.

例 3　如图 10, $ABCD$ 为正方形. 四个小圆与这个正方形的边的延长线相切,又与大圆相切, A_1、B_1、C_1、D_1 为切点. 求证:直线 AA_1、BB_1、CC_1、DD_1 共点.

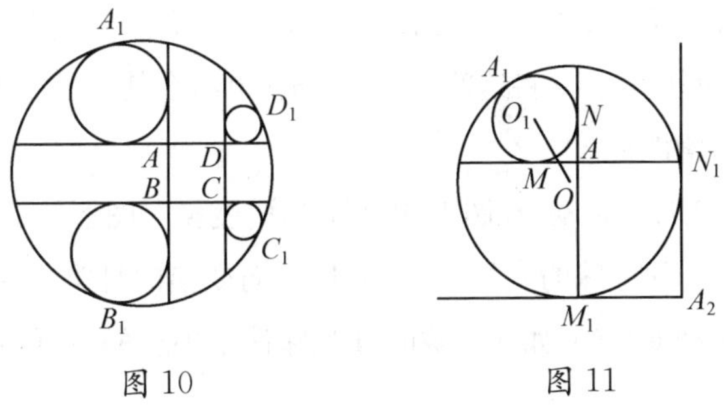

图 10　　　　图 11

图中有 1 个大圆与 4 个小圆,4 个小圆的地位是平等的,我们可以先选定过 A_1 的那个圆来研究,暂时将其他 3 个圆隐去(以免干扰我们的视线),得到图 11.

图 11 中两个圆内切于 A_1. 这个切点有很多性质. 例如它与两个圆的圆心 O_1、O 共线,并且 $A_1O_1:A_1O=R_1:R$, 其中 R_1、R 分别为$\odot O_1$、$\odot O$ 的半径. 观点更高一些,应当看出 A_1 是这两个圆的位似中心.

如果以 A_1 为位似中心, $R_1:R$ 为位似比,那么 O_1 将变为 O, $\odot O_1$ 将变为$\odot O$,而$\odot O_1$ 的切线将变为$\odot O$ 的切线,并与原来的切线平行(除非原来的切线过 A_1). $\odot O_1$ 的过 A 的两条切线变为$\odot O$ 的两条切线,分别与原来的两条切线平行,因而也是互相垂直的. 并且设这两条切线的交点是 A_2,则 A 与 A_2 是一对对应点,直线 AA_2 过位似中心 A_1,换句话说,直线 AA_1 就是直线 A_1A_2.

对其他 3 个圆作同样的处理,可以得出 A_2、B_2、C_2、D_2 四点, A_2B_2、B_2C_2、C_2D_2、D_2A_2 分别是与 AB、BC、CD、DA 平行的、$\odot O$ 的切线. 它们围成一个矩形,边长都等于$\odot O$ 的直径,因而是一个正方形.

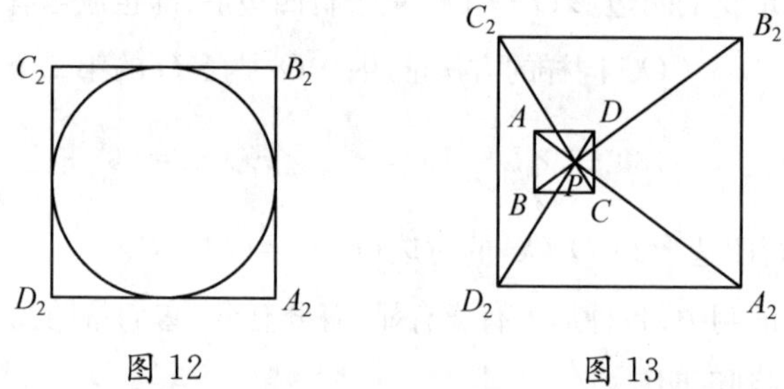

图 12　　　　图 13

问题化为 AA_2、BB_2、CC_2、DD_2 四条线是否共点(图 13),这是不难证明的.

在本例中,原来的图形比较复杂,我们先分出其中的一部分(隐去其余部分),使形象更加鲜明突出. 将这部分研究透彻,再结合其余部分研究. 这也就是华罗庚先生所说的“退”.

在例 1 中,图形在运动. 在图 11 中,图形作“变换”. 图形的变换比运动更为一般. 运动只是一种特殊的变换(全等变换),而变换的种类繁多,有位似变换(例如图 11)、相似变换(例如同一城市的比例不同的地图)、仿射变换(例如将圆压成椭圆的压缩变换)、射影变换、拓扑变换等.

有了变换的观点,就可以更好地进行形象思维.看到相切的圆就会想到位似变换,看到图 13 也会想到两个正方形 $ABCD$、$A_2B_2C_2D_2$ 是位似的(AA_2、BB_2 的交点 P 就是位似中心).观点提高了,就能看到很多原来未看到的东西.

在以上三个例题中,我们对几何图形进行处理,有显有隐,有分有合,有静有动(变换),以使形象更加鲜明,特点更加突出,联系更加紧密,也就是更便于进行形象思维.

平面几何的一个功能就是提供很多形象,培养学生形象思维的能力,如果过于削弱平面几何,而又没有很好的内容代替,对培养学生形象思维的能力是很不利的,一些国家的前车之鉴值得我们警惕.

参考文献

[1]　[美]G·波利亚.怎样解题[M].上海:上海科技教育出版社,2002.

[2]　[美]G·波利亚.数学与猜想[M].北京:科学出版社,2001.

[3]　单墫.解题研究[M].南京:南京大学出版社,2002.

[4]　单墫.数学是思维的科学[J].数学通报,2001(6).

解题漫谈

面对一道难题,从何着手? 这里举一个例子. 1993 年国际数学竞赛的第 5 题:是否存在函数 f: **N**→**N**,满足

(1) $f(f(n))=f(n)+n$;

(2) $f(1)=2$;

(3) $f(n+1)>f(n)$.

这是一个需要判断的问题:满足要求的函数是否存在? 当然,如果存在,这个函数是什么? 有没有简单的表达式? 如果不存在,为什么不存在? 这些都需要进一步说明.

解决这个问题,只有“尝试”(有一位哲人说过“自古成功在尝试”).

尝试(或者叫做探索)应当从简单的情况入手.

最简单的函数(也是我们最熟悉的函数)莫过于线性函数(一次函数) $f(n)=an+b$,在 $b=0$ 时,即正比例函数 $f(n)=an$(自变量只在 **N** 上取值,因此,我们用 n 表示). 我们先看看它能否满足条件.

由于(3),函数递增,所以 $a>0$.

函数值为整数,为方便起见,取 a 为整数. $a=1$ 时, $f(n)=n$, 这时 $f(f(n))=f(n)=n(=f(n)+0\cdot n)$ 不合要求(1)(也不合要求(2)). $a=2$ 时, $f(n)=2n$,这时 $f(f(n))=2f(n)=f(n)+f(n)=f(n)+2n$, 也不合要求(1).

更大的 a,导致(比 $f(n)+2n$) 更大的 $f(f(n))$,因此没有使(1)成立的 a.

$f(n)=an+b$ 也不能使(1)成立. 因为 b 是常数,当变数 n 很大时,重要的是一次项(n 的系数大小是关键),而不是常数项. b 对于一次项系数毫无影响,因而在 $f(f(n))$中是微不足道的.

线性函数不能满足要求,尝试失败了. 能否就此断言满足要求的函数不存在呢? 不能! 线性函数不满足要求,其他的函数(如二次函数)还可能满足要求.

需要进一步尝试. 如果将次数升高, $f(f(n))$的次数就比 $f(n)$大(在 $f(n)$

为二次时，$f(f(n))$为四次)，因此要求(1)也不能成立.

还是回到一次好. 但一次函数又不满足要求，真是“山重水复疑无路”了.

仔细回顾一下，$a=1$ 太小，$a=2$ 太大，a 应当在1、2之间，这样就必须放弃 a 为整数的限制(这一限制是我们自己加的，我们可以放弃，不必坚持).

对于 $f(n)=an$，有 $f(f(n))=af(n)=a^2n$ 要使(1)成立，即

$$a^2n=an+n. \tag{4}$$

我们应取 $a=\frac{\sqrt{5}+1}{2}$. (5)

遗憾地是 $f(n)=\frac{\sqrt{5}+1}{2}n$ 不是整值函数，它的值是无理数，虽然 $f(n)=\frac{\sqrt{5}+1}{2}n$ 能使(1)成立.

这就好像爱丽丝漫游奇境记中的情况：顾了这头，就顾不了那头.

四个要求(1)、(2)、(3)及整值中，(1)是关键(最难满足)，既然 $f(n)=\frac{\sqrt{5}+1}{2}n$ 能使(1)成立，千万不要轻易放弃.

要使 $f(n)$为整值，这并不难，我们加一个取整符号：用$[x]$表示 x 的整数部分. 并令

$$f(n)=\left[\frac{\sqrt{5}+1}{2}n\right]. \tag{6}$$

(6)是整值函数，而且满足(3)，但不满足(2).

只需作一些修正.

令 $f(n)=\left[\frac{\sqrt{5}+1}{2}n\right]+1$ 即可满足(2)、(3). 但是否满足(1)呢?

$$\begin{aligned}f(f(n))&=\left[\frac{\sqrt{5}+1}{2}f(n)\right]+1=\left[\frac{\sqrt{5}-1}{2}f(n)\right]+f(n)+1\\&=\left[\frac{\sqrt{5}-1}{2}\left[\frac{\sqrt{5}+1}{2}n\right]+\frac{\sqrt{5}-1}{2}\right]+f(n)+1\\&=\left[n-\frac{\sqrt{5}-1}{2}\left\{\frac{\sqrt{5}+1}{2}n\right\}+\frac{\sqrt{5}-1}{2}\right]+f(n)+1=f(n)+n+1,\end{aligned}$$

与条件(1)差一个1.

再作修正,令 $f(n)=\left[\frac{\sqrt{5}+1}{2}n+b\right]$,其中 $0<b<1$ 是一个待定(在下面确定)的常数.这时

$$
\begin{aligned}
f(f(n)) &= \left[\frac{\sqrt{5}+1}{2}f(n)+b\right]=\left[\frac{\sqrt{5}-1}{2}f(n)+b\right]+f(n) \\
&= \left[\frac{\sqrt{5}-1}{2}\left[\frac{\sqrt{5}+1}{2}n+b\right]+b\right]+f(n) \qquad (7) \\
&= f(n)+n+\left[b+\frac{\sqrt{5}-1}{2}b-\frac{\sqrt{5}-1}{2}\left\{\frac{\sqrt{5}+1}{2}n+b\right\}\right].
\end{aligned}
$$

我们希望上式中[ ]为0,即

$$
0<\frac{\sqrt{5}+1}{2}b-\frac{\sqrt{5}-1}{2}\left\{\frac{\sqrt{5}+1}{2}n+b\right\}<1. \qquad (8)
$$

为此,取 $b=\frac{1}{\frac{\sqrt{5}+1}{2}}=\frac{\sqrt{5}-1}{2}$,因为 $\frac{\sqrt{5}+1}{2}n+b=\frac{\sqrt{5}+1}{2}(n+1)-1$ 是无理数,所以 $0<\left\{\frac{\sqrt{5}+1}{2}n+b\right\}<1$,则

$$
0<1-\frac{\sqrt{5}-1}{2}\left\{\frac{\sqrt{5}+1}{2}n+b\right\}<1
$$

即(8)成立.

因此 $f(n)=\left[\frac{\sqrt{5}+1}{2}n+\frac{\sqrt{5}-1}{2}\right]$ 满足所有要求,它就是所求的函数.

满足要求的函数并不唯一.我们可以作出其他满足要求的函数,参见《数学竞赛研究教程》第24讲(单墫著,江苏教育出版社1993年出版)

解题面对面（一）

甲：有些题目不会做，找解答看，更加糊里糊涂.

师：请举个例子.

乙：例如这道题：

设 p 是给定的质数，$a_1, a_2, \cdots, a_k$ 是 $k\ (k \geqslant 3)$ 个整数，均不被 p 整除，且模 p 互不同余. 记

$$S=\{n \mid 1 \leqslant n \leqslant p-1,\ (na_1)_p<(na_2)_p<\cdots<(na_k)_p\},$$

其中，$(b)_p$ 表示整数 b 被 p 除的余数. 证明：

$$|S|<\frac{2p}{k+1}. \qquad ①$$

甲：解答的第一句话就是"需要一个引理".

引理　设 $b_0=p-a_k$，$b_i=a_i-a_{i-1}$，$i=1, 2, \cdots, k$，其中 $a_0=0$. 令

$$S'=\{n \mid 1 \leqslant n \leqslant p-1,\ (b_0n)_p+(b_1n)_p+\cdots+(b_kn)_p=p\}.$$

则 $|S'|=|S|$. ②

我不知道这个引理从何而来(乙：好像是从天上掉下来的). 符号越来越多，不但有 S，又来了个 S'，不但有 a_i，又增加了 b_i. 太玄了.

师：玄，就是抽象. 抽象性是数学的一个特点. 但抽象，首先得有具体直观的像. 学数学，应当先搞清具体直观的像. 有了具体直观的像，抽象也就易于理解了.

乙：您说说这道题怎么解.

师：看不懂的解答先不要管它. 原题中，$(b)_p$ 表示整数 b 被 p 除的余数. 这句话不很清楚. 确切地说，是指区间 $[0, p)$ 中，与 $b\pmod p$ 同余的整数，即 $0 \leqslant (b)_p<p$，且

$$(b)_p \equiv b \pmod p,$$

在 b 是正整数时,就是通常作除法 $b \div p$ 时,所得的余数.

不妨先设 $a_1, a_2, \cdots, a_k$ 都是正整数,且

$$0 < a_1 < a_2 < \cdots < a_k < p. \quad ③$$

此时, $n=1 \in S$ (S 中至少有一个很具体的元素), $(a_1)_p$, $(a_2)_p$, $\cdots$, $(a_k)_p$ 就是$a_1, a_2, \cdots, a_k$.

$a_1, a_2, \cdots, a_k$ 可以用数轴上的点直观地表示. 更方便一些,它们可以用区间$[0, p)$中的整点表示,如图 1,其中 0 与 a_i 的距离就是a_i.

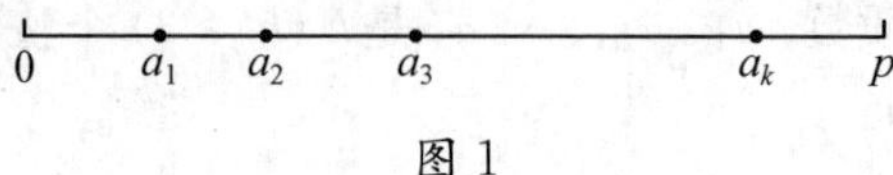

图 1

甲:这张图有什么用?

师:图上的点 $a_1, a_2, \cdots, a_k$ 将区间$(0, p)$分成 $k+1$ 条(互不重叠的)线段. 你们看看这些线段有什么关系?

乙:这 $k+1$ 条线段的长度相加,正好是 p.

师:说得很好. 如果 n 是S 中另一个元素,那么……

甲:$(na_1)_p$, $(na_2)_p$, $\cdots$, $(na_k)_p$ 也是$(0, p)$中的 k 个整点,将$(0, p)$分成 $k+1$ 条线段. 这些线段相加,总长度也是 p.

师:设 $t=|S|$. 对 S 中每个元素n,画一张图(画在一起看不清楚). 这样就有 t 张图,如图 2.

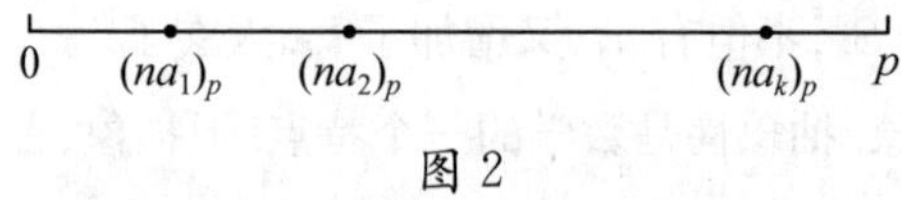

图 2

每一张图中, $k+1$ 条线段的总长是 p. 因此 t 张图中, $t(k+1)$ 条线段的总长是 tp.

这 $t(k+1)$ 条线段的总长,有没有其他的方法计算或估算呢?

乙:您的意思是否是将每张图上的第一条线段的长先加起来,再将每张图上的第二条线段的长加起来……最后,算出这些和的和?

甲:这就是您说过的方法:算两次.

乙：可是，第一条线段的长，我们并不知道，怎样加呢？

师：设 $S=\{n_1(n_1=1),\ n_2,\ \cdots,\ n_t\}$. 看看

$$(n_1a_1)_p,\ (n_2a_1)_p,\ \cdots,\ (n_ta_1)_p \qquad ④$$

（每张图中第一条线段的长）中，是否有相同的？

甲：如果 $1\leqslant i<j\leqslant t$，而

$$(n_ia_1)_p=(n_ja_1)_p, \qquad ⑤$$

那么，$n_ia_1\equiv n_ja_1(\bmod p)$. 从而，

$$(n_j-n_i)a_1\equiv 0(\bmod p). \qquad ⑥$$

但 $1\leqslant n_i,\ n_j\leqslant p-1$ 且 $n_i\neq n_j$，a_1 不被 p 整除，所以，式⑥不成立，即式⑤不成立. 式④中没有相同的数，它们又都是正整数，因此，它们的长度至少是 $1,\ 2,\ \cdots,\ t$（顺序不一定这样）.

乙：因此，各图中第一条线段长度的和大于或等于

$$1+2+\cdots+t=\frac{t(t+1)}{2}.$$

甲：第二条线段的长度是

$$(n_1a_2)_p-(n_1a_1)_p,\ (n_2a_2)_p-(n_2a_1)_p,\ \cdots,\ (n_ta_2)_p-(n_ta_1)_p. \qquad ⑦$$

如果有 $1\leqslant i<j\leqslant t$，而

$$(n_ja_2)_p-(n_ja_1)_p=(n_ia_2)_p-(n_ia_1)_p,$$

那么，

$$n_j(a_2-a_1)\equiv n_i(a_2-a_1)(\bmod p),$$

$$(n_j-n_i)(a_2-a_1)\equiv 0(\bmod p). \qquad ⑧$$

由于 a_1、a_2 模 p 不同余，同样式⑧不成立. 所以，⑦中没有相同的数. 各图中第二条线段长度的和大于或等于 $\frac{t(t+1)}{2}$.

乙：依此类推，$t(k+1)$ 条线段的长度的总和大于或等于 $(k+1)\frac{t(t+1)}{2}$.

甲:因此,$tp \geqslant (k+1)\dfrac{t(t+1)}{2}$.

从而,$t+1 \leqslant \dfrac{2p}{k+1}$.

故 $|S| = t < \dfrac{2p}{k+1}$.

乙:假设③没有多少作用,可以取消.

师:是的.

甲:引理中的 b_i 就是图 1 中线段的长. S'是什么呢?

师:S'是什么,不重要. 既然我们有了自己的解答,那就不必跟着别人的脚印走.

乙:的确,现在的解答直观,而且容易理解. 不明白,那本书上的解答为什么那样写?

师:大概写解答的人自己也没有弄得很清楚吧.

解题面对面（二）

甲：2008 年中国数学奥林匹克的最后一题是一道数论题：

试确定所有同时满足

$$q^{n+2} \equiv 3^{n+2} (\bmod p^n), \quad ①$$

$$p^{n+2} \equiv 3^{n+2} (\bmod q^n) \quad ②$$

的三元数组(p, q, n)，其中 p、q 为奇质数，n 为大于 1 的整数.

乙：可以先求出一些特殊解. 显然 $q=3$ 是式①的解，代入式②得 $p=3$. 所以，$(3, 3, n)(n=2, 3, \cdots)$ 都是解.

甲：如果 $p=q$，那么，由式①得到 $p=3$. 还是上面所说的解.

师：因此，可以设 $q>p\geqslant 5$.

乙：很多不定方程(组)，除了显而易见的那些解外，没有其他的解. 是不是？

甲：但我们并不清楚现在的方程组有没有其他解. 往下该怎么做呢？

师：要充分利用 p、q 是质数这一条件，适当地估计. 先看看最简单的情况：$n=2$.

乙：$n=2$ 时，$p^4 \equiv 3^4 (\bmod q^2)$，即

$$q^2 \mid (p^4-3^4)=(p^2+3^2)(p^2-3^2).$$

又 $(p^2+3^2, p^2-3^2)=(p^2+3^2, 2p^2)=2(3^2, p^2)=2$，

而 q 为奇质数，所以，

$$q^2 \left| \frac{p^2+3^2}{2} \text{ 或 } q^2 \right| \frac{p^2-3^2}{2}.$$

但 $q^2>\dfrac{p^2+3^2}{2}$，故以上两式均不成立.

所以，$n\geqslant 3$.

师：式①即 $p^n \mid (q^{n+2}-3^{n+2})$. 从而，

$$p^n \mid (q^{n+2}+p^{n+2}-3^{n+2}).\tag{③}$$

同理，$q^n \mid (q^{n+2}+p^{n+2}-3^{n+2})$. ④

又 p、q 是不同的质数，故由式③、④得

$$p^n q^n \mid (q^{n+2}+p^{n+2}-3^{n+2}).\tag{⑤}$$

这样就有

$$1 \leqslant \frac{q^{n+2}+p^{n+2}-3^{n+2}}{p^n q^n} < \frac{q^2}{p^n}+\frac{p^2}{q^n}.$$

于是，只须证明反向不等式

$$\frac{q^2}{p^n}+\frac{p^2}{q^n} \leqslant 1.\tag{⑥}$$

那么，本题就没有其他解了.

甲：因为 $q^n \mid (p^{n+2}-3^{n+2})$，所以，

$$q^n < p^{n+2},\ p > q^{\frac{n}{n+2}}.$$

当 $n \geqslant 4$ 时，$\dfrac{n^2}{n+2}-2 \geqslant \dfrac{2}{3}$，

$$\frac{q^2}{p^n}+\frac{p^2}{q^n} < \frac{1}{q^{\frac{2}{3}}}+\frac{1}{q^2} \leqslant \frac{1}{7^{\frac{2}{3}}}+\frac{1}{7^2} < \frac{1}{2}+\frac{1}{2}=1.$$

但当 $n=3$ 时，只得到

$$\frac{q^2}{p^3}+\frac{p^2}{q^3} < q^{\frac{1}{5}}+\frac{1}{q},$$

$q^{\frac{1}{5}}$ 不小于 1.

师：很好的尝试. 现在只剩下 $n=3$ 的情况，即只须证明

$$\frac{q^2}{p^3}+\frac{p^2}{q^3} \leqslant 1.\tag{⑦}$$

估计可以精确一些.

首先将 $q^3 \mid (p^5-3^5)$ 中 p 的方次减少 1，即由于 q 是大于 p 的质数，$q \nmid$

$(p-3)$.

从而，$q^3 \left| \frac{p^5-3^5}{p-3} \right.$.

乙：于是，当 $p \geqslant 5$ 时，

$$q^3 \leqslant \frac{p^5-3^5}{p-3} < \frac{p^5}{p-3} = \frac{p^4}{1-\frac{3}{p}} \leqslant \frac{5}{2}p^4. \tag{⑧}$$

甲：再代入式⑦得

$$\frac{q^2}{p^3}+\frac{p^2}{q^3} < \left(\frac{5}{2}\right)^{\frac{3}{4}}\left(\frac{1}{q}\right)^{\frac{1}{4}}+\frac{1}{q} \leqslant \left(\frac{5}{2}\right)^{\frac{3}{4}}\left(\frac{1}{7}\right)^{\frac{1}{4}}+\frac{1}{7},$$

可惜 $\left(\frac{5}{2}\right)^{\frac{3}{4}}\left(\frac{1}{7}\right)^{\frac{1}{4}} > 1$. 还是不成.

师：不要紧. p 的值可以增大.

因为 $5^5-3^5=2\times 11\times 131$ 不被质数的平方整除，所以，$q^3 \mid (5^5-3^5)$ 不成立，即必须 $p \geqslant 7$, $q \geqslant 11$.

乙：此时，式⑧变成

$$q^3 < \left(1-\frac{3}{7}\right)^{-1} p^4 = \frac{7}{4}p^4.$$

再代入式⑦得

$$\frac{q^2}{p^3}+\frac{p^2}{q^3} < \left(\frac{7}{4}\right)^{\frac{3}{4}}\left(\frac{1}{11}\right)^{\frac{1}{4}}+\frac{1}{11} < 1.$$

甲：于是，本题的解只有 $(3, 3, n)(n=2, 3, \cdots)$.

乙：本题的关键应当是式⑤，将式③、④中的 p^n、q^n 变成 p^nq^n.

师：是的. p^n、q^n 变成 p^nq^n，“阶”(也就是次数)升高了很多. 至于后面的估计，已经不是最重要的. 但“小处也不可随便”，要学会根据需要将估计逐步精细化，直至达到目标(式⑥、⑦).

解题面对面(三)

师:现在讨论一个不等式:

给定正整数 n 及实数 $x_1 \leqslant x_2 \leqslant \cdots \leqslant x_n$, $y_1 \geqslant y_2 \geqslant \cdots \geqslant y_n$, 满足

$$\sum_{i=1}^{n} i x_i = \sum_{i=1}^{n} i y_i. \quad ①$$

证明:对任意实数 α,有

$$\sum_{i=1}^{n} [i\alpha] x_i \geqslant \sum_{i=1}^{n} [i\alpha] y_i, \quad ②$$

其中,$[\beta]$表示不超过实数 β 的最大整数.

(2008 年中国数学奥林匹克)

甲:我想设

$$z_i = y_i - x_i (i = 1, 2, \cdots, n),$$

这样式①就成为 $\sum_{i=1}^{n} i z_i = 0$, 于是,只须证明

$$\sum_{i=1}^{n} [i\alpha] z_i \leqslant 0.$$

师:很好的想法,将问题简化了. 注意

$$z_1 \geqslant z_2 \geqslant \cdots \geqslant z_n.$$

乙:形如 $\sum_{i=1}^{n} a_i b_i$ 的和,使我想到阿贝尔求和法 $(B_i = \sum_{j=1}^{i} b_j,\ i = 1, 2, \cdots, n)$:

$$\sum_{i=1}^{n} a_i b_i = \sum_{i=1}^{n-1} (a_i - a_{i+1}) B_i + a_n B_n.$$

取 $a_i = z_i$, $b_i = i$, 就有

$$0=\sum_{i=1}^{n}z_i i=\sum_{i=1}^{n-1}(z_i-z_{i+1})B_i+z_nB_n. \quad ③$$

甲:令 $C_i=\sum_{j=1}^{i}[j\alpha]$. 同样有

$$\sum_{i=1}^{n}z_i[i\alpha]=\sum_{i=1}^{n-1}(z_i-z_{i+1})C_i+z_nC_n. \quad ④$$

如果 $\frac{C_i}{B_i}\leqslant\frac{C_n}{B_n}$, ⑤

那么,由式④、③及 $z_i-z_{i+1}\geqslant 0\ (i=1,\ 2,\ \cdots,\ n-1)$, 就有

$$\sum_{i=1}^{n}z_i[i\alpha]\leqslant\sum_{i=1}^{n-1}(z_i-z_{i+1})B_i\cdot\frac{C_n}{B_n}+z_nC_n=\frac{C_n}{B_n}\left[\sum_{i=1}^{n-1}(z_i-z_{i+1})B_i+z_nB_n\right]=0.$$

因此,只须证式⑤.

师:可证稍强的结论:$\frac{C_i}{B_i}$随 i 递增,即

$$\frac{C_i}{B_i}\leqslant\frac{C_{i+1}}{B_{i+1}}(i=1,\ 2,\ \cdots,\ n-1). \quad ⑥$$

乙:式⑥就是

$$\begin{aligned}&(i+1)([\alpha]+[2\alpha]+\cdots+[i\alpha])\leqslant(1+2+\cdots+i)[(i+1)\alpha]\\ \Leftrightarrow\ &2([\alpha]+[2\alpha]+\cdots+[i\alpha])\leqslant i[(i+1)\alpha]. \quad ⑦\end{aligned}$$

甲:因为 $[x]+[y]\leqslant[x+y]$, 所以,

$$[\alpha]+[i\alpha]\leqslant[(i+1)\alpha],$$
$$[2\alpha]+[(i-1)\alpha]\leqslant[(i+1)\alpha],$$
$$\cdots$$
$$[i\alpha]+[\alpha]\leqslant[(i+1)\alpha].$$

将以上 i 个不等式相加即得式⑦.

师:本题有三个关键的解题步骤:一是化简;二是阿贝尔求和法的运用;三是$\frac{C_i}{B_i}$的递增性,即式⑥.

对每一道题的解法,应当明确它有哪几个关键步骤. 关键步骤清楚了,解题的脉络也就分明了. 通过总结、回顾,弄清关键步骤,有助于解题能力的提高.

从简单的做起

解题经验丰富的人都知道:一个复杂的数学问题A做不出来,多半是由于一个比问题A简单的问题B未做好.反过来说,简单的问题B做好了,那么,问题A的难度就会大大下降,甚至迎刃而解.

本文举几个图论的问题作为说明.

1 问题A

下面的五个问题作为问题A,都有一定的难度.

A1. 将完全图K_v的边任意二染色(染上红色或蓝色).图中单色三角形(三边同色的三角形)的个数最少是多少?

A2. 将完全图K_8的边二染色,必有两个独立的(即没有公共点的)单色三角形.而对完全图K_7,相应的结论不成立.

A3. 求最小的正整数v,使得完全图K_v的边任意二染色时,都有m个独立的单色三角形.

A4. 求最小的正整数v,使得完全图K_v的边任意二染色时,都有两个单色三角形恰有一个公共点.

A5. 能将完全图K_{25}的边四染色,不出现单色三角形吗?

2 问题B及其解

问题B 完全图K_6的边染上红色或蓝色.则图中一定有一个单色三角形.

问题B是常见的,证法很多.下面就是一种证明.

证明 完全图K_6的每个点引出的5条边中有3条同色.

设x_1x_2、x_1x_3、x_1x_4都是红色.

如果$\triangle x_2x_3x_4$不是单色三角形,则必有一条边为红色(设x_2x_3为红色).

故$\triangle x_1x_2x_3$是单色三角形.

问题B已经证完.但还不能说问题B已经做好了.因为还有一些与之有关的问题值得讨论,也应当讨论.

首先,完全图K_6中的单色三角形是否只有一个呢?

不是的.完全图K_6中至少有两个同色三角形.

这比原来的结果要强.证法也有多种.

证法1　设$\triangle x_1x_2x_3$为红色三角形.

如果$\triangle x_4x_5x_6$不是红色三角形,则必有一边为蓝色(设x_4x_5为蓝色,如图1,蓝色边用虚线表示,以下全同).如果x_1x_4、x_1x_5均为蓝色,则$\triangle x_1x_4x_5$是蓝色三角形.否则,设x_1至少向x_4或x_5引一条边为红色.同样,设x_2、x_3也是如此.

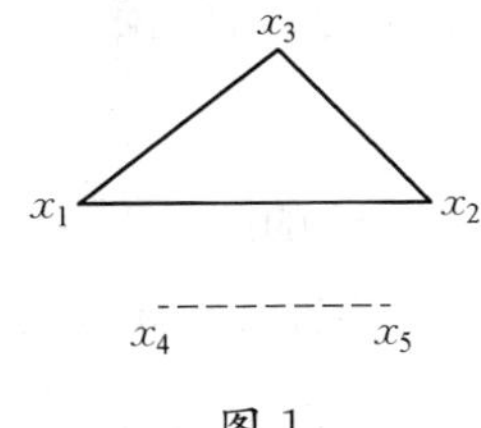

图 1

于是,x_1、x_2、x_3向x_4、x_5至少引出三条红边.因此,x_4、x_5中有一个引出两条红边(设x_4x_1、x_4x_2为红色).故$\triangle x_1x_2x_4$又是一个单色三角形.

证法2　如果一点引出的两条边颜色不同,则称它们组成一个"异色角".

注意到点x引出r条红边,b条蓝边$(b+r=5)$.

则以x为顶点的异色角有rb个.而

$$rb \leqslant 2\times 3=6,$$

于是,异色角的总数小于或等于$6\times 6=36$.

另一方面,单色三角形的角都不是异色角.而不是单色的三角形中有两个异色角,因此,不是单色的三角形至多有$36\div 2=18$个.

故单色三角形至少有

$$\mathrm{C}_6^3-18=20-18=2(\text{个}).$$

其次,两个是不能再加强的结果.

例如,两个红色$\triangle x_1x_2x_3$与$\triangle x_4x_5x_6$之间的边全为蓝色.这样的二染色,完全图K_6只有两个单色三角形.

再次,完全图K_5的二染色,不一定有单色三角形,如图2.

由一开始对完全图K_6的证明,在完全图K_5没有单色三角形时,每一点引出的四条边中,不能有三条同色.因而,一定是两条红边、两条蓝边.

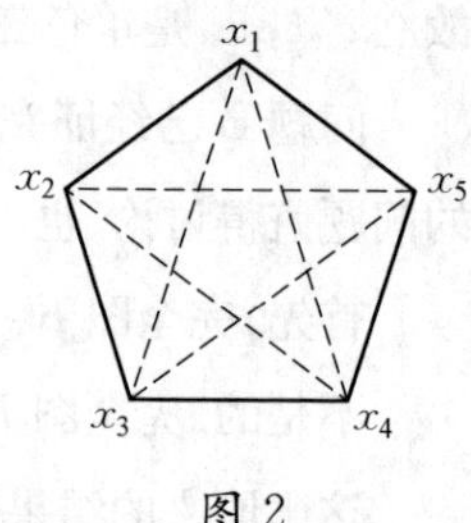

图 2

因为每一点有两条红边,所以,红边形成一个圈,蓝边也形成一个圈.

故完全图 K_5 没有单色三角形时,染色方式一定是图 2.

经过这一彻底的讨论,问题 B 已经做好了,可以动手解问题 A 了.

3 问题 A 的解

A1 的解 采用问题 B 的讨论中异色角的计算方法.

先设 v 为偶数 $(v=2n)$.

顶点 x 引出 r 条红边, b 条蓝边,则以 x 为顶点的异色角个数为

$$br \leqslant n(n-1).$$

故异色角总数小于或等于 $2n \times n(n-1)$.

不是单色的三角形的个数小于或等于

$$2n \times n(n-1) \div 2 = n^2(n-1).$$

单色三角形的个数大于或等于

$$C_{2n}^3 - n^2(n-1) = \frac{1}{3}n(n-1)(n-2) = 2C_n^3.$$

另一方面,取集合

$$X = \{1, 2, \cdots, n\},$$
$$Y = \{n+1, n+2, \cdots, 2n\}.$$

X、Y 之间每两点连蓝线. X 内部任两点均连红线, Y 也是如此. 这样得到的二染色的完全图 K_{2n} 恰有 $2C_n^3$ 个单色(全为红色)三角形.

所以,单色三角形的个数最少是 $2C_n^3$.

再设 v 为奇数 $(v=2n+1)$. 此时,

$$br \leqslant n \times n.$$

故异色角总数小于或等于 $(2n+1)n^2$.

单色三角形的个数大于或等于

$$C_{2n+1}^3-\frac{1}{2}(2n+1)n^2=\frac{(2n+1)n(n-2)}{6}=C_n^3+C_{n+1}^3-\frac{n}{2}.$$

从而,单色三角形的个数大于或等于

$$C_n^3+C_{n+1}^3-\left[\frac{n}{2}\right],$$

其中, $[x]$表示不超过实数 x 的最大整数.

取两个集合

$$X=\{1,\ 2,\ \cdots,\ n\},$$
$$Y=\{n+1,\ n+2,\ \cdots,\ 2n+1\}.$$

当 n 为奇数时, X 内部的边均染红色, Y 内部

$$(n+1,\ n+2),\ (n+3,\ n+4),\ \cdots,\ (2n-2,\ 2n-1)$$

染蓝色,其余边染红色;X、Y 之间

$$(1,\ n+1),\ (2,\ n+2),\ \cdots,\ (n-1,\ 2n-1)$$

染红色,其余边染蓝色.

此时, X 内部有 C_n^3 个红色三角形, Y 内部有 $C_{n+1}^3-\frac{n-1}{2}(n-1)$ 个红色三角形, X、Y 之间有 $\frac{n-1}{2}(n-2)$ 个蓝色三角形.

故共有 $C_n^3+C_{n+1}^3-\frac{n-1}{2}$ 个单色三角形.

当 n 为偶数时, X 内部同上, Y 内部

$$(n+1,\ n+2),\ (n+3,\ n+4),\ \cdots,\ (2n-1,\ 2n)$$

染蓝色,其余边染红色; X、Y 之间

$$(1,\ n+1),\ (2,\ n+2),\ \cdots,\ (n,\ 2n)$$

染红色,其余边染蓝色.

此时，X 内部有 C_n^3 个红色三角形，Y 内部有 $C_{n+1}^3-\frac{n}{2}(n-1)$ 个红色三角形，X、Y 之间有 $\frac{n}{2}(n-2)$ 个蓝色三角形.

故共有 $C_n^3+C_{n+1}^3-\frac{n}{2}$ 个单色三角形.

因此，单色三角形的个数最少是

$$\begin{cases} 2C_n^3, & v=2n; \\ C_n^3+C_{n+1}^3-\left[\frac{n}{2}\right], & v=2n+1. \end{cases}$$

【评注】 有些书上说图中至少有 $\frac{v(v-1)(v-5)}{24}$ 个三角形. 这在 $v\equiv 1 \pmod 4$ 时，与上面的结果一致. 其他情况则稍弱，不是最大值.

文[5]中用矩阵方法证明上述结果. 对于这种方法，该书作者认为"也许不是最短和最快捷的".

A2 的解 图中有一个单色三角形(设$\triangle x_6x_7x_8$ 是红色三角形).

假设结论不成立. 于是，另 5 个点所组成的完全图 K_5 中无单色三角形，因而是图 2.

而将 x_6 加入完全图 K_5 中，得到的完全图 K_6 有两个同色三角形，当然都是以 x_6 为顶点的三角形.

此时，有两种情况.

(1) 有一个蓝色的$\triangle x_6x_4x_1$(如图 3).

考虑由 x_2、x_3、x_5、x_7、x_8 组成的完全图 K_5. 这个图也是图 2 那样，x_5 应引出两条蓝边，两条红边. 所以，$\triangle x_5x_7x_8$ 是红色三角形，与$\triangle x_6x_4x_1$ 独立，矛盾.

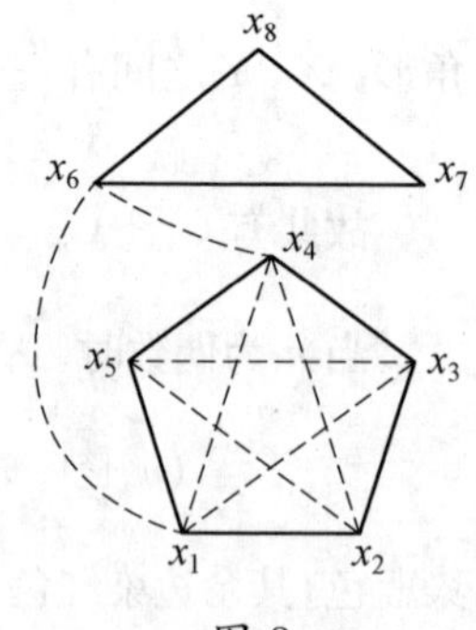

图 3

(2) 两个单色三角形都是红色的(设$\triangle x_4x_5x_6$ 是红色的)(如图 4).

由对称性，可设另一个红色三角形有一条边是 x_6x_1.

对于 x_7、x_8 情况类似，即各与 x_1、x_2、x_3、x_4、x_5 中的点组成两个红色三

角形，这些红色三角形只能是$\triangle x_7x_4x_5$、$\triangle x_7x_5x_1$、$\triangle x_8x_4x_5$、$\triangle x_8x_5x_1$，否则，结论已经成立. 但此时$\triangle x_4x_6x_7$、$\triangle x_8x_5x_1$是两个独立的红色三角形，矛盾.

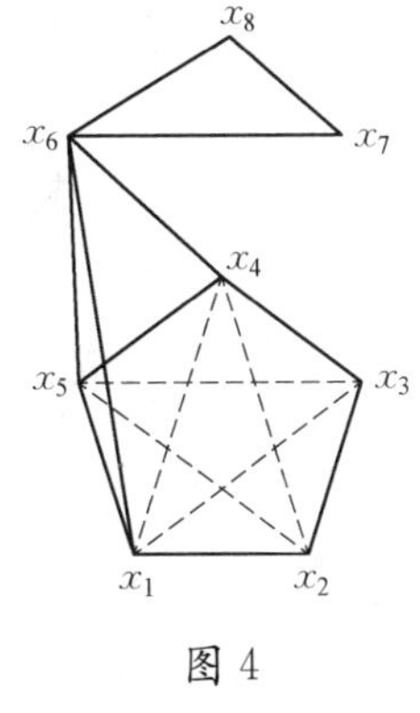

图 4

另一方面，取一个红色的完全图K_5及一个红色的完全图K_2，两者之间的边染成蓝色. 于是，所得的完全图K_7中没有蓝色三角形，也没有两个独立的红色三角形.

A3 的解　当$m=1$时，$v=6$(问题 B).

当$m=2$时，$v=8$(问题 A2).

设$m\geqslant 2$时，完全图K_{3m+2}的任意二染色都有m个独立的单色三角形. 对于完全图$K_{3(m+1)+2}$的二染色，当然有一个单色三角形. 去掉它，剩下的完全图K_{3m+2}由归纳假设存在m个独立的单色三角形. 于是，对一切自然数$m(m\geqslant 2)$，均有$v\leqslant 3m+2$.

另一方面，对于完全图K_{3m+1}，将一条边x_1x_2染成红色，剩下的完全图K_{3m-1}的边也全染成红色. x_1x_2与完全图K_{3m-1}之间的边染成蓝色. 此时，完全图K_{3m+1}无蓝色三角形，而且任m个红色三角形的$3m$个顶点不可能全不相同(因为$3m>3m-1$).

于是，$v=\begin{cases}6, & m=1;\\ 3m+2, & m\geqslant 2.\end{cases}$

A4 的解　$v=9$.

一方面，取两个红色的完全图K_4，在它们之间的边全为蓝色. 所得完全图K_8没有两个单色三角形恰有一个公共点.

另一方面，考虑完全图K_9的任意二染色. 由问题 B，有一个单色三角形(设$\triangle x_1x_2x_3$为红色). 另外六个点组成的完全图K_6，边全为红色时结论显然.

假设结论不成立. 设有边v_4v_5为蓝色. 与问题 B 的论证相同(图 1)，可以设又有一个红色$\triangle x_1x_2x_4$.

而另外五个点x_5、x_6、x_7、x_8、x_9组成的完全图K_5，边全为红色时结论显然.

设边x_8x_9为蓝色(如图 5).

在结论不成立时，x_4x_8、x_4x_9不全为蓝色.

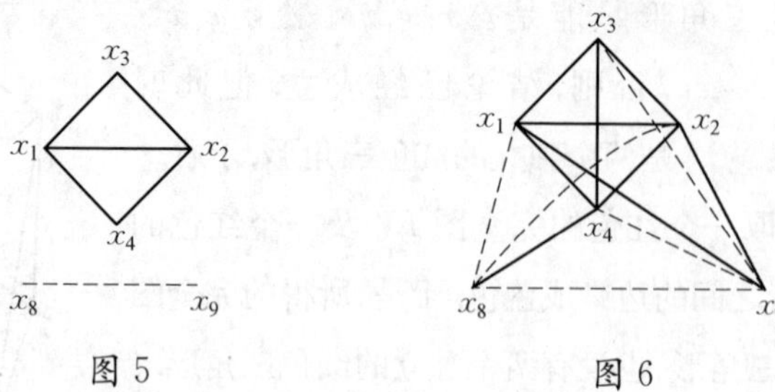

图 5　　　　图 6

设 x_4x_8 为红色(图 6). 则 x_1x_8、x_2x_8 为蓝色, x_9x_1、x_9x_2 为红色, x_9x_4、x_9x_3 为蓝色, x_3x_4 为红色, $\triangle x_2x_3x_4$ 与$\triangle x_2x_1x_9$ 是两个红色三角形恰有一个公共点.

于是,结论成立.

A5 的解　将 25 个点分为五组,每组 5 个点,构成五个完全图 G_i $(1 \leqslant i \leqslant 5)$, 每个染色均如图 2(用 1、2 两种颜色).

再将每个完全图 G_i 当作一个点,所得的完全图 K_5 仍按图 2 染成另外两种颜色,即完全图 G_i 与 G_{i+1} $(i=1, 2, 3, 4, 5, G_6=G_1)$ 之间的边染第三种颜色,完全图 G_i 与 G_{i+2} $(G_7=G_2)$ 之间的边染第四种颜色.

这样得到的完全图 K_{25} 的四染色没有单色三角形.

可见,问题 A1～A5(Advanced problems)的解都利用了问题 B(Basic problem)的解. 为此,解好基本问题十分重要. 最后,强调一句,解决问题应从简单的做起.

参考文献

[1] Graph Theory—An Introductory Course, B. Bollobás, Springer-Verlag, 1979.
[2] Graph Theory with Applications, Bondy and Murty. Macmillan Press, 1976.
[3] Introduction to Graph Theory, R. J. Wilson, Longman Group, 1979.
[4] 单墫. 趣味的图论问题[M]. 上海教育出版社,1980.
[5] 任韩. 图论[M]. 上海科技教育出版社,2009.

第四章　　数学普及(一)

数学科普与我　江苏教育报,1996 年 7 月 15 日

从实数到向量　湖南数学年刊,1995(4)

一个错误的例题　数学通报,1980(1)

从 3 到 n　国内外中学数学,1984(9)

The Gaps Between Consecutive Binomial Coefficients(与 Edward T. H. Wang 合作)　Mathematics Magazine,1990(2)

也谈取棋子游戏(笔名肖韧吾)　数学通报,2009(12)

一个古老等式的组合证明(与宋艳红合作)　大学数学,2005(3)

1994 年全国理科试验班招生统一考试(数学)　初中生数学学习,1994(12

不等式的证明(一)　中学教研(数学),2007(1)

不等式的证明(二)　中学教研(数学),2007(2)

利用导数证明不等式　中等数学,2006(2)

介绍一个几何不等式(与刘亚强合作)　中等数学,1989(6)

一个角平分线不等式的推广(与陈计合作)　数学通讯,1995(11)

三角不等式的一种证法　中学教研(数学),1982(6)

游戏与数学归纳法　中学生数学,1982(2)

什么时候用归纳法　中学生数学,1989(3)

数学归纳法在整数问题中应用一例　中学生数学,1985(6)

数学归纳法一例　中学生数学,1985(5)

方法一例:数学归纳法一例　中学生数学,1985(1)

三角中的归纳法　中学生数学,1985(2)

处处留心皆学问　中学生,1991(6)

一个作图问题(笔名肖韧吾)　数学通讯,1983(2)

几个尺规作图问题(载《初等数学研究在中国(第 2 辑)》)　哈尔滨工业大学出版社,2019

存在性问题的一种证法 数学通讯,1983(5)
从“挖坑”谈起 数学教学通讯,1985(1)
几何变换与证题 数学教学,1982(3)
一个几何命题的探讨与推广 中学数学教学,1984(4)
一道几何题的推广 中等数学,1985(5)
填满空间的正多面体 中等数学研究,2007(10)
熊的颜色 中学数学研究,2006(8)
莫莱定理 中学数学研究,2002(1)
一点补充 中等数学,2007(12)
利用图形解决问题 数学通讯,1985(12)
过三角形外心与内心的直线 中学数学研究,2007(2)
关于匹窦不等式的讨论 安徽教育,1980(1)
一类恒等式的证明 数学教学,1982(6)
三角形中一些量的关系 数学教学,1983(4)
利用面积证题一例 数学教学通讯,1985(5)
几个容易混淆的极值问题 中等数学,1983(6)
方程组的巧解 中学数学教学,1983(3)
$y=2x^2$ 与 $y=\frac{1}{2}x^2$ 的图像形状一样吗? 安徽教育,1981(12)
韦达定理的一个应用 安徽教育,1982(10)
正方体的分解 科学,1998(6)
十三个球的问题 科学,1997(5)

数学科普与我

科学昌明,不仅需要科学家筚路蓝缕,作开拓性的研究,还应当有更多的普及工作者播种耕耘,提高全人类的水平.

不少老一辈的科学工作者,例如华罗庚先生,既是一流的科学家,又写了很多优秀的科普作品. 我做学生的时候,就读过华先生的《从杨辉三角谈起》《从祖冲之的圆周率谈起》等著作. 当时流行的普及读物还有中国青年出版社翻译的苏联的数学小丛书,许莼舫先生的《几何定理和证题》等. 正是这些众多的科普读物,使我喜爱数学,并且立下了一个志愿:将来也要写一些这样的书,奉献给数学爱好者,特别是年轻的中小学生.

1978 年,上海教育出版社王文才和赵斌两位先生到中国科学技术大学组稿,找到我. 我便报了三个选题,他们都采纳了. 从这以后,我自己定了一个不成文的规划:每年写一篇论文,一本小册子. 十多年来,我先后为十多家出版社完成了近 20 本书:《几何不等式》《趣味的图论问题》《覆盖》《棋盘上的数学》《组合数学的问题与方法》《趣味数论》《解析几何的技巧》《对应》《解题思路训练》《巧解应用题》《帮你学因式分解》《数学竞赛史话》《国际数学竞赛解题方法》《数学趣题巧解 100 例》《不定方程》《算两次》《数学竞赛研究教程》《组合几何》《集合及其子集》(其中有几种是合著),并主编了《奥数教程》(发行数百万套)、《数学奥林匹克题典》等. 有些书,台湾的九章出版社等已用繁体字在港澳台发行.

1989 年,我带队在德国参加国际数学竞赛时,遇到新加坡领队许其明教授. 他一见面就说:“早就知道你的大名,新加坡的书店里有很多你的书. ”新加坡出的大概是盗版(从未与我联系过),不过,科普著作的目的就在于普及科学,正如鲁迅所说,“有人翻印,功德无量. ”

好的科普读物,我以为,应当做到四个字:准、新、浅、趣.

准:内容选择得当,有意义,论说正确,史实无讹.

新:有新意,不雷同. 材料新颖或用新的观点处理熟悉的问题,令人耳目为之一新.

浅:通俗易懂,生动活泼,可读性强.浅,并不是肤浅,而是深入浅出,它需要作者对写的内容有透彻的理解,就像未受污染的水潭,深,然而澄澈见底.

趣:幽默风趣,而不流于油滑,使人能轻松愉快地学习一些有用的知识.

做到以上四点,当然不很容易,但国内做得好的却也不乏其人.例如胡作玄先生,他写的《布尔巴基学派的兴衰》,与赵斌先生合写的《菲尔兹奖获得者传》等,都是介绍现代数学与数学史的上乘之作.任宏硕先生写的《奥林匹克金钥匙》,也是给低年级小学生的一份好礼物,对培养数学的思维大为有益,尤为难得的是这本书图文并茂,适合儿童特点.还有上海第二军医大学的谈祥柏先生,20 世纪 50 年代就在《科学画报》上发表文章,至今笔耕不辍.谈老先生文学修养好,他的科普文章妙趣横生,是这一领域中的奇葩.

我自己写的书,虽然也受到一些读者的称赞,却有一个明显的缺点:数学太多.由于是搞数学的,又喜欢做题,一写书总想多放点数学问题进去,结果太浓腻了,化不开.

或许科普工作需要形成一个系列,从科研的前沿开始,逐步普及.起初层次较高,受众较少;随着范围的扩大,不断"加水稀释",直至一般的群众也都能明白.

例如,英国的 A. Wiles 与他的学生 R. Taylor 在 1994 年证明了 Fermat 大定理(文章正式发表在 1995 年 *Annals of Mathematics* 第三期),这是轰动数学界的一件大事.开始,能够读懂证明的只有寥寥数人,但很快就有了许多讨论班来学习他们的工作,这一过程正在继续下去,已经有通俗的报告.将来一定会有人写一本普及的读物,介绍 Fermat 大定理及其证明.

这几年,出版社约稿,我均婉言谢绝.主要原因是精力不够,写不动了.

初写小册子,由于积累多,可以连续写几本;写了十几本后,"库存"渐渐空了,年龄也大了,杂事又多,写起来很吃力.例如写《组合几何》,断断续续写了 3 年,查阅了 30 多种外文资料,还花了大量时间想问题、做问题,比写一篇论文困难得多.现在,我已经写了 20 本书,内容涉及不少数学分支,对象从小学直到大学.继续写,很难找到合适的题材,也很难再有新意.

"长江后浪推前浪,一代新人胜旧人."希望年轻人能够关心科普领域,在这片土地上大显身手.

从实数到向量

第 20 届(1994)全俄数学竞赛中有这样一道题:“在直线上标出 n 个不同的蓝色点和 n 个不同的红色点.证明同色点之间两两距离之和不超过异色点之间两两距离之和.”

证法很多,如果以这条直线为数轴,设蓝色点的坐标为 $x_1<x_2<\cdots<x_n$,红色点的坐标为 $y_1<y_2<\cdots<y_n$.

当 $n=1$ 时,结论显然.考虑 $n>1$.因为

$$\begin{aligned}\sum_{i<n}|y_n-y_i|+\sum_{i<n}|x_n-x_i|&=\sum_{i<n}(y_n-y_i)+\sum_{i<n}(x_n-x_i)\\&=\sum_{i<n}(y_n-x_i)+\sum_{i<n}(x_n-y_i)\\&\leqslant\sum_{i<n}|y_n-x_i|+\sum_{i<n}|x_n-y_i|,\end{aligned}$$

如果结论对于 $n-1$ 已经成立,即

$$\sum_{1\leqslant i<j\leqslant n-1}|y_j-y_i|+\sum_{1\leqslant i<j\leqslant n-1}|x_j-x_i|\leqslant\sum_{\substack{1\leqslant j\leqslant n-1\\1\leqslant j\leqslant n-1}}|x_i-y_j|$$

那么将以上二式相加便得结论对 n 成立.

清华附中理科实验班一位同学给出稍有不同的另一种证法:设同前,则

$$\begin{aligned}\sum_{\substack{1\leqslant j\leqslant n-1\\1\leqslant j\leqslant n-1}}|x_i-y_j|&\geqslant\sum_{1\leqslant i<j\leqslant n}(|x_i-y_j|+|x_j-y_i|)\\&\geqslant\sum_{1\leqslant i<j\leqslant n}(-x_i+x_j+y_j-y_i)\\&=\sum_{1\leqslant i<j\leqslant n}|x_i-x_j|+\sum_{1\leqslant i<j\leqslant n}|y_j-y_i|\end{aligned}$$

很自然地,我们会问:如果这 $2n$ 个点不在同一条直线上,那么结论是否仍然成立?

设这 $2n$ 个点中,A_1,A_2,…,A_n 为蓝色点,B_1,B_2,…,B_n 为红色点.简单地利用三角形不等式,得

$$A_iA_j \leqslant A_iB_k + B_kA_j,$$

对 k 求和得 $$n \cdot A_iA_j \leqslant \sum_{1\leqslant k\leqslant n} A_iB_k + \sum_{1\leqslant k\leqslant n} B_kA_j,$$

再对 i、j 求和得 $$n\sum_{\substack{1\leqslant i\leqslant n\\1\leqslant j\leqslant n}} A_iA_j \leqslant n\sum_{\substack{1\leqslant k\leqslant n\\1\leqslant i\leqslant n}} A_iB_k + n\sum_{\substack{1\leqslant k\leqslant n\\1\leqslant j\leqslant n}} B_kA_j,$$

即 $$\sum_{\substack{1\leqslant i\leqslant n\\1\leqslant j\leqslant n}} A_iA_j \leqslant 2\sum_{\substack{1\leqslant k\leqslant n\\1\leqslant i\leqslant n}} A_iB_k. \tag{1}$$

(1)比预计的结果多出一个系数 2,能否将这个系数 2 改成 1 或较 2 小的数. 1994 年 10 月的全国几何不等式会议上,笔者向与会者提出上面的问题,引起不少人的兴趣. 会后有些朋友在通信中讨论过这个问题,发现即使是平面上的六个点(三红三蓝)也不很容易,空间中的六个点就更困难了(大概都可以作数学竞赛的试题).

最近,笔者解决了这个问题.

首先考虑同一平面的 $2n$ 个点(n 红 n 蓝). 这 $2n$ 个点可以用复数或向量表示为 $x_1, x_2, \cdots, x_n$(蓝点)与 $y_1, y_2, \cdots, y_n$(红点).

如果用 a 表示 x 轴上的单位向量,那么 x_i 与 a 的数量积(内积) $x_i \cdot a$ 就是 x_i 在 x 轴上的射影(复数 x_i 的实部),根据上面所证

$$\sum_{\substack{1\leqslant i\leqslant n\\1\leqslant j\leqslant n}} |x_i \cdot a - y_j \cdot a| \geqslant \sum_{1\leqslant i<j\leqslant n} |x_i \cdot a - x_j \cdot a| + \sum_{1\leqslant i<j\leqslant n} |y_i \cdot a - y_j \cdot a|, \tag{2}$$

即

$$\sum_{\substack{1\leqslant i\leqslant n\\1\leqslant j\leqslant n}} |x_i - y_j| \cdot |\cos\psi_{ij}| \geqslant \sum_{1\leqslant i<j\leqslant n} |x_i - x_j| \cdot |\cos\varphi_{ij}| + \sum_{1\leqslant i<j\leqslant n} |y_i - y_j| \cdot |\cos\eta_{ij}|, \tag{3}$$

其中 ψ_{ij}, φ_{ij}, η_{ij} 分别为 $x_i - y_j$, $x_i - x_j$, $y_i - y_j$ 与 x 轴所成的角.

类似于(2)、(3)的结论对 $x_1, \cdots, x_n, y_1, \cdots, y_n$ 在任一条直线上的射影成立,即设 a 为单位向量,与 x 轴夹角为 θ,则(2)仍成立. 而由数量积定义,

$$|(x_i - y_j) \cdot a| = |x_i - y_j| \cdot |\cos(\theta - \psi_{ij})|$$

等等,所以(2)即

$$\sum_{\substack{1\leqslant i\leqslant n\\1\leqslant j\leqslant n}}|x_i-y_j|\cdot|\cos(\theta-\psi_{ij})|\geqslant\sum_{1\leqslant i<j\leqslant n}|x_i-x_j|\cdot$$
$$|\cos(\theta-\varphi_{ij})|+\sum_{1\leqslant i<j\leqslant n}|y_i-y_j|\cdot|\cos(\theta-\eta_{ij})|,\tag{4}$$

将(4)两边对 θ 从 0 到 2π 积分. 由于

$$\int_0^{2\pi}|\cos(\theta-\psi_{ij})|\,\mathrm{d}\theta=\int_0^{2\pi}|\cos\theta|\,\mathrm{d}\theta=4,$$
$$\int_0^{2\pi}|\cos(\theta-\varphi_{ij})|\,\mathrm{d}\theta=\int_0^{2\pi}|\cos(\theta-\eta_{ij})|\,\mathrm{d}\theta=\int_0^{2\pi}|\cos\theta|\,\mathrm{d}\theta,$$

所以

$$\sum_{\substack{1\leqslant i\leqslant n\\1\leqslant j\leqslant n}}|x_i-y_j|\int_0^{2\pi}|\cos\theta|\,\mathrm{d}\theta\geqslant\sum_{1\leqslant i<j\leqslant n}|x_i-x_j|\int_0^{2\pi}|\cos\theta|\,\mathrm{d}\theta+$$
$$\sum_{1\leqslant i<j\leqslant n}|y_i-y_j|\int_0^{2\pi}|\cos\theta|\,\mathrm{d}\theta,$$

即

$$\sum_{\substack{1\leqslant i\leqslant n\\1\leqslant j\leqslant n}}|x_i-y_j|\geqslant\sum_{1\leqslant i<j\leqslant n}|x_i-x_j|+\sum_{1\leqslant i<j\leqslant n}|y_i-y_j|,\tag{5}$$

因此,将实数 x_1, $\cdots$, x_n, y_1, $\cdots$, y_n 换成复数(同一平面的向量),结论依然成立.

对于更一般的 n 维空间的向量 x_1, $\cdots$, x_n, y_1, $\cdots$, y_n 证明完全类似,只是得出(2)式后采用 n 维单位球面 S(圆周是二维球面)上的积分(a 为过原点与 S 上一点的任一向量),由于球面的对称性,

$$\int_S|(x_i-y_j)\cdot a|\,\mathrm{d}S=k\int_S|x_i-y_j|\,\mathrm{d}S,$$
$$\int_S|(x_i-x_j)\cdot a|\,\mathrm{d}S=k\int_S|x_i-x_j|\,\mathrm{d}S,$$
$$\int_S|(x_i-x_j)\cdot a|\,\mathrm{d}S=k\int_S|y_i-y_j|\,\mathrm{d}S,$$

其中 k 为正数,可以算出而不必算出. (2)积分后再在两边约去 k 便得到(5).

于是,设 M、N 都是 α_1, α_2, $\cdots$, α_n 的若干个线性式的绝对值的和. 如果在 α_1, $\cdots$, α_n 为实数时 $M\geqslant N$, 那么在 α_1, $\cdots$, α_n 为复数或 n 维向量时,仍有 $M\geqslant N$.

一个错误的例题

邦德列雅金著《连续群》第六章有一个如下的例题.

例 65 设 $\alpha_1, \cdots, \alpha_r$ 为线性无关的无理数的有限组(以整数 $n_1, \cdots, n_r$ 为系数的和 $n_1\alpha_1+\cdots+n_r\alpha_r$, 只当所有系数为零时才为零). 我们来证,对任何正数 ε 及实数组 $d_1, \cdots, d_r$ 总存在整数组 $n_1, \cdots, n_r$ 及整数 m,使得

$$|m\alpha_i-d_i-n_i|<\varepsilon,\ i=1,\ \cdots,\ r.$$

原书证明如下:

"设 G 为具有 r 个线性无关的生成元素 $\alpha_1, \cdots, \alpha_r$ 的离散群. 把每一整数 m 用下列方法对应于群 G 的特征标 β_m. 假如 $x=n_1\alpha_1+\cdots+n_r\alpha_r$,那么 $\beta_m(x)=k(m(n_1\alpha_1+\cdots+n_r\alpha_r))$. ($k$ 定义见该书 260 页). 容易看出 $\beta_m+\beta_n=\beta_{m+n}$. 因此,形式 β_m 的所有特征标的集合 B 是群. 以 X 表群 G 的特征标群;B 是 X 的子群. 以 Φ 表集合 B 在 X 中的闭包,容易看出,假如对每一 m 有 $\beta_m(x)=0$,则 $x=0$(*); 这是由于数 α_1 线性无关的缘故. 由此得出, $(G, \Phi)=0$; 由于零化子的相互性,由此得出 $\Phi=X$. 因此集合 B 在 X 中到处稠密,即群 G 的每一特征标可以任意精确地用形式 β_m 的特征标来逼近. 由此就可直接得出所要证的叙说. 事实上,假如 $d_1, \cdots, d_r$ 为所给的数,那么用 $\beta_{(\alpha i)}=k(d_i)(i=1, 2, \cdots, r)$ 来决定群 G 的特征标 β. 用特征标 β_m 来逼近特征标 β,我们即得所需要的关系式."

这个证明有一个错误,由 α_1 线性无关及 $\beta_m(x)=0$ 也就是 $\sum ml_i\alpha_i=$ 整数(注意这个整数不一定是零),并不能得出(*)式 $x=0$.

假如: $r=2$, $\alpha_1=\sqrt{2}$, $\alpha_2=1-\sqrt{2}$,显然对于 $x=\alpha_1+\alpha_2$ 有 $\beta_m(x)=0$, 而且不难看出 α_1 与 α_2(在有理数域上)线性无关,但是 $x\neq 0$.

事实上,例题本身就是错的. 例如上面举出的 α_1 与 α_2 在有理数域上是线性无关的,但对 $d_1=d_2=\frac{1}{4}$,并不存在整数 m, n_1, n_2 使 $|m\alpha_i-n_i-d_i|<\frac{1}{5}(i=1, 2)$ 同时成立,因为如果这两个等式同时成立,那么

$$\frac{2}{5} > |m\alpha_1 - n_1 - d_1| + |m\alpha_2 - n_2 - d_2|$$
$$\geqslant |m\alpha_1 - n_1 - d_1 + m\alpha_2 - n_2 - d_2| = \left|m - n_1 - n_2 - \frac{1}{2}\right| \geqslant \frac{1}{2},$$

这就导致矛盾.

正确的叙述应当是:

定理Ⅰ　如果 $1, \alpha_1, \alpha_2, \cdots, \alpha_r$(在有理数域上)线性无关,那么对任意的正数 ε 及实数组 $d_1, \cdots, d_r$,总存在整数组 $n_1, n_2, \cdots, n_r$ 及整数 m,使得

$$|m\alpha_i - d_i - n_i| < \varepsilon,\ i = 1, 2, \cdots, r.$$

这里的条件加强了,不难看出由 $1, \alpha_1, \cdots, \alpha_r$ 在有理数域上线性无关可以推出 $\alpha_1, \alpha_2, \cdots, \alpha_r$ 为无理数并且在有理数域上线性无关,但反之不然,上面的 $\alpha_1 = \sqrt{2}$, $\alpha_2 = 1 - \sqrt{2}$ 仍然可以充当反例.

定理Ⅰ的证明可以照搬上面所引的证明,由于条件已经加强,(*)式是成立的.

实际上,我们还可以证得更多一些,即有:

定理Ⅱ　如果 $1, \alpha_1, \cdots, \alpha_r$ 在有理数域上线性无关,那么 $(m\alpha_1, \cdots, m\alpha_r)$ 在 $\mathbf{R}^r$ 中一致分布(mod1).

证明主要利用下面的 Weyl 判别法:

1. 序列 $\{x_m\} \cdot m = 1, 2, \cdots$ 在 $\mathbf{R}$ 中一致分布(mod1)

$$\Leftrightarrow \text{对所有整数 } k \neq 0,\ \lim_{N \to \infty} \frac{1}{N} \sum_{m=1}^{N} \mathrm{e}^{2\pi i k x_m} = 0.$$

2. 序列 $\{X_m\} = \{(x_m^{(1)}, x_m^{(2)}, \cdots, x_m^{(r)})\}$ 在 $\mathbf{R}^r$ 中一致分布(mod1)$\Leftrightarrow$ 对所有整点 $K = (k^{(1)}, k^{(2)}, \cdots, k^{(r)}) \neq (O)$, $\lim\limits_{N \to \infty} \frac{1}{n} \sum\limits_{m=1}^{N} \mathrm{e}^{2\pi i \langle K, X_m \rangle}$. 其中 $\langle K, X_m \rangle = k^{(1)} x_m^{(1)} + k^{(2)} x_m^{(2)} + \cdots + k^{(r)} x_m^{(r)}$ 为 K 与 X_m 的内积.

由判别法 1 与 2 立即得到:

引理　$\{X_m\}$ 在 R^r 中一致分布(mod1)$\Leftrightarrow\{\langle K, X_m \rangle\}$ 在 $\mathbf{R}$ 中一致分布(mod1),其中 K 为 $\mathbf{R}$ 中的所有$\neq O$ 的整点.

现在我们来证明定理Ⅱ,记 $(\alpha) = (\alpha_1, \alpha_2, \cdots, \alpha_r)$,因为 $1, \alpha_1, \cdots, \alpha_r$ 在

有理数域上线性无关,对任一整点 $K=(k^{(1)}, k^{(2)}, \cdots, k^{(r)})\neq 0$, $\langle K, \alpha\rangle=k_1\alpha_1+k_2\alpha_2+\cdots+k_r\alpha_r$ 与1(在有理数域上)线性无关,所以$\langle \mathbf{R}, \alpha\rangle$是无理数. 根据 Чебышев 的一个定理(华罗庚著《数论导引》289 页), $\{\langle K, m\alpha\rangle\}=\{m\langle K, \alpha\rangle\}$ 在 $\mathbf{R}$ 中一致分布(mod1). 根据引理, $\{m\alpha\}=\{(m\alpha_1, \cdots, m\alpha_r)\}$ 在 $\mathbf{R}^r$ 中一致分布(mod1).

从 3 到 n

大家都知道，结论越是一般，应用的范围也越是广泛. 因此，人们决不满足于已有的结果，而是不断地寻求更一般、更普遍的结论. 这种一般化、普遍化的倾向是数学的一大特点. 即使在古老的欧几里得几何中，我们也可以看到很多"从 3 到 n"的推广. 本文的目的就是举几个这样的例子.

例 1　正三角形中任一点 P 到各边的距离之和为定值.

推广一下就可以变成下面的两个结论：

边相等的凸 n 边形内一点 P 到各边的距离之和为定值.

角相等的凸 n 边形内一点 P 到各边的距离之和为定值.

后一个结论的证明比前一个要难一点.

例 2　设 P 为$\triangle ABC$ 内一点，我们知道 $y = PA + PB + PC$ 有最小值也有最大值. 如果$\triangle ABC$ 的每个角都小于 $120°$，那么$\triangle ABC$ 内有一点 F，满足 $\angle AFB = \angle BFC = \angle CFA = 120°$，在这点 y 取得最小值(F 称为 Fermat 点). 否则的话，设 $\angle A \geqslant 120°$，则 y 在 A 点取得最小值.

推而广之，我们可以得出：

如果在凸 n 边形 $A_1A_2\cdots A_n$ 内有一点 F，满足

$$\angle A_1FA_2 = \angle A_2FA_3 = \cdots = \angle A_{n-1}FA_n = \angle A_nFA_1, \tag{1}$$

那么 $y = PA_1 + PA_2 + \cdots + PA_n$，其中 P 为 n 边形 $A_1A_2\cdots A_n$ 内一点，在 P 点与 F 点重合时取得最小值.

证明可见《初等数学论丛》第 2 辑 66—81 页(上海教育出版社出版).

但是，在 n 边形内未必有满足(1)式的点. 这时 y 在什么位置取得最小值，是一个到现在还没有彻底解决的问题.

另一方面，我们也知道，对于$\triangle ABC$，y 在 A、B、C 三点中的某一个取得最大值.

推广到凸 n 边形就有：

“设 P 为凸 n 边形 $A_1A_2\cdots A_n$ 内一点，则 $y=PA_1+PA_2+\cdots+PA_n$ 在某个顶点处取得最大值.”

最近中国科学技术大学陈计同学(后为宁波大学教授)对这个结论作了一个简短的证明(待发表).

例 3 如果 $\triangle A_1A_2A_3\backsim\triangle A_1'A_2'A_3'$，$A_1''$、$A_2''$、$A_3''$ 分别为 A_1A_1'、A_2A_2'、A_3A_3'的中点，那么

$$\triangle A_1''A_2''A_3''\backsim\triangle A_1A_2A_3.$$

推广到 n 边形就有：

如果 n 边形 $A_1A_2\cdots A_n\backsim n$ 边形 $A_1'A_2'\cdots A_n'$，A_1''、A_2''、$\cdots$、A_n'' 分别在 A_1A_1'、A_2A_2'、$\cdots$、A_nA_n'上，并且

$$\frac{A_1A_1''}{A_1A_1'}=\frac{A_2A_2''}{A_2A_2'}=\cdots=\frac{A_nA_n''}{A_nA_n'},$$

那么 n 边形 $A_1''A_2''\cdots A_n''\backsim n$ 边形 $A_1A_2\cdots A_n$.

证明可见《数学通讯》1983 年第 11 期《复数的一个应用》.

例 4 1935 年 Erdös 提出了一个几何不等式：

设 P 为$\triangle ABC$ 内一点，P 到三边距离为 PD、PE、PF，则

$$PA+PB+PC\geqslant 2(PD+PE+PF). \tag{2}$$

1937 年 Mordell 给出了一个简单的证明，此后(2)就被称为 Erdös-Mordell 不等式.

1981 年上海复兴中学李伟同学提出一个猜测：

设 P 为凸 n 边形 $A_1A_2\cdots A_n$ 内一点，P 到各边距离为 PD_1，PD_2，$\cdots$，PD_n，则

$$PA_1+PA_2+\cdots+PA_n\geqslant\sec\frac{\pi}{n}(PD_1+PD_2+\cdots+PD_n). \tag{3}$$

他证明了(3)在 $n=4$ 时成立. 后来，陈计证明了(3)在 $n=5$、9、17 时均成立. 作者(单墫)证明了(3)式对于一切 $n(\geqslant 3)$均成立(待发表).

例 5 自三角形的外接圆上一点 P 向三边引垂线，则三个垂足共线. 这条

线称为 Simson 线.

这也是一个众所周知的定理.

现在考虑圆内接 n 边形 $A_1A_2\cdots A_n$，自它的外接圆上一点 P 向各边引垂线，垂足记为 A_1'，A_2'，$\cdots$，A_n'.

A_1'，A_2'，$\cdots$，A_n'是否一定共线呢？不是的，这只要画一个图就清楚了. 这样，上面关于 Simson 线的定理就不能简单地推广了. 但是，如果自 P 点向 n 边形 $A_1'A_2'\cdots A_n'$的各边作垂线，连结垂足得 n 边形 $A_1''A_2''\cdots A_n''$；自 P 向 n 边形 $A_1''A_2''\cdots A_n''$ 的各边作垂线，连结垂足得 n 边形 $A_1'''A_2'''\cdots A_n'''$；……；这样继续下去，至第 $n-2$ 次，就会发现 n 个垂足 $A_1^{(n-2)}$，$A_2^{(n-2)}$，$\cdots$，$A_n^{(n-2)}$ 恰好在一条直线上（请读者以 $n=4$ 或 $n=5$ 为例，绘图验证）.

Simson 定理就是 $n=3$ 的特例. 一般的结论是 Zylbertrest 首先发现并证明的.

例 6　分别以$\triangle ABC$ 的边 BC、CA、AB 为边向外作正三角形 BCA'、CAB'、ABC'，则$\triangle BCA'$、$\triangle CAB'$、$\triangle ABC'$的重心 O_1、O_2、O_3 构成一个正三角形.

这是众所周知的拿破仑(Napoleon)定理.

这也可以换成另一种等价的说法：

分别以$\triangle ABC$ 的边 BC、CA、AB 为底边向外作顶角为 $120^\circ\left(=\dfrac{2\pi}{3}\right)$ 的等腰三角形 BCO_1、CAO_2、ABO_3，则$\triangle O_1O_2O_3$ 是正三角形.

1940 年，J. Douglus 把上面的结论推广到 n 边形，1941 年 B・H・Neumann 给出了一个漂亮的证明. 具体见常庚哲著《复数计算与几何证题》（上海教育出版社出版）.

除了将关于三角形的结果推广到 n 边形外，几何中还有一种常见的推广，即将平面（二维）几何的结果推广到三维空间，再进一步推广到 n 维空间. 这里就不多谈了.

最后留下两个与 n 有关的几何问题供读者考虑：

1. 试将关于三角形的梅涅劳斯定理（一条直线交$\triangle ABC$ 的边 BC 于 D、CA 于 E、AB 于 F，则 $\dfrac{DC}{DB}\times\dfrac{EA}{EC}\times\dfrac{FB}{FA}=1$）推广到 n 边形.

2. 已知$\odot O_1$ 半径为 r_1、$\odot O_2$ 半径为 $r_2(r_2<r_1)$，两圆内切. 陆续作$\odot C_0$、$\odot C_1\cdots$，每一个$\odot C_i$ 与$\odot O_1$、$\odot O_2$ 相切，并且与前面的一个圆相切 ($i=0, 1, \cdots$)，如图 1 所示. 试求出$\odot C_i$ 的半径.

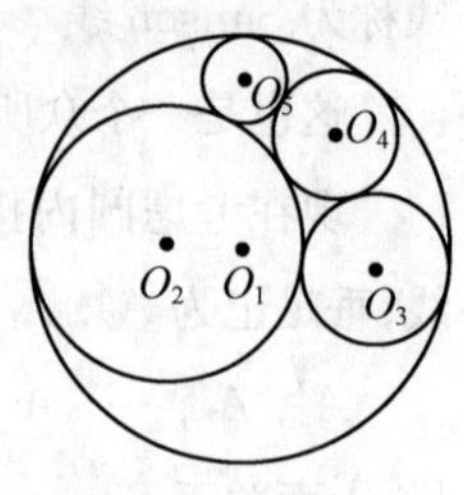

图 1

(答案：$\odot C_i$ 的半径为 $\dfrac{r_1 r_2(r_1-r_2)}{r_1 r_2+i^2(r_1-r_2)^2}$)

The Gaps Between Consecutive Binomial Coefficients

Among all combinatorial quantities, the binomial coefficients are unique. They are simple in concept and derivation, yet they appear in almost all the combinatorial identities and seem to give rise to an endless variety of problems. In this note we investigate one such problem which we believe to be new. Put in a simple way, the problem is to determine the smallest and largest "gaps" between consecutive binomial coefficients in any row of Pascal's triangle. More precisely, for positive integers n and k with $n \geqslant 2$ and $1 \leqslant k \leqslant n$, we define

$$d_k(n) = \left| \binom{n}{k} - \binom{n}{k-1} \right|$$

and we let $d_m(n) = \min_{1 \leqslant k \leqslant n} d_k(n)$, and $d_M(n) = \max_{1 \leqslant k \leqslant n} d_k(n)$. Our purpose is to solve the following problems:

P1. Determine $d_m(n)$ and all k such that $d_k(n) = d_m(n)$.

P2. Determine $d_M(n)$ and all k such that $d_k(n) = d_M(n)$.

We start out with P1, which is the easier one. Due to symmetry, it clearly suffices to consider $d_k(n)$ for $1 \leqslant k \leqslant [(n+1)/2]$. Since

$$\binom{n}{k} - \binom{n}{k-1} = \frac{n!\,(n-2k+1)}{k!\,(n-k+1)!} = 0$$

if and only if $n - 2k + 1 = 0$, we see that $d_m(n) = 0$ if and only if n is odd and $d_k(n) = 0$ if and only if $k = (n+1)/2$. When n is even we shall show that $d_m(n) = n - 1$ except when $n = 4$ in which case direct inspection reveals that $d_m(4) = 2$. In fact, we shall show that in general if we define $d'_m(n) = \min\{d_k(n) \mid 1 \leqslant k \leqslant n, k \neq (n+1)/2\}$ (so that the aforementioned trivial fact that $d_m(n) = 0$ when n is odd is excluded), then $d'_m(n) = n - 1$ holds for all $n \neq 4$.

THEOREM 1. *For all $n \geqslant 2$, except when $n = 4$, $d'_m(n) = n - 1$.*

Furthermore, if $n \neq 6$, *then* $d_k(n) = n-1$ *if and only if* $k=1$ *or* n, *and if* $n=6$, *then* $d_k(6)=5$ *if and only if* $k=1, 3, 4$, *or* 6.

Proof. Since the cases when $n \leqslant 6$ can be verified directly by inspection, we assume that $n > 6$ and use induction. Examining the appropriate rows in Pascal's triangle reveals that the theorem holds for $n=7$ and 8. Due to symmetry and the fact that $d_1(n) = d_n(n) = n-1$ we need only to show that $d_k(n) > n-1$ for all k such that $2 \leqslant k \leqslant [n/2]$ where $n \geqslant 9$. By Pascal's identity, we have:

$$\begin{aligned} \binom{n}{k} - \binom{n}{k-1} &= \left\{\binom{n-1}{k} + \binom{n-1}{k-1}\right\} - \left\{\binom{n-1}{k-1} + \binom{n-1}{k-2}\right\} \\ &= \left\{\binom{n-1}{k} - \binom{n-1}{k-1}\right\} + \left\{\binom{n-1}{k-1} - \binom{n-1}{k-2}\right\}. \end{aligned} \tag{1}$$

If n is odd, then by the induction hypothesis, we have

$$\binom{n-1}{k} - \binom{n-1}{k-1} \geqslant n-2 \quad \text{and} \quad \binom{n-1}{k-1} - \binom{n-1}{k-2} \geqslant n-2.$$

Hence from (1) we obtain

$$\binom{n}{k} - \binom{n}{k-1} \geqslant 2(n-2) = (n-1) + (n-3) > n-1. \tag{2}$$

If n is even then the above argument still holds except when $k = [n/2]$ in which case

$$\binom{n-1}{k} - \binom{n-1}{k-1} = 0.$$

By a second application of Pascal's identity we obtain from (1)

$$\begin{aligned} \binom{n}{k} - \binom{n}{k-1} &= \binom{n-1}{k-1} - \binom{n-1}{k-2} \\ &= \left\{\binom{n-2}{k-1} - \binom{n-2}{k-2}\right\} + \left\{\binom{n-2}{k-2} - \binom{n-2}{k-3}\right\}. \end{aligned} \tag{3}$$

Since $n-2$ is even we have, by the induction hypothesis,

$$\binom{n-2}{k-1}-\binom{n-2}{k-2}\geqslant n-3 \quad \text{and} \quad \binom{n-2}{k-2}-\binom{n-2}{k-3}\geqslant n-3.$$

Hence from (3), we obtain

$$\binom{n}{k}-\binom{n}{k-1}\geqslant 2(n-3)=(n-1)+(n-5)>n-1. \tag{4}$$

By (2) and (4) our proof is complete.

Now we turn to P2 the answer of which is somewhat less obvious than that of P1. The complete answer to this problem is contained in the next theorem.

THEOREM 2. *Let* $\tau=(1/2)(n+2-\sqrt{n+2})$. *Then*

$$d_M(n)=\binom{n}{[\tau]}-\binom{n}{[\tau]-1}.$$

Furthermore, $d_k(n)=d_M(n)$ *if and only if*

$$k=\begin{cases}[\tau] \text{ or } n-[\tau]+1 \text{ if } \tau\notin \mathbf{Z}\\ \tau-1,\ \tau,\ n-\tau+1 \text{ or } n-\tau+2 \text{ if } \tau\in\mathbf{Z}\end{cases}$$

where $\mathbf{Z}$ *denotes the set of integers.*

Proof. Due to symmetry we may again assume that $k\leqslant[n/2]$. Direct computations show that

$$\left\{\binom{n}{k}-\binom{n}{k-1}\right\}-\left\{\binom{n}{k-1}-\binom{n}{k-2}\right\}=\frac{n!\,\sigma}{k!\,(n-k+2)!}$$

where

$$\begin{aligned}\sigma&=(n-k+2)(n-2k+1)-k(n-2k+3)\\&=(n-2k+1)(n-2k+2)-2k\\&=(2k)^2-(2n+4)(2k)+(n+1)(n+2)\\&=\{2k-(n+2)\}^2-(n+2)\\&=\{2k-(n+2)+\sqrt{n+2}\}\{2k-(n+2)-\sqrt{n+2}\}\\&=4(k-\tau)\left\{k-\frac{1}{2}(n+2+\sqrt{n+2})\right\},\end{aligned}$$

where $\tau=(1/2)(n+2-\sqrt{n+2})$. Since $k\leqslant[n/2]<(1/2)(n+2+\sqrt{n+2})$ we conclude that

$$\binom{n}{k}-\binom{n}{k-1}\geqslant\binom{n}{k-1}-\binom{n}{k-2}$$

if and only if $k\leqslant\tau$ with equality holding if and only if $\tau\in\mathbf{Z}$. The statement of our theorem now follows immediately.

To illustrate Theorem 2, we consider an example.

Example. When $n=13$, $\tau=\frac{1}{2}(15-\sqrt{15})\notin\mathbf{Z}$ and $[\tau]=5$. Thus the largest gap in the 13th row of Pascal's triangle occurs twice: once between $\binom{13}{4}$ and $\binom{13}{5}$, and the other between $\binom{13}{8}$ and $\binom{13}{9}$. These gaps have absolute value $\binom{13}{5}-\binom{13}{4}=1287-715=572$. When $n=14$, $\tau=6\in\mathbf{Z}$. Thus the largest gaps in the 14th row of Pascal's triangle occur four times: between $\binom{14}{4}$ and $\binom{14}{5}$; $\binom{14}{5}$ and $\binom{14}{6}$; $\binom{14}{8}$ and $\binom{14}{9}$; and between $\binom{14}{9}$ and $\binom{14}{10}$. The absolute value of these gaps is $\binom{14}{5}-\binom{14}{4}=2002-1001=1001$.

Theorem 2 also indicates an interesting fact regarding consecutive binomial coefficients which form an arithmetic progression. It is well known [1, p. 54] that no four consecutive binomial coefficients can form an arithmetic progression. Our result in Theorem 2 implies that there are infinitely many triples of consecutive binomial coefficients which form an arithmetic progression. In fact, we have:

COROLLARY. *If* $n>2$ *such that* $n+2$ *is a perfect square, then* $\binom{n}{\tau-2}$,

$\binom{n}{\tau-1}$, *and* $\binom{n}{\tau}$ *form an arithmetic progression where* $\tau = (1/2)(n+2-\sqrt{n+2})$. *Furthermore, the common difference of this arithmetic progression yields the largest gap in the nth row of Pascal's triangle.*

In closing, we point out that if, in the left half of each row in Pascal's triangle, we use a dot to represent the larger one of the binomial coefficients whenever a largest gap occurs and connect the dots in consecutive rows with line segments, then we would obtain a chain which has some kind of zigzag pattern and which has a "diamond" whenever $n > 2$ is such that $n + 2$ is a perfect square. We leave it to the readers to do the actual drawing.

Acknowledgement. This article was written while the first author was visiting Wilfrid Laurier University, July 1987 – April 1988. The hospitality of WLU is greatly appreciated. This research was supported by the Natural Sciences and Engineering Research Council of Canada under grant A9121.

Reference

[1] Daniel I. A. Cohen, *Basic Techniques in Combinatorial Theory*, John Wiley & Sons, New York, 1978.

也谈取棋子游戏

张慧欣在文[1]中提出一个取棋子的游戏：

游戏 A 有若干堆棋子，两人轮流取棋子，每人每次只能从 1 堆中取 1—4 个棋子，取到最后一个棋子者获胜.

文中提出这个游戏是否有必胜策略的问题，后来被牛伟强[2]解决.

游戏 A 中每次取 1—4 个棋子的规定过于特殊，自然地应该改为“每次取 1—k 个棋子”，其中 k 是一个可以任意预先商定的正整数. 本文将证明，这样的游戏的必胜策略问题仍是可以解决的，特别地这也给出文[2]中结果的另一个证明.

首先我们来说明下面一个游戏的必胜策略：

游戏 B 有若干堆棋子，甲乙两人轮流取棋子（第一次由甲取棋子），每人每次只能从 1 堆中取至少 1 个棋子（取子数没有上限），取到最后一个棋子者获胜.

这个游戏的必胜策略问题早已解决，至少可以在 60 余年前的杂志上查到. 下面我们用现代的语言叙述这个问题的答案并给出证明，这样比较简单些. 我们需要用到二进制数，对二进制数我们用 $\dot{+}$ 记半加法，即对任意两个二进制数 b 和 b'，$b \dot{+} b'$ 为将 b 和 b' 的各位数字分别相加，但不进位，例如 $10101 \dot{+} 1111 = 11010$. 显然半加法满足交换律和结合律，此外对任意二进制数 b 有 $b \dot{+} b = 0$.

设游戏中某时刻共有 n 堆棋子，各堆棋子的个数分别为 $a_1, \cdots, a_n$，将棋子个数用二进制表出，为清楚起见将 a_i 对应的二进制数记为 $b_i (1 \leqslant i \leqslant n)$. 令

$$b = b_i \dot{+} \cdots \dot{+} b_n, \tag{1}$$

称为所有棋子的总半和.

命题 1 在游戏 B 的进程中，如果遇到棋子的总半和为 0 的情形，则无论怎样取棋子，余下的棋子的总半和一定不为 0；反之，如果遇到棋子的总半和不为 0 的情形，则总有办法取棋子使得余下的棋子的总半和为 0.

因此,如果开始时总半和为 0,则乙有必胜的策略;反之则甲有必胜的策略.

证明　设进程中棋子的总半和为 b,此时如果从第 i 堆中取棋子使得第 i 堆还剩 $a'_i<a_i$ 个棋子(其余各堆棋子的个数当然不变),记 b'_i 为 a'_i 所对应的二进制数,则余下的棋下的总半和为

$$b'=b_1\dot{+}\cdots\dot{+}b_{i-1}\dot{+}b'_i\dot{+}b_{i+1}\dot{+}\cdots\dot{+}b_n=b_1\dot{+}\cdots\dot{+}b_n\dot{+}b_i\dot{+}b'_i=b\dot{+}b_i\dot{+}b'_i,\tag{2}$$

由于 $a_i>a'_i$,有 $b_i\neq b'_i$,故 $b_i\dot{+}b'_i\neq 0$,这说明 $b'\neq b$,因此若 $b=0$ 则 $b'\neq 0$.

若 $b\neq 0$,设 b 为 r 位数,则必有一个 b_i 的(右数)第 r 位数字为 1,从而 $b'_i=b_i\dot{+}b$ 的第 r 位数字为 0 而更高位数字均与 b_i 相同,故 b'_i 对应的整数 $a'_i<a_i$. 在第 i 堆中取 $a_i-a'_i$ 个棋子,则由(2)可见余下棋子的总半和为 $b\dot{+}b_i\dot{+}b_i\dot{+}b=0$.

最后,如果开始时总半和为 0,则由上所述,乙可以保持每次取子后余下棋子的总半和仍为 0,直到所有棋子取完为止,注意甲取子后余下棋子的总半和非 0,而总半和非 0 时棋子数必非零,所以甲无论何时也不可能将棋子取尽,换言之最后一次必为乙取子;反之,如果开始时总半和非零,则甲第一次可取子使得余下棋子的总半和为 0,故与上同理甲可保证取到最后一个棋子. 证毕.

有趣的是,如果将游戏 B 中的胜负规则改为相反,即规定取最后一个棋子者输,则上面的必胜策略仍然适用! 为方便起见,记:

游戏$\bar{\mathrm{B}}$　有若干堆棋子,甲乙两人轮流取棋子(第一次由甲取棋子),每人每次只能从 1 堆中取至少 1 个棋子(取子数没有上限),取最后一个棋子者输.

则有:

推论 1　对游戏$\bar{\mathrm{B}}$假设开始时至少有 1 堆棋子的个数 >1. 如果开始时总半和为 0,则乙有必胜的策略;反之则甲有必胜的策略.

证明　不妨设开始时总半和为 0,否则由命题 1 可以类似地证明.

乙的必胜策略与游戏 B 基本相同,即保持总半和为 0,所不同的是在游戏接近尾声时,即只剩下一堆棋子个数 >1 时,乙取该堆棋子使得余下棋子的总半和为 1,这总是可以做到的:设此时共有 n 堆棋子,其中第 i 堆的棋子个数 >1 而其余各堆棋子的个数都是 1,则在 n 为奇数时取第 i 堆的棋子留下 1 个,而在 n 为偶数时取尽第 i 堆的棋子. 这样轮到甲取子时有奇数堆棋子每堆 1 个,当然甲必

取最后一个棋子无疑. 证毕.

下面我们来讨论开始时所说的游戏,即:

游戏C 设k为正整数. 有若干堆棋子,甲乙两人轮流取棋子(第一次由甲取棋子),每人每次只能从1堆中取$1\sim k$个棋子,取到最后一个棋子者获胜.

设游戏中某一时刻共有n堆棋子,各堆棋子的个数分别为$a_1, \cdots, a_n$,令$\bar{a}_i$为a_i除以$k+1$所得的余数,将$\bar{a}_i$对应的二进制数记为$\bar{b}_i(1\leqslant i\leqslant n)$. 令

$$\bar{b}=\bar{b}_1\dot{+}\cdots\dot{+}\bar{b}_n, \tag{3}$$

称为所有棋子的余半和.

命题2 在游戏C的进程中,如果遇到棋子的余半和为0的情形,则无论怎样取棋子,余下的棋子的余半和一定不为0;反之,如果遇到棋子的余半和不为0的情形,则总有办法取棋子使得余下的棋子的余半和为0.

因此,如果开始时余半和为0,则乙有必胜的策略;反之则甲有必胜的策略.

证明 设进程中棋子的余半和为$\bar{b}$,此时如果从第i堆中取棋子使得第i堆还剩$a_i'<a_i$个棋子(其余各堆棋子的个数不变),令$a_i=c(k+1)+\bar{a}_i$, $a_i'=c'(k+1)+\bar{a}_i'$, 且令$\bar{b}_i'$为$\bar{a}_i'$所对应的二进制数$(1\leqslant i\leqslant n)$,

$$b'=\bar{b}_1\dot{+}\cdots\dot{+}\bar{b}_{i-1}\dot{+}\bar{b}_i'\dot{+}\bar{b}_{i+1}\dot{+}\cdots\dot{+}\bar{b}_n, \tag{4}$$

由游戏规则有$1\leqslant a_i-a_i'\leqslant k$,故$c\geqslant c'\geqslant c-1$,这样就有两种可能的情形:

情形1: $c'=c$,此时有$\bar{a}_i'<\bar{a}_i$,故$\bar{b}_i\neq\bar{b}_i'$, 与(2)类似地有$\bar{b}'=\bar{b}\dot{+}\bar{b}_i\dot{+}\bar{b}_i'\neq\bar{b}$.

情形2: $c'=c-1$,此时有$\bar{a}_i'>\bar{a}_i$,故仍有$\bar{b}_i\neq\bar{b}_i'$,从而$\bar{b}'=\bar{b}\dot{+}\bar{b}_i\dot{+}\bar{b}_i'\neq\bar{b}$.

因此,若$\bar{b}=0$则$\bar{b}'\neq0$.

如果$\bar{b}\neq0$, 设$\bar{b}$为r位数,则必有一个$\bar{b}_i$的(右数)第r位数字为1,从而$\bar{b}_i'=\bar{b}_i\dot{+}\bar{b}$的第$r$位数字为0而更高位数字均与$\bar{b}_i$相同,故$\bar{b}_i'$所对应的整数$\bar{a}_i'<\bar{a}_i\leqslant k$. 在第$i$堆中取$d=\bar{a}_i-\bar{a}_i'$个棋子(注意$1\leqslant d\leqslant k$),则与(2)类似地可见余下的棋子的余半和为$\bar{b}\dot{+}\bar{b}_i\dot{+}\bar{b}_i\dot{+}\bar{b}=0$.

最后,如果开始时余半和为0,则由上所述乙可以保持每次取子后余下棋子

的余半和仍为 0,直到所有棋子取完为止,注意甲取子后余下棋子的余半和非零,而余半和非零时棋子数必非零,所以甲无论何时也不可能将棋子取尽,换言之最后一次必为乙取子;反之,如果开始时余半和非零,则甲第一次可取子使得余下棋子的余半和为 0,故与上同理,甲可保证取到最后一个棋子. 证毕.

注　命题 2 的证明中已具体给出一种必胜策略,但这不是唯一的必胜策略. 对于实际遇到的 $\bar{b}\neq 0$ 的情形,经常有不止一种取法可将余半和化为 0,上面只是给出一种而已. 例如在 $k>1$ 时,有两堆棋子,一堆有 $k+1$ 个另一堆有 2 个,可以将 2 个一堆的取尽,也可以将 $k+1$ 个一堆的取出 $k-1$ 个.

如果甲乙两人只有一人懂得必胜策略,则在实际比赛中懂得必胜策略的一方基本上可保必胜,因为在开始时如果有很多堆棋子每堆个数很多,则取法非常多,不懂必胜策略的一方"碰巧" 每次都按必胜策略取子的概率实在太小了,总会让懂得必胜策略的一方得到机会的.

在 $k=4$ 的情形,命题 2 给出的结果和取法与[2] 一致.

与游戏$\bar{\text{B}}$类似地可建立:

游戏$\bar{\textbf{C}}$　设 k 为正整数. 有若干堆棋子,甲乙两人轮流取棋子(第一次由甲取棋子),每人每次只能从 1 堆中取 $1\sim k$ 个棋子,取最后一个棋子者输.

推论 2　对游戏$\bar{\text{C}}$,假设开始时至少有 1 堆棋子的个数模 $k+1$ 的余数 >1. 如果开始时余半和为 0,则乙有必胜的策略;反之则甲有必胜的策略.

证明　不妨设开始时余半和为 0,否则由命题 2 可以类似地证明.

乙的必胜策略与游戏 C 基本相同,即保持余半和为 0,所不同的是在只有一堆棋子的个数模 $k+1$ 的余数 >1 时,乙取该堆棋子使得余下棋子的余半和为 1,这总是可以做到的:设此时共有 n 堆棋子,其中第 i 堆棋子的个数模 $k+1$ 的余数 >1,而其余各堆棋子的个数模 $k+1$ 的余数都是 1 或 0,则余半和 $\bar{b}>1$,注意 $\bar{b}'=\bar{b}\dotplus\bar{b}_i$ 等于 0 或 1,令 $\bar{b}_i'\dotplus 1$ 所对应的整数为 $\bar{a}_i'$,则 $\bar{a}_i'<\bar{a}_i$ 且 $d=\bar{a}_i-\bar{a}_i'\leqslant k$, 从第 i 堆取 d 个棋子就使得余下棋子的余半和为 1. 此后乙可保持在取子后余下棋子的余半和为 1. 注意余半和为 1 等价于有奇数堆棋子的个数模 $k+1$ 余 1 而其余各堆棋子的个数均为 $k+1$ 的倍数,此时任一堆棋子的个数若 >1,则 $\geqslant k+1$, 甲的取子有下列几种可能的情形:

情形 1:甲在某堆取 d 个棋子而该堆棋子的个数 >1(特别地当 $d>1$ 时总

是如此),则乙可在同一堆取 $k+1-d$ 个棋子;

情形2:甲在某堆取1个棋子,而该堆只有1个棋子,但还有另一堆,此时若另一堆棋子的个数模 $k+1$ 余1,则乙可在该堆取1个棋子;而若另一堆棋子的个数是 $k+1$ 的倍数,则乙可在该堆取 k 个棋子;

情形3:甲在某堆取1个棋子而该堆只有1个棋子,但没有另一堆,此时游戏已经结束.

总之,乙可继续保持余半和为1状态,直到甲取最后一个棋子.证毕.

顺便指出,有另一种游戏的数学原理与上面的取棋子游戏等价,就是所谓"自由棋"(参看[3]),这种棋的棋盘只有一排格子(格数不限),棋子只有一种(个数不限),对弈时任取一些棋子放在一些格中(一个格里可放多枚棋子),然后"弈者双方递行一着",每人每次取一枚棋子按指定的方向前进至少1格,直到所有棋子都走到最后一格为止,走最后一步者获胜.如果每次走的格数没有上限,则这种自由棋的数学原理与前面的游戏B等价;如果规定每次至多走 k 格,则数学原理与游戏C等价.若将胜负规则改为"走最后一步者输",则在每次走的格数没有上限时这种自由棋的数学原理与游戏 $\bar{B}$ 等价;而在每次至多走 k 格时数学原理与游戏 $\bar{C}$ 等价.道理很简单:如果将棋盘的格子从最后一格开始往前依次标号0, 1, 2, ….(见下表),那么在开始放棋子时就可以将第 i 格中的一枚棋子对应于游戏B或C中的一堆 i 个棋子(格中放了几个棋子就对应于游戏B或C中有几堆棋子),而将一个棋子前进 d 格对应于相应的那堆棋子取出 d 个,不难看出这样就建立了自由棋与取棋子游戏之间的数学等价.因此,命题1和命题2也给出了自由棋的必胜策略.

n	$n-1$		1	0

参考文献

[1] 张慧欣.数学中的几个小游戏[J].数学通报,2008(11).
[2] 王跃进,牛伟强.数学中的几个小游戏遗留问题的探索[J].数学通报,2009(11).
[3] 尹裕.自由棋[N].小学生数学报,1996.

一个古老等式的组合证明

摘要　采用组合的方法,对 $\sum_{k=0}^{n}\binom{n+k}{k}\binom{n}{k}(m-1)^{n-k}=\sum_{k=0}^{n}\binom{n}{k}^{2}m^{k}$ 这一等式提供了一种全新的证明. 此外,还提供了一种完全不用微积分的代数证明.

在国际象棋棋盘上,从点$(0,\ 0)$到点$(n,\ n)$,如果只允许横向的单步移动(比如从$(x,\ y)$移到$(x+1,\ y)$)和纵向的单步移动(比如从$(x,\ y)$移到$(x,\ y+1)$),那么所有可能路径的总和应该是 $a_n=\binom{2n}{n}$. 如果除了横向和纵向的单步移动之外,还允许斜线移动(比如从$(x,\ y)$移到$(x+1,\ y+1)$),那么 $b_n=\sum_{k=0}^{n}\binom{n+k}{k}\binom{n}{k}$ 将是所有这些路径的总和.

这两个古老组合的证明非常有趣:假定从点$(0,\ 0)$移到点$(n,\ n)$,有一种路径总共包括 k 步水平和 1 步垂直以及 m 步斜线移动,那么 $k+m=1+m=n$. 即 $1=k,\ m=n-k$.

因此,这条路径一共有 $k+k+(n-k)=n+k$ 次移动. 在这 $n+k$ 次移动中包含有 k 次水平方向的移动,而其余的 n 次移动中包含有 k 次垂直方向的移动. 于是 $a_k=\binom{n+k}{k}\binom{n}{k}$,所有可能的路径为 $\sum_{k=0}^{n}a_k=b_n$.

显然,如果没有斜线方向的移动, $a_k=\binom{n+k}{k}\binom{n}{k}=\binom{2n}{n}$ 就是所有这些路径的总和.

下面是有关 b_n 的两个定理:

(Ⅰ) $\sum_{k=0}^{n}\binom{n+k}{k}\binom{n}{k}(x-1)^{n-k}=\sum_{k=0}^{n}\binom{n}{k}^{2}x^{k}$;

(Ⅱ) $b_n=\sum_{k=0}^{n}\binom{n}{k}^2 2^k$.

当 $x=2$ 时,由(Ⅰ)可以直接得出(Ⅱ).

文献[5]用代数方法对上述定理进行了证明,下面是其简要的证明过程.

证 设 $p_n(x)=\sum_{k=0}^{n}\binom{n}{k}^2 x^k$, $p_n(x)$的 k 阶导数为 $p_n^{(k)}(x)$. 由于 $(x^m)^{(k)}=k!\binom{m}{k}x^{m-k}$,从而有

$$p_n^{(k)}(x)=\sum_{j=k}^{n}\binom{n}{j}^2 k!\binom{j}{k}x^{j-k}.$$

于是

$$\begin{aligned}\frac{p_n^{(k)}(1)}{k!}&=\sum_{j=k}^{n}\binom{n}{j}\binom{n}{j}\binom{j}{k}=\sum_{j=k}^{n}\binom{n}{j}\binom{n}{k}\binom{n-k}{n-j}\\&=\binom{n}{k}\sum_{j=k}^{n}\binom{n}{j}\binom{n-k}{n-j}=\binom{n}{k}\binom{2n-k}{n}.\quad \text{(Vandermonde 定理)}\end{aligned}$$

从而

$$\frac{p_n^{(n-k)}(1)}{(n-k)!}=\binom{n}{n-k}\binom{n+k}{n}=\binom{n}{k}\binom{n+k}{n}.\tag{1}$$

另一方面,由 Taylor 展开公式,可得

$$p_n(x)=\sum_{k=0}^{n}\frac{p_n^{(k)}(1)}{k!}(x-1)^k=\sum_{k=0}^{n}\frac{p_n^{(n-k)}(1)}{(n-k)!}(x-1)^{n-k}.$$

将(1)式代入上式,于是(Ⅰ)式得证.

显然,当 $x=1$ 时,(Ⅰ)式变为 $\binom{2n}{n}=p_n(1)$,即 $a_n=p_n(1)=\sum_{k=0}^{n}\binom{n}{k}^2$;当 $x=2$ 时, $b_n=p_n(2)=\sum_{k=0}^{n}\binom{n}{k}^2 2^k$. 多项式 $p_n(x)$神奇地将 a_n 和 b_n 联系在了一起.

文献[5]在文末希望用组合的方法对(Ⅱ)进行说明,本文就此给出一个生

动的组合证法,并对(Ⅰ)给出一种完全不用微积分的代数证明.此外,还分别对这两个等式做了进一步扩展.

构造　从 n 个男生和 n 个女生中选出 n 人,并有若干人担任委员,但男生不得充任委员.选法共有多少种呢?

组合证明　考虑先从 n 个女生中选出 j 人,再从 n 个男生中选出 $n-j$ 人.由于每个女生都有当委员和不当委员两种选择,所以选法应该是 $\sum\limits_{j=0}^{n}\binom{n}{n-j}\binom{n}{j}2^j=\sum\limits_{j=0}^{n}\binom{n}{j}^2 2^j$,即 $\sum\limits_{k=0}^{n}\binom{n}{k}^2 2^k$ 种.

另一方面,先从 n 女中选 k 人充当委员,再在余下的 $2n-k$ 人中选 $n-k$ 人,这种选法应该是 $\sum\limits_{k=0}^{n}\binom{n}{k}\binom{2n-k}{n-k}$ 种.

因为

$$\sum_{k=0}^{n}\binom{n+k}{k}\binom{n}{k}=\sum_{k=0}^{n}\binom{2n-k}{n-k}\binom{n}{n-k}=\sum_{k=0}^{n}\binom{2n-k}{n-k}\binom{n}{k},$$

故(Ⅱ)式成立.

推广 1　对所有的整数 m,

(Ⅲ)　$$\sum_{k=0}^{n}\binom{n+k}{k}\binom{n}{k}(m-1)^{n-k}=\sum_{k=0}^{n}\binom{n}{k}^2 m^k.$$

证　当 $m=1$ 时,考虑从 n 女 n 男中选 n 人的选法.先从 n 个女生中选出 k 人,再从 n 个男生中选出 $n-k$ 人.选法应该是 $\sum\limits_{k=0}^{n}\binom{n}{n-k}\binom{n}{k}=\sum\limits_{k=0}^{n}\binom{n}{k}^2$ 种.

另一方面,先从 n 个女生中选出 k 人,再在余下的 $2n-k$ 人中选出 $n-k$ 人.选法应该是 $\sum\limits_{k=0}^{n}\binom{2n-k}{n-k}\binom{n}{k}=\sum\limits_{k=0}^{n}\binom{n+k}{k}\binom{n}{n-k}$ 种.因此上式成立.

当 $m>1$ 时,考虑从 n 女 n 男中选 n 人,并有若干人分别担任第一委员,第二委员,……,第 $m-1$ 委员,但男生不得充任委员.

一方面,先从 n 个女生中选出 j 人,再从 n 个男生中选出 $n-j$ 人.由于每个女生都有不当委员,当第一委员,第二委员,……,第 $m-1$ 委员共 m 种可能,所

以选法应该是 $\sum_{j=0}^{n}\binom{n}{n-j}\binom{n}{j}m^j=\sum_{j=0}^{n}\binom{n}{j}^2 m^j$，即 $\sum_{k=0}^{n}\binom{n}{k}^2 m^k$ 种.

另一方面，先从 n 个女生中选出 k 人充当委员，再在余下的 $2n-k$ 人中选 $n-k$ 个人. 由于这 k 个女生都有充当第一委员，第二委员，…，第 $m-1$ 委员共 $m-1$ 种可能，所以选法应该是 $\sum_{k=0}^{n}\binom{2n-k}{n-k}\binom{n}{k}(m-1)^k$ 种.

而

$$\begin{aligned}\sum_{k=0}^{n}\binom{2n-k}{n-k}\binom{n}{k}(m-1)^k&=\sum_{k=0}^{n}\binom{2n-k}{n-k}\binom{n}{n-k}(m-1)^k\\&=\sum_{k=0}^{n}\binom{n+k}{k}\binom{n}{k}(m-1)^{n-k},\end{aligned}$$

所以对一切整数 m，(Ⅲ)式都成立.

引理 $\sum_{k=0}^{n-h}(-1)^{n-h-k}\binom{n-h}{k}\binom{n+k}{k}=\binom{n}{h}$.

证 等式的左边为 $\sum_{k=0}^{n-h}(-1)^{n-h-k}\binom{n-h}{k}(1+x)^{n+k}$ 中 x^n 的系数，即 $((1+x)-1)^{n-h}(1+x)^n$ 中 x^n 的系数. 因而是 $(1+x)^n$ 中 x^h 的系数 $\binom{n}{h}$.

推广 2 将整数 m 进一步扩展，(Ⅲ)式就推广到了(Ⅰ)式.

下面对(Ⅰ)式给出一个完全不用微积分的新证明.

证 左边 $=\sum_{k=0}^{n}\binom{n+k}{k}\binom{n}{k}(x-1)^{n-k}=\sum_{k=0}^{n}\binom{n+k}{k}\binom{n}{k}\sum_{h=0}^{n-k}(-1)^{n-k-h}\binom{n-k}{h}x^h$

$$\begin{aligned}&=\sum_{h=0}^{n}x^h\sum_{k=0}^{n-h}\binom{n+k}{k}(-1)^{n-k-h}\binom{n-k}{h}\binom{n}{k}\\&=\sum_{h=0}^{n}x^h\binom{n}{h}\sum_{k=0}^{n-h}(-1)^{n-k-h}\binom{n+k}{k}\binom{n-h}{k}=\sum_{h=0}^{n}\binom{n}{h}^2x^h\\&=\text{右边}.\end{aligned}$$

最后一步利用了引理.

推广 3 当 $x=\theta=\mathrm{e}^{\frac{\pi}{3}}$ 时，有 $\theta^2-\theta+1=0$.

在 $3|n$ 时,因为 $(\theta-1)^{n-k}=\theta^{2(n-k)}=\theta^{-2k}=(-\theta)^k$,有

$$(\text{IV})\qquad \sum_{k=0}^{n}\binom{n+k}{k}\binom{n}{k}(-\theta)^k=\sum_{k=0}^{n}\binom{n}{k}^2\theta^k.$$

参考文献

[1] Lawden D F. On the solution of linear difference equations [J]. Math. Gaz. XXXVI 1952: 193-196.

[2] Moser L. King paths on a chessboard [J]. Math. Gaz. XXXIX 1955: 54.

[3] Grassl R, Mingus T. Equivalence classes and a familiar combinatorial identity [J]. Math. Gaz. 1998, (82): 98-100.

[4] Simonsl S. A curious identity [J]. Math. Gaz, 2001, (85): 296-298.

[5] Bataille M. A curious identity [J]. Math. Gaz, 2003, (87): 144-148.

A Combinatorial Method for an Old Equation

Abstract The aim of this note is to obtain by combinatorial method the equation $\sum_{k=0}^{n}\binom{n+k}{k}\binom{n}{k}(m-1)^{n-k}=\sum_{k=0}^{n}\binom{n}{k}^2 m^k$. Furthermore, this note also provides a new algebraic proof other than complex calculus proof for the sum.

1994年全国理科试验班招生统一考试（数学）

第一试试卷（1994年7月17日，100分钟）

一、选择题：

1. 直线 $y=\frac{1}{2}x+k$ 与 x 轴、y 轴的交点分别是 A、B. 如果 $\triangle AOB$ 的面积 $S_{\triangle AOB}\leqslant 1$, 那么 k 的取值范围是（　）

(A) $k\geqslant -1$　　(B) $k\leqslant 1$

(C) $-1\leqslant k\leqslant 1$　　(D) $k\geqslant 1$ 或 $\leqslant -1$

2. 如图1，已知边长为4的正方形截去一角成为五边形 $ABCDE$，其中 $AF=2$, $BF=1$. 在 AB 上的一点 P，使矩形 $PNDM$ 有最大面积，则矩形 $PNDM$ 的面积的最大值是（　）

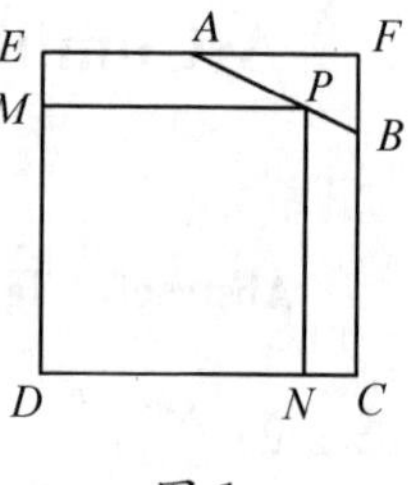

图 1

(A) 8　　(B) 12

(C) $\frac{25}{2}$　　(D) 14

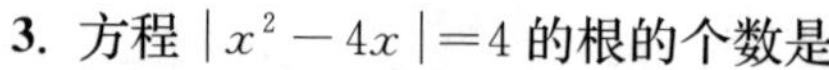

3. 方程 $|x^2-4x|=4$ 的根的个数是（　）

(A) 1　　(B) 2

(C) 3　　(D) 4

4. 设有含铜百分率不同的两块合金，甲重40千克，乙重60千克，从这两块合金上切下重量相等的一块，并把所切下的每块与另一种剩余的合金加在一起，熔炼后两者含铜的百分率相等，则切下来的合金的重量是（　）

(A) 12千克　　(B) 15千克

(C) 18千克　　(D) 24千克

5. 如图 2,已知四边形 $ABCD$ 的对角线 AC、BD 交于 O,且 $S_{\triangle ABC}=5$, $S_{\triangle BCD}=9$, $S_{\triangle CDA}=10$, $S_{\triangle ADB}=6$,则 $S_{\triangle ABO}=$ (　　)

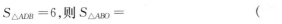

(A) 2　　(B) 4

(C) 6　　(D) 8

图 2

6. 若方程 $x^2-2ax+4a-3=0$ 的两根均大于 1,则实数 a 的取值范围是 (　　)

(A) $a\geqslant 1$　　(B) $0<a\leqslant 3$

(C) $a\leqslant 3$　　(D) $a\geqslant 3$

7. $\triangle ABC$ 中, E、F 将 BC 三等分, BM 为中线,连接 AE、AF 分别交 BM 于 G、H, G 在 B、H 之间,则 $BG:GH:HM=$ (　　)

(A) $4:3:2$　　(B) $5:3:2$　　(C) $4:2:3$　　(D) $3:3:2$

8. 有序数对的运算 $*$ 定义为:

$$(a,\ b)*(c,\ d)=(ac+bd,\ ad+bc).$$

如果对所有的 $(a,\ b)$ 均有 $(a,\ b)*(x,\ y)=(a,\ b)$, 那么 $(x,\ y)$ 是 (　　)

(A) $(0,\ 0)$　　(B) $(1,\ 0)$　　(C) $(0,\ 1)$　　(D) $(1,\ 1)$

二、填空题:

1. 当 $x=1-\sqrt{2}$ 时,代数式

$$\frac{x}{x^2+a^2-x\sqrt{x^2+a^2}}+\frac{2x-\sqrt{x^2+a^2}}{x^2-x\sqrt{x^2+a^2}}+\frac{1}{\sqrt{x^2+a^2}}$$

的值是________.

2. 若方程 $x^2+(m-2)x+(m^2-7)=0$ 的两实根平方和最大,则实数 m 的值为________.

3. 若 $\sqrt{x}=\dfrac{1-a}{2}$(a 是不大于 1 的实数),且化简 $\sqrt{x+a}-\sqrt{x-a+2}$ 的结果为 -2,则字母 a 的取值范围是________.

4. 已知 $\triangle ABC$ 中, $\angle B=45^\circ$, D 是 BC 上一点, $AD=5$, $AC=7$, $DC=3$, 则 AB 的长为________.

5. 当 $x=\frac{2}{\sqrt{3}-1}$ 时，$y=\frac{1}{2}x^3-x^2-x+1$ 的值是________.

6. 如图 3,扇形 OAB 的半径为 2，$\angle AOB$ 为直角，M 是以 OB 为直径的半圆的圆心，$MP \parallel OA$，MP 与半圆相交于 N 点,则图中阴影部分的面积为________.

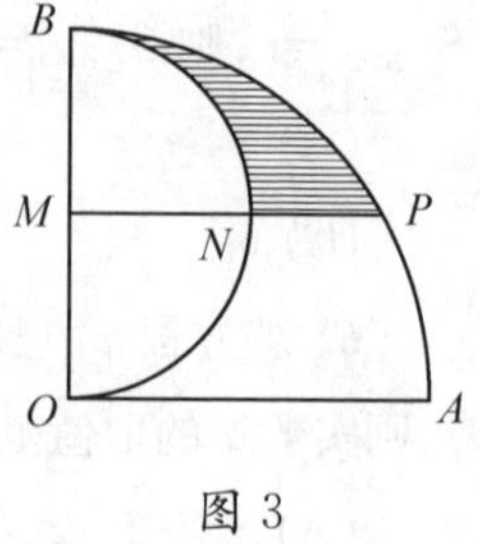

图 3

7. 设 $\sqrt{27-10\sqrt{2}}=a+b$,其中 a 为正整数，b 在 0，1 之间,则 $\frac{a+b}{a-b}=$________.

8. 已知 a 为非负整数,若关于 x 的方程 $2x-a\sqrt{1-x}-a+4=0$ 至少有一个整数根,则 a 可取的值为________.

三、解答题:

1. 在周长为 300 cm 的圆周上有甲、乙两球,以大小不等的速度作匀速圆周运动,甲球从 A 点出发,按顺时针方向运动,乙球同时从 B 点出发,按逆时针方向运动,两球相遇于 C 点,相遇后,两球各自反向作匀速圆周运动,但返时甲球速度是原来的 2 倍,乙球速度是原来的一半,它们第二次相遇于 D 点(如图 4). 已知 $\overset{\frown}{AMC}=40\,\text{cm}$，$\overset{\frown}{BND}=20\,\text{cm}$，求 $\overset{\frown}{ACB}$ 的长度.

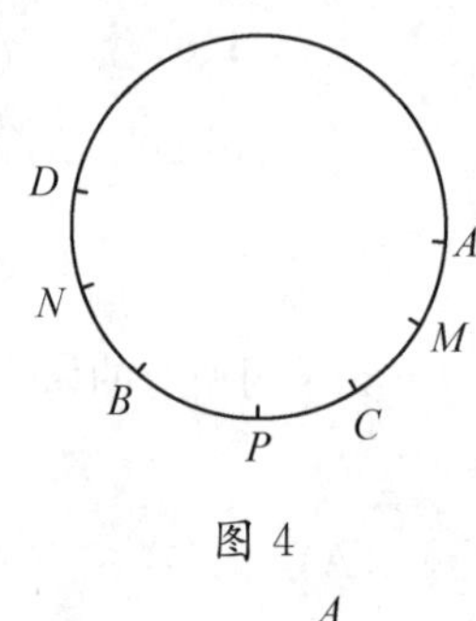

图 4

2. 如图 5,已知 $ABCD$ 是圆内接四边形,对角线 AC、BD 交于一点 E,且 $BE=DE$. 求证:$AB^2+BC^2+CD^2+DA^2=2AC^2$.

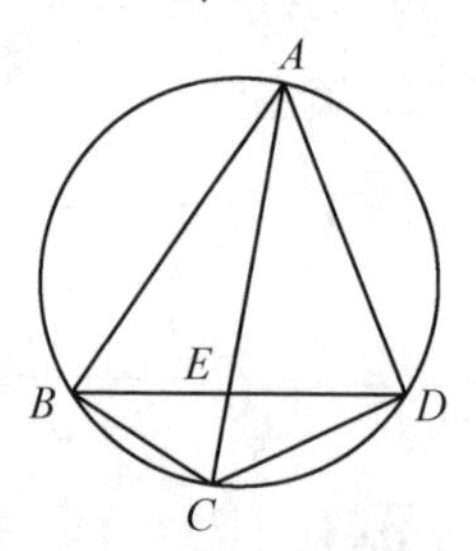

图 5

第二试试卷（1994 年 7 月 17 日，80 分钟）

1. 用 d_k 表示抛物线 $y=k(k+1)x^2-(2k+1)x+1$ 在 x 轴上截得的线段的长,求 $d_1+d_2+\cdots+d_{999}$ 的值.

2. 如果对于自然数 n，7^n 的个位数字用 a_n 表示,那么:(1) 求 a_{1994} 的值;(2) 当 n 为什么数时，$-n^2+2na_n$ 取得最大值? 并求出这个最大值.

3. 在 $n\times n$ 的正方形表格中,写上非负整数,如果在某一行和某一列的交汇处的数是 0,那么该行和该列上各数的和不小于 n. 证明:表中所有数的和不小于$\frac{n^2}{2}$.

4. $\odot O$与$\odot O'$外切于F,直线AB切$\odot O$于A,切$\odot O'$于B,直线$CE\parallel AB$,切$\odot O'$于C,交$\odot O$于D、E,证明:(1) A、F、C三点共线;(2) $\triangle ABC$的外接圆与$\triangle BDE$的外接圆的公共弦通过点F.

答　案

一试

一、**1.** 直线在x轴、y轴的截距分别为$-2k$、k.所以$S_{\triangle AOC}=\frac{1}{2}\cdot|-2k|\cdot|k|=k^2\leqslant 1$. 答案为(C).

2. 以D为原点,直线DC、DE为x、y轴建立直角坐标系. A、B坐标分别为$(2, 4)$、$(4, 3)$. AB方程为$\frac{y-4}{-1}=\frac{x-2}{2}$,即$x=10-2y$. $S_{PNDM}=xy=2y(5-y)$,在$y\geqslant\frac{5}{2}$时递减.由于$3\leqslant y\leqslant 4$,所以在$y=3$即P与B重合时,S_{PNDM}取最大值12.答案为(B).

3. 由$x^2-4x=4$得$x=2\pm 2\sqrt{2}$.由$x^2-4x=-4$得$x=2$. 答案为(C).

4. 设切下合金重x千克.又设甲、乙两种合金含铜率分别为a、b, $a\neq b$. 由题意,$\frac{ax+b(60-x)}{60}=\frac{a(40-x)+bx}{40}$.化简得$100-(a-b)x=40\cdot 60(a-b)$.由于$a\neq b$,所以$x=24$. 答案为(D).

5. $\frac{BO}{OD}=\frac{S_{\triangle ABC}}{S_{\triangle CDA}}=\frac{5}{10}=\frac{1}{2}$. $S_{\triangle ABO}=\frac{1}{1+2}S_{\triangle ABD}=\frac{1}{3}\times 6=2$. 答案为(A).

6. 由判别式$\geqslant 0$得$(a-1)(a-3)\geqslant 0$(*).又两根之和$2a>2$,所以$a>1$. 结合(*)得$a\geqslant 3$. 在$a\geqslant 3$时,小根$a-\sqrt{(a-1)(a-3)}=1+\sqrt{a-1}(\sqrt{a-1}-\sqrt{a-3})>1$. 答案为(D).

7. 过M作BC的平行线,分别交AE、AF于P、Q. $\frac{BG}{GM}=\frac{BE}{PM}=\frac{BE}{EC}\cdot\frac{EC}{PM}=\frac{BE}{EC}\cdot\frac{AC}{AM}=1$, $\frac{BH}{HM}=\frac{BF}{HM}=\frac{BF}{FC}\cdot\frac{FC}{HM}=\frac{BF}{FC}\cdot\frac{AC}{AM}=4$,所以$BG=4HM-GH$

$=GH+HM=GM$, $2GH=3HM$. 答案为(B).

8. 答案为(B).

二、**1.** 原式$=\left(\dfrac{x}{(\sqrt{x^2+a^2}-x)\sqrt{x^2+a^2}}+\dfrac{1}{\sqrt{x^2+a^2}}\right)+\dfrac{x}{x^2-x\sqrt{x^2+a^2}}$ $+\dfrac{x-\sqrt{x^2+a^2}}{x^2-x\sqrt{x^2+a^2}}=\dfrac{1}{\sqrt{x^2+a^2}-x}+\dfrac{1}{x-\sqrt{x^2+a^2}}+\dfrac{1}{x}=\dfrac{1}{x}=\dfrac{1}{1-\sqrt{2}}=$ $-(1+\sqrt{2})$.

2. $x_1^2+x_2^2=(x_1+x_2)^2-2x_1x_2=(m-2)^2-2(m^2-7)=-(m+2)^2+22$. 当 $m=-2$ 时,方程成为 $x^2-4x-3=0$,有两个实数根,并且 $x_1^2+x_2^2=22$ 为最大.

3. $x=\dfrac{(1-a)^2}{4}$, $\sqrt{x+a}=\sqrt{\dfrac{(1+a)^2}{4}}=\dfrac{|1+a|}{2}$, $\sqrt{x-a+2}=\sqrt{\dfrac{(a-3)^2}{4}}=\dfrac{3-a}{2}$. 由 $3-a-|1+a|=4$ 得 $a\leqslant-1$.

4. 由余弦定理得 $\cos ADC=\dfrac{3^2+5^2-7^2}{2\times3\times5}=-\dfrac{1}{2}$,所以 $\angle ADC=120^\circ$, $\angle ADB=60^\circ$. 由正弦定理得 $AB=\dfrac{5\sin60^\circ}{\sin45^\circ}=\dfrac{5}{2}\sqrt{6}$.

5. $x=\sqrt{3}+1$, $(x-1)^2=3$,即 $x^2-2x-2=0$. $y=\dfrac{1}{2}\times(x^2-2x-2)+1=1$.

6. $OP=2\cdot OM$, $\angle POA=\angle MPO=30^\circ$,所求面积$=\dfrac{\pi\times2^2}{4}-\dfrac{\pi\times1^2}{4}-\dfrac{1}{2}\times1\times\sqrt{2^2-1^2}-\dfrac{1}{3}\times\dfrac{\pi\times2^2}{4}=\dfrac{5\pi}{12}-\dfrac{\sqrt{3}}{2}$.

7. $\sqrt{27-10\sqrt{2}}=\sqrt{27-2\sqrt{25\times2}}=\sqrt{25}-\sqrt{2}=5-\sqrt{2}=3+(2-\sqrt{2})$, $a=3$, $b=2-\sqrt{2}$, $\dfrac{a+b}{a-b}=\dfrac{5-\sqrt{2}}{1+\sqrt{2}}=(5-\sqrt{2})(\sqrt{2}-1)=6\sqrt{2}-7$.

8. 显然 $x\leqslant1$. 又 $a(\sqrt{1-x}+1)=2x+4\geqslant0$,所以 $x\geqslant-2$. 当 $x=1$, 0, -2 时,a 分别为 6, 2, 0. 当 $x=-1$ 时, a 不是整数. 所以 a 可取 6, 2, 0.

三、**1.** 设 $\overset{\frown}{ACB}$ 的长度为 x cm. 又设甲、乙原来的速度分别为 v_1、v_2($v_1\neq$

v_2). 由题意

$$\frac{40}{v_1}=\frac{x-40}{v_2}, \tag{1}$$

$$\frac{300-20-(x-40)}{2v_1}=\frac{x-40+20}{\frac{1}{2}v_2}, \tag{2}$$

从而 $\frac{40}{x-40}=\frac{v_1}{v_2}=\frac{320-x}{4(x-20)}$.

化简得 $x^2-200x+9600=0$, $x_1=120$, $x_2=80$.

x_2 导致 $v_1=v_2$ 不合题意. 所以 $\overparen{ACB}$ 为 120 cm.

2. 由中线公式, $AB^2+AD^2=\frac{1}{2}BD^2+2AE^2=2BE^2+2AE^2$, $BC^2+CD^2=2BE^2+2EC^2$. 又 $BE^2=BE\times ED=AE\times EC$, 所以

$$AB^2+BC^2+CD^2+DA^2=2AE^2+4AE^2\times EC+2EC^2=2AC^2.$$

二试

1. $k(k+1)x^2-(2k+1)x+1=0$ 的根为$\frac{1}{k}$, $\frac{1}{k+1}$,所以

$$d_k=\frac{1}{k}-\frac{1}{k+1}\cdot d_1+d_2+\cdots+d_{999}=\frac{1}{1}-\frac{1}{999+1}=0.999.$$

2. (1) $a_1=7$, $a_2=9$, $a_3=3$, $a_4=1$ 且 $a_{n+4}=a_n$,所以 $a_{1994}=a_{94}=a_2=9$.

(2) $-n^2+2na_n=-(n-a_n)^2+a_n^2$. 在 $n\neq 4k+2$ 时, $-n^2+2na_n\leqslant a_n^2\leqslant 7^2<80$. 在 $n=4k+2$ 时, $-n^2+2na_n=-(n-9)^2+81\leqslant -1+81=80$. 等号仅在 $n=10$ 时成立. 所以 $n=10$ 时, $-n^2+2na_n$ 取最大值 80.

3. 设行和、列和中最小的为第一行,其和为 k,则第一行至少有 $n-k$ 个 0. 这 $n-k$ 个 0 所在的 $n-k$ 列,每列的和至少为 $n-k$. 其他 k 列,每列的和至少为 k. 所以表中所有数的和 $\geqslant(n-k)^2+k^2\geqslant\frac{1}{2}[(n-k)+k]^2=\frac{n^2}{2}$.

又解:设至多有 k 个 0 每两个不在同一行也不在同一列,并且它们分别在第 i 行第 i 列 ($i=1, 2, \cdots, k$). 对于 $1\leqslant i\leqslant k$,第 i 行与第 i 列的和 $\geqslant n$. 对于 $k+1\leqslant i\leqslant n$, 第 i 行上若有零,则这个零必在前 k 列. 设在第 j 列 ($1\leqslant j\leqslant$

k)，则第 j 行第 i 列的数非零(否则将第 j 行第 j 列的零用第 i 行第 j 列的零与第 j 行第 i 列的零代替，得到 $k+1$ 个 0，每两个不在同一行也不在同一列). 于是第 i 行与第 i 列的和仍然 $\geqslant n$. 从而 n 个行与 n 个列的和 $\geqslant n^2$. 表中所有数的和 $\geqslant \frac{n^2}{2}$.

4. (1) 如图 6，过 F 作两圆的公切线交 AB 于 G、交 CD 于 H，因为 $AB \parallel CD$，所以 $\angle AGF=\angle CHF$. 因为 GF、GA 与 $\odot O$ 相切，所以

$$\angle GFA=\angle GAF=\frac{1}{2}(180^\circ-\angle AGF).$$

同理，$\angle CFH=\angle FCH=\frac{1}{2}(180^\circ-\angle CHF)$.

所以 $\angle GFA=\angle CFH$，A、F、C 三点共线.

(2) 易知 CB 是 $\odot O'$ 的直径，所以 $CB \perp AB$，$BF \perp AC$，$CB^2=CF\times CA=CE\times CD$. 因而 CB 是 $\triangle BDE$ 的外接圆的切线. 这三角形的外心在 CB 的垂线 BA 上，又在 DE 的垂直平分线上. 易知 A 在 DE 的垂直平分线上. 因而 A 是 $\triangle BDE$ 的外心.

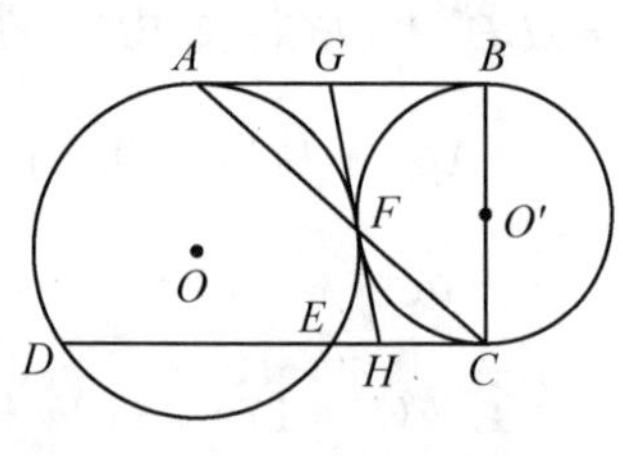

图 6

熟知直角三角形 ABC 的外心为 AC 的中点. 因此 $\triangle ABC$ 的外接圆与 $\triangle BDE$ 的外接圆的连心线为 AC. B 为这两个圆的公共点，$BF \perp AC$，所以两圆公共弦所在直线通过点 F.

又解：$\angle AEF=\angle GFA=\angle CFH=\angle FCH$，所以 $\triangle AEF \backsim \triangle ACE$，$AE^2=AF\times AC=AB^2$，$AE=AB$，又显然 $AE=AD$. 所以 A 是 $\triangle BDE$ 的外心. 以下与上面的解法相同.

不等式的证明（一）

已知 x, y, z 为正实数,求证:

$$(xy+yz+zx)\left[\frac{1}{(x+y)^2}+\frac{1}{(y+z)^2}+\frac{1}{(z+x)^2}\right]\geqslant\frac{9}{4}. \tag{1}$$

甲:我在一本书上看到这题的解答,看不懂,太复杂了.老师有没有简单的做法?

师:左边式子很复杂,我也得试一试.

乙:是不是可以设 $x+y+z=1$?

师:可以这样设,但未必有什么好处,因为 $\sum xy$ 是比较小的,常见的不等式都是它的上界估计,而现在要找它的下界.

甲:用调和平均小于等于几何平均这个性质,可以得出 $\sum\frac{1}{(x+y)^2}$ 的下界.

师:将 $\sum xy$ 与 $\sum\frac{1}{(x+y)^2}$ 分开处理恐怕难以奏效.我想将它们乘起来,在不等式证明中,和往往比积容易处理.

乙:那将有 9 项.

师:$(xy+yz+zx)\cdot\frac{1}{(x+y)^2}=\frac{xy}{(x+y)^2}+\frac{z(x+y)}{(x+y)^2}=\frac{xy}{(x+y)^2}+\frac{z}{x+y}$,所以式(1)即为

$$\sum\frac{xy}{(x+y)^2}+\sum\frac{z}{x+y}\geqslant\frac{9}{4}. \tag{2}$$

甲:这样式(2)的左边只有 6 项.不妨设 $x\geqslant y\geqslant z$,由排序不等式可得

$$\frac{x}{y+z}+\frac{y}{z+x}+\frac{z}{x+y}\geqslant\frac{y}{y+z}+\frac{z}{z+x}+\frac{x}{x+y}, \tag{3}$$

$$\frac{x}{y+z}+\frac{y}{z+x}+\frac{z}{x+y}\geqslant\frac{z}{y+z}+\frac{x}{z+x}+\frac{y}{x+y},\tag{4}$$

所以式(3)、式(4)两边对应相加得

$$\frac{x}{y+z}+\frac{y}{z+x}+\frac{z}{x+y}\geqslant\frac{3}{2}.\tag{5}$$

乙:这是我做过的不等式,可是

$$\sum\frac{xy}{(x+y)^2}\leqslant\sum\frac{1}{4}=\frac{3}{4},\tag{6}$$

式(2)与式(6)不等号方向相反,失败了!

师:不要轻易放弃,我们看到式(2)的左边分成两部分,一部分有下界估计式(5),另一部分却是上界估计式(6),但如果能证明差

$$\sum\frac{x}{y+z}-\frac{3}{2}\geqslant\frac{3}{4}-\sum\frac{xy}{(x+y)^2},\tag{7}$$

那不就大功告成了!

乙:是啊! 式(7)右边可化为

$$\sum\left(\frac{1}{4}-\frac{xy}{(x+y)^2}\right)=\frac{(x-y)^2}{4(x+y)^2}+\frac{(x-z)^2}{4(x+z)^2}+\frac{(y-z)^2}{4(y+z)^2}.\tag{8}$$

甲:由式(3)、式(4)得,式(7)左边为

$$\frac{(x-y)^2}{2(y+z)(z+x)}+\frac{(x-z)^2}{2(y+z)(x+y)}+\frac{(y-z)^2}{2(x+y)(x+z)}.\tag{9}$$

乙:式(9)的第1项显然大于等于式(8)的第1项,式(9)的第2项也大于等于式(8)的第2项,因为

$$2(x+z)^2-(y+z)(x+y)=(2x^2-xy-y^2)+(4xz-xz-yz)+2z^2\geqslant 0,\tag{10}$$

但第3项就不一定了,因为 $4(y+z)^2$ 可以很小,小于 $2(x+y)(x+z)$ 还是不好办啊!

师:设法证明式(9)、式(8)第2项的差大于等于式(8)、式(9)第3项的差.

甲:去掉分母,也就是要证明

$$(x-z)^2[2(x+z)^2-(x+y)(y+z)](y+z)$$
$$\geqslant (y-z)^2[(x+y)(x+z)-2(y+z)^2](x+z). \tag{11}$$

乙:式子仍然很繁啊!

师: $x-z \geqslant y-z$,而且

$$(x-z)(y+z)-(y-z)(x+z)=2(xz-yz) \geqslant 0, \tag{12}$$

所以式(11)的左边-右边

$$\geqslant (y-z)^2(x+z)[2(x+z)^2-(x+y)(y+z)$$
$$-(x+y)(x+z)+2(y+z)^2]$$
$$=(y-z)^2(x+z)(x^2+y^2-2xy+2xz+2yz+4z^2) \geqslant 0.$$

甲:这个证明可以看懂,而且知道为什么要这样做.

乙:为什么把式(9)和式(8)的第1项去掉?又为什么把式(11)左边的 $(x-z)^2(y+z)$ 减少为 $(y-z)^2(x+z)$?万一不成立怎么办?

师:证明不等式,胆子要大、心要细.要舍得去掉一些东西,使运算简化而不等式仍能成立.万一不成立也没有关系,从头再来就是了.对于大小,要有感觉,要在混乱中发现秩序,在复杂中发现简化的办法,这当然要多探索,多总结.

不等式的证明（二）

设 $a_1, a_2, \cdots, a_n$ 与 $b_1, b_2, \cdots, b_n$ 是两组不成比例的实数，实数 $x_1, x_2, \cdots, x_n$ 满足

$$\sum_{i=1}^{n} a_i x_i = 0, \tag{1}$$

$$\sum_{i=1}^{n} b_i x_i = 1, \tag{2}$$

证明
$$\sum_{i=1}^{n} x_i^2 \geqslant \frac{\sum_{i=1}^{n} a_i^2}{\left(\sum_{i=1}^{n} a_i^2\right)\left(\sum_{i=1}^{n} b_i^2\right) - \left(\sum_{i=1}^{n} a_i b_i\right)^2}. \tag{3}$$

题中的条件"$a_1, a_2, \cdots, a_n$ 与 $b_1, b_2, \cdots, b_n$ 不成比例"可以省去，因为若 $a_1, a_2, \cdots, a_n$ 与 $b_1, b_2, \cdots, b_n$ 成比例，则由式(1)可得 $\sum_{i=1}^{n} b_i x_i = 0$，与式(2)矛盾，所以条件(1)和条件(2)已隐含此意.

熟悉 Lagrange 恒等式的人立即可以看出式(3)的分母

$$\left(\sum_{i=1}^{n} a_i^2\right)\left(\sum_{i=1}^{n} b_i^2\right) - \left(\sum_{i=1}^{n} a_i b_i\right)^2 = \frac{1}{2}\sum_{ij}(a_i b_j - a_j b_i)^2, \tag{4}$$

而
$$\begin{aligned}
\sum_{i=1}^{n} x_i^2 \sum_{j=1}^{n}\sum_{i=1}^{n}(a_i b_j - a_j b_i)^2 &= \sum_{j=1}^{n}\left[\sum_{i=1}^{n} x_i^2 \sum_{i=1}^{n}(a_i b_j - a_j b_i)^2\right] \\
&\geqslant \sum_{j=1}^{n}\left[\sum_{i=1}^{n} x_i (a_i b_j - a_j b_i)\right]^2 \quad \text{（此处利用了 Cauchy 不等式）} \\
&= \sum_{j=1}^{n}\left(b_j \sum_{i=1}^{n} a_i x_i - a_j \sum_{i=1}^{n} x_i b_i\right)^2 = \sum_{j=1}^{n} a_j^2, \quad \text{（此处利用了条件(1) 和(2)）}
\end{aligned} \tag{5}$$

即
$$\sum_{i=1}^{n} x_i^2 \geqslant \frac{\sum_{j=1}^{n} a_j^2}{\sum_{j=1}^{n}\sum_{i=1}^{n}(a_ib_j-a_jb_i)^2}. \tag{6}$$

这与式(3)已经很接近了,遗憾的是式(4)右边为 $\frac{1}{2}\sum_{j=1}^{n}\sum_{i=1}^{n}(a_ib_j-a_jb_i)^2$, 所以由式(6)得到

$$\sum_{i=1}^{n} x_i^2 \geqslant \frac{\frac{1}{2}\sum_{j=1}^{n} a_j^2}{\left(\sum_{i=1}^{n} a_i^2\right)\left(\sum_{i=1}^{n} b_i^2\right)-\left(\sum_{i=1}^{n} a_i \cdot b_i\right)^2}, \tag{7}$$

多出因子$\frac{1}{2}$而无法消去,所以这种方法只能得出较弱的式(7). 做错了,只好重头再来.

先要弄清题意. $\boldsymbol{a}=(a_1, a_2, \cdots, a_n)$, $\boldsymbol{b}=(b_1, b_2, \cdots, b_n)$, $\boldsymbol{x}=(x_1, x_2, \cdots, x_n)$ 都是 n 维(实)向量,两个向量的和、差即将相应的分量(坐标)相加、减得到的向量,而两个向量的数量积是一个数,即

$$\boldsymbol{a}\cdot\boldsymbol{b}=\sum_{i=1}^{n} a_ib_i. \tag{8}$$

在数量积为 0 时,称两个向量互相垂直,式(1)、式(2)即

$$\boldsymbol{x}\cdot\boldsymbol{a}=0(\boldsymbol{x}\text{ 与 }\boldsymbol{a}\text{ 垂直}),\ \boldsymbol{x}\cdot\boldsymbol{b}=1. \tag{9}$$

满足式(9)的向量很多,我们先找最简单的,即与 $\boldsymbol{a}$, $\boldsymbol{b}$ 共面的向量(在 $\boldsymbol{a}$, $\boldsymbol{b}$ 所成平面上的向量),这种向量 $\boldsymbol{y}$ 可写成

$$\boldsymbol{y}=\alpha\boldsymbol{a}+\beta\boldsymbol{b},\ \alpha,\ \beta\in\mathbf{R}, \tag{10}$$

记
$$A=\boldsymbol{a}\cdot\boldsymbol{a}=\sum a_i^2,\ B=\boldsymbol{b}\cdot\boldsymbol{b}=\sum b_i^2,\ C=\boldsymbol{a}\cdot\boldsymbol{b}=\sum a_ib_i, \tag{11}$$

$\boldsymbol{y}$ 满足式(9)即

$$(\alpha\boldsymbol{a}+\beta\boldsymbol{b})\cdot\boldsymbol{a}=\alpha A+\beta C=0, \tag{12}$$

$$(\alpha\boldsymbol{a}+\beta\boldsymbol{b})\cdot\boldsymbol{b}=\alpha C+\beta B=1. \tag{13}$$

由式(12)、式(13)易得 $\alpha=\dfrac{-C}{AB-C^2},\ \beta=\dfrac{A}{AB-C^2}$, (14)

$$\boldsymbol{y}^2=\boldsymbol{y}\cdot(\alpha\boldsymbol{a}+\beta\boldsymbol{b})=\beta\boldsymbol{y}\cdot\boldsymbol{b}=\beta, \tag{15}$$

式(15)即
$$\sum_{i=1}^n y_i^2=\frac{\sum\limits_{i=1}^n a_i^2}{\left(\sum\limits_{i=1}^n a_i^2\right)\left(\sum\limits_{i=1}^n b_i^2\right)-\left(\sum\limits_{i=1}^n a_i\cdot b_i\right)^2}. \tag{16}$$

所以在 $\boldsymbol{x}=\boldsymbol{y}$ 时,式(3)成为等式.

特殊情况既然成立(而且是等式),一般情形当然也应当正确,这时 $\boldsymbol{x}=\boldsymbol{y}+(\boldsymbol{x}-\boldsymbol{y})$, 而由于 $\boldsymbol{x}$, $\boldsymbol{y}$ 均是满足式(9)的向量,所以有

$$(\boldsymbol{x}-\boldsymbol{y})\cdot\boldsymbol{a}=\boldsymbol{x}\cdot\boldsymbol{a}-\boldsymbol{y}\cdot\boldsymbol{a}=0, \tag{17}$$

$$(\boldsymbol{x}-\boldsymbol{y})\cdot\boldsymbol{b}=\boldsymbol{x}\cdot\boldsymbol{b}-\boldsymbol{y}\cdot\boldsymbol{b}=1-1=0, \tag{18}$$

从而
$$(\boldsymbol{x}-\boldsymbol{y})\cdot\boldsymbol{y}=(\boldsymbol{x}-\boldsymbol{y})\cdot(\alpha\boldsymbol{a}+\beta\boldsymbol{b})=0, \tag{19}$$

(几何意义即为向量 $\boldsymbol{x}-\boldsymbol{y}$ 与 $\boldsymbol{a}$, $\boldsymbol{b}$ 均垂直,所以 $\boldsymbol{x}-\boldsymbol{y}$ 与 $\boldsymbol{a}$, $\boldsymbol{b}$ 张成的平面垂直. 特别地,与 $\boldsymbol{y}$ 垂直.)

$$\boldsymbol{x}^2=[\boldsymbol{y}+(\boldsymbol{x}-\boldsymbol{y})]^2=\boldsymbol{y}^2+2\boldsymbol{y}\cdot(\boldsymbol{x}-\boldsymbol{y})+(\boldsymbol{x}-\boldsymbol{y})^2=\boldsymbol{y}^2+(\boldsymbol{x}-\boldsymbol{y})^2\geqslant\boldsymbol{y}^2, \tag{20}$$

由式(20),式(15)即得式(3).

另有一种证法是模仿 Cauchy 不等式的一种证法.

对任意实数 λ,有

$$\lambda^2=\left[\sum_{i=1}^n x_i(a_i+\lambda b_i)\right]^2\leqslant\sum_{i=1}^n x_i^2-\sum_{i=1}^n(a_i+\lambda b_i)^2, \tag{21}$$

即
$$\lambda^2\left(\sum_{i=1}^n x_i^2\sum_{i=1}^n b_i^2-1\right)+2\lambda\sum_{i=1}^n x_i^2\sum_{i=1}^n a_i\cdot b_i+\sum_{i=1}^n x_i^2\sum_{i=1}^n a_i^2\geqslant 0. \tag{22}$$

对 Cauchy 不等式

$$\sum_{i=1}^n x_i^2\cdot\sum_{i=1}^n b_i^2\geqslant\left(\sum_{i=1}^n x_i\cdot b_i\right)^2=1, \tag{23}$$

（Ⅰ）当式(23)等号不成立时，式(22)左边作为 λ 的二次式，判别式 $\Delta \leqslant 0$，即

$$\left(\sum_{i=1}^{n} x_i^2 \cdot \sum_{i=1}^{n} a_i \cdot b_i\right)^2 \leqslant \left(\sum_{i=1}^{n} x_i^2 \cdot \sum_{i=1}^{n} b_i^2 - 1\right) \sum_{i=1}^{n} x_i^2 \cdot \sum_{i=1}^{n} a_i^2. \tag{24}$$

仍用式(11)中记号，式(24)可化简得

$$\sum_{i=1}^{n} x_i^2 (AB - C^2) \geqslant A, \tag{25}$$

由 Cauchy 不等式得，$AB > C^2$，所以式(3)成立.

（Ⅱ）当式(23)中等号成立时，$x_1, x_2, \cdots, x_n$ 与 $b_1, b_2, \cdots, b_n$ 成比例. 由式(1)得 $C=0$，式(3)即为式(23)，显然成立.

第二种证法亦颇有技巧，但其背景不如第一种证法清楚.

利用导数证明不等式

导数是研究函数的重要工具，在证明不等式时也极为有用. 本文拟对此作一些介绍.

1 预备知识

一般地，设函数 $y=f(x)$ 在某个区间 I 内可导.

如果 $f'(x)>0\ (x\in I)$，则 $f(x)$ 严格递增 $(x\in I)$；如果 $f'(x)<0\ (x\in I)$，则 $f(x)$ 严格递减 $(x\in I)$.

上面结论中的"$>$"或"$<$"也可以改为"$\geqslant$"或"$\leqslant$"，只要等号成立的点不形成一个区间.

2 无条件的不等式

对于无条件的不等式，通常运用上面的预备知识即可解决.

例 1 求 $\dfrac{\sin x}{2}+\dfrac{2}{\sin x}\ (x\in(0,\ \pi))$ 的最小值.

解 由于函数 $f(t)=\dfrac{t}{2}+\dfrac{2}{t}\ (t\in(0,\ 1])$ 的导数

$$f'(t)=\frac{1}{2}-\frac{2}{t^2}=\frac{t^2-4}{2t^2}<0,$$

所以，$f(t)$ 严格递减.

因此，$f(t)$ 的最小值为 $f(1)=\dfrac{5}{2}$.

从而，$\dfrac{\sin x}{2}+\dfrac{2}{\sin x}$ 的最小值为 $\dfrac{5}{2}$，且在 $x=\dfrac{\pi}{2}$ 时取得.

本题若不用导数，可用算术—几何平均不等式求解，但技巧性较高. 其他解

法则比较麻烦.

例 2　已知 x、y、$z \in \mathbf{R}_+$. 求证:

$$\frac{xyz}{(1+5x)(4x+3y)(5y+6z)(z+18)} \leqslant \frac{1}{5120}. \tag{1}$$

解　对函数

$$f(t)=\frac{t}{(at+b)(ct+d)},\ t \in \mathbf{R}_+,\ a、b、c、d>0$$

求导得

$$f'(t)=p[(at+b)(ct+d)-t(2act+bc+ad)]=p(bd-act^2).$$

其中 $p=(at+b)^{-2}(ct+d)^{-2}$ 表示正的量.

当 $t^2<\dfrac{bd}{ac}$ 时, $f'(t)>0$, 所以, $f(t)$严格递增;当 $t^2>\dfrac{bd}{ac}$ 时, $f'(t)<0$,

所以, $f(t)$严格递减. 因此, $f(t)$在 $t^2=\dfrac{bd}{ac}$ 时取得最大值

$$f\left(\sqrt{\frac{bd}{ac}}\right)=\frac{1}{(\sqrt{ad}+\sqrt{bc})^2}.$$

固定 y,对 x 的函数

$$g(x)=\frac{x}{(1+5x)(4x+3y)}$$

用上面的结果 ($a=5$, $b=1$, $c=4$, $d=3y$) 得

$$\frac{x}{(1+5x)(4x+3y)} \leqslant \frac{1}{(\sqrt{15y}+2)^2}. \tag{2}$$

同样,对 z 的函数有

$$\frac{z}{(5y+6z)(z+18)} \leqslant \frac{1}{(\sqrt{5y}+6\sqrt{3})^2}. \tag{3}$$

所以,由式(2)、(3)并对 y 的函数再一次用上面的结果得

$$\text{式①左边} \leqslant \frac{y}{(\sqrt{15y}+2)^2(\sqrt{5y}+6\sqrt{3})^2} = \left[\frac{\sqrt{y}}{(\sqrt{15y}+2)(\sqrt{5y}+6\sqrt{3})}\right]^2$$

$$\leqslant \left[\frac{1}{(\sqrt{18\sqrt{5}}+\sqrt{2\sqrt{5}})^2}\right]^2 = \frac{1}{5120}.$$

3 条件不等式

对于某些条件不等式,可以固定其中一些变数,化为一元函数,再经过调整得出所需要的结果.这种方法称为"导数调整法".

例3 已知 a、b、c 为非负实数,且 $a+b+c=1$. 求证:

$$(1-a^2)^2+(1-b^2)^2+(1-c^2)^2 \geqslant 2.$$

证明 注意到,在 $a=1$, $b=c=0$ 时,所证不等式成立.不妨设 $a \geqslant b \geqslant c$. 固定 b,则 $c=1-b-a$ 是 a 的函数.

考虑 a 的函数

$$f(a)=(1-a^2)^2+(1-b^2)^2+(1-c^2)^2,$$

有 $f(a)$ 递减 $\Leftrightarrow f'(a) \leqslant 0$

$\Leftrightarrow -a(1-a^2)+c(1-c^2) \leqslant 0$

$\Leftrightarrow (a-c)(a^2+ac+c^2-1) \leqslant 0$

$\Leftrightarrow a^2+ac+c^2-1 \leqslant 0$

$\Leftarrow (a+c)^2 \leqslant 1.$

最后的不等式显然成立.所以, $f(a)$ 递减.可以使 a 增大 c 减小 ($a+c$ 保持不变),而 $f(a)$ 一直减小,直至 c 减小至0.于是,

$$(1-a^2)^2+(1-b^2)^2+(1-c^2)^2 \geqslant (1-a^2)^2+(1-b^2)^2+1,$$

其中 $a+b=1$.

对函数 $\varphi(a)=(1-a^2)^2+(1-b^2)^2+1(a+b=1)$ 采用同样的做法(前面的 c 换成 b),可知 $\varphi(a)$ 递减.所以,

$$\varphi(a) \geqslant (1-1)^2+(1-0^2)^2+1=2.$$

因此,所证不等式成立.

由例 3 可以看出,导数调整法的步骤是:

(1) 选定目标,也就是不等式中等号成立(函数取最大值或最小值)的情况;

(2) 固定一些变量,使函数成为一元函数;

(3) 确定调整方向,也就是应证明该一元函数递增或递减;

(4) 利用导数证明函数递增或递减,从而使一个变量调整到所需要的值;

(5) 继续采取上面的做法,直至使各个变量都调整到所需要的值.

例 4 已知 $a \geqslant b \geqslant c \geqslant 0$,且 $a+b+c=3$. 求证:$ab^2+bc^2+ca^2 \leqslant \frac{27}{8}$.

证明 在 $a=b=\frac{3}{2}$, $c=0$ 时,所证不等式的等号成立. 固定 c,则 $b=3-c-a$.

考虑函数 $f(a)=ab^2+bc^2+ca^2$,有

$$\begin{aligned} f(a)\text{ 递减} &\Leftrightarrow f'(a) \leqslant 0 \\ &\Leftrightarrow b^2-2ba+2ac-c^2 \leqslant 0 \\ &\Leftrightarrow (b-c)(c+b-2a) \leqslant 0. \end{aligned}$$

最后的不等式显然成立.

所以, $f(a)$递减. 可以使 a 减小到 b (c 保持不变),而 $f(a)$一直增加到 $b^3+bc^2+cb^2(c=3-2b)$.

$\varphi(b)=b^3+bc^2+cb^2$ 递增 $\Leftrightarrow \varphi'(b) \geqslant 0$

$\Leftrightarrow 3b^2+c^2-4bc+2bc-2b^2 \geqslant 0$

$\Leftrightarrow (b-c)^2 \geqslant 0$.

最后一个不等式显然成立.

所以, $\varphi(b)$递增,

$$\varphi(b) \leqslant \varphi\left(\frac{3}{2}\right)=\frac{27}{8}.$$

因此,所证不等式成立.

例 4 的第一步不是将 a 调整至$\frac{3}{2}$或将 c 调整至 0,而是将 a 调整至 b(然后再继续调整).

4 更复杂的例子

例 5 已知 x、y、$z\in\mathbf{R}_+\cup\{0\}$，且 $x+y+z=\frac{1}{2}$. 求

$$\frac{\sqrt{x}}{4x+1}+\frac{\sqrt{y}}{4y+1}+\frac{\sqrt{z}}{4z+1} \tag{1}$$

的最大值.

解 易猜到 $x=y=z=\frac{1}{6}$ 时，式(1)取最大值 $\frac{3}{5}\sqrt{\frac{3}{2}}$（至少这是最大值的一个"候选者". 另一个候选者是 $x=\frac{1}{2}$，$y=z=0$ 时，函数的取值 $\frac{1}{3}\sqrt{\frac{1}{2}}$. 这些特殊的值值得注意）.

不妨设 $x>\frac{1}{6}>z$. 固定 y，则

$$z=\frac{1}{2}-y-x.$$

考虑 x 的函数 $f(x)=\sum\frac{\sqrt{x}}{4x+1}$，有

$f(x)$严格递减 $\Leftrightarrow f'(x)<0$

$$\Leftrightarrow\frac{1-4x}{\sqrt{x}(4x+1)^2}<\frac{1-4z}{\sqrt{z}(4z+1)^2}. \tag{2}$$

当 $x\geqslant\frac{1}{4}$ 时，式(2)显然成立（左边负，右边正）.

当 $x<\frac{1}{4}$ 时，$\frac{1-4x}{\sqrt{x}(4x+1)^2}$ 是 x 的减函数，所以，式(2)也成立.

因此，$f(x)$递减. 可将 x 调小 z 调大（$x+z$ 保持不变），而 $f(x)$增加. 从而，可使 x 减至 $\frac{1}{6}$ 或 z 增至 $\frac{1}{6}$（根据 y 大于或小于 $\frac{1}{6}$ 而定）.

不妨设 $z=\frac{1}{6}$，而 $x>y$. 对

$$\varphi(x)=\frac{\sqrt{x}}{4x+1}+\frac{\sqrt{y}}{4y+1}+\frac{\sqrt{\frac{1}{6}}}{4\times\frac{1}{6}+1}$$

（其中 $y=\frac{1}{3}-x$）同样处理，得出当 $x=y=z=\frac{1}{6}$ 时，$\sum\frac{\sqrt{x}}{4x+1}$ 取得最大值 $\frac{3}{5}\sqrt{\frac{3}{2}}$.

上面的解法同时得出，当 $y=z=0$，$x=\frac{1}{2}$ 时，$\sum\frac{\sqrt{x}}{4x+1}$ 取得最小值 $\frac{1}{3}\sqrt{\frac{1}{2}}$.

例 6　已知 x、y、$z\in\mathbf{R}_+$，且 $x+y+z=1$. 求

$$\frac{\sqrt{x}}{4x+1}+\frac{\sqrt{y}}{4y+1}+\frac{\sqrt{z}}{4z+1}$$

的最大值.

此题与例 5 不同的地方是，条件 $x+y+z=\frac{1}{2}$ 换为 $x+y+z=1$.

解　不妨设 $x\geqslant\frac{1}{3}\geqslant z$. 固定 y，有

$$f(x)=\sum\frac{\sqrt{x}}{4x+1}\text{ 严格递减}\Leftrightarrow f'(x)<0$$

$$\Leftrightarrow\frac{1-4x}{\sqrt{x}(4x+1)^2}<\frac{1-4z}{\sqrt{z}(4z+1)^2}. \tag{1}$$

当 $z\leqslant\frac{1}{4}$ 时，式(1)显然成立(左边负，右边正). 此时，$f(x)$递减. 可将 x 调小 z 调大（$x+z$ 保持不变），而 $f(x)$增加. 所以，可使 $x=\frac{1}{3}$ 或 $z=\frac{1}{4}$. 前者已有一个变量为$\frac{1}{3}$. 否则 $z=\frac{1}{4}$.

设 $z\geqslant\frac{1}{4}$. 同理可设 $y\geqslant\frac{1}{4}$. 这时，$x\leqslant\frac{1}{2}$.

要证式(1)成立,只须证

$$\varphi(x)=\frac{4x-1}{\sqrt{x}(4x+1)^2}$$

在$\left[\frac{1}{4},\frac{1}{2}\right]$上单调递增.

$\varphi(x)$严格递增, $x\in\left[\frac{1}{4},\frac{1}{2}\right]$

$\Leftrightarrow\varphi'(x)>0,\ x\in\left[\frac{1}{4},\frac{1}{2}\right]$

$\Leftrightarrow 8x(4x+1)-(4x+1)(4x-1)-16x(4x-1)>0,\ x\in\left[\frac{1}{4},\frac{1}{2}\right]$

$\Leftrightarrow 1+24x-48x^2>0,\ x\in\left[\frac{1}{4},\frac{1}{2}\right]$

$\Leftrightarrow 1+6t-3t^2>0,\ t\in[1,2].$

不难看出,最后的不等式成立.

所以,式(1)成立.

因此, $f(x)$递减. 可将 x 调小 z 调大 ($x+z$ 保持不变),而 $f(x)$增加. 直至 x 或 z 中有一个成为$\frac{1}{3}$,不妨设 $z=\frac{1}{3}$, $x\geqslant y$. 对

$$g(x)=\frac{\sqrt{x}}{4x+1}+\frac{\sqrt{y}}{4y+1}+\frac{\sqrt{\frac{1}{3}}}{4\times\frac{1}{3}+1}$$

同样处理,得出 $x=y=z=\frac{1}{3}$ 时, $\sum\frac{\sqrt{x}}{4x+1}$ 取得最大值$\frac{3\sqrt{3}}{7}$.

例 6 与例 5 的函数相同,只是条件 $x+y+z=\frac{1}{2}$ 换成了 $x+y+z=1$, 其调整就比例 5 复杂一些.

顺便指出,很多人喜欢用琴生(Jensen)不等式来处理此类问题. 但使用琴生不等式是有条件的,即函数应为凸(凹)函数. 要证明函数是凸(凹)的,需要求二阶导数,并证明二阶导数恒负(正)(除少数已知的凸(凹)函数,如 $y=\lg x$, $y=ax^2+bx+c$, $y=\sin x\ (x\in(0,\pi))$ 等).

例 7　已知 x、y、$z \in \mathbf{R}_+$，且 $xyz=1$. 求证：

$$\frac{1}{\sqrt{1+x}}+\frac{1}{\sqrt{1+y}}+\frac{1}{\sqrt{1+z}} \leqslant \frac{3\sqrt{2}}{2}. \tag{1}$$

证明　不妨设 $x \geqslant y \geqslant z$.

当 $y \leqslant 2$ 时，固定 x，则 $z=\frac{1}{xy}$.

函数 $f(y)=\frac{1}{\sqrt{1+x}}+\frac{1}{\sqrt{1+y}}+\frac{1}{\sqrt{1+z}}$ 递减

$\Leftrightarrow f'(y) \leqslant 0$

$\Leftrightarrow -\frac{1}{(1+y)^{\frac{3}{2}}}+\frac{1}{(1+z)^{\frac{3}{2}}} \cdot \frac{1}{xy^2} \leqslant 0$

$\Leftrightarrow \frac{y^2}{(1+y)^3} \geqslant \frac{z^2}{(1+z)^3}$

$\Leftarrow \frac{y^2}{(1+y)^3}$ 递增

$\Leftrightarrow \left(\frac{y^2}{(1+y)^3}\right)' \geqslant 0$

$\Leftrightarrow 2(1+y)-3y=2-y \geqslant 0$.

最后一个不等式显然成立.

因此，$f(y)$递减. 此时，可增大 z 减小 y，直至 $y=z$.

当 $y>2$ 时，固定 z，则 $x=\frac{1}{yz}$.

$\varphi(y)=\frac{1}{\sqrt{1+x}}+\frac{1}{\sqrt{1+y}}+\frac{1}{\sqrt{1+z}}$ 递减

$\Leftrightarrow \varphi'(y) \leqslant 0$

$\Leftrightarrow$ ……（前面的 z 换成 x）

$\Leftrightarrow \frac{y^2}{(1+y)^3} \geqslant \frac{x^2}{(1+x)^3}$

$\Leftarrow \frac{y^2}{(1+y)^3}$ 递减

$\Leftrightarrow \left(\frac{y^2}{(1+y)^3}\right)' \leqslant 0$

$\Leftrightarrow 2-y \leqslant 0.$

最后一个不等式显然成立.因此，$\varphi(y)$递减.此时,可减小 y 增大 x,直至 $y=2$. 再用前面的方法调整使 $y=z$. 于是,式(1)化为

$$\frac{1}{\sqrt{1+x}}+\frac{2}{\sqrt{1+y}} \leqslant \frac{3\sqrt{2}}{2}\left(x y^{2}=1,\ y \leqslant x\right).$$

$h(y)=\dfrac{1}{\sqrt{1+x}}+\dfrac{2}{\sqrt{1+y}}$ 递增

$\Leftrightarrow h'(y) \geqslant 0$

$\Leftrightarrow \dfrac{1}{(1+x)^{\frac{3}{2}}} \cdot \dfrac{1}{y^{3}}-\dfrac{1}{(1+y)^{\frac{3}{2}}} \geqslant 0$

$\Leftrightarrow y^{-2}(1+y) \geqslant 1+x$

$\Leftrightarrow 1+y \geqslant \dfrac{1+x}{x} \Leftrightarrow x y \geqslant 1.$

因为 $y \leqslant x$, $x y^{2}=1$,则 $y \leqslant 1$, $x y \geqslant 1$. 所以, $h(y)$递增.可使 y 增大 x 减小,直至 $x=y=1$. 在这个过程中, $h(y)$一直增加,直至取得最大值$\dfrac{3\sqrt{2}}{2}$.

例 8 已知 x、y、$z \in \mathbf{R}_{+} \cup\{0\}$,且 $x+y+z=1$. 求证:

$$\sum\left(\frac{20}{15-9 x^{2}}-9 x^{2}\right) \leqslant \frac{9}{7}.$$

证明 当 $x=y=z=\dfrac{1}{3}$ 时,等号成立.

$\sum\left(-9 x^{2}\right)$ 在 $x=y=z=\dfrac{1}{3}$ 时,取得最大值,但$\sum \dfrac{20}{15-9 x^{2}}$ 却不是在 $x=y=z=\dfrac{1}{3}$ 时,取得最大值. 先说明如下:

不妨设 $x \geqslant y \geqslant z$. 固定 y,则

$$z=1-y-x.$$

$f(x)=\sum \dfrac{20}{15-9 x^{2}}$ 递增 $\Leftrightarrow f'(x) \geqslant 0$

$\Leftrightarrow \dfrac{x}{\left(15-9 x^{2}\right)^{2}}-\dfrac{z}{\left(15-9 z^{2}\right)^{2}} \geqslant 0.$

由于 $x \geqslant z$，$15-9x^2 \leqslant 15-9z^2$，所以，最后一个不等式成立. 从而，$f(x)$ 递增. 因此，可使 x 增大 z 减小，直至 z 变为 0. 在这个过程中，$\sum \frac{20}{15-9x^2}$ 一直增加. 同样，可使 x 增大 y 减小，直至 y 也变为 0，x 变为 1. 所以，

$$\sum \frac{20}{15-9x^2} \leqslant \frac{20}{15-9}+\frac{20}{15}+\frac{20}{15}=6,$$

最大值 6 在 $x=1$，$y=z=0$ 时取得.

因为 $\sum \frac{20}{15-9x^2} \leqslant 6$，所以，在 $9\sum x^2 \geqslant \frac{33}{7}=6-\frac{9}{7}$ 时，所证不等式成立.

设 $9\sum x^2 < \frac{33}{7}(<5)$.

不妨设 $x \geqslant \frac{1}{3} \geqslant z$. 固定 y，则

$$z=1-y-x.$$

$F(x)=\sum\left(\frac{20}{15-9x^2}-9x^2\right)$ 递减

$\Leftrightarrow F'(x)<0$

$\Leftrightarrow \frac{20x}{(15-9x^2)^2}-x-\frac{20z}{(15-9z^2)^2}+z<0$

$\Leftarrow \left(\frac{20x}{(15-9x^2)^2}-x\right)'<0$

$\Leftrightarrow 20(15+27x^2)-(15-9x^2)^3<0.$

因为 $15-9x^2>15-5=10$，所以，

$$\begin{aligned}&(15-9x^2)^3-20(15+27x^2)\\>&100(15-9x^2)-20(15+27x^2)\\=&20(60-72x^2)\\>&20(60-8\times 5)>0.\end{aligned}$$

于是，$F(x)$ 递减. 可谓小 x 增大 z，直至 x 或 z 成为 $\frac{1}{3}$. 在此过程中，$F(x)$ 一直增加.

不妨设 $z=\frac{1}{3}$，又设 $x\geqslant\frac{1}{3}\geqslant y$. 继续采用上面的做法，便知所证不等式成立，其中等号仅在 $x=y=z=\frac{1}{3}$ 时成立.

本例也不适合用琴生不等式证明. 因为函数 $g(x)=\frac{20}{15-9x^2}-9x^2$ 并非凸函数，它的二阶导数与上面的 $20(15+27x^2)-(15-9x^2)^3$ 只差一个正因子. 而 $20(15+27x^2)-(15-9x^2)^3$ 并不恒负，例如，$x=1$ 或接近 1 时，即取正值.

例 9 给定正整数 n. 求最小的正数 λ，使对任何 $\theta_i\in\left(0,\frac{\pi}{2}\right)(i=1,2,\cdots,n)$，只要

$$\tan\theta_1\cdot\tan\theta_2\cdot\cdots\cdot\tan\theta_n=2^{\frac{n}{2}}, \tag{1}$$

就有

$$\cos\theta_1+\cos\theta_2+\cdots+\cos\theta_n\leqslant\lambda. \tag{2}$$

解 当 $n=1$ 时，

$$\cos\theta_1=(1+\tan^2\theta_1)^{-\frac{1}{2}}=\frac{\sqrt{3}}{3},\ \lambda=\frac{\sqrt{3}}{3}.$$

设 $n\geqslant 2$. 令 $x_i=\tan^2\theta_i(1\leqslant i\leqslant n)$，则题设条件变为

$$x_1x_2\cdots x_n=2^n. \tag{3}$$

要求最小的 λ，使

$$\sum\frac{1}{\sqrt{1+x_i}}\leqslant\lambda.$$

不妨设 $x_1,x_2,\cdots,x_n$ 中 x_1 最大，则由式(3)有 $x_1\geqslant 2$.

若 $x_2\geqslant 2$，则与例 7 中 $y\geqslant 2$ 的情况完全一样(x 即 x_1，y 即 x_2)，可以得出 $\sum\frac{1}{\sqrt{1+x_i}}=\varphi(x_2)$ 递减. 从而，可调小 x_2 直至 $x_2=2$，这时，$\varphi(x_2)$ 一直增加. 同样，可设 $x_3\leqslant 2,x_4\leqslant 2,\cdots,x_n\leqslant 2$. 再由例 7 中 $y\leqslant 2$ 的情况，可逐步调整，使得 $x_2,x_3,\cdots,x_n$ 都调成 $\left(\frac{2^n}{x_1}\right)^{\frac{1}{n-1}}$. 于是，只须求最小的 λ，使

$$g(y)=\frac{1}{\sqrt{1+x}}+\frac{n-1}{\sqrt{1+y}}\leqslant\lambda,$$

其中 $x\geqslant 2\geqslant y$，$xy^{n-1}=2^n$.

$$g'(y)=\frac{-(n-1)}{2(1+y)^{\frac{3}{2}}}+\frac{1}{2(1+x)^{\frac{3}{2}}}\cdot\frac{(n-1)2^n}{y^n}=\frac{n-1}{2y}\left[\frac{x}{\sqrt{(1+x)^3}}-\frac{y}{\sqrt{(1+y)^3}}\right].$$

$g(y)$递增 $\Leftrightarrow g'(y)\geqslant 0$

$\Leftrightarrow \dfrac{x^2}{(1+x)^3}\geqslant\dfrac{y^2}{(1+y)^3}$

$\Leftrightarrow x^2(1+y)^3-y^2(1+x)^3\geqslant 0.$

当 $n=2$ 时，$xy=2^2$，

$$y^2[x^2(1+y)^3-y^2(1+x)^3]=4^2(1+y)^3-y(y+4)^3=(2+y)(2-y)^3\geqslant 0.$$

所以，$g(y)$递增.

因此，最大值为 $g(2)=\dfrac{2}{\sqrt{3}}$，即 $\lambda=\dfrac{2}{\sqrt{3}}$.

当 $n\geqslant 3$ 时，$xy^{n-1}=2^n$，

$$\begin{aligned}&y^{3n-5}[x^2(1+y)^3-y^2(1+x)^3]\\&=2^{2n}y^{n-3}(1+y)^3-(y^{n-1}+2^n)^3\\&=(2^n-y^n)(y^{2n-3}+2^ny^{n-3}+3\times 2^ny^{n-2}-2^{2n})\\&\leqslant(2^n-y^n)(2^{2n-3}+2^{2n-3}+3\times 2^{2n-2}-2^{2n})=0.\end{aligned}$$

所以，$g(y)$递减.

当 $y\to 0$ 时，$g(y)\to n-1$，即 $\lambda=n-1$.

介绍一个几何不等式

1 问题的提出

Walther Janous 1986 年在 Crux Mathemalicorum, Vol. 12, No. 4 上提出一个几何不等式：

命题 1 设三角形的三边为 a, b, c, $s=\dfrac{a+b+c}{2}$, 中线为 m_a, m_b, m_c, 则

$$\frac{1}{m_a}+\frac{1}{m_b}+\frac{1}{m_c}\geqslant\frac{3\sqrt{3}}{s}. \tag{1}$$

这样的不等式并不难构造,实际上只要取一个关于 a, b, c 对称的零次齐次函数 $F(a, b, c)$,算出它在 $a=b=c$ (即三角形为正三角形)时的值 $F(a, a, a)$,便可提出形如

$$F(a, b, c)\underset{(\leqslant)}{\geqslant}F(a, a, a)$$

的不等式. 例如,令 $F(a, b, c)=s\left(\dfrac{1}{m_a}+\dfrac{1}{m_b}+\dfrac{1}{m_c}\right)$, 则有 $F(a, a, a)=3\sqrt{3}$, 进而得到命题 1.

2 反例

这样提出的不等式并非永远正确(尽管正三角形具有多种极端性质). 实际上命题 1 就是错误的. 当 $b=c$, 而 $a\to 0$ 时, $m_c=m_b\to\dfrac{b}{2}$, $m_a\to b$,从而

$$s\left(\frac{1}{m_a}+\frac{1}{m_b}+\frac{1}{m_c}\right)\to 5<3\sqrt{3},$$

这已足以说明问题.

3　修正

命题 1 虽然不正确,但我们不希望问题就此结束. 可以对(1)式进行修正,式中的 $3\sqrt{3}$ 须改为一个较小的数,这个较小的数当然至多是 5. 因此可提出:

命题 2　设三角形的三边为 a, b, c, $s=\dfrac{a+b+c}{2}$, 中线为 m_a, m_b, m_c,则

$$\frac{1}{m_a}+\frac{1}{m_b}+\frac{1}{m_c}>\frac{5}{s}. \tag{2}$$

本文的目的就是证明命题 2.

4　转化

引理(Klamkin 中线对偶定理) 设三角形的三边为 a, b, c,三条中线为 m_a, m_b, m_c,又设 $\Phi(a, b, c, m_a, m_b, m_c)$是关于 a, b, c 对称的零次齐次函数,则

$$\Phi(a, b, c, m_a, m_b, m_c)\geqslant k \tag{a}$$

对任意三角形成立与

$$\Phi\left(\frac{2}{3}m_a, \frac{2}{3}m_b, \frac{2}{3}m_c, \frac{1}{2}a, \frac{1}{2}b, \frac{1}{2}c\right)\geqslant k \tag{b}$$

对任意三角形成立等价.

证明　设(a)式对任意$\triangle ABC$ 成立,$\triangle ABC$ 的三边为 a, b, c,三条中线为 $AD=m_a$, $BE=m_b$, $CF=m_c$, 重心为 G. 延长中线 CF 至 H,使 $FH=FG$ (如图 1),则易证$\triangle AGH$ 三边的长为$\triangle ABC$ 三条中线的$\dfrac{2}{3}$,$\triangle AGH$ 三条中线的长为$\triangle ABC$ 三边的$\dfrac{1}{2}$. 因(a)式对任意三角形成立,故在$\triangle AGH$ 中有

$$\Phi\left(\frac{2}{3}m_a, \frac{2}{3}m_b, \frac{2}{3}m_c, \frac{1}{2}a, \frac{1}{2}b, \frac{1}{2}c\right)\geqslant k.$$

即(b)式对$\triangle ABC$成立.

另一方面,若(b)式对任意$\triangle AGH$成立,$\triangle AGH$的三边为a, b, c,三条中线为$AF=m_a$, $GI=m_b$, $HJ=m_c$,按图1所示的方法不难得到$\triangle ABC$,使$\triangle ABC$的三边是$\triangle AGH$三条中线的2倍,$\triangle ABC$的三条中线是$\triangle AGH$三边的$\frac{3}{2}$.由于在$\triangle ABC$中仍有(b)式成立,可知,(a)式对$\triangle AGH$成立.引理得证.

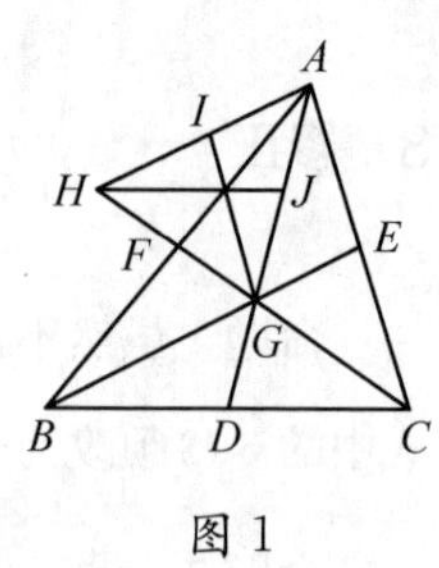

图 1

据引理,(2)式与

$$\frac{1}{\frac{1}{2}a}+\frac{1}{\frac{1}{2}b}+\frac{1}{\frac{1}{2}c}>\frac{5}{\frac{1}{2}\left(\frac{2}{3}m_a+\frac{2}{3}m_b+\frac{2}{3}m_c\right)},$$

即

$$\frac{2}{3}(m_a+m_b+m_c)\left(\frac{1}{a}+\frac{1}{b}+\frac{1}{c}\right)>5$$

等价.

因此,要证命题2只要证:

命题3 设三角形的三边为a, b, c,中线为m_a, m_b, m_c,则

$$\frac{2}{3}(m_a+m_b+m_c)\left(\frac{1}{a}+\frac{1}{b}+\frac{1}{c}\right)>5. \tag{3}$$

5 推广

命题4 设三角形的三边为a, b, c, P为任意一点, P到三顶点的距离分别为x, y, z,则

$$(x+y+z)\left(\frac{1}{a}+\frac{1}{b}+\frac{1}{c}\right)>5. \tag{4}$$

命题4是命题3的推广,事实上,在命题4中, P是三角形的重心时, P到三顶点的距离为$\frac{2}{3}m_a$, $\frac{2}{3}m_b$, $\frac{2}{3}m_c$,即为命题3.下面证明命题4成立.

6　两种情况

我们知道,如果$\triangle ABC$有一个角,比如说$\angle A \geqslant 120^{\circ}$,那么在$P$点与$A$重合时,$x+y+z$为最小,这时

$$(x+y+z)\left(\frac{1}{a}+\frac{1}{b}+\frac{1}{c}\right)=(b+c)\left(\frac{1}{a}+\frac{1}{b}+\frac{1}{c}\right)$$
$$=\frac{b+c}{a}+(b+c)\left(\frac{1}{b}+\frac{1}{c}\right)>1+2^{2}=5.$$

命题4成立.

如果$\triangle ABC$的内角均$<120^{\circ}$,那么当P为$\triangle ABC$的Fermat-Torricelli点时,$x+y+z$为最小.这时,$\angle APB=\angle BPC=\angle CPA=120^{\circ}$,由余弦定理易知

$$a=\sqrt{y^{2}+yz+z^{2}},\ b=\sqrt{x^{2}+xz+z^{2}},\ c=\sqrt{x^{2}+xy+y^{2}}.$$

在此种情形下证明命题4即要证:

命题5　设x,y,z为正数,则

$$(x+y+z)\left(\frac{1}{\sqrt{y^{2}+yz+z^{2}}}+\frac{1}{\sqrt{x^{2}+xz+z^{2}}}+\frac{1}{\sqrt{x^{2}+xy+y^{2}}}\right)>5. \tag{5}$$

7　加强

命题5虽然只是一个普通的三元函数的极值问题,但证明并不容易.Wolfgang Gmeiner与问题的提出者Walther Janous在这一步卡住了.当然人们有理由怀疑命题5是否正确,但用计算机搜索的结果表明命题5不但成立,而且还可以加强为:

命题6　设x,y,z为正数,则

$$(x+y+z)\left(\frac{1}{\sqrt{y^{2}+yz+z^{2}}}+\frac{1}{\sqrt{x^{2}+xz+z^{2}}}+\frac{1}{\sqrt{x^{2}+xy+y^{2}}}\right)>4+\frac{2}{\sqrt{3}}. \tag{6}$$

这里的 $4+\frac{2}{\sqrt{3}}$ 是最佳值,因为在 $x=\frac{1}{2}$, $y=\frac{1}{2}$, $z=0$ 时两边相等.

8 证明的完成

现在我们证明命题 6.

1° 将(6)式的左边记为 $F(x, y, z)$,不妨设 $x\geqslant y\geqslant z$. 又令 $u=x+y$, 则

$$F(x, y, z)=F(x, u-x, z),\ x\geqslant \frac{u}{2}\geqslant z.$$

$$\begin{aligned}
&\frac{\partial}{\partial x}F(x, u-x, z)\\
&=(x+y+z)\cdot\left(-\frac{2x-(x+y)}{2\sqrt{(x^2+xy+y^2)^3}}-\frac{2x+z}{2\sqrt{(x^2+xz+z^2)^3}}\right.\\
&\qquad\left.+\frac{2y+z}{2\sqrt{(y^2+yz+z^2)^3}}\right)\\
&=\frac{x+y+z}{2}\left(\frac{y-x}{\sqrt{(x^2+xy+y^2)^3}}-\frac{2x}{\sqrt{(x^2+xz+z^2)^3}}\right.\\
&\qquad\left.+\frac{2y}{\sqrt{(y^2+yz+z^2)^3}}+\frac{z}{\sqrt{(y^2+yz+z^2)^3}}-\frac{z}{\sqrt{(x^2+xz+z^2)^3}}\right)\\
&\geqslant\frac{x+y+z}{2}\left(\frac{y-x}{\sqrt{(x^2+xy+y^2)^3}}-\frac{2x}{\sqrt{(x^2+xz+z^2)^3}}\right.\\
&\qquad\left.+\frac{2y}{\sqrt{(y^2+yz+z^2)^3}}\right).
\end{aligned}$$

由于

$$\begin{aligned}
&\frac{2y}{\sqrt{(y^2+yz+z^2)^3}}-\frac{2x}{\sqrt{(x^2+xz+z^2)^3}}\\
&=2\int_x^y\frac{\mathrm{d}}{\mathrm{d}t}\left[\frac{t}{\sqrt{(t^2+tz+z^2)^3}}\right]\mathrm{d}t\\
&=2\int_x^y\frac{(t^2+tz+z^2)-\frac{3}{2}t(2t+z)}{\sqrt{(t^2+tz+z^2)^5}}\mathrm{d}t=\int_y^x\frac{4t^2+tz-2z^2}{\sqrt{(t^2+tz+z^2)^5}}\mathrm{d}t
\end{aligned}$$

$$
\begin{aligned}
&>\int_y^x \frac{t^2+tz+z^2}{\sqrt{(t^2+tz+z^2)^5}}\mathrm{d}t\\
&=\int_y^x \frac{1}{\sqrt{(t^2+tz+z^2)^3}}\mathrm{d}t\\
&>\frac{x-y}{\sqrt{(x^2+xy+y^2)^3}},
\end{aligned}
$$

所以

$$\frac{\partial}{\partial x}F(x,\ u-x,\ z)>0.$$

从而在 $x\geqslant\frac{u}{2}\geqslant z$ 时,

$$F(x,\ u-x,\ z)\geqslant F\left(\frac{u}{2},\ \frac{u}{2},\ z\right)=F\left(\frac{x+y}{2},\ \frac{x+y}{2},\ z\right).$$

2° 由 $F(x,\ y,\ z)$的齐次性,不妨设 $\frac{x+y}{2}=1\geqslant z$,我们只需证明

$$F(1,\ 1,\ z)=(2+z)\cdot\left(\frac{1}{\sqrt{3}}+\frac{2}{\sqrt{z^2+z+1}}\right)>4+\frac{2}{\sqrt{3}},$$

即

$$\frac{z}{\sqrt{3}}-\frac{2z+4}{\sqrt{z^2+z+1}}>4. \tag{7}$$

3° (7)式的证明如下:

$$
\begin{aligned}
&\frac{z}{\sqrt{3}}+\frac{2z+4}{\sqrt{z^2+z+1}}>4\\
&\Leftrightarrow \frac{4\sqrt{3}-z}{\sqrt{3}}\leqslant\frac{2z+4}{\sqrt{z^2+z+1}}\\
&\Leftrightarrow (4\sqrt{3}-z)^2(z^2+z+1)\leqslant 12(z+2)^2\\
&\Leftrightarrow -z^3+(8\sqrt{3}-1)z^2+(8\sqrt{3}-37)z+8\sqrt{3}\geqslant 0\\
&\Leftarrow (8\sqrt{3}-2)z^2+(8\sqrt{3}-37)z+8\sqrt{3}\geqslant 0\\
&\Leftrightarrow 4(8\sqrt{3}-2)\cdot 8\sqrt{3}\geqslant(8\sqrt{3}-37)^2\\
&\Leftrightarrow 3(8\sqrt{3})^2-8^2\sqrt{3}\geqslant 37^2-16\times 37\sqrt{3}\\
&\Leftrightarrow 9\times 64+16\times 33\sqrt{3}\geqslant 37^2.
\end{aligned}
$$

最后一式显然成立.至此,命题 2、3、4、5、6 均已得到证明.

9 两点附注

1° 为了证明不等式(6),我们用了微积分.是否有完全初等的、不用微积分的证明?我们期待这种证明的出现.

2° Klamkin 对偶定理尚有其他应用.例如下面的不等式(m_a, m_b, m_c 分别为a, b, c 边的中线)

$$\frac{a^2}{m_a}+\frac{b^2}{m_b}+\frac{c^2}{m_c}\geqslant\frac{4}{3}(m_a+m_b+m_c),$$

直接证颇不容易.由对偶定理,只需证

$$\frac{m_a^2}{a}+\frac{m_b^2}{b}+\frac{m_c^2}{c}\geqslant\frac{3}{4}(a+b+c).$$

后者用中线公式可化简为

$$(a^2+b^2+c^2)\left(\frac{1}{a}+\frac{1}{b}+\frac{1}{c}\right)\geqslant 3(a+b+c),$$

这可由 Cauchy 不等式(或用其他方法)导出.

一个角平分线不等式的推广

1

文献[1]提出猜想：设 w_a, w_b, w_c 是$\triangle ABC$ 的角平分线，则

$$\sum w_a^{-1} > \sum a^{-1}. \tag{1}$$

匡继昌在文献[2]第 772 页中将它列为 100 个未解决的问题(问题 42). 1994 年，成萱[3]首先发表了这一猜测的证明. 我们考虑(1)的一个指数推广.

定理 1　在$\triangle ABC$ 中，

$$\sum \frac{1}{w_a^2} \geqslant \sum \frac{1}{a^2} + \frac{1}{3}\sum \frac{1}{bc}, \tag{2}$$

等号成立当且仅当$\triangle ABC$ 是正三角形.

证　设 s, R, r 分别为$\triangle ABC$ 的半周长、外接圆半径、内切圆半径，则由三角形中的恒等式

$$\sum \frac{1}{w_a^2} = \frac{2R+r}{8Rr^2} + \frac{4R+r}{8Rs^2}, \tag{3}$$

$$\sum \frac{1}{a^2} = \left(\frac{s^2+4Rr+r^2}{4Rrs}\right)^2 - \frac{1}{Rr}, \tag{4}$$

以及

$$\sum \frac{1}{bc} = \frac{1}{2Rr}, \tag{5}$$

(2)等价于

$$\frac{2R+r}{8Rr^2} + \frac{4R+r}{8Rs^2} \geqslant \left(\frac{s^2+4Rr+r^2}{4Rrs}\right)^2 - \frac{5}{6Rr}, \tag{6}$$

即

$$\begin{aligned} H(s^2) &\equiv 3(s^2+4Rr+r^2)^2 - 40Rrs^2 - 6Rs^2(2R+r) - 6Rr^2(4R+r) \\ &= 3s^4 - 2s^2(6R^2+11Rr-3r^2) + 3r^2(8R^2+6Rr+r^2) \leqslant 0. \end{aligned} \tag{7}$$

由 Gerretsen 不等式：

$$16Rr-5r^2\leqslant s^2\leqslant 4R^2+4Rr+3r^2, \tag{8}$$

以及

$$\begin{aligned}&H(16Rr-5r^2)\\&=-192R^3r+500R^2r^2-256Rr^3+48r^4\\&=-4r(R-2r)(48R^2-29Rr+6r^2)\leqslant 0,\end{aligned} \tag{9}$$

$$\begin{aligned}&H(4R^2+4Rr+3r^2)\\&=-40R^3r+44R^2r^2+48Rr^3+48r^4\\&=-4r(R-2r)(10R^2+9Rr+6r^2)\leqslant 0,\end{aligned} \tag{10}$$

即知(7)成立，从而(6)、(2)得证.

2

1994 年，刘健[4]、杨学枝[5]、王振与陈计[6]分别将不等式(1)加强成

$$\sum\frac{1}{w_a}\geqslant 3\left(\sum\cos\frac{A}{2}\right)\cdot\sum\frac{1}{a}\geqslant\frac{2}{\sqrt{3}}\sum\frac{1}{a}, \tag{11}$$

$$\sum\frac{1}{w_a}\geqslant\frac{1}{3}\left(\sum\sec\frac{A}{2}\right)\sum\frac{1}{a}\geqslant\frac{2}{\sqrt{3}}\sum\frac{1}{a}, \tag{12}$$

$$\sum\frac{1}{w_a}\geqslant\left(3\sum\frac{1}{w_aw_b}\right)^{\frac{1}{2}}\geqslant\frac{2}{\sqrt{3}}\sum\frac{1}{a}. \tag{13}$$

本节中，我们给出(1)的一种新的加强：

定理 2 在△ABC 中，有

$$\left(\sum\frac{1}{w_a}\right)^2\geqslant\sum\left(\frac{1}{b}+\frac{1}{c}\right)^2\geqslant\frac{4}{3}\left(\sum\frac{1}{a}\right)^2, \tag{14}$$

等号成立当且仅当△ABC 是正三角形.

证 由定理 1，要证(14)的第一个不等式，只需证

$$2\sum\frac{1}{w_bw_c}\geqslant\sum\frac{1}{a^2}+\frac{5}{3}\sum\frac{1}{bc}, \tag{15}$$

由三角形恒等式

$$w_aw_bw_c=\frac{4abc(a+b+c)S_\triangle}{(b+c)(c+a)(a+b)}, \tag{16}$$

以及 $w_a\geqslant h_a$ 等,得

$$\begin{aligned}2\sum\frac{1}{w_bw_c}&\geqslant\frac{(b+c)(c+a)(a+b)\sum h_a}{2abc(a+b+c)S_\triangle}\\&=\frac{(b+c)(c+a)(a+b)}{abc(a+b+c)}\sum\frac{1}{a},\end{aligned} \tag{17}$$

所以要证(15),只需证

$$\frac{(b+c)(c+a)(a+b)}{abc(a+b+c)}\sum\frac{1}{a}\geqslant\sum\frac{1}{a^2}+\frac{5}{3}\sum\frac{1}{bc}, \tag{18}$$

由于 $3\prod(b+c)\cdot\sum bc-\left(3\sum b^2c^2+5abc\sum a\right)\sum a=abc\sum a^2-abc\sum bc=\frac{abc}{2}\cdot\sum(b-c)^2\geqslant 0$,所以(17)成立,从而(15)及(14)的第一个不等式得证.又因为

$$3\sum\left(\frac{1}{b}+\frac{1}{c}\right)^2-4\left(\sum\frac{1}{a}\right)^2=\sum\left(\frac{1}{b}-\frac{1}{c}\right)^2\geqslant 0, \tag{19}$$

所以(14)的第二个不等式也成立.

注记:上述不等式(15)是(13)第二个不等式

$$\sum\frac{1}{w_bw_c}\geqslant\frac{4}{9}\left(\sum\frac{1}{a}\right)^2 \tag{20}$$

的加强.

参考文献

[1] Koz. Mat. Lapok, 1984, No. 3, 126.
[2] 匡继昌.常用不等式(第二版)[M].湖南教育出版社,1993:772.
[3] 成萱.关于三角形角平分线的一个猜想的证明[J].中学数学(武汉),1994(5):31.
[4] 刘健.一些新的三角形不等式[J].中学数学(苏州),1994(5):9-12.
[5] 杨学枝.关于三角形角平分线的两个不等式[J].中学数学(武汉),1994(7):30-31.
[6] 王振,陈计.两个猜想不等式的加强及其它[J].中学教研(数学),1994(7-8):51-53.
[7] D. S. Mitrinović.几何不等式的新进展[M].陈计等译.北京大学出版社,1995.

三角不等式的一种证法

对于三角不等式，中学生往往感到比较棘手. 据了解，有两个原因：第一，没有掌握一种普遍的解法，遇到题目不知如何下手. 看到解答，也只知其然，不知其所以然，即一步一步地看下去知道解答是正确的，却不知道这个解答是怎样想出来的，因而遇到其他题目仍然解不出. 第二，有些书籍或刊物上介绍的解答，用到不少其他的知识，这就增加了难度. 例如某一刊物上有一篇文章为了证明$\triangle ABC$中，$\sin A+\sin B+\sin C\leqslant\frac{3\sqrt{3}}{2}$，先去证另一个不等式$a^2+b^2+c^2\leqslant 9R^2$，而后一个不等式并不比原先的不等式简单.

为了帮助同学解决这两方面的困难，本文介绍一种简单的、容易依循的方法，这种证法并不涉及三角以外的知识.

牛顿说过，"在数学中，例子比定律更为重要."我们先举一个例子来说明这一方法的要点是什么.

例1 证明：在$\triangle ABC$中，

$$\sin A+\sin B+\sin C\leqslant\frac{3\sqrt{3}}{2}. \tag{1}$$

解 为证明(1)式成立，先看一看(1)中的等号能不能成立？什么时候成立？

只要熟悉几个特殊角的三角函数值，立即就知道在$A=B=C=60^\circ$时，(1)是成立的. 因而我们要证的不等式(1)也就是

$$\sin A+\sin B+\sin C\leqslant\sin 60^\circ+\sin 60^\circ+\sin 60^\circ. \tag{2}$$

不妨设$A\geqslant B\geqslant C$（以下各题均如此，不再申明），于是$A\geqslant 60^\circ$，$C\leqslant 60^\circ$. 把A缩小，C增加，保持和$A+C$不变，使得A、C这两个角中，一个被调整为60°，另一个为$A+C-60^\circ$. 我们证明这时$\sin A+\sin C$的值不会减少，即证明

$$\sin A+\sin C\leqslant\sin 60^\circ+\sin(A+C-60^\circ). \tag{3}$$

由于 $\sin A+\sin C=2\sin\dfrac{A+C}{2}\cos\dfrac{A-C}{2}$，$\sin 60^\circ+\sin(A+C-60^\circ)=2\sin\dfrac{A+C}{2}\cos\dfrac{A+C-120^\circ}{2}$，而 $A\geqslant 60^\circ\Rightarrow A-C\geqslant 120^\circ-A-C$，$C\leqslant 60^\circ\Rightarrow A-C\geqslant A+C-120^\circ$，$A-C\geqslant|120^\circ-A-C|$，所以

$$\cos\frac{A-C}{2}\leqslant\cos\frac{A+C-120^\circ}{2},\tag{4}$$

从而(3)成立.

现在再来调整 B 与 $A+C-60^\circ$. 我们证明在保持和 $B+A+C-60^\circ=120^\circ$ 不变,将 B 调整为 60°(从而另一角也等于 60)时，$\sin B+\sin(A+C-60^\circ)$ 的值不减少,即

$$\sin B+\sin(A+C-60^\circ)\leqslant\sin 60^\circ+\sin 60^\circ.\tag{5}$$

由于 $\sin B+\sin(A+C-60^\circ)=2\sin\dfrac{180^\circ-60^\circ}{2}\cos\dfrac{A+C-B-60^\circ}{2}=2\sin 60^\circ\cos\dfrac{A+C-B-60^\circ}{2}\leqslant 2\sin 60^\circ$，故(5)式成立.

由(3)、(5)即得

$$\sin A+\sin B+\sin C\leqslant\sin 60^\circ+\sin(A+C-60^\circ)+\sin B\leqslant 3\sin 60^\circ=\frac{3\sqrt{3}}{2}.$$

证毕.

由例题 1 的解法可以看出,我们的方法就是保持 $A+B+C$ 不变,逐步地将 A、B、C 调整为适当的角(60°),证明在每一步调整时,和 $\sin A+\sin B+\sin C$ 均不减少,因而最后(经过两步调整)得出(1)式.

注 1 仔细分析一下上面的解法,就可以发现(1)中等号仅在 $A=B=C=60^\circ$ 时成立.

注 2 同时,我们解答了一个极值问题:在$\triangle ABC$ 中，$\sin A+\sin B+\sin C$ 的最大值为$\dfrac{3\sqrt{3}}{2}$,并且在 $A=B=C=60^\circ$ 时,取得最大值. 如果运用正弦定理,在(1)的两边同乘 $2R$,就得出:圆内接三角形中,以正三角形的周长为最大,最大值为 $3\sqrt{3}R$.

例 2 证明:在$\triangle ABC$中,

$$\cos A+\cos B+\cos C\leqslant\frac{3}{2}. \tag{6}$$

解 等号仍然是在$A=B=C=60^\circ$时取得,因此,我们的方针是将A、B、C逐步地调整为60°. 与上面的证明相仿,

$$\cos A+\cos C=2\cos\frac{A+C}{2}\cos\frac{A-C}{2}$$

$$\leqslant 2\cos\frac{A+C}{2}\cos\frac{A+C-120^\circ}{2}=\cos 60^\circ+\cos(A+C-60^\circ). \tag{7}$$

又 $\cos B+\cos(A+C-60^\circ)=2\cos 60^\circ\cos\frac{A+C-60^\circ-B}{2}\leqslant 2\cos 60^\circ$, (8)

由(7)、(8)得

$$\cos A+\cos B+\cos C\leqslant\cos 60^\circ+\cos(A+C-60^\circ)+\cos B\leqslant\cos 60^\circ+2\cos 60^\circ=\frac{3}{2}.$$

这里的调整法是完全初等的. 在有些刊物上,采用另一种"调整法",即由

$$\cos A+\cos C=2\cos\frac{A+C}{2}\cos\frac{A-C}{2}\leqslant 2\cos\frac{A+C}{2},$$

得出在$\cos A+\cos B+\cos C$取最大值时,必有$A=C$,然后依据同理得$B=C$,从而导出$A=B=C=60^\circ$时,$\cos A+\cos B+\cos C$为最大. 这种"调整法"有很大的局限性(不适用于下面的例题3),并且需要预先假定最大值存在,而严格说来,这一点是要证明的,证明常常用到连续性,因此不宜在中学生中提倡.

例 3 证明:在锐角三角形ABC中,

$$\sin A+\sin B+\sin C>2.$$

解 这里没有等号成立的情况,但是如果考虑"退化"的三角形,即在A、B、C分别为90°、90°、0°时,$\sin A+\sin B+\sin C$等于2. 因此(9)式即

$$\sin A+\sin B+\sin C>\sin 90^\circ+\sin 90^\circ+\sin 0^\circ. \tag{10}$$

我们的方针是将三个角逐步调整为两个90°,一个0°并且保证在调整中,$\sin A+\sin B+\sin C$的值不增加. 为此先证明

$$\sin A+\sin C>\sin 90^\circ+\sin(A+C-90^\circ). \tag{11}$$

因为$\triangle ABC$为锐角三角形,$\angle A<90^\circ$,所以

$$A-C<180^\circ-A-C,$$

$$\cos\frac{A-C}{2}>\cos\frac{180^\circ-A-C}{2}, \tag{12}$$

$$\sin A+\sin C=2\sin\frac{A+C}{2}\cos\frac{A-C}{2}>$$

$$2\sin\frac{A+C}{2}\cos\frac{180^\circ-A-C}{2}=\sin 90^\circ+\sin(A+C-90^\circ).$$

即(11)成立. 又

$$\sin B+\sin(A+C-90^\circ)=\sin B+\cos B>1, \tag{13}$$

由(11)、(13)即得

$$\sin A+\sin B+\sin C>\sin 90^\circ+\sin(A+C-90^\circ)+\sin B>\sin 90^\circ+1=2.$$

仿照例题 3,不难证明在锐角三角形ABC中,$\cos A+\cos B+\cos C>1$.

例 4 证明:在$\triangle ABC$中

$$\sin A+\sin B+\sin C-\cos A-\cos B-\cos C\leqslant\frac{2\sqrt{3}-3}{2}, \tag{14}$$

并且在$\triangle ABC$为锐角三角形时,

$$\sin A+\sin B+\sin C-\cos A-\cos B-\cos C>1. \tag{15}$$

解 (14)即

$$\sin A+\sin B+\sin C-\cos A-\cos B-\cos C\leqslant 3(\sin 60^\circ-\cos 60^\circ). \tag{16}$$

我们保持和$A+C$不变,将A、C中一个角调整为60°(另一个$A+C-60^\circ$),这时,

$$\sin A+\sin C-\cos A-\cos C=2\cos\frac{A-C}{2}\left(\sin\frac{A+C}{2}-\cos\frac{A+C}{2}\right),$$

$$\sin 60^\circ+\sin(A+C-60^\circ)-\cos 60^\circ-\cos(A+C-60^\circ)$$

$$=2\cos\frac{A+C-120^\circ}{2}\left(\sin\frac{A+C}{2}-\cos\frac{A+C}{2}\right),$$

由于 $A+C>B$,所以 $A+C>\frac{180^\circ}{2}=90^\circ$, $\frac{A+C}{2}>45^\circ$,又显然 $90^\circ>\frac{A+C}{2}$,所以

$$\sin\frac{A+C}{2}-\cos\frac{A+C}{2}>0. \tag{17}$$

由(4)及(17)得

$$\begin{aligned}&2\cos\frac{A-C}{2}\left(\sin\frac{A+C}{2}-\cos\frac{A+C}{2}\right)\\ \leqslant{}&2\cos\frac{A+C-120^\circ}{2}\left(\sin\frac{A+C}{2}-\cos\frac{A+C}{2}\right),\end{aligned}$$

即

$$\begin{aligned}&\sin A+\sin C-\cos A-\cos C\\ \leqslant{}&\sin 60^\circ+\sin(A+C-60^\circ)-\cos 60^\circ-\cos(A+C-60^\circ).\end{aligned} \tag{18}$$

再调整 B 与 $A+C-60^\circ$, 这时

$$\begin{aligned}&\sin B+\sin(A+C-60^\circ)-\cos B-\cos(A+C-60^\circ)\\ ={}&2\sin 60^\circ\cos(B-60^\circ)-2\cos 60^\circ\cos(B-60^\circ)\\ ={}&2(\sin 60^\circ-\cos 60^\circ)\cos(B-60^\circ)\leqslant 2(\sin 60^\circ-\cos 60^\circ).\end{aligned} \tag{19}$$

由(18)、(19)即得(14).

不等式(14)即

$$\sin A+\sin B+\sin C-\cos A-\cos B-\cos C>2\sin 90^\circ+\sin 0^\circ-2\cos 90^\circ-\cos 0^\circ. \tag{20}$$

由于

$$\begin{aligned}&\sin A+\sin C-\cos A-\cos C\\ ={}&2\sin\frac{A+C}{2}\cos\frac{A-C}{2}-2\cos\frac{A+C}{2}\cos\frac{A-C}{2}\\ ={}&2\cos\frac{A-C}{2}\left(\sin\frac{A+C}{2}-\cos\frac{A+C}{2}\right),\end{aligned}$$

$$\begin{aligned}&\sin 90^\circ+\sin(A+C-90^\circ)-\cos 90^\circ-\cos(A+C-90^\circ)\\ ={}&2\cos\frac{A+C-180^\circ}{2}\left(\sin\frac{A+C}{2}-\cos\frac{A+C}{2}\right),\end{aligned}$$

由(17),(12)得

$$\cos\frac{A-C}{2}\left(\sin\frac{A+C}{2}-\cos\frac{A+C}{2}\right)>\cos\frac{A+C-180^\circ}{2}\left(\sin\frac{A+C}{2}-\cos\frac{A+C}{2}\right),$$

即

$$\sin A+\sin C-\cos A-\cos C$$
$$>\sin 90^\circ+\sin(A+C-90^\circ)-\cos 90^\circ-\cos(A+C-90^\circ). \tag{21}$$

又

$$\sin B+\sin(A+C-90^\circ)-\cos B-\cos(A+C-90^\circ)$$
$$=\sin B+\cos B-\cos B-\sin B=0, \tag{22}$$

由(21)、(22)即得(15).

例 5　证明:在$\triangle ABC$中,

$$\sin\frac{A}{2}\sin\frac{B}{2}\sin\frac{C}{2}\leqslant\frac{1}{8}. \tag{23}$$

解　只要证明

$$\sin\frac{A}{2}\sin\frac{B}{2}\sin\frac{C}{2}\leqslant\left(\sin\frac{60^\circ}{2}\right)^3. \tag{24}$$

由于(4)

$$\sin\frac{A}{2}\sin\frac{C}{2}=\frac{1}{2}\left(\cos\frac{A-C}{2}-\cos\frac{A+C}{2}\right)$$
$$\leqslant\frac{1}{2}\left(\cos\frac{A+C-120^\circ}{2}-\cos\frac{A+C}{2}\right)=\sin\frac{60^\circ}{2}\sin\frac{A+C-60^\circ}{2},$$

又

$$\sin\frac{B}{2}\sin\frac{A+C-60^\circ}{2}$$
$$=\frac{1}{2}\left(\cos\frac{A+C-60^\circ-B}{2}-\cos\frac{A+B+C-60^\circ}{2}\right)$$
$$=\frac{1}{2}\left(\cos\frac{A+C-60^\circ-B}{2}-\cos 60^\circ\right)$$
$$\leqslant\frac{1}{2}(1-\cos 60^\circ)=\left(\sin\frac{60^\circ}{2}\right)^2,$$

由(25)、(26)即得(24). 证毕.

以上方法不仅限于三角形的内角,对于满足$A+B+C=2\pi$之类的角也同样适用,还可以推广至n $(n\geqslant 3)$个角.

游戏与数学归纳法

很多游戏与数学归纳法有关系,我们举几个例子作为说明.

例 1 “抢三十”.

这是很多人熟悉的游戏,由两个人从一开始轮流往下报数,每次报一个、两个或三个数,谁先报到 30 就算胜.

显然,如果甲能“抢”到 26,那么无论对方报 27,或报 27、28,或报 27、28、29,甲总能“抢”到 30. 这样一来,问题就化为“抢 26”. 而依照这样理由可知“抢 26”又可以化为“抢 22”,这样逐步往前推,问题归结为“抢 2”,因而先说的人只要依次抢 2, 6, 10, …,最后一定能抢到 30.

这个游戏也可以用更一般的形式叙述出来,即甲、乙二人从 1 开始往下报数,每次可报 1, 2, …, k 个数,谁报到 N 为胜(或谁报到 N 为负).

解决一个与自然数有关的问题,我们往往先退一步,或退若干步,退到最简单的、容易解决的场合(比如例 1 中,退到“抢 2”的问题),解决这个容易的问题后再进行原来的问题,甚至更为一般的问题. 这一思想方法,与数学归纳法有着重要的联系.

数学归纳法的基本思想是:为了证明一个与自然数有关的命题 $P_{(n)}$,我们先证明 $P_{(1)}$(或者 $P_{(n_0)}$,其中 n_0 是一个较小的自然数),然后再证明对每个自然数 k,由 $P_{(k)}$ 的正确性可以推出 $P_{(k+1)}$ 的正确性,这两步合起来就完成了命题 $P_{(n)}$ 的证明.

例 2 桌上有 111 根火柴,甲、乙两人轮流取,每次取 1 根或素数根,取最后 1 根的为胜,问甲应当如何取?

我们先从简单的情况出发,当火柴根数为 1, 2 或 3 的时候,显然甲胜(一次全部取完),在根数为 4 的时候,不论甲怎样取,总是乙胜(乙把剩下的全部取完).

在火柴根数大于 4 的时候,如果甲能取走若干根火柴,使得剩下的火柴根数为 4,那么甲就能胜,因此在火柴根数为 5、6 或 7 的时候,甲应先取 1、2 或 3

根. 但是,在火柴根数为 8 的时候,无论甲怎样取(当然只能取 1 根或素数根),剩下的火柴根数总不是 4,因此乙总能取得胜利.

归纳起来,在火柴根数为 4, 8 的时候甲负,在火柴根数为 1, 2, 3, 5, 6, 7 的时候甲胜(双方均按正确的取法),因此我们可以产生一个猜测,在火柴根数为 $4n$ 的时候,乙能胜,而在火柴根数为 $4n-1$, $4n-2$, $4n-3$ 的时候,甲能胜. 这个猜测在 $n=1$(1, 2, 3, 4 根)与 $n=2$(5, 6, 7, 8 根)的时候已经证实,逐步推下去可知对于 $n=3, 4, \cdots$ 也成立. 一般地假如对于 $n=1, 2, \cdots, k$,猜测均成立,那么对于 $4k+1$, $4k+2$ 或 $4k+3$ 根火柴,甲可以先取 1, 2 或 3 根火柴,使乙面临 $4k$ 根火柴的情况,从而甲能胜. 而对于 $4(k+1)$ 根火柴,由于甲只能取 1 根或素数根,剩下的火柴根数总不是 4 的倍数,因而乙能胜. 这样,我们的猜测对于任意的 n 都是正确的. 特别地,在根数为 111 的时候,甲能胜,只要注意每次留下的根数都是 4 的倍数就行了.

例 3　14 根火柴如图 1 放置,每根火柴可以跳过两根火柴与另一根火柴并在一起(但仍算两根),如图 2 所示,你能不能把这 14 根火柴,按照上面的"跳法",两两并在一起?

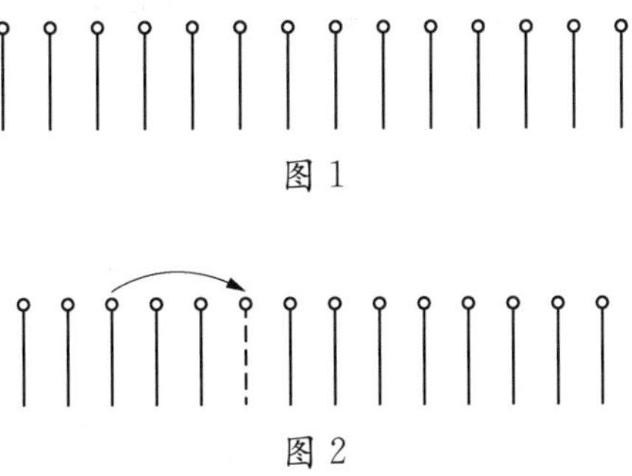

图 1

图 2

我们还是先退到最简单的情况. 容易验证 4 根或 6 根火柴是不可能两两并在一起的,8 根火柴则不难用下面的方法并起来(图 3):

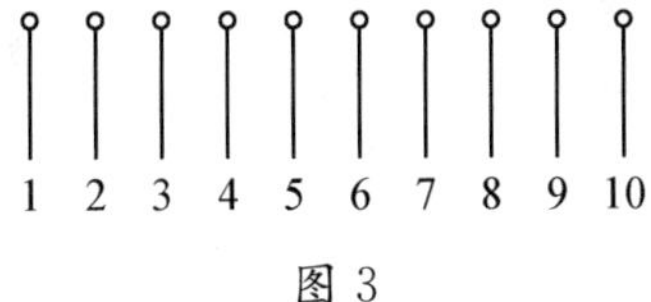

图 3

5 跳到 2, 3 跳到 7, 4 跳到 1, 6 跳到 8.

对于 10 根火柴,可以先将 7 跳到 10,从而化为 8 根的情况. 依此办理,可以处理一般的 $2n$ $(n \geqslant 4)$ 根火柴的情况.

例 4 有三个同样的塔,1 号塔上有 100 个环,自上而下,一个比一个大,现在要把这 100 个环移到 2 号塔上,但每次只能移动 1 个环,放到其他两个塔上,并且每一个环只能放在比它大的环上,不能放在比它小的环上. 问应如何移动?至少要移动多少次?

如果不加思索,贸然就动手去移,那是办不成功的. 我们仍然从最简单的情况着手,一个环的情形显然只要移一次. 只有两个环的情形,可以先将小环移到 3 号塔,将大环移到号 2 塔,最后再将小环移到 2 号塔,共需 $3=2^2-1$ 次.

假如对于 k 个环,经过 2^k-1 次移动可以把它们移到 2 号塔上,那么对于 $k+1$ 个环,我们可以先将上面的 k 个环经过 2^k-1 次移动,移到 3 号塔上,然后将最大的环移到 2 号塔上,最后再将 2 号塔上的环全部移到 3 号塔上 (2^k-1 次). 这样共需 $2^k-1+1+2^k-1=2^{k+1}-1$ 次,这也就是次数最少的移法. 因此,对于 n 个环,至少要经过 2^n-1 次移动,才能全部移到 2 号塔上. 对于 $n=100$,需要经过 $2^{100}-1$ 次移动,才能把这 100 个环移到 2 号塔上,这是一个大得惊人的数字.

什么时候用归纳法?

这个问题,因为情况千差万别,不宜笼统回答.但我们可以说,凡是与自然数有关的命题,都应当想到采用归纳法去尝试一下(当然,有时归纳法未必能够奏效或者未必给出最简单的解法).

下面就是一个与自然数有关的问题:

例题　设在圆周上写有 n ($n \geqslant 3$) 个自然数(允许重复),每一个数的相邻两数的和与这个数的比是整数,证明这 n 个比的和$\leqslant 3n$.

我们应当想到采用归纳法.首先奠基.在 $n=3$ 时,设这三个数中最大的为 M,另两个分别为 a, b.由于 $M \geqslant a$, $M \geqslant b$, 所以

$$a+b=2M$$

或

$$a+b=M.$$

如果 $a+b=2M$,那么 $a=b=M$,这时三个比的和 $=2+2+2=6<9$.

如果 $a+b=M$, 那么

$$a+M=2a+b$$

是 b 的倍数,所以 $2a$ 是 b 的倍数.同理 $2b$ 是 a 的倍数,于是有以下三种情况:

(1) $a=b$.

(2) $a>b$,这时由于 $2a>2b$,并且 $2b$ 是 a 的倍数,所以 $a=2b$.

(3) $b>a$,这时 $b=2a$.

不论是哪一种情况,三个比的和

$$\frac{a+b}{M}+\frac{a+M}{b}+\frac{b+M}{a} \leqslant 8<9.$$

因此,命题在 $n=3$ 时成立.

假定命题对于 $n-1$ 成立,现在考虑 n 个数的情况.设其中最大的为 M,与 M 相邻的为 a、b,与 b 相邻的另一个数为 c,与 a 相邻的另一个数为 d(如图 1).

去掉 M，这样圆周上只剩下 $n-1$ 个数，由于 $a+b=2M$ 或 M，我们仍分两种情况讨论.

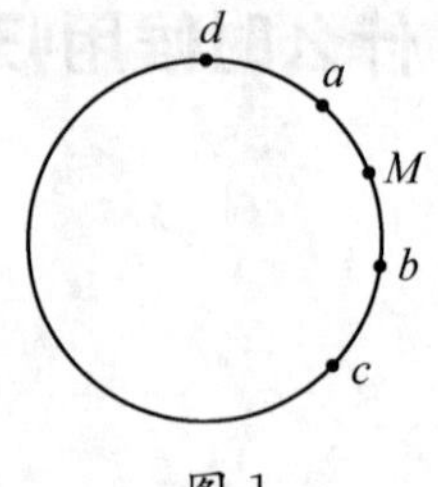

图 1

(1) 如果 $a+b=2M$，那么 $a=b=M$. 这时比

$$\frac{a+c}{b}=\frac{M+c}{b},\ \frac{d+b}{a}=\frac{d+M}{a}$$

都是自然数. 于是根据对 $n-1$ 个数所作归纳假设，相应的 $n-1$ 个比的和 $S'\leqslant 3(n-1)$. 回到原来的 n 个数的情况，这时 n 个比的和

$$S=S'+\frac{a+b}{M}=S'+2\leqslant 3(n-1)+2<3n.$$

(2) 如果 $a+b=M$，那么比

$$\frac{a+c}{b}=\frac{M-b+c}{b}=\frac{M+c}{b}-1$$

是自然数，比

$$\frac{d+b}{a}=\frac{M+d}{a}-1$$

也是自然数，仍然化为 $n-1$ 个数的情况. 由归纳假设，我们得到 $n-1$ 个比的和 $S'\leqslant 3(n-1)$. 回到 n 个数的情况，由于 $\frac{M+c}{b}$、$\frac{M+d}{a}$ 分别比 $\frac{a+c}{b}$、$\frac{d+b}{a}$ 大 1，比 $\frac{a+b}{M}=1$，这时 n 个比的和

$$S=S'+3\leqslant 3(n-1)+3=3n.$$

于是命题对一切自然数 $n\geqslant 3$ 成立.

在例题中有字母 n 出现，容易想到归纳法. 还有许多与自然数有关的命题，虽然没有明显地写出字母 n，我们也应当想到归纳法.

下面给出一道这样的习题，供大家思考练习.

练习题 在圆周上写有 64 个非零整数，将每个数改成与它相邻的两个数的乘积，得到 64 个新的数，再照这样继续处理，证明：经过有限多步，最后得出 64 个正数.

数学归纳法在整数问题中应用一例

数学归纳法是一种十分重要的方法，无论是三角、几何、代数，都需要用到归纳法. 当然在应用归纳法时，也或多或少地需要一些有关的知识，需要一些技巧. 这里举一个与整数有关的问题：

例　设正整数 $a_0 < a_1 < a_2 < \cdots < a_n$，证明

$$\frac{1}{[a_0, a_1]} + \frac{1}{[a_1, a_2]} + \cdots + \frac{1}{[a_{n-1}, a_n]} \leqslant 1 - \frac{1}{2^n}.$$

其中$[a, b]$表示 a 与 b 的最小公倍数.

我们先将解这题所需要的知识介绍一下：

(1) 最小公倍数是大家熟悉的，例如 $12 = 2^2 \times 3$ 与 $40 = 2^3 \times 5$ 的最小公倍数是 $[12, 40] = 2^3 \times 3 \times 5 = 120$.

(2) 最大公约数也是大家熟悉的，例如 12 与 40 的最大公约数是 $(12, 40) = 2^2 = 4$.

(3) 注意 $120 \times 4 = 12 \times 40$，一般地，$[a, b] \times (a, b) = ab$. 我们要利用这个结论.

(4) 还要指明一个简单的事实：如果正整数 $b < a$，那么 $(a, b) \leqslant a - b$.

这是很容易证明的(因为(a, b)整除 a，又整除 b，所以也整除正整数 $a-b$，从而 $(a, b) \leqslant a - b$).

现在来解上面的例题.

解　当 $n = 1$ 时，由于 $a_0 < a_1$，并且 a_0，a_1 都是正整数，所以 $a_1 \geqslant 2$，更有 $[a_0, a_1] \geqslant a_1 \geqslant 2$. 于是 $\frac{1}{[a_0, a_1]} \leqslant \frac{1}{2} = 1 - \frac{1}{2}$，即 $n = 1$ 时结论成立.

假设当 $n = k$ 时，结论成立. 考虑 $n = k + 1$ 的情况，这时有两种可能：

(1) $a_{k+1} \geqslant 2^{k+1}$. 这时有

$$\frac{1}{[a_k, a_{k+1}]} \leqslant \frac{1}{a_{k+1}} \leqslant \frac{1}{2^{k+1}},$$

又由归纳假设

$$\frac{1}{[a_0, a_1]}+\frac{1}{[a_1, a_2]}+\cdots+\frac{1}{[a_{k-1}, a_k]}\leqslant 1-\frac{1}{2^k},$$

两式相加得

$$\frac{1}{[a_0, a_1]}+\frac{1}{[a_1, a_2]}+\cdots+\frac{1}{[a_k, a_{k+1}]}\leqslant 1-\frac{1}{2^k}+\frac{1}{2^{k+1}}=1-\frac{1}{2^{k+1}}.$$

(2) $a_{k+1}<2^{k+1}$. 这时利用上面所说的知识,

$$\frac{1}{[a_i, a_{i+1}]}=\frac{(a_i, a_{i+1})}{a_i a_{i+1}}\leqslant\frac{a_{i+1}-a_i}{a_i a_{i+1}}=\frac{1}{a_i}-\frac{1}{a_{i+1}}, i=0, 1, \cdots, k.$$

将这 k 个不等式相加便得到

$$\frac{1}{[a_0, a_1]}+\frac{1}{[a_1, a_2]}+\cdots+\frac{1}{[a_k, a_{k+1}]}$$
$$\leqslant\left(\frac{1}{a_0}-\frac{1}{a_1}\right)+\left(\frac{1}{a_1}-\frac{1}{a_2}\right)+\cdots+\left(\frac{1}{a_k}-\frac{1}{a_{k+1}}\right)=\frac{1}{a_0}-\frac{1}{a_{k+1}}.$$

由于 $a_0\geqslant 1$, $a_{k+1}<2^{k+1}$, 因此

$$\frac{1}{a_0}-\frac{1}{a_{k+1}}\leqslant 1-\frac{1}{2^{k+1}}.$$

从而

$$\frac{1}{[a_0, a_1]}+\frac{1}{[a_1, a_2]}+\cdots+\frac{1}{[a_k, a_{k+1}]}\leqslant 1-\frac{1}{2^{k+1}}.$$

综合以上两种情况可知 $n=k+1$ 时,结论仍然成立. 于是对于一切自然数 n,结论成立.

最后赘述一句,上面的解法中所用的一个技巧是根据 a_{k+1} 的大小,分为两种情况(或更多种情况),用不同的方法处理. 这虽然简单,可是中学同学对这一点还是比较生疏的. 请读者仔细体会.

数学归纳法一例

在用归纳法解题时，有一个值得注意的现象：一个较强的结论可以用归纳法来证明，而较弱的结论却不能用归纳法直接证. 我们先举一个例子来说明这个现象.

例 1　设 n 为自然数，$x>-1$，证明

$$(1+x)^n \geqslant nx. \tag{1}$$

如果采用归纳法，$n=1$ 的情况是显然的，假定

$$(1+x)^k \geqslant kx. \tag{2}$$

却无法推出

$$(1+x)^{k+1} \geqslant (k+1)x.$$

（只能推出 $(1+x)^{k+1} \geqslant kx(1+x)=kx+kx^2$）.

要采用归纳法来解这道题，得把(1)换为更强的结论

$$(1+x)^n \geqslant 1+nx \tag{3}$$

（如果(3)成立，那么(1)当然也成立. 这就是说(3)比(1)强）.

现在 $n=1$ 的情况也是显然的. 假定

$$(1+x)^k \geqslant 1+kx, \tag{4}$$

那么

$$(1+x)^{k+1} \geqslant (1+x)(1+kx)=1+x+kx+kx^2 \geqslant 1+(k+1)x,$$

即(3)在 $n=k+1$ 时也成立，这就证明了(3)对一切自然数 n 成立. 从而(1)也成立.

为什么更强的结论(3)能用归纳法来证明，而较弱的结论(1)反而不能用归纳法直接证呢？

这是由于结论越强,那么我们的归纳假设也就越强. 在例 1 中,要证明(3),可以假定(4)成立,而要证明(1)却只有假定(2)成立,(4)这个归纳假设比(1)强,因而也就容易导出(3)(从(2)不能导出(1)).

因此,在考虑使用归纳法时,有时还需要这样的思路:"把结论加强到便于使用归纳法". 下面再举一例说明.

例 2 设 $a_n=\dfrac{1\cdot 3\cdot\cdots\cdot(2n-1)}{2\cdot 4\cdot\cdots\cdot(2n)}$, 证明

$$a_n<\frac{1}{\sqrt{3n}}. \tag{5}$$

首先,我们把结论(5)加强为

$$a_n\leqslant\frac{1}{\sqrt{3n+1}}. \tag{6}$$

由于 $a_1=\dfrac{1}{2}\leqslant\dfrac{1}{\sqrt{4}}$, 所以在 $n=1$ 时,(6)式成立. 假定

$$a_k\leqslant\frac{1}{\sqrt{3k+1}}, \tag{7}$$

那么

$$a_{k+1}=a_k\cdot\frac{(2k+1)}{2(k+1)}\leqslant\frac{1}{\sqrt{3k+1}}\cdot\frac{2k+1}{2(k+1)},$$

不难看出

$$\begin{aligned}\left(\sqrt{\frac{3k+4}{3k+1}}\right)^2&=\frac{3k+4}{3k+1}=1+\frac{3}{3k+1}\\&=1+\frac{2}{2k+1}+\frac{3}{3k+1}-\frac{2}{2k+1}\\&=1+\frac{2}{2k+1}+\frac{1}{(2k+1)(3k+1)}\\&<1+\frac{2}{2k+1}+\frac{1}{(2k+1)^2}=\left(\frac{2k+2}{2k+1}\right)^2,\end{aligned}$$

从而

$$a_{k+1} < \frac{1}{\sqrt{3k+4}} = \frac{1}{\sqrt{3(k+1)+1}},$$

即在 $n = k+1$ 时，(6)式也成立. 因此(6)式对一切自然数 n 成立.

由于(6)成立，较弱的结论(5)当然也成立.

方法一例：数学归纳法一例

编者按：数学归纳法是高二教材的内容，本文涉及的知识仅限于初二的范围，因此初中同学也可以阅读.

请大家先考虑一个几何命题：

“在$\triangle ABC$中，若$AB=2AC$，则$\angle C>2\angle B$.”

这一命题可以证明如下：

如图1，延长BC到B'，使$CB'=AC$，则$\angle B'=\angle CAB'=\frac{1}{2}\angle ACB$，

$$AB'<AC+CB'=2AC=AB.$$

于是$\angle B<\angle B'$，从而$\angle ACB=2\angle B'>2\angle B$. 证毕.

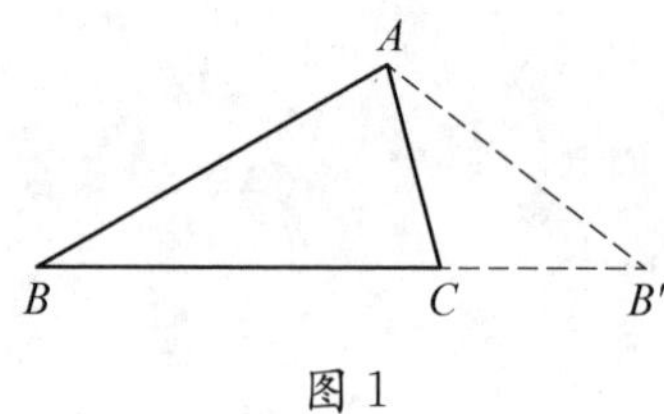

图1

上面命题有一个系数2，我们称这一命题为P_2. 如果将命题中系数2都改为3或都改为4(称为P_3或P_4)，命题是否成立呢？请大家考虑一下.

事实上，设n是大于1的整数，命题P_n也是成立的.

命题P_n 在$\triangle ABC$中，如果$AB=nAC$，那么$\angle C>n\angle B$.

下面我们来证明P_n也是正确的. 它的证明与对P_2的证明有点类似(可参照图1或另绘一图)：

延长BC到B'，使$CB'=(n-1)AC$，则

$$AB'<AC+CB'=nAC=AB.$$

于是$\angle B<\angle B'$，

$$\angle ACB=\angle CAB'+\angle B'>\angle CAB'+\angle B.$$

如果能够证明

$$\angle CAB'>(n-1)\angle B', \tag{1}$$

那么 $\angle C>n\angle B$ 就成立了.

但在$\triangle ACB'$中,有

$$CB'=(n-1)CA, \tag{2}$$

因此要证明命题 P_n,就归结为在(2)式假定下证明(1),这恰好是命题 P_{n-1}. 从而命题 P_{n-1} 成立,就可推出命题 P_n 成立.

同理,在 $n-2>1$ 时,由命题 P_{n-2} 成立可推出命题 P_{n-1} 成立. 即要证明命题 P_{n-1},只要证明命题 P_{n-2}.

依此类推,经过有限多步,命题 P_n 的证明归结为命题 P_2 的证明. 由于我们已经证明了命题 P_2 是正确的,于是逐步推得命题 P_3、P_4、…、P_{n-2}、P_{n-1}、P_n 都是正确的.

以上的证法就是数学归纳法. 在书写时可以将它分为两个部分:

1° 证明命题在 $n=2$ 时是成立的,这一步称为奠基. 在许多问题中,奠基是证明命题在 $n=1$ 时成立,但也可能是对某个整数 n_0 进行奠基(即证明命题在 $n=n_0$ 时成立).

2° 证明由命题 P_k 成立可推出命题 P_{k+1} 成立,这里 k 是任一$\geqslant 2$(或$\geqslant 1$,或$\geqslant n_0$)的整数(这样由命题 P_2 成立可推出 P_3、P_4、…P_n、…均成立).

我们再举一个相近的例子:

设 n 为自然数,在$\triangle ABC$ 中,如果

$$\angle C=n\angle B, \tag{3}$$

那么

$$AB\leqslant nAC. \tag{4}$$

我们采用数学归纳法来证明这个命题成立.

首先进行奠基. 当 $n=1$ 时,命题显然成立(这时(4)是等式).

其次,假设命题在 $n=k$ 时成立,这称为归纳假设,我们要证明由这个归纳

假设可以推出命题在 $n=k+1$ 时成立. 为此, 设在图 2 的 $\triangle ABC$ 中, $\angle C=(k+1)\angle B$.

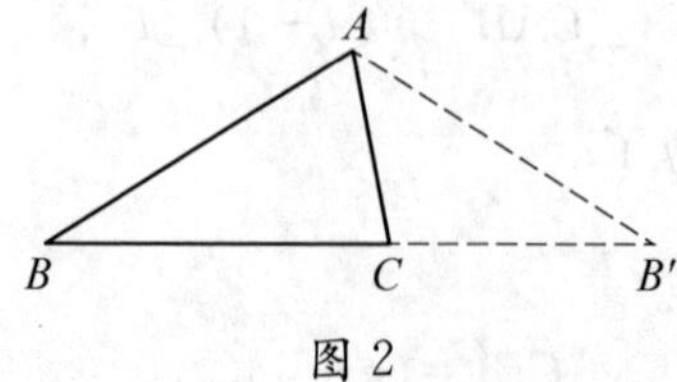

图 2

作 $\angle CAB'=k\angle B$, 它的边 AB' 与 BC 的延长线交于 B', 则

$$\angle B'=\angle ACB-\angle CAB'=(k+1)\angle B-k\angle B=\angle B,$$

从而 $AB=AB'$.

在 $\triangle ACB'$ 中, $\angle CAB'=k\angle B=k\angle B'$, 根据归纳假设 $CB'\leqslant kAC$, 从而

$$AB=AB'<AC+CB'\leqslant AC+kAC=(k+1)AC.$$

证毕.

如果 $n>1$, 那么(4)中不等式也可以改为严格不等式, 从上面的证明也可以看出这一点. 当然, 证明时要从 $n=2$ 进行奠基.

三角中的归纳法

在三角中也有很多应用归纳法的问题，例如1965年全国高考的第4题：

(1) 证明 $|\sin 2x| \leqslant 2|\sin x|$ (x 为任意值).

(2) 已知 n 为任意正整数，用数学归纳法证明

$$|\sin nx| \leqslant n|\sin x|. \ (x \text{ 为任意值}) \qquad (*)$$

(1)的证明是不难的. 因为 $|\cos x| \leqslant 1$，所以

$$|\sin 2x| = |2\sin x\cos x| \leqslant 2|\sin x|.$$

(2)的奠基部分即 $n=1$ 时($*$)是显然的等式.

$n=2$ 时即(1). 假设有 $|\sin(n-1)x| \leqslant (n-1)|\sin x|$，那么

$$\begin{aligned}
&|\sin nx| \\
=&|\sin(n-1)x\cos x + \cos(n-1)x\sin x| \\
\leqslant&|\sin(n-1)x\cos x| + |\cos(n-1)x\sin x| \\
\leqslant&|\sin(n-1)x| + |\sin x| \\
\leqslant&(n-1)|\sin x| + |\sin x| \\
=&n|\sin x|.
\end{aligned}$$

因此($*$)对任意正整数 n 成立.

有趣的是这道高考题与上一篇文章的几何问题密切相关.

事实上，若在$\triangle ABC$中，$c=nB$，则由($*$)式得 $\sin C = \sin B \leqslant n\sin B$.

再由正弦定理，得

$$c = \frac{b\sin C}{\sin B} \leqslant nb.$$

上一篇文章的另一个几何问题也可以这样处理：

由 $b=nc$ 及正弦定理，得 $\sin C = n\sin B$.

所以由(*)得 $n\sin\frac{C}{n}\geqslant\sin C$.

所以 $\sin\frac{C}{n}\geqslant\sin B$.

由于$\frac{C}{n}$与 B 都是锐角,所以 $\frac{C}{n}\geqslant B$, 即 $C\geqslant nB$. (注意如果用 $\sin nB\leqslant n\sin B=\sin C$ 不能立即得出 $nB\leqslant C$, 因为 nB 未必是锐角).

下面的一些恒等式都可以用归纳法证明:

$$\frac{1}{2}+\cos\theta+\cos2\theta+\cdots+\cos n\theta=\frac{\sin\frac{2n+1}{2}\theta}{2\sin\frac{\theta}{2}},$$

$$\cos\theta+\cos2\theta+\cdots+\cos n\theta=\frac{\sin\frac{n}{2}\theta\cos\frac{n+1}{2}\theta}{\sin\frac{\theta}{2}},$$

$$\cos\theta+\cos3\theta+\cdots+\cos(2n-1)\theta=\frac{\sin2n\theta}{2\sin\theta},$$

$$\sin\theta+\sin2\theta+\cdots+\sin n\theta=\frac{\sin\frac{n}{2}\theta\sin\frac{n+1}{2}\theta}{\sin\frac{\theta}{2}},$$

$$\frac{\sin\theta}{\sin\theta}+\frac{\sin3\theta}{\sin\theta}+\frac{\sin5\theta}{\sin\theta}+\cdots+\frac{\sin(2n-1)\theta}{\sin\theta}=\left(\frac{\sin n\theta}{\sin\theta}\right)^2,$$

$$\underbrace{\sqrt{2+\sqrt{2+\sqrt{2+\cdots+\sqrt{2}}}}}_{n\text{个根号}}=2\sin\left(1+\frac{1}{2}+\frac{1}{4}+\cdots+\frac{1}{2^n}\right)\frac{\pi}{4}.$$

供有兴趣的读者练习.

处处留心皆学问

学习，应当不断地提出问题，“于无疑处有疑”，切勿轻易放过那些可以动动脑筋的机会.

这里举两个简单有趣的几何作图中的问题.

例 1　过直线 l 外一点 A 作 l 的平行线.

这样的作图，初中同学人人学过. 我们应该再进一步，提出新的要求——“至少要画几条线，就可以完成上述作图？”

这里的线指直线或圆（弧），并且作图工具只有圆规与（没有刻度的）直尺，不允许用三角板推.

答案是 3 条线.（请同学们先想一想，试一试，然后再看下面的解答）作法如下：

如图 1，在 l 上任取两点 B、C. 分别以 A、C 为圆心，BC、BA 为半径作圆，两圆（与 A 在 l 同侧）的交点为 D，过 A、D 作直线.

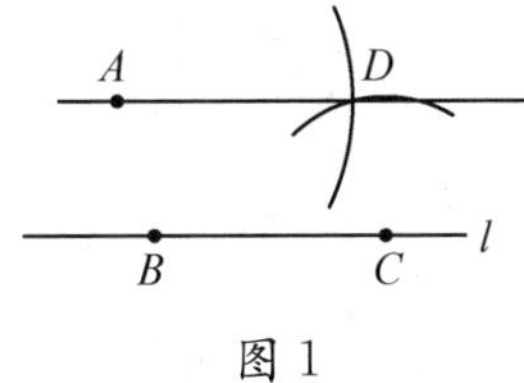

图 1

易知四边形 $ABCD$ 是平行四边形，所以 AD 就是所求的平行线.

我们一共作了三条线：两个圆，一条直线. 能不能减少一条线呢？不能！因为最后总要作一条直线，它由 A 及另一点 D 确定；而定出 D 点必须两条线，所以至少要作三条线.

类似地，还可以考虑：

例 2　过点 A 作直线 l 的垂线，至少要作几条线（只允许用圆规、直尺，不允许用三角板）？

解答分两种情况:

(1) A 在 l 外

通常的作法需要 4 条线. 但是,只要我们多想一想,可以知道用 3 条线就足够了. 如图 2,在 l 上任取两点 B、C,分别以 B、C 为圆心,BA、CA 为半径作圆,相交于(异于 A 的)点 D,则 AD 就是 l 的垂线.

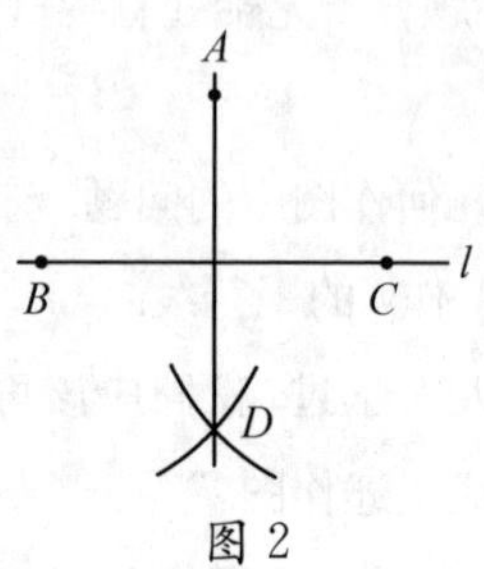

图 2

(2) A 在 l 上

上面的作法不再适用,必须另辟蹊径.

在 l 外任取一点 B,以 B 为圆心,BA 为半径作圆,交 l 于(异于 A 的)点 C(如果这圆与 l 相切于 A,那么 BA 就是 l 的垂线). 过 C 作⊙B 的直径 CD,则直线 DA 即为所求(图 3).

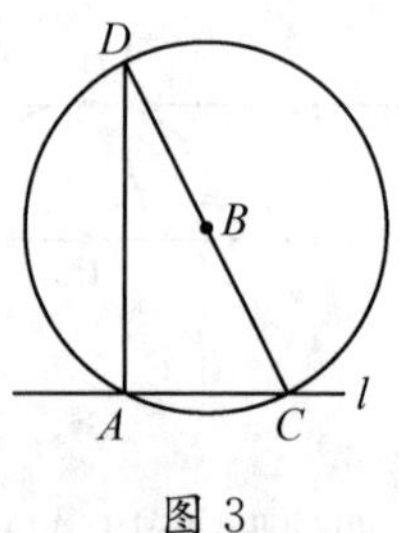

图 3

与例 1 的道理相同,3 条线是最少的.

上面的两道作图题虽不困难,但也不一定很快就能想到正确的答案. 在学习的过程中应当多思,多问几个为什么. 坚持这样做,你的学问必将大有长进.

一个作图问题

几何学家 Pedöe 曾经在一家印度的数学刊物上提出一个问题：用一个张脚固定为 r 的圆规(即只能作半径为 r 的圆)，能否作出一点 A，使它与两个已知点 B、C 构成等边三角形的三顶点？

这个问题一直未得到解答. 据 Pedöe 本人的意见，如果线段 $BC \leqslant 2r$ 或者线段 BC 已经画出，那么问题都不难解决，但一般情况，似乎是不能解决的.

其实，这个问题是可以解决的. 我们采用归纳法来作图.

首先，当 $BC \leqslant 2r$ 时可以用下面的方法作出点 A：

分别以 B、C 为圆心作半径为 r 的圆，两圆相交于 D. 再以 D 为圆心作半径为 r 的圆，与 $\odot B$、$\odot C$ 分别交于 E、F(如图 1). 最后以 E、F 为圆心，作两个径为 r 的圆相交于 A，则点 A 即为所求.

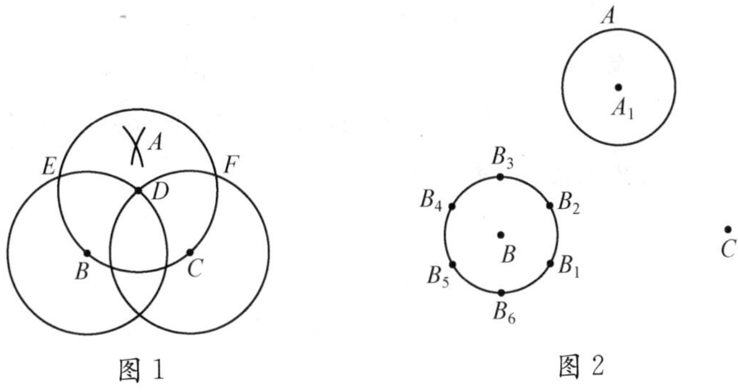

图 1　　图 2

证明甚易故从略. 我们指出这时常有两解，图 1 中 A、B、C 三点是依逆时针方向排列的. A 点关于 BC 的对称点 A' 也是问题的解，但 A'、B、C 三点是依顺时针方向排列的.

现在我们设在 $BC \leqslant \frac{n}{2}r$ 时可以作出点 A，使 $\triangle ABC$ 为等边三角形，并且 A、B、C 三点是依逆时针方向排列的. 则当 $BC \leqslant \frac{n+1}{2}r$ 时，如图 2，以 B 为圆

心作半径为 r 的圆，在$\odot B$ 的圆周上任取一点 B_1，然后在$\odot B$ 的圆周上作出点 B_2、B_3、B_4、B_5、B_6，使 $B_1B_2=B_2B_3=B_3B_4=B_4B_5=B_5B_6=r$. 由于$\angle B_iBC$ 中必有一个 $\leqslant\frac{60^\circ}{2}=30^\circ(i=1,2,3,4,5,6)$，不妨设 $\angle B_1BC\leqslant 30^\circ$，那么根据余弦定理，得

$$CB_1^2=CB^2+r^2-2CB\cdot r\cos\angle B_1BC\leqslant\left(\frac{n+1}{2}r\right)^2+r^2-(n+1)r^2\cdot\frac{\sqrt{3}}{2},$$

所以

$$CB_1\leqslant\frac{n+1}{2}r-\frac{1}{2}r=\frac{n}{2}r,$$

根据归纳假设，可以作出一点 A_1，使$\triangle A_1B_1C$ 为等边三角形，并且 A_1、B_1、C 三点是依逆时针排列的.

考虑绕点 C 的旋转，旋转角为 60°(顺时针方向旋转)，这时 B_1 变为 A_1，B 变为所求出的点 A，因此 $AA_1=BB_1=r$，即以 A_1 为圆心 r 为半径作圆，则点 A 必在这个圆上.

我们在$\odot B$ 上再取点 B_1'、B_2'、B_3'、B_4'、B_5'、B_6'，使 $B'_1B'_2=B'_2B'_3=B'_3B'_4=B'_4B'_5=B'_5B'_6=r$，设 $\angle B_1'BC\leqslant 30^\circ$，根据归纳假设作出点 A_1'，使$\triangle A_1'B_1'C$ 为等边三角形，并且 A_1'、B_1'、C 三点是依逆时针顺序排列的，以 A_1'为圆心 r 为半径作圆. $\odot A_1'$与$\odot A_1$ 的交点就是所求的点 A.

几个尺规作图问题

直尺、圆规是欧氏平面几何的作图工具. 今天,有了几何画板,不以规矩也能成方圆,但是尺规仍然是常用的作图工具. 不仅如此,尺规作图在培养思维能力方面极具作用,这或许正是希腊人当时限定仅用尺规作图的初衷.

苏联数学家曾写过小册子《直尺作图》《圆规作图》.《直尺作图》讨论了使用直尺的作图,其中最重要的一个作图是下面的命题.

命题 1 已知直线 $AB \parallel CD$, 那么可仅用直尺作出线段 AB 的中点.

作法 如图 1,在直线 AB、CD 外任取一点 P,联结 PA、PB,分别交直线 CD 于 E、F,联结 BE、AF,相交于 N,过 P、N 作一条直线,交 AB 于 M,则 M 为线段 AB 的中点.

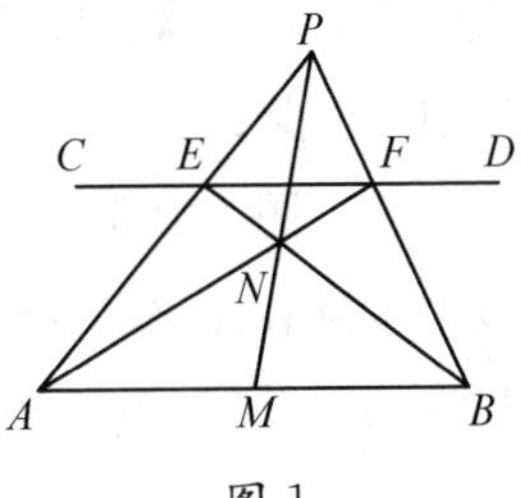

图 1

证明不难,曾被用作 1978 年第一届全国高中数学联赛的试题,这里从略.

仅用直尺可以完成许多尺规作图,但不能完成全部的尺规作图. 理由很简单:直尺作出的直线都满足一次方程(在建立直角坐标系后),所得到的点都与原有的点(可假定为有理点)的坐标及原有的直线方程的系数(也可假定为有理数)在同一个数域中;而圆满足二次方程,用尺规作图会得到一些点,坐标在有理数域的二次(乃至 2^n 次)扩张中;因此仅用直尺不能完成全部的尺规作图. 但如果给出一个圆及其圆心,那么用直尺就可以完成全部的尺规作图.

仅用圆规可完成全部的尺规作图,这是有名的马斯开龙尼定理. 当然, 这里的"过两点 A、B 作直线"不可能真的仅用圆规画出来,而是仅用圆规可以完成直线 AB 的全部功能,即:

(1) 可作出直线 AB 上任意多个点.

(2) 可作出直线 AB 与另一条(也是已知两个点的)直线 CD 的交点.

(3) 可作出直线 AB 与任一个圆的交点.

这此都可以参看上面提到的那两本小册子.

本文主要讨论几个尺规作图,其中做如下的一些限定:

(1) 限定尺、规使用的总次数.

(2) 限定直尺使用的次数.

(3) 限定圆规使用的次数.

作图 1 已知直线 l 及直线 l 外一点 P,过 P 作 l 的垂线.

通常的作法是以 P 为圆心,任作一个圆交 l 于 A、B,分别以 A、B 为圆心,过 P 作圆,两圆又相交于点 Q,过 P、Q 作一条直线,则直线 PQ 为所求(图 2).

这里圆规用了 3 次,直尺用了 1 次,共计 4 次.

能不能使总次数减少?最少是几次?这个问题不难.在 l 上任取两点 A、B,分别以 A、B 为圆心,过点 P 作圆,又相交于点 Q,再过 P、Q 作一条直线,PQ 即为所求(图 3).尺规使用的总次数为 3,而且 3 是最少的.因为最后需要作直线 PQ,用一次直尺,而定出点 Q,必须作两条线(直线或圆)相交,所以尺规必须使用至少 3 次.

图 2、图 3 中都用了两次圆规,如果限定只能用一次圆规呢?可以如图 4,先以 P 为圆心,任作一个圆与 l 相交于 A、B, PB、PA 分别再交圆于 C、D 两点,这时 $CD \parallel AB$,由命题 1,只用直尺可作出 AB 的中点 M,而直线 $PM \perp AB$.

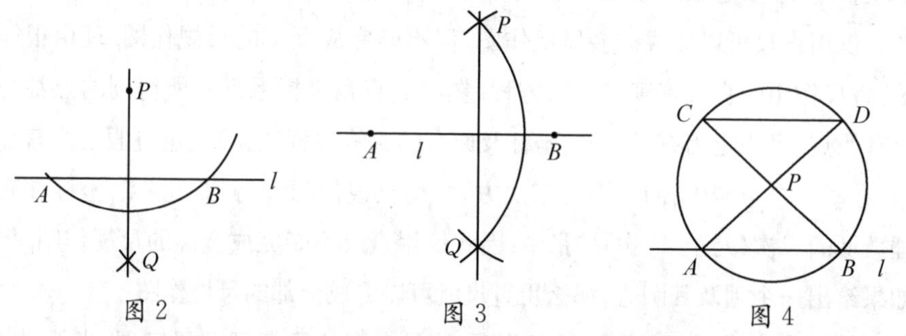

图 2　　图 3　　图 4

不用圆规能作出 l 的垂线 PQ 吗?这个问题留给读者想一想.

作图 2 已知直线 l 及 l 上一点 P,过 P 作 l 的垂线.

通常的作法,尺规使用的总次数为 4(与图 2 类似).

怎样减少为 3?(由作图 1 所说的理由,3 是最少的)在 l 外任取一点 O,以 O 为圆心,过 P 作圆,又交 l 于 Q,过 O、A 作一条直线又交圆于 Q.过 P、Q 作一条直线,则 PQ 即为所求(图 5).

这里圆规只用了一次，而直尺用了两次，能否只用一次直尺呢？可以的.在图5中得到点A后，以A为圆心过O作圆，交圆O于B(图6)，再以B为圆心过O作圆，交圆O于C，以C为圆心过O作圆，交圆O于Q，则直线PQ即为所求(如不需要实际作出直线PQ，按上面《圆规作图》一书的说法，不用直尺就完成了作图).

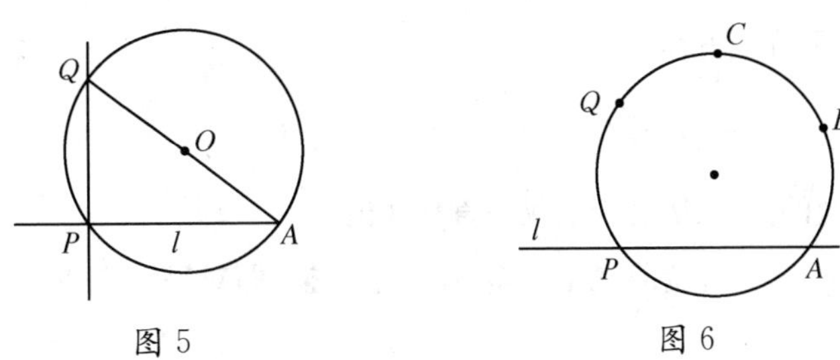

图5　　图6

作图3　已知$\angle AOB$，作$\angle AOB$的角平分线.

通常的作法是以O为圆心，任作一个圆，分别交OA、OB于C、D.再分别以C、D为圆心，过O作圆，两圆交于E(图7)，射线OE即为所求.

其中圆规用了3次.能不能减少为2次？

如图8，以O为圆心作圆，分别交OA、OB于C、D，再以O为圆心，以与OC不同的长度为半径作圆，分别交OA、OB于E、F，联结CF、DE，相交于G，则OG即为所求.

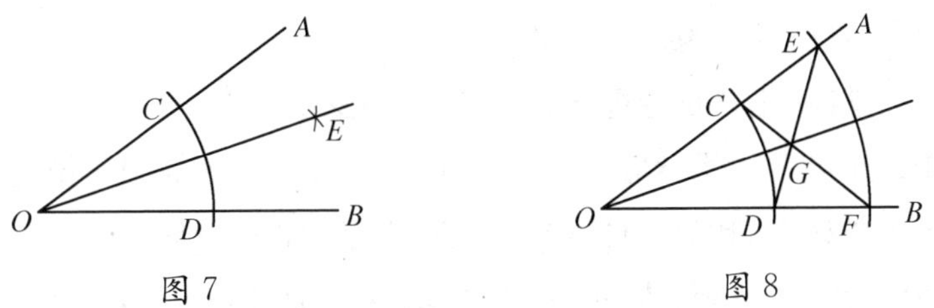

图7　　图8

能不能只用一次圆规？如图4(将其中点P改为点O)，以O为圆心，作一个圆分别交OA、OB及它们的延长线于A、B、C、D.再用直尺作出线段AB的中点M，过O、M作直线即可.

能不能完全不用圆规？不能.

下面介绍两种证法.

第一种证法.证明$45°$的$\angle AOB$不能仅用直尺平分.

以 O 为原点，OA 为 x 轴，OC 为 y 轴建立直角坐标系(图 9).

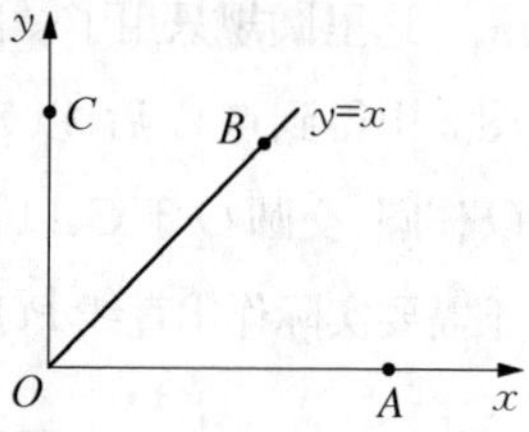

OB 的方程是

$$y=x, \tag{1}$$

而$\angle AOB$ 的平分线方程为

$$y=(\tan 22.5^\circ)x, \tag{2}$$

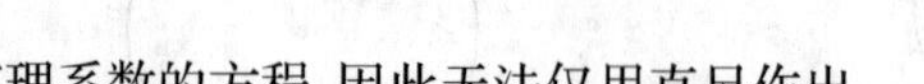

图 9

不是有理系数的方程，因此无法仅用直尺作出.

第二种证法. 仍建立直角坐标系. 作压缩变换，即对每一点 $M(x,\ y)$，定义它的像为 $M'\left(x,\ \frac{y}{2}\right)$.

容易看出压缩变换将直线变为直线，点与直线的从属关系保持不变.

如果仅用直尺就能平分一个角，那么作压缩变换，所得的像采用同样作法，得到平分线的像应当是像的平分线. 但事实并非如此，例如在图 9 中，直角$\angle AOC$ 压缩后仍是自身，它的平分线 OB 压缩后却不是 OB(方程变为 $y=\frac{x}{2}$)，不是$\angle AOC$ 的平分线.

注 直角经过压缩，通常并不是直角，仅在直角边平行于坐标轴时才是这样. 由此也可得出为什么仅用直尺不能作垂线. 一般地，在压缩变换下不能保持的性质，仅用直尺不能完成有关作图.

还可以举许多作图问题进行研讨，不过本文不想写长，留给读者去考虑吧！

再赘述几句. 初等数学历史悠久，因此重大的研究几乎都已被做完. 我以为初等数学的普及或许比研究更为重要(有时普及中也蕴含研究). 当然，初等数学中也可能会发现一些新结果(例如叶中豪就在平面几何中发现不少新问题)，但大多只是一些问题，不能形成庞大的理论体系. 我国初等数学研究最大的成果或许就是组合数学中陆家羲的工作. 如果打个不很妥帖的比方，钻研这些难题好比啃骨头，而从事新理论的发现与研究好比吃肉. 我国数学界的研究，过去也是啃骨头多(如哥德巴赫问题就是一根坚硬的骨头). 但现在注意到并且能够吃肉的新一代数学家也多起来了(如恽之伟、孙斌勇、许晨阳、田野等)，他们大多经过奥数的训练或熏陶.

存在性问题的一种证法

在数学中，常常出现这样的问题：要证明具有某种性质 A（性质 A 的具体内容由具体问题给出）的数学对象是存在的．这类存在问题有的是很难的，有的是至今尚未解决的，但也有不少是属于初等数学的范畴，常常在各种数学竞赛中出现．本文的目的就是介绍一种方法，用以解决这类初等的存在性问题．在很多场合，这一方法是可以奏效的（当然，决不是万能的），这一方法的要点就是在所讨论的数学对象（我们可以称它为元素或元）的全体所组成的集合 M 中，选取具有某种极端性质（最大或最小）的那一个元，然后再证明这个元恰好具有性质 A．有时集合 M 中的元素并不具备这种极端性质（或具有极端性质的元素并不具备性质 A），我们也可以考虑与 M 有关的另一个集合 N，在 N 中选取某个具有极端性质的元，再证明 M 中与它相关联的一个元具有性质 A．

我们通过下面的一些例题来阐述这一方法．其中集合 M（或 N）并无必要指明，所以我们也不一一指出，以免繁琐冗长，令人生厌．

例 1　已知平面上 100 个点，任意两点间的距离 $\leqslant 1$，任意三点组成一个钝角三角形．证明存在一个半径为 $\frac{1}{2}$ 的圆，将这 100 个点全部盖住．

解　以这 100 个点中每两点的连线为直径作圆，这些圆的半径都不大于 $\frac{1}{2}$，我们希望在这些圆中找出一个圆，将这 100 个点全部盖住．

设 $\odot O$ 为其中一圆，以已知点 A、B 的连线为直径．对任一已知点 C（C 不同于 A、B），如果 $\angle ACB$ 为钝角，即 AB 为 $\triangle ABC$ 的最大边，那么 C 就被 $\odot O$ 盖住．因此，如果 $\odot O$ 是上述圆中直径最大的圆，那么这个圆就把 100 个已知点全部盖住．

在这个例题中，集合 M 与 N 相同，即以每两个已知点的连线为直径作圆，这些圆的全体就是集合 M．

例 2　证明任一四面体 $ABCD$ 中总可以找到一个顶点，由这点引出的三条棱可以组成一个三角形.

解 设四面体 $ABCD$ 的六条棱中最长的一条为 AB,我们证明 A、B 两点中必有一点符合要求,不然的话,将有

$$AB \geqslant AC + AD,$$
$$AB \geqslant BC + BD,$$

相加得

$$2AB \geqslant (AC + BC) + (AD + BD) > AB + AB = 2AB,$$

矛盾!

在本例中,集合 M 由四个顶点组成,集合 N 由六条棱组成. 在下面的例题中,我们不再指明集合 M、N.

例 3 已知正多边形 $A_1A_2\cdots A_n$ 内接于⊙O, P 为⊙O 内一点,证明必定存在两个顶点 A_h、A_k,使得

$$180^\circ\left(1-\frac{1}{n}\right) \leqslant \angle A_hPA_k \leqslant 180^\circ.$$

解 设在 A_1、A_2、…、A_n 中, A_i 距点 P 最近. 这时有两种情况:

(1) 如果在直线 A_jP 上还有一个顶点 $A_j\,(1 \leqslant j \leqslant n)$,那么显然有

$$180^\circ\left(1-\frac{1}{n}\right) < \angle A_iPA_j = 180^\circ.$$

(2) 如果在直线 A_iP 上无其他顶点,那么其他顶点(全部在 A_iP 的一侧时,命题显然正确)分居在直线 A_iP 的两侧,因而其中有两个相邻的顶点 A_j、$A_{j+1}\,(1 \leqslant j \leqslant n$,约定 A_{n+1} 即 $A_1)$,分别在直线 A_iP 的两侧(图 1). 于是由于

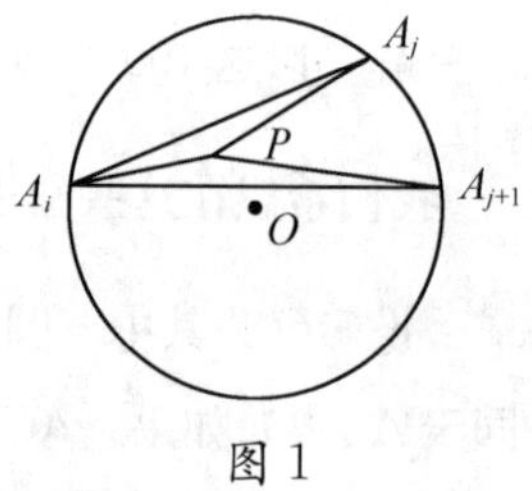

图 1

$$PA_j \geqslant PA_i,\ PA_{j+1} \geqslant PA_i,$$

所以

$$\angle PA_iA_j \geqslant \angle PA_jA_i,\ \angle PA_iA_{j+1} \geqslant \angle PA_{j+1}A_i,$$

从而

$$\angle PA_jA_i+\angle PA_{j+1}A_i\leqslant\angle PA_iA_j+\angle PA_iA_{j+1}=\angle A_jA_iA_{j+1}=180^\circ\times\frac{1}{n},$$

$$\angle A_iPA_j+\angle A_iPA_{j+1}\geqslant 360^\circ-2\times180^\circ\times\frac{1}{n}=360^\circ\left(1-\frac{1}{n}\right),$$

因此 $\angle A_iPA_j\geqslant180^\circ\left(1-\frac{1}{n}\right)$ 与 $\angle A_iPA_{j+1}\geqslant180^\circ\left(1-\frac{1}{n}\right)$ 这两个不等式至少有一个成立，证毕.

例 4　有限多个圆覆盖着面积为 S 的区域，证明可以从中找出一组没有公共点的圆，它们所覆盖的区域的面积$\geqslant\frac{1}{9}S$.

解　从这些圆中取出最大的一个，设这圆为$\odot(O_1, r_1)$(即圆心为 O_1 半径为 r_1)，其面积 $S_1=\pi r_1^2$. 将$\odot(O_1, r_1)$及所有与$\odot(O_1, r_1)$有公共点的圆全部去掉，由于这些圆的半径均不大于 r_1，所以这些圆都在$\odot(O_1, 3r_1)$内，因而覆盖着面积$\leqslant 9S_1$ 的区域 P_1(图 2)，换句话说，$S_1\geqslant P_1$ 的面积的$\frac{1}{9}$.

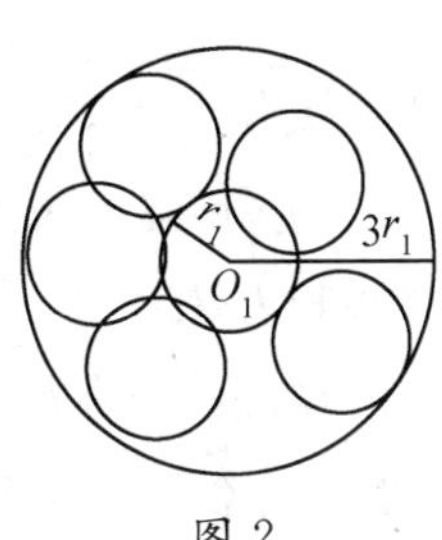

图 2

对剩下的圆进行同样的处理，即再取出一个最大的圆$\odot O_2$，将这圆及所有与它有公共点的圆全部去掉，这样继续下去. 不难看出$\odot O_1$、$\odot O_2$、…就是所求的圆.

例 5　n 个点 $(n>4)$，每三个点不共线，证明以这些点为顶点至少可以构成 C_{n-3}^2 个凸四边形.

解　以其中任三点为顶点作三角形，设其中面积最大的为$\triangle ABC$，过 A、B、C 分别作对边的平行线得到$\triangle A'B'C'$(图 3).

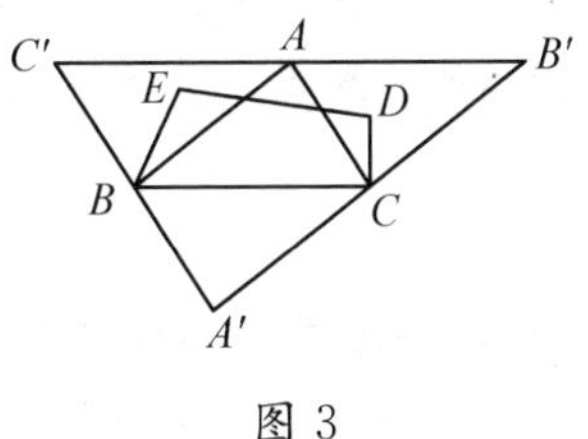

图 3

因为$\triangle ABC$ 的面积最大，所以其余的 $n-3$ 个点均在$\triangle A'B'C'$中，过其中任意两点 D、E 的直线只能与$\triangle ABC$ 的两条边相交，不妨设直线 DE 与边(线段)BC 不相交，则 B、C 在直线 DE 同侧，D、E 在直线 BC 同侧，从而四边形 $DEBC$ 为凸四边形. 因此至少有 C_{n-3}^2 个凸四边形.

从“挖坑”谈起

一条路上，挖了些坑. 走路时，如果不当心，就会落在坑里.

路上有坑，不是好事. 但是，挖坑的思想却可以用来解题.

例 1 a、d 都是正整数，证明在等差数列

$$a,\ a+d,\ a+2d,\ a+3d,\ \cdots \tag{1}$$

中一定有一个数的首位数字为 9.

为了解这道题，我们设想有一位机器人站着数轴的正方向，从 a 点出发，每步步长为 d.

在机器人前进的路上挖一个坑，也就是在数轴上取一个半开区间 $[9\times10^n,\ 10^{n+1})$，这个坑(区间)的长为 10^n.

只要 n 足够大(比如说 $n>\lg d$)，那么坑长 $10^n>$步长 d. 于是机器人必然要落到坑里，即存在 $a+kd\in[9\times10^n,\ 10^{n+1})$. 这时 $a+kd$ 的首位数字当然是 9.

实际上，我们用同样的方法可以证明(1)中存在着一个数，它的前若干位数字组成预先给定的数.

例 2 证明存在一个平方数，它的数字中有 0, 1, 2, 3, 4, 5, 6, 7, 8, 9.

如果我们设 $b=1234567890$，那么只需要证明有一个平方数，它的前十位数字组成的数为 b. 于是，设想有一位机器人走过平方数数列

$$1^2,\ 2^2,\ 3^2,\ \cdots,\ k^2,\ (k+1)^2,\ \cdots, \tag{2}$$

虽然这里机器人的步长是递增的，但我们仍然可以挖一个坑 $[b\times10^n,\ (b+1)\times10^n)$ 使机器人落进去. 这只要取 $n>12$，这时

$$\sqrt{b\times10^n}<\sqrt{10^{2n-2}}=10^{n-1},$$

当机器人走到坑前，即

$$k^2\leqslant b\times10^n<(k+1)^2$$

时，由于步长

$$2k+1\leqslant 2\sqrt{b\times 10^n}+1<2\times 10^{n-1}+1<10^n$$

（坑长），所以再向前一步，机器人就落进坑内.

例 3 设机器人从原点出发，步长为固定正实数 d. 如果每个整数都看成是一个坑（坑长退化为 0），问 d 是什么数时机器人总要落进坑里？

答案是当且仅当 d 为有理数时，机器人迟早落进坑里.

因为 d 为有理数 $\frac{n}{m}$ 时（m、n 正整数），有

$$md=n, \tag{3}$$

即在第 m 步，落入坑 n 里.

反之，如果在第 m 步，落入坑 n 里，那么(3)成立，从而 $d=\frac{n}{m}$ 为有理数.

在例题 3 中，d 为正无理数时，机器人不会落入坑 n 里，但是如果坑长不是零，情况就大不相同了.

对于任意的正数 ε（ε 可以取得很小，例如 $\varepsilon=0.000000001$），在数轴上挖坑

$$(m-\varepsilon,\ m+\varepsilon),\ m=0,\ 1,\ 2,\ \cdots,$$

如果 d 是正无理数，那么从原点出发，步长为 d 的机器人迟早要落入坑里.

这是一个很重要的结论，称为**迪里赫勒**(Dirichlet)**定理**，通常叙述成：

如果 d 是无理数，那么对任意正数 ε，存在着整数 m、n，使得

$$|nd-m|<\varepsilon. \tag{4}$$

现在我们来证明这个定理. 将区间[0, 1]分成若干等份，每份的长$<\varepsilon$. 显然只要份数 $k>\frac{1}{\varepsilon}$，就能达到这一要求.

考虑 $nd-[nd]$，这里 $[x]$ 表示 x 的整数部分，也就是不超过 x 的最大整数. 显然，$nd-[nd]$ $(n=1,\ 2,\ 3,\ \cdots)$ 都在区间[0, 1]中. 由于这种形状的数有无限多个，而[0, 1]只分为 k 个等份. 所以必有两个数 $n_1d-[n_1d]$ 与 $n_2d-[n_2d]$ $(n_2\neq n_1)$ 落入同一个等份里（这个简单的原理称为**迪里赫勒原则**或**抽屉原则**，有着广泛的应用），即有

$$|(n_1d-[n_1d])-(n_2d-[n_2d])|<\varepsilon, \tag{5}$$

改记 $n_2-n_1=n$, $[n_2d]-[n_1d]=m$, 则上式即成为(4). 证毕.

迪里赫勒定理有许多应用.

例 4 证明存在自然数 n,使

$$|\sin n|<0.0000001.$$

我们取 $\varepsilon=\arcsin 0.0000001$. 根据迪里赫勒定理,存在整数 m、n 满足

$$|n-2m\pi|<\varepsilon, \tag{6}$$

于是

$$|\sin n|=|\sin(2m\pi+n-2m\pi)|=\sin(n-2m\pi)|<\sin\varepsilon=0.0000001.$$

在迪里赫勒定理中所挖的坑 $(m-\varepsilon, m+\varepsilon)$ 是以整数 m 为中心的,其实并不需要如此. 对于任意的 a、b: $0<a<b<1$, 可以证明如果机器人从原点出发,步长为正无理数 d,那么它迟早要落到形如 $(m+a, m+b)$ $(m=0, 1, 2, \cdots)$ 的坑里.

为了证明这个事实,我们在迪里赫勒定理中取 $\varepsilon=b-a$. 则有整数 n、m 满足

$$|nd-m|<b-a \tag{7}$$

(不妨假定(7)中 $n>0$, 否则用 $-n$ 及 $-m$ 代替 n 及 m).

考虑数

$$nd-m, 2(nd-m), 3(nd-m), 4(nd-m), \cdots,$$

也就是考虑一个自原点出发、步长为 $|nd-m|$ 的机器人. 由于(7)式,它迟早要落到长为 $b-a$ 的某个形如

$$(l+a, l+b), l=0, \pm 1, \pm 2, \cdots$$

的坑中,即有正整数 k 及整数 l 使

$$l+a<k(nd-m)<l+b,$$

改记 kn 为 n, $km+l$ 为 m,则

$$m + a < nd < m + b,$$

这就是要证明的.

刚刚证明的结论称为**克朗涅克**(Kronecker)**定理**,这个定理也有很多应用.

例 5　证明存在正整数 m,使得 7^m 的前 N 位数字组成的数为一个给定的 N 位数 b(例如 b 为四位数 8541).

所要证明的结论也就是存在正整数 m 及 n,使得 7^m 落入 $[b \times 10^n, (b+1) \times 10^n)$ 内,即

$$b \times 10^n \leqslant 7^m < (b+1) \times 10^n, \tag{8}$$

(8)等价于

$$n + \lg b \leqslant m\lg 7 < n + \lg(b+1),$$

即 $m\lg 7$ 落入区间 $(n + \lg b, n + \lg(b+1))$ 中. 由于 $\lg 7$ 是无理数,因此由克朗涅克定理知道有这样的 m 及 n 存在. 证毕.

显然 7 可以换成任意一个正整数,只要这个数不是 10 的非负整数幂.

迪里赫勒定理与克朗涅克定理都与有理数逼近无理数的问题有关. 它们还有许多推广(例如联立逼近的问题),这里就不赘述了.

注　通常的**克朗涅克定理**表述为:

如果 d 是无理数,c 是任意实数,那么对于任给正数 ε,有整数 m、n 存在,使

$$|nd - m - c| < \varepsilon.$$

这与我们上面的说法实质上是一致的 $(b = c + \varepsilon, a = c - \varepsilon)$.

几何变换与证题

几何学中所研究的图形并非孤立静止的，它们之间可能存在着各种各样的联系，通过变换可以把一个几何图形变为另一个几何图形.

虽然在中学课程里不可能(也没有必要)采用变换群的观点，但是注意培养学生用变换的思想来考察问题还是十分必要的.

本文的目的就是通过一些例题说明如何利用变换来证题.

例 1 如图 1，$\triangle ABC \cong \triangle A'B'C'$，$AD$ 为 $\angle BAC$ 的平分线，$A'D'$ 为 $\angle B'A'C'$ 的平分线，证明 $AD = A'D'$.

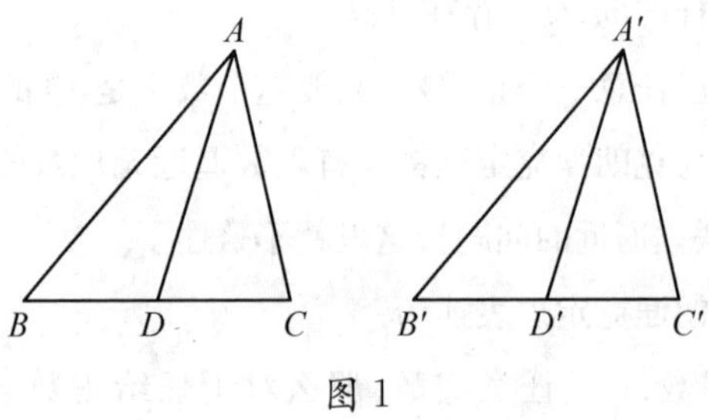

图 1

这个问题，如果不采取变换的观点，证起来还不很容易(当然也不难). 但是，采取变换的观点，问题就成为显然了. 我们把 $\triangle ABC$ 放在 $\triangle A'B'C'$ 上，由于 $\triangle ABC \cong \triangle A'B'C'$，可以使对应的顶点与边全部重合，这时 $\angle BAC$ 的平分线与 $\angle B'A'C'$ 的平分线重合，因此 $AD = A'D'$.

这里采用的变换就是全等变换或者叫做合同变换、运动.

采用合同变换的观点，两个全等的图形都可以看成是一个图形，因此两个全等的图形中对应的线段一定相等.

可以证明平面上的合同变换都可以由反射(轴对称)、平移及(绕一个定点的)旋转组成.

我们先来看看反射在证题中的应用.

例 2 在正三角形 ABC 的三条边上各取一点 D、E、F，证明这个内接三角

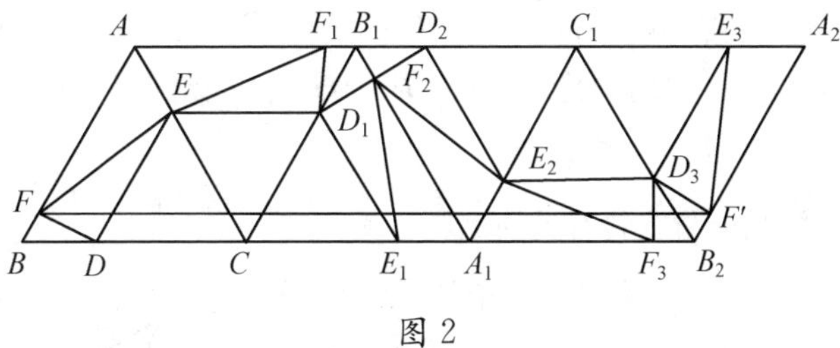

图 2

形 DEF 的周长不小于$\triangle ABC$ 的周长的一半.

为了解这问题,我们将$\triangle ABC$ 接连作 5 次反射(轴为 AC、CB_1、B_1A_1、A_1C_1、C_1B_2),不难看出图 2 中折线 $FED_1F_2E_2D_3F' \geqslant FF'$,而折线 $FED_1F_2E_2D_3F'$是$\triangle DEF$ 的周长的两倍,$FF'=AA_2$ 即$\triangle ABC$ 的周长.因此$\triangle DEF$ 的周长$\geqslant \frac{1}{2}\triangle ABC$ 的周长.

这里的正三角形也可改为正 n 边形. 1981 年 25 省市数学竞赛题中有这样的问题:"一张台球桌形状是正六边形 $ABCDEF$. 一个球从 AB 中点 P 击出,击中 BC 边上某点 Q,并且依次碰击 CD、DE、EF、FA 各边,最后击中 AB 边上的某一点. 设 $\angle BPQ=\theta$,求 θ 的取值范围."由于入射角等于反射角,这个问题也可以仿照例题 2,用轴对称(反射)来解决.

在几何作图中常常采用平行移动法. 在证明题中利用平行移动的例子也不少.

例 3　如图 3,$\square ABCD$ 中有一点 P,已知$\angle PAD=\angle PCD$,求证$\angle PBC=\angle PDC$.

这个问题的已知条件中有两个角相等,要证明的结论是另两个角相等,这就启发我们去利用共圆点的性质,但通常共圆的四点构成图 4. 将图 3 与图 4 比较就不难发现,只要把$\triangle PAB$ 沿 BC 方向平行移动,使 B 变成 C,则因为 $AD \mathrel{\underline{\parallel}} BC$,$A$ 变成 D. 设这时 P 点变为 P',则$\triangle PAB$ 变为$\triangle P'DC$(参见图 5).

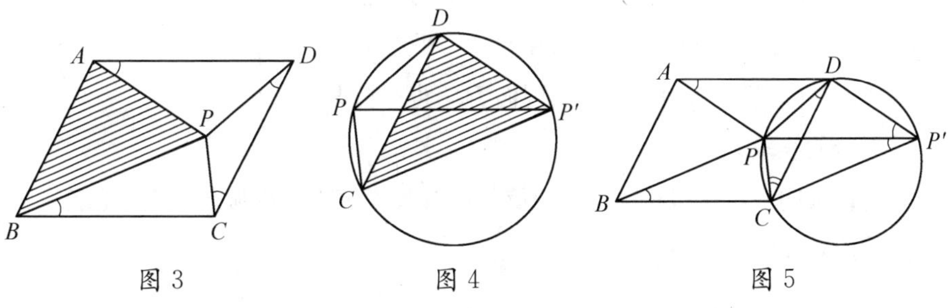

图 3　　图 4　　图 5

由于 $PP' \mathrel{\underline{\parallel}} BC \mathrel{\underline{\parallel}} AD$，易知 $\angle PP'C = \angle PBC$，$\angle PAD = \angle PP'D$. 从而 $\angle PCD = \angle PAD = \angle PP'D$，$P$、$C$、$P'$、$D$ 四点共圆，$\angle PDC = \angle PP'C = \angle PBC$.

现在我们举几个与旋转有关的例题.

例 4 如图 6，在$\triangle ABC$ 的三条边上向外各作一个正三角形：$\triangle ABC'$、$\triangle BCA'$、$\triangle CAB'$. 则 $AA' = BB' = CC'$，并且每两条线的夹角都是 $60°$.

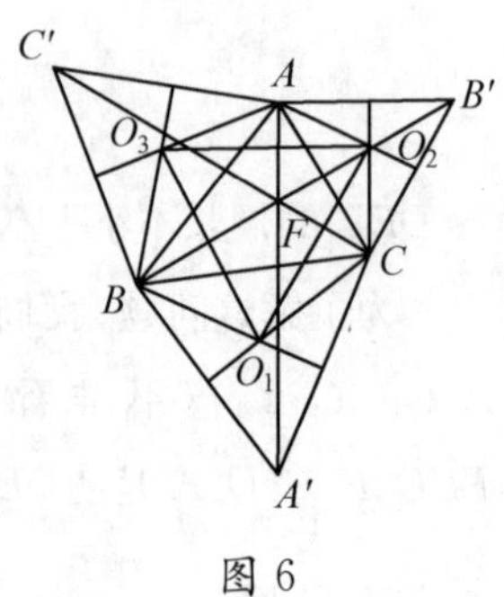

图 6

为了证明这两个结论，只要绕 A 点作 $60°$的(顺时针方向的)旋转. 由于 $\angle B'AC = 60°$，$AB' = AC$，所以 B'变成 C，同理 B 变成 C'，从而 BB'经过这个旋转变成 CC'. 因此 $BB' = CC'$，并且 BB'与 CC'的夹角为 $60°$.

读者不难进一步证明 AA'、BB'、CC'交于一点 F，这一点也是$\odot ABC'$、$\odot BCA'$、$\odot CAB'$的交点. 当 F 在$\triangle ABC$ 内部时，称为 Fermat 点，它到三个顶点 A、B、C 的距离的和为最小.

类似地，如果在$\triangle ABC$ 的边 AB、AC 上向外作正方形 $ABDE$、$ACFG$，则 $BG = CE$，并且 $BG \perp CE$，这只要绕 A 点作 $90°$的旋转就可以证实了.

例 5 如图 7，过 P 点作直线分别与$\triangle ABC$ 的边 AB、AC 及中线 AE 垂直，这些垂线分别与高 AD 相交于 L、M、N，证明 $LN = MN$.

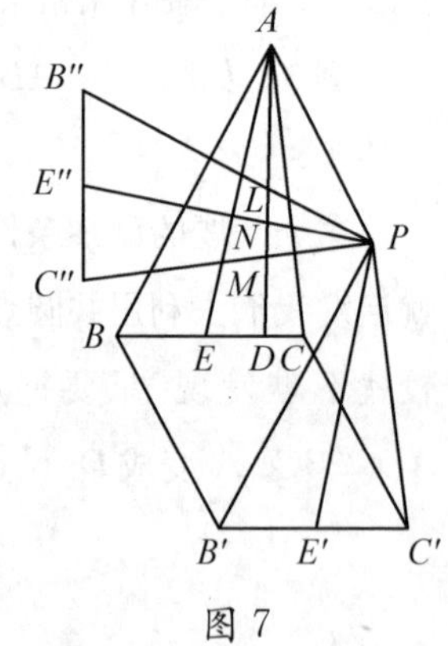

图 7

这次我们先将$\triangle ABC$ 平行移动，使得 A 点与 P 点重合(即图中$\triangle PB'C'$)，再绕 P 点作 $90°$的旋转(顺时针方向)，成为$\triangle PB''C''$. 经过这样变换后，直线 AB、AC、AE 分别与它们的垂线 PL、PM、PN(也就是 PB''，PC''，PE'')重合，并且直线 BC 的新位置 $B''C''$与它原来的垂线 AD 平行，由于 AE 平分 BC，PE''平分 $B''C''$，即 PN 平分 $B''C''$，因而 PN 也平分与 $B''C''$平行的线段 LM.

可以证明一次平行移动与一个旋转结合起来等于一次旋转(当然是绕另一个点的旋转).

除了合同变换，中学里最常见的变换是相似变换，它可以分成一次位似变换与一次合同变换，因此我们主要讨论位似变换.

例 6　三个全等的圆有一个公共点 K，并且都在一个已知的三角形内，每一个圆与三角形的两条边相切. 证明三角形的内心 I，外心 O 与 K 点共线.

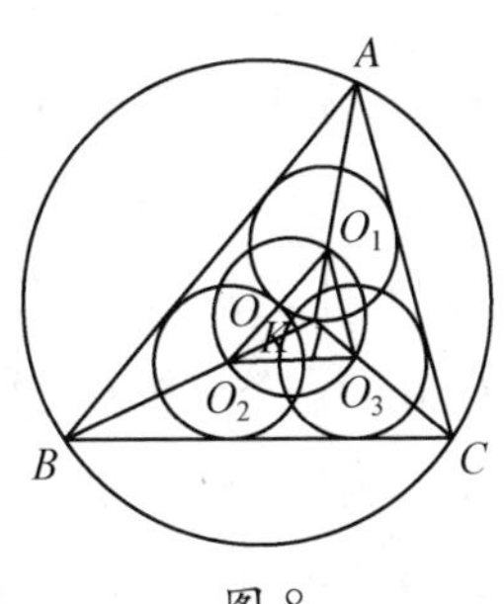

图 8

本题是 22 届国际数学竞赛题的第 5 题. 如果熟知位似变换，就不难证明. 首先，由于 O_1、O_2 到 AB 的距离相等(等于$\odot O_1$ 的半径)，所以 $O_1O_2 \parallel AB$. 同理 $O_2O_3 \parallel BC$，$O_3O_1 \parallel CA$，$\triangle O_1O_2O_3 \backsim \triangle ABC$. 又由于 AB、AC 均与$\odot O_1$ 相切，O_1A 平分$\angle BAC$. 同理 O_2B 平分$\angle ABC$，O_3C 平分$\angle BCA$. 因此$\triangle ABC$ 与$\triangle O_1O_2O_3$ 的对应顶点的连线 O_1A、O_2B、O_3C 相交于$\triangle ABC$ 的内心 I，即$\triangle O_1O_2O_3$ 与$\triangle ABC$ 位似，位似中心为 I.

因为 K 到 O_1、O_2、O_3 的距离相等(等于$\odot O_1$ 的半径)，所以 K 为$\triangle O_1O_2O_3$ 的外心，由于每一对对应点的连线都过位似中心 I，所以 K 与$\triangle ABC$ 的外心 O 的连线过 I，即 O、K、I 三点共线.

例 7　(Napoleon 定理)图 6 中，O_1、O_2、O_3 分别为三个正三角形的中心，证明$\triangle O_1O_2O_3$ 也是正三角形.

我们将$\triangle AO_2O_3$ 绕 A 点旋转 30°(顺时针方向)，则 O_3 落到 AC'上，O_2 落到 AC 上. 不难算出

$$\frac{AO_3}{AC'}=\frac{AO_2}{AC}=\frac{2}{3}\times\frac{\sqrt{3}}{2}=\frac{\sqrt{3}}{3},$$

所以$\triangle AO_2O_3 \backsim \triangle AC'C$，并且 $O_2O_3=\frac{\sqrt{3}}{3}C'C$. 同理 $O_3O_1=\frac{\sqrt{3}}{3}A'A$，$O_1O_2=\frac{\sqrt{3}}{3}B'B$. 而例题 4 中已经证明 $A'A=B'B=C'C$，所以 $O_2O_3=O_3O_1=O_1O_2$，$\triangle O_1O_2O_3$ 为正三角形.

例 8　如图 9，自$\triangle ABC$ 的顶点 A 向$\angle ABC$、$\angle ACB$ 及其外角的角平分线作垂线，证明四个垂足 D、E、F、G 共线.

延长 AD 交 BC 于 D'. 由于 $\angle 1=\angle 2$，$\angle ADB=\angle D'DB=90°$，$BD=BD$，

所以 $\triangle ADB \cong \triangle D'DB$，$AD = D'D = \frac{1}{2}AD'$. 换句话说，以 A 为位似中心，$1:2$ 为相似比作位似变换，则 D 点变为直线 BC 上的一点 D'. 同理，在这变换下，E、F、G 分别变为 BC 上的点 E'、F'、G'. 由于 D'、E'、F'、G' 在同一条直线 BC 上，所以在变换前，D、E、F、G 共线.

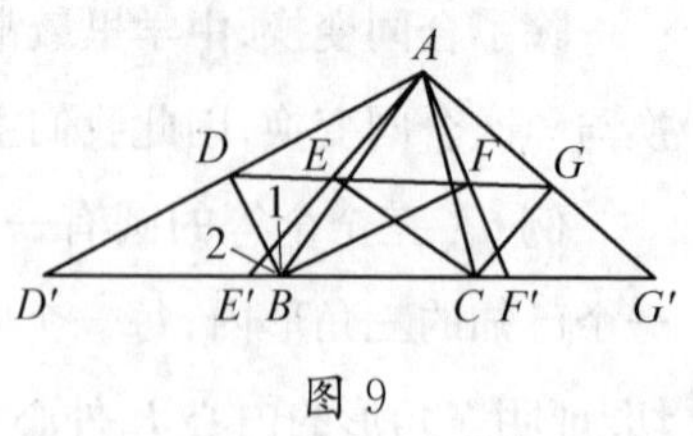

图 9

例 9　图 10 中，$\triangle ABC$ 为正三角形，$MP \parallel BC$，M 在 AB 上，P 在 AC 上，D 为 $\triangle AMP$ 的外心，E 为 BP 的中点，求 $\angle DEC$ 与 $\angle CDE$.

显然 $\triangle AMP$ 为正三角形. 设 AP 中点为 N，MP 中点为 F，易知 N、F、E 共线，并且 $NE \parallel AB$，$\frac{FE}{PC} = \frac{FE}{MB} = \frac{1}{2}$.

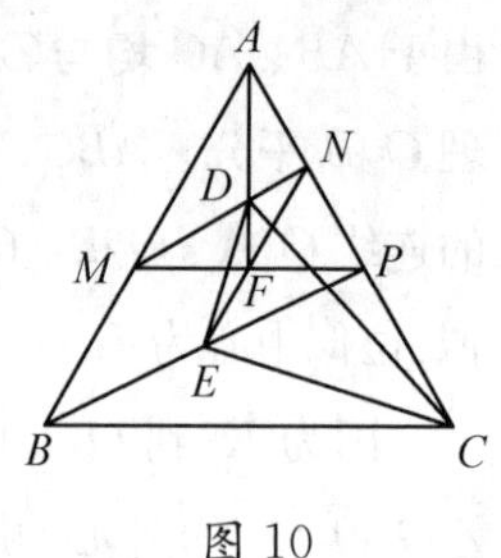

图 10

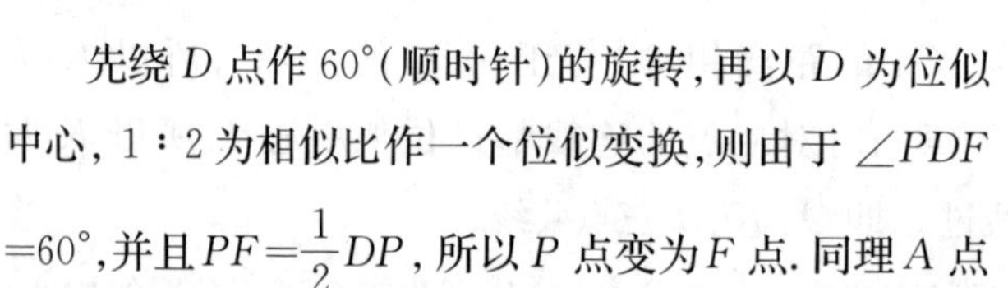

先绕 D 点作 $60°$（顺时针）的旋转，再以 D 为位似中心，$1:2$ 为相似比作一个位似变换，则由于 $\angle PDF = 60°$，并且 $PF = \frac{1}{2}DP$，所以 P 点变为 F 点. 同理 A 点变为 N 点. 于是直线 PA 变为直线 NF. 由于 $\frac{FE}{PC} = \frac{1}{2}$，所以在上述变换下，直线 PA 上的点 C 变为直线 NF 上的 E 点. 于是 $\angle CDE = 60°$，并且 $\frac{DE}{DC} = \frac{1}{2}$，从而 $\angle DEC = 90°$.

一个几何命题的探讨与推广

去年我曾与一位同志讨论了一个几何问题.

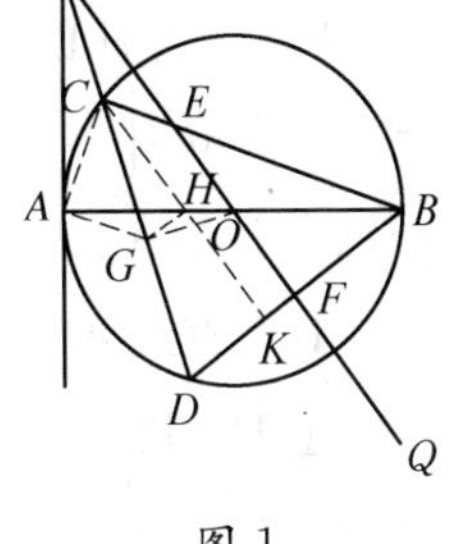

图 1

"设 AB 为⊙O 的直径，P 在过 A 的切线上，过 P 作割线交⊙O 于 C、D，直线 BC、BD 分别与 PO 相交于 E、F，则 $OE=OF$". (图 1)

证明并不太困难. 设 G 为 CD 中点，则 $OG\perp CD$. 过 C 作直线平行于 EF，分别交 AB、BD 于 H、K. 连 HG、AG、AC. 因为

$$\angle OGP=\angle OAP=90°,$$

所以 O、G、A、P 四点共圆，从而

$$\angle GPE=\angle GAH.$$

又由于 $CH\ /\!/\ EF$，$\angle GCH=\angle GPE$，所以

$$\angle GCH=\angle GAH.$$

从而 G、H、C、A 四点共圆，$\angle HGC=\angle HAC$.

但同弧上的圆周角相等，

$$\angle HAC=\angle CDB,$$

从而 $\angle HGC=\angle CDB$，于是 $HG\ /\!/\ KD$.

因为 G 是 CD 的中点，所以 H 是 CK 的中点.

又由于 $CK\ /\!/\ EF$，易知 O 是 EF 的中点.

上述命题与结论能不能作进一步的推广？下面就来看一看可以怎样推广.

如果直线 PO 与过 B 的切线相交于 Q，那么 $PO=OQ$. CD 是过 P 的割线与⊙O 的交点，B 是过 B 的切线与⊙O 的切点，这个切点可以看成是两个交点重合为一，E、F 分别是 BC、BD 与 PQ 的交点(图 1). 于是，更一般地，我们应

当有:

“设⊙O 的圆心 O 在过 P、Q 两点的直线上,并且 $OP=OQ$. 过 P、Q 各作一条直线,分别交⊙O 于 C、D 及 A、B, AC、BD 分别交 PQ 于 E、F,那么 $OE=OF$.”(图 2)

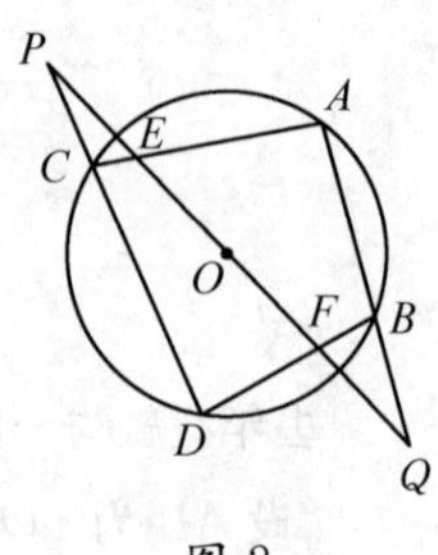

图 2

当 A、B 重合时,就是开始提出的问题.

这个结论还可以进一步推广到有心的二次曲线:

“设 V 为有心二次曲线,过中心 O 任作一条直线,在这直线上取 P、Q,使 $PO=OQ$. 过 P、Q 各作一条直线,分别交 V 于 A、B 及 C、D,直线 AC、BD 分别交 PQ 于 E、F,则 $EO=OF$.”

其实直线 AC、BD 还可以换成任一条过 A、B、C、D 的二次曲线. 即有:

定理 设 V 为有心二次曲线. 过中心 O 任作一条直线,在这直线上取 P、Q,使 $PO=OQ$. 过 P、Q 各作一条直线,分别交 V 于 A、B 及 C、D. 过 A、B、C、D 四点的二次曲线 V_1 交直线 PQ 于 E、F,则 $EO=OF$.

证明如下:

以直线 PQ 为 x 轴,O 为原点,建立直角坐标系. 设 P、Q 的坐标分别为 $(-x_0, 0)$、$(x_0, 0)$,曲线 V 方程为

$$ax^2+bxy+cy^2-d=0. \tag{1}$$

又设直线 AB、CD 的方程分别为

$$y-k_1(x+x_0)=0 \text{ 与 } y-k_2(x-x_0)=0,$$

或者用一个二次方程

$$[y-k_1(x+x_0)][y-k_2(x-x_0)]=0 \tag{2}$$

表示这两条直线(退化的二次曲线).

过曲线(1)与曲线(2)的交点 A、B、C、D 的二次曲线组成一个曲线族,其中任一条曲线 V_1 的方程可表为

$$\lambda_1(ax^2+bxy+cy^2-d)+\lambda_2[y-k_1(x+x_0)][y-k_2(x-x_0)]=0, \tag{3}$$

因此 V_1 与 x 轴的交点 E、F 的横坐标满足一个方程,这个方程就是在(3)中令

$y=0$ 而得到的

$$\lambda_1(ax^2-d)+\lambda_2k_1k_2(x^2-x_0^2)=0. \tag{4}$$

注意(4)的一次项系数为零(这就是证明的关键!),所以(4)的两根之和

$$x_1+x_2=0,$$

即 $EO=OF$.

定理中的 V 与 V_1 既可以是非退化的、也可以是退化的二次曲线,直线 AB、(1)也可以是切线(A 与 B 重合或 C 与 D 重合),这样就有许许多多的特例,与图 1 与图 2 相关联的两个命题就是定理的特例(这时有心曲线 V 为圆).不仅如此,如果注意到上述证明的关键是"(4)的一次项系数为 0",那么在 V 为圆时,只要圆心 O_1 在 y 轴上,即使它不与线段 PQ 的中点 O 重合(这时 V 的方程为 $x^2+(y-y_0)^2=1$),结论也成立. 即有:

定理　设⊙O_1 的圆心在线段 PQ 的垂直平分线上,过 P、Q 任作两条直线分别交⊙O_1 于 A、B 及 C、D, AC、BD 分别交直线 PQ 于 E、F,那么 $PE=FQ$.

特别地,在 P 与 Q 重合时,这个定理成为如下的"蝴蝶定理":

过点 O 任作两条直线交⊙O_1 于 A、B 及 C、D,作直线 $GH\perp OO_1$, AC、BD 分别交 GH 于 E、F,则 $EO=OF$.

从上述定理还可以得到许多结论,例如:

设⊙O 的圆心 O 为线段 PQ 的中点,过 P、O 任作一圆交⊙O 于 A、B,过 Q、O 任作一圆交⊙O 于 C、D. 过 A、C、O 的圆交 PQ 于 E,过 B、D、O 的圆交 PQ 于 F,则 $EO=OF$.

一道几何题的推广

在数学中常常见到各种各样的推广.将特殊的结论推广为一般的结论,这当然是一种有意义的工作.不仅如此,一般的结论揭示了普遍的规律,把握了事物的本质.而在特殊问题中,这种本质往往被掩盖在各种特殊的性质之下,不易发现.所以有时导出一个一般的结论并不比导出一个特殊的结论更难,甚至反倒简单得多,下面的一道几何题即是一例.

这道题是梁绍鸿著《初等数学复习及研究(平面几何)》一书中的习题,原题如下:

"设 AB 为⊙O 的直径,P 在过 A 的切线上,过 P 作割线交⊙O 于 C、D,直线 BC、BD 分别与 PO 相交于 E、F,则 $OE=OF$."(图1)

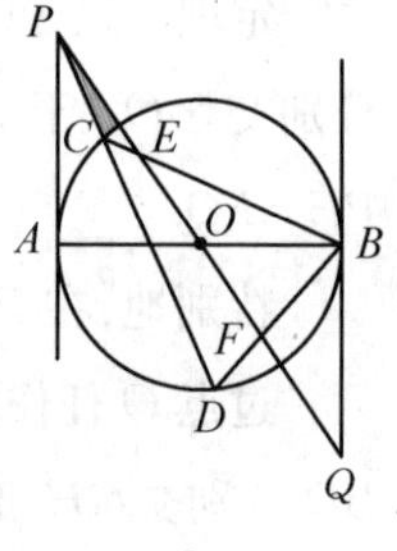

图1

这道题的纯几何证法可见《中学数学教学》1984年第4期拙作《一个几何命题的探讨与推广》,这里不再赘述.本文的目的是将上述命题推广(比上面所提的文章更进一步).

首先,设过 B 的切线交 PO 于 Q,则 $PO=OQ$.因此,上面的问题也可改述成等价的形式:

"设 P、Q 在过圆心 O 的直线上,且 $PO=OQ$.过 P 作割线交圆于 C、D,过 Q 作切线切圆于 B,BC、BD 分别交 PQ 于 E、F,则 $EO=OF$."

这里 PC 是任一割线,而 QB 却是极特殊的"割线"——切线.因此,很自然地会想到,命题可推广成:

"P、Q 在过圆心 O 的直线上,且 $PO=OQ$.过 P 作割线交圆于 C、D,过 Q 作割线交圆于 A、B,AC、BD 分别交 PQ 于 E、F,则 $EO=OF$."(图2)

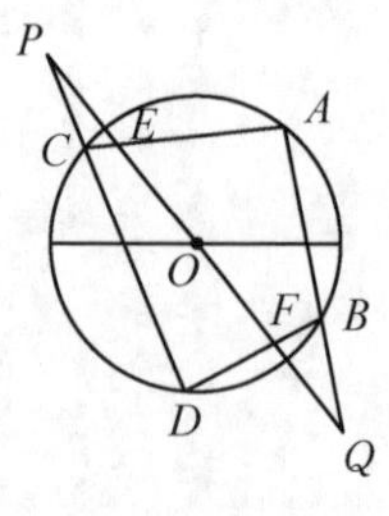

图2

这里 PQ 的中点是圆心 O,更进一步,可把圆心(直径的中点)O 改为一条弦的中点 O_1,即:

"设 O_1 为圆 O 一条弦的中点,P、Q 在这弦所在直线上，并且 $PO_1=O_1Q$. 过 P 作割线交圆于 C、D，过 Q 作割线交圆于 A、B，AC、BD 分别交 PQ 于 E、F，则 $EO_1=O_1F$.”（图 3）

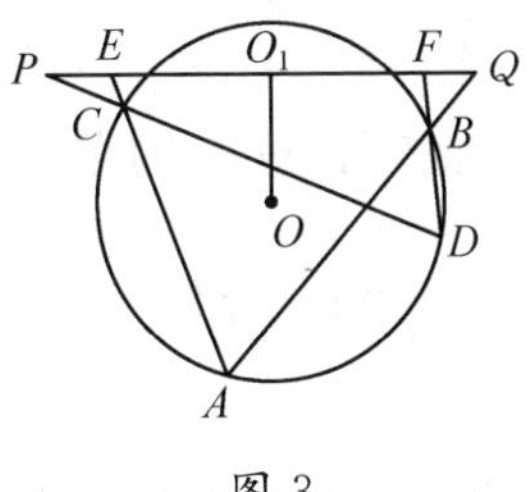

图 3

以上命题是否正确，尚有待证明. 不过，为了更清楚地揭示出内在的规律，我们宁愿再推广一步，即把圆改为一般的圆锥曲线：

“设 O 为圆锥曲线 Γ 的一条弦的中点，P、Q 在这弦所在直线上，并且 $PO=OQ$. 过 P 作直线交曲线 Γ 于 C、D，过 Q 作直线交曲线 Γ 于 A、B，AC、BD 分别交 PQ 于 E、F，则 $EO=OF$.”

直线 AC 与 BD 可看成是一条退化的圆锥曲线，直线 PC、QA 也可看成是一条退化的圆锥曲线，因而更一般地有：

“设 O 为圆锥曲线 Γ 的一条弦的中点，P、Q 在这弦所在直线上，并且 $PO=OQ$. 过 P、Q 的圆锥曲线 Γ_1 与曲线 Γ 相交于 A、B、C、D，过 A、B、C、D 的任一条圆锥曲线 Γ_2 与 PQ 相交于 E、F，则 $EO=OF$.”

现在我们来证明这个命题. 首先建立坐标系. 以直线 PQ 为 x 轴，O 为原点，建立直角坐标系，如图 4 所示. 设这时曲线 Γ 的方程为

$$ax^2+bxy+cy^2+dx+ey+f=0.$$

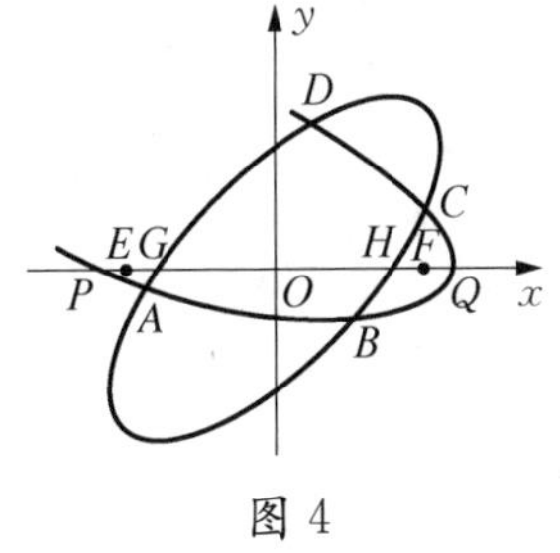

图 4

曲线 Γ 与 x 轴有两个交点 G、H，这两点的横坐标满足方程

$$ax^2+dx+f=0.$$

由于 O 为弦 GH 的中点，所以 G、H 的横坐标互为相反数，故由韦达定理（根与系数的关系）有 $d=0$. 因此曲线 Γ 的方程为

$$ax^2+bxy+cy^2+ey+f=0. \tag{1}$$

其中 x 的一次项的系数为 0. 反之，形如(1)的曲线与 x 轴的交点的横坐标恰好相差一个符号（两者之和为 0）.

由于 O 也是 PQ 的中点，所以曲线 Γ_1 的方程中 x 的一次项系数也为 0，即它的方程为

$$a'x^2+b'xy+c'y^2+e'y+f'=0. \tag{2}$$

过曲线 Γ 与 Γ_1 的交点 A、B、C、D 的二次曲线为

$$\lambda_1(ax^2+bxy+cy^2+ey+f)+\lambda_2(a'x^2+b'xy+c'y^2+e'y+f')=0 \tag{3}$$

的形式(这与直线束或圆束相当),在(3)中 x 的一次项系数仍然为 0. 因此,这曲线与 x 轴相交于 E、F 时,必有 $EO=OF$. 这就是要证明的结论.

在上述证明中并未用到圆的特殊性质(例如圆弧上的圆周角相等),而只用到(1)、(2)、(3)中 x 的一次项系数为 0,这才是问题的本质.

将一般的结论"特殊化",又可导出许多结论. 例如设曲线 Γ 为圆, P、Q 与 O 点重合,那么就得到:

"设 O 为圆内一弦 GH 的中点,过 O 作弦 AB、CD,AC、DB 分别交 GH 于 E、F,则 $EO=OF$." (图 5)

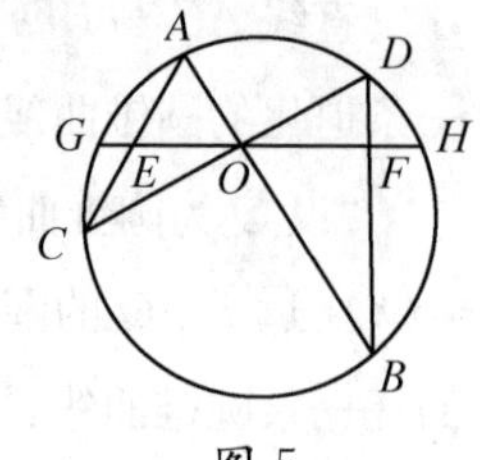

图 5

图 5 的形状像一只翩翩起舞的蝴蝶,所以常被称为蝴蝶定理.

填满空间的正多面体

众所周知,正多边形中正三角形、正方形、正六边形可以填满平面,而且也只有这三种正多边形可以填满平面.哪些正多面体可以填满空间呢?

显然正方体可以填满空间.但是否只有正方体,其他 4 种正多面体能填满空间吗?如果不能,为什么不能呢?这个问题很有趣.

为了将空间情况与平面对照、类比,我们先回忆一下平面的情况是如何解决的.

如果平面一点 O 是(填满平面的)正多边形的顶点,那么在点 O 的周角,也就是以 O 为圆心、1 为半径的圆周长是 2π 个弧度($360°$).因此,在 O 点的各正多边形的内角和应等于 2π.这些内角又都相等,所以应当是 2π 除以一个正整数.而正 n 边形的角内是 $(n-2)\pi/n$,所以 $2n=2(n-2)+4$ 是 $n-2$ 的倍数,从而 $n=3$、4、6.

对于三维空间,类似地,如果一点 O 是正多面体的顶点,以 O 为心、1 为半径作一个球面.球面的面积是 4π(它相当于平面中圆的周长 2π).以 O 为顶点的正多面体,在 O 处的棱(作为射线)与球面相交,这些交点形成一个球面多边形,这些射线形成一个以 O 为顶点的"立体角",而这球面多边形的面积就表示这个"立体角"的大小.类似地,这个立体的大小应当是 4π 或 $720°$除以一个整数.

现在我们列举 5 种正多面体的(各个面上的)面角与(每两个相邻面的)二面角 θ(这是一道很好的计算题):

正多面体面数	面角	二面角 θ	$\sin\theta$
4	$60°$	$70°31'43.6''$	$\frac{2\sqrt{2}}{3}$
6	$90°$	$90°$	1
8	$60°$	$109°28'16.4''$	$\frac{2\sqrt{2}}{3}$
12	$108°$	$116°33'54.2''$	$\frac{2}{\sqrt{5}}$
20	$60°$	$138°11'22.8''$	$\frac{2}{3}$

最后一列表明实际上是先算出 $\sin\theta$,再求出 θ .

球面多边形的面积等于“角盈”,也就是各个角(即上述二面角)的和减去 π. 为方便起见,就用度数表示(乘以$\frac{2\pi}{360^\circ}$即化为弧度),并列出下表:

正多面体面数	内角和	角盈
4	$3\theta=211^\circ33'$	$31^\circ33'$
6	$3\theta=270^\circ$	90°
8	$4\theta=437^\circ52'$	$257^\circ52'$
12	$3\theta=349^\circ39'$	$169^\circ39'$
20	$5\theta=690^\circ55'$	$510^\circ55'$

(其中 θ 的系数是 O 点引出的棱数. 正 4、6、12 面体,每个顶点引出 3 条棱,而正 8、20 面体,每个顶点分别引出 4、5 条棱).

表中只有正方体(即正 6 面体)的角盈 90°乘以整数等于球面面积 720° (4π),所以只有正方体可以填满空间. 而且,各个正多面体角盈的组合也无法凑成 720°,除非全是正方体. 所以只有全用正方体才能填满整个空间.

熊的颜色

波利亚的名著《怎样解题》的第四篇提供了一些问题. 其中第一个就是:

一头熊从 P 点出发,向正南走 1 千米[①],然后改变方向向正东走 1 千米,再向左转,往正北走 1 千米. 这时它正好回到它出发的 P 点. 这头熊是什么颜色?

熊的颜色分黑、棕、白. 白熊,也就是北极熊,产在北极一带. 因而容易猜想到 P 点是北极,而熊的颜色是白的.

不难验证 P 为北极合乎题意:当熊从北极出发,向正南走 1 千米,即沿一条经线向南走 1 千米;改向东走 1 千米,即沿一条纬线向东走 1 千米;最后向北走 1 千米,即沿一条经线向北走 1 千米. 无论这条经线是否开始走过的经线,由于同一纬线上的点到北极的(球面)距离都相同,所以熊刚好回到出发点北极.

但是,P 是否一定是北极,有没有其他合乎要求的点呢? 本文将指出,这样的点很多,有无穷多个,它们形成无穷多个纬圈(完整的纬线,我们简称为纬圈).

如果 P 在北半球,但不是北极 N,那么熊先沿经线向南行 1 千米到达 Q,然后沿 Q 处的纬线行 1 千米. 它必须先回到 Q,才能再向北走回到 P. 因此, Q 处的纬线的周长应当是$\frac{1}{n}$千米,其中 n 是自然数 1, 2, 3, ….

设地球的中心为 O, $\angle NOQ=\alpha$ (弧度),如图 1(这里,我们将地球当作一个数学中的球,半径为 $R=6400$ 千米,图中的 $\overset{\frown}{PQ}$、α 都应当很小,但为了看得清楚,我们将它们"放大"了). Q 所在纬圈的半径

$$AQ=\frac{1}{2\pi n}\ (n=1,\ 2,\ \cdots),\tag{1}$$

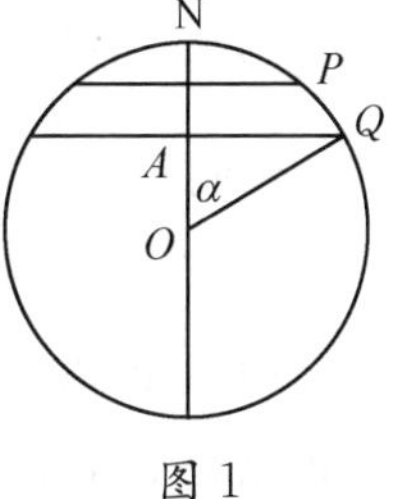

图 1

① 原文为 1 英里,即 1.609 千米. 我们改为 1 千米,不影响题意与解答. 以下的单位均为千米,不再注出.

$$\sin\alpha = \frac{AQ}{OQ} = \frac{1}{2\pi nR}. \tag{2}$$

$\overset{\frown}{NQ}$ 的长是 $R\alpha$，而

$$R\alpha < R \cdot \frac{\pi}{2}\sin\alpha = \frac{1}{4n} < 1. \tag{3}$$

这与 $\overset{\frown}{PQ}$ 的长为 1 矛盾. 因此在北半球，除了北极 N，没有符合要求的点.

但如 P 在南半球，如图 2，P 先沿经线向南到 Q，同样(1)、(2)成立. 这时，设 $\angle POQ = \beta$，则 $\overset{\frown}{PQ}$ 的长为

$$R\beta = 1. \tag{4}$$

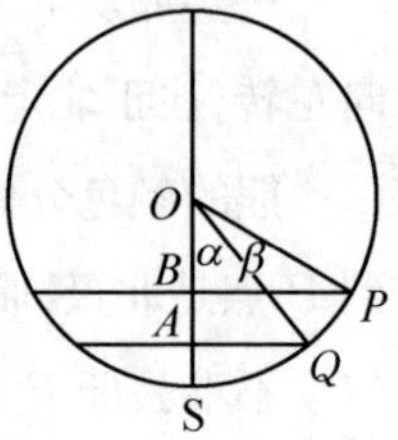

图 2

P 所在纬圈的半径

$$\begin{aligned} BP &= R\sin(\alpha+\beta) = R\sin\alpha\cos\beta + R\sin\beta\cos\alpha \\ &< R\sin\alpha + R\beta = \frac{1}{2\pi n} + 1. \end{aligned} \tag{5}$$

由于 R 很大，所以由(2)得到的 α 与由(4)得到的 β 都很接近于 0，$\cos\alpha$ 与 $\cos\beta$ 都很接近于 1，$\frac{\sin\beta}{\beta}$ 也很接近于 1. 因此可以认为(5)是一个等式，即

$$BP \approx \frac{1}{2\pi n} + 1. \tag{6}$$

$$P \text{ 所在纬圈的周长} \approx \frac{1}{n} + 2\pi. \tag{7}$$

因此符合要求的点 P 有无穷多个，它们形成周长约为(略小于) $2\pi + \frac{1}{n}$ 千米的纬圈 ($n=1, 2, \cdots$). (注：1982 年科学出版社的《怎样解题》译本，解答写成 $2\pi + 1$. 2002 年上海科技教育出版社的译本已更正).

但是，南极一带没有熊，只有企鹅(北极一带没有企鹅)，所以虽然数学上 P 点可以有无穷多个，但熊仍只能从北极出发，它的颜色是白的.

本文系看到左再思先生的文章《零星的回忆》(《中学数学研究》2006 年第 1 期)后而写成的，谨对左先生表示衷心的感谢.

莫莱定理

莫莱(Frank Morley, 1860—1937)在讨论平面上的“n——线”时,给出了几个一般性的定理,其中的一个特例即为著名的**莫莱定理**:一个三角形的角的三等分线的、分别靠近三边的三个交点,构成正三角形.

莫莱定理的证明很多. 1908 年, T. Delahaye 与 H. Lez 所作的证明可能是最早在印刷物上出现的证明. 他们的证明很优雅,要点是计算莫莱三角形的边长.

如图 1,设$\triangle ABC$ 中各角的三等分线构成三角形 DEF(莫莱三角形),又设$\triangle ABC$ 的三个内角分别为 3α、3β、3γ,三边边长分别为 a、b、c,外接圆半径为 R,则可证

$$EF=8R\sin\alpha\sin\beta\sin\gamma. \tag{1}$$

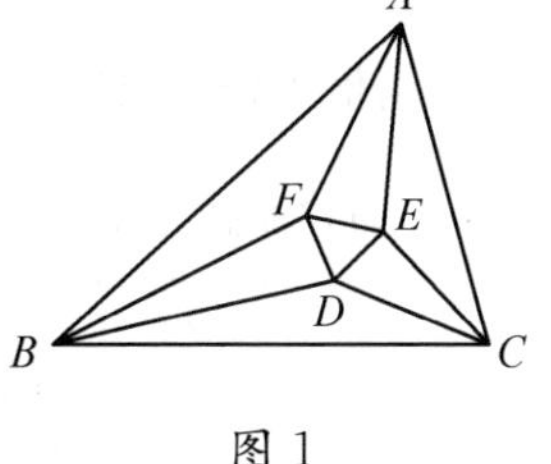

图 1

由于(1)是关于 α、β、γ 对称的,同样 DE、FD 也由(1)表出,所以$\triangle DEF$ 是正三角形.

我们分两步来证明(1).

第一步,在$\triangle ABF$ 中,由正弦定理,

$$AF=\frac{c\sin\beta}{\sin(\alpha+\beta)}=\frac{2R\sin\beta\sin 3\gamma}{\sin\left(\frac{\pi}{3}-\gamma\right)}. \tag{2}$$

而由积化和差公式,

$$2\sin\left(\frac{\pi}{3}-\gamma\right)\sin\left(\frac{\pi}{3}+\gamma\right)=\cos 2\gamma-\cos\frac{2\pi}{3}=\cos 2\gamma+\frac{1}{2},$$

$$2\sin\gamma\left(\cos 2\gamma+\frac{1}{2}\right)=\sin 3\gamma-\sin\gamma+\sin\gamma=\sin 3\gamma,$$

所以

$$AF=\frac{2R\sin\beta\sin 3\gamma}{\sin\left(\frac{\pi}{3}-\gamma\right)}=\frac{4R\sin\beta\sin\gamma\left(\cos 2\gamma+\frac{1}{2}\right)}{\sin\left(\frac{\pi}{3}-\gamma\right)}=8R\sin\beta\sin\gamma\sin\left(\frac{\pi}{3}+\gamma\right).\tag{3}$$

同理，
$$AE=8R\sin\beta\sin\gamma\sin\left(\frac{\pi}{3}+\beta\right).\tag{4}$$

第二步，在$\triangle AEF$中，由余弦定理，

$$EF^2=AE^2+AF^2-2\cdot AE\cdot AF\cos\alpha.\tag{5}$$

因此要证(1)，只需证明

$$\sin^2\left(\frac{\pi}{3}+\beta\right)+\sin^2\left(\frac{\pi}{3}+\gamma\right)-2\sin\left(\frac{\pi}{3}+\beta\right)\sin\left(\frac{\pi}{3}+\gamma\right)\cos\alpha=\sin^2\alpha.\tag{6}$$

由倍角公式及三角恒等变形，

$$\begin{aligned}(6)式左边&=\cos^2\left(\frac{\pi}{6}-\beta\right)+\cos^2\left(\frac{\pi}{6}-\gamma\right)-2\cos\left(\frac{\pi}{6}-\beta\right)\cos\left(\frac{\pi}{6}-\gamma\right)\cos\alpha\\&=\frac{1+\cos\left(\frac{\pi}{3}-2\beta\right)}{2}+\frac{1+\cos\left(\frac{\pi}{3}-2\gamma\right)}{2}\\&\quad-\cos\alpha(\cos(\beta-\gamma)+\cos\alpha)\\&=1+\frac{1}{2}\left(\cos\left(\frac{\pi}{3}-2\beta\right)+\cos\left(\frac{\pi}{3}-2\gamma\right)\right)\\&\quad-\cos^2\alpha-\cos\alpha\cos(\beta-\gamma)=\sin^2\alpha.\end{aligned}$$

因此(6)式成立，从而(1)成立，$\triangle DEF$是等边三角形.

莫莱定理中的角，也可以是三角形的外角，请读者自己去验证.

一点补充

文[1]对下面的问题进行了详细、深入的讨论.

题 1　证明:对任意三角形,一定存在两条边,它们的长 u、v,满足

$$1 \leqslant \frac{u}{v} < \frac{1+\sqrt{5}}{2}. \tag{1}$$

在给出上界 $\frac{1+\sqrt{5}}{2}$ 时,文[1]已有多种富有启发性的证明. 本文意在说明 $\frac{1+\sqrt{5}}{2}$ 是怎样得来的. 为此提出下面的问题,作为文[1]的一点补充.

题 2　设$\triangle ABC$ 中, $a \geqslant b \geqslant c$,

$$q = \min\left\{\frac{a}{b}, \frac{b}{c}\right\}. \tag{2}$$

对所有的$\triangle ABC$,求 q 的最小的上界.

我们知道,在$\triangle ABC$ 中,

$$a < b + c, \tag{3}$$

而 $\frac{a}{b} \geqslant q$, $\frac{b}{c} \geqslant q$, 即

$$a \geqslant qb,\ c \leqslant \frac{b}{q}. \tag{4}$$

将式(4)中的 a、c"代入"式(3),"消去"a、c 得 $qb < b + \frac{b}{q}$,即 $q^2 - q - 1 < 0$.

所以,

$$q < \frac{1+\sqrt{5}}{2}. \tag{5}$$

$\frac{1+\sqrt{5}}{2}$ 就是这样得来的.

上面的式(3)、(4)都是不等式,不是方程.我们借用解方程中的术语代入、消去,但要注意不等号的方向不可搞错.

式(5)表明,$\frac{1+\sqrt{5}}{2}$ 是 q 的上界.它是不是最小的上界呢?

确实是最小的上界.为此,取 a 满足

$$1<a<\frac{1+\sqrt{5}}{2}, \tag{6}$$

而 $b=1$, $c=\frac{1}{a}$,则 $a<b+c$.

因此,a、b、c 构成三角形,且 $q=a$.而 a 可以任意接近 $\frac{1+\sqrt{5}}{2}$(只要满足式(6),a、b、c 就能构成三角形),所以任意一个比 $\frac{1+\sqrt{5}}{2}$ 小的数,都不能充作 q 的上界.

但 q 不能等于 $\frac{1+\sqrt{5}}{2}$,所以$\frac{1+\sqrt{5}}{2}$ 不是 q 的最大值,而只是"可望而不可及"的上确界——最小的上界.

对文[1]中的问题本身,本文也可以算作一种新的证明.

参考文献

[1] 罗增儒.探索、发现、论证(上)[J].中等数学,2007(9):2-4,14.

利用图形解决问题

借助图形,一些难以下手的问题可以迎刃而解,这里举几个例子.

例1 用 d_k 表示 N 市中住户不少于 k 人的住宅个数,又用 c_k 表示 N 市中人数为第 k 位多的那种住宅里的人数,显然 $c_1 \geqslant c_2 \geqslant c_3 \cdots$, $d_1 \geqslant d_2 \geqslant d_3 \geqslant \cdots$. 证明:

(1) $c_1 + c_2 + c_3 + \cdots = d_1 + d_2 + d_3 + \cdots$

(2) $c_1^2 + c_2^2 + c_3^2 + \cdots = d_1 + 3d_2 + 5d_3 + \cdots$

(3) $d_1^2 + d_2^2 + d_3^2 + \cdots = c_1 + 3c_2 + 5c_3 + \cdots$

解 为便于理解,先举一个简单例子. 设有 7 户住宅,每户住宅里的人数分别为 2, 3, 4, 4, 5, 6, 8. 于是,

$$d_1 = 7,\ d_2 = 7,\ d_3 = 6,\ d_4 = 5,\ d_5 = 3,\ d_6 = 2,\ d_7 = 1,\ d_8 = 1;$$

$$c_1 = 8,\ c_2 = 6,\ c_3 = 5,\ c_4 = 4,\ c_5 = 4,\ c_6 = 3,\ c_7 = 2.$$

画一个点阵图(图 1),使第 k 列的点数为 c_k,由检验即知第 k 行的点数恰为 d_k.

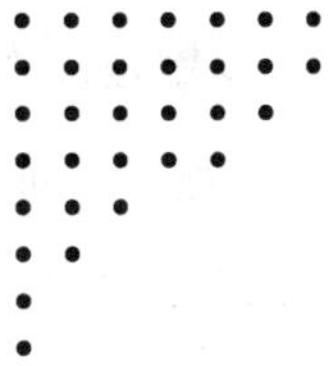

图 1

一般地,像右面那样画一个图(图 1),使第一列为 c_1 个点,第二列为 c_2 个点,……,也就是第 k 列表示 N 市中人数为第 k 位多的那种住宅里的人数. 这时,第一行恰好有 d_1 个点——它表示 N 市中住户不少于 1 人的住宅数,第二行恰好有 d_2 个点,…….

图中总的点数,如果先算每一列,然后将各列相加便得到

$$c_1 + c_2 + c_3 + \cdots,$$

如果先算行,再将各行相加便得到

$$d_1 + d_2 + d_3 + \cdots,$$

不论先横先竖,算出来的总和一样,所以

$$c_1+c_2+c_3+\cdots=d_1+d_2+d_3+\cdots,$$

即(1)成立.

为了证明(2),我们将图中的点“加权”,然后再相加,即把第一行的点乘 1,第二行乘 3,第三行乘 5,等等.

如果先算行再相加,得到

$$d_1+3d_2+5d_3+\cdots,$$

如果先算列,每列和为

$$1+3+5+(2c_k-1)=c_k^2,$$

因此各列总和为

$$c_1^2+c_2^2+c_3^2+\cdots,$$

从而(2)成立,同样可证(3)成立.

例 2 用$[x]$表示不超过 x 的最大整数,证明对任一正整数 n,

$$[\sqrt{n}]+[\sqrt[3]{n}]+\cdots+[\sqrt[n]{n}]=[\log_2 n]+[\log_3 n]+\cdots+[\log_n n].$$

证明 考虑满足 $y^x\leqslant n$, $x\geqslant 2$, $y\geqslant 2$ 的整点(x, y)的个数(整点即坐标 x、y 都是整数的点).

如果一列一列地数, $x=2$ 时,有$[\sqrt[2]{n}]$ 个; $x=3$ 时,有$[\sqrt[3]{n}]$ 个; ……;共 $[\sqrt{n}]+[\sqrt[3]{n}]+\cdots$ 个.

如果一行一行地数, $y=2$ 时,有$[\log_2 n]$个; $y=3$ 时,有$[\log_3 n]$ 个;……,共有

$$[\log_2 n]+[\log_3 n]+\cdots$$

个,所以原式成立.

例 3 如果正数 a_1, a_2, a_3, $\cdots$, a_{100} 满足

$$a_1+a_2+\cdots+a_{100}=300,$$
$$a_1^2+a_2^2+\cdots+a_{100}^2>10000,$$

证明:一定存在三个数 a_i, a_j, a_k,满足

$$a_i + a_j + a_k \geqslant 100.$$

解　不妨设 $a_1 \geqslant a_2 \geqslant a_3 \geqslant a_4 \geqslant \cdots$. 只要证

$$a_1 + a_2 + a_3 \geqslant 100,$$

假设

$$a_1 + a_2 + a_3 < 100. \tag{4}$$

考虑图 2,100 个小正方形(边长分别为 a_1, a_2, a_3, …),可放入三个边长为 100 的正方形中,这三个正方形是并立着的,一个挨着一个,小正方形靠上方的边顺次排列,它们占据的面积不超过图中阴影部分.

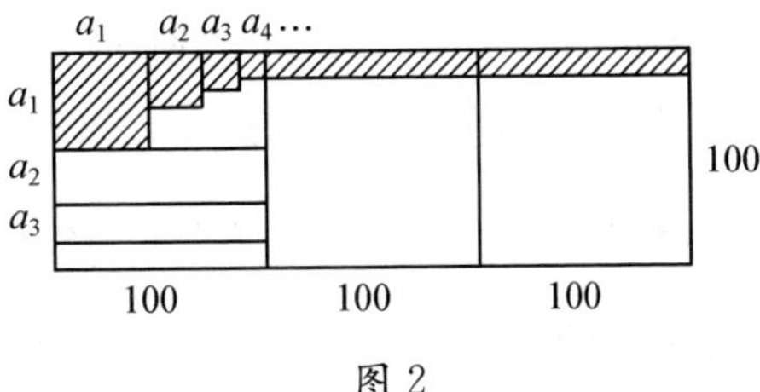

图 2

如果(4)成立,那么可将第一正方形剖为 4 个长方形,每个长为 100,而宽分别为 a_1, a_2, a_3 及 $100 - a_1 - a_2 - a_3$.

由于 $a_2 \geqslant a_3 \geqslant a_4 \geqslant \cdots$,所以第二个大正方形中的阴影部分可纳入宽为 a_2 的长方形中,第三个大正方形中的阴影部分可纳入宽为 a_3 的长方形中,于是 100 个小正方形的总面积小于第一个大正方形的面积,这与已知

$$a_1^2 + a_2^2 + a_3^2 + \cdots + a_{100}^2 > 10\,000$$

矛盾,因此(4)不成立,即

$$a_1 + a_2 + a_3 > 100.$$

过三角形外心与内心的直线

三角形的外心与重心的连线称为欧拉线，这是人所周知的. 三角形的外心，如果与内心相连，这条直线有什么性质呢？在文献中未见有人说起. 我的一位亲戚东南大学土木建筑系的单建教授最近发现：

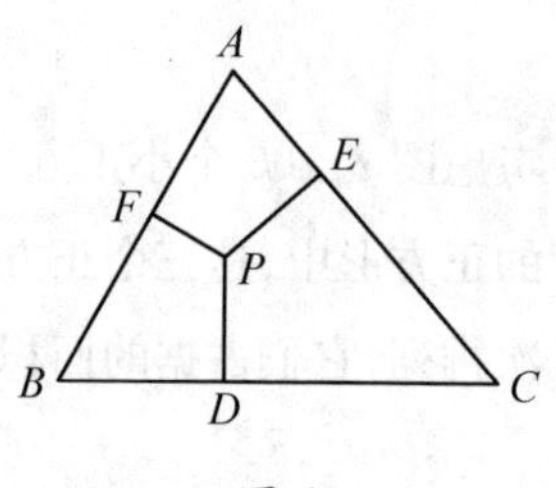

图 1

如图 1，设$\triangle ABC$不是正三角形. 自一点 P 向$\triangle ABC$三边作垂线，垂足分别为 D、E、F，那么

$$BD+CE+AF=DC+EA+FB \tag{1}$$

成立的充分必要条件是 P 在直线 OI 上，其中 O、I 分别为$\triangle ABC$的外心与内心. 在$\triangle ABC$为正三角形时，平面上任一点 P 均使(1)成立.

这个命题颇为有趣. 单建先生有一个几何的证明，本文拟给出一个不同的解析几何的证明.

以$\triangle ABC$的外心 O 为坐标原点，设 A、B、C、P 的坐标分别为$(x_A,\ y_A)$，$(x_B,\ y_B)$，$(x_C,\ y_C)$，$(x,\ y)$，又设 $\overrightarrow{BC}$、$\overrightarrow{CA}$、$\overrightarrow{AB}$ 上的单位向量分别为

$$(\cos\alpha,\ \sin\alpha),\ (\cos\beta,\ \sin\beta),\ (\cos\gamma,\ \sin\gamma),$$

则
$$BD=\overrightarrow{BP}\cdot(\cos\alpha,\ \sin\alpha)=(x-x_B)\cos\alpha+(y-y_B)\sin\alpha \tag{2}$$

为 x 与 y 的线性式(一次式).

同理 CE、AF 以及 $DC=a-BD$(其中 $a=BC$)、EA、FB 也都是如此，因此(1)是 x 与 y 的一次方程，即

$$\left(2\sum\cos\alpha\right)x+\left(2\sum\sin\alpha\right)y-2\sum x_B\cos\alpha-2\sum y_B\sin\alpha=a+b+c, \tag{3}$$

其中 x、y 的系数分别为

$$2\sum\cos\alpha, \ 2\sum\sin\alpha. \tag{4}$$

如果(4)中二式同时为 0,即 $\overrightarrow{BC}$、$\overrightarrow{CA}$、$\overrightarrow{AB}$ 上的单位向量构成三角形(向量 $\left(\sum\cos\alpha, \ \sum\sin\alpha\right)$ 为零向量),这时△ABC 是正三角形.

因此,在△ABC 不是正三角形时,(3)是真正的一次方程,表示一条直线,而显然外心 O 满足(1)(因 D、E、F 分别是 BC、CA、AB 的中点),即 O 在这条直线上. 内心 I 也满足(1)(因 $BD=FB$, $DC=CE$, $EA=AF$),即 I 也在这条直线上,所以这条直线就是直线 OI. 注意,我们选择 O 为原点,而 O 点坐标适合(3),所以将(0, 0)代入(3)得

$$-2\sum x_B\cos\alpha-2\sum y_B\sin\alpha=a+b+c, \tag{5}$$

因此 OI 的方程(3)即

$$x\sum\cos\alpha+y\sum\sin\alpha=0. \tag{6}$$

在△ABC 是正三角形时,由于 $\overrightarrow{OA}+\overrightarrow{OB}+\overrightarrow{OC}=\vec{0}$, 故

$$\sum\cos\alpha=0, \ \sum\sin\alpha=0. \tag{7}$$

因此(6)也就是(3)恒成立,即平面上任一点 P 都满足(1).

还需指出,上面的 BD、DC 等均为有向线段,即当 D 在线段 BC 内部时,BD、DC 均为正;当 D 在 BC 延长线上时, BD 为正,而 DC 为负,它是 CD 的相反数;当 D 在 CB 延长线上时, BD 为负,而 DC 为正. 上面的结果对于双心 n 边形(包括正 n 边形)同样成立.

关于匹窦不等式的讨论

设$\triangle ABC$和$\triangle A'B'C'$的边长分别为a、b、c和a'、b'、c'，它们的面积记为$\triangle$和$\triangle'$，则

$$a'^2(-a^2+b^2+c^2)+b'^2(a^2-b^2+c^2)+c'^2(a^2+b^2-c^2)\geqslant 16\triangle\triangle'. \tag{1}$$

这个不等式称为匹窦(Pedoe)不等式.

匹窦不等式引起了人们的注意，围绕着匹窦不等式开展了不少讨论，《安徽教育》1979年第11期，程龙同志给出了匹窦定理的三角证明，并将匹窦不等式加强为

$$\begin{aligned}&a'^2(-a^2+b^2+c^2)+b'^2(a^2-b^2+c^2)+c'^2(a^2+b^2-c^2)\\&\geqslant 16\triangle\triangle'+\frac{2}{3}[(ab'-a'b)^2+(bc'-b'c)^2+(ca'-c'a)^2],\end{aligned} \tag{2}$$

并且进一步猜测$\frac{2}{3}$可改进为1，即

$$\begin{aligned}&a'^2(-a^2+b^2+c^2)+b'^2(a^2-b^2+c^2)+c'^2(a^2+b^2-c^2)\\&\geqslant 16\triangle\triangle'+(ab'-a'b)^2+(bc'-b'c)^2+(ca'-c'a)^2.\end{aligned} \tag{3}$$

本文指出所述猜测(3)是不成立的. 实际上，$\frac{2}{3}$不能换成更大的数K，即当$K>\frac{2}{3}$时，

$$\begin{aligned}&a'^2(-a^2+b^2+c^2)+b'^2(a^2-b^2+c^2)+c'^2(a^2+b^2-c^2)\\&\geqslant 16\triangle\triangle'+K[(ab'-a'b)^2+(bc'-b'c)^2+(ca'-c'a)^2]\end{aligned} \tag{4}$$

不成立，证明如下：

令

$$\begin{aligned}Q=&16\triangle\triangle'+K[(ab'-ba')^2+(bc'-cb')^2+(ca'-ac')^2]\\&-[a'^2(-a^2+b^2+c^2)+b'^2(a^2-b^2+c^2)+c'^2(a^2+b^2-c^2)],\end{aligned}$$

取

$$a=a'=1,\ A=A'=90^\circ,\ b=c'=\sin B,\ c=b'=\cos B,$$

则 $Q=16\triangle^2+K[(b-c)^2+(b^2-c^2)^2+(b-c)^2]-[(-1+b^2+c^2)$

$\quad +c^2(1-b^2+c^2)+b^2(1+b^2-c^2)]$

$=4b^2c^2+(K-1)(b^2-c^2)^2+2K-4Kbc-1$

$=\sin^2 2B+(K-1)\cos^2 2B+2K-2K\sin 2B-1$

$=(K-2)\cos^2 2B+2K(1-\sin 2B)$

$=(1-\sin 2B)[2K+(K-2)(1+\sin 2B)]$

$=(1-\sin 2B)[(3K-2)+(K-2)\sin 2B].$

在 $\frac{2}{3}<K\leqslant 1$ 时，$0<\frac{3K-2}{2-K}\leqslant 1.$

因此存在 α，满足 $0<\alpha\leqslant\frac{\pi}{2}$，且

$$\sin\alpha=\frac{3K-2}{2-K}.$$

取 $0<B<\frac{\alpha}{2}$，因 $\sin\alpha>\sin 2B$，则

$$Q=(1-\sin 2B)(\sin\alpha-\sin 2B)\cdot(2-K)>0,$$

即(4)式不成立.

在 $K>1$ 时，$\frac{3K-2}{2-K}>1>\sin 2B$，当然(4)式也不成立.

由此可见，(2)中系数 $\frac{2}{3}$ 已经不能改进，即程龙同志文中所得结果已经是最好的了.

一类恒等式的证明

在一些课本,如著名的《初等数学教程》(法国布尔勒著,有中译本)或习题集中,要求证明下面的一类恒等式:

已知 $a+b+c=0$, 求证:

$$\frac{a^5+b^5+c^5}{5}=\frac{a^2+b^2+c^2}{2}\cdot\frac{a^3+b^3+c^3}{3};\tag{1}$$

$$\frac{a^4+b^4+c^4}{2}=\frac{a^2+b^2+c^2}{2}\cdot\frac{a^2+b^2+c^2}{2};\tag{2}$$

$$\frac{a^7+b^7+c^7}{7}=\frac{a^2+b^2+c^2}{2}\cdot\frac{a^5+b^5+c^5}{5};\tag{3}$$

$$\frac{a^5+b^5+c^5}{5}\cdot\frac{a^2+b^2+c^2}{2}=\frac{a^3+b^3+c^3}{3}\cdot\frac{a^4+b^4+c^4}{2};\tag{4}$$

稍加推广,如已知 $a_1+a_2+\cdots+a_n=0$, 求证:

$$\frac{a_1^5+a_2^5+\cdots+a_n^5}{5}=\frac{a_1^2+a_2^2+\cdots+a_n^2}{2}\cdot\frac{a_1^3+a_2^3+\cdots+a_n^3}{3};\tag{5}$$

等等.

这一类恒等式如何证明?最容易想到的是直接将 $c=-(a+b)$ 代入等式两边去验证,但这样做是非常麻烦的.

其实,这类恒等式形式整齐,可以有简便的证法.

在学过微积分以后,利用对数函数的幂级数的展开式

$$\ln(1+x)=x-\frac{x^2}{2}+\frac{x^3}{3}-\cdots+\frac{(-1)^{n-1}}{n}x^n+\cdots,\ |x|<1$$

这个问题就很容易解决.

由于 $a+b+c=0$, 所以有恒等式

$$(1+ax)(1+bx)(1+cx)=1+qx^2+rx^3,$$

其中 $q=ab+bc+ca$, $r=abc$.

取 $|x|$ 充分小,可使 $|ax|<1$, $|bx|<1$, $|qx+rx^2|<1$, 对上恒等式两边取对数得

$$\ln(1+qx^2+rx^3)=\ln(1+ax)+\ln(1+bx)+\ln(1+cx),$$

展开得

$$\begin{aligned}&(qx^2+rx^3)-\frac{1}{2}(qx^2+rx^3)^2+\frac{1}{3}(qx^2+rx^3)^3-\cdots\\=&\left[ax-\frac{1}{2}(ax)^2+\frac{1}{3}(ax)^3-\cdots\right]+\left[bx-\frac{1}{2}(bx)^2+\frac{1}{3}(bx)^3+\cdots\right]\\&+\left[cx-\frac{1}{2}(cx)^2+\frac{1}{3}(cx)^3+\cdots\right]\\=&-\frac{a^2+b^2+c^2}{2}x^2+\frac{a^3+b^3+c^3}{3}x^3-\frac{a^4+b^4+c^4}{4}x^4+\frac{a^5+b^5+c^5}{4}x^5+\cdots.\end{aligned}$$

比较两边 x 各次幂的系数,得

$$-q=+\frac{a^2+b^2+c^2}{2},$$

$$r=\frac{c^3+b^3+c^3}{3},$$

$$+\frac{q^2}{2}=\frac{a^4+b^4+c^4}{4},$$

$$-qr=\frac{a^5+b^5+c^5}{5},$$

$$\frac{r^2}{2}-\frac{q^3}{3}=\frac{a^6+b^6+c^6}{6},$$

$$q^2r=\frac{a^7+b^7+c^7}{7}\cdots.$$

因此,

$$\frac{a^5+b^5+c^5}{5}=\frac{a^2+b^2+c^2}{2}\cdot\frac{a^3+b^3+c^3}{3}=-qr,$$

$$\frac{a^4+b^4+c^4}{2}=\frac{a^2+b^2+c^2}{2}\cdot\frac{a^2+b^2+c^2}{2}=q^2,$$

$$\frac{a^7+b^7+c^7}{7}=\frac{a^2+b^2+c^2}{2}\cdot\frac{a^5+b^5+c^5}{5}=q^2r,$$

$$\frac{a^5+b^5+c^5}{5}\cdot\frac{a^2+b^2+c^2}{2}=\frac{a^3+b^3+c^3}{3}\cdot\frac{a^4+b^4+c^4}{2}=q^2r.$$

即(1)、(2)、(3)、(4)成立.同样可以证明(5).

由上面的证明,我们还可以导出许多恒等式来.例如,已知 $a+b+c=0$,则

$$a^5+b^5+c^5=-5abc(bc+ca+ab)(=-5qr); \tag{6}$$

$$a^6+b^6+c^6=3a^2b^2c^2-2(bc+ca+ab)^3(=3r^2-2q^3); \tag{7}$$

等等.

如果令 $x-y=a$, $y-z=b$, $z-x=c$,那么显然有 $a+b+c=0$,因此运用上面的方法,可知对于任意的 x、y、z,有以下恒等式:

$$\begin{aligned}&\frac{(x-y)^5+(y-z)^5+(z-x)^5}{5}\\&=\frac{(x-y)^2+(y-z)^2+(z-x)^2}{2}\cdot\frac{(x-y)^3+(y-z)^3+(z-x)^3}{3};\end{aligned} \tag{8}$$

$$\begin{aligned}&25\{(y-z)^7+(z-x)^7+(x-y)^7\}\{(y-z)^3+(z-x)^3+(x-y)^3\}\\&=21\{(y-z)^5+(z-x)^5+(x-y)^5\}^2;\end{aligned} \tag{9}$$

$$\begin{aligned}&\{(y-z)^2+(z-x)^2+(x-y)^2\}^3-54(y-z)^2(z-x)^2(x-y)^2\\&=2(y+z-2x)^2(z+x-2y)^2(x+y-2z)^2;\end{aligned} \tag{10}$$

$$\begin{aligned}&(y-z)^6+(z-x)^6+(x-y)^6-3(y-z)^2(z-x)^2(x-y)^2\\&=2(x^2+y^2+z^2-yz-zx-xy)^3;\end{aligned} \tag{11}$$

$$\begin{aligned}&(y-z)^7+(z-x)^7+(x-y)^7\\&=7(y-z)(x-x)(x-y)(x^2+y^2+z^2-yz-zx-xy)^2;\end{aligned} \tag{12}$$

等等.

实际上,运用前面的记号,可知(8)、(9)、(10)、(11)、(12)的两边分别等于 $-qr$、$525q^2r^2$、$-2(4q^3+27r^2)$、$-2q^3$、$7q^2r$.

当然还可以导出更多的恒等式来.

三角形中一些量的关系

设$\triangle ABC$的边长为a、b、c，面积为$\triangle$，半周长为s，即

$$a+b+c=2s. \tag{1}$$

又设外接圆的半径为R，内切圆的半径为r，有不少关于这些量的等式，例如

$$\frac{abc}{4\triangle}=R, \tag{2}$$

$$abc=4Rrs, \tag{3}$$

$$R(\cos A+\cos B+\cos C)=R+r, \tag{4}$$

等等.

(2)式不难证明，只要应用面积公式$\triangle=\frac{1}{2}bc\sin A$及正弦定理$a=2R\sin A$就可以推出，在(2)中将$\triangle$换为相等的量$sr$，就导得(3).

(4)式似乎比较困难，当然解决这个问题的方法还是很多的：几何的、三角的、代数的方法都有. 但是，在着手解决一个问题之前，我们最好先把这个问题的意义思考一下. 由余弦定理可知，$\cos A$、$\cos B$、$\cos C$都可以用a、b、c来表示，因此(4)式的意义就是将a、b、c的一个有理式用R、r来表示，更准确一点地说，是将a、b、c的一个对称有理式(将字母a、b、c中任意两个互换时，保持不变的有理式称为a、b、c的对称有理式)用R、r来表示. 于是，可以提出一个进一步的问题：是否a、b、c的任意一个对称有理式都可以用R、r来表示？稍作思考就知道这是办不到的，因为a、b、c是三个独立的量(仅受两者之和大于第三者的约束)，仅用两个量R、r来表示是不够的，必须再加入一个量. 最简单的，当然是加入s(当然也可以选$\triangle$或者其他的有关量)我们希望能证明下面的命题.

命题　a、b、c的对称有理式一定可以用s、R、r来表示.

证明　因为每个对称有理式是两个对称多项式的商,所以只要证明命题对 a、b、c 的对称多项式成立. 在高等代数中有一个熟知的结论:a、b、c 的每个对称多项式可以用它们的基本对称函数即 $a+b+c$, $ab+bc+ca$, abc 来表示,因此命题的证明就化为将 $a+b+c$, $ab+bc+ca$, abc 用 s、R、r 来表示了,我们已经有

$$a+b+c=2s, \tag{1}$$

及

$$abc=4sRr, \tag{3}$$

剩下的问题就是用 s、R、r 来表示 $ab+bc+ca$. 将这个量记为 σ.

由根和系数的关系(或者由直接的乘法运算来验证),我们知道

$$(x-a)(x-b)(x-c)=x^3-2sx^2+\sigma x-4sRr.$$

在上式中令 $x=s$ 得到

$$(s-a)(s-b)(s-c)=s^3-2s^3+\sigma s-4sRr=s(-s^2+\sigma-4Rr).$$

由面积公式

$$(s-a)(s-b)(s-c)=\triangle^2/s=sr^2,$$

因此

$$sr^2=s(-s^2+\sigma-4Rr),$$

从而

$$\sigma=s^2+r^2+4Rr,$$

即

$$ab+bc+ca=s^2+r^2+4Rr. \tag{5}$$

于是命题成立.

现在举几个例子来说明命题的应用. 通过这些例子可以具体地掌握如何将 a、b、c 的对称有理式表为 R、r、s 的函数(也是有理式).

例1　将 $a^2+b^2+c^2$ 用 s、R、r 来表示.

解　$a^2+b^2+c^2=(a+b+c)^2-2(ab+bc+ca)=4s^2-2\sigma=2(s^2-4Rr-r^2)$(利用了(1)、(5)).

例2　证明:$a^3+b^3+c^3=2s(s^2-6Rr-3r^2)$.

解　$a^3+b^3+c^3=(a^3+b^3+c^3-3abc)+3abc$

$$=(a+b+c)(a^2+b^2+c^2-ab-bc-ca)+3abc$$
$$=2s[2(s^2-r^2-4Rr)-(s^2+r^2+4Rr)+12sRr]$$
$$=2s(s^2-6Rr-3r^2).$$（利用了上题及(1)、(3)、(5)）.

最后回到我们原来的问题，即：

例 3 证明：$R(\cos A+\cos B+\cos C)=R+r$.

解
$$2abc(\cos A+\cos B+\cos C)$$
$$=a(b^2+c^2-a^2)+b(c^2+a^2-b^2)+c(a^2+b^2-c^2)$$
$$=a(b^2+c^2+a^2)+b(c^2+a^2+b^2)+c(a^2+b^2+c^2)-2(a^3+b^3+c^3)$$
$$=(a+b+c)(a^2+b^2+c^2)-2(a^3+b^3+c^3)$$
$$=4s(s^2-4Rr-r^2)-4s(s^2-6Rr-3r^2)$$
$$=4s(2Rr+2r^2)=8sr(R+r)$$（利用了例 2 及例 3）.

由(3)，

$$8srR(\cos A+\cos B+\cos C)=8sr(R+r),$$

所以 $$R(\cos A+\cos B+\cos C)=R+r.$$

还可以举出更多的例子，如将 $a^4+b^4+c^4$ 表为 s、R、r 的多项式等. 一般说来，只要先将 a、b、c 的对称有理式用 $a+b+c$，$ab+bc+ca$，abc 来表示，再利用(1)、(3)、(5)便可以将这对称有理式用 s、R、r 来表示.

数学问题并非孤立的，一个问题可能只是冰山的一个尖，如果能看到这个“冰山”的整体，那么就有可能解决一类问题，这往往比零打碎敲地解决一个个的孤立的问题省力得多.

注：利用本文中所用的方程 $(x-a)(x-b)(x-c)=0$ 的根与系数的关系及基本对称多项式 $a+b+c$，$ab+bc+ca$，abc，同样可以解决笔者在《一类恒等式的证明》（《数学教学》1982 年第 6 期）中所说的问题，而不必利用对数函数的展开式.

利用面积证题一例

1984年笔者在贵州参加全国数学竞赛的命题工作，会议期间曾讨论过一道题：

如图1,设在$\triangle ABC$中，D为BC的中点，G为重心,过G任作直线分别交AB、AC于E、F. 设$\frac{AE}{AB}=h$，$\frac{AF}{AC}=k$，求证：

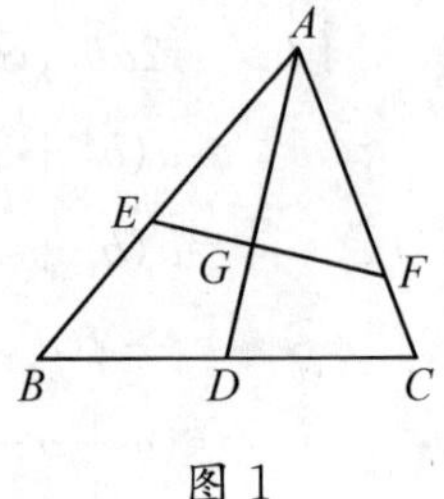

图 1

$$\frac{1}{h}+\frac{1}{k}=3. \tag{1}$$

我们不想局限于这一个问题,所以先作一些推广. 考虑一下，D为BC上任意一点，G为AD上任意一点,这时的结论是什么?

设$\frac{BD}{DC}=\frac{\lambda_1}{\lambda_2}$，$\lambda_1+\lambda_2=1$，$\frac{AG}{AD}=t$，我们断言有

$$\frac{\lambda_1}{k}+\frac{\lambda_2}{h}=\frac{1}{t}. \tag{2}$$

为了证明(2),我们借助于三角形的面积. 设$\triangle AEG$，$\triangle AGF$，$\triangle ABC$的面积分别为S_1，S_2，S. 则由于$\triangle ABD$的高与$\triangle ABC$的高相同,而$\frac{BD}{BC}=\frac{\lambda_1}{\lambda_1+\lambda_2}=\lambda_1$，所以$\triangle ABD$的面积为$\lambda_1 S$,同理$\triangle ADC$面积为$\lambda_2 S$. 由于$\triangle AEG$与$\triangle ABD$有一个角相同,所以

$$\frac{S_1}{\lambda_1 S}=\frac{AE\times AG}{AB\times AD}=ht. \tag{3}$$

同样

$$\frac{S_2}{\lambda_2 S}=kt. \tag{4}$$

$\lambda_1\cdot(3)+\lambda_2\cdot(4)$得

$$\frac{S_1+S_2}{S}=\lambda_1 ht+\lambda_2 kt. \tag{5}$$

由 S_1+S_2 就是$\triangle AEF$ 的面积,所以

$$\frac{S_1+S_2}{S}=\frac{AE\cdot AF}{AB\cdot AC}=hk. \tag{6}$$

从而由(5)、(6)得 $\lambda_1 ht+\lambda_2 kt=hk$, 两边同时除以 hkt 就得到(2).

如果 D 为中点,那么 $\lambda_1=\lambda_2=\frac{1}{2}$, 这时(2)成为

$$\frac{1}{k}+\frac{1}{h}=\frac{2}{t}. \tag{7}$$

(7)有个有趣的应用,计算并联电路的电阻.

如果 G 是重心,那么 $t=\frac{2}{3}$. 于是由(7)就得到(1).

(2)还有种种特殊的情况,例如 G 是$\triangle ABC$ 的内心,这时我们有 $\frac{BD}{DC}=\frac{AB}{AC}=\frac{c}{b}$, 所以 $\lambda_1=\frac{c}{b+c}$, $\lambda_2=\frac{b}{b+c}$.

而 $\frac{AG}{GD}=\frac{AB}{BD}=\frac{c}{\frac{c}{b+c}\cdot a}=\frac{b+c}{a}$, 所以

$$t=\frac{AG}{AD}=\frac{b+c}{a+b+c}.$$

于是(2)成为

$$\frac{c}{k(b+c)}+\frac{b}{h(b+c)}=\frac{a+b+c}{b+c}.$$

即

$$\frac{c}{k}+\frac{b}{h}=a+b+c. \tag{8}$$

以上推导出的(2)式对于 D 在 BC(或 CB)的延长线上,或 G 在 AD(或 DA)的延长线上仍然适用,例如 G 为$\triangle ABC$ 的一个旁心,这时也可以从(2)导出一个与(8)类似的公式.

几个容易混淆的极值问题

下面的三个极值问题貌同而实异,不可把它们混淆起来.

问题 1 n 个正数的和为 1983,问它们的积在什么时候最大,最大值是多少?

解 由于 n 个正数的几何平均数不大于它们的算术平均数,所以在 n 个正数 $a_1, a_2, \cdots, a_n$ 的和为 1983 时,

$$a_1a_2\cdots a_n \leqslant \left(\frac{a_1+a_2+\cdots+a_n}{n}\right)^n = \left(\frac{1983}{n}\right)^n.$$

其中等号当且仅当 $a_1=a_2=\cdots=a_n=\frac{1983}{n}$ 时成立.

因此,所求的最大值为 $\left(\frac{1983}{n}\right)^n$,并且在 n 个数都等于 $\frac{1983}{n}$ 时取得.

显然,1983 可以换成任意一个正数 S,类似的结论仍然成立.

问题 2 若干个正数的和为 1983,问它们的积在什么时候最大,最大值是多少?

解 由上题,如果 n 个正数的和为 1983,那么它们的积在 n 个数都相等时取得最大值,最大值 M_n 为 $\left(\frac{1983}{n}\right)^n$. 但与问题 1 不同的是,本题中 n 并非一个预先确定的数,我们需要确定 n,使得 $\left(\frac{1983}{n}\right)^n$ 为最大.

取对数(我们用自然对数), $\log\left(\frac{1983}{n}\right)^n = n\log 1983 - n\log n$. 我们面临的问题就是要求出在 n 为什么值时, $n\log 1983 - n\log n$ 为最大.

考虑函数 $y=x\log 1983 - x\log x$. 由于它的导数 $y'=\log 1983-1-\log x=\log\frac{1983}{\mathrm{e}x}$, 在 $x<\frac{1983}{\mathrm{e}}$ 时, $y'>0$;在 $x=\frac{1983}{\mathrm{e}}$ 时, $y'=0$;在 $x>\frac{1983}{\mathrm{e}}$ 时, $y'<0$,所以在 $x<\frac{1983}{\mathrm{e}}$ 时,函数 y 递增;在 $x>\frac{1983}{\mathrm{e}}$ 时,函数 y 递减;在 $x=\frac{1983}{\mathrm{e}}$

时，函数 y 取得最大值$\frac{1983}{\mathrm{e}}$.

但$\frac{1983}{\mathrm{e}}$不是整数(众所周知，自然对数的底 $\mathrm{e}=2.71828\cdots$)，所以函数 $y=x\log 1983-x\log x$ 的最大值并非 $n\log 1983-n\log n$ 的最大值！怎样求 $n\log 1983-n\log n$ 的最大值呢？我们需要利用函数 $y=x\log 1983-x\log x$ 的增减性.

由于在 $x<\frac{1983}{\mathrm{e}}$ 时，y 递增，所以在 $n<\frac{1983}{\mathrm{e}}$ 时，$n\log 1983-n\log n$ 在 $n=\left[\frac{1983}{\mathrm{e}}\right]$ 时为最大. 这里$[x]$表示 x 的整数部分，例如 $\left[\frac{1983}{\mathrm{e}}\right]=[729.\cdots]=729$. 同样，在 $n>\frac{1983}{\mathrm{e}}$ 时，$n\log 1983-n\log n$ 在 $n=\left[\frac{1983}{\mathrm{e}}\right]+1$ 时为最大.

经过计算知道

$$\left[\frac{1983}{\mathrm{e}}\right]\log 1983-\left[\frac{1983}{\mathrm{e}}\right]\log\left[\frac{1983}{\mathrm{e}}\right]$$
$$<\left(\left[\frac{1983}{\mathrm{e}}\right]+1\right)\log 1983-\left(\left[\frac{1983}{\mathrm{e}}\right]+1\right)\log\left(\left[\frac{1983}{\mathrm{e}}\right]+1\right).$$

所以 $n\log 1983-n\log n$ 在 $n=\left[\frac{1983}{\mathrm{e}}\right]+1=730$ 时为最大，最大值为

$$\left(\left[\frac{1983}{\mathrm{e}}\right]+1\right)\log 1983-\left(\left[\frac{1983}{\mathrm{e}}\right]+1\right)\log\left(\left[\frac{1983}{\mathrm{e}}\right]+1\right)$$
$$=730\log 1983-730\log 730.$$

从而本题的答案为：积的最大值等于$\left(\frac{1983}{730}\right)^{730}$，并且在加数的个数为 730，各个加数都等于$\frac{1983}{730}$时取得.

一般地，如果若干个正数的和等于固定值 m. 则它们的积的最大值是 $\left(\frac{m}{\left[\frac{m}{\mathrm{e}}\right]}\right)^{\frac{m}{\left[\frac{m}{\mathrm{e}}\right]}}$ 与 $\left(\frac{m}{\left[\frac{m}{\mathrm{e}}\right]+1}\right)^{\frac{m}{\left[\frac{m}{\mathrm{e}}\right]+1}}$ 中较大的那个值，并且在这些数都等于$\frac{m}{\left[\frac{m}{\mathrm{e}}\right]}$或都等于 $\frac{m}{\left[\frac{m}{\mathrm{e}}\right]+1}$ 时积取得最大值.

值得注意的是函数 $y=f(x)$ 与 $f(n)$ (n 为自然数)的最大值是不一定相同的. 读者可再考虑一个例子:二次函数 $y=2x^2-9x+1$ 与 $2n^2-9n+1$ 的最小值,前者在 $x=\frac{9}{4}$ 时取得,而后者在 $n=2$ 时取得.

问题 3 若干个自然数的和等于 1983,问它们的积在什么时候最大,最大值是多少?

解 由于 $\frac{1983}{\left[\frac{1983}{\mathrm{e}}\right]}$ 与 $\frac{1983}{\left[\frac{1983}{\mathrm{e}}\right]+1}$ 都不是整数,所以不能套用上题的解法. 需要另辟途径,采用完全不同的方法.

假定自然数 $a_1, a_2, \cdots, a_n$(当然这里的 n 也是未知的)的和等于 1983,它们的积 P 为最大. 由于 $x \geqslant 4$ 时, $2(x-2) \geqslant x$, 所以可以假定 $a_i<4\ (i=1, 2, \cdots, n)$, 否则可将 a_i 换成两个数: 2 与 a_i-2, 使得总和保持不变,而积 P 增大. 同样,由于 $1 \cdot x<x+1$, 所以可假定 $a_i>1\ (i=1, 2, \cdots, n)$, 否则可将两个数 $a_i(=1)$, a_j 换为一个数 $1+a_j$, 总和仍为 1983,而积 P 增大.

于是 a_i 只可能为 2 或 $3(i=1, 2, \cdots, n)$. 注意 $2\times2\times2<3\times3$, 所以 $a_1, a_2, \cdots, a_n$ 中至多有两个 2 出现,否则可将 2、2、2 换为 3、3,和仍为 1983,而积 P 增大. 这样,所求的最大乘积 $P=2^x \cdot 3^y$,其中 y 为非负整数, $x=0, 1, 2$. 并且 $2x+3y=1983$. 如果 $x=1$ 或 2,该方程无整数解. 在 $x=0$ 时,由该方程得 $y=661$. 因此乘积的最大值为 3^{661},并且在各个加数都等于 3 时取得.

上面的解法也适用于一般的情况. 要注意,各个加数并不一定是全相等的. 例如,将 1983 换为 1984,答案则是:积的最大值为 $2^2\times3^{660}$, 在 660 个加数为 3, 2 个加数为 2(或 1 个加数为 4)时取得. 由此更可以看出,套用问题 1、2 的解法来解问题 3 是错误的.

方程组的巧解

方程组通常是用消元法来解的，但在未知数个数较多(三个以上)，系数是字母或代数式时，用消元法往往是比较麻烦的．在方程组的形状比较整齐时，常常有一些巧妙的解法，我们仅就一次方程组举几个例子说明．

例1　a、b、c 互不相等，试解方程组

$$\begin{cases} x+y+z=0, & (1) \\ ax+by+cz=0, & (2) \\ a^2x+b^2y+c^2z=k. & (3) \end{cases}$$

解　我们取 $p=-(b+c)$，$q=bc$，那么由韦达定理(根与系数的关系)，b、c 是二次方程

$$u^2+pu+q=0 \tag{4}$$

的两个根，即

$$b^2+pb+q=0,\ c^2+pc+q=0, \tag{5}$$

并且有恒等式

$$u^2+pu+q=(u-b)(u-c), \tag{6}$$

在(1)的两边同乘以 q，在(2)的两边同乘以 p，然后再与(3)相加，那么由于(5)，得到

$$(a^2+ap+q)x=k,$$

因此

$$x=\frac{k}{a^2+ap+q}.$$

由于(6)，得

$$x=\frac{k}{(a-b)(a-c)}. \tag{7}$$

原方程组 x、y、z(及 a、b、c)的地位是完全平等的(对称的),因此由同样理由可得(将(7)式中 x 换为 y, a、b、c 换为 b、c、a)

$$y=\frac{k}{(b-c)(b-a)}$$

及(将(7)式中将 x 换为 z, a、b、c 换为 c、a、b)

$$z=\frac{k}{(c-a)(c-b)},$$

故方程组的解为

$$\begin{cases} x=\dfrac{k}{(a-b)(a-c)}, \\ y=\dfrac{k}{(b-c)(b-a)}, \\ z=\dfrac{k}{(c-a)(c-b)}. \end{cases}$$

这一方法可以推广到更多个未知数的方程组,例如解方程

$$\begin{cases} x_1+x_2+\cdots+x_n=1, \\ a_1x_1+a_2x_2+\cdots+a_nx_n=k, \\ \qquad\cdots \\ a_1^{n-1}x_1+a_2^{n-1}x_2+\cdots+a_n^{n-1}x_n=k^{n-1}, \end{cases}$$

其中 a_1, a_2, $\cdots$, a_n 互不相等.

这只要用 σ_{n-1}, σ_{n-2}, $\cdots$, σ_1 分别与方程组中第一个、第二个、……、第 $n-1$ 个方程相乘,然后再与第一个方程相加,其中 σ_{n-1}, σ_{n-2}, $\cdots$, σ_1 是由恒等式

$$u^{n-1}+\sigma_1u^{n-2}+\cdots+\sigma_{n-1}\equiv(u-a_2)(u-a_3)\cdots(u-a_n)$$

确定的,换句话说, a_2, a_3, $\cdots$, a_n 是方程

$$u^{n-1}+\sigma_1u^{n-2}+\cdots+\sigma_{n-1}=0$$

的 $n-1$ 个根. 于是从原方程组导出

$$(a_1^{n-1}+\sigma_1a_1^{n-2}+\cdots+\sigma_{n-1})x_1=k^{n-1}+\sigma_1k^{n-2}+\cdots+\sigma_{n-1},$$

即
$$x_1=\frac{(k-a_2)(k-a_3)\cdots(k-a_n)}{(a_1-a_2)(a_1-a_3)\cdots(a_1-a_n)}$$

同样可得 x_2, …, x_n.

例 2　解方程组

$$\begin{cases}\dfrac{x}{a+\lambda}+\dfrac{y}{b+\lambda}+\dfrac{z}{c+\lambda}=1,\\ \dfrac{x}{a+\mu}+\dfrac{y}{b+\mu}+\dfrac{z}{c+\mu}=1,\\ \dfrac{x}{a+\gamma}+\dfrac{y}{b+\gamma}+\dfrac{z}{c+\gamma}=1,\end{cases}$$

其中 λ、μ、γ 互不相等，a、b、c 也互不相等.

解　设 x、y、z 是方程组的解，考虑 θ 的二次多项式

$$(a+\theta)(b+\theta)(c+\theta)\left[\frac{x}{a+\theta}+\frac{y}{b+\theta}+\frac{z}{c+\theta}\right],$$

在 $\theta=\lambda$、μ、γ 时，它的值分别为

$$(a+\lambda)(b+\lambda)(c+\lambda),\ (a+\mu)(b+\mu)(c+\mu),\ (a+\gamma)(b+\gamma)(c+\gamma),$$

因为中括号中的值为 1.

另一方面，θ 的二次多项式

$$(a+\theta)(b+\theta)(c+\theta)\left[1-\frac{(\theta-\lambda)(\theta-\mu)(\theta-\gamma)}{(a+\theta)(b+\theta)(c+\theta)}\right]$$

在 $\theta=\lambda$、μ、γ 时的值也分别为

$$(a+\lambda)(b+\lambda)(c+\lambda),\ (a+\mu)(b+\mu)(c+\mu),\ (a+\gamma)(b+\gamma)(c+\gamma),$$

即这两个 θ 的二次多项式在 $\theta=\lambda$、μ、γ 时的值对应相等，根据多项式恒等定理，这两个多项式恒等，则有

$$(a+\theta)(b+\theta)(c+\theta)\left[\frac{x}{a+\theta}+\frac{y}{b+\theta}+\frac{z}{z+\theta}\right]$$
$$\equiv(a+\theta)(b+\theta)(c+\theta)\left[1-\frac{(\theta-\lambda)(\theta-\mu)(\theta-\gamma)}{(a+\theta)(b+\theta)(c+\theta)}\right],$$

从而有

$$x+(a+\theta)\left[\frac{y}{b+\theta}+\frac{z}{c+\theta}\right]\equiv a+\theta-\frac{(\theta-\lambda)(\theta-\mu)(\theta-\gamma)}{(b+\theta)(c+\theta)},$$

令 $\theta=-a$，便得

$$x=\frac{-(-a-\lambda)(-a-\mu)(-a-\gamma)}{(b-a)(c-a)}=\frac{(a+\lambda)(a+\mu)(a+\gamma)}{(b-a)(c-a)},$$

同样可得 y、z.

本题也可以推广到 n 个未知数的方程组. 例如 $a_1, a_2, \cdots, a_n$ 互不相等，$b_1, b_2, \cdots, b_n$ 互不相等时，用完全一样的方法可得方程组

$$\begin{cases}\dfrac{x_1}{a_1-b_1}+\dfrac{x_2}{a_1-b_2}+\cdots+\dfrac{x_n}{a_1-b_n}=1,\\ \dfrac{x_1}{a_2-b_1}+\dfrac{x_2}{a_2-b_2}+\cdots+\dfrac{x_n}{a_2-b_n}=1,\\ \qquad\qquad\cdots\\ \dfrac{x_1}{a_n-b_1}+\dfrac{x_2}{a_n-b_2}+\cdots+\dfrac{x_n}{a_n-b_n}=1\end{cases}$$

的解为

$$x_1=-\frac{(b_1-a_1)(b_1-a_2)\cdots(b_1-a_n)}{(b_1-b_2)(b_1-b_3)\cdots(b_1-b_n)},$$

等等.

上面所说的方程组当然也可以用行列式来解，但这里行列式的计算也不很容易.

$y=2x^2$ 与 $y=\frac{1}{2}x^2$ 的图像形状一样吗?

在讨论二次函数的图像时，常常有人说“抛物线 $y=2x^2$ 与 $y=\frac{1}{2}x^2$ 的形状是不一样的”. 其实这样说是有疑问的.

通常所说的两个图形 F_1 与 F_2 形状相同是什么意思呢? 对于三角形，这是人们所熟知的，所谓两个三角形的形状相同就是指两个三角形相似，即两个三角形的点之间可以建立起一一对应，使得连结对应点的对应线段的比均相等. 推而广之，两个图形 C_1 与 C_2 形状相同即相似，也就是指在这两个图形的点之间可以建立起这样的一一对应，使得连结对应点的对应线段的比均相等. 例如所有的正方形(或正 n 边形)，形状都是相同的. 所有的圆，形状也都是相同的. 这些在直观上都是显然的.

有趣的是所有的抛物线的形状都相同，即所有抛物线都是彼此相似的. 这个结论乍看起来很奇怪，难道“矮而胖”的抛物线 $y=2x^2$ 与“瘦而长”的抛物线 $y=\frac{1}{2}x^2$ 是相似的吗? 是的，根据上面的定义，确实如此. 不仅这样，我们还可以证明更一般的结论:

“如果两条圆锥曲线的离心率相同，那么这两条圆锥曲线相似.”

证明　经过运动(平移、旋转或反射)使两条圆锥曲线 C_1 与 C_2 的焦点(在椭圆或双曲线的情况，使左焦点与左焦点)及过焦点的对称轴重合，并且在抛物线的情况，使两条曲线的开口的方向成为相同.

按照课本上的办法，以共同的焦点为极点，以过焦点的对称轴为极轴建立极坐标，这时在每一条射线 $\theta=\theta_0$ 上，圆锥曲线 C_1 与 C_2 各有一个点，其极半径分别为

$$\rho_1=\frac{p_1}{1-e\cos\theta_0},$$

$$\rho_2=\frac{p_2}{1-e\cos\theta_0},$$

其中 p_1、p_2 分别为两条曲线的半焦正弦(即曲线过焦点的与轴垂直的弦的一半).

显然

$$\rho_1 : \rho_2 = p_1 : p_2,$$

因此在所述位置上的两条曲线是位似的,以极点为位似中心, $p_1 : p_2$ 为相似比. 从而在移动前两条曲线是相似的. 证毕.

由于一切抛物线的离心率均为 1,因此我们有以下推论:

"一切抛物线都是相似的".

由此看来,我们不应该说抛物线 $y=2x^2$ 与 $y=\frac{1}{2}x^2$ 的形状是不一样的. 如果要指出抛物线 $y=2x^2$ 与 $y=\frac{1}{2}x^2$ 之间的差别,可以说它们是不全等的,或者说它们的焦正弦(跨度)不相同,或者说它们的焦点与顶点之间的距离(矢)不同.

韦达定理的一个应用

在解析几何中，常常需要从两个代数方程里消去一个参数，即消元. 用代入法消元在一个方程(对其中某个元)为一次时最为便利. 如果方程的次数高于二，初等的方法往往需要一些技巧，其中韦达定理(即根与系数的关系)是一个有力的工具，特别是方程的形状十分整齐、对于各个元是对称时，利用这一定理来消元常常是方便的.

下面举两个例题来说明如何利用韦达定理来消元，这两个例题是与今年(指 1982 年)高考第八题相关联的问题.

例 1　已知 y_1、y_2、y_3 满足下列方程

$$y_1y_2(y_1+y_2)=-2p^2q, \tag{1}$$

$$y_1y_3(y_1+y_3)=-2p^2q, \tag{2}$$

并且 $y_2\neq y_3$，证明

$$y_2y_3(y_2+y_3)=-2p^2q. \tag{3}$$

解　问题的实质就是从(1)、(2)两式消去参数 y_1.

将(1)写成

$$y_1y_2^2+y_1^2y_2+2p^2q=0,$$

我们看出 y_2 是方程

$$y_1u^2+y_1^2u+2p^2q=0 \tag{4}$$

的一个根. 同样，y_3 也是(4)的一个根，因此由韦达定理，得

$$\begin{cases} y_2+y_3=-y_1, & (5)\\ y_2y_3=\dfrac{2p^2q}{y_1}. & (6)\end{cases}$$

(5)、(6)二式相乘得

$$y_2 y_3(y_2+y_3)=-2p^2 q,$$

这就是所要证明的结论.

例 2 已知 $x_2 \neq x_3$，试从

$$x_1^2 x_2^2=k(x_1+x_2),\ x_1^2 x_3^2=k(x_1+x_3)$$

中消去 x_1.

解 x_2、x_3 是方程

$$x_1^2 u^2-ku-kx_1=0$$

的两个根,因此,由韦达定理,得

$$\begin{cases} x_2+x_3=\dfrac{k}{x_1^2}, & (7)\\ x_2 x_3=-\dfrac{k}{x_1}. & (8)\end{cases}$$

从(7)、(8)极易消去 x_1,得

$$x_2^2 x_3^2=k(x_2+x_3). \tag{9}$$

在两个方程关于 x_2、x_3 为对称时(即将 x_2、x_3 互换,两个方程不变),都可以用上面的方法来消元. 如果,我们预先知道三个参数 x_1、x_2、x_3(或 y_1、y_2、y_3)是对称的,即在所讨论的问题中,三个参数处于同样的地位,那么我们还可以预先知道消去 x_1(或 y_1)后所得到的方程的形状是和已有的两个方程相同或相似.

现在我们来看一看今年(指 1982 年)高考试题的第八题,这一题的主要困难在于消元,而利用上面的方法是很方便的(当然也可以用其他方法消元). 原题如下:

抛物线 $y^2=2px$ 的内接三角形有两边与抛物线 $x^2=2qy$ 相切,证明这个三角形的第三边也与 $x^2=2qy$ 相切.

解法 1 设这三角形的顶点为 $A_i(x_i,\ y_i)$ $(i=1,\ 2,\ 3)$，则

$$y_i^2=2px_i(i=1,\ 2,\ 3), \tag{10}$$

直线 A_1A_2 的方程为

$$y-y_1=\frac{y_2-y_1}{x_2-x_1}(x-x_1), \tag{11}$$

由(11)与方程 $x^2=2qy$ 得

$$x^2=2q\left[y_1+\frac{y_2-y_1}{x_2-x_1}(x-x_1)\right],$$

即

$$x^2-\frac{2q(y_2-y_1)}{x_2-x_1}\cdot x+\frac{x_1y_2-y_1x_2}{x_2-x_1}\cdot 2q=0. \tag{12}$$

因为 A_1A_2 与抛物线 $x^2=2qy$ 相切,所以

$$\left[\frac{2q(y_2-y_1)}{x_2-x_1}\right]^2=\frac{4(x_1y_2-y_1x_2)}{x_2-x_1}\cdot 2q.$$

利用(10)式将上式化简得

$$y_1y_2(y_1+y_2)=-2p^2q.$$

同理

$$y_1y_3(y_1+y_3)=-2p^2q.$$

由于 A_1、A_2、A_3 是不同点,所以 $y_2\neq y_3$(否则由(10)得 $x_2=x_3$, A_2 与 A_3 重合). 由例 1 得到

$$y_2y_3(y_2+y_3)=-2p^2q,$$

这就表明直线 A_2A_3 与抛物线 $x^2=2qy$ 也相切.

解法 2　考虑 $x^2=2qy$ 的三条切线

$$x_ix=qy+qy_i \quad (i=1,\ 2,\ 3),$$

其中$(x_i,\ y_i)$是切点 $(i=1,\ 2,\ 3)$,由

$$\begin{cases}x_1x=qy+qy_1,\\ x_2x=qy+qy_2,\end{cases}$$

得出第一、二两条切线的交点 B_3 的坐标为

$$x=\frac{q(y_1-y_2)}{x_1-x_2},\ y=\frac{x_1y_2-x_2y_1}{x_2-x_1}.$$

因为 $x_i^2=2qy_i(i=1,2,3)$，所以 B_3 的坐标为

$$x=\frac{q\left(\frac{x_1^2}{2q}-\frac{x_2^2}{2q}\right)}{x_1-x_2}=\frac{x_1+x_2}{2q},\ y=-\frac{x_1x_2}{2q}. \tag{13}$$

同样第一、三两条切线的交点 B_2 坐标满足

$$x=\frac{x_1+x_3}{2q},\ y=-\frac{x_1x_3}{2q}. \tag{14}$$

第二、三两条切线的交点 B_1 坐标满足

$$x=\frac{x_2+x_3}{2q},\ y=-\frac{x_2x_3}{2q}. \tag{15}$$

我们假定 B_2、B_3 在抛物线 $y^2=2px$ 上，要证明 B_1 也在这条抛物线上.

由(13)知

$$\left(-\frac{x_1x_2}{2q}\right)^2=2p\left(\frac{x_1+x_2}{2q}\right),$$

即

$$x_1^2x_2^2=4pq(x_1+x_2).$$

同理由(14)知

$$x_1^2x_3^2=4pq(x_1+x_3).$$

由于 $x_2\neq x_3$（否则第二、三两条切线重合），所以从例 2 推出

$$x_2^2x_3^2=4pq(x_2+x_3).$$

这也就证明了 B_1 点也在抛物线 $y^2=2px$ 上. 从而命题成立.

正方体的分解

一个正方形可以分解成 4 个正方形，这是众所周知的事. 稍加考虑便知道，一个正方形也可以分解成 6、7、8 个正方形(图 1). 其中 7 个正方形的分解法是先将正方形分解成 4 个小正方形，再将其中一个小正方形分成 4 个更小的正方形，从而将一个正方形分成 $4+3=7$ 个正方形，用这个方法也可以将正方形分成 9、10、11…个正方形. 因此可将正方形分成 n 个正方形，这里 $n=4$ 或大于等于 6 的整数. 另一方面，稍作推理便可知道，一个正方形不能分成 2、3 或 5 个正方形. 因此，正方形的分解问题已彻底解决.

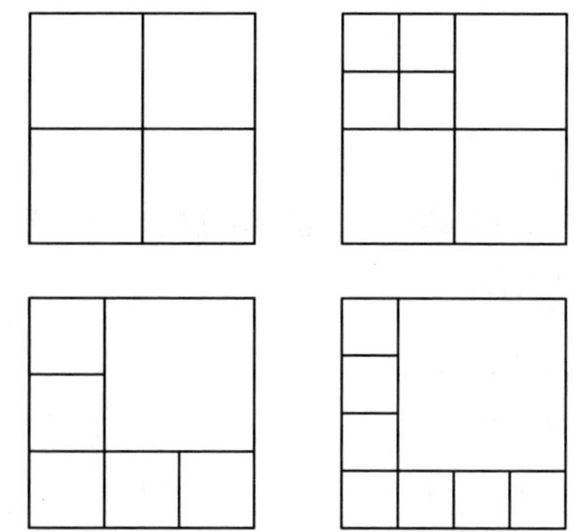

图 1　把正方形分解成 4、6、7、8 个小正方形

1　正方体分解的提出

很自然地，人们会想到正方体的分解.

显然，一个正方体可以分成 8 个正方体，这 8 个正方体大小相等，边长是原正方体的$\frac{1}{2}$. 一个正方体也可以分成 27 个正方体，这 27 个正方体的边长都是

原正方体的$\frac{1}{3}$.

如果正方体可以分解成 n 个正方体，那么将其中一个小正方体再分成 8 个或 27 个，这样原正方体就可分成 $n+7$ 或 $n+26$ 个正方体.

于是，一个正方体可以分解成 15、22、29、36、43、50…及 34、41、48…个正方体.

1946 年，斯科特(William Scott)证明了：一个正方体可以分解成 n 个正方体，这里 $n=1$、8、15、20、22、27、29、34、36、38、39、41、43、45、46、48、49、50、51、52、53 及大于 54 的任一整数.

伊夫斯(Howard Eves)在他的著作《几何学概观》(*A Survey of Geometry*)中介绍了上述结果. 一位数学爱好者蒂尔(Von Christian Thiel)读了此书后，在 1969 年告诉伊夫斯，n 也可以为 54. 于是在该书 1972 年的修订版中，已改成 $n=$ 1、8、15、20、22、27、29、34、36、38、39、41、43、45、46 及大于 47 的任一整数.

2 具体的分法

伊夫斯的书中并未给出具体的分法，但这些分法只要多想一想就不难得出. 下面我们就来谈谈具体的分法.

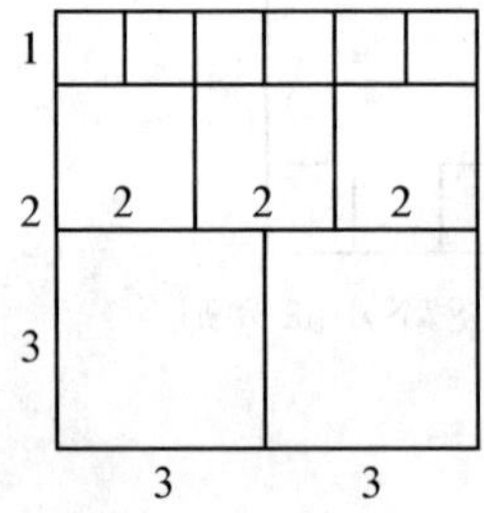

图 2 分解成 49 个小正方体的俯视图. 假定正方体的边长为 6，从图中可以看出，边长为 1、2、3 的小正方体分别有 6^2、3^2、2^2 个，从而一共有 $36+9+4=49$ 个小正方体.

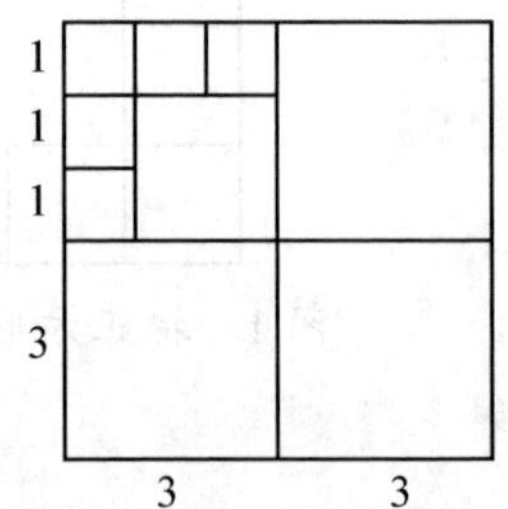

图 3 分解成 51 个小正方体的上层俯视图(下层的俯视图仍为图 2). 其中边长为 3 的正方体有 5 个，边长为 2 的正方体有 5 个(下层 3 个，上层 1 个，还有 1 个在两层之间). 至于边长为 1 的正方体，由图可知一共有 $6\times2+5\times4+9=41$ 个.

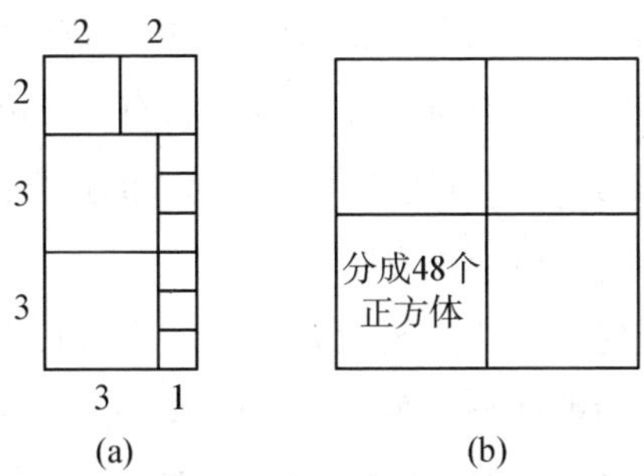

图 4　分解成 54 个正方体的示意图. 如图(a)，先将两个边长为 4 的正方体叠在一起，将其分解为 48 个小正方体，其中边长为 1、2、3 的正方体分别有 42、4、2 个. 如图(b)，假定一个正方体边长为 8，先将其分成 8 个边长为 4 的正方体，再选择两个叠在一起的正方体，按照图(a)的方法分成 48 个小正方体. 这样就完成了分解成 $6+48=54$ 个正方体的工作.

首先，$20=3^3-2^3+1$. 这就是说，如果将正方体分成 27 个小正方体，再将其中 8 个(比如在左上方的 8 个)并做 1 个，那么正方体就分成了 20 个小正方体. 于是，如果一个正方体可以分解成 n 个正方体，那么它也可以分解成 $n+19$ 个正方体.

同样，$38=4^3-3^3+1$. 于是，一个正方体可以分解成 38 个正方体，而且根据上面所说，也可以分解成 45、52…个正方体.

又因为 $27=20+7$，$39=20+19$，$46=20+26$，$53=34+19$，所以正方体可以分解为 27、39、46、53 个正方体.

在斯科特所给出的 n 中，小于 54 的数只剩下 49 与 51 了. 这两种分解均比较困难.

由于 $49=6^3-4\times(3^3-1)-9\times(2^3-1)$，这表明先将正方体分解成 6^3 个边长为原正方体$\dfrac{1}{6}$的正方体，再将正前面的 4×3^3 个正方体合成 4 个正方体，中间 9×2^3 个正方体合成 9 个正方体，就可以得出 49 个正方体.

同样，等式 $51=6^3-5\times(3^3-1)-5\times(2^3-1)$ 表明正方体可以分解为 51 个正方体. 做法仍是先将正方体分解为 6^3 个正方体，再适当合并. 假定原正方体边长为 6，其中边长为 3 的正方体有 5 个，边长为 2 的正方体有 5 个(下层 3 个，上层 1 个，还有 1 个在两层之间).

为了将正方体分成 54 个小正方体，首先将底面为正方形、高为底面边长 2

倍的长方体(实际上就是两个相同的正方体叠在一起)分成48个正方体.不妨设底面边长是4个单位.从正前面看去(主视图),有2个边长为3的正方体.最上面是4个边长为2的正方体,其余都是边长为1的正方体,共有$2\times4^3-2\times(3^3-1)-4\times(2^3-1)=48$个正方体.

一个正方体,假定它的边长为8,可以分成8个边长为4的正方体.其中6个保持不动,而将左前方的两个垒在一起的正方体,作为一个上面所说的$4\times4\times8$的长方体,分成48个正方体.这样原正方体便分成了$48+6=54$个正方体.或许这就是蒂尔的分解方法.

3　结论与进一步问题

总结一下,我们已经给出将正方体分解为n个正方体的方法,这里$n=1$、8、15、20、22、27、29、34、36、38、39、41、43、45、46、48、49、50、51、52、53、54.

注意48、49、50、51、52、53、54是7个连续整数,而如果正方体能分成n个正方体,那么它也能分成$n+7$个正方体,所以对于大于47的整数n,正方体都可以分成n个正方体.

这样,本文就得到了伊夫斯书中的全部结论.当然,在能够分解时,分解的种数未必只有一种.要确定分解种数,除了少数情况,看来不是一件轻而易举的事.

正方体能不能分解成47个正方体?这是遗留下的一个问题,或许某一位读者能够解决它.

十三个球的问题

1694 年,英国牛津的一位天文学家格雷戈里(David Gregory)与他的朋友、大名鼎鼎的牛顿(Isaac Newton),讨论体积不同的星星在天空中如何分布,引出了一个问题:一个单位球能否与 13 个(互不相交的)单位球相切? 牛顿认为不可能,而格雷戈里则猜测:一个单位球能够与 13 个单位球相切. 他们的讨论记录在格雷戈里的一本笔记本中,没有发表,保存在牛津的一所教堂里.

1　问题的等价形式与推广

在单位球 A 与单位球 O 相切时,点 O 到球 A 的切线形成一个圆锥. 这个圆锥含有球面 O 的一个球冠,切点 A_1 就是球冠的极(顶点)(见图 1). A_1 与球冠上任一点的球面距离 $\leqslant\frac{\pi}{6}$(即 $30°$). $\frac{\pi}{6}$ 称为这个球冠的半径.

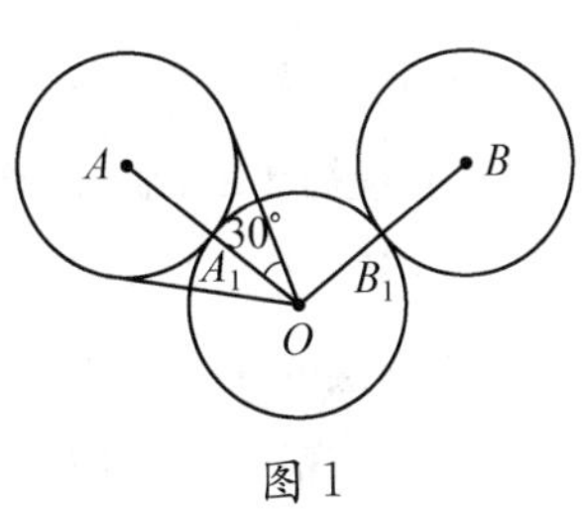

图 1

同样,对另一个与球 O 相切的单位球 B,也有一个以切点 B_1 为极、$\frac{\pi}{6}$ 为半径的球冠.

于是,格雷戈里的猜测等价于下面的问题:球面上能否有 13 个半径为 $\frac{\pi}{6}$ 的球冠,互不重叠?

由于 $\overset{\frown}{A_1B_1}\geqslant 2\times\frac{\pi}{6}=\frac{\pi}{3}$,所以 $A_1B_1\geqslant OA_1=1$,即每两个切点之间的(直线)距离 $\geqslant 1$. 格雷戈里的猜测还可以再换一个等价的说法:

单位球面上能否有 13 个点,每两个点之间的距离 $\geqslant 1$?

另外,问题可以不限于三维空间. 高维的我们后面再谈. 至于在平面上,相应的问题简单得多. 二维的"球"就是圆. 设 $\odot(O,\ 1)$ 是圆心为 O、半径为 1 的圆.至多有多少个互不重叠的单位圆与⊙$(O, 1)$相切?

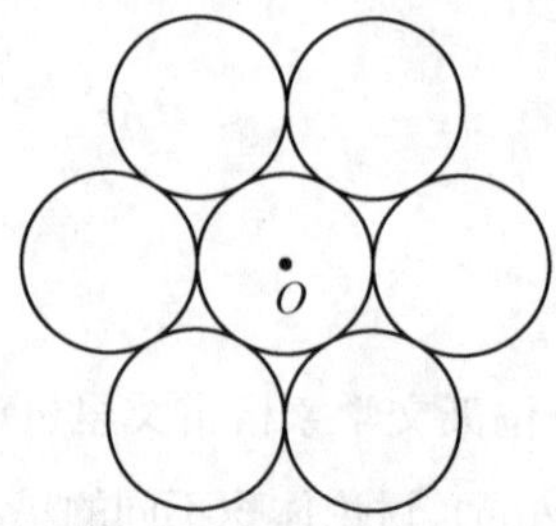

图2　6个互不相交的单位圆和⊙$(O, 1)$相切.

图2表明,可以有6个这样的单位圆与⊙$(O, 1)$相切,这6个圆"挤"得很紧,容易证明,第7个圆没有"立足之地".

2　12≤相切球数≤14

回到三维空间,不难举出一个例子,说明可以有12个单位球与同一个单位球相切.最简单的例子是将球一层一层地堆起来.最上面1个,第2层4个,第3层9个,第4层16个.那么,在第3层核心的那个球既与上一层的4个球相切,又与下一层的4个球相切,还与同一层的4个球相切.

另一个很自然的例子,就是球O的内接正二十面体.它有12个顶点,20个面,每个面为正三角形,不难算出这正二十面体的每条棱长为

$$2\sin\left(\arccos\frac{1}{2\sin\frac{\pi}{5}}\right)=\frac{1}{5}\sqrt{50-2\sqrt{125}}=1.0514\cdots>1.$$

即12个顶点(它们可以作为切点),两两的距离大于1.

这12个点的距离大于1,而不是等于1.因此,它们还可以挪近一些.格雷戈里猜测还可以再放进一个点,并不是毫无道理的.

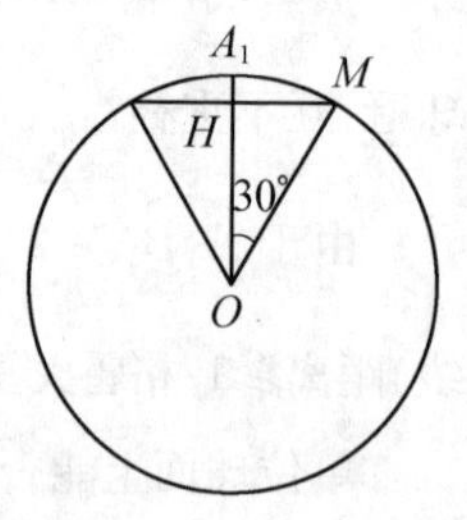

图3　半径为$\frac{\pi}{6}$的球冠平面图

另一方面,我们可以证明与单位球O相切、互不重叠的单位球不超过14个,也就是在球面O上,两两距离≥1的点,个数≤14.为此,考虑半径为$\frac{\pi}{6}$的球冠,如图3,球冠

的高 $h = A_1H = 1 - OH = 1 - \frac{\sqrt{3}}{2}$. 所以球冠的面积为 $2\pi h = \pi(2-\sqrt{3})$. 球面积为 4π，所以互不重叠的、半径为 $\frac{\pi}{6}$ 的球冠，其个数 $\leqslant \frac{4\pi}{2\pi h} = 4(2+\sqrt{3}) = 14.928\cdots$. 因此，15 个半径为 $\frac{\pi}{6}$ 的球冠必有重叠.

3　相切球数≤12

上节的结果有差距，我们还应当进一步证明相切球数不仅≤13，而且≤12，即格雷戈里的猜测不成立.

这一证明是困难的. 虽然大多数人倾向于牛顿的观点，认为至多 12 个球与一个单位球相切，但是严格的证明却姗姗来迟，直至将近 260 年以后的 1953 年，才由许特(K. Schütte)与范・德・瓦尔登(B. L. van der Waerden，出生于 1903 年的著名代数学家)给出. 一说是 1874 年霍佩(R. Hoppe)已有证明，但我们未见到原始文献. 1956 年，利奇(John Leech)给出一个较为简单的证明. 为了介绍这一证明的梗概，需要了解一些球面几何的基本知识.

4　球面几何

对球面上每两个点 A、B，过 A、B 与球心 O 作一个平面，这平面与球面的交线是一个圆，半径与球半径相等，称为大圆.

大圆，在球面几何中也称为"直线"，与平面上的直线不同，球面上的直线，长度是有界的(即 2π)，但却是"无涯"的，即没有起点与终点. 球面上任意两条直线必有两个公共点. 所以球面几何不是欧几里得几何，通常也称它为黎曼几何.

大圆上，连结 A、B 两点的弧有两条，其中较短的一条称为连结 A、B 的"线段". 它是在球面上从 A 到 B 的最短途径(这也正是我们称大圆为直线的一个原因).

不过球心的平面，与球面相交得到的圆，半径小于球半径，称为小圆或简称圆.

对球面上任意三点 A、B、C，将每两点用“线段”(大圆弧)连结起来，组成球面三角形，其边长即线段(大圆弧)的长，与平面几何类似，通常记为 a、b、c. 球面三角形的角，例如，$\angle A$ 是平面 OAB 与 OAC 所成的二面角. 球面三角形与平面三角形有不少类似之处(如三角形全等的定理)，但也有很多的不同：

球面三角形的三个角的和一定大于 π，即差 $E=A+B+C-\pi$ 是一个正数. E 称为角盈，对于单位球，它正好是球面三角形 ABC 的面积.

球面三角形有以下基本公式：

(1) 正弦定理：$\dfrac{\sin a}{\sin A}=\dfrac{\sin b}{\sin B}=\dfrac{\sin c}{\sin C}$；

(2) 余弦定理：$\cos a=\cos b\cos c+\sin b\sin c\cos A$；

(3) 面积公式：$\tan\dfrac{E}{4}=\sqrt{\tan\dfrac{s}{2}\tan\dfrac{s-a}{2}\tan\dfrac{s-b}{2}\tan\dfrac{s-c}{2}}$，

其中 $s=\dfrac{a+b+c}{2}$.

有了这些公式，便可以进行球面几何的计算，例如：

例 1 设球面三角形的边长为 $\dfrac{\pi}{3}$，$\arccos\dfrac{1}{7}$，$\dfrac{\pi}{3}$，则由(2)可以算出它的角为 $\dfrac{\pi}{3}$，$\arccos\left(-\dfrac{1}{7}\right)$，$\dfrac{\pi}{3}$；

例 2 边长为 $\dfrac{\pi}{3}$ 的正三角形，由(2)可算出角为 $\arccos\dfrac{1}{3}$，面积为 $3\arccos\dfrac{1}{3}-\pi=0.55128\cdots$.

例 3 边长均为 $\dfrac{\pi}{3}$、角均为 $\arccos\left(-\dfrac{1}{3}\right)$ 的正四边形，它的对角线互相垂直平分，长均为 $\dfrac{\pi}{2}$.

例 4 边长均为 $\dfrac{\pi}{3}$，一条对角线为 $\arccos\dfrac{1}{7}$ 的四边形，它被这条对角线分为两个全等的三角形，如例 1，所以它的面积是 $2\left(\arccos\left(-\dfrac{1}{7}\right)+\dfrac{2\pi}{3}-\pi\right)=1.334\cdots$.

例 5 边长均为$\frac{\pi}{3}$. 自某顶点引出的两条对角线均为 $\arccos\frac{1}{7}$，这种五边形的面积为

$$1.334\cdots+\left(\arccos\frac{47}{96}+2\arccos\frac{1}{12}-\pi\right)$$
$$=1.334\cdots+0.892\cdots=2.226\cdots.$$

5 证明梗概

假设单位球面上有 13 个点，两两的距离（本节距离均指球面距离）$\geqslant\frac{\pi}{3}$.

如果其中某两个点的距离$<\arccos\frac{1}{7}$（$\approx 1.4274\cdots$），就用一条线段将它们连结起来. 这样，13 个点与它们之间的一些连线形成球面上的一个网络.

如果有一个点与其他 12 个点的距离均$\geqslant\arccos\frac{1}{7}$，那么可以移动这个点，使它至少与一个点的距离$<\arccos\frac{1}{7}$，同时与其他点的距离仍$\geqslant\frac{\pi}{3}$. 于是没有孤立点，即每一个点至少与其他一点相连. 采用同样手法，可以假定每点至少与其他两点相连.

自同一点引出的两条线 AB、AC，在$\triangle ABC$ 成例 1 中的三角形时，$\angle A$ 取得下确界$\frac{\pi}{3}$，此时 AB、AC 一为$\frac{\pi}{3}$，一为 $\arccos\frac{1}{7}$，于是恒有 $\angle BAC>\frac{\pi}{3}$，由于 A 点处的周角为 2π，故 A 点引出的线$<\frac{2\pi}{\frac{\pi}{3}}=6$ 条，即每一点至多引出 5 条线.

利用余弦定理还可以证明，在上述网络中，每两条线除了端点外没有公共点. 因此对于它，有欧拉定理 $v-e+f=2$，其中 $v=13$ 是点数，e 是线的条数，$f=f_3+f_4+f_5+\cdots$ 是面数，f_3、f_4、$f_5\cdots$分别表示三角形、四边形、五边形$\cdots$的个数，又易知 $2e=3f_3+4f_4+5f_5+\cdots$，所以 $2v-4=2(e-f)=f_3+2f_4+3f_5+\cdots$.

网络中,三角形的面积以例 2 的正三角形为最小,四边形的面积以例 4 的为最小,五边形的以例 5 为最小. $n(\geqslant 5)$边形可化作 $n-2$ 个三角形,其中 $n-3$ 个的面积至少是例 2 所说的面积,剩下的一个至少是例 5 里两条对角线与一边所围的 $0.892\cdots$,因此有

$$\begin{aligned}4\pi &= \text{网络总面积} \geqslant 0.55128\cdots \times f_3 + 1.334\cdots \times f_4 + 2.226\cdots \times f_5 + \cdots \\ &= 0.55128\cdots \times (f_3 + 2f_4 + 3f_5 + \cdots) + 0.231\cdots \times f_4 \\ &\quad + 0.572\cdots \times (f_5 + \cdots) \\ &= 0.55128\cdots \times (2v-4) + 0.231\cdots \times f_4 + 0.572 \times (f_5 + \cdots).\end{aligned}$$

因为 $2v-4=22$, 所以由上式得

$$0.438 \geqslant 0.231 f_4 + 0.572 \times (f_5 + \cdots).$$

从而 $f_5 = f_6 = \cdots = 0$, $f_4 = 0$ 或 1.

如果 $f_4 = 0$,那么 $f_3 = f = \frac{2e}{3}$,结合欧拉定理得 $13 - e + \frac{2e}{3} = 2$, $e = 33$. 平均每点引出$\frac{2e}{v} = \frac{66}{13} > 5$ 条线,与上面所述每点至多引 5 条线矛盾. 因此 $f_4 = 1$. 由 $13 - e + f_3 + 1 = 2$ 及 $2e = 3f_3 + 4$ 解得 $f_3 = 20$, $e = 32$. 由于 13 个点所引线数之和为 $2e = 64$, 每点至多引出 5 条线,所以必有一个点引出 4 条线,其余 12 个点各引出 5 条线.

这样的网络(13 个点,1 点引 4 条线,其他每点各引 5 条线,组成 1 个四边形,20 个三角形)如果存在,应当能在球面或平面上画出来. 我们先画一个四边形,将其余的 9 个点均装在这个四边形内(在球面上,这个四边形的“外面”才是球面四边形). 这里长度、角度都无关紧要,只需注意点与线的从属关系及各面都是三角形.

如果四边形的四个顶点都引出 5 条线,由于同一点引出的两条相邻的线必须构成三角形,所以如图 4 产生出一个八边形 $ABCDEFGH$. 不妨设除 G、H 外,其余每点均引出 5 条线,这样就至少再产生 P、Q、R 三个点,得出矛盾. 如果四边形有一个顶点只引出 4 条线,那么其余的点均引出 5 条线. 如图 5 产生一个七边形 $ABCDEFG$,进而又产生三个点 P、Q、R,总点数仍超过 13.

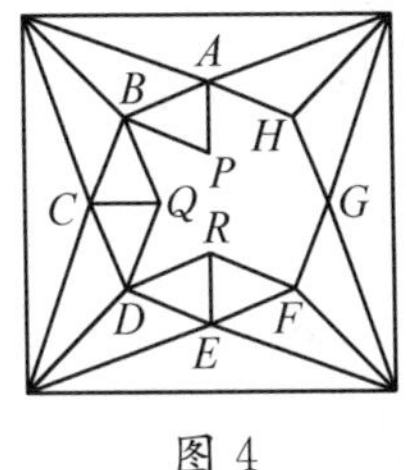

图 4

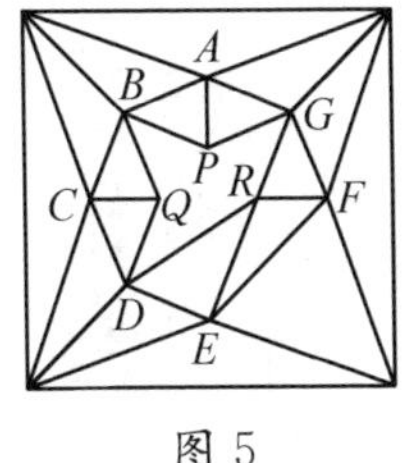

图 5

于是,所述的网络不存在,从而格雷戈里的猜测不成立.

6　有关问题

与 13 个球的问题有关联的问题甚多.

首先是生物学家塔莫斯(Tammes)在 1930 年提出的问题:在单位球上放 n 个半径均为 a_n 的球冠,互不重叠, a_n 的最大值是多少? 这个问题到目前为止只知道 $n \leqslant 12$ 与 $n = 24$ 的答案. 其中 $n = 12$ 的答案就是 12 个球冠的极构成正 20 面体的 12 个顶点,从而 $2\sin a_n$ 就是前面所算的棱长 $1.0514\cdots$, $a_n \approx \dfrac{63°26'}{2}$.

其次是球的装箱问题. 这个著名问题我们将另撰文介绍.

还有高维的推广,例如在四维空间中,类似的相切个数已经知道是 24 或 25,但究竟是哪一个却还不能确定. 当 $d > 3$ 时,在 d 维空间中,能确定的现在只有两种,即: $d = 8$, 相切个数为 240; $d = 24$, 相切个数为 196 560(这一结果也是利奇获得的).

参考文献

[1] Fejes Toth L. On the number of equal discs that can touch another of the same kind. *Studia Sci Math Hunger*, 1967, 2: 363.

[2] Leech J. The problem of the thirteen spheres, *Math Gaz*, 1956, 40: 22.

[3] ibid. Some sphere packings in higher space. *Canadian J Math*, 1964, 16: 657.

[4] Schütte K, van der Waerden B L. Das Problem der dreizehn Kugeln. *Math Ann*, 1953, 125: 325.

[5] Gardner M. *Mathematical circus*. Vantage Books, New York, 1992: 35; 264.

[6] Croft H T, Falconer K J, Guy R K. *Unsolved problems in geometry*. Springer-Verlag, New York, 1991: 114; 121.

第五章　数学普及（二）

细毛虫问题	中学生,1990(10)
不要急于“分母有理化”	中学生数理化(初中),1986(1)
一个不等式的证明	中学生数理化(高中),1986(1)
关于二次三项式的几个问题	安徽教育,1984(2)
“黄金矩形”的一些性质	中学生数学,1981(1)
谈一道竞赛题	中学生数学,2002(3)
走进美妙的数学花园	中学数学研究,2006(10)
做生意,还是做数学?	中学数学研究,2006(6)
拉佛阁兄弟	初中生数学学习,2002(31)
恒等式的证明(一)	初中生数学学习,2002(35)
恒等式的证明(二)	初中生数学学习,2003(1)
恒等式的证明(三)	初中生数学学习,2003(1)
条件等式的证明(一)	初中生数学学习,2003(3)
条件等式的证明(二)	初中生数学学习,2003(4)
平方差公式的应用(一)	初中生世界,2004(1/2)
平方差公式的应用(二)	初中生世界,2004(8)
孙庞斗智	时代数学学习,2002(C2)
图形帮助思考	时代数学学习,1999(C1)
算术更好	时代数学学习,1999(9)
输送信息	时代数学学习,2000(23)
乘与除	时代数学学习,2001(25)
人与球	时代数学学习,2001(10)
质数与合数(一)	时代数学学习,2000(7)
质数与合数(二)	时代数学学习,2000(9)
居中和	时代数学学习,1999(13)
逆序	时代数学学习,1999(15)

行程问题	时代数学学习,2001(34)
行程问题(一)	时代数学学习,2000(17)
行程问题(二)	时代数学学习,2000(19)
相遇几次	时代数学学习,2000(21)
染圆上的点	初中生数学学习,2002(10)
“四连”(一)	初中生数学学习,2004(11)
“四连”(二)	初中生数学学习,2004(12)
解题漫谈(二)	数理天地,1995(5)
棋盘上的一个组合问题(1)	数理天地,2000(4)
棋盘上的一个组合问题(2)	数理天地,2000(9)
自动扶梯	时代数学学习,2004(9)
数学探索无止境	时代数学学习
达·芬奇密码	时代数学学习
数学就是动脑	时代数学学习
风速的影响	时代数学学习
探索:从最简单的入手	时代数学学习
“乌龟”在谁手上	时代数学学习
不用分数解工程问题	时代数学学习
如履平地	时代数学学习
晴天和雨天	时代数学学习
颠倒乾坤	时代数学学习
哲学家的年龄问题	时代数学学习
两全其美	时代数学学习
切饼与切瓜	时代数学学习
走出迷宫	时代数学学习

细毛虫问题

桌上有一条长为 1 的细毛虫(宽度忽略不计)，蜷成任意的形状．用一张圆形的纸片把这条虫完全覆盖，这个圆的直径最小是多少？

回答这个问题并不困难．当细毛虫伸直时，长为 1，所以圆的直径必须$\geqslant 1$，才能把虫覆盖．反过来，我们可以证明直径$\geqslant 1$ 的圆形纸片可以把虫完全覆盖．

怎样证明呢？

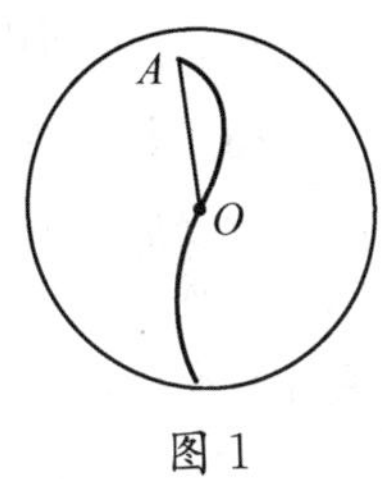

图 1

说出来非常简单．如图 1，这条虫身上有一点 O，从 O 沿虫身到头或到尾的长都是$\frac{1}{2}$．将圆纸片的中心对准 O，这时，对虫身上任一点 A，线段 $AO\leqslant$从 O 沿虫身到 A 的长$\leqslant\frac{1}{2}$．A 到圆心 O 的距离$\leqslant\frac{1}{2}$，所以 A 被圆纸片盖住．整个细毛虫也就被这圆纸片盖住．

圆纸片的面积是

$$\pi r^2=\pi\times\left(\frac{1}{2}\right)^2=0.78539\cdots.$$

如果纸片的形状不再是圆，能不能用面积更小的纸片覆盖这条细毛虫呢？

能．对角线的长为 1 的正方形纸片，可以覆盖这条细毛虫，这张纸片的面积为 0.5．

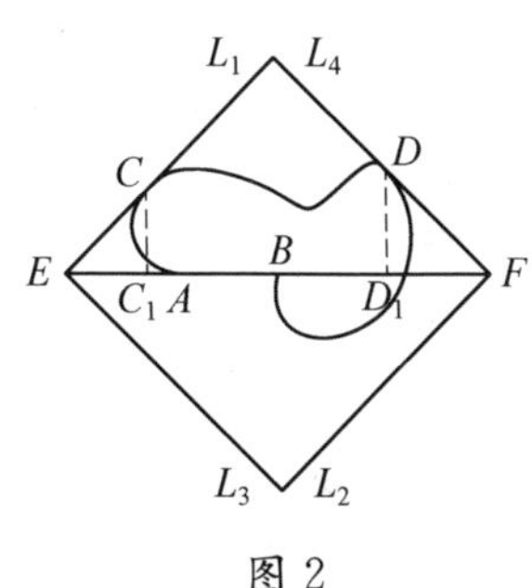

图 2

为了用正方形纸片覆盖细毛虫，我们设虫头为 A，虫尾为 B(如图 2)．作直线 AB，再作两组平行线：$L_1\,/\!/\,L_2$，$L_3\,/\!/\,L_4$，L_1 与 AB 的夹角为 45°，L_3 与 L_1 垂直．这样，虫在 L_1、L_2 之间，也在 L_3 与 L_4 之间．将 L_1、L_2、L_3、L_4 适当平移，总可以构成一个正方形，它的对角线通过 A、B，并且，四条边中至少有两条边与虫相接触(即有公共点).

不妨设 C 是虫与 L_1 的公共点, D 是虫与 L_4 的公共点(其他情况与此类似),正方形的对角线 EF 过 A、B.

从图中容易看出

$EF = EC_1 + C_1D_1 + D_1F = CC_1 + C_1D_1 + DD_1 \leqslant AC + CD + DB \leqslant$ 虫长,

即

$$EF \leqslant 1.$$

换句话说,对角线为 1 的正方形覆盖这条细毛虫.

上面所说的面积还可以减少. 实际上,直径为 1 的半圆形纸片就可以覆盖细毛虫,它的面积为 0.39269….

覆盖细毛虫的纸片,面积最小是多少? 形状是什么? 这是尚未解决的问题. 目前知道的最好结果是,可以用一个面积为 0.28863 的纸片覆盖细毛虫,这纸片的形状是对角线长分别为 1 和$\frac{\sqrt{3}}{3}$的菱形切去一个角.

边长为 1 的正三角形,不一定能覆盖这条小虫. 这是许多人料想不到的.

不要急于“分母有理化”

在根式的加减法中，为了合并同类根式常常需要分母有理化.但是，并不是在所有的场合都必须立即进行分母有理化.尤其在计算过程中，“静观待变”，不急于进行分母有理化，反倒是好办法.这里举一个简单的例子.

如图 1，在 Rt$\triangle HAB$ 中，两条直角边 AH、AB 的长分别为 1、6，AN 是斜边 HB 上的高，$AF \perp AN$，并且 $AF=2$，求 $\tan\angle ANF$、NF 及 $\cos\angle ANF$.

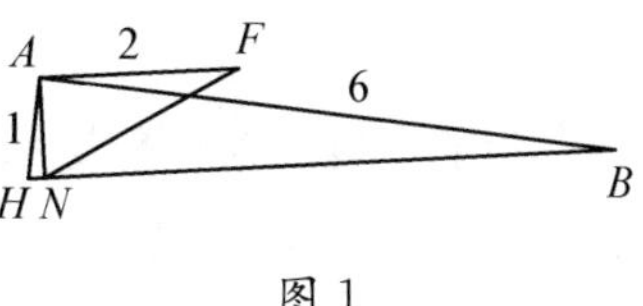

图 1

解法是很简单的.首先由勾股定理求出 $BH=\sqrt{37}$，再由$\triangle AHB$ 的面积可知$AN=\dfrac{AH \cdot AB}{HB}=\dfrac{6}{\sqrt{37}}$.不要急于分母有理化！下一步就可以求出

$$\tan\angle ANF=\frac{AF}{AN}=\frac{2}{\frac{6}{\sqrt{37}}}=\frac{\sqrt{37}}{3},$$

$$NF=\sqrt{AF^2+AN^2}=\sqrt{2^2+\frac{6^2}{37}}=2\sqrt{\frac{37+9}{37}}=2\sqrt{\frac{46}{37}},$$

$$\cos\angle ANF=\frac{AN}{NF}=\frac{\frac{6}{\sqrt{37}}}{2\sqrt{\frac{46}{37}}}=\frac{3}{\sqrt{46}}.$$

如果一定要将最后的结果分母有理化，那么不难得到

$$NF=2\sqrt{\frac{46}{37}}=\frac{2}{37}\sqrt{37\times 46}=\frac{2}{37}\sqrt{1702},\ \cos\angle ANF=\frac{3}{46}\sqrt{46}.$$

请急于分母有理化的同学自己算一算这道题，比较一下，对于“在计算过程中不要急于分母有理化”这句话就更有体会了.当然，计算的最终结果中如果分母含有根式时，则应将分母有理化.

一个不等式的证明

不等式的证明较等式难,但也不是完全没有可循的方法. 下面举一道题,略见一斑.

"设$\triangle ABC$的周长为2, a、b、c为三边的长,证明:

$$a^2+b^2+c^2+2abc<2.\text{"} \tag{1}$$

为了证明这个不等式,我们设法将字母的个数减少,如果一个不等式(或等式)中仅含一个字母,那么这个不等式(等式)的证明就大有希望. 所以应当尽量**消元**,减少字母的个数. 由于

$$a+b+c=2, \tag{2}$$

所以(1)可改成等价的形式

$$(2-c)^2+c^2-2ab+2abc<2,$$

即

$$2(1-c)^2<2ab(1-c). \tag{3}$$

由于$c<\dfrac{a+b+c}{2}=1$,所以$1-c>0$, (3)又等价于

$$2(1-c)<2ab. \tag{4}$$

这已经是一个形式很简单的不等式了,可惜的是ab不能表成c的代数式. 但是,在c固定时, $a+b$也固定, ab的最小值是可以求得的,即由于

$$4ab=(a+b)^2-(a-b)^2,$$

所以在$|a-b|$最大时, ab的值最小,因为

$$|a-b|<c,$$

所以

$$4ab>(a+b)^2-c^2=(2-c)^2-c^2=4(1-c). \tag{5}$$

(5)就是(4)式,因此原不等式成立.

关于二次三项式的几个问题

二次三项式 $ax^2+bx+c\ (a\neq 0)$ 是中学生最熟悉的函数,这里举几个问题,供复习讨论之用.

例1　已知 $ax^2+bx+c=0$ 有实数根,$k\neq 0$,证明 $ax^2+bx+c+k(2ax+b)=0$ 有两个不相等的实数根,并且第二个方程恰有一个根在第一个方程的两根之间.

证明　先将 $ax^2+bx+c=0$ 改写为 $\left(x+\dfrac{b}{2a}\right)^2-d=0$,其中 d 是一个常数(不难算出 $d=\dfrac{b^2-4ac}{4a^2}$,不过我们并不需要 d 的表达式),即

$$y^2-d=0, \tag{1}$$

其中 $y=x+\dfrac{b}{2a}$.

同样,$ax^2+bx+c+k(2ax+b)=0$ 可改写为

$$y^2+2ky-d=0. \tag{2}$$

现在只要证明(2)有两个不相等的实数根,并且其中恰有一个在(1)的两个根之间.

由于(1)有实数根,所以 $d\geqslant 0$,从而

$$(2k)^2-4\cdot(-d)=4(k^2+d)>0,$$

即(2)有两个不相等的实数根.

在 $d=0$ 时,方程(1)的两个根均为 0,而(2)的一根为 0,另一根 $y=-2k\neq 0$,即只有一个根($y=0$)在(1)的两根之间.

在 $d>0$ 时怎样证明(2)恰有一个根在(1)的两根之间呢?我们注意函数 $u=y^2-d$ 的图像是开口向上的抛物线,数 y' 在 $y^2-d=0$ 的两根 y_1、y_2 之间的充分必要条件是 $y'^2-d\leqslant 0$(即函数 u 在 y' 处的值是负数或零). 对于 y^2+

$2ky-d=0$ 的两个根 y_1'、y_2'，由韦达定理有

$$y_1'\cdot y_2'=-d<0,$$

即 y_1'、y_2'的符号相反，从而$-2ky_1'$与$-2ky_2'$的符号相反.

另一方面，又有

$$y_i'^2-d=-2ky_i' \quad (i=1,\ 2),$$

因此 $y_1'^2-d$ 与 $y_2'^2-d$ 的符号相反，这就是说(2)的两个根 y_1'、y_2'恰有一个在(1)的两个根之间.

例 2 已知在 $|x|\leqslant 1$ 时，$|ax^2+bx+c|\leqslant 1$，证明 $|x|\leqslant 1$ 时，$|2ax+b|\leqslant 4$.

证明 不妨假定 $a>0$. 由于 x 轴上的区间$[-1,\ 1]$关于原点对称，所以可以假定 $b\geqslant 0$(否则用$-x$ 代替 x，$-b$ 代替 b).

由于 $|x|\leqslant 1$ 时，$|ax^2+bx+c|\leqslant 1$，所以

$$|a+b+c|\leqslant 1 \quad (x=1),$$
$$|c|\leqslant 1 \qquad (x=0),$$

从而 $|x|\leqslant 1$ 时，

$$|2ax+b|\leqslant 2a+b\leqslant 2(a+b)\leqslant 2(|a+b+c|+|c|)\leqslant 2(1+1)=4.$$

例 3 a、b、c 都是正数，ax^2+bx+c 有实数根，证明

$$\max\{a,\ b,\ c\}\geqslant \frac{4}{9}(a+b+c).$$

证明 不妨设 $a+b+c=1$. 如果 $b\geqslant \frac{4}{9}$，则结论成立. 因此，我们设 $b<\frac{4}{9}$，这时

$$a+c=1-b>\frac{5}{9}. \tag{1}$$

另一方面，由于 ax^2+bx+c 有实数根，所以

$$b^2-4ac\geqslant 0,$$

但 $b<\frac{4}{9}$，所以

$$4ac\leqslant b^2<\left(\frac{4}{9}\right)^2,$$

即

$$ac<\frac{4}{81}. \tag{2}$$

由(1)得

$$\left(\frac{5}{9}-c\right)c<ac<\frac{4}{81},$$

化简得

$$c^2-\frac{5}{9}c+\frac{4}{81}>0. \tag{3}$$

由(3)得

$$c>\frac{4}{9}\text{ 或 }c<\frac{1}{9}.$$

在 $c<\frac{1}{9}$ 时，由(1)得

$$a>\frac{5}{9}-c>\frac{4}{9}.$$

因此，总有

$$\max\{a,\ b,\ c\}\geqslant\frac{4}{9}(a+b+c).$$

第 2、3 题的结论都可以推广到 n 次多项式.

"黄金矩形"的一些性质

所谓"黄金分割"是指下述的问题:

"分已知线段为两部分,使其中一部分是线段本身和另一部分的比例中项."

就是在线段 AB 上求一点 P,使 $AP^2=AB\cdot BP$(图 1).

如何求出 P 点? 分析如下:

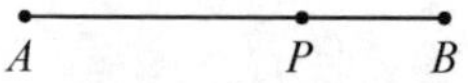

A P B

图 1

设 AB 的长为 1, AP 的长为 x,则 BP 的长为 $1-x$. 于是 $1\cdot(1-x)=x^2$,即 $x^2+x-1=0$.

解方程得 $$x=\frac{-1\pm\sqrt{5}}{2}.$$

舍去负值,得 $$x=\frac{\sqrt{5}-1}{2}\approx 0.618.$$

几何中的具体求法是:(1) 作 $BC\perp AB$,且使 $BC=\frac{1}{2}AB$, 连 AC; (2) 在 AC 上截取 $CQ=CB$; (3) 在 AB 上截取 $AP=AQ$; 则 P 点就是所求的分点.(图 2,证略)

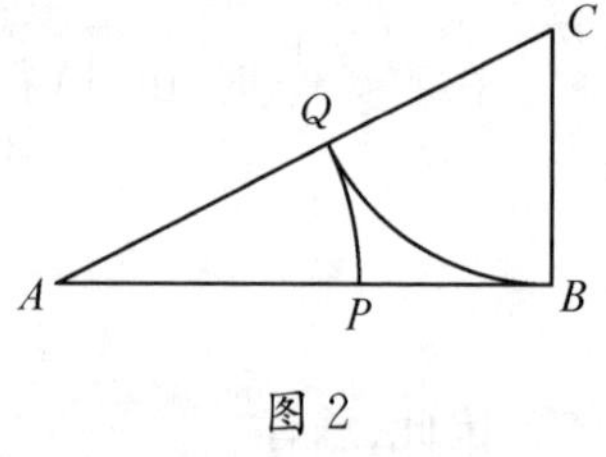

图 2

这样的分点 P 称为"黄金分割点".

在美术家看来,最美观的矩形是长宽比为 $1:0.618$(即在上图中,把 APB 折成直角 APB 所作出的矩形). 在绘图、建筑、舞台设计等过程中,这种分割法是经常遇见的.

如果一个矩形的长与宽的比为 $1:\omega$,这里 $\omega=\frac{\sqrt{5}-1}{2}\approx 0.618$, 我们就称它为黄金矩形.

黄金矩形有很多有趣的性质,列举如下,证明都很简单,读者不难得出.

如图 3, $AD=1$, $AB=\omega$, 矩形 $ABCD$ 是一个黄金矩形.

(1) CD 是 AD 与 $AD-CD$ 的比例中项,即 $\frac{AD}{CD}=\frac{CD}{AD-CD}$.

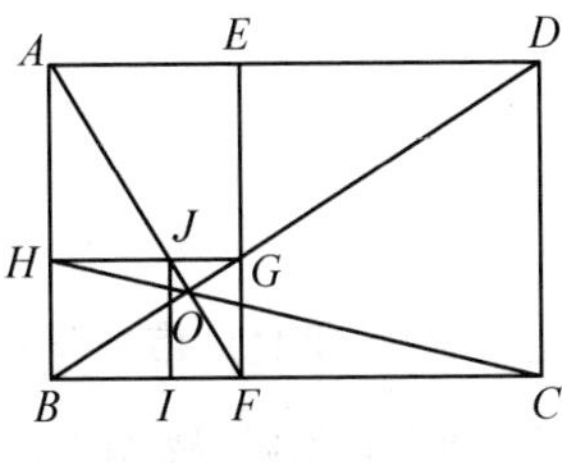

图 3

(2) 作正方形 $CDEF$,则得到的矩形 $BFEA$ 仍是黄金矩形.

如果再作正方形 $AEGH$,又得一个黄金矩形 $FGHB$,如此继续下去,可得一串正方形与一串黄金矩形.

(3) 上面得到的正方形 $CDEF$, $AEGH$, $BIJH$, …的边长组成一个等比数列 ω, ω^2, ω^3, ….

(4) 上面得到的黄金矩形 $ABCD$, $BFEA$, $FGHB$, …是相似的,每两个相邻的矩形的相似比为 ω.

(5) 这些黄金矩形一个套住一个,全体黄金矩形有一个唯一的公共点 O,点 O 是这些黄金矩形的相似中心.

(6) D、G、B 三点共线, A、J、F 三点共线.

(7) 直线 AF、GD、CH 共点,这个公共点就是 O 点.

(8) $AF\perp GD$.

(9) $\frac{OA}{OD}=\frac{OB}{OA}=\frac{OF}{OB}=\frac{OG}{OF}=\omega$, $\frac{OH}{OC}=\omega^2$.

(10) 如果以 B 为原点,直线 BC 为 x 轴, BA 为 y 轴,则 O 点的坐标为 $\left(\frac{\omega^2}{1+\omega^2}, \frac{\omega^3}{1+\omega^2}\right)$.

(11) 点 D、A、B、F、G、J、…在同一条对数螺线上.

谈一道竞赛题

2001 年北京市数学竞赛初二年级决赛题四是一道很好的几何题. 看似简单,却不容易. 原题如下:

如图 1,在等腰三角形 ABC 中,延长边 AB 到点 D,延长边 CA 到 E,连结 DE,恰有

$$AD = BC = CE = DE.$$

求证: $\angle BAC = 100°$.

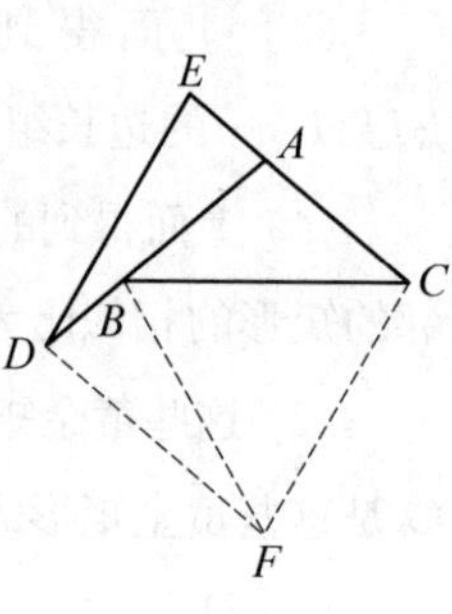

图 1

除去《中学生数学》2001 年第 6 期公布的答案外,这里再提供两个解答.

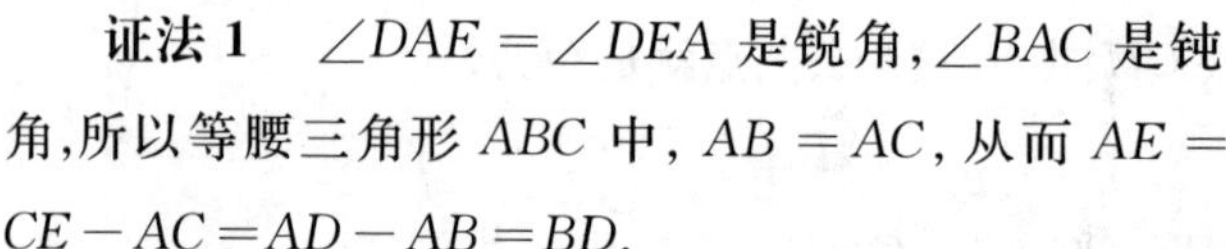

证法 1 $\angle DAE = \angle DEA$ 是锐角, $\angle BAC$ 是钝角,所以等腰三角形 ABC 中, $AB = AC$, 从而 $AE = CE - AC = AD - AB = BD$.

如图 1, 以 BD 为底, 作与 $\triangle ADE$ 全等的等腰三角形 BDF, 连 CF, 则 $\angle FDB = \angle EAD$, 所以 $DF \underline{\underline{/\!/}} CE$, 四边形 $CEDF$ 是平行四边形.

$CF = DE = BF = BC$, $\triangle BCF$ 是正三角形.

设 $\angle ABC = \alpha$, 则 $\angle FBD = \angle EAD = 2\alpha$, $\alpha + 60° + 2\alpha = 180°$.

所以 $\alpha = 40°$, $\angle BAC = 100°$.

证法 2 如图 2, 连 CD. 同解法 1, 易知 $AB = AC$, 设 $\angle ABC = \alpha$, 则 $\angle DEA = \angle DAE = 2\alpha$.

在等腰三角形 DEC 中, 作高 EF, 交 AB 于 P, 连 CP, 则 $\angle DEF = \angle FEC = \alpha$, $\angle DPF = \angle CPF$.

在等腰三角形 CEB 中, $\angle CEB = \angle CBE = \dfrac{180° - \alpha}{2}$, $\angle PEB = \angle CEB - \angle FEC = \dfrac{180° - \alpha}{2} - \alpha = 90° - \dfrac{3}{2}\alpha$.

$$\angle PBE=180^\circ-\angle CEB-\angle BAE=180^\circ-\frac{180^\circ-\alpha}{2}-2\alpha=90^\circ-\frac{3}{2}\alpha.$$

由于 $\angle PEB=\angle PBE$，所以 P 在$\triangle CEB$ 的对称轴上，即 CP 是顶角的平分线，$\angle APC=\angle CPF$. 所以 $\angle DPF=\angle CPF=\angle APC=\frac{180^\circ}{3}=60^\circ$.

从而 $90^\circ-\frac{3}{2}\alpha=\angle PBE=\frac{1}{2}\angle BPF=\frac{60^\circ}{2}=30^\circ$，$\alpha=40^\circ$，$\angle BAC=100^\circ$.

对于特殊角 30°、45°、60°、90°，大家很有兴趣. 但 80°（或 100°）的角也有不少特点. 图 2 的$\triangle EDC$ 就是一个顶角为 80°的等腰三角形. 我们还可以举出几个与它有关的问题.

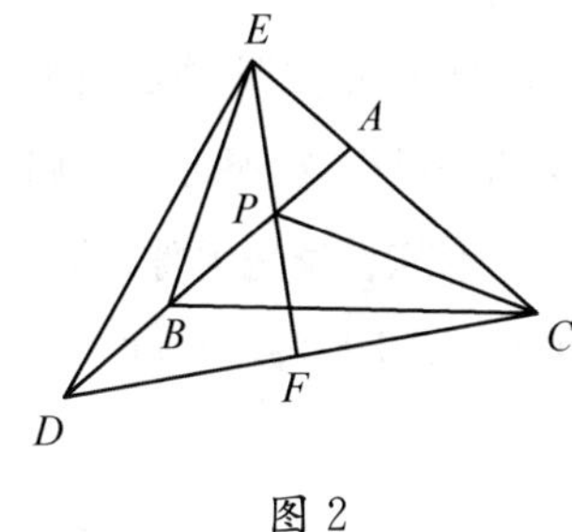

图 2

1. 设$\triangle EDC$ 中，$ED=EC$，$\angle DEC=80^\circ$，B 为$\triangle EDC$ 内一点，$\angle BCD=10^\circ$，$\angle BDC=30^\circ$. 求$\angle EBD$.

2. 设$\triangle EDC$ 中，$ED=EC$，$\angle DEC=80^\circ$，B 为$\triangle EDC$ 内一点，$BC=EC$，$\angle BCE=40^\circ$. 求$\angle EDB$.

3. 设$\triangle ABC$ 中，有一点 P，$\angle PAB=10^\circ$，$\angle PBA=20^\circ$，$\angle PCA=30^\circ$，$\angle PAC=40^\circ$. 证明$\triangle ABC$ 是等腰三角形（1996 年第 25 届美国数学奥林匹克第 5 题）.

2001 年国际数学竞赛的第 5 题也是一道涉及 80°的题. 原题如下：

在$\triangle ABC$ 中，AP 平分$\angle BAC$，P 在 BC 上，BQ 平分$\angle ABC$，Q 在 CA 上. 已知 $\angle BAC=60^\circ$，$AB+BP=AQ+QB$，$\triangle ABC$ 的角的可能值是什么？

答案是 $\angle ABC=80^\circ$，$\angle ACB=40^\circ$.

走进美妙的数学花园

“走进美妙的数学花园”是由中国青少年发展服务中心、中国少年科学院举办的一项数学活动.

2002 年 8 月,第二十四届国际数学家大会在中国北京举办的同时,作为大会的卫星会议,举办了首届“走进美妙的数学花园”中国青少年数学论坛. 著名数学家陈省身先生曾亲笔题词,对这项活动给予极大的支持. 此后,“走进美妙的数学花园”每年举办一届.

走进美妙的
数学花园
陈省身题

第四届活动于 2006 年 7 月 31 日至 8 月 3 日在江苏省南京市南京师范大学附属实验学校举行.

参加本届活动的有来自全国各地(包括台湾、香港、澳门)的 600 多名选手,分三、四、五、六、七、八年级参加四项活动:**“趣味数学解题技能展示”**(其中七年级的试题及解答附后);**“数学建模小论文答辩”**;**“数学益智游戏”**(分个人赛和团体赛);(不分年级的)**“发现之旅”**.

“小论文答辩”历届均为展示选手数学专业水准的好机会.

本届**“发现之旅”**要求选手用正多边形做成一个球.

“益智游戏”的个人赛是甲、乙二人轮流在一个五角星(如图 1)的 10 个顶点($A-J$)上放白子与黑子. 谁放的棋子是图 1 中某个三角形的 3 个顶点就算输(例如 A、G、D 三点都放白子,则甲输). 这是一个很有趣的游戏,读者不妨试一试.

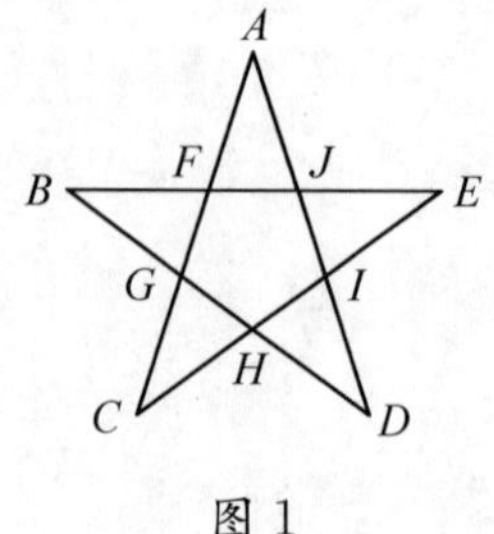

图 1

“益智游戏”的团体赛是每方各出二人，分别坐在如图 2 的棋盘的东、西、南、北，并依照这个次序轮流将棋盘中的虚线（一个小正方形的边）改成红线与蓝线. 东、西方希望能连成一条贯穿东西的红色路线，而南、北方希望能连成一条贯穿南北的蓝色路线. 先连成的一方为赢，并规定如双方不能连成时算东西方输.

台湾队夺得“发现之旅”与“益智游戏”团体赛的第一名.“益智游戏”个人赛的各年级第一名分别是北京郑祎琳（三年级），南京陈劭瀚（四年级），澳门曾鹰（五年级），北京王宇辰（六年级），张家口张东（七年级），河北过云龙（八年级）.

北

西　　东

南

图 2

最后，根据各项总成绩的高低，评选出 8 个队进行**决赛**（团体对抗赛）. 浙江金华代表队夺得决赛团体第一名.

这一活动的主试委员会由陈平、刘守军、刘来福、胡大同、吴康、单墫、段扬、葛军、张玉香等人组成.

第四届“走进美妙的数学花园”中国青少年数学论坛“趣味数学解题技能展示”决赛七年级试题和解答

注意事项：不允许使用计算器.

一、填空题（共 10 题，第 1～5 题每题 8 分，第 6～10 题每题 10 分）

1. 计算：$(-2)\times 3-(-1)\times(-4)=$________.

2. 由六个棱长为 1 的小正方体拼成如图 3 所示立体图形，它的表面积是________.

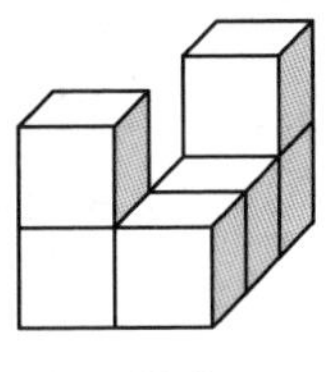

图 3

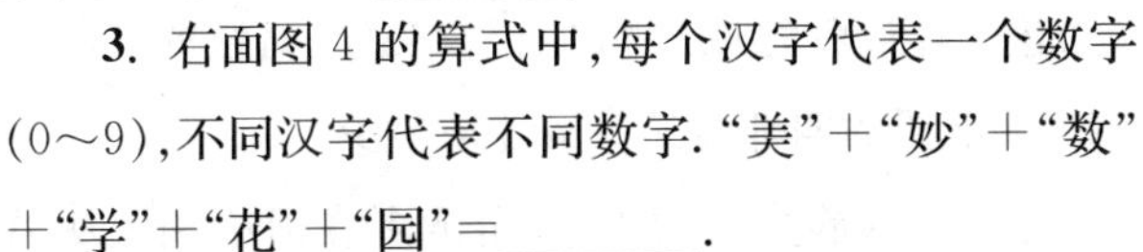

3. 右面图 4 的算式中，每个汉字代表一个数字（0～9），不同汉字代表不同数字.“美”+“妙”+“数”+“学”+“花”+“园”=________.

		美	妙	数	学
×				花	园
	数	学	真	美	妙
4	2	3	8	0	
5	好	好	好	美	妙

图 4

4. 规定：$A\bigcirc B$ 表示 A、B 中较大的数，$A\triangle B$ 表示 A、B 中较小的数. 若 $(A\bigcirc B+B\triangle 3)\times(B\bigcirc 5$

$+A\triangle 3)=96$，且 A、B 均为大于 0 的自然数，$A\times B$ 的所有取值为________.

5. 五次测验平均成绩是 90，中位数(即 5 个成绩按大小次序排列，居中的那个数)是 91，众数(即 5 个成绩中，出现次数最多的那个数)是 94. 最低两次测验的成绩之和是________.

6. 在 3×3 的棋盘上共有 24 条长为 1 的小线段. 甲、乙二人轮流将小线段标数，每次标一条，甲标 0，乙标 1. 甲的目的是可以沿标 0 的线段从南到北，乙的目的是可以沿标 1 的线段从东到西，谁先实现目的为胜. 现已有 6 条线段标好，甲下一条怎样标才可不败？(在图 5 上标出)

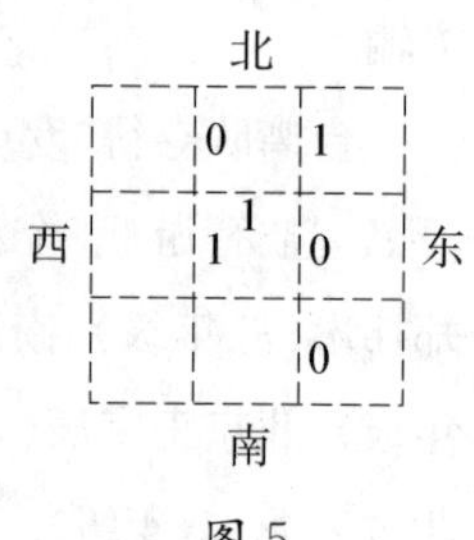

图 5

7. 已知 $\dfrac{(b-c)(d-a)}{(a-b)(c-d)}=3$，那么 $\dfrac{(a-c)(b-d)}{(a-b)(c-d)}=$________.

8. 有 20 堆石子，每堆都有 2006 粒石子. 从任意 19 堆中各取一粒放入另一堆，称为一次操作. 经过不足 20 次操作后，某一堆中有石子 1990 粒，另一堆石子数在 2080 到 2100 之间. 这一堆石子有________粒.

9. 甲、乙二人同时分别从 A、B 两地出发相向而行，到达 B、A 立即返回(假设他们速度都保持不变). 若第一次相遇点距 A 的距离与第二次相遇点距 B 的距离之比为 $6:7$，则甲、乙的速度之比为________.

10. 如图 6，一个 3×3 表格中的两个方格已经被染成黑色. 用红、黄、蓝、绿四种颜色对其余 7 个方格染色，使得每行、每列以及两条对角线上各个方格所染颜色都各不相同. 共有________种不同的染色方式.

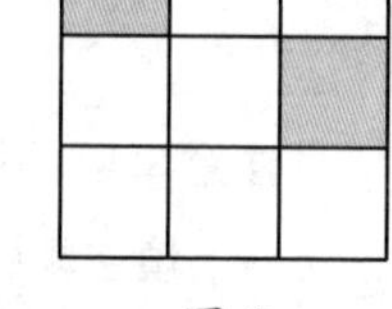
图 6

二、解答题(共 2 题，每题 15 分)

11. 某商店为了找零方便，每一件商品的价格均定为 0.5 元的整数倍. 李明在此商店里购买了四件商品，心算了一下，共需付 9 元. 在收银台结帐时，发现收银员在计算器上按的不是加法键，而都是乘法键. 他正要交涉，奇怪的是收银员算出的总价也是 9 元. 问：李明购买的四件商品价格各是多少？

12. 河水是流动的，在 B 点处流入静止的湖中. 一游泳者在河中顺流从 A 点到 B 点，然后穿过湖到 C 点，共用 3 小时. 而由 C 到 B 再到 A，共 6 小时. 如果湖水也是流动的，从 B 流向 C，速度等于河水速度，那么，这名游泳者从 A 到 B

再到 C 只需 2.5 小时. 问在这样的条件下,他由 C 到 B 再到 A,共需多少小时?

解　答

1. -10.

2. 上、下的面有 $2\times4=8$ 个,前、后的面有 $2\times4=8$ 个,左、右的面有 $2\times5=10$ 个. 表面积是 $8+8+10=26$.

3. $42380=5\times8476$, 且 42380 不被 6、7、8、9 整除,故

$$\text{“花”}=5,\ \text{“美妙数学”}=8476.$$

“园”×“学”的个位为“妙”,即 4,所以“园”$=9$.

$$\text{“美”}+\text{“妙”}+\text{“数”}+\text{“学”}+\text{“花”}+\text{“园”}=39.$$

4. 不妨设 $A\geqslant B$. ① 如果 $A\leqslant5$,那么

$$(A\bigcirc5+B\triangle3)\times(B\bigcirc5+A\triangle3)\leqslant(5+3)(5+3)<96.$$

② 如果 $A>5$, $B\leqslant5$, 那么

$$(A\bigcirc5+B\triangle3)\times(B\bigcirc5+A\triangle3)=8(A+B\triangle3).$$

所以

$$A+B\triangle3=12.$$

$$A=12-B\triangle3\geqslant12-3=9,\text{且}A\leqslant12-1=11.$$

故 $A=9$, $B=5$, 4, 3. $A=10$, $B=2$. $A=11$, $B=1$.

③ 如果 $A>5$, $B>5$, 那么

$$(A\bigcirc5+B\triangle3)\times(B\bigcirc5+A\triangle3)=(A+3)(B+3).$$

$A+3\geqslant B+3>8$, 而 96 不能分成两个大于 8 的因数相乘. 因此,总结得 $A\times B$ 的所有取值是 $9\times3=27$, $9\times4=36$, $9\times5=45$, $10\times2=20$, $11\times1=11$.

5. $90\times5-91-94\times2=171$.

6. 甲必须标 EF. 如图 7,如果甲不标 EF,那么在甲也不标 AB、CD 时,乙标 EF. 然后对 AB、CD 两条边,乙总可标其中的一条,从而获胜;而在甲标

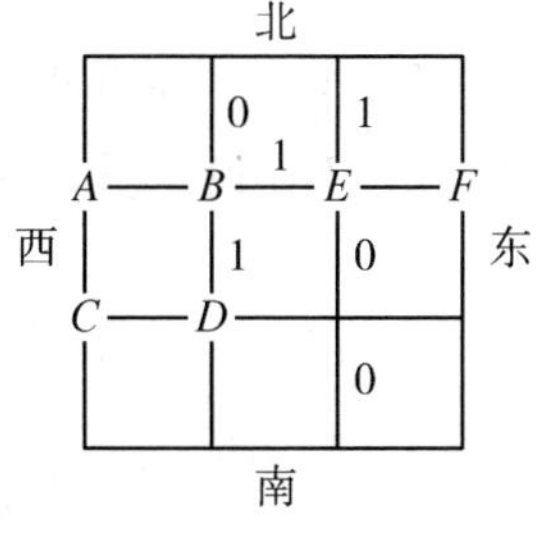

图 7

AB 或 CD 时，乙标两条中剩下的一条，下一步即可取胜.

7. 由已知得

$$(bd+ac)-(ab+cd)=3(bd+ac)-3(ad+bc),$$

即

$$3(ad+bc)-(ab+cd)=2(bd+ac).$$

所以

$$\begin{aligned}\frac{(a-c)(b-d)}{(a-b)(c-d)}&=\frac{(ab+cd)-(bc+ad)}{(ac+bd)-(bc+ad)}\\&=\frac{-2(bd+ac)+2(ad+bc)}{(ac+bd)-(bc+ad)}=-2.\end{aligned}$$

8. 1990 比 2006 少 16 粒. 如果 1990 粒这一堆曾有一次增加 19 粒，那么经过不足 20 次操作，石子数不会少于 2006. 所以这一堆，每次操作都是取走 1 粒. 操作次数为 16.

因为 $19=20-1$，所以每一堆石子经过 16 次操作，数量都减少 16 再加上若干个(包括 0 个)20，即是 1990 加上若干个 20. 因此石子数在 2080 与 2100 之间的这堆的数量应为

$$1990+20\times 5=2090.$$

9. 从出发到第一次相遇，两人合起来走完一个全程. 从出发到第二次相遇，两人合起来走完三个全程. 因此，甲从出发到第二次相遇，所走路程是他从出发到第一次相遇的 3 倍. 于是全程与他从出发到第一次相遇所走路程之比是 $(3\times 6-7):6=11:6$. 从出发到第一次相遇，乙所走路程与甲所走路程的比是

$$(11-6):6=5:6,$$

即甲、乙二人速度之比是 $6:5$.

10. 共有 $4\times 3\times 2\times 2\times 2=96$ 种染法.

11. 设四件商品价格分别为 $0.5a$, $0.5b$, $0.5c$, $0.5d$(元)，a、b、c、d 为正整数，不妨设 $a\geqslant b\geqslant c\geqslant d$，则

$$\begin{cases}a+b+c+d=18, & (1)\\ abcd=2^4\times 9. & (2)\end{cases}$$

由(1)，

$$\frac{18}{4}\leqslant a\leqslant 18-1-1-1=15,$$

所以 $2^4\times 9$ 的约数 a 只能取以下的值：6、8、9、12. 将 $2^4\times 9$ 作分解，分成 4 个数的连乘积，并且最大的约数 a 取上述值，只有 $9\times 4\times 4\times 1$ 与 $8\times 6\times 3\times 1$ 两种满足(1).

因此 $$a=9,\ b=c=4,\ d=1;$$

或 $$a=8,\ b=6,\ c=3,\ d=1.$$

四种商品价格分别为 0.5, 2, 2, 4.5 或 0.5, 1.5, 3, 4(元).

12. 设游泳者的速度为每小时 v_1 千米，河水速度为每小时 v_2 千米，又设游泳者从 A 到 B 顺水所用时间为 t 小时，则从 B 到 A 逆水所用时间为 $t+(6-3)=t+3$ 小时，从 B 到 C 静水所用时间为 $3-t$ 小时，从 B 到 C 顺水所用时间为 $2.5-t$ (小时). 我们有

$$\begin{cases}\dfrac{v_1+v_2}{v_1-v_2}=\dfrac{t+3}{t}, & (1)\\[2ex] \dfrac{v_1+v_2}{v_1}=\dfrac{3-t}{2.5-t}. & (2)\end{cases}$$

由(2)得 $$\frac{v_1+v_2}{v_1-v_2}=\frac{3-t}{2(2.5-t)-(3-t)}=\frac{3-t}{2-t}. \tag{3}$$

由(1), (3)得 $$\frac{v_1+v_2}{v_1-v_2}=\frac{(t+3)+(3-t)}{t+(2-t)}=3.$$

即顺水速度为逆水的 3 倍. 所以逆水由 C 到 B 再到 A 共需 $2.5\times 3=7.5$(小时).

做生意，还是做数学？

在华人数学家中,周炜良(1911—1995)的名气不算最大,但他的贡献肯定是最大的几位之一.翻一翻岩波数学辞典,就会见到"周定理""周—小平定理""周坐标""周簇""周环"等以他命名的定理、定义与方法.陈省身先生曾说过,周炜良未被评为美国科学院院士,"我想这对科学院来说是一大损失"(引自王元《陈省身文集》读后感).

周炜良先生的经历很奇特.他几乎未读过中小学,全靠自学.18岁到美国芝加哥大学主修经济学,后来才逐步转向数学,并在德国汉堡大学听课时认识了同龄的陈省身.周炜良认为"只要愿意学习,在数学上缺乏某些重要学科的知识并不重要".

由于纳粹排犹,周炜良的岳父一家被迫离开德国,几乎一文不名.周炜良得负担妻子与两个孩子,还要负担岳父母,不得不放弃数学研究回国去经商.将近十年完全脱离了数学活动.

1946年春,周炜良重新见到陈省身.陈鼓励他研究数学.于是周面临一项根本性的决定,也是他一生中最重要的决定:是否应该停掉生意,回到数学中去?

当时周炜良的生意已做得不错,且年已35岁,取得博士学位已经10年,之后几乎没有做过任何学问.重回数学是否太晚?经济问题又如何解决?他感到这是在冒一次大的风险."因为在一生中的这个阶段做一次这样新的努力,哪怕是只想取得一般的成功,也是完全没有把握的",但他又觉得"生活中有时必须要采取一些大胆的行动".于是他下定决心,结束了生意,来到美国普林斯顿高等研究所,后来又到霍普金斯大学,不久便担任教授、系主任,直至退休.在此期间,他取得了众多的成就.陈省身评价他是一位富有创见的数学家.

周炜良的抉择是正确的.否则,商界或许多了一位精明的生意人,数学界却少了一位杰出的数学家,数学中少了许多重要的结果.

2005年是周先生逝世十周年,谨以此短文表示纪念.(文中引文出自周炜良先生的文章《一份永久的感激之情》)

拉佛阁兄弟

在 2002 年的国际数学家大会上，有一对法国兄弟十分引人注目.

哥哥是 36 岁的洛朗 · 拉佛阁，他是这届菲尔兹奖的两名得主之一. 弟弟是 28 岁的文森特 · 拉佛阁，应邀在会上作了 45 分钟的大报告. 兄弟俩都毕业于著名的巴黎高等师范学院，又都是法国国家科学研究中心的终身研究员.

洛朗很早就专攻“朗兰兹纲领”. 这是美国普林斯顿大学罗伯特 · 朗兰兹教授在 1967 年给著名数论专家安德烈 · 韦依的一封长信中提出的宏伟规划. 在很多人的眼中，数学的各个分支（如代数、几何、分析）是各自独立的，互不相关，就像十几个岛国，各用各的语言，各有各的习惯，而朗兰兹要将这些分支统一起来. 为了实现这种“统一性”，需要解决一组意义深远的猜想. 当然每一个猜想的解决都是非常艰巨的工作.

洛朗由于专攻“朗兰兹纲领”，好几年都没有拿出什么成果. 幸而法国国家科学研究中心的政策十分宽厚，并未将他扫地出门. 这种鼓励数学家钻研重大问题的政策，免除了洛朗的后顾之忧. 而辛勤的劳动终于结出了硕果. 1999 年，洛朗在普林斯顿高等研究院，关于“函数域上的相应的朗兰兹猜想”，作了一系列的报告. 他用 6 年多时间写成的长达 300 页的手稿在数学圈中广为流传. 朗兰兹本人对洛朗的工作给予高度称赞，很多人认为洛朗可望获得菲尔兹奖.

荣誉接踵而来，洛朗连续获得法国国家科学研究中心奖、法国柯莱研究奖，今年终于在北京获得数学界的最高荣誉——菲尔兹奖.

文森特虽然比哥哥小了 8 岁，却早在 1990 年就已经到过北京. 他当时是作为法国中学生的代表来参加 31 届国际数学奥林匹克的. 当时，他完全是抱着对数学的兴趣而来的，并未考虑能否获得金牌，结果他不但得到了金牌，而且成为获得满分的四名选手之一（另三名是俄罗斯的玛林列科娃与中国的汪建华、周彤）. 文森特在算子代数 K 理论的研究中取得重要进展，荣获 2000 年欧洲数学会青年科学家奖. 如果他能在下一届国际数学家大会上获得菲尔兹奖，那么将在数学史上留下兄弟同获殊奖的佳话. 但记者和他谈起这件事时，他的回答是：“我从事数学研究是因为我对数学有兴趣，而不是为了什么奖.”

恒等式的证明(一)

恒等式的证明是初中代数的一个重要内容.请看下面的问题:

已知 $xy\neq-1$, $yz\neq-1$, $zx\neq-1$,求证:

$$\frac{x-y}{1+xy}+\frac{y-z}{1+yz}+\frac{z-x}{1+zx}=\frac{x-y}{1+xy}\cdot\frac{y-z}{1+yz}\cdot\frac{z-x}{1+zx}. \tag{1}$$

我们可以在左边进行运算,使所得的结果等于右边.所以恒等式的证明实际上是已知答案的计算题,但必须老老实实、一步一步地算,不能简单地写:"显然,左边=右边."

在本题中,左边先将前两项相加得

$$\begin{aligned}
\text{左边}&=\frac{(x-y)(1+yz)+(y-z)(1+xy)}{(1+xy)(1+yz)}+\frac{z-x}{1+zx}\\
&=\frac{x-y^2z+xy^2-z}{(1+xy)(1+yz)}+\frac{z-x}{1+zx}\\
&=\frac{(x-z)(1+y^2)}{(1+xy)(1+yz)}+\frac{z-x}{1+zx}\\
&=\frac{(x-z)[(1+y^2)(1+zx)-(1+xy)(1+yz)]}{(1+xy)(1+yz)(1+zx)}\\
&=\frac{(x-z)(y^2+zx-xy-yz)}{(1+xy)(1+yz)(1+zx)}\\
&=\frac{(x-z)(y-x)(y-z)}{(1+xy)(1+yz)(1+zx)}\\
&=\text{右边}.
\end{aligned}$$

如果一开始就将左边三项加在一起(公分母为 $(1+xy)(1+yz)(1+zx)$),比较麻烦,不如上面先加两项简单.

还可以先将左边第三项移到右边,即证明:

$$\frac{x-y}{1+xy}+\frac{y-z}{1+yz}=\frac{x-y}{1+xy}\cdot\frac{y-z}{1+yz}\cdot\frac{z-x}{1+zx}-\frac{z-x}{1+zx}. \tag{2}$$

(1)与(2)是等价的,即如果(1)成立,那么(2)也成立. 反过来,如果(2)成立,那么(1)也成立. 证明(2)就是第二种证明(1)成立的方法.

上面已经得出(2)的左边$=\dfrac{(x-z)(1+y^2)}{(1+xy)(1+yz)}$.

$$而(2)的右边=\frac{(z-x)[(x-y)(y-z)-(1+xy)(1+yz)]}{(1+xy)(1+yz)(1+zx)}$$

$$=\frac{(z-x)[-y^2-xz-1-xy^2z]}{(1+xy)(1+yz)(1+zx)}$$

$$=\frac{(x-z)(1+y^2)(1+xz)}{(1+xy)(1+yz)(1+zx)}$$

$$=\frac{(x-z)(1+y^2)}{(1+xy)(1+yz)}.$$

所以(2)的两边相等,(2)成立,(1)也成立.

第二种证明是将(2)的两边分别进行运算,证明所得结果相同. 这也是证明恒等式的常用方法.

恒等式的证明（二）

已知 $a+b$、$b+c$、$c+a$ 都不为 0，求证：

$$\frac{ab}{(b+c)(c+a)}+\frac{ac}{(b+c)(a+b)}+\frac{bc}{(a+b)(a+c)}$$
$$=\frac{c^2}{(b+c)(c+a)}+\frac{b^2}{(b+c)(a+b)}+\frac{a^2}{(a+b)(a+c)}. \tag{1}$$

(1)的两边都是分式，运算中肯定要通分，不如索性一开始就去分母，即将(1)的两边同乘 $(b+c)(c+a)(a+b)$，得出等价的

$$ab(a+b)+ac(a+c)+bc(b+c)$$
$$=c^2(a+b)+b^2(a+c)+a^2(b+c). \tag{2}$$

(2)的左边 $=a^2b+b^2a+a^2c+c^2a+b^2c+c^2b$，而右边也等于这一结果，所以(2)成立，从而(1)也成立.

(1)的证明不难，但乍一看却很难想到它的两边会是相等的，这是我在证明一个不等式时偶然发现的.

恒等式的证明（三）

已知 $s=\frac{a+b+c}{2}$，求证：$s(a+2s)(b-c)+b(a+2s)(c-s)\left(1-\frac{c^2}{2}\right)-c(a+2s)(b-s)\left(1-\frac{b^2}{2}\right)+bs(b-c)\left(1-\frac{c^2}{2}\right)+cs(b-c)\left(1-\frac{b^2}{2}\right)+b^2(c-s)-c^2(b-s)-bc(b-s)\left(1-\frac{a^2}{2}\right)+bc(c-s)\left(1-\frac{a^2}{2}\right)=0.$

这个恒等式的左边比较复杂，正是锻炼我们运算能力的好材料. 已知 $s=\frac{a+b+c}{2}$，可以将式中的 s 换作$\frac{a+b+c}{2}$，但不必一开始就换，因为那样写起来较为麻烦，只要心里记住 s 就是 $\frac{a+b+c}{2}$，然后在适当的时候再将它代换进去.

式子很繁，需要有条不紊地进行运算. 首先注意式子中有 1，从而（如果将式子全乘出来）其中的单项式的次数有两种，一种是 3 次的，另一种是 5 次的，将它们分开处理为好. 即分别证明 3 次单项式的代数和为 0，5 次的单项式的代数和也为 0. 具体地说，就是要去证明

$$s(a+2s)(b-c)+b(a+2s)(c-s)-c(a+2s)(b-s)+bs(b-c)+cs(b-c)+b^2(c-s)-c^2(b-s)-bc(b-s)+bc(c-s)=0 \tag{1}$$

与

$$-b(a+2s)(c-s)c^2+c(a+2s)(b-s)b^2-bs(b-c)c^2-cs(b-c)b^2+a^2bc(b-s)-a^2bc(c-s)=0. \tag{2}$$

(1)的左边前三项的和为 0（请你心算一下），后六项的和也为 0，所以(1)成立.

(2)的左边提出 bc 后剩下部分成为

$$(a+2s)(b(b-s)-c(c-s))-s(b-c)(b+c)+a^2(b-c),$$

其中 $b(b-s)-c(c-s)=(b-c)(b+c-s)$,所以再提出 $b-c$ 后,左边剩下部分成为

$$\begin{aligned}&(a+2s)(b+c-s)-s(b+c)+a^2\\=&a(b+c-s)+s(b+c)-2s^2+a^2\\=&a(a+b+c-s)+s(b+c)-2s^2\\=&as+s(b+c)-2s^2\\=&s(a+b+c)-2s^2=0.\end{aligned}$$

于是(2)成立,原式也成立.

上面的运算,最好不用草稿纸,所有过程全部写在正式的卷面上. 如果依赖草稿纸,不但心算能力不能提高,而且誊来誊去,很容易出错.

条件等式的证明（一）

已知：

$$x=cy+bz, \tag{1}$$

$$y=az+cx, \tag{2}$$

$$z=bx+ay. \tag{3}$$

求证：

$$\frac{x^2}{1-a^2}=\frac{y}{1-b^2}=\frac{z^2}{1-c^2}\text{(假定分母均不为 0)}. \tag{4}$$

(1)、(2)、(3)组成一个方程组. 显然

$$x=y=z=0$$

是这个方程组的解，它使(4)成立.

如果方程组仅有这一组解，那么问题已经完全解决，但我们并不知道方程组是否仅有这一组解(事实是在 $a^2+b^2+c^2+2abc=1$ 时方程组并不只有这一组解). 如果去解方程组，将是比较麻烦的，而且最后并不能得到所希望的、对我们有利的结论. 所以，我们稍作变通. 具体做法如下：

$c\cdot(2)+(1)$，消去 y 得

$$(1-c^2)x=(ac+b)z,$$

所以

$$\frac{x}{ac+b}=\frac{z}{1-c^2}.$$

同样消去 x，得

$$\frac{y}{bc+a}=\frac{z}{1-c^2},$$

即

$$\frac{x}{ac+b}=\frac{y}{bc+a}=\frac{z}{1-c^2}. \tag{5}$$

这里已出现一个我们希望见到的，在(4)中有用的分母 $1-c^2$.

再由(1)、(3)得

$$\frac{x}{ab+c}=\frac{y}{1-b^2}=\frac{z}{bc+a}. \tag{6}$$

将(5)、(6)的后一半相乘得

$$\frac{y^2}{(bc+a)(1-b^2)}=\frac{z^2}{(bc+a)(1-c^2)},$$

所以

$$\frac{y^2}{1-b^2}=\frac{z^2}{1-c^2}.$$

同理可得(4)的前半部(要利用方程(2)、(3)).

条件等式的证明(二)

已知：
$$a_1x+b_1y+c_1z=0, \tag{1}$$
$$a_2x+b_2y+c_2z=0, \tag{2}$$

求证：$\dfrac{x}{b_1c_2-b_2c_1}=\dfrac{y}{c_1a_2-c_2a_1}=\dfrac{z}{a_1b_2-a_2b_1}$(假定分母均不为0).　(3)

从(1)、(2)可以消去 z 得出 x、y 的关系,即 $c_2\cdot(1)-c_1\cdot(2)$,得
$$(a_1c_2-a_2c_1)x+(b_1c_2-b_2c_1)y=0.$$
所以　$(a_1c_2-a_2c_1)x=-(b_1c_2-b_2c_1)y$, $\dfrac{x}{b_1c_2-b_2c_1}=\dfrac{y}{c_1a_2-c_2a_1}$.

同理(消去 x)可得(3)的另一半.

本题的结论很有用,值得记忆.记忆的办法是采用“行列式”,即约定
$$\begin{vmatrix} a & b \\ c & d \end{vmatrix}=ad-bc,$$
这样(3)就可以写成 $\dfrac{x}{\begin{vmatrix} b_1 & c_1 \\ b_2 & c_2 \end{vmatrix}}=\dfrac{y}{\begin{vmatrix} c_1 & a_1 \\ c_2 & a_2 \end{vmatrix}}=\dfrac{z}{\begin{vmatrix} a_1 & b_1 \\ a_2 & b_2 \end{vmatrix}}$,

即 x 的分母是将数表(也称为矩阵)$\begin{vmatrix} a_1 & b_1 & c_1 \\ a_2 & b_2 & c_2 \end{vmatrix}$ 的第一列划去所得到的行列式,z 的分母是将第三列划去所得到的行列式.而 y 的分母是将第二列划去后,并将剩下两列交换位置所得到的行列式,它与不交换二行位置的行列式 $\begin{vmatrix} a_1 & c_1 \\ a_2 & c_2 \end{vmatrix}=a_1c_2-a_2c_1$ 正好差一个符号.这是要特别当心的,不要搞错.

平方差公式的应用（一）

师：今天看到一个与平方差有关的问题：

证明 2003 可以写成两个整数 x、y 的平方差，即有整数 x、y，使得

$$x^2-y^2=2003. \tag{1}$$

生：这可以用因式分解的方法. 由上式得 $(x+y)(x-y)=2003$，所以

$$\begin{cases}x+y=2003,\\x-y=1.\end{cases}$$

$$x=\frac{2003+1}{2}=1002,\ y=\frac{2003-1}{2}=1001.$$

师：你得到一组使(1)成立的解. 还有一组解是 $x=-1002$，$y=-1001$. 不过，这道题只要求找出一组使(1)成立的整数解. 所以你的解答是很好的.

进一步，可以考虑更一般的问题：设 $2n+1$ 是奇数，证明 $2n+1$ 可以写成两个整数的平方差.

生：我还是用因式分解的方法.

师：因式分解的方法可行. 不过，也可以直接猜一下：

$(\quad)^2-(\quad)^2=2n+1$ 中，$(\quad)$里可以填什么？

生：可以填 $n+1$ 与 n. 因为$(n+1)^2-n^2=n^2+2n+1-n^2=2n+1$.

师：很好. 所以我们得到更一般的结论：每个奇数都可以写成平方差.

但是，能写成平方差的整数不一定是奇数，4 的倍数 $4n$(n 为整数)，也能写成平方差.

生：$(n+1)^2-(n-1)^2=4n$，所以 $4n$ 能写成平方差.

平方差公式的应用（二）

师:今天讨论两个与平方差有关的问题. 第一个问题是:

例 1　如果 a 是正整数,证明有两个整数 x、y,使得 $x^2-y^2=a^3$.

生:我还是进行分解:

$$(x+y)(x-y)=a^3,$$

所以 $\begin{cases}x+y=a^2,\\ x-y=a.\end{cases}$　解得 $\begin{cases}x=\dfrac{a^2+a}{2},\\ y=\dfrac{a^2-a}{2}.\end{cases}$

师:你得到的 x、y 是不是整数?

生:如果 a 是偶数, $a^2\pm a$ 是偶数. 如果 a 是奇数, a^2 与 a 同是奇数, $a^2\pm a$ 是偶数,所以上面得到的 x、y 都是整数,而且

$$\left(\frac{a^2+a}{2}\right)^2-\left(\frac{a^2-a}{2}\right)^2=\frac{a^2+a+a^2-a}{2}\cdot\frac{a^2+a-(a^2-a)}{2}=a^3,$$

所以 a^3 是平方差.

师:能否利用上次得到的结论来证?(前一篇文章《平方差公式的应用(一)》中的结论为:每个奇数都可以写成平方差,是 4 的倍数的偶数也能写成平方差.)

生:如果 a 是奇数,那么 a^3 也是奇数,它能写成平方差.

如果 a 是偶数,那么 a^3 是 4 的倍数,它能写成平方差.

这种证法更容易一些.

师:于是第一道题就有两种不同的证法. 现在看第二道题:

例 2　a、b、c、d 是四个自然数, $a<b<c<d$,并且后一个减去前一个,所得的差都相同(称这四个数组成等差数列). 证明 $abcd$ 是平方差.

生:例如 1、3、5、7 这四个数,它们的积是奇数,所以也是平方差.

一般地,如果 a 是奇数, $abcd$ 是不是奇数? 好像不一定.

师:不一定. 你可以设差 $b-a=t$,将 4 个数表成 a, $a+t$, $a+2t$, $a+3t$,再考虑能不能用上一次的结论.

生:如果 a 是奇数, t 是偶数,那么 $a+t$, $a+2t$, $a+3t$ 都是奇数, $a(a+t)(a+2t)(a+3t)$ 也是奇数,所以能写成平方差.

如果 a 是奇数, t 是奇数,那么 $a+t$, $a+3t$ 都是偶数,积 $a(a+t)(a+2t)(a+3t)$ 是 4 的倍数,所以能写成平方差.

如果 a 是偶数,那么 a, $a+2t$ 都是偶数,积 $a(a+t)(a+2t)(a+3t)$ 是 4 的倍数,所以能写成平方差.

师:证得很好. 如果不用上次的结论. 直接做也可以:

$$\begin{aligned}&a(a+t)(a+2t)(a+3t)\\=&(a^2+3at)(a^2+3at+2t^2)\\=&(a^2+3at+t^2-t^2)(a^2+3at+t^2+t^2)\\=&(a^2+3at+t^2)^2-t^4.\end{aligned}$$

生:将 a 与 $a+3t$ 相乘, $a+t$ 与 $a+2t$ 相乘,得出的 a^2+3at 与 $a^2+3at+2t^2$ 不仅第一项相同,而且第二项也相同. 这就可以用平方差公式了,很巧妙.

师:不过,就本题而言,还是利用上次的结论更简单,基本上不需要计算.

孙庞斗智

鬼谷子先生有两个绝顶聪明的门徒，一个叫孙宾(后来改名孙膑)，一个叫庞涓.

有一天鬼谷子对他们说："两个大于 1 而小于 100 的自然数相加得 x，相乘得 y. 谁能猜出我说的这两个自然数是多少?"

庞涓说："条件太少了. x、y 都不知道，我没有办法做."

"好，你附耳过来."鬼谷子悄悄地告诉庞涓 x 是多少. 然后又悄悄地告诉孙宾 y 是多少.

"现在你们分别知道 x、y，谁能说出答案?"

庞涓想了一会，得意地说："虽然我不知道答案是哪两个自然数，但是我知道孙宾也不知道这两个数是多少!"

孙宾淡淡一笑："听你一说，现在我倒知道这两个数是多少了."

庞涓大吃一惊："慢，你先别说答案."他又想了一想说："师傅，我也知道这两个数是多少了."

亲爱的读者，你能知道鬼谷子说的自然数是多少吗?

这是一道难题，做不出来不要紧，后面有神算子写的解答.

孙宾的法宝是**分解因数**，即将 y 分成两个数的积. 如果 $y=33$，那么孙宾由 $33=3\times 11$ 立即得出这两个数是 3、11. 如果 $y=24$，那么由于 $24=4\times 6=3\times 8$，所以孙宾无法确定这两个数.

庞涓肯定孙宾不知道这两个数，这表明两个数不全是质数(否则孙宾立即能说出这两个数)，也表明 x 不能写成两个质数的和(否则庞涓不能肯定孙宾不知道这两个数).

不难验证每个大于 6、小于 200 的偶数都能写成两个不同的质数的和.

又容易知道 x 不小于 5(否则 x 不是两个不同的、大于 1 的自然数的和). 而 $x=6$ 时庞涓立即得出两数是 2、4，所以 $x\neq 6$(因为庞涓不知道两个数).

因此，x 不能是偶数.

2 也是质数. 根据同样的理由(x 不是两个质数的和), x 不是下列各数:

$$2+3,\ 2+5,\ 2+7,\ 2+11,\ 2+13,\ 2+17,\ 2+19,$$
$$2+23,\ 2+29,\ 2+31,\ 2+37,\ 2+41,\ 2+43,\ 2+47.$$

如果 x 大于 54,那么 x 是 53 与一个大于 1 的自然数 a 的和. 而 $53\times a$ 分解成两个小于 100、大于 1 的自然数的积时,一个因数必定是 53(53 的 2 倍已经大于 100),另一个因数当然也只能是 a. 这样,由于孙宾知道积 $y=53\times a$,他立即能说出答案. 既然庞涓断定孙宾不知道答案, x 一定小于 54.

综合上面所说,庞涓的一句话泄漏了不少信息,孙宾可以断定 x 只能是下列各数:

$$11,\ 17,\ 23,\ 27,\ 29,\ 35,\ 37,\ 41,\ 47,\ 51,\ 53. \qquad (1)$$

现在孙宾可以得出答案了! 因为他知道 y. 例如 $y=24$,将 24 分解成 3×8(不分解成 2×12), $3+8=11$ 是(1)中的数($2+12$ 不是(1)中的数),并且 24 只有这一种分解,能使所得的两个因数的和在(1)中. 这样,所求的两个数就是 3、8. 同样, $y=28$ 时, $28=4\times 7$, $4+7=11$, 并且 28 只有这一种分解满足要求(即两个因数的和在(1)中),所求的两个数就是 4 与 7.

我们不知道 y,所以到这一步,还无法说出答案. 但听了孙宾的话,我们可以知道(庞涓也知道)y 不等于 72,因为 $72=8\times 9=24\times 3$,而 $8+9=17$, $24+3=27$ 都在(1)中,所以 72 的两种分解均满足要求,这时孙宾无法确定所求的数究竟是 8 与 9,还是 24 与 3. 既然孙宾能确定出答案,那么 y 一定不是 72.

类似地, $30=2\times 15=5\times 6$, $3\times 14=2\times 21$, $5\times 12=3\times 20$, $6\times 11=2\times 33$, $7\times 10=35\times 2$ 等数均有两种以上的分解满足要求,它们均不能作为 y.

现在,我们可以列出一些可能的 y 值(不是全部),并将这些 y 分解后所确定的两个自然数的和与 y 用线相连,得到下面的(2):

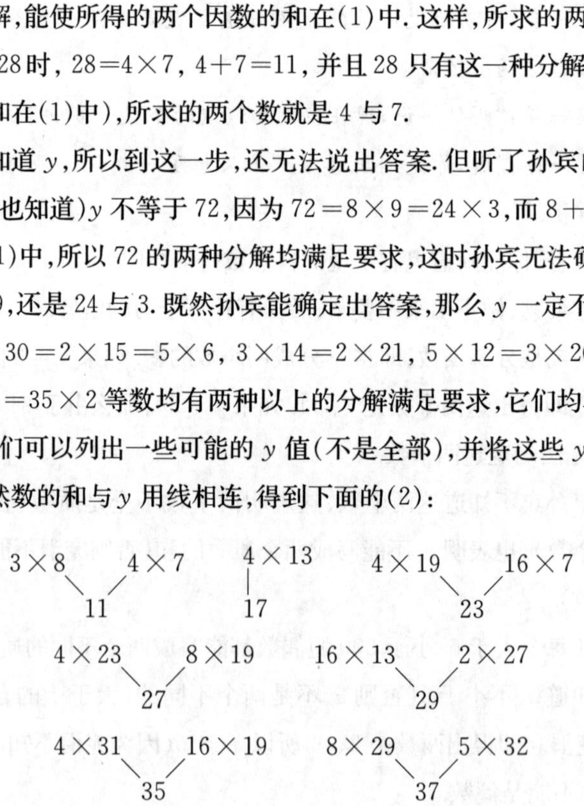

```
3×8   4×7     4×13      4×19   16×7
   \  /         |           \  /
    11          17           23

   4×23   8×19      16×13   2×27
       \  /              \  /
        27                29

   4×31   16×19     8×29   5×32
       \  /             \  /
        35               37
```

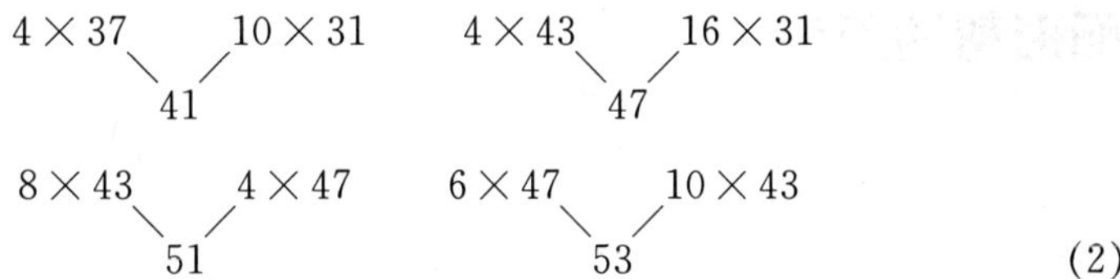

(2)

上面列出的 y 均只有一种满足要求的分解. 而且请留意, 17 上面只有一根"辫子".

听了孙宾的话, 庞涓便可以得出答案. 因为他知道 x. 例如 $x=17$ 时, 他可以将 17 分拆为两个自然数的和:

$$17=2+15=3+14=4+13=5+12=6+11=7+10=8+9.$$

但上面已经说过 2×15, 3×14, 5×12, 6×11, 7×10, 8×9 均不能作为 y, 所求的两个自然数只能是 4 与 13.

我们虽然不知道 x, 但听到庞涓说他知道所求的两数时, 便可以断定 $x=17$. 因为在(2)中, 其他的 x(例如 $x=11$) 均至少有两种分拆 $(11=3+8=4+7)$, 产生两个 $y(3\times8,\ 4\times7)$. 这种"头上有两根以上辫子"的 x 均不能定出唯一的答案. 只有 17"头上一根辫子", 才能定出答案为 4 与 13.

以上就是神算子所作的解答.

图形帮助思考

“华罗庚金杯赛”(以下简称“华杯赛”)是为了纪念我国数学家华罗庚先生而举行的数学竞赛,由小学六年级与初中一年级的优秀学生参加,每两年一届,自 1985 年起,已经举行了 7 届.“华杯赛”的问题较难,有不少问题很有新意,可以开拓同学们的眼界,提高学习兴趣,而且饶有余味,可以反复琢磨,甚至高中同学也能从中获得益处.

好的问题应当有好的解法.在本文与后面几篇文章中,我们介绍一些“华杯赛”的问题与解答.

例 (1997 年第六届“华杯赛”决赛初一组第一试第 5 题)一批大小略有不同的长方体盒子,高都等于 6 厘米,长和宽都大于 5 厘米,长、宽的比不小于 2.若在任一盒子中放一层边长为 5 厘米的小立方体,无论怎样放,放完后被小立方体所覆盖的底面积都不超过原底面积的 40%.现向盒子中注水,问:

(1) 要使最小的盒子不往外溢,最多能注多少立方厘米的水?

(2) 要使最大的盒子开始外溢,最少要注进去多少立方厘米的水?

解 将盒子的底面画出来,如图 1.设放入尽可能多的小立方体,经移动后全靠在左上部,横数 a 个,竖数 b 个.长剩下的一段长 p(厘米),宽剩下的一段长 q(厘米).

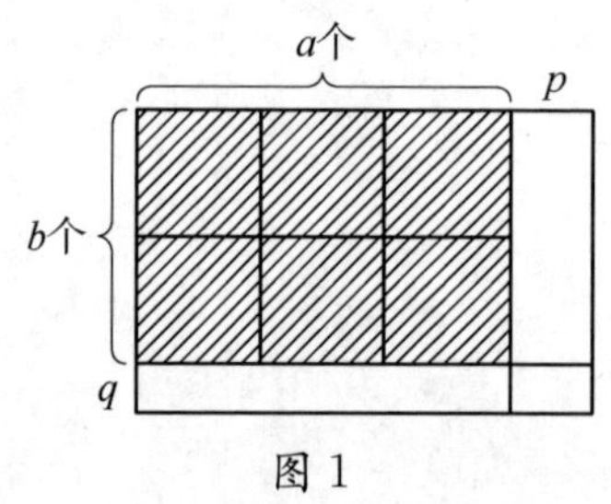

图 1

由题意 $a \geqslant b \geqslant 1$(因为长、宽都大于 5 厘米),$p$、$q$ 均小于 5(否则还能再放一些小立方体).

从图 1 看出在 $b \geqslant 2$ 时,阴影部分的面积大于它下方(以 q 为宽的)矩形面积的 2 倍,也大于它右方(以 p 为宽的)矩形面积的 2 倍,并且大于右下角($p \times q$ 的)矩形面积的 4 倍.如果阴影部分不超过底面积的 0.4(即 40%),那么其余部分面积小于底面积的 $0.4\times\frac{1}{2}+0.4\times\frac{1}{2}+0.4\times\frac{1}{4}=0.5$,这是不可能的(至少应当占 0.6). 所以 $b=1$, 即小立方体只排成一行.

如果 $a\geqslant 4$, 用同样的理由(请读者自己画图)可以得出阴影部分面积不超过底面积的 0.4 时, 其余部分面积小于底面积的 $0.4\times 1+0.4\times\frac{1}{4}+0.4\times\frac{1}{4}=0.6$. 所以 $a\leqslant 3$.

如果 $a=1$. 那么底面的长为 $5+p$, 宽为 $5+q$. 长显然小于宽的 2 倍($2\times(5+q)\geqslant 10>5+p$).

如果 $a=2$, 那么底面的长为 $10+p$, 宽为 $5+q$. 因为长、宽的比不小于 2, 所以 $p\geqslant 2q$, 从而 $q<\frac{5}{2}$. 如图 2, 阴影部分的面积大于下方面积的 2 倍, 也大于右方面积的 2 倍, 并且大于右下角面积的 2 倍. 如果阴影部分只占底面积的 0.4, 那么其余部分的面积小于底面积的 $0.4\times\frac{1}{2}+0.4\times\frac{1}{2}+0.4\times\frac{1}{2}=0.6$. 所以只能 $a=3$.

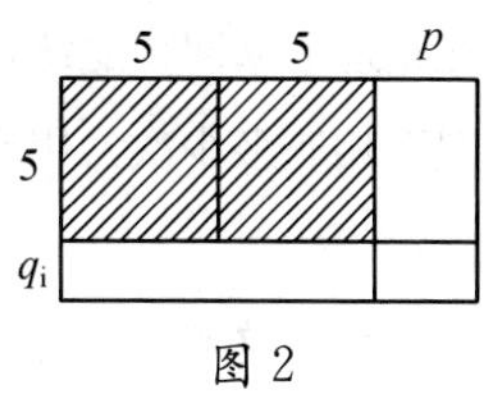

图 2

于是, 底面如图 3 所示, 其中可放 3 个边长为 5 的正方形, 这部分的面积为 $5\times 5\times 3=75$(平方厘米), 所以底面面积至少为 $75\div 40\%=\frac{375}{2}$(平方厘米).

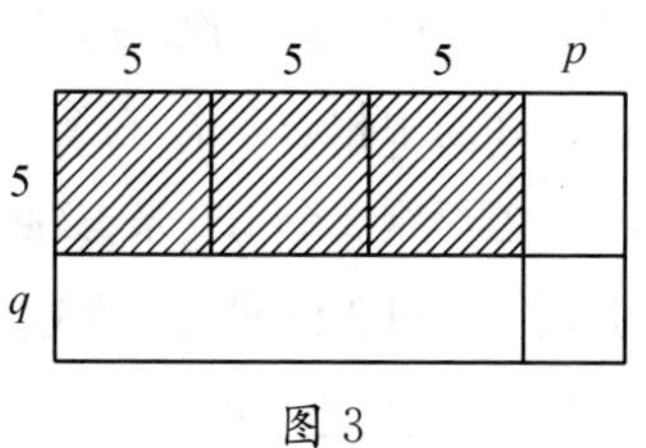

图 3

同时, 由于 p、q 都小于 5, 底面面积小于 $(5\times 3+5)(5+5)=200$(平方厘米).

于是问题(1)的答案是 $1125\left(=6\times\frac{375}{2}\right)$ 立方厘米, 问题(2)的答案是 $1200(=6\times 200)$ 立方厘米.

本题的关键是确定盒中可放几个边长为 5 的小立方体, 即定出 $b=1$, $a=3$. 利用图形较为方便. 其中阴影部分面积≤底面积的 0.4 是极重要的条件, 在证明 $b=1$ 及 $a\leqslant 3$ 时均需用到. 证明 $a\neq 2$ 时, 除了这个条件, 还要用到长、宽的比不小于 2. 如果不利用图形, 纯粹从式子上分析, 虽然也能得出结论, 但不如利用图形清晰、自然.

算术更好

例 (1997 年第六届"华杯赛"决赛初一组第一试第 3 题)一段跑道长 100 米,甲、乙分别从 A、B 端点同时相向出发,各以每秒 6 米和每秒 4.5 米的速度在跑道上来回往返练习跑步. 问:在 10 分钟内(包括第 10 分钟),甲和乙在途中迎面相遇多少次? 甲在途中追上乙多少次? 甲和乙在 A、B 两端点共相遇多少次?

解 这道题用算术的方法比用代数方法更简单.

甲、乙两人速度的比是 6∶4.5 即 4∶3,所以甲跑 4 个 100 米时,乙跑 3 个 100 米. 在这个过程中,乙先由 B 到 A(第 1 个 100 米),与甲迎面相遇 1 次. 乙再由 A 到 B(第 2 个 100 米),又与甲迎面相遇 1 次. 最后乙由 B 到 A(第 3 个 100 米),又与甲相遇 1 次. 并且甲最后在 A 处追上乙(即甲比乙恰好多跑 100 米). 因此在途中相遇 3 次,在途中甲追上乙 0 次,在端点相遇 1 次.

乙再跑 3 个 100 米,又回到 B 点,这时甲跑了 4 个 100 米,回到 A 点(正好是前一个过程倒回去). 在这个过程中,两人在途中相遇 3 次,在途中甲追上乙 0 次,在端点相遇 0 次.

以后,无非是重复这一过程(即乙每跑 6 个 100 米,相当于 1 个周期,每跑 3 个 100 米,相当于半个周期).

由于在 10 分钟内,乙跑 27 个 100 米 ($4.5\times60\times10=2700$),甲跑 36 个 100 米,所以甲、乙共相遇 $3\times\frac{27}{3}=27$(次),在途中甲追上乙 0 次,在端点相遇 $\frac{27-3}{6}+1=5$(次).

算术的思想、方法很有用处,它是今后学习(特别是学习数论与组合)的基础,也是我国传统数学的一种特色. 算术与代数解法各有千秋,不可因为学习代数而摒弃算术.

输送信息

某计算机接收信息的速度为每秒 2.8 千字节;发送信息的速度为每秒 3.8 千字节.现在要从 A 处接收,往 B 处发送,还要将机内储存的 58 千字节的信息也发送至 B 处.如果发送、接收轮流进行,每次发与收各 10 秒.问:

(1) 若先发送,经过多少秒恰好将机内储存的信息送完?

(2) 若先接收,经过多少秒恰好将机内储存的信息送完?(答案保留分数)

这是第七届"华杯赛"试题.

(1) 先发送 10 秒,发出

$$10 \times 3.8 = 38(\text{千字节}),$$

还剩

$$58 - 38 = 20(\text{千字节}).$$

以后每 20 秒(收、发各 10 秒),可发机内储存的

$$10 \times (3.8 - 2.8) = 10(\text{千字节}).$$

因此,将机内储存的信息送完需要

$$10 + 2 \times 20 = 50(\text{秒}).$$

(2) 每 20 秒(收、发各 10 秒),可发机内储存的 10 千字节.100 秒可发机内储存的 50 千字节.还剩

$$58 - 50 = 8(\text{千字节}).$$

再过 10 秒,又输入 28(=2.8×10) 千字节,共有

$$8 + 28 = 36(\text{千字节}).$$

需用

$$\frac{36}{3.8} = \frac{36}{38} \times 10 = 9\frac{9}{19}(\text{秒}).$$

因此,将机内储存的信息送完需要

$$100 + 10 + 9\frac{9}{10} = 119\frac{9}{19}(\text{秒}).$$

乘与除

a 是自然数,且 $17a=\underbrace{111\cdots111}_{n\text{个}1}$, 求 a 的最小值.

这是第七届“华杯赛”的试题.

问题不难,千万别把问题想得太难. 17 乘 a 的结果是 n 个 1 组成的数,所以 n 个 1 组成的数除以 17 便是 a,当然这里的 n 要使商 a 为整数. 因为 n 越小,商 a 也越小,所以做竖式除法一直除到 n 个 1 能被 17 整除,这时的 a 就是所求的最小值.

具体算式如下(略写部分,请自行补上):

```
           6 5 3 5 9 4 7 7 1 2 4 1 8 3
    ______________________________________
 17 ) 1 1 1 1 1 1 1 1 1 1 1 1 1 1 1 1
      1 0 2
      -----
          9 1
          8 5
          ---
            6 1
            5 1
            ---
            1 0 1
              …
                                …
                            -------
                            1 4 1
                            1 3 6
                            -------
                                5 1
                                5 1
                                ---
                                  0
```

a 的最小值是 65 359 477 124 183.

一般地,只要自然数 m(例如 $m=17$) 与 10 互质,一定有一个 a,使得 ma 是全由 1 组成的数.

人与球

每个男生有 k 个白球、没有花球,每个女生有 n 个花球、没有白球. A 组有男生 7 人,女生 6 人;B 组有男生 8 人,女生 7 人. A 组的白球比花球多;B 组的白球比花球少. 如果 A 组男生每人拿出一个白球给 B 组,那么这时 A 组的白球就不比花球多了;而 B 组的白球也不比花球少了. 求:

(1) 最大的 n 是几? 相应的 k 是几?

(2) 最小的 n 是几? 相应的 k 是几?

这是第七届"华杯赛"的试题.

从已知条件容易得出

$$7k > 6n,\ 8k < 7n, \tag{1}$$

$$7k - 7 \leqslant 6n,\ 8k + 7 \geqslant 7n, \tag{2}$$

于是由(2)得

$$\frac{6}{7}n + 1 \geqslant k \geqslant \frac{7}{8}n - \frac{7}{8}, \tag{3}$$

即

$$1 + \frac{7}{8} \geqslant \left(\frac{7}{8} - \frac{6}{7}\right)n,\ n \leqslant 105.$$

于是 n 的最大值是 105. 在 $n = 105$ 时,由(3)得 $k = 91$.

不难验证 $n = 105$, $k = 91$ 符合要求. 因此,这就是第(1)小题的答案.

第(2)小题我们分两种情况讨论.

(i) 如果 n 是 7 的倍数,设 $n = 7m$, 则由(1)、(2)的第一个不等式得

$$6m < k \leqslant 6m + 1.$$

因为 k 是整数,所以 $k = 6m + 1$.

再由(1)的第二个不等式得

$$8(6m+1)<49m.$$

即 $m>8$. 所以最小的 $m=9$, $n=63$.

(ii) 如果 n 不是 7 的倍数,设 $n=7m+l$,其中 $1\leqslant l\leqslant 6$, 则由(1)、(2)的第一个不等式得

$$6m+\frac{6}{7}l<k\leqslant 6m+\frac{6}{7}l+1,$$

即

$$6m+l-\frac{l}{7}<k\leqslant 6m+l+1-\frac{l}{7}<6m+l+1.$$

因为 k 是整数,所以从上式得

$$k=6m+l. \tag{4}$$

再由(1)的第二个不等式得

$$8(6m+l)<49m+7l,\ m>l.$$

l 的最小值是 1,所以 m 的最小值是 2, n 的最小值是 15,相应的 $k=13$.

不难验证 $n=15$, $k=13$ 符合要求.

所以第(2)小题的答案是 $n=15$, $k=13$.

本题需要解不等式,还要特别注意整数的性质(如(4)的导出及设 $n=7m+l$).

质数与合数（一）

b 与 p 是大于 1 的自然数，且

$$p+2b,\ p+4b,\ p+6b,\ p+8b,\ p+10b$$

都是质数. 求 $p+b$ 的最小值.

先看看 p 的最小值是多少.

$p=2$ 时，$p+2b$ 是 2 的倍数，并且大于 2，所以 $p+2b$ 不是质数. 同理 p 是正偶数时，$p+2b$ 也不是质数. 因此 p 是大于 1 的奇数.

$p=3$ 时，$p+6b$ 有真因数 3，它不是质数. 因此 p 是大于 3 的奇数.

$p=5$ 时，$p+10b$ 有真因数 5，它不是质数. 因此 $p\geqslant 7$.

再看看 b 的最小值.

注意 2、4、6 这三个数，除以 3 余数互不相同. 如果 b 不是 3 的倍数，那么 $p+2b$，$p+4b$，$p+6b$ 除以 3 余数也互不相同(因为它们两两的差是 $2b$ 或 $4b$，不被 3 整除). 因此，这三个数除以 3 的余数恰好是不同的三个数 0、1、2，其中必有一个能被 3 整除，因而它不是质数. 所以 b 一定是 3 的倍数.

同理，如果 b 不是 5 的倍数，那么 $p+2b$，$p+4b$，$p+6b$，$p+8b$，$p+10b$ 这五个数，除以 5 余数互不相同，即余数恰好是不同的五个数 0，1，2，3，4，其中必有一个能被 5 整除，因而它不是质数. 所以 b 一定是 5 的倍数.

综上所述，b 一定是 $15(=3\times 5)$ 的倍数，从而 $b\geqslant 15$.

在 $p=7$，$b=15$ 时，$p+2b$，$p+4b$，$p+6b$，$p+8b$，$p+10b$ 依次为：

$$37,\ 67,\ 97,\ 127,\ 157.$$

它们都是质数.

因此，$p+b$ 的最小值为 $7+15=22$.

本题由第七届“华杯赛”的一道试题改编而成. “华杯赛”的原题还要难一些，我们下一篇文章再说.

质数与合数(二)

b 与 p 是大于 1 的自然数,且

$$p+2b,\ p+4b,\ p+6b,\ p+8b,\ p+10b,\ p+12b$$

都是质数. 求 $p+b$ 的最小值.

这是第七届"华杯赛"的一道试题. 与前一篇文章的问题相比,现在要求 $p+12b$ 也是质数.

对上次所给答案 $p=7$, $b=15$, $p+12b=187=11\times 17$ 不是质数. 因此,上次的答案不再正确. 但其中的推导论证仍然有效. 例如 $p\geqslant 7$, p 是奇数,而且(从推导中可以看出)p 不是 3、5 的倍数. b 是 15 的倍数.

注意 2、4、6、8、10、12 这六个数,除以 7 余数互不相同而且都不为 0,如果 b 不是 7 的倍数,那么 $2b$、$4b$、$6b$、$8b$、$10b$、$12b$ 除以 7 余数也互不相同而且都不为 0,即余数恰好是 1、2、3、4、5、6 这六个数. 如果 p 也不是 7 的倍数,那么 $p+2b$, $p+4b$, $p+6b$, $p+8b$, $p+10b$, $p+12b$ 这六个数除以 7,其中必有一个余数是 0,因而不是质数. 所以 b、p 中至少有一个是 7 的倍数.

如果 $p=7$, 那么 b 的可能值是

$$15,\ 30,\ 45,\ 60,\ 75,\ \cdots$$

$b=15$ 时, $p+12b=187$ 不是质数. $b=30$ 时, $p+6b=187$. $b=45$ 时, $p+4b=187$. $b=60$ 时, $p+4b=247$,不是质数. 所以 $b\geqslant 75$. 在 $b=75$ 时, $p+2b$, $p+4b$, $p+6b$, $p+8b$, $p+10b$, $p+12b$ 依次为

$$157,\ 307,\ 457,\ 607,\ 757,\ 907,$$

都是质数. 因此, $p+b$ 的最小值 $\leqslant 7+75=82$.

如果 b 是 7 的倍数,那么 b 是 $105(=3\times 5\times 7)$ 的倍数. $p+b>105>82$. 因此 $p+b$ 最小时, b 不是 7 的倍数. 从而 p 一定是 7 的倍数. p 的可能值是(依从小到大的顺序排列)

$$7, 49, 77, \cdots$$

在 $p=49$ 时，b 的可能值是(依从小到大的顺序排列)

$$15, 30, 45, \cdots$$

其中 $b \geqslant 45$ 使 $p+b \geqslant 49+45>82$,不可能为最小值. $b=15$ 时,

$$p+8b=49+120=169=13\times 13$$

不是质数. $b=30$ 时，$p+4b=169$.

因此，$p=49$ 时，$p+b$ 不可能最小. 在 $p \geqslant 77$ 时，$p+b \geqslant 77+15>82$ 也不可能最小.

综上所述，$p+b$ 的最小值是 82,其中 $p=7$, $b=75$.

本题首先要知道质数的定义及判别法. 判别的方法是试除. 例如 907,可用小于 31 的质数去试除,不难发现 3、5、7、11、13、17、19、23、29 都不是 907 的因数(约数),所以 907 是质数(这里 31 是第一个平方超过 907 的正整数).

其次,在上一篇文章已经说到整数除以 3,余数有三个不同值;除以 5,余数有五个不同值. 这一讲用到整数除以 7,余数有七个不同值,其中非零的有六个. 通过余数的讨论,我们发现 p、b 中至少有一个是 7 的倍数. 这是解决本题的关键.

居中和

例　(1997年第六届"华杯赛"决赛小学组第一试第3题)将1, 2, 3, …, 49, 50任意分成10组,每组5个数.每一组中,数值居中的那个数称为"居中数",10个"居中数"的和称为"居中和".对每一种分法有相应的"居中和".问:最大的与最小的居中和各是多少?

解　设10个"居中数"依从小到大的顺序是a_1, a_2, …, a_{10}.

显然a_{10}应小于同组中两个比它大的数,所以$a_{10} \leqslant 50-2=50-3+1$.

而a_9不但小于同组中的两个比它大的数,还小于a_{10}那一组中的三个数,所以$a_9 \leqslant 50-2-3=50-2\times 3+1$.

类似地,$a_8 \leqslant 50-3\times 3+1$, $a_7 \leqslant 50-4\times 3+1$, …$a_1 \leqslant 50-10\times 3+1$.

将上述不等式相加得

$$a_1+a_2+\cdots+a_{10} \leqslant 10\times 50-3\times(1+2+\cdots+10)+10=345. \quad (*)$$

即最大的"居中和"不超过345.

另一方面,"居中和"可以等于345.例如:(1, 2, 21, 22, 23), (3, 4, 24, 25, 26), (5, 6, 27, 28, 29), (7, 8, 30, 31, 32), (9, 10, 33, 34, 35), (11, 12, 36, 37, 38), (13, 14, 39, 40, 41), (15, 16, 42, 43, 44), (17, 18, 45, 46, 47), (19, 20, 48, 49, 50)这10组数的"居中和"恰好为$48+45+42+\cdots+24+21=345$.

同样可以证明$a_1 \geqslant 3$, $a_2 \geqslant 3\times 2$, …, $a_{10} \geqslant 3\times 10$,所以"居中和"最小为$3+3\times 2+\cdots+3\times 10=165$(请读者举例说明"居中和"可以取这个值).

"居中数",在统计学中称为"中位数",今后还会学到.

在本题中,举例说明(*)中等号可以成立,是不可缺少的部分.举例,在数学中也是非常重要的.使(*)中等号成立的例子不只一个,使"居中和"最小(等于165)的例子也不只一个.

逆序

例　(1997 年第六届“华杯赛”决赛初一组第二试第 2 题)用 1, 2, …, 99, 100 共一百个数排成一个数列

$$a_1,\ a_2,\ \cdots,\ a_{99},\ a_{100}. \tag{1}$$

已知数列中第 6 个是 $a_6=60$,第 94 个是 $a_{94}=98$. 如果相邻两个数 $a_i>a_{i+1}$,就将它们交换位置. 如此操作,直到左边的数都小于右边的数为止. 请回答最少需实行多少次交换? 最多需实行多少次交换?

解　如果在上述排列(1)中, $j>i$,而 $a_i>a_j$,我们就称$(a_i,\ a_j)$是一个逆序. 例如在

$$\begin{aligned}&1,\ 2,\ 3,\ 4,\ 5,\ \mathbf{60},\ 6,\ 7,\ \cdots,\ 58,\ 59,\ 61,\ \cdots,\\&92,\ 93,\ \mathbf{98},\ 94,\ 95,\ 96,\ 97,\ 99,\ 100\end{aligned} \tag{2}$$

中, (60, 6), (60, 7), …, (60, 59), (98, 94), (98, 95), (98, 96), (98, 97)都是逆序,共有 $54+4=58$ 个逆序. 而在排列

$$\begin{aligned}&100,\ 99,\ 97,\ 96,\ 95,\ \mathbf{60},\ 94,\ 93,\ \cdots,\\&8,\ 7,\ \mathbf{98},\ 6,\ 5,\ 4,\ 3,\ 2,\ 1\end{aligned} \tag{3}$$

中,如果去掉 60 与 98 两个数,共有

$$97+96+\cdots+2+1=\frac{97\times 98}{2}=4753$$

个逆序. 60 放入后,增加 $5+59=64$ 个逆序. 98 放入后,又增加 $2+6=8$ 个逆序. 所以(3)中共有 $4753+64+8=4825$ 个逆序.

每交换一次,恰好减少一个逆序,所以(2)调整成自然排列(即从小到大的排列)

$$1,\ 2,\ 3,\ \cdots,\ 97,\ 98,\ 99,\ 100 \tag{4}$$

需实行 58 次交换. 而(3)调整成(4)需实行 4825 次交换.

由于(1)中, $a_6=60$, 所以 1, 2, …, 59 中至少有 $59-5=54$ 个在 a_6 后面,即 60 至少产生 54 个逆序. 又(1)中 $a_{94}=98$, 所以在 a_{94} 后面的 6 项中,至少有 $6-2=4$ 个数比 a_{94} 小,即 98 至少产生 4 个逆序. 所以(1)中至少有 $54+4=58$ 个逆序,至少需交换 58 次才能变成(4). 而(2)就是交换 58 次变成(4)的实例.

由于(1)中 $a_6=60$, 所以在 a_6 前面至多有 5 个数比 60 大,即 60 至多产生 $59+5=64$ 个逆序. 又在 $a_{94}=98$ 前面至多有 2 个数比 98 大,即 98 至多产生 $6+2=8$ 个逆序. 去掉 60 与 98 后,其余的 98 个数至多产生 4753 个逆序(即全部从大到小排列). 所以(1)中至多有 4825 个逆序,至多需交换 4825 次才能变成(4). 而(3)就是交换 4825 次变成(4)的实例.

本题的答案是 58 与 4825.

逆序是一个重要的概念. 本题不仅要计算(2)、(3)中逆序的个数,而且还应证明(2)、(3)提供了两个实例,分别使(1)中逆序的个数达到最小与最大. 这种推理能力,需要逐步培养,并非一日之功. 揣摩已有的解法,就是培养这种能力的必由之路.

行程问题

早上八点钟，甲、乙、丙三人从东往西直行.乙在甲前 400 米，丙在乙前 400 米.甲、乙、丙三人的速度分别为每分钟 120 米、100 米、90 米.问什么时刻甲和乙、丙的距离相等？

本题是 2001 年"华杯赛"中学组第一试第 2 题.

由于甲的速度最快，所以甲先追上乙，而乙尚未追上丙.具体地说，甲追上乙需

$$400 \div (120-100)=20(\text{分}),$$

而此时乙在丙后面

$$400-(100-90)\times 20=200(\text{米}).$$

在这 20 分内，甲与丙的距离始终大于甲与乙的距离.

设再过 t 分钟，甲到达乙、丙的中间，与乙、丙的距离相等，则这时乙、丙的距离是

$$200-(100-90)t=200-10t(\text{米}),$$

而甲、丙的距离是

$$200-(120-90)t=200-30t(\text{米}),$$

并且

$$200-10t=2(200-30t),$$

即

$$50t=200,$$

$$t=4.$$

此后，甲与丙的距离比甲与乙的距离小，直到乙追上丙，这时共行

$$400 \div (100-90)=40(\text{分}).$$

在这以后，甲与丙的距离比甲与乙的距离大.

因此，在 8∶24 与 8∶40 时，甲与乙、丙的距离相等.

行程问题（一）

C、D 两地相距 45 千米.甲、乙二人骑自行车分别从 C、D 两地同时出发，相向而行.甲的速度是每小时 9 千米，乙的速度是每小时 7 千米.丙骑摩托车，每小时行 63 千米，与甲同时从 C 地出发，在甲、乙二人间来回穿梭(与乙相遇立即返回，与甲相遇也立即返回).问：甲、乙二人相距 20 千米时，甲与丙相距多少千米?

这道行程问题比较复杂，需要仔细弄清题意，并尽可能做得简单一点.

如图 1，设在甲、丙第一次相遇时，甲行了 $9s$ 千米到达 E，这时乙行了 $7s$ 千米到达 F.丙应当行 $63s$ 千米，从 C 到达 G 与乙相遇，再折回与甲在 E 相遇，因此 EG 这段长是(单位千米均省略不写)

$$(63s-9s)\div 2=27s.$$

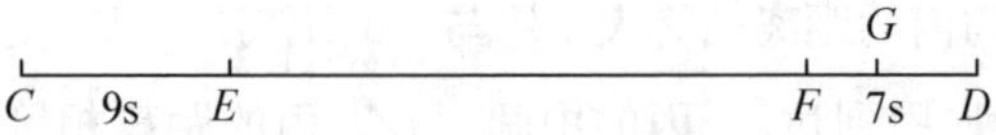

图 1

CG 这段是

$$9s+27s=36s.$$

乙行的路程是丙的 $\frac{1}{9}$，所以在乙与丙相遇时，乙行

$$36s\times\frac{1}{9}=4s.$$

即 $DG=4s$. 从而

$$GF=7s-4s=3s,\ EF=27s-3s=24s.$$

由于 $CD=9s+27s+4s=40s$，所以 EF 是 CD 的 $\frac{24}{40}$，即 $\frac{3}{5}$.

于是，在甲、丙第一次相遇时，甲、乙相距

$$45\times\frac{3}{5}=27.$$

同样,将 C、D 两点换成 E、F 来讨论.我们得到甲、丙再相遇时,甲、乙的距离是

$$27\times\frac{3}{5}=\frac{81}{5}.$$

因为 $\frac{81}{5}<20$,所以在甲、乙相距 20 千米时,甲与丙尚未相遇.或者说,甲、乙从相距$\frac{81}{5}$的两地各自向后退回一些,这时丙也相应地退回一些.这样就可算出甲、乙相距 20 千米时,甲、丙的间距.由于

$$20-\frac{81}{5}=\frac{19}{5},$$

而甲、乙速度之比是 $9:7$,所以甲应退

$$\frac{19}{5}\times\frac{9}{9+7},$$

乙应退

$$\frac{19}{5}\times\frac{7}{9+7},$$

而丙的速度是甲的 7 倍,所以丙应退

$$\frac{19}{5}\times\frac{9}{9+7}\times 7.$$

于是,在甲、乙相距 20 千米时,甲、丙相距

$$\frac{19}{5}\times\frac{9}{9+7}\times(1+7)=\frac{19}{5}\times\frac{9}{2}=17\frac{1}{10}(\text{千米}).$$

上面的解法,主要利用速度的比,设出一个未知数 s 就可以解决问题.其中设甲的行程为 $9s$,而不是 s,目的是尽量利用整数,少用分数,使计算尽量简便.

这道题由第七届“华杯赛”的一道题改制而成,“华杯赛”的原题还要复杂些,我们下一篇文章再谈.

行程问题（二）

A、B 两地相距 125 千米. 甲、乙二人骑自行车分别从 A、B 两地同时出发，相向而行. 丙骑摩托车每小时行 63 千米，与甲同时从 A 出发，在甲、乙之间来回穿梭(与乙相遇立即返回，与甲相遇也立即返回). 若甲速度为每小时 9 千米，且当丙第二次回到甲处时(甲、丙同时出发的那一次为丙第 0 次回到甲处)，甲、乙二人相距 45 千米. 问甲、乙二人相距 20 千米时，甲与丙相距多少千米?

这是第七届“华杯赛”的试题.

首先要求出乙的速度，然后根据上一篇文章就可以得出答案.

如图 1，设乙的速度为甲的 k 倍，又丙与乙相遇时甲行了 s 千米，则这时丙行了 $7s$ 千米，乙行了 ks 千米，于是

$$7s+ks=125(\text{千米}). \tag{1}$$

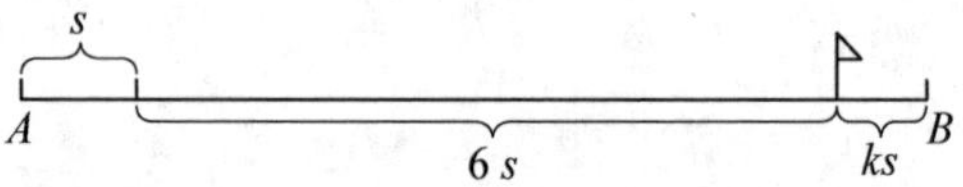

图 1

这时甲、丙相距 $6s(=7s-s)$ 千米. 如图 2，丙第一次回到甲处时，甲又向前行

$$6s\div(7+1)=\frac{3}{4}s(\text{千米}),$$

丙行 $\frac{3}{4}s\times 7$ 千米，乙行 $\frac{3}{4}s\times k$ 千米，所以甲、乙相距

$$\frac{3}{4}s\times 7-\frac{3}{4}s\times k=\frac{3}{4}s(7-k)(\text{千米}), \tag{2}$$

即[将(1)代入(2)消去 s]

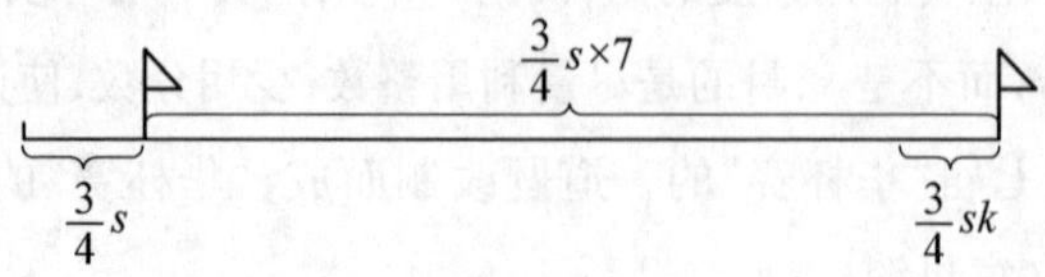

图 2

$$\frac{3}{4}\left(\frac{7-k}{7+k}\right)\times 125(\text{千米}). \tag{3}$$

注意(3)中的 125,如果改成其他数(例如 A、B 两地原来相距 250 千米),推导完全一样,于是,在丙第二次回到甲处时,甲、乙相距

$$\frac{3}{4}\left(\frac{7-k}{7+k}\right)\times\frac{3}{4}\left(\frac{7-k}{7+k}\right)\times 125(\text{千米}). \tag{4}$$

(推导与上面完全一样,只是 125 千米换成了 $\frac{3}{4}\left(\frac{7-k}{7+k}\right)\times 125$ 千米)

根据已知条件,得

$$\frac{3}{4}\left(\frac{7-k}{7+k}\right)\times\frac{3}{4}\left(\frac{7-k}{7+k}\right)\times 125=45,$$

即

$$\left(\frac{7-k}{7+k}\right)^2=\frac{16}{25}.$$

于是(只取正值)

$$\frac{7-k}{7+k}=\frac{4}{5},$$

从而

$$k=\frac{7}{9},$$

即乙的速度是每小时

$$\frac{7}{9}\times 9=7(\text{千米}).$$

根据上一篇文章,甲、乙二人相距 20 千米时,甲与丙相距 $17\frac{1}{10}$千米.

相遇几次

上午 8 点,快车与慢车同时从 A 站出发.慢车环行一次用 43 分钟,到 A 站休息 5 分钟;快车环行一次用 37 分钟,到 A 站休息 4 分钟.求 22 点以前,两车在 A 站相遇几次?

这是第七届"华杯赛"试题.

$$22-8=14,\ 14\times 60=840=41\times 20+20=48\times 17+24.$$

慢车每隔 $48(=43+5)$ 分钟,又从 A 站出发.快车每隔 $41(=37+4)$ 分钟,又从 A 站出发.因此,慢车在 8 点后,$(48h-5)$ 分 $\sim 48h$ 分$(h=1,2,\cdots,17)$ 在 A 站,快车在 8 点后,$(41k-4)$ 分 $\sim 41k$ 分$(k=1,2,\cdots,20)$ 在 A 站,两车相遇的条件是:

$$41k-4\leqslant 48h,\ 48h-5\leqslant 41k,$$

即

$$-4\leqslant 48h-41k\leqslant 5. \tag{1}$$

显然

$$48\times 41-41\times 48=0. \tag{2}$$

又经试验可以知道 $48-1,\ 48\times 2-1,\ 48\times 3-1,\ 48\times 4-1,\ 48\times 5-1,\ 48\times 6-1,\ \cdots$ 中,第一个被 41 整除的是:

$$48\times 6-1=41\times 7,$$

即

$$48\times 6-41\times 7=1. \tag{3}$$

将(3)分别乘 2,乘 3,…,得

$$48\times 12-41\times 14=2, \tag{4}$$

$$48 \times 18 - 41 \times 21 = 3, \tag{5}$$

$$48 \times 24 - 41 \times 28 = 4, \tag{6}$$

$$48 \times 30 - 41 \times 35 = 5. \tag{7}$$

又用(2)分别减去(3)、(4)、(5)、(6)得

$$48 \times 35 - 41 \times 14 = -1, \tag{8}$$

$$48 \times 29 - 41 \times 34 = -2, \tag{9}$$

$$48 \times 23 - 41 \times 27 = -3, \tag{10}$$

$$48 \times 17 - 41 \times 20 = -4, \tag{11}$$

其中(3)、(4)、(11)符合要求，(h, k)分别为

$$(6, 7), (12, 14), (17, 20) \tag{12}$$

其他各式的 h 均已超出 17，k 均已超出 20，不符合要求.

如果还有 h'、k'满足(1)，那么它们与(2)至(11)中某式相减导出

$$48h'' - 41k'' = 0,$$

从而 h''是 41 的倍数，h'等于h''与(2)至(11)中某个 h 的和，即 41 的倍数与(2)与(11)中某个 h 的和. 这个数当然不在 1～17 之间.

因此，满足要求的(h, k)只有(12)中所说的 3 组，即两车在 A 站相遇 3 次.

染圆上的点

圆上有 100×1024^{10} 个点,顺次编号为 $1, 2, 3, \cdots, 100\times1024^{10}$. 按下列规则染色:(1) 先将 1 号点染色;(2) 若上次染 n 号点,则沿编号方向数 n 个点并将最后数到的点染色. 如此继续下去,问共染多少个点?

本题是 2001 年"华杯赛"中学组第一试第 6 题.

根据题意,所染的点号码是

$$1, 2, 2^2, 2^3, 2^4, \cdots, 2^{102}, \cdots.$$

染到 16×2^{102} 时,共染 $102+4+1=107$ 个点. 注意 $100\times1024^{10}=25\times2^{102}$, 接下去染的点是 7×2^{102}($2\times16-25=7$,已经绕过一圈),再接下去是 14×2^{102}, 3×2^{102}, 6×2^{102}, 12×2^{102}, 24×2^{102}, 23×2^{102}, 21×2^{102}, 17×2^{102}, 9×2^{102}, 18×2^{102}, 11×2^{102}, 22×2^{102}, 19×2^{102}, 13×2^{102},最后仍回到 2^{102}(已染过). 所以共染 122 个点.

如果自然数 n 的质因数分解是

$$n=p_1^{\alpha_1}p_2^{\alpha_2}\cdots p_k^{\alpha_k}$$

(α_i 是正整数, $i=1, 2, \cdots, k$, $p_1<p_2<\cdots<p_k$ 是质数),那么

$$\varphi(n)=p_1^{\alpha_1-1}p_2^{\alpha_2-1}\cdots p_k^{\alpha_k-1}(p_1-1)(p_2-1)\cdots(p_k-1)$$

称为欧拉函数. 例如 $n=25=5^2$ 时, $\varphi(n)=5(5-1)=20$.

数论中有一个欧拉定理,即在 a 与 n 互质时, $a^{\varphi(n)}-1$ 能被 n 整除. 所以 $2^{20}-1$ 能被 25 整除,也就是 2^{20} 除以 25 余 1. 因此,在染了点 2^{102} 后,再染 20 个点又回到 2^{102}.

“四连”（一）

一个 6×6 的棋盘，有 36 个小方格，在方格里放棋子. 如果在一条直线（竖线、横线或斜线）上有 4 个同色的棋子相连，就称为一个“四连”.

甲放白棋，乙放黑棋. 如果允许甲先放，他至少要放多少个棋子，才能使乙随后放的棋子不可能构成“四连”？

最好自己画一个棋盘先试一试.

图 1 是一种放法.

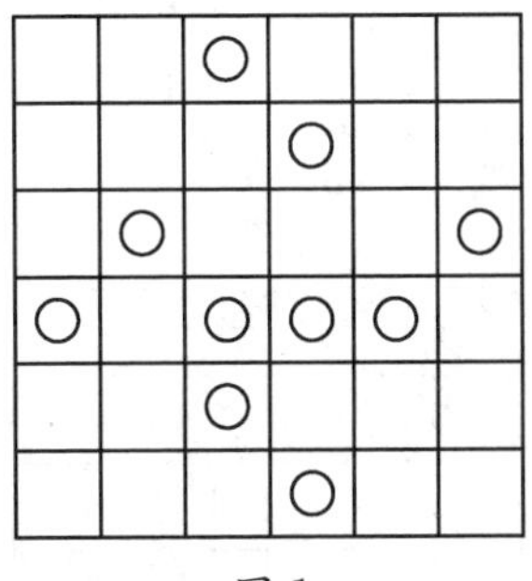

图 1

这种放法共用了 10 个白子，不难检验它符合要求（当然这不是唯一的放法）.

要证明“至少要 10 个白子”，不是一件容易的事.

首先注意每个 1×4（由同一行或同一列的 4 个相连的方格组成）的块，至少要放 1 个白子，所以边上的每个 2×4 的块（图 2 中阴影部分）至少要放 2 个白子. 如果中央（A、B、C、D 四个方格）放的白子数大于或等于 2，那么白子总数大于或等于 $4\times2+2=10$.

如果中央不放白子，那么中间的两列（第三、四列），每列必须放两个白子，中间两行及两条主对角线（从左上到右下，从左下到右上的两条斜线）也是如此，而且这些白子互不相同，白子总数大于或等于 $6\times2=12$.

如果中央恰有 1 个白子，不妨设放在 C 处，那么由于 A、B 为空格，第三行

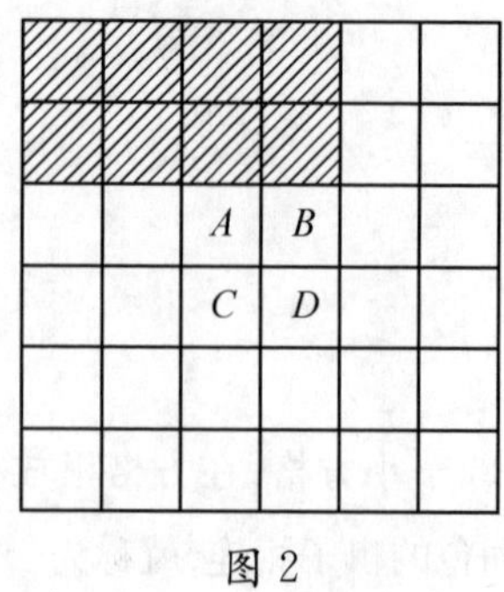

图 2

必有 1 个白子放在阴影中(图 3). 同样,第四列有 1 个白子在阴影中; A、D 所在主对角线上有两个白子在阴影中;B 所在的、与 AD 平行的斜线也有 1 个白子在阴影中. 这 5 个白子互不相同,所以图中两块 2×4 的阴影中,有一块至少放 3 个白子. 其他 3 个 2×4 的块(与它组成棋盘的"边"),每个块中至少放 2 个白子,白子总数大于或等于 $1+3+3\times2=10$.

因此,10 个白子是最少的.

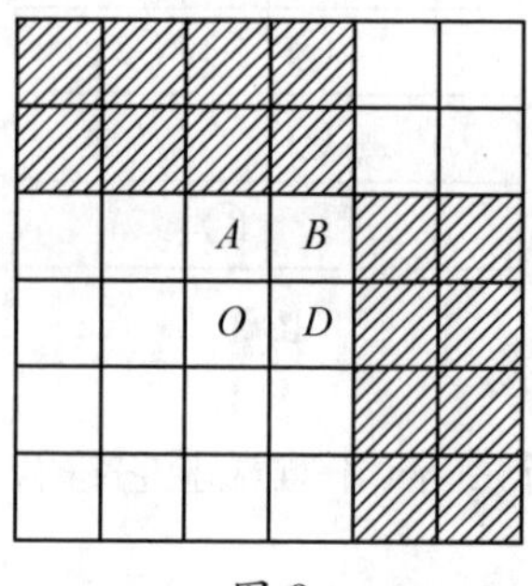

图 3

“四连”(二)

上一篇文章说过,在 6×6 的棋盘上放白子,使得后放的黑子不能形成“四连”,白子至少要用 10 个.这 10 个白子使黑子不能形成“四连”,但它自己能不能形成“四连”?至多形成多少个?

我们探讨一下这个问题.

如果某列有 4 个白子形成“四连”,不妨设第三列的前 4 行是“四连”(图 1),这时其他列各需 1 个白子放在前 4 行,5、6 行也各需 1 个白子,共需白子

$$4+5+2=11(个).$$

如果主对角线上前 4 个方格成“四连”(图 2),那么边上 3 个 2×4 的块(阴影部分)共有 6 个白子,A、B 两格中没有白子.同理,C、D 两格中没有白子,这样 4 个空格 A、B、C、D 在一条线上.所以这种情况不会发生.

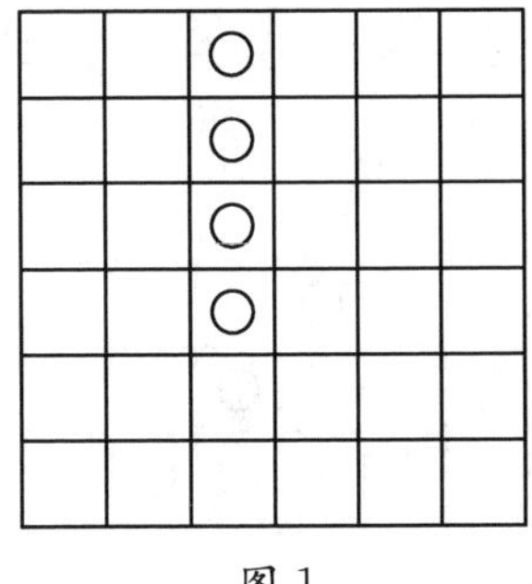

图 1

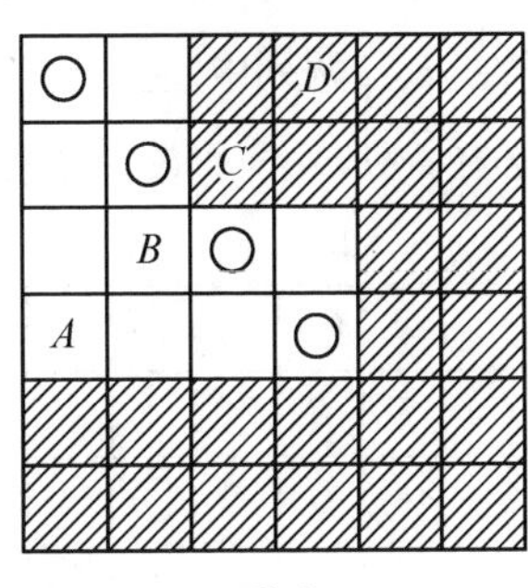

图 2

如果主对角线上中间 4 个方格成“四连”(图 3),那么由于中央已有 2 个白子,边上每个 2×4 的块中恰有 2 个白子.斜线 AB 上应有白子,不妨设白子在 A 或 B.由于 B 这列已有白子,所以 B 无白子,A 放白子(而且左上角的 2×4 的块中再没有白子了).同理 C 无白子.这样 BC 所在斜线的前 4 个方格均无白子,所以这种情况也不会发生.

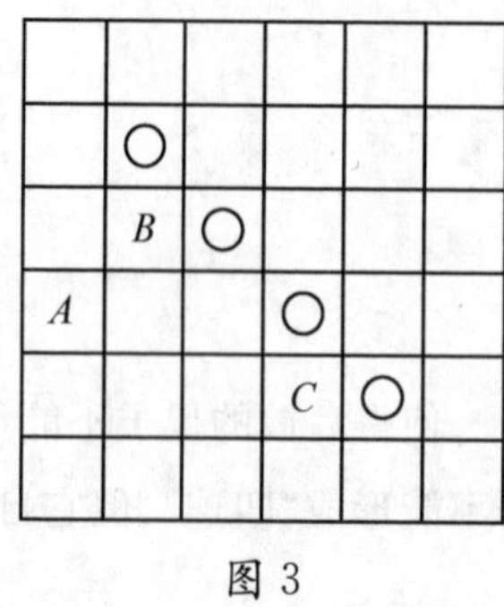

图 3

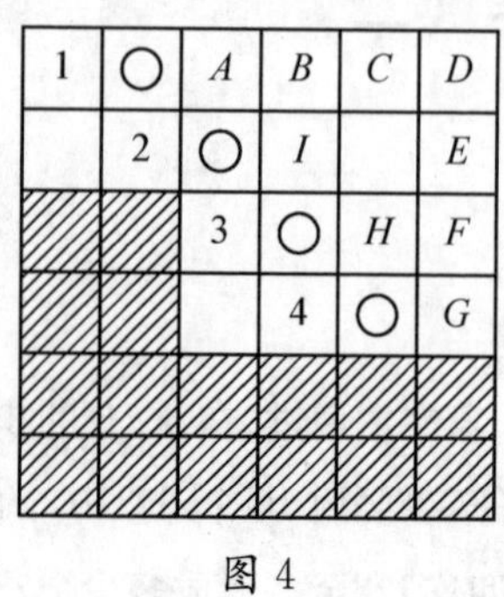

图 4

如果“四连”如图 4,那么右上角的三角形 ADG 的每条边上至少有 1 个白子,3 条边上至少有 2 个白子,剩下 4 个白子在阴影(2 个 2×4 的块)中. 1、2、3、4 中无白子,这种情况也不会发生.

最后,“四连”如图 5,这时 A、B、C、D 中必有一个白子, E、F、G、H 中也必有一个白子,其余 4 个白子在阴影中,所以 I 处无白子. 这样一来, C 处必有白子. G、H 处不会全有白子,不妨设 G 处没有白子. B、G、I 均无白子,所以 2 处必有白子. 右下的 2×4 的块中,另一个白子在最后一行,又在过 D 的斜线上,因此必在 3 处. 主对角线的后 4 个方格中必有白子,它在 H 处. 最后还剩两个白子在一、二两列中. 它们应当在第三、四行,4 和 I 所在的这条斜线上应有白子,它必在 4 处. 第一列的白子可在 5 处或 6 处.

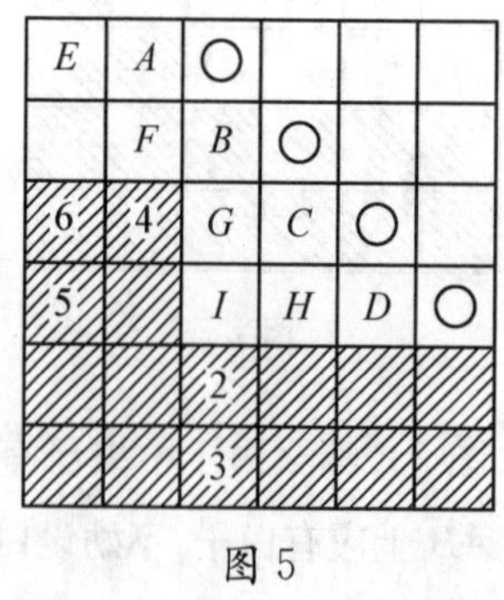

图 5

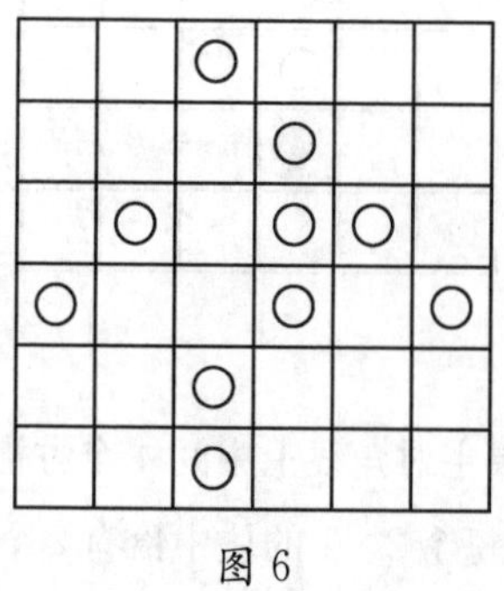
图 6

于是,有一个“四连”的情况只有两种,一种如图 6,另一种是将图 6 的第一列的白子上移 1 格. 除去对称的图形外,其他图形均不形成“四连”. 所以至多只能形成 1 个“四连”.

解题漫谈（二）

如图1所示,圆中两条互相垂直的弦将圆分成4个部分,两个阴影部分的面积之和为S_1,另两个部分的面积之和为S_2. S_1与S_2谁大? 大多少?

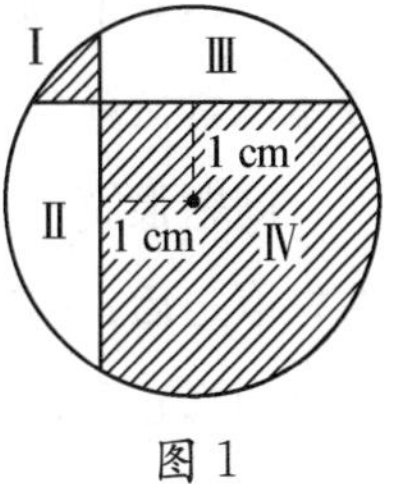

图1

如果知道圆的半径,可以算出S_1、S_2,但相当麻烦,不如直接比较,现在不知道半径,当然只能直接比较.

先猜一猜S_1与S_2谁大? 我相信稍有感觉的人都可以猜出S_1大;因为图2中直径CD将图分成两个对称的部分,阴影部分显然多占了一些地方(超过一半).

至于大多少,需要细致一点的分析. 主要思想仍然是对称.

如图2,作弦$A'B'$与AB(关与轴CD)对称. 面积Ⅰ与Ⅰ′相等,Ⅱ与Ⅱ′相等,所以只需看一看Ⅵ比Ⅴ大多少.

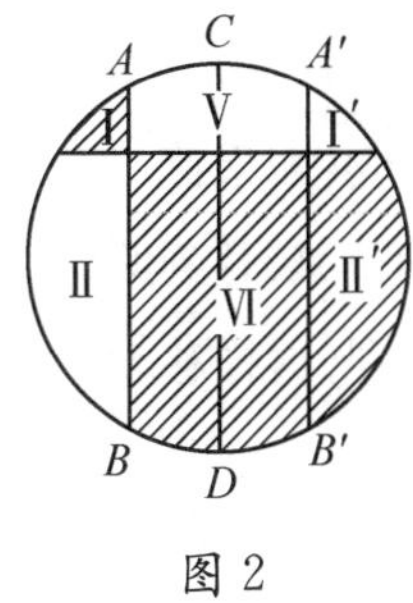

图2

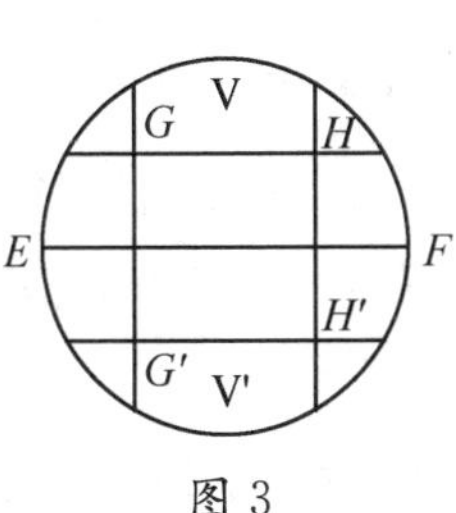

图3

如图3,作$G'H'$与GH关于轴(直径)EF对称. 面积Ⅴ′与Ⅴ相等. 所以

$$S_1-S_2=\text{矩形 } GG'H'H \text{ 的面积}=2\times2=4 \text{ 平方厘米}.$$

本题如果给出圆的半径,这个条件会是多余的,还有可能将人诱入歧途:分别计算S_1、S_2的值. 其实,利用对称性,直接比较,远为简单.

棋盘上的一个组合问题(1)

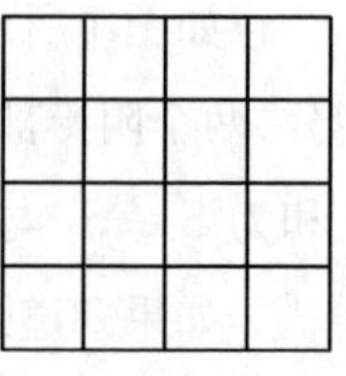
图 1

在一个 4×4 的正方形棋盘(图 1)中,取 4 个方格,每两个方格既不同行,也不同列,有多种取法?

问题很简单:第一行取第一个方格,有 4 种取法,第一个方格取定后,在第二行取第二个方格,有 3 种取法(不能与第一个同列).第二个方格取定后,在第三行取第三个方格,有 2 种取法(不能与第一个、第二个同列).最后,第 4 个方格只有唯一的取法.因此,由乘法原理,共有 $4\times3\times2\times1=24$ 种取法.

当然也可以直接用全排列得出取法为 $4!$.

我们称这样的 4 个方格(每两个既不同行也不同列)为一个"好组".可以提出一个更进一步的问题:

取两个好组,它们没有公共的方格,有多少种取法?

首先,取第一个好组有 $4!$ 种取法.在第一个好组取定后,我们看看第二个好组有多少种取法.

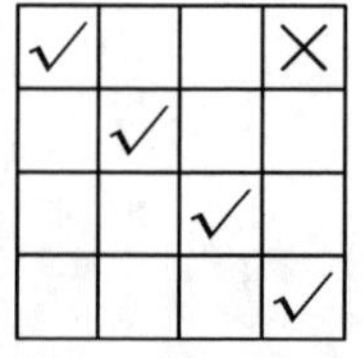
图 2

不妨设第一个好组由图 2 中打√的方格组成(如果第一行打√的不在第一列,我们可以将√所在的列剪下来,移到第一列.由于好组只要求每两个方格不同行、不同列,这样移动不影响论证).

第二个好组的方格,我们打上×,第一行有 3 处可以打×,不妨设(理由与前相同)在右上角打上×.

第一行打好×后,第二行有 2 处可以打×.这两种情况不完全相同,需要分开讨论.

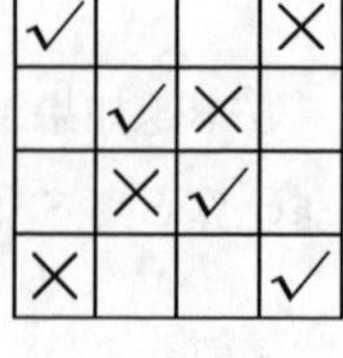
图 3

(1) 如果第二行的×在图 3 中的位置,那么第三行有两处可以打×,第三行打好×后,第四行的×位置也就唯一确定了.因此第二个好组有两种,如图 3、图 4.

(2) 如果第二行的×在图 5 中的位置,那么第三行只有一处可以打×，第三行打好×后，第四行的×，位置也就唯一确定了. 因此第二个好组有一种，如图 5.

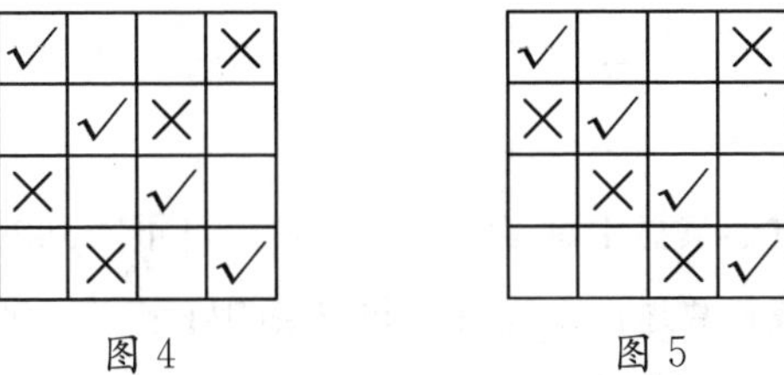

图 4　　　　图 5

于是，第一行打好×后，第二个好组有 $2+1=3$ 种，由于第一行的×有 3 个位置，由乘法原理，第二个好组有 $3\times3=9$ 种. 由于第一个好组有 $4!$ 种，再根据乘法原理，没有公共方格的取两个好组，有 $4!\times9\div2!=108$ 种取法，其中除以 $2!$ 是由于不考虑两组之间的顺序.

注意：我们多次用到乘法原理，也用到加法原理. 特别要弄清“不妨设在右上角打×”. 这个“不妨”导致用 3 去乘后面算出的好组的种数. 而后面计算时，却需要分情况讨论，应当用加法原理，如果对这个“不妨”不放心，最好的办法是自己把 9 种情况都画一画.

棋盘上的一个组合问题(2)

在 4×4 的正方形棋盘中,4 个两两不同行,也不同列的方格,称为一个好组.上一篇文章中我们算过两个没有公共方格的好组,有 108 种取法,现在我们算一算,取三个好组,每两个没有公共方格?有多少种不同的取法?

首先,第一个好组有 4!种取法,设第一个好组已经取定,并用图 1 中的√表示它的四个方格.

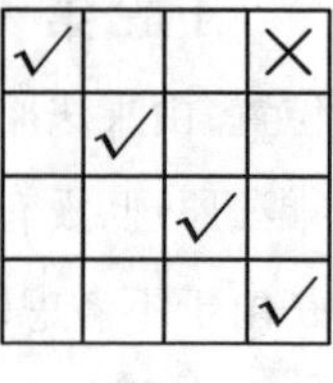

图 1

在第二个好组的方格打上×,第一行打×的方格有 3 种取法,不妨在右上角打×,于是就有图 2、3、4 三种情况,在每一个图中,取第 3 个好组,好组的方格打○.

在图 2 中,第一行打○有 2 个地方,不妨设为图 3,第一行打○后,第二行打○的方法有 2 种,第一、二行打○后,第三、四行打○的方法只有 1 种,如图 3、4 所示,因此,第三个好组的取法有 2×2=4 种.

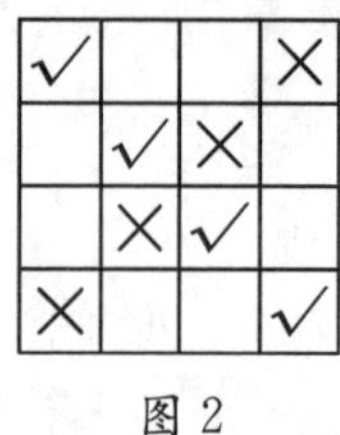

图 2

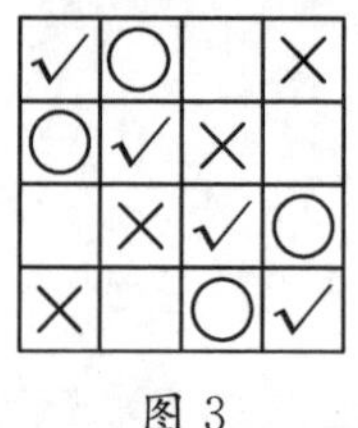

图 3

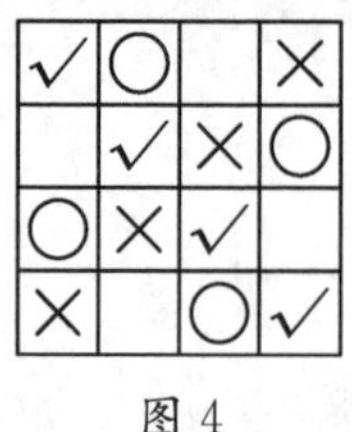

图 4

在图 5 中,第一行打○有 2 个地方,第一行打○后,第三行打○只有 1 种方法.第三行打○后,第二行打○也只有 1 种方法.最后,第四行打○的地方也唯一确定(如图 6).因此,第三个好组的取法有 2 种.

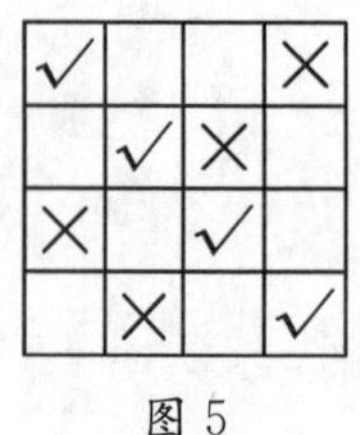

图 5

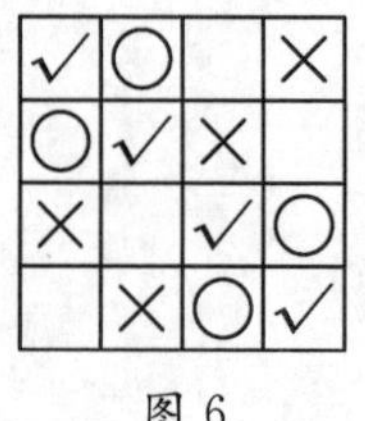

图 6

同理，在图 7 中，第三个好组的取法有 2 种.

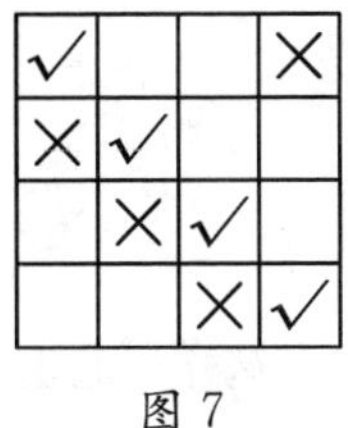

图 7

于是，取三个没有公共方格的好组，有

$$4! \times 3 \times (4+2+2) \div 3! = 96 \text{ 种}.$$

其中除以 3! 是由于不考虑三组之间的顺序.

自动扶梯

在我主编的某套奥数教材中有这样一道题：

"自动扶梯以均匀速度往上行驶着，两位性急的孩子从扶梯走上楼，男孩每分钟走20级梯，女孩每分钟走15级梯，结果男孩用了5分钟到达梯顶，女孩用了6分钟到达梯顶，问扶梯共有多少级."

解答是 $(20-15)\times 6\times 5=150$(级).

深圳市吴乃华先生指出："150级，这个数据脱离实际，如按一级0.2米来计算，150级将高达 $0.2\times 150=30$(米)."

吴先生批评的意见很对，一般商厦两层楼之间很少有高达30米的，自动扶梯也就20多级. 我见到的最长也是最宽的自动扶梯在香港科技大学，24小时不停地转动，蔚为壮观. 但究竟有多少级却未曾数过. 即使超过150级，也只是特殊的例子(或许飞机场的平地的自动扶梯能达到150级). 此外，一分钟走20级，也太慢了. 走得快些，20级应当只用几秒钟.

因此，题中的数据需要修改.

怎样修改呢？先采用字母表示数.

设男孩每秒钟走 a 级梯，女孩每秒钟走 b 级梯，结果男孩用了 s 秒钟，女孩用了 t 秒钟到达梯顶(其他条件不变). 问扶梯共有多少级.

本题中，应当有 $a>b$, $s<t$. 此外, a、b、s、t 尽可能取整数值，以便计算.

解法如下：

设电梯有 m 级，每秒行 n 级，则有

$$\begin{cases} m=s(n+a), & (1)\\ m=t(n+b). & (2)\end{cases}$$

消去 n，即将(1)$\times t$ 减去(2)$\times s$ 得

$$(t-s)m=st(a-b),$$

所以

$$m=\frac{st(a-b)}{t-s}. \tag{3}$$

希望 m 是整数,最简单的办法是令 $t-s=1$. 例如 $t=5$, $s=4$, 这时

$$m=20(a-b).$$

m 不能太大. 只好也令 $a-b=1$. 例如 $a=3$, $b=2$,这样 $m=20$, 即扶梯共 20 级.

又如令 $s=4$, $t=6$,则 $t-s=2$, $m=12(a-b)$,也是整数,如再取 $a=4$, $b=3$,则 $m=12$. 但这组数却是不可以取的. 因为 $as=16$, 已经超过 12(除非扶梯是下行的. 虽然可以这样解释,毕竟不太符合实际).

如令 $s=4$, $t=6$, $a=3$, $b=2$,则 $m=12$. 恰好与 sa、tb 相等,这时扶梯停了.

所以在取 s、t、a、b 的数值时,不但要使 m 为整数,不很大,还要使 $m\geqslant sa$, 读者利用(3)不难发现这一要求等价于 $sa\geqslant tb$.

满足这些要求的 s、t、a、b 的值并不是很多,稍一不慎便会出现不符实际的情况,命题时千万要注意.

数学探索无止境

3, 7, 11, 15, 19, …这列数,从第二个起,每一个减去前一个的差,都相等:7－3＝11－7＝15－11＝19－15＝…

这样的一列数就称为等差数列,这相等的差称为公差.上面这列数的公差就是4.

3, 7, 11这三个数,不但成等差数列,而且都是质数,可惜的是,再往后面的一个数15,却不是质数了.

能不能找4个质数成等差数列?

这并不难找.比如5, 11, 17, 23, 29,这5个质数(比4个还多1个)就成等差数列,公差是6.

能不能再长一点,比如说,找6个质数组成的等差数列?

也能够找到.不过,要注意检验你找的6个数确实是质数.

继续往下找,还可以找到10个成等差数列的质数,199, 409, 619, 829, 1039,1249, 1459, 1669, 1879, 2089,公差为210.

再多一些,若从107 928 278 317开始,每次加99 227 882 870,可以有18个成等差数列的质数;从8 297 644 381开始,每次加4 180 566 390,可以有19个成等差数列的质数.

数学探索无止境.著名数学家厄尔迪什大胆猜想:可以找到任意长的等差数列,其中每一个数都是质数.这"任意长"的意思就是要多长就有多长.当时,大家都认为这个猜想很难得到证明.

2006年获得菲尔兹奖的华裔数学家陶哲轩成功证明了上述猜想.这是他众多成果中的一个.

达·芬奇密码

前些日子(指 2006 年),根据同名小说改编的美国电影大片《达·芬奇密码》在全国各地上映,反响热烈.你看过了吗?

在小说(或电影)中,卢浮宫美术博物馆的馆长雅克·索尼埃在临死前挣扎着写下了一串数:

13, 3, 2, 21, 1, 1, 8, 5.

这串数当然是十分重要的密码.这密码肯定有规律,否则馆长要告诉的人便无法破解,密码也就失去了作用.但馆长为了不使别人,尤其是他的敌人掌握密码,又故意将密码打乱.需要根据一定的规律重新整理,才能得出所需要的正确信息.

这串密码究竟有什么规律呢?

数学家对数字有着特殊的敏感.只需将上面的数按照从小到大的顺序重排一下,变成:

1, 1, 2, 3, 5, 8, 13, 21.

现在你们看出其中的规律了吗?从第三个数开始,每个数都是前两个的和.这下你恍然大悟了吧,只要知道开始两个数是 1,就可以不断地往下写.我们一起继续写下去,就是:

34, 55, 89, 144, …

这种数叫做斐波那契(Fibonacci)数.它是文艺复兴时期意大利数学家比隆迪·列昂纳多(斐波那契是他的笔名)首先发现的.斐波那契数的应用十分广泛.国外还有一本杂志《斐波那契季刊》,专门刊登研究斐波那契数的文章呢!

数学就是动脑

数学,就是动脑.只要你肯动脑,会动脑,爱动脑,你就一定能学好数学.遇到数学题,不要害怕,应当想:“啊哈,这是一个动脑的机会哎,不能放过它!”

比如说,下面这道题:

李大爷用一批化肥给承包的麦田施肥.若每亩施6千克,则缺少化肥300千克;若每亩施5千克,则余下化肥200千克.那么,李大爷共承包了麦田________亩,这批化肥有________千克.

拿到题目,首先要做的是分析题意.已知的条件是:(1) 每亩施6千克,缺少化肥300千克;(2) 每亩施5千克,余下化肥200千克.

假如你就是李大爷,你会怎样施肥呢?先每亩施5千克吧,然后将多出来的200千克也加到田里,每亩增加1千克,这样就有200亩田中施的化肥变成6千克了.还有多少亩田仍然只施了5千克化肥呢?

比照条件(1),显然还有300亩田中仍然只施了5千克化肥,这就表明,李大爷承包的麦田一共有:$300+200=500$(亩),这批化肥总共有:$5\times 500+200=2700$(千克).

怎么样,轻松解决了吧!告诉你,这是今年“华罗庚金杯赛”的决赛试题.开动脑筋,“华杯赛”的试题也不可怕.

题做完了,还应该总结、回顾一下解题的过程,想想还有没有其他方法.假如先每亩施6千克化肥,会怎样呢?

孔子说:“学而时习之,不亦说乎.”习就是反复练习.一个方法,只有反复练习才能熟练掌握,才能由此及彼,由表入里,举一反三,闻一知十.这时,你就会变得自信.

开动脑筋,你就会越来越自信;越有自信,你就越来越愿意自己动脑!

风速的影响

飞机往返于甲、乙两地. 如果飞机的速度不变，那么在风速增大时，飞机在甲、乙两地间往返一次所用的时间是增加，还是减少呢？

不妨设从甲地到乙地是顺风，则从乙地到甲地是逆风.

在风速增大时，从甲到乙的时间比原来少，但从乙到甲的时间比原来多. 因此，往返一次所用的总时间是增加，还是减少，还需要进一步分析.

在没有风的时候，飞机从甲飞到乙，与从乙飞到甲，所用的时间相等，不妨设都是 t 小时.

在有风的时候，如果飞机从甲往乙飞仍用同样的时间（t 小时），那么由于顺风飞行，飞机将飞过乙，到达丙处（如图 1）. 其中乙到丙这段距离为：风速 $\times t$.

甲　　　　　　乙　丙

图 1

如果飞机从乙往甲飞也用 t 小时，那么由于逆风飞行，飞机飞不到甲地，只能飞到丁处（如图 2）. 其中丁到甲这段距离为：风速 $\times t$.

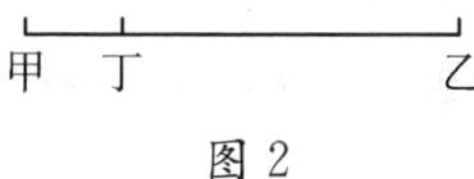

图 2

由此看出乙到丙这一段与丁到甲这一段的距离，是相等的.

飞机实际飞行时，从甲飞到乙后即往回飞，省下从乙到丙这一段时间. 但这一段时间是顺风飞行，所用时间少于飞机逆风飞行时飞过同样距离的丁到甲这一段所用时间.

因此，有风时飞机往返甲、乙一次，所用时间超过 $2t$，即比无风时所用的时间要长.

探索：从最简单的入手

在 1，2，…，2006 中，有多少个数不能被 2、3、5 整除？

2006 显然是个较大的数，如果把这些数全都写出来，再逐个检查，那就太麻烦了．这时，我们应该先考虑最简单的情况，即 1 到 30 中有多少个数不能被 2、3、5 整除：

1 ~~2~~ ~~3~~ ~~4~~ ~~5~~ ~~6~~ 7 ~~8~~ ~~9~~ ~~10~~ 11 ~~12~~ 13 ~~14~~ ~~15~~ ~~16~~ 17 ~~18~~ 19 ~~20~~ ~~21~~ ~~22~~ 23 ~~24~~ ~~25~~ ~~26~~ ~~27~~ ~~28~~ 29 ~~30~~

把能被 2、3、5 整除的数都划去后，还剩下 8 个数 1，7，11，13，17，19，23，29．

进一步，在 31 到 60 中又有多少个这样的数呢？继续用上面的方法，仍剩下 8 个数，即 31，37，…，59．

同样，在 61 到 90 中，也有 8 个数不能被 2、3、5 整除．

……

其实，完全不用这么麻烦，因为一个数如果不能被 2、3、5 整除，那么这个数加上 30 的倍数后，所得的和也不能被 2、3、5 整除．

因为 $2006 \div 30 = 66 \cdots 26$，所以在 1 到 $30 \times 66 = 1980$ 这 1980 个数中，共有 $8 \times 66 = 528$(个)数不能被 2、3、5 整除．

再看，从 1981 到 2006 这 26 个数中，满足条件的数共有 7 个．于是，所求的答案是 $528 + 7 = 535$(个)．

问题解决了，你有没有回过头来想一想，为什么选 1 到 30，而且是加上 30 的倍数呢？这是因为，30 是 2、3、5 的最小公倍数！

从最简单的情况入手，寻找规律，往往能帮助你找到解题的捷径．

“乌龟”在谁手上

54 张扑克，点数相同的 2 张算一对(不论花色)，大王、小王算一对. 预先藏起一张，然后甲、乙两人各拿一些牌，比如一人 27 张，一人 26 张.

每个人将手中的对子放到一边，最后甲手上剩下 7 张，乙剩下 6 张. 再互相抽对方的牌，每次抽 1 张，抽牌后，如果手中的牌能成对子，也放到一边去. 如此继续下去，直到最后剩下 1 张，称为“乌龟”.

问在开始抽牌时，“乌龟”在谁手上？为什么？

猜到“乌龟”在谁手上，不难. 重要的是把理由说清楚.

除了“乌龟”外，两人手中剩下的 $7+6-1=12$(张)牌中，每 2 张配成一对，并且 1 张在甲手上，1 张在乙手上(如果在同一个人手上，事先就应当扔掉了). 因此，甲、乙手中应各有 6 张牌. 甲手中却有 7 张牌，多出的 1 张就是“乌龟”，与它算对子的就是预先藏起的那张牌.

不用分数解工程问题

一项工程，若甲、乙两人合作，8 天完成；若甲单独做，12 天完成. 现在，甲、乙两人合作几天后，余下的工程由乙独自完成. 已知乙前后两段所用时间的比为 $1:3$，问这个工程实际上多少天完成？

这是一个工程问题，解决工程问题通常要用到分数. 但这道题有比较简单的解法，而且完全不需要用分数，你能想到吗？

由题意，甲、乙合作需 8 天完成，甲单独做需 12 天完成，这就表明，甲做 $12-8=4$(天)，相当于乙做 8 天，即甲做 1 天相当于乙做 2 天.

现在，设甲、乙合作(前一段)的时间分别是 1 份，那么后一段乙单独做的时间就是 3 份，即乙做的时间共 4 份. 乙的 4 份时间相当于甲的 2 份时间，如果将乙做的时间转换为甲做的时间，甲单独做这项工就需要 $1+2=3$(份)时间. 因此，每份时间是：$12\div3=4$(天)，实际工期是：$4\times4=16$(天).

还有一种简便的解法是：

$$8+1\times2\times2+4=16(\text{天}).$$

你能说出这样列式的理由吗？

如履平地

汽车从 A 地到 B 地，既有平路，也有上坡与下坡，速度分别为每小时 63 千米、56 千米、72 千米. 汽车从 A 地到 B 地用了 8 小时，返回用了 10 小时. 问：从 A 地到 B 地有多少千米？

汽车经过平路部分，往返都是平路；而汽车在去的路途中的上坡，在回来时就变成下坡；对于下坡，在回来时就变成了上坡.

这样，可以算出：平路每千米往返需用

$$\frac{1}{63}+\frac{1}{63}=\frac{2}{63}(\text{时});$$

上坡每千米往返需用

$$\frac{1}{56}+\frac{1}{72}=\frac{9+7}{7\times8\times9}=\frac{2}{63}(\text{时});$$

下坡每千米往返同样也需用$\frac{2}{63}$小时.

因此，在本题中，有往有返时，三种路段每千米所用时间实际上是一样的. 由于往返这段路程共用 $8+10=18$(时)，所以两地距离为：

$$18\div\frac{2}{63}=567(\text{千米}).$$

晴天和雨天

甲、乙两人各做一项工程. 如果全是晴天,甲需 12 天,乙需 15 天完成. 雨天甲的工作效率比晴天低 40%,乙降低 10%. 两人同时开工,恰好同时完成. 问工作中有多少个雨天?

本题有很多种解法. 应当找一种尽量简单的解法.

如果全是晴天,乙比甲多用(15－12)天. 现在两人同时开工,同时完成,可见其中必有雨天.

本题的关键就是将雨天换成晴天.

甲 1 个雨天相当于(1－40%)个晴天.

乙 1 个雨天相当于(1－10%)个晴天.

因此,每将 1 个雨天换为晴天,乙比甲多工作 (1 － 10%) － (1 － 40%) ＝ (40% － 10%) (天). 最后,雨天全换成晴天,乙比甲多工作(15－12)天. 所以共有 (15 － 12) ÷ (40% － 10%) ＝10(个)雨天被换成晴天,也就是工作中有 10 个雨天.

在推理过程中,遇到一些算式,如 15 － 12, 40% － 10%, 可以计算,也可以不计算,而留到列出总的算式以后再一并计算.

颠倒乾坤

一个 I 字型的部件，如图 1，将它放置成图 2，问图 2 中两个打“?”号的点，对应在图 1 中标的字母是什么？

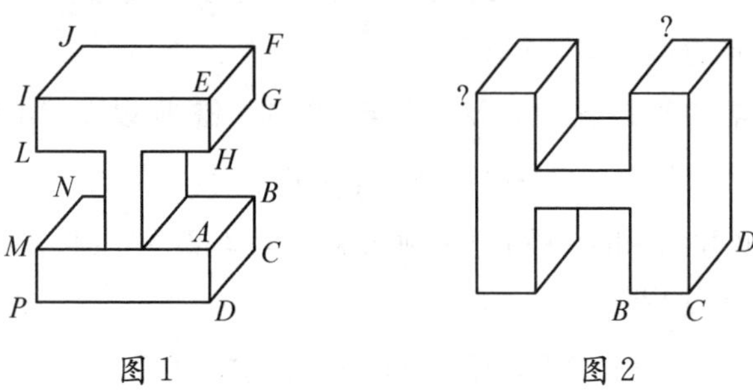

图 1　　　　图 2

如果在现实中，我们只需将如图 1 形状的部件转转、翻翻，放置成图 2，就很容易得到正确答案.

现在只能在纸上观察比较图 1、图 2，其实只要认真观察，这个问题并不难，但需要一点对空间图形的想象能力.

如图 2，右边那个问号的正下方应当是 A（图 2 中只标出 B、C、D，未标 A）. 而在图 1 中，在 A 左方的顶点是 M，将部件放置成图 2 时，四边形 $ABCD$ 在底部，M 在 A 的正上方，所以这个标“?”的点对应为字母 M.

同样，可以知道左边标“?”的正下方是 F（图 1 中 B 的上方），所以这个标“?”的点对应为字母 J.

按照上述的方法，同学们不难对照图 1，将图 2 中所有的点标上字母，自己试一试吧！

哲学家的年龄问题

《献给非哲学家的小哲学》是法国哲学家阿尔贝·雅卡尔的一本畅销哲学书.书中标为“X”的一节,有两个算术问题,都与年龄有关.

问题1 15年后,汤姆的年龄是去年的3倍,汤姆今年多大?

先考虑去年的年龄.

去年的年龄加上15再加1,就是去年年龄的3倍,所以去年年龄的$3-1=2$倍就是16岁,去年是$16\div2=8$(岁),今年是$8+1=9$(岁).

问题2 我和父亲的年龄和是我俩年龄差的2倍.5年前,父亲年龄是我的4倍,现在我俩各几岁?

这个问题稍难一些.我们要仔细分析题中的数量关系.

因为两人的年龄和减去年龄差,一定是我年龄的2倍,而题目中已知年龄和是年龄差的2倍,所以年龄差的$2-1=1$倍正好是我年龄的2倍.于是,现在父亲的年龄是我的$2+1=3$倍.

5年前,就是把现在我的年龄减少5岁,父亲的年龄也减少5岁,父亲年龄就变成我的4倍.假如父亲年龄减少$5\times3=15$岁,那么父亲年龄仍是我的3倍,这少减少的$15-5=10$(岁)便是那年我的年龄的$4-3=1$倍,即5年前我的年龄是$10\div1=10$(岁),现在我年龄是$10+5=15$(岁),父亲是$15\times3=45$(岁).

此题若列方程解,可以轻松求解,同学们可以动手比较一下.

两全其美

爱丽丝跌进兔子洞里，发现一个门，门很小，进不去．她喝了一瓶饮料，人就变小了．但是开门需要钥匙，钥匙放在桌上，而桌子却太高了（因为爱丽丝现在只有十英寸高），她拿不到钥匙．幸而，她又看到一块饼．吃了饼，她又变大了，能够拿到钥匙开门，却又无法走进那个小门．

怎样才能两全其美呢？数学中也常有这样的问题．

如图 1，在 8×8 的棋盘中，填 64 个自然数，每个小格一个数，要使得整个棋盘中，64 个数的平均数小于 2；而每个 5×5 的子棋盘中，25 个数的平均数大于 3，怎样填？

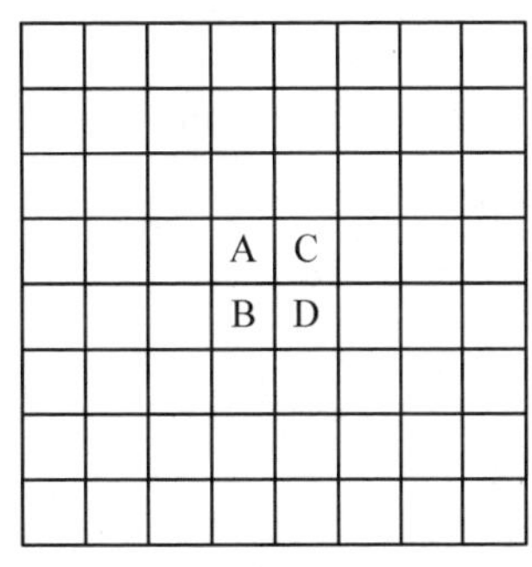

图 1

如果每个方格都填 1，那么整个棋盘的平均数是 1，小于 2；但 5×5 的子棋盘（图中有 16 个）中，25 个数的平均数也是 1，1 小于 3．不符合条件．

如果每个方格都填 4，那么 5×5 的子棋盘中，25 个数的平均数是 4，大于 3；但整个棋盘中的平均数也是 4，大于 2．也不符合条件．

怎样才能两全其美，都符合条件呢？

首先，64 个数不能全部相等．其次，注意 16 个 5×5 的子棋盘都会有中间的 A、B、C、D 这 4 个方格．

如果 A（或 B、C、D）中的数比较大，那么，5×5 的子棋盘中，25 个数的平均数就比较大．如果其他的数比较小，那么，整个 8×8 的棋盘，平均数就比较小．

比如说，A 格中填 $1+25=26$，其余方格中都填 1，那么每个 5×5 的子棋盘中，25 个数的平均数就是 $(26+24)\div 25=2$．如果 A 格中填 $1+2\times 25=51$，其余方格中都填 1，那么每个 5×5 的子棋盘中，25 个数的平均数就是 $(51+24)\div 25=3$．

继续下去，只要 A 格中填 $1+51=52$，每个 5×5 的子棋盘中 25 个数的平均数就大于 3 了. 而 8×8 棋盘中，因 $52+63<64\times2$，64 个数的平均数便小于 2.

当然，填法不只一种. 只要确定了 A(或 B、C、D)，其他数很快就能填出. 同学们可以试一试.

切饼与切瓜

一块大饼，一刀切成 2 块. 切两刀，可以切成 4 块. 切三刀，最多能切成几块? 切十刀，最多能切成几块呢?

我们可以在纸上先画出一个圆，代表一块饼.

一刀切下去，变成两块，也就是比原来增加了 1 块 $(1+1=2)$.

切两刀，又可增加 2 块，只要这一刀(一条直线)与前一刀(也是一条直线)相交，就可以将前面切成的两块中的每块中的每一块都切成两块. 于是变成 $1+1+2=4$(块).

切三刀，最多可以切成 7 块(如图 1)，即比原先增加 3 块.

只要这一刀(一条直线)与前两刀(两条直线)相交，那么这一条直线就被前两条直线分成 3 段，每一段将原先的一块分成两块. 因此，每一段线增加 1 块，3 段共增加 3 块.

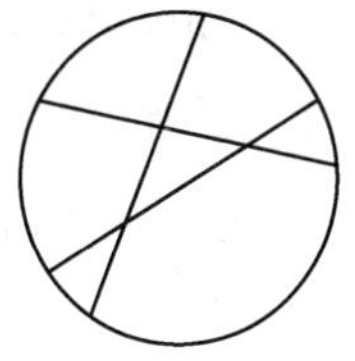

图 1

如果再切一刀，这一刀，也就是图上的一条直线，它最多与前 3 条直线都相交，被前 3 条直线分成 4 段，每一段将原先的一块分成两块. 因此，每一段增加 1 块，4 段共增加 4 块. 变成 $1+2+3+4=11$(块).

依次类推，切十刀最多比上一次增加 10 块. 所以切十刀，最多有 $1+1+2+3+4+5+6+7+8+9+10=56$(块).

同样的，一个西瓜，一刀切成 2 块，那么切两刀、切三刀，直至切十刀，最多又能切成几块呢?

参考上面说的切饼方法，第一刀将西瓜切成两半后，每半个西瓜(或者说每半个瓜皮)都可以想象成一块大饼.

再切一刀，将“每块大饼”切成$(1+1)$块. 因此，西瓜变为 $2\times(1+1)=4$(块).

切三刀，西瓜将变为 $2\times(1+1+2)=8$(块).

切十刀，西瓜最多被切成 $2\times(1+1+2+3+4+5+6+7+8+9)=92$(块).

当然，这只是理论上的推断，如果真的实际操作，那大饼与西瓜多半没有切完，就凌乱不堪了.

走出迷宫

图1是一个“迷宫”.你能找到一条从入口A到出口B的路吗?

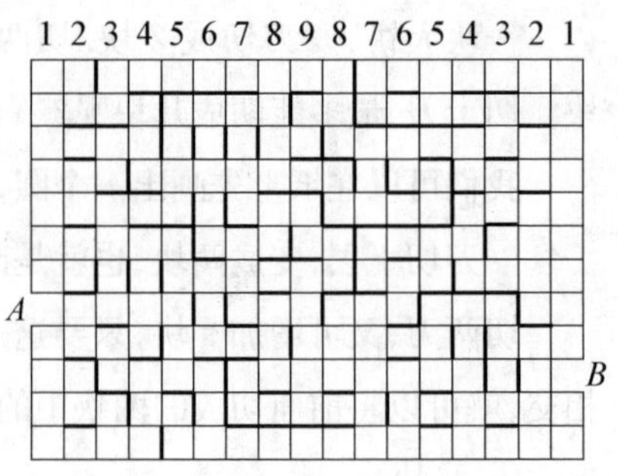

图1

这很容易,从入口A进来,然后右拐沿第1列走到底,再沿最下面一行走.走到第4列遇到阻挡(粗线表示阻挡),向上走1路,再向右、向下沿最下面一行走到头.最后向上2格就到出口B了.

如果要你最快走出迷宫,走一条最短的路呢?瞎摸乱撞可行不通,这时需要数学来帮忙了.刚才的这条路是不是最短的?计算一下,共走了$5+3+2+13+2=25$(格).

如果从入口A进来,一直向前走到第5列,向下1格;再向右沿水平方向到第8列,再向下1格;然后继续向右沿水平方向到倒数第4列,向下1格;最后沿水平方向走到头,向上1格,便到出口B了.这条路的长是$5+4+7+4+1=21$(格),比刚才那条路少了4格.

还有更短的路线吗?观察迷宫,从A到B共有17列,每列都必须要走到;A在B上面2行,所以从A到B至少要走$17+2=19$(格).

如果正好是19格,那么这条路线不能超过A所在的行,也不能低于B所在的行.但在这2行及所夹的1行中有许多障碍.如果仅在这3行中走,就必须有时向上走(从倒数第5列至倒数第3列,就说明不向上走是走不通的除非走到B所在行的下一行),而向上一步就必定要向下一步,所以从A到B的路至少要走$19+2=21$(格).21格是最短的路程.

第六章　数学师友

高山仰止(载《怀念华罗庚》)　　中国大百科全书出版社,2004

捧着一颗心来,不带半根草去——怀念陈景润先生

南京周末,1996 年 3 月 30 日;南京师大报,1996 年 4 月 10 日

悼念元老　　2021 年 5 月 14 日

怀念常庚哲老师　　2018 年 11 月 18 日

悼念张筑生教授　　中等数学,2002(2)

忆肖刚　　2015 年 5 月

学习谈老　　2019 年 9 月

叶中豪五十初度　　2016 年 8 月 22 日

高山仰止

华罗庚先生，是我国数学界的喜马拉雅山，仰之弥高. 虽然很难有人能够达到华先生那样的水平，但他的精神给每个从事数学工作的人以巨大的鼓舞，他的教诲永远是我们大家的座右铭.

（一）

华先生感人至深的是“精勤不倦，自强不息”.

华先生的正规学历是初中. 他完全依靠自学，不仅读了当时视为很难的 Hall 与 Knight 的《大代数》，而且还学习了解析几何、微积分、代数方程等大学课程. 1929 年 19 岁时就发表论文《Sturm 氏定理的研究》，次年又发表文章《苏家驹之代数的五次方程式解法不能成立的理由》，不仅表明他已经熟悉阿贝尔关于五次方程不可能有代数解的理论，而且还直接指出苏文中一个 12 阶行列式计算有误. 正是这篇文章引起了当时的最高学府清华大学算学系主任熊庆来的重视，邀请他到清华担任助理员.

华先生到清华后，眼界大开，告别了过去初级水平的“研究”，如饥似渴地学习高等数学，特别是跟杨武之教授学习数论. 他在两年内学完了大学课程并开始做文章. 他的研究跨上了一个台阶. 3 年内连续发表了 21 篇论文，大多刊登在国外一流杂志上，包括日本的《东北数学杂志》以及当时世界上最重要的数学杂志——德国的《数学年鉴》（华先生是第二个在这个杂志上发表文章的中国人）.

正确地说，华罗庚先生自学最成功的一段并不是他在家乡的时候. 在清华这个时期才是最重要最成功的一段. 正如王元先生在《华罗庚》一书中所说：

“因为一个人是容易在自己熟悉及习惯的学术领域里停步的，即不再更新知识，尤其是自学的人，更易如此. 华罗庚却能在两三年内，改弦更张，更上一层楼，这是最宝贵的.”

在清华期间,他还接触到来中国讲学的、世界一流的数学家阿达玛与维纳,从他们那里了解到数学的最新动向.维纳热心地写信将华罗庚推荐给英国剑桥大学的著名数学家哈代.

为了在数学上有更大的作用,获得更高的成就,华罗庚放弃了在清华晋升讲师的机会,到英国剑桥去学习.

当时剑桥大学是世界数学中心之一,分析与解析数论尤强,那里及附近集中了一批朝气蓬勃、才华横溢的青年数学家,如达文坡特、海尔布伦、埃斯特曼、赖特、蒂奇马什等.华罗庚尽量利用这个良好的学术环境.他风趣地说:"有人去英国,先补习英文,再听一门课,写一篇文章,然后得一个学位.我听七八门课,记了一厚叠笔记,回国后又重新整理了一遍,仔细地加以消化.在剑桥时,我写了10多篇文章."他还说:"我只有两年的研究时间,自然要多学点东西,多写点有意思的文章,念博士不免有些繁文缛节,太浪费时间了.我不想念博士学位,我只要求做一个访问学者.我来剑桥是为了求学问,不是为了学位."

在剑桥大学,华罗庚又告别了他在清华大学时期的研究.在清华大学时他的工作虽然是国内数学界的佼佼者,但研究的课题仍然较为零散,也不属于数论主流或重大课题.到剑桥后,他真正做出了世界第一流的工作(如关于完整三角和的估计、华罗庚不等式),引起了国际上的重视,达到了他一生中的第一个创作高峰.与清华时的工作相比,又跃上了更新的台阶.

这时,华罗庚先生已经形成了自己的学术观点.他无论搞哪一门数学,总是抓住中心问题,力求在方法上创新.我国有句成语"班门弄斧",形容一些人不自量力.可华先生却认为"弄斧必到班门".他还说:"人说不到黄河心不死,我说到了黄河志更高."

华先生不仅"专""精",而且"漫",他能从一个研究领域转入另一个研究领域.除了数论,他在代数、多复变函数、矩阵几何等领域也都有杰出的贡献.尤其是华先生特有的"直接法",即用尽量简单的初等的数学工具,单刀直入地处理数学中的一些重要问题.这种风格,备受国内外同行的赞赏.

华先生曾经题词:"努力不计年,自强永不息,学习数学是一辈子的事."他实践了这句话,一生奉献给数学,鞠躬尽瘁,直到倒在讲台上.

（二）

华罗庚先生非常重视培养人才与数学的普及工作.他对我国数学界影响之大,恐怕没有人能与之相比.在某种程度上可以说没有华罗庚就没有现代的中国数学.

1950年,华先生回国后不久即参加数学研究所的筹备工作.1952年出任数学所所长.他广泛搜罗人才(其中包括1956年进所的陈景润先生),并亲自领导数论组与代数组的讨论班,培养研究人才.

他还在清华大学、中国科学技术大学任教,亲自给大学生上基础课并撰写讲义(即后来出版的《高等数学引论》).

华先生特别热心数学的普及工作,经常给中学师生与数学爱好者作通俗讲座,并写作一些科普读物,如《从杨辉三角谈起》《从祖冲之的圆周率谈起》《从孙子的神奇妙算谈起》《数学归纳法》《统筹方法平话》《优选法平话》等.

华先生非常重视自己所做的普及工作.在日本东京大学报告的提纲中,就将一生的工作分为“理论”与“普及”两块,相提并论.

华先生还说:“深入浅出是真功夫.”他的报告就是深入浅出的典范.1964年,我第一次听到华先生的报告.华先生特别注意听众的要求与接受能力.报告一开始就说:“今天来的人很多,我一定要把‘音’定好,不能太高,也不能太低.”他定的“音”果然恰到好处.那天他先举了茶杯与杯盖的例子.圆口的茶杯,杯盖也是圆,直径稍大一些.杯盖无论怎么放,也不会落到杯中.但是,如果杯口是正方形呢?即使杯盖是稍大的正方形,依然会落入杯中.这个通俗的问题立即引起了大家的兴趣.接着,他又举了当时苏联发射洲际导弹宣布的禁入区域.他说,“从这个四边形区域的四个顶点,立即可以推出发射塔是在乌拉尔山的某处.”这样的例子雅俗共赏,充分说明生活中处处有数学,生活离不开数学.

1966年,华老来南京推广“两法”,在人民大学堂作报告.他的普通话非常标准,不疾不徐,语言幽默生动,例如说到工厂任务月头松月尾紧,他就用了“天女散花”“夜战马超”“萧何月下追韩信”等比方,引得哄堂大笑.

10年浩劫之后,华老决心振兴我国数学.1981年,71岁高龄的华老率庞大的讲学队伍来到合肥中国科学技术大学,同行的有王元、吴方、杨乐、张广厚等,又邀请复旦大学的夏道行、谷超豪、胡和生，南京大学的周伯勋、叶彦谦，还有科技大学本校的龚升、石钟慈、彭家贵等. 国内有 20 多所大学，派近百人前来听讲. 华老自己做了“矩阵几何与狭义相对论”“微积分方程的几何理论”“普及数学方法的若干个人体会”与“国民经济中所用到的数学方法”等 4 次报告. 当时人如潮水般涌向报告厅，听众对华老的报告热烈鼓掌欢迎，反响极为强烈.

华老还认真听取其他人的报告. 有一次吴方先生的报告中用到一个三角不等式，华老立即说：“这个不等式，我来证明.”说着便起身拄着拐杖上了讲台，当场进行演算. 可见华老虽已年逾古稀，思维仍然十分敏捷，而且童心未泯，很喜欢“露一手”给大家看看.

华老特别重视数学竞赛. 早在 1946 年访苏期间，就专门考察了数学竞赛活动，并听了柯尔莫多洛夫与阿历山德罗夫对参加数学竞赛活动的部分师生所作的讲演(题目分别是“对数”与“复数”)，在心中埋藏了中国倡办数学竞赛活动的种子.

1956 年，华先生著文欢呼“我们也要搞数学竞赛了”，并亲自倡导在北京、上海、天津与武汉四大城市举办了中学生的数学竞赛活动. 后来由于“四人帮”的干扰，数学竞赛被迫停止. “文化大革命”结束后，1978 年华老又亲自出任竞赛委员会主任，主持规模空前的北京、上海、天津、山西、安徽、辽宁、四川、广东 8 省市的中学数学竞赛，并主持出版了《全国中学数学竞赛题解》一书，华老专门写了长达 7000 字的《前言》.

可以告慰华老的是，在华老逝世的次年，我国首次派出正式的 6 人代表队参加国际数学奥林匹克(IMO)，并取得了三金一银一铜、总分第四的好成绩. 1989 年，又在 IMO 取得总分第一. 1990 年，在北京成功地主办了第 31 届 IMO，并再次取得总分第一.

华老对年轻一代寄予厚望. 他写了很多介绍学习方法的文章，把自己的心得传授给大家. 他勉励青年人努力学习，反复强调“聪明在于学习，天才由于积累”. 在给母校金坛中学题词时，华老语重心长地写了 4 个字：“后来居上.”

华老是江苏人. 虽然长期在北京居住，但仍爱吃南方的大米. 当时北京的大米是按比例分配的. 华老是全国政协常委(后来是副主席)，可以另外分到一些大米. 但全家都吃大米还是不够，而且北方大米比较贵，又总不及家乡的米好吃. 1981 年，华老到合肥讲学时，发现合肥的米价廉物美，便买了一袋. 但华老年事已高,又有残疾,显然无法自己将米带回去.华老便委托吴方先生办理.吴先生虽然乐意为这斗米折腰,但他也年届半百,感到这个任务的沉重.正好我与吴先生同车去北京,这件事理所当然地由我承担了.我把大米扛上火车,又从火车运上汽车,再从汽车上搬下来送到华老住所,圆满地完成了任务.

现在已到开镰时节,凉风习习,仿佛送来阵阵稻谷的清香.可是,华罗庚先生已经离开我们14年了.

1999年9月9日

捧着一颗心来，不带半根草去——怀念陈景润先生

陈先生走了,真没有想到!

陈先生有恩于我,我的博士答辩就是先生主持的. 那一天,报告厅挤满了人,各系的学生都来了,都想亲眼看一看这位仰慕已久的数学家. 后面的人看不到,就站在折叠椅上,椅子踩坏了很多.

陈先生没有惊人的外貌,永远是那么简朴,那么平常. 平顶头,戴一副极普通的眼镜,眼黑显得略大些,转动得比较少,老是呈沉思状,神游在数学王国里.

陈先生为人非常谦恭. 1984 年夏天,先生住在贵州民族学院. 我们上山拜望,陈先生一定要亲自沏茶,我们连声说不必费事,先生坚持要沏,水瓶空了,又提了水壶,装水,点火,忙个不停. 我们告辞出来,又一定要送. 因为是晚上,又是山路,大家一再请先生留步,先生才很勉强地答应了,站在月光下,目送我们离去.

我们沿着蜿蜒的山路走了一刻多钟,到了山脚,正要上公路,忽然后面跑出一个人,扑到大家身上,大笑:“我一直跟在后面,你们都没有发现.”原来竟是陈先生!

1984 年后,陈先生身体一直不好,患了帕金森氏综合症,在友谊医院住了一阵,出院后行走仍然不便,又不幸被车撞了一次. 1988 年我从加、美回来,特地去问候. 陈先生一见就伸出手握我的手,嘴里还说了一句听不清楚的话. 数学所陪伴的秘书,懂得先生的语言,立即“译”给我听:“问单教授好.”接着先生又告诉我:“因为行走不便,吃了药,走路好些,却影响了说话.”这段话当然又是秘书翻译的.

真没想到这就是最后一次见到陈先生. 后来只听说先生的身体时好时坏,常常住院. 我总以为天佑善人,陈先生的身体能够好起来. 没想到先生走得这么快.

陈先生的一生实在太刻苦了.

1957 年,由于对塔锐问题的研究受到华罗庚先生的赞赏,陈先生被调入数

学所. 那时先生还很年轻. 10 年时间(1957—1966)做了很多出色的工作:解决了华林问题的 g(5);对圆内整点问题的研究和对 g(4)的估计也都创造了“世界纪录”,并保持了很久. 当然,最著名的还是创造了陈氏筛法,解决了“1+2”. 这项成果的摘要发表于 1966 年的《科学通报》. 由于“文革”,全文延迟到 1974 年才在《中国科学》刊出.

这么多的成果,全是在 6 平方米的小屋中,啃着干馒头完成的.

“文革”后,陈先生已经誉满天下. 一次,我奉校方之命请先生去中国科学技术大学讲学(当时我在科大). 这时先生已搬过家,但也只有两小间,仍然单身. 先生执意留我吃饭,亲手做了自诩的“盖浇饭”,就是一碗米饭一个荷包蛋加上若干片胡萝卜,大概先生平时自奉的伙食还没有这样“丰盛”. 先生还一再提出“不要买软卧,硬卧、硬座都可以”,并向我要票:“把票给我,我自己去换成硬卧.”

先生就是这样的俭朴,付出的那么多,享用的又是这样少,真正“捧得一颗心来,不带半根草去”. 如果先生能早点注意身体,为了国家,为了自己,为了孩子,或许……

陈先生留下自己的业绩,匆匆地走了. 真没想到陈先生走得这样急.

悼念元老

惊悉元老逝世.

王元先生是我的恩师,大家都尊称他为元老.

1978 年 4 月我到中国科学技术大学读研. 此前,1964 年我从扬州师院数学系毕业,并在中学任教 14 年,扬州师院数学系的大学基础课与中学数学研究抓得极为扎实,但当时数学研究在全国高校并未形成风气. 我到中科大,跟常庚哲老师学习样条函数,又参加陆鸣皋老师主持的解析数论讨论班与冯克勤老师的代数数论讨论班. 如何进行科研,实在是门外汉. 元老从北京来合肥指导,对陆鸣皋老师与我做的工作大加奖励,并在校系领导面前表扬. 于是我就被定为首批博士学位的一个候选者. 元老还多次谈他自己学习与治学的心得,他说读中学时并不很努力,外国电影倒看了不少. 考大学时未能考上最好的大学,读了英士大学. 但后来院系调整,合并成立浙江大学. 有一次复变函数考试,考 $e^{-\frac{1}{x}}$ 的奇点属性,只有他一个人答对,从此得到系里青睐,并选定以数学研究为终身方向.

他在中科院数学所主要研究解析数论,华罗庚先生擅长圆法. 而筛法在国际上刚刚起步,元老便研究这新的方法,并在 26 岁时就用这新方法证明了哥德巴赫问题的"3+4",取得当时最佳结果. 元老还用这方法研究拉丁方的个数,也获得长期领先的结果. 他自己说那时年轻,什么都能做.

元老曾说现在他不急于陷入某个问题之中,大部分时间在观察动态,看看有什么值得做的有意义的问题(他多次对我说太小的问题不值得做),有什么能够做的问题.

他还说:"我的学生,我都要和他合作一篇文章."他给了我一个题目,但对我说:"不一定能做出来,做不出来也不要紧,慢慢想."

但这个问题,我一直未做出来(觉得无从下手,想了想就放下了). 1983 年,我去北京参加首批博士学位的大会. 想到要去见元老,觉得不好交代,便在火车上连夜想，居然想出来了(幸亏当时火车要开 10 个小时以上). 见到元老便将这事告诉他，他也很高兴. 后来又经过修改，合作完成了这篇关于哥德巴赫数的文章(《*A Conditional Result on Goldbach Problem*》，发表在《数学学报(英文版)》1985 年第 1 期).

元老任中科院数学所所长时，倡导改革，邀请一些人去访问. 我也躬逢其盛，在访问期间读了一些文章，做了一些结果.

我在南京师范大学数学系任系主任时，元老来南京访问，我与陈永高一同陪元老故地重游，寻访了石鼓路的教堂与当时的国立六中，但现在已毫无踪影，只好去看白下路的六中，当时学校还请元老题了字. 又一同去看了东南大学，即原中央大学. 元老那时的家即在成贤街口，可惜在三天前刚刚拆得精光，已无踪迹可以凭吊.

1992 年我应香港数学会岑嘉评主席之邀去访问，正好元老也在那边，共同讨论了一个问题，岑教授写了一篇文章，由元老推荐在《中国科学》发表.

有一年到北京，正好元老家的小世兄升初中，要报考某校，招我去辅导. 多年后，小世兄也在美国大学担任数学教授了.

元老喜爱游泳，有一年在山东开数论会议，由山东大学潘承洞先生任东道主. 元老是常常下水的.

元老晚年喜爱书法，曾送我一个手卷，写的是辛稼轩的词《沁园春 · 灵山齐庵赋时筑偃湖未成》，笔力遒劲而又工整，远胜当下一些"书法家".

我今天上午到学校(南师大)，忽然觉得身体不适，眼一黑，坐到地上，幸有李君华老师与物理系的谭老师等在旁，由学校用救护车将我送仙林鼓楼医院. 我无大问题，没想到元老去世了，或许是某种心灵感应吧.

心中悲哀，匆匆写下上述文字，以作纪念.

2021 年 5 月 14 日晚

怀念常庚哲老师

（一）

一九七七年　刚将左毒变
常彭两位来　告我招考研
次日即考试　心态放坦然
考好好处大[①]　考差亦无玷
上午考分析　下午代数篇
除一附加题　其他皆完卷
克正[②]拔头筹　常公笑逐颜
我亦附骥尾　都过标准线
常公极得意　勉勗多良言
音容与笑貌　历历在眼前

（二）

奥数与普及　常公着先鞭
复数解几何　抽屉幻万变
著述遍海内　甘棠留人间
今日升遐去　能不常怀念

2018 年 11 月 18 日

① 常公原话.
② 克正:李克正教授,当时是一工厂工人.

悼念张筑生教授

张筑生教授于2002年2月6日去世.对数学竞赛来说,这是一个难以弥补的损失.

张教授长期参与冬令营(中国数学奥林匹克)及国家集训队选拔的命题工作.他眼界开阔,又与科研前沿保持密切的联系,所命的题大多新颖而又有深刻的背景.如第一届中国数学奥林匹克的第六题:

"MO牌足球由若干多边形皮块用三种不同颜色的丝线缝制而成,有以下特点:

(1) 任一多边形皮块的一条边恰与另一多边形皮块同样长的一条边用一种颜色的丝线缝合;

(2) 足球上每一结点恰好是三个多边形的顶点,每一结点的三条缝线的颜色不同.

求证:可以在这MO牌足球的每一结点上放置一个不等于1的复数,使得每一多边形块的所有顶点上放置的复数的乘积都等于1."

再如,2000年中国数学奥林匹克的第三题:

"某乒乓球俱乐部组织交流活动,安排符合以下规则的双打赛程表.规则为:

(1) 每名参加者至多属于两个对子;

(2) 任意两个不同对子之间至多进行一次双打;

(3) 凡表中同属一对的两人就不在任何双打中作为对手相遇;

统计各人参加的双打次数,约定将所有不同的次数组成的集合称为'赛次集'.

给定由不同的正整数组成的集合 $A=\{a_1, a_2, \cdots, a_k\}$,其中每个数都能被6整除.试问最少必须有多少人参加活动,才可以安排符合上述规则的赛程表,使得相应的赛次集恰为 A? 请证明你的结论."

这些问题是张教授留下的一笔宝贵的精神财富.

张教授多才多艺,不仅有深厚的文学功底,而且懂得医理. 1988 年,他根据病情,查看医书,断定自己患了鼻咽癌,随后去医院检查,果真如此.

十多年来,他虽身患绝症,仍以惊人的毅力参与领导我国的数学竞赛事业. 1995 年作为领队兼主教练去加拿大参加第 36 届 IMO,取得总分第一. 1997、1999、2000、2001(其中 1998 年 IMO 在台湾举行,未派队参加)年,他都作为主教练主持国家队的训练工作,均取得总分第一.

张筑生教授将自己的后半生完全奉献给数学竞赛. 直至 2002 年年初才被迫停止自己喜爱的工作,第一次没有参加冬令营的命题. 而后病情急转直下,但直到癌细胞已向全身扩散,他还与自己的学生、前来探视的第 35 届 IMO 金牌得主姚健钢谈竞赛的事情,真正做到了鞠躬尽瘁,死而后已.

忆肖刚

年龄小我近十岁　才力胜我逾十倍

竟然先我去极乐　能不令人长深悲

玄武湖上荡轻舟　锡惠山前赏红梅

思想已存天地间　莫嗟生命似芦苇

1977 年，复旦大学与中国科学技术大学分别被教育部与中国科学院批准提前招收研究生. 科大招了三名，都在数学系. 最早的一位就是肖刚，上半年就到了合肥. 十月份，李克正与我一同被科大的常庚哲、彭家贵老师考核(笔试加面试). 此前李克正还参加了复旦的考试. 结果复旦、科大都录取了李克正，他选择了科大. 我由于种种原因，1978 年 4 月份才去科大. 全国正式招收研究生在 1978 年(杨劲根兄文中说查建国兄 1977 年读研，似应为 1978 年).

中科大在文革中被迫南迁. 行到河南，河南不要. 再到合肥，当时安徽革委会主任宋佩璋同意接收，将两所小的学校合肥师范学院、银行干校拨给科大安身. 1977 年一招生，顿时觉得人多屋少. 教学区、生活区混在一堆. 我们住的那座楼，既有教室、办公室，又有学生宿舍、教工宿舍. 肖刚、李克正和我三个人住在楼下进门的第一间. 室内除了两张床(有上下铺)，只能再在中间放一张办公桌当三人共用的书桌. 夏天热，没有空调，只好把门开着. 一天中午饭后，躺在床上，刚有睡意，进来一位小偷，偷了肖刚晾的衬衫. 我与李克正奋力去追，在黄山路邮局前合力将小偷捉住送派出所. 这位小偷年纪轻轻，已是惯犯，派出所的警察一见就认识.

肖、李两位是曾肯成先生的得意门生. 当时有人送了曾先生两盆牡丹，一盆叫照粉，一盆叫洛阳红. 曾先生很高兴，拟了一付对联：

肖刚李克正

照粉洛阳红

我见过不少聪明人. 数学界不像政界，没有特别愚蠢的，但说到天才，恐怕只有肖刚才当得起. 他小我八岁(原先以为他小我十岁)，但才力的确超过我十倍.不用说他在代数几何方面的卓越贡献,就说初等数学吧.我文革前已在中学工作两年,可以算得上初等数学的解题高手.但肖、李二人常有非常独特、优雅的解法,令人赞佩.例如 Polya 的名著《数学的发现》中有一道题:证明 11, 111, 1111,…中没有平方数.

原书的解法比较麻烦.肖刚张了一眼就说:“mod8”(后来我改为 mod4).这个解法现在广为流传,它就源自肖刚.

肖刚做题往往就是这样,话不多,一两句就击中要害.

肖刚与曾先生一样,平时考虑问题,几乎不用草稿.我问他:“遇到复杂的计算怎么办?”他笑笑说:“我不会算.”但有一次我看他做一道四点共双曲线的问题,其中有多个正切,又有复杂的行列式,他算得很快,也拿了一张纸,但写得很少,大部分都用心算.可见他的算功非常之好,只是不屑做那些繁琐而无趣的计算.

他评论范德瓦尔登的《代数学》时说:“这书初版有很多的计算,后来删了.因为时尚变了,能不算的尽量不算.”

他还说:“代数是一种解释.”

他用下面同调代数中最常见的图(图 1)解释解决问题的两种方式:

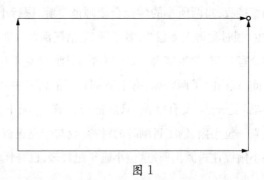

图 1

(从图中左下角到右上角)一种方式是先在平地上向前,然后艰难地向上攀登.另一种是先将理念(观点)上升到一定高度,然后在天上行走,如履平地.

肖刚的观点极高,所以做任何事情都很容易.

他看数学书就像读小说,非常之快.为什么能这么快呢?他说:“我先看这本书想要解决什么问题.再看作者为了解决这个问题,引入哪些工具、概念.然后就自己思考应当如何展开,得产生或需要哪些定理?再翻翻书验证一下,果然如此.”

我问他:“定理的证明看不看?”

他说:“一般不看,没有必要为作者做校对工作. 有些定理,如果有趣,也会想一想如何去证明.”

“想不出来怎么办?”

“那就算了.”

肖刚对书与文章,有自己的评价. 如说 Shafarevich 的《Basic Algebraic Geometry》“没有什么东西”,而 Hartshorne 的《Algebraic Geometry》“应当细读”. 他做了后一本书的所有习题,还写了不少自己的心得.

肖刚的学习效率极高. 他说:“我看书的连续时间从不超过一小时. 到一小时我就休息. 否则头昏脑涨,没有效率.”

科大不少人认为肖刚聪明但不用功,甚至说他“老在校园里晃荡.”的确,他每隔一小时就在校园里晃荡一次. 外面的人常看到他晃荡,而与他同一宿舍的我,总看到他在看书想问题.

肖刚文革时只上到初一,后来插队. 他的数学完全是自学的. 他的英语也完全是自学的. 虽然上了江苏师院(现在的苏州大学)外语系,课上念的都是外国人不说的 Chinese English,如什么“The foreigh language is a weapon of the class struggle”. 所以肖刚常常逃课,自己听 BBC,看原版的小说. 到科大读研后,因为曾老师建议他与李克正分开,李去美国,肖去法国. 所以肖刚又自学法语. 他真的背字典. 那时录音机还很罕见,而且是磁带绕在很大的盘子上. 他一个字一个字地读,用录音机录下后,再反复回放给自己听. 他说:“单词的遗忘率在三天内最高,所以得在三天里复习一次,抢在遗忘之前巩固.”

有人说肖刚把字典一页一页地撕掉,我没有看见,他的字典都是好好的.

肖刚是个很幽默的人. 有一次在图书馆看书. 他对我说:“你注意那边那个人吗?”我说:“没有啊.”“你过去看看他在看哪一页.”我看了回来告诉他. 肖刚笑了:“那个人一早就在那里,正襟危坐,书放在桌上. 我去看了一次,他一小时一页都未动过. 你去看,又过了一小时,还是一页未动. 思想不知开到哪里去了.”

肖刚也爱玩. 合肥的景点,逍遥津、包公祠、城隍庙、教弩台,他全去过. 他到过南京,和李克正、我三个人在玄武湖用木浆划船. 他还邀李克正和我去无锡梅园. 那天杨劲根兄也从上海来. 杨兄的母亲大寿,他到无锡买寿桃. 我第一次看到大如面盆的桃子.

肖刚很重友情. 每次李克正或我离开合肥,他都送我们到火车站. 1987 年,应陈省身先生之邀,我从加拿大去旧金山,他到机场接我. 还驾车带我在旧金山沿海岸玩了整整一天. 他的夫人陈馨也陪同,好像是她在海滩放风筝,人美如画(也可能是看别人放,记不清了).

肖刚在法国拿到国家博士回国后,决定在上海华东师范大学工作. 科大方面听到消息很着急,派常庚哲老师去劝肖刚回合肥. 行前系里头头想到找一个与肖刚有交情的一道去. 当时李克正还在美国未回. 于是就想到与肖刚同宿舍的我,要我也去游说. 我问系里:"能开出什么条件?"系里说:"准备给肖刚报副教授,并考虑给他分房."我说:"听说华东师大已给肖刚分了房,报了正教授. 如果科大在上海,华东师大在合肥,或许肖刚还可以考虑科大的条件. 现在一切都是华东师大的条件好,怎么可能把肖刚动员回来?"头头也只好说:"你们去试试看."

我与常老师到了上海,在肖刚丈人家见到肖刚. 他的岳父园林专家陈从周先生午睡刚起,也见到了. 聊了一会,上面派的任务当然无法完成. 不过,我跑了一趟,倒有收获. 系里有人说:"你们许愿给肖刚分房,可单某在合肥,也不给他分房."于是我也得到半套房子,可谓不虚此行.

在科大读研时,肖刚对数学的普及工作也饶有兴致. 他和李克正支持我写通俗的小册子. 给我出主意,提供材料. 他还建议我们三人用一个共同的笔名轮流写普及的文章. 这笔名一人提一个字,他提"肖"字,李克正提"韧"字,我提"吾"字,合成"肖韧吾"(小人物). 可惜我刚用这笔名写了第一篇"生锈圆规的作图",他二人就出国了.

读研时,肖刚、李克正、我三个人通过不少次信. 后来,肖刚就改手写为电脑打字了. 那时电脑刚刚出现,他用的是自创的汉字输入法,都是繁体字. 肖刚到法国后,我与他直接联系不是很多. 因为也没有什么事需要打扰他,而且他的研究越来越深,我觉得他走得越来越快,越来越远,完全跟不上. 他的信息我大多从李克正那里间接获得.

我未见过肖刚锻炼身体. 他的身体似乎也还可以. 没想到一下就走进天国了,连个招呼也没跟我打. 但这样的天才,他的思想一定永远存留在世间,供大家学习,研究.

学习谈老

数学发展快
一日数千里
前贤开大道
希望后人继
希望真理解
希望能普及

数学作普及
实际大不易
数学重抽象
普及得具体
数学似枯燥
普及要风趣

外国加德纳
为此费大力
中国有谈老
堪与相匹敌
誉满科普界
众口赞第一

文章似海多
著作如山积
凤凰出版人
编成十册集

有益爱好者
功德莫大矣

谈老多创造
时常发新义
文史功底醇
文字富真趣
妙笔著妙文
尤妙想法奇

此次十册出
各地需求急
先睹自为快
一读手难释
唯祈谈老寿
再出新文集

注:2019年谈祥柏先生(1930年5月生于上海)九十大寿.江苏凤凰出版集团推出《谈祥柏趣味数学详谈》丛书十册,我于2019年9月作此诗遥贺.

叶中豪五十初度

当年号小疯
如今成封翁
声名震海内
豪情传域中
烟士披里纯
酒必饮大钟
永远乐呵呵
不忌增臃肿
乘兴去希腊
会见老欧公
四壁尽图形
变化妙无穷
勗勉讲学勤
多播花草种
既已知天命
立言当为重
应有名山作
不必藏山峰
早日呈世人
亦是一大功
说罢水陆献
弦歌似春风
不觉酩酊醉
醉眠美人丛
红尘多纷扰

佛说色即空
蝴蝶化庄生
庄生蝴蝶梦
美梦了无痕
梦觉如生龙
身在大上海
满座皆高朋

2016 年 8 月 22 日

附录

一、单墩著作（独著及合著类）

序号	书名	出版时间	出版社	备注
1	1—20 届国际数学奥林匹克题解	1978 年		安徽省教育厅教材编写室资料第九期，与李克正、杜锡录合作
2	几何不等式	1980 年 2 月	上海教育出版社	
3	趣味的图论问题	1980 年 9 月	上海教育出版社	
4	覆盖	1983 年 7 月	上海教育出版社	
5	解题思路训练	1987 年 2 月	中国少年儿童出版社	
6	棋盘上的数学	1987 年 11 月	上海教育出版社	与程龙合作
7	趣味数论	1987 年 12 月	中国青年出版社	
8	帮你学因式分解	1988 年 2 月	中国少年儿童出版社	与段扬合作
9	巧解应用题	1988 年 7 月	中国少年儿童出版社	
10	解析几何的技巧	1989 年 6 月	中国科学技术大学出版社	与程龙合作
11	对应	1989 年 10 月	科学技术文献出版社	与王子侠合作
12	组合数学的问题与方法	1989 年 12 月	人民教育出版社	
13	数学竞赛史话	1990 年 5 月	广西教育出版社	
14	国际数学竞赛解题方法	1990 年 6 月	中国少年儿童出版社	与葛军合作

（续表）

序号	书名	出版时间	出版社	备注
15	不定方程	1991 年 9 月	上海教育出版社	与余红兵合作
16	数学趣题巧解 100 例	1991 年 12 月	中国少年儿童出版社	
17	算两次	1992 年 3 月	中国科学技术大学出版社	
18	数学竞赛研究教程	1993 年 4 月	江苏教育出版社	
19	组合几何	1996 年 6 月	上海教育出版社	
20	趣味数学百题解答	1996 年 9 月	中国少年儿童出版社	
21	十个有趣的数学问题	1999 年 3 月	上海教育出版社	
22	华罗庚数学奥林匹克教材(初一年级)	2000 年 1 月	知识出版社	与王晓琴合作
23	集合及其子集	2001 年 7 月	上海教育出版社	
24	平面几何中的小花	2002 年 5 月	上海教育出版社	
25	解题研究	2002 年 6 月	南京师范大学出版社	
26	概率与期望	2005 年 4 月	华东师范大学出版社	数学奥林匹克小丛书
27	因式分解技巧	2005 年 4 月	华东师范大学出版社	数学奥林匹克小丛书
28	单墫数学科普著作选	2006 年 10 月	江苏教育出版社	包括《快乐的数学》《数学什锦》《趣题巧解》《思路训练》《巧解应用题》等
29	数列与数学归纳法	2009 年 1 月	上海科技教育出版社	
30	集合与对应	2009 年 1 月	上海科技教育出版社	
31	初等数论的知识与问题	2011 年 3 月	哈尔滨工业大学出版社	
32	我怎样解题	2013 年 1 月	哈尔滨工业大学出版社	

（续表）

序号	书名	出版时间	出版社	备注
33	单墫初中数学指津：数的故事与灵活策略	2014 年 7 月	上海辞书出版社	
34	单墫初中数学指津：代数的魅力与技巧	2014 年 6 月	上海辞书出版社	
35	单墫初中数学指津：平面几何的知识与问题	2014 年 7 月	上海辞书出版社	
36	平面几何 100 例	2015 年 5 月	中国科学技术大学出版社	
37	三角函数	2016 年 6 月	中国科学技术大学出版社	
38	数列与极限	2016 年 10 月	中国科学技术大学出版社	
39	Probability and Expectation	2016 年 9 月	East China Normal University Press	
40	解题研究	2016 年 12 月	上海教育出版社	新版
41	解题漫谈	2016 年 12 月	上海教育出版社	
42	代数不等式的证明	2017 年 3 月	中国科学技术大学出版社	
43	我怎样解题	2017 年 5 月	上海教育出版社	新版
44	数学竞赛研究教程（上下）	2018 年 6 月	上海教育出版社	新版
45	国际数学竞赛解题方法・数学竞赛史话	2019 年 10 月	上海教育出版社	新版，与葛军合作
46	平面几何的知识与问题	2019 年 4 月	中国科学技术大学出版社	新版
47	代数的魅力与技巧	2020 年 11 月	中国科学技术大学出版社	新版
48	数学随笔	2021 年 3 月	哈尔滨工业大学出版社	

二、单墫著作（翻译类）

序号	书名	出版时间	出版社	备注
1	几何不等式	1991 年 9 月	北京大学出版社	
2	近代欧氏几何学	1999 年 8 月	上海教育出版社	
3	近代欧氏几何学	2012 年 3 月	哈尔滨工业大学出版社	新版
4	近代的三角形几何学	2012 年 7 月	哈尔滨工业大学出版社	

三、单墫著作（主编类）

序号	书名	出版时间	出版社	备注
1	数学奥林匹克（1987—1988)(高中版)	1990 年 8 月	北京大学出版社	与胡大同合作
2	数学奥林匹克(1989)——30 届国际数学竞赛预选题	1990 年 10 月	北京大学出版社	与刘亚强、葛军合作
3	数学奥林匹克——第 31 届国家集训队资料(1990)	1991 年 6 月	北京大学出版社	与葛军合作
4	数学奥林匹克(小学版)(修订版)	1991 年 6 月	北京大学出版社	包括四至六年级共 3 分册
5	数学奥林匹克(初中版)(修订版)	1991 年 6 月	北京大学出版社	包括初一至初三共 3 分册
6	数学奥林匹克基础训练(小学版)	1992 年 10 月	海洋出版社	
7	数学奥林匹克(小学版新版)	1992 年 10 月	北京大学出版社	包括启蒙篇、基础篇、提高篇三册
8	数学奥林匹克(初中版新版)	1992 年 12 月	北京大学出版社	包括基础篇、知识篇、提高篇共 3 册

(续表)

序号	书名	出版时间	出版社	备注
9	数学奥林匹克(高中版新版)	1992 年 12 月	北京大学出版社	包括基础篇、知识篇、竞赛篇共 3 册
10	数学奥林匹克题典	1995 年 2 月	南京大学出版社	
11	初中课本	1995 年 12 月	江苏科学技术出版社	包括代数 4 册、几何 2 册
12	数学奥林匹克	1999 年 6 月	南京大学出版社	包括小学三至六年级共 4 册
13	几何不等式在中国	1996 年 9 月	江苏教育出版社	
14	数学奥林匹克竞赛题解精编	1999 年 10 月	南京大学出版社	
15	华罗庚数学奥林匹克教材	2000 年 1 月	知识出版社	包括小学二至六年级共 5 册
16	初等数论	2000 年 7 月	南京大学出版社	高校教材
17	华罗庚金杯少年数学邀请赛集训题典	2000 年 8 月	中国大百科全书出版社	
18	奥数教程	2000 年 10 月	华东师范大学出版社	与熊斌共同担任总主编,包括小学、初中、高中所有年级
19	数学名题词典	2002 年 7 月	江苏教育出版社	
20	全国高中数学联赛模拟试题	2002 年 9 月	河海大学出版社	
21	小学数学奥林匹克水平测试卷	2003 年 6 月	江苏教育出版社	包括三至六年级分册
22	小学数学奥林匹克教程	2004 年 6 月	江苏教育出版社	包括三至六年级共 4 册

（续表）

序号	书名	出版时间	出版社	备注
22	华数奥赛集训指南（初中组）	2005 年 5 月	中国大百科全书出版社，北方妇女儿童出版社	
23	初中数学奥林匹克教程	2004 年 6 月	江苏教育出版社	包括初一至初三共 3 册
24	华数奥赛集训模拟（高中组）	2005 年 5 月	中国大百科全书出版社，北方妇女儿童出版社	
25	苏教版高中数学教材	2004 年 6 月	江苏教育出版社	与高中数学课程标准配套
26	多功能题典（高中数学竞赛）	2008 年 8 月	华东师范大学出版社	与熊斌共同主编
27	数学奥林匹克命题人讲座	2009 年 1 月	上海科技教育出版社	主编该丛书，并编写了其中几册
28	初等数论	2020 年 1 月	南京大学出版社	新版
29	走进美妙的数学花园	2010 年 1 月	南京师范大学出版社	包括三至八年级共 6 册
30	强基计划深度学习丛书	2020 年 10 月	湖南科学技术出版社